QUANTITATIVE REASONING

MATHEMATICS FOR CITIZENS IN THE 21st CENTURY

PRELIMINARY EDITION

Jeffrey O. Bennett
University of Colorado at Boulder

William L. Briggs
University of Colorado at Denver

Cherilynn A. Morrow
University of Colorado at Boulder

Addison-Wesley Publishing Company

Reading, Massachusetts • Menlo Park, California • New York • Don Mills, Ontario
Harlow, United Kingdom • Amsterdam • Bonn • Sydney • Singapore • Tokyo
Madrid • San Juan • Milan • Paris

Credits

Pages 1, 4, 12, 15, 18, 19, 94, 338 — Reprinted with permission from *Everybody Counts: A Report to the Nation on the Future of Mathematics Education*. Copyright 1989 by the National Academy of Sciences. Courtesy of the National Academy Press, Washington, D.C. Page 5 — From *On the Pulse of the Morning* by Maya Angelou. Copyright © 1993 by Maya Angelou. Reprinted by permission of Random House, Inc. Page 7 — From "Education: The Knowledge Gap," supplement to *The Wall Street Journal*, 2/9/90. Reprinted by permission of The Wall Street Journal, © 1990 Dow Jones & Company, Inc. All Rights Reserved Worldwide. Page 13 — From P. J. Davis and R. Hersh, "The Limits of Mathematics" in *The World Treasury of Physics, Astronomy, and Mathematics*, Timothy Ferris, Ed., 1991, Little, Brown, & Co., p. 559. Reprinted with permission. Page 13 — From John Allen Paulos, *A Mathematician Reads the Newspaper*, 1995, HarperCollins, p. 166. Reprinted with permission. Page 72 — From James Gleick, *Chaos: Making a New Science*, 1987, Penguin Books. Reprinted with permission. Pages 54, 75 — Molly Ivins. Reprinted with permission of the Fort Worth Star-Telegram. Page 135 — From *A Brief History of Time* by Stephen W. Hawking. Copyright © 1988 by Stephen W. Hawking. Used by permission of Bantam Books, division of Bantam Doubleday Dell Publishing Group, Inc. Page 179 — Found in Douglas Hofstadter, *Scientific American*, May 1982. Page 204 — From "The S&L Tab" by John Allen Paulos, *The New York Times*, June 28, 1990, A,25:1. Page 204 — "The Biggest Robbery in History — You're the Victim" by Michael Gartner, August 9, 1990. Reprinted with permission of The Wall Street Journal, © 1990 Dow Jones & Company, Inc. All rights reserved. Page 232 — Reprinted from *Beyond the Limits* copyright © 1992 by Meadows, Meadows, and Randers. With permission from Chelsea Green Publishing Co., White River Junction, Vermont and the Canadian publishers, McClelland & Stewart, Toronto. Page 235 — "Passenger Pigeons" by Larry Bleiberg, *The Dallas Morning News*, November 21, 1993, p. 11G. Reprinted with permission of The Dallas Morning News. Pages 238, 248 — From A. K. Dewdney, *200% of Nothing*, 1993, John Wiley & Sons. Reprinted with permission. Page 254 — Jessica Mathews, "The 'Waste, Fraud, and Abuse' Fraud," *The Washington Post*, February 13, 1995. © The Washington Post. Pages 278, 290 — From Thomas Levenson, *Measure for Measure: A Musical History of Science*, 1994, Simon & Schuster. Page 350 — From John Neihardt, *Black Elk Speaks*, 1932. Published in 1982 by Pocket Books, New York. Reprinted with permission. Page 489 — From Darrell Huff, *How to Lie with Statistics*, © 1954. New York: W.W. Norton, p. 8. Reprinted with permission. Page 533 — From Lester R. Brown and Hal Kane, *Full House*, © 1994 Worldwatch Institute. New York: W.W. Norton. Reprinted with permission. Page 575 — From John L. Casti, *Complexification*, 1994, New York: HarperCollins, p. 3. Reprinted with permission.

ISBN 0-201-76554-3
1 2 3 4 5 6 7 8 9 10-CRS-99989796

Quantitative Reasoning on the Internet

Thank you for choosing to use this Preliminary Edition of *Quantitative Reasoning — Mathematics for Citizens in the 21st Century*, a textbook based on a course developed by the authors at the University of Colorado. To facilitate communication between the authors, instructors, and students, Addison-Wesley has established a *Quantitative Reasoning* discussion site on the World Wide Web. To reach this site:

1. Connect to the Addison-Wesley Higher Education home page at

   http://www.aw.com/he

2. Select *Mathematics* from the options offered
3. Select *Quantitative Reasoning* from the options offered

You will find three main categories to choose from at the *Quantitative Reasoning* site (subject to change):

- *Quantitative Reasoning in the News:* Select this option to see suggestions for teaching or studying quantitative reasoning based on current events. Suggestions will be posted by the authors.
- *Quantitative Reasoning Users Discussion Group:* Select this option to join a dialog with other users of the Preliminary Edition. Read ideas submitted by others, and submit your own ideas for teaching, suggestions on how to use the book, and other items likely to be of interest to other users of the textbook.
- *Textbook Comments and Criticism:* Select this option to submit your comments, criticisms, or suggestions regarding the Preliminary Edition of this textbook directly to the authors. We will incorporate all comments into our plans for the First Edition, and will respond to your comments on-line as needed.

If you do not have access to the World Wide Web, write to the authors: c/o Jeff Bennett, Campus Box 389, University of Colorado, Boulder, CO 80309.

PREFACE

Why did we write a book called *Quantitative Reasoning — Mathematics for Citizens in the 21st Century*? We wrote it because we believe that the quantitative concepts and skills covered in this book are crucial to meeting the future challenges facing individuals, societies, and the human species.

The challenges are many and difficult. In the rapidly changing technological world, individuals must have flexibility and knowledge well beyond what was necessary in the past. For example, we as individuals are faced with increasingly complex decisions in everything from personal finance to voting; in the home there are endless choices such as which foods to eat, which computers to purchase, and whether to lease or buy a car; and most people today (including the authors!) go through numerous career changes during their lives. As a society, we face the challenges of adjusting to a rapidly changing global economy while still providing opportunity for all citizens. As a species, our challenges include accommodating a population growing by almost 100 million people per year; alleviating environmental threats posed by global climate change, pollution, and deforestation; and dealing with the risks posed by terrifying, modern weapons.

In the face of such challenges, it is no exaggeration to state that the survival of the nation, the civilization, and perhaps even the species, depend upon the future decisions we must make. Yet challenges can be met only if they are thoroughly understood by the public that must face them. Education therefore is the critical ingredient to survival and prosperity, and an essential component of that education is the ability to reason with quantitative information. The writer H. G. Wells realized this 75 years ago when he said "Human history becomes more and more a race between education and catastrophe" (*The Outline of History*, 1920). Thus our ultimate purpose in writing this book is to present the quantitative skills and concepts that are necessary to meet the challenges of the modern world.

This book is designed primarily for those of you who are not planning additional course work in mathematics. The selection of topics, examples, and problems is driven by what we believe is necessary to make a person *quantitatively literate*. In particular, if you work carefully through this book, you should reach the following four goals:

1. You will be able to think critically about *any* quantitative issue covered in the news. For example, you should be able to understand and critically interpret *any* article in your local newspaper, and in more widely distributed publications such as *The New York Times* or *The Wall Street Journal*.
2. You will be able to make informed decisions on quantitative issues that confront you both in your personal life and in your life as a voting citizen. The former includes issues such as personal finance and planning; the latter includes public policy issues on topics such as the environment, economics, and public health.
3. You will be prepared with the quantitative skills needed for your subsequent college course work, particularly for courses in the natural and social sciences.
4. You will develop the ability to reason quantitatively, and to clearly explain and present your reasoning, so that you are prepared for a modern career. In the modern job market, both employees and employers must be prepared to use advanced technology, to work with quantitative information, and to respond flexibly to a rapidly changing economy.

As you can see, the emphasis in all these goals is the practical use of mathematics. We want you to become confident in applying mathematics to problems and issues in the world around you. You will learn about many topics that are covered in traditional mathematics classes, but you will study them in the context of issues that are personally and socially relevant. We hope that by the time you finish the book, you will view mathematics as exciting, relevant, and as a subject that is essential for a complete understanding of many other disciplines.

FEATURES OF THE BOOK

Your reading will be most effective if you take advantage of several special features of the book.

Chapter Preview

The *Chapter Preview* provides a brief "road map" to each chapter. Read the introductory paragraph and the preview to know what to expect from the chapter.

Examples

The book is heavily supported by examples and solutions designed to provide specific illustrations of skills and concepts. There are at least two ways to use the examples. Most of the time you probably will read the example/solution as a part of the text. Occasionally, however, you might want to treat them as exercises: read the example, try to solve the given problem, and then check your answer with the given solution.

Time-Out to Think

Throughout the text you will find short interludes called *Time-Out to Think*, which should prompt you to step back from the text and think about what you have read. Use them as one-minute exercises to help consolidate your understanding or think about the importance of the concepts under discussion.

Thinking About...

Occasionally there are topics that are not central to the main flow of the text, but which can help amplify or provide perspective on topics covered in the text. These topics, ranging from the plight of passenger pigeons to the history of numerology, appear in the *Thinking About* boxes. We urge you to take these fascinating excursions that give mathematics another dimension.

Chapter Conclusions

Each chapter ends with a short *conclusion* section, summarizing those intellectual threads that tie the chapter together.

Starred Sections

A few optional sections are included. These starred sections (*) help to complete the theme of the chapter, but may be skipped without hindering study in the remainder of the book. If you want to see where mathematical ideas can lead, read these sections.

Problems

Every chapter ends with a section devoted to problems, and each problem has a title that summarizes its general theme. Although there are more problems than anyone could work in a single course, reading through the problem section will help you to appreciate the many diverse applications of just a few basic mathematical ideas.

The problems appear in the following three categories:

1. **Reflection and Review.** This category of problems closely follows the flow of the chapter, and the problems are subdivided according to the section with which they are associated. Studying these problems will help you test yourself on the material presented in the chapter. Noting the section number with which a particular problem is associated can help you find the relevant discussion and examples in the text of the chapter.
2. **Further Topics and Applications.** These problems extend the ideas covered in the chapter to related applications not discussed in the text. Although their order corresponds roughly to the order in which chapter material is presented, they may require concepts from several sections or a synthesis of ideas from the entire chapter.
3. **Projects.** These problems generally require reading or research beyond the textbook; sometimes they require group work. The projects are more involved and open-ended than problems from the prior two categories.

Answers to Problems

Answers to selected problems appear in the back of the book. Answers are included for at least a few problems from every chapter and to every other problem from the Appendices. Because many problems have several possible answers, we generally provide only the straightforward portions of answers, and do not answer the discussion portions of these problems. We encourage you to attempt the discussion portions of all problems, including those to which we provide answers, because it will help you develop your understanding of the mathematical concepts and improve your critical reasoning skills.

HOW TO SUCCEED IN YOUR QUANTITATIVE REASONING COURSE

We have found that most students find this book very different from the textbooks they used in prior mathematics courses. Therefore, one of the keys to success in your quantitative reasoning course is to approach it with an open and optimistic frame of mind. We urge you to read Chapter 1 carefully, as it will help you to understand the philosophy of this book. In addition, the following hints may help you to succeed in this and other courses.

How Much Time Will Your Class Require?

In any college class, as a rule of thumb, you should expect to study about 2 to 3 hours per week *outside* of class for each unit of credit. For example, in a 3-credit course, you should expect to spend 6 to 9 hours per week studying outside of class. If you are spending fewer hours than the rule of thumb suggests, the course probably is too easy for you or you are not learning as much as you could. If you are spending more hours than the rule of thumb suggests, you may be studying inefficiently.

Note that a student taking 15 credit hours should expect to spend 30 to 45 hours each week studying outside of class. Combined with time in class, this works out to a total of 45 to 60 hours spent on academic work, which is not much more than the time required of a typical job (except that in college you more or less get to choose your own hours). Of course, if you are working while you attend school, you will need to budget your time carefully.

As a rough guideline, your 6 to 9 hours per week studying for your quantitative reasoning course might be divided as follows:

- 1 to 2 hours reading the text material assigned by your instructor.
- 3 to 5 hours completing the homework assignments.
- An average of about 2 hours for additional study (such as preparing for exams).

Everyone enrolled in college is capable of learning the material in this book. If you put in the time, and make efficient use of it, you will succeed in your quantitative reasoning course.

General Strategies for Studying

- Don't miss class. Listening to lectures and participating in discussions is much more effective than reading someone else's notes. Active participation will help you retain what you are learning.
- Budget your time effectively. An hour or two each day is more effective, and far less painful, than studying all night before homework is due or before exams. Also, if you procrastinate, it will be too late for you to ask for help from your instructor or friends.
- If a concept gives you trouble, do additional reading or problem solving beyond what has been assigned. Don't give up: keep working at it until you understand and overcome your difficulties.
- If you still have trouble, ask for help! Don't be embarrassed; you surely can find friends, colleagues, or teachers who will be glad to help you learn.
- Don't highlight — underline! Many students use highlight pens to mark portions of the text for later study. Unfortunately, highlighting can easily become "mindless," especially if you are tired. Using a pen or pencil to underline, instead of highlighting, requires greater care and therefore helps to keep you alert as you study.

- Working together with friends can be valuable in helping you to solve difficult problems. However, be sure that you learn *with* your friends and do not become dependent on them.
- Keep a journal to help you understand and appreciate what you are learning. See the section that follows later in this preface called *Keeping a Journal*.

Strategies for Using This Textbook

- On your first pass, read through the entire chapter quickly to gain a "feel" for the material and concepts presented.
- Reread the chapter in greater detail on your second pass, concentrating on the sections emphasized by your instructor.
- Use the margins! The wide left margins of this textbook are designed so that you can make notes to yourself as you study. Jot down your thoughts or insights so that you can easily return to them when it is time to study for an exam.
- Take advantage of the features of the textbook described earlier, such as *Time-Out to Think* and *Thinking About* boxes.
- Work through the examples to be sure you can obtain the answers on your own.
- Once you understand the examples, *then* do assigned problems.
- If you still have trouble with the assigned problems and feel you need additional practice, try a few more of the *Reflection and Review* problems that were described in the "Features" section above. If necessary, seek help from friends, colleagues, or teachers.

Preparing for Exams

- Rework problems and other assignments; try additional problems to be sure you understand the concepts. Study your performance on quizzes or exams from earlier in the semester.
- Study your notes from lectures and discussions. Pay attention to what your instructor expects you to know for an exam.
- Reread the relevant sections in the textbook, paying special attention to notes you have made in the margins.
- Study individually before joining a study group with friends. Study groups are efficient only if *every* individual comes prepared to contribute.
- Don't stay up too late before an exam. Don't eat a big meal within an hour of the exam (thinking is more difficult when blood is being diverted to the digestive system).
- Try to relax before and during the exam. If you have studied effectively, you are capable of doing well. Staying relaxed will help you think clearly.

Presenting Problem Solutions

We urge students in all classes to submit homework of *collegiate quality*: neat and easy to read, well-organized, and make good use of the English language (e.g., grammar, spelling) as well as mastery of the subject matter. Future employers and instructors will expect this quality of work. Moreover, although submitting homework of collegiate quality requires a great deal of effort, it serves two important purposes directly related to learning:

1. The effort you expend in clearly explaining your work solidifies your learning. In particular, research has shown that writing and speaking trigger different areas of your brain. By writing something down — even when you think you already understand it — your learning is reinforced by involving other areas of your brain.

2. By making your work clear and self-contained (that is, making it a document that you can read without referring to the questions in the text), it will be a much more useful study guide when you review for a quiz or exam.

The following guidelines will help ensure that your homework meets the standards of collegiate quality.

- All answers should be explained clearly in a form that can be understood without reference to the question. This doesn't mean you should rewrite the questions. Rather, it means that your solutions should be self-contained. Study the sample questions that follow and use your judgment.
- Be sure to *show your work* clearly in all calculations. By doing so, both you and your instructor can follow the process you used to obtain an answer. Without a step-by-step presentation, it may be impossible to determine where you went wrong if your answer is incorrect, and your homework would be useless as a study guide.
- Word problems should have word answers. That is, after you have completed any necessary calculations, any problem stated in words should be answered with one or more *complete sentences*. The purpose of these complete sentence answers is to demonstrate that you understand the point of the problem and the meaning of your solution.
- Your solutions should be well organized and easy to read. Always use proper grammar, proper sentence and paragraph structure, and proper spelling.
- Pay attention to details that will make your homework *look* good. For example,

  use standard 8.5 by 11 inch paper (never turn in pages torn from notebooks with ragged edges),

  staple all pages together so that your work will not become disordered or lost (don't use paper clips or folded corners that tend to come apart),

  use a ruler when you want to make straight lines,

  draw graphs on graph paper, and

  use illustrations whenever they can help to explain your answer.

- If possible, make your work look professional by using a word processor and by creating graphs and illustrations with a spreadsheet or other software. Many word processors allow easy creation of equations, but if you don't have access to a computer, print your work neatly.
- If you study with friends, be sure that you turn in your own work stated in your own words — it is important for you to avoid any possible *appearance* of academic dishonesty.

Sample Format for Problem Solutions

The following two sample questions are answered according to the preceding guidelines, followed by notes explaining the format of the solution.

Sample Problem 1 If you travel at 55 miles per hour for three hours, how far will you go?

Solution:

$$\left(55\ \frac{\text{mi}}{\text{hr}}\right)\times(3\ \text{hr}) = 165\ \text{mi}.$$

Traveling at 55 miles per hour, you would travel 165 miles in three hours.

Notes: The calculation shows how the answer was obtained; this is what we mean by "showing work." Note that units accompany each number, making it is easy to see the reasoning that led to the final answer. Also, note that the final answer is stated in words with a complete sentence.

Sample Problem 2 Explain the reasons why many colleges are creating courses in quantitative reasoning.

Solution: Over the past decade, colleges throughout the United States have wrestled with the question of how to teach mathematical ideas to students who do not plan to major in mathematics, engineering, or science. Traditional mathematics courses, such as algebra or geometry, are no longer considered adequate preparation for the kinds of issues that students will face in their future courses, careers, or lives as citizens. Instead, colleges seek new courses that will meet student needs in a rapidly changing world. Courses which present mathematics in the context of personal and societal issues, often called courses in *quantitative reasoning*, represent one common solution to the problem of meeting student needs in mathematics.

Note: This answer can be understood without referring to the question, even though the question has not been repeated verbatim. The answer is concise but still gives a clear and readable explanation.

Keeping a Journal

The work associated with this text involves more than just solving mathematical problems; it involves interpretation of quantitative concepts, often through discussion or writing. Devoting a bit of time each day to learning about quantitative issues in society, perhaps by reading newspapers, will help you understand and appreciate the relevance of the course material.

Keeping a journal can be an effective way to collect and record what you learn. Writing in a journal can help you clarify your thinking about new concepts, and the process of collecting ideas and information in your journal can help you keep organized. In addition, a written journal allows you to go back later and see how your thinking has evolved over time. All you need is a book with blank pages or a computer and disk, plus the time that you devote to journal entries. We suggest that you concentrate on three types of entries in a quantitative reasoning journal:

- Write short essays, personal reactions, or opinions about topics presented in the textbook or class. By placing these entries in the journal you will be assured of easy access to them (especially when it is time to study for an exam).
- Include news items, accompanied by a brief comment or calculation, that help you see the relevance of what you are learning. State your opinion of whether the news items are reported fairly or unfairly. Possible sources for entries include newspapers, magazines, television, radio, or the inventiveness of your own mind. For major news stories, editorials, and science articles we recommend *The New York Times* (especially the Science Times section that appears on Tuesdays). Advertisements can be quantitatively interpreted or critiqued for their logic. Even comic strips are fair game, as they sometimes require quantitative literacy to be fully appreciated.
- Personal entries documenting your attitudes toward the course and connections between your quantitative reasoning course and other classes may improve your motivation for learning.

Finally, creative freedom is the name of the game in your journal. Almost any topic can be molded into a good journal entry. You can share your journal with others, or you can keep it private. The more you use your journal as a learning tool, the greater the benefits you are likely to realize.

ACKNOWLEDGMENTS

Development of our quantitative reasoning course, and this textbook, began in 1987. Along the way we have been fortunate to have help from many people. The concept for the quantitative reasoning course was developed through the efforts of an interdisciplinary faculty committee at the University of Colorado chaired by John Williamson of the Department of Mathematics. Other committee members included Otomar Bartos (Sociology), Steve Bernstein (Biology), Gary Bradshaw (Psychology), Michael Breed (Biology), Greg Carey (Psychology), Gary Gaile (Geography), Andrew Hamilton (Astronomy), John Hodges (Mathematics), Robert MacRae (Mathematics), Richard Roth (Mathematics), and Tom Swain (Biology, University Learning Center). The first class was taught at

Colorado in fall 1988 by the author Jeffrey O. Bennett with the help of Steve Bernstein and Tom Swain. Following that successful trial, the program grew rapidly through the efforts of Dean Charles Middleton and Associate Dean J. Michael Shull of the College of Arts and Sciences. Dozens of undergraduate and graduate teaching assistants also helped us to develop this course. We particularly thank Hal Huntsman, John Supra, David Wilson, Megan Donahue, Mark Anderson, Debbie Segal, Bev Boydston, Jon Goldberg, Laura Loughry, Jon Dowling, Karen Herendeen, and Joe Kunches. In addition, Hal Huntsman, John Supra, and David Theobald devoted countless hours during their graduate student careers to helping us develop the course and working on an as-yet-unpublished computer laboratory curriculum.

We thank Jim Lightbourne and others at the National Science Foundation, which provided a grant through the *Undergraduate Course and Curriculum Development* program (NSF Grant USE-915060) that helped us develop a computer laboratory to accompany the course (not yet published) and the general pedagogy for the course.

Special thanks to our editors and friends at Addison-Wesley, especially Bill Poole, Elka Block, Christine O'Brien, Andy Fisher, Mary Clare McEwing, Greg Tobin, Linda Davis, Karen Guardino, Patsy DuMoulin, Sally Simpson, Susan Dainis, and Jenny Bagdigian.

Most of the illustrations in this book were created by Hal Huntsman, a graduate student at the University of Colorado who double majored in Mathematics and English. Huntsman, along with Rhodes Scholar David Wilson, also contributed numerous ideas for problems. Thanks also to Andy Danielson, who reviewed many of the problems and prepared many of the solutions that appear in the solutions manual for Instructors.

We greatly appreciate the tremendous efforts of our copy editor Jerry Moore. We thank all of the reviewers who have helped us revise and improve the manuscript including: Bob Bernhardt — East Carolina University, Walter Czarnec — Framingham State College, Marsha J. Driskill — Aims Community College, John Emert — Ball State University, Lynn R. Hun — Dixie College, Jim Koehler — University of Colorado at Denver, Timothy C. Swyter — Frederick Community College, Emily Whaley — DeKalb College, and Donald J. Zielke — Concordia Lutheran College. In particular we appreciate the efforts of the instructors who have agreed to class test the project at California State University at Monterey Bay, Essex Community College, La Guardia Community College, Marshall University, Millersville University, the University of Hawaii, the University of Texas at San Antonio, and the University of Wisconsin at Madison.

Finally, we thank the people who have put up with us while we have worked for years on this project, including our colleagues and students at the University of Colorado Boulder and Denver campuses, and most especially, Lisa, Julie, and Katy.

Because you are using a preliminary edition, you may find typos and other errors. We apologize for any inconvenience these errors may cause, and hope they will not significantly interfere with your enjoyment of the book.

Jeffrey O. Bennett
William L. Briggs
Cherilynn A. Morrow

Contents

Part 1 Thinking Logically

Part 2 Thinking Quantitatively

Part 3 Thinking Mathematically

PART 1

THINKING LOGICALLY

1 LITERACY FOR THE MODERN WORLD

What does Literacy for the Modern World mean? In short, literacy in today's world demands not only reading and writing, but also an understanding of fundamental concepts of logic, mathematics, and science. This book focuses on *quantitative* literacy — literacy in terms of information involving mathematical ideas or numbers. This aspect of literacy is fundamental to nearly every discipline of study, as well as to the complex issues that citizens routinely face in a modern democracy. In this chapter we help you begin your quest for quantitative literacy by describing its meaning and importance, both to your personal life and to society.

CHAPTER 1 PREVIEW:

1.1 INTERDISCIPLINARY THINKING. Modern issues can be understood only when examined from various perspectives. We examine this idea of interdisciplinary thinking in the first section.

1.2 WHAT IS QUANTITATIVE LITERACY? In this section we define quantitative literacy and discuss its importance to personal life, society, culture, and jobs.

1.3 CHALLENGING MISCONCEPTIONS ABOUT MATHEMATICS. The difficulties that many people experience with mathematics are rooted in several unfortunate misconceptions. We discuss some common misconceptions in this section.

1.4 WHAT IS MATHEMATICS? Having discussed what mathematics is not, we turn our attention to what it is. We look at mathematics in three ways: as the sum of its branches, as a way of modeling the world, and as a language.

1.5 THE ROAD TO QUANTITATIVE LITERACY. We end the chapter with a few suggestions that may help you achieve quantitative literacy.

Mathematics is the key to opportunity. No longer just the language of science, mathematics now contributes in direct and fundamental ways to business, finance, health, and defense. For students, it opens doors to careers. For citizens, it enables informed decisions. For nations, it provides knowledge to compete in a technological economy. To participate fully in the world of the future, America must tap the power of mathematics.

— *Everybody Counts, A Report to the Nation on the Future of Mathematics Education* (National Academy Press, Washington, D.C., 1989)

1.1 INTERDISCIPLINARY THINKING

Virtually every major issue in the modern world involves mathematical ideas. Environmental issues are studied with quantitative measurements and mathematical modeling on computers. Conflicts between nations are analyzed in terms of resource bases, economic policies, and the power of weapons. Economic decisions involve millions, billions, or trillions of dollars and require analysis of relative benefits and costs. Even personal decisions involve mathematics, as in choosing a health insurance policy or financing a new car.

We could argue similarly about disciplines other than mathematics. Historical context, fundamental ideas from psychology and sociology, and principles of physics also are essential to understanding current issues. The reality of the modern world is that important issues, whether personal or societal, are *interdisciplinary* in nature.

In fact, $E = mc^2$ can be derived with only high school algebra. We do so in Chapter 16.

An example of the interdisciplinary nature of human knowledge is Einstein's famous equation, $E = mc^2$ (where E stands for energy, m for mass, and c for the speed of light) and its social importance. Einstein derived this equation in 1905 as part of his *special theory of relativity*. The concept behind this particular arrangement of symbols is that mass, the amount of material in an object, is a form of energy. Thus, under certain circumstances, mass can be converted into energy and energy into mass.

Nuclear weapons present a dramatic and destructive example of the conversion of mass into energy, in accordance with Einstein's equation $E = mc^2$.

Einstein's equation helped scientists discover how the Sun produces energy and, soon thereafter, how to release nuclear energy. This discovery led to the invention of the atomic bomb. Einstein, an admirer of Ghandi who strongly believed in pacifism, was devastated that his simple discovery in 1905 led to so many deaths 40 years later when the atomic bombs were dropped on Hiroshima and Nagasaki. Yet the project to develop the atomic bomb started, at least in part, because Einstein argued for it in letters to President Roosevelt. Why would a pacifist argue in favor of developing a weapon of mass destruction? Einstein feared that Nazi Germany, with whom he had first-hand experience (he had lived in Germany until 1933), might develop the bomb first. Thus even in Einstein's own life, the mathematics and science he loved were irrevocably tied up with history, psychology, sociology, and politics.

Although nuclear tensions have diminished since the demise of the Soviet Union in 1991, the possibility of nuclear warfare remains one of the world's greatest social and political concerns. It dominates much of the political landscape as policymakers seek to prevent nuclear war, halt nuclear proliferation, and safeguard nuclear arsenals against terrorist attacks. The existence of weapons of mass destruction psychologically affects children everywhere. Many works of literature, film, music, and art from the second half of the twentieth century were influenced by fears of nuclear annihilation. The peaceful use of Einstein's equation in nuclear power plants also has an impact on the world. Nuclear power provides a significant fraction of the world's current energy, even while its continued use and expansion are the focus of many environmental and political battles.

Time-Out to Think: Do you support the use of nuclear power as an energy source? Why or why not? Do you feel that you understand enough about nuclear issues to clearly defend your opinion? Explain.

We cannot simply separate disciplines when discussing the impact of Einstein's discovery. Nevertheless, mathematics classes that teach algebra rarely make a connection to the physics that Einstein studied. Physics classes that teach $E = mc^2$ usually avoid discussing the political implications of nuclear weapons or nuclear energy. Political scientists who teach about arms control generally do not discuss the psychological impacts of the nuclear threat on children. English professors who teach about bomb-inspired works of literature virtually never discuss how simply Einstein was able to derive his equation or the beauty of that equation in the scheme of nature. To understand the importance of $E = mc^2$ requires a different approach to learning, an approach that recognizes how the myriad parts of human knowledge are interconnected. One of our tasks in this book is to help you develop such an integrated approach to knowledge. We hope that you will find it valuable and that it will help you integrate what you learn in other courses into a single, coherent, and personal understanding of the world in which you live.

THINKING ABOUT ... *MAKING EDUCATION INTERDISCIPLINARY*

In a democracy, which depends on an educated populace for strength, educational systems must offer formal education to everyone. For practical reasons, the teaching of so many people makes inevitable the compartmentalization of education. Two common methods of the compartmentalizing knowledge are by subject and by vocation (Figure 1-1). Despite their artificial nature, such subdivisions of knowledge can be remarkably effective; indeed, modern democracies have the highest overall literacy rates in human history.

The only danger in subdividing human knowledge for the sake of education occurs when this system of convenience is mistaken for reality. People falling into this trap might not grasp the importance of writing in a mathematics class, understanding history in a science class, or understanding science in an business class. The result is a lack of perspective, or inability to see the "big picture," which often leads to disastrous consequences: business failures when managers cannot anticipate future needs or environmental destruction when long-term benefits are overlooked for short-term gain.

The solution to these problems, fortunately, is quite simple. Interdisciplinary connections should be emphasized in every course of study. We encourage you, as a student, to seek such connections for yourself. And we hope that the myriad connections between mathematics and other disciplines will become clear as you read this book.

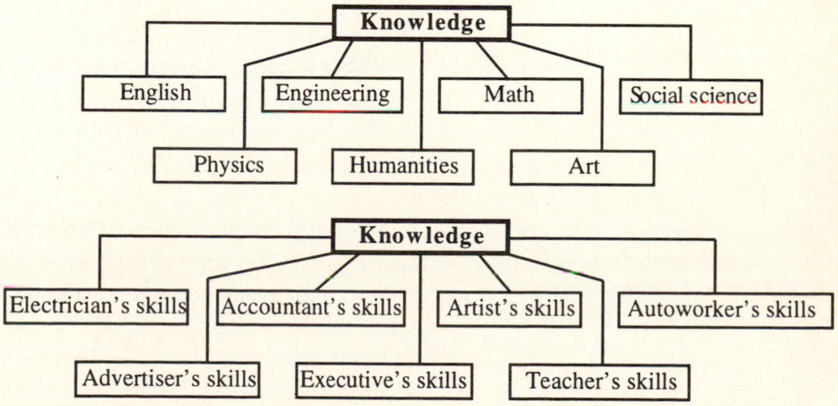

Figure 1-1. The job of education is to pass knowledge from one generation to the next. Here we show knowledge subdivided by two schemes: one based on traditional subjects and the other based on vocational skills. Can you think of another way to do it?

1.2 WHAT IS QUANTITATIVE LITERACY?

***Quantitative literacy* is a crucial aspect of literacy for the modern world. It entails the abilities to interpret and reason with information involving mathematics or numbers.**

***Quantitative reasoning* is the process of interpreting and reasoning with quantitative information.**

Literacy is the ability to read and write, and it comes in varying degrees. Some people can recognize only a few words and write only their names; others read and write in many languages. A primary goal of the educational system is to provide citizens with a level of literacy sufficient to read and write about the important issues of our time. In the modern world, the abilities to interpret and reason with **quantitative information** — information that involves mathematical ideas or numbers — are crucial aspects of this literacy. This so-called **quantitative literacy** is essential to comprehending modern issues that appear in the news everyday. The process of interpreting and reasoning with quantitative information is called **quantitative reasoning.**

1.2.1 Who Needs It?

Who needs quantitative literacy? Let's begin by thinking at the level of an individual, where quantitative literacy is a survival skill. Personal finance, for example, is increasingly complex in a world that requires many choices — whether to rent or buy, which telephone company to use, which health insurance policy to use, and which features to get in a personal computer.

The term *innumeracy* was popularized by the mathematician and writer John Allen Paulos in his book *Innumeracy, Mathematical Illiteracy and its Consequences* (Hill and Wang Publishers, New York, 1988).

A *lack* of quantitative literacy, or **innumeracy,** can therefore lead to both financial trouble and personal problems. A marriage may be strained if, for example, a couple purchases a home that is too expensive for their budget. For single college students, personal debt in the form of student loans or credit card balances can lead to stress or poor performance in school. Casino gambling and state lotteries can senselessly drain the resources of those who do not understand the underlying probabilities.

Time-Out to Think: Describe an instance in which a poor financial decision led to stress for you or your family.

At the societal level, quantitative literacy is crucial in a world where governments face decisions that influence not only the general well-being of their citizens, but also the chances of survival of civilization. Consider, for example, decisions about the use of nonrenewable resources, limiting environmental damage, devoting money to space exploration, or intervening in regional wars. These decisions inevitably involve quantitative information. If politicians and the people who elect them lack quantitative literacy, how can sensible policies be formulated and achieved?

Time-Out to Think: Think of a current issue facing the United States that will have a major impact on your life. How is quantitative information involved in this issue? Do you think that politicians, or the public, understand the issue well enough to make a wise decision? Why or why not?

Mathematical illiteracy is both a personal loss and a national debt. — from *Everybody Counts* (National Academy Press, Washington, D.C., 1989, p. 18)

A related consequence of innumeracy — usually involving a misunderstanding of logic, probability, or statistics — is an inability to distinguish between legitimate science and pseudoscience or fraudulent science. A great deal of study is not needed for someone to realize that astrology has no scientific basis, that evolution is a theory supported by an abundance of evidence, and that experimental evidence for telepathy is nonexistent. Nevertheless, many people lose a great deal of time and money to charlatans and practitioners of pseudoscience. Society also pays a heavy price when, for example, school biology curricula present incomplete or inaccurate discussions of concepts of evolution, health, or nutrition. So who needs quantitative literacy? Everyone!

1.2.2 Quantitative Literacy and Culture

A ROCK, A RIVER, A TREE
Hosts to species long since departed
Marked the mastodon,
The dinosaur, who left dry tokens
Of their sojourn here
On our planet floor,
Any broad alarm of their hastening doom
Is lost in the gloom of dust and ages.

— from *On the Pulse of the Morning,* by Maya Angelou

Quantitative literacy enriches the appreciation of both ancient and modern culture. The historical record shows that all great cultures devoted substantial energy to mathematics and to science (or to observational and theoretical methods that predated modern science). Without a sense of how quantitative concepts are used in art, architecture, and science, you cannot fully appreciate the incredible achievements of the Mayan civilization of Central America, the builders of the great city of Zimbabwe in Africa, the ancient Egyptian and Greek civilizations, the ancient Polynesians with their navigational expertise, or countless other cultures.

Quantitative literacy also can help you appreciate literature. Writer Mark Twain frequently wrote about scientific innovations and often used deliberately invalid quantitative extrapolations to make humorous points in his essays and novels. Poets Maya Angelou, Walt Whitman, Robert Frost, T. S. Eliot, John Milton, J. W. Goethe, William Wordsworth, and W. H. Auden, among others, allude to quantitative and scientific ideas in their literary works.

Time-Out to Think: How can quantitative understanding enhance appreciation of the excerpt from a poem by Maya Angelou? For example, does knowing that dinosaurs became extinct 65 million years ago affect how a person responds to the poem? Does putting 65 million years into perspective make a difference? Why or why not?

Many historical figures known best for their achievements in art, literature, or politics made substantial contributions to mathematics, science, and technology. For example, the great Renaissance painter Leonardo Da Vinci made important discoveries in science and engineering, and incorporated observations of nature and mathematical concepts (especially geometry) into his artistic masterpieces.

Benjamin Franklin, best known as a statesman, conducted important experiments with electricity and lightning, and wrote the founding document of the first scientific society in America. Another well-known statesman, Thomas Jefferson, was an ardent student of science and deeply involved in scientific farming, meteorological measurements, and the collection and classification of fossils. Theodore Roosevelt used a break between presidential campaigns to write a scientific paper about the ecology of the Amazon.

Theodore Roosevelt was president of the United States from 1901 to 1909 and a presidential candidate in 1912 and 1916. In 1914, he led an expedition along the Amazon river in Brazil and wrote a paper on the coloration of birds and mammals.

History reveals that most great contributors to Western culture were versed in mathematics and that quantitative literacy is beneficial to personal achievement in any field of endeavor.

Most of the names and information on this list are from an article by Steven G. Buyske ("Famous Nonmathematicians," *American Mathematical Monthly,* November 1993).

THINKING ABOUT ... LOOK WHO STUDIED MATHEMATICS!

The training in critical thinking provided by the study of mathematics is valuable in many careers. The following list is only a small sample of people who studied mathematics, but became famous for work in other fields.

Corazon Aquino, former president of the Philippines. A mathematics minor.

Harry Blackmun, Supreme Court justice. Summa cum laude in mathematics, Harvard University.

Lewis Carroll, author of *Alice in Wonderland*. A mathematician.

David Dinkins, former mayor of New York City. BA in mathematics, Howard University.

Alberto Fujimori, president of Peru. MS in mathematics, University of Wisconsin.

Art Garfunkel, musician. MA in mathematics, Columbia University.

Mae Jemison, first African-American woman in space. Studied mathematics as part of her degree in chemical engineering from Stanford University.

Omar Khayyam, author of *The Rubaiyat*. Also wrote on algebra and geometry.

John Maynard Keynes, economist. MA in mathematics, Cambridge University.

Lee Hsien Loong, politician in Singapore. BA in mathematics, Cambridge University.

Edwin Moses, three-time Olympic Champion in the 400-meter hurdles. Studied mathematics as part of his degree in physics from Morehouse College.

Florence Nightingale, pioneer in nursing. Studied mathematics and applied it to her work.

William Perry, secretary of defense. Ph.D. in mathematics, Pennsylvania State University.

Sally Ride, first American woman in space. Studied mathematics as part of her Ph.D. in physics from Stanford University.

David Robinson, basketball star. Bachelor's degree in mathematics, U.S. Naval Academy.

Alexander Solzhenitsyn, Nobel prize–winning Russian author. Degrees in mathematics and physics from the University of Rostov.

Bram Stoker, author of *Dracula*. Studied mathematics at Trinity University, Dublin.

Laurence Tribe, Harvard law professor. Summa cum laude in mathematics, Harvard University.

Virginia Wade, Wimbledon champion. Bachelor's degree in mathematics, Sussex University.

1.2.3 Quantitative Literacy in the Work Force

Quantitative literacy also is important in the work force, as a lack of sufficient quantitative skills closes off many of the most challenging and highest paying jobs. It is especially important today, as satisfying and secure employment is

Jobs are becoming more demanding, more complex. But our schools don't seem up to the task. They are producing students who lack the skills that business so desperately needs to compete in today's global economy. And in doing so, they are condemning students to a life devoid of meaningful employment. — from *Education: The Knowledge Gap*, supplement to the *Wall Street Journal*, February 9, 1990.

increasingly difficult to find. A few years ago, a college degree in any major field of study virtually guaranteed a well-paying job. Today, however, only a small fraction (as small as 10 to 20%) of college graduates find work in their major fields. Many college graduates are forced to take low-paying jobs in service industries or are unemployed after graduation.

Even if you are fortunate enough to land a good job in your major field, the odds are high that you will not keep that job for life. In the future the average American worker is expected to change jobs a dozen times and make a complete career change four to five times. This outlook applies even to the most talented individuals. Consider, for example, business executives who switch to completely different types of companies (in 1993, a former chief executive of a tobacco and food company was chosen to head the computer company IBM).

Because of the changing nature of work and the prospect of not finding a job in your chosen field, limiting yourself to a narrow line of study is risky. The broader your knowledge and your skills, the more likely you are to find satisfying employment. Indeed, today's world is so interdisciplinary that most businesses seek employees with broad training. Quantitative literacy can be an important discriminator in a competitive field of applicants.

To appreciate the skills required for various jobs, study Tables 1-1 through 1-3 on pages 8 and 9. Table 1-1 defines a set of language and mathematics skill levels. The skills required for selected jobs are shown in Table 1-2. Table 1-3 compares the skill levels of jobs in 1990 with those expected for jobs in 2000.

Time-Out to Think: What are your current skill levels, as defined in Table 1-1? What do you hope your skill levels will be when you graduate from college?

In this text we cover all the mathematics skills for levels 1 through 3, most of the skills for level 4, and some of the skills for level 5 (including an introduction to statistics and some of the concepts of calculus). In addition, the training you will gain in quantitative reasoning should help your critical thinking skills and thereby improve your language skills as well.

Calvin and Hobbes **by Bill Watterson**

Tables 1-1 through 1-3 are adapted from a chart in *Education: The Knowledge Gap*, supplement to the *Wall Street Journal*, February 9, 1990.

Table 1-1. Skill Levels

Level	Language Skills	Math Skills
1	Recognizes 2500 two- or three-syllable words. Reads at a rate of 95-120 words per minute. Writes and speaks simple sentences.	Adds and subtracts two-digit numbers. Does simple calculations with money, volume, length, and weight.
2	Recognizes 5000-6000 words. Reads 190-215 words per minute. Reads adventure stories and comic books, as well as instructions for assembling model cars. Writes compound and complex sentences with proper grammar and punctuation.	Adds, subtracts, multiplies, and divides all units of measure. Computes ratio, rate, and percentage. Draws and interprets bar graphs.
3	Reads novels and magazines, as well as safety rules and equipment instructions. Writes reports with proper format and punctuation. Speaks well before an audience.	Understands basic geometry and algebra. Calculates discount, interest, profit and loss, markup, and commissions.
4	Reads novels, poems, newspapers, and manuals. Prepares business letters, summaries, and reports. Participates in panel discussions and debates. Speaks extemporaneously on a variety of subjects.	Deals with complex algebra and geometry, including linear and quadratic equations, logarithmic functions, and axiomatic geometry.
5	Reads literature, book and play reviews, scientific and technical journals, financial reports, and legal documents. Can write editorials, speeches, and critiques.	Knows calculus and statistics; able to deal with econometrics.
6	Same types of skills as level 5, but more advanced.	Works with advanced calculus, modern algebra, and statistics.

In Table 1-2, note that the occupations requiring higher skill levels are generally the most prestigious and highest paying. Data from "Education: The Knowledge Gap," supplement to *The Wall Street Journal*, 2/9/90.

In addition, note that most of these occupations require high skill levels in both language and math, refuting the myth that if you're good at language you don't have to be good at mathematics, and vice versa.

Table 1-2. Skill-Level Requirements (1990)

Occupation	Language Level	Math Level	Occupation	Language Level	Math Level
Biochemist	6	6	Corporate executive	4	5
Computer engineer	6	6	Computer sales agent	4	4
Mathematician	6	6	Management trainee	4	4
Cardiologist	6	5	Insurance sales agent	3	4
Social psychologist	6	5	Retail store manager	3	4
Lawyer	6	4	Cement mason	3	3
Tax attorney	6	4	Dairy farm manager	3	3
Newspaper editor	6	4	Poultry farmer	3	3
Accountant	5	5	Tile setter	3	3
Personnel manager	5	5	Travel agent	3	3
Corporate president	5	5	Telephone operator	3	2
Weather forecaster	5	5	Janitor	3	2
Secondary teacher	5	5	Short-order cook	3	2
Elementary teacher	5	3	Assembly-line worker	2	2
Disk jockey	5	3	Toll collector	2	2
Financial analyst	4	5	Laundry worker	1	1

Table 1-3 assumes the same level for both language and math skills. Column totals do not add to 100% because each category was rounded to the nearest percent.

Note that 41% of jobs in 2000 are expected to require skills at level 4 or above. Further, the percentage of jobs available to unskilled workers (levels 1 and 2) will be much smaller than in 1990.

Table 1-3. Present and Future Job Availability by Skill Level

Skill Level	Percentage of Jobs in 1990	Percentage of Jobs in 2000
1	9	4
2	31	23
3	35	34
4	21	28
5	5	11
6	1	2

1.2.4 Learning to "Drive" Mathematics

How much mathematics do you need to know to qualify as quantitatively literate and, in particular, as someone who can use quantitative reasoning? Clearly, we can't expect everyone to use mathematics like a professional mathematician, just as we wouldn't expect everyone to play music like a concert musician, or to write award-winning poetry and prose. However, every citizen should enjoy some familiarity with mathematics, music, and literature. An analogy between working with mathematics and working with cars may help clarify the meaning of quantitative literacy.

For the purpose of this analogy, we call the highest level of working with mathematics **theoretical mathematics.** People working at this level discover new principles and advance the frontiers of knowledge. Analogous to working with cars, this level corresponds to discovering entirely new principles for automobile design, such as new aerodynamic shapes or new ideas for generating power. The people working with cars at this level usually are Ph.D. engineers or scientists working in research labs for the government or industry.

We call the next level in our analogy **applied mathematics.** At this level, known mathematical tools are applied to problems of immediate interest, such as designing efficient quality control methods for manufacturing, analyzing risk for insurance policies, developing mathematical models to assess human impact on the environment, or teaching mathematics to the next generation. Analogous to working with cars, applied mathematics corresponds to the design and building of a car, where advanced technologies are applied in developing a new model automobile. The people working with cars at this level usually are engineers who work as members of an automobile design team.

We call the next level in our analogy **vocational mathematics.** Vocational mathematicians use mathematical tools routinely. For example, a computer programmer must translate mathematical ideas into computer languages. Accountants and bankers use mathematical methods to analyze financial records and investment strategies. Any successful businessperson depends on mathematics to analyze inventories, devise purchasing schemes or modify pricing strategies. Analogous to working with cars, vocational mathematics corresponds to creatively identifying problems and repairing automobiles. The people working with cars at this level are professional auto mechanics. They generally do not discover new principles or apply principles in new ways, but they understand the construction of cars and can repair them very well.

The fourth level of mathematical understanding, quantitative literacy, is necessary for everyone. Analogous to working with cars, quantitative literacy is all the things you must know about driving. It involves operating the accelerator and brakes, steering, being aware of others on the road, being able to do simple repairs, and knowing the laws and courtesy of the road. Just as these basic driving skills are needed for survival on the highway, quantitative literacy is a survival skill in today's technological society. Table 1-4 summarizes the analogy.

Table 1-4. Working with Mathematics: An Analogy to Working with Cars

Mathematics	Car Analogy
Theoretical mathematics (mathematician or theoretical scientists)	Fundamental principles that apply to cars (Ph.D. engineer or scientist in research lab)
Applied mathematics (engineer, scientist, teacher, statistician, or business analyst)	Design and construction of cars (engineers on automobile design teams)
Vocational mathematics (computer programmer, accountant, or statistician)	Professional auto mechanic
Quantitative literacy	Driving (and all the skills and concepts required to drive well)

THINKING ABOUT ... APPLYING MATHEMATICS

Applied mathematics, in which mathematics is developed or used in practical applications, sometimes is distinguished from theoretical mathematics, or mathematics for its own sake. In ancient times, the development of mathematics was driven by practical needs in agriculture, architecture, engineering, navigation, and taxation. Religious and spiritual concerns also drove mathematics; many mathematical techniques of geometry and trigonometry were developed to aid in charting the motions of the heavens for astronomy and astrology. During the European Renaissance, mathematics became the primary tool of science and engineering. It also played a prominent role in art: Artists including Leonardo Da Vinci and Michelangelo made extensive use of geometry in their work.

Today, mathematics is used in virtually every field of study (see Figure 1-2), including those outside the traditional sciences. For example,

- World War II gave birth to the mathematics of operations research, in which optimum strategies are sought for business, policy, and management decisions;
- economic theories have become increasingly quantitative, and many now take the form of complex mathematical models;
- mathematical statistics provides powerful descriptive and predictive tools in public health, sociology, and psychology;
- mathematics provides the language for the design and use of computers, which in turn are used to enlarge the boundaries of mathematics;
- developments in computer science have led to robotics and artificial intelligence;
- mathematical modeling methods, originally developed for the natural sciences, now are applied to problems as diverse as the spread of epidemics, the management of ecosystems, and the study of development of civilizations;
- mathematical theories of data management are crucial to consumer applications such as credit card billing systems and airline and rental car reservation systems;
- the mathematics of networks (see Chapter 12) is used to analyze telecommunications systems and design complex routing systems for airlines and delivery services.

Of course, some mathematicians always are working with no specific motivation other than to expand mathematical knowledge. However, in every case yet encountered, such work eventually has found practical use. One of the best examples is the use of nineteenth century ideas in group theory, or abstract algebra. Once thought to have no practical applications, group theory is now used by physicists to explain fundamental properties of subatomic particles. More recently, number theory, which we discuss briefly in Chapter 6, has been used to design codes and security systems; topology, the mathematical study of shapes, has been used to model the structure of large protein molecules; and mathematical theories of sets and logic have led to "fuzzy logic" with applications in artificial intelligence, and the operation of car brakes and video cassette recorders. This transfer of abstract mathematics to practical problems is likely to continue at an accelerated rate in the future, further blurring the boundaries between theoretical and applied mathematics.

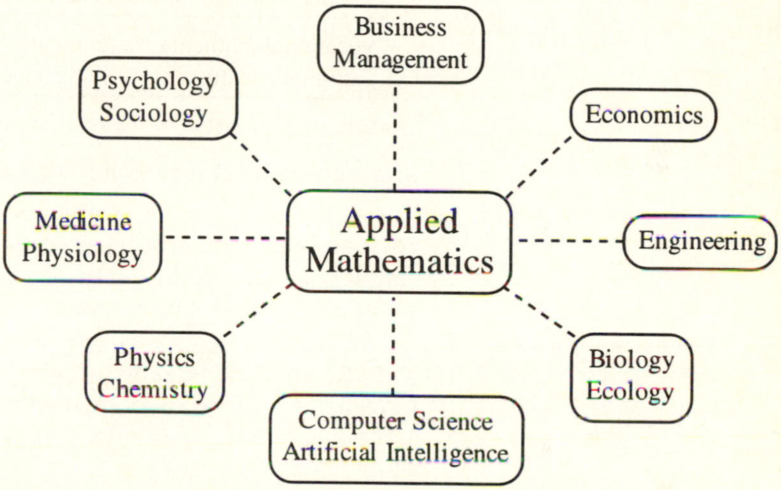

Figure 1-2. Applied mathematics can be viewed as a central resource for addressing and solving problems in a wide and growing variety of disciplines.

1.3 CHALLENGING MISCONCEPTIONS ABOUT MATHEMATICS

You definitely aren't alone if you consider yourself to be "math phobic" or "math loathing." Many people hold these attitudes and unfortunately, the fear and loathing are often reinforced by classes that present mathematics as an obscure and sterile subject. We hope that as you study this text you will learn that mathematics is neither especially difficult nor dull. Instead, you may find that it can be enjoyable, interesting, and relevant to your life.

THE FAR SIDE By GARY LARSON

Math phobic's nightmare

Indeed, experience with teaching thousands of students suggests that barriers to learning mathematics are more often psychological than intellectual.

Time-Out to Think: How would you describe your current attitude toward mathematics? Do you consider yourself to be either "math phobic" (fear of mathematics) or "math loathing" (dislike of mathematics)? If so, how do you think your attitude towards mathematics formed, and what will it take to change it?

As a result, the first step in learning to "drive" mathematics involves confronting these psychological barriers. We begin this task by challenging several common misconceptions about mathematics — the very misconceptions that often lead to mental barriers against mathematics.

Misconception One: Math Requires a Special Brain

One of the most pervasive misconceptions is that some people just aren't good at mathematics because learning mathematics requires special and rare abilities. The reality is that nearly everyone can do mathematics, although to do so may require self-confidence and hard work.

Equity for all requires excellence for all; both thrive when expectations are high. — from *Everybody Counts* (National Academy Press, Washington, D.C., 1989, p. 29)

Why should anyone think it otherwise? Years of work are required to learn to read, master a musical instrument, or become skilled at a sport. Indeed, the belief that mathematics requires special innate talent, and is accessible to only a few elite people, is peculiar to the United States. Elsewhere, particularly in European and Asian school systems, all students are expected to become highly proficient in mathematics. Quantitative literacy is but the first essential step.

Of course, different people learn mathematics at different rates and in different ways — for example, some people by concentrating on concrete problems, others by thinking visually, and still others by thinking abstractly. No matter what type of thinking style you prefer, you can succeed in mathematics.

Time-Out to Think: Do you believe that men and women tend to hold different attitudes toward mathematics? If so, why? Do you think misconceptions play a role in those differences? If not, how would you refute the argument of a friend who claims that differences *do* exist?

The large number of successful women mathematicians should be proof enough that gender doesn't determine mathematical ability; the same can be said for race and culture, as successful mathematicians come from every ethnic and cultural background. Any apparent differences in mathematical proficiency between ethnic or gender groups can be attributed largely to differences in opportunity and training, not inherent ability.

Misconception Two: Math in Modern Issues Is Too Complex

A common belief is that the advanced mathematical concepts underlying many modern issues are too complex for the average person to understand. Although only a few people receive the training needed to work with or discover advanced mathematical concepts, most people are capable of understanding enough about the mathematical basis of important issues to develop informed and reasoned opinions.

You may already recognize this difference in other fields. For example, years of study and practice are required to become a proficient, professional writer; but most people can read a book. To become a lawyer requires hard work and law school, but many people can understand how the law affects them in a particular situation. And though few have the musical talent of a Mozart, anyone can learn to appreciate his music. Mathematics is no different. If you can read this book, you can understand enough mathematics to succeed as an individual and a concerned citizen. Don't let anyone tell you otherwise!

Emotions, beliefs, attitudes, dreams, intentions, jealousy, envy, yearning, regret, longing, anger, compassion, and many others. These things—the inner world of human life — can never be mathematized. — from "The Limits of Mathematics" by P. J. Davis and R. Hersh in *The World Treasury of Physics, Astronomy,* and *Mathematics,* edited by Timothy Ferris (Little, Brown & Co., Boston, 1991)

Misconception Three: Math Makes You Less Sensitive

Some people believe that learning mathematics will somehow make them less sensitive to the romantic and aesthetic aspects of life. Although a person may be indifferent to a beautiful sunset or ignorant of artistic beauty, the study of mathematics can't be blamed. Indeed, understanding the underlying mathematics that produces the colors of the sunset or the geometric beauty in a work of art can actually enhance aesthetic appreciation. Furthermore, many people find beauty and elegance in mathematics itself. It is no accident that, as we have already mentioned, many people trained in mathematics have made important contributions to art, music, and many other fields.

Misconception Four: Math Makes No Allowance for Creativity

The "turn the crank" nature of the problems in many textbooks may give the impression that mathematics stifles creativity. Some of the facts, formalisms, and skills required for mathematical proficiency are fairly cut and dried; but using these mathematical tools demands creativity. Consider designing and building a home. The task demands specific skills to lay the foundation, frame in the structure, install plumbing and wiring, and paint walls. But building the home involves much more: It requires creativity to develop the architectural design, to respond to on-the-spot problems during construction, and to factor in constraints based on budgets and building codes. The mathematical skills you've learned in school are like the skills of carpentry or plumbing. But the essence of applying mathematics is like the creative process of building a home.

Probably the most harmful misconception is that mathematics is essentially a matter of computation. Believing this is roughly equivalent to believing that writing essays is the same as typing them. — from *A Mathematician Reads the Newspaper* by John Allen Paulos (HarperCollins, New York, 1995)

Misconception Five: Math Provides Exact Answers

A mathematical formula may yield a specific result, and in school that result may be marked right or wrong. But when you use mathematics in real situations, answers are never so clear-cut. For example:

> A bank offers simple interest of 5%, paid at the end of one year (that is, after one year the bank pays you 5% of your account balance). If you deposit $1000 today and make no further deposits or withdrawals, how much will you have in your account after one year?

A straight mathematical calculation *seems* simple enough: 5% of $1000 is $50; so you should have $1050 at the end of a year. But will you? How will your balance be affected by service charges or by taxes on interest earned? What if the bank fails? What if the bank is located in a country in which the currency collapses during the year? Choosing a bank in which to invest your money is a *real* mathematics problem. It is a problem that doesn't necessarily have a simple or definitive solution.

Another aspect of this misconception is the belief that numbers must always be exact in mathematics. For example, if asked to calculate 100,000 + 1, you will probably answer 100,001. But:

> According to the census your hometown population is 100,000. A friend of yours moves to town. What is the new population?

Answering 100,001 would be incorrect. Why? Because any population census must be an approximation. Making an exact count of 100,000 people in a city is impractical, if not impossible. Aside from the difficulties of finding and counting everyone, people will be born, will die, and will move in or out of town during the time needed to conduct a census. Real mathematics problems require creative strategies for making useful estimates and approximations.

Misconception Six: Math Is Irrelevant to My Life

By now, we hope that enough has been said to dispel the myth that mathematics is useless in everyday life. Understanding and applying mathematics, at least at the level of quantitative literacy, is important to personal enrichment, satisfying employment, and good citizenship.

> **Time-Out to Think:** Do you harbor any of these misconceptions? If so, how did they develop? Can you think of any other common misconceptions about mathematics?

1.4 WHAT *IS* MATHEMATICS?

The word *mathematics* is derived from the Greek word *mathematikos*, which means "inclined to learn."

In discussing misconceptions we identified what mathematics is not; now, let's describe what mathematics is. The word *mathematics* is derived from the Greek word *mathematikos*, which means "inclined to learn." Thus, literally speaking, to be mathematical is to be curious, open-minded, and interested in always learning more. A root as appealing as this one makes especially tragic the fear or boredom that so many people associate with mathematics.

Of course, in the modern world the word *mathematics* has other connotations. In this section, we look at mathematics in three different ways: as the sum of its branches, as a way to model the world, and as a language.

1.4.1 Mathematics as the Sum of its Branches

As you progressed through school, you undoubtedly learned to associate mathematics with some of its branches. Among the better known branches of mathematics are:

- **logic** — the study of principles of reasoning;
- **arithmetic** — methods for operating on numbers;
- **algebra** — methods for working with unknown quantities;
- **geometry** — the study of size and shape;

- **trigonometry** — the study of triangles and their uses;
- **probability** — the study of chance;
- **statistics** — methods for analyzing data; and
- **calculus** — the study of quantities that change.

One way to view mathematics is as the sum of its branches. Indeed, most "traditional" mathematics books focus on one branch of mathematics at a time. In this book, however, we do not do so except to use a tool or technique from one of the branches to help you understand or solve a problem. Nevertheless, familiarity with the various branches of mathematics is helpful, even if only to know when you are using them.

1.4.2 Mathematics as a Way to Model the World

Mathematics also may be viewed as a tool for creating **models,** or representations that allow us to study real phenomena. Modeling is not unique to mathematics. A road map, for example, is a model that represents the roads in some region.

Mathematical models can be as simple as a single equation that predicts how the money in your bank account will grow or as complex as a set of thousands of interrelated equations and parameters used to represent the global climate. By studying models, you can gain insight into otherwise intractable problems. A global climate model, for example, can help you understand weather systems and allows you to ask "what if" questions about how human activity may affect the climate. Further, when a model is used to make a prediction that does not come true, it can point to areas where further research is needed.

Mathematical models take many different forms, and you are probably familiar with many of them: Tables, graphs, and equations are just a few examples. In this text, we emphasize recognizing when and how mathematics is used to make models. We also cover numerous techniques for building mathematical models and using them to study meaningful problems. Of course, just as "a map is not the territory," models are not the real thing. Models are a valuable tool for studying a problem, but they are only as good as the equations and observations from which they are made. We therefore focus on both the value and limitations of mathematical models.

Skills are to mathematics what scales are to music or spelling is to writing. The objective of learning is to write, to play music, or to solve problems — not just to master skills. — from *Everybody Counts* (National Academy Press, Washington, D.C., 1989, p. 57)

1.4.3 Mathematics as a Language

A third way of looking at mathematics is as a language, with its own vocabulary and grammar. Indeed, mathematics often is called "the language of nature" because it is so useful for modeling the natural world. Like any language, different degrees of fluency are possible. Again, quantitative literacy is the level of fluency required for success in today's world.

The idea of mathematics as a language also is useful in thinking about how to *learn* mathematics. Table 1-5 on the next page shows a useful analogy between learning a language and learning mathematics, along with a comparison to learning art.

1.5 THE ROAD TO QUANTITATIVE LITERACY

We've spent most of this first chapter discussing the meaning of quantitative literacy and the nature of the mathematics that lies at its heart. We spend the rest of the book helping you become quantitatively literate by learning to reason quantitatively. Before we begin, let's think a bit about how best to proceed.

Table 1-5. Learning Mathematics: An Analogy to Language and Art

Learning a language	Learning the language of art	Learning the language of mathematics
Become familiar with many styles of speaking and writing such as essays, poetry, and drama.	Become familiar with many styles of art such as classical, renaissance, impressionist, and modern.	Become familiar with techniques from many branches of mathematics such as arithmetic, algebra, and geometry.
Place literature in context by studying its history and the social conditions in which it was created.	Place art in context by studying its history and the social conditions under which it was created.	Place mathematics in context by studying its history, purposes, and applications.
Learn the elements of language — such as words, parts of speech (nouns, verbs, etc.), and rules of grammar — and practice their problems.	Learn the elements of visual form — such as lines, shapes, colors, and textures — and practice using them in your own art work.	Learn the elements of mathematics — such as numbers, variables, and operations — and practice using them to solve simple problems.
Develop the ability to analyze language in complex forms critically — novels, short stories, essays, poems, speeches, debates, and similar works.	Develop the ability to analyze works of art critically — painting, sculpture, architecture, photography, and similar works.	Develop the ability to analyze quantitative information critically — mathematical models, statistical studies, economic forecasts, investment strategies, and similar works.
Use language creatively for your own purposes, such as writing a term paper or story, or engaging in debate.	Use your sense of art creatively, such as in designing your house, taking a photograph, or making a sculpture.	Use mathematics creatively to solve important problems or to create models for understanding some aspect of reality such as the climate or economic trends.

Table 1-5 draws on and extends to mathematics an analogy between language and art suggested by Edmund Burke Feldman in *Thinking About Art* (Prentice-Hall, Englewood Cliffs, New Jersey, 1985)

1.5.1 Identify Your Personal Goals

You can clarify your approach to quantitative reasoning by thinking about how it fits your personal goals. Ask yourself what you hope to gain from this course and how it can improve your life.

Goal: an end toward which effort is directed.

Strategy: a plan or method for achieving a goal.

It is important to distinguish between *goals* and *strategies*. For example, you might set a goal of developing confidence in your ability to do mathematics. Your strategy for achieving this goal might include attending all classes, studying the text, completing homework, and allowing plenty of time to prepare for exams.

Let your strategy, and even your goals, evolve as you learn more about the course and your own abilities. Remember that failure of a particular strategy need not prevent you from achieving a goal. Many students achieve far more than they once thought possible. Achieving quantitative literacy may even lead you in new study and career directions. Perhaps you have denied yourself a career path you might otherwise enjoy because you doubted your ability to develop the needed mathematical skills. Believe in yourself, constantly re-examine your goals and strategies, and see how far you can go.

THINKING ABOUT ... PERSONAL GOALS AND STRATEGIES

Mathematics will feel far more relevant if you first examine how it fits your personal goals. Because distinguishing between goals and strategies is important, here are a few examples.

- The goal of attending college is usually to gain an education. The strategy involves selecting courses and a major. A problem may arise if the goal and strategy are confused. For instance, some students focus only on learning isolated pieces of a subject (often the pieces that will be on an exam), losing sight of the goal of gaining knowledge.
- An individual may have a goal of helping children grow up healthy. As a strategy, that person may decide to become a pediatrician (medical doctor for children). However, if the individual is not accepted to medical school, the

goal is not necessarily lost. The person might choose an alternative strategy to reach the overall goal, such as becoming a social worker to help parents learn to care for their children or a police officer to help combat violence.

- Many students say that their goal in college is to graduate. In fact, college graduation actually is a strategy for achieving larger goals. Those goals might include becoming an educated person, learning about a variety of subject areas in order to find what interests you most, acquiring the training you need for a chosen career path, or demonstrating to yourself that you are capable of hard work and achievement. If you have such a goal, the act of graduating from college is simply a strategy for achieving that goal.

1.5.2 Break Down Your Psychological Barriers

Perhaps the single most important step that you can take is to break down any psychological barriers that get in the way. As we have already discussed, nearly everyone can achieve quantitative literacy. Nevertheless, it takes courage — you will be working with mathematical skills that may have tormented you in the past, and everyone inevitably makes mistakes along the way. If you persevere and stay focused on your goals, you will succeed.

Also you're not alone on this road. Your classmates will face the same challenges; work with them to develop your skills and understanding together. In addition, try to work with your instructors, not against them — they share the goal of seeing you achieve quantitative literacy. Finally, put a little faith in this book. It is the product of many years of experience working with students like you. If you follow the basic plan, study the examples, and work the problems, you will be amazed at what you learn and at how much fun you can have in the process.

1.5.3 Set Your Course

The last step before embarking on your journey is briefly to chart the course. This book is organized into four broad parts, each consisting of several chapters. Regardless of how your instructor chooses to move through them, they provide a useful framework for success.

Part 1 focuses on logical thinking, which is necessary to understanding and evaluating quantitative information, as well as to problem solving. In Part 2, we focus on different ways in which numbers are used in real life; there, you will begin to discover the true meaning of quantitative literacy. In Part 3, we extend our mathematical tools and thereby broaden the scope of problems that you can understand and solve. Finally, in Part 4, we devote special attention to some of the most important quantitative topics in modern society, including probability, statistics, and the growth of populations.

Within the context of this general structure, you will encounter many different applications along the way. To be conversant about the environment or the economy can be empowering and enjoyable, as can being confident of your ability to make wise decisions in the voting booth or knowing that you can become a better bridge or poker player. Or you might simply be satisfied to be free from fear when a child asks for help in figuring out a mathematics problem brought home from school. Good luck!

1.6 CONCLUSION

At this point in a chapter, it is worthwhile to pause and reflect on what we have covered. We therefore end every chapter with a Conclusion. Its purpose is not to restate the chapter material — rather, it is to reflect on the intellectual threads that tie the chapter together and to think about how the ideas presented promote quantitative reasoning.

In this chapter, you studied the meaning and importance of quantitative literacy. As you move forward, keep in mind the following key ideas.

- Even though you will choose a discipline of emphasis in your studies, remember that real issues and problems are interdisciplinary and best understood through examination from a variety of perspectives.
- Quantitative literacy is a crucial aspect of literacy for the modern world. It is necessary to making wise personal decisions, exercising citizenship through social discourse and voting, and appreciating culture fully. In addition, quantitative literacy is crucial to remaining competitive in a rapidly changing job market. The many skills of quantitative reasoning presented in this book form the cornerstone of quantitativel literacy.
- If you are one of the many people who have a particular fear of mathematics, confronting your attitudes may be the most important step in learning to reason quantitatively. We urge you to challenge constantly your attitudes and any misconceptions you may hold.
- Perhaps the most insidious of all misconceptions about mathematics is the one that holds that some otherwise intelligent people are incapable of understanding the subject. As you continue your studies, you should constantly remind yourself that you are fully capable of understanding everything presented in this book. It may require a great deal of effort, but success is well within your reach.

Mathematics, of course, is not the only cornerstone in today's world. Reading is even more fundamental as a basis for learning and life. What is different today is the great increase in the importance of mathematics to so many areas of education, citizenship, and careers.
— from *Everybody Counts* (National Academy Press, Washington, D.C., 1989, p. 3)

PROBLEMS

Reflection and Review

Section 1.1

1. **The Interdisciplinary Nature of Modern Issues.** List at least one component of each issue that involves mathematics, science, political science, sociology, and economics. Add other subject areas if you want to. *Example*: AIDS. *Solution:* The following list is but one of the many possible ways to answer this question.
   - Mathematics is used to study the probability of contracting AIDS.
   - Science is used to study the biology of the AIDS virus.
   - Political science is used to analyze the political battles in Congress over funding for AIDS research.
   - Sociology is used to examine societal attitudes about allowing children with AIDS in public schools.
   - Economics is used to estimate the cost of AIDS treatment to society.
   - Art allows us to critique literature and films about AIDS.
   - History provides the framework for studying the causes and effects of past epidemics.

   a. The long-term viability of the Social Security system.
   b. The appropriate level for the federal gasoline tax.
   c. Health care reform.
   d. Job discrimination against women or ethnic minorities.
   e. Effects of population growth (or decline) on your community.
   f. Possible bias in standardized tests (e.g., the SAT).
   g. The degree of risk posed by carbon dioxide emissions.
   h. Immigration policy of the United States.
   i. Violence in public schools.
   j. Pick an issue of your choice from today's newspaper.

2. **Quantitative Concepts in the News.** Identify the major unresolved issue discussed on the front page of today's newspaper. List at least three areas in which quantitative concepts play a role in the policy considerations of this issue.

Section 1.2

3. **Personal Consequences of Innumeracy.** In one or two paragraphs, give your own example of how innumeracy can lead to difficulties in personal life.

4. **Business Consequences of Innumeracy.** Search the Business section of a newspaper from the past week for an article about a company that is having financial problems (e.g., falling sales, layoffs, dropping stock price, or bankruptcy). Briefly describe any decisions that led the company to trouble that might have been caused by innumeracy on the part of employees or management.

5. **Social Consequences of Innumeracy.** Choose an important social, economic, or environmental issue (e.g., welfare, tax breaks for business, or pollution controls) in which you believe the current policy should be changed or improved. Describe at least one way in which you think the current policy stems from innumeracy.

6. **Quantitative Literacy and the Arts.** Choose a well-known, historical figure in a field of art in which you have a personal interest (e.g., a painter, sculptor, musician, or architect). Briefly describe how mathematics played a role in or influenced that person's work.

7. **Quantitative Literature.** Choose a favorite work of literature (poem, play, short story, or novel). Describe one or more instances in which quantitative literacy is helpful to understanding the subtleties intended by the author.

8. **Your Quantitative Major.** What is your major field of study? Identify ways in which quantitative reasoning is important within that field. (If you haven't yet chosen a major, pick a field that you are considering for your major.)

9. **Preparing for the Future.** Realizing that most Americans will change careers several times during their lives, identify at least three occupations in Table 1-2 that interest you. Do you have the necessary skills for them at this time? If not, outline a plan for your remaining college semesters to ensure that you acquire the necessary skills before you graduate. (Note: Many of these occupations require training in graduate school or extensive job experience. For such cases, explain how you will prepare for graduate school or prerequisite jobs.)

10. **Reopening Career Options.** Have you ever ruled out any careers because they require too much mathematical skill? If so, describe a plan by which you could, if you chose, catch up and learn the skills necessary to reopen that career option. If not, explain why you never considered such a career.

Section 1.3

11. **Examples of Misconceptions.** For each misconception described in Section 1.3, give another example showing why the misconception doesn't reflect reality.

12. **Sources of Misconceptions.** Why do you think so many people hold misconceptions about mathematics?

13. **Merit Badge?** The following quote is from *Everybody Counts* (National Academy Press, Washington, D.C., 1989, p. 76).

    > Too many Americans seem to believe that it does not really matter whether or not one learns mathematics. Only in America do adults openly proclaim their ignorance of mathematics ("I never was very good at math") as if it were some sort of merit badge. Parents and students in other countries know that mathematics matters.

    Suppose that you met someone who wears ignorance of mathematics as a "merit badge." Write a short letter to try to convince that person to change his or her attitude.

14. **Teaching Without Misconceptions.** Imagine that you are an elementary school teacher. Describe a few techniques you might use to help your students learn to enjoy mathematics and to avoid developing misconceptions.

Section 1.4

15. **Looking at Mathematics.** Consider the three ways of looking at mathematics described in Section 1.4. Do you prefer one way over the others? If so, explain why. If not, explain how each way can contribute to understanding the question "what is mathematics?"

16. **Thinking About Models.** List three models that you encounter in everyday life that represent some underlying reality. In each case,
    a. explain how the model differs from the underlying reality (possibly in size, shape, texture, appearance), and
    b. explain how the model resembles the underlying reality, thereby making it useful.

Section 1.5

17. **Attitudes Toward Mathematics.** Most children have a natural affinity for mathematics; they take pride in their counting skills and enjoy puzzles, building blocks, and computers. Unfortunately, this natural interest seems to be snuffed out in most people by the time they reach adulthood. What is your attitude toward mathematics? If you have a negative attitude, can you identify when in your childhood that attitude developed? Was your attitude affected by something said by a teacher, friend, or relative? What might have been done differently in your education to have kept you interested in mathematics? Write a short essay responding to these questions.

18. **Personal Goals and Strategies.** The following questions are designed to help you think about your goals in this course, in college, and in your life. Write a short (two or three paragraph) response to each.
    a. Identify a personal goal that you can achieve more easily by improving your quantitative reasoning.
    b. What is the goal that you hope a college education will help you reach? Do you believe that college is the best strategy, at this time in your life, for working toward your goal? If yes, how can keeping focused on your goal help you have a better college experience? If no, explain why you believe that you should or should not remain in college now.
    c. Imagine that you are 100 years old and looking back at your life. What would make you feel that life had been worthwhile?

Further Topics and Applications

19. **The Mathematics of Abortion.** To show how virtually every issue involves concepts in mathematics and science, consider the emotionally charged issue of abortion. Without revealing your position on this issue, list several areas in which mathematical or scientific concepts are important to the debate over this issue.

20. **The Mathematics of Recycling.** As another example of the interdisciplinary nature of issues, consider recycling. Briefly list some of the impacts that recycling policies have on industry, government, the economy, the environment, and individual behavior. Discuss the quantitative concepts needed to understand each impact.

21. **Are Lottery Purchases Quantitatively Informed?**
    a. Do you ever purchase lottery tickets? Why or why not?
    b. In your decision to buy or not buy lottery tickets, do you consider the probability of winning? Explain.
    c. State-sponsored lottery games have consequences beyond individual gambling; for example, lottery revenues often are an important source of state funding of programs. Considering all the factors involved in state-sponsored lotteries, do you believe it reasonable for people to spend 1% of their income on the lottery? What about 5%? 10%? Is there a level at which lottery spending becomes foolish and, if so, what is that level? Do you think that state-sponsored lotteries are a good idea for society?

22. **Innumeracy and Pseudosciences: Astrology.** Does your local newspaper carry an astrological horoscope, and do you ever read it? Why or why not? Suppose that you were convinced (if you aren't already) that astrological predictions have no scientific basis; would you still read your horoscope? Why or why not? Can reliance on a horoscope pose any danger to an individual or to society? Explain.

23. **Quantitative Reasoning and Cartoons.** Find a favorite cartoon that includes a quantitative concept. Place a copy of the cartoon in your journal or assignment and explain the quantitative concept covered.

24. **More Misconceptions?** Can you think of any other common misconceptions about mathematics, beyond those described in Section 1.3? Explain.

Projects

25. **Mathematics and Culture.** Identify and describe several great achievements of your cultural ancestors that required mathematical or scientific skills.

26. **Career Skills.** What do you want to be when you "grow up"? What skills are required for that career? Make a list from the skill levels in Table 1-1 and add any other skills you think you might need. Discuss your list with at least one person who works in that career at present; modify your list according to her or his recommendations.

27. **Improving Quantitative Education.** What steps (if any) do you think the educational system should take to ensure that everyone is quantitatively literate? For example, should a certain level of quantitative literacy be required for graduation from high school? from college? If so, how would you ensure that graduates have achieved the required level of proficiency? Write your answer in the form of a one-page editorial entitled "Quantitative Literacy and Our Schools."

28. **Looking Into the Future.** The process of extrapolation — inferring the unknown based on things that are known — is important in mathematics. To practice extrapolation, write a brief essay about what you expect the United States to be like in 20 years. Extrapolate from the trends you see taking place today. For example, will the standard of living be higher or lower? What kinds of jobs will be available and how much unemployment will there be? What will have happened with problems such as AIDS, drugs, crime, traffic, and the cost of health care? Will the United States be at peace or involved in wars? How will education have changed? Use numbers wherever you can to make your predictions quantitative. Save your essay to reread after you complete this course and after you complete college; seeing how your predictions turned out will be fun.

2 PRINCIPLES OF REASONING

Before you can learn to reason with mathematical or quantitative information, you must first understand the general principles of reasoning. Cultivating skills of reasoning can offer many satisfactions in everyday life, including confidence in your ability to learn, to ask good questions, and to figure things out. Reasoning can help you understand the myriad forces of persuasion and fallacy that bombard you from television and other media every day. Reasoning also is an important ingredient in sound decision making in all walks of life. In this chapter we investigate the principles of reasoning, focusing primarily on problems that do not involve mathematics. This proficiency with reasoning and organized thinking will serve you well in later chapters, both in learning new mathematical skills and in applying them confidently to quantitative problems.

CHAPTER 2 PREVIEW:

2.1 ***LOGIC: THE STUDY OF REASONING.*** We begin the chapter with a discussion of the role of logic and reasoning in our lives.

2.2 ***THE VALUE OF LOGICAL ARGUMENT.*** A logical argument follows a more rigid structure than typical everyday arguments. In this section we discuss the value of logical argument.

2.3 ***PROPOSITIONS — BUILDING BLOCKS OF ARGUMENTS.*** In this section we examine propositions and how they can be represented with Venn diagrams. We then cover compound and conditional (if ... then) propositions, which allow investigation of computer database applications such as key word searches.

2.4 ***DEDUCTIVE ARGUMENTS.*** The primary characteristic of a valid deductive argument is that its conclusion necessarily follows from its premises. In this section we present methods for assessing deductive validity and distinguishing validity from truth.

2.5 ***INDUCTIVE ARGUMENTS.*** The second broad category of argument is inductive. In this section we investigate inductive arguments, showing that the concept of validity doesn't apply; instead, inductive arguments are evaluated subjectively according to their strength. We also compare the roles of deductive and inductive argument in mathematics.

2.6 ***ARGUMENTS IN THE REAL WORLD.*** We close the chapter by demonstrating how to analyze arguments that aren't clearly structured, like most arguments in everyday life! This analysis leads to the concepts of assumed premises and intermediate conclusions, as well as to methods of flow charting arguments.

Civilized life depends upon the success of reason in social intercourse, the prevalence of logic over violence in interpersonal conflict.

— Juliana Geran Pilon

In a republican nation, whose citizens are to be led by reason and persuasion and not by force, the art of reasoning becomes of first importance.

— Thomas Jefferson

2.1 LOGIC: THE STUDY OF REASONING

Life is full of decisions: big ones, such as what to study in school or whether and whom to marry; and little ones, such as what color of shirt to put on in the morning. How do we make such decisions? How do we develop our beliefs and opinions? How can we evaluate the opinions or decisions of others?

No single answer to these questions exists. Everyone makes decisions for many different reasons. To take a simple example, consider how you might make a decision regarding the question: "Should I do the assigned reading in this book tonight?" You might make an *emotional* decision, going ahead with the reading because you enjoy the material or skipping it because you are distraught over relationship problems. In contrast you might make a decision based on **reasoning,** in which you carefully analyze the potential impacts of your decision on your personal goals. In fact, most people base their decisions on a combination of emotions and reasoning.

Logic is the study of the methods and principles involved in reasoning.

Reasoning plays an important role in much more than just a person's own decision making. Beliefs and opinions tend to be formed by a combination of personal background, emotional state, and reasoning ability. Furthermore, the ability to reason well is crucial to the individual's evaluation of the beliefs, opinions, and decisions of others. It therefore is a necessary aspect of citizenship (see the Pilon and Jefferson quotes at the beginning of the chapter).

In this book, we are concerned primarily with developing **quantitative reasoning**—the ability to reason with quantitative or mathematical information. However, because reasoning without quantitative information is prerequisite to reasoning with it, we first cover the methods and principles involved in reasoning, or **logic**.

In this chapter, we concentrate on the construction and analysis of logical arguments. Although we apply these skills to a few quantitative situations, most of the work in this chapter is nonmathematical. This approach should allow you to build your confidence at thinking logically before moving on to more quantitative situations. Then, in Chapters 3–5, we continue to develop logic and apply it both to understanding quantitative arguments and to problem solving.

"Contrariwise," continued Tweedledee, "if it was so, it might be; and if it were so, it would be; but as it isn't, it ain't. That's logic."—from Lewis Carroll, *Through the Looking Glass*

Time-Out to Think: In *Star Trek* (original cast), Mr. Spock believed that all decisions should be based on logic. Captain Kirk argued that logic should be only one ingredient in decision making. Which side do you take? When should logic be used, and when should other factors be considered?

2.2 THE VALUE OF LOGICAL ARGUMENT

In logic the word **argument** has a more precise meaning than it does in ordinary English usage. Consider the following "argument" that might be heard between two people on opposing sides of the abortion debate.

Jack: "Abortion is immoral."

Jill: "No it isn't."

Jack: "Yes it is! Doctors who perform abortions should go to jail."

Jill: "You don't even know anything about abortion."

Jack: "I know a lot more than you know! I've watched lots of films about it."

Jill: "I can't talk to you, you're an idiot!"

People generally quarrel because they cannot argue. — G.K. Chesterton

Although this heated conversation may be typical of many arguments heard daily, it is not a *logical* argument. It has virtually no structure, and neither Jack's conclusion (abortion is immoral) nor Jill's conclusion (Jack is an idiot) is established in a compelling way. Neither Jack nor Jill is likely to sway the other with this type of argument; in fact, such arguments often have the effect of making the opposing parties more belligerent.

Can Jack and Jill possibly argue about abortion without becoming angry? Jack's goal apparently is to persuade Jill that abortion is immoral, and Jill's apparently is to persuade Jack that it isn't. Let's consider how they might argue their respective positions logically.

A proposition is a statement that makes a distinct claim, such as an assertion or denial.

A logical argument is made of building blocks called **propositions** — statements that make a distinct claim, such as an assertion or denial. If you look carefully at Jack and Jill's argument, you can find several examples of propositions, including:

"Abortion is immoral."

"Abortion is not immoral."

"Jack knows a lot about abortion."

"Jack is an idiot."

We call such statements propositions because they *propose* something to be true or false. A logical argument begins with a set of propositions that describe the ideas, facts, or assumptions on which the argument is based, called the **premises** of the argument. Arguments end with one or more propositions that represent the **conclusions** of the argument.

Suppose that, instead of simply proclaiming abortion to be immoral, Jack used the following structured argument.

Jack's opening argument

Premise:	The killing of a human being is immoral.
Premise:	Abortion is the killing of a human being.
Conclusion:	Abortion is immoral.

Note that both the premises and the conclusion are propositions: each makes a distinct claim that, depending on your viewpoint, is either true or false.

As Jack's conclusion follows logically from his premises, Jill cannot simply attack the truth of his conclusion. Instead, Jill might argue that Jack's premises are flawed, which in turn makes his conclusion flawed. For example, Jill might argue against Jack's two premises in the following manner.

Jill's argument against Jack's first premise

Premise:	The killing of a human being in self-defense is not immoral.
Premise:	Killing in circumstances of war is not immoral.
Premise:	The death penalty for murderers is not immoral.
Conclusion:	The killing of a human being is not necessarily immoral.

Jill's argument against Jack's second premise

Premise:	If a fetus is not viable outside the womb, then it is not yet a human being.
Premise:	Most abortions occur before the fetus is viable.
Conclusion:	Most abortions do not involve killing a human being.

Like Jack's argument, Jill's arguments are clear and well-conceived: her premises are propositions that lead logically to her conclusions. As a result, if Jack wants to pursue his position, he must seek flaws in Jill's premises. The argument could continue for a long time and, given the controversy over abortion, Jack and Jill aren't likely to resolve their differences. Nevertheless, logical argument has great value: through this process each side may better understand the other, which someday might open the way to an acceptable compromise.

Arguments, at least when they are thoughtfully and logically conceived, are generally used for any of four main purposes:

1. to *discover* the truth of a proposition. For example, a scientific study may seek to establish the truth of a proposition whose truth is currently unknown, such as "continued use of fossil fuels will lead to global warming."
2. to *justify* belief in a proposition. For example, you might construct an argument to justify to yourself the proposition, "studying harder will make my grades better."
3. to *explain* the truth of a proposition. For example, you might construct an argument to explain to your child the truth of the proposition, "eating broccoli is good for you."
4. to *persuade* someone of the truth of a proposition. For example, you might create an argument designed to persuade people that "my new cola is better than either Coke or Pepsi."

With these purposes in mind, we devote the rest of the chapter to studying arguments in greater detail. We begin by studying propositions because they are the building blocks of arguments.

2.3 PROPOSITIONS — BUILDING BLOCKS OF ARGUMENTS

Propositions — statements that *propose* something to be true or false—are the basic building blocks of arguments. Like any English sentence, a proposition must have a *subject* and *predicate.*

- "Joan is sitting in the chair" is a proposition because it makes an assertion. The subject is "Joan" and the predicate is "is sitting in the chair."
- "I did not steal the car" is a proposition because it makes a denial. The subject is "I" and the predicate is "did not steal the car."
- "Are you going to the store?" is not a proposition because it makes neither an assertion nor a denial.
- "Three miles south of here" is not a proposition because it does not make any distinct claim and does not have the structure of a sentence.

A proposition must be *capable* of being either true or false, though we may not know which it is. With regard to the preceding two propositions, Joan either is or is not sitting in the chair, and I either did or did not steal the car. We are now ready to investigate the structure of propositions and how propositions can be connected to form more complex statements.

2.3.1 Categorical Propositions

A particularly important type of proposition in logic is one that expresses a relationship between two categories, or **sets.** For example, the proposition

All politicians are liars

compares the *subject set* (politicians) to the *predicate set* (liars). Because they compare two categories, propositions of this type are often called **categorical propositions.** If we use the letter S for the subject set, and the letter P for the predicate set, we can rewrite the preceding proposition as

All S are P (where S = politicians; P = liars).

Table 2-1 identifies and describes the four standard categorical propositions.

Traditionally each type of categorical proposition is designated by a single letter.

A: All S are P.
E: No S is P.
I: Some S are P.
O: Some S are not P.

The letters come from the Latin words *affirmo* (**A**ff**I**rmo), which means "I affirm," as the **A** and **I** statements are affirmative; and *nego* (n**E**g**O**), which means "I deny," as the **E** and **O** statements are negative.

Table 2-1. The Four Standard Categorical Propositions

Form	Example	Subject (S)	Predicate (P)
All S are P.	All apes are mammals.	apes	mammals
No S are P.	No fish are mammals.	fish	mammals
Some S are P.	Some doctors are women.	doctors	women
Some S are not P.	Some teachers are not men.	teachers	men

Rephrasing into Standard Form

Many propositions, when expressed in ordinary English, do not look like any of the four standard forms shown Table 2-1. Because analyzing propositions in standard form is easiest, rephrasing a proposition so that it fits one of the four forms and *retains its original meaning* often is useful. To do so usually requires looking at a proposition carefully to determine its subject and predicate sets. For example, note how each of the following propositions is rephrased into standard form.

- All diamonds are valuable. ⇒ All diamonds are things of value. This proposition has the standard form *all S are P,* where S = diamonds and P = things of value.
- Some birds do not fly. ⇒ Some birds are not flying animals. This proposition has the standard form *some S are not P,* where S = birds and P = flying animals.
- Some people never learn. ⇒ Some people are people who never learn. This proposition has the standard form *some S are P,* where S = people and P = people who never learn.
- Elephants never forget. ⇒ No elephants are creatures that forget. This proposition has the standard form *no S are P,* where S = elephants and P = creatures that forget.

Time-Out to Think: Rephrase the proposition "toxic waste sites are always dangerous" so that it fits one of the four standard forms for categorical propositions. Identify the subject and predicate sets.

Sometimes propositions refer to a set that has only one member, that is, to a **singular set.** We can translate them directly into standard form, as long as we recognize that the single member represents all the members of a singular set as, for example, the following.

- *Jane is a doctor* has the standard form *all S are P,* if we identify *S = the set consisting of Jane* and *P = doctors.*
- *My dinner is good* has the standard form *all S are P,* if we identify *S = the set of all dinners that I am eating right now* and *P = good dinners.*

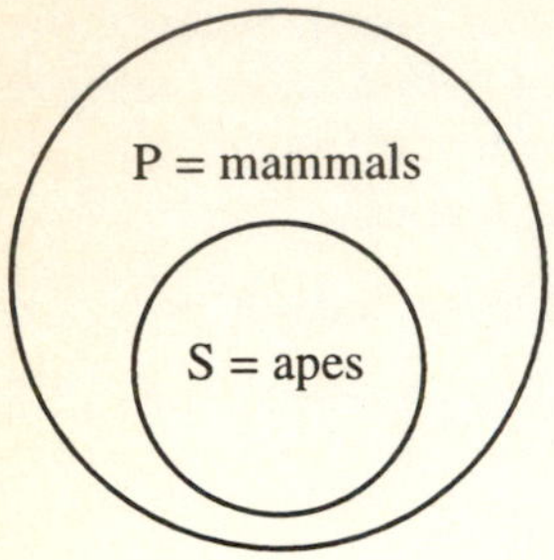

Figure 2-1. A Venn diagram for the proposition *all apes are mammals* shows the subject circle S inside the predicate circle P.

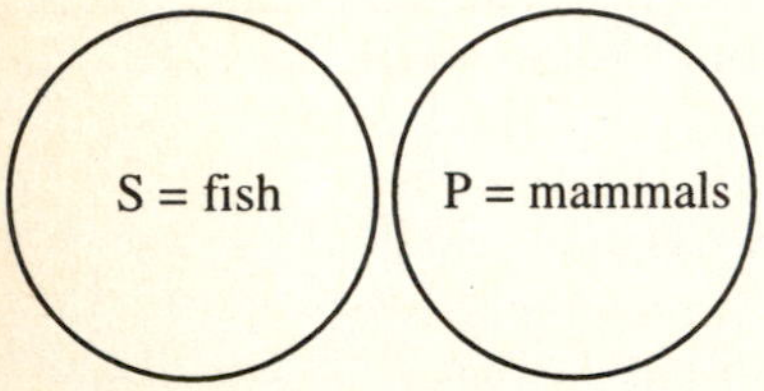

Figure 2-2. A Venn diagram for the proposition *no fish are mammals* shows that two separate circles are needed for propositions of the form *no S are P*.

Venn Diagrams

You will soon begin analyzing arguments that consist of one or more propositions. One visual way of doing so is by drawing circles to represent the sets compared in each of the propositions. Diagrams consisting of such circles are called **Venn diagrams,** after the English logician John Venn (1834–1923).

Figure 2-1 shows a Venn diagram for the proposition *all apes are mammals,* which has the form *all S are P.* Because all members of the subject set (S = apes) also are members of the predicate set (P = mammals), the subject circle is drawn *inside* the predicate circle. The fact that one circle is inside the other is the *only* important feature of the Venn diagram; the size of the circles doesn't matter.

Figure 2-2 shows a Venn diagram for the proposition *no fish are mammals.* In this case, the subject set (S = fish) and predicate set (P = mammals) are completely distinct; we therefore draw the two circles separately.

To draw the Venn diagram for propositions of the form *some S are P*, we consider the proposition *some politicians are liars.* We can draw a Venn diagram for this proposition with two overlapping circles, one for S = politicians, and one for P = liars (Figure 2-3a). However, this proposition introduces some ambiguity: it makes clear that at least *some* politicians are liars, but it doesn't say whether any politicians are *not* liars. To avoid confusion, we add labeled arrows to identify what is stated and what is unstated by the proposition (Figure 2-3b).

Similarly, the Venn diagram for propositions of the form *some S are not P*, such as *some princes are not charming,* also consists of two overlapping circles. As shown in Figure 2-4, ambiguity arises because the proposition doesn't state whether any princes *are* charming. Again, we can avoid possible confusion by adding labeled arrows to indicate what is stated and what is unstated.

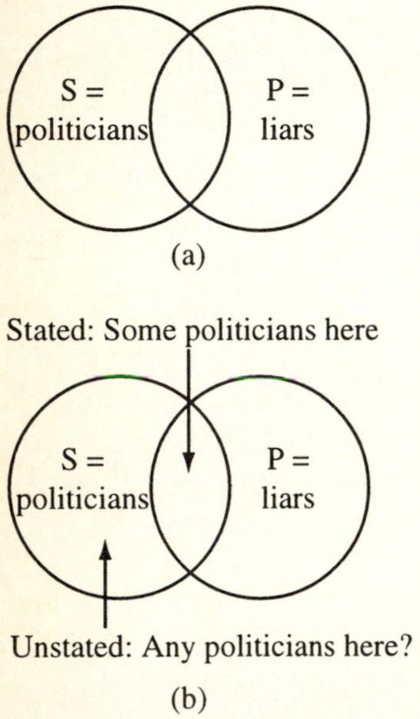

Figure 2-3. (a) The Venn diagram for *some politicians are liars* shows that propositions of the form *some S are P* consist of two overlapping circles. (b) We can avoid confusion by adding labeled arrows that indicate what is stated and what is unstated by the proposition.

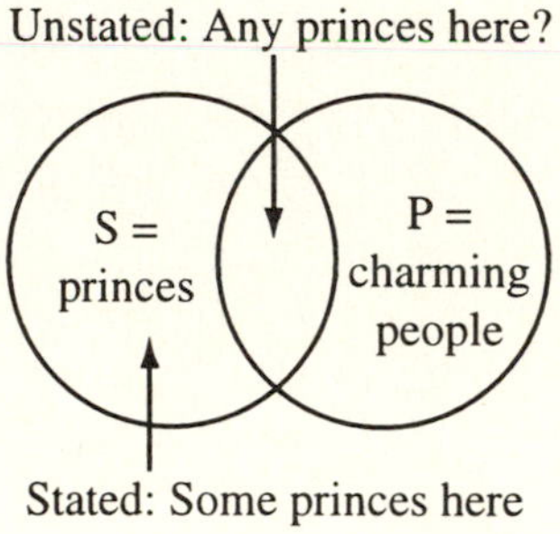

Figure 2-4. The Venn diagram for *some princes are not charming* (people) shows that propositions of the form *some S are not P* also consist of two overlapping circles. Again, labeled arrows that show what is stated and unstated can be helpful.

Claims of Truth

A categorical proposition always makes a *claim* of truth. However, this claim is not necessarily true. And, although a proposition must be *capable* of being either true or false, determining which it is may not be possible. Generally speaking, propositions fall into one of three broad classes.

1. A proposition is *unambiguous* if no person could reasonably disagree with its truth or falsehood. For example, we can unambiguously determine whether the propositions "Joan is sitting in the chair" and "there are 145 students in the graduating class" are true or false.
2. A proposition is *unverifiable* if it would require impossible or impractical procedures to show whether it is true or false. Consider the proposition "continued use of fossil fuels will cause the Earth's climate to change dramatically." If use of fossil fuels continues, we will eventually find out whether this proposition is true. At present, however, we can only argue over the likelihood that it is true.

3. A proposition also may be a *matter of opinion*. The truth of propositions such as "computers have improved the quality of our lives" and "lawyers are paid too much" can be argued endlessly.

Time-Out to Think: In terms of the three types of truth claims, how would you classify *abortion is immoral*? Why? Would everyone agree with you? Does this classification tell you anything about why the abortion issue is subject to so much emotion and debate? Explain.

2.3.2 Compound Propositions

Two or more propositions can be joined to form a **compound proposition.** The words that join them are called **logical connectors;** the two most common are the words *and* and *or*.

If both propositions joined by the connector *and* are true, then the entire statement is true.

The logical use of *and* usually is unambiguous. The statement "the capital of France is Paris and Antarctica is cold" is a true statement because both propositions in the statement are true. However, the statement "the capital of France is Paris and the capital of America is Madrid" is false because one of the propositions is false. A reliable rule is that if both propositions joined by the connector *and* are true, then the entire statement is true.

***Or* is inclusive when it means "either or both." *Or* is exclusive when it means "either but not both."**

In contrast, the meaning of *or* isn't always clear. If a health insurance policy covers hospitalization in cases of illness *or* injury, it probably means that it covers either illness *or* injury, or both. Thus the **inclusive** use of the connector *or* means "either or both." But when a restaurant offers a choice of soup *or* salad, the general understanding is that you may not choose both. Thus the **exclusive** use of the connector *or* means "one or the other." Further, the word *or* sometimes involves implicit qualifications. For example, a tire warranty that covers "5 years *or* 50,000 miles" usually means "these tires are guaranteed for 5 years *or* 50,000 miles, whichever comes first." A reliable rule is that if at least one of the two propositions joined by the connector *or* is true, then the entire statement is true.

Because the word *or* can have different meanings, considering its context is important when you confront it in ordinary English. In logic, however, the situation is simpler: *Or* is assumed to be *inclusive,* unless stated otherwise.

Time-Out to Think: Imagine that your house is robbed during looting after an earthquake. Your insurance policy states that your house is insured for earthquakes, fire, *or* robbery. Should you expect your insurance to cover both the earthquake damage and the loss from the robbery? Why or why not?

Connectors, Computers, and Key Word Searches

The Internet links computers around the world and facilitates information sharing. Most colleges have computers that are connected to the Internet. You can also gain Internet access for a personal computer by subscribing to a commercial service. If you have never tried the Internet, you don't know what you're missing. Get online, and use key word searches to find something that interests you.

The use of logical connectors is more than a minor grammatical matter. Computerized databases, such as "online" library catalogs, encyclopedias, and directories of information available on the Internet, hold enormous amounts of information. Looking for a specific book, fact, or Internet site in such databases can be like trying to find the proverbial "needle in a haystack." Fortunately, the use of logical connectors and **key words** — author names, subject areas, words from titles, and the like — makes the task less daunting.

Imagine that you are searching a library catalog for a book on impressionist art by an author named Jones. If you do a search on the subject *impressionist art* you are likely to find hundreds of books listed. Similarly, you probably would get a list of hundreds of books by searching on the author name *Jones*. A more efficient search would be on the combination *impressionist art* AND *Jones*. The logical connector *and* tells the computer to list only those books whose data files contain both the subject area *impressionist art* and the author name *Jones*. The resulting list will be far more manageable.

> **Time-Out to Think:** What would happen if you searched on "impressionist art *or* Jones"? Don't forget that a logical *or* is assumed to be inclusive!

In general, use of the connector *and* makes a search more specific, and use of the connector *or* makes a search more general.

Example 2-1 *Key Word Search.* Suppose that you are searching a database for articles on the federal deficit that appeared in *Time* or *Newsweek* between 1993 and 1995. How would you structure the search?

Solution: Let's assume that the database allows for three separate search parameters: subject or title, magazine, and date. For the date, you would use the key word search *1993* OR *1994* OR *1995*. The use of the word *or* ensures that the list will include articles from all three years. For the magazine, you would use *Time* OR *Newsweek*, to ensure that both magazines are included on the list.

The subject is a bit more difficult. Searching on *federal* OR *deficit* would lead to a list with many articles unrelated to the federal deficit, such as articles on federal crimes or attention deficit disorder. Searching on *federal* AND *deficit* would certainly limit the search to articles about the federal deficit; however, it might still miss articles that are cataloged under different key words. You might, for example, search on (*federal* OR *United States* OR *government*) AND (*deficit* OR *budget*). This search has the effect of looking for any article cataloged under federal deficit, United States deficit, government deficit, federal budget, United States budget, or government budget. If this search generates too many items on the list, you can always narrow the scope further.

❖

2.3.3 Conditional (If ... Then) Propositions

Another common way of connecting propositions is with the words "if ... then." Consider the following examples.

If the president signs that bill, then she is courageous.
If all politicians are liars, then Representative Smith is a liar.
If it is Tuesday, then this must be Belgium.

Statements of this type are called **conditional propositions** because they propose something to be true (the *then* part of the statement) only on the *condition* that something else is true (the *if* part of the statement). Note that conditional propositions have the general form "if p, then q" where both p and q are categorical propositions.

Conditional propositions have the form "if p, then q."

Many conditional statements in English are not stated in standard form, but we can always rephrase them if necessary. For example, note how each of the following statements has been rephrased into the standard form "if p, then q."

- If I leave, I'm not coming back. ⇒ If I leave, then I am not coming back.
- Further rain will cause a flood. ⇒ If it rains further, then there will be a flood.

- He's a nurse, so he must know CPR. ⇒ If a person is a nurse, then the person must know CPR.
- Your ad implies that the price should be only $20. ⇒ If your advertisement is true, then the price is only $20.

Besides their importance in logical arguments "if ... then" statements are fundamental in computer programming and in many computer applications.

2.4 DEDUCTIVE ARGUMENTS

We are now ready to investigate how arguments actually proceed from premises to conclusions. This process is often called **inference** because we try to *infer* the conclusion from the premises. The two basic types of inferential processes are

- **deductive inference**, in which a specific conclusion is *deduced* from more general premises; and
- **inductive inference**, in which a conclusion is formed by *generalizing* from more specific premises.

In this section, we consider deductive arguments, saving inductive arguments for Section 2.5.

All the deductive arguments we present are short, consisting of no more than three propositions (two premises and one conclusion). Nevertheless, they convey nearly all the essential features of deduction. Thus, even though deductive arguments can be far longer than those discussed here, we develop all the basic concepts needed for their analysis.

2.4.1 Deductive Arguments with Categorical Propositions

Consider the following argument.

Argument 1

Premise:	All politicians are crooks.
Premise:	All crooks are liars.
Conclusion:	All politicians are liars.

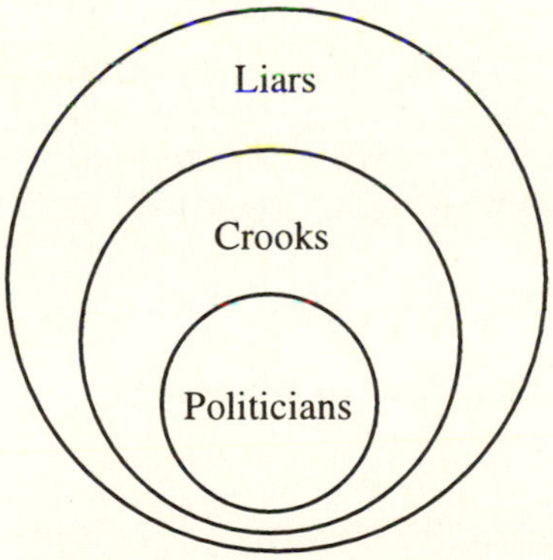

Figure 2-5. The first premise of argument 1 puts the circle for *politicians* inside the circle for *crooks*; and the second premise puts the circle for *crooks* inside the circle for *liars*. Note that the circle for *politicians* therefore is within the circle for *liars*, just as the conclusion claims.

Either of the two premises of this argument probably could provoke a long and heated debate. Nevertheless, *if* the premises of this argument are true, its conclusion automatically follows. We can show this result visually with Venn diagrams. The Venn diagram for the first premise has the circle for *politicians* inside the circle for *crooks*, and the second premise has the circle for *crooks* inside the circle for *liars*. The Venn diagram for the two premises together is shown in Figure 2-5. The circle for *politicians* is clearly within the circle for *liars*, so the conclusion follows from the premises.

Because the conclusion of argument 1 follows directly from the premises, argument 1 is a **valid argument**—even though we may not agree that its premises are true or that its conclusion is true!

This condition illustrates one of the most important distinctions in logic: the difference between validity and truth. Validity is concerned only with the *logical structure* of an argument, not the truth of its premises or conclusions. A few more examples should make the distinction clear for the four basic cases can arise.

A deductive argument is valid if its conclusion necessarily follows from its premises — regardless of whether the premises or conclusion are true.

An argument is sound if it is valid *and* its premises are true.

- If an argument is valid *and* its premises are true, its conclusion also must be true. Such an argument is a **sound argument** (Example 2-2). Of course, disagreement over whether a premise is true will lead to disagreement over whether the argument is sound.
- If an argument is *valid* but its premises are false, its conclusion also must be false. The argument is valid but not sound (Example 2-3).
- If an argument's premises are true but its conclusion is false, a flaw in its logical structure must exist; that is, it is an **invalid argument.** Because any argument that involves incorrect reasoning is fallacious, we say that this argument suffers from a **formal fallacy:** that is, a fallacy of structure or form (Example 2-4).
- The final case occurs when an argument has a true conclusion but is invalid because of a flaw in its logical structure (Example 2-5). That is, the argument suffers from a formal fallacy because its conclusion does not follow from its premises; the truth of the conclusion might be attributed to a "lucky guess."

Finally, the examples show that the method of testing an argument with Venn diagrams proceeds in three basic steps.

Step 1: Before you can draw Venn diagrams, the premises and conclusion must be in standard form for categorical propositions so that you can clearly identify the sets being compared.
Step 2: Draw a Venn diagram that combines the information from all the premises.
Step 3: Think about the Venn diagram for the conclusion; if the diagram for the premises implies the diagram for the conclusion, the argument is valid. Otherwise, the argument is invalid.

Time-Out to Think: Look again at argument 1. We have determined that it is valid. Is it also sound? Would everyone else agree with your assessment? Why or why not?

Example 2-2 *A Sound Argument.* Evaluate the following argument.

Premise:	All narcotics are habit-forming.
Premise:	Some drugs are narcotics.
Conclusion:	Some drugs are habit-forming.

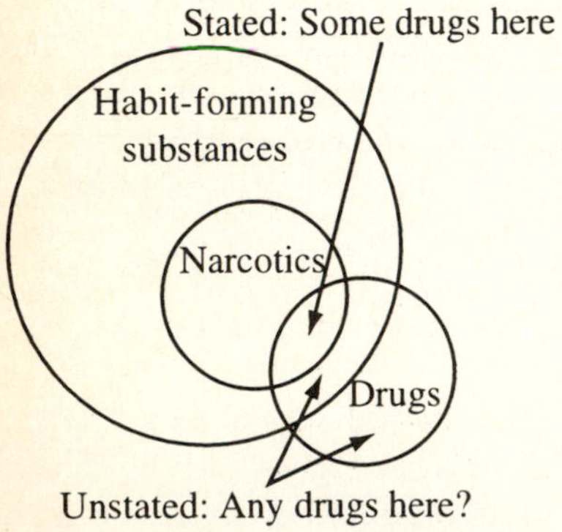

Figure 2-6. The Venn diagram for the argument in Example 2-2 confirms that the conclusion (some drugs are habit-forming) follows from the premises. The argument therefore is valid. *Habit-forming substances* is abbreviated *hfs.*

Solution: First, note that all three propositions (both premises and the conclusion) are true statements. Next, determine whether the conclusion follows necessarily from the premises. As usual, common sense may have led you to the answer already. However, let's use Venn diagrams to check the argument visually.

Step 1: We first need to express the propositions in standard form to clearly identify the sets. We do so by recognizing that *habit-forming* refers to the set of *habit-forming substances*. The three sets in the argument are: *narcotics, drugs,* and *habit-forming substances.*

Step 2: The first premise tells us that the circle for *narcotics* should be inside the circle for *habit-forming substances* in the Venn diagram. The second premise tells us that the circle for *drugs* overlaps the circle for *narcotics*. However, the premises say nothing about whether any drugs are *not* narcotics. To include all possible cases, we draw the *drugs* circle so that it overlaps both of the other circles (Figure 2-6) and then indicate what is stated and what is unstated by the premises.

Step 3: Finally, we ask whether the conclusion can be inferred from the premises. The conclusion requires that some *drugs* be inside the circle for *habit-forming substances*. Because the Venn diagram for the premises clearly shows that to be the case, the conclusion follows from the premises.

The conclusion follows from the premises, so the argument is valid. And, because its premises also are true, the argument is sound.

❖

Example 2-3 ***Valid But Not Sound.*** Evaluate the following argument.

Premise:	All fish are mammals.
Premise:	All mammals are human beings.
Conclusion:	All fish are human beings.

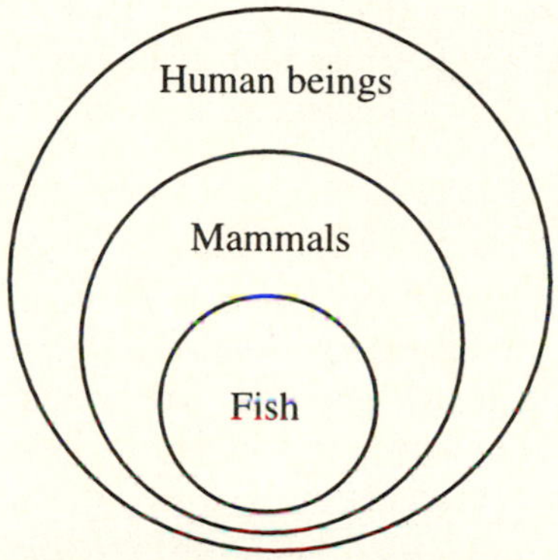

Figure 2-7. The Venn diagram for the argument in Example 2-3 shows that the conclusion (all fish are human beings) follows from the premises. The argument is therefore valid, even though the premises and conclusion are all false.

Solution: Note that all three propositions (two premises and one conclusion) are blatantly false. The argument therefore cannot be sound. But is it valid? Let's check it with a Venn diagram.

Step 1: The three propositions are already in standard form; the three sets involved are: *fish*, *mammals*, and *human beings*.

Step 2: The first premise tell us that the circle for *fish* belongs inside the circle for *mammals*. The second premise tells us that the circle for *mammals* should be inside the circle for *human beings*. The Venn diagram for the two premises together is shown in Figure 2-7.

Step 3: The conclusion requires that the circle for *fish* be inside the circle for *human beings*. That is the case in the diagram for the premises, so the conclusion follows. The argument therefore is valid, even though its conclusion is blatantly false.

❖

Example 2-4 ***True Premises, Invalid.*** Evaluate the following argument.

Premise:	Fish live in the water.
Premise:	Frogs are not fish.
Conclusion:	Frogs do not live in the water.

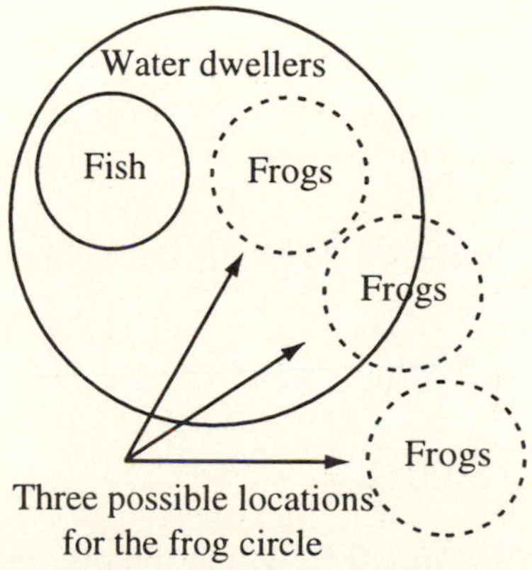

Figure 2-8. The Venn diagram for the argument in Example 2-4 shows that frogs are not fish, but it doesn't necessarily show that frogs are not water dwellers. The argument is invalid.

Solution: There is a quick way to evaluate this argument. Clearly, both of its premises are true, yet its conclusion is false. This outcome can occur only when the structure of the argument contains a logical flaw; the argument therefore is invalid. We can also evaluate this argument with a Venn diagram.

Step 1: Rewrite the argument with its propositions in standard form. Note that the three sets in the argument are *fish*, *frogs*, and *water dwellers* (creatures that live in the water).

Premise:	All fish are water dwellers.
Premise:	No frog is a fish.
Conclusion:	No frog is a water dweller.

Step 2: The first premise tells us that the circle for *fish* should be inside the circle for *water dwellers*. The second premise tells us that the circle for *frogs* should be separate from the circle for *fish*. However, it does not indicate whether the *frog* circle is inside, outside, or overlapping the *water dweller*

circle; we therefore draw all three possibilities, using dotted lines to show that they are possible locations for the circle (Figure 2-8).

Step 3: The conclusion requires that the *frog* circle be separate from the *water dweller* circle. However, because that is only one of three possibilities shown in the diagram for the premises, we cannot say that the conclusion necessarily follows. The argument therefore is structurally flawed, or invalid.

❖

Example 2-5 ***Invalid But True Conclusion.*** Evaluate the following argument.

Premise:	All narcotics are habit-forming.
Premise:	Some drugs are habit-forming.
Conclusion:	Some drugs are narcotics.

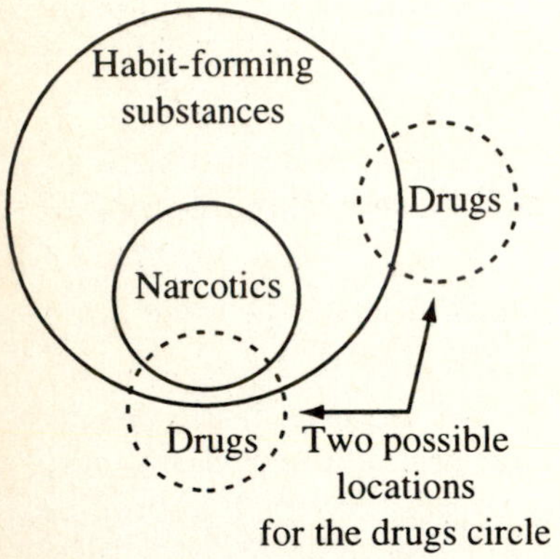

Figure 2-9. Although the conclusion of the argument in Example 2-5 (some drugs are narcotics) is true, the Venn diagram shows that it is not supported by the premises. The argument therefore is invalid.

Solution: In this argument note that the two premises and the conclusion are all true. Yet, the argument is not valid. We can demonstrate why with a Venn diagram.

Step 1: As in Example 2-2, we identify the words *habit-forming* with the set of *habit-forming substances.*

Step 2: The first premise puts the circle for *narcotics* inside the circle for *habit-forming substances*. The second premise tells us that the circle for *drugs* overlaps the circle for *habit-forming substances*. However, the premises say nothing about whether the *drugs* circle overlaps the *narcotics* circle. We therefore have two possible locations for the *drugs* circle, as shown in Figure 2-9.

Step 3: The conclusion requires overlap between the *drugs* circle and the *narcotics* circle. However, because this overlap is only one of the two possibilities shown in the diagram, the conclusion doesn't *necessarily* follow. The argument therefore is invalid. That is, its conclusion is not established logically; rather, this is a case of either a lucky guess or knowing the answer ahead of time!

If you are having difficulty understanding why this argument is invalid, try substituting the set *sports* for the set *drugs*. The argument now reads as follows.

Premise:	All narcotics are habit-forming.
Premise:	Some sports are habit-forming.
Conclusion:	Some sports are narcotics.

This substitution does not affect the structure of the argument. Nevertheless, even though its premises still may be true, the conclusion now is clearly false.

❖

Example 2-6 ***Argument with a Singular Proposition.*** Evaluate the following argument.

Premise:	All U.S. presidents have been male.
Premise:	Harry Truman was a male.
Conclusion:	Harry Truman was a U.S. president.

Solution: Again, this case presents an invalid argument in which the premises and conclusion are all true.

Step 1: We rephrase the propositions into standard form. Recall how singular propositions are rephrased: Harry Truman becomes the single member of *the set consisting of Harry Truman.*

Premise:	All U.S. presidents are males.
Premise:	All members of *the set consisting of Harry Truman* are males.
Conclusion:	All members of *the set consisting of Harry Truman* are U.S. presidents.

Step 2: The first premise puts the circle for *U.S. presidents* inside the circle for *males*. The second premise states that the circle for *the set consisting of Harry Truman* (*HT* for short) also is inside the circle for *males*. However, the premises say nothing about whether the circle for *U.S. presidents* and the circle *HT* overlap. Figure 2-10 shows two possible ways to position the circle for *HT*.

Step 3: The conclusion requires the *HT* circle to be inside the *U.S. presidents* circle. Because this condition does not necessarily reflect the premises, the conclusion does not follow. The argument is invalid. You can spot the invalidity simply by substituting another name (e.g., Mahatma Ghandi) for Harry Truman. Structurally the argument remains the same, but the conclusion is then clearly false.

❖

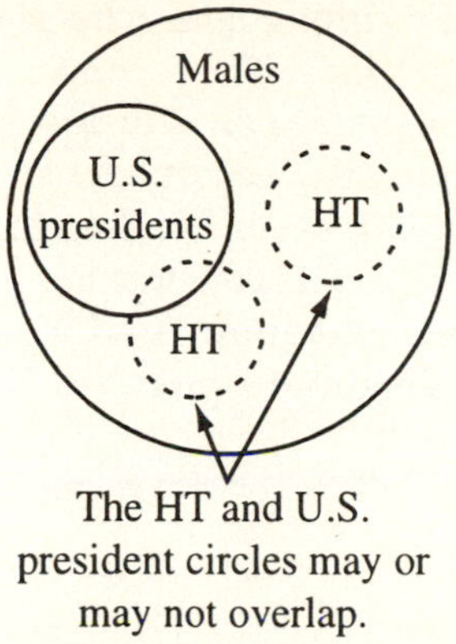

Figure 2-10. The Venn diagram shows that the conclusion of the argument in Example 2-6 (Harry Truman was a U.S. president) does not follow from the premises, even though it is true. The argument therefore is invalid.

2.4.2 Deductive Arguments with One Conditional Premise

Another important type of deductive argument is formed by starting with a conditional proposition. For example, consider the conditional proposition

If it is Tuesday, then this must be Belgium.

If It's Tuesday, It Must Be Belgium is the title of a 1969 movie starring Ian McShane and Suzanne Pleshette.

This proposition has the standard form:

If p, then q, where p = *it is Tuesday* and q = *this is Belgium*.

Suppose that an argument begins with this proposition as its first premise. Further, suppose that it *is* Tuesday. Clearly, we can then conclude that *this is Belgium*. That is, the following argument is valid.

Premise:	If it is Tuesday, then this is Belgium.
Premise:	It is Tuesday.
Conclusion:	This is Belgium.

Time-Out to Think: The preceding argument is valid, but is it sound? Recall that, to be sound, it must not only be valid but its premises must also be true. Can you think of any circumstances under which the argument would be sound? Can you think of any circumstances under which it would not be sound? Explain.

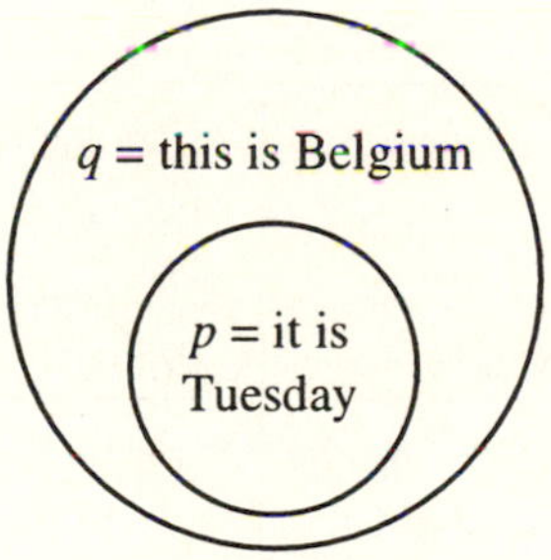

(a)

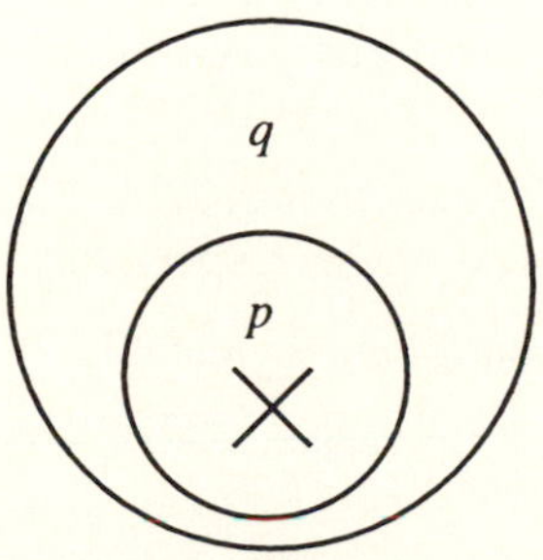

(b)

Figure 2-11. (a) A conditional proposition of the form *if p, then q* can be represented by placing the circle for p inside the circle for q. In this case, p = *it is Tuesday* and q = *this is Belgium*. (b) We now place an X in the p circle to represent the second premise (it is Tuesday). Because the X also is inside the q circle, we conclude that q also must be true (it really is Belgium).

We can also establish the validity of this argument with a modification of the Venn diagram technique. Instead of using circles to represent sets, as we have been doing, we use circles to represent the p and q statements. The conditional proposition *if p, then q* is represented by placing the p circle inside the q circle. In this case, we place the circle for *it is Tuesday* inside the circle for *this is Belgium*, as shown in Figure 2-11a. The second premise in argument 2 asserts that p is true — that is, it really is Tuesday. To indicate this on the diagram, we place an X in the p circle (Figure 2-11b). Because the X also lies within the q circle, we can validly conclude that q is true — this *is* Belgium.

Next, let's examine the validity of a slightly different argument:

Premise:	If it is Tuesday, then this is Belgium.
Premise:	This is Belgium.
Conclusion:	It is Tuesday.

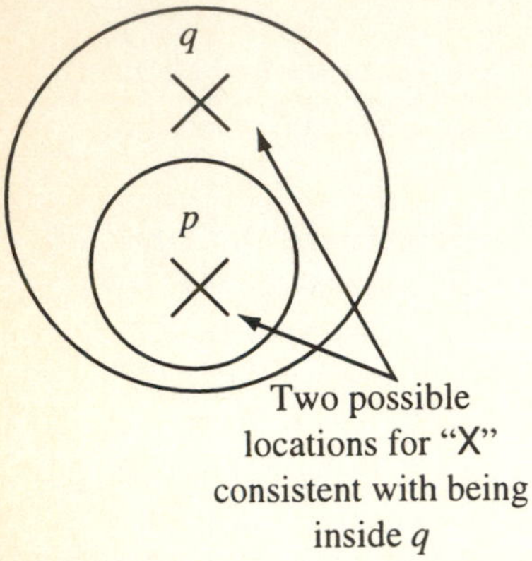

Figure 2-12. The second premise of the argument asserts the truth of q (*this is Belgium*); we therefore place an X inside the q circle. However, we do not know whether the X also belongs within the p circle. The conclusion *it is Tuesday* does not follow.

As with the previous argument, the first premise simply tells us to place the circle for *it is Tuesday* inside the circle for *this is Belgium.* The second premise asserts the truth of *this is Belgium,* so we must place an X inside the circle for *this is Belgium.* Note, however, that the premise does not tell us whether the X should also be within the circle for *it is Tuesday.* We therefore indicate two possible locations for the X in Figure 2-12. The conclusion asserts the truth of the p statement (*it is Tuesday*), which requires an X in the p circle. Because this is not necessarily the case according to the premises, the conclusion does *not* follow and the argument is invalid.

Note that the preceding argument takes the following general form.

Premise:	If p, then q.
Premise:	q.
Conclusion:	p.

The first part of the conditional proposition, p, is called the **antecedent.** The second part, q, is called the **consequent.** Because the second premise asserts, or affirms, the truth of the consequent q, it is said to affirm the consequent. Because the conclusion p does not follow, this argument suffers the **fallacy of affirming the consequent.**

In fact, a simple conditional argument may be made in only four ways: we affirm or deny the truth of the p statement or affirm or deny the truth of the q statement. We have already covered what happens when we affirm p and when we affirm q. In Examples 2-7 and 2-8 we show what happens when we deny either p or q. Table 2-2 summarizes the validity of the four arguments.

The two valid arguments in Table 2-2 have Latin names. The valid argument that affirms the antecedent is called *modus ponens.* The valid argument that denies the consequent is called *modus tollen.*

Table 2-2. Four Basic Conditional Arguments

Structure of Argument	If p, then q. p ——— q	If p, then q. Not p ——— Not q	If p, then q. q ——— p	If p, then q. Not q ——— Not p
Validity	Valid	Invalid	Invalid	Valid
Name of Argument	Affirming the Antecedent	Denying the Antecedent	Affirming the Consequent	Denying the Consequent

Example 2-7 ***Denying the Antecedent.*** Using p = *you liked the book* and q = *you'll love the movie,* evaluate the validity of a conditional argument that denies the antecedent.

Solution: Find the general structure of denying the antecedent in Table 2-2. Replace p and q with *you liked the book* and *you'll love the movie,* respectively.

If p, then q.		If *you liked the book*, then *you'll love the movie.*
Not p	$\Rightarrow$	*You did NOT like the book.*
Not q		*You will NOT love the movie.*

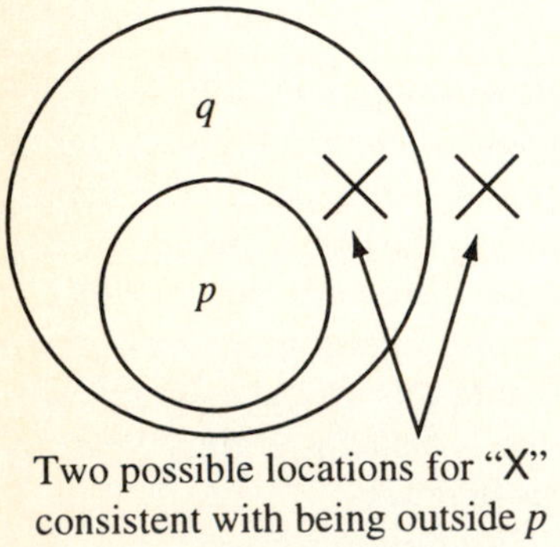

Figure 2-13. The Venn diagram shows that denying the antecedent is invalid and therefore a formal fallacy.

Represent the first premise by placing the circle for *you liked the book* inside the circle for *you'll love the movie.* The second premise denies the p statement; therefore place an X outside the circle for p (*you liked the book*). The premise does not tell us whether this X also is outside the circle for q (*you'll love the movie*); Figure 2-13 therefore shows both possibilities. The conclusion (*you will NOT love the movie*) requires that the X be outside the q circle; this assertion is not necessarily the case, so the conclusion does not follow, and the argument is invalid.

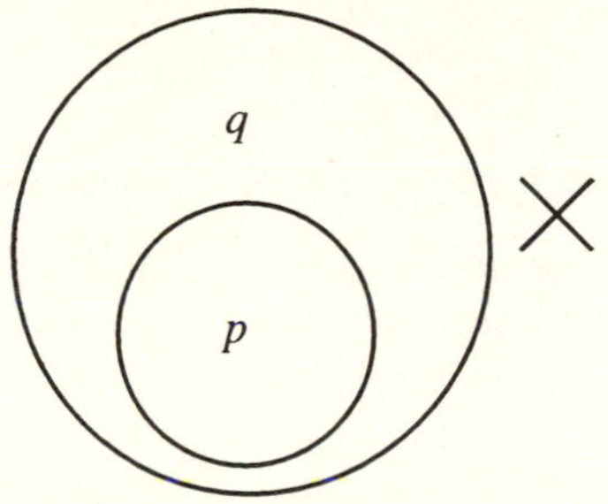

Figure 2-14. The Venn diagram shows that denying the consequent is a valid form of argument.

Example 2-8 ***Denying the Consequent.*** Using p = *we are eating at home* and q = *I need to go shopping*, evaluate the validity of the conditional argument that denies the consequent.

Solution: Find the general structure of denying the consequent in Table 2-2. Replace p and q with *we are eating at home* and *I need to go shopping*, respectively.

If p, then q.		If *we are eating at home*, then *I need to go shopping*.
Not q	$\Rightarrow$	*I do NOT need to go shopping.*
Not p		*We are NOT eating at home.*

In this case, the second premise has the form of *NOT q*; therefore the X is placed outside the circle for q (*I need to go shopping*), as shown in Figure 2-14. The conclusion (*We are NOT eating at home*) requires that the X be outside the p circle; this assertion is the case, so the argument is valid.

❖

Example 2-9 ***The Fallacy of Affirming the Consequent.*** Evaluate the argument: "I knew that, if I got a B on the final, then I'd pass the course. Well, the professor told me that I passed! Therefore I must have gotten a B on the final."

Solution: First, put this statement into the standard form of a conditional argument.

Premise:	If I got a B on the final, then I passed the course.
Premise:	I passed the course.
Conclusion:	I got a B on the final.

The first statement has the form *if p, then q*, where p = *I got a B on the final* and q = *I passed the course*. The second premise, *I passed the course*, asserts the truth of q. That is, it affirms the consequent. The argument therefore is invalid because affirming the consequent is a formal fallacy. Common sense tells you that the argument is invalid because the student might have passed by getting an A on the final!

❖

2.4.3 Deductive Arguments with a Chain of Conditionals

We need to discuss one more form of deductive argument, which is both easy to identify and quite common. Consider the following argument.

Premise:	Frost in Florida means a shortage of oranges.
Premise:	A shortage of oranges means high prices for oranges.
Conclusion:	Frost in Florida means high prices for oranges.

The first step in analyzing this argument is to put the statements into standard form. The first premise can be written as the conditional proposition "if there is frost in Florida, then there will be a shortage of oranges." Treating the other two statements similarly yields the following argument.

Premise:	If there is frost in Florida, then there will be a shortage of oranges.
Premise:	If there is a shortage of oranges, then there will be high orange prices.
Conclusion:	If there is frost in Florida, then there will be high orange prices.

Note that this argument consists of *three* conditional propositions: two conditional premises and a conditional conclusion. This argument takes the following form.

If p, then q.	where:	p = *there is frost in Florida;*
If q, then r.		q = *there will be a shortage of oranges;* and
If p, then r.		r = *there will be high orange prices.*

In this form, the argument is a chain of inference from p to q to r. It therefore is a valid argument. This type of argument is particularly important because it suggests cause and effect relationships: Establishing cause and effect (which is covered in Chapter 3) is fundamental to all science and statistics.

> **Time-Out to Think:** Taking this argument a step further, suppose that orange prices *are* high. Can you conclude that the high prices were caused by a frost in Florida? Why or why not?

Example 2-10 *A Chain of Conditionals.* Analyze the argument: "Maria Lopez advocates high academic standards in the school district, which will benefit my children's education. Therefore, if I vote for Maria Lopez, my children will get a better education."

Solution: This argument can be rephrased in the standard form for a chain of conditionals.

Premise:	If I vote for Maria Lopez, then the school district will have higher academic standards.
Premise:	If the school district has higher academic standards, then my children will benefit.
Conclusion:	If I vote for Maria Lopez, then my children will benefit.

Cast in this form, the validity of the argument is clear: It represents a chain of inference from p (I vote for Maria Lopez) to q (the school district will have higher standards) to r (my children will benefit). Is the argument sound? The first premise has a couple of hidden assumptions, namely, that voting for Maria Lopez will get her elected and that once elected she really will be able to implement higher standards. The second premise is more plausible if you believe that higher standards really do benefit children. Thus the truth of the premises seems to be a matter of opinion, so the soundness of the argument also is a matter of opinion.

❖

2.5 INDUCTIVE ARGUMENTS

We now turn to **inductive arguments.** Consider the following example of an inductive argument.

Premise:	Birds fly up into the air but eventually come back down.
Premise:	People or animals that jump into the air fall back down.
Premise:	Rocks thrown into the air come back down.
Premise:	Balls thrown into the air come back down.
Conclusion:	What goes up must come down.

Note that each premise represents a specific case of something that goes up and then comes back down. The conclusion represents a **generalization** of these specific cases.

What do you think about the preceding argument? At first, it might seem quite compelling because the premises seem to offer plenty of evidence for the conclusion. But the argument involves generalization, so it can be ruined by a single counterexample. In this case, the two *Voyager* spacecraft launched by NASA went up but will never come back down. They are on their way out of the solar system — which invalidates the conclusion.

Voyager 1 and Voyager 2 were launched in 1977. Both passed by the planets Jupiter (1979) and Saturn (1981); Voyager 2 also passed by Uranus (1986) and Neptune (1989). Both spacecraft are on their way out of the solar system and will never return to Earth.

A fundamental distinction between deductive and inductive arguments is that the conclusion of a valid deductive argument follows necessarily from its premises, whereas that can never be the case for an inductive argument. The process of generalization always introduces the possibility that a new fact will show the generalization to be incorrect. As a result, concepts of validity and soundness do not apply to inductive arguments. Instead, we discuss the **strength** of inductive arguments; that is, we make a subjective judgment about how well the premises support the generalization in the conclusion.

Even for the strongest inductive arguments, the best we can say is that their conclusions probably are true. Table 2-3 compares the major features of deductive and inductive arguments.

Table 2-3. A Comparison of Deductive and Inductive Arguments

Deductive Arguments	Inductive Arguments
The conclusion usually is more specific than the premises.	The conclusion usually is more general than the premises.
In a *valid* deductive argument, the conclusion *necessarily* follows from the premises.	There is no such thing as a valid inductive argument. Inductive arguments can be analyzed only in terms of their *strength*. The conclusion of a strong inductive argument seems likely to follow from its premises, but it does not necessarily do so.
Validity concerns only logical structure—an argument can be valid even when its conclusion is blatantly false.	Although an inductive argument cannot be valid, it can be *invalidated* by a single premise that counters its conclusion.
A deductive argument is *sound* if it is valid *and* its premises are true. Because people may disagree about the truth of the premises, they may also disagree about the soundness of a deductive argument.	Because an inductive argument cannot be valid, neither can it be sound. At best, the conclusion of a strong inductive argument *probably* is true.

2.5.1 Induction and Deduction in Everyday Life

People usually form reasoned opinions and decisions through inductive processes. For example, we might decide to vote for a particular candidate because we usually agree with her views, or we might believe that democracy is the best form of government because of the freedoms allowed its citizens.

When we need *proof*, however, induction will not do. Proof requires deduction, in which a conclusion is necessarily established from a set of premises. Thus induction and deduction appear to play distinct roles in logic: induction helps a person to organize knowledge and suggest possible truths, whereas deduction allows a person to prove or disprove those possible truths.

However, in real life deduction and induction are intertwined. For example, look back at the abortion argument between Jack and Jill in Section 2.2. All three structured arguments (Jack's opening argument and Jill's two arguments in response) are deductively valid: their conclusions follow necessarily from their premises. Yet, clearly neither Jack nor Jill has proven anything.

Time-Out to Think: Study Jill's arguments against each of Jack's premises. Develop an argument against each of Jill's premises. Are your arguments deductive or inductive? Can you then create other arguments that counter these?

I think, therefore I am. — from René Descartes (1596–1650), French philosopher and mathematician

The truth of any logical conclusion, whether reached through induction or deduction, rests on the truth of its premises. Yet, if we reach deeply enough into any premise, we can find reason to doubt its truth. Ultimately, all arguments start from certain basic premises that have not been deductively established; instead, the truth of such premises is either taken on faith or argued inductively. The question of whether any absolute truths exist — or even whether there are any truths that can be universally agreed upon — is one of the most vexing in philosophy. We return to this question in Chapter 5.

2.5.2 Induction and Deduction in Mathematics

Perhaps more so than any other subject, mathematics relies on the idea of proof. Mathematicians suggest possible **theorems** (statements of mathematical truth) and then seek to prove them. As proof is possible only through deductive logic, the process of mathematical proof always is deductive. Of course, proofs must start somewhere! The starting points for mathematical proofs are called **axioms**; these are the "givens" that mathematicians assume to be true without proof.

Although mathematical proof relies on deduction, induction also plays an important role: ideas for theorems usually come through inductive reasoning. An interesting example is the "Goldbach conjecture," first proposed in 1742. To understand the Goldbach conjecture, recall that a **prime number** is a number whose only divisors are 1 and itself. The first few primes are 2, 3, 5, 7, 11, 13, 17, 19, and 23 (we discuss prime numbers further in Chapter 6). Considering even numbers (except the number 2), we find that they always can be expressed as the sum of two prime numbers:

$$4 = 2 + 2,\ 6 = 3 + 3,\ 8 = 5 + 3,\ 10 = 7 + 3,\ 12 = 7 + 5,$$

and so on. The Goldbach conjecture asserts that every even number (except 2) can be expressed as the sum of two prime numbers.

No one has ever found an even number that violates the Goldbach conjecture. Computers have checked even numbers of astronomical size and verified that they always can be expressed as the sum of two prime numbers. Nevertheless, all lists have an end, so checking lists of numbers provides only *inductive* evidence for the Goldbach conjecture. Mathematicians are still searching for a deductive proof; only when one is found will the Goldbach conjecture be elevated to the stature of a proven theorem.

Although the Goldbach conjecture may be a bit esoteric, the principle of seeking inductive evidence can be useful to everyone. In Examples 2-11 and 2-12, we demonstrate how a few test cases can be used to check a mathematical rule. Although test cases constitute inductive evidence only, and not proof, they often are enough to satisfy yourself of a rule's truth. Moreover, if even one test case fails, the rule cannot be true.

Example 2-11 ***Inductively Testing a Mathematical Rule.*** Use induction to test the rule: For all numbers a and b, it is true that $a \times b = b \times a$.

Solution: Try some test cases.

Is $7 \times 6 = 6 \times 7$?	$\Rightarrow$	you verify that it is true	$\Rightarrow$	Yes!
Is $(-23.8) \times 9.2 = 9.2 \times (-23.8)$?	$\Rightarrow$	you try both on a calculator	$\Rightarrow$	Yes!
Is $\sqrt{2} \times 1/3 = 1/3 \times \sqrt{2}$?	$\Rightarrow$	you try both on a calculator	$\Rightarrow$	Yes!

The three test cases are somewhat different (mixing fractions, decimals, and negative numbers), yet the rule works in each case. This outcome offers strong inductive argument in favor of the rule. Although you have not proved the rule $a \times b = b \times a$, you now have good reason to believe that it is true. Your belief would be strengthened by additional test cases that confirm the rule.

❖

Example 2-12 ***Invalidating a Proposed Rule.*** Suppose that you cannot recall whether adding the same amount to both the numerator and denominator (top and bottom) of a fraction such as 2/3 is legitimate. That is, you are wondering whether, for any number a, it is true that

$$\frac{2}{3} = \frac{2+a}{3+a}.$$

Solution: Try some test cases.

Suppose that $a = 0$. Is it true that $\frac{2}{3} = \frac{2+0}{3+0}$? $\Rightarrow$ Yes!

Suppose that $a = 1$. Is it true that $\frac{2}{3} = \frac{2+1}{3+1}$? $\Rightarrow$ No!

Although the rule worked in the first test case, it failed in the second. Thus you may conclude that it is *not* generally legitimate to add the same value to the top and bottom of the fraction 2/3. Generalizing (inductively), you could conclude that adding the same value to the top and bottom of *any* fraction is not legitimate. By doing tests like this one before applying a "rule" from memory, you can avoid mistakes caused by doing illegitimate operations.

❖

THINKING ABOUT ... THE MYSTERY OF FERMAT'S LAST THEOREM

An extraordinary example of the interplay between deductive and inductive reasoning in mathematics is offered by Fermat's last theorem. Pierre Fermat (1601–1665) spent most of his life in Toulouse, France. A lawyer and politician, he studied mathematics for leisure and became one of the most prolific mathematicians of all time.

A good way to understand Fermat's last theorem is to recall the theorem of Pythagoras: If a, b, and c are the lengths of the sides of a right triangle and c is the length of the longest side (the hypotenuse), then $a^2 + b^2 = c^2$. Sets of three numbers that satisfy the Pythagorean theorem are called Pythagorean triples, which include the sets:

$a = 3$, $b = 4$, $c = 5$, as $3^2 + 4^2 = 5^2$;
$a = 5$, $b = 12$, $c = 13$, as $5^2 + 12^2 = 13^2$;
$a = 6$, $b = 8$, $c = 10$, as $6^2 + 8^2 = 10^2$.

Fermat raised the following question: If the exponent 2 is replaced by some other natural number (say, 3 or 4 or 5), can similar triples be found that satisfy the relationship? In other words, are there natural numbers a, b, and c that satisfy

$$a^3 + b^3 = c^3 \quad \text{or} \quad a^4 + b^4 = c^4 \quad \text{or} \quad a^n + b^n = c^n?$$

In 1621, Fermat concluded that for any natural number n besides 1 or 2, it is impossible to find natural numbers a, b, and c that satisfy the relationship $a^n + b^n = c^n$. In the margin of his notebook he left one of the most tantalizing sentences in mathematical history: "I have assuredly found a wonderful proof of this, but the margin is too narrow to contain it." When Fermat died in 1665, his notebook was discovered and the margin note became a challenge to mathematicians around the world. For more than three centuries, mathematicians have searched in vain for the proof of what became known as Fermat's last theorem.

Today, the general belief is that the theorem must be true. This belief is based on experimental inductive evidence. With the help of computers, an exponent (say, $n = 3$ or $n = 4$) may be selected and millions of triples a, b, and c tested. Except when $n = 1$ or $n = 2$, no triple of natural numbers has ever been found to satisfy the relationship $a^n + b^n = c^n$. However, all the computer trials in the world do not constitute a proof of the theorem. An acceptable proof of the truth of Fermat's last theorem must be careful, rigorous, and deductive.

The story isn't finished, however. In the summer of 1993, an English mathematician, Andrew Wiles, quietly concluded a lecture in Cambridge, England, with the announcement that he had proved Fermat's last theorem. Wiles's claim was taken seriously because of his reputation and the apparent rigor of his 200-page proof. Unfortunately, after several months of scrutiny, other mathematicians found a flaw in the proof. Wiles now claims to have corrected the flaw, and other experts are examining the revised proof. Mathematicians generally agree on two matters: First, based on inductive evidence, Fermat's last theorem probably is true; second, Fermat himself could not have used a proof as complex as Wiles's (much of the mathematics in Wiles's proof was unknown in Fermat's time). Thus the question remains as to whether Fermat actually proved his theorem. Unless someone else someday finds a simple, clever, and valid proof—one that Fermat himself might have found—the world will never know for sure.

2.6 ARGUMENTS IN THE REAL WORLD

Arguments in everyday life rarely are as clean and simple as the arguments presented so far. Nevertheless, the concepts, skills, and techniques discussed can be used for the analysis of complex real-life arguments.

The most important skill in argument analysis is clear, organized thinking. When you encounter an argument, you need to "pick it apart," and figure out how it all fits together. One method of doing so is to rephrase the argument in terms of premises and conclusions. Another is to rephrase each proposition in standard form for a categorical or conditional proposition. However, as arguments become more complex, visualizing how all the pieces fit together becomes more difficult. Further, some pieces (e.g., unstated premises or conclusions) often are missing from real arguments. In this section, we describe a few techniques for analyzing the more complex arguments that you are likely to encounter in the real world.

2.6.1 Mapping the Flow of an Argument

A useful technique for examining an argument's construction is to map its flow from its premises to its conclusions. We call this technique making a **flow chart** for an argument. The basic idea is to identify all the distinct propositions in an argument and to decide which are premises and which are conclusions. In the process, we can also identify any unstated assumptions.

Independent and Additive Premises

Consider the following inductive argument, in which propositions 1–4 are premises and proposition 5 is the conclusion.

(1) Birds are animals and they are mortal.
(2) Fish are animals and they are mortal.
(3) Spiders are animals and they are mortal.
(4) Humans beings are animals and they are mortal.
———
(5) All animals are mortal.

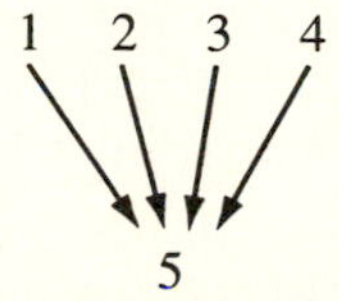

Figure 2-15. In an inductive argument, each premise independently supports the conclusion.

To make a flow chart, we draw arrows to show the flow of the argument from its premises to its conclusion. In this argument, no single premise is essential to reaching the conclusion, but each additional premise strengthens the argument. We therefore say that each premise *independently* support the conclusion. Thus the flow chart shows an arrow pointing from each premise to the conclusion (Figure 2-15).

In contrast consider the following deductive argument.

(1) All carcinogens (cancer-causing materials) are dangerous.
(2) Asbestos is a carcinogen.
———
(3) Asbestos is dangerous.

Figure 2-16. In deductive arguments the premises additively lead to the conclusion.

Again, we want to map the flow of the argument from its premises to its conclusion. In this case, however, neither premise provides independent support to the conclusion. That is, the statement *all carcinogens are dangerous* does *not* suggest that asbestos is dangerous, unless we also know that asbestos is a carcinogen. Similarly, the premise that *asbestos is a carcinogen* implies that asbestos is dangerous only if we also know that carcinogens are dangerous. The two premises support the conclusion only when they are linked. We therefore say that the two premises support the conclusion only when they are combined *additively*. We show this situation on a flow chart by linking the two premises and drawing only one arrow to the conclusion (Figure 2-16).

The flow charts shown in Figures 2-15 and 2-16 suggest a general rule for determining whether premises provide independent or additive support to a conclusion:

- inductive arguments involve premises that *independently* support a conclusion; and
- deductive arguments involve premises that must be combined *additively* to support the conclusion.

Time-Out to Think: Look back at earlier arguments in this chapter. Can you confirm that deductive arguments depend on premises that combine additively, whereas inductive arguments involve independent premises? Explain how.

Example 2-13 Make a flow chart for the following argument and then evaluate it.

Buying the bicycle was a good idea because I needed transportation and the price was reasonable.

Solution: Before you can chart the argument, you need to identify each of its distinct propositions. They are numbered in the order they appear in the argument.

(1) Buying the bicycle was a good idea.
(2) I needed transportation.
(3) The price of the bicycle was reasonable.

Figure 2-17. The flow chart for the argument in Example 2-13 shows two premises leading independently to the conclusion.

Next, recognize that proposition (1) is the conclusion of the argument and that (2) and (3) are premises. The only remaining question is whether the premises independently support the conclusion or provide support only when they are combined. Does statement (2), by itself, offer a reason for buying a bicycle? Yes, the need for transportation certainly is a reason for doing so. Similarly, statement (3) also independently supports the conclusion. Thus the flow chart shows separate arrows pointing from each premise to the conclusion (Figure 2-17).

With the diagram drawn, you can now evaluate the argument. It is not a deductive argument because the conclusion doesn't necessarily follow from the premises; it therefore is inductive (as suggested by the independent premises). Is the argument strong? A need for transportation can argue for buying a bike, but it might also argue for buying a bus pass! A reasonable price also supports the idea of buying a bike, but many things are reasonably priced. However, each premise lends some independent support to the conclusion, and together they make the argument reasonably strong. We could strengthen the argument further with additional independent premises.

Assumed Premises

Consider the argument: "We should support the new Forest Service regulations because they will protect old-growth forests." We can identify and number two propositions in this argument.

(1) We should support the new Forest Service regulations.
(2) The new Forest Service regulations will protect old-growth forests.

Clearly, (1) is the conclusion and (2) is a premise. However, this premise *by itself* doesn't support the conclusion. It supports the conclusion only if it is combined with two unstated assumptions: that protecting old-growth forests is a good idea and that we should support good ideas.

Assumptions that are not explicitly stated but are crucial to the conclusion of an argument are called **assumed premises.** We label assumed premises with an A; hence the two assumed premises in the forest service argument become A1 and A2. The complete list of propositions, including those stated and those assumed, then is

(1) We should support the new Forest Service regulations.
(2) The new Forest Service regulations will protect old-growth forests.
(A1) Protecting old-growth forests is a good idea.
(A2) We should support good ideas.

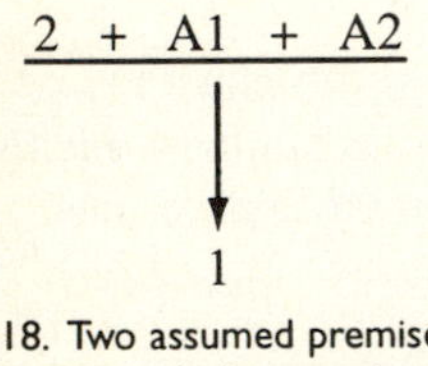

Figure 2-18. Two assumed premises combined additively with the stated premise lead to the conclusion.

Proposition (1) is the conclusion, and (2), (A1), and (A2) are the premises. Note that all three premises must be combined to support the conclusion. We therefore show all three premises to be additive in the flow chart (Figure 2-18). With the assumed premises, the argument is deductively valid because the conclusion follows necessarily from the premises.

As this example suggests, assumed premises usually are additive. We identify them by recognizing that an argument is not logically consistent without them. Further, assumed premises usually are statements that the speaker (or writer) considers so obvious that they need not be stated. Of course, not everyone will always agree!

> **Time-Out to Think:** Would everyone agree that the assumed premises in the forest service argument are "obvious"? Consider, for example, a group of logging executives. Would you agree that assumed premises often are the real points of contention in debates? Why or why not?

Intermediate Conclusions

So far, we have dealt with arguments in which a set of premises leads directly to a conclusion. Often, however, arguments contain many layers of premises and conclusions, as in the following argument.

> I was driving 55 mph (miles per hour) in a 40 mph zone when the cop gave me a ticket. The penalty is $10 for each mile per hour over the speed limit, so I had to pay $150.

We can identify and number five distinct propositions in this argument.

(1) I was driving 55 mph.
(2) The speed limit was 40 mph.
(3) The cop gave me a ticket.
(4) The state penalty is $10 for each mile per hour over the speed limit.
(5) I had to pay $150.

The conclusion of this argument is the proposition *I had to pay $150*. This conclusion is based on three facts: first, that the driver got a ticket, which is proposition (3); second, that the penalty is $10 for each mile per hour over the speed limit, which is proposition (4); and third, that the driver was going 15 mph over the speed limit. Then, multiplying 15 × $10 = $150 yields the conclusion.

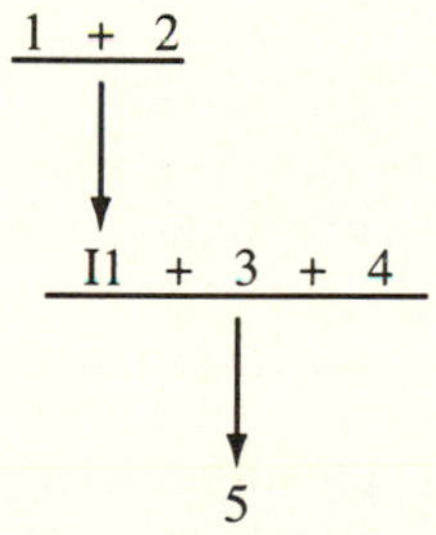

Figure 2-19. An unstated intermediate conclusion labeled (I1) is needed to reach the final conclusion.

However, the fact that the driver was going 15 mph over the speed limit is not explicitly stated in the argument. Instead, we deduced this fact from the first two propositions: that the driver was going 55 mph and that the speed limit was 40 mph. This fact therefore represents an unstated **intermediate conclusion**, which is necessary for proceeding to the final conclusion. We label intermediate conclusions with an I, so this one is denoted I1:

(I1) I was driving 15 mph over the speed limit.

The complete flow chart for this argument is shown in Figure 2-19. Note that propositions (1) and (2) additively make a deductive argument for the intermediate conclusion (I1). It then combines with propositions (3) and (4) to make a deductive argument for the final conclusion (5).

> **Time-Out to Think:** In the preceding argument, we called the unstated fact that the driver was going 15 mph over the speed limit an intermediate conclusion. Why didn't we call it an assumed premise? How do assumed premises differ from unstated intermediate conclusions?

Example 2-14 Make a flow chart and evaluate the following argument.

> I need to take three social science courses to graduate. History courses count as social science courses. Therefore, if I take the course *The American West*, which is being offered this semester, I will satisfy a graduation requirement.

Solution: Begin by identifying and numbering the distinct propositions. To make your work easier, express (2) and (4) as conditional propositions.

(1) Three social science courses are required for graduation.
(2) If a course is a history course, then it counts as a social science course.
(3) *The American West* is offered this semester.
(4) If I take *The American West* this semester, then I will satisfy a graduation requirement.

Proposition (4) is the conclusion of the argument. To reach this conclusion, you need to know the unstated fact that *The American West* is a social science course. This intermediate conclusion (I1), in turn, depends on an assumed premise—that *The American West* is a history course (A1). That is, you reach the intermediate conclusion with the following deductive argument.

(2) If a course is a history course, then it is a social science course.
(A1) *The American West* is a history course.

(I1) *The American West* is a social science course.

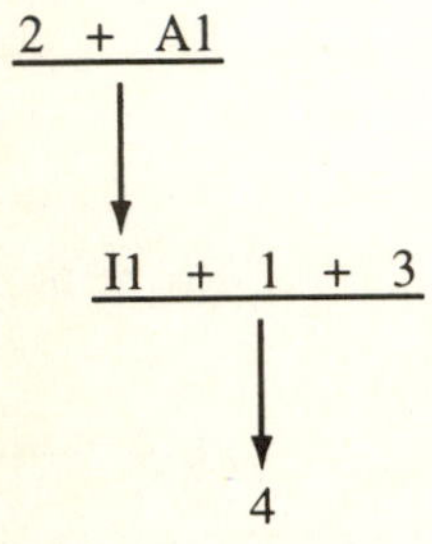

Figure 2-20. This flow chart is based on both an assumed premise and an intermediate conclusion.

The intermediate conclusion, in combination (additively) with propositions (1) and (3), leads to the final conclusion (Figure 2-20). With the assumed premise and intermediate conclusion, the argument is deductively valid because its conclusion follows necessarily from the premises.

❖

2.6.2 Putting It All Together

We now have introduced all the components of ordinary arguments: stated premises and conclusions, assumed premises, and unstated intermediate conclusions. Therefore we're ready to address arguments that are more typical of those encountered in real life. Keep in mind that the process of evaluating an argument consists of four basic steps.

Step 1: Identify all the stated propositions and determine which are premises and which are conclusions.
Step 2: Look for and identify any assumed premises or intermediate conclusions.
Step 3: Analyze the overall flow of the argument; a flow chart can be a useful tool for this purpose but is not required.
Step 4: Evaluate the argument. If it is deductive, is it valid? If it is inductive, how strong is it?

Examples 2-15 and 2-16 illustrate these principles and should solidify your understanding of the process.

Example 2-15 Analyze the following argument.

We should buy a house, as interest rates have declined to reasonable levels and we have enough money for a down payment. Also houses in this part of town have appreciated in value during the last two years, and we could use another bedroom.

Solution: Begin by identifying and numbering the propositions.

(1) We should buy a house.
(2) Interest rates are affordable at our income level.
(3) We have enough money for a down payment.
(4) Houses are appreciating in value.
(5) We could use another bedroom.

The conclusion is proposition (1). Next look for assumed premises and intermediate conclusions. Premises (2) and (3) together seem to imply that *we can afford a house*. However, this implication is based on an assumption, namely, that the combination of money for a down payment and reasonable interest rates makes a house affordable. Thus the argument takes the following form.

(A1) If interest rates are affordable *and* we have enough money for a down payment, then we can afford to buy a house.
(2) Interest rates are affordable.
(3) We have enough money for a down payment.

(I1) We can afford to buy a house.

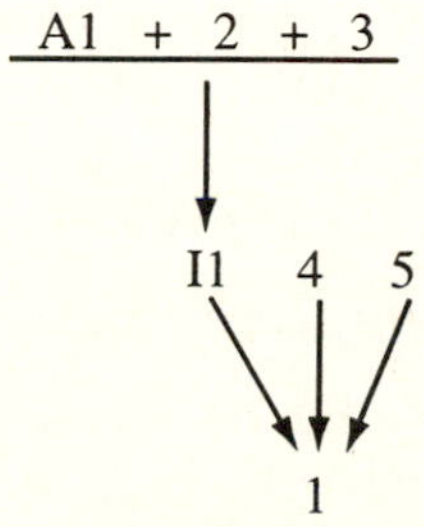

Figure 2-21. This flow chart shows that both an assumed premise and an intermediate conclusion are needed for the argument; the intermediate argument is deductive and the final argument is inductive.

With the assumed premise stated as a conditional proposition, this deductive argument leads to the intermediate conclusion that *"we can afford to buy a house."*

The intermediate conclusion *"we can afford to buy a house"* offers one reason to support the final conclusion *"we should buy a house."* Propositions (4) and (5) independently support the same conclusion. The final conclusion therefore is inductive. By itself none of (I1), (4), or (5) is a very strong reason to buy a house; together they provide a fairly strong inductive argument.

Figure 2-21 shows a flow chart for the argument. Note that the intermediate argument is deductive and hence has additive premises; the final inductive argument has independent premises.

❖

Example 2-16 ***Arguing for Prisons.*** Even apparently simple statements can hide logical complexity. Analyze the following argument.

> We should build more prisons because incarcerating more criminals will reduce the crime rate.

Solution: Begin by identifying the two stated propositions in this argument.

(1) We should build more prisons.
(2) If we incarcerate more criminals, then the crime rate will be reduced.

Proposition (1) is the conclusion and (2) is a conditional premise. However, (2) by itself says nothing about building prisons. Clearly, there are hidden assumptions. You can begin to make sense of the argument by identifying two assumed premises, which additively with premise (2) support an intermediate conclusion.

(A1) If we build more prisons, then more prisoners can be incarcerated.
(2) If we incarcerate more criminals, then the crime rate will be reduced.
(A2) If the crime rate is reduced, then we will have a more desirable society.

(I1) If we build more prisons, then we will have a more desirable society.

Note that this intermediate argument is deductive, because it is a chain of conditionals. The intermediate conclusion can now be combined with a third assumed premise to reach the final conclusion.

(I1) If we build more prisons, then we will have a more desirable society.
(A3) All policies that lead to a more desirable society should be enacted.

---

(1) We should build more prisons.

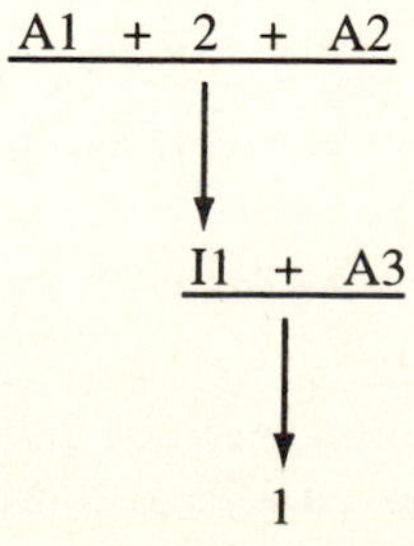

Figure 2-22. This flow chart reveals complexity hidden in an apparently simple statement.

Figure 2-22 shows the complete flow chart. Note how much complexity lies in this apparently simple argument.

Now, let's evaluate the argument. With the inclusion of the assumed premises and intermediate conclusion, the argument is deductively valid. However, of the four premises and one intermediate conclusion necessary to reach the final conclusion, the original argument stated only one. Moreover, having identified all the premises (stated and assumed) you can now evaluate their truth. Assumed premises (A1) and (A2) are fairly reasonable, and most people probably would agree that they are true. Assumed premise (A3) is much more a matter of opinion; for example, some people might argue that a good policy should not be enacted if it has a particularly high cost or if another policy could achieve the same goal at lower cost. Finally, premise (2) is highly contentious; some people believe that there is little correlation between crime rate and the number of people in prison and that locking up more prisoners might not reduce the crime rate at all. Indeed, this premise probably falls into the category of unverifiable propositions. It can't be verified unless "more" prisons are built, and it does not explain how many constitute "more."

In summary, this argument is extremely weak; if closely scrutinized, it is unlikely to persuade anyone to build more prisons.

❖

2.6.3 Onward to Quantitative Reasoning

We have developed most of the basic principles of reasoning. With a few minor exceptions, however, we have not yet done any *quantitative* reasoning. Therefore now is a good time to add some numbers to give you a "taste" of the types of problems you will learn to work in the rest of this book.

In fact, the only difference between the reasoning we introduced in this chapter and *quantitative* reasoning is that the latter adds mathematical or quantitative information. To illustrate the idea, we close the chapter with an example that involves numbers: analyzing airline ticket prices. Although this analysis doesn't make use of a Venn diagram or a flow chart, it does use the most important analytic skill: clear, logical thinking. It also demonstrates the importance of logical analysis, as it portrays a realistic situation with hundreds of dollars on the line.

Example 2-17 *Out-Thinking the Airlines.* If you ever fly, you know that airlines typically offer many different prices for the same trip. Suppose that you are planning a trip six months in advance and discover that you have two choices in purchasing an airline ticket: (A) the lowest fare is $1100, but 25% of the fare is nonrefundable if you change or cancel the ticket; or (B) a fully refundable ticket is available for $1900. Analyze this situation and explain how you would make a decision.

Solution: The first step is to understand the options presented. In fact, you can think of them as a set of conditional propositions. Option (A) may be described by two conditional propositions.

(1) If you purchase ticket A and you go on the trip, then you will pay $1100.
(2) If you purchase ticket A and you cancel the trip, then you will pay $275.

The $275 in the latter case represents the nonrefundable portion of the ticket: 25% of $1100 is $275. Similarly, option B may be expressed as two conditional propositions; however, in this case the ticket is fully refunded if you cancel.

(3) If you purchase ticket B and you go on the trip, then you will pay $1900.
(4) If you purchase ticket B and you cancel the trip, then you will pay $0.

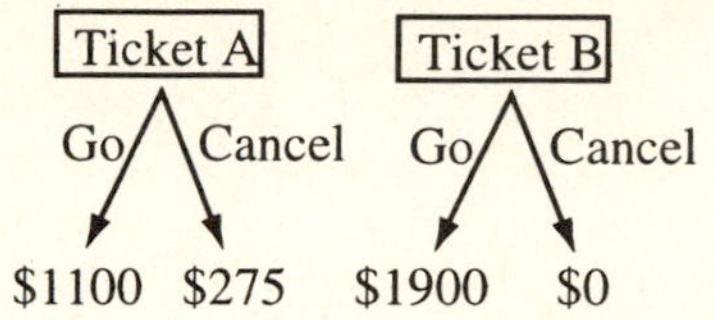

Figure 2-23. A simple flow chart clearly shows the four possible costs of the two ticket options.

Figure 2-23 shows the flow chart representing the four possibilities. Clearly, option A is the better buy if you go on the trip, and option B is the better buy if you end up canceling your trip. Because you are planning six months in advance of the trip, knowing whether you will cancel your trip is difficult. You are therefore interested in the *difference* between the two tickets under the two possibilities (going on the trip or canceling). Combining propositions (1) and (3) yields the intermediate conclusion:

(I1) If you go on the trip, then ticket B costs $800 more than ticket A.

Similarly, propositions (2) and (4) may be combined to conclude:

(I2) If you cancel the trip, then ticket A costs $275 more than ticket B.

With all the possibilities analyzed, you may now weigh the relative benefits of each ticket. In effect, you are balancing the risk of spending an extra $800 if you go on the trip, versus spending an extra $275 if you cancel. Unless you think the probability is high that you will cancel your trip, ticket A is the way to go. (You can quantify this probability by recognizing that $800 is about three times larger than $275. In that case, unless you think the probability of cancellation is at least *three times* higher than the probability of going, ticket A makes more sense. Of course, you may have no way to know the probability that you will cancel the trip!)

If you are familiar with real ticket pricing, perhaps an alternative strategy has occurred to you: You could buy ticket B six months in advance. Then, as the departure approaches (perhaps a month in advance) and you are more certain that you will not need to cancel, ask for a refund on ticket B and buy ticket A. This plan is a good one, but it carries some risk. Air fares may rise and make ticket A more expensive. In addition, airlines typically offer only a limited number of seats at the lower fare, and they may be sold out for the flight(s) you want.

Let's look at the alternatives from the point of view of the airlines. An airline is motivated by profit, and empty seats on a flight mean lost profit. A cancellation penalty reduces the likelihood of cancellations; therefore the airline has a better idea of how many seats will be filled and can accept a smaller profit per ticket (hence the lower price). With fully refundable tickets, the probability is higher that some passengers will cancel — perhaps by not showing up and leaving empty seats. To offset this risk, the airline seeks more profit per ticket (hence the higher price).

2.7 CONCLUSION

In this chapter we introduced the basic principles of reasoning, which are fundamental to developing the ability to reason quantitatively. Key ideas to keep in mind from this chapter include the following.

- Although decision making often involves more than logical reasoning, the ability to organize your thinking through logic will provide you with insight into your own beliefs, as well as those of others.
- Though we do not always explicitly label types of propositions and arguments, understanding the principles of reasoning is fundamental to all the work in the remainder of this book.
- Validity and truth are very different. Validity concerns only logical structure. Truth is nearly always subjective.
- Real-world arguments rarely are clear and simple. Nevertheless, the principles developed in this chapter can help you analyze arguments in depth, thereby better understanding their strengths and weaknesses.

PROBLEMS

Reflection and Review

Section 2.1

1. **How Do *You* Make Decisions?** Make a list of at least five decisions, major or minor, that you made today. Explain the process by which you made each decision. For example, if you based a decision on an emotion, state the emotion involved and how it led to your decision. If you based a decision on logic, explain the premises and reasoning that led to your logical conclusion. If the decision involved more than one process (e.g., logic *and* emotion), explain how each was involved and how you balanced them.

2. **When Logic Fails You.** Undoubtedly you have been involved in disputes in which you were sure you had logic on your side, but you failed to convince the person with whom you disagreed. Relate a personal experience in such a situation and explain why the other person did not agree with your conclusion. Was the other person equally logical, even though he or she came to a different conclusion? If not, on what nonlogical basis did he or she disagree with you?

3. **Do You Trust Your Logic?** Relate a personal situation in which your logic told you to make one choice, but other factors (emotions, etc.) told you to make another. Which choice did you make (the logical or the nonlogical)? In retrospect, was it the right decision? Why or why not?

Section 2.2

4. **Can Logic Settle Debates?** The purpose of logic is to help people distinguish between legitimate and fallacious reasoning. Some believe that applying logic can reveal unquestionably which side of a debate has the better argument. Do you agree? Explain.

5. **Arguments in the Media.** Look back at the four purposes of arguments listed in Section 2.2. Find and describe an example of an argument in the news media that is used for each of the four purposes.

Section 2.3

6. **Working with Categorical Propositions.** For each of the following propositions: (i) if necessary, rephrase the proposition into one of the four standard forms for categorical propositions; (ii) identify the subject set (S) and the predicate set (P); (iii) draw a Venn diagram for the proposition; and (iv) classify the claim of truth made by the proposition as unambiguous, unverifiable, or a matter of opinion; be sure to explain your classification.
   - **a.** All biology courses are science courses.
   - **b.** Some police officers are women.
   - **c.** Some police officers are not tall people.
   - **d.** No Republican is a socialist.
   - **e.** Some states have no coastlines.
   - **f.** No bachelors are married.
   - **g.** Every nurse knows CPR.
   - **h.** Ronald Reagan was a great president.
   - **i.** I do not like broccoli.
   - **j.** Tom Hanks is my favorite actor.

7. **More on Claims of Truth.** For each of the following propositions, explain whether you think its truth or falsity could be established unambiguously, whether it is unverifiable, or whether it is a matter of opinion.
   - **a.** The large federal debt is a drain on the U.S. economy.
   - **b.** Euthanasia is immoral.
   - **c.** The Serbian general was personally responsible for the murder of Bosnian civilians.
   - **d.** Social Security taxes place an unfair burden on the young.
   - **e.** The Social Security System will collapse by the year 2020.
   - **f.** Chinese is the most widely spoken language in the world.
   - **g.** The widespread use of automobiles was the most significant development of the twentieth century.
   - **h.** The average salary of public school teachers is less than 2% of the average salary of a major league baseball player.
   - **i.** Public school teacher salaries are too low.
   - **j.** Hawaii was the fiftieth state to join the Union.

8. **Compound Propositions.** Each of the following statements consists of two propositions joined by either *and* or *or*. State whether each individual proposition in the statement is true or false and then state whether the compound proposition is true or false. Explain your logic.
   - **a.** Washington, D.C., is the capital of the United States, and Beijing is the capital of China.
   - **b.** Washington, D.C., is the capital of the United States, and Shanghai is the capital of China.
   - **c.** Washington, D.C., is the capital of the United States, or Shanghai is the capital of China.

9. **Interpreting *or*.** For each of the following statements interpret the use of the connector *or*. Is it used in the exclusive or inclusive sense? Should the statement be further qualified to clarify it?
   - **a.** The menu offers a choice of appetizer or dessert.
   - **b.** The car warranty covers parts for 3 years or 36,000 miles.
   - **c.** If I win the lottery, I will go to Brazil or Nepal.
   - **d.** I will wear either boots or shoes.
   - **e.** The road will be of made asphalt or concrete.
   - **f.** The biscuits cost 35 cents or three for a dollar.
   - **g.** The insurance policy covers fire or theft.
   - **h.** A discount is offered for children or seniors.

10. **Interpreting Connectors.** A consumer survey of popcorn preferences found 24 people who eat WonderCorn only, 32 people who eat PrimePop only, and 18 people who eat both brands.
    - **a.** How many people were surveyed?
    - **b.** How many people eat WonderCorn only? PrimePop only?

c. How many people eat WonderCorn?
d. How many people eat PrimePop?
e. How many people eat WonderCorn and PrimePop?
f. How many people eat WonderCorn or PrimePop?

11. **Interpreting a Survey.** A survey of newspaper readers asked which of the following three newspapers they read daily: *New York Times (NYT), Washington Post (WP),* and *Wall Street Journal (WSJ).* The results were as follows:

Paper(s)	Readers	Paper(s)	Readers
NYT only	24	*NYT* and *WSJ* only	14
WSJ only	27	*NYT* and *WP* only	16
WP only	26	*WP* and *WSJ* only	13
None	15	All three	8

a. How many people read the *WSJ* only?
b. How many people read the *NYT* and *WSJ* only?
c. How many people read *NYT*?
d. How many people read the *WP* and the *WSJ*?
e. How many people read the *NYT* and *WSJ* and *WP*?
e. How many people read the *NYT* or *WSJ* or *WP*?
f. How many people read the *NYT*, but not the *WP*?

12. **Library Search.** Suppose that you wanted to find the call number of the book *Finnegan's Wake* by James Joyce. For each of the following key word combinations, state whether the book would be on the list. Explain.
a. Finnegan
b. Finnegan and Joyce
c. Joyce or Yeats
d. Joyce and James
e. Joyce and Yeats
f. Finnegan and (Joyce or Yeats)

13. **Conditional Propositions.** Each of the following statements is a conditional proposition. For each: (i) if necessary, rephrase it in the standard form "if p, then q"; (ii) identify p and q; and (iii) explain why you think the conditional proposition is true or false.
a. If it rains today, then I will not need to water the garden.
b. If she is a member of Congress, then she must be a lawyer.
c. Reducing the tax rate will cause our schools to close.
d. He must be fit if he finished that marathon.
e. Graduating from college implies an ability to write.

14. **"If ... Then" to "All S are P."** Conditional (if ... then) propositions often can be interpreted as "all S are P" propositions and vice versa. For example, the proposition "if the animal is a mammal, then it is warm-blooded" can be written as "all mammals are warm-blooded." Rewrite the following conditional propositions in the form of "all S are P" and make a Venn diagram for each proposition.
a. If the candidate is running for governor, then the candidate must be rich.
b. If a person is a nurse, then the person must know how to draw blood.
c. If the flower is a columbine, then it must be blue.
d. If it rains today, then I am staying home.

Section 2.4

15. **Deductive Arguments with Categorical Propositions.** For each of the following arguments: (i) if necessary, rephrase the argument so that all propositions are in a standard categorical form; (ii) draw a Venn diagram from the premises and use it to determine whether the argument is valid; and (iii) discuss the truth of the premises and whether the argument is sound.

a. Premise: All islands are tropical.
Premise: All tropical lands have jungles.
Conclusion: All islands have jungles.

b. Premise: All horses are mammals.
Premise: All horses have long tails.
Conclusion: All mammals have long tails.

c. Premise: All dairy products contain protein.
Premise: No soft drinks contain protein.
Conclusion: No soft drinks are dairy products.

d. Premise: All salty foods cause high blood pressure.
Premise: Some snack foods are salty.
Conclusion: All snack foods cause high blood pressure.

e. Premise: No men are secretaries.
Premise: Some secretaries are tall.
Conclusion: Some tall people are not men.

f. Premise: Some reptiles are snakes.
Premise: All snakes can fly.
Conclusion: Some flying animals are reptiles.

g. Premise: Some lobbyists work for the oil industry.
Premise: All lobbyists are persuasive people.
Conclusion: Some persuasive people work for the oil industry.

h. Premise: No horses have wings.
Premise: All animals with wings can breathe.
Conclusion: No horses can breathe.

i. Premise: No one can get medical treatment without health insurance.
Premise: Some people do not have health insurance.
Conclusion: Some people cannot get medical treatment.

j. Premise: All U.S. presidents have been men.
Premise: George Washington was a man.
Conclusion: George Washington was a U.S. president.

k. Premise: States in the Eastern Standard Time Zone are east of the Mississippi River.
Premise: Maine is in the Eastern Standard Time Zone.
Conclusion: Maine is east of the Mississippi River.

l. Premise: All literature courses are classified as humanities courses.
Premise: "Shakespeare" is a literature course.
Conclusion: "Shakespeare" is a humanities course.

m. Premise: Prime numbers are numbers that have only two factors — themselves and 1.
Premise: All even numbers have 2 as a factor.
Conclusion: No even number greater than 2 is prime.

16. **Deductive Arguments with One Conditional.** For each of the following arguments: (i) if necessary, rephrase the argument so that it fits the standard form of one of the arguments in Table 2-2; (ii) draw a Venn diagram from the premises and use it to determine whether the argument is valid; and (iii) if the argument is invalid, explain the fallacy.

a. Premise: If I don't eat breakfast, then I eat lunch.
Premise: I didn't eat breakfast.
Conclusion: I ate lunch.

b. Premise: If I don't eat breakfast, then I eat lunch.
Premise: I didn't eat lunch.
Conclusion: I ate breakfast.

c. Premise: If I don't eat breakfast, then I eat lunch.
Premise: I ate breakfast.
Conclusion: I didn't eat lunch.

d. Premise: If we can put a man on the Moon, we can build a VCR that works.
Premise: We can build a VCR that works.
Conclusion: We can put a man on the Moon.

e. Premise: When interest rates decline, the bond market improves.
Premise: Last week the bond market improved.
Conclusion: Interest rates must have declined.

f. Premise: When it rains, it pours.
Premise: It is pouring.
Conclusion: It is raining.

g. Premise: Nurses must know CPR.
Premise: Tom is a nurse.
Conclusion: Tom knows CPR.

h. Premise: Knowledge implies power.
Premise: The president is powerful.
Conclusion: The president is knowledgeable.

17. **Chains of Conditionals.** Each of the following arguments may be rephrased as a chain of conditionals. For each: (i) rephrase the argument as a chain of conditionals; (ii) evaluate its validity; and (iii) discuss the truth of its premises and whether the argument is sound.

a. Premise: In the United States, we have the right to say anything at any time.
Premise: Yelling "fire!" in a theater is saying something.
Conclusion: In the United States, we have the right to yell "fire!" in a theater.

b. Premise: If taxes are cut, then taxpayers will have more disposable income.
Premise: With more disposable income, taxpayer spending will fuel the economy.
Conclusion: A tax cut will fuel the economy.

c. Premise: If taxes are cut, the U.S. government will have less revenue.
Premise: If there is less revenue, then the deficit will be larger.
Conclusion: Tax cuts will lead to a larger deficit.

Section 2.5

18. **Analyzing Inductive Arguments.** For each of the following arguments: (i) discuss the truth of each of the premises; (ii) discuss the strength of the argument; and (iii) discuss the truth of the conclusion.

a. Premise: Cows have four limbs, and they are mammals.
Premise: Monkeys have four limbs, and they are mammals.
Premise: Lions have four limbs, and they are mammals.
Conclusion: All animals with four limbs are mammals.

b. Premise: $(-6) \times (-4) = 24$
Premise: $(-2) \times (-1) = 2$
Premise: $(-27) \times (-3) = 81$
Conclusion: Whenever we multiply two negative numbers, the result is a positive number.

c. Premise: Michael Jordan wears Nike shoes, and he is a great basketball player.
Premise: Charles Barkley wears Nike shoes, and he is a great basketball player.
Premise: Larry Bird wears Nike shoes, and he is a great basketball player.
Conclusion: If I wear Nike shoes, I will be a great basketball player.

d. Premise: Bach, Buxtehude, Beethoven, Brahms, Berlioz, and Britten are all great composers.
Conclusion: All great composers have names that begin with B.

e. Premise: Sparrows are birds and they fly.
Premise: Eagles are birds and they fly.
Premise: Hawks are birds and they fly.
Premise: Larks are birds and they fly.
Conclusion: All birds fly.

19. **Testing Mathematical Rules with Inductive Arguments.** As shown in the examples in Section 2.4, testing mathematical rules with an inductive argument is possible. Remember that a single instance in which the rule fails is sufficient to prove that it is not true but that several instances with different types of numbers are needed to make a strong inductive argument in support of a rule. Test the following statements with as many different numbers as you think necessary and decide whether you think the statement is true.

a. Is it true for all numbers a and b that $a \times b = b \times a$?
b. Is it true for all numbers a and b that $a \div b = b \div a$?
c. Is it true for numbers a, b, and c that $a^2 + b^2 = c^2$?
d. Is it true for all positive integers n that

$$1+2+3+\cdots+n=\frac{n\times(n-1)}{2}?$$

Section 2.6

20. Flow Charts for Arguments. Evaluate each of the following arguments according to the four-step process described in Subsection 2.6.2 and prepare a flow chart for each case.

a. This school should be closed because its walls contain asbestos, a known cancer-causing material.
b. Charlie obviously has something to hide. He pleaded the Fifth Amendment in court last week; only people with things to hide plead the Fifth.
c. I saw Jenny in a limousine, so she must be rich.
d. The national inspiration afforded by the Moon landings in 1969–1972 was justification enough for the cost of the program.
e. The Soviet Union lost the war in Afghanistan because the United States provided weapons to the Afghani rebels.
f. Every American has a right to adequate medical treatment. Therefore I can only support a health care reform package that guarantees insurance coverage for all U.S. citizens.
g. Statistics show that in the U.S. a criminal offense occurs every 2 seconds, violent crimes occur every 16 seconds, and robberies occur every 48 seconds. Clearly we need to increase the conviction rate of offenders and strengthen police forces.
h. Good advice in investing is to "buy low and sell high." By the same reasoning, as the number of business degrees awarded by universities has been increasing for 20 years and the number of mathematics degrees has been decreasing for 20 years, becoming a mathematics major would be wise.
i. In a recent poll, 65% of Americans claim to have had at least one prayer answered sometime in their lives. Even if you are not a religious person, trying a few prayers would be a good idea.

21. Quantitative Reasoning. Analyze each of the following situations and explain how you would make a decision.

a. You are planning a trip to visit your family two months in advance. The airline offers you two ticket choices: (1) you may purchase a ticket for $350, and the ticket price can be refunded for a $75 penalty; or (2) you may purchase a fully refundable ticket for $600.
b. You need three sticks of butter for baking. Your local store sells individual sticks for 45¢ each and packages of four sticks for $1.25.
c. You own a small business and need to be in another city on Monday for a meeting. You may fly to the meeting on Monday morning and return Monday evening on a ticket priced at $750. However, if you include a Saturday night stay in your trip, the ticket price is only $335. A hotel will cost $105 per night, and you estimate that meals away from home will cost you $55 per day.
d. You are married and expecting a baby. Your current health insurance costs $115 per month, but doesn't cover prenatal care or delivery. You can upgrade to a policy that covers prenatal care and delivery, but your new premium will be $275 per month. The cost of prenatal care and delivery is approximately $4000.
e. You fly frequently between two cities 1500 miles apart. Airline A offers the trip for an average round-trip cost of $350. Airline B offers the same trip for only $325. However, Airline A has a frequent flyer program by which you earn a free round-trip ticket after you fly 25,000 miles.

Further Topics and Applications

22. Complete Argument Analysis. Use any of the logical tools at your disposal to evaluate the following arguments. Explain the details of your evaluation clearly and be sure to identify assumed premises and intermediate conclusions. Discuss the validity and soundness of any deductive arguments (or parts of arguments) and the strength of any inductive arguments.

a. Amoebas are not plants because they are capable of locomotion, and no plant has that capacity.
b. The athletic program is given more money than any academic department, so this university must value athletics over academics. Also, the football coach is the highest paid university employee.
c. Dependency on foreign oil has put the U.S. economy at great risk, yet the nation has at least a decade's worth of oil untouched in wilderness areas and offshore. The federal government should immediately open all potential oil fields to drilling, regardless of the environmental consequences.
d. During the next decade the number of Americans aged 5–15 will increase. Furthermore the number of Americans aged 25–35 will decline. Therefore the demand for teachers will increase during the next decade.
e. On average, Americans save 4.6% of their disposable incomes, whereas Japanese people save 14.6% of their incomes. Moreover, the personal debt of Americans has increased sixfold in 20 years, and the consumer confidence index has dropped since 1990. These trends show that, overall, Americans have adopted an increasingly "live for the moment" attitude.
f. That we have a health care crisis is clear: Large numbers of Americans are uninsured and the cost of health care to our nation is skyrocketing. I believe that only the government has the power to step in and solve this crisis, so I urge you to support candidates who will make health care reform their top priority.
g. The Federal Reserve Board should lower interest rates because the economy is slowing down, and when the economy slows down there is less investment.
h. The populations of China and India dwarf that of the United States, yet they use far less energy. In the future, the people of China and India no doubt will

demand a standard of living comparable to that in the United States. To achieve it, the governments of China and India are likely to exploit their enormous reserves of coal as an energy source. Their burning of so much fossil fuel undoubtedly will cause a worldwide environmental catastrophe.

i. Overpopulation is not a real problem. Modern technology, especially in bioengineering, will enable scientists to develop far more efficient agriculture. In addition, advances in irrigation technology, along with the development of crops that can grow in salt water, will enable the conversion of much of the world's desert wastelands into productive farms. As a result, agribusinesses will be able to produce enough food for at least 50 billion people, or more than eight times the current world population. Some people claim that such large-scale agriculture would cause environmental damage, but this result is unlikely: the human ingenuity of a much larger population undoubtedly will prevent such damage.

j. Fourth grade students should use pencil and paper to learn to add fractions and strengthen their mental skills. Drill problems also induce good discipline. Therefore fourth grade students should not be allowed to use calculators. Anyway, their teachers and parents never used calculators when they were learning.

k. Word processors with spell checkers allow students to grow up without any spelling skills. Now, word processors have grammar checkers, so students will not learn grammar. What's next? Word processors should not be allowed in the schools.

l. I hate movies based on comics, and *Batman* is based on a comic. And, even if I didn't hate comic-based movies, I never liked the *Batman* comic strip anyway. But my friend Bill said he'd never speak to me again unless I went with him to see the movie because he loves one of the actresses. So that's why I went to see a movie that I knew I would hate.

23. A Financial Decision. Consider the following situation: "I need a special computer for a project I will be working on for the next three months; after that, I will no longer need this computer. I can lease the computer for $350 per month, or I can buy it for $2100. Further, if I resell it after the three months, I can expect to get $1200. Oops, I almost forgot, sales tax is 5%."

a. Identify all the propositions and make a flow chart for the argument.

b. What should you do if you want to make the most economical choice for acquiring a computer?

c. Identify those premises that are "hard facts" and those that are estimates that could affect the outcome of the argument.

d. What would you do in this situation? Why?

24. Airline Overbooking. Most airlines regularly "overbook" flights; that is, they sell tickets for more seats than available for particular flights. By law, the airline has to compensate any passengers denied seats because of overbooking. Explain why overbooking might be a reasonable strategy for an airline trying to maximize profit.

25. Poetry and Mathematics. Consider the following poem, which was written by the English classical scholar and poet A. E. Housman (1859–1936).

Loveliest of Trees

Loveliest of trees, the cherry now
Is hung with bloom along the bough
And stands about the woodland ride
Wearing white for Eastertide.
Now, of my threescore years and ten,
Twenty will not come again,
And take from seventy springs a score,
It only leaves me fifty more.
And since to look at things in bloom
Fifty springs are little room,
About the woodlands I will go
To see the cherry hung with snow.

a. How old was the poet at the time he wrote this poem? (Hint: a score is 20.)

b. Describe how the poem indicates subtraction. What do you think the poet means by "it only leaves me fifty more"?

c. How much longer does the poet expect to live? Did he live that long?

d. Does analysis of the numbers in the poem enhance appreciation of the poem? Defend your opinion.

Projects

26. Analysis of Higher Education. The following statement is a collection of facts and observations concerning American higher education. First analyze the structure of the argument and diagram its flow. Then determine whether the cited facts are true of the college or university that you attend. Explain whether the argument is sound or strong. Can you bring other facts to the argument that might alter the conclusion?

American public universities are approaching a crisis. State funding is declining as priorities shift to public school systems and prisons. At the same time, population projections indicate a 30% increase in high school graduates in the next 15 years. Many universities are at capacity and cannot expand even if funds were available. The solution is to raise both admission standards and tuition at four-year colleges and universities, thereby limiting the number of students admitted to these institutions. Even if the number of university graduates remains level, the U.S. would have one of the highest university graduation rates of any country in the world.

27. Fuzzy Logic. The system of logic that you studied in this chapter is inherited from the ancient Greeks and is sometimes called *two-valued*, or *binary* logic. It based on the assumption that a proposition must be capable of being either true or false, but not both and certainly nothing in-between. Recently, other systems of logic have been devised in which other "truth values" are possible; some of these systems allow for shades of doubt and uncertainty. One form of logic, called **fuzzy logic**, allows for a continuous range of values between absolutely true and absolutely false.

a. Discuss situations in which these multiple-valued systems of logic would be useful and more realistic than the traditional two-valued system.

b. Many engineers are attempting to improve the performance of modern computers and appliances by incorporating fuzzy logic in their designs. Investigate and report on at least one use of fuzzy logic in modern technology. Be sure to explain the advantages offered by the use of fuzzy logic instead of two-valued logic.

28. Animal Reasoning. Do you believe that animals are capable of reasoning, as defined in this chapter? Defend your opinion. Be sure to cite research into animal behavior to support your view.

29. Library Search. Identify a project (for class, work, or personal interest) for which you expect to need materials from the library soon. Go to the library and find the materials you need by using key word searches in the library's computerized catalog. Describe the key word searches you used and the material you obtained.

30. Government Materials on the Internet. Many government materials, including speeches by the president, recent legislation, proposed bills in Congress, and reports by government agencies, is now available on the Internet. Choose a policy issue that is of interest to you. Use the Internet and find recently enacted and proposed legislation that concerns this issue. Describe the process by which you found the legislation (key word searches, etc.).

31. Mail Merge. Many word processors are capable of generating personalized versions of a standard or form letter. This function is usually called a *mail merge.* Most mail merges rely on the use of logical connectors to generate the personalized letters. Study the manual that comes with a word processor you use, and learn how to do a mail merge. Create a list of 5 people (real or imaginary) and write a short form letter. Use the mail merge to generate a personalized version of the letter for each of the 5 people on your list. After you are finished, write a short description of the role of logical connectors in your mail merge.

32. Who Has a Right to March? In July 1995, the U.S. Supreme Court ruled (unanimously) that the private organizers of Boston's St. Patrick's Day parade have the First Amendment right to exclude gay rights activists from marching in the parade. This decision surprised many people, because the march occurs on public land and everyone has the right to march on public land. To understand the decision, you could obtain and study its full text. However, most people get their information from short news stories. Most news stories about this decision included a quote from Justice David Souter's concurring opinion that "one important manifestation of the principle of free speech is that one who chooses to speak may also decide what not to say." Based on this quote, make an educated guess about the chain of reasoning that led to this decision. Do you agree with it? Why or why not?

3 REASONING WITH CARE

In a perfect world, all arguments would be carefully constructed according to the general principles of reasoning presented in Chapter 2. In the real world, however, people often encounter incorrect, or fallacious, reasoning. Sometimes, particularly in advertisements and political campaigns, clever but fallacious arguments are offered in a deliberate attempt to deceive. At other times, mistakes in reasoning are inadvertent or are the result of hidden biases. In this chapter we describe a few of the most common types of errors in reasoning. By studying them, you will learn to evaluate critically the flood of information being generated today and to exercise care in your own reasoning.

CHAPTER 3 PREVIEW:

3.1 ***THE FORCES OF PERSUASION.*** A short introduction sets the stage for a detailed analysis of common errors in reasoning.

3.2 ***FALLACIES OF RELEVANCE.*** Fallacies in which an argument ends in a conclusion that is not relevant to its premises are common. In this section we discuss a set of general fallacies of relevance that are primarily nonquantitative.

3.3 ***FALLACIES OF NUMBERS AND STATISTICS.*** In this section we examine fallacies that are more closely related to numbers, particularly those that are common in the use of statistics.

3.4 ***CAUSAL CONNECTIONS.*** Perhaps the most serious error of reasoning involves claiming that one thing causes another when the evidence is inadequate. In this section we describe the concept of causality and explain how to establish causal connections.

3.5 ***CHAOS.*** In this section we explore the new and exciting field of chaos theory, which holds that some systems, like the weather or the economy, are so complex that establishing clear causal connections may be impossible.

3.6 ***FALLACIES INVOLVING PERCENTAGES.*** Percentages are commonly used in everyday life, yet calculations involving percentages are particularly prone to errors of logic or arithmetic. In this concluding section we demonstrate common fallacies involving percentages.

Here comes the orator! with his flood of words and his drop of reason.

— Benjamin Franklin

...practically no one knows what they're talking about when it comes to numbers in the newspapers. And that's because we're always quoting other people who don't know what they're talking about, like politicians and stock-market analysts.

— Molly Ivins, syndicated columnist

3.1 THE FORCES OF PERSUASION

In the late 1980s, the R. J. Reynolds Tobacco Company began an advertising campaign for Camel cigarettes that featured a cartoon character named Joe Camel. A 1991 study of 229 preschool children by researchers at the Medical College of Georgia concluded that, by age 6, children recognized Joe Camel as readily as Mickey Mouse. According to studies published in the *Journal of the American Medical Association*, illegal sales of Camels to minors skyrocketed from $6 million to $476 million a year between 1988 and 1991. Further, a study of 131 teenaged smokers found that the popularity of Camel cigarettes had increased in parallel with the buildup of the Joe Camel advertising campaign.

Why did R. J. Reynolds think a cartoon character would persuade anyone to smoke its cigarettes? Did it really work? How did the researchers determine that children recognized Joe Camel as readily as Mickey Mouse? How were illegal sales to minors estimated? On a more fundamental level, is there anything wrong with selling cigarettes? Executives the company claimed that no link between smoking and lung cancer had ever been established and that cigarettes are merely a relaxing and enjoyable diversion. Opponents labeled the executives "merchants of death," claiming that they seek to hook young children on cigarettes to replace the customers they kill with their product.

Clearly, the powers of persuasion are being wielded by those on all sides of this issue. To uncover the truth, you must be able to discard arguments based on fallacies while listening to and learning from truly persuasive arguments. In theory, this ability merely requires analyzing the arguments according to the principles of reasoning presented in chapter 2. In practice, doing so can be far more difficult. You may have only a relatively short time to evaluate a question that others have studied for years. Further, those who want to persuade you can call upon a vast array of argument styles — some legitimate and some fallacious — developed during years of research.

In this chapter, we describe some of the fundamental concepts behind the forces of persuasion that are used to influence you. We identify common types of fallacies, including some that are nonquantitative and some that are quantitative in nature. We also address the important question of causality. How, for example, can you decide whether cigarettes cause cancer? You also will encounter one of the most exciting developments in contemporary science: a new tool called *chaos theory* which provides insight into the dynamics of complex systems and places fundamental limits on the ability to assess cause and effect.

3.2 FALLACIES OF RELEVANCE

The word *fallacy* comes from the Latin *fallacia,* meaning "deceit or trick." Fallacies are deceptive because they may lead people to form an opinion or draw a conclusion based on mistaken logic or false notions. In Chapter 2, we presented numerous examples of *formal fallacies,* wherein logical errors occur through a flaw in the *form* (or structure) of an argument. Here, we consider **informal fallacies,** in which an argument is deficient because of its content.

Many common informal fallacies fall into a broad category known as **fallacies of relevance.** A fallacy of relevance is committed when a premise is irrelevant to the conclusion of an argument. Studying fallacies of relevance is important for at least four reasons.

- They are particularly common in arguments over important political, economic, legal, scientific, and technological issues, as well as in political and commercial advertisements.
- They tend to be psychologically persuasive despite their incorrect reasoning and therefore may be difficult to recognize as fallacious.
- Because many people can be fooled by fallacies of relevance, they often are deliberately invoked to sway individual or public opinion.
- Understanding fallacies can help you to understand and evaluate your beliefs and opinions.

Time-Out to Think: More than $100 billion per year is spent on advertising in the United States. Even though some advertisements offer strong and persuasive arguments, many involve fallacies of relevance. What impact do you think advertising has on your ability to detect and understand fallacies?

Fallacies of relevance are categorized and described in many texts. The names and categories used here may differ in some cases from those in other texts. The descriptions here are brief; for more detailed discussion of fallacies, consult a textbook on logic.

Although a poor argument can be created in numerous ways, certain characteristics are particularly common in fallacies of relevance. In this section we categorize and describe some of them, focusing on those that are non-quantitative in nature. In the next section, we move on to consider fallacies that involve numerical information. Keep in mind that a particular fallacious argument will not always fall neatly into one category.

3.2.1 Subjectivism

Consider the following argument.

> I don't care what the Supreme Court says. I was brought up to believe that prayer is an important part of every day, so I am sure that our Constitution cannot prohibit prayer in the public schools.

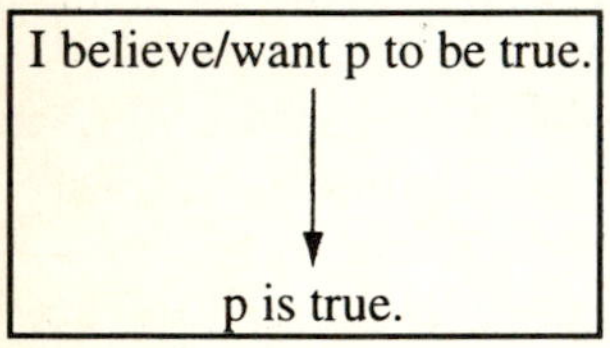

Figure 3-1. The fallacy of subjectivism has the form "I believe/want p to be true, therefore p is true."

The conclusion of this argument is that prayer in public schools is constitutional. However, the premises offered in support are personal opinions. The speaker is certainly entitled to believe that prayer in the schools is a good idea, but this belief is irrelevant to the question of constitutionality. Thus the argument suffers from a fallacy of relevance. More specifically, the fallacy in this argument is called **subjectivism,** because its premises are *subjective* (personal) beliefs or desires.

The general form of the fallacy of subjectivism is: "I believe/want something to be true; therefore it is true." We can draw a flowchart for this fallacy (Figure 3-1), using the letter p to represent any proposition.

Subjectivism is common in advertising. Many advertisements feature some person (often a celebrity) simply stating, "I love this product, and you will too." One person's opinion about a product is not a logical basis for deciding that others also will like it.

Subjectivism also is commonly used to *evade* logical conclusions. Consider the statement, "I'll go on believing that this tax deduction is legal until I'm told otherwise." That simply is another way of saying, "I want this tax deduction to be legal; therefore I assume it is legal." The restatement makes clear the fallacy of subjectivism.

Time-Out to Think: Legally, ignorance is not an excuse for breaking the law. For example, you cannot avoid a penalty by pleading that you didn't know you were taking an illegal tax deduction. What would happen in society if ignorance of the law *did* allow you to avoid punishment? Why would such avoidance make the entire legal system *subjective?*

Subjectivism sometimes takes more extreme forms. In 1610, Galileo used his telescope to discover four moons orbiting the planet Jupiter. Because the moons were *not* orbiting the Earth, this discovery contradicted the prevailing belief that everything circles the Earth. Many of Galileo's contemporaries believed so strongly in an Earth-centered universe that they refused to consider any evidence to the contrary and even refused to look through Galileo's telescope themselves. That is, they took the Earth-centered universe to be an **irrefutable hypothesis** that can never be disproved.

A refusal to consider contradictory evidence is only one way of preserving belief in an irrefutable hypothesis. Another is to invent new, or **ad hoc,** explanations to "explain away" any contradictory evidence that comes to light. An example is provided by experiments in telepathy, in which a subject is supposed to use mental powers to determine which card (from a deck of cards) the experimenter holds. Proponents originally claimed that such experiments would prove the existence of telepathy. When an experiment failed to reveal evidence of telepathy, the ad hoc explanation was that the experimental conditions somehow suppressed the subject's mental powers. When several experiments showed that supposed telepaths guessed the correct card *less often* than would be expected by chance, the proponents claimed this "negative effect" proved the existence of telepathy. When all the experiments, taken together, were shown to occur as chance would predict, the entire setup was said to be flawed. The existence of telepathy appears to be an irrefutable hypothesis to many of its proponents.

The Latin words *ad hoc* mean "for this," and are generally taken to mean "for this special purpose." An *ad hoc explanation* is one created for the special purpose of "explaining away" some piece of evidence.

Time-Out to Think: Ad hoc explanations are particularly prevalent in conspiracy theories. For example, proponents of the theory that the Kennedy assassination was a conspiracy simply assume that any contradictory "evidence" is misinformation put out by the conspirators. Can a true believer be convinced that a conspiracy does not exist? If so, how? If not, why not?

3.2.2 Appeal to Ignorance and Limited Choice

Consider the following argument.

> There is no evidence that any of the predicted consequences of global warming are occurring. The dire warnings of environmentalists about the consequences of global warming clearly are bunk.

The *lack* of evidence for global warming does not necessarily imply that global warming is *not* occurring or that it won't occur in the future. The argument suffers from the fallacy of **appeal to ignorance**, in which ignorance about something is used to support an opposing proposition, which is another fallacy of relevance. The absence of evidence for one proposition doesn't necessarily imply the truth of the opposing proposition (Figure 3-2).

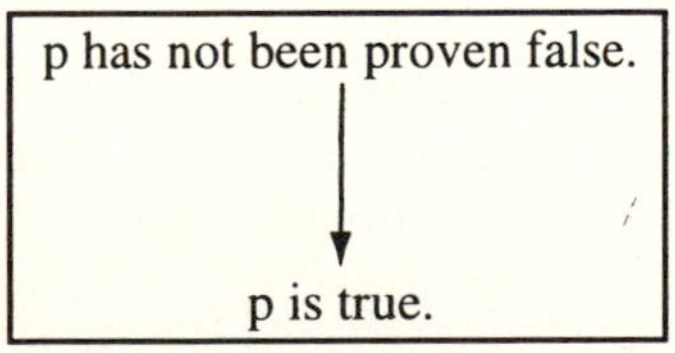

p has not been proven false.
p is true.

Figure 3-2. The fallacy of appeal to ignorance has the form "*p* has not been proven false, therefore *p* is true."

Appeal to ignorance takes the general form: "if something has not been proven false, then it must be true," or "if something has not been proven true, then it must be false." A common example of this fallacy is in statements like, "Can you prove it doesn't exist? No? Then it does!"

Closely related to appeal to ignorance is the fallacy of **limited choice** (or **false choice**), in which an argument precludes choices that should be considered (Figure 3-3). For example, the argument, "if you don't support the president then you are anti-American," ignores (fails to offer) many reasonable choices, such as being patriotic while disliking a particular president. Another example of the limited-choice fallacy is the argument, "You're wrong, so I must be right."

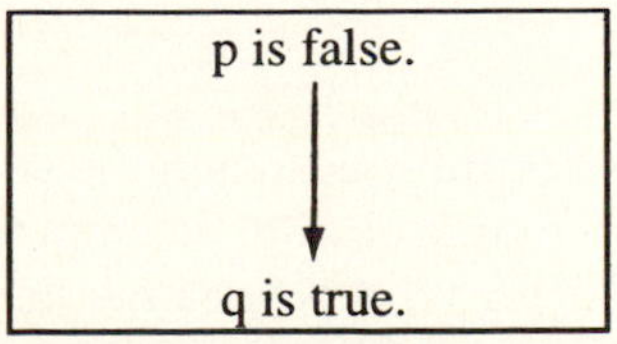

Figure 3-3. The limited choice fallacy has the form "*p* is false, therefore, *q* is true."

3.2.3 Appeal to Emotion

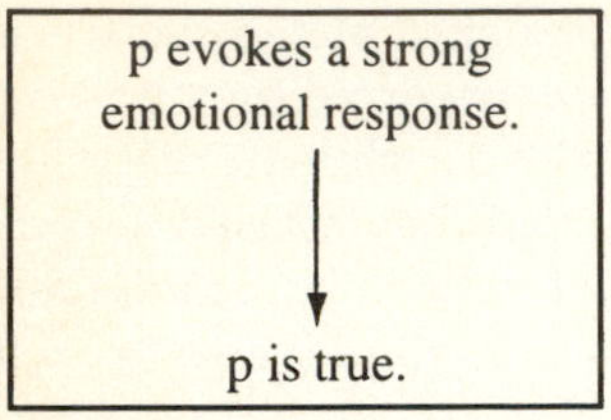

Figure 3-4. The appeal to emotion fallacy has the form "*p* evokes a strong emotional response, therefore *p* is true."

The fallacy of **appeal to emotion** evokes an emotional response in an attempt to persuade someone to accept the truth of a proposition. For example:

> I deserve an A in this class, Professor Bennett! I've been sick for a week, my roommate keeps me up all night, football practice takes three hours a day, I'm always tired, and I try really hard.

No logical argument is offered for the requested grade; instead, the speaker seeks sympathy from the professor. Unfortunately for the student, feelings of sympathy are logically irrelevant to the grade that is deserved (Figure 3-4). Appeal to emotion is one of the most common techniques of persuasion.

Recall a perfume commercial featuring an attractive couple in love or a beer commercial in which beer drinkers are having a great time at the beach. In neither case does the advertiser offer any logical reason for choosing its product. Instead, the advertiser hopes that pleasing images will become associated in your mind with its product.

Time-Out to Think: Turn on the television. Count the number of commercials you watch before you find one that makes an appeal to emotion. How often do commercials use the fallacy of appeal to emotion?

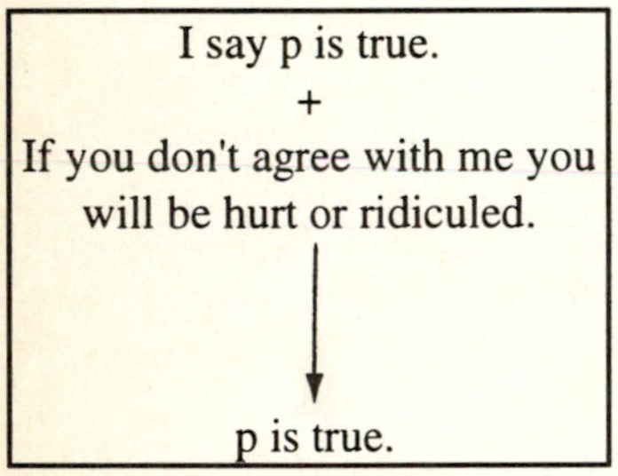

Figure 3-5. Appeal to force has the form "I say *p* is true and if you don't agree with me you will be hurt or ridiculed; therefore *p* is true."

Appeal to emotion also is common in politics. For example, supporters of nuclear power might offer pleasing pictures of families living happily with electricity generated by nuclear power plants, while showing disgusting pictures of pollution produced by non-nuclear sources such as coal-burning power plants. Opponents might show gruesome pictures of people suffering from radiation sickness, or the disastrous consequences of the Chernobyl nuclear accident. As with television advertisements, these appeals offer no logic. However, because public relations experts have found that appealing to emotion often is a successful technique of persuasion, it is sure to remain commonplace.

A variation on the appeal to emotion fallacy, sometimes called **appeal to force,** involves the use of intimidation, pressure, threats of physical or emotional abuse, or other forms of coercion (Figure 3-5). Blackmail is one of the more obvious forms of this fallacy, as reflected in:

> I'm sure you can see that the disturbance at the Alpha Beta House has no news value, Mr. Editor. After all, you must be aware that the members of our fraternity place large amounts of advertising in your newspaper.

The only "argument" advanced for overlooking the disturbance at the Alpha Beta House is an implied threat to stop advertising.

Advertisers sometimes use an appeal to force by threatening ridicule, as in "only nerds don't use our product." Appeal to force is particularly common in negative political advertisements. For example, a candidate might say, "If my opponent is elected, taxes will rise." This argument is an appeal to force because it carries an implied threat: "Vote for me or you will suffer."

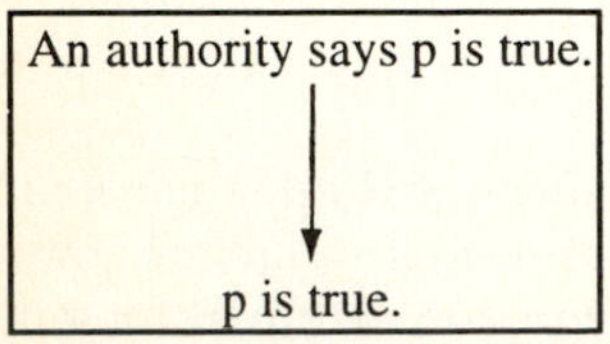

Figure 3-6. An inappropriate appeal to authority has the form "an authority says *p* is true, therefore *p* is true."

3.2.4 Inappropriate Appeal to Authority

The fallacy of **inappropriate appeal to authority** is committed when the support for a proposition relies on the testimony of an inappropriate or unqualified authority (Figure 3-6). For example, when an actor who plays a medical doctor on a popular TV show endorses a particular painkiller, there is no logical reason to think that the actor could provide authoritative advice. But viewers captivated by the actor's role may be charmed into following her advice. The key word in this fallacy is *inappropriate*. Because many modern issues are complex, you often have no choice but to consider the opinions of experts who have studied the issues in depth. Thus, an appeal to authority isn't *necessarily* a fallacy. It may be appropriate to trust expert advice when

1. the supposed authority is a credible expert in the subject at issue,
2. the supposed authority is free of bias or conflict of interest on the particular issue, and
3. the issue is so complex that it requires the specialized knowledge of an authority.

The first criterion demands that a person truly be an expert. The truth may be difficult to assess, and you often must base a judgment on credentials and a proven record. In addition, the first criterion warns that an expert in one field may not be an authority in another. For example, a scientist who studies the Earth's atmosphere may be a legitimate authority on the question of how human activity can affect the climate but probably isn't an authority on the *economic* impact of climatic change.

The second criterion insists on objectivity. If the authority has a financial, emotional, or other special attachment to the issue or product, that person's commentary must be taken with a grain of salt. For example, if an expert is paid for her testimony, as is the case in most product endorsements, she may have an incentive to be less than objective.

Time-Out to Think: In recent years expert witnesses at criminal or legal trials have commonly received payment for their testimony. Can someone who is paid by one side in a trial be considered an objective witness? Why or why not?

Finally, you should not rely on an authority in lieu of thinking for yourself. On complex issues such as global warming, where you have no choice but to consider expert opinions, the authorities themselves often disagree. For example, some experts believe that global warming will produce only minor changes if it is a problem at all; others believe that it may cause catastrophic consequences to civilization. Indeed, the most important benefit of quantitative reasoning skills may be that they enable you to evaluate and judge the opinions of experts.

3.2.5 Personal Attack (Ad Hominem)

The Latin words *ad hominem* mean "to the person." The **ad hominem,** or **personal attack,** fallacy involves attacking the character, circumstances, or motives of a person making an argument (Figure 3-7). For example, consider the statement, "We cannot trust Mr. Roberts's view on health care, for he is an unhappy, bitter, and neurotic man." Even if the claims about Mr. Roberts's character are true, they aren't relevant to the issue of whether his reasoning on health care is correct. Instead, the argument offers only a personal attack on Mr. Roberts's character.

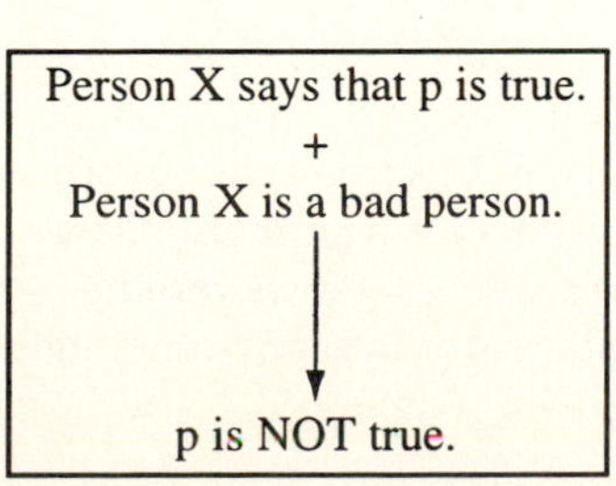

Figure 3-7. The ad hominem fallacy has the form "Person X says that *p* is true and person X is a bad person; therefore *p* is NOT true."

An example of a circumstantial attack is the following exchange:

Gwen: "You should stop drinking because it is hurting your grades, endangering people when you drink and drive, and destroying your relationship with your family."

Merle: "I've seen you drink a few too many on occasion yourself!"

Gwen's argument is well reasoned, with premises offering strong support for her conclusion that Merle should stop drinking. Merle rejects this argument by citing circumstances in which Gwen sometimes drinks too much herself. That may be true, but it isn't a logical response to Gwen's argument.

Another type of personal attack tries to dismiss an argument because of a person's or group's motives. For example:

> This new bill will be an environmental disaster because, as we've shown, its sponsors received large campaign contributions from heavy industrial polluters.

This argument is fallacious because it doesn't challenge the bill; instead, it questions the sponsor's motives. However, consideration of a person's character, circumstances, or motives *may be* logically relevant in one type of situation: when assessing testimonial evidence. In that case, character, circumstances, and motives may determine the witness's credibility. Still, concluding that a witness is unreliable doesn't mean that *all* the person's statements are false.

3.2.6 Begging the Question (Circular Reasoning)

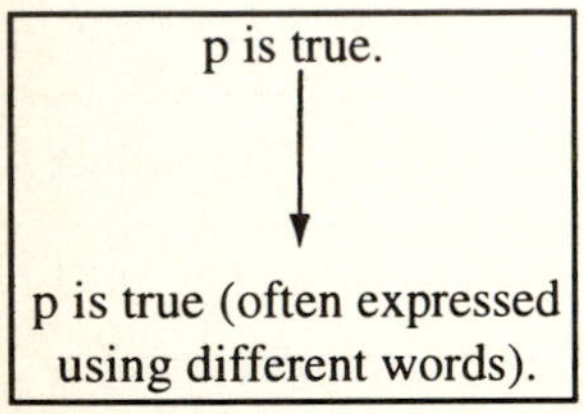

Figure 3-8. Begging the question has the form "*p* is true, therefore *p* is true" (often expressed using different words).

In the fallacy of **begging the question,** or **circular reasoning,** the premises of an argument "beg"the listener to accept the conclusion by asserting it, often in a disguised form (Figure 3-8). For example, consider the statement:

> Society has an obligation to shelter the homeless because the needy have a right to the resources of the community.

Both the premise (the needy have a right to resources) and the conclusion (society has an obligation to shelter the homeless) express the same concept in different ways.

Circular reasoning also is involved anytime the conclusion of an argument is presupposed by its premises, as in:

> Property owners should be allowed to do what they want with their property. You therefore have no right to object to our construction plans because we are the owners of this land.

A variation on circular reasoning is the **complex question,** in which a question is worded so that a particular conclusion is inevitable. A classic example of a complex question is: "Have you stopped abusing your child?" The question presupposes that you have abused your child in the past. Another question, "Why are private services so much more efficient than governmental services?" presupposes the conclusion that private enterprise is more efficient than government. Judges generally rule complex questions out of order in court for attempting to "lead the witness."

3.2.7 Non Sequitur

The Latin words *non sequitur* mean "does not follow." In some sense every fallacy is a non sequitur, because by definition a fallacious argument has a conclusion that does not logically follow from its premises. However, the term is commonly used when a fallacy doesn't fall neatly into any other category, as in:

> Talking on a car phone increases the risk of accident. There should be a law against it.

This argument might at first sound compelling. It can be made deductively valid by adding the assumed premise that "government should make laws against dangerous behavior." But without the assumed premise the argument is a non sequitur: The fact that something is risky or dangerous is a separate issue from the question of what the government should legislate.

Two types of non sequiturs deserve special mention because they are so common. The first is a **diversion,** or **red herring** (herring is a fish that turns red when rotten). It occurs when attention is diverted from the real issue to another issue, as for example:

> We should not continue to fund genetic research because there are so many ethical issues involved. Ethics is at the heart of our society, and we cannot afford to have too many ethical loose ends.

Note that the statement argues about ethics but that its conclusion is about funding genetic research. The argument seeks to divert attention from funding for genetics by instead focusing on ethics.

The term *straw man* comes from the idea that we are arguing against something of our own invention (as we can build a man made out of straw), rather than against a real person or issue.

A second special type of non sequitur is the **straw man,** in which an argument is made against a distortion of someone's idea or position. For example, columnist William F. Buckley has written many articles in which he advocates legalization of drugs. A straw man argument against him might begin: "Buckley thinks that drugs are not a major problem, so...." In fact, none of Buckley's argument's has ever claimed that drugs are a minor problem; he argues for legalization for other reasons. Thus the argument is against a caricature of Buckley's beliefs, rather than against what he actually believes.

A straw man also is frequently invoked when people misunderstand science. For example:

> You cannot tell me that I am wrong, because Einstein's theory of relativity proved that everything is relative.

Einstein's theory does *not* say that everything is relative; the argument uses a straw man distortion of Einstein's theory.

3.3 FALLACIES OF NUMBERS AND STATISTICS

We now turn to a few common fallacies that involve numbers or the collection and analysis of statistical data. These fallacies are particularly important to your work in this text, and you will see them arise again and again.

3.3.1 Appeal to Popularity and Appeal to Numbers

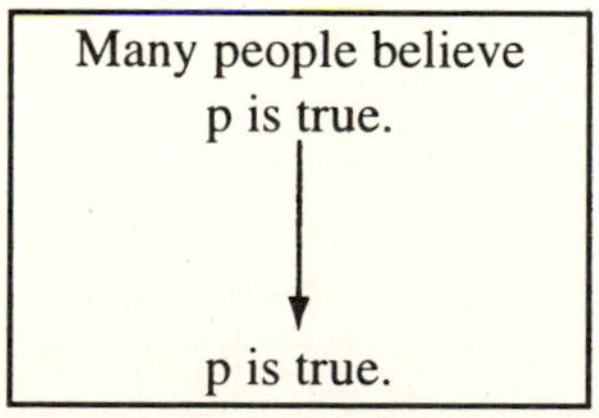

Figure 3-9. The fallacy of appeal to popularity has the form "many people believe *p* is true, therefore *p* is true."

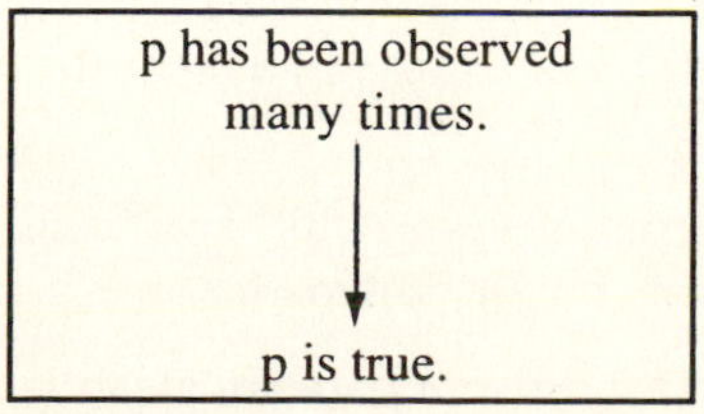

Figure 3-10. The appeal to numbers fallacy has the form "*p* has been observed many times, therefore *p* is true."

The following statement paraphrases a television commercial:

> Ford makes the best automobile in America; after all, more people drive Fords than any other American car.

Aside from claiming that Fords are very popular, no other reason is given to support the conclusion that Fords are the best American-made cars. This argument suffers the fallacy of **appeal to popularity** (also called **appeal to majority**), in which the fact that large numbers of people believe a proposition is used as evidence of its truth (Figure 3-9). This is a fallacy of relevance because the number of adherents to some belief is irrelevant to its truth. After all, only a few hundred years ago, most people believed the Earth to be flat and at the center of the universe. Similarly, the truth of a particular religious belief cannot be determined simply by counting the number of believers!

The fallacy of **appeal to numbers,** in which a conclusion is drawn solely on the basis of quantity, is a variation on appeal to popularity (Figure 3-10). Consider the argument:

> There is no question that Earth has been visited by flying saucers. There have been too many eyewitness reports to believe otherwise.

A large number of "eyewitness reports" cannot, alone, constitute proof. The reports must be analyzed for their quality and reliability. However, a large number of reports might suggest that something is worthy of further study. In the case of flying saucers, enough sightings have been reported to justify scientific investigation. Unfortunately for proponents of flying saucers, these studies have turned up no credible evidence: Pictures are always fuzzy, witnesses are unreliable, or a simple alternative explanation exists for what was seen. This lack of credible evidence doesn't preclude the existence of flying saucers, but many proponents take flying saucers to be an irrefutable hypothesis. Every time a piece of "evidence" of flying saucers is discredited, an ad hoc explanation is invented, usually involving a conspiracy or cover-up by the government.

Many forms of Government have been tried, and will be tried in this world of sin and woe. No one pretends that democracy is perfect or all-wise. Indeed, it has been said that democracy is the worst form of Government except all those other forms that have been tried from time to time. — Winston Churchill

Democracy substitutes election by the incompetent many for appointment by the corrupt few. — George Bernard Shaw

THINKING ABOUT ... IS DEMOCRACY BASED ON A FALLACY?

Although appeal to popularity is not a logical basis for drawing conclusions, it can be an excellent basis for suggesting *possible* truths that can then be tested. Democracy is founded on this very idea. Citizens decide who should hold governmental offices by voting. The candidate who appeals to the majority of voters wins. This method doesn't ensure that those elected are the best choices; it means only that, in a democracy, the people have agreed that there is no better way of making the choice. Democracy allows the winning candidates a chance to prove their worthiness through their actions in office; if the voters don't like they way elected officials perform, they can vote them out in the next election. Of course, a successful democracy depends on the electorate being willing to study candidates and issues in sufficient depth to make reasonable judgments.

An understanding of the fallacy of appeal to popularity also explains why the founders chose to create a *constitutional democracy,* rather than a "pure" democracy (in which *all* decisions are subject to popular vote). The idea behind the Constitution and the rule of law is that some truths must be established independently of popular opinion. For example, the legal system (since the 1960s, at least) recognizes that people of all races are entitled to equal rights under the law. This recognition is based on carefully considered and well-reasoned logical arguments that led to the adoption of numerous statutes, constitutional amendments, and court opinions. Thus, even if the majority of people in some community vote to deny equal rights to members of some race, the law will not recognize that majority opinion. That is because the majority opinion, by itself, cannot logically overrule the weight of evidence and argument that led to the established law.

3.3.2 Hasty Generalization

Subtly different from appeals to popularity and numbers, the fallacy of **hasty generalization** supports a proposition with an inadequate number of instances or instances that are atypical (Figure 3-11). For example, a person might claim:

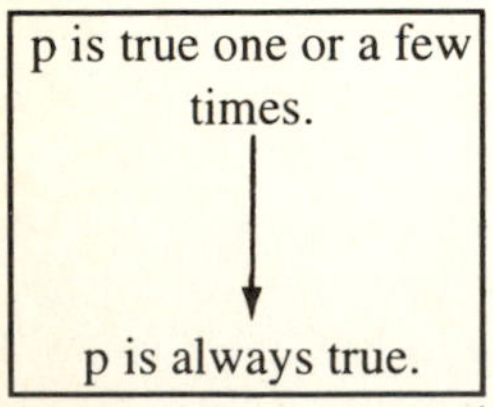

Figure 3-11. The fallacy of hasty generalization has the form "*p* is true one or a few times, therefore *p* is always true."

> I am sure that the chemical company is contaminating our soil. Two local children have already died of leukemia since its factory opened.

Two supporting cases certainly are not enough to establish a pattern. Hasty generalization with atypical instances is illustrated by:

> Look at that woman buying steak with food stamps and driving off in a Cadillac. See? Government programs just create welfare queens.

Not only is a single instance used to support a general conclusion about government programs, but the instance is atypical in that few welfare recipients drive Cadillacs.

Hasty generalization doesn't arise only from a small number of instances. For example, throwing many objects upward shows that "what goes up must come down." In this case, the problem is that a sufficient *variety* of test cases wasn't chosen: If an object is "thrown" with great enough velocity — as is a rocket bound for outer space — it doesn't come back down. Thus avoiding the fallacy of hasty generalization requires observation not only of many distinct cases, but also of a variety of different types of cases.

In statistics, the fallacy of hasty generalization often manifests itself through inadequate sample sizes. For example, suppose that you want to take a poll to learn whether the citizens of a town are likely to vote for a particular candidate. If you were to draw a conclusion based on talking to a few people in a local bar you would have committed the fallacy of hasty generalization. Not only would you have too few people in your sample but, because they all like the same bar, they probably are not representative of the variety of opinions in the town.

3.3.3 Bias and the Availability Error

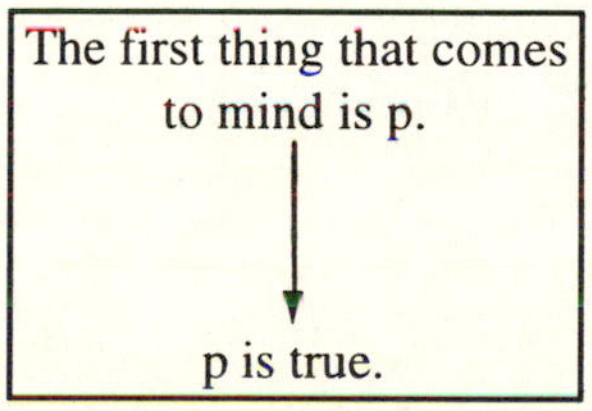

Figure 3-12. The availability error takes the form "the first thing that comes to mind is *p*, therefore *p* is true."

Sound reasoning must be based on objectivity, rather than on subjective bias or prejudice. But to reason without bias is incredibly challenging! One common manifestation of unconscious bias is in what psychologists call the **availability error**, or the human tendency to make judgments based on what is *available* in the mind (Figure 3-12). For example, suppose that you are asked to name a restaurant right after reading an article about tacos, burritos, and enchiladas. A Mexican restaurant is likely to come to mind first. Salespeople, news reporters, pollsters, politicians, con artists, or anyone else trying to persuade you of something can use the availability error in attempting to bias your decision.

The following example, from John Allen Paulos's book *A Mathematician Reads the Newspaper*, shows how opinion poll questions can be biased. According to Paulos, the Yankelovich polling organization asked a sample of respondents a question taken from a questionnaire by Ross Perot's organization:

> Should laws be passed to eliminate all possibilities of special interests giving huge sums of money to candidates?

Of the people polled, 80% responded *yes* and 18% responded *no*. Then the question was rephrased and asked again:

> Should laws be passed to prohibit interest groups from contributing to campaigns, or do groups have a right to contribute to the candidate they support?

The results were dramatically different: 40% *yes*, and 55% *no*. The likely explanation for the change in results is that the original question did not make the opposing argument psychologically *available* to respondents. That is, the original question biased respondents toward a *yes* answer.

> **Time-Out to Think:** Try the following experiment on a few friends. Spell out the following words: MacDonald, macintosh, MacWorld, machine. In each case, spell the word slowly like "m-a-c-(pause)-d-o-n-a-l-d," and ask your friends to pronounce it. How many people get hung up trying to say "Mac–Hine" instead of "machine?" How does this instance illustrate the availability error?

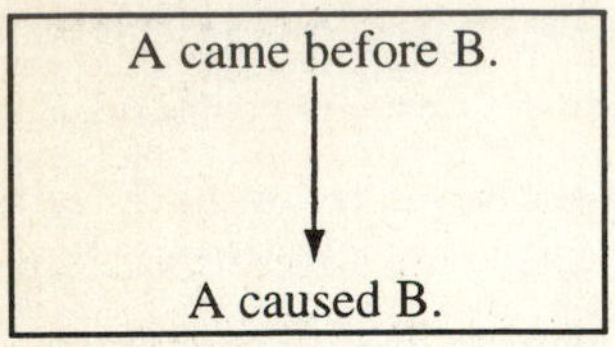

Figure 3-13. The false cause fallacy has the form "A came before B, therefore A caused B."

The effectiveness of a drug often is tested by giving it to some patients, while other patients receive a **placebo:** something that looks just like the drug but contains no medication. If the drug is effective, patients receiving it improve, but patients receiving the placebo do not. Often, however, even patients receiving the placebo improve. This phenomenon, called the *placebo effect*, presumably is a result of the interaction of psychological and physiological processes in the human body; it is a subject of ongoing research.

3.3.4 False Cause

When the fact that one event came before another is incorrectly taken as evidence that the first event *caused* the second event, the fallacy is called **false cause** (Figure 3-13). It is also called the **post hoc** fallacy, from the Latin phrase *post hoc, ergo propter hoc,* meaning "after this, therefore because of this." For example:

> I placed the quartz crystal on my forehead and in five minutes my headache was gone. The crystal alleviated my pain.

The fact that the crystal was on the forehead before the pain went away certainly does not mean the crystal *caused* the headache to go away. It may have been a mere coincidence. Alternatively, it may have been an example of the **placebo effect,** in which *belief* in the benefit of a "treatment" seems to make it work. The fallacy of false cause probably is the most common error in the interpretation of statistical data. By itself, data can only suggest a correlation between two events (e.g., the placement of the crystal appears to be related to the headache). Establishing causality is a much more difficult process, and so important that we devote the next section entirely to it.

3.4 CAUSAL CONNECTIONS

Consider how the answers to the following questions affect decision making.

- Will an extra hour of study improve your test grade or just make you sleepy during the test?
- Does smoking cause lung cancer?
- Will continued use of fossil fuels cause an environmental catastrophe?

If the extra study will not help your grade, there is no point in doing it. If smoking causes lung cancer, smoking might not be a good idea. If continued use of fossil fuels will cause an environmental catastrophe, significant changes in the generation and use of energy are needed. Unfortunately, none of these questions is easy to answer because they are questions of **causality:** They ask whether one thing causes another.

A **causal connection,** or **cause and effect relationship,** may be suggested in many ways. For example, some of the many ways to suggest a relationship in which drunk driving is the cause and auto accidents are the effect include:

- Drunk driving causes auto accidents.
- Driving while drunk often leads to accidents.
- Higher accident rates occur when drunk drivers are on the road.
- At the root of many accidents, we find a drunk driver.

3.4.1 Correlation Versus Cause

A **correlation** exists between two different events when the incidence of one event is related in some way to the incidence of the other. A **positive correlation** means that the incidence of both events rises (or falls) together. For example, air temperature and ice cream consumption are positively correlated: People eat more ice cream during hot weather, and eat less during cold weather. A **negative correlation** means that the incidence of one event increases while the other decreases. For example, smoking and life expectancy are negatively correlated: The more a person smokes, the shorter is that person's life expectancy.

It is tempting to assume a causal connection whenever we find a correlation. However, a correlation may have at least three possible explanations:

1. The correlation may be merely a coincidence.
2. The correlated effects may have a common underlying cause.
3. One of the correlated effects may be the cause of the other.

An example of coincidence is found in a reported correlation between the performance of the stock market (as measured by the Dow–Jones Industrial Average) and the winner of the Super Bowl. The correlation showed that when the stock market rose during the period from November until Super Bowl Sunday in January, the winner of the Super Bowl usually was the team whose city name comes later in the alphabet. That is, if the teams were Denver and San Francisco, a rising market would suggest San Francisco as the winning team because S comes after D. This correlation successfully matched the Super Bowl result in all but 3 of the first 24 Super Bowl games. Nevertheless, it seems preposterous to believe that this is a real cause and effect, and few people would bet on the Super Bowl by checking the stock market and the alphabet.

An example of common underlying cause is found in the correlation: Over a period of several decades the number of ministers in Detroit rose at roughly the same rate as beer sales. Does this correlation imply that going to church causes people to drink more beer? Of course not. Instead, both effects are attributable to an underlying cause: increasing population. More people mean higher beer sales and increased church attendance.

The third explanation for a correlation is the most important, but it is also the most problematic: *Proving* that one event is the cause of another is exceedingly difficult. Let's investigate it.

3.4.2 Causal Factors

A first step in investigating causality is to determine whether one event is a **casual factor,** or **causal condition,** for another. An event is a **necessary condition** if the effect *cannot* happen in its absence. An event is a **sufficient condition** if the effect *always* happens whenever the event occurs. We can illustrate these concepts with some examples.

If an event is a *necessary* condition, the effect cannot happen without it. If an event is a *sufficient* condition, the effect always happens when it occurs.

- Generally, sexual intercourse is *necessary* to cause pregnancy (disregarding artificial insemination and test-tube conception). However, it is not *sufficient* to cause pregnancy: You will not automatically get pregnant if you have intercourse (particularly if you are a male!).
- Starvation is *sufficient* to cause death. It is not *necessary*; there are many other ways to die.
- Having a motive usually is *necessary* to commit a crime, but it is not *sufficient*. Many people might have a motive to commit a particular crime, but still not do it.
- Receiving AIDS-infected blood in a transfusion is *sufficient* to infect the recipient, but it is not *necessary*. The AIDS virus may be contracted in other ways.

Sometimes a causal factor is neither necessary nor sufficient, at least in a single case. For example, smoking is not a necessary condition for lung cancer because some nonsmokers contract lung cancer. Nor is smoking a sufficient condition, because many smokers do not get lung cancer. Instead, smoking is a **probabilistic causal factor** for lung cancer. In other words, smokers have a higher probability than nonsmokers of contracting the disease.

Smoking is necessary to explain the high incidence of lung cancer among smokers. Further, scientific understanding of how tobacco smoke can create cancerous mutations in lung cells shows that it is sufficient to explain the high incidence. Thus, although smoking cannot be established as the cause of any *particular* case of lung cancer, it can be established as the cause of many cases of lung cancer.

Time-Out to Think: In a civil suit, the plaintiffs must show that they have been damaged by the defendant. Suppose that Mr. Smith is ill with lung cancer, and sues the tobacco company whose cigarettes he smoked. Why is proving damage by the tobacco company (the defendant) so difficult for Mr. Smith (the plaintiff)?

3.4.3 Establishing Causality

How is a suspected causal factor truly established as a cause of some effect? Unfortunately, there is no general procedure. Nevertheless, a few general guidelines are available for establishing causality.

Mill's Methods

Mill's methods are somewhat more elaborate than the brief synopsis presented here; for more detail consult any textbook in logic. In addition, we describe only four of Mill's five methods; the fifth, called the *joint method of agreement and difference,* involves using the method of agreement and the method of difference simultaneously.

Nineteenth-century philosopher John Stuart Mill (1806–1873) developed a set of guidelines for establishing causality known as **Mill's methods.** Although Mill's methods have fancy names, they are simple techniques that you probably use in everyday reasoning. Their underlying principles also are used in designing scientific experiments and observations.

The **method of agreement** seeks a common factor when a particular effect is present. For example, suppose that your stomach is frequently upset. The method of agreement suggests looking for a common factor among your cases of upset stomach. Perhaps you realize that an upset stomach always occurs on days that you drink coffee. The *agreement* is that every case of upset stomach is associated with consumption of coffee. Although this agreement doesn't *prove* that coffee is a causal factor, it at least suggests that further investigation is warranted.

The **method of difference** looks for factors whose absence eliminates the effect. Continuing the upset stomach example, perhaps you drink coffee sometimes but do not get an upset stomach. What is the *difference* between your good days and your bad days? You might eventually realize that on the good days you drank decaffeinated coffee. This difference leads you to suspect that the causal factor is caffeine, rather than the coffee itself.

The method of concomitant variation is particularly useful when you cannot use the methods of agreement or difference. Mill applied this method to proving that the Moon is a cause of tides. The methods of agreement and difference cannot be used, because the Moon is always present and cannot be removed. However, noting that high tide occurs when a place on Earth is along the line of sight to the Moon, and low tide occurs when the place is farthest from the line of sight, demonstrates concomitant variation between the tides and the Moon.

The **method of concomitant variation** looks for a *quantitative* relationship between the suspected causal factor and the effect. Continuing the example, you look for effects of *varying* your caffeine consumption. If the amount of caffeine correlates with the degree of your upset stomach, you have shown concomitant variation: When the suspected causal factor (caffeine) increases or decreases, the effect (upset stomach) also increases or decreases. The existence of concomitant variation provides strong evidence that the suspected casual factor is at least partially responsible for the observed effect.

The **method of residues** involves subtracting *known* effects in order to test what is left (the "residue") as a possible cause. Buying salad by the pound for lunch illustrates the idea of a residue. You put salad on a plate and the cashier weighs the salad and plate together. After subtracting the weight of the plate, the remaining weight, or residue, is the weight of the salad. To apply the method of residues to the caffeine example, you need prior knowledge of the precise effects of *everything* else that you eat (besides caffeine) and the knowledge that these effects do not explain your upset stomach. In that case, you could conclude that your upset stomach — the residue after accounting for all other effects — must be caused by caffeine. In this case, the method of residues would be very difficult to apply because of the large number of variables involved and the knowledge of them required.

Going Astray

Suppose that you always put sugar in your coffee but don't add sugar to anything else that you eat. In the upset stomach example, you might therefore have used the method of agreement to conclude that the common factor in all your upset stomachs was the *sugar*, rather than the coffee. The method of difference would substantiate your claim: On days that you did not use sugar your stomach felt fine (no sugar means no coffee). The method of concomitant variation provides further evidence for your belief that when you use more sugar (because you had more coffee) you are sicker. However, blind application of Mill's methods can easily lead you astray, and these methods must be used with care.

Time-Out to Think: Think of an experiment that would prove that sugar is *not* the cause of your upset stomach.

A Physical Model

The mistaken belief that sugar might be the cause of your upset stomach shows that appearances can be deceiving. To be sure that you have discovered a necessary causal factor, you must have tested *every* other possible factor in its absence to prove that the effect cannot occur without it. Similarly, you can be sure that a factor is sufficient only if the effect still occurs after your experiments have eliminated *every* other associated factor. Unfortunately, it is very difficult to be sure that all possibilities have been considered and tested.

As a result, you should not accept a causal link until you understand *how* the suspected factor causes the observed effect. That is, you should seek a **physical model** to explain the cause and effect relationship. Caffeine generally is accepted as a probabilistic causal factor for upset stomachs because there *is* a physical model that explains the causal connection: Caffeine inhibits certain nerve signals in the human body, which can lead to secretions that irritate the stomach wall. Still, although this model shows that caffeine *can* cause an upset stomach, you cannot be certain that this model applies to *your* particular case.

Another example is provided by the question of what causes the planets to orbit the Sun. After discovering the laws of planetary motion in the early 1600s, Johannes Kepler suggested that the laws might be explained by a force emanating from the Sun. However, this idea gained acceptance only after Newton's discovery of the law of gravity made possible the creation of a physical model explaining *how* the planets remain in orbit about the Sun.

3.4.4 Confidence in Causality

The difficulty of proving causality has led some philosophers, most notably David Hume, to conclude that cause and effect don't exist. We do not take such an extreme view.

As we have shown, establishing causality rarely is straightforward. In many cases, proving causality beyond all doubt seems *impossible*. Thinking about a *level of confidence* in a causal link is a more realistic approach. Our level of confidence can range from nil to near certainty, but it is useful to think in terms of three broad levels:

- **Possible Cause:** An *apparent* linkage exists between two events, such as a correlation, but no other evidence suggests causality.
- **Probable Cause:** A good reason to suspect causality exists, perhaps because one or more of Mill's methods substantiates a belief in a causal factor. For example, the legal standard for obtaining a warrant for a search, wiretap, or other method of seeking evidence by the police must be probable cause to suspect that a crime was committed.
- **Cause Beyond Reasonable Doubt:** A model is so successful in explaining the linkage between events that it seems unreasonable to doubt the causal connection. For example, scientists seek a physical model to explain the action of the causal factor, and in court the prosecution presents a model designed to show that the defendant committed the crime.

THINKING ABOUT *... REASONABLE DOUBT*

When a crime has been committed, the prosecution seeks to prove that the defendant was the *cause* of the crime. The standard for conviction in a criminal case is that the prosecution must prove guilt beyond a "reasonable doubt." Unfortunately, the term *reasonable doubt* is rather vague. A 1994 Supreme Court decision illustrates the difficulty.

Traditionally, judges instruct juries on how to interpret reasonable doubt. Until the 1994 decision, a common instruction involved the term *moral certainty.* This instruction dated from an 1850 decision in which the Supreme Court defined reasonable doubt as a mental state in which a juror "cannot say he or she feels an abiding conviction to a *moral certainty* of the truth of the charge." The author of the 1994 decision, Justice Sandra Day O'Connor, traced how the common meaning of the phrase *moral certainty* had changed since 1850 and concluded that the term today conveys a *lesser* degree of certainty than it did in 1850. The 1994 decision therefore warned against continued use of the phrase.

Although the Court agreed that the term should no longer be used, the justices were unable to agree on a replacement definition. Justice Ruth Bader Ginsburg endorsed a set of jury instructions on reasonable doubt proposed by the Federal Judicial Center. Those instructions read:

> *Proof beyond a reasonable doubt is proof that leaves you firmly convinced of the defendant's guilt. There are very few things in this world that we know with absolute certainty, and in criminal cases the law does not require proof that overcomes every possible doubt. If, based on your consideration of the evidence, you are firmly convinced that the defendant is guilty of the crime charged, you must find him guilty. If on the other hand, you think there is a real possibility that he is not guilty, you must give him the benefit of the doubt and find him not guilty.*

Justice Ginsburg stated that this instruction "surpasses others I have seen in stating the reasonable doubt standard succinctly and comprehensibly." Nevertheless, the instructions still seem circular. For example, how firm is "firmly convinced"? What is a "real possibility that he is not guilty"? Some courts, as noted by Justice Ginsburg, have avoided the problem entirely by not providing any definition of reasonable doubt. Like all questions of causality, determining guilt always will involve at least some level of subjectivity.

3.4.5 A Few Case Studies

The best way to learn about establishing causality is through case studies. Each of the following case studies offers a slightly different twist on the problem of causality.

Fluoride and Tooth Decay

Even though correlation does not show causality, it often points the way to research in which a causal factor is discovered. In the 1940s researchers discovered that people in some cities averaged fewer tooth cavities than people in other cities. In other words, the number of cavities in different populations was *correlated* with the cities in which the populations lived. Further investigation of this correlation found that the common factor (using Mill's method of agreement) among the populations with fewer cavities was a high natural level of fluoride in the water supply. As a result, fluoride is now an ingredient in most dental products.

The Milwaukee Water Contamination Incident

In 1993, thousands of people in Milwaukee mysteriously began to suffer from acute diarrhea. Researchers soon discovered that an infectious organism called *Cryptosporidium* had contaminated the water supply. That finding may be regarded as an example of the method of residues. The normal water supply didn't cause illness and the only known change in the water supply was the presence of *Cryptosporidium.* So the *Cryptosporidium* represents the residue believed to cause the illnesses.

The Milwaukee AIDS Project blamed the water contamination for 88 deaths of people with weakened immune systems. However, this conclusion was disputed. *Cryptosporidium* can cause the symptoms observed during the outbreak, but it is not necessary; other organisms produce similar symptoms. Thus, blaming *Cryptosporidium* for a *particular* case of illness is far more difficult than showing causality statistically.

Time-Out to Think: Suppose that an individual with AIDS died during the outbreak and an autopsy found that *Cryptosporidium* was present in the bloodstream. Would you then say that the organism was the cause of death? Why or why not?

What Caused the Challenger Explosion?

In January 1986, the Space Shuttle Challenger exploded shortly after liftoff. Seven astronauts were killed, and the remaining space shuttle fleet was grounded for the next three years. Because the explosion was a single event, Mill's methods weren't applicable — they generally require contrasting situations. Hence a commission was established to determine cause by developing a physical model. Their model indicated that a small hole had formed in one of the O-rings used to seal joints in the shuttle's solid rocket boosters. When the hole formed, a jet of hot material shot through the side of one booster. That, in turn, ignited the shuttle's large liquid fuel tank, causing the catastrophic explosion.

This physical model accounts for all known aspects of the Challenger explosion, and the O-ring failure was therefore considered to be the immediate cause of the accident. But was it the ultimate cause? Further study found a basic design flaw in the joints containing the O-rings; the design was changed after Challenger. Thus it might be argued that a faulty design caused the failure. However, some of the engineers involved in building the solid rocket boosters apparently knew about the problem and claimed that they warned of potential failure in cold weather. In that case, because the temperature was below freezing at the time of the accident, it might be argued that the failure was caused by a negligent launch decision. Another possibility was lack of communication between the engineers and the NASA team responsible for the launch decision, who may not have been aware of the engineers' concerns. In that case, blame might be assigned to a poor management structure or to the people who created the management structure. Who or what would you blame?

Time-Out to Think: Are you old enough to remember the Challenger accident? Did it affect you personally in any way? What effect do you think it had on the nation?

Asbestos and Lung Cancer

Although the level of risk posed by exposure to asbestos, a mineral used in many building materials, is subject to controversy, researchers generally agree that asbestos is a probabilistic causal factor for lung cancer. This conclusion originally was established through work reflecting three of Mill's methods.

The first indications of causality came from the finding that an unusually large number of workers in the asbestos industry had lung cancer. Thus the method of agreement suggested asbestos as the common factor. Further agreement was found when more detailed study showed that the individuals who contracted lung cancer had been exposed to airborne asbestos (dust containing asbestos) in their work. The method of difference was applied by studying workers from the same companies who were *not* exposed to airborne asbestos. Their lung cancer rates were not unusually high, lending further support to the conclusion that asbestos was a causal factor. The method of concomitant variations showed that groups of workers with higher degrees of exposure to airborne asbestos had a higher incidence of lung cancer.

Although the evidence strongly suggested that asbestos is a causal factor, asbestos workers may have been exposed to some *other* material in their work. The asbestos link was generally accepted only after a physical model was developed to explain how it causes lung cancer. According to the model, the microscopic size and shape of asbestos fibers allows them to lodge in the lungs. Once there, they can cause cell damage that can lead to cancer.

The model also explains why asbestos is a *probabilistic* causal factor: Any particular asbestos fiber lodged in the lungs has only a small probability of causing the specific type of cell damage that leads to cancer. The greater the exposure to asbestos, the more fibers lodge in the lungs and the higher is the probability that the fibers will damage a cell and cause it to become cancerous.

The Cosmic Background Radiation

In the mid 1960s two scientists at Bell Labs, Arno Penzias and Robert Wilson, sought to calibrate a radio antenna by pointing it toward parts of the sky with no known sources of radio noise. To their surprise, they found an annoying "hiss" in their receiver no matter where or when they pointed the antenna. Because the hiss was always present, and never seemed to vary, the method of residues was the only practical way to identify the causal factor.

During two years of careful work, Penzias and Wilson accounted for every source of radio noise they could imagine: atmospheric storms, ground interference, the circuitry of the antenna itself, and even pigeon droppings that had fallen into the antenna. In the end, the only residue remaining was the sky itself. They concluded that the hiss represented some kind of radio noise that is found everywhere in space. Because they had no way of explaining this result, Penzias and Wilson did not make it known widely.

At about that time, Penzias sat next to another astronomer, Bernard Burke, on a flight home from an astronomical conference. Penzias told Burke of the troubles with the radio antenna. Burke had recently read a paper discussing the idea that the universe began in a cosmic explosion called the **Big Bang.** The paper predicted that the universe would be filled with remnant radiation from the Big Bang and that this radiation would be observable with radio telescopes. Burke suggested that Penzias might want to talk to the authors, who worked just a few miles down the road at Princeton. Shortly thereafter, the Princeton group joined Penzias and Wilson in announcing the discovery of **cosmic background radiation** — presumed to be the leftover radiation from the Big Bang. Penzias and Wilson received the Nobel Prize for their work.

The Ozone Hole

Our final case study involves an instance when a physical model, at least initially, did not work properly. **Ozone** is a gas that naturally exists high in the atmosphere, where it absorbs harmful ultraviolet radiation from the Sun. During the 1980s, researchers discovered a major reduction in the concentration of ozone occurring each spring over Antarctica — the so-called **ozone hole.**

CFCs have been used in many products including Styrofoam, air conditioners, refrigerators, and industrial solvents. Because they are chemically inert, they were once considered ideal: safe, nonburning, and nontoxic. However, this same inertness keeps them intact for a long time. When air conditioners go to the dump, for example, the CFCs eventually escape and rise into the atmosphere. Once they reach the ozone layer, solar ultraviolet light breaks the CFCs apart, leaving by-products that destroy ozone. After the discovery of the ozone hole, an international treaty was negotiated that calls for eliminating production of CFCs by the year 2000.

A debate quickly arose over the cause of the ozone hole. Existing models of ozone chemistry suggested that human-made chemicals called **chlorofluorocarbons (CFCs)** can destroy ozone. These light gases can rise high into the atmosphere where ultraviolet light from the Sun breaks down their molecules. According to the models, the by-products of CFC breakdown can lead to ozone depletion. However, no existing model could explain the enormous, localized depletion represented by the ozone hole.

The first step in establishing causality involved the discovery of a negative correlation between the atmospheric concentration of the chemical *chlorine monoxide* and ozone; that is, more chlorine monoxide meant less ozone in the ozone hole. Chlorine monoxide is a known by-product of CFC breakdown, but it also can come from other sources. However, the model that explains the breakdown of CFCs also predicts the presence of fluorine products. Finding these fluorine products added strong evidence that CFCs are a causal factor in the ozone hole.

However, because CFCs are distributed globally throughout the atmosphere at all times of the year, they could not be the only causal factor; the ozone hole was found only in the Antarctic spring. Eventually, scientists developed a model to explain the existence of the ozone hole by invoking another causal factor: ice crystals that form in the swirling, springtime atmosphere over Antarctica. According to this model, CFCs are trapped on these ice crystals, which thereby facilitates the chemical reactions that destroy ozone.

3.5 CHAOS

We have shown that the best way to establish causality is by finding a physical model that explains an observed causal connection. For example, models explained how caffeine can cause an upset stomach, how asbestos can cause lung cancer, how an O-ring failure caused the Challenger explosion, how the Big Bang caused the cosmic background radiation, and how CFCs caused the ozone hole. In each case, the causal link seems to be strong and, for this reason, these models are useful for making *predictions.*

What happens when models are used to describe more complex systems like the weather, the human body, the criminal justice system, or rain forest ecology? Can causal links still be identified, and do the models still have predictive value?

3.5.1 The Butterfly Effect

Suppose that you want to predict the location of a car one minute from now. You need two basic pieces of information: where the car is now and how it is moving. For example, if the car is passing Exit 17 on Highway 95 and traveling due north at 1 mile per minute, it will be one mile north of Exit 17 in one minute.

Now suppose that you want to predict the weather. Again, you need two basic types of information for this task:

1. the current weather, and
2. how weather changes from one moment to the next.

Imagine creating a "model world" with which to predict the weather. You could overlay a globe with a graph paper. At each point where lines intersect, you specify the current weather with a set of numbers representing temperature, pressure, and wind. These are the starting points, or **initial conditions,** for predicting weather conditions. Next, you input all these data into a computer, along with a set of equations (physical laws) that describe how the weather changes from one moment to the next.

Suppose that the initial conditions represent the weather around the world *at this very moment*, and you run this model (on the computer) to predict the weather for the next month in New York City. The model might tell you that tomorrow will be warm and sunny, cooling during the week, with a major storm passing through a month from now. Suppose that you run the model again but make one minor change in the initial conditions — say, a tiny change in the wind speed over Brazil. For tomorrow's weather, you will find that this slightly different initial condition will yield a slightly different weather prediction for New York City. For next week's weather, the model may yield a much different prediction. And for next month's weather, the two predictions are unlikely to agree at all!

Those studying chaotic dynamics discovered that the disorderly behavior of simple systems acted as creative process. It generated complexity; richly organized patterns, sometimes stable and sometimes unstable, sometime finite and sometimes infinite, but always with the fascination of living things. — from *Chaos — Making a New Science*, by James Gleick

Why? Because the laws governing change in weather can cause very tiny changes in initial conditions to be magnified with time, leading to huge changes down the line. This extreme sensitivity to the initial conditions has come to be called the **butterfly effect:** If initial conditions change by as much as the flap of a butterfly's wings, the resulting prediction may be very different.

Time-Out to Think: In the 1940s and 1950s many scientists believed that not only would predicting weather with precision months ahead be possible, but also *controlling* the weather would be possible. Explain why the butterfly effect proves that neither of these hopes can be realized.

3.5.2 Chaotic Systems

The butterfly effect is one of the hallmarks of what are called **chaotic systems.** The second defining feature of chaotic systems is the presence of apparently random behavior. Taken together, the two effects impose fundamental limits on making predictions in chaotic systems. They also mean that, in chaotic systems, identifying causal connections is extremely difficult; indeed, the idea of cause and effect may not be relevant at all!

We discuss linear equations in detail in Chapter 10. Generally, the mathematics of nonlinear equations is beyond what we cover in this book, though we do work with a few nonlinear formulas in Chapter 11 and beyond.

What is it about the weather that makes it chaotic and inherently unpredictable? Simple systems are described by **linear equations** in which, for example, increasing a cause produces a corresponding increase in an effect. In contrast, chaotic systems are described by **nonlinear equations,** which allow for more subtle and intricate interactions.

Many nonlinear systems exhibit chaotic behavior. For example, the criminal justice system is nonlinear because increasing the sentence for a certain crime does not automatically cause a corresponding decrease in the crime rate. The economy is nonlinear because a rise in interest rates does not automatically produce a corresponding change in consumer spending. Other examples of nonlinear systems include the stock market and the beating of the human heart, as depicted in Figure 3-14. Despite the name *chaos,* chaotic systems are not completely random. In fact, many chaotic systems have a kind of underlying order that explains the general features of their behavior, even while details at any particular moment remain unpredictable. In a sense, many chaotic systems are *predictably unpredictable.* The understanding of chaotic systems is increasing at a tremendous rate, but much remains to be learned about them.

(a)

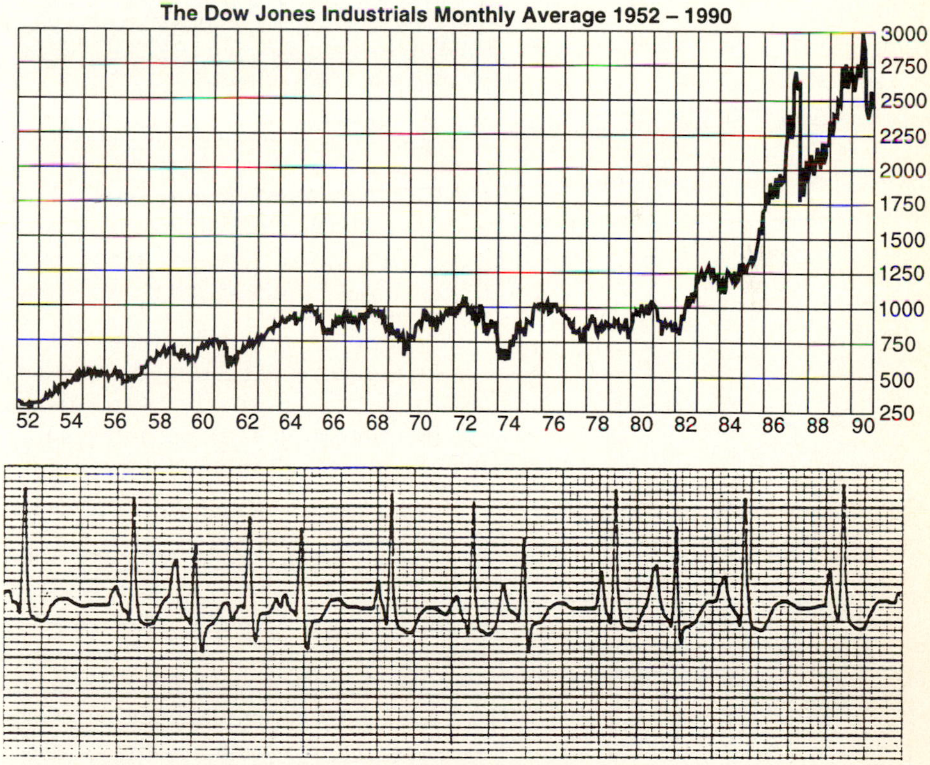

(b)

Figure 3-14. (a) The fluctuations in the stock market show erratic and unpredictable behavior. (b) An electrocardiogram (heart scan) shows a human heart going into fibrillation, in which the regular heartbeat is suddenly replaced by apparently random fluctuations. Both patterns appear to be random, and neither can be precisely predicted, chaos theory can be used to study the underlying complexity of these systems. (b) from Dale Davis, *Differencial Diagnosis of Arrhythmias*, 1991, W. B. Saunders Co. Reprinted with permission.

3.5.3 The Complexity of Global Warming

We began our discussion of causality (Section 3.4) by posing the question, "Will continued use of fossil fuels cause an environmental catastrophe?" This question illustrates the difficulty in reasoning with care about chaotic systems.

The average temperature near the Earth's surface is maintained at livable levels by the **greenhouse effect.** Molecules of carbon dioxide and other **greenhouse gases** in the atmosphere cause this effect by absorbing infrared energy. As shown in Figure 3-15, this action traps excess energy in the atmosphere, which warms the Earth. Without the insulation provided by the greenhouse effect, the average temperature on Earth would be 0°F — inhospitable to life as we know it.

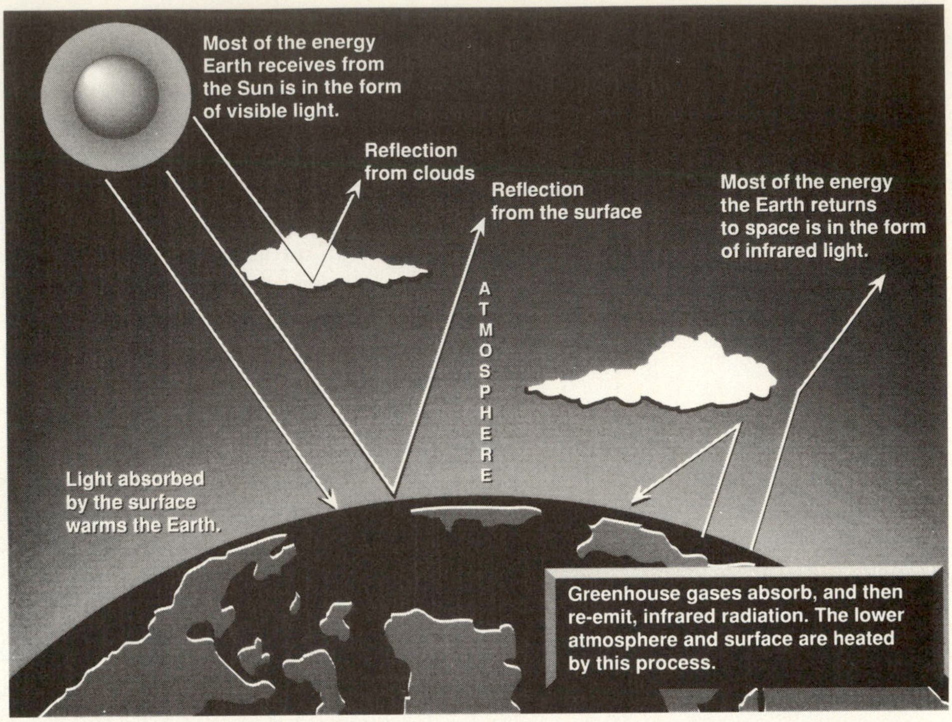

Figure 3-15. The greenhouse effect occurs because carbon dioxide and other "greenhouse gases" in the atmosphere absorb infrared radiation.

Thanks to the naturally occurring greenhouse effect, the global average temperature is a more pleasant 60°F. However, combustion of fossil fuels (coal, oil, and natural gas) releases additional carbon dioxide into the Earth's atmosphere. Thus human activity may enhance the greenhouse effect and increase the global average temperature, which is known as **global warming.** It is a simple argument, which we can write in the following manner.

Premise 1:	Burning fossil fuels adds carbon dioxide to the atmosphere.
Premise 2:	If carbon dioxide is added to the atmosphere, then more energy (heat) will be trapped in the atmosphere.
Premise 3:	If more energy is trapped in the atmosphere, then the Earth will get warmer.
Conclusion:	The burning of fossil fuels will make the Earth warmer.

Premise 1 is a simple statement of fact, easily proved by measuring the carbon dioxide released when fossil fuels are burned. However, both premises 2 and 3 involve statements of causality about the Earth's atmosphere, which is a nonlinear system. Therefore you should not be surprised that premises 2 and 3 are less obvious than they appear!

For example, premise 3 isn't necessarily correct. Adding energy to the atmosphere will not necessarily increase the temperature, at least not in the short term. Instead, the energy might be expended in melting the polar ice caps, increasing evaporation from the oceans, or generating more and bigger storms.

Time-Out to Think: All storms involve atmospheric energy, whether they are winter storms or summer storms. Explain why this condition yields the paradoxical suggestion that "global warming" might mean more frequent and severe winter blizzards or summer hurricanes.

Increased evaporation from the oceans could lead to more clouds, which might lead to less sunlight reaching the ground, which might counteract the warming effects. Some people have used this possible **negative feedback** mechanism to argue that the greenhouse effect will be self-correcting and therefore does not pose a threat to our civilization.

However, because water vapor also is a greenhouse gas, increased evaporation may *further* enhance the greenhouse effect. Could this **positive feedback** make the greenhouse effect far worse than generally feared? Could a collection of unforeseen positive feedbacks cause a runaway greenhouse effect like that which bakes the surface of Venus to 800°F?

Given the uncertainty in predicting whether the burning of fossil fuels will warm the Earth, you can appreciate the problem of trying to predict the precise consequences of an enhanced greenhouse effect. Indeed, the question "Will continued use of fossil fuels cause an environmental catastrophe?" has only one reasonable answer: No one knows because the system is inherently unpredictable. One of the primary challenges of policymaking is to act sensibly in dealing with environmental issues, the possible consequences of which range from minor to catastrophic.

3.5.4 Chaos in Human Behavior

Here we sit, debating the difference between the Republican plan and the Democratic plan as though anyone knows what the economy will be doing seven to ten years from now. It is not only profoundly silly, but the hilarious certitude that our polls invest in this nonsensical debate makes it deliciously funny as well.— Molly Ivins, syndicated columnist

Perhaps the effects of chaos are most evident in human behavior. The human brain is a nonlinear system and often acts in unpredictable ways. Does a child clean up a room when asked? Do smokers quit when they learn that smoking causes lung cancer?

Anytime that human behavior is involved in a system, the system is chaotic. Whether we are talking about a tax cut, a new welfare policy, a crackdown on drugs, or a cut in interest rates, the long-term consequences cannot be accurately predicted. Anyone who claims otherwise is engaged in fallacious reasoning and unaware of the realities of nonlinear systems.

So how can you make decisions that involve chaotic systems? First, remember that although specific consequences can't be predicted, a model of a chaotic system can lead to an understanding of its complexity and general features. In this way you can make an *informed* decision, even though precise predictions are unavailable. Most important, you should not be surprised if a new policy has unintended consequences. Thus you should be flexible in decision making when chaos is present.

Chaos theory has been called the third great revolution in science in the twentieth century (after the theories of relativity and quantum mechanics). Even if that judgment is premature, clearly this exciting new subject has fundamentally altered the way in which we view the world.

3.6 FALLACIES INVOLVING PERCENTAGES

Up to this point in the chapter, we have been concerned primarily with reasoning that involves relevance. We described common fallacies of relevance in Sections 3.2 and 3.3. We then expanded on the important concept of when we can legitimately claim causality and thereby avoid the fallacy of false cause.

We now consider a few common fallacies involving percentages. Unlike fallacies of relevance, these fallacies involve a misunderstanding of how to work with percentages.

Before investigating misuses, let's briefly review the meaning of percentages. The words *per cent* mean *divided by 100.* Thus 10% = 10/100 = 1/10 = 0.1, which shows that percentages are simply fractions.

Percentages are often called *rates;* for example, a sales tax of 6% is called a tax rate of 6%. At this rate, a \$5 purchase will have a tax of

$$\$5 \times 6\% = \$5 \times 0.06 = \$0.30$$

and the total price will be \$5.30. Note that the total price can also be found by multiplying the pre-tax price by 1.06: \$5 × 1.06 = \$5.30. This observation suggests the general rule:

$$\text{total price} = (\text{pre-tax price}) \times (1 + \text{tax rate}).$$

Example 3-1 *A Taste Test.* A taste test asks 125 people to choose between Clear Cola and Cloudy Cola. Forty-nine people prefer Clear Cola, and the rest prefer Cloudy Cola. Express the results as percentages.

Solution: Of the 125 people polled, 49 people preferred Clear Cola; that is, the fraction of the people who preferred Clear Cola is 49/125. To write this as a percentage, first convert the fraction to decimal form and then to a percentage:

$$\frac{49}{125} = 0.392 = 0.392 \times 100\% = 39.2\%.$$

Thus 39.2% of the people sampled preferred the taste of Clear Cola. As the total of the percentages favoring the two colas must be 100%, the percentage favoring Cloudy Cola was 100% – 39.2% = 60.8%

❖

Example 3-2 *Tax Calculations.* Suppose that the local sales tax rate is 5%. (a) If you purchase a shirt with a pre-tax price of \$17, what is the total price? (b) Suppose the total price of a compact disk is \$18.40. What was the pre-tax price?

Solution: **a**. With a 5% tax rate, multiply the pre-tax price by 1.05 to find the total price:

$$\text{total price} = (\text{pre-tax price}) \times (1 + \text{tax rate}) = \$17 \times 1.05 = \$17.85.$$

The total price of the shirt with tax is \$17.85.

b. In this case, the total price is given and the pre-tax price must be determined. Recall that

$$\begin{aligned} \text{total price} &= (\text{pre-tax price}) \times (1 + \text{tax rate}) \\ \$18.40 &= (\text{pre-tax price}) \times 1.05. \end{aligned}$$

Solve for the pre-tax price by dividing both sides by 1.05:

$$\frac{\$18.40}{1.05} = \frac{(\text{pre-tax price}) \times 1.05}{1.05} \Rightarrow \text{pre-tax price} = \$17.52$$

The pre-tax price of the compact disk is \$17.52.

❖

3.6.1 Absolute and Relative Change

Most fallacies involving percentages may be traced to the fact that percentages are used to describe relative, rather than absolute change. We can investigate the difference by using Table 3-1, which lists the estimated world population, by decade since 1950 (from the U.S. Bureau of the Census). The **absolute change** in a pair of numbers is their difference. For example, the absolute change in the world's population from 1950 to 1960 was

$$\begin{aligned} \text{absolute change} &= (\text{population in 1960}) - (\text{population in 1950}) \\ &= 3{,}050{,}000{,}000 - 2{,}570{,}000{,}000 \\ &= 480{,}000{,}000. \end{aligned}$$

Table 3-1

Year	World Population
1950	2,570,000,000
1960	3,050,000,000
1970	3,720,000,000
1980	4,480,000,000
1990	5,320,000,000

The population increased by 480 million people from 1950 to 1960. Because we are comparing populations at two different times, we think of the population in 1950 as the **previous value** and the population in 1960 the **new value.** Then, the definition of absolute change is

absolute change = (new value) – (previous value).

To calculate the **relative change** in the population between 1950 and 1960, we divide the absolute change by the previous value, or

$$\text{relative change} = \frac{\text{absolute change}}{\text{previous value}}$$

In this case, the previous value is the population in 1950, and the absolute change is the increase from 1950 to 1960. Therefore

$$\text{relative change} = \frac{\text{absolute change}}{\text{previous value}} = \frac{480{,}000{,}000}{2{,}570{,}000{,}000} = 0.187 = 18.7\%.$$

The relative change in population from 1950 to 1960 was an increase of 18.7%.

Relative changes often are used to describe prices, populations, or investments. Because relative changes most often are stated in terms of percentages, we sometimes refer to relative change as the **percent change.** In general, if a number changes from a previous value to a new value, we calculate the percent change with the following formula:

$$\text{percent change} = \frac{\text{new value} - \text{previous value}}{\text{previous value}} \times 100\%.$$

Relative change often is more meaningful than absolute change. A few examples should illustrate the value of using relative change.

Example 3-3 ***Stock Gain.*** Suppose that you invest $800 in the stock market. At the end of one year you sell the stock receiving, after commissions, $875. What was your absolute gain from your investment? What was your relative gain, as a percentage?

Solution: The previous value of your stock was the price you paid, or $800; the new value was the price you received, or $875. Your absolute gain was $875 – $800 = $75, because you gained $75 from your sale of the stock. Therefore, your relative gain was

$$\frac{\$75}{\$800} = 0.09375 \quad \text{or} \quad 9.375\%.$$

You gained 9.375% on your investment.

❖

Example 3-4 ***Absolutely Large and Relatively Small.*** The president of the United States earns a salary of $200,000 per year. Suppose that the president decided to donate this salary, which is large by the salary standards of most Americans, to help pay off the federal debt. How much would that help, given that the federal debt is approximately $5 trillion ($5,000,000,000,000)?

Solution: The president's contribution would *reduce* the federal debt by $200,000; that is, it represents an absolute change in the federal debt of *negative* $200,000 (negative changes are decreases). The relative change in the debt would be

$$\frac{-\$200{,}000}{\$5{,}000{,}000{,}000{,}000} = -\ 0.00000004 \quad \text{or} \quad -0.000004\%$$

The relative change in the federal debt would be only 4 one-millionths of 1 percent! Thus, although $200,000 is a lot of money in absolute terms, relative to the salaries of most Americans, it is insignificant compared to the federal debt.

❖

Example 3-5 ***Did Iraq Cause a Stock Crash?*** In July 1990, shortly before the Iraqi invasion of Kuwait, the Dow Jones Industrial Average (DJIA) reached a record value for a day-end closing of 2999.75. By August 22, 1990, the DJIA had fallen to 2603.96. What was the relative change in the DJIA during that month? Do you think the invasion of Kuwait *caused* the decline in the DJIA?

Solution: Use the August 22 average as the new value and the July record as the previous value. The percent change was

$$\text{percent change in DJIA} = \frac{2603.96 - 2999.75}{2999.75} \times 100 = -13.19\%.$$

The negative result indicates that the DJIA *lost* 13.19% of its value during the month. The Iraqi invasion of Kuwait precipitated a major foreign policy crisis for the United States, which eventually led to the Gulf war in 1991. The ensuing decline in the DJIA suggests a possible cause and effect. Perhaps investors sold, fearing that a disruption of oil supplies from oil-rich Kuwait would harm the economy and lower corporate profits. Of course, for every person who sold stock, someone else bought it, presumably thinking that it was a good investment at the new lower price. So, did the Iraqi invasion cause the decline? More than likely the decline was the result of many factors, including the psychology of investors. Indeed, because the stock market is a nonlinear system, attributing causality probably isn't possible.

❖

3.6.2 Fallacies of Relative Change

Let's now look at some of the common ways that calculations with percentages, particularly with relative change, can be misleading. The first lesson, applicable to all the examples that follow, is that percentages often are stated in vague or misleading ways.

Consider, for example, the change in the world's population from 1 billion in about 1850 to about 6 billion today. That is, the population today is *a factor of six* larger than it was in 1850. Because a factor of six is the same as 600%, we can equivalently say that today's population is 600% of the population in 1850. How much has the population increased? It increased by 5 billion, from 1 billion to 6 billion, which is *five* times the 1850 population. Thus we might say that the current population is a factor of five, or 500%, more than the 1850 population. Hence we may describe the relative change using percentages in two different ways:

- today's population is 600% of the 1850 population, and
- today's population is 500% more than the 1850 population.

Both statements are correct, even though they present two different percentages. Note how a subtle change in wording makes an important difference in the use of percentage. Not surprisingly, percentages often are stated incorrectly or ambiguously.

Time-Out to Think: How would you interpret the statement, "The population has increased by a factor of 500%?" Is this statement more ambiguous than the preceding statements?

Example 3-6 *How Deep is the Snow Pack?* On the same day, the following statements were offered on the news about the amount of snow in the mountains: (1) "the snow pack is currently 300% of normal," and (2) "the snow pack is 300% above normal." Are these statements equivalent? Suppose that the normal snow pack is 2 feet deep. What does each statement imply?

Solution: Statement 1 suggests the snow pack is 300% of its normal depth, or 3 times its normal depth. If the normal depth is 2 feet, statement 1 implies that the snow pack is 6 feet deep. Statement 2 suggests that the snow pack is 300% above normal. That means it is 6 feet more than the normal depth of 2 feet, making the total snow pack 8 feet deep. Clearly, one of the two news reports was wrong. Unfortunately, without more information, there is no way of determining which one is wrong.

❖

More Is Less: Shifts in the Previous Value

Imagine that your employer is experiencing a temporary slowdown in business. To avoid layoffs, the employer asks you and other employees to accept a 10% pay cut. However, the employer promises that within six months after the pay cut, you will be given a 10% pay increase. Will the increase restore your original salary?

At first glance, you might think that you will recover your original salary. After all, 10% is 10%, right? Wrong, and this is the heart of the fallacy. To see why, suppose that your pay is $100 per week. Because 10% of $100 is $10, a 10% cut lowers your salary from $100 to $90 per week. When you later receive a 10% raise, the raise is 10% of $90, or $9. Thus your salary after the raise will be only $99 per week, or $1 *less* than your original salary!

Beware of shifts in the previous (or base) value in successive changes.

Overall, the 10% pay cut followed by the 10% pay raise leaves you short of where you began. The reason is that the *previous value*, used to calculate percent change, shifts during the problem from $100 to $90. Any time one percent change follows another, the previous value shifts.

Example 3-7 *Shifting Investment Value.* A stockbroker is confronted by angry investors and offers the following defense: "I admit that the value of your investments under my management fell by 60% during my first year on the job. This year, however, their value has increased by 75%! So stop complaining." Evaluate the stockbroker's defense.

Solution: The stockbroker's defense is clever, because it shows that your percentage gain in the second year was larger than your percentage loss in the first year. In absolute terms, however, the 75% second-year gain leaves you well short of recouping your 60% first-year loss. The broker has tried to conceal a significant loss by using percentages deceptively.

To understand why, imagine that you began with an investment of $1000. During the first year, your investment lost 60% of its value, or $600. Thus your investment of $1000 was worth only $400 at the end of the first year. During the second year, your investment gained 75% in value. But, because it was worth only $400 at the start of the year, its increase in value is only 75% of $400, or $300. Your second year gain brings your investment value only to $700.

Thus at the end of the two years your $1000 investment is worth only $700. Overall, the value of your investment has fallen by $300, or 30%. Despite the large relative gain in the second year, your investment has lost value overall.

❖

Example 3-8 *Just 1%.* Suppose that you invest some money, say $10,000, through a stockbroker who charges a fixed fee of $100 to manage your investments. The broker boasts that his fee amounts to only 1% of your investment. At the end of the year your broker tells you that the value of your investment has decreased slightly and that the fixed fee of $100 is now 2% of your investment's new value. How has your investment performed?

Solution: Because the $100 is now 2% of your investment's value and 2% = 0.02 = 1/50, your investment must be worth 50 times $100, or $5000. Thus your investment has lost half its original value of $10,000! By talking only about the 1% change in his fee, the broker implies that you've lost only a little, when actually you've lost a lot.

❖

Compounding

Compounding involves a series of shifts in previous value. Suppose that you begin with an investment of $100 that earns 10% per year for five consecutive years. You might, at first, expect that five years' worth of 10% returns would mean a total return of 50%, bringing your investment value to $150. Thanks to compounding, however, your actual return is much *better!* We can summarize what happens with a table.

We discuss compounding as an example of exponential growth in Chapter 15, and present the compound interest formula in Chapter 16.

Year	Investment value at beginning of year	10% of investment value	New value at end of year
1	$100.00	$10.00	$110.00
2	$110.00	$11.00	$121.00
3	$121.00	$12.10	$133.10
4	$133.10	$13.31	$146.41
5	$146.41	$14.64	$161.05

Note that, even though the *percentage* increase remains constant at 10% every year, the *absolute* increase grows each year. The reason is that the *new value* at the end of one year becomes the *previous value* for the beginning of the next year. Over the five years, your gain is $61.05, or 61.05% of your original investment.

Example 3-9 *Tax Cuts.* A politician promises, "If elected, I will cut your taxes by 20% for each of the first three years of my term. None of my opponents is willing to promise of a 60% tax cut in three years!" Evaluate the politician's statement.

Solution: The politician neglected the effects of compounding. A cut of 20% in each of three years will *not* make an overall cut of 60%. To see what really happens, suppose that you currently pay $1000 in taxes. The following table shows how your taxes change over the three years.

Year	Tax paid in previous year	20% of tax from previous year	New tax this year
1	$1000	$200	$800
2	$800	$160	$640
3	$640	$128	$512

Over three years, your taxes decline by \$1000 – \$512 = \$488. But \$488 is only 48.8% of \$1000, so your tax bill declines by 48.8% of its original value, not by 60%! (That also means that the new tax bill is 51.2% of its old value.)

❖

Testing Accuracy

Another common fallacy with percentages occurs in statements of testing accuracy, such as a claim like "this test is 90% accurate." The claim might well be true, yet cause very misleading interpretations.

A **polygraph** is a machine that measures a variety of bodily functions, including heart rate, skin temperature, and blood pressure. Widely used by government, law enforcement, and businesses, the polygraph often is called a *lie detector* because many people who are telling lies exhibit subtle changes in bodily functions that can be detected by the polygraph. Unfortunately, lie detection by polygraph test is notoriously inaccurate: As a result, polygraph results are not admissible as evidence in legal proceedings.

The inaccuracies in lie detection by polygraphs stem from a variety of factors: Some people are able to lie without exhibiting changes in bodily functions; some people who are telling the truth become nervous; and the polygraph results must be interpreted by a human operator. The overall accuracy of polygraphs as lie detectors is hotly debated, but even among their strongest proponents few claim an accuracy of greater than 90%.

The fallacy with accuracy is illustrated by the following scenario. A company with 1000 employees has a policy of firing anyone who uses illegal drugs. To show that it is serious, the company decides to give a polygraph test to all employees, asking whether they have used illegal drugs since being hired. Knowing that an admission of drug use would mean getting fired, not a single employee admits to drug use. After testing, company officials announce that 108 employees were lying and fires them. Confronted by the angry former employees, a company official explains: "The test is 90% accurate. We are sorry that there may be a few of you who were fired wrongly, but the vast majority of you are guilty of drug use."

Unfortunately, the official has committed a significant fallacy. Even if the test is 90% accurate, it is *not true* that 90% of the fired employees were guilty! For example, suppose that only 1% of the employees (10 of the 1000 employees) actually were drug users who lied on the polygraph. A 90% accurate polygraph test will catch 9 of these 10 liars, and one will go free.

How, then, does the company end up firing 108 people? Consider what happens to the other 99%, or 990 employees, who were truthful in saying they had not used drugs. Again assuming 90% accuracy, the polygraph test correctly will find that 90% of the 990 people are telling the truth. In 10% of the cases, however, the polygraph test mistakenly concludes that the employees were lying. Because 10% of 990 employees is 99 employees, the polygraph test will lead to 99 *false accusations* of lying.

Thus, the polygraph test catches 9 of the 10 liars, but also falsely accuses 99 of 990 honest employees, making the total number of accused employees 108 (Figure 3-16). Nearly 92% (99 ÷ 108 = 0.917) of the fired employees were innocent; only about 8% actually committed the crime of which they were accused. The company official's statement that "the vast majority of you are guilty of drug use" was far off the mark.

Invalid Adding of Percentages

Percentages often are used to compare different characteristics of the individuals in a particular population. For example, imagine a study of high school dropouts that finds 40% of them came from single-parent families; 20% had suffered physical abuse by an adult; and 25% were alcoholics. You might be

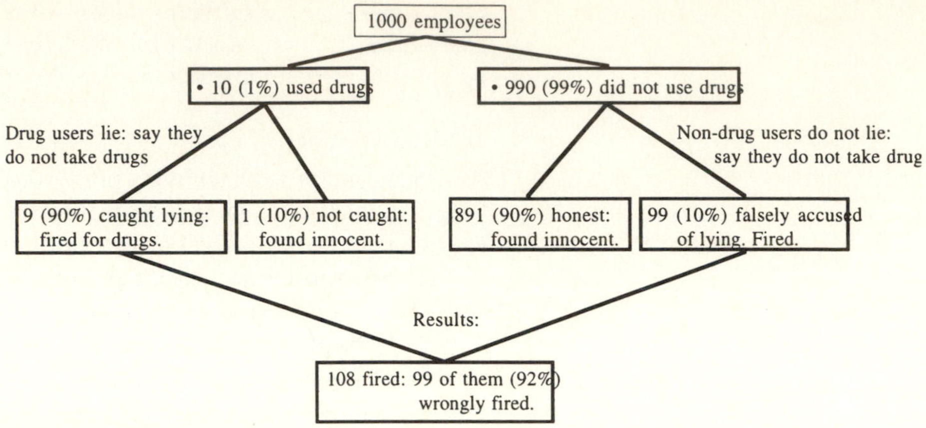

Figure 3-16. A schematic flow chart shows how a lie detector test that is 90% accurate nevertheless leads to a situation in which the vast majority of those accused of lying are falsely accused.

tempted to conclude that 40% + 20% + 25% = 85% of the dropouts were from single-parent families, victims of abuse, *or* alcoholics.

Such a conclusion would be incorrect because an individual dropout might have more than one of the three characteristics. For example, some dropouts might be both from a single-parent family and an alcoholic. The percentages cannot necessarily be added simply.

Time-Out to Think: For the preceding data, what is the range of possible answers to the question: What percentage of dropouts were at least one of the following: from single-parent families, victims of abuse, or alcoholics?

Just Plain Mistakes

So far, we have considered fallacies in which percentages were correctly calculated, but interpreted incorrectly. Unfortunately, many fallacies occur simply because percentages are calculated incorrectly. For example, a store might promise "one-third off," but then mark an item down from $1.50 to $1.10; a one-third discount should result in a price of $1.00.

Another common error is in miscalculating relative change. In his wonderful book, *200% of Nothing,* A. K. Dewdney relates the story of a utility company that promised savings of "200 percent on energy" with new efficient lighting. Because a savings of 100% on energy would mean that no energy is needed at all for the lighting, a savings of 200% implies that the lighting actually produces energy when it is used!

The utility company example is taken from *200% of Nothing*, by A. K. Dewdney (John Wiley & Sons, New York, 1993).

Dewdney traces the problem to the use of an incorrect definition for percent change. Imagine that the new lighting actually reduces energy use by two-thirds, so that lighting that initially required 75 watts of power (the previous value) now requires only 25 watts (the new value). Using the definition of percent change, we find the savings is

$$\frac{25-75}{75}\times 100\% = \frac{-50}{75}\times 100\% = -66.67\%.$$

Note that the percent change is negative, meaning that the new lighting uses about 67% *less* energy than the old lighting. So where did the utility company get its claim of 200%? Look what happens if we "accidentally" switch the new value and previous value in the formula:

$$\frac{75-25}{25}\times 100\% = \frac{50}{25}\times 100\% = 200\%.$$

The utility company found its "200% savings" by miscalculating the change.

3.7 CONCLUSION

In this chapter, we explored common fallacies and how to avoid them, with the aim of helping you learn to reason with care. Key lessons that you should remember from this chapter include the following.

- The forces of persuasion have become extremely sophisticated. Nevertheless, you can avoid being misled by learning to recognize and challenge the many fallacies you encounter.
- Establishing causality is essential to science, courtroom justice, and many other decision-making situations. Yet it is rarely straightforward and is best regarded in terms of a level of confidence in a causal connection.
- Whenever you encounter an analysis of a social, economic, or environmental system, you should consider the possibility of nonlinear or chaotic behavior. If it is present, then the system is inherently unpredictable.
- Calculations involving percentages are particularly prone to errors. Skill with calculating percentages and the ability to recognize fallacies with percentages are essential in today's world.

PROBLEMS

Reflection and Review

Section 3.1

1. **Persuaded.** Identify and describe three instances in which you were persuaded of something that you later decided was untrue. In each case, explain how you were persuaded, and why you later changed your mind.

Section 3.2

2. **Finding Fallacies.** For each type of fallacy listed, find and describe at least one example from an advertisement, a news report, or a political campaign.
   - **a.** Subjectivism
   - **b.** Appeal to ignorance
   - **c.** Limited choice
   - **d.** Appeal to emotion
   - **e.** Inappropriate appeal to authority
   - **f.** Personal attack
   - **g.** Circular reasoning
   - **h.** Diversion
   - **g.** Straw man
3. **Oklahoma City Conspiracy Theory.** In April 1995, terrorists detonated a bomb that killed 167 people and destroyed the Federal office building in Oklahoma City. The terrorists were apparently associated with an anti-government movement that believes the U.S. government is "out of control," and is conspiring with the United Nations to take over the United States. Members of the movement claim that the government is spying on them and seeking to destroy their way of life. Some reporters suggested that, if the claims were true, the government should have known about the planned bombing and prevented it. In response, members of the movement claimed that the government *itself* blew up the building, in order to discredit their group. Explain how this conspiracy theory is an irrefutable hypothesis, and why the response to the suggestion from reporters is ad hoc.

Section 3.3

4. **Finding Fallacies.** For each type of fallacy listed, find and describe at least one example from either an advertisement, a news report, or a political campaign.
   - **a.** Appeal to popularity
   - **b.** Appeal to numbers
   - **c.** Hasty generalization
   - **d.** Availability Error
   - **e.** False Cause
5. **Opinion Poll Bias.** Find a recent news report of a single-question opinion poll. State the exact words of the question and the results of the poll.
   - **a.** Find a way to reword the question in a way that you think would change the poll results.
   - **b.** How do you think your new wording would affect the poll results? Why?
   - **c.** Do you think that the original question was biased? Explain.

Section 3.4

6. **Causal Connections in the News.** Search newspapers from the past week and find at least three instances in which a causal connection is claimed. For each case, briefly explain the claimed causal link. Then, write a paragraph or more explaining whether you believe a cause and effect relationship is involved.

7. **Considering Correlations.** Each of the following exercises describes a correlation. State whether you believe that the correlation occurs because of mere coincidence, because of an underlying cause for both phenomena, or because one is the cause of the other. Explain your answers. Propose a cause when you believe one is involved.
   - **a.** The crime rate in the United States has been rising at the same time that the number of people in prison has increased. That is, the crime rate is correlated with the number of people sent to prison.
   - **b.** The incidence of AIDS is correlated with sexual promiscuity. That is, the more sexual partners a person has had, the more likely he or she is to be infected with the AIDS virus.
   - **c.** Over the past three decades, the number of freeways and freeway lanes in Los Angeles has grown, and traffic congestion has worsened. That is, there is a correlation between increased availability of freeways and increased traffic congestion.
   - **d.** Astronomers have discovered that, outside the Milky Way's relatively small neighborhood of the universe, all galaxies are moving away from it. Moreover, the farther the galaxy, the faster it is moving away. That is, there is a correlation between the distance to a galaxy and the speed at which it is receding.
   - **e.** In one section of a class grades correlated with the age of the students' mothers. That is, the older the mother, the higher was the grade.
   - **f.** In some studies, the incidence of melanoma (the most dangerous form of skin cancer) increases with latitude. That is, skin cancer is correlated with living in the generally cooler climates of more northerly latitudes.
   - **g.** When gasoline prices rise, attendance at National Parks declines. That is, there is a correlation between gasoline price and the number of visitors to National Parks.
   - **h.** A large number of classical composers had last names beginning with the letter B (Beethoven, Bach, Brahms, Bartok, Berlioz, and Bruckner). That is, talent at classical composition is correlated with the first letter of last names.
   - **i.** Sales of ice tea in a local restaurant are positively correlated with ticket sales at the local swimming pool.
   - **j.** When a ban on deer hunting was imposed in a mountainous region, the incidence of mountain lion sightings increased.
   - **k.** As the volume of water flowing through mountain streams increases, the sale of snow shovels in nearby communities decreases.
   - **l.** A grocer notices that if she raises the price of peaches, fewer peaches are sold in a single day.
   - **m.** Among the clients of a particular astrologer, people born between April 21 and May 20 seem to be stubborn and strong-willed.

8. **Causal Conditions.** Each of the following exercises lists a cause and an effect. State whether the cause is a necessary causal condition, a sufficient causal condition, or a probabilistic causal condition for the effect described. Explain your answers.
   - **a.** Cause: poor grades. Effect: expelled from school.
   - **b.** Cause: getting caught for speeding. Effect: paying a fine.
   - **c.** Cause: a high fat diet. Effect: coronary disease.
   - **d.** Cause: being a famous movie star. Effect: getting the lead role in a movie whose script you love.
   - **e.** Cause: temperature above freezing. Effect: snow is melting.
   - **f.** Cause: it is winter. Effect: it is cold.
   - **g.** Cause: air bags in cars. Effect: people surviving head on collisions.
   - **h.** Cause: ozone depletion. Effect: skin cancer.
   - **i.** Cause: sexual intercourse with an AIDS-infected person. Effect: contraction of AIDS.
   - **j.** Cause: infection with the AIDS virus. Effect: death.
   - **k.** Cause: suspect found present at the scene of the murder. Effect: suspect found guilty of murder.
   - **l.** Cause: rising interest rates. Effect: falling stock market.
   - **m.** Cause: always have trouble being on time. Effect: being unemployed.
   - **n.** Cause: poor mathematical skills. Effect: job applications for managerial positions are routinely rejected.

9. **What Was the Cause of Death?** A person who was on a hunger strike died. Can you conclude that starvation was the cause of death? Why or why not?

10. **Smoking and Lung Cancer.** There is a strong correlation between tobacco smoking and incidence of lung cancer, and most physicians believe that tobacco smoking causes lung cancer. Yet, not everyone who smokes gets lung cancer. Briefly describe how smoking could cause cancer when not all smokers get cancer.

11. **Suing the Tobacco Companies.** Many of the families of smokers who died of lung cancer have sued tobacco companies, claiming that actions of the tobacco companies were the cause of death. Despite the strong evidence showing that smoking is a probabilistic causal factor in lung cancer, few (if any) of the plaintiffs have won such cases. Why do you think that winning such a case is so difficult? What kinds of evidence might lead a jury to rule in favor of the plaintiffs?

12. **Upset Stomach.** Suppose that your frequent upset stomachs are correlated with coffee consumption. Imagine that you suspect the underlying causal factor is the *dark color* of coffee. You try a series of experiments in which you test the effects of other drinks such as water, fruit juice, milk, coffee, tea, cola, and lemon–lime soda. You find that an upset stomach occurs whenever you consume coffee, tea,

or cola. You are fine when you drink water, fruit juice, milk, or lemon–lime soda.

a. Does the experiment provide further evidence for your belief that the dark color of the drink causes your stomach aches? Explain.
b. Describe an experiment to show that dark color is *not* a causal factor in your upset stomachs.

13. **Identifying Causes: Headaches.** Suppose that you are trying to identify the cause of late-afternoon headaches that plague you several days each week. For each of the following tests and observations, explain which of Mill's methods you used and what you concluded.
    a. The headaches occur only on days that you go to work.
    b. If you stop drinking Coke at lunch, the headaches persist.
    c. In the summer, the headaches occur less frequently if you open the windows of your office slightly. They occur even less often if you open the windows of your office fully.
    d. Having made all these observations, what reasonable conclusion can you reach about the cause of the headaches?

14. **Identifying Causes: Car Trouble.** Suppose that your car is stalled on the side of the road. For each of the following tests and observations, explain which of Mill's methods you used and what you concluded.
    a. The car still has all four tires securely attached.
    b. The oil is at the correct level.
    c. The starter is getting an electrical charge, but the engine will not "turn over."
    d. The gas gauge reads empty (remember the car is not running).
    e. You last put gas in your car three weeks ago.
    f. Having made all these observations, what reasonable conclusion can you reach about the cause of the breakdown?

15. **Physical Models.** Each of the following statements describes a generally accepted causal connection. Suggest a physical model that explains each connection.
    a. Running out of gas causes a car to stop.
    b. Being unable to breathe for half an hour causes death.
    c. Dropping a book causes it to fall.
    d. Inflating a balloon and then letting it go (without tying the end) causes it to fly about the room.
    e. Lack of rain creates a forest fire hazard.

Section 3.5

16. **Chaotic Systems Around You.** Characterizing chaotic systems as those that exhibit irregular behavior or sensitive dependence on the initial state, identify three news stories or everyday situations that seem to characterize chaotic behavior. Explain your choices.

17. **Sensitive Dependence.** Imagine that you are driving around town and approach a traffic light. Now imagine two scenarios: In the first, the traffic light is already yellow and you decide to stop as it turns red. In the second, the light is just turning yellow and you drive through. These two scenes differ only in the split second between the traffic light turning yellow and already being yellow.
    a. Is this split second difference *likely* to cause any major change in your life? Why or why not?
    b. Describe a way in which this split second difference *could* drastically change your life.
    c. Explain how this example illustrates that your life is like a chaotic system, with extreme sensitivity to initial conditions.

18. **Weather Forecast.** Look at a five-day-old newspaper and find a weather report that forecasts the local weather five days ahead.
    a. For each of the five days, describe whether or not the forecast was accurate.
    b. In general, do you think a forecast five days ahead is as likely to be accurate as tomorrow's forecast? Explain.

Section 3.6

19. **Practice with Change.** Express the following changes in terms of absolute change and relative change (with a percentage). Don't forget that decreases are indicated by a negative change.
    a. a change from 50 to 55
    b. a change from 0.1 to 0.5
    c. a change from 1000 to 1001
    d. a change from 23 to 20
    e. a change from 50,000 to 45,000
    f. a change from 0.0001 to 0.001

20. **Working with Percentages.** In the following problems, you are told that an amount is some percentage of a total. In each case, find the total. *Example:* 25 is 45% of the total. *Solution:* The total is 25/0.45 = 55.55.
    a. 23 is 3% of the total.
    b. 100 is 15% of the total.
    c. 150 is 89% of the total.

21. **Sales Tax.** Suppose that your local sales tax is 8.3% on nonfood purchases. You purchase a TV set and a videodisk player for $495 and $429 (pretax prices), respectively. What is your total bill, tax included?

22. **Decimals, Fractions, and Percentages—All in One!** Suppose that you buy a $250 CD player at a "⅓-off" sale. The sales tax is 5.5%. How much do you actually pay for the CD player?

23. **Change Everywhere!** Look at any newspaper or news magazine, listen to any TV or radio newscast, and notice how often you are told of change. Quite frequently, change is described in terms of percentages. Translate each of the following (factual) statements into a statement about percent change.
    a. The U.S. population has changed from 220 million people in 1974 to 265 million people in 1995.

b. The number of deaths from heart disease has decreased from 250,000 in 1970 to 140,000 in 1993.
c. Defense spending by the U.S. government increased from $90 billion in 1974 to $390 billion in 1989.
d. Consumer debt in the United States increased from $160 billion in 1974 to $850 billion in 1995.
e. The average student–teacher ratio in U.S. public schools decreased from 21 in 1974 to 17 in 1988.
f. The percentage of bachelor's degrees awarded to women increased from 44% in 1972 to 54% in 1992. The percentage of doctoral degrees awarded to women changed from 18% to 38%. (Hint: Don't be confused by the fact that the quantities changing are themselves percentages; you can still use the relative change formulas.)
g. The percentage of single-parent households in the United States increased from 12% (of all households) in 1970 to 30% (of all households) in 1995.

24. Watch the Words! The average annual precipitation on Mt. Washington, New Hampshire, is 90 inches. During one particularly wet year, more rain and snow fell than usual. The news carried both of the following statements.

- The precipitation this year is 200% of normal.
- The precipitation this year is 200% above normal.

What do each of these statements imply about the precipitation during this year? Do the two statements have the same meaning? Explain.

25. Pay Cut. Suppose that you are earning a salary of $1000 per week.
a. Your company experiences a slowdown in earnings, and asks all workers to take a 20% pay cut. What is your salary after the cut?
b. Six months later, the company has recovered. They offer you a 20% increase on your salary at that time (i.e., the salary you have been earning since the time of the pay cut). How much will you be earning after the increase?
c. Explain *why* your salary after the increase is not back to its original level.
d. What percentage pay raise would you need to restore your original salary after a 20% cut?

26. Profitable Company? In defense of apparent losses, the president of a company states: "I realize that last year our profits dropped by 25%. Fortunately, we increased our profits by 45% this year over last year. So I'd say we are doing pretty well!" Evaluate the president's analysis.

27. Changing Test Scores. A high school reports that their students' SAT scores were down by 20% for one year. The next year, however, they rose by 30%. As a result, the high school principal announced at the PTA meeting, "Overall, test scores have improved by 10% over the past two years."
a. Is the principal's announcement correct? Describe what the principal has done.
b. How much have test scores actually changed over the two-year period?

28. Shifting Percentages. Suppose that you pay a flat $10 monthly service charge on your checking account. In March, that fee is 1% of your balance. In April, the (same) fee is 4% of your balance.
a. What was your balance in March?
b. What was your balance in April?
c. By what percent did your balance change from March to April?

29. Compounding Interest. On January 1, 1990, you deposited $500 in a checking account that earns 5% interest compounded at the end of each year. Make a table that shows the balance in your account on January 1, 1991, through 2000. By how much has your balance increased during the decade, in absolute and relative terms?

30. Drug Test Accuracy. Suppose that 1000 people are given a drug test that is 98% accurate and that 50 of the people actually are drug users.
a. If the test is 98% accurate, we can expect approximately 98% of the 50 drug users to test positive. How many of the 50 drug users can we expect to test positive?
b. Similarly, we can expect 98% of the 950 nonusers to test negative. How many of the nonusers can we expect to test positive?
c. Using your answers from parts (a) and (b), how many of the 1000 people can we expect to test positive? Of these, how many are users?
d. What percentage of the positive tests are, in fact, false positives (nonusers who test positive because of the inherent inaccuracy of the test)?

31. Polls and Percentages. Political pollster John Farsight pitches the following argument to his campaign staff: "35% of the voters in the district are Union members, and 40% of the voters in the district are African-American. We can capture virtually all the votes in each of these two groups. That will give us 75% of the votes in the district and an easy win." Suppose all of Farsight's premises are true. Explain, including a numerical example, how Farsight's conclusion could be false and how his candidate could, in fact, lose the election in this district.

32. Be Prepared! On a certain Boy Scout outing, 70% of the scouts carried a compass, 75% carried a knife, 80% carried a map, and 85% carried a flashlight. What percentage of scouts, at a maximum, carried all four items? (Extra credit: what percentage of scouts, at a *minimum*, carried all four items?)

33. Car Deal of the Century. A car dealer advertises a great deal. If you trade in your old car, he says that he will offer a 100% discount on a select line of new cars that originally cost $10,000.
a. How much should the new car cost you?
b. When you actually try to take advantage of the deal, the dealer tells you the car will cost $5000. What error has he made in his use of percentages? Explain.

34. Bargain Hunting. In Niwot, you pay $5 in tax for a lawn mower, which is 5% of the purchase price. In Longmont, you pay $5 in tax for the same lawn mower, but it is 7% of the purchase price. In which town do you get a better price for the lawn mower and why?

Further Topics and Applications

More on Fallacies

35. Advertisers Hoping You'll See A Causal Connection. A common, if somewhat subtle, form of advertising involves simply associating a product with something that many people desire. For example, a beer commercial might show young, attractive people having fun at the beach. Nothing is explicitly stated about the beer; it simply is associated with the fun. Because it makes no special claim about the beer, the commercial itself may not be committing any fallacy. Nevertheless, the advertiser clearly hopes that you will commit the false cause fallacy. That is, the advertiser hopes that you will see the association between the beer and fun and conclude that the beer caused the fun. Describe at least three other advertisements that use this same technique and state what the advertiser wants you to see a causal relationship. For each of the three cases, state whether you believe a cause and effect relation actually exists and explain your reasoning.

36. Editorial Analysis. Find a newspaper editorial on a topic of interest to you. Attach a copy of the editorial so that your instructor can see it when grading this question.

a. Analyze the arguments made in the editorial.
b. Identify and describe any fallacies involved.
c. Overall, do you agree with the editorial? Discuss.

37. Johnson Versus Goldwater. In 1964, President Lyndon Johnson ran for reelection against Barry Goldwater. During the campaign, Goldwater made statements about Vietnam and the Soviet Union that led some people to conclude that he would lead the nation deeper into the war in Vietnam and perhaps into other wars as well. Despite his denials, Goldwater fell behind in the campaign. Then Johnson's campaign aired the following television ad:

A little girl, strolling through a field, picks a daisy and plucks its petals while counting, "one, two, three...." In the background, a gruff male voice begins a countdown, "ten, nine, eight...," slowly drowning out the little girl's voice. As the countdown ends, the screen lights up with an atomic explosion. It is followed by the voice of President Lyndon Johnson, who begins, "These are the stakes: to make a world in which all of God's children can live, or go into the dark."

a. What implication does the ad make about Goldwater? What did Johnson's campaign hope to accomplish with the ad?
b. What fallacy is committed in Johnson's narration at the end (these are the stakes...)? Discuss.
c. Identify and discuss any other fallacies committed in the ad's attempt to get people to vote for Johnson.
d. Briefly research the 1964 presidential campaign. How significant was this ad to the outcome of the election?

38. Mathematical Battle of the Sexes. Often a claim is made that boys inherently are more capable at mathematics than girls. The following arguments are commonly used to make this point:

- Nearly all the great mathematicians have been men.
- Males outnumber females in fields requiring mathmatical expertise, such as the physical sciences and engineering.
- Boys score higher, on average, than girls on standarized mathematics tests such as the GRE and SAT.
- Even in preschool and kindergarten, boys show more interest in mechanical things than girls do.

a. Each of these arguments suffers one or more fallacies. Identify and analyze the fallacies involved.
b. A conclusion can be true even if the argument supporting it is fallacious. Do you think that boys inherently are more capable at mathematics than girls? Defend your opinion.
c. Have you ever experienced different treatment in a math class because of your gender? Describe the incident and its impact on you. If you cannot recall a personal experience, you may have witnessed it happening to someone else; if so, describe the incident and its impact on the person.
d. In light of your answers to parts (a)–(c) discuss the following quotation from Mary Wollstonecraft.

It would be an endless task to trace the variety of meanness, cares, and sorrows into which women are plunged by the prevailing opinion that they were created rather to feel than to reason, and that all the power they obtain must be obtained by their charms and weakness.

— Mary Wollstonecraft (1759–1797)

39. Fallacies in Advertising: Why? Fallacies are extremely common in advertising. Why do you think this is so? Which types of fallacies are the most common in commercial advertising? Do you believe that fallacies fool many people into buying products for which there is no legitimate evidence of special quality? Why or why not?

40. Fallacies in Politics. Fallacies also are common in politics. Recall or research a recent major election (e.g., for president, governor, mayor), and describe the role that fallacies played in the campaign. Were particular types of fallacies more common than others? Do you believe that the fallacies influenced the outcome of the vote?

41. Are Fallacies Dangerous? What do you think about the preponderance of fallacious logic in our society? Do you believe that it is harmful? Can anything be done to reduce the incidence of fallacies? Explain.

42. **Teaching About Fallacies.** Imagine that you are a fourth-grade teacher. Is this a good time to teach your children about recognizing fallacies and how to deal with fallacies they encounter? Why or why not? Describe how you would present a lesson on fallacies to your class (or to your own children).

43. **Identifying Fallacies.** Identify the fallacies in each of the following arguments and explain the reasoning for your selection. If you think that more than one answer is possible and you can't decide which is best, explain the reason for the ambiguity.
   a. Dave says that I would be better off buying the Subaru, but I'm not going to because Dave's an idiot.
   b. Following President Reagan's defense buildup, the Soviet Union began the process of democratization that ultimately led to its breakup. Therefore Reagan is responsible for the changes that led to the demise of the Soviet Union.
   c. The Golden Rule is a sound ethical principle because it is basic to every system of ethics in every culture.
   d. Lawyer to defendant: "Have you stopped abusing your wife?"
   e. Of the four candidates for mayor, the former dog catcher is leading in the polls, so I will vote for her.
   f. My mom says that I should never smoke, but I'm not going to pay any attention because she smokes at least a pack a day.
   g. If you love me, you'll loan me the money.
   h. I believe in telepathy, because no one has ever proven that it doesn't exist.
   i. Can you prove that a fetus isn't a human being? No? Then abortion should be illegal.
   j. The preponderance of male doctors and lawyers tells me that males have an aptitude for these careers.
   k. I will not allow my children to attend school with kids who have AIDS. I don't care if the experts say that it's perfectly safe. I remember going to school thinking I would not catch a cold, but I did. The cold turned into pneumonia. We can never know when a person will catch a disease.
   l. Pepsi — it's the choice of the new generation.
   m. There's no way that Senator Smith's bill can help the cause of gun control because he is one of the biggest recipients of campaign contributions from the National Rifle Association.
   n. Our data show that the global average temperature has risen more than 1°C during the past 50 years. During that time, the concentration of carbon dioxide in the atmosphere rose dramatically. Thus carbon dioxide emissions must be the cause of rising average global temperatures.
   o. Hundreds of lives have been saved by air bags, so air bags have undoubtedly improved auto safety.
   p. Violent crime by youth has risen in virtual lockstep with increased violence on television. Can there be any doubt that television violence leads to real violence?
   q. During the last ten years, the number of ministers in Peaceful Valley has doubled. At the same time the number of arrests for drunkenness has increased 20%. The ministers must be driving people to drink.
   r. Since 1960, the percentage of the population over 18 that smokes has decreased from 40% to about 20%. During the same period, the percentage of overweight people has increased from 25% to 35%. Clearly, quitting smoking leads to overeating.

44. **Arguments and Fallacies.** Each of the following exercises presents a quotation making a simple argument. Evaluate each argument, and describe any fallacies committed.
   a. "I think, therefore I am." — Rene Descartes (1596–1650)
   b. "A vain man can never be utterly ruthless: he wants to win applause and therefore he accommodates himself to others." — Johann Wolfgang von Goethe (1749–1832)
   c. "The newspaper is the second hand in the clock of history; and it is not only made of baser metal than those which point to the minute and the hour, but it seldom goes right." — Arthur Schopenhauer (1788–1860)
   d. "It is as absurd to say that a man can't love one woman all the time as it is to say that a violinist needs several violins to play the same piece of music." — Honoré Balzac (1799–1850)
   e. "Training is everything. The peach was once a bitter almond; cauliflower is nothing but cabbage with a college education." — Mark Twain (Samuel Clemens) (1835–1910)
   f. "The study of mathematics is apt to commence in disappointment.... We are told that by its aid the stars are weighed and the billions of molecules in a drop of water are counted. Yet, like the ghost of Hamlet's father, this great science eludes the efforts of our mental weapons to grasp it." — Alfred North Whitehead (1861–1947)
   g. "The most remarkable thing about socialist competition is that it creates a basic change in people's view of labor, since it changes the labor from a shameful and heavy burden into a matter of honor, matter of fame, matter of valor and heroism." — Joseph Stalin (1879–1953)
   h. "Cowards die many times before their deaths; the valiant never taste of death but once. Of all the wonders I have yet heard, it seems to me most strange that men should fear, seeing that death, a necessary end, will come when it will come." — William Shakespeare (1564–1616)

Causality and Chaos

45. **Other Causal Factors for Lung Cancer.** Several things besides smoking have been shown to be probabilistic causal factors in lung cancer. For example, exposure to asbestos and exposure to radon gas, both of which are

found in many homes, can cause lung cancer. Suppose that you meet a person who lives in a home that has a high radon level and insulation that contains asbestos. The person tells you, "I smoke, too, because I figure I'm doomed to lung cancer anyway." What would you say in response?

46. Establishing Smoking as a Cause of Lung Cancer. Clearly, the correlation between smoking and lung cancer isn't sufficient to establish smoking as a cause of lung cancer. Why, then, do people generally accept that smoking causes lung cancer? What kinds of evidence led researchers to establish the causal relationship?

47. High Voltage Power Lines. Suppose that people living near a particular high-voltage power line have a higher incidence of cancer than people living farther from the power line. Can you conclude that the high-voltage power line is the cause of the elevated cancer rate? If not, what other explanations might there be for it? What other types of research would you like to see before you conclude that high-voltage power lines cause cancer?

48. The Gulf War and Oil Prices. The U.S. economy went into recession at nearly the same time that Iraq invaded Kuwait in 1990, precipitating a short-lived jump in the price of oil. Can you conclude that the Iraqi invasion of Kuwait was a cause of the recession? Did the Iraqi invasion cause the increase in oil prices? Explain.

49. Do Guns Cause a Higher Murder Rate? Those who favor gun control often point to a correlation between the availability of handguns and murder rates to support their position that gun control would save lives. Does this correlation, by itself, indicate that handgun availability causes a higher murder rate? Suggest some other factors that might support or weaken this conclusion.

50. Math and Heart Rates. In 1983, a graduate student in psychology found a correlation between heart rate (in beats per minute) and the type of mental activity in which her subjects were involved. Specifically, she found that, when subjects were asked to do tasks that involved concentrated mental activity (e.g., doing a math word problem), their heart rates increased (above each individual's resting heartbeat). Moreover, when her subjects were asked to perform tasks that didn't involve intense concentration, their heart rates decreased. Do you think that a causal connection exists between heart rate and behavioral activity? Can you think of other possible explanations for the results of this study? What further research might you suggest to clarify the issue?

51. Seasonal Affective Disorder. In a study published in *Scientific American*, several prominent medical professionals suggested that there is a correlation between where people live in the U.S. and seasonal affective disorder (SAD). This disorder is characterized by mood swings and a generally depressed and lethargic state. The researchers found that in the northern states (those above a certain latitude) 100 of 100,000 individuals suffered form SAD, whereas only 6 of 100,000 in the southern states suffered from this disorder. They concluded that lack of sunlight causes the disorder. What do you think of this conclusion? What further information might be needed before you would accept it?

52. The Cricket and the Scientist. Here is a macabre parable that carries a lesson about causality. A scientist studying the behavior of crickets commands "Jump!" Seeing the cricket jump, he notes in his book that "crickets with six legs can jump." The scientist then removes a leg of the cricket and commands "Jump!" Seeing the cricket jump, he notes that "crickets with only five legs can jump." The scientist then removes a leg of the cricket and commands "Jump!" Seeing the cricket jump, he notes that "crickets with only four legs can jump." The scientist continues the experiment until the cricket has no legs. The scientist commands the legless cricket to jump. Seeing the cricket remain still, the scientist notes that "crickets with no legs cannot hear." Comment on which of Mill's methods the scientist used in this experiment and whether his conclusion is well founded.

53. Traffic Jam. The average "rush hour" speed on major Los Angeles freeways was 65 miles per hour in 1963. By 1993, the average speed during the same hours was approximately 17 miles per hour.

a. What was the percentage decline in average rush hour speed during the 30-year period?

b. Over this 30-year period, the capacity of Los Angeles freeways dramatically increased: New roads were built, and lanes were added to existing freeways. Analyze the causality claimed in the following statement: "Over a period of many years, the effect of building more roads is inevitably an increase in traffic problems."

Problems Involving Percentages

54. Lottery Chances. Suppose that you buy two lottery tickets, for $1 each, every week. Over the course of a year, you get a $10 winner once, a $5 winner twice, and $2 winners fifteen times. (Note that, for example, a $5 winner means that you will be given $5 when you turn in your ticket, for which you had paid $1.)

a. How much money have you spent, and how much have you won back?

b. What percentage of your original money did you lose?

c. If your winning tickets are distributed evenly over the year, in what percentage of weeks did you get some kind of winning ticket?

d. Comment on the psychological effects of the game; for example, comment on the fact that you win every few weeks, yet lose overall.

55. Inflated Chocolate. Suppose that, between 1990 and 1994, the price of a chocolate chip cookie at your student union rose from 45¢ to 60¢. What was the percentage increase in price? Based on a national inflation rate during this period of about 3% per year, was the price increase for the cookies reasonable?

56. **Health Care Costs.** Using inflation-adjusted dollars (that is, inflation has already been taken into account in these costs), health care spending in the United States rose from $53 billion in 1970 to $600 billion in 1990. During the same period, the U.S. population rose from 210 million to 250 million. What were the percentage increases in health care spending and population? Comment on the differing percentages.

57. **Distribution of Air Time.** Over the five days spanning April 11–15, 1994, Rocky Mountain Media Watch (RMMW) recorded the evening newscasts of the four major local Denver TV stations. Analysis of these newscasts showed how much air time each of the following categories received.

Category	Hours of air time
Commercials	4.5
News	5.08
Sports	1.57
Public service announcements	0.09
Weather	1.02
Previews	0.88
Chatter	0.28

a. How many hours of air time did RMMW study?
b. What percentage of air time did each category get during these newscasts?
c. If *you* could decide how the air time was distributed, what percentage of time would you give to each category? Explain your answer.

58. **The Subject Matter of News.** In the same study (see problem 57), RMMW analyzed the subject matter of news and obtained the following results.

Story category	Hours of air time	Story category	Hours of air time
War	3.04	Education	0.12
Children	0.11	Environment	0.63
Public works	0.41	Government	0.90
Crime	4.09	Health	1.43
Disaster	1.04	Other	1.35
Economy	0.34		

a. What percentage of air time did each category get during these newscasts?
b. What changes would you like to see in what is shown on the news? Explain.

59. **Murder for Murder.** Sectarian violence in Northern Ireland has taken many lives. From 1970 to 1994, the IRA killed 2000 people and Ulster Loyalists killed 900. This violence has occurred within a Catholic population of 600,000 and a Protestant unionist population of 900,000. In comparison, consider the number of Americans killed by guns each year. In 1991, 40,000 were killed by firearms. The U.S. population is 260 million.

a. How many people died each year in Northern Ireland because of sectarian violence, if approximately the same number of people were killed each year between 1970 and 1994?
b. What percentage of the population of Northern Ireland does your answer to part (a) represent?
c. What percentage of the U.S. population was killed by handguns in 1991?
d. How do the percentages found in parts (b) and (c) compare? What do they suggest about the difference in the levels of violence in Northern Ireland and the United States?

60. **Oil Crises of the 1970s.** Two major oil crises occurred during the 1970s. Both arose from dramatic increases in oil prices.

a. The first crisis occurred in late 1973. Between October 1973 and January 1974, the price of a barrel of crude oil from the Middle East rose from $3.01 to $10.95. What was the absolute and relative change in the crude oil price during this time period?
b. The second crisis occurred between January and April of 1979. The price of crude oil rose from $13.34 to $14.54 per barrel. What was the absolute and relative change in the crude oil price during this period?
c. Which crisis do you think was more dramatic? If you wanted to capture the large quantitative difference between these two events, would you use absolute or relative change to convey the point? Explain.

61. **Global Coal Consumption.** Between 1970 and 1989, global coal consumption rose from 1543 million tons to 2218 million tons (units are based on oil equivalents, rather than actual weights).

a. What was the absolute and relative change during these 19 years?
b. With growing evidence that coal consumption causes significant environmental damage, ranging from the strip mining used to retrieve coal to the pollution it causes, consumption peaked in 1989. By 1992, consumption had dropped to 2180 million tons. What was the absolute and relative change over this three-year period?
c. Suppose that the relative change found in part (b) continued for six more three-year periods (18 years total). What would the coal consumption be at the end of that time? Would coal use have dropped back to 1970 levels?

(Note: Although total global coal use declined from 1989 through at least 1993, the source of this decline was reduced use of coal in industrialized nations, such as the United States, Western Europe, and Japan, and especially in nations of Eastern Europe, such as Russia, Ukraine, and Poland. However, coal use in developing nations, particularly in India and China, continued to increase. If these trends continue, global coal use again will begin to increase as developing nations increase their energy and coal consumption.)

62. **CFC Percentages.** Following the discovery of an ozone hole over the Antarctic in 1985, the United Nations sponsored an international conference in Montreal to work to decrease chlorofluorocarbons (CFCs) emissions by half over the following ten years. In response to the Montreal protocol, a company that makes refrigerators containing CFCs announced that it will reduce its CFC production by 25% over the next five years, followed by another 25% reduction over the subsequent five years. The CEO publicly commented that "With a 25% reduction per five-year period, we will cut our CFC production by 50% in ten years."
   a. What is wrong with the CEO's use of percentages?
   b. By what percentage will the company actually cut its CFC production during the next ten years, assuming it accomplishes a 25% reduction in five years, followed by another 25% reduction in the subsequent five years?

63. **Military Spending: The Top 10.** The following table lists the top ten countries in the world in military spending during 1993; values are billions of US dollars:

Country	Rank	1993 Military Spending
United States	1	$291
Japan	2	40
France	3	36
United Kingdom	4	35
Germany	5	31
Russia	6	29
China	7	22
Italy	8	17
Saudi Arabia	9	16
South Korea	10	12

   a. How much did these ten countries combined spend on the military during 1993?
   b. Calculate the percentage of the total (from part (a)) spent by each country.
   c. Suggest several reasons for the differences in spending from country to country.

Projects

64. **The Methods of Concomitant Variations and Residues in the San Francisco Earthquake.** Many of the deaths in the 1989 San Francisco earthquake occurred because of the collapse of the double-decked Nimitz freeway. Because future earthquakes are likely to occur in the area, it understanding the reasons for the failure of the structure is important in order to prevent loss of life the next time. Mill's methods of agreement and difference rely on the examination of many cases to establish causality. Because earthquakes are rare we cannot examine many cases of the Nimitz freeway failure, so these methods are of little use in determining its cause. Instead, engineers use the methods of concomitant variations and residues to try to establish the cause of the failure.
   a. The method of concomitant variations looks for an association of quantitative changes in one factor with quantitative changes in another factor. Suggest a way that this method could be used to help identify the cause of the Nimitz freeway collapse. (Hint: Consider the possibility of studying parts of the structure that did not collapse, or other similar structures in the area.)
   b. The method of residues also can be used to try and establish a cause for the collapse of the freeway. Because only one section of the structure collapsed, some residual factor — a factor not present in other parts of the structure — obviously was involved in the collapse. Engineers might search for this factor by identifying something in the construction that was different in the collapsed sections from that in other sections. Still, they might not find such a factor in the construction. Suggest some possible factors besides the construction that might have caused only one section of the freeway to collapse. Do additional research as needed.

65. **Mill's Methods of Agreement and Difference Lead to a Nobel Prize.** The 1989 Nobel Prize in chemistry was awarded to Professor Thomas Cech (University of Colorado) for his work in proving that RNA can act as a catalyst. Prior to Cech's work, researchers believed that only proteins could act as catalysts. Cech's work began with an experiment designed to identify the catalyst for a particular biochemical reaction. Under natural conditions, the reaction takes place in the presence of numerous proteins, as well as in the presence of RNA and other substances. Cech's experiment involved removing proteins, one by one, to determine whether the reaction still occurred. To Cech's surprise, the reaction continued normally even after all the proteins had been removed. Continuing, he found that the reaction ceased only when he removed the RNA. He then went back and tried removing the RNA in mixtures that still contained proteins. He found that, without the RNA, the reaction could not proceed. He concluded that RNA was necessary for this particular catalytic reaction. Moreover, the RNA apparently was sufficient to cause the reaction, because the reaction proceeded in the presence of RNA even when no other substances were present.
   a. Cech's work proved that, in the cases under study, the presence of the RNA is necessary for the catalytic action. Briefly explain how he used the method of difference to establish this fact.
   b. The method of agreement looks for factors common to every case in which an effect occurs. Thus the method can establish that the factor is sufficient to cause the effect. Describe how the method of agreement led Cech to conclude that RNA is sufficient to cause the reaction.

c. As Cech continued his work, however, he found that the could add certain substances to the RNA-containing mixture that would cause the reaction to cease. That is, the reaction would be proceeding normally, but then cease when he added those substances. Suggest why the RNA might be a sufficient condition for catalytic action when it is alone but not sufficient when certain other substances are present. What do these other substances do?

d. One of the outstanding questions about the origin of life on Earth is how the first living cells were able to reproduce themselves. Modern cells pass their genes along to subsequent generations through the genetic code, which is encoded in either RNA or DNA (all higher forms of life use DNA as their genetic material, but some viruses use RNA); biologists generally assume that early organisms relied on RNA. However, catalysts are required for the reproduction reactions to proceed. Prior to Cech's work, researcher thought that all catalysts were proteins and that the instructions for making these proteins were encoded in the RNA (or DNA). Thus we have a "chicken and egg" problem: Which came first, the catalysts that help the RNA to replicate or the RNA that contains the instructions for making the catalysts? Explain how Cech's discovery of catalysis by RNA might help explain some of the gaps in our understanding of the origin of life. Do additional research as needed.

66. **Marie Curie and the Methods of Agreement, Concomitant Variations, and Residue.** In 1898, a young physicist named Marie Curie began experimenting with uranium and thorium to learn about the curious emissions from those elements. Curie measured the amount of energy emitted by lumps of pure uranium and thorium. She then tried larger "lumps" that contained the same amounts of uranium or thorium as in her earlier experiment but was mixed with other materials. She found that the added materials had no effect on the emitted energy. She did the experiment again, this time heating the uranium and thorium and then cooling them. She did many more variations of the experiment. Each time, no matter the conditions, she found that the emitted energy was affected only by the amounts of uranium and thorium in the lumps of material. This led her to assert that the energy emitted by a substance is proportional only to the amount of uranium and thorium it contains.

Next, Curie tested a substance called pitchblende, a shiny, black-brown mineral. Knowing how much uranium and thorium were present in pitchblende, she predicted (based on her earlier results) how much energy it would emit. To her surprise, her measurements showed a much higher level of energy emission than she had expected. She repeated the experiment several times but always obtained the same result. Finally, to explain the discrepancy between the energy emission she had expected and what she measured, Curie concluded that the pitchblende must contain at least one more, previously unknown, energy-emitting element besides uranium and thorium. She eventually found that the pitchblende contained two previously unidentified elements. She named one of them Polonium, after her native Poland, and the other Radium, after the property radioactivity (a term she also coined).

a. Explain how Curie used the method of agreement to support her assertion that the energy emitted by a substance is proportional only to the amount of uranium or thorium that it contains.

b. Explain how the method of concomitant variations also helped support her assertion that the energy emitted by a substance is proportional only to the amount of uranium or thorium that it contains.

c. Suggest a simple experiment that would have represented the method of difference that might have further supported her assertion about the energy-emitting properties of uranium and thorium.

d. Explain how Curie used the method of residue in her work with the pitchblende to propose the existence of at least one new element.

e. Do you think that Curie's initial conclusion that the pitchblende contained unidentified elements was reasonable? What other explanations could there have been for the unexpectedly high energy emissions she measured?

f. Research Curie's career. In particular, discuss obstacles that she faced and how she overcame them, as a woman in science at a time when women scientists were rare.

67. **More on CFCs.** The CFC production figures from 1990 to 1993 were 820,000 tons in 1990, 720,000 tons in 1991, 630,000 tons in 1992, and 510,000 tons in 1993.

a. Calculate the absolute and relative change between 1990 and 1991, between 1991 and 1992, and between 1992 and 1993.

b. If you were helping decide how CFC levels should be reduced, would you advocate an annual change at a fixed absolute rate or a fixed relative rate? Why?

c. Determine the current level of global CFC production. Discuss whether CFCs will soon be eliminated.

d. Despite the drop in global CFC production, the concentration of chlorine in the atmosphere (from CFCs) has continued to rise. Find out why. Will ozone depletion continue to be a problem if all CFC production is halted? If so, why and for how long?

68. **Trees Cause Pollution?** During his 1980 presidential campaign, Ronald Reagan argued against many environmental controls by claiming that "eighty percent of air pollution comes from plants and trees." Briefly research the 1980 presidential campaign and discuss the controversy that surrounded Reagan's statement. How did Reagan make this "just plain mistake" with percentages?

69. The Nazi Propaganda Machine. When Hitler assumed power in Germany in 1933, many outsiders considered him to be a buffoon and his threats to be bluffs. As late as 1939, the Allies attempted to appease Hitler without war. By the end of the war in 1945, some 50 million people had died, including 6 million Jews — more than 2⁄3 of Europe's Jewish population — killed in the Nazi period of extermination known as the Holocaust.

How did such an evil regime hold power in a country of well-educated citizens like Germany? How was Hitler able to make other leaders believe that his threat wasn't real? At least part of the answer to both questions lies in the elaborate propaganda industry that Hitler created to garner support for his government. Research and report on the Nazi propaganda machine: how it worked and how much it influenced events prior to and during World War II. Could such a thing ever happen in the United States? Explain your answer.

70. Causality in Law. Legal proceedings involve many examples of attempts to establish causal connections. Watch a set of legal proceedings for a couple of hours (either by going to court or on TV). Analyze what you observe. Describe any causal connections sought and how evidence for the causal connections is introduced and used.

4 PROBLEM-SOLVING STRATEGIES

In the first three chapters we explored ideas and examples intended to help you think critically about important personal and social issues. In this chapter, we ask you to collect your critical thinking skills and apply them to the solution of quantitative problems. Because nearly every important issue involves mathematical or quantitative information at some level, you inevitably will be confronted with quantitative problems. Sometimes the problems are routine, such as figuring out how much carpet is needed for a bedroom. Other problems are far more complex, such as choosing the best health care plan for your needs or arranging your class schedule in order to graduate in four years. Although no general techniques will solve all problems, in this chapter we explore several approaches that often are useful.

CHAPTER 4 PREVIEW

4.1 ***THE ART OF PROBLEM SOLVING.*** Before developing problem-solving strategies, we introduce the chapter by discussing how problem solving is more of an art than a science.

4.2 ***PROBLEM SOLVING THROUGH UNIT ANALYSIS.*** One of the most powerful problem-solving tools involves keeping track of units in calculations. In this section we show how to identify and work with units in problems, as well as how to convert between different sets of equivalent units.

4.3 ***THE PROCESS OF PROBLEM SOLVING.*** Problem solving requires both creativity and organization. In this section we present a four-step process that will help you stay organized while tapping your creative powers.

4.4 ***STRATEGIC HINTS FOR PROBLEM SOLVING.*** We end the chapter by presenting a series of hints, along with examples of how to use them, that you may find useful in the future. By working through the examples, you will also gain valuable problem-solving experience.

Today's world is more mathematical than yesterday's, and tomorrow's world will be more mathematical than today's. As computers increase in power, some parts of mathematics become less important while other parts become more important. While arithmetic proficiency may have been "good enough" for many in the middle of the century, anyone whose mathematical skills are limited to computation has little to offer today's society that is not done better by an inexpensive machine.

— *Everybody Counts* (National Academy Press, Washington, D.C., 1989, p. 45)

4.1 THE ART OF PROBLEM SOLVING

In many ways, problem solving is more of an art than a science. No fixed rules are prescribed and no set of tools will always work. Problems may be practical or abstract, and may involve many different creative processes.

To begin with a nonmathematical example, consider the types of problems a plumber faces. Confronted with a leaky pipe, the plumber must determine the source and cause of the leak and the best way to repair it. If the source is clear and the pipe accessible, the plumber might solve the problem quickly and easily. However, if the pipe is in a difficult location, the plumber may need to devise a unique method for access and repair. The process may also be *iterative;* if water still leaks after the first attempt at repair, another strategy may be called for.

Similarly, the quantitative problems associated with modern issues may be as straightforward as calculating a sales tax or as complex as assessing the risks associated with global warming. Many different ways may be available to address the same problem, and the first strategy chosen may or may not work.

Further, the process of solving a quantitative problem — whether posed in a mathematics book or one confronted in real life — in essence is no different from the process used by a plumber. It requires ingenuity, experience, and a bag of tools. The only real difference lies in the character of the tools to be used. Whereas a plumber may call up mechanical tools like wrenches and saws, quantitative problem solving calls for tools such as adding and subtracting, algebraically solving equations, or using statistics.

Of course, some problems require tools or experience that go beyond what we can cover in this book. Nevertheless, knowledge and experience always is helpful. Returning to the plumbing analogy, a bit of knowledge can help you ensure that an unscrupulous plumber does not overcharge or make unnecessary repairs, even if you could not do the repairs yourself. Similarly, experience with problem solving and the use of mathematical tools will help you make intelligent decisions on personal and social issues, and ensure that you are not fooled by so-called experts who use poor quantitative reasoning.

In this chapter we use mathematical tools of arithmetic. If you need to review of these tools, study Appendix 1.

This chapter is devoted to giving you problem-solving experience with familiar tools only — logical tools developed in Chapters 2 and 3 and mathematical tools you learned in high school or earlier. In later chapters, we add new tools to the "tool box," expanding the range of problems that you can solve. As you progress, you will learn to use mathematics to understand both the natural world and the social world around you. In the process, we hope that you will gain great confidence in your ability to understand any issue that you might encounter.

THINKING ABOUT ... *DREADED WORD PROBLEMS*

Unfortunately, many people look back with dread at the occasional attempts made to show them the practical utility of mathematics. You may remember from your earliest grade school days the exercise sections in mathematics books. They probably contained many drill problems at the top of the page, and then, the print became dense and the rest of the page was filled with problems consisting of nothing but words. You may still remember (or be unable to forget) problems like:

> *Bob is three years older than his sister Jill will be when she is twice the present age of her niece Sarah who was born when Bob's dad was 30. Bob's dad is five years older than his mom, who is now 29. How old was Bob's mom when he was born?*

(Although you might be tempted to answer "why not ask Bob's mom?" the daring among you can verify that the answer is 18.)

Undoubtedly, these problems seemed perverse at the time and, indeed, their relevance to daily life is hard to appreciate. Perhaps as a result, few students excelled at these problems and few teachers emphasized them.

The irony in this state of affairs is striking. A common misconception is that mathematics is abstract and irrelevant and has nothing to do with "real" issues. Surely, the *purpose* of word problems is to change that perception and to introduce some relevance into mathematics courses. Yet, with rare exceptions, word problems have had the opposite effect.

We hope that you will find the problems in *this* book to be much more interesting and worthy of your time than those that led Gary Larson to draw the "Hell's Library" cartoon. Nevertheless, even though we have tried mightily to write problems that are relevant, problems in a book are necessarily somewhat artificial compared to the problems encountered in real life. You will therefore better appreciate the time spent working these problems if you can connect them, in your own mind, to problems that you care about in the real world. This connection is the key to overcoming the dread of word problems and learning to appreciate the deep connections between mathematics and real life.

Hell's library

4.2 PROBLEM SOLVING THROUGH UNIT ANALYSIS

If we add five apples and three apples we get eight apples. But if we add five apples and three oranges, we get ... five apples and three oranges! This simple illustration of the old saying that you can't compare apples and oranges points to one of the most powerful techniques for problem solving.

Numbers in real problems almost always represent a quantity of *something,* such as the number of apples in a basket, the length of a room in feet, or the distance in miles that a car can travel in an hour. Terms that describe a quantity — *apples, feet, or miles* — are called its **units,** or **dimensions.** The lesson illustrated by the apples and oranges is that you must keep careful track of units as you work through a problem. We now explore this technique, often called **unit** (or **dimensional**) **analysis.**

4.2.1 Identifying the Units

The first step in unit analysis is to identify the units involved in a problem. The easiest units to identify are **simple units,** which embody only a single concept. For example, the quantities 45 minutes, 75 dollars, and 12 buckets of apples, respectively, have the simple units of *minutes, dollars,* and *buckets of apples.* As the last of these three units shows, sometimes more than one word is required to describe a simple unit.

Many quantities are associated with **compound units,** in which two or more simple units are multiplied, divided, or raised to powers. Working with compound units requires identification of the operations (multiplication, division, or powers) involved. Because the operations usually are not explicit when compound units are stated in English, you need to learn a few simple ways of making the identification.

Division of Units: *Per*

***per* means "divided by."**

Compound units formed by division usually are identifiable by the key word *per,* which simply means "divided by." When you work with such units in problems, writing the division of units as a fraction is helpful, as shown in the following examples.

- A speed of 65 miles per hour $\Rightarrow$ $65\dfrac{\text{miles}}{\text{hour}}$ or $\dfrac{65 \text{ miles}}{1 \text{ hour}}$.
- A price of \$1.22 per pound $\Rightarrow$ $1.22\dfrac{\$}{\text{pound}}$ or $\dfrac{\$1.22}{1 \text{ pound}}$.
- There are 12 inches in a foot $\Rightarrow$ $12\dfrac{\text{inches}}{\text{foot}}$ or $\dfrac{12 \text{ inches}}{1 \text{ foot}}$.

In the last example, note that equivalent expressions, such as *in a,* sometimes are used instead of *per.*

The use of abbreviations for units is especially helpful in dealing with compound units. For example, we write 12 inches per foot, as 12 in./ft. Some units have standard abbreviations, such as "m" for meters, "in." for inches, or the more mysterious "lb" for pounds. In other cases, you may make up your own abbreviation if it is helpful. In general, you should use unit abbreviations only when they are unambiguous in the context.

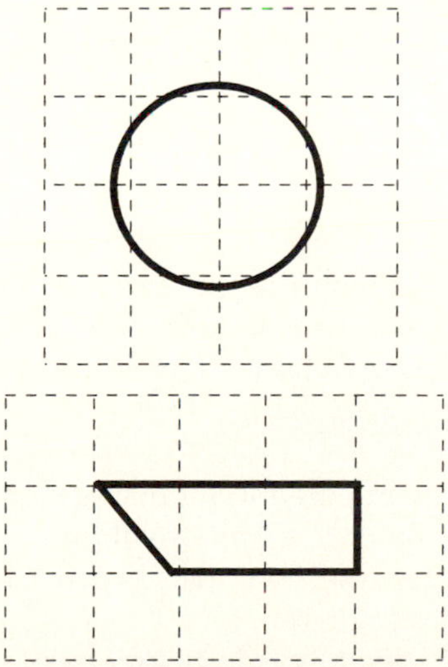

Figure 4-1. The most basic way to measure an area is to count the number of squares of known size that fit within it. If each small square represents one square centimeter, the areas of the two figures are about 4 square centimeters and 2.5 square centimeters, respectively.

Units Raised to Powers: Area and Volume

Sometimes compound units are formed by raising a simple unit to a power. The most common examples of units raised to powers — and the ones with the most direct physical interpretations — occur with area and volume. **Area** describes the size of a surface such as a floor, a wall, or a piece of cloth. The most basic way of measuring area is by counting the number of squares of a known size that fit within the region (Figure 4-1). For example, if each of the squares is one centimeter on a side, the units of area will be square centimeters; if the squares are a foot on a side, the area units are square feet.

To understand why units like square centimeters or square feet involve a unit raised to a power, recall that the area of a rectangle is the product of its length and

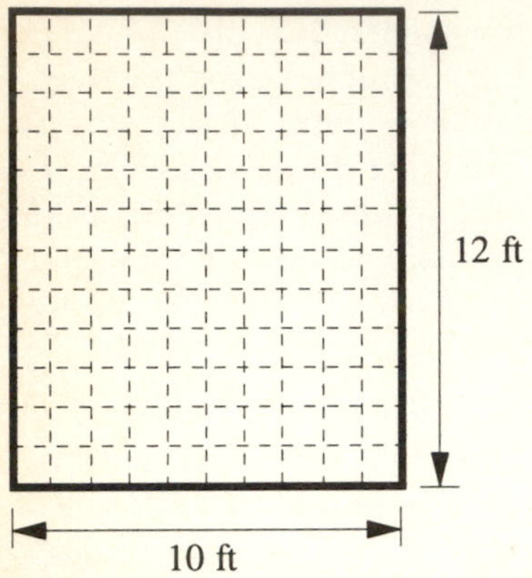

Figure 4-2. The dotted lines marking square feet show that a rectangular room measuring 10 feet by 12 feet has an area of 120 square feet.

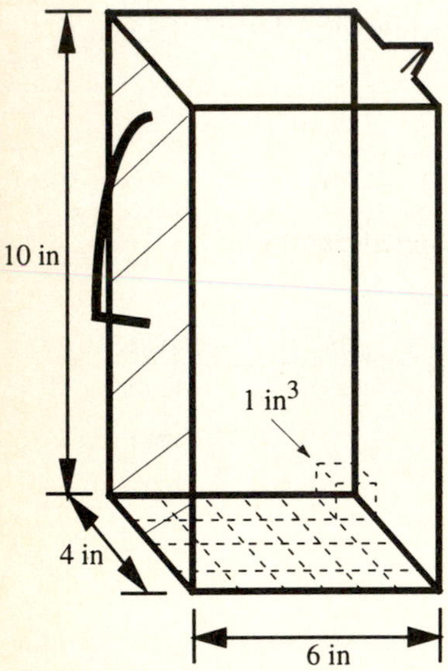

Figure 4-3. A pitcher 4 inches wide, 6 inches long, and 10 inches tall has a volume of 240 cubic inches (240 in^3).

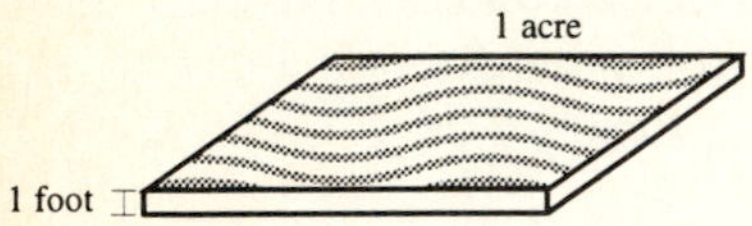

Figure 4-4. An acre-foot is a volume that covers an area of one acre to a depth of one foot.

1 acre = 43,250 square feet.

width. The area of a room that measures 10 feet by 12 feet (Figure 4-2) is

$$10 \text{ ft} \times 12 \text{ ft} = 120 \text{ ft} \times \text{ft} = 120 \text{ ft}^2 = 120 \text{ square feet.}$$

Thus the term *square* in units of area refers to the second power. Square meters can be written as meter^2 (m^2), square yards as yards^2 (yd^2), and so on.

Volume is used to describe the capacity of any container or space. The most basic way of measuring volume is by counting cubes of a known size that fit in the container. Thus volume is measured in units such as cubic inches or cubic meters. The term *cubic* refers to the third power, which always appears when calculating volumes. Consider the pitcher in Figure 4-3. Because volume is the product of length, width, and height, the volume of the pitcher is

$$4 \text{ in.} \times 6 \text{ in.} \times 10 \text{ in.} = (4 \times 6 \times 10)\ (\text{in.} \times \text{in.} \times \text{in.}) = 240 \text{ in}^3.$$

That is, cubic inches are the same thing as in^3. Similarly, cubic meters are m^3, cubic centimeters are cm^3 (sometimes called "cc's"), and so on.

Units raised to powers often occur in combination with other units. The following examples illustrate the idea.

- Cloth priced at \$2 per square yard $\Rightarrow \dfrac{\$2}{yd^2}$ or $2\dfrac{\$}{yd^2}$.
- A mineral with a density of 3 grams per cubic centimeter $\Rightarrow 3\dfrac{g}{cm^3}$.
- A stream flowing at a rate of 200 cubic feet per second $\Rightarrow 200\dfrac{ft^3}{s}$.

Compound Units with Simple Names

One complication in identifying units occurs when compound units have simple-sounding names. For example, an **acre** sounds like a simple unit because it is a single word. However, an acre is a unit of area, so it has the underlying dimensions of a length to the second power: an acre is equal to 43,250 square feet. For practical or historical reasons, many other compound units have been given simple names. You will encounter a few of them, such as *watts* and *calories*, later in the book.

Multiplication of Units

Another way of forming compound units is to multiply two *different* units together. Such units usually are used as a matter of convenience. For example, suppose that we want to know the volume of a reservoir that covers an area of 5 acres to an average depth of 10 feet. The volume of the reservoir can be calculated by multiplying its area by its average depth, so the volume of water is

$$5 \text{ acres} \times 10 \text{ feet} = 50 \text{ acres} \times \text{feet.}$$

Although "acres × feet" might look a bit strange compared to more common volume units such as cubic feet, it is simply an alternative unit of volume (Figure 4-4). Moreover, for engineers who measure reservoir areas in acres, these units may be the most convenient ones to work with.

In English, multiplication in compound units usually is indicated by a hyphen. For example,

- a reservoir with a volume of 50 acre-feet ⇒ 50 acres × feet;
- an electric bill for 100 kilowatt-hours of energy ⇒ 100 kilowatts × hours;
- an engine that produces a torque of 75 foot-pounds ⇒ 75 feet × pounds.

4.2.2 Working with Units

After identifying the units involved in a problem, you can operate on the units along with the numbers in a problem. Keep in mind several valuable principles which will aid you in working with units:

1. You can add or subtract numerical values only if they have the same units. That is, 5 apples + 3 apples = 8 apples; but if you add 5 apples and 3 oranges, the answer must be stated as 5 apples + 3 oranges.
2. When you multiply, divide, or raise units to powers, we form new compound units. For example, if you divide a distance of 60 miles by a time of one hour, the answer has the compound unit miles per hour, or mi/hr.
3. Sometimes you must translate words like *per* (which suggests division) and *of* (which suggests multiplication) to know which operation to use.
4. Units that appear in both the numerator and denominator can be canceled leaving the answer with the desired units.
5. Mathematically, it does not matter whether a unit is singular (e.g., *bucket)* or plural (e.g., *buckets*). For this reason, the abbreviation *ft* can stand for either *foot* or *feet*.
6. Finding the reciprocal of units is similar to finding reciprocals of numbers. For example, the reciprocal of 3 hr is $\frac{1}{3\text{ hr}}$, and the reciprocal of $\frac{2\text{ ft}}{1\text{ s}}$ is $\frac{1\text{ s}}{2\text{ ft}}$.

Recall that the **reciprocal** is formed simply by interchanging the numerator (top) and denominator (bottom) of a fraction. See Appendix A for review.

A few examples should clarify the process of working with units. In each case, note the preceding principles in action.

Example 4-1 Suppose that you have 3 buckets of 12 apples each. How many apples do you have total?

Solution: To identify the units, you recognize that "12 apples each" means 12 apples *per* bucket, or 12 apples/bucket. The word *of* indicates that you must multiply the 3 buckets by the 12 apples per bucket to find that

$$3 \text{ buckets} \times 12\frac{\text{apples}}{\text{bucket}} = 36 \text{ apples}.$$

The three buckets contain a total of 36 apples.

❖

Example 4-2 You purchase a home with a lot 70 feet wide and 100 feet long. What is the area of the lot?

Solution: You simply multiply the length and width of the lot, forming the compound unit of square feet (or feet to the second power):

$$70 \text{ ft} \times 100 \text{ ft} = 7000 \text{ ft}^2.$$

The lot has an area of 7000 square feet.

❖

Example 4-3 Suppose that a car travels 25 miles every half-hour. How fast is it going?

Solution: The problem can be rephrased to state that the car is covering 25 miles *per* half-hour, indicating that you need to divide. Note, however, that dividing makes canceling units more difficult. You therefore should replace division with multiplication by the reciprocal, which makes it easier to see the final units:

$$25 \text{ mi} \div \frac{1}{2}\text{hr} = 25 \text{ mi} \times \frac{2}{1 \text{ hr}} = 50\frac{\text{mi}}{\text{hr}}.$$

The car is traveling at a speed of 50 miles per hour.

❖

Example 4-4 Suppose that a storm leaves 6 inches of rain. If a farmer has 100 acres of land, how much water does the land receive from the storm?

Solution: The 6 inches of rain delivered by the storm refers to the depth of water that falls on the land. To find the total volume of water that falls, you multiply the area (100 acres) by the depth (6 inches, or ½ foot):

$$100 \text{ acres} \times \frac{1}{2}\text{ft} = 50 \text{ acre} \times \text{feet}.$$

The farmer's land received a total volume of 50 acre-feet of water from the storm.

❖

Units Can Help You Find a Solution

Analyzing the units of a problem often helps you approach a problem that might otherwise give you trouble. To illustrate the technique, suppose that a car can go 25 miles per gallon of gas. How much gas is needed for it to go 90 miles?

The problem asks "how much gas?" so the answer should have units of gallons. This means you must *divide* the distance by the gas mileage:

$$90 \text{ mi} \div 25\frac{\text{mi}}{\text{gal}} = 90 \text{ mi} \times \frac{1 \text{ gal}}{25 \text{ mi}}.$$

Although that's all there is to it, we can demonstrate the power of unit analysis by imagining that we were completely lost on how to approach this problem. We would begin by observing that there only two quantities are involved in the problem: a distance of 90 miles and the car's gas mileage of 25 miles per gallon. With just two quantities involved, there are only four possible ways to combine them multiplicatively:

1. $25\frac{\text{mi}}{\text{gal}} \div 90 \text{ mi} = 25\frac{\text{mi}}{\text{gal}} \times \frac{1}{90 \text{ mi}} \quad \Rightarrow \quad \text{units of } \frac{1}{\text{gal}}.$

2. $25\frac{\text{mi}}{\text{gal}} \times 90 \text{ mi} \quad \Rightarrow \quad \text{units of } \frac{\text{mi}^2}{\text{gal}}.$

3. $90 \text{ mi} \div 25\frac{\text{mi}}{\text{gal}} = 90 \text{ mi} \times \frac{\text{gal}}{25 \text{ mi}} \quad \Rightarrow \quad \text{units of gal}.$

4. $90 \text{ mi} \times 25\frac{\text{mi}}{\text{gal}} \quad \Rightarrow \quad \text{units of } \frac{\text{mi}^2}{\text{gal}}.$

For each of these four possible combinations, we identified the units of the answer. Because the answer must have units of gallons, combination (3) must be the correct approach to the problem.

Example 4-5 A jet traveling at an average speed of 830 kilometers per hour made a particular trip in 3.7 hours. How far did it travel?

Solution: You begin by recognizing that the answer should have units of kilometers because the question asks "how far?" Then identify the two quantities

Why did we express our final answer as "about 3100 kilometers," rather than as 3071 kilometers? Like most numbers in real problems, the speed and travel time given for the airplane could not have been exact. Rounding the answer shows the degree of "exactness" that we think is justified. In Chapter 8, where you will study uncertainty in detail, you will see that this choice was made because 3100 km has two *significant* digits — just like the two numbers (830 km/hr and 3.7 hours) in the problem statement.

involved in the problem: a speed of 830 kilometers per hour and a time of 3.7 hours. Multiplying these quantities leads to the expected units of kilometers:

$$830 \text{ km/hr} \times 3.7 \text{ hr} = 3071 \text{ km}.$$

The jet traveled about 3100 kilometers during its trip.

❖

Example 4-6 Imagine that you were hired as a grader for a math course. An exam question reads: "Cheryl purchased 5 pounds of apples at a price of 50 cents per pound. How much did she pay for the apples?" Suppose that a student has written the answer: "50 ÷ 5 = 10. She paid 10 cents." Write a note to the student explaining where he went wrong.

Solution: Dear student: You can see your error by keeping track of units. The number 50 in this problem carries units of *cents per pound* and the number 5 carries units of *pounds*. Tracking the units in your calculation gives

$$50\frac{¢}{\text{lb}} \div 5 \text{ lb} = 50\frac{¢}{\text{lb}} \times \frac{1}{5 \text{ lb}} = 10\frac{¢}{\text{lb}^2}$$

In writing your sentence ("she paid 10 cents"), you correctly realized that the units of the answer should be cents. But your calculation yields units of *cents per square pound*. To get an answer with units of cents, you should have multiplied:

$$50¢ \times 5 \text{ lb} = 250¢.$$

Thus the correct answer is that the 5 pounds of apples cost 250¢, or $2.50. In the future, keep track of your units so that you don't make similar mistakes!

❖

4.2.3 Unit Conversions

Many problems involve converting one set of units to another. For example, you might want to convert a distance in kilometers to miles, or you might need to convert a measurement from quarts to cups.

Summary Box 4.1: Using Unit Analysis in Problem Solving

(1) Always keep track of units in calculations, and be sure that your answer has the units you expect.

(2) To make working with the units easier, always replace division with multiplication by the reciprocal.

(3) State your final answer clearly in words, using complete sentences. Be sure that your statement makes sense as the answer to the original question.

Multiplying by 1

The basic "trick" of unit conversions is to devise an appropriate way of multiplying by 1. For example, the following are all different ways of writing 1.

$$1 = \frac{1}{1} = \frac{8}{8} = \frac{1/4}{1/4} = \frac{\sqrt[3]{\pi}}{\sqrt[3]{\pi}} = \frac{1 \text{ kilogram}}{1 \text{ kilogram}} = \frac{12 \text{ inches}}{1 \text{ foot}}.$$

The last expression shows the necessity of stating units: 12 ÷ 1 is *not* 1, but 12 inches ÷ 1 foot *is* 1. The key to unit conversions is simply to write 1 in a form that helps you change from one set of units to another.

Example 4-7 Convert a distance of 7 feet into inches.

Solution: Multiply by 1 in the form that equates 12 inches to 1 foot:

$$7 \text{ ft} \times \underbrace{\frac{12 \text{ in.}}{1 \text{ ft}}}_{1} = 84 \text{ in.}$$

Seven feet is the same as 84 inches.

❖

Conversion Factors

To convert from one set of units to another requires knowing a **conversion factor.** The conversion factor between feet and inches, for example, is

12 inches = 1 foot. Note that we can write this conversion factor in three equivalent ways:

$$12 \text{ in.} = 1 \text{ ft} \quad \text{or} \quad \frac{12 \text{ in.}}{1 \text{ ft}} = 1 \quad \text{or} \quad \frac{1 \text{ ft}}{12 \text{ in.}} = 1.$$

As Example 4-7 demonstrated, one of the last two forms generally is the most useful because it allows multiplication by 1. In solving a problem, you should simply choose the form of 1 that works best.

> **Time-Out to Think:** Starting with the conversion factor 12 inches = 1 foot, show how you can write it in the other two forms by dividing both sides of the equation either by 12 inches or by 1 foot.

Example 4-8 Convert a length of 102 inches to feet.

Solution: Multiply by the form of 1 in which *inches* is in the denominator, so *inches* will cancel:

$$102 \text{ in.} \times \underbrace{\frac{1 \text{ ft}}{12 \text{ in.}}}_{1} = 8.5 \text{ ft}.$$

We have shown that 102 inches is equivalent to 8.5 feet.

❖

Example 4-9 If you have four quarts of milk, how many pints do you have?

Solution: The problem asks you to convert from quarts to pints. One quart contains two pints, so the conversion factor between quarts and pints is:

$$1 \text{ quart} = 2 \text{ pints} \quad \text{or} \quad \frac{1 \text{ quart}}{2 \text{ pints}} = 1 \quad \text{or} \quad \frac{2 \text{ pints}}{1 \text{ quart}} = 1.$$

Because you want the answer in pints, we choose the conversion factor with *quarts* in the denominator so *quarts* will cancel:

$$4 \text{ quarts} \times \frac{2 \text{ pints}}{1 \text{ quart}} = 8 \text{ pints}.$$

Four quarts of milk is the same as eight pints of milk.

❖

Example 4-10 Convert a distance of 2.5 miles into feet.

Solution: You need the conversion factor between miles and feet: recall that there are 5280 feet in one mile. You can write this conversion factor as:

$$1 \text{ mi} = 5280 \text{ ft} \quad \text{or} \quad \frac{1 \text{ mi}}{5280 \text{ ft}} = 1 \quad \text{or} \quad \frac{5280 \text{ ft}}{1 \text{ mi}} = 1.$$

Because you are converting miles to feet, the most useful form is the one that has miles in the denominator. Again, note how the units of miles cancel when you multiply by 1 in the form of 5280 ft/1 mi:

$$2.5 \text{ mi} \times \frac{5280 \text{ ft}}{1 \text{ mi}} = 13{,}200 \text{ ft}.$$

A distance of 2.5 miles is equivalent to 13,200 feet.

❖

Note that

$$\frac{24\text{ hr}}{1\text{ day}} = \frac{60\text{ min}}{1\text{ hr}} = \frac{60\text{ s}}{1\text{ min}} = 1.$$

Example 4-11 *A Chain of Unit Conversions.* How many seconds are there in one day?

Solution: This problem requires several conversion factors. Because you probably don't know the conversion factor between days and seconds, you must use more familiar conversion factors:

$$1\text{ day} = 24\text{ hours},\quad 1\text{ hour} = 60\text{ minutes},\quad \text{and}\quad 1\text{ minute} = 60\text{ seconds}.$$

Using these three conversion factors, you can solve the problem by setting up a *chain* of unit conversions. Because the answer should have units of seconds, you should write the chain of conversions in a form in which all units except for seconds cancel:

$$1\text{ day} \times \frac{24\text{ hr}}{1\text{ day}} \times \frac{60\text{ min}}{1\text{ hr}} \times \frac{60\text{ s}}{1\text{ min}} = 86,400\text{ s}.$$

There are 86,400 seconds in one day.

❖

Units Raised to Powers

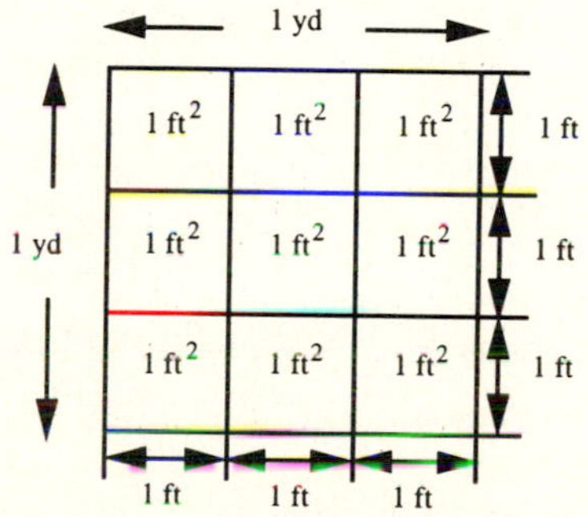

Figure 4-5. The area of the large square is one square yard because it is one yard long and one yard wide. Note that one square yard contains nine square feet.

Special care is required in converting units that are raised to a power. For example, suppose that you want to know the number of square feet in a square yard. You may not know the conversion factor between square yards (yd^2) and square feet (ft^2); however, you do know that 1 yard = 3 feet.

To find the conversion factor between square yards and square feet, you can square both sides of the yards-to-feet conversion factor:

$$1\text{ yd} = 3\text{ ft} \xrightarrow{\text{square both sides}} (1\text{ yd})^2 = (3\text{ ft})^2,\text{ or } 1\text{ yd}^2 = 9\text{ ft}^2.$$

Thus one square yard is the same as nine square feet. To see why, study Figure 4-5, which represents an area of one square yard. Each side of the square yard is one yard, or three feet long. Each of the smaller squares is one foot long and one foot wide. Because nine of these square feet fit in the square yard, we have visual confirmation of the conversion factor $1\text{ yd}^2 = 9\text{ ft}^2$.

Example 4-12 *The Cost of Carpet.* The area of a room in a house usually is measured in units of square feet. However, carpet usually is sold in units of square yards. Suppose that you have a room that measures 12 feet by 15 feet and you wish to install carpet that costs $15 per square yard. How much will the carpet cost?

Solution: The area of the room is $12\text{ ft} \times 15\text{ ft} = 180\text{ ft}^2$. Because you will purchase carpet in units of square yards, you must convert from square feet to square yards. Use the conversion factor that has square feet in the denominator so that you can cancel the ft^2 units:

From $1\text{ yd}^2 = 9\text{ ft}^2$ we get the equivalent conversion factor

$$\frac{1\text{ yd}^2}{9\text{ ft}^2} = 1.$$

$$180\text{ ft}^2 \times \underbrace{\frac{1\text{ yd}^2}{9\text{ ft}^2}}_{1} = 20\text{ yd}^2.$$

The room needs 20 square yards of carpet, and at a price of $15 per square yard, the total cost of the carpet will be

$$20\text{ yd}^2 \times \$15/\text{yd}^2 = \$300.$$

The carpet for the room will cost $300.

❖

Example 4-13 How many square feet are in a square mile? How many acres are in a square mile?

Solution: You can find the conversion factor between square miles and square feet by squaring both sides of the miles-to-feet conversion factor:

$$1 \text{ mi} = 5280 \text{ ft} \xrightarrow{\text{square both sides}} (1 \text{ mi})^2 = (5280 \text{ ft})^2, \text{ or } 1 \text{ mi}^2 = 27{,}878{,}400 \text{ ft}^2.$$

To convert into acres, use the conversion factor of 1 acre = 43,250 square feet:

$$27{,}878{,}400 \text{ ft}^2 \times \frac{1 \text{ acre}}{43{,}250 \text{ ft}^2} = 644.6 \text{ acre}.$$

A square mile is equivalent to almost 28 million square feet, or about 640 acres.

❖

Example 4-14 Convert a volume of 12 cubic yards to cubic feet.

Solution: To find the conversion factor from cubic yards to cubic feet, simply cube the yards-to-feet conversion:

$$1 \text{ yd} = 3 \text{ ft} \xrightarrow{\text{cube both sides}} (1 \text{ yd})^3 = (3 \text{ ft})^3, \text{ or } 1 \text{ yd}^3 = 27 \text{ ft}^3.$$

Writing the conversion in the form that has cubic yards in the denominator gives

$$12 \text{ yd}^3 \times \underbrace{\frac{27 \text{ ft}^3}{1 \text{ yd}^3}}_{1} = 324 \text{ ft}^3.$$

Note that you can solve this problem in a single step, by cubing the yards-to-feet conversion as you work:

As with squaring, cubing both sides of a conversion factor gives an equivalent conversion factor.

$$12 \text{ yd}^3 \times \left(\frac{3 \text{ ft}}{1 \text{ yd}}\right)^3 = 12 \text{ yd}^3 \times \left(\frac{27 \text{ ft}^3}{1 \text{ yd}^3}\right) = 324 \text{ ft}^3.$$

Either way, we see that 12 cubic yards is the same as 324 cubic feet.

❖

Time-Out to Think: Example 4-14 shows that one cubic yard contains 27 cubic feet. Make a mental picture or a sketch to explain why this is true.

4.3 THE PROCESS OF PROBLEM SOLVING

Perhaps you noticed that the strategy used to solve the problems in Section 4.2 was not always the same. Instead, we suggested a strategy that seemed to lend itself to each particular problem. This approach demonstrates one of the most important principles of problem solving: Be flexible, because no particular strategy always works!

Problem solving demands creativity. Indeed, the more complex the problem, the more creative you will need to be in devising an appropriate strategy. At the same time, being organized and logical is important, so that your creativity will be focused on the problem at hand.

How to Solve It (Princeton University Press, Princeton, N.J., 1957), by George Polya (1887–1985), was first published in 1945. More than 1 million copies have been sold and it is available in at least 17 different languages.

The process of problem solving can be thought of in terms of four basic steps. The steps shown here are modified from those described by George Polya in his classic book about problem solving, *How to Solve It*.

1. **Understand the problem.** As baseball great Yogi Berra once said, "If you don't know where you're going, you'll probably end up someplace else." The first step therefore is to determine where you are going. Be sure that you understand what the problem is asking, and form a general idea of what a solution will look like. In particular, you may find it helpful to
   - think about the context of the problem (that is, how it relates to other problems in the real world) to gain insight into its purpose;
   - make a list or table of the specific information given in the problem;
   - draw a picture or diagram to help you make sense of the problem;
   - restate the problem in different words to clarify its question; or
   - make a mental or written model of the solution, into which you will fill in details as you work through the problem.
2. **Devise a strategy for solving the problem.** Once you understand the problem, the next step is to decide how to go about solving it. This step is the most difficult, and is the one that requires creativity, organization, and experience. In devising a strategy, you may find it worthwhile to
   - identify and obtain any information needed to solve the problem that was not provided in the problem statement, using recall, estimation, or research;
   - make a list of possible strategies and hints that will help you select your overall strategy;
   - map out your strategy with a flow chart or diagram; or
   - if you have a mental model of the solution (see step 1), think of your strategy as an argument whose conclusion is the solution; then, use your tools for constructing arguments to help formulate the strategy.
3. **Carry out your strategy, and revise it if necessary.** After you have selected a strategy, the next step is to carry it out to find the result. In this step in which you are most likely to call on analytical and computational tools as you work through the mathematical details of the problem. You may find it valuable to
   - keep an organized, neat, and written record of your work, which will be helpful if you later need to review or study your solution;
   - double-check each step you take so that you do not risk carrying errors through to the end of your solution; or
   - constantly reevaluate your strategy as you work; if you find a flaw in your strategy, return to step 2 and create a revised strategy.
4. **Look back to check, interpret, and explain your result.** Although you may be tempted to feel like you have finished after you find a result in step 3, this final step is the most important. After all, a result is useless if it is wrong, misinterpreted, or cannot be explained to others. In this final step you should
   - be sure that your result makes sense; for example, be sure that it has the expected units, that its numerical value is reasonable, and that it is a reasonable answer to the original problem;

- once you are sure that your result is reasonable, check its specifics; this task can be as simple as rechecking your calculations in step 3, but finding an independent way of checking the result is preferable;
- identify and understand potential sources of uncertainty in your result;
- write your solution clearly and concisely, including discussion of any relevant uncertainties or assumptions; and
- consider and discuss any pertinent implications of your result.

If you look back at the examples presented already — and ahead to the examples in the remainder of this book — you will see that we rarely enumerate the four steps. Nevertheless, the four-step process is implicit in every problem we solve. Likewise, you are not likely to enumerate the steps explicitly, except when asked to do so by an instructor! Still, as you confront more complex problems in this book and face them in your life, you should keep the four-step process in mind.

Example 4-15 Without actually solving the problem, we describe how to apply the four-step process to the question: How much sales tax revenue is generated by textbook sales at your school each year?

Solution: Although we follow the four-step process, note that only some of its specific details are relevant to this problem. We make use of those that are helpful.

Step 1: We must first understand the problem. It is about tax revenue (the context), so we can assume that the tax rate is set by the state or locality. In addition, when it asks "how much?" we presume that it is asking for an amount of money. We therefore expect the solution to have a general form like (a model solution): Textbook sales generate *x* dollars in sales tax revenue each year.

Step 2: Next, we must devise a strategy. We note that at least two pieces of information are needed but not given in the problem.

(1) We need to know the tax rate.

(2) Because sales tax is calculated as a percentage of total (pretax) sales, we need to know the total amount of money spent on textbooks at your school.

The tax rate should be relatively easy to find by asking at any local store or contacting the local government or chamber of commerce. Finding the total amount of money spent on textbooks poses a greater challenge. We might ask the bookstore management about total sales; however, this number probably includes merchandise and books other than textbooks. A better method might be to make an estimate based on the number of students at the school and a survey to estimate the average amount of money that each student spends on books. Based on all these factors, the following is a potential step-by-step strategy.

- Find the tax rate.
- Find out how many students attend the school.
- Conduct a small survey, trying to include students of different majors and different years, to estimate the average amount of money spent on textbooks each year.
- Multiply the number of students at the school by the average amount spent on textbooks to find the total amount of money spent on textbooks per year.
- Multiply the total amount of money spent on textbooks by the tax rate to find the total sales tax revenue per year.

Step 3: Now, we carry out the strategy. To illustrate the process, suppose that your school has 20,000 students; they spend an average of $300 per year on textbooks; and the local sales tax rate is 6.5%. The total sales tax revenue is

$$\underbrace{\text{sales tax revenue}}_{\text{units of }\frac{\$}{\text{yr}}} = \underbrace{\text{number of students}}_{\text{units of students}} \times \underbrace{\text{spending per student}}_{\text{units of }\frac{\$}{\text{student}\times\text{yr}}} \times \underbrace{\text{tax rate}}_{\substack{\text{no units}\\ \text{percentage}}}$$

$$= 20,000 \text{ students} \times \frac{\$300}{\text{student} \times \text{yr}} \times 0.065 = \$390,000\,/\,\text{yr}.$$

Note that we included years in the denominator of both the sales tax revenue and the per student spending units because both these numbers are annual values.

Step 4: We need to do several things to complete the process. First, we ask whether the answer is reasonable. The $390,000 in sales tax revenue means that each of the 20,000 students pays about $20 in tax; for a spending level of $300 per student on books, this result agrees with the tax rate of 6.5%. Second, we check our work carefully to be sure that the units work out properly and that we've made no arithmetic errors. Third, we recognize that, because we used estimates for both the number of students and average textbook spending, our solution involves some uncertainty. We might make an educated guess that the solution could be off by as much as 10%; as a result, rounding $390,000 to $400,000 is reasonable. Finally, we look for other relevant factors: for example, we might note that the answer is likely to change from year to year. Combining all this information into a clear and concise statement, we might write:

"The total sales tax revenue generated by textbook sales at my school is in the neighborhood of $400,000 per year. This number may be off by as much as 10% because it was based on the number of students and the average annual textbook spending per student, both of which were estimates. In addition, because both of these numbers and the sales tax rate may change from year to year, the amount given is good for this year only."

❖

4.4 STRATEGIC HINTS FOR PROBLEM SOLVING

Although the four-step process may help you become better organized and better able to tap your creativity, the only sure way to improve at problem solving is by doing it. This section is designed to help you gain some of that experience.

In the following subsections we offer hints about problem solving. The meaning and utility of each hint is illustrated with one or more examples. Of course, no list of hints can be exhaustive — we can't anticipate every type of problem that may arise! Rather, the hints and examples will demonstrate a few common features of problems and problem solving. Keeping these hints and examples in mind should help you broaden your skills in devising problem-solving strategies.

The hints themselves relate more to *attitudes* about problem solving than to actual techniques. Indeed, their primary purpose is to help you develop a "mind set" that is conducive to enjoyable and successful problem solving. The better you become at problem solving now, the better you will be at it in the remainder of your schoolwork, your career, and your life generally.

4.4.1 There May Be More Than One Answer

Greenhouse gases include carbon dioxide, methane, and water vapor; these gases can trap heat in the atmosphere. Many people are concerned that human activity is causing the concentration of greenhouse gases in the atmosphere to rise, which may in turn alter the Earth's climate.

How can society best reduce the total amount of greenhouse gases emitted into the atmosphere? We won't even attempt to answer this question, but it should make the point that no *single* best answer may be available. Indeed, many different political and economic strategies could yield similar reductions in greenhouse gas emissions.

Many people recognize that policy questions do not have unique answers. Perhaps surprisingly, the same is true of many mathematical problems. For example, 20, 40/2, and 60/3 are equally correct answers to 10 + 10. Somewhat more substantial are cases where mathematical operations or equations yield two or more *different,* but valid answers; for example, both $x = 4$ and $x = -4$ are solutions to the equation $x^2 = 16$. There are many examples in which proof of a mathematical theorem is found one way and, later, a shorter or more ingenious proof is found. The later proof does not invalidate the first proof; there simply are two "right" answers. Thus our first strategic hint is to recognize that real problems often have more than one answer.

Nonunique Solutions

In problems of everyday life, nonunique solutions may occur because not enough information is available to distinguish among a variety of possibilities. Example 4-16 demonstrates this point by showing that the data from a traffic counter generally is not enough to determine the precise number of vehicles that have passed.

Example 4-16 *Traffic Counters.* A traffic counter is a device designed to count the number of vehicles passing along a street. It usually is a thin black tube stretched across a street or highway and connected to a "brain box" at the side of the road. Usually pressure activated, the device registers one "count" each time a set of wheels (that is, wheels on a single axle) rolls over the tube. A normal automobile registers two counts: one for the front wheels and one for the rear wheels. A light truck with three axles (front wheels plus a double set of rear wheels) registers three counts. A large semitrailer truck might register four or five counts.

Suppose that, during a one-hour period, a particular traffic counter registers 35 counts on a residential street on which heavy trucks are prohibited; that is, only two-axle vehicles (cars) and three-axle vehicles (light trucks) are allowed. How many cars and light trucks passed over the traffic counter?

Solution: We might begin by trial and error. For example, we can rule out the possibility of 12 light trucks, because 12 light trucks would yield 36 counts (3 counts per truck) — more than the 35 registered. However, 11 light trucks would yield only 33 counts. We then realize that, because a car registers 2 counts, 11 light trucks and 1 car would produce the 35 counts. But is this solution the only possible one?

Continuing by trial and error, we find that 9 light trucks and 4 cars also is a solution: the 9 trucks yield 27 counts, and the four cars yield 8 counts, for a total of 35. In fact, six different combinations of cars and light trucks will produce a total count of 35 (we leave it for you to find the remaining four solutions in the end-of-chapter problems). Yet, during a particular hour on a particular street, only one combination of cars and trucks actually passed that point. Unfortunately, with only the traffic counter information, we cannot determine which of the six solutions represents the actual traffic flow during the particular hour.

❖

Algorithms

Many important problems are posed as "how to" rather than "how many" problems. That is, they ask for a method or procedure for doing some task. A mathematical recipe or procedure is called an **algorithm.** As you might guess, many different algorithms may accomplish the same task, though with different levels of efficiency. In Example 4-17 we investigate sorting algorithms, which enable computers to sort lists of numbers (in ascending or descending order) or words (in alphabetical order). Note that the example offers two solutions. However, the second algorithm requires fewer calculations, and is therefore more *efficient;* that is, a computer program using the second algorithm will run considerably faster than one using the first algorithm.

Example 4-17 *Sorting Algorithms.* Find an algorithm that sorts a list of 10 numbers into ascending order. To keep the example specific, we begin with a worst-case list in the order opposite from what we desire: 10, 9, 8, 7, 6, 5, 4, 3, 2, 1.

Solution — Algorithm 1: A straightforward algorithm compares each number with every other in the list, interchanging pairs of numbers when they are in the wrong order. To begin this algorithm we compare 10 (the first number) to the next number on the list, 9. As 10 is larger, we interchange its place with the 9. Next we compare the 10 to the 8, again finding that the smaller number should be moved ahead on the list. We continue on down the list, eventually moving the 10 to where it belongs at the end of the list:

$$(\underbrace{10,9,}_{\text{compare}} 8,7,6,5,4,3,2,1) \Rightarrow (9, \underbrace{10,8,}_{\text{compare}} 7,6,5,4,3,2,1) \Rightarrow \cdots \Rightarrow \underbrace{(9,8,7,6,5,4,3,2,1,10)}_{\substack{\text{"10" at end of list} \\ \text{after 9 comparisons}}}.$$

Note that moving the 10 to its correct place at the end of the list requires a total of nine comparisons: First we compared it to 9, then to 8, and so on down to 1. But we're still not done! We repeat the process, moving the 9, comparing it to each of the nine other numbers on the list, and interchanging places when necessary. We then compare the 8 to each of the other nine numbers on the list, and so on. This method requires 9 comparisons for each of the 10 numbers, or a total of 90 comparisons:

$$(\underbrace{9,8,}_{\text{compare}} 7,6,5,4,3,2,1,10) \Rightarrow (8, \underbrace{9,7,}_{\text{compare}} 6,5,4,3,2,1) \Rightarrow \cdots \Rightarrow \underbrace{(8,7,6,5,4,3,2,1,9,10)}_{\substack{\text{"9" in correct place} \\ \text{after 9 comparisons}}}.$$

Solution — Algorithm 2: Perhaps you can already see a better way! Once the 10 has been moved to the end of the list, there is no reason to consider it again while we sort the remaining numbers. Thus we actually need only eight comparisons, rather than nine, to move the 9 to its correct location. Similarly, with the 9 and 10 properly placed, the 8 needs to be compared with only seven other numbers. Continuing on, we can place the 7 with only six comparisons, and so on. In this case, the total number of comparisons needed is

$$9 + 8 + 7 + 6 + 5 + 4 + 3 + 2 + 1 = 45.$$

This second algorithm, in which we skip unnecessary comparisons, reduces by half the total number of comparisons needed (from 90 in the first algorithm to 45 in this one). It is therefore a much more efficient algorithm — and there are other sorting algorithms that are more efficient still! Indeed, because sorting is so important in computer applications, finding more efficient sorting algorithms is an active field of research.

❖

4.4.2 There May Be More Than One Method

Just as we must often admit to more than one right answer, we should often expect more than one strategy for finding an answer. Mathematics and the human mind are far too rich and diverse to expect that everyone will follow the same path to a solution. As illustrated Example 4-18, choosing the more efficient strategy can save a lot of time and work.

Example 4-18 ***Jill and Jack's Race.*** Jill and Jack ran a 100-meter race. When Jill crossed the finish line Jack had run only 95 meters. They decide to race again, but this time Jill handicaps herself by starting 5 meters behind the starting line. Assuming that both runners maintain their respective paces, who wins the second race?

Solution — Method 1: One approach to this problem is analytical, in which we analyze each race quantitatively. We were not told how fast either Jill or Jack ran, so we can choose some reasonable numbers. For example, we might assume that Jill completed the 100 meters in the first race in 20 seconds. In that case, her pace was: 100 m ÷ 20 s = 5 m/s, or 5 meters per second. Because Jack ran only 95 meters in the same 20 seconds, his pace was: 100 m ÷ 20 s = 4.75 m/s, or 4.75 meters per second.

Using these numbers, we can analyze the second race. With Jill starting 5 meters behind the starting line, she needs to run 105 meters while Jack needs to run only 100 meters. Using the paces found above, their times for the second race would be

$$\text{Jill}: \ 105\text{ m} \div 5\frac{\text{m}}{\text{s}} = 105\text{ m} \times \frac{1\text{ s}}{5\text{ m}} = 21\text{ s};$$

$$\text{Jack}: \ 10\text{ m} \div 4.75\frac{\text{m}}{\text{s}} = 10\text{ m} \times \frac{1\text{ s}}{4.75\text{ m}} = 21.05\text{ s}.$$

Jill will win the second race by 0.05 seconds.

Solution — Method 2: Although the analytical method works, we can use a much more intuitive and direct solution. We simply note that Jill runs 100 meters in the same time that Jack runs 95 meters. Therefore Jill will pull even with Jack 95 meters from the starting line in the second race. In the remaining 5 meters, Jill's faster speed will allow her to pull away and win. Note how this insight allows us to avoid the calculations needed in method 1!

❖

4.4.3 Use Appropriate Tools

Mastery requires more than simply learning how to work with problem-solving tools. It is equally important to know when to use them. You don't need a computer to check your tab in a restaurant, and you don't need calculus to find the area of a rectangular room! For any given task there is an appropriate level of power that is needed, and it is a matter of style and efficiency to neither underestimate nor overestimate that level.

Thus our third hint concerns use of your logical, analytical, and computational tools. Usually you will have a choice of tools to use in addressing any problem; be sure that you choose the ones that are most suited to the job. The next example shows a somewhat unrealistic application of this hint. However, despite its frivolous nature, it is mathematically similar to an important problem regarding China's population that could have worldwide repercussions.

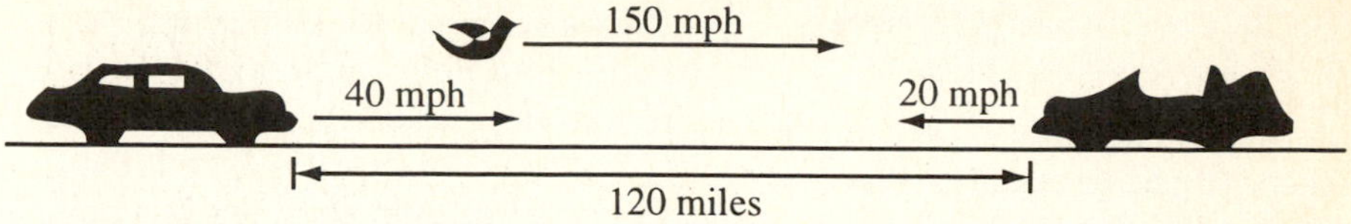

Figure 4-6. A diagram can help you understand the problem described in Example 4-19.

Example 4-19. *The Cars and the Canary.* Two cars, 120 miles apart, begin driving toward each other on a long straight highway. One car travels 20 miles per hour and the other 40 miles per hour (Figure 4-6). At the same time a canary, starting on one car, flies back and forth between the two cars as they approach each other. If the canary flies 150 miles per hour and spends no time to turn around at each car, how far has it flown when the cars collide?

Solution: Because the problem asks "how far?" we might be tempted to approach it by calculating the sum of the *distances* traveled by the canary on each trip back and forth between the cars. Note, however, that as the cars approach one another each trip is shorter than the previous one. Thus we have to find the distances covered by the canary during many separate trips back and forth, and add them. In principle, we would need to add up an *infinite* number of ever-smaller distances! A sum of an infinite number of numbers is called an **infinite series.** We could therefore solve this problem using the analytical tools for infinite series.

Incidentally, you may be perplexed by the fact that an *infinite* series can still lead to the *finite* result that the canary flies 300 kilometers. We will investigate this apparent contradiction further in Chapter 5, with Zeno's paradox.

However, a much easier way to solve this problem requires only simple arithmetic. Instead of focusing on distance, we focus on how much *time* the canary spends in flight. Note that, because one car is traveling at 20 mi/hr and the other at 40 mi/hr, they are approaching each other at a combined rate of 60 mi/hr. Initially they are 120 miles apart, are approaching each other at 60 mi/hr, and will collide after 120 mi ÷ 60 mi/hr = 2 hours. The canary is flying at a speed of 150 mi/hr, so it will fly 300 miles in the two hours before the cars collide. Solving this problem intuitively with arithmetic is far easier than with an infinite series!

❖

China's Population Policy

Faced with a daunting overpopulation problem, the government of China has officially allowed only one child per family since 1978. The stated goal of this policy is to reduce China's population from more than 1.1 billion today to about 700 million by 2050. Because of sometimes lax enforcement, the one-child policy has been only partially successful to date. It also has led to many unintended social consequences.

Perhaps the most serious unintended consequence concerns an apparent shortage of girls in China. Under normal circumstances, approximately equal numbers of boys and girls should be born. In China, however, surveys indicate that boys considerably outnumber girls. The reason for this shortage of girls is thought to be rooted in cultural traditions that stress the importance of a family having at least one son. As a result, many Chinese families apparently determine the sex of fetuses by ultrasound, and abort those that are female. There are even disputed reports of baby girls being killed after birth.

Statistical studies show that boys and girls are not born in precisely equal numbers. Instead, approximately 106 boys are born for every 100 girls. However, because males have higher mortality rates than females the numbers even out in adulthood and women outnumber men in old age.

Time-Out to Think: Many people are concerned that this shortage of girls could lead to severe consequences, both within China and worldwide, as the current generation of children matures. Think of some of these potentially severe consequences. Can you think of any potentially positive consequences?

How did China become so overpopulated? The culprit was previous government policies. The Communist government of Mao Zedong encouraged large families from the late 1940s through the 1960s. Mao believed that population growth was beneficial; to those who argued that an additional person represented "one more mouth to feed," Mao was said to have replied that it also meant "two more hands to work." China's population more than doubled from the time of the Communist takeover to the time of the one-child policy.

One proposal for stopping these practices asks the government to replace its one-child policy with a one-son policy. Families would be allowed to have children until a son is born. That is, if the first child is a boy, a family has met its limit of children. However, if the first child is a girl, a family can have additional children until one is a boy. Such a proposed policy change immediately raises two questions:

- Would a one-son policy lead to more girls than boys, more boys than girls, or equal numbers of boys and girls?
- How would this policy change affect the overall birth rate in China?

In fact, this problem is mathematically similar to the car and the canary problem. One way to address it is by counting the number of children that would be born. We could count the number of boys and girls born as a first child. Then, under the assumption that only those families with a girl have a second child, we could count the number of boys and girls born as a second child. Families having two girls would be allowed to have a third child, so we would count the number of boys and girls born as third children, and so on. We would end up with a long list of numbers to add.

However, a moment of insight allows us to answer the two questions quickly. The issue of when a family decides to stop having children cannot affect the natural probability that any child will be a boy or a girl. As long as the policy stops the practice of abortion or infanticide based on a child's sex, the result must be approximately equal numbers of boys and girls. Moreover, because *every* family would have one boy under this policy — and there are equal numbers of boys and girls overall — the *average* number of children per family must be two. Thus a one-son policy would lead to twice as many children as the one-child policy.

4.4.4 Consider Simpler, Similar Problems

Sometimes we are confronted with a problem that at first may seem daunting. Our fourth hint is that sometimes it can be approached by solving a simpler but similar problem. The insight gained from solving the easier problem may then help understand the original problem.

Example 4-20 *Coffee and Milk.* You have two cups in front of you: One holds coffee and one holds milk. You take a teaspoon of milk from the milk cup and stir it into the coffee cup. Next, you take a teaspoon of the mixture in the coffee cup and put it back into the milk cup. After the two transfers, is there more milk in the coffee cup or more coffee in the milk cup?

Before reading the solution, stop and think about this question carefully! What does your intuition tell you?

Solution: Clearly, only three answers are possible: (1) more coffee in the milk cup than milk in the coffee cup; (2) less coffee in the milk cup than milk in the coffee cup; or (3), equal amounts of coffee in the milk cup and milk in the coffee cup. Unfortunately, visualizing the transfer of fluids is very difficult, making it difficult to decide among the three possibilities.

However, if you note that the key idea in this problem is the *mixing* of two things, you can try a similar mixing problem that is much easier: mixing two bowls of marbles.

Imagine that you have a pile of 10 black marbles (the black pile) and another of 10 white marbles (the white pile). Now do two transfers analogous to the coffee–milk transfers. That is, first take 2 white marbles and put them in the

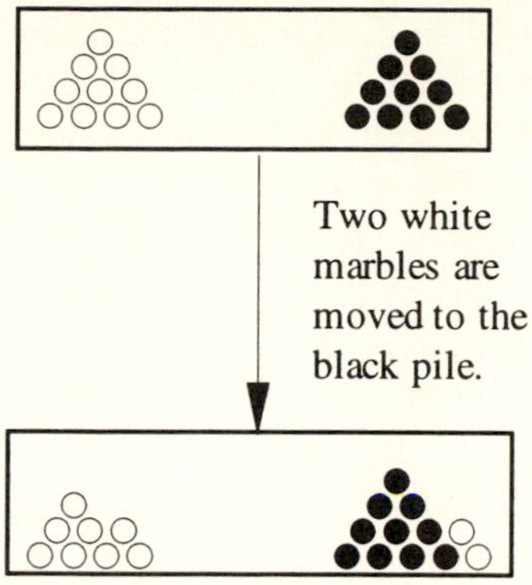

Figure 4-7. The first marble transfer leaves the "white pile" with 8 white marbles, and the "black pile" with 10 black and 2 white marbles.

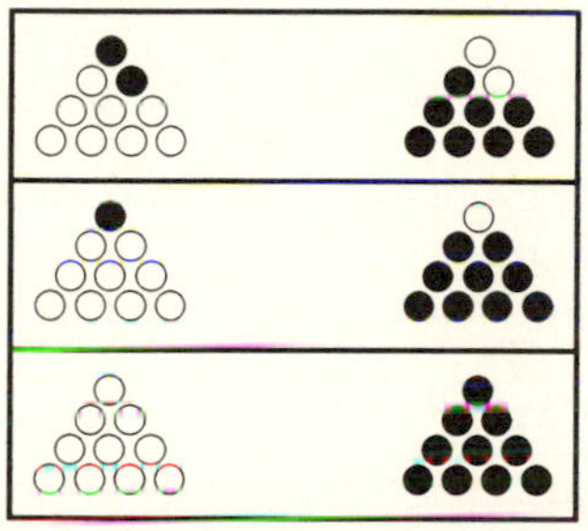

Figure 4-8. In the second transfer two marbles are moved to the white pile. The three possibilities are top to bottom: two black marbles are transferred to the white pile; one white and one black marble are transferred; and two white marbles are transferred. Note that all three cases yield the same number of black marbles in the white pile as white marbles in the black pile.

black pile; this step is like transferring the teaspoon of milk into the coffee. The white pile now has 8 white marbles, and the black pile has 10 black and 2 white marbles (Figure 4-7).

For the second transfer, take *any* two marbles from the black pile and put them in the white pile; this step is analogous to taking the teaspoon from the mixture in the coffee cup to the milk cup. Analogous to our original coffee–milk question, we now ask: Are there more black marbles in the white pile or white marbles in the black pile?

The marble version of the problem allows you to carry out the transfers, either mentally or with real marbles. Because you are transferring only 2 marbles from the black pile, you have only three possible cases to consider: the 2 marbles transferred can be either both black, both white, or 1 of each. After the second transfer, each pile again has 10 marbles. Further, Figure 4-8 shows that all three cases yield the same number of black marbles in the white pile as white marbles in the black pile

The choice of transferring two marbles each time was arbitrary. Transferring, say, one, three, or seven marbles would not have changed the essential result: You always end up with the same number of black marbles in the white pile as white marbles in the black pile. Starting with 10 marbles in each pile also was arbitrary; the conclusion would remain the same if you started with 20, 50, or a trillion trillion marbles.

The last statement holds the key to the coffee–milk question. Ultimately, the coffee and milk are made out of molecules which, for the purposes of this problem, can be viewed as miniature marbles. Thus the answer to the coffee–milk question must be the same as the answer to the marble question: After the two transfers, the amounts of coffee in the milk and milk in the coffee are equal.

Most people are surprised by this result. Are you? Before leaving this problem, we note that the marble analogy can lead to further surprising generalizations. For example, the conclusion is the same even if we start with *different* amounts of coffee and milk in the two cups. It also remains the same no matter how much liquid is transferred (but it must be the *same* amount in each of the two transfers) and regardless of whether or how the coffee cup is stirred. Finally, what the cups contain — regular, decaf, or something entirely different — doesn't matter!

❖

4.4.5 Consider Equivalent Problems with Simpler Solutions

All the hints so far involve finding innovative ways to address problems that, at first, seem difficult. In Example 4-20, we replaced a difficult problem with a simpler, similar problem. This process can help you gain valuable insights, but it generally doesn't provide a numerical solution; fortunately, a numerical solution was not called for.

When a numerical solution is needed, sometimes you can apply a similar strategy. Rather than replacing a difficult problem with a simpler one, you can try to replace it with an *equivalent* problem — a problem that will have the same numerical solution but will be easier to solve.

Example 4-21 *A Coiled Wire.* Let's consider a problem of a type often encountered by plumbers, electricians, and engineers: measuring or wrapping a wire around cylindrical pipe. Suppose that eight turns of a wire are wrapped around a pipe with a length of 20 centimeters and a circumference of 6 centimeters. What is the length of the wire?

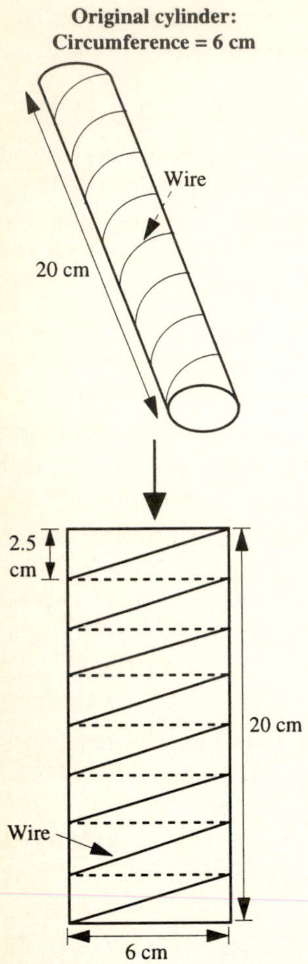

Figure 4-9. A wire is coiled around a cylindrical pipe eight times. If we imagine cutting the pipe (lengthwise) and pressing it flat, it would take the shape of a rectangle. Its length remains 20 cm, and its width is the 6-cm pipe circumference. The wire would then lie in eight straight diagonal pieces.

Solution: At first, this problem appears difficult, as though it might require tools from calculus and three-dimensional geometry. However, one crucial observation transforms the problem to one that can be solved with relative ease. Imagine that the pipe is made out of a soft plastic material.

Now, imagine cutting the pipe along its length and laying it out flat. This operation transforms the three-dimensional cylinder into a flat rectangle. As shown in Figure 4-9, it leaves the wire cut into eight diagonal segments (assume that the wire is glued to the pipe so that it does not fall off as the pipe is cut).

Although transforming the cylinder into a flat rectangle changes the problem, it does not change the solution. Remember that we are seeking the length of the wire; although it is now lying in pieces, its total length remains unchanged. We need only to find the length of each of the eight pieces.

Because each wire piece lies diagonally, it forms the hypotenuse of a right triangle. You may recall the **Pythagorean theorem**, which states that for a right triangle:

$$\text{base}^2 + \text{height}^2 = \text{hypotenuse}^2 \quad \text{or} \quad \text{hypotenuse} = \sqrt{\text{base}^2 + \text{height}^2}\,.$$

The height of each triangle formed by a wire segment is the 6-cm width of the rectangle. The base of each triangle is one-eighth of the rectangle's length, or 20 cm ÷ 8 = 2.5 cm. Substituting these values into the Pythagorean theorem yields

$$\text{wire segment length} = \text{hypotenuse} = \sqrt{(2.5\text{ cm})^2 + (6\text{ cm})^2} = 6.5\text{ cm}.$$

For the eight segments, the total wire length lying in the rectangle is 8 × 6.5 cm = 52 cm. Because the length of wire in the rectangle is the same as that coiled around the cylinder, the solution to the original problem is that the coiled wire has a length of 52 centimeters.

4.4.6 Do Not Be Reluctant to Use Approximations

The next hint is to recognize that approximations can be extremely valuable — either to find a preliminary solution or to reach an acceptable final solution. As long as you know that you are making an approximation, and assess the uncertainties introduced, they can save a lot of work.

Some approximations come from mathematical results. For example, one mathematical theorem states that, for small values (much less than 1) of a number x,

Recall the notation ≈ means "approximately equal to."

$$\sqrt{1+x} \approx 1 + \frac{1}{2}x\,.$$

Even without knowing how the approximation was derived, we can confirm its validity with a few test cases. For example, if we try a value of $x = 0.05$ (a calculator is needed for the first calculation),

$$\sqrt{1+0.05} = \sqrt{1.05} = 1.024695 \quad \text{and} \quad 1 + \frac{1}{2}(0.05) = 1.025\,.$$

The two values are quite close, confirming the validity of the approximation in this case.

Time-Out to Think: Try the preceding approximation formula with smaller numbers (say, 0.01, or 0.0002). Convince yourself that the approximation gets better as x gets smaller. Then try it with some larger values. Does the formula still work with $x = 0.1$? $x = 0.5$? When does it begin to fail?

We also need to consider another type of approximation — those made in setting up a problem. The next example illustrates the point.

Example 4-22 ***A Bowed Rail.*** Imagine a long bar of metal, such as a rail along a mile-long length of railroad track. Suppose that the rail is anchored on both ends (a mile apart) and that, on a hot day, its length expands by 1 foot. If the added length causes the rail to bow upward (as shown in Figure 4-10a), about how high would the center of the rail rise above the ground?

Solution: Seeking an exact solution would be difficult; aside from involving trigonometry and complicated algebra, we would also need to determine the exact shape of the bow in the rail. Instead, let's make an approximation.

As shown in Figure 4-10(b), we can approximate the height of the center of the rail by replacing the bow with two straight lines. We now have two right triangles, which means that we can use the Pythagorean theorem. The lengths of the base and hypotenuse are labeled — be sure you understand where these numbers come from. We can now solve for the height of the triangle (a mile is 5280 feet, so a half-mile is 2640 feet):

$$\text{Height of triangle} = \sqrt{(2640.5\text{ ft})^2 - (2640\text{ ft})^2} = 51.4\text{ ft.}$$

Based on our approximation, the center of the rail would rise more than 50 feet off the ground! This result is surprisingly large (to most people). Nevertheless, a review of our method turns up nothing unreasonable and, in fact, it is quite accurate. Of course, if we look back to step 4 of the problem-solving process, we realize that something is unrealistic about the problem posed — rails are not generally a mile long and do not rise 50 feet off the ground.

❖

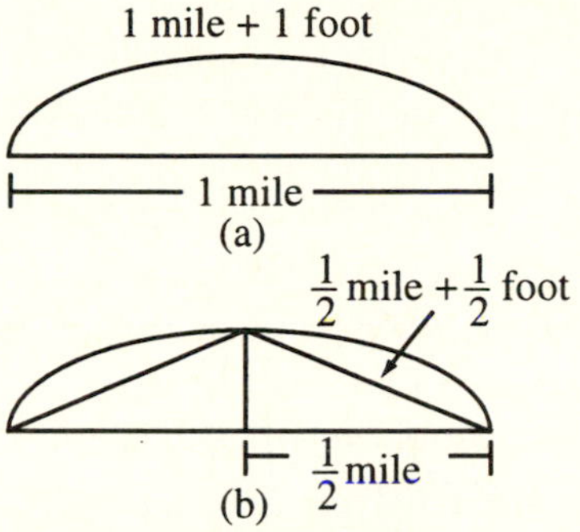

Figure 4-10. On a hot day a rail that is originally 1 mile in length expands by 1 foot. (a) The rail, which is now one mile plus one foot in length, bows upward. (b) The problem of finding the height of the center of the rail is simplified if the curved rail is replaced by two straight pieces of rail.

4.4.7 Try Alternative Patterns of Thought

When solving problems, you should try to avoid rigid patterns of thought that tend to suggest the same ideas and methods over and over again. Instead, you should approach every problem with an openness and freshness that allows innovative ideas to percolate.

In its most wondrous form, this approach is typified by what Martin Gardner, a well-known popularizer of mathematics, calls "aha!" problems, after the common term for exclaiming insight. These are problems whose best solution involves a penetrating insight that reduces the problem to its essential parts. Once that insight is gained, the solution usually appears immediately.

Example 4-23 ***The Monk and the Mountain.*** A monk sets out from a monastery in the valley at dawn. He walks all day up a winding path, stopping for lunch and taking a nap along the way. At dusk he arrives at a temple on the mountaintop. Several days later the monk makes the return walk to the valley, leaving the temple at dawn, walking the same path for the entire day and arriving at the monastery in the evening. Must there be one point along the path that the monk occupies at the same time of day on both the ascent and descent?

This problem is adapted from Arthur Koestler, *The Act of Creation*, 1964, Dell Publishing. Reprinted with permission.

Solution: The temptation for a mathematically inclined person might be to devise functions to describe the ascent and descent and then try to prove that they have an intersection point. The more insightful approach is to abandon the sequential thinking that the problem imposes and imagine the monk and an imaginary twin making the ascent and the descent at the same time. No matter how fast or slowly the monk and his imaginary twin walk, they must pass each other somewhere along the path. At this passing point, the monk and his twin would be at the same point at the same time. This description captures all the essential elements of the original problem, so the conclusion must be the same: the monk *does* occupy a particular point along the path at the same time of day on both his ascent and descent.

❖

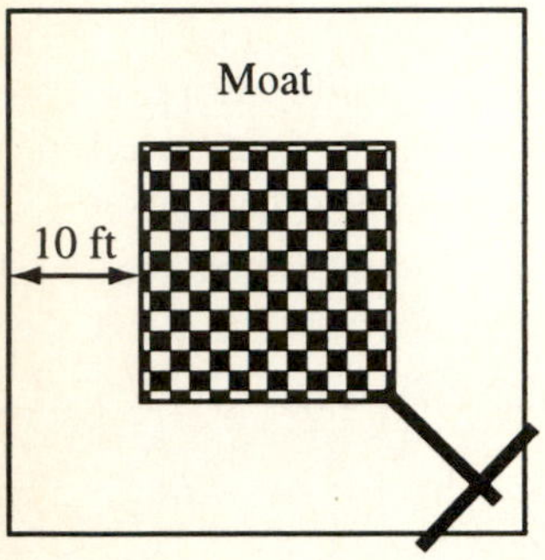

Figure 4-11. The "aha!" solution to the moat problem.

Example 4-24 *The Moat.* A castle is surrounded by a deep rectangular moat 10 feet across. A knight on a rescue mission must cross the moat, but he has only two 9 1/2-foot planks. How does he do it?

Solution: As with most "aha!" problems, the solution to this problem is undeniable once it is seen. One look at the solution in Figure 4-11 should convince you of the "aha!" solution.

❖

4.4.8 Do Not Spin Your Wheels

Finally, everyone has had the experience of getting "bogged down" with a problem. When your wheels are spinning, let up on the gas! Often the best strategy in problem solving is to put a problem aside for a few hours or days. You will be amazed at what you might see (and what you overlooked) when you return to it.

4.5 CONCLUSION

In this chapter we focused on the problem-solving process. With the experience you have gained in this chapter, you will be able to apply new mathematical tools as you learn them in this book and in the future. This chapter includes the following key lessons about problem solving.

- Working with the units of a problem, as well as with the numbers, is one of the best problem-solving techniques. Always consider the units of any problem you face. Take advantage of unit analysis both to help you in setting up problems and in checking your answers.
- Problem solving is more of an art than a science, and requires both creativity and organization. Implicit use of the four-step problem solving process (Section 4.3) will help keep you organized while tapping your creativity.
- The only sure way to improve at problem solving is by doing it.

PROBLEMS

Reflection and Review

Section 4.1

1. **Your Problem-Solving Experience.** Write two or three paragraphs reflecting on your own experience at problem solving in prior mathematics classes. Have you enjoyed "word problems," or do you relate to the "Hell's Library" cartoon? How do you think your attitudes towards word problems formed? What steps do you think you need to take to improve your problem-solving ability?

Section 4.2

2. **Identifying the Units.** Rewrite each statement by explicitly identifying the units. *Example:* Detergent sells for \$2 per ounce. *Solution:* Detergent sells for \$2/oz.
   a. Apples sell for 50 cents per pound.
   b. A mile is 1760 yards long.

c. The tile installation will cost \$3.50 per square foot.
d. The river is flowing at a rate of 5000 cubic feet per second.
e. An acre contains 43,250 square feet.
f. A liter holds 1000 cubic centimeters.
g. Atmospheric pressure is 14.7 pounds per square inch.
h. The reservoir volume is 750 acre-feet.
i. In an hour, this light bulb uses 100 watt-hours of energy.
j. Earth's average density is 5.5 grams per cubic centimeter.
k. The rocket accelerated at 7 meters per second per second.
l. The ore contains 5 ounces of gold in every cubic foot.

3. **The Key Words *Of* and *Per*.** Solve the following problems. Recall that *of* generally suggests multiplication and *per* generally suggests division.
   a. Sam used 1/3 of a dozen eggs for his Sunday morning omelet. How many eggs did he use?
   b. The 25 scouts in Troop 11 raised \$200. How much was raised (on average) per scout?
   c. Two fifths of the 25,000 people in Midway are Republicans, and 1/3 of the Republicans are women. How many Republican women are there in Midway?
   d. A total of 550 apples were picked and put into 11 crates. How many apples were there per crate?

4. **Units of Area.** The United States, excluding Alaska and Hawaii, is shaped roughly like a rectangle with a length of about 3500 miles and width of about 1000 miles. In contrast, the state of Colorado is shaped like a rectangle, with a length of about 400 miles and a width of about 300 miles.
   a. What is the approximate area of the United States in square miles? in square feet? in acres?
   b. What is the area of Colorado in square miles? in square feet? in acres?
   c. What fraction of the area of the United States does Colorado represent? State your answer as a percentage.

5. **Units of Volume.** Calculate the volumes of the following objects.
   a. A swimming pool that is 3 meters deep, 10 meters long, and 5 meters deep.
   b. A package measuring 22 inches by 15 inches by 12 inches.
   c. A skyscraper that is 1000 feet high with a 25,000 square foot base.
   d. A human being shaped roughly like a box, with a height of 5.5 feet, width of 1.25 feet, and a depth by a 0.5 feet.

6. **Working with Units.** Solve each of the following problems, being sure to show your work with units clearly.
   a. Suppose that you buy 3 pounds of apples priced at 50 cents per pound. How much will you pay?
   b. There are 100 centimeters in a meter and 1000 meters in a kilometer. How many centimeters are in a kilometer?
   c. There are 3 feet in a yard, and 5280 feet in a mile. How many yards are in a mile?
   d. Through a stroke of great luck, you find a 252-milligram ruby. With 200 milligrams per carat (for precious stones), what is its weight in carats?
   e. Suppose that you work 40 hours per week and are paid \$6 per hour. How much money will you make in a week?
   f. Suppose that you take a trip driving 1200 miles in 20 hours. What is your average speed for the trip?
   g. If you sleep an average of 7.5 hours each night, how many hours do you sleep in a year?
   h. An average human heart beats 60 times per minute. If an average human being lives to the age of 75, how many times does the average heart beat in a lifetime?

7. **Unit Analysis: What Went Wrong?** Imagine that you are grading an exam. Each of the following problems states an exam question and a solution given by a student. State whether each solution is right or wrong. If it is wrong, write a note to the student explaining why the answer is wrong and how to solve it correctly. (See Example 4-6.)
   a. *Exam Question:* A candy store sells chocolate for \$7.70 per pound. The piece you want to buy weighs 0.11 pounds. How much will it cost, to the nearest cent? (Neglect sales tax.) *Student Solution:* 0.11 ÷ 7.70 = 0.014. It will cost 1.4 ¢.
   b. *Exam Question:* You bought a copy of *The Economist* magazine (weekly) for five straight weeks. Each copy costs \$3.50. How much did you spend? (Neglect sales tax.) *Student Solution:* \$3.50 × 5 = \$17.5. I spent \$17.50.
   c. *Exam Question:* You ride your bike up a steep mountain road at 5 miles per hour. How far do you go in 3 hours? *Student Solution:* 5 ÷ 3 = 1.7. I rode 1.7 miles.
   d. *Exam Question:* You can buy a 50-pound bag of flour for \$11 or you can buy a 1-pound bag for \$0.39. Compare the per pound cost for the large and smal bags. *Student Solution:* The large bag price is 50 ÷ \$11 = \$4.50 per pound, which is much more than the 39 ¢ per pound price of the small bag.
   e. *Exam Question:* The average person needs 1500 calories a day. A can of Coke contains 140 calories. How many Cokes would you need to drink to fill your daily caloric needs? (Note: this diet may not meet other nutritional needs!) *Student Solution:* 1500 ÷ 140 = 60,000. You would need to drink 60,000 Cokes to meet your daily caloric needs.

8. **Unit Conversions.**
   a. There are 8 ounces in a cup, 4 cups in a quart, and 4 quarts in a gallon. How many ounces are in a gallon?
   b. How many minutes are in a century?
   c. A car is driving at 55 miles per hour. What is this speed in feet per hour? feet per second?

9. **Conversions with Units Raised to Powers.**
   a. Find a conversion factor between cubic inches and cubic feet.
   b. Find a conversion factor between cubic inches and cubic yards.
   c. A kilometer is 1000 meters long. How many square meters are in one square kilometer?
   d. How many cubic meters are in one cubic kilometer?
   e. A football field is 100 yards long and 55 yards wide. Find its area in square yards, square feet, and acres.

10. **New Flooring.** Imagine that you own a house and want to install new flooring. The rooms to be covered are the kitchen, which measures 10 feet by 12 feet; the master bedroom, which is 14 feet by 15 feet; the living room, which is 16 feet by 12 feet; the bathroom, in which the floor area to be covered is 3 feet by 5 feet; and the hallway, which is

3 feet by 8 feet. You plan to cover the kitchen with linoleum, at a price of $4 per square foot. The bathroom will be tiled, at $7 per square foot. The living room and hallway will be carpeted, at $18 per square yard. And the bedroom will be carpeted at $14.25 per square yard. What is the total price for your planned flooring?

Section 4.3

11. **Sales Tax from Textbooks.** How much sales tax revenue is generated by textbook sales at your school? Follow the four-step process in Example 4-15, but use numbers that you find for *your* school.

12. **Four-Step Problem Solving.** In each of the following problems, describe how the four-step problem-solving process could be used to obtain a solution. Following the scheme of Example 4-15, you may use reasonable numbers that will allow you to obtain a sample answer.
   a. How much sales tax revenue is generated by the sale of automobiles in your state?
   b. How much sales tax revenue is generated by the sale of automobiles nationwide?
   c. How much revenue does the National Football League earn from ticket sales?
   d. Suppose that you want to drive from New York to Los Angeles with one other person. How much will you spend on gas if the two of you split the cost evenly?

Section 4.4

13. **Traffic Counters.**
   a. Find, by trial-and-error, all six solutions to the traffic counter problem described in Example 4-16.
   b. Suppose that the same traffic counter described in the example registers 41 counts during another period of time. Find all possible combinations of two- and three-axle vehicles that might have passed. Look for patterns in your answers to simplify the process.
   c. A similar traffic counter is placed on a major highway, where vehicles with 2, 3, and 4 axles can travel (that is, vehicles that register 2, 3, or 4 counts). Suppose that, during a 5-minute period late at night, the counter registers 10 counts. Find all possible combinations of two-, three-, and four-axle vehicles that might have passed that point.
   d. Suppose that the traffic counter from part (c) registers 66 counts during a 1-minute period of heavy traffic. Find all possible combinations of two-, three-, and four-axle vehicles that might have passed. Be sure to look for patterns to simplify your work.

14. **Your Own Nonunique Solution.** Describe another problem that, like the traffic counter example, does not have a unique solution. You may describe any problem, subject to the following constraints: first, your problem should represent something real (just as traffic counters actually exist and are used); second, it should be a problem that you can describe clearly and concisely.

15. **Sorting 20 Numbers.** For each sorting algorithm given in Example 4-17, find how many comparisons would be needed to sort a list of 20 numbers.

16. **Alphabetic Sorting.** Describe an algorithm for sorting a list of names in alphabetic order. How many comparisons would be needed to sort a list of 10 names using your algorithm? To keep the problem relatively simple, assume that the names all have different first letters.

17. **Dictionary Algorithm.** Although you may not realize it, you probably use mental algorithms every day. For example, consider the process of looking up a word in a dictionary. One possible, though highly inefficient, algorithm would be to start at the beginning of the dictionary, and compare the word you are looking for to every word in the dictionary, stopping only when you find a match.
   a. Why is this algorithm inefficient? How many comparisons would be needed to look up a typical word?
   b. Describe the algorithm that you follow when you look up a word in a dictionary.
   c. In words, compare your algorithm to the one described in the problem. Why is yours more efficient? Can you think of an even more efficient algorithm?

18. **More on Jack and Jill.**
   a. In Example 4-18, suppose that Jill's time was 10 seconds in the first race. Following solution method 1 in that example, by how much would she win the second race?
   b. Hack and Quill race 200 meters and Hack wins by 10 meters. They race a second time, with Hack starting 10 meters behind the starting line. Who wins the second race?

19. **The Cars and Canary Revisited.** Two cars, 150 kilometers apart, begin driving toward each other on a long, straight highway. One car travels 80 kilometers per hour and the other 100 kilometers per hour. At the same time a canary, starting on one car, flies back and forth between the two cars as they approach each other. If the canary flies 120 kilometers per hour and spends no time to turn around at each car, how far has it flown when the cars collide?

20. **Mixing Marbles.** To show in Example 4-20 that the number of marbles transferred or the number of marbles we start with does not matter, do the following. Consider the case in which each pile initially has 15 marbles. Suppose that on the first transfer *three* black marbles are moved to the white pile. On the second transfer, any three marbles are taken from the white pile and put into the black pile. Demonstrate, preferably with the aid of diagrams in addition to words, that you will always end up with as many white marbles in the black pile as black marbles in the white pile.

21. **The Coiled Wire Revisited.** Suppose that 10 turns of a wire are wrapped around a pipe having a length of 20 centimeters and a circumference of 6 centimeters. What is the length of the wire?

22. **The Bowed Rail Revisited.** Consider a 1-kilometer-long rail, anchored on both ends. On a hot day, its length expands by one centimeter, causing the rail to bow upward. About how high does the center of the rail rise above the ground? Comment on the realism of this problem.

Further Topics and Applications

More Unit Analysis Problems

23. **Full of Hot Air.** The average person breathes 15 times per minute (at rest), inhaling and exhaling half a liter of air each time. How much "hot air" (the air is warmed by the body) does the average person exhale each day?

24. **Degrees of Longitude.** One degree of longitude at the equator corresponds to 69.171 miles. Use the fact that there are 360° of longitude around the Earth to find the circumference of the Earth.

25. **Water in a Reservoir.** Suppose that a water reservoir covers an area one mile wide by two miles long and has an average depth of 250 feet.
    - **a.** What is the area of the reservoir in acres?
    - **b.** What is the volume of the reservoir in acre-feet?
    - **c.** What is the volume of the reservoir in cubic feet?
    - **d.** Given that a cubic foot of water is the same as 7.5 gallons, how many gallons of water are in the reservoir?
    - **e.** Given that a cubic foot of water weighs 62.4 pounds, how much does all the water in the reservoir weigh?

26. **Dog Years.** Sometimes the age of dogs is described in a unit called "dog years." A commonly used conversion is that 1 real year equals 7 dog years.
    - **a.** If your dog is 15 real years old, what is her age in dog years?
    - **b.** People often refer to the third year in the life of a human child as the "terrible twos" stage. If dogs have a terrible twos stage in their third dog year, how old are they, in real years, during this stage?
    - **c.** Why do you think that anyone ever started using the unit of dog years? Do you think the common conversion of 1 real year to 7 dog years is reasonable? Explain.
    - **d.** What is the physical meaning of a year (that is, in terms of the Earth and Sun)? Does a dog year have a similar physical meaning? Explain.

27. **Explain This One!** A Goodyear tire commercial shown during the 1995 NFL playoffs began by stating that the Goodyear Aquatread tire can "channel away" 1 gallon of water per second. The announcer then goes on to state: "one gallon per second — that's 396 gallons per mile." What's wrong with this statement? Hypothesize about how the advertising agency was able to come up with such a bizarre statement.

Four-Step Problem Solving

Use the four-step process on each of the following problems.

28. **Finding a Good Deal on Tax.** Suppose that you live in a city where the sales tax is 7.8%. You plan to purchase a car from a dealer offering a pretax price of $11,345. However, you find that a nearby city with a tax rate of 4.7% has a dealer offering the same car for a pretax price of $11,450. Which is the better deal, and by how much?

29. **Buying a Newspaper.** Suppose that the newsstand price of a newspaper is $1 per issue on Monday through Saturday, and $4.50 on Sunday. A home delivery subscription special offers the newspaper for $250 for an entire year. How much would you save if you subscribed? Answer in both absolute (dollars and cents) and relative (percentage) terms, and be sure to discuss uncertainties when you look back (step 4 of the problem-solving process).

30. **Gas Tax Comparison.** In 1988, the average gasoline sales tax in the United States was about 45%. In contrast, the gasoline sales tax in Great Britain was 178%. The price of a gallon of gas in the United States was approximately $1.20. Assuming that the pretax cost of gasoline was the same in both countries, how much did a gallon of gas cost in Great Britain at the time?

31. **Incoming Sophomore.** Advance placement exams are offered in many disciplines. In 1994, each test cost $72, and a passed test counted for an average as 3.5 college credits. Suppose that a highly motivated student at your school received enough advance placement credit to enter college as a sophomore and thereby graduated in three years rather than four. Considering the cost of the tests and the cost of tuition, how much does this student save on her college education?

32. **Computer Stored Books.** Computer memory is measured in units of "bytes," where one byte is enough memory to store one character (a letter in the alphabet or a number). How many typical pages of text can be stored on a 500-megabyte hard drive? (A megabyte is one million bytes.)

33. **Budgeting for a New Car.** After graduating and getting a job paying $24,000 a year, you decide it's time to park your bike and buy a brand new car. You find a great deal on a new car with monthly payments of $300. Insurance adds another $100 per month.
    - **a.** Suppose that federal income tax takes 12% of your income, state income tax takes 5%, and FICA (Social Security and Medicare) takes another 7.5%. What is your monthly take-home pay (after deducting taxes)? What percentage of your take-home pay goes toward paying for your car?
    - **b.** If you work 8 hours a day, 5 days a week, how many hours of each day do you spend working to pay for your car?
    - **c.** Suppose that you are a parent of this recent graduate. Based on the results in parts (a) and (b), would you advise your child to go ahead with the car purchase? Why or why not?

34. **Cars and Bikes.** Consider the following data about cars and bikes in 1990:
    - There were 900 million bicycles worldwide.
    - There were 400 million cars worldwide.
    - The global population was 5.5 billion people.
    - China's population was about 1 billion.
    - The U.S. population was about 260 million.
    - There were 100 million bikes and 140 million cars in the United States.
    - There were 300 million bikes and 1.2 million cars in China.

Contrast and discuss the typical numbers of cars and bikes per person for the United States, China, and the rest of the world in 1990.

35. Global Energy Consumption at the U.S. Rate? In 1994, the average American consumed fossil fuels at a rate estimated to be equivalent to 40 barrels of oil per year. The estimated global reserve of recoverable fossil fuels was the equivalent of 10 trillion barrels of oil. If everyone in the world consumed fossil fuels at the same rate as Americans, how long would the reserves last?

36. Implicit Four-Step Problem Solving. The four-step problem-solving process is rarely enumerated but usually implicit. In particular, you could rework any of the examples in this chapter by explicitly enumerating the four-step process. Although that means extra work for an already known solution, it is good practice for future problem solving.

a. Do so for one example of your choice from section 4.2.
b. Do so for one example of your choice from section 4.4.

Just for Fun

37. More "Aha!" Problems. Each problem has an "aha!" type solution (Subsection 4.4.7). Although they have little practical relevance, you might find them fun.

a. Twelve blue socks and 12 white socks are loose and mixed up in a drawer. If you try to select them in the dark, how many socks do you need to take from the drawer to be sure that you have a matching pair?
b. Each of 10 large barrels is filled with golf balls that all look alike. The balls in nine of the barrels weigh one ounce and the balls in one of the barrels weigh two ounces. With only *one* weighing on a scale, how can you determine which barrel contains the heavy golf balls?
c. A woman is traveling with a wolf, a goose, and a mouse. She must cross a river in a boat that will hold only herself and one other animal. If left to themselves, the wolf will eat the goose and the goose will eat the mouse. How many crossings are required to get all four creatures across the river alive?
d. Three boxes are labeled APPLES, ORANGES, and APPLES AND ORANGES. Each label is incorrect. Can you select one fruit from one box and determine the correct labels? If so, explain how.
e. How do you measure nine minutes with a seven-minute and a four-minute hourglass?
f. A 150-foot rope is suspended at its two ends from the tops of two 100-foot flagpoles. The lowest point of the rope is 25 feet from the ground. What is the distance between the flagpoles?
g. If a clock takes 5 seconds to strike 5:00, how long does it take to strike 10:00?
h. A man arrives in a small Nevada town in need of a haircut. He discovers that the town has two barbers. One is well-groomed, with splendidly cut hair; the other is unkempt, with atrocious and unattractive hair. Which barber should the visitor patronize and why?
i. You are considering buying 12 gold coins that look alike but have been told that one of them is a heavy counterfeit. How can you find the heavy coin in three weighings on a balance scale?

Projects

38. Analyzing Supercomputer Problems. Difficult problems in many fields today are routinely addressed with the aid of powerful computers, or supercomputers. But just because a problem requires the use of a supercomputer, doesn't mean that it is beyond the comprehension of the average person. One way of making sense of problems studied with supercomputers is to analyze the research in terms of the four-step problem solving process.

Choose a problem of interest to you that has recently been studied with the aid of a supercomputer. Good sources of information on such problems include popular magazines on science or mathematics, such as *Scientific American, Discover, Science News,* or the weekly *Science Times* section of the *New York Times*. After choosing the problem, write a short essay in which you describe the research in terms of the four-step process, without actually going through any calculations. That is, describe the problem clearly, explain the strategy used to address it, describe how the strategy was carried out with the aid of a supercomputer, and look back at the meaning of the results.

39. Textbook Analysis: Word Problems for Kids. Although research shows that most adults today have difficulty with "word problems," we might hope that the next generation will have far less difficulty with them. Find a current textbook in mathematics that is used at the upper elementary school level (grades 4–6). Read through the "word problems" in the textbook. Write a critical analysis of the word problems and conclude with an opinion as to whether use of this textbook should be continued. *Extra Credit:* Identify

a teacher who is currently using the textbook you have analyzed. Interview the teacher about the attitudes of her or his students towards word problems. Include the results of your interview in your evaluation of the textbook.

40. Reflecting on China's One-Child Policy.

a. Suppose that, as the current generation matures, China's population of young adults has more men than women by a ratio of 118 to 100. With 400 million young adults in China, how many men will be unable to find a spouse?

b. One surprising fact about China's population is that, even if the one-child policy is strictly enforced, China's population will continue to *grow* for a short time before it begins to fall. Explain how that is possible.

c. In 1978, China's policy planners calculated that a one-child policy would reduce the population to 700 million by 2050. However, the problem of "missing girls" already demonstrates that China's population in 2050 is likely to be very different in size or makeup (male–female ratio) than intended by the policy planners. Discuss any general lessons that policy planners should learn from this prospect.

d. Imagine that you are a Chinese leader. How would you address the problem of "missing girls"? Explain.

e. To convince yourself that a one-son policy would lead to an average of two children per family, with equal numbers of boys and girls, do the following. Suppose that 100,000 families are having children according to this policy. Describe the general makeup of all of the families (that is, start with the fact that 50,000 families have a boy as their first and therefore only child, and continue on). Use this process to show that the average number of children is two and that boys and girls are equal in number.

41. Watering the Grass. The most popular lawn grass in the arid southwestern United States is Kentucky bluegrass. Unfortunately, keeping this grass healthy requires a tremendous amount of water: Each square foot of grass needs 20 gallons of water over the 20-week summer season. In contrast, a grass that is native to the region, such as buffalo grass, needs only about 1.5 gallons per square foot per 20-week season (of course, it doesn't look quite as green).

a. Contrast the total amount of water needed for quarter-acre lawns of Kentucky bluegrass and buffalo grass.

b. Choose a particular arid state, such as California, Nevada, New Mexico, Arizona, Utah, Wyoming, or Colorado. Determine the total amount of water that goes to nonagricultural use in the state and the percentage of that total devoted to watering lawns. How much water could be saved if all lawns were replaced with native grasses with low water requirements, such as buffalo grass? Comment on the monetary and environmental impacts of such a conversion.

42. The Stock Market Page. With a few terms defined, your problem-solving prowess should enable you to interpret the newspaper stock market pages. They typically offer the following information for each listed stock.

- 365-Day High/Low: These are the high and low prices for the stock over the past 365 days.
- Stock: The name of the stock is given, usually as an abbreviation.
- Div: The current annual dividend payment is listed, in dollars per share.
- Yld %: The yield percentage is the ratio of the annual dividend to the share price, expressed as a percentage.
- P/E: The price/earnings ratio is the share price of the stock divided by earnings per share over the past year. (The earnings per share is the total company earnings divided by the total number of shares.)
- Sales or Volume: The number of shares traded during the day is shown (sometimes expressed in blocks of 100 shares).
- High, Low: The highest and lowest prices, respectively, at which shares traded during the day are reported.
- Last, Change: The closing price of the stock (price of the last trade for day) and the change from the previous closing price, respectively, are listed.

a. Choose three stocks that you are interested in for any reason. Find a stock market page that gives all the preceding information and record it for your three stocks. Be sure that at least one of your stocks (preferably all three) pays a dividend.

b. How does the most recent closing price compare to the 365-day high and low? Answer in both absolute and relative terms. Find prices for your stocks at the end of each of the past 10 weeks (by checking old newspapers, or using a service that lists changes in stock prices). If you had purchased shares in these three stocks 10 weeks ago, how would you be doing now?

c. Using the dividend and closing price, calculate the yield percentage for each stock. Compare it to the value printed as "Yld" and discuss any discrepancies. What does the listing show if a stock does not pay any dividend?

d. Convert the yield percentage to a decimal fraction. The reciprocal of this value is the share price divided by the dividend. Calculate this value for your stocks. Is it the same as the P/E ratio for the stocks? Explain.

e. Not all stocks have a P/E ratio listed. Why? In general, are stocks with high or low price/earnings ratios better buys? How can you make a decision if no P/E ratio is available? Explain your answer.

f. If a stock doesn't pay a dividend, can a yield percentage be calculated? Explain. Why don't all stocks pay dividends? How can you decide whether a stock without a dividend is a good buy?

g. The yield percentage is comparable in meaning to the interest (annual yield) on money in a savings account. Compare the yield percentages for the companies you chose to the interest available through savings accounts at local banks. Discuss the advantages and disadvantages of investing money in these stocks against putting money in a savings account.

5 THE SEARCH FOR KNOWLEDGE

In Chapters 1–4, we explored the meaning and role of quantitative literacy, discussed the basic principles of reasoning, and showed you how to apply the principles of logic to the task of solving quantitative problems. All of this material was organized under *Thinking Logically,* the title of Part 1 of this book. We now conclude Part 1 with an in-depth investigation into how knowledge is acquired. We describe the interrelationships among logic, mathematics, and science, which open the way to understanding the scientific method — the principle means by which knowledge is acquired today. This understanding, in turn, will help you learn to be skeptical about the untested claims of pseudoscience. Finally, in studying the search for knowledge, you will also have the opportunity to investigate the concept of truth, with some surprising results from mathematics.

CHAPTER 5 PREVIEW:

5.1 ***THE SEARCH FOR TRUTH.*** We begin the chapter by recounting the history of the search for truth. In the process, we demonstrate how logic, mathematics, and science are interconnected.

5.2 ***LOGIC AND SCIENCE.*** In this section we explore the relationship between logic and science in the scientific method. Understanding this relationship is crucial to understanding the role of science in society.

5.3 ***PARADOXES.*** A primary step in the search for knowledge is identifying unsolved problems. The study of paradoxes often points the way, and in this section we investigate several important and well-known paradoxes.

5.4 ****THE LIMITATIONS OF LOGIC.*** In this final section we explore the limitations of logic and the theorem of Gödel, which is surely one of the most astonishing discoveries of the twentieth century.

If we could find characters or signs appropriate for expressing all our thoughts as definitely and as exactly as arithmetic expresses numbers or geometric analysis expresses lines, we could in all subjects in so far as they are amenable to reasoning accomplish what is done in Arithmetic and Geometry. For all inquiries which depend on reasoning would be performed by the transposition of characters and by a kind of calculus.... And if someone would doubt my results, I should say to him: "let us calculate, Sir," and thus by taking to pen and ink, we should soon settle the question.

— Gottfried Wilhelm von Leibniz (1646–1716)

5.1 THE SEARCH FOR TRUTH

Can the question of the existence of a God, or Creator, be settled through logic? Consider the argument of the nineteenth century biologist and philosopher William Paley, which begins with the claim that a watch implies the existence of a watchmaker:

Paley's argument in favor of the existence of a Creator is from his book, *Natural Theology*, first published in 1802.

> *In crossing a heath, suppose I pitch my foot against a stone, and were asked how this stone came to be there; I might possibly answer, that for anything I knew to the contrary, it had lain there forever.... But suppose I had found a watch upon the ground, and it should be inquired how the watch happened to be in the place; I should hardly think of the answer which I had given before, that, for anything I knew, the watch might have always been there.... The inference, we think, is inevitable; that the watch must have had a maker.*

Paley continued his argument by noting that biological structures, such as the human eye, are far more complex than a watch and are perfectly suited to their functions. He thereby concluded that such structures must have been designed by a Creator. In succinct form, Paley's argument is as follows:

Premise:	The existence of something as complex and functional as a watch implies the existence of a watchmaker.
Premise:	Biological systems are more complex and functional than a watch.
Conclusion:	Biological systems imply the existence of a Creator.

Time-Out to Think: What do you think of Paley's argument? Is it convincing? How might you counter Paley's logic to argue *against* the existence of a Creator?

Paley's "proof" of the existence of a Creator provides an example of how philosophers, since time immemorial, have tried to use logic to ascertain ultimate truths. It also demonstrates the interrelationships among logic, mathematics, and science. The role of logic is clear in Paley's attempt to make a deductive argument for a Creator. Science also plays a clear role, as an understanding of biology is invoked in the second premise. The role of mathematics in Paley's argument is more subtle but is essential in at least two ways. First, mathematics is implicit in the concept of complexity; the idea that an eye is more complex than a watch is quantitative, depending on some notion of the number of interconnections among the parts of the eye compared to the parts of a watch. Second, mathematics is involved through the long, intertwined history of logic and mathematics.

In this chapter, we delve into the concept of truth and, in the process, investigate how knowledge is acquired. In doing so, we explore science and the scientific method in some detail. By the time you finish this chapter, you will have developed a much clearer understanding of how science works, why it is so important in the modern world, and how it is related to logic and mathematics.

5.1.1 Classical Logic and Mathematics

Mathematics and logic always have been intertwined. Indeed, the study of logic, like algebra or geometry, is considered to be one of the many branches of mathematics. Just as logic is used to test the validity of arguments, a primary

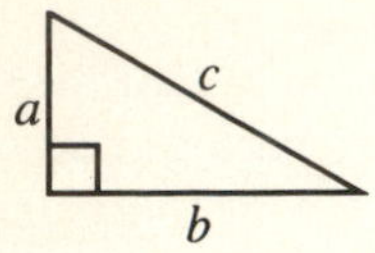

Figure 5-1. The Pythagorean Theorem states that $a^2 + b^2 = c^2$ for a right triangle.

goal of mathematics is to establish the truth of mathematical propositions. A good example is provided by the Pythagorean theorem, which states the relationship between the side lengths of a right triangle (Figure 5-1). Mathematicians establish the truth of a theorem by constructing a proof, which essentially is an argument wherein established mathematical facts serve as premises and, in this case, the Pythagorean theorem is the conclusion (see the discussion of mathematical proof in subsection 2.5.2).

The similarity between arguments and proofs explains why many great philosophers also were mathematicians. Indeed, this relationship between logic and mathematics extends back to ancient times. Aristotle (384–322 B.C.), the Greek philosopher well-known for his contributions to many fields of science and as the tutor of Alexander the Great, probably was the first person to attempt to give logic a rigorous foundation. Aristotle believed that truth could be established from three basic laws.

The law of identity:	A thing is itself.
The law of the excluded middle:	A statement is either true or false.
The law of noncontradiction:	No statement is both true and false.

Time-Out to Think: What do you think of Aristotle's three laws? Do they seem to be correct? Can you imagine any statements that violate his laws? If so, identify and explain them.

Aristotle's laws found almost immediate application in mathematics, where they formed the basis of the logic used by the Greek mathematician Euclid to establish the foundations of geometry. We cover Euclid's logic in more detail in Chapter 11; for now, we need to say only that Euclid began with a set of five postulates, or premises, from which he then derived all of classical geometry.

Perhaps because of Euclid's astonishing success, philosophers and mathematicians placed great trust in the validity of **classical**, or **Aristotelian, logic**. This trust continued to be validated, as it led directly to the development of the modern scientific method and the accompanying advances in human knowledge and technology. Indeed, this interplay of logic and mathematics may have been the single greatest factor in the rise of the Western world, beginning in the Renaissance, as the center of scientific, industrial, and technological development. However, even great ideas can be carried too far, as you will soon discover.

***THINKING ABOUT ...** PROOF OF THE PYTHAGOREAN THEOREM*

Despite its name, the Pythagorean theorem was known in many cultures long before Pythagoras (c. 580–500 B.C.) was born. Babylonians used the theorem nearly a thousand years earlier, and it is also clearly stated in ancient Hindu and Chinese texts.

The Pythagorean theorem can be illustrated by drawing squares along each side of a right triangle (Figure 5-2). Simply measure the area of each square; the sum of the areas of the squares along the two smaller sides (sides a and b) is the same as the area of the square along the **hypotenuse**, or longest side (side c), confirming that $a^2 + b^2 = c^2$.

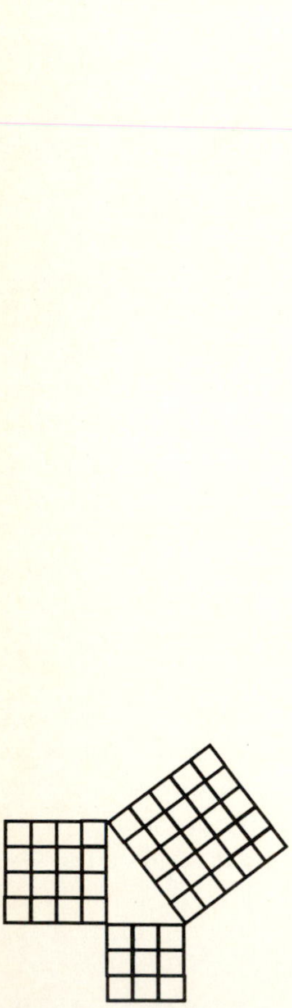

Figure 5-2. The Pythagorean theorem is illustrated by showing the area of the square on each side of a right triangle.

Although this illustration demonstrates the Pythagorean theorem for the particular right triangle shown, it doesn't prove that the theorem is true in general. For the most part, we aren't concerned with mathematical proofs in this book. However, the proof of the Pythagorean theorem is elegant and offers a good example of how mathematicians prove theorems.

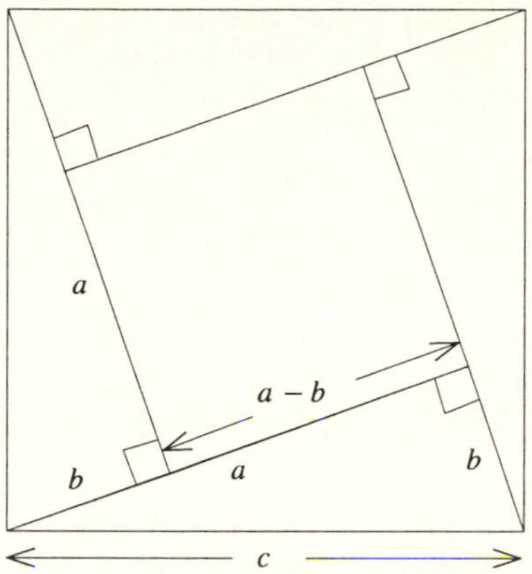

Figure 5-3. This diagram shows the construction used in Bhaskara's proof of the Pythagorean theorem.

Although many different proofs of the Pythagorean theorem have been given, one of the simplest is attributed to a twelfth-century Hindu mathematician named Bhaskara. His proof begins with a diagram of a large square, inside of which is a smaller square surrounded by four right triangles (Figure 5-3). To follow the proof, first note that the

area of *large* square = (area of *small* square) + (area of the four right triangles).

The side length of the *large* square is c, so its area is c^2. The side length of the *small* square is $a - b$, so its area is $(a - b)^2$. Because the area of a triangle is one-half its base multiplied by its height, the area of each of the four right triangles is $\frac{a \times b}{2}$. Substituting all these areas into the preceding equation gives

$$c^2 = (a - b)^2 + 4 \times \left(\frac{a \times b}{2}\right) = (a - b)^2 + 2ab.$$

Finally, applying a bit of algebra yields $(a - b)^2 = a^2 - 2ab + b^2$. Substituting this expression into the previous equation we get

$$\begin{aligned} c^2 = (a - b)^2 + 2ab &= a^2 - 2ab + b^2 + 2ab \\ &= a^2 + b^2. \end{aligned}$$

This result is the Pythagorean theorem. Legend has it that, when Bhaskara showed his proof to others, he accompanied it with just a single word: "Behold!"

5.1.2 Leibniz's Dream: A Calculus of Reasoning

Recognizing that logic could be used to establish mathematical truths, scholars eventually began to wonder whether logic could be used to determine other truths. For that matter, could logic be used to determine "universal truth," including truth in matters of ethics and morality?

In the seventeenth century the philosopher Gottfried Wilhelm von Leibniz (1646–1716) attempted to answer these questions. Leibniz is better known as one of two inventors of the branch of mathematics called *calculus* (the other inventor was Isaac Newton). Leibniz believed that formulating a *calculus of reasoning*, analogous to his calculus of mathematics, should be possible. This calculus of reasoning would be used to decide all arguments and would be the logical analog of using mathematics to answer mathematical questions.

Leibniz suggested that an international symbolic language for logic be developed. With **symbolic logic,** he believed that general equations of logic could be written in the same way that symbols are used to write equations in algebra. Those equations of logic would be used to calculate a "solution" to any argument.

Time-Out to Think: Reread the quote from Leibniz on the first page of this chapter. What do you think of Leibniz's dream of finding a calculus of reasoning?

Leibniz himself made little progress beyond outlining his ideas. Real work on creating a symbolic logic had to wait nearly two centuries, until George Boole published a book called *The Laws of Thought* in 1854. Boole tried to treat logic as a mechanical process akin to algebra and developed the fundamental ideas for using mathematical symbols and operations to represent statements and to solve problems in logic. Boole's work created what is known today as the **algebra of sets**, or **Boolean algebra**, which has many applications, particularly in computer science.

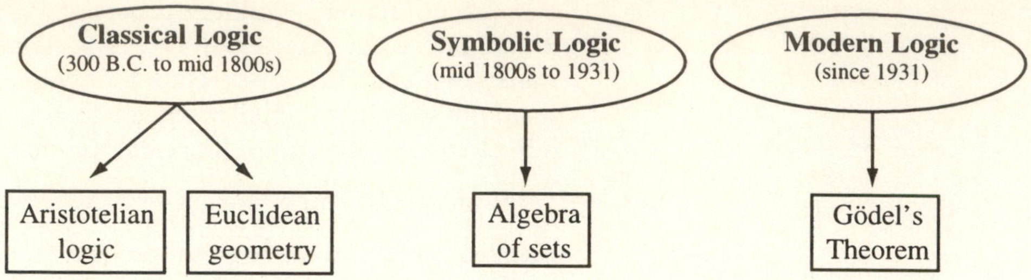

Figure 5-4. The history of logic may be divided into three periods, reflecting the relationship between logic and mathematics.

During his work on *Principia*, Russell himself became concerned that his hope was unattainable when he encountered a paradox with no apparent resolution. In brief, the paradox goes like this: A set is a collection of things. Now, consider a coin collection; the set is the collection, and its members are the individual coins. Note that, in this case, the set is not a member of itself — the collection is not an individual coin. Next, consider the set of all sets that are not members of themselves. Is this set a member or itself or not? This question seems to be unanswerable and caused Russell great distress. As Gödel later proved, Russell's distress was fully warranted.

The success of Boole's work led other philosophers and mathematicians to continue the development of symbolic logic. The culmination of this period was Bertrand Russell and Alfred North Whitehead's three-volume work, entitled *Principia Mathematica*, published during 1910–1913. In *Principia*, Russell and Whitehead sought to put all of mathematics into a standard logical form by attempting to derive all known mathematics from symbolic laws of thought. Russell hoped that putting mathematics onto firm logical ground would lead to Leibniz's dream of creating a system of logic in which *all* truth could be derived from a few basic principles.

Time-Out to Think: Do you believe in the notion of universal or absolute truth? Why or why not?

So what became of Leibniz's dream? Did mathematicians find a way to settle all arguments unquestionably and objectively? Alas, the quest for Leibniz's dream was not only unsuccessful but, in 1931, mathematician Kurt Gödel discovered a proof that the dream could *never* be achieved. We will return to Gödel's theorem and its implications in section 5.4. For now, we note that the shattering of Leibniz's dream ushered in a new period in the relationship between logic and mathematics, often termed the period of **modern logic** (Figure 5-4).

5.2 LOGIC AND SCIENCE

Science: knowledge acquired through careful observation and study; knowledge as opposed to ignorance or misunderstanding.

Scientific method: a set of principles and procedures, based on logic, for the systematic pursuit of knowledge.

What is **science**? The word comes from the Latin *scientia*, which means "having knowledge" or "to know." The words *conscience* (related to the idea of self-knowledge or self-awareness) and *omniscience* (all knowing) also have this root. Today, science connotes a special kind of knowledge: knowledge acquired through careful observation and study.

Note that the term *science* applies to much more than the traditional school sciences of biology, chemistry, and physics. Indeed, *any* subject can be studied scientifically if knowledge is systematically pursued according to the principles and procedures known as the **scientific method**.

The scientific method developed slowly in human history, and the modern idea of the scientific method dates back only a few hundred years. Nevertheless, it is a clear outgrowth of principles of experimentation and observation developed, to some degree, by every human culture. The modern scientific method depends on logical analysis both in determining how to pursue knowledge and in testing and analyzing proposed theories. It also depends on mathematics, not only in the close historical ties between mathematics and science, but also in the demand for quantitative measures of the success, just as theories are subjected to rigorous testing and evaluation.

Because logic, mathematics, and science are interconnected, the study of science is crucial to quantitative reasoning.

Human knowledge has increased dramatically in the few hundred years since the scientific method came into widespread use. Hardly any aspect of life today is not deeply influenced by science; for example, science contributes to advances in energy use, understanding of environmental issues, making business operations more efficient, understanding the interplay of social and economic forces, developing modern weaponry, and exploring the universe. An understanding of science and the scientific method therefore is critical to understanding human history, and present and future challenges.

> **Time-Out to Think:** What do you consider to be the most important change in the human condition over the past several hundred years? What role did science or technology play in that change?

5.2.1 Fact, Law, Hypothesis, and Theory

The study of science and the scientific method requires some terminology — in particular, the terms *fact*, *law*, *hypothesis*, and *theory*. Unfortunately, no standard, generally agreed upon definitions exist for these terms; in fact, the terms often are used in unclear or even contradictory ways, and distinguishing among them can be difficult. Nevertheless, for the purposes of discussion, we offer a set of definitions for them. Keep in mind that the definitions presented here may be different from those offered by others who write or speak on science.

Fact: something that is indisputably true, or as close as possible to being so.

Law: a statement of a particular pattern or order in nature; this term may overlap with the others defined here.

Hypothesis: a *tentative* explanation for some set of natural phenomena.

Scientific theory: an extensively tested and verified model that explains a broad range of phenomena.

A **fact** is a simple statement that is objectively true. For example, we consider it a fact that the Sun rises each morning, that the planet Mars appeared in a particular place in our sky last night, and that the Earth rotates. However, facts are not always obvious. For most of human history people did not recognize that the Earth orbits the Sun; rather, people thought that the Sun orbited the Earth. A fact is supposed to be indisputable but, because it may be difficult to establish (e.g., the orbiting of the Earth), someone usually will dispute any fact.

A **law** usually is named for historical reasons, rather than scientific reasons. For example, *Kepler's laws* of planetary motion, which describe how planets move in their orbits about the Sun, were not called "laws" until more than 100 years after their discovery by Kepler. By then scientists knew that Kepler's laws are only approximations, albeit very good ones, to the precise motions of the planets. In general, the term *law* is applied to a statement about a particular pattern of order in nature. It is probably the most difficult of the terms considered here because it may overlap with all of the others.

A **hypothesis** is a tentative explanation, sometimes called an "educated guess." A hypothesis represents a model by which many phenomena can be explained. Because a hypothesis has not yet been thoroughly tested, it later may be rejected. A good hypothesis suggests observations or experiments that could lead to its own confirmation or demise. Kepler, for example, hypothesized that the explanation for his laws was a force emanating from the Sun, which he thought might be magnetic. This hypothesis turned out to be partially correct: a force *does* emanate from the Sun, but it is gravity, not magnetism.

A **scientific theory** is an *accepted* model that explains a broad range of phenomena. A sound scientific theory must not only agree with facts found in prior observations and experiments, but it must also offer additional predictions that can be tested and studied further. For a hypothesis to be elevated to the status of theory it must have survived numerous tests without failure. For example, Isaac Newton's equations describing gravity constitute a model that meets the criteria for a scientific theory. His model explains facts such as apples falling to the ground when dropped, the orbit of the Moon around the Earth, and the existence of tides; it also shows gravity to be the force that causes the planets to obey Kepler's laws. In addition, Newton's model made verifiable predictions: Sir Edmond Halley, for example, used it to predict the return of a comet (which we now call *Halley's comet*).

Note that, by this definition, a theory is a model that has survived rigorous and repeated attempts to find its flaws. Thus, when we say that something is "only a theory," we mean that it has been tested and retested and that it has withstood a multitude of challenges. Of course, this process does not guarantee that a scientific theory is infallible or that it will never be modified or abandoned. Because a scientific theory makes predictions, it remains dynamic by being continually tested and refined. A sound theory explains a great many facts but may be replaced later with another theory that explains even more.

Newton's theory of gravity is a special case of Einstein's theory of general relativity. It works well in weak gravitational fields (as on Earth), but it yields inaccurate predictions in strong gravity (as near the Sun or a black hole). Einstein's theory also has some known failings, and physicists today are working to develop an even more complete theory, sometimes called ***supergravity***, of which Einstein's theory will be only a special case.

For example, despite the incredible success of Newton's gravitational theory, eventually experiments and observations were found that it could not explain. Thus Newton's theory came to be recognized as limited in its range of applicability. Today it is known to be only a special case of a broader theory of gravity, Einstein's *General Theory of Relativity*. Note that, even though a better theory replaced Newton's, it was not discarded. The many tests that Newton's theory passed have not been nullified, it is still useful in explaining many facts. Indeed, Newton's theory of gravity is all that is required to plot trajectories for space travel and to put people on the Moon!

Time-Out to Think: Look up *theory* in a dictionary. How does its definition in ordinary English differ from that given here for a scientific theory? If you hear someone talking about a "theory," how can you decide whether it is a mere hypothesis or a tested and verified scientific theory?

Interestingly, an idea very much like Darwin's idea of natural selection was proposed by a Greek scientist, Empedocles, in about 450 B.C. Empedocles developed his ideas in an attempt to explain the diversity of life, as well as the existence of fossils. Noting that many fossils matched no living organisms, he concluded they represented extinct species. But, like many of the great discoveries of the ancients, most of Empedocles's ideas were lost.

THINKING ABOUT *... THE THEORY OF EVOLUTION*

Another example of a scientific theory, but one that has been greatly misunderstood, is the **theory of evolution**. Charles Darwin first enunciated modern ideas of evolution. Darwin was aware of fossil records that showed changes in the species living on the Earth over time, and he made detailed observations of similarities and differences in existing species. Based on many observations, Darwin hypothesized that species evolve over time through a mechanism he called **natural selection**.

Unfortunately, public debate has been intense about the status of evolution, with frequent arguments over whether evolution is a "fact" or a "theory." In part, the confusion arises from misunderstanding of the term *theory*. Recall that a scientific theory has been substantially tested and verified. In addition, the theory of evolution has been confused with the facts that it is designed to explain. For example, the following are considered facts by nearly all scientists who have studied them.

- The Earth is approximately four and one-half billion years old.
- Early in the history of the Earth life forms were simple.
- Increasingly complex forms of life appeared over time.
- Many species from the past are now extinct.
- Tens of millions of distinct species are alive today.

The theory of evolution seeks to explain these facts. Darwin's original ideas about natural selection have been substantially tested and refined since his time, but they still do not explain all the facts. For example, no one has yet established how the first living cells came into existence.

Because not all the facts about the history of life on Earth have been explained, the theory of evolution is incomplete; hence biologists are constantly proposing modifications to the theory. Until each proposed modification is tested and verified or rejected disagreement about its validity can be substantial. Thus, like any scientific theory, the theory of evolution is subject to continuing debate. Nevertheless, its underlying basis, as represented by the facts about the history of life and the basic idea of natural selection, is firmly established.

5.2.2 The Scientific Method

The discovery of new science or new knowledge involves carefully planned observations and experiments. However, it also involves intuition, collaboration with others, moments of insight, and luck. Indeed, no set rules exist for making discoveries, and individual scientists go about their work in unique ways. Nevertheless, the scientific method, which consists of certain logical principles and procedures, represents an *idealization* of the process used to discover or construct new knowledge.

The following list presents a few basic characteristics of the scientific method.

- Observations or experiments reveal new facts, which lead to the recognition and formulation of an unsolved problem.
- A hypothesis is constructed that attempts to explain the observed or experimental facts. Note that the construction of a hypothesis is *inductive*, because it involves generalization in an attempt to find a general model for many distinct facts.
- The hypothesis is analyzed logically in search of deductions that follow from it. Ideally, these deductions represent predictions of the hypothesis that haven't previously been observed or tested experimentally.
- Observational or experimental methods are developed to test each new prediction. Tests are supposed to be unbiased and reproducible. That is, they should be independent of the testing conditions and the people conducting the test. And, most importantly, the tests must be designed with the potential to prove the prediction wrong.
- If the tests confirm the prediction, the hypothesis is strengthened and further tests are proposed. If the predictions fail, the hypothesis must be modified or discarded.
- If a hypothesis passes many tests without failure, it achieves the status of a scientific theory.
- The theory is challenged continually with new tests and repeated checking of old tests; it is refined, expanded, or replaced as needed.

A flow chart for the scientific method is shown in Figure 5-5.

Again, keep in mind that the scientific method is an idealization; in practice, it is rarely followed precisely. Indeed, some philosophers have gone so far as to state that the scientific method is a myth. Support for this idea comes from studying the processes that have led to important scientific discoveries: In most, if not all, cases the actual process deviates from the idealization of the scientific method.

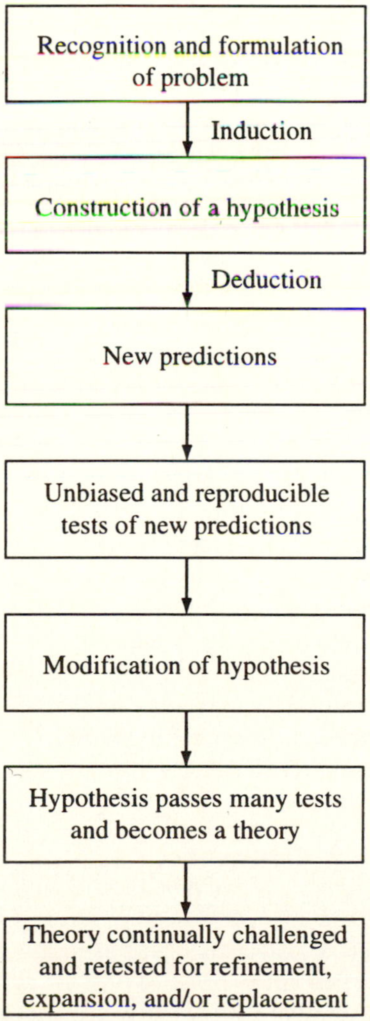

Figure 5-5. The scientific method involves both induction and deduction.

5.2.3 Science, Nonscience and Pseudoscience

Curiosity and the quest for knowledge seem almost instinctual in human beings. The scientific method, with its logical and systematic approach, has been enormously successful in helping humans discover new knowledge. In this century, alone, science has led to nuclear weapons and nuclear energy; to the creation of transistors, television, and computers; and to humans walking on the Moon. Clearly, the scientific method works.

Nonscience: any attempt to search for knowledge that knowingly does not follow the scientific method.

Nevertheless, people seek knowledge in many ways that do not follow the basic tenets of the scientific method, and therefore are not science. We can categorize such methods as either nonscience or pseudoscience. By **nonscience**, we mean any attempt to search for knowledge that *knowingly* does not follow the scientific method. The many forms of nonscience include any knowledge sought through faith, emotion, religion, or other processes that deliberately do not follow the scientific method.

Time-Out to Think: Can the scientific method be applied to questions of faith, prayer, emotion, or religion? Defend your view.

Pseudoscience: that which purports to be science but, under careful examination, fails tests conducted by the scientific method.

By **pseudoscience** we mean attempts to search for knowledge that *claim* to be scientific, but are not; the prefix *pseudo* means "false." Examples of pseudosciences include most of astrology, parapsychology (e.g., mental telepathy), UFOlogy (studies of "aliens" visiting Earth in UFOs), and many conspiracy theories. Of course, each of these subjects may be studied scientifically, and many people have done so. Despite this scientific study we think of them as pseudosciences because their central claims have never been borne out in scientific tests. For example, many practitioners of astrology claim to be able to predict your future by charting the positions of the stars and planets at the moment of your birth. However, in scientific tests in which astrologers attempt to make predictions about people they have never met, their predictions turn out no better than would be expected by pure chance. Similarly, many parapsychologists claim to be able to demonstrate mental telepathy, but in controlled experiments mental telepathy has never been conclusively demonstrated.

The key ingredient in the scientific method, which differentiates science from nonscience or pseudoscience, is unbiased and reproducible testing. Any person, with the necessary background knowledge, can test any hypothesis or theory. In nonscience such testing is unimportant; beliefs are based on ideas such as faith and on nonreproducible events such as miracles. Pseudosciences may fail to meet the scientific method in any number of ways, but the most common is that testing is biased. That is, when practitioners of pseudosciences test their craft, they generally do so in a way that others cannot replicate. However, as the scientific method is an idealization of how to acquire new knowledge, the boundaries between science, nonscience, and pseudoscience are not always clear.

THINKING ABOUT ... THE VALUE AND VALIDITY OF NONSCIENCE AND PSEUDOSCIENCE

Can we say anything about the value or validity of nonscience and pseudoscience? Let's start with nonscience . Even scientists often practice nonscience, as when they follow personal biases in choosing a topic for research or when they practice their personal religions. Religion is an example of nonscience because it seeks knowledge through non-scientific means such as faith, prayer, or meditation. Historically, science and religion often have been portrayed as being in conflict. Such shouldn't be the case, however; science and religion simply operate in different realms of human thought. In particular, science generally answers "how" questions, whereas religion attempts to answer more fundamental "why" questions. Religion generally makes no claim about following scientific method, so science can say little about the validity of "truths" revealed through religion. Even with the scientific method, if some aspect of a particular religion is logically proved to be false, the followers of that religious faith are under no obligation to accept that logic.

The crucial question is: Are religious beliefs and scientific knowledge equally valid? This question lies at the heart of debates over what should be taught in school and other issues of separation of church and state. Although not easily answered, it can be useful to distinguish between validity for individuals and validity for societies. Individuals are likely to consider their own religious beliefs to have validity equal to or greater than that of science. That is a key reason why the Constitution does not allow the government to regulate individual religious beliefs. Society, however, must recognize that different religions often answer the same questions differently. Thus different people will disagree about the validity of particular religious ideas and, because they cannot be tested scientifically, society has no clear basis upon which to decide what beliefs should be taught in school. In contrast, knowledge acquired by science may be tested or checked by anyone, regardless of religious belief. Indeed, important scientific discoveries have been made by people of virtually every religion. Thus scientific knowledge may be considered to have validity for everyone in a society, and understanding science therefore is important for all citizens.

What about pseudosciences? The difficulty with pseudosciences is that they claim to be scientific when they are not. Pseudosciences thus place themselves in direct conflict with science. Scientific study shows the claims of pseudosciences to be unsubstantiated and in some cases outright false. Thus the pseudosciences have no validity and, in some cases, they may even damage society. Nevertheless, many people hold deep beliefs in some pseudosciences, and trying to ban pseudoscientific practices would be counterproductive. Instead, individuals should learn about science and the scientific method, so that they will be able to make their own decisions about the validity of scientific claims.

5.2.4 Is Science Objective?

By their very definition, scientific theories must be objective. That is, they are not subject to individual interpretation or biases because they can be tested and confirmed by anyone.

However, individual scientists are always biased. This bias shows up in many ways. A hypothesis, for example, often reflects the biases of the person who proposes it. Because a hypothesis hasn't yet been thoroughly tested, plenty of room for opinion is allowed in suggesting one. In particular, some ideas may not even be considered because they fall too far outside the general scientific thought, or **paradigm**, of the time. An example was the development of Einstein's theory of relativity: Many other scientists had hints of this theory in the decades before Einstein, but did not investigate them at least in part because they seemed too outlandish.

Biases also can show up in the testing of a hypothesis. Individual scientists have been known to "cheat" in order to obtain results they desire. Some cases of deliberate cheating, or **scientific fraud,** have been documented, but in even more cases the "cheating" probably is subconscious or unintentional.

The beauty of the scientific method is that it allows continued testing by many people. Thus, even if personal biases affect some results, tests by others eventually will discover the mistakes. Undoubtedly the fallacy of *subjectivism* is common in the practice of individual scientists. However, the collective action of many scientists, following the basic tenets of the scientific method, makes science, *as a whole*, objective.

5.3 PARADOXES

The first step in the scientific method involves recognizing an unsolved problem. One of the most important ways by which problems are recognized is through the discovery of a **paradox**: a situation or statement that seems to violate common sense or to contradict itself.

The study of a paradox forces people to look deeply into their understandings and beliefs. In the effort to explain a paradox, they may discover new principles, new facts, or be led down the path to a new scientific theory. They might even be forced to change their "common sense"!

The value of using paradoxes to point toward new knowledge is illustrated by a simple example all young children encounter. At a very young age, you learned "common sense" meanings for *up* and *down*: *up* was above your head and *down* was toward your feet. Thus you knew exactly what would happen to a person suspended upside-down from the ceiling (Figure 5-6a).

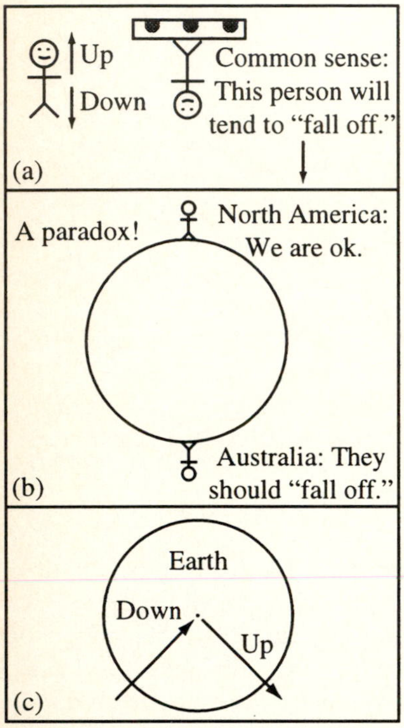

Figure 5-6. (a) Childhood common sense about *up* and *down* leads to (b) a paradox when the child learns that the Earth is round. (c) To resolve the paradox requires developing a new "common sense."

One day, however, we learned that the Earth is round. Moreover, globes generally show the Northern Hemisphere on the top and the Southern Hemisphere on the bottom. You immediately were confronted with a paradox: Your common sense told you that people should fall off the Earth if they tried to live in Australia (Figure 5-6b), yet photographs proved that they didn't.

To resolve this paradox you were forced to accept that your "common sense" understanding of *up* and *down* was incorrect. Rather than being the direction pointing above your head, *up* is away from the center of the Earth and *down* is toward the center of the Earth (Figure 5-6c). With this new understanding, the paradox no longer seems surprising. Your new common sense recognized that *up* and *down* are determined relative to the center of the Earth and consequently that no such thing as *up* and *down* exists in space.

> **Time-Out to Think:** Why do you suppose that most maps and globes show the Northern Hemisphere on the top, and the Southern Hemisphere on the bottom?

Of course, not all paradoxes are so easily resolved. Consider the statement "I never tell the truth." It clearly is a paradox: If the person who makes this statement never tells the truth, the statement cannot be true either; but if the statement is false, the person sometimes tells the truth! Logicians never have found a satisfactory way of dealing with such statements.

These two brief examples show that paradoxes sometimes point toward new understanding and sometimes identify fundamental problems that might never be answered. In this section we examine three paradoxes of great historical importance. The first, Zeno's paradox, stumped mathematicians for two thousand years but has a clear resolution today. The second, the paradox of light, is now scientifically understood, yet it remains baffling to our "common sense." The third, the paradox of creation, has not yet been resolved, and may never be.

5.3.1 Zeno's Paradox

None of Zeno's original writings has survived. However, reference to his work is found in the surviving writings of Aristotle, who lived about 100 years later.

One of the most famous paradoxes also is one of the oldest. It is one of several attributed to a Greek philosopher Zeno of Elea (c. 460 B.C.). The paradox begins by imagining a race between the warrior Achilles and a slow-moving tortoise. The tortoise is given a small head start. As Achilles is much faster, he will soon overtake the tortoise and win the race.

Or will he? Figure 5-7 illustrates Zeno's paradox. To understand it, imagine that Achilles runs twice as fast as the tortoise and that the tortoise is given a headstart of 10 meters. To catch up with the tortoise, Achilles must first run 10 meters to reach the point where the tortoise began, labeled P1. During this time, however, the tortoise will move ahead 5 meters to P2, because it runs half as fast as Achilles. Achilles must therefore run an additional 5 meters to P2 before he can catch the tortoise. Again, however, during the time Achilles covers these five meters, the tortoise moves half as far, putting it 2.5 meters ahead at P3.

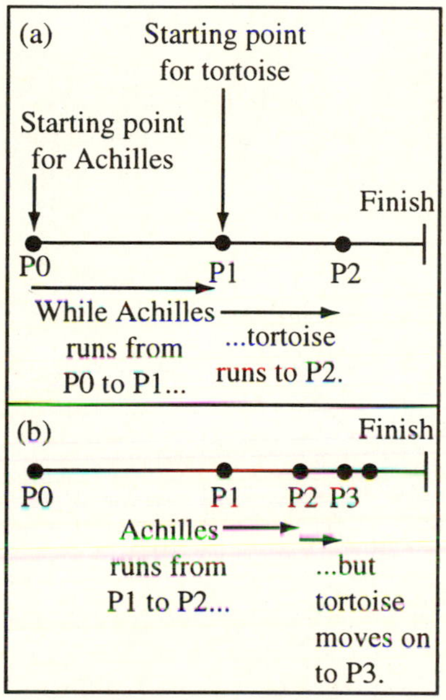

Figure 5-7. Zeno's paradox: Achilles runs faster but the tortoise is given a head start. (a) While Achilles runs to the point P1 (where the tortoise started) the tortoise moves ahead to P2. (b) To catch the tortoise, Achilles must keep running to P2; meanwhile the tortoise moves on to P3. Continuing this chain of logic suggests that Achilles can never catch the tortoise.

The sequence of such events is never-ending. To catch the tortoise, Achilles must run the 2.5 meters to P3; while he does so, the tortoise will move ahead (1.25 meters) to P4. And so on. Each time Achilles reaches the previous position of the tortoise, the tortoise has moved ahead. Thus the paradox is that Achilles seemingly will never catch the tortoise. To summarize, the following are basic elements of Zeno's paradox.

- In each stage of the race, as viewed by Zeno, Achilles makes up only half the distance to the tortoise.
- Thus an infinite number of such stages is required for Achilles to catch the tortoise. However, the distance that Achilles must cover in each stage is only half as long as the distance in the previous stage.
- The infinitely many stages required for Achilles to catch the tortoise must require an infinite amount of time. Thus Achilles will never catch the tortoise.

Zeno's paradox puzzled philosophers and mathematicians for more than 2000 years. The resolution is that Zeno's argument about the amount of time required to cover the infinite set of distances was wrong. An infinite set of distances can be covered in a finite amount of time, so long as each is sufficiently smaller than the previous one. To illustrate let's consider what happens if we add an infinite number of fractions together, each half as large as the previous one. That is, imagine doing the following sum, which is called a **series**:

$$1+\frac{1}{2}+\frac{1}{4}+\frac{1}{8}+\frac{1}{16}+\frac{1}{32}+\frac{1}{64}+\frac{1}{128}+\frac{1}{256}+\frac{1}{512}+\frac{1}{1024}+\frac{1}{2048}+\cdots.$$

Time-Out to Think: Using your calculator, add the first four terms of the series. Then add the first 8 terms. Then add the first 12 terms. What do you think would happen if you kept adding additional terms of this series forever?

The unexpected property of this series is that, although the sum increases each time another term is added, it *does not grow to infinity*. In fact, we can prove mathematically that the sum of the series is 2. Although Zeno's paradox shows that Achilles must cover an infinite number of distances (stages) to catch the tortoise, it *does not follow* that this run will require an infinite amount of time: Achilles *will* pass the tortoise and win the race.

Why did mathematicians take more than 2000 years to resolve Zeno's paradox? The reason is that the necessary mathematics wasn't invented until the seventeenth century. Like the series of fractions, resolution of Zeno's paradox requires the concept of a *limit*. This particular series of fractions is said to have a value of 2 because, as more and more terms are added, the sum gets closer and closer to 2 but never exceeds it. That is, 2 is the **limit**, or target value, of this sum. The concept of limits is fundamental to calculus. Thus the resolution of Zeno's paradox leads directly to the ideas of calculus, upon which much of modern mathematics and science is based.

If the ideas of calculus are needed to resolve Zeno's paradox, why weren't they developed much earlier? Apparently, humans were ready intellectually for calculus more than 2000 years ago; writings of the Greek scientist and mathematician Archimedes (287–212 B.C.) suggest that he came very close to inventing calculus. Some historians therefore argue that humans weren't politically or sociologically ready for such advances. According to this belief, the result of the disparity between intellectual development and sociopolitical development was the fall of ancient Greece, the decline of the Roman Empire, and the ensuing period known as the Dark Ages. That explanation, of course, brings us to yet another paradox: The scientific and technological advances of the twentieth century offer the power both to dramatically improve human life and to catastrophically destroy civilization. Is humanity prepared politically and sociologically to deal rationally with *these* advances?

5.3.2 The Light Paradox — Particle or Wave?

An example of light acting like a wave is the phenomenon of *diffraction*, in which light "waves" can bend around corners. An example of light acting like a particle is its influence on your eye: A single piece, or *photon*, of light can cause a change in a molecule in your retina; the change is registered by your brain and interpreted as vision.

Prior to the twentieth century, a scientific debate raged over whether light is a wave or a particle. The paradox is that in some scientific experiments light acts like a stream of particles, but in others it acts like a wave. To understand this phenomenon better, think of particles and waves as two different ways of sending information.

Suppose that you wanted to send some information to another person. One method would be to write all the information on a piece of paper and then take the paper to the other person. The paper, in this sense, is a particle: It is transferred physically from one person to the other. An alternative method of sending the information would be through waves, such as waves on a string or sound waves. For example, you could connect a string from yourself to the other person. When you shake your end of the string, a wave moves along the string to the other end. If you and the other person agree on a code (e.g., Morse code, in which each letter is represented by a series of short or long shakes of the string), you could communicate your message via the waves on the string. Note that nothing physically passes from you to the other person: The string stays in the same place (except for its shaking up and down) the entire time.

As this analogy shows, particles and waves behave quite differently. How, then, do we explain the paradoxical behavior of light sometimes acting like a particle and sometimes like a wave? The answer, demonstrated by physicists early in the twentieth century, is that light is *both* a wave and a particle! In other words, our common sense about differences between waves and particles turns out to be incorrect, as did our common sense about up and down. Nature, on its most fundamental level, makes no such distinction.

The resolution to the paradox of the nature of light may seem unsatisfying, and even the greatest physicists have difficulty creating a "mental picture" of something that is both a wave and a particle. Nevertheless, the discovery of the dual nature of light opened a vast new area of physics known as **quantum mechanics**. It led to the discovery that not only is light both a particle and wave but that the same is true of electrons and other subatomic particles. That is, an electron also is both a particle and a wave. Strange as that may seem, scientists and engineers depend on this fact to create the electronic components of all modern computers. Thus, although the wave–particle paradox remains baffling to our "common sense," the fact that it can be understood scientifically is directly responsible for much of our modern technology.

5.3.3 The Paradox of Creation

Perhaps the most fascinating of all paradoxes is the paradox of creation. We exist, and the universe exists. Thus to assume that the universe must have been created at some time in the past seems natural. This creation may have been deliberate — the action of an omniscient God or Creator — or it may have been "accidental," a consequence of some random shuffling of the laws of nature and of matter.

According to present scientific understanding, the universe was created in the cosmic explosion called the Big Bang, which occurred between 7 and 20 billion years ago. At the moment of the Big Bang the universe was infinitely tiny and dense; it has been expanding ever since.

The Big Bang meets the criteria of a scientific theory. It explains a great deal about how the universe grew from its dense and tiny beginning to its wondrous diversity today. It also makes many predictions that have been tested and verified. However, it does not explain *why* the Big Bang occurred, nor does it reveal what existed before, or "outside," the Big Bang. Herein lies the paradox: If we ever learn what created the universe, we will be confronted with a new question: What created the material that created the universe? Or, in terms of an omniscient Creator: Who or what created the Creator?

Logically, these questions must have answers since the universe does, in fact, exist. Nevertheless, seeing a way out of this paradox is difficult; seemingly another question about creation always will arise from the latest one answered. The current system of logic may be inadequate for addressing the paradox of creation. Perhaps, as with the paradoxes discussed earlier, the paradox of creation will someday be resolved. Or perhaps creation will remain a mystery forever.

According to present scientific understanding, the Big Bang was the origin of *both* time and space. In this sense, time did not exist until it was created in the Big Bang. Thus to ask about "before" the Big Bang is meaningless because the concept of "before" didn't exist until time was created. Nevertheless, logic demands that something must be "outside" the Big Bang and hence "outside" the universe, even if that something cannot be understood in the context of time. Thus, the question of "where did the Big Bang come from?" remains to represent the paradox of creation.

A well-known scientist ... once gave a public lecture on astronomy. He described how the earth orbits around the sun and how the sun, in turn, orbits around the center of a vast collection of stars called our galaxy. At the end of the lecture, a little old lady at the back of the room got up and said: "What you have told us is rubbish. The world is really a flat plate supported on the back of a giant tortoise." The scientist gave a superior smile before replying, "What is the tortoise standing on?" "You're very clever, young man, very clever," said the old lady. "But it's turtles all the way down!" — from *A Brief History of Time* by Steven Hawking (Bantam Books, 1988)

Time-Out to Think: Do you believe that the paradox of creation will ever be resolved? Why or why not?

5.4* THE LIMITATIONS OF LOGIC

Having examined how logic helps in the acquisition of knowledge through science, we now return to Leibniz's dream of creating a system of logic from which all truth can be derived. If such a system were possible, all paradoxes would be resolvable.

5.4.1 Gödel's Theorem

Following the publication of *Principia Mathematica* by Bertrand Russell and Alfred North Whitehead in 1913 (recall the discussion in subsection 5.1.2), other mathematicians sought to fill in the remaining gaps and finally realize the ultimate system of logic. They realized that a first step required showing that *mathematics* could be wholly understood as a system of logic; only then could mathematical logic be developed into Leibniz's dream of a calculus of reasoning.

One of the key players in this effort was the mathematician David Hilbert, who sought to show that mathematics could be constructed by starting with a few basic assumptions, called **axioms**, and then applying rules of logic. The idea was that mathematics should be *formalized* as a system in which all other mathematical truths, or theorems, could be derived from the basic axioms.

Although Hilbert did not actually create such a formalized system of mathematics, he and other mathematicians showed that such a system must have three properties.

1. It must be *finitely describable*. That is, the number of basic axioms should be limited so that, at least in principle, writing them all down is possible.
2. It must be *consistent*. That is, it should have no internal contradictions. If the basic axioms are used to derive other statements (or theorems), they must never lead to two that are mutually contradictory; in other words, they should never produce a situation in which a statement (or theorem) is both true and false.
3. It must be *complete*. That is, the basic axioms should allow analysis of every possible situation. If someone proposes a statement or theorem, the basic axioms should allow determining whether it is true or false.

By the late 1920s, with these required properties identified, the dream of Leibniz and the symbolic logicians was clearly enunciated. It seemed only a matter of time before a system for explaining *everything* might be developed. Then, suddenly, the dream came crashing down.

In 1931, the Austrian mathematician Kurt Gödel shook the mathematical world to its foundations: He proved that no formal system of logic can possess all three required properties. That is, he *proved* that no system can be simultaneously complete, consistent, and finitely describable. The actual proof is beyond the scope of this book, but the consequences of **Gödel's theorem** (as it is called) have astonishing mathematical and philosophical implications. Gödel's theorem spawned entirely new branches of mathematics and philosophy. Without going into detail, we can offer a taste of some its consequences.

- Some true mathematical theorems can never be proven.
- Some mathematical problems can never be solved.
- No systematic approach to mathematics can answer all mathematical questions.

Interestingly, Gödel's theorem does not address the question of the existence of absolute truth. It states that the concept of truth cannot be absolutely *defined* and that all truth can never be discovered, but it says nothing about whether truth exists. Consider an analogy to time. Many properties of time can be described, but precisely defining the concept of time may be impossible. Nevertheless, time exists, as evidenced by its passage while you read this page.

By extension, a philosophical consequence of Gödel's theorem is that no absolute way exists to define the concept of *truth*. Gödel's theorem therefore represents the final end of Leibniz's dream; a calculus of reasoning, resolving all arguments of all kinds, will never be found. Both mathematically and philosophically, Gödel's theorem may well be one of the most important discoveries in human history. For those interested in learning more, numerous popular books on Gödel's theorem have been written at a level comprehensible to anyone who has completed this chapter.

5.4.2 The Value of Logic

If no system of logic can be perfect, as Gödel's theorem states, what good is logic? The answer lies in its effectiveness. Logic works: Using logic allows the discovery of new knowledge and the development of new technology. Logic provides ways to address disputes, even if it cannot always ensure their resolution.

Recall that the basic purpose of logic is to help you draw conclusions from premises. Logic does, in fact, help you to do that. Through logic, you can study your personal beliefs and societal issues; you can even explore the nature of truth. What Gödel's theorem says, however, is that logic cannot ultimately answer *all* questions.

The value of logic should not be understated: Logical reasoning is an excellent tool for understanding and acquiring knowledge. Nor should it be overstated: Under some circumstances logic alone will fail. Finding the proper balance between logic and other processes of decision making is one of the greatest challenges of being human.

5.5 CONCLUSION

In this chapter we explored the roles and interrelationships among logic, mathematics, and science in the search for knowledge. Key ideas to remember from this chapter include the following.

- The process of discovering new knowledge rarely is as straightforward as suggested by the idealized notion of the scientific method. As a result, claimed scientific results often are biased or otherwise inaccurate. Because nearly every major societal issue relies to some extent on scientific knowledge or assertions you need to learn how to examine and evaluate scientific claims in order to make intelligent decisions and choices.
- The process of resolving a paradox can provide needed insight into an idea or even lead to a new discovery. Therefore you should not only study paradoxes when you encounter them, but also actively search for paradoxical consequences of any hypothesis or theory.
- The concept of truth is complex and has been studied for thousands of years. Modern mathematics, by means of Gödel's theorem, has proven that the ability to uncover truths has some basic limitations. If appreciated by everyone, this lesson might well lead to less acrimony between people who hold different beliefs.

PROBLEMS

Reflection and Review

Section 5.1

1. **The Pythagorean Theorem.** The Pythagorean theorem is one of the most important ideas in mathematics, and we return to it again and again throughout this book. To be sure you understand it, try the following.
   a. Find the length of the hypotenuse of a right triangle whose other two sides have lengths of $a = 6$ and $b = 8$.
   b. Find the length of the hypotenuse of a right triangle whose other two sides have lengths of $a = 7$ and $b = 10$.
   c. A right triangle has a hypotenuse of length $c = 15$ and another side with a length of $a = 3$. Find the length of the third side.
   d. In New York City, most streets are aligned either north–south or east–west. Suppose you need to walk three blocks north and four blocks west to get from your hotel to a particular restaurant. Without calculating, explain how you would use the Pythagorean theorem to compare the distance you will walk along the streets to the straight-line distance ("as the crow flies") between the hotel and the restaurant.
   e. Suppose that a staircase leads from the ground floor of a building to the second floor. Assume that you know the height of each floor of the building and the length of the staircase. Without calculating, explain how you would use the Pythagorean theorem to determine how far the staircase extends along its base.
   f. The distance between bases along the base paths of a baseball diamond is 90 feet. Knowing that the baseball diamond actually is a square, explain (without calculating)

how you would use the Pythagorean theorem to calculate the distance from home plate to second base.

2. **A Calculus of Reasoning.** Suppose that Leibniz's dream of a calculus of reasoning had been realized. Write a short essay describing how you think it might have changed the world.

Section 5.2

3. **Fact, Law, Hypothesis, or Theory.** Categorize each of the following as either a fact, law, hypothesis, or theory. Because others might categorize them differently, be sure to defend your categorization.
   a. The population of the United States is about 250 million.
   b. Democracy is the best form of government.
   c. AIDS is caused by a virus that undergoes frequent mutations, which is why the disease is difficult to cure or prevent.
   d. No device can be built that creates more energy than it uses.
   e. The Sun is one of many billions of stars in the universe.
   f. The universe began with an event called the Big Bang, which occurred between 7 and 20 billion years ago.
   g. Adult depression is caused by childhood memories that can be addressed only through psychotherapy.
   h. The average number of pupils per teacher in U.S. public schools fell from 21 in 1974 to 17.5 in 1994.
   i. The average number of pupils per teacher in American public schools decreased from 21 in 1974 to 17.5 in 1994 because there were fewer students in 1994 than in 1974.
   j. When the supply of a product increases, the demand for that product decreases.
   k. Students learn mathematics most effectively when they are actively engaged in creating their own understanding.
   l. The fact that 20% of U.S. students watch more than five hours of TV a day, compared to 11% of Korean students, explains why U.S. students score significantly lower than Korean students on standardized mathematics exams.

4. **Hypotheses in the News.** Identify a recent hypothesis reported in the news: for example, a new idea about how to lose weight; a claim about who committed a crime; the cause of a regional war; or why a species is vanishing. Explain how the hypothesis was developed. Did the development follow the scientific method? Briefly comment on whether you believe that the hypothesis will hold up under further testing and verification.

5. **The Meaning of *Theory*.** Each of the following statements uses the word *theory* in ordinary English; for each, explain whether the usage refers to a tested and verified theory, or whether it is more akin to a hypothesis.
   a. According to the theory of supply and demand, prices rise when supply becomes limited and demand remains constant.
   b. The theory behind building large numbers of nuclear weapons is that the threat of total annihilation deters war.
   c. The police are investigating a new theory in which the victim was killed because he was blackmailing the Mafia boss.
   d. The theory of the *greenhouse effect* suggests that continued use of fossil fuels may eventually cause catastrophic damage to our environment.
   e. The Republican gains in the 1994 congressional elections are best explained by the theory of the "angry white male."
   f. The theory of gravity is used to calculate spacecraft orbits and trajectories, and thus is the basis for ensuring that a space probe like *Cassini* successfully reaches its destination of Saturn.
   g. My theory is that she broke up with me because she couldn't handle my superiority.
   h. My theory is that she broke up with him because he's a stuck-up jerk.
   i. The fact that a 5-year-old cannot answer your question is well explained by Piaget's theory of child development.

6. **Scientific Testing of Astrology.** Astrology holds that the positions of the planets among the constellations at the moment of your birth affect your entire life. Suggest some ways in which this idea could be tested with the scientific method.

Section 5.3

7. **Zeno's Paradox.** Explain Zeno's paradox if Achilles is running 10 times as fast as the tortoise. Does this alter fundamentally anything in the paradox?

8. **Series Sum.** Look again at the series,

$$1+\frac{1}{2}+\frac{1}{4}+\frac{1}{8}+\frac{1}{16}+\frac{1}{32}+\frac{1}{64}+\frac{1}{128}+\frac{1}{256}+\frac{1}{512}+\frac{1}{1024}+\cdots,$$

   in which each denominator is twice the preceding one.
   a. Add the first four terms.
   b. Add the next four terms. Compare this value to the fourth term. What is the total for the first eight terms?
   c. Add the next four terms. Compare this value to the eighth term in the sum. What is the total for the first 12 terms? Compare the value of the total for the fifth through twelfth terms to the value of the fourth term.
   d. Can you generalize to state a rule about how the sum increases as you add more terms? What do you think will be the total value of the sum? Why?

Section 5.4

9. **Russell's Paradox.** One of many versions of Russell's paradox (named after the English philosopher and logician Bertrand Russell) goes like this: Suppose that, in a certain library, every book has a list of references. Some books include themselves in their list of references and some do not. A master book in the library lists every book in the library that does not reference itself. Question: Does the master book list itself? Discuss this paradox.

10. **Another Paradox.** Imagine opening a book and reading the sentence "This statement is false." Can the statement be true? Can the statement be false? Explain.

Further Topics and Applications

11. **The Faith of Leibniz.** Comment on the great faith that Leibniz held in the ability of logic to discern truth in any situation. Where do you think this faith came from? Do you think this same faith helped him in creating the mathematical subject known as calculus? Explain.

12. **Is Absolute Truth a Western Idea?** Some people have argued that the notion of a logic that could discover absolute truth is unique to Western thought. That is, they argue that mathematicians in other cultures would not have believed these same notions. Write a short essay arguing either for or against this proposition.

13. **Evolution and the Schools.** In the United States teaching the theory of evolution in the schools continues to be controversial. Why do you think so many people object to the teaching of evolution? Should an understanding of the theory of evolution be required as part of courses in science? Defend your opinion.

14. **The Earth-Centered Universe.** Through most of human history (until the 1600s), most people believed that the Earth was at the center of the universe. List some of the observed facts upon which this belief was based. Was the idea of an Earth-centered universe a hypothesis or a theory? Explain your answer.

15. **The Scientific Method and the Legal Profession.** Briefly describe how the scientific method is important in legal work, especially in court cases.

16. **Not Quite Zeno's Paradox.** Consider the following situation, which might at first seem similar to Zeno's paradox. Imagine that you start some distance away from home. Magically, every minute you cover half the distance to your home. That is, in the first minute you are halfway home. In the next minute you cover half the remaining distance. And so on. Explain how this scenario differs from Zeno's paradox. In particular, explain why (neglecting the finite size of atoms) you would never reach home if you traveled half the distance each minute.

17. **A Logical Puzzle.** An old puzzle goes like this: You are standing at the entrance of two doorways. One leads to heaven and the other to hell, but you do not know which is which. A demon and an angel are present. The demon intends to send you to hell and always tells lies, and the angel intends to send you to heaven and always tells the truth. To help you decide which door to walk through you are allowed to ask only one question, and only one spirit will answer. However, as both the demon and angel are invisible, you will not know *which* spirit answers. Is there a question that you can ask from which you will be able to choose the door to heaven, regardless of which spirit answers? Explain.

Projects

18. **Researching a Pseudoscience.** Research scientific studies of one of the pseudosciences. Report on why the claims of the pseudoscience do not meet the criteria for scientific validity.

19. **How Scientific Discoveries Really Are Made.** Choose a particular, significant scientific discovery that is of interest to you. Research how the discovery was made. Report on the processes that led to the discovery. How closely did the discoverers follow the idealized scientific method? Explain.

PART 2

THINKING QUANTITATIVELY

6 CONCEPTS OF NUMBER

In this chapter we help you begin to integrate quantitative reasoning with logical reasoning by investigating the concept of numbers: a concept that we use every day, yet which is difficult to define. We introduce the history of numbers, how numbers are used, and the modern system of numbers. In a sense this chapter describes the underlying basis upon which mathematics is built. You must understand it before you can move on to the study of more advanced quantitative concepts.

CHAPTER 6 PREVIEW:

6.1 ***THE LANGUAGE OF NATURE.*** Number systems and mathematics developed in tandem with human understanding of nature. In this opening section we discuss mathematics as the language of nature.

6.2 ***A BRIEF HISTORY OF NUMBERS.*** In this section we briefly discuss the development of number systems and various ways of writing numbers.

6.3 ***BUILDING THE MODERN NUMBER SYSTEM.*** Numbers may be categorized according to certain properties. In this section we explore the construction of the number system, from natural numbers through complex numbers.

6.4 ***SYSTEMS OF STANDARDIZED UNITS.*** In real problems numbers almost always are associated with units. We therefore explore the two commonly used systems of units to help you understand the meaning of numbers in real contexts.

6.5 ***EXACT AND APPROXIMATE NUMBERS: ROUNDING.*** Another important concept is the difference between exact and approximate numbers. We briefly discuss rounding numbers to a desired level of approximation.

6.6 ****PRIME NUMBERS: MYSTERIES AND APPLICATIONS.*** The study of prime numbers has a long and fascinating history. In this section we consider some of the mysteries and applications of prime numbers.

6.7 ****INFINITY.*** We conclude the chapter by exploring briefly the tantalizing idea of infinity.

The concept of number is the obvious distinction between the beast and man. Thanks to number, the cry becomes song, noise acquires rhythm, the spring is transformed into a dance, force becomes dynamic, and outlines figures.

— Joseph de Maistre, nineteenth century French philosopher

Philosophy is written in this grand book — I mean the universe — which stands continuously open to our gaze, but it cannot be understood unless one first learns to comprehend the language and interpret the characters in which it is written. It is written in the language of mathematics.

— Galileo Galilei, *Il Saggiatore* (1623)

6.1 THE LANGUAGE OF NATURE

The power of mathematics to model and describe natural phenomena has amazed mathematicians, scientists, and philosophers for centuries. In the twentieth century, mathematics also has proven valuable in modeling phenomena of human nature, including many economic and social interactions. Not surprisingly, then, mathematics often is called *the language of nature.*

Numbers are fundamental to mathematics in the same way that words are fundamental to spoken languages. Moreover, just as spoken languages can express similar ideas with different alphabets, vocabularies, and grammatical rules, numbers may be written and used in many different ways. Historically, many different systems have been developed for writing numbers. Today, however, modern mathematics is written and used in the same way everywhere. If you pick up a mathematics text in Chinese or Hindi or Spanish or Swahili, the words will look vastly different but the equations will look the same.

Anyone who understands algebraic notation reads at a glance in an equation results reached arithmetically only with great labor and pains. — from *Theory of Wealth,* by A. Cournot, 1897

As you delve more deeply into the language of mathematics, you will encounter more terminology and greater use of symbols and equations. Until you become very familiar with mathematics, its symbols may seem abstract or foreign. In fact, the notation of mathematics is *not* that of ordinary English. The numerals have Hindu–Arabic origins, and the letters used as mathematical symbols often come from other alphabets, such as Greek and Hebrew.

In fact, learning mathematics is much like learning a language. Abstraction is common to both mathematics and languages; the word *animal,* which represents a wide range of biological organisms, is no less abstract than x in an equation. In addition, like any language, mathematics has its own vocabulary and its own grammatical rules. Most important, mathematical statements always can be translated into English (or other languages). Herein lies a key to learning mathematics. No matter what strange-looking symbols or abstract concepts a mathematical statement contains, always stop and think: What does the statement mean in English? If you can learn to do this, without panicking or skipping mathematical statements when you come to them, you will find that mathematics is no more difficult than any other language.

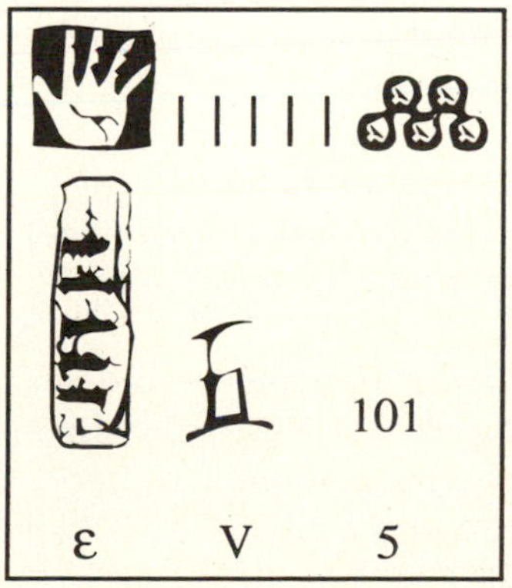

Figure 6-1. The concept of five can be represented in many different ways. From top to bottom and left to right, five is represented as: digits (fingers); tally marks; a pile of stones; notches on wood; a Chinese numeral; a binary numeral; a Greek numeral; a Roman numeral; and a Hindu–Arabic numeral. Note that a *numeral* is a written representation of a number.

6.2 A BRIEF HISTORY OF NUMBERS

What do we mean when we say the number 5? Five what? Does 5 have the same meaning when we refer to "five people" as when we refer to "five laws"? If we say "five people," is the meaning always the same? That is, does it convey anything about the people, whether they are young or old, short or tall, and so on? Indeed, does the *concept* of the number 5 have a reality independent of things being counted or measured?

No single, formal definition of the concept of **numbers** exists. The term *numbers* is used to describe many different ideas, and the concept of numbers has evolved over time. For example, the number 0 was unknown to the ancient Greeks and Egyptians, and negative numbers wouldn't have been considered "real" to most ancient cultures. Today, mathematicians and scientists routinely work with *imaginary numbers,* which despite their name are no more abstract or less useful than other numbers.

Historically, the concept of numbers developed in parallel with methods for writing **numerals** — symbols that represent numbers. For example, the symbol 5 is a numeral, whereas the concept of five is a number (Figure 6-1). In this section we explore the historical development of numbers and numerals.

6.2.1 The Origin of Modern Numerals

For most of human history, people used numbers only for simple counting, and numeral systems relied on *tallies* with fingers or toes, piles of stones, or notches cut on a bone or a piece of wood. These systems offered no way to write (or even conceptualize) large numbers, which would take too long to count. To simplify the process of counting, many early tribes grouped their counts by 2s, 3s, or other amounts. Eventually, the most common group sizes became 5, 10, and 20, probably because these numbers correspond to the digits on hands and feet.

In about 3000 B.C., the Egyptians and the Babylonians independently introduced the first numeral systems to go beyond tallying. Egyptian hieroglyphics used separate symbols for 1, 10, and each successive power of 10 up to 1 million; for example, to write the number 30, the Egyptians repeated the symbol for 10 three times. The Babylonians also used separate symbols for 1 and 10, but didn't invent symbols for larger numbers. Instead, when they reached 60, they began writing numbers in a column to the left, just as we now use different columns to represent 1s, 10s, 100s, and so on.

Modern numerals trace directly to the work of Hindu mathematicians in India during the first few centuries A.D. In about A.D. 800, when major Hindu works on astronomy were translated into Arabic, the Hindu numerals became part of the Arab culture. The Arabs led the development of mathematics during the next several centuries, and the written shapes of the Hindu numerals slowly changed over time. As a result, modern numerals are called **Hindu–Arabic numerals** (many people drop the credit to the Hindus and refer to the numerals simply as Arabic).

Despite their origins, the numerals with which you are familiar — made by combining the symbols 0, 1, 2, 3, 4, 5, 6, 7, 8, and 9 — look very different from the numerals used in the modern Hindi and Arabic languages. The final transformation into their current form occurred in Western Europe after the great works of the Arab mathematicians were translated into Latin in about A.D. 1200.

THINKING ABOUT *... USES OF NUMBERS*

Although we tend to think of numbers only in terms of their use in mathematics, numbers actually serve three distinct purposes: counting, ordering, and labeling.

Numbers associated with counting, or with answers to a question of "how many?," are called **cardinal numbers**. Note that the cardinal numbers do not include negative numbers because counting a set of physical objects and getting a negative answer isn't possible.

The **ordinal numbers** are used to indicate the *order* of members in a set. For example, if we speak of a person's "second-born child" we are referring to the order in which the child was born. That is, the ordinal number "second" answers the question of "where in the birth order was this child?" Note that it is *not* the answer to a question of "how many?"

Finally, numbers used as labels or names are called **nominal numbers**. Examples of nominal numbers include Social Security numbers, license plate numbers, and numbers on football uniforms. In these cases the numbers serve as identifiers but do not answer questions of either counting or order. Nominal numbers sometimes include symbols other than Hindu–Arabic numerals; license plate "numbers," for example, often include both letters and numbers.

6.2.2 Roman Numerals

Along the way from ancient Egyptian and Babylonian numerals to Hindu–Arabic numerals, many other systems of numerals came into use. Of the greatest historical interest are the systems used in ancient Greece and Rome, in which letters of the alphabet (Greek and Latin, respectively) were used to represent numbers. **Roman numerals**, developed in about 500 B.C., are a remnant of that era. The seven basic symbols in the Roman numeral system are

I = 1, V = 5, X = 10, L = 50, C = 100, D = 500, and M = 1000.

In general, Roman numerals are read from left to right, starting with the symbol representing the largest number; the value of the numeral is the sum of the values of the symbols. Thus, for example,

III	= 1 + 1 + 1	= 3;
XX	= 10 + 10	= 20;
XXVII	= 10 + 10 + 5 + 1 + 1	= 27;
MDCCLXXXVII	= 1000 + 500 + 100 + 100 + 50 + 10 + 10 + 10 + 5 + 1 + 1	= 1787.

The original Roman numeral system did not use subtraction. Thus the number 4 was written as IIII, rather than as IV. Subtraction was introduced in the sixteenth or seventeenth century but, by then, Hindu–Arabic numerals were far more common than Roman numerals.

An exception to left-to-right reading occurs when a symbol is followed immediately by a symbol of greater value; then, the smaller value is subtracted from the larger. For example,

IV	= 5 – 1	= 4;
IX	= 10 – 1	= 9;
XL	= 50 – 10	= 40;
XLVI	= (50 – 10) + 5 + 1	= 46;
MCMXCIX	= 1000 + (1000 – 100) + (100 – 10) + (10 – 1)	= 1999.

Despite the fact that Roman numerals were invented more than two thousand years after Babylonian numerals, in many ways they were less useful. Writing large numbers is extremely difficult with Roman numerals, and they offer no convenient way to represent fractions. Despite these drawbacks, however, the Roman numeral system was dominant in Europe for more than a thousand years. Even today, Roman numerals are used for decorative or artistic purposes, such as on clock faces or monuments.

> **Time-Out to Think:** Why would writing very large numbers with Roman numerals be difficult? What about writing fractions?

6.2.3 The Decimal System

Roman numerals are an example of an **additive numeral system** in which values are determined by adding the values of individual symbols. Consider, for example, the value of the Roman numeral XXX:

XXX = 10 + 10 + 10 = 30.

Note that the position of a symbol doesn't affect its value: All three X's have the same value of 10. In contrast, consider the Hindu–Arabic numeral 444:

$$444 = (4 \times 100) + (4 \times 10) + (4 \times 1) = \text{four hundred, forty-four.}$$

The value of each 4 is different, depending on its column, or **place**: The first (leftmost) 4 is in the *hundreds* column, representing the number 400; the middle 4 is in the *tens* column, representing the number 40; and the rightmost 4 is in the *ones* column, representing the number 4. Because the column, or place, of a symbol affects its value, the Hindu–Arabic numeral system is an example of a **place-value system**.

As an interesting aside, the word *decimate* also comes from the Latin *decimus* and comes from a Roman army practice in which every tenth man of a platoon was killed.

Moreover, because each successive column (moving left) is larger in magnitude by a factor of 10, Hindu–Arabic numerals are a **decimal**, or **base-10**, place-value system. The word *decimal* comes from the Latin *decimus*, meaning "tenth." The 10 symbols used in Hindu–Arabic numerals (0, 1, 2, 3, 4, 5, 6, 7, 8, and 9) are called **digits**, from the Latin *digitus*, meaning "finger."

Additive numeral systems, such as the Roman numeral system, do not have a symbol for **zero**. Place-value systems, in contrast, require a symbol for zero. To see why, try writing the number *twenty*. In Roman numerals, it is simply

$$XX = 10 + 10 = \text{"twenty."}$$

With Hindu–Arabic numerals, writing *twenty* requires a 2 in the *tens* column. However, the only way to identify the *tens* column is to see the *ones* column to its right. The number 20 has only two 10s, and no 1s, so a 0 is needed to serve as a **place-holder** for the ones column.

Thus the idea of *zero* was crucial to development of the modern number system. Although the Babylonians (c. 3000 B.C.) had a rudimentary idea of zero, it didn't appear as a meaningful number until about A.D. 600, when Hindu mathematicians introduced it. However, the Mayan civilization in America had independently developed the idea of zero as much as five hundred years earlier. Unfortunately, most of the great mathematical achievements of the Mayans and other native American cultures were lost after the European conquest of the Americas.

Names and Values of Decimal Columns

Each column in the decimal system has a name and a value. The values and names for the first nine columns in the decimal system are

Hundred millions	Ten millions	Millions	Hundred thousands	Ten thousands	Thousands	Hundreds	Tens	Ones
100,000,000	10,000,000	1,000,000	100,000	10,000	1000	100	10	1

To express the meaning of a decimal numeral, we can write it in **expanded form**; for example,

$$\begin{aligned} 403{,}598{,}123 &= (4 \times 100{,}000{,}000) + (0 \times 10{,}000{,}000) + (3 \times 1{,}000{,}000) + (5 \times 100{,}000) \\ &\quad + (9 \times 10{,}000) + (8 \times 1000) + (1 \times 100) + (2 \times 10) + (3 \times 1) \\ &= \text{four hundred three million, five hundred ninety-eight thousand,} \\ &\quad \text{one hundred and twenty-three.} \end{aligned}$$

The columns beyond (to the left of) the hundred millions also have common names. However, care must be used because the same names have different meanings in the United States and Great Britain; a summary is provided in Table 6-1. In the United States every group of three columns has a new name; the British group six columns before using a new name. The differences can cause great confusion. For example, what Americans call a "billion dollars" ($1,000,000,000), the British call a "thousand million dollars." Similarly, what the British call a "billion dollars" is what Americans call a "trillion dollars." When in doubt, ask how many zeros are involved.

Table 6-1. Names and Values of Numbers

Name	U.S. Meaning	British Meaning
Million	1,000,000	1,000,000 (same as United States)
Billion	1,000,000,000 (9 zeros)	1,000,000,000,000 (12 zeros)
Trillion	1,000,000,000,000 (12 zeros)	1,000,000,000,000,000,000 (18 zeros)
Quadrillion	1 followed by 15 zeros	1 followed by 24 zeros
Quintillion	1 followed by 18 zeros	1 followed by 30 zeros
Sextillion	1 followed by 21 zeros	1 followed by 36 zeros
Septillion	1 followed by 24 zeros	1 followed by 42 zeros
Octillion	1 followed by 27 zeros	1 followed by 48 zeros

Decimal Fractions

Fractions historically presented special problems in notation, even though merchants and traders commonly used them. Interestingly, the earliest known method for writing fractions, invented by the Babylonians in about 2000 B.C., was nearly identical to the modern method of writing decimal fractions. However, the Babylonian number system was based on powers of 60 rather than powers of 10. This ancient Babylonian idea was reintroduced in 1585 by a Flemish mathematician, Simon Stevin, in his book *The Tenth*. The method of writing fractions with a numerator and denominator (e.g., ½) probably was developed by Hindu mathematicians.

Each successive column to the right of the decimal point has one-tenth the value of the previous column. The values and names for each of the first six columns right of the decimal point, both as common fractions and in decimal form, are

Ones	Decimal point	Tenths	Hundredths	Thousandths	Ten thousandths	Hundred thousandths	Millionths
1	.	0.1	0.01	0.001	0.0001	0.00001	0.000001
		1/10	1/100	1/1000	1/10,000	1/100,000	1/1,000,000

As with other decimal numerals, writing a decimal fraction in expanded form can help clarify its meaning. For example,

$$8.375 = (8 \times 1) + (3 \times 0.1) + (7 \times 0.01) + (5 \times 0.001)$$

= eight and three hundred seventy-five one-thousandths.

6.2.4 Numbers in Other Bases

The decimal system is only one of many possible place-value systems, each having a different relationship among the values of columns. To understand how numbers can be represented in other bases, you need only think about writing them in expanded form. As we have shown, each column in the base-10 system represents a change in value by a factor of 10; for this reason, 10 different symbols are needed to write base-10 numerals: 0, 1, 2, 3, 4, 5, 6, 7, 8, and 9. The same principle applies to other bases.

In **base-2**, or the **binary** system, only two symbols are needed: 0 and 1. These two symbols often are called **bits**, for *binary digits*. The rightmost column in base-2 is the *ones*, as in base-10. However, from left to right the values of base-2 columns increase by factors of two: 2, 4, 8, 16, and so on. We can convert a base-2 numeral into a base-10 numeral by writing it in expanded form. For example,

The subscript 2 on 1101_2 means base-2.

$$1101_2 = (1 \times 8) + (1 \times 4) + (0 \times 2) + (1 \times 1) = 8 + 4 + 0 + 1 = 13.$$

In other words, the base-2 numeral 1101 is just another way of writing the decimal numeral 13.

A numeral in any base can be represented in any other base. The familiar decimal world in which you live, work, and balance your checkbook is only one of many possible numeric worlds: The choice is one of convenience and custom. Indeed, the base-10 system in common use today is a historical accident. Past civilizations have used many other bases.

For example, each successive column in the ancient Babylonian system increased in value by a factor of 60. Vestiges of the Babylonian base-60 system remain today in our systems of time keeping (an hour is 60 minutes and a minute is 60 seconds) and angle measurement (a degree is 60 minutes of arc and a minute is 60 seconds of arc). The Mayans used a base-20 system to achieve very advanced mathematics for their time. Indeed, a study of Native American cultures by W. C. Eels in 1913 found that only about a third used a decimal system; nearly another third used base-5 systems, and the remainder used a variety of systems, including base-2, base-3, and base-20.

Nevertheless, why confuse matters with bases other than the base-10 system that works today? This point is well taken, but it is a bit anthropocentric (human-centered). If we attribute the choice of a decimal number system to the fact that humans have 10 fingers, no profound science fiction is needed to imagine extraterrestrials whose anatomy might result in other bases. Far less fanciful is the fact that computers do not have 10 fingers; in the depths of its hardware, a computer senses only whether electricity is flowing or not flowing through a circuit. The natural base for computers therefore is base-2: electricity flowing corresponds to the symbol 1, and not flowing corresponds to the symbol 0. All of the miraculous speed and versatility of any computer, from hand-held calculators to supercomputers, is built upon simple, ingenious manipulations of the binary digits 0 and 1.

Example 6-1 *Counting in Base-2.* Write the first eight numbers in base-2.

Solution: To write the first eight numbers in base-2, you need know only the values of each column.

$1 = 1_2$

$2 = (1 \times 2) + (0 \times 1) = 10_2$

$3 = (1 \times 2) + (1 \times 1) = 11_2$

$4 = (1 \times 4) + (0 \times 2) + (0 \times 1) = 100_2$

$5 = (1 \times 4) + (0 \times 2) + (1 \times 1) = 101_2$

$6 = (1 \times 4) + (1 \times 2) + (0 \times 1) = 110_2$

$7 = (1 \times 4) + (1 \times 2) + (1 \times 1) = 111_2$

$8 = (1 \times 8) + (0 \times 4) + (0 \times 2) + (0 \times 1) = 1000_2$

That is, the numbers 1–8 in base-2 are 1, 10, 11, 100, 101, 110, 111, 1000.

❖

Example 6-2. *More Base-2.* Convert the numeral 10101_2 to base-10.

Solution: You convert by writing the numeral in expanded form; remember that the column to the left of the *eights* column is twice as large, or *sixteens*:

$$10101_2 = (1 \times 16) + (0 \times 8) + (1 \times 4) + (0 \times 2) + (1 \times 1) = 16 + 0 + 4 + 0 + 1 = 21.$$

The base-2 numeral 10101_2 represents the decimal numeral 21.

❖

Example 6-3. *A Base-5 Numeral.* Convert the numeral 431_5 from base-5 to base-10. How many symbols are needed to write numbers in base-5?

Solution: In base-5, the values of columns as you move leftward increase by factors of 5. That is, from right to left, the values are 1, 5, 25, 125, 625, and so on. Thus

$$431_5 = (4 \times 25) + (3 \times 5) + (1 \times 1) = 100 + 15 + 1 = 116.$$

The numeral 431, in base-5 is the same as 116 in base-10. Because each column to the left is larger by a factor of five, base-5 requires only five symbols: 0, 1, 2, 3, and 4.

❖

6.3 BUILDING THE MODERN NUMBER SYSTEM

The modern number system can be defined according to how we add, subtract, multiply, or divide with numbers. In this section we build the modern number system, beginning with the numbers used for counting.

6.3.1 Natural Numbers

Figure 6-2. The set of natural numbers is shown on a number line; the set continues to the right forever.

Because counting always begins at 1, the counting numbers, or **natural numbers**, comprise the set {1, 2, 3, 4,...}. The natural numbers are represented on a number line with a series of equally spaced dots (Figure 6-2). Natural numbers are further categorized according to their **factors** (or **divisors**). To understand factors, let's consider the natural number 6 and try dividing 6 by all the smaller natural numbers:

$$6 \div 1 = 6,\ \ 6 \div 2 = 3,\ \ 6 \div 3 = 2,\ \ 6 \div 4 = 1.5,\ \ 6 \div 5 = 1.2,\ \text{ and }\ 6 \div 6 = 1.$$

When we divide 6 by 1, 2, 3, or 6, the result is another natural number; therefore 1, 2, 3, and 6 are the factors of 6. We do not get another natural number when we divide 6 by 4 or 5; therefore 4 and 5 are *not* factors of 6.

The number 1 is considered to be neither prime nor composite because its only factor is itself.

A number such as 7, which has only itself and 1 as factors, is called a **prime number**. Numbers such as 6, which have factors besides themselves and 1, are called **composite numbers**.

Time-Out to Think: By looking for other factors, convince yourself that the first 10 prime numbers are 2, 3, 5, 7, 11, 13, 17, 19, 23, and 29. Note that, except for 2, all of these numbers are odd. Explain why 2 is the only even number that is prime.

The idea that every composite number can be uniquely expressed as a product of prime numbers is called the **fundamental theorem of arithmetic**.

Many composite numbers can be factored in several ways. For example, 8 can be factored three ways: $8 = 1 \times 8$, $8 = 2 \times 4$, and $8 = 2 \times 2 \times 2 = 2^3$; because the latter product involves only prime numbers, it is called the **prime factorization**.

Essentially, finding the factors of a number is a process of trial and error, as shown in Examples 6–4 to 6–6. Fortunately, there are some rules that can help you determine divisibility; a few of the simplest rules are listed in Table 6-2 on the next page.

Table 6-2. Divisibility Rules for 2, 3, 4, 5, 9, and 10

Is 2 a factor?	All even numbers are divisible by 2. For example, $126 \div 2 = 113$.
Is 3 a factor?	A number is divisible by 3 if the *sum of its digits* is divisible by 3. For example, the sum of the digits in the number 654 is 15, which is divisible by 3; thus 654 is divisible by 3 ($654 \div 3 = 218$).
Is 4 a factor?	A number is divisible by 4 if the *number formed by its last two digits* is divisible by 4. For example, 1736 is divisible by 4 because its last two digits (36) are divisible by 4 ($1736 \div 4 = 434$).
Is 5 a factor?	A number is divisible by 5 if its *last digit* is 5 or 0. For example, the numbers 75, 90, and 20,060 all are divisible by 5 ($75 \div 5 = 15$; $90 \div 5 = 18$; $20{,}060 \div 5 = 40{,}112$).
Is 9 a factor?	A number is divisible by 9 if the *sum of its digits* is divisible by 9. For example, the sum of the digits in the number 2772 is 18, which is divisible by 9; thus 2772 is divisible by 9 ($2772 \div 9 = 308$).
Is 10 a factor?	A number is divisible by 10 if its *last digit* is 0. For example, the number 310 is divisible by 10 because it ends in a 0 ($310 \div 10 = 31$).

Example 6-4. Find all the ways of factoring of 24. Which one is its prime factorization?

Solution: Look for all the ways of multiplying numbers together to get 24:

$$1 \times 24 = 24, \quad 3 \times 8 = 24, \quad 2 \times 3 \times 4 = 24, \quad 2 \times 2 \times 2 \times 3 = 24,$$
$$2 \times 12 = 24, \quad 4 \times 6 = 24, \quad 2 \times 2 \times 6 = 24.$$

The factors of 24 are: 1, 2, 3, 4, 6, 8, 12, and 24. Its prime factorization is $24 = 2^3 \times 3$.

❖

Example 6-5. Find the prime factorization of 100.

Solution: First, recognize that $10 \times 10 = 100$. Next, realize that each 10 can be factored as $10 = 2 \times 5$. As 2 and 5 are both prime, the prime factorization of 100 is $100 = 2 \times 5 \times 2 \times 5 = 2^2 \times 5^2$.

❖

Example 6-6.
Find the prime factorization of 702.

Solution: Because factoring is a matter of trial and error, you can approach this problem in various ways. One way is to try dividing by primes in order. For each prime chosen, divide by that prime as many times as possible before choosing the next prime. To start, first try dividing by 2.

$702 = 2 \times 351$	Can't divide 351 by 2, so try 3.
$= 2 \times 3 \times 117$	Try 3 again.
$= 2 \times 3 \times 3 \times 39$	Try 3 again.
$= 2 \times 3 \times 3 \times 3 \times 13$	Because 13 is prime, the factorization is done.

The prime factorization of 702 is $702 = 3^3 \times 2 \times 13$.

❖

6.3.2 Integers

Adding two (or more) natural numbers always results in another natural number; for example, $5 + 3 = 8$. Subtracting two natural numbers, however, results in three possible general outcomes.

1. Subtracting a smaller natural number from a larger one gives another natural number; for example, $10 - 6 = 4$.
2. Subtracting a natural number from itself yields *zero;* for example, $9 - 9 = 0$.
3. Subtracting a larger natural number from a smaller one does *not* give a natural number; instead, it is a negative number: for example, $6 - 10 = -4$.

The last outcome shows that negative numbers are an almost "natural" outgrowth of the natural numbers. Indeed, negative numbers arise naturally in many applications. Their first use probably was in commerce, where debts or losses can be expressed as negative numbers. For example, if your business has income of \$1000 and expenses of \$1500, your net income is –\$500; that is, you have a loss of \$500. Other common uses of negative numbers are in temperature and elevation measurements. On the Celsius scale, 0° is the temperature at which water freezes to ice; the temperature in your freezer must be below 0°C to keep ice cubes frozen. Elevations usually are measured with sea level as 0 elevation. Thus the elevation of a place below sea level, such as Death Valley, California, is negative.

The set of all numbers that we can make by adding or subtracting pairs of natural numbers is called the **integers**. The integers therefore include all the natural numbers, which are also called the **positive integers**; zero; and the negatives of all the natural numbers, or **negative integers**. In other words, the integers are the set: {..., –5, –4, –3, –2 –1, 0, 1, 2, 3, 4, 5,...}. The set of integers is shown on the number line in Figure 6-3. Also shown is the subset of the integers called the **whole numbers**, which includes only the positive integers *and* zero; that is, whole numbers are the set {0, 1, 2, 3,...}.

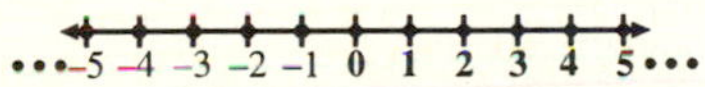

Figure 6-3. The integers are shown on a number line. The whole numbers include only the positive integers and 0; they are indicated with bold labels.

The integers have two basic properties:

- Every integer except 0 has a **sign** which indicates whether it is positive (+) or negative (–); 0 is neither positive nor negative.
- Every integer has a **magnitude** (also called **absolute value**) which indicates how far it lies from 0 on the number line.

Multiplying and Dividing Integers

Working with negative numbers requires special care with signs. A more complete review is provided in Appendix 1. The following is a brief summary of the results of multiplication and division with positive and negative numbers.

Results of Integer Multiplication or Division

× or ÷	*Positive*	*Negative*
Positive	Positive	Negative
Negative	Negative	Positive

- Multiplying or dividing two positive numbers yields another positive number: (positive) × (positive) = positive, and (positive) ÷ (positive) = positive.
- Multiplying or dividing a positive and negative number yields a negative number: (positive) × (negative) = negative, and (positive) ÷ (negative) = negative.
- Multiplying or dividing two negative numbers yields a positive result: (negative) × (negative) = positive, and (negative) ÷ (negative) = positive

Time-Out to Think: If a negative times a negative gives a positive, why don't "two wrongs make a right"? (Hint: Are "two wrongs" added or multiplied?)

Alternative Notations for Negative Numbers

In mathematics, negative numbers commonly carry the minus (–) sign. However, other methods of writing negative numbers may be used, particularly in financial work. Accountants, for example, often write negative numbers in *red ink* to ensure they cannot be accidentally confused with positive numbers, which they write in black ink. Hence the expression "in the red" means being in debt; while "in the black" means having net assets. Another alternative notation is to place negative numbers in parentheses. Tax forms, for example, usually show losses in that way.

6.3.3 Rational and Real Numbers

The answer obtained by adding, subtracting, or multiplying two or more integers always is another integer. For example, $5 \times (-4) = -20$, $7 + 3 = 10$, and $(-4) - 5 = (-9)$. Dividing integers, however, results in three possible general outcomes.

1. Dividing an integer by one of its factors gives another integer; for example, $24 \div 24 = 1$ or $24 \div 6 = 4$.
2. Dividing an integer by 0 is "undefined" (see Appendix 1).
3. Dividing an integer by any nonzero integer that is *not* one of its factors gives a fraction, *not* another integer; for example: $24 \div 7 = 24/7$ or $24 \div 48 = 1/2$.

The set of all possible outcomes of dividing integers (except dividing by 0) is called the **rational numbers**. The name *rational* comes from the word *ratio*, which refers to the division of two numbers. A more formal definition of the rational numbers is: the set of all numbers that can be expressed in the form x/y, where both x and y are integers and $y \neq 0$. The integers therefore are a *subset* of the rational numbers because any integer x is the same as $x/1$.

The word *planet* comes from the Greek word meaning "wandering star," and originally referred to any object that moves through the fixed constellations of stars. The ancient Greeks recognized seven "planets": the Sun, the Moon, Mercury, Venus, Mars, Jupiter, and Saturn. Besides feeding Pythagorean mysticism, the seven Greek planets also led to the names of the seven days of the week.

At one time in ancient Greece *all* numbers were believed to be rational numbers. The famous philosopher Pythagoras (c. 500 B.C.) inspired a group of followers who formed their own "secret society." The Pythagoreans believed that numbers had special and mystical meanings. For example, the number 1 was considered divine; even numbers were considered to be feminine, and odd numbers (besides 1) were considered to be masculine. The number 5, sum of the first feminine and masculine numbers, represented marriage. The origin of the common belief that 7 is a "lucky number" probably goes back to the Pythagoreans: It had special significance because it represented the seven "planets" known to the Greeks. Indeed, the motto of the Pythagoreans was "all is number."

One of the most sacred beliefs of the Pythagoreans was that all numbers were either "whole," by which they meant the natural numbers (they did not recognize zero or negative numbers), or fractions made by the division of these "whole" numbers. Unfortunately, by using the theorem of Pythagoras, the Pythagoreans soon realized that a right triangle with two sides of length 1 has a third side with a length of $\sqrt{2}$ (Figure 6-4).

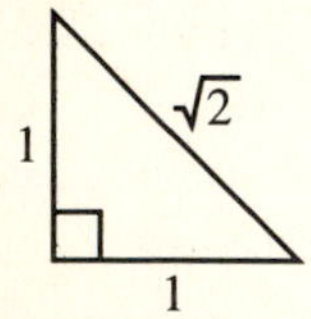

Figure 6-4. According to the Pythagorean theorem, a right triangle with two sides of length 1 has a hypotenuse of length $\sqrt{2}$. Although $\sqrt{2}$ is irrational, and cannot be written exactly in decimal form, you can use your calculator to verify that $\sqrt{2} \approx 1.41421$.

Try as they might, the Pythagoreans could not express $\sqrt{2}$ as a fraction. Eventually, they proved that it could not be done and hence that is an **irrational** number: It cannot be expressed by dividing two whole numbers. Legend has it that this discovery was so devastating to the Pythagoreans that they attempted to keep it secret, lest their fundamental doctrines be challenged. Supposedly they even killed one of their members, Hippasus, for telling others of the discovery of irrational numbers.

Time-Out to Think: The word *irrational*, literally, means "inexpressible as a ratio." Considering the story of the Pythagoreans, how do you think the word *irrational* came to mean strange or unreasonable thinking or behavior?

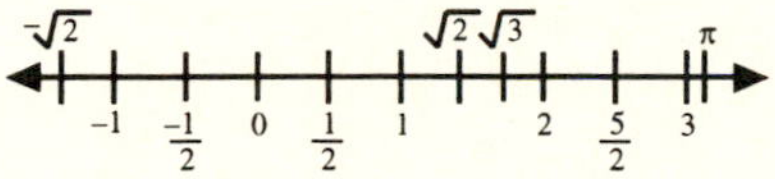
$-\sqrt{2}$
$\sqrt{2}\,\sqrt{3}$
π
-1 $\frac{-1}{2}$ 0 $\frac{1}{2}$ 1 2 $\frac{5}{2}$ 3

Figure 6-5. The real numbers consist of both the rational and irrational numbers. Here, we show on a number line a few rational numbers along with the irrational numbers $-\sqrt{2}$, $\sqrt{2}$, $\sqrt{3}$, and π.

We now know that the complete number line contains two types of numbers: rational numbers (or fractions) that can be expressed as the ratio of integers and irrational numbers that cannot be expressed as ratios. The combination of the rational and irrational numbers is called the **real numbers**. Another way to describe the real numbers is as the rational numbers and "everything in between." Each point on the number line has a corresponding real number, and each real number has a corresponding point on the number line (Figure 6-5).

THINKING ABOUT *... THE NUMBER OF THE BEAST*

The Pythagorean belief that numbers held special meanings or power is a form of what is called **numerology**. Practitioners of numerology claim to find both special meaning in numbers and numerical meaning in names. Although numerology is recognized as a pseudoscience today, it was widely practiced in many ancient cultures.

Often mystical, numerology also may have been used for deliberate coding in some cases. A famous example is found in the last book of the New Testament: The Revelation of St. John the Divine. In chapter 13, verse 18, he says:

> *Here is wisdom. Let him that hath understanding count the number of the beast: for it is the number of a man; and his number is Six hundred three-score and six.*

If you recall that a *score* is 20, you will realize that this verse is the origin of the number 666 representing the number of the beast, or devil. However, many biblical scholars believe that it was in fact intended to represent a particular, real man. Who? Revelation was written shortly after (within a few decades) the first great persecution of Christians under the Roman emperor Nero. If Nero's name is written in Hebrew, the sum of the numbers represented by the corresponding Hebrew letters is 666. Thus some scholars have suggested that Revelation is a political commentary couched in coded language: It would have been understood by John's intended audience of Christians but would have been meaningless to the Roman authorities.

A more recent example of numerology concerns four great composers — Beethoven, Shubert, Bruckner, and Mahler. Each man died shortly after completing his ninth (and last) symphony. Speculating that nine symphonies must be some sort of natural limit, twentieth-century composer Arnold Schönberg suggested that nine symphonies must bring a composer "too close to the hereafter." Numerology, despite its origins in ancient mysticism, remains with us today.

DILBERT® by Scott Adams

I'M PREDICTING THAT THE WORLD WILL END IN THE YEAR 2000.
THE CREATOR OF THE UNIVERSE WORKS IN MYSTERIOUS WAYS. BUT HE USES A BASE TEN COUNTING SYSTEM AND LIKES ROUND NUMBERS.
SO YOU REALLY WANT TO AVOID BEING, LET'S SAY, IN MOBILE HOME NUMBER 1,000,000 IN THE YEAR 2000.
I'M FEELING ANXIETY.

DILBERT reprinted by permission of United Feature Syndicate, Inc.

6.3.4 Imaginary and Complex Numbers

The operations addition, subtraction, multiplication, or division on real numbers always give another real number. Consider what happens, though, when you take the square root of a real number. If the number is positive, there is no problem; for example,

$$\sqrt{9} = 3 \text{ because } 3 \times 3 = 9;$$
$$\sqrt{0.25} = 0.5 \text{ because } 0.5 \times 0.5 = 0.25.$$

However, finding a real number that is the square root of a negative number is impossible. Why? Because the only way to get a negative product is to multiply a negative number by a positive number; thus a real number can never be multiplied *by itself* to yield a negative result. For example, no real number can be multiplied by itself to yield –4, so $\sqrt{-4}$ is not a real number.

Time-Out to Think: Convince yourself that $\sqrt{-4}$ cannot be a real number and that the same is true for the square roots of all other negative numbers. What happens if you try to take the square root of a negative number on your calculator?

To solve the problem of finding square roots of negative numbers, mathematicians invented **imaginary numbers,** or numbers that represent the square roots of negative numbers. A special number called ***i*** (for "imaginary") is defined to be the square root of negative 1. That is,

$$i = \sqrt{-1} \text{ or, equivalently, } i^2 = \sqrt{-1} \times \sqrt{-1} = -1.$$

Using i, you can find the square root of any negative number. For example,

$$\sqrt{-4} = 2i \text{ because } 2i \times 2i = 4 \times i^2 = 4 \times (-1) = -4.$$

The use of imaginary numbers makes it possible to take square roots of all real numbers, including the negative numbers. However, imaginary numbers cannot be shown on a real number line because they are *not* real numbers. The **complex numbers** are the set of numbers that include all the real numbers *and* all the imaginary numbers. We do not work much with either imaginary or complex numbers in this book, but they are very important and useful in mathematics, science, and engineering.

Time-Out to Think: Think about how the imaginary numbers are defined, and about the concept of numbers. Do you think that imaginary numbers are any less "real" than other numbers? Why or why not?

6.3.5 Summarizing the Modern Number System

Let's reflect for a moment on what we accomplished in this section. We began with the ancient concept of natural numbers, which are used for counting. We then showed, by observing the effects of arithmetic operations, how to broaden the concept of numbers. We defined the complex numbers, which encompass both real and imaginary numbers. We demonstrated that the real numbers can be either rational or irrational. The rational numbers include the integers and other fractions that can be expressed in the form of x/y. The integers include the natural numbers, 0, and the opposites (negatives) of the natural numbers. Finally, we showed that the natural numbers can be categorized according to whether they are prime, composite, or 1. Figure 6-6 summarizes the relationships among these sets of numbers.

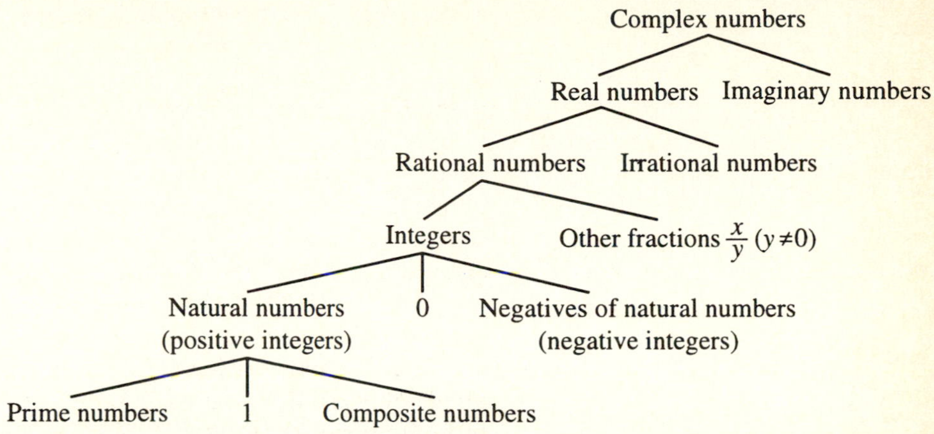

Figure 6-6. The relationships among the different sets of numbers are summarized.

6.4 SYSTEMS OF STANDARDIZED UNITS

As presented so far in this chapter, the concept of numbers is abstract. However, in most real situations numbers have a specific meaning; for example, the number 5 might mean 5 feet, or 5 apples, or 5 hours. Thus the meaning of a number depends on the units associated with it. Indeed, as we showed in Chapter 4, working with units along with numbers is essential in problem solving.

The most appropriate units differ in different situations. For example, you probably purchase apples in units of *apples* because you eat them one at a time. But the store probably charges for them in units of *pounds.* Note that pounds are an example of a **standardized unit**: *Two pounds* has the same (standard) meaning at all stores. *Apples*, in contrast, is not a standardized unit because apples come in different sizes and shapes. In this section, we describe the two systems of standardized measurement in common use today.

6.4.1 The U.S. Customary System of Measurement

The U.S. customary system is sometimes referred to as the "English system" because it is very similar to the customary system used for a long time in Great Britain. However, since Great Britain converted to the use of metric units, this system is now used only in the United States. In addition, there are a few subtle differences between the units used in the United States and those in the system that was used in Great Britain.

The **U.S. customary system** (USCS) represents units — feet, yards, miles, pounds, quarts, and so on — customarily used in the United States. The origin of U.S. customary units goes back to ancient systems of measurement used by Middle Eastern civilizations, including the Egyptians, Sumerians, Babylonians, and Hebrews. These systems of measurement were further developed in ancient Greece and Rome and in Europe during the Middle Ages. Eventually, the English standardized many of these ancient measures and brought them to the colonies whose revolution created the United States.

Because these units were developed over thousands of years and over large geographical regions, and because merchants in various trades tended to define their own units, many different units came into use. The result, unfortunately, is that the USCS is almost hopelessly complicated. Nevertheless, we will help you try to make some sense of it because these units are still used in the United States. If you find them confusing, you aren't alone! Indeed, except in the United States, these ancient units have been virtually abandoned in favor of metric units.

Measures of Length

Lengths in the U.S. customary system are derived from units originally based on body parts. For example, the **cubit** was the length of the forearm from the elbow to the tip of the middle finger. Distance was paced by the **foot**. The

Romans considered the foot to be equivalent to 12 *thumb-widths*, called *uncia*, which is the origin of our word *inch* and our 12 inches in 1 foot. They measured distances by pacing: A distance of a thousand paces was called *milia passum* (Latin for "one thousand paces") from which we get our word *mile*.

Because not everyone is the same size, lengths based on body parts vary from person to person. Many civilizations therefore standardized their units of length, often based on body parts of royalty. In England, for a long time, the foot was the length of the King's foot, which changed every time the King changed! The first true standardization of the English system began when King Henry I (1100–1135) set the *yard* as the measurement from the tip of *his* nose to the tip of *his* thumb on *his* outstretched arm.

Table 6-3 summarizes USCS units of length. Inches, feet, yards, and miles are the most common units. Fathoms are used to measure depth in oceans; rods and furlongs are used in horse racing. Leagues are rarely used today but were a common measure in the recent past; you can find distances in leagues in many historical documents, as well as in fictional works such as Jules Verne's *20,000 Leagues Under the Sea*. The only USCS unit of length that gets much use outside the United States is the **nautical mile**, commonly used to measure distances at sea; a related unit, the **knot**, measures speed in nautical miles per hour.

A nautical mile is equivalent to the distance along one minute of arc on the Earth's surface; that is, 60 nautical miles correspond to one degree of arc (e.g., one degree of latitude, or one degree of longitude along the equator). Thus the nautical mile is a convenient unit for navigation.

Time-Out to Think: The deepest trench in the oceans of Earth is about 35,000 feet below sea level. Thus, is a *depth* of 20,000 leagues possible? What did Jules Verne mean in the title of his novel?

The official definitions of USCS units are now based on metric standards, so there really is no "basic unit" of length; the inch is listed as the basic unit because it is the shortest. The official definition of the inch is 2.54 centimeters.

Table 6-3. Lengths in the U.S. Customary System of Units

(abbreviations for commonly used units are listed in parentheses)

1 inch (in.) (basic unit of length)	1 furlong = 40 rods (= 1/8 mile)
1 foot (ft) = 12 inches	1 statute mile (mi) = 1760 yards = 5280 feet
1 yard (yd) = 3 feet	1 nautical mile = 1.1516 statute miles = 6080.20 feet
1 rod = 5.5 yards	1 league on land = 3 statute miles
1 fathom = 6 feet	1 marine league = 3 nautical miles

Measures of Weight

USCS measures of weight evolved from an ancient unit called the **grain**, which probably referred to the weight of a typical grain of wheat. However, measures of weight became far more complicated than measures of length; in fact, three distinct sets of measures of weight are recognized within the U.S. customary system.

The word *avoirdupois* comes from the old French words meaning "goods of weight."

The first, the **avoirdupois** system, was adopted in about 1300 by London merchants and is still commonly used in the United States. It is based on a **pound**, defined as a weight of 7000 grains, which in turn is divided into 16 ounces; each avoirdupois ounce (oz) is 437.5 grains. The unit for measuring very heavy weights is the **ton**: An ordinary ton is 2000 avoirdupois pounds and a British ton, or long ton, is 2240 pounds.

The second system, called the **troy** system (probably named for the town of Troyes, France), was established during the fifteenth century. It is still sometimes used for weighing precious metals and stones. A **troy ounce** weighs 480 grains. A **troy pound** consists of 12 troy ounces, which makes it lighter than an avoirdupois pound.

A **carat** originally was a measure of weight in the troy system but today is defined to be 200 milligrams. For example, a "5 carat diamond" weighs 5×200 mg = 1 gram.

The **karat** is a term describing the purity of gold. Pure gold is defined to be 24 karats. Thus, for example, a necklace that is 50% gold (and 50% other materials) is 12-karat gold because 50% × 24 karats = 12 karats.

The third system was developed by medieval apothecaries (drug stores) and is called the **apothecary** weight system. It was used for weighing drugs but is no longer in common use anywhere; pharmacists today use metric units. The apothecary ounce and pound are the same as the troy units, but the apothecary used different units for smaller weights. Table 6-4 summarizes USCS measures of weight.

Table 6-4. Weights in the U.S. Customary System of Units
(abbreviations for commonly used units in parentheses)

1 grain (basic unit of weight for all three systems: approximately 0.0648 grams = 64.8 milligrams)		
Avoirdupois Measures	**Troy Measures**	**Apothecary Measures**
1 ounce (oz) = 437.5 grains	1 ounce = 480 grains	1 ounce = 480 grains
1 pound (lb) = 16 oz = 7000 grains	1 pound = 12 ounces = 5760 grains	1 pound = 12 ounces = 5760 grains
1 ton = 2000 lb	1 pennyweight = 24 grains	1 dram = 3 scruples = 60 grains
1 British, or long, ton = 2240 lb	1 carat = 3.086 grains = 0.2 gram	1 scruple = 20 grains

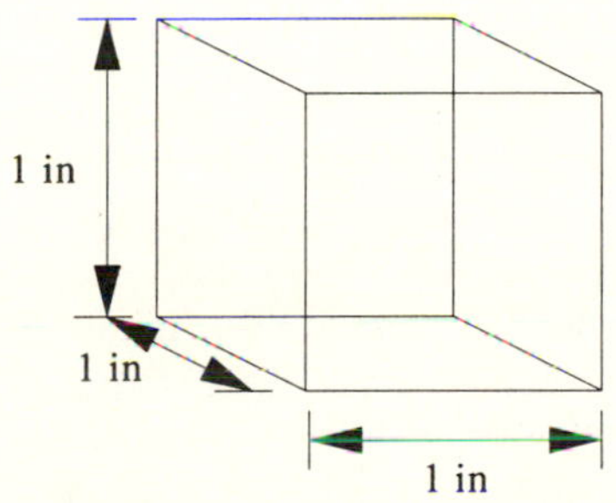

Figure 6-7. A cubic inch is the volume of a cube one inch on a side.

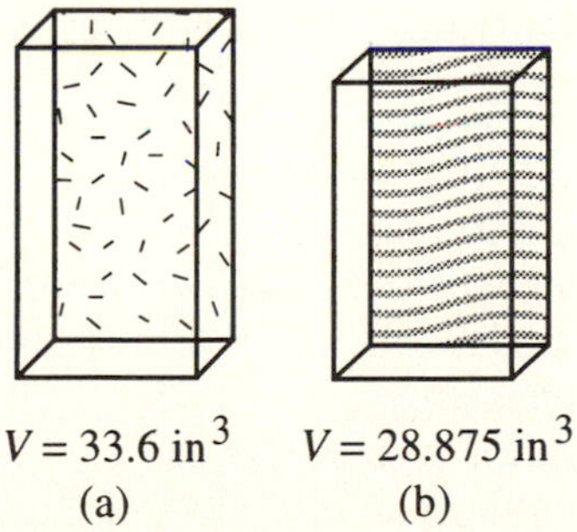

Figure 6-8. Volumes representing (a) a dry pint and (b) a liquid pint are compared. The figure is drawn to scale.

Measures of Volume

Units of volume, or capacity, measure the amount of material that can fit in a container. The basic USCS unit of volume is the cubic inch (Figure 6-7). Unfortunately, as with all USCS measures, the story of volume quickly becomes complicated.

First, many different units are used to measure volume, making it very difficult to keep track of all of them. Further, units of volume have different meanings depending on whether dry or liquid materials are being measured. For example, a **dry pint** is 33.60 cubic inches, but a **liquid pint** is only 28.875 cubic inches (Figure 6-8). Thus a container that holds one pint of water is too small for one pint of flour!

A further complication is that liquid measures with the same name have different definitions in the U.S. and British customary systems. For example, the British fluid ounce (1.734 cubic inches) is slightly smaller than the U.S. fluid ounce (1.804 cubic inches); and the British pint holds 20 British fluid ounces, whereas the U.S. pint has only 16 ounces. Beer drinkers may be familiar with this difference, as a British pint is about 20% larger than a U.S. pint.

$$1 \text{ liquid pint (U.S.)} = (16 \text{ oz}) \times \left(\frac{1.804 \text{ in}^3}{1 \text{ oz}} \right) = 28.87 \text{ in}^3$$

$$1 \text{ liquid pint (British)} = (20 \text{ oz}) \times \left(\frac{1.734 \text{ in}^3}{1 \text{ oz}} \right) = 34.68 \text{ in}^3.$$

A particularly confusing, but often used, measure of volume is the **barrel.** The definition of a barrel depends on what is being measured, and on state law. For example, a barrel of petroleum is 42 gallons, but a barrel of liquor (e.g., beer or wine) is 31 gallons. For other substances, the definition of a "barrel of liquids" varies from state to state, with values ranging from 31 to 42 gallons.

Table 6-5 on the next page summarizes USCS measures of volume and some measures from the old British customary system. A few additional units, commonly used in cooking, include the **teaspoon** (six teaspoons to one fluid ounce); the **tablespoon** (3 teaspoons to one tablespoon); and the **cup** (8 fluid ounces).

Table 6-5. Volume in the U.S. and British Customary Systems of Units
(abbreviations for commonly used units in parentheses)

1 cubic inch (in^3) = 16.387 cubic centimeters		(basic unit of volume)
U.S. Liquid Measures	**U.S. Dry Measures**	**Old British Measures**
1 fluid dram = 0.225 in^3	—	1 fluid dram = 0.217 in^3
1 fluid ounce (fl oz) = 8 fluid drams = 1.804 in^3	—	1 fluid ounce (fl oz) = 8 fluid drams = 1.734 in^3
1 gill = 4 fluid ounces	—	1 gill = 5 fluid ounces
1 pint (pt) = 16 fluid ounces = 28.87 in^3	1 dry pint (pt) = 33.60 in^3	1 pint (pt) = 20 fluid ounces = 34.68 in^3
1 quart (qt) = 2 pints = 57.75 in^3	1 dry quart (qt) = 2 dry pints = 67.2 in^3	1 quart (qt) = 2 pints = 69.36 in^3
1 gallon (gal) = 4 quarts = 231 in^3	—	1 imperial gallon = 4 quarts = 277.4 in^3
1 barrel: 31 to 42 gallons	1 peck = 8 dry quarts = 538.6 in^3	1 peck = 2 gallons = 554.8 in^3
(depending on usage and state law)	1 bushel = 4 pecks = 2150.4 in^3	1 bushel = 4 pecks = 2219.4 in^3

6.4.2 Basic SI (Metric) Units

After studying customary units carefully, you almost certainly will conclude that "there's got to be a better way!" Fortunately, there is. Recognizing the difficulties with customary systems of measurement, French politicians and scientists got together to invent the **metric system** in the late 1700s. Adopted for use by the Republic of France in 1795, the metric system was a product of the French Revolution of 1789. The two basic ideas behind the creation of the metric system were:

- to create a coherent and sensible set of standardized units to replace the customary systems in use around the world; and
- to simplify conversions and calculations by organizing relationships among the units by factors of 10 (decimal relationships).

Although the metric system was adopted for use in France in 1795, Napoleon abandoned its use in 1812. It was re-adopted by the French in 1840.

Use of the metric system slowly spread internationally throughout the 1800s. In 1875, the French government convened a conference to consider metric standards that was attended by delegates from 20 nations. The conference produced the *Treaty of the Meter*, which created mechanisms for refining the metric system and encouraging its international use.

The modern version of the metric system is known as the *Systeme Internationale d'Unites* (French for the International System of Units), or **SI**, which was formally established in 1960. By 1975, SI had been adopted for everyday use by every nation except Burma (now Myanmar), Liberia, and the United States. Even in the United States, SI units are legally established and widely used in science, manufacturing, and commerce.

Time-Out to Think: Who was the first U.S. president to propose that the United States adopt the metric system? Make an educated guess; you can find the answer on page 161 in *Thinking About ... Will the United States Go Metric?*

Although the present SI has seven basic units, for now we are concerned with only three of them:

- the **meter** for length, abbreviated "m";
- the **kilogram** for mass, abbreviated "kg"; and
- the **second** for time, abbreviated "s."

In addition, we need one *derived* unit:

- the **liter** for volume, abbreviated "ℓ," is the same as 1000 cubic centimeters, or 0.001 cubic meter.

The liter is a derived unit because its definition is derived from the definition of the meter.

Decimal-Valued Prefixes

Multiples of metric units are formed by powers of 10, as shown in Table 6-6. For example, the prefix *kilo* means 10^3, or 1000, so a kilometer is 1000 meters; similarly, a microgram is one millionth of a gram because the prefix *micro* means 10^{-6}, or one millionth.

To understand the table, recall how to work with powers of 10. For example,

$$10^2 = 10 \times 10 = 100, \text{ and}$$

$$10^{-2} = \frac{1}{10^2} = \frac{1}{100} = 0.01.$$

A more detailed review of powers of 10 is provided in Section 7.2, where we cover scientific notation.

Table 6-6. SI (Metric) Prefixes

Small Values			Large Values		
Prefix	*Abbrev.*	*Value*	*Prefix*	*Abbrev.*	*Value*
Deci	d	10^{-1}	Deca	da	10^1
Centi	c	10^{-2}	Hecto	h	10^2
Milli	m	10^{-3}	Kilo	k	10^3
Micro	μ	10^{-6}	Mega	M	10^6
Nano	n	10^{-9}	Giga	G	10^9
Pico	p	10^{-12}	Tera	T	10^{12}
Femto	f	10^{-15}	Peta	P	10^{15}
Atto	a	10^{-18}	Exa	E	10^{18}
Zepto	z	10^{-21}	Zetta	Z	10^{21}
Yocto	y	10^{-24}	Yotta	Y	10^{24}

Example 6-7 How many nanoseconds are in a microsecond?

Solution: A microsecond is a millionth of a second and a nanosecond is a billionth of a second, so a microsecond is 1000 times longer than a nanosecond, or

$$\frac{1\ \mu s}{1\ \text{ns}} = \frac{0.000001\ \text{s}}{0.000000001\ \text{s}} = 1000.$$

Thus 1000 nanoseconds make a microsecond.

❖

Time-Out to Think: Popular usage has adapted the prefix *mega* to mean "a lot." For example, people say that expensive things cost "megabucks," or that something really fun is "megafun." What do these statements mean literally? Should the federal debt of several trillion dollars be called "terabucks"? Do you think you can start a new trend by calling things, say, "gigafun" or even "yottafun"?

Definitions of Basic Units

Precise definitions of SI units are established by the *International Bureau of Weights and Measures*, headquartered in Sèvres, France. To the extent possible, definitions are based on properties of atoms. The advantage of using atomic properties is that they can be measured, in principle, by anyone in any location — even an alien civilization, if we ever encounter one!

As an example of atomic-based standards, consider the definition of the second. Until 1967, the second was defined as 1/86,400 of a day. The problem with this definition, however, was that the precise length of a day changes slowly over time. An atomic standard, adopted in 1967, solved this problem. All atoms emit light under certain circumstances, and the emitted light has particular colors (or frequencies) that are characteristic of the type of atom. As light is a wave (it is also a particle — see subsection 5.3.2), a particular frequency represents a particular number of vibrations in a specific amount of time. The atomic definition of one second is the amount of time required for 9,192,631,770 vibrations of a particular frequency of light emitted by atoms of the element cesium-133. Thus the most precise time-keeping devices available are **atomic clocks**, which work essentially by counting vibrations of light from cesium atoms (Figure 6-9).

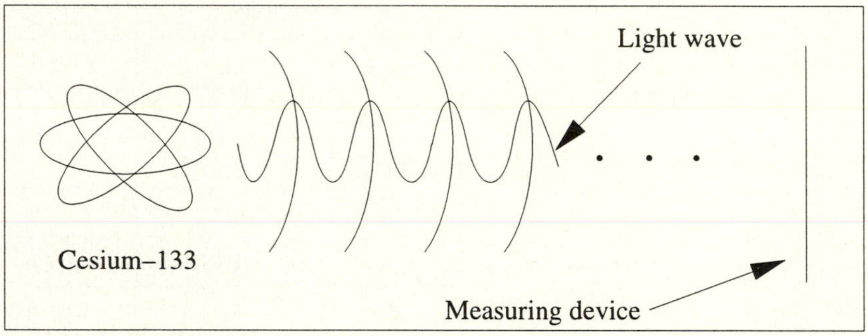

Figure 6-9. Essentially, an atomic clock counts the number of light-wave crests passing through a measuring device. One second is the time required for 9,192,631,770 wave crests emitted by cesium-133 atoms.

Many metric definitions have changed over time. The meter, for example, was originally defined as *one ten millionth* of the distance from the equator to the north pole. Based on this measurement, a metal rod was cut to a length of 1 meter and stored in Sèvres, France; this particular bar was the official standard for the meter. Later, the definition of a meter was changed to an atomic standard based on light emitted by the element krypton. In 1983, the meter was redefined again, this time in terms of the speed of light: 1 meter is the distance traveled by light (in a vacuum) during a time of 1/299,792,458 second. The advantage of this most recent definition is that it relies on two other established definitions: the definitions of the second and the speed of light, which are the same everywhere in the universe.

Unlike the second and the meter, the present definition of the kilogram is not based on atomic standards. Instead, it is defined as the mass of a particular block, made of platinum and iridium, housed in Sèvres, France. Clearly, this definition is problematic: The only way that scientists can precisely measure mass is by comparison to this one particular block. A new definition of the kilogram is currently being sought.

THINKING ABOUT ... WILL THE UNITED STATES GO METRIC?
Throughout most of the world, speed limits are posted in kilometers per hour, milk and gasoline are sold by the liter, and meat and fruit are sold by the kilogram. Will the United States ever join the rest of the world in the use of SI units?

In fact, the metric system has a long history in the United States. Several early leaders, including Thomas Jefferson and Benjamin Franklin, spent time in France and were familiar with the scientific discussions that led to the invention of the metric system. Jefferson was especially fond of the metric system reliance on *decimal* relationships; indeed, Jefferson had used decimal relationships himself in devising the U.S. system of currency based on the dollar, which was adopted in 1785.

George Washington also was concerned about the confusion caused by many different systems of measurement and, early in his first term as president, he urged action to ensure a uniform system of measurement for all the states. In 1790, while serving as secretary of state to President Washington, Thomas Jefferson formally proposed the adoption of the metric system to Congress. Had Jefferson's proposal been accepted, the United States would have been the second nation in the world, after France, to adopt the metric system. To Jefferson's disappointment, no legislation was enacted.

Nevertheless, the move toward metric never ceased. In 1866, the newly organized National Academy of Sciences urged Congress to authorize use of the metric system. The result was legislation legalizing (but not mandating) use of the metric system, which was signed into law by President Andrew Johnson on July 20, 1866. In 1875, the United States became one of the 17 signatories of the original *Treaty of the Meter*. In 1893, after receiving copies of the official standards for the meter and kilogram from France, Congress adopted metric units as legal standards. Thus, since 1893, all U.S. customary units have been defined by their metric equivalents; for example, the inch is defined as 2.54 centimeters. In 1901, the United States established the *National Bureau of Standards* (renamed the *National Institute for Standards and Technology* in 1992) for the purpose of developing measurement standards for commerce and science.

Since then, Congress has made several attempts to spur conversion to the metric system. In 1988, for example, Congress passed legislation requiring most federal agencies to use metric units in their procurements, grants, and business activities beginning in 1992. Nevertheless, most people in the United States continue to rely on customary units.

Despite public resistance, the complete conversion of the United States to SI units probably is inevitable. Besides the fact that SI units are far easier to learn and work with, increasing global trade, travel, and communication will necessitate falling in line with the rest of the world. The push toward metric is likely to be especially strong from industry and business. Because U.S. businesses hope to prosper through exports, they must conform to international standards. Indeed, the European Community has threatened to restrict imports that do not conform to SI standards. Today, nearly all motor vehicles, farm machinery, and computer equipment is built to metric specifications, and the labels on most containers of food and drink state metric equivalents.

6.4.3 Metric–USCS Conversions

Conversions between SI and USCS units are carried out like any other unit conversions (see Chapter 4): you need only to know the conversion factors. Although Table 6-7 on the next page lists only a few basic conversions, it should be sufficient for most of the conversions you need in your everyday life or in other college classes.

If you plan to do any traveling or if you commonly work with metric units in sports or business, *memorizing* these conversions, at least in rough terms, is a good idea. For example, if you recall that a kilometer is about 0.6 mile, you will know that a 10-kilometer road race is about 6 miles. If you recall that a gallon is a little less than 4 liters, you will know that a gasoline price of $1 per liter is a little less than $4 per gallon.

Table 6-7. USCS ⇔ SI Conversions

USCS to SI	SI to USCS
1 inch = 2.540 cm	1 cm = 0.3937 inch
1 inch = 0.02540 m	1 m = 39.37 inch
1 mile = 1.609 km	1 km = 0.6214 mile
1 pound = 0.4536 kg	1 kg = 2.205 pound
1 gal = 3.785 ℓ	1 ℓ = 0.2642 gal

Example 6-8. International athletic competitions generally use metric distances. Compare the length of a 100-meter race to a 100-yard race.

Solution: Table 6-7 shows that 1 meter is 39.37 inches. Convert 100 meters to 100 yards with a chain of conversion factors:

$$(100\text{ m})\times\left(\frac{39.37\text{ in}}{\text{m}}\right)\times\left(\frac{1\text{ yd}}{36\text{ in}}\right)=109.4\text{ yd}.$$

Note that 100 meters is almost 110 yards; a good "rule of thumb" to remember is that distances in meters are about 10% longer than the corresponding number of yards.

❖

Example 6-9. You go to a gas station and find that the price of gasoline is stated as $0.275 (27.5 cents) per liter. What is the equivalent cost per gallon?

Solution: Simply convert from price per liter to price per gallon. In Table 6-7 note that there are 3.785 liters in a gallon. Thus

$$\frac{\$0.275}{1\,\ell}\times\frac{3.785\,\ell}{1\text{ gal}}=\frac{\$1.041}{1\text{ gal}}.$$

The price of the gasoline is $1.041 (or 104.1 cents) per gallon.

❖

Example 6-10. How many square kilometers are in 1 square mile?

Solution: Simply square the conversion factor of 1.609 kilometers to a mile:

$$\left(1\text{ mi}^2\right)\times\left(\frac{1.609\text{ km}}{1\text{ mi}}\right)^2=\left(1\text{ mi}^2\right)\times\left(1.609^2\,\frac{\text{km}^2}{\text{mi}^2}\right)=2.589\text{ km}^2.$$

Therefore 1 square mile is about 2.589 square kilometers.

❖

6.4.4 Standardized Units of Temperature

Three temperature scales are commonly used today. People in the United States usually measure temperature on the **Fahrenheit** scale, which is defined so that water freezes at 32°F and boils at 212°F. Internationally, temperature is

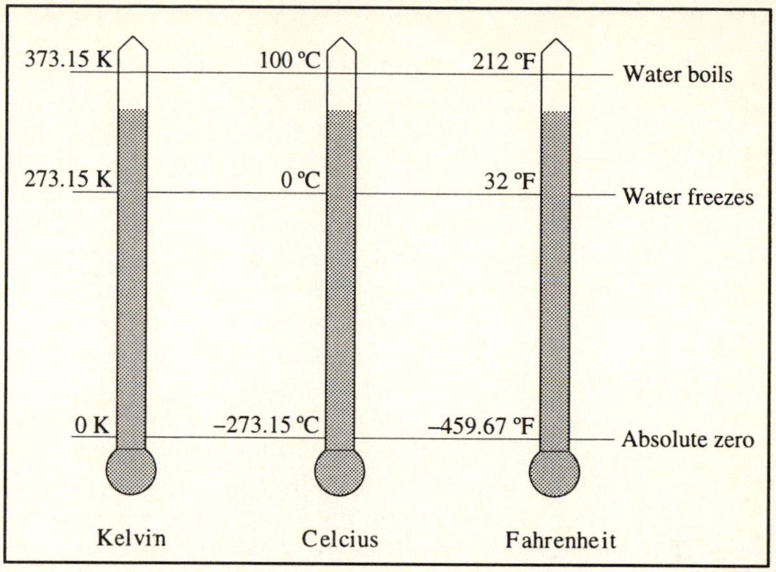

Figure 6-10. Celsius, Kelvin, and Fahrenheit temperature scales are compared.

usually measured on the **Celsius** scale, which places the freezing point of water at 0°C and the boiling point at 100°C.

The degree symbol (°) is not necessary when writing temperatures on the Kelvin scale.

The SI unit of temperature is called the **Kelvin**. The Kelvin scale is the same as the Celsius scale except in its zero point. A temperature of 0 K is **absolute zero**, which is the coldest possible temperature; 0 K is equivalent to –273.15°C.

As shown in Figure 6-10, any particular temperature has a Kelvin value that is numerically 273.15 larger than its Celsius value. That is,

$$\text{Temperature (Kelvin)} = \text{Temperature (°C)} + 273.15, \text{ or}$$

$$\text{Temperature (°C)} = \text{Temperature (Kelvin)} - 273.15.$$

To find the conversion between Fahrenheit and Celsius, note that the Fahrenheit scale has 180 degrees (212°F – 32°F = 180°F) between the freezing and boiling points of water, whereas the Celsius scale has only 100 degrees between these points. Each Celsius degree therefore represents a temperature change equivalent to 1.8 Fahrenheit degrees. Furthermore, the freezing point of water is numerically 32 larger on the Fahrenheit scale than the Celsius scale. Combining these two facts, the conversions between Fahrenheit and Celsius are

$$\text{Temperature (°C)} = \frac{[\text{Temperature (°F)}] - 32\text{°F}}{1.8\frac{\text{°F}}{\text{°C}}}, \text{ or}$$

$$\text{Temperature (°F)} = 32\text{°F} + \left(1.8\frac{\text{°F}}{\text{°C}}\right) \times [\text{Temperature (°C)}].$$

Time-Out to Think: Using a temperature of 32°F, confirm that the first formula above yields the freezing point of water on the Celsius scale (0°C). Use a temperature of 100°C to confirm that the second formula yields the equivalent 212°F.

6.5 EXACT AND APPROXIMATE NUMBERS: ROUNDING

Most real-life problems involve measurements, and measured numbers are only approximate. For example, your weight fluctuates, and different scales often give slightly different readings. Thus you probably *know* your weight only to within a couple of pounds. Even if a scale reads 122.32 pounds, saying "I weigh about 122 pounds" would be more honest. Moreover, even when people *do* know numbers exactly, they often aren't interested in anything more than an approximation. Imagine, for example, a store offering sale prices that are "one third off." If the regular price of an item is \$1, one third off means that you should get a discount of 33.3333...¢ (33⅓¢). As the smallest unit of currency is the cent, however, you probably won't care if the discount is approximated as 33¢.

Learning to recognize when numbers are approximate and estimate their level of uncertainty is one of the most important aspects of the concept of numbers. In fact, it is so important that we devote Chapter 8 entirely to this topic. For the moment, we consider only the important process of rounding, which we've already used in some of our examples. The basic process of rounding is straightforward.

Some statisticians use a more complex rounding rule if the value in the next column is exactly 5: They round up if the last digit being kept is odd and down if it is even. We don't worry about this subtlety in this book, but you may encounter it if you take a course in statistics.

- First, decide which decimal column (e.g., tens, ones, tenths, or hundredths) is the smallest that should be kept.
- Then, look at the number in the next column to the *right* (e.g., if rounding to tenths, look at hundredths). If the value in the next column is *less than 5*, round *down*; if it is *5 or greater*, round *up*.

For example, the number 382.2593 is given to the nearest ten thousandth. It can be rounded in the following ways.

382.2593 rounded to the nearest thousandth is 382.259.

382.2593 rounded to the nearest hundredth is 382.26.

382.2593 rounded to the nearest tenth is 382.3.

382.2593 rounded to the nearest one is 382.

382.2593 rounded to the nearest ten is 380.

382.2593 rounded to the nearest hundred is 400.

How do you know when to round a number? When numbers are exact, as in the "one third off" sale, it is a matter of personal taste or convenience. Perhaps you care only about the nearest penny; or, if the price is higher, you might care only about the nearest dollar. However, when measurements or estimates are involved, rounding is important so that you don't imply more certainty than you actually have. Until we discuss this topic in Chapter 8, we round by "feel." For example, suppose that you walk around a circle with a diameter of 1 kilometer. How far have you walked? The formula for the distance around a circle (the circumference) is the diameter multiplied by π; using a calculator, you obtain an answer of 3.141592654 km (the calculator's approximation to π km). Because the diameter of the circle was a measurement, however, it is not exactly 1 km; further, you could not have walked exactly on the edge of the circle. Thus, in describing how far you walked, you should use your "feel" for the level of certainty; in this case, you might say that you walked approximately 3.1 kilometers.

> **Time-Out to Think:** Electronic timers for sports such as swimming or track can record times to the nearest thousandth of a second, or better. Why do you think that world records are recorded only to the nearest hundredth of a second?

6.6* PRIME NUMBERS: MYSTERIES AND APPLICATIONS

Recall that a prime number is any natural number whose only factors are 1 and itself. With such a concise definition, it might seem that generating the sequence of prime numbers — or at least testing whether a given number is prime — should be easy. However, neither task is easy, which leads to both the mystery and the utility of primes.

Perhaps the most basic question about primes is: How many are there? If a number is very large, it seemingly ought to have factors besides 1 and itself. However, that isn't the case. As first proven by Euclid (c. 300 B.C.), the list of prime numbers has no end; there are infinitely many primes.

Eratosthenes is also well known as the first person to make an accurate estimate of the circumference of the Earth (see Chapter 11).

How can a list of prime numbers be generated? Perhaps the most systematic method is attributed to another Greek mathematician, Eratosthenes, who lived in the third century B.C. His method, called **the sieve of Eratosthenes,** uses a list of natural numbers starting at 1. For example, we might start with the numbers 1 through 10. To find the prime numbers on the list we begin by crossing out the 1 because it is neither prime nor composite. The next number, 2, is prime. Therefore, we cross out all subsequent multiples of 2 because they have 2 as a factor and so they must be composite. We call 2 a **sieve number** because it helps us "sift through," or remove, other numbers on the list. At this point, the list is

~~1~~ ② 3 ~~4~~ 5 ~~6~~ 7 ~~8~~ 9 ~~10~~

The next number, 3, is prime. We continue by using 3 as a sieve number, crossing out all subsequent multiples of 3. When we cross out 9 (it is a multiple of 3) the list becomes

~~1~~ ② ③ ~~4~~ 5 ~~6~~ 7 ~~8~~ ~~9~~ ~~10~~

We move to the next number that hasn't already been crossed out. The number, 5 in this case, is prime. Therefore we use 5 as a sieve number and cross out all multiples of 5 that haven't already been eliminated. We continue this process until we reach the end of our list of numbers. The numbers that remain after all the "crossings out" are the primes on the original list. For a list that goes only to 10, the primes are 2, 3, 5, and 7.

In principle, we could use the sieve of Eratosthenes on a list of numbers of any length. In practice, however, the method is extremely tedious. Consider the plight of a nineteenth-century Austrian astronomer named Kulik, who spent 20 years using this method to find all the prime numbers between 1 and 100 million! Perhaps unappreciative of the magnitude of his effort, the library to which he gave his manuscripts lost the sections containing the primes between 12,642,000 and 22,852,800.

> **Time-Out to Think:** Some time-saving tricks help reduce the effort involved in using the sieve of Eratosthenes. For example, on our list of 10 numbers we did not need to use the 5 as a sieve number because its only multiple (10) had already been crossed out. If you used the method on a list of the first 100 numbers, how high would you have to go in searching for multiples? Why?

The sieve of Eratosthenes is an *algorithm* (procedure) that generates prime numbers. A *formula* to produce primes could generate lists with much less effort. Alas, mathematicians have searched in vain for such a formula for more than two thousand years. A few formulas work over a limited range of numbers before failing. For example, the formula $n^2 - n + 41$ successfully produces primes for small values of n; let's try it for $n = 1, 2, 3,$ and 4:

$$n = 1 \quad \Rightarrow \quad n^2 - n + 41 = 1^2 - 1 + 41 = 41;$$

$$n = 2 \quad \Rightarrow \quad n^2 - n + 41 = 2^2 - 2 + 41 = 43;$$

$$n = 3 \quad \Rightarrow \quad n^2 - n + 41 = 3^2 - 3 + 41 = 47;$$

$$n = 4 \quad \Rightarrow \quad n^2 - n + 41 = 4^2 - 4 + 41 = 53.$$

Each of the numbers generated is prime. However, this formula fails after $n = 41$; that is, it successfully generates only 41 prime numbers before it produces a composite number. In addition, the formula misses many prime numbers along the way; note, for example, that it missed all the primes less than 41. Other formulas similarly fail after generating a relatively short list of primes. Indeed, most mathematicians now believe that a formula for reliably generating primes simply doesn't exist.

A visual way to explore prime numbers is shown in Figure 6-11, which represents factors of the numbers 1 through 80. The black squares in each column represent the factors of the column number. For example, the four black squares in column 10 indicate that 10 has four factors (1, 2, 5, and 10). The prime numbers are distinguished by having only two black squares — representing the factors 1 and themselves — in their respective columns. The way in which the number of factors varies from column to column follows no apparent pattern and remains an intriguing mystery.

Time-Out to Think: Lying above the diagonal in Figure 6-11 is a slanted line of squares in which only every other column has a filled square. Above that lies a line with only every third column having a filled square. What do these lines represent?

Careful study of Figure 6-11 reveals that several pairs of primes, called **twin primes**, are separated only by a single even number. The first six twin primes are

(3, 5), (5, 7), (11, 13), (17, 19), (29, 31), and (41, 43).

How often do twin primes occur? Is their appearance predictable? Do they go on forever? These unanswered questions will continue to inspire and haunt prime number enthusiasts for many years to come.

Prime numbers also have practical utility. Some of the most interesting applications are in the field of **cryptography**, or the writing of messages in code. In the modern age, a vast amount of information is available electronically, and money can be exchanged through electronic transactions. Codes therefore are extremely important in protecting privacy and maintaining security.

One strategy for electronic security relies on the difficulty of finding the factors of large numbers and testing whether certain numbers are prime. For

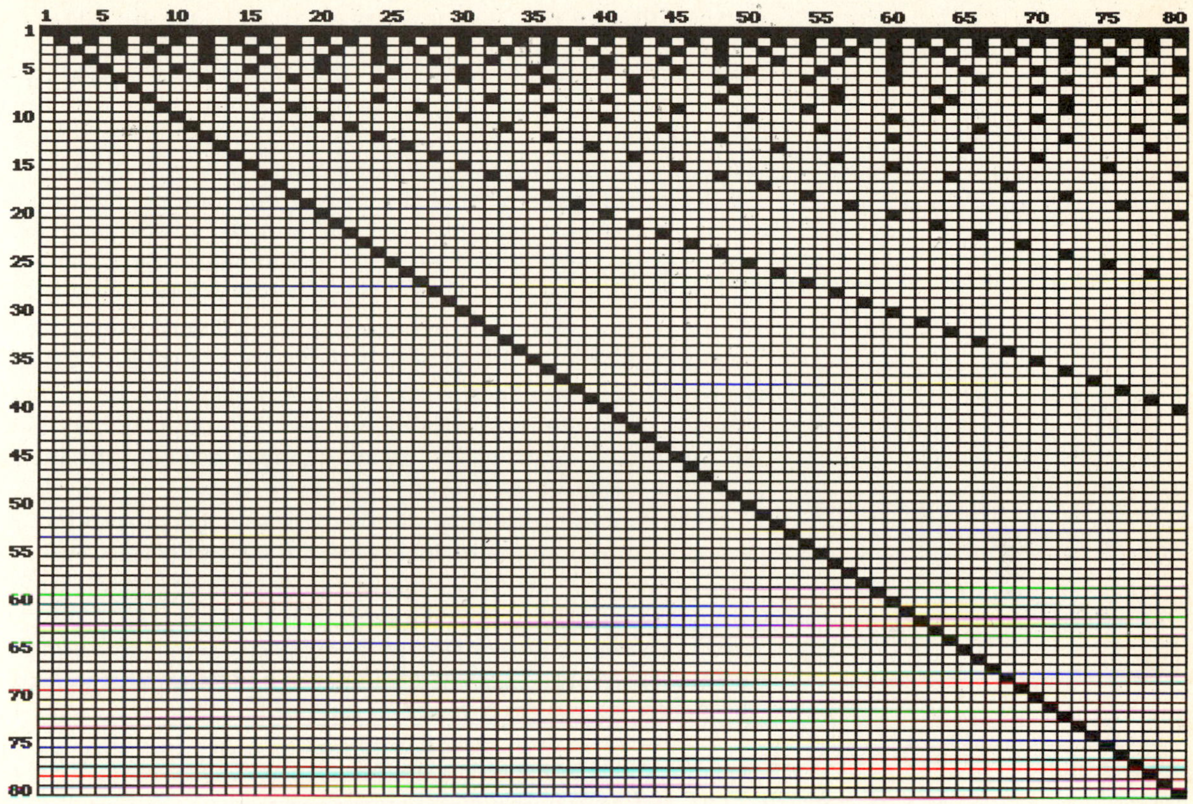

Figure 6-11. In each column of this grid, a square is filled only if a row number is a factor of the column number. For example, in column 8 the squares of rows 1, 2, 4, 8 are filled because these numbers are factors of 8. Note that row 1 is entirely filled because 1 is a factor of every number and the diagonal is filled because every number is a factor of itself. The column corresponding to a prime number such as 41 has only the squares in rows 1 and 41 filled.

example, both 65,537 and 6,700,417 are prime numbers. Finding their product is a simple:

$$65{,}537 \times 6{,}700{,}417 = 439{,}125{,}228{,}929.$$

However, if you were given only the number 439,125,228,929, finding its two prime factors it would be extremely difficult. This dramatic difference between the *ease* of multiplication and the *difficulty* of factoring allows reliable security systems to be devised. When extremely large prime numbers are multiplied the resulting composite number can be so much larger that even supercomputers can't find their prime factorizations. A security system can utilize this large composite number as a *lock*. The two primes that were multiplied to make the composite represent the *keys*.

Because there is no efficient way to find the prime factorization, the lock can be opened only by people who hold the keys; that is, by those who know the primes that were multiplied to create the composite number. Of course, research is seeking more efficient methods of factoring large numbers, and computers are getting faster. Thus a lock that is impenetrable today may be vulnerable in the future. But, as larger and larger prime numbers are found, more and more inviolable locks can be designed.

6.7* INFINITY

Infinity may be the most astonishing aspect of the concept of numbers. Serious study of infinity began just over a century ago, with the work of German mathematician Georg Cantor. His results shocked the mathematical world at the time, and the waves have not yet fully subsided.

Before we begin, recall that a *set* is a collection of objects, such as the cards in a standard deck, the even numbers from 1 through 100, or the leaves of an oak tree. A *subset* is some collection of objects from the original set. Subsets of the aforementioned sets include, respectively, the spades in a deck of cards, the even numbers between 1 and 50, and the dead leaves on the oak.

A fancy term for the number of elements in a set is its **cardinality**. For example, a standard deck has 52 cards, so we say that its cardinality is 52. Similarly, there are 50 even numbers from 1 through 100, so the cardinality of this set is 50. The cardinality of the set of leaves on an oak tree might be 23,568, or whatever the number of leaves.

How can you determine whether two sets have the same cardinality? One way is to count the members of each set and check to see whether the count is the same for both sets. For example, the cardinality of the set of cards in a standard deck is the same as the cardinality of the set of 52 whole weeks in a year. Another way to determine whether two sets have the same cardinality involves *matching* rather than counting. Consider the two sets of numbers

$$\{2, 4, 6, 8, 10, 12, 14, 16, 18\} \text{ and } \{1, 3, 5, 7, 9, 11, 13, 15, 17\}.$$

We can pair the members of each set and then check for any "leftovers." If there are no leftovers, the sets must have the same number of members, or cardinality. Let's try matching those two sets:

$$\begin{array}{ccccccccc} 2 & 4 & 6 & 8 & 10 & 12 & 14 & 16 & 18 \\ \updownarrow & \updownarrow & \updownarrow & \updownarrow & \updownarrow & \updownarrow & \updownarrow & \updownarrow & \updownarrow \\ 1 & 3 & 5 & 7 & 9 & 11 & 13 & 15 & 17 \end{array}$$

Each member of the first set has only one counterpart in the second set. We therefore conclude that the sets have the same cardinality, or the same number of members, without even knowing what the cardinality is. In mathematical terminology, we found a **one-to-one correspondence** between the members of the two sets.

6.7.1 The Paradox of Infinite Sets

Now let's step over the brink and try the same approach on two sets with *infinite* numbers of members. We choose one set to be the natural numbers:

$$\{1, 2, 3, 4, 5, 6, \ldots\}$$

The natural numbers form an **infinite set** because, as indicated by the ellipsis (...), the numbers continue endlessly. We choose a second infinite set to be the even natural numbers

$$\{2, 4, 6, 8, 10, 12, \ldots\}$$

We now ask an apparently absurd question: Do these sets have the same number of members, or the same cardinality? In any *finite* range, such as from 1 to 100, there are only half as many even numbers as natural numbers. As a result, like most people, you probably are tempted to answer that the second set has fewer members than the first. But let's try the matching test.

As before, we start pairing numbers from each set:

1	2	3	4	5	6	7	8	9	...
$\updownarrow$	$\updownarrow$	$\updownarrow$	$\updownarrow$	$\updownarrow$	$\updownarrow$	$\updownarrow$	$\updownarrow$	$\updownarrow$	...
2	4	6	8	10	12	14	16	18	...

Clearly, we could continue this matching process indefinitely. Each member of the first set always will have a corresponding member of the second set. We can even state a rule for the pairs: The number n from the first set always is paired with the number $2n$ from the second set (where n is any natural number). What are we to conclude? If we abide by the rule that we established for finite sets, we have no choice but to conclude that these two infinite sets, one a subset of the other, have the same number of members!

A similar argument can be applied to the set of natural numbers and any of its infinite subsets, and the same conclusion is inescapable. For example, by matching members we find a one-to-one correspondence that shows there are

- as many natural numbers as odd numbers,
- as many natural numbers as multiples of 3 ({3, 6, 9, 12, 15,...}), and
- as many natural numbers as multiples of 5 ({5, 10, 15, 20, 25,...}).

Time-Out to Think: Think of a one-to-one correspondence to verify each of the preceding claims. Can you state a rule for the pairing in each case? Can you show a one-to-one correspondence between the natural numbers and the *squares* of the numbers?

If you find these ideas unsettling, you aren't alone. Galileo first recognized them in 1638, but he considered them to be unexplainable paradoxes and therefore chose not to work with infinity any further. For the next 250 years, mathematicians followed Galileo's lead in ignoring infinity. However, Georg Cantor took these paradoxes as starting points for further work, and invented a new arithmetic, called **transfinite arithmetic**, that applies to infinity.

Cantor's work did not sit well with most mathematicians of the late nineteenth century. They branded him a heretic and ridiculed him. One principal rival, Leopold Kronecker (who believed that infinity could not be considered mathematically), used his influence over hiring at many universities and as editor of one of the most important mathematical journals to carry out a personal vendetta against Cantor. As a result, Cantor's career was blocked, and he was unable to publish the results of his work in mathematical journals.

My theory stands as firm as rock; every arrow directed against it will return quickly to its archer. How do I know this? Because I have studied it from all sides for many years; because I have examined all objections which have ever been made against the infinite numbers; and above all because I have followed its roots, so to speak, to the first infallible cause of all created things. — Georg Cantor

Cantor became embittered and paranoid, alienating even the mathematicians who supported his work. He found greater appreciation among philosophers, especially Catholic theologians who believed his work to be of fundamental importance to their conception of God and the universe. Cantor therefore was able to continue his work and publish in philosophical journals. Nevertheless, Cantor suffered a series of mental breakdowns and personal tragedies; he died in a mental hospital in 1918.

6.7.2 The Arithmetic of Infinity

The paradoxical result that the natural numbers and the even natural numbers have the same cardinality is just the beginning of the surprises and paradoxes

$\aleph_0$ is made up of the first letter of the Hebrew alphabet, ℵ (aleph), with a subscript zero.

in the arithmetic of the infinite. How do we designate the cardinality of the natural numbers? You may be familiar with the symbol ∞ for infinity, but for reasons that will become clear, this symbol is inadequate for describing the cardinality of infinite sets. Instead, the standard symbol for the cardinality of the set of natural numbers is $\aleph_0$, usually pronounced "aleph-naught."

Because the set of even natural numbers has the same number of members as the set of natural numbers, it also has a cardinality of $\aleph_0$. Now, let's do some transfinite arithmetic. What happens if we add one member to the set of even numbers? We do so by starting the set from 0 instead of 2:

	1	2	3	4	5	6	7	8	9	...
	↕	↕	↕	↕	↕	↕	↕	↕	↕	...
0	2	4	6	8	10	12	14	16	18	...

The set on the bottom row now has one more member than the top set. Or does it? Note what happens if we simply *shift* the bottom row to the right:

1	2	3	4	5	6	7	8	9	...
↕	↕	↕	↕	↕	↕	↕	↕	↕	...
0	2	4	6	8	10	12	14	16	...

The bottom row again has a one-to-one correspondence with the top row; that is, both sets have the same number of members and cardinality $\aleph_0$. Yet, we formed the bottom row by adding one member to a set which *already* had cardinality of $\aleph_0$. Thus we have one of the curious results of transfinite arithmetic:

$$\aleph_0 + 1 = \aleph_0.$$

What happens if we add two infinite sets together? Just as we have shown that the even numbers have a cardinality of $\aleph_0$, we could also show that the odd numbers have a cardinality of $\aleph_0$. If we add the odd and even numbers together, we get the complete set of natural numbers. However, this set also has a cardinality $\aleph_0$, so we have another strange result:

$$\aleph_0 + \aleph_0 = \aleph_0.$$

You might wonder whether any sets have *more* members than the set of natural numbers; that is, do all infinite sets have the same cardinality $\aleph_0$? In search of a "larger" infinite set, let's consider the set of positive rational numbers. As before, we seek a system of matching between the two sets. To begin we imagine a square array with infinite numbers of columns and rows. Along the top of the array we label the columns with all the natural numbers; we do the same to label the rows along the side. Then we fill the array with fractions in which the numerator is the row number, and the denominator is

DILBERT reprinted by permission of United Feature Syndicate, Inc.

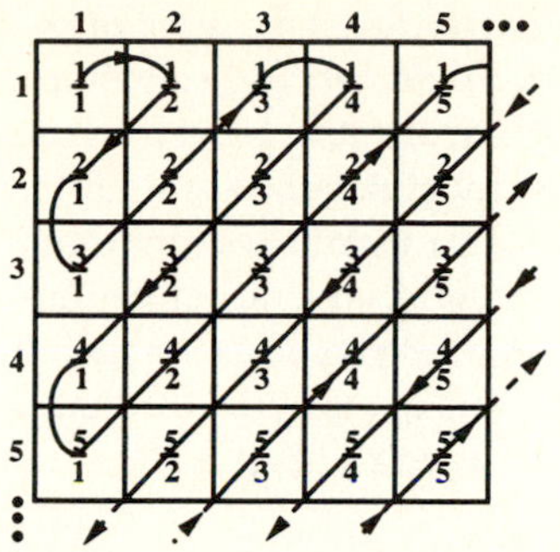

Figure 6-12. The rational numbers are displayed in an array in which the fraction *n*/*m* appears in the *n*th row and *m*th column. By following the path through the array shown by the arrows, the rational numbers can be matched one-to-one with the natural numbers.

the column number. Figure 6-12 shows the first five columns and rows. Because the array extends to infinity along both the top and the side, it has a place for *every* rational number. The first column, for example, contains all rational numbers with a denominator of 1; the second column contains all rational numbers with a denominator of 2; and so on.

How many rational numbers are there? We can devise a counting scheme simply by passing back and forth through the array of rational numbers along ascending, then descending, diagonals. The beginning of this path is shown in Figure 6-12. We start at $1/1$, which we match with the natural number 1. Then we move to $1/2$ which we match with the natural number 2. We continue along the path, matching each fraction with the next natural number; the chosen path ensures that we will pass through *every* rational number as we move through the array, following the pattern

$$\begin{array}{cccccccccc} 1 & 2 & 3 & 4 & 5 & 6 & 7 & 8 & 9 & \ldots \\ \updownarrow & \updownarrow & \updownarrow & \updownarrow & \updownarrow & \updownarrow & \updownarrow & \updownarrow & \updownarrow & \ldots \\ \frac{1}{1} & \frac{1}{2} & \frac{2}{1} & \frac{3}{1} & \frac{2}{2} & \frac{1}{3} & \frac{1}{4} & \frac{2}{3} & \frac{3}{2} & \ldots \end{array}$$

We can never run out of natural numbers, so we can match every fraction in one-to-one correspondence. Our astonishing conclusion is that there are as many natural numbers as there are fractions, or rational numbers!

What does this outcome mean for transfinite arithmetic? Because the columns in the array are labeled by natural numbers, the number of columns can be designated $\aleph_0$. The same is true for the number of rows. But we have also shown that the total number of fractions in the array is $\aleph_0$. Thus

$$\aleph_0 \times \aleph_0 = \aleph_0, \qquad \text{or} \qquad (\aleph_0)^2 = \aleph_0.$$

Let's look at one more example of transfinite arithmetic. We've already shown that $(\aleph_0)^2 = \aleph_0$, so we can follow through to obtain

$$\aleph_0 \times \aleph_0 \times \aleph_0 = \aleph_0 \times (\aleph_0)^2 = \aleph_0 \times \aleph_0 = \aleph_0, \quad \text{or} \quad (\aleph_0)^3 = \aleph_0.$$

Following similar lines of reasoning we can show, for example, that

$$100 + \aleph_0 = \aleph_0; \quad 2 \times \aleph_0 = \aleph_0; \quad 1{,}000{,}000 \times \aleph_0 = \aleph_0;$$

$$(\aleph_0)^4 = \aleph_0; \quad \text{and } (\aleph_0)^{1{,}000{,}000} = \aleph_0.$$

The arithmetic of infinity is very different from ordinary, finite arithmetic!

6.7.3 Bane of the Pythagoreans

With a clever argument, Cantor was able to show that some infinite sets *do* have cardinality greater than $\aleph_0$. In particular, he showed that the *real* numbers, including both rational and irrational numbers, *cannot* be put into one-to-one correspondence with the natural numbers. As we have already shown that the cardinality of the rational numbers is the same as that of the natural numbers, he therefore proved that there are more irrational than rational numbers. In fact, his proof showed that between *any* two points on the number line, the number of irrational numbers is greater than the number of all rational numbers. Thus, for example, there are more irrational numbers between 0.0000000000000001 and 0.0000000000000002 than the total of all rational numbers. The cardinality of this new, higher infinity is designated $\aleph_1$ ("aleph-one"). You can now understand that the symbol ∞ could not be used because infinity has more than one "level"!

Recall the Pythagorean belief that all numbers should be rational and how they tried to hide their discovery of the irrationality of $\sqrt{2}$. Imagine their reaction if they had been alive to see Cantor's work and to learn that the irrational numbers infinitely outnumber all the rational numbers.

Although it is subtle, we can outline Cantor's argument showing that there are more irrational than rational numbers. Recall that irrational numbers cannot be written exactly in decimal form because they are decimals that never end. Suppose that someone claims to have found a scheme for matching the irrational numbers to the natural numbers, as in the list:

$$1 \Rightarrow 0.142678435\ldots$$
$$2 \Rightarrow 0.383902892\ldots$$
$$3 \Rightarrow 0.293758778\ldots$$
$$4 \Rightarrow 0.563856365\ldots$$
$$5 \Rightarrow 0.486793597\ldots$$
$$\vdots \Rightarrow \quad \vdots$$

Regardless of the method used for matching, we will always be able to write another irrational number that is *not* already on the list. To do so, for the first digit of our new number, we choose something other than the first digit of the first number on the list; that is, anything other than 1. For the second digit, we choose something other than the second digit of the second number on the list, or something other than 8. And so on to infinity. The resulting irrational number will differ in at least one digit from every number on the list. In other words we will have found a number that was "missed" by the matching scheme. This shows that the natural numbers cannot be put in one-to-one correspondence with the irrational numbers. Our conclusion: The cardinality of the irrational numbers is greater than that of either the natural or rational numbers.

Time-Out to Think: Describe a method for finding another irrational number that would have been missed in the original matching scheme. Do you see why *infinitely many* irrational numbers are missed by *any* attempt to match them with the natural numbers? Explain.

6.7.4 Higher Orders of Infinity

We have said that the cardinality of the natural and rational numbers is $\aleph_0$ and that of the real and irrational numbers is $\aleph_1$. Does a level of infinity exist *between* $\aleph_0$ and $\aleph_1$? The answer is unknown, but a set with such a cardinality has never been found. Most mathematicians would bet on what is called the **continuum hypothesis:** that no set with cardinality between $\aleph_0$ and $\aleph_1$ exists.

The final mind-bending step is to ask whether any infinities have higher cardinality than the real numbers. Again, Cantor was able to answer this question with a "yes." In fact, he showed that an infinite number of higher levels of infinity exist and that their cardinality might be designated

$$\aleph_0, \aleph_1, \aleph_2, \aleph_3, \ldots.$$

Although Cantor's transfinite arithmetic provides mathematical rules for constructing these higher orders of infinity, no one has ever been able to describe a set with an infinity higher than $\aleph_2$.

6.8 CONCLUSION

In this chapter we have developed three principal aspects of the concept of numbers: mathematics from its ancient origins to the esoteric nature of infinity; units associated with numbers; and approximate as opposed to exact numbers. Key lessons from this chapter that will help you in your further work include the following.

- Our modern number system, language, and culture are all intertwined. Thus an appreciation of the history and mechanics of our number system will enhance your understanding and enjoyment of language and culture.
- In most real situations, numbers are not abstract mathematical ideas. Instead, they refer to concrete measures or objects through their association with units. Understanding common units therefore is fundamental to understanding the use of numbers in modern society.
- Numbers in real situations tend to be approximate rather than exact. Recognizing when numbers are approximate and learning to estimate their uncertainties is crucial to interpreting numbers in the real world.
- Although people often think of arithmetic as something learned in elementary school, in-depth study of arithmetic is a rich and dynamic field of mathematical research. Anyone can appreciate the strange and wonderful results found in such fields as the study of prime numbers and the study of infinity.

PROBLEMS

Reflection and Review

Section 6.1

1. **A Language of Symbols.** Mathematics uses symbols as abstractions, enabling its users to generalize rules. For practice with symbols, rewrite each of the following statements in symbolic form, using the symbol indicated.

   Example: Any number s, except 0, raised to the fifth power.

   Solution: $s^5, s \neq 0$.

   a. The quotient of 10 divided by any number x except 0.
   b. The product of 5 times any even number y.
   c. The difference when 12 is subtracted from any number a.
   d. The sum of 13 and the square root of any number b except 0.

2. **Mathematical Patterns.** A common character in mathematics is the *ellipsis* (…), which indicates the continuation of a pattern. In each of the following problems describe the pattern. If the pattern continues only briefly write out the entire statement described in "shorthand" with the ellipsis; if the pattern continues forever, list the next three entries that would follow the last entry before the ellipsis.

   Example: $1 + 2 + 3 + \dots + 10$. *Solution:* The pattern is adding a set of consecutive numbers. The pattern continues only from 1 through 10, so the entire statement would read: $1 + 2 + 3 + 4 + 5 + 6 + 7 + 8 + 9 + 10$. Note that the sum is 55.

   Example: 2, 4, 6, 8,… *Solution:* The pattern is counting by 2, or listing the even natural numbers. No stopping point is indicated, so the pattern continues forever; the next three entries following 8 would be 10, 12, 14.

   a. $1 + 2 + 3 + \dots + 6$
   b. 3, 6, 9, 12, …
   c. $1 + 10 + 100 + 1000 + \dots + 1{,}000{,}000$
   d. $1 + 3 + 5 + 7 + 9 + 11 + \dots$
   e. $2^0 + 2^1 + 2^2 + 2^3 + \dots + 2^{10}$
   f. 2, 3, 5, 7, 11, 13, 17, 19, 23, 29, 31, …
   g. 2, 4, 8, … , 1024
   h. $1^2 + 2^2 + 3^2 + 4^2 + 5^2 + \dots$
   i. $1 + 5 + 14 + 30 + 55 + \dots$
   j. 0, 1, 1, 2, 3, 5, 8, 13, 21, 34, 55, 89, … (Hint: look at the *two* previous entries for each number. This pattern is called the *Fibonacci sequence*.)

3. **Learning a Language.** Have you ever studied a foreign language? If so, write a short essay about how that experience compares to your past experience with learning mathematics. If you have not studied a foreign language, write a short essay about how your experience in learning

mathematics compares with your experience in studying English vocabulary and grammar.

Section 6.2

4. **Reading Roman Numerals.** Find the value for each of the following Roman numerals.

a. LIII	b. LXVI	c. CXXI
d. CLXXXI	e. MCDXLIII	f. XLIV
g. XXXVII	h. XCIX	i. CCCXLVI
j. CMLIII	k. CV	l. MMMCCXLIV
m. MMCMIII	n. CLXXII	o. CDLXXVIII

5. **Writing Roman Numerals.** Write each of the following as a Roman numeral.

a. 4	b. 37	c. 41
d. 49	e. 106	f. 334
g. 445	h. 490	i. 499
j. 972	k. 1995	l. 2001
m. 2462	n. 3789	o. 3540

6. **Roman Numerals in Use.** The ability to read Roman numerals can help you determine when something was created.
   a. On the back of a U.S. one dollar bill is a picture of a pyramid with an eye. What date is written along the base of the pyramid? What is the significance of that date?
   b. Walking through Oxford University, England, you discover a building with MCCCLXXIX chiseled on the wall. What do you think it means? Would you expect to find a building with this Roman numeral in the United States? Why or why not?
   c. The band *Enigma* produced a recording called "MCMXC a.D." What do you think the title means?
   d. In every film that shows a Roman numeral date in its credits, the numeral always begins with MCM. Why?

7. **Numbers in Expanded Form.** Write each number in expanded form and in words (as expressed in the United States, not Great Britain).

a. 75	b. 106	c. 227
d. 687	e. 432,067	f. 1,001,001
g. 42,329,000	h. 854,040,332	i. 3,635,000,000
j. 17,950,000,000,000	k. 0.003	l. 1000.7
m. 98.24816	n. 1.0099	o. 65.15
p. 99.999999	q. 0.000000011	r. 1000.0001

8. **Number Name Confusion.** Each of the following numbers is expressed as it would be expressed in Great Britain. Write each number using Hindu–Arabic numerals and in words as it would be expressed in the United States.
   a. Ten thousand million.
   b. Four hundred twenty-two thousand million.
   c. Three billion.
   d. Seventeen billion, six hundred thousand million.
   e. One thousand billion.
   f. Ten thousand billion, two thousand million

9. **Base-2 to Base-10.** Express the following base-2 numbers in base-10.
   a. 1001_2 b. 1010_2 c. 10000_2 d. 11111_2

10. **Base-10 to Base-2.** Express the following base-10 numbers in base-2:
    a. 4 b. 12 c. 23 d. 32

11. **Base-8 numbers.** How many different symbols are needed in the base-8 system of numbers? Give the values of the first four columns of a base-8 number.

12. **Uses of Numbers.** For the numbers expressed in each of the following statements indicate whether they are being used as cardinal, ordinal, or nominal numbers. Explain your reasoning.
    a. "I've got three boxes here for a Ms. Jones."
    b. Deli worker to crowd: "Next up is number 45!"
    c. "The price of the car is $19,200."
    d. "Go left at the next light; you'll see signs for Highway 66."
    e. "This is the fourth time I've told you to clean up your room!"
    f. National Football League guidelines: Any player wearing number 80 is supposed to be a tight end.
    g. Joe came in second in the 10-kilometer race.

Section 6.3

13. **Divisibility Rules.** In the pairs of numbers shown determine whether the first is a factor of the second by using the divisibility rules of Table 6-2.

a. 10; 2140	b. 4; 1272	c. 3; 831	d. 5; 5554
e. 9; 5436	f. 2; 7532	g. 4; 446	h. 3; 7452
i. 10; 255	j. 9; 813		

14. **Prime Numbers?** Decide whether the following numbers are prime or composite. Explain.
    a. 23 b. 49 c. 67 d. 101 e. 143

15. **Factoring.** Find all the factors of each of the following numbers and identify its prime factorization.

a. 10	b. 12	c. 16
d. 25	e. 32	f. 36
g. 48	h. 100	i. 200

16. **Prime Factorizations.** Find the prime factorization of each of the following numbers.

a. 75	b. 390	c. 400
d. 1320	e. 625	f. 1000
g. 242	h. 800,000	i. 6,000,000

17. **Classifying Numbers.** Consider this list of sets: natural numbers, integers, rational numbers, and real numbers. For each of the following numbers identify from the list the *first* set that describes the number. Explain.

a. 2.3	b. $-3/2$	c. 3	d. -5
e. 100.1	f. -6.1	g. $5/3$	h. π

18. **How Operations Affect Numbers.** Each part describes a numerical operation. For each part (i) state whether the result would be negative, positive, or zero; and (ii) state whether the result would be integer, rational, or irrational.
    a. multiplying three negative integers
    b. dividing a positive integer by a negative rational number
    c. multiplying a positive rational number by a positive irrational number

d. adding two negative irrational numbers
e. subtracting a negative rational number from a positive irrational number
f. dividing a positive integer by a negative integer and multiplying the result by a negative integer
g. adding two negative rational numbers and dividing the result by a negative integer

19. **Irrational to Rational?** Starting with an irrational number, is it possible to obtain a *rational* result by adding, subtracting, multiplying, or dividing by any other number? Explain.

20. **Imaginary Numbers.** Use the definition of i to show the following.
a. $i^3 = -i$ b. $i^4 = 1$ c. $i^5 = i$ d. $\sqrt{-9} = 3i$

Section 6.4

21. **Everyday Metric.** Describe three ways that you use metric units in your everyday life.

22. **Grocery Metric.** Describe three examples of the use of metric units in the grocery store.

23. **USCS Lengths.** Make the following conversions within the U.S. customary system of measurement.
a. Convert your own height in feet to inches. If you don't know your height, measure it.
b. A basketball player might be 6.75 feet tall. Express this height in inches and yards.
c. The Kentucky Derby horse race is 10.2 furlongs in length. What is this distance in miles? How does it compare to a road race of 6.2 miles or a marathon of 26 miles?
d. How far is Jules Verne's 20,000 leagues under the sea in feet? in miles?
e. Deep sea trenches can reach a depth of 6000 fathoms. How deep is that in feet? in miles? in leagues?
f. If a boat is moving at 30 knots, how fast is it going in statute miles per hour? As a percentage, how much faster or slower is 30 knots than 30 statute miles per hour?

24. **USCS Weights.** Make the following conversions within the U.S. customary system of measurement.
a. Convert your own weight in pounds to ounces using avoirdupois measures. How much do you weigh in tons?
b. A gallon of water weighs about 128 ounces. How many pounds is that?
c. You can send a letter overseas with a weight of up to 154 grains for the lowest postal rate. How many ounces is this in avoirdupois measures?

25. **USCS Volumes.** Make the following conversions within U.S. customary system of measurement.
a. Most soda cans contain 12 fluid ounces. How many cubic inches is that?
b. At many microbreweries, you can buy 1/2 gallon jugs of beer to take home. How many pints is this? How many British pints is it?
c. A large car's gas tank might hold 20 gallons. How many cubic inches is 20 gallons? How many barrels (of oil)?
d. Agricultural products such as corn and wheat often are traded in units of 5000 bushels. How many cubic inches does this quantity represent? If 150 million bushels of wheat are traded on one day, how many cubic inches is this? How many cubic feet is that? Describe, in words, the size of a building that would hold 5000 bushels of wheat.

26. **Metric Prefixes.** For each of the following, state how much larger or smaller the first unit is than the second. Remember to be especially careful if units are raised to powers (e.g., squared or cubed).
a. centimeter, millimeter
b. femtometer, nanometer
c. cubic meter, cubic centimeter
d. teragram, megagram
e. gigasecond, yottasecond
f. zeptosecond, attosecond
g. kilometer, micrometer
h. picogram, microgram
i. square millimeter, square kilometer
j. milliliter, deciliter
k. decaliter, centiliter
l. exaliter, yottaliter

27. **Comparing Units.** For each pair of measurements determine which is larger and state how much larger, as a percentage.
a. 1 kilometer, 1 mile.
b. 100 kilograms, 100 pounds.
c. 1 quart, 1 liter.
d. 1500 meters, 1 mile.

28. **USCS–SI Conversions.** Convert each measurement to the units specified.
a. 10 meters to feet.
b. 880 yards to kilometers.
c. 20 gallons to liters.
d. 5 milliliters to cubic inches.
e. 150 pounds to kilograms.
f. 1200 square feet to square meters.
g. 100 kilometers per hour to miles per hour.
h. 5.5 grams per cm^3 to pounds per cubic foot.
i. 25 miles per hour to kilometers per hour.
j. 105 centimeters to yards.

29. **Sensible or Ridiculous?** Determine whether each of the following statements is sensible or patently ridiculous. Explain why.

Example: I ate 2 meters of apples at lunch. *Solution:* The statement is ridiculous. A meter is a unit of length, so talking about "meters of apples" makes no sense.

Example: My brother is 4 meters tall. *Solution:* The statement is ridiculous. A meter is slightly longer than a yard, so 4 meters is slightly more than 12 feet; no one is 12 feet tall.
a. I drank 2 liters of water today.
b. A professional football player weighs 300 kilograms.
c. Bill drove along the interstate at 100 kilometers per hour.
d. Fred ran 35 liters per second.
e. The world record high jump for men is 7 meters.

f. Sue ran 10,000 meters in less than an hour.
g. The book I sent you weighs 3 milligrams.
h. An infant eats 2500 grams of food each day.

30. **Celsius–Kelvin Conversions.** In each of the following convert, as appropriate, Kelvin into Celsius or Celsius into Kelvin. State answers to the nearest degree.

a. 50 K	b. 240 K	c. 500,00 K
d. 10°C	e. 100°C	f. 320 K

31. **Celsius–Fahrenheit Conversions.** In each of the following convert, as appropriate, Fahrenheit into Celsius or Celsius into Fahrenheit. State answers to the nearest tenth of a degree.

a. 0°F	b. 200°C	c. 100°C
d. 10,000°C	e. 70°F	f. –273.15°C
g. 415°F	h. 15°C	i. 98.6°F
j. 0°C	k. –40°F	l. 37°C

Section 6.5

32. **Rounding Practice.** Round the following numbers to the nearest whole number:

a. 2.4	b. 14.500	c. 779.49	d. 13.500
e. 4.1999	f. 13.5001	g. 234.5	h. 1.099
i. 1999.5	j. 88.71		

33. **Converting Common Fractions to Decimals.** Convert the following common fractions to decimal form. If the decimal continues beyond the hundred thousandths column, round to the nearest hundred thousandth.

a. 3/4	b. 4/11	c. 5/8	d. 1/6
e. 4/7	f. 11/20	g. 37/60	h. 96/137

34. **More Rounding.** Round each of the following numbers to the nearest thousandth, tenth, ten, and hundred.

a. 2365.98521	b. 322354.09005	c. 6000
d. 34/3	e. 78.555	f. 0.45232768
g. –12.1	h. –850.7654	

35. **Real-World Rounding.** Find at least three different examples of numbers that are rounded in news stories. Explain how and why each number is rounded. (Hint: you might consider baseball batting averages, times in athletic events, or the stock market.)

Section 6.6

36. **Consecutive Primes?** Can two consecutive natural numbers be prime? Explain your answer.

37. **The Sieve of Eratosthenes.** Write the list of natural numbers between 1 and 100, and carry out the sieving process. Based on this process do the following.
    a. List the prime numbers less than 100.
    b. Use your list to identify the twin primes less than 100.
    c. You should have found that you needed to use sieve numbers only up to 10 before all of the work was done. That is, using 11, 12, or any higher number to find the primes on your list of numbers up to 100 isn't necessary. Why?
    d. Suppose that you used the sieve of Eratosthenes on a list of numbers up to 1000; how high would you need to go with the sieve numbers? What about on a list of numbers up to 10,000? Can you state a general rule?

38. **Number of Factors.** Make a table showing the number of factors possessed by each number less than 30 (e.g., 2 has two factors, 3 has two factors, 4 has three factors). Identify the primes as those numbers with only two factors. Is there a discernible pattern in the number of factors? Explain.

Section 6.7

39. **Cardinality of the Integers.** Describe a method by which the natural numbers can be matched one-to-one with the integers, thereby showing that the cardinalities of the natural numbers and the integers are equal.

40. **Cardinality of the Prime Numbers.** The prime numbers also form an infinite set. What is the cardinality of the set of prime numbers? Explain.

Further Topics and Applications

41. **Thinking About Numbers.** For each part that describes a particular use of number, do the following.
    (i) Identify the first set that could always describe this number: natural numbers (positive integers), negative integers, integers, rational numbers, irrational numbers, or real numbers.
    (ii) State whether the number is likely to be exact or approximate. Explain.

    Example: The balance in your checking account. **Solution:**
    (i) Rational numbers describe the account balance because it is expressed in dollars and fractions of dollars (cents).
    (ii) The balance is approximate, rounded to the nearest cent.

    a. The number of students in your English class.
    b. The number of people in Brazil.
    c. The temperature outside your kitchen.
    d. Your supply of chips in a poker game (you may not always be winning!).
    e. The fraction of students at your school who are women.
    f. The national debt.
    g. The rate at which your elevation changes after you jump from an airplane (with a parachute).

42. **Tourist Conversions.** Suppose that you are traveling in Europe. Make the following common conversions, assuming that the monetary exchange rates are

    1 British pound = $1.60
    1 French franc = $0.21
    1 German mark = $0.72
    1580 Italian lire = $1.00

    a. Convert 25 francs to dollars.
    b. Convert 17 marks to dollars.
    c. Convert a price of 4 francs per liter to dollars per gallon.
    d. Convert a price of 2 marks per kg to dollars per pound.
    e. Convert 2000 lire/kg to dollars per pound.
    f. Convert 0.5 British pounds per liter to dollars per gallon.
    g. Convert 20 francs to British pounds.

43. Repeating Decimals. Some rational numbers cannot be written exactly in decimal form. Instead, they are repeating decimals, in which a pattern repeats forever. For each common fraction shown:

(i) use your calculator to convert the fraction to decimal form;
(ii) identify the repeating pattern of digits; and
(iii) round the decimal form to the nearest thousandth.

a. 2/3	**b.** 1/7	**c.** 3/13	**d.** 32/6
e. 2/11	**f.** 7/9	**g.** 1/11	**h.** 7/15

44. Body Measurements. How tall are you? How much do you weigh? State your answers in SI units.

45. A Pint of Beer. A U.S. pint contains 16 fluid ounces, whereas a British pint contains 20 British fluid ounces. As a percentage, how much more beer do you get when you order a pint in England than in the United States?

46. Pacing a Mile. The word *mile* comes from the Latin words *milia passum,* which mean "one thousand paces." The Romans considered a pace to be two steps.

a. A statute mile is 5280 feet. Is 1000 paces per mile reasonable? If so, what kind of paces (e.g., little steps, running steps, etc.)?
b. The Romans considered a pace to have a length of 5 feet. How many feet were in a Roman mile? How does the Roman mile compare, by percentage, to a statute mile? to a nautical mile?

47. Mountains and Trenches.

a. Mt. Elbert, the tallest mountain in Colorado and also second-tallest in the continental United States, is 14,462 feet above sea level. How high is that in miles? in meters? in kilometers?
b. The tallest mountain in the world, Mt. Everest, rises 29,023 feet above sea level. How high is that in miles? in meters? in kilometers?
c. Mauna Kea, the highest mountain on the island of Hawaii, rises 13,796 ft above sea level. It extends an additional 18,200 ft from sea level to its base on the ocean floor. How tall is Mauna Kea from its base to its peak, in feet, miles, meters, and kilometers? Compare its total extent to the height of Mt. Everest. Would it be fair to call Mauna Kea the highest mountain in the world? Why or why not?
d. The deepest point in the oceans is a gorge called the *challenger deep* that lies within the Marianas trench in the western Pacific ocean. It reaches a depth of 36,201 feet. How deep is that in miles, meters, and kilometers? Compare the depths of the oceans to the heights of the tallest mountains on Earth.

48. The Metric Mile. In track and field, the 1500-meter race is sometimes called the *metric mile.*

a. Compare the metric mile to a statute mile. How much longer or shorter is it, by percentage?
b. Look up the current men's and women's world records for the (statute) mile. If you assume that the runners maintain the same pace for the metric mile, what should their times be for the metric mile?
c. Look up the current world records for the metric mile. Do they agree with the expected times you calculated in part (b)? If not, why do you think they differ?

49. Carats and Karats. A *carat* is a measure of weight: 1 carat = 0.2 gram. A *karat* is a measure of the purity of gold: Pure gold is defined to be 24 karats; a mixture containing 50% gold is 12-karat gold; a mixture containing 10% gold is 2.4 karats; and so on. Use the definitions of carat and karat to answer the following questions.

a. If you find a nugget that is 75% gold, what is its purity in karats?
b. Suppose that you purchase a 14-karat gold chain that weighs 15 grams. How much gold have you purchased?
c. Find the price of gold today and use it to calculate the value of the gold chain from part (b).
d. Suppose that you have a 4-gram pendant made of gold that is 50% pure. Explain each of the following facts about the pendant: (i) the pendant weighs 20 carats; (ii) it is made of 12-karat gold; (iii) it contains 10 carats of gold.
e. Is it possible to have jewelry made of 30-karat gold? Why or why not? Is it possible to have a 30-carat gold nugget? Explain.
f. Diamonds are sold according to their weight in carats. How much does a 23-carat diamond weigh in grams?
g. Can diamonds be sold in units of karats? Why or why not?

50. The Cullinan Diamond and the Star of Africa. The largest single rough diamond ever found, the Cullinan diamond, weighed 3106 carats; it was used to cut the world's largest diamond gem, the Star of Africa (530.2 carats), which is part of the British crown jewels collection. How much did the Cullinan diamond weigh in milligrams? in pounds? How much does the Star of Africa weigh in milligrams? in pounds?

51. Metric Tools. Many tools come in both USCS and SI standards. In a standard socket set, the smallest USCS subdivision is $\frac{1}{16}$ inch; the smallest SI subdivision is 0.5 millimeters. Which of these subdivisions is smaller? Suggest some advantages and disadvantages of using these two standards.

52. Temperature Formulas. Starting from the conversion formula from Fahrenheit to Celsius,

$$\text{Temperature } (°\text{C}) = \frac{[\text{Temperature } (°\text{F})] - 32°\text{F}}{1.8\dfrac{°\text{F}}{°\text{C}}},$$

find the conversion formula for temperature in Celsius to temperature in Fahrenheit. Show all the required steps clearly.

53. The Goldbach Conjecture. The Goldbach conjecture states that every even number (except 2) can be expressed as the sum of two primes (see subsection 2.5.2).

a. Show that this conjecture holds for all the even numbers from 4 to 20.
b. Would the conjecture hold if you changed the word *even* to *composite*? Explain.

54. **A Divisibility Trick.** Pick a natural number between 1 and 9. Multiply your number by 9. Add the two digits of your number together. Subtract 5 from this number. Now match this number with a letter of the alphabet (1 ⇒ A, 2 ⇒ B, 3 ⇒ C, etc.). Think of the name of a country that starts with this letter. Think of the name of an animal that starts with the second letter of this country's name. Think of a color associated with this animal. You should now have a country, an animal, and a color. Are you thinking of a gray elephant from Denmark? Explain. (Try this trick on a large group of people; you will be surprised how often it works.)

Projects

55. **The Price of Gasoline.** In September 1993, the price of gasoline in England was 50 pence per liter.
   a. In September 1993, one U.S. dollar was worth 0.690 British pounds (the British pound, or pound sterling, is the basic unit of British currency; its symbol is £). Note that there are 100 pence in 1£. Use this information to find a conversion factor from pence to dollars.
   b. Find a conversion factor to convert the price of gasoline from pence per liter to dollars per gallon.
   c. Use the conversion factor from part (b) to find the price of gasoline in England (September 1993) in dollars per gallon.
   d. In September 1993, the price of gasoline in the United States was about $1.20 per gallon. Compare this price to the price at the same time in England. Identify possible reasons for the price difference.
   e. What is the approximate price of a gallon of gas in the United States today? What is the current exchange rate between dollars and British pounds? (You should be able to obtain both pieces of information easily; exchange rates are published daily in many newspapers.) Convert the current price of gasoline in the United States to pence per liter.
   f. Find the current price of gasoline in England. How does it compare to the current price of gasoline in the United States? Have the comparative prices changed since September 1993? If so, explain the change and suggest possible explanations.
   g. Do enough research to find the real reasons for the difference in fuel prices between the United States and Great Britain. Also compare gasoline prices in other European countries, Japan, and Canada. Write a short essay explaining the reasons for those differences. Conclude your essay by stating your own opinion about whether the United States, overall, is benefiting or suffering because of its low gasoline prices.

56. **Making Metric Products.** By going metric, the standard sizes of many products might need modification to simplify their presentation. For example, a standard 12-fluid-ounce soda can contains 355 milliliters of liquid. Producers might want to change this volume to 350 or 400 milliliters.
   a. Suppose that producers decided to change the volume of soda cans to either 350 or 400 milliliters. How much more or less would these new cans hold than a 12-fluid-ounce can? Give your answers as percentages.
   b. Choose two other products that are sold in volumes that are whole numbers in U.S. customary units. How might you suggest the sizes be changed to make the metric units simpler but still be comparable in size?
   c. Should companies change the sizes of their products when the United States completely converts to metric? Why or why not?
   d. Research product sizes used in other countries. For example, how much do cans of soda hold in other countries? Write a short summary of your findings.

57. **Going Metric?** Do some further research and write a short paper on some aspect of the history of the metric system in the United States. For example, you might write on why Congress didn't act on Jefferson's proposal to adopt the metric system in 1790, the role of the United States in the adoption of the Treaty of the Meter, or the recent history of metric conversion in the United States.

58. **Accounting Ledgers.** The "books" that small businesses use to keep track of finances provide an excellent example of the practical use of simple arithmetic. Find a small business that is willing to let you look at some of its accounting ledgers. Learn how the financial data is entered and interpreted. Then carefully go through the portions of their books provided and check for errors or omissions. Write a short summary of your efforts. The business may thank you for helping them with their accounting!

59. **More Transfinite Arithmetic.** $\aleph_0$ is called a *countable* infinity because members of a set with cardinality $\aleph_0$ can be counted with the natural numbers. In contrast, $\aleph_1$ is *uncountable* because members of a set with cardinality $\aleph_1$ cannot be put into one-to-one correspondence with the natural numbers. The real numbers, with cardinality $\aleph_1$, are said to form a *continuum* because the real numbers encompass every point on the number line; that is, the number line has no gaps in the real numbers. In contrast, the set of points on the number line that forms the rational numbers isn't continuous: Between any two rational numbers a gap always exists in which only irrational numbers lie.
   a. In your own words, explain *infinitely many* irrational numbers must lie between any two rational numbers on the number line. Then explain why this condition means that, in total, infinitely more irrational numbers than rational numbers exist.
   b. Explain why the cardinality of the irrational numbers is the same as the cardinality of the real numbers. Use this argument to prove that $\aleph_1 + \aleph_1 = \aleph_1$.
   c. The real numbers include both the rational and irrational numbers. Use this definition to explain the following result of transfinite arithmetic: $\aleph_1 + \aleph_0 = \aleph_1$.
   d. Explain what you expect from the following operations in transfinite arithmetic: $\aleph_1 \times \aleph_1$ and $\aleph_1 \times \aleph_0$.
   e. Do some further reading or research on transfinite arithmetic. Find an example of a set that has cardinality $\aleph_2$ and explain its cardinality.

7 NUMBERS LARGE AND SMALL

In ancient times, people had no way to express extremely large and small numbers; in fact, doing so was unnecessary. Today, however, life is filled with numbers that may at first seem incomprehensible: a population measured in billions of people, a national debt measured in trillions of dollars, and distances ranging from nanometers to light-years. In this chapter, we continue to focus on thinking quantitatively by developing methods for interpreting such numbers. We hope that, by the end of this chapter, numbers that once seemed incomprehensibly large or small will have new meaning for you, and that you will be confident in dealing with such numbers wherever you encounter them.

CHAPTER 7 PREVIEW

7.1 ***PUTTING NUMBERS IN PERSPECTIVE.*** We begin the chapter by explaining briefly why putting numbers in perspective is important.

7.2 ***WRITING LARGE AND SMALL NUMBERS.*** Dealing with large and small numbers is much easier with a special (scientific) notation, involving powers of 10, which we cover in this section.

7.3 ***ESTIMATION.*** One of the best ways to put large or small numbers in perspective is by comparing them to quantities that are easier to understand. Estimation, which we discuss in this section, helps in making reasonable comparisons.

7.4 ***SCALING AND SCALING LAWS.*** Another method of putting numbers in perspective is through scaling laws, which we use in this section to illuminate ideas from various fields including biology and astronomy.

7.5 ***A FEW CASE STUDIES.*** Ultimately, the only way to learn to put numbers in perspective is through practice. We therefore conclude the chapter with a set of case studies involving important numbers from everyday life.

John F. Kennedy told a story about the famous French soldier, Marshall Lyautey. One day the Marshall asked his gardener to plant a row of trees of a certain rare variety in his garden the next morning. The gardener said he would gladly do so, but he cautioned the Marshall that trees of this kind take a century to grow to full size. "In that case," replied Lyautey, "plant them this afternoon."

— Douglas Hofstadter, *Scientific American*, May 1982

A billion here, a billion there; soon you're talking real money.

— Senator Everett Dirksen

7.1 PUTTING NUMBERS IN PERSPECTIVE

At the polling booth, you choose among candidates who talk about spending and taxation in units of tens or hundreds of billions of dollars. Do you understand what these candidates are saying?

Some of the most important policy issues of our time involve human impact on the environment, that is, the collective impact of almost six billion people. Do you understand the meaning of a world population of that size or, for that matter, a population that is growing by almost 100 million people each year?

Despite the reduction in global tensions since the end of the Cold War, thousands of nuclear weapons still exist. Do you know what a nuclear weapon with one megaton of explosive power means?

To make sense of a large or small number, we must relate it to something with which we already are familiar.

At a more personal level, the purchase of a computer involves choosing among models featuring megabytes or gigabytes of memory and processing times measured in nanoseconds, microseconds, or milliseconds. Can you realistically assess their relative values?

As these questions show, survival and prosperity in the modern world depend on decisions that involve numbers that may, at first, seem incomprehensibly large or small. To make wise decisions, you must find ways of putting such numbers into perspective. That is the task in this chapter: learning how to make extremely large or small numbers comprehensible by relating them to numbers with which you already are familiar.

7.2 WRITING LARGE AND SMALL NUMBERS

Consider the numbers in the following statements.

- The diameter of the Galaxy is about 1,000,000,000,000,000,000 kilometers.
- The nucleus of a hydrogen atom has a diameter of about 0.000000000000001 meters.
- The U.S. federal debt is about $5,000,000,000,000.

The many zeros make the numbers so difficult to read that most people simply skip right over them. Thus one of the most difficult things about extremely large or small numbers is simply finding a way to say or write them in a meaningful way.

Fortunately, a better way of expressing them is available. Note how much easier the preceding statements are to read when rewritten using powers of 10.

- The diameter of the Galaxy is about 10^{18} kilometers.
- The nucleus of a hydrogen atom has a diameter of about 10^{-15} meters.
- The U.S. federal debt is about 5×10^{12} dollars.

In this section, we demonstrate how to write numbers by using powers of 10, along with the advantages and disadvantages of this technique.

7.2.1 Powers of 10

Before getting into details, recall that powers of 10 simply indicate how many times to multiply by 10. For example,

$$10^2 = 10 \times 10 = 100, \text{ and } 10^6 = 10 \times 10 \times 10 \times 10 \times 10 \times 10 = 1{,}000{,}000.$$

Negative powers are the reciprocal of the corresponding positive powers. For example,

$$10^{-2} = \frac{1}{10^2} = \frac{1}{100} = 0.01, \quad \text{and} \quad 10^{-6} = \frac{1}{10^6} = \frac{1}{1,000,000} = 0.000001.$$

Table 7-1 lists powers of 10, along with their values and names, for powers between –12 and 12. Note that powers of 10 follow two basic rules.

A positive exponent tells how many 0s follow the 1.

A negative exponent tells how many places are to the right of the decimal point; all places are filled with 0s except the last, which is a 1.

1. A positive exponent tells how many 0s follow the 1. For example, 10^0 is a 1 followed by no 0s, and 10^8 is a 1 followed by eight 0s.
2. A negative exponent tells how many places are to the right of the decimal point; all places are filled with 0s except the last, which is a 1. For example, 10^{-1} has just one place to the right of the decimal point, so it is written as 0.1; 10^{-6} has six places to the right of the decimal, so it is written as 0.000001 (the six places contain five 0s and the 1).

Table 7-1. Powers of 10

Zero and Positive Powers			Negative Powers		
Power	*Value*	*Name*	*Power*	*Value*	*Name*
10^0	1	One			
10^1	10	Ten	10^{-1}	0.1	Tenth
10^2	100	Hundred	10^{-2}	0.01	Hundredth
10^3	1000	Thousand	10^{-3}	0.001	Thousandth
10^4	10,000	Ten thousand	10^{-4}	0.0001	Ten thousandth
10^5	100,000	Hundred thousand	10^{-5}	0.00001	Hundred thousandth
10^6	1,000,000	Million	10^{-6}	0.000001	Millionth
10^7	10,000,000	Ten million	10^{-7}	0.0000001	Ten millionth
10^8	100,000,000	Hundred million	10^{-8}	0.00000001	Hundred millionth
10^9	1,000,000,000	Billion	10^{-9}	0.000000001	Billionth
10^{10}	10,000,000,000	Ten billion	10^{-10}	0.0000000001	Ten billionth
10^{11}	100,000,000,000	Hundred billion	10^{-11}	0.00000000001	Hundred billionth
10^{12}	1,000,000,000,000	Trillion	10^{-12}	0.000000000001	Trillionth

Multiplying and Dividing Powers of 10

Multiplying and dividing powers of 10 simply requires adding exponents (multiplication) or subtracting exponents (division). A few examples should clarify this process:

To multiply powers of 10, *add* exponents:
$$10^n \times 10^m = 10^{n+m}.$$

$$10^3 \times 10^2 = \underbrace{1000}_{10^3} \times \underbrace{100}_{10^2} = \underbrace{100,000}_{10^5}$$ (add exponents $10^3 \times 10^2 = 10^5$);

$$10^4 \times 10^7 = \underbrace{10,000}_{10^4} \times \underbrace{10,000,000}_{10^7} = \underbrace{100,000,000,000}_{10^{11}}$$ (add exponents $10^4 \times 10^7 = 10^{11}$);

$$10^5 \times 10^{-3} = \underbrace{100,000}_{10^5} \times \underbrace{0.001}_{10^{-3}} = \underbrace{100}_{10^2}$$ (add exponents $10^5 \times 10^{-3} = 10^2$);

$$\frac{10^5}{10^3} = \underbrace{100,000}_{10^5} \div \underbrace{1000}_{10^3} = \frac{100,000}{1000} = \underbrace{100}_{10^2}$$ (subtract exponents $10^5 \times 10^3 = 10^2$);

To divide powers of 10, *subtract* exponents:
$10^n \div 10^m = 10^{n-m}$.

$$\frac{10^3}{10^7} = \underbrace{1000}_{10^3} \div \underbrace{10,000,000}_{10^7} = \frac{1000}{10,000,000} = \underbrace{0.0001}_{10^{-4}}$$ (subtract exponents $10^3 \times 10^7 = 10^{-4}$).

We can generalize these rules using n and m to represent any numbers.

- To multiply powers of 10, *add* exponents: $10^n \times 10^m = 10^{n+m}$.
- To divide powers of 10, *subtract* exponents: $10^n \div 10^m = 10^{n-m}$.

Time-Out to Think: Use the rules for multiplication and division to confirm that $10^{-4} \times 10^9 = 10^5$ and $10^{-18} \div 10^{-20} = 100$.

Adding and Subtracting Powers of 10

Unlike multiplication and division, no shortcut is available for adding or subtracting powers of 10. The values must be written in longhand notation. For example,

$10^6 + 10^2 = 1,000,000 + 100 = 1,000,100$;

$10^8 + 10^{-3} = 100,000,000 + 0.001 = 100,000,000.001$;

$10^7 - 10^3 = 10,000,000 - 1000 = 9,999,000$.

Time-Out to Think: Suppose that someone told you that $10^6 + 10^2 \approx 10^6$. Although this answer looks strange — it has completely ignored the second term in the summation — is it a good approximation to the correct answer? Why? What can you say in general about adding small powers of 10 to large powers of 10?

7.2.2 Scientific Notation

A number is in scientific notation when it is written in the form
$a \times 10^n$,
where a must be *between* 1 and 10 and n must be an integer.

The number four hundred million can be written as 400×10^6, 4×10^8, 40×10^7, 0.4×10^9, or in many other ways. All these expressions are valid, but only 4×10^8 clearly reveals that the number involves *hundred millions*. Similarly, 5.6×10^4 immediately shows that the number involves *ten thousands* (56,000), as does 3.2×10^{-3} for *thousandths* (0.0032). Numbers written with a number *between* 1 and 10 multiplied by a power of 10, are said to be in **scientific notation.**

Despite its name, scientific notation is useful in many nonscientific applications. Indeed, scientific notation is so useful that you should think of the word *science* according to its Latin root, which means "knowledge." Proficiency in scientific notation makes knowledge about numbers easier to acquire.

A number written in scientific notation can be quickly converted to ordinary notation by remembering the following simple rules.

- The power of 10 indicates how many places to move the decimal point; move it to the *right* if the power of 10 is positive and to the *left* if it is negative.
- If moving the decimal point creates any open places, fill them with 0s.

For example, note the following conversions.

$$4.01 \times 10^2 \xrightarrow[\text{2 places to right}]{\text{move decimal}} 401$$

$$3.6 \times 10^6 \xrightarrow[\text{6 places to right}]{\text{move decimal}} 3,600,000$$

$$5.7 \times 10^{-3} \xrightarrow[\text{3 places to right}]{\text{move decimal}} 0.0057$$

Converting a number into scientific notation simply involves the reverse process.

- Move the decimal point so that it appears after the *first* nonzero digit.
- The number of places the decimal point moves tells you the power of 10; the power is *positive* if the decimal point moves to the left and *negative* if it moves to the right.

Note the conversions.

$$3042 \xrightarrow[\text{3 places to left}]{\text{decimal needs to move}} 3.042 \times 10^3$$

$$0.00012 \xrightarrow[\text{4 places to right}]{\text{decimal needs to move}} 1.2 \times 10^{-4}$$

$$226 \times 10^2 \xrightarrow[\text{2 places to left}]{\text{decimal needs to move}} \left(2.26 \times 10^2\right) \times 10^2 = 2.26 \times 10^4$$

Multiplying or dividing numbers given in scientific notation also is straightforward: Simply operate on the powers of 10 and the other parts of the number separately. For example,

$$\left(6 \times 10^2\right) \times \left(4 \times 10^5\right) = (6 \times 4) \times \left(10^2 \times 10^5\right) = 24 \times 10^7 = \left(2.4 \times 10^1\right) \times 10^7 = 2.4 \times 10^8$$

$$\frac{4.2 \times 10^{-2}}{8.4 \times 10^{-5}} = \frac{4.2}{8.4} \times \frac{10^{-2}}{10^{-5}} = \frac{1}{2} \times 10^{-2-(-5)} = 0.5 \times 10^3 = \left(5 \times 10^{-1}\right) \times 10^3 = 5 \times 10^2.$$

Note that, in both these examples, we first got a leading number that was *not* between 1 and 10. We therefore followed the earlier rules to put the final answer in scientific notation.

Addition and Subtraction with Scientific Notation

As with powers of 10, scientific notation offers no shortcuts to addition and subtraction. In general, you must write numbers in longhand notation before adding or subtracting. For example:

$(3 \times 10^6) + (5 \times 10^2) = 3,000,000 + 500 = 3,000,500 = 3.0005 \times 10^6$;

$(4.6 \times 10^9) - (5 \times 10^8) = 4,600,000,000 - 500,000,000 = 4,100,000,000 = 4.1 \times 10^9$.

The only exception occurs when both numbers have the same power of 10; in that case, we can add or subtract directly. For example:

$$(7 \times 10^{10}) + (4 \times 10^{10}) = (7 + 4) \times 10^{10} = 11 \times 10^{10} = 1.1 \times 10^{11};$$

$$(2.3 \times 10^{-22}) - (1.6 \times 10^{-22}) = (2.3 - 1.6) \times 10^{-22} = 0.7 \times 10^{-22} = 7.0 \times 10^{-23}.$$

Scientific Notation on a Calculator

Further details of how to use scientific notation on your calculator are presented in Appendix D.

Most calculators have a special key for entering scientific notation. The key, usually labeled either *exp* (short for "exponent") or *EE* (short for "enter exponent"), is used to enter a power of 10. For example, to enter the number 3.5×10^6 use the key sequence

<3.5> <EE> <6>.

Think of the *EE* (or *exp*) as "times 10 to the power that I enter next." Note that you must be especially careful in entering "pure" powers of 10. For example, to enter the number 10^5 you would press

<1> <EE> <5>.

> **Time-Out to Think:** Find the key on *your* calculator for entering scientific notation. Try entering the number 10^5. What happens if you use the *EE* key without first entering 1? Another common mistake is to press <10> <EE> <5> instead of <1> <EE> <5>; why is that incorrect? To be sure you know how to use your calculator with scientific notation, check each of the examples in this subsection.

7.2.3 Advantages and "Dangers" of Scientific Notation

Rounding and expressing numbers in scientific notation allow quick approximations to the exact answers.

We have demonstrated that scientific notation simplifies writing extremely large and small numbers. A second advantage of scientific notation is that it lets you *approximate* answers "in your head" easily.

For example, consider multiplying 5795×326 in your head — a difficult task for most people. However, if you round 5795 to one decimal place it becomes 6000, or 6×10^3; similarly, 326 rounds to 300, or 3×10^2. Now, you can easily multiply $(6 \times 10^3) \times (3 \times 10^2)$ to get 18×10^5, or 1.8×10^6. Because mistakes can occur in calculator work, this approximation technique provides a useful check on your results.

> **Time-Out to Think:** Use this approximation technique to estimate the answers to 3962×0.17; $4{,}785{,}378 \times 3{,}997{,}321$; and $0.0022 \div 0.00019$. In each case, compare your estimate to the exact answer obtained on your calculator.

Despite its advantages, working with scientific notation has one potential danger: It makes extremely large or small numbers deceptively easy to write. In particular, numbers that hardly *look* any different on paper can have vastly different meanings. For example, 10^{26} doesn't look much different from, say, 10^{20}, but it is *a million times* larger (10^{26} is 10^6 times larger than 10^{20}). Or consider the number 10^{80}, which doesn't look so incredibly large on paper — yet it is a number larger than the total number of atoms in the entire universe!

7.3 ESTIMATION

The actual height of the Empire State Building is about 1250 feet; including antennas on its top its height is about 1470 feet.

How high is 1000 feet? For most people, the quantity "1000 feet" has little meaning by itself. However, if we say that 1000 feet is about the height of a 100-story building (e.g., the Empire State Building), it becomes much easier to visualize. In this section, we discuss methods of estimation, which will help you put large and small numbers into perspective.

7.3.1 The Process of Estimation

How did we know that 1000 feet is about the height of a 100-story building? In this case, we simply recognize that a story is typically about 10 feet from floor to ceiling. Thus a 100-story building has a height of about

$$(100 \text{ stories}) \times (10 \text{ ft/story}) = 1000 \text{ ft.}$$

Like problem solving, estimation is more an art than a science. Indeed, the following basic estimation process is closely related to the problem–solving process described in Section 4.3.

1. *Decide what you want to estimate, and why.* We chose to estimate the height of a 100-story building as a way of putting 1000 feet in perspective. In other cases, you may have other reasons for making estimates. Nevertheless, deciding what you plan to estimate and understanding why you are doing it may be the most important step in the estimation process.
2. *Devise a strategy for making the estimate.* In estimating the height of the building, the strategy required knowing the number of stories and the typical height of each story. Thus the strategy itself involved making an estimate: the typical height of a story.
3. *Carry out the strategy and revise it if necessary.* The building estimate simply involved multiplication. More steps may be required in other cases.
4. *Look back to check, interpret, and explain your estimate.* As with any form of problem solving, looking back and being sure that you understand the meaning and implications of your result is essential.

Simple estimates, such as educated guesses at things like the height of each story in a building or the number of drinks that students consume each week, sometimes are called *back of the envelope* calculations. This expression reflects the idea that with a small amount of writing space, such as the back of an envelope, you can do calculations that may provide insights into such things as the height of a building, the amount of money spent on alcohol, or the laws of nature.

Example 7-1 ***Student Spending on Alcoholic Beverages.*** Describe the amount of money spent on alcoholic beverages by students at a large university.

Solution: Step 1. in this case, estimating the total amount of money spent on alcohol by the students *each year* seems to be a reasonable approach.

Step 2. A strategy for making the estimate is needed. One method is to estimate the average amount spent per student and multiply by the number of students. That is,

$$\text{total spending} = (\text{spending per student}) \times (\text{number of students}).$$

You now need to know each of these other numbers. For a large university, simply choose a reasonable number of students, say, 25,000. However, finding a reasonable value for the typical spending per student requires further thought. For example, you might estimate spending per student from the typical number of drinks per week, the typical cost of a drink, and the number of weeks in a year. The following equation shows the product; note the units:

Recall that units of "per student per year" mean *divided by* students and *divided by* years. Thus per student per year means 1 ÷ (student × year).

$$\underbrace{\text{annual spending per student}}_{\frac{\$}{\text{student}\times\text{year}}} = \underbrace{\begin{array}{c}\text{number of drinks}\\ \text{per student per week}\end{array}}_{\frac{\text{drinks}}{\text{student}\times\text{week}}} \times \underbrace{\begin{array}{c}\text{cost per}\\ \text{drink}\end{array}}_{\frac{\$}{\text{drink}}} \times \underbrace{\begin{array}{c}\text{52 weeks}\\ \text{per year}\end{array}}_{\frac{\text{weeks}}{\text{year}}}.$$

Step 3. One way to estimate the number of drinks per student per week is by taking a poll of friends or a class. Suppose that the average student (accounting for both heavy drinkers and nondrinkers) has about five alcoholic drinks per week. A survey of local bars indicates that the typical cost of a drink is \$2. You can now make the estimate:

$$\text{annual spending per student} = \frac{5\text{ drinks}}{\text{student} \times \text{week}} \times \frac{\$2}{\text{drink}} \times \frac{52\text{ weeks}}{\text{year}} = \frac{\$520}{\text{student} \times \text{year}};$$

$$\text{total annual spending} = \frac{\$520}{\text{student} \times \text{year}} \times (25{,}000\text{ students}) = \frac{\$1.3 \times 10^7}{\text{year}}.$$

Step 4. Your estimate is that students at a large university probably spend about $\$1.3 \times 10^7$, or \$13 million, on alcoholic beverages each year. Of course, a fair amount of uncertainty is involved in that result. If students drink more or less than you assumed or if the price of a drink is much different from your finding, the estimate could be significantly off. Nevertheless, a fair conclusion is that students at that university spend about ten million dollars annually on alcoholic beverages. This conclusion certainly meets the goal of describing this amount of money.

❖

Time-Out to Think: How many students attend your school? How much do you think typical students at your school spend on alcoholic beverages? Do a quick estimate of the amount of money spent each year on alcoholic beverages by students at your school. Does the result surprise you?

Example 7-2 ***The New York Marathon.*** Put the total amount of running done by participants in the New York Marathon in perspective.

Solution: The first step is to choose one of the many ways of answering this question. Let's estimate the total mileage run by marathon participants (in a single race). That will allow comparison to other known distances, which should provide some perspective.

If you know how many people run the race, you can multiply by the distance covered by each runner to find the total distance run, or

$$\text{total distance} = (\text{distance per runner}) \times (\text{number of runners}).$$

Thus the first part of the strategy involves knowing the marathon distance of about 42 kilometers (26 miles) and an estimate of the number of runners in the race. Newspaper accounts of the New York Marathon relate that, typically, about 20,000 runners finish the race each year. Substituting these numbers and units into the preceding equation yields

$$\text{total distance} = \frac{42\text{ km}}{\text{runner}} \times \left(2 \times 10^4\text{ runners}\right) = 84 \times 10^4\text{ km} = 8.4 \times 10^5\text{ km}.$$

The total distance covered by the 20,000 runners is about 840,000 kilometers. Comparing this distance to the circumference of the Earth (about 40,000 kilometers) gives

$$\frac{\text{total distance}}{\text{circumference of Earth}} = \frac{8.4 \times 10^5\text{ km}}{4 \times 10^4\text{ km}} = 2.1 \times 10^1 = 21.$$

In other words, run as a relay, the marathon runners could circle the Earth approximately 21 times. This outcome might lead to further thoughts on putting the race in perspective; for example, how long would running such a relay take? If the runners take an average of 4 hours to finish the marathon, running the relay would take

$$\text{total time} = \frac{4\text{ hr}}{\text{runner}} \times \left(2 \times 10^4 \text{ runners}\right) = 8 \times 10^4 \text{ hr}.$$

Converting from hours to years yields

$$8 \times 10^4 \text{ hr} \times \left(\frac{1\text{ day}}{24\text{ hr}}\right) \times \left(\frac{1\text{ yr}}{365\text{ day}}\right) = 9.1\text{ yr}.$$

In summary, the New York Marathon run as a relay would circle the Earth 21 times and take more than nine years to complete! Of course, the estimates of the number of runners and the average time to complete the race introduce some uncertainty into the results; nevertheless, we have succeeded in putting the race in perspective.

❖

7.3.2 Order of Magnitude Estimation

In Example 7-1, we estimated that the amount of money spent on alcoholic beverages by students at a large university is about $10 million. Because of the uncertainties in the assumptions used, the actual number could be substantially different: At a school where students consume half as much alcohol, the amount would be half as much; if drinks cost twice as much as estimated the entire amount would be twice as large. Nevertheless, the number $10 million provides a "feel" for the typical amount of money spent, even if it may be off by millions of dollars!

Estimates that involve large uncertainties, but still give a feel for their size, are called **order of magnitude** estimates. An order of magnitude is a power of 10. For example,

- 100 is an order of magnitude larger than 10;
- 10,000 is two orders of magnitude (10^2, or 100 times) larger than 100; and
- 10^{23} is five orders of magnitude (10^5, or 100,000 times) larger than 10^{18}.

The order of magnitude of a number can tell you a lot. Knowing that the amount of money spent on alcoholic beverages is on the order of $10 million, rather than on the order of $100,000 or $100 million, makes a big difference. Similarly, suppose that you wanted to tell a distant friend about the size of the town or city in which you live. Towns and cities range in population from tiny ones with only a few residents to megalopolises with millions of people. Simply stating the order of magnitude of the population of your city will tell your friend a lot. A city with a population on the order of 10,000 will have a much different character than one with a population on the order of 1 million.

Or, consider the age of the universe. Until this century, scholarly guesses about the age of the universe ranged from just a few thousand years to infinity! Scientists now estimate that it is on the order of 10 billion years — a significant advance in human knowledge, even though the actual age may lie somewhere in the range of 7 to 20 billion years.

Aside from the fact that order of magnitude estimates require less precision, the process of making them is just like that for any other estimate. To illustrate the method, let's make a quick order of magnitude estimate of the total amount of money spent on ice cream in the United States each year. We simply need to multiply the population of the United States by the average per capita spending on ice cream, or

total spending = (population of United States) × (average spending per person).

The U.S. population is available from many sources; it is about 250 million. Determining the average annual spending on ice cream is much more difficult. Because we are concerned only with an order of magnitude, however, we seek a reasonable "guess" by asking: Is \$1 per person per year a reasonable estimate? Since a typical ice cream cone costs about \$1, that would mean that the average American has ice cream only once a year, which seems too low. Next, we try \$10 per person; does this amount seem reasonable? That would mean getting an ice cream cone about 10 times per year, or almost once a month. How about \$100 per person per year? This amount would mean getting ice cream about 100 times a year, or almost once every three days, which seems too high. Thus \$10 per person per year seems like a reasonable order of magnitude estimate. Multiplying by the population of the United States, we find that

$$\text{total spending} = (2.5 \times 10^8 \text{ persons}) \times \frac{\$10}{\text{person} \times \text{yr}} = \$2.5 \times 10^9 \text{ per year.}$$

That is, annual spending on ice cream in the United States is on the order of \$2 billion per year. As with any order of magnitude estimate, it could be off by as much as a factor of 10 either way. But at least we know that the spending is on the order of billions of dollars, not millions or trillions.

Example 7-3 Make an order of magnitude estimate of the total value of all the "stuff" (e.g., clothing, books, compact disks, televisions, and cars) — except real estate — owned by a typical adult in the United States.

Solution: You are concerned only with orders of magnitude, so you might begin with a guess of \$100. Clearly, this amount seems too small because the value of the clothes you are wearing and the textbooks you are using is likely to be close to \$100. Next, you could try \$1000, which seems more reasonable. However, most adults own a car, and the average value of a car is several thousand dollars; thus \$1000 also seems too small for the total value of an adult's stuff. Skip to \$100,000, which seems too large because few people own \$100,000 worth of stuff (not including real estate). As \$1000 seems too small, and \$100,000 too big, you could reasonably settle on \$10,000 as an *order of magnitude* estimate of the value of an average person's "stuff."

❖

The Number of Atoms in the Human Body

Another good example of order of magnitude estimation is the number of atoms in the human body. In devising an appropriate strategy, we need to recall a few facts from chemistry.

- The human body is made mostly of water, H_2O.
- The atomic weight of hydrogen is 1, and the atomic weight of oxygen is 16. The total of the atomic weights for the three atoms in a molecule of water, H_2O, therefore is 18. Thus the *average* atomic weight of the three atoms in water is 6.
- The average atomic weight of the atoms in water, 6, reminds us that 6 grams of water contains 6×10^{23} (**Avogadro's number**) atoms.

Avogadro's number, 6×10^{23}, relates the number of atoms in an object to their atomic weight. For example, because the atomic weight of hydrogen is 1, one gram of hydrogen contains 6×10^{23} atoms. Similarly, because the atomic weight of carbon is 12, twelve grams of carbon contains 6×10^{23} atoms.

Because 6 grams of water contains 6×10^{23} atoms, 1 gram of water contains 1×10^{23} atoms. Because a kilogram is 10^3 grams, 1 kilogram of water contains $10^{23} \times 10^3 = 10^{26}$ atoms. As the human body is made mostly of water, we can reasonably assume that each kilogram of the human body contains about 10^{26} atoms.

How much does a person weigh? A relatively small child weighs 10 kilograms, or about 22 pounds. A relatively large adult weights 100 kilograms, or about 220 pounds. For an order of magnitude estimate, we are safe in assuming that "a person" weighs on the order of 10 to 100 kilograms:

Number of atoms in human body = $(10^{26}$ atoms/kg$) \times (10$ to 100 kg$) = 10^{27}$ to 10^{28} atoms.

That is, the human body contains on the order of 10^{27} to 10^{28} atoms.

Time-Out to Think: Take a moment to consider the remarkable nature of this last result. Combining a few simple facts from high school chemistry with logical thinking about estimation, we roughly determined the number of atoms in the human body!

Example 7-4 Approximately how many atoms are there among all living humans?

Solution: Currently the world population is about 6 billion people. To an order of magnitude that is 10 billion, or 10^{10}, people. Based on the estimate of 10^{27} to 10^{28} atoms in the human body, the number of atoms in all living humans is

total atoms = $(10^{27}$ to 10^{28} atoms/person$) \times (10^{10}$ persons$) = 10^{37}$ to 10^{38} atoms.

That is, the total number of atoms in all living humans is on the order of 10^{37} to 10^{38}.

❖

7.4 SCALING AND SCALING LAWS

Another technique for giving meaning to numbers that might otherwise seem incomprehensible is through the process of scaling: comparing numbers to other things that you already know, or creating maps or scale models. In addition, scaling has many other uses, which we briefly examine in this section.

7.4.1 Scale Factors

You probably are familiar with ideas of scaling from reading maps and working with scale models. Scales usually are expressed in one of three ways.

- *Verbally:* A scale can be described in words such as "one centimeter on the map represents one kilometer on the ground." Sometimes, this description is written simply as 1 cm = 1 km.
- *Graphically:* A scale can be shown with a "miniruler" marked to show the represented distances (Figure 7-1).
- *As Fractions:* Because a kilometer is 100,000 larger than a centimeter (there are 100 centimeters in a meter and 1000 meters in a kilometer), the scale at which 1 centimeter represents 1 kilometer can be written "1 to 100,000," or as the fraction 1/100,000. In this case 100,000 is the **scale factor.**

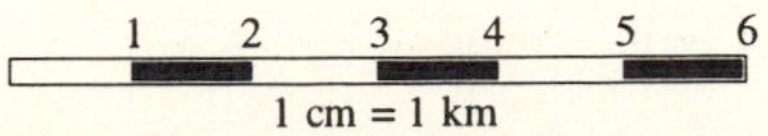

Figure 7-1. According to this map scale, one centimeter on the map equals one kilometer on the ground.

The scale factor shows directly how distances on the scale compare to actual distances. For example, a scale factor of 100,000 on a map means that actual distances are 100,000 times larger than distances on the map. That is,

$$\text{scale factor} = \frac{\text{actual distance}}{\text{map distance}}.$$

Example 7-5 Suppose that you have a map with a scale expressed as "one inch on the map represents one mile on the ground." Write this scale as a fraction, and state the scale factor.

Solution: The map scale can be written as

$$1 \text{ inch (map)} = 1 \text{ mile (actual)}.$$

To find the scale, simply divide the map distance by the real distance. However, to simplify the division first convert 1 mile to inches so that both distances have the same units.

$$1 \text{ mi} = 1 \text{ mi } \times 5280 \frac{\text{ft}}{\text{mi}} \times 12 \frac{\text{in}}{\text{ft}} = 6.34 \times 10^4 \text{ in}$$

The division is now easy.

$$\text{scale} = \frac{\text{map size}}{\text{actual size}} = \frac{1 \text{ in}}{1 \text{ mi}} = \frac{1 \text{ in}}{6.34 \times 10^4 \text{ in}} = \frac{1}{63,400}.$$

The *scale* of the map is 1 to 63,400. The *scale factor* is 63,400 (the reciprocal of the fraction defining the scale).

❖

Example 7-6 Suppose that you have a map of the United States on which the scale factor is 10 million; that is, the map scale is 1 to 10 million. If two towns are actually 600 kilometers apart, how far apart will they be on the map?

Solution: The scale of 1 to 10 million means that distances on the map are one ten millionth of actual distances. Thus to find the distance on the map simply divide the actual distance by the scale factor of 10 million (10^7) to obtain

$$\text{map distance } = \frac{\text{actual distance}}{\text{scale factor}} = \frac{6 \times 10^2 \text{ km}}{10^7} = 6 \times 10^{-5} \text{ km}.$$

Always express numbers in the units that will be the most meaningful to others or the easiest to interpret.

Thus, on the map, the towns are separated by 0.00006 km. However, for most people such a tiny fraction of a kilometer has little meaning. To make the answer more meaningful, convert it to a unit that is easier to understand; in this case, to centimeters:

$$6 \times 10^{-5} \text{ km } \times \left(10^3 \frac{\text{m}}{\text{km}}\right) \times \left(10^2 \frac{\text{cm}}{\text{m}}\right) = 6 \text{ cm}.$$

The two towns are separated by 6 centimeters on the map.

❖

Example 7-7 Suppose that, using the same map as in Example 7-6 (scale of 1 to 10 million), two towns lie 7 inches apart. How far apart actually are the towns?

Solution: With a scale of 1 to 10 million, the actual distance is 10 million times the map distance, so the

$$\text{actual distance} = (\text{map distance}) \times (\text{scale factor}) = (7 \text{ in}) \times (10^7) = 7 \times 10^7 \text{ in}.$$

The distance is 70 million inches, which is meaningless. Convert this distance to units that are more meaningful, such as miles or kilometers:

$$7 \times 10^7 \text{ in} = 7 \times 10^7 \text{ in} \times \left(\frac{1 \text{ ft}}{12 \text{ in}}\right) \times \left(\frac{1 \text{ mi}}{5280 \text{ ft}}\right) = 1105 \text{ mi};$$

$$1105 \text{ mi} = 1105 \text{ mi} \times (1.62 \text{ km/mi}) = 1790 \text{ km}.$$

The actual distance between the two towns is about 1100 miles, or 1800 kilometers. Note that we rounded the answer to reflect its approximate nature. ❖

Example 7-8 Express a scale of 1 to 10 million verbally and graphically.

Solution: Verbally, the statement could be: "One centimeter (or one inch) on the map represents __?__ in the real world." On a scale of 1 to 10 million, 1 centimeter on the map is the same as 10 million centimeters in the real world, and 1 inch on the map represents 10 million inches. To make these numbers more meaningful, convert them to kilometers and miles, respectively:

$$10^7 \text{ cm} = 10^7 \text{ cm} \times \left(\frac{1 \text{ m}}{100 \text{ cm}}\right) \times \left(\frac{1 \text{ km}}{1000 \text{ m}}\right) = 100 \text{ km}$$

$$10^7 \text{ in} = 10^7 \text{ in} \times \left(\frac{1 \text{ ft}}{12 \text{ in}}\right) \times \left(\frac{1 \text{ mi}}{5280 \text{ ft}}\right) = 158 \text{ mi}.$$

Verbally then the scale is: "One centimeter on the map represents 100 kilometers in the real world," or "One inch on the map represents 158 miles in the real world." Graphically, we can represent this scale by drawing a short ruler marked with the appropriate distances:

❖

Uses of Scales

So far, we have discussed scales only for distances on maps. But scales have many other uses. Architects and engineers often build scale models to show what a proposed building, car, or airplane might look like.

Another common use of scales is with time. History often is represented on **timelines,** on which each centimeter along the timeline represents a certain number of years of history. Time also can be scaled against other times: for example, **time-lapse photography** allows movies and videos to show events in fast motion. Many TV stations use time-lapse photography to show short videos of an entire day's weather; for example, 1 second of video might represent 1 hour of real time. At this scale, a full day's weather is presented in just 24 seconds.

One of the most important uses of scaling today is with computers. On a computer, researchers can create a scale for size or time, or many other physical measurements, which they can change at will. For example, astronomers used a computer to create scale models of two large galaxies and then scaled time so that a few billion years could be represented in just a few minutes. This **computer model** simulates interactions that take billions of years in nature before our very eyes.

(a)

(b)

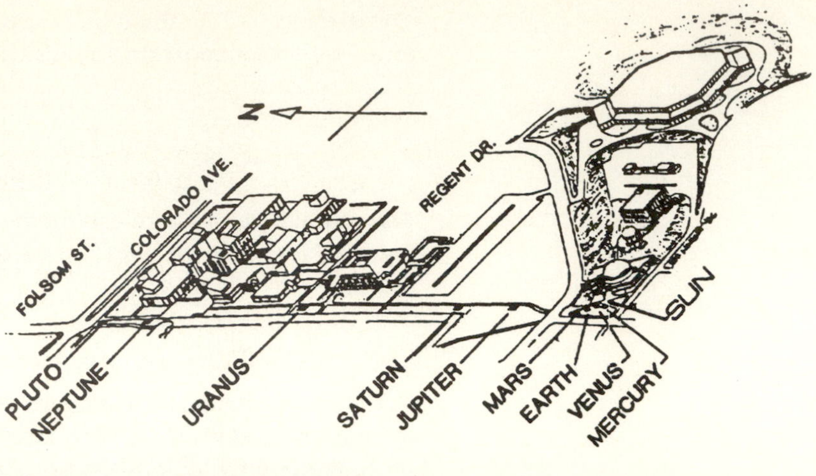

Figure 7- 2. A scale model of the solar system is a good way of making sense of the scale of space. (a) The model Sun sits atop a two-meter tall black granite pyramid in the Colorado scale model solar system; the scale is 1 to 10 billion. (b) The map shows the layout of the model solar system on the University of Colorado at Boulder campus. (a) and (b) Jeff Bennett, Center for Astrophysics & Space Astronomy at the University of Colorado.

7.4.2 The Scale of Space and Time

How big is the universe? How old is the universe? These are among the most ancient of all human questions. Remarkably, during the past century we finally began to find answers to these incredible questions. The answers themselves are large numbers; indeed, they are literally astronomical! Yet, through the use of scaling, we can begin to appreciate their meaning.

The Solar System

Each day, the Earth spins once on its axis; as it does, the Sun, Moon, planets, and stars appear to rise and set in the sky. The Earth, traveling together with the Moon, makes one orbit of the Sun each year. Eight other planets, numerous moons, asteroids, and comets also orbit the Sun.

One of the best ways to visualize the solar system is to build or imagine a scale model of it. A convenient scale to use is 1 to 10 billion; on this scale the Sun is about the size of a grapefruit, the pinhead-sized Earth orbits the Sun at a distance of about 15 meters, and the entire solar system lies within a short walk (about 3/4 kilometer) of the Sun (Figure 7-2).

Example 7-9 The planet Jupiter has a diameter of about 140,000 kilometers, and its average distance from the Sun is 7.8×10^8 kilometers. What is its diameter and distance from the Sun on a scale of 1 to 10 billion?

Solution: To find the scaled diameter and distance divide by the scale factor of 10 billion (10^{10}). Then convert the answers to more meaningful units: centimeters for the diameter and meters for the distance.

$$\text{scale diameter} = \frac{\text{actual diameter}}{10^{10}} = \frac{1.4\times10^5\text{ km}}{10^{10}} = 1.4\times10^{-5}\text{ km }\times10^3\frac{\text{m}}{\text{km}}\times10^2\frac{\text{cm}}{\text{m}} = 1.4\text{ cm}$$

$$\text{scale distance} = \frac{\text{actual distance}}{10^{10}} = \frac{7.8\times10^8\text{ km}}{10^{10}} = 7.8\times10^{-2}\text{ km }\times10^3\frac{\text{m}}{\text{km}} = 78\text{ m}.$$

On a 1 to 10 billion scale Jupiter has a diameter of 1.4 centimeters, about the size of a marble, and orbits about 78 meters from a model Sun.

❖

The Stars

For distances beyond the solar system, units of miles or kilometers are too small to work with easily. Instead, we use a special unit to express distances among the stars, called the **light-year,** or the *distance* that light can travel in one year. For example, 10 light-years is the distance that light can travel in ten years; 7500 light-years is the distance that light can travel in 7500 years. The light-year is a convenient unit for three important reasons:

1. The speed of light always is the same, regardless of who is measuring it and where the light comes from. Thus the definition of a light-year contains no ambiguity. The speed of light is approximately 300,000 kilometers per second.
2. As first proved by Einstein, the speed of light is the fastest possible speed. Nothing can exceed the speed of light, and no physical object can even reach it. Thus, for example, light takes two years to travel a distance of two light-years, so anything else (e.g., a spaceship) takes *longer* than two years to travel the same distance.
3. A light-year is an extremely long distance; it is far enough to make it a convenient unit for cosmic scales.

How far is a light-year? It is simply the speed of light multiplied by one year. The only trick is in keeping the units straight:

A light-year is a unit of distance, *not* a unit of time. It is about 9.5 trillion kilometers.

$$\begin{aligned}1 \text{ light-year} &= (\text{speed of light}) \times (1 \text{ yr}) \\ &= \left(3\times10^5 \frac{\text{km}}{\text{s}}\right) \times \left(1 \text{ yr} \times \frac{365 \text{ day}}{1 \text{ yr}} \times \frac{24 \text{ hr}}{1 \text{ day}} \times \frac{60 \text{ min}}{1 \text{ hr}} \times \frac{60 \text{ s}}{1 \text{ min}}\right) \\ &= 9.5\times10^{12} \text{ km}.\end{aligned}$$

That is, "one light-year" is just an easy way of saying "9.5 trillion kilometers" or "almost 10 trillion kilometers."

> **Time-Out to Think:** Suppose that you hear someone say: "It will take humanity light-years to evolve to that point." What is wrong with this sentence? Explain why it makes no sense, based on the definition of a light-year.

The distances between stars are enormous compared to the distances within the solar system. Consider again the 1 to 10 billion scale model of the solar system, in which the distance from the Sun to the outer planets is a short walk. How far would you have to walk to reach "nearby" stars on this scale? The nearest star besides our Sun is called *proxima centauri,* one of three stars in the *alpha centauri* star system. *Proxima centauri* is about 4.3 light-years from the solar system. To find its distance on the 1 to 10 billion scale, first convert its distance from light-years to kilometers:

$$4.3 \text{ light-years} \times \frac{9.5\times10^{12} \text{ km}}{1 \text{ light-year}} = 4.1\times10^{13} \text{ km}.$$

Next divide by the scale factor of 10 billion to find the scale distance, or

$$\text{scale distance} = \frac{\text{actual distance}}{10^{10}} = \frac{4.1\times10^{13} \text{ km}}{10^{10}} = 4.1\times10^3 \text{ km} = 4100 \text{ km}.$$

The distance to even the nearest star on this scale is more than 4000 kilometers, or approximately the same as the east–west distance across the entire United States!

Time-Out to Think: At the 1 to 10 billion scale you can walk the distance from the Sun to Pluto in just a few minutes. About how long do you think it would take to walk the 4100 kilometers to the nearest star at this scale?

The Milky Way Galaxy

The Sun is but one of *several hundred billion* stars that make up the disk-shaped system of stars called the **Milky Way galaxy.** Stretching 100,000 light-years in diameter, the entire galaxy is rotating slowly. Our solar system, located in the outskirts about 25,000 light-years from the galactic center, completes one orbit every 200 million years. In the region of the Sun, each star is separated by enormous distances: On the 1 to 10 billion scale, the stars are like grapefruits spaced thousands of kilometers apart. In contrast, on the scale of Figure 7-3, representing the Milky Way, the period at the end of this sentence would contain a million star systems!

Example 7-10 One thousand light-years is only 1% of the diameter of the Milky Way galaxy. Imagine trying to travel a thousand light-years with present technology, whereby spacecraft can travel at a maximum speed of about 100,000 kilometers per hour. How long would the trip take?

Solution: A thousand light-years is a *distance*; to find the time required to travel that far, divide the distance by the speed. The distance is given in light-years and the speed in kilometers per hour, so you need to convert from light-years to kilometers. You can do all that at the same time:

$$\text{travel time} = \frac{\text{distance}}{\text{speed}} = \frac{10^3 \text{ light-year}}{10^5 \frac{\text{km}}{\text{hr}}} \times \frac{9.5 \times 10^{12} \text{ km}}{1 \text{ light-year}} = 9.5 \times 10^{10} \text{ hr}.$$

Next, convert the answer to units that are more meaningful; in this case from hours to years:

$$9.5 \times 10^{10} \text{ hr} \times \left(\frac{1 \text{ day}}{24 \text{ hr}}\right) \times \left(\frac{1 \text{ yr}}{365 \text{ day}}\right) = 1.1 \times 10^7 \text{ yr}.$$

Using present technology, to travel just 1% of the distance across the galaxy would take more than 10 million years.

❖

Figure 7-3. An artist's conception shows the Milky Way galaxy (NASA).

Figure 7-4. An artist's conception of giant archipelagos of galaxies stretching across the universe; each dot represents an entire galaxy of stars (NASA).

The uncertainty about the age of the universe — between 7 and 20 billion years — comes from difficulty in measuring precisely the expansion rate.

The Universe

The Milky Way galaxy is only one of some *ten billion galaxies* in the universe. Galaxies are like islands of stars in space. Most galaxies exist in **clusters** with as many as a few thousand galaxies. The Milky Way galaxy is part a small cluster of about 30 galaxies located within 3 million light-years of one another. Clusters of galaxies, in turn, are parts of **superclusters** that stretch like giant archipelagos for hundreds of millions of light-years across the universe (Figure 7-4).

Telescopic observations made since the 1920s have proven that the entire universe is expanding; that is, the distances between groups of galaxies are getting larger over time. Thus believing that the universe must have been smaller in the past seems reasonable. Extrapolating the expansion backward in time, the entire universe would have been compressed into a single point somewhere between 7 and 20 billion years ago. This conjecture, along with many other pieces of observational and theoretical evidence, led scientists to conclude that the Universe must have begun between 7 and 20 billion years ago with the **Big Bang,** or the beginning of the expansion.

Example 7-11 *Universal Timeline.* Suppose that the universe is 15 billion years old. Imagine that you make a timeline 100 meters long. How far along the timeline represents 1 billion years? Written human history extends back only about 10,000 years. How far would that be on the timeline?

Solution: The scale of the timeline is

$$15 \text{ billion years} = 100 \text{ meters.}$$

One billion years must therefore be 1/15 of the timeline:

$$1 \text{ billion years} = 100 \text{ meters} \div 15 = 6.7 \text{ meters.}$$

To find the distance on the timeline that represents 10,000 years use the same basic procedure. As 1 billion years is 10^9 years and 10,000 years is 10^4 years, 1 billion years is 10^5 times longer than 10,000 years. Thus the length for 10,000 years must be $1/10^5$, or 10^{-5}, of the length for a billion years:

$$10^9 \text{ years} = 6.7 \text{ meters} \Rightarrow 10^4 \text{ years} = 6.7 \times 10^{-5} \text{ meters.}$$

To make this result more meaningful, convert it to millimeters because they are familiar from metric rulers:

$$(6.7 \times 10^{-5} \text{ m}) \times (10^3 \text{ mm/m}) = 6.7 \times 10^{-2} \text{ mm} = 0.067 \text{ mm.}$$

On a timeline that represents the age of the universe as the approximate length of a football field, all of written human history occupies only a tiny fraction (about 7%) of a millimeter!

❖

THINKING ABOUT ... THE COSMIC PERSPECTIVE

What is the value of knowing the scale of the solar system or the scale of the universe? After all, the times and distances involved are hardly the kinds of things encountered in everyday life. Nevertheless, understanding the scale of the universe provides a certain perspective — a "cosmic perspective" — on human existence and the Earth.

In subtle ways, the cosmic perspective can significantly influence lives. Most people conduct their daily business with reference to a personal "worldview" developed through a combination of education, religious training, and personal thought. This worldview shapes both beliefs and actions. For example, consider that for most of human history people assumed that the Earth — and, by implication, humanity — was at the center of the universe.

About 400 years ago the work of Copernicus, Kepler, and Galileo proved otherwise: The Earth is but one small planet orbiting the Sun. Shortly thereafter, Dutch scientist Christiaan Huygens made the first reasonable estimate of distances to the stars. In a sense, Huygens may have been the first person in history truly to understand that the planets and the stars constitute real *worlds*, in many cases far larger than the Earth. Huygens believed that this knowledge should fundamentally change the way people thought about themselves. He wrote (c. 1690):

> *How vast those Orbs must be, and how inconsiderable this Earth, the Theatre upon which all our mighty Designs, all our Navigations, and all our Wars are transacted, is when compared to them. A very fit consideration, and matter of Reflection, for those Kings and Princes who sacrifice the Lives of so many People, only to flatter their Ambition in being Masters of some pitiful corner of this small Spot.*

To Huygens, wars and other evils reflected a misplaced sense of self-importance. In his quote, he seems to be arguing that war makes no sense when viewed from the perspective of the Earth as a tiny planet in a vast universe. Perhaps he hoped that the discovery that the Earth is *not* the center of the universe would change human behavior for the better.

If a worldview can so fundamentally affect the way that people live their lives, worldviews should be based on realistic knowledge about the universe, space, and time. That is, developing a cosmic perspective based on modern knowledge of the universe is important. Even though such a perspective won't answer "ultimate questions" about the nature of existence or the purpose of life, it provides a factual basis upon which to build a personal worldview. This worldview, in turn, can help people to make their own judgments about life and its meaning.

Thus, although questions about the size, age, and nature of the universe may seem remote from everyday life, they actually may be among the most important questions that anyone can study. Perhaps that is why every human culture, in every epoch of history, has attempted to learn as much as possible about answers to these questions. We are very fortunate to live at a time when we understand far more than ever before.

7.4.3 Scaling Laws for Area and Volume

In each of the previous examples of scaling, we worked with units such as centimeters, kilometers, miles, or seconds that are **linear;** that is, they are raised only to the first power. However, if we work with areas or volumes the units are **nonlinear;** that is, they are raised to some power. For example, areas are measured in units of square meters (m^2) or square feet (ft^2), while volumes are measured in units of cubic meters (m^3) or cubic feet (ft^3).

Suppose that an architect creates a scale model of a building using a scale of 1 to 20, meaning that one meter in the model represents 20 meters in the building. If the actual building is 10 meters tall, the model is 0.5 meter, or 50 centimeters tall:

$$\text{model height} = \frac{\text{actual height}}{\text{scale factor}} = \frac{10 \text{ m}}{20} = \frac{1}{2} \text{ m} = 50 \text{ cm}.$$

How much paint would be needed to paint the model compared to the amount needed to paint the actual building? Paint covers walls and ceilings, which represent areas. To keep things simple, let's consider just a single wall:

$$\text{area of actual wall} = (\text{actual length}) \times (\text{actual height}).$$

In the 1 to 20 scale model, the wall length is 1/20 of the actual length, and the wall height is 1/20 of the actual height. Thus

$$\begin{aligned}\text{area of model wall} &= (\text{model length}) \times (\text{model height}) \\ &= \frac{(\text{actual length})}{20} \times \frac{(\text{actual height})}{20} \\ &= \frac{(\text{actual length}) \times (\text{actual height})}{20^2} \\ &= \frac{\text{area of actual wall}}{400}.\end{aligned}$$

The area of the model wall is smaller than the actual wall by a factor of 20^2, or 400, so only 1/400 as much paint is needed to cover the model wall as the actual wall. (Although we did this calculation for a rectangular wall, the shape of the area doesn't matter.) For 1 to 20 scale, 20 is the *scale factor;* thus 400 is the *square* of the scale factor. We can therefore summarize the **scaling law for area** as

$$\text{scaled area} = \frac{\text{actual area}}{(\text{scale factor})^2}.$$

Time-Out to Think: To convince yourself that the shape of the area doesn't matter, imagine that the building has a circular floor. Its area is πr^2, where r is the radius of the circle. On a 1 to 20 scale, the radius in the model is 1/20 of the actual radius. Following a procedure similar to that just used, convince yourself that the area of the model floor will be 1/400 of the area of the actual floor.

Next, consider how the interior *volume* of the model compares to the volume of the actual building. The volume of the building is the product of its length, width, and height. Each dimension is a factor of 20 smaller in the model than in the actual building, so the volume of the model is smaller than the volume of the actual building by a factor of 20^3, or 8000:

$$\begin{aligned}\text{model volume} &= (\text{model length}) \times (\text{model width}) \times (\text{model height}) \\ &= \frac{(\text{actual length})}{20} \times \frac{(\text{actual width})}{20} \times \frac{(\text{actual height})}{20} \\ &= \frac{\text{actual volume}}{20^3} = \frac{\text{actual volume}}{8000}.\end{aligned}$$

Generalizing, we can write a **scaling law for volume** (again, the shape of the volume does not matter) as

$$\text{scaled volume} = \frac{\text{actual volume}}{(\text{scale factor})^3}$$

Pressure Scaling

Suppose that, somehow, your size magically doubled; that is, all your linear dimensions like height, width, and depth suddenly doubled. For example, if you were 5 feet tall before, you now are 10 feet tall. Let's see what happens to some of your other dimensions.

Question 1: By what factor has your waist size increased? To answer this question you must recognize that waist size is a linear dimension because it is measured in units of centimeters or inches. Therefore, just as your height doubled, your new waist size doubled.

Question 2: With your sudden growth, you will need to replace your wardrobe. How much more material will be required for your new set of clothes? In this case, you must recognize that clothing covers surface *area* and so scales as the *square* of the linear scale factor. Thus the amount of material required for clothing increases by a factor of 2^2, or four times.

Question 3: By what factor has your weight changed? Your weight depends on your *volume,* which scales as the *cube* of the linear scale factor. Your new volume, and new weight, are therefore 2^3, or 8 times their old values.

Using these results, we can investigate one more interesting quantity. Your new size will result in a change in the **pressure** on your weight-bearing joints (e.g., your ankles, knees, and hips). Pressure is defined as force per unit area, which in this case, comes from your weight:

$$\text{pressure on a weight-bearing joint} = \frac{\text{weight}}{\text{surface area of joint}}.$$

Your weight has increased by a factor of 8 and your surface area has increased only by a factor of 4, so the pressure on a weight-bearing joint has increased by a factor of 8/4, or 2. That is, by doubling your size, the pressure on your weight-bearing joints also has doubled.

Figure 7-5. Why are an elephant's legs thicker than a deer's legs? Because of scaling laws for area and volume.

Pressure scaling explains a lot about the appearance of animals. For example, you have probably noted that, in proportion to their bodies, an elephant's legs are much thicker than a deer's legs. Why? If a deer were simply "scaled up" to the size of an elephant, the pressure on its joints would be so great that its legs would collapse. With thick legs, the surface area of the weight-bearing joints is increased: the added surface area reduces the pressure (Figure 7-5).

Time-Out to Think: The blue whale is the largest animal ever to live on the Earth. An average blue whale is 35 meters long, and weighs 120 tons. Using what you have just learned about pressure on weight-bearing joints, explain why the largest animals live in the ocean.

7.4.4 The Surface Area to Volume Ratio

Why do squirrels eat almost incessantly, whereas people typically eat only three times a day? Why is the Earth, but not the Moon, volcanically active? Why does crushed ice cool your drink faster than ice cubes?

Remarkably, the answers to all three questions involve relative scaling laws for area and volume. Dividing the surface area of any object by the object's volume yields the **surface area to volume ratio;** that is,

$$\text{surface area to volume ratio} = \frac{\text{surface area}}{\text{volume}}.$$

How does the surface area to volume ratio vary with the size of an object? Consider again the architectural model with a 1 to 20 scale. Because the area of the actual building is larger than that of the model by a factor of 20^2 and the volume is greater by a factor of 20^3, the actual surface area to volume ratio is *smaller* by a factor of 20 than the model ratio, or

$$\text{scaling of surface area to volume ratio} = \frac{\text{surface area scaling}}{\text{volume scaling}} = \frac{20^2}{20^3} = \frac{1}{20}.$$

Generalizing, whenever an object is "scaled up", its surface area to volume ratio *decreases* by the scale factor; when an object is "scaled down," its surface area to volume ratio increases. That is:

- larger objects have *smaller* surface area to volume ratios than similarly proportioned small objects; and
- smaller objects have *larger* surface area to volume ratios than similarly proportioned large objects.

All mammals maintain a warm body temperature through metabolism, which takes place in the body *volume*. However, mammals lose heat through their skin, which represents the body *surface area*. Because squirrels are *smaller* than humans, they have *larger* surface area to volume ratios. Hence they have relatively more surface area through which to lose heat, compared to their heat-generating volume. Thus, compared to humans, squirrels must eat more food in proportion to their body weights each day just to stay warm.

How about volcanic activity? Both the Earth and Moon are roughly spherical but, because the Moon is much *smaller* (the Moon's diameter is about one fourth the Earth's), its surface area to volume ratio is *larger*. That is, the Moon has relatively more surface area than the Earth through which to lose internal heat. Thus, 4.6 billion years after both the Earth and Moon formed with hot interiors, the Moon has cooled significantly. In contrast, the Earth remains volcanically active because it still retains much of its internal heat.

Finally, a small piece of ice has a *larger* surface area to volume ratio than a large ice cube; that is, smaller pieces of ice have relatively more surface area. Consequently, if you crush an ice cube into small pieces, the total volume of ice is unchanged but the total surface area of the ice increases. Because a drink is cooled by contact between the liquid and the ice surface, the greater surface area of the crushed ice means that your drink cools more rapidly.

Time-Out to Think: Find two ice cubes of roughly the same size and crush one of them. Pour two glasses of water and place the crushed ice in one and the ice cube in the other. Does the glass of water with the crushed ice cool faster, as claimed here?

7.5 A FEW CASE STUDIES

We have presented two useful techniques for putting numbers in perspective: estimation and scaling. However, as we have shown, the problem of putting a number in perspective rarely has a single solution. Instead, you must examine each situation and use your creativity to come up with a way of relating a large or small number to something more familiar. Thus, as with any problem-solving approach, the key to becoming good at putting numbers in perspective is practice. We therefore conclude this chapter with some case studies involving numbers from daily life.

7.5.1 The National Debt

The federal government typically describes money in units of millions, billions, or trillions of dollars. Let's begin by thinking about a billion dollars.

How many people can you employ with a billion dollars per year? Typical starting salaries for college graduates range from about $20,000 to about $30,000, so let's take $25,000 as a reasonable estimate. To that, we must add the cost of office space, computers, benefits, and other costs of employment. Businesses typically figure this *overhead* to be about the same as a person's salary. Thus the typical cost for an employee who is a recent college graduate is about $50,000 per year. The information provided is

- amount available = 1 billion dollars per year = $\$10^9$/yr;
- cost of typical employee = $50,000 per employee per year
  = $\$5 \times 10^4$/employee/yr.

To determine how many people you could employ for a billion dollars, simply divide:

$$\frac{\$10^9}{\text{yr}} \div \frac{\$5\times 10^4}{1\text{ employee}\times\text{yr}} = \frac{\$10^9}{\text{yr}} \times \frac{1\text{ employee}\times\text{yr}}{\$5\times 10^4} = 2\times 10^4\text{ employee.}$$

Thus, one billion dollars per year could support a work force of 20,000 employees at a cost of $50,000 (salary and overhead) per employee.

Estimating the cost of a single B-2 stealth bomber or a single space shuttle mission involves many factors, and different people make widely varying estimates. The $1 billion estimate for the shuttle comes from dividing NASA's total budget for space shuttle operations by the typical number of launches per year.

Time-Out to Think: The cost of each B-2 stealth bomber, one of the most advanced military airplanes, is estimated at about $1 billion. The cost of a space shuttle mission also is estimated to be about $1 billion. Are these worthwhile expenses? Defend your view. Be sure to keep in mind that much of the money spent is used to pay employees working on these projects.

Another way to put a billion dollars in perspective also points out how different numbers can be, even when they sound similar (like million, billion, and trillion). Suppose, for example, that you become a sports star and earn a salary of a million dollars per year. How long would it take you to earn a billion dollars?

$$\frac{\$1\text{ billion}}{\$1\text{ million/yr}} = \frac{\$10^9}{\$10^6/\text{yr}} = 10^3\text{ yr} = 1000\text{ yr.}$$

We show how revenues and expenditures are tracked in more detail in Chapter 9.

Even at the rate of a million dollars per year, earning a billion dollars would take a thousand years. Alas, most athletic careers do not last that long.

Now that we have put a billion dollars in perspective, let's consider a much larger number. Over the past several decades, the U.S. government has consistently spent more money than it has received from taxes, fees, and other revenue sources. The result is that, each year, the government runs a budget **deficit.** For example, if in a given year the government receives $1.1 trillion in revenues but spends $1.4 trillion, then the deficit is

$$\$1.4\text{ trillion} - \$1.1\text{ trillion} = (\$1.4\times 10^{12}) - (\$1.1\times 10^{12}) = \$3\times 10^{11} = \$300\text{ billion.}$$

Where does the government get the money to make up this deficit? It borrows it. Private investors loan money to the government and in return receive a promise that it will be paid back with interest. The mechanisms through which this lending is accomplished are government bonds and other securities backed by a government promise of repayment.

Time-Out to Think: Most citizens believe that the government spends much of its money wastefully. Yet, investors say that government-backed securities are the safest of all investments: that is, you are less likely to lose your investment in government-backed securities than in any other investment. Why are government-backed securities considered to be so safe?

Each year, the government borrows money to cover that year's deficit. Over the years and decades, the amounts borrowed each year add up. The total amount of money owed by the government is called the **national debt.** At the end of 1995, the U.S. national debt was approximately $5 trillion, and rising. Can we make sense of a huge number like $5 trillion?

Suppose that the president and Congress suddenly decided that the national debt needed to be paid off. To raise the needed money, they decide to ask everyone in the United States to pay a one-time tax. If everyone were asked to pay the same amount, how much would each woman, man, and child have to pay?

All things considered, the '90's were a pretty good decade. For one thing, Congress finally got a handle on the pesky federal budget deficit. That was in 1995, when the House of Representatives, after weeks of heated debate over the issue of what comes after "trillion," hit on the idea of using exponents. This led to the historic Setting a Definite Serious Limit on the Deficit of No More Than 10^{15} Act *of 1996, which held federal spending firmly and historically in check until it was replaced by the historic* Let's Make It 10^{17} But That Is Really the Absolute Final Amount Act *of 1997.*
— from a satirical article by Dave Barry. From *Newsweek*, January 3, 1994, © 1994 Newsweek, Inc. All rights reserved. Reprinted by permission.

We obtain the answer simply by dividing the total debt by the population of the United States, which is approximately 250 million people:

$$\frac{\$5\times 10^{12}}{2.5\times 10^{8}\text{ persons}} = \frac{\$2\times 10^{4}}{\text{person}}.$$

Thus to retire the national debt a U.S. citizen would have to pay about $20,000 to the federal government. A family of four would owe $80,000!

Another way to think about the debt is to visualize it in $1 bills. If you piled them up, how high would the pile reach? Doing this calculation requires knowing that the thickness of a $1 bill is approximately 0.2 millimeters.

Time-Out to Think: Measuring the thickness of a single dollar bill is difficult. A better way to determine the thickness of a $1 bill is to make a stack of, say, ten $1 bills and measure the thickness of the stack. Then, to get the thickness of a single bill, divide by 10. Try this process; does your result agree with the 0.2 millimeter thickness stated?

How high would the pile of 5 trillion $1 bills reach? We simply multiply the number of bills by the thickness of each bill:

$$\left(5\times 10^{12}\text{ bills}\right)\times\left(\frac{0.2\text{ mm}}{\text{bill}}\right) = 1\times 10^{12}\text{ mm}.$$

The stack of bills would rise a trillion millimeters. Converting to a more meaningful unit, in this case to kilometers, we get

$$10^{12}\text{ mm}\times\frac{1\text{ m}}{10^{3}\text{ mm}}\times\frac{1\text{ km}}{10^{3}\text{ m}} = 10^{6}\text{ km}.$$

A stack of $1 bills equaling the national debt would rise to a height of 1 million kilometers, or more than twice the 400,000 kilometer distance from the Earth to the Moon!

Let's carry this exercise a little further. Suppose that you could drive your car in space, along the pile of 5 trillion $1 bills. How long would driving the length of the entire national debt take you? A typical highway cruising speed is about 100 kilometers per hour (62 miles per hour). Thus to drive 1,000,000 kilometers (the height of the stacked bills) would take

$$\text{time} = \frac{\text{distance}}{\text{speed}} = \frac{10^{6}\text{ km}}{100\text{ km/hr}} = 10^{4}\text{ hr}\times\frac{1\text{ day}}{24\text{ hr}}\times\frac{1\text{ month}}{30\text{ day}} \approx 14\text{ months},$$

or more than a year to "visit" the national debt (Figure 7-6).

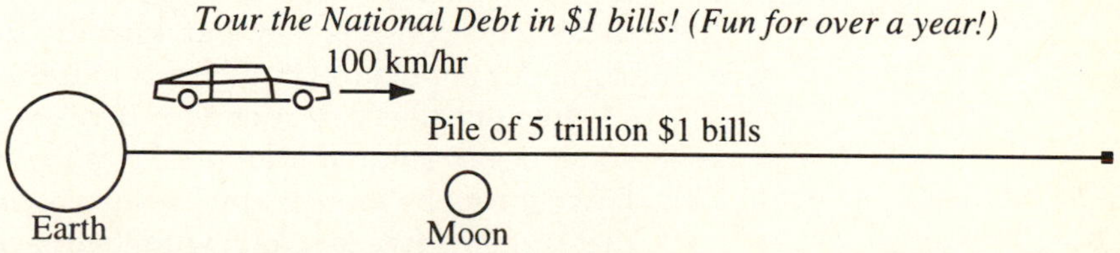

Figure 7-6. A stack of $1 bills equaling the federal debt would stretch more than twice the distance to the Moon.

7.5.2 Counting the Stars

The Sun is only one of at least one hundred billion stars in the Milky Way galaxy. How large is the number 100 billion?

Imagine that, tonight, you are having difficulty falling asleep. Instead of counting sheep, you decide to count stars. You begin counting as quickly as you can; at first, you count so fast that the numbers almost run together:

"onetwothreefourfive...." After a while, however, you need more time to count clearly: "four hundred sixty-two thousand, nine hundred and seventy-six... four hundred sixty-two thousand, nine hundred and seventy-seven...." Let's assume that, on average, you are able to count about one star each second. How long will counting 100 billion stars take?

Clearly, the answer, is one hundred billion seconds. But how long is that? We need to convert 100 billion (10^{11}) seconds into something more meaningful:

$$10^{11}\ \text{s} \times \left(\frac{1\ \text{min}}{60\ \text{s}}\right) \times \left(\frac{1\ \text{hr}}{60\ \text{min}}\right) \times \left(\frac{1\ \text{day}}{24\ \text{hr}}\right) \times \left(\frac{1\ \text{yr}}{365\ \text{day}}\right) = 3171\ \text{yr}.$$

You would need *more than three thousand years* just to count to one hundred billion. And, that assumes that you never take a break: no sleeping, no eating, and absolutely no dying!

Another way to put the number of stars in perspective is by thinking about the possibility of alien civilizations. Astronomers believe that most stars probably have planets and that some of these planets might be Earth-like. If so, perhaps some harbor life and are home to civilizations like our own. Huge uncertainties surround guesses about how common, or uncommon, such civilizations might be. Perhaps civilizations arise around 10% of stars; or, civilizations might be so rare that ours is the only one in the entire galaxy. For the sake of argument, suppose that the odds of finding a civilization around some star are about the same as the odds of becoming a big winner in the lottery: about one in a million. If that were the case, how many civilizations would exist in the Milky Way galaxy?

If there are one hundred billion stars in the Milky Way and one in a million have a planet harboring a civilization, the number of civilizations is

$$10^{11}\ \text{stars} \times \left(\frac{1\ \text{civilization}}{10^{6}\ \text{stars}}\right) = 10^{5}\ \text{civilizations}.$$

With civilizations around only one in a million stars, 100,000 civilizations would exist in the Milky Way galaxy alone!

Time-Out to Think: Take a moment to ponder a hundred thousand other worlds where students, like you, might learn about numbers and think about the stars in their sky....

7.5.3 Meet Your Constituents

How many people go to your school? How many people live in your community? How many people live in the United States? Finding numerical answers to these questions is easy, but what do the numbers mean? One way of thinking about population numbers is to think about personal contact.

Imagine that you attend (or perhaps you do attend) a large university of about 20,000 students. Suppose that the university hires a new president, Mr. Bowtie, who believes that the only way he can truly understand the problems that students face is to talk with them personally. Thus, President Bowtie decides to spend a little time with each of the 20,000 students at his university. He thinks hard about how best to accomplish this task, and decides the answer is: "Let's do lunch!" Mr. Bowtie asks every student to sign up for a lunch date with him; to ensure lively discussions, he suggests that the students sign up in groups of four for each lunch. Is Mr. Bowtie quantitatively literate?

If President Bowtie has lunch with four students per day, how long will he need to have lunch with all 20,000 students? Let's do the calculation:

$$20{,}000 \text{ students} \times \frac{1 \text{ day}}{4 \text{ students}} = 5000 \text{ days} \times \frac{1 \text{ yr}}{365 \text{ day}} = 13.7 \text{ yr}.$$

Even "doing lunch" every day, including weekends and holidays, he would need almost 14 years to meet all 20,000 students. Because new students enter the university each year and many students graduate in less than 14 years, President Bowtie's plan is hopelessly flawed: He will never be able to "do lunch" with all the students as he had planned.

Perhaps you are interested in politics but believe that election campaigns should involve more personal contact. So you decide to run for the U.S. House of Representatives. As part of your campaign, you intend to shake hands with every one of your future constituents. How long will that take?

We can estimate how many constituents you will represent by dividing the population of the United States (250 million) by the 435 members in the House of Representatives:

$$\text{average number of constituents} = \frac{2.5 \times 10^8 \text{ constituents}}{435 \text{ representatives}} = \frac{5.7 \times 10^5 \text{ constituents}}{\text{representatives}}.$$

Each member of the House of Representatives has about 570,000 constituents. Suppose that you can shake a person's hand, say hello, and move on to the next person in about ten seconds. Thus, if you're fortunate enough to have all your constituents waiting in line to meet you, shaking all their hands would take

$$570{,}000 \text{ handshakes} \times 10 \text{ s/handshake} = 5.7 \times 10^6 \text{ s},$$

or, equivalently,

$$5.7 \times 10^6 \text{ s} \times \left(\frac{1 \text{ min}}{60 \text{ s}}\right) \times \left(\frac{1 \text{ hr}}{60 \text{ min}}\right) \times \left(\frac{1 \text{ day}}{24 \text{ hr}}\right) \times \left(\frac{1 \text{ wk}}{7 \text{ day}}\right) \approx 9\frac{1}{2} \text{ wk}.$$

That's nine and a half weeks to shake all their hands, assuming no breaks for eating or sleeping! You *do* need to eat, sleep, and plan campaign strategy, so let's assume that you shake hands only for six hours each day, or one fourth of a day. Then, instead of nine and half weeks, you will need four times as long: 38 weeks, or about nine months. Perhaps that's why campaigns seem to drag out for such a long time....

In 1993, President Clinton became the first president to have his own electronic mail (**e-mail**) address. Using the international computer network called the **Internet,** anyone can send an e-mail message directly to the president. By mid-1995, an estimated 25 million Americans had access to the Internet. If, each day, one out of a thousand people sends the president a message, how many messages would he receive? Note that we can write the latter assumption as one message per 1000 Internet users per day. We multiply to obtain

$$\left(2.5 \times 10^7 \text{ Internet users}\right) \times \frac{1 \text{ message}}{(1000 \text{ Internet users}) \times (1 \text{ day})} = \frac{2.5 \times 10^4 \text{ message}}{\text{day}}.$$

The president would receive about 25,000 messages per day, which is about 750,000 messages per month, or about nine million messages per year! When do you think you'll receive a personal response to *your* message?

The banking industry in the United States consists of several types of institutions differentiated by regulatory law and their base of customers. The most basic differentiation is between *commercial banks* and *savings institutions,* which include the S&Ls. The differences between commercial and savings institutions in recent years have been reduced by deregulation and competition.

7.5.4 The Savings and Loan Bailout

During the 1980s and early 1990s a large number of banking institutions failed, particularly those known as *savings and loan associations* (S&Ls). To understand what happened, let's explore briefly how S&Ls and other banking institutions work. Essentially, a bank earns profits by charging interest on money that it lends to **borrowers.** Where does the bank get the money to lend? From **depositors** who give their money to the bank for safekeeping.

Because banks are dependent for their survival on recovering the money they lend, they try to ensure that loans go only to "good credit risks"; that is, they try to screen borrowers to make sure that the borrower will be able to repay the loan. If a borrower defaults — fails to repay a loan — the bank loses money. If enough borrowers default, the bank could "fail"; that is, it will have lost some or all the money entrusted to it by depositors. Most people keep a large portion of their savings in banking institutions, so a failure can be catastrophic; for example, retired individuals might suddenly lose their lifetime savings.

Time-Out to Think: Have you ever applied for a loan? What type of information did the lender ask for? Why did the lender request this information? Do you think you are a "good credit risk"?

Originally, the FDIC served commercial banks and the FSLIC served savings and loan associations (S&Ls). In 1989, in the wake of the S&L crisis, Congress eliminated the insolvent FSLIC. Today, S&Ls are covered by a subsidiary of the FDIC.

During Great Depression in the 1930s, nearly two thousand savings institutions failed, taking the savings of their depositors with them. In response, the U.S. government created several organizations to insure depositor's accounts including the **Federal Deposit Insurance Corporation** (FDIC) and the **Federal Savings and Loan Insurance Corporation** (FSLIC). These corporations collect insurance payments from banks or S&Ls, and in return insure depositors' accounts of up to $100,000.

Although the underlying causes of the S&L crisis remain hotly debated, the immediate cause of the failures was bad loans: the S&Ls had loaned money to many individuals, groups, and businesses that defaulted. The amount of depositors' money lost by the S&Ls was so great that the FSLIC didn't have the money to cover all its insurance obligations. As a result, Congress stepped in and provided the necessary money to cover those losses, which became know as the **savings and loan bailout.**

Where did the money go? The depositors' money went primarily to two sources: to the borrowers who defaulted on their loans and to salaries and profits for individuals who operated the S&Ls. Who repaid the depositors? The losses were repaid by the U.S. government, so ultimately the taxpayers reimbursed the depositors.

Because of uncertainties caused by different accounting practices, the final cost of the bailout may never be known precisely; estimates, however, range as high as $500 billion. Mathematician John Allen Paulos, in an article in the *New York Times*, offered this interesting perspective (among others) on $500 billion:

> *Given that gold sells for roughly $350 an ounce, and that the distance from the east coast to the west coast of the United States is about 15 million feet, $500 billion could buy a transcontinental gold bar weighing about five and half pounds a foot.... If we were to stretch the gold bar into a rainbow extending from Capitol Hill to 1,500 miles above the Kansas prairies and ending over the Phoenix headquarters of Charles Keating's failed savings and loan empire, this golden arch would weigh in at almost four pounds a foot.*

***THINKING ABOUT ...** THE BIGGEST ROBBERY IN HISTORY*

The following article, titled "The Biggest Robbery in History — You're the Victim," appeared in the *Wall Street Journal* (August 9, 1990). The author, Michael Gartner, is editor and co-owner of the *Daily Tribune* in Ames, Iowa, and president of NBC News in New York.

"A man was murdered in my neighborhood in New York the other day. He was a 33-year old advertising man. The phones in his apartment building were out of order, so, late in the evening, he stepped out to the corner to call a colleague. A would-be robber came by and shot him dead.

"Murders in New York are hardly unusual — last year, there were 1,905, one every four and a half hours — but the murder in my generally quiet neighborhood was different. It wasn't a domestic dispute. It wasn't a drug deal. Rather, it was a robbery in which the holdup man, according to police, was a homeless person, one of several who have taken over a little nearby park where mothers and children used to gather. The *Times* ran his picture last week. I've seen him often.

"I read about the murder the same day I read yet another story about another increase in the amount of money we're going to have to pay to keep the savings and loan industry afloat. Now, its up to around $500 billion, including interest. I'm sure there will be higher estimates in coming months. But what difference does it make? The figures became numbing months ago.

"Yet, I wondered: Would my neighbor be alive if just some of that money had been earmarked to help the homeless? How much money is the federal government spending on the poor, deranged people who have no place to live?

"The answer, nationwide, is about $598 million. That's just a little more than one-tenth of 1% of the amount we'll spend on the savings and loan scandal. That puts in perspective, a little, the enormity and enormousness of the scandal. Here is the government spending half a trillion dollars to rectify bad decisions — sometimes just dumb, sometimes criminal — made by executives and directors who, it turns out, were accountable to no one. Why isn't anyone upset by this?

"Here are some other comparisons with the $500 billion bailout: The entire cost of World War II, in current dollars and including service-connected veterans' benefits, is about $460 billion. That's $40 billion less than the bailout.

"The cost of the Vietnam War, including benefits, stands at $172 billion. Korea was $70 billion. World War I was $63 billion. The Civil War was $7 billion — 1.5% of the cost of the savings and loan debacle.

"The National Cancer Institute spent $1.3 billion on cancer research last year. The National Institutes of Health spent $582 million on heart disease, $171 million on lung disease, $170 million on blood disease. Add them up: less than one-half of 1% of the S&L money. Why isn't anyone upset?

"Across the country, corporations, foundations and individuals give away about $100 billion a year. In other words, if every charitable gift from every company, every foundation, every living human and the will of everyone who died this year were earmarked to bail out the savings and loan industry, it wouldn't be enough to handle the cost even without worrying about the interest on that cost.

"If you add up the budget of every State, you get total expenditures of about $450 billion, or at least you do if you add up the fiscal 1987 figures, the latest I could find. Again, less than the savings and loan bailout.

"You get the idea. It's hard to find anything in this country — or anything in history — as costly as the savings and loan scandal. No war, no defense program, no social program, no other scandal has ever cost what this will cost. Why isn't anyone upset?

"All of the ads in all of the newspapers, on all of the television stations and networks, on all of the billboards and everywhere else cost the advertisers about $100 billion a year. If you add up the assets of Prudential and Metropolitan Life and Equitable Life and Aetna and Teachers Insurance and

New York Life and Connecticut General and Travelers and John Hancock and Northwestern Mutual you still don't get to $500 billion. The 1988 profits of all the companies on the Fortune 500 list added up to just $115 billion. Why isn't anyone upset?

"Probably because the savings and loan horror isn't a crime like the murder of the man in my neighborhood. There is no one, identifiable victim, other than the taxpayer. There is no one, identifiable perpetrator, other than the system.

"Yet, again, you wonder: if just a piece of that money could be directed elsewhere, how many murders could be prevented, how many homeless could be helped, how many diseases could be conquered, how many police could be hired, how many parks could be built?

"For it just isn't right. The advertising man in my neighborhood is dead. The homeless man in my neighborhood is in jail. But the people — officers and directors, mainly — who are costing us $500 billion, robbing us of money to cure diseases or educate the poor or just pay off our debt, are still free to enjoy the riches of life. Why?"

7.5.5 Until the Sun Dies

From *Ozymandias*
by Percy Bysshe Shelley

I met a traveller from an antique land
Who said: Two vast and trunkless legs of stone
Stand in the desert... Near them, on the sand,
Half sunk, a shattered visage lies, whose frown,
And wrinkled lip, and sneer of cold command,
Tell that its sculptor well those passions read
Which yet survive, stamped on these lifeless things,
The hand that mocked them, and the heart that fed:
And on the pedestal these words appear:
'My name is Ozymandias, king of kings:
Look on my works, ye Mighty, and despair!'
Nothing beside remains. Round the decay
Of that colossal wreck, boundless and bare,
The lone and level sands stretch far away.

Life on Earth flourishes with energy that ultimately comes from the Sun. Thus we can expect that living things will continue to inhabit the Earth until the Sun dies, which astronomers estimate will be in about 5 billion years. How long is five billion years?

First, let's think about human history. The pyramids in Egypt were constructed between about 2700 and 2100 B.C.; that makes the oldest of them a little less than 5000 years old. Although many people think of the pyramids as "eternal," they are slowly eroding. The main source of erosion, today, is tourists walking on them and chipping off souvenirs, but pollution, wind, and rain also take a toll. In fact, the pyramids have eroded substantially since they were first built. Suppose that the pyramids have lost about one tenth of 1% of their original mass to erosion since they were constructed. How long will they last?

As one tenth of 1% is one thousandth (0.001), the pyramids would last 1000 times their present age, or

$$5000 \text{ years} \times 1000 = 5 \text{ million years.}$$

At an erosion rate of only one tenth of 1% every 5000 years, the pyramids will completely disappear in 5 million years — a long time, but 5 *billion* years is a thousand times longer still!

How does a human lifetime compare to 5 billion years? If we assume that a human lifetime is about 100 years, we can compare through division:

$$\frac{100 \text{ yr}}{5 \times 10^9 \text{ yr}} = 2 \times 10^{-8}.$$

A human life is about two *one-hundred millionths* of 5 billion years. What is two one-hundred millionths of a human lifetime? Let's convert it to minutes for a better perspective:

$$\left(2 \times 10^{-8}\right) \times 100 \text{ yr} = 2 \times 10^{-6} \text{ yr} \times \frac{365 \text{ day}}{1 \text{ yr}} \times \frac{24 \text{ hr}}{1 \text{ day}} \times \frac{60 \text{ min}}{1 \text{ hr}} = 1 \text{ min}.$$

Thus a human lifetime in comparison to the life expectancy of the Sun is roughly the same as *one minute* (or about 60 heartbeats) in comparison to a human lifetime.

On a more somber note, our species is at a crucial juncture in history. During the past century, humanity has acquired sufficient technology and power to destroy human life totally. If that happens, no humans will be alive during the next 5 billion years. Life on Earth, however, is likely to continue and to evolve even if humans are extinct. Would another intelligent species ever emerge on the Earth? Of course, there is no way to know for sure. However, some scientists believe that some species of dinosaurs might have evolved intelligence if they hadn't been wiped out about 65 million years ago. At the same time the dinosaurs died, so did all other large land species of animals. Nevertheless, 65 million years later, humans dominate the Earth. Suppose that another intelligent species could evolve 65 million years after humans disappear. If they also destroyed themselves, another species could evolve 65 million years after that, and so on. How many more chances would the Earth have to produce intelligent species, if new ones evolve every 65 million years? The answer is

$$\frac{5 \text{ billion years}}{65 \text{ million years}} = \frac{5 \times 10^9 \text{ years}}{6.5 \times 10^7 \text{ years}} = 77.$$

The Earth would have some 80 more chances! Perhaps one of those species will not destroy itself, and might move on to other star systems by the time the Sun finally dies. Perhaps it will be humans who survive.

7.6 CONCLUSION

In this chapter we focused on methods of putting large and small numbers in perspective. In the process, we also developed techniques for estimation and discussed scaling and scaling laws. Key ideas for you to leave this chapter with include the following.

- Numbers that sound alike — million, billion, and trillion — may have very different meanings. Be aware of the differences among numbers. Scientific notation can help you to analyze, compare, and work with large and small numbers.
- Estimation is an important problem-solving process. In many cases, an order of magnitude estimate is enough to give you a "feel" for the size of some quantity.
- Understanding scaling and scaling laws will help you not only make sense of many large and small quantities, but also explain many physical phenomena related to the important surface area to volume ratio.
- Perhaps the most important lesson in this chapter, and one of the most important in this entire book, is that numbers have real meanings. Just because a number is easy to say — a billion years or a trillion dollars — doesn't mean that it is easy to comprehend. Any number can be put into some perspective.

PROBLEMS

Reflection and Review

Section 7.1

1. **Large Numbers in the News.** Search today's newspaper for as many instances of numbers larger than 100,000 you can find. Briefly explain the context within which each large number is used.

Section 7.2

2. **Reading Powers of 10.** Convert each of the following numbers from scientific notation and write its name. *Example:* 2×10^3. *Solution:* $2 \times 10^3 = 2000$ = two thousand.

a. 5×10^6 b. 7×10^9 c. -2×10^{-2}
d. 8×10^{11} e. 1×10^{-7} f. 9×10^{-4}
g. 4.6×10^9 h. 3.95×10^{-6} i. -6.02×10^{-10}
j. 9.9863×10^3 k. 3.0001×10^{-2} l. 7.7869×10^4

3. **Writing Powers of 10.** Write each of the following numbers in scientific notation.
a. 600 b. 0.9 c. 50,000
d. 0.003 e. 0.0005 f. 70,000,000,000
g. 0.00000002 h. 7 million i. 9 billionths
j. 4 trillion k. 80 billion l. 600 million

4. **Converting to Scientific Notation.** Convert each of the following numbers to scientific notation.
a. 1,000,000 b. 150,000 c. 45×10^{-1}
d. 18 hundredths e. 540×10^6 f. 530×10^{23}
g. 645 h. 0.92 i. 500.098
j. 0.002×10^6 k. 250 million l. $99 \div 10^{-2}$

5. **Converting from Scientific Notation.** Convert each of the following numbers from scientific notation.
a. 2.2×10^{-4} b. 2.2×10^{-4} c. 2×10^{-1}
d. 6.667×10^1 e. 3.5×10^4 f. 6.2×10^{26}
g. 7.0×10^3 h. 1.5×10^9 i. 3.0906×10^3
j. 9.828×10^7 k. 1.501×10^{-10} l. 1.01×10^{-34}

6. **Large and Small Numbers Everywhere.** Rewrite each statement with a number in scientific notation.
a. The U.S. national debt in 1993 was about four and a half trillion dollars.
b. Corporate profits in the United States in 1993 were 450 billion dollars.
c. Consumer debt (loans and credit cards) in the United States in 1994 was 800 billion dollars.
d. In 1993, heart disease caused 739,860 deaths in the United States.
e. The area of the Earth's surface is 509,600,000 square kilometers.
f. In 1990, Americans used 2,626,165 gigawatt-hours of electricity; express this quantity in watt-hours. (Hint: Recall that the prefix "giga" means 1 billion).

7. **Approximation with Powers of 10.**
a. Suppose that you add $10^{26} + 10^7$. What, approximately, is the answer? Explain.
b. Suppose that you subtract $10^{81} - 10^{62}$. What, approximately, is the answer? Explain.

8. **Practice with Scientific Notation.** Do each of the following operations without a calculator and show your work clearly. Be sure to express the final answers in scientific notation. You may round your answers by writing only two decimal columns (as in 3.2×10^5 or 7.6×10^{-27}).
a. $(3 \times 10^4) \times (8 \times 10^5)$ b. $(6.3 \times 10^2) + (1.5 \times 10^1)$
c. $(9 \times 10^3) \times (5 \times 10^{-7})$ d. $(4.4 \times 10^{99}) \div (2.0 \times 10^{11})$
e. $(8 \times 10^{12}) \div (4 \times 10^9)$ f. $(7.5 \times 10^{21}) \div (1.5 \times 10^{13})$
g. $(3.2 \times 10^{22}) \div (1.6 \times 10^{-14})$ h. $(6 \times 10^{10}) - (5 \times 10^9)$
i. $(9 \times 10^3) + (5 \times 10^9)$ j. $(8.1 \times 10^{30}) + (9 \times 10^{15})$
k. $(2.5 \times 10^{-4}) \times (3 \times 10^{-4})$ l. (8.1×10^{14}) ((3×10^{-7})
m. $(6.6 \times 10^{-2}) \div (4.4 \times 10^{-3})$ n. $(2.1 \times 10^4) - (1.5 \times 10^5)$

9. **Approximation with Scientific Notation.** For each of the following problems: (i) make an "in your head" estimate of the answer; explain your process in words (you may want to follow the pattern of the explanation provided in subsection 7.2.3); (ii) do the exact calculation (with a calculator if necessary); (iii) compare the approximation to the exact answer; how well did your approximation technique work?
a. $20{,}000 \times 100$ b. $9{,}642 \div 31$
c. $-12.5 \times 11{,}890$ d. 250 million $\times$ 40
e. 7.453×291 f. $6{,}570{,}0999 \div 32.7$
g. 5.6 billion $\div$ 200 h. 4 trillion $\div$ 260 million
i. $9{,}000 \times 54{,}986$ j. 3 billion $\div$ 25,000
k. $5{,}987 \times 341$ l. $43 \div 765$

10. **They Don't Look Very Different!** Compare each pair of numbers. ***Example:*** 10^6, 10^4. ***Solution:*** 10^6 is 10^2, or 100, times larger than 10^4.
a. 10^{35}, 10^{26} b. 10^{17}, 10^{27}
c. 1 billion, 1 million d. 7 trillion, 7 thousand
e. 2×10^{-6}, 2×10^{-9} f. 6.1×10^{27}, 6.1×10^{29}
g. 4×10^{15}, 2×10^{-3} h. 8×10^{15}, 1.6×10^{13}
i. 250 million , 5 billion j. 9.3×10^2, 3.1×10^{-2}
k. 10^{-8}, 2×10^{-13} l. 3.5×10^{-2}, 7×10^{-8}

Section 7.3

11. **Estimation Practice.**
a. Estimate total annual spending on alcoholic beverages by students at *your* school.
b. Estimate the total amount of money spent each year by students at your school on newspapers and news magazines. Compare that to the amount spent on alcoholic beverages.
c. Estimate the total number of words in this textbook.

12. **Comparisons Through Estimation.** Make estimates as needed to answer clearly each of the following questions.
a. Which is bigger, the height of a 10-story apartment building or the length of a football field? By how much?
b. Could a person walk across the United States (New York to California) in a year? If not, about how long would it take?
c. Which is more, the number of miles Americans fly each year or the number of miles Americans drive each year?
d. Which weighs more, a pound of dirt or a pound of feathers?
e. Which holds more people, a football stadium or 10 movie theaters?

13. **Orders of Magnitude.** In each of the following problems, compare the two quantities by order of magnitude. *Example:* A small car (weight about 1500 to 2000 pounds) is about one order of magnitude heavier than a typical person (weight about 150 pounds).
a. The population of China (1.1 billion) and the population of Japan (125 million)
b. The population of the United States (260 million) and the population of Iceland (250,000)
c. The area of Canada (10 million square kilometers) and the area of Lebanon (10,500 square kilometers)
d. The mass of the Sun (2×10^{30} kg) and the mass of the Earth (6×10^{24} kg)
e. The diameter of a cell (10^{-6} m) and the diameter of a proton (10^{-15} m)

14. A Bolder Relay. One of the premiere running events in the world, the Bolder Boulder 10-kilometer race, is held each year in Boulder, Colorado. In 1995, approximately 35,000 runners completed the 10 kilometer race.

a. What was the total distance run by the 35,000 runners?
b. If the race were a relay, and could be run around the Earth, how many times would the relay circle the Earth?
c. Suppose that the average runner completes the race in one hour. How long would the relay take, in hours? in years?

15. Order of Magnitude Estimates. Make an order of magnitude estimate for each of the following. Explain your reasoning.

a. Estimate the average amount of money spent on going to the movies by an American family each year.
b. Estimate the average amount of money spent on food by an American each year.
c. Estimate the average price of a house in the United States.
d. Estimate the average amount of gasoline purchased by an American automobile owner each year.

16. Numbers of Atoms. Estimate each of the following.

a. How many atoms are in a liter of water (a liter of water weighs one kilogram).
b. How many atoms are there in all the people living in the United States?
c. Gold has an atomic weight of 197; that is, 197 grams of gold contain Avogadro's number (6×10^{23}) of atoms. How many atoms are in 20 grams of pure gold?

Section 7.4

17. Scale Factors. What is the scale factor for each of the following?

a. 1 cm on the map represents 1 km on the ground.
b. 2 inches on the map represents 0.5 mile on the ground.
c. 5 cm (map) = 100 km (actual).
d. 1 foot (map) = 100 meters (actual).

18. Topographic Map Scale. A common scale on topographic maps from the United States Geological Survey is 1 to 24,000.

a. If you measure the distance between two locations as 7.5 cm, how far apart are they actually?
b. If you knew that two towns on the map were 5 km apart, how far would they be separated on this map?
c. If the smallest reasonably measurable distance on the map is 2 mm, what is this distance on the ground?

19. Imagining Scale. For the following scaling factors, express each verbally and graphically. Give an example in which each scale might be useful.

a. 1 to 100
b. 1 to 1,000,000
c. 1 to 24,000
d. 1 to 10^8

20. A Scale Model Solar System. The following table lists equatorial diameters for the Sun and nine planets, along with average distances from the Sun. Note that the column heading on the distance data indicates that the units are 10^6 km. In other words, every number in the column should be multiplied by 10^6 km. For example, the average distance to Mercury is 57.9×10^6 km.

Object	Diameter (km)	Average Distance from Sun (10^6 km)
Sun	1,392,500	—
Mercury	4,880	57.9
Venus	12,100	108.2
Earth	12,760	149.6
Mars	6,790	227.9
Jupiter	143,000	778.3
Saturn	120,000	1,427
Uranus	52,000	2,870
Neptune	48,400	4,497
Pluto	2,260	5,900

a. Calculate the scaled diameters and distances for all the planets and the Sun on a scale of 1 to 10 billion. Be sure to answer in meaningful units: diameters in millimeters or centimeters and distances in meters.
b. In words, describe the overall layout of the scale model solar system. For example, you could compare the scaled planets to familiar objects (e.g., raisins, grapes, peppercorns). You could describe distances in relation to some well-known local distances (e.g., how the planets would lay out on your campus).
c. In the chapter we stated that the distance to the nearest star on this scale would be about 4000 kilometers. A typical walking speed is about 3 miles per hour. How long would walking this distance take? Compare that to how long walking to the model of Pluto on this scale would take.
d. Based on your findings, describe in words why detecting planets around other stars is so difficult.

21. Shining Light on Pluto. The greatest distance between the planet Pluto and the Sun is about 7.4×10^9 kilometers. How far is this distance in light-years? How long does light take to reach Pluto from the Sun at this distance?

22. Scaling Laws with a Model Airplane. Suppose that you build a scale model airplane using a scale of 1 to 50.

a. How much longer are the actual airplane wings than the model airplane wings?
b. How much more paint is needed to paint the actual airplane than the model airplane?
c. How does the volume of the cabin space on the actual airplane compare to that of the model?

23. Tripling Your Size. Suppose that you magically tripled in size (while your density remained unchanged); that is, your linear dimensions all triple from their current values.

a. By what factor has your height increased?
b. By what factor has your waist size increased?
c. By what factor has your weight increased?
d. How much more material is required for your new set of clothes compared to your old set?
e. By what factor has the pressure on your weight-bearing joints increased?

24. **Comparing People.** Consider a person, let's call him or her Sam, who is 10% taller than you, but proportioned in approximately the same way. (In other words, Sam looks just like a larger version of you.)
  a. By what factor is Sam taller than you? How tall are you? How tall is Sam?
  b. By what factor is Sam's waist size larger than yours? What is your waist size? What is Sam's?
  c. By what factor is Sam's weight greater? How much do you weigh? How much does Sam weigh?
  d. How much more material is required for Sam's clothes than for yours? Imagine that you make your own clothes and that the cost of the material (sold by the square foot) for a shirt is $5.50. How much would similar material cost to make a shirt for Sam?
  e. By what factor is the pressure on Sam's weight-bearing joints greater than yours?
  f. Assuming that Sam is proportioned in the same way as you, who is more likely to have joint injuries? Why?

25. **Areas and Volumes in a Scale Model Solar System.** Consider again a scale model of the solar system, built at a scale of 1 to 10 billion.
  a. How does the circumference (i.e., the distance around the equator) of an actual planet compare with the circumference of its model planet?
  b. How does the surface area of an actual planet compare with the surface area of its model?
  c. How does the volume of an actual planet compare to the volume of its model?
  d. Suppose that you compare the *ratio* of two planet volumes, such as Jupiter and the Earth. That is, you divide the volume of Jupiter by the volume of the Earth. How will the ratio of the volumes of the model planets compare with the ratio of volumes for the actual planets? Explain.

26. **The Surface Area to Volume Ratio Explains a Lot!** Use the concept of the surface area to volume ratio to answer each of the following questions. Explain your reasoning.
  a. Suppose that you have an ice cube that you want to use to cool a glass of water. What should you do with the ice cube to cool the water as rapidly as possible?
  b. Are you able to digest your food more easily if you chew it into small pieces?
  c. Suppose that two people are proportioned roughly the same and engage in roughly the same amount of physical activity but that one person weighs 20% more than the other. Would you expect the larger person to eat 20% more food than the smaller person?
  d. Why do you stay warmer if you curl up into a ball-like position?
  e. Why can flies walk on the ceiling?
  f. Why can smaller animals fall from greater heights with less physical damage than humans?

27. **Understanding Scaling to Prevent Minor Disasters.** Suppose that a heavy piece of furniture, such as a water bed, is on the second floor of an apartment building. By thinking about how pressure scales with size, explain why the water bed is more likely to crash through the floor if it is supported on four posts (one at each corner of the bed) than if it sits on a large, flat platform.

Section 7.5

28. **A Billion Dollars Worth of Homes.** The average price of a family home in the United States is about $100,000. How many such homes could you buy with a billion dollars?

29. **Counting Your Gigabuck of Cash.** Suppose that you were given $1 billion, in $1 bills. How long would you need to count your fortune? Explain any assumptions you make.

30. **A Billion Dollars Worth of Stuff.** In Example 7-3, we estimated that the total value of a person's "stuff" (not including real estate) is on the order of $10,000.
  a. Suppose that you had a billion dollars to spend. How many people's "stuff" could you buy? Compare this number to the population of the town or city in which you live.
  b. Suppose that you decided to buy all this stuff by telephoning individuals and negotiating to buy all of their stuff. If you spend an average of 15 minutes on each call (you're a good negotiator!), how long would completing all your negotiations take?

31. **You're Going Further into Debt.** By 2002, according to some estimates, the national debt will reach $7 trillion. If the U.S. population has grown to 280 million by then, how much will be the debt per person?

32. **Paving with Dollar Bills.** Measure the length and width of a $1 bill and use your result to find its area.
  a. Suppose that you began laying $1 bills to cover the ground. If you had the 1995 debt of $5 trillion in $1 bills, how much total area could you cover?
  b. The total land area of the United States is about 10 million square kilometers. Could you cover the United States with the national debt in $1 bills?
  c. The surface area of the entire Earth is about 510 million square kilometers. Could you cover the surface area of the entire Earth with those $1 bills?

33. **Finding the Right Partner.** Suppose that you attend a school with 20,000 students in which half are men and half are women. Not wanting to get married too quickly, you decide you should do some dating before settling on a person for a permanent relationship. You decide that you should go on a date with each of the 10,000 students of the opposite sex. Assuming that you're such an incredible person that no one could possibly turn down your request for a date and that you limit yourself to one date per day, how many years would you need for the 10,000 dates?

34. **Going to Congress by Kissing Babies?** Suppose that you are running for election to the House of Representatives. Recognizing that shaking all your constituent's hands would take too long, you decide to restrict yourself to kissing babies. If 3% of your 570,000 constituents are babies, how many babies will you be kissing? If you can kiss each baby and tell the baby how cute she or he is in about 30 seconds, how long will you need to kiss them all?

35. **The President Reads Your E-Mail Personally.** Suppose that the president wanted personally to read each of the 25,000 e-mail messages received in a typical day. If he can quickly read each message in about 10 seconds, could he read them all in a day? Explain.

36. The Disappearance of the Pyramids. Suppose that the rate of erosion for the pyramids is 1% every 5000 years. How many years will elapse before the pyramids are erased?

Further Topics and Applications

37. Are Billions Spent Wisely? Look through this week's newspapers for at least two instances in which expenditures (by government, corporations, or other groups) of at least a billion dollars are mentioned. Describe each instance and write a paragraph explaining whether you think the money is being spent for a good purpose.

38. Employing a City. Suppose that you live in a city of 100,000 people. Assume that about half the population is working and that keeping a working person employed costs about $50,000 per year. What is the cost of keeping this city employed? Compare that to NASA's annual budget of about $13 billion. How many people do you think NASA employs?

39. Zipper Money. Suppose that you invented and patented a useful, inexpensive product such as the zipper. Imagine that, on average, every person in the United States buys about 10 items using your product each year. Further imagine that you earn a royalty of 1¢ on each item. How much money will you earn each year from your invention?

(The first zipper, called the Hookless Fastener, was invented in 1893 by Whitcomb L. Judson. It was used on boots and shoes and consisted of two thin metal chains that could be locked together with a metal slider. In 1910, Judson developed the C-Curity Fastener, for trousers and skirts. B. F. Goodrich bought Judson's company in 1923 and used the zipper for its rubber galoshes.)

40. Interest on the National Debt. Because the government finances the national debt by borrowing money, it must pay interest on the debt. Assume that the government is paying interest on $5 trillion at a rate of 3% per year (i.e., the interest payment is an amount equal to 3% of the total debt). What is the annual interest payment? What would be the interest payment, assuming a 3% rate, when the debt reaches $7 trillion? If interest rates rose and government had to pay at a rate of 5%, what would be the interest payment?

41. Printing Money. Suppose that the government decides to pay off the national debt simply by printing more money and decides to print it in $1 bills. Assume that the printing starts when the national debt is $5 trillion. How many dollars will need to be printed each second in order to print the entire $5 trillion within 1 year? How many bills would have to be printed each second if $100 dollar bills, instead of $1 bills were printed? Comment briefly on the economic repercussions if the government were to pay off the debt by printing money.

42. What Would You Do with a Billion Dollars? Imagine that you suddenly inherited a billion dollars. List, in detail, what you would do with the money during the first year. Divide your list into investments, purchases, and charities, and then itemize proposed disbursements within each group. (Don't just say that you will give $100 million to charity or invest $500 million in the stock market; list the particular charities you would donate to and the particular stocks or mutual funds you would invest in.) Explain each item on your list; for example, if you plan to give some amount to a particular charity, be sure to explain what you think the charity can do with that money. Note that you must be careful with your investments, in particular, or you risk losing your money! Putting it all in a bank account, for example, wouldn't be wise because bank accounts are insured only up to $100,000.

43. What Would *She* Do with a Billion Dollars? Suppose that two large companies merge. In the transaction, the chief executive of one of the companies gains a personal fortune worth $1 billion, through stock options and other mechanisms. Imagine that, instead of keeping the billion dollars personally, the executive decides to use it to invest in new ideas. She decides to hire some of our nation's best and brightest for a period of a year, offering them the opportunity to work creatively to come up with new technologies that could benefit society and generate profits for the company. She offers salaries for the year of $100,000 per person and incurs overhead of the same amount per person.

a. Under the assumptions stated, how many people could the executive hire with the billion dollars?

b. Write a paragraph or two contrasting the value to stockholders of having an executive earn a billion dollars personally versus using the money to hire new employees.

c. For one person to earn a billion dollars might seem excessive, but some people argue that the system that allows such huge personal gain keeps the entire economic system functioning better. However, others argue that such large salaries are a burden on the overall economic system. Count the number of letters in your last name. If the number of letters is even, write a few paragraphs defending large personal gains as reasonable and beneficial. If the number of letters in your last name is odd, argue the opposite.

44. The Cost of Everything. We estimated that the typical American owns on the order of $10,000 worth of "stuff" (not including real estate). Use this estimate and the U.S. population (about 250 million) to determine the cost of buying *everything* (except real estate) owned by everyone in the United States. The United States owns an estimated one fourth of the world's wealth. How much would buying everything owned by everyone in the world (again, except real estate) cost? Compare these numbers to the U.S. national debt.

45. More on the S&L Bailout.

a. The $500 billion in losses from the S&L crisis occurred over about a decade. If that amount was lost at a constant rate, how much money was lost each year? each day? each second?

b. In 1990, the budget of the U.S. Department of Education was about $25 billion and NASA's budget was about $13 billion. Compare these amounts to $500 billion.

c. Suppose that, instead of bailing out the S&Ls, the government had decided to give $500 billion to individual citizens by sending checks to every person in the United States. If everyone received the same amount, what would be the value of each check? Suppose, instead, that the government decided to divide this amount among the entire world population. How much would each check be worth in that case?

d. Reread the John Allen Paulos quote about the cost of the S&L crisis at the end of subsection 7.5.4. Explain how he arrived at the price of the transcontinental gold bar.

e. Write a paragraph or two stating your own opinion of the damages resulting from the S&L crisis.

46. **One of the Very Best.** Imagine that you are very good at a particular sport; so good, in fact, that you are better in your sport than 99.9% of all people. This means that only 0.1% of the population can compete in this sport at your level. How many of the 250 million people in the United States can perform the sport as well as you?

47. **Scaling Pizza.** A pizza shop sells a small and a large pizza. The small pizza is 12 inches in diameter, and the large pizza is 16 inches in diameter. The prices, without toppings, are $8.40 and $11.20, respectively.
   a. By what factor is the diameter of the large pizza larger than that of the small one?
   b. Compare the areas of the large and small pizzas.
   c. Is one of the pizzas a better deal than the other? Explain.
   d. You can calculate the actual area of each pizza by recalling that the area of a circle is π times the square of the radius, and that the radius is half of the diameter. That is: area (circle) = $\pi \times (\text{radius})^2$. What is the cost of each pizza, in dollars per square inch?

48. **Photo Enlargements.** Suppose that you are printing photographs from negatives of 35-millimeter film. Note that "35-millimeter" film actually measures 24 by 36 millimeters.
   a. Commercial prints are enlarged to an effective size of 4 inches by 6 inches ("standard" prints, which are 3.5 by 5 inches, are cropped from this larger size). How do the linear dimensions of the negative and the 4 by 6–inch print compare? How do their areas compare?
   b. One reason that commercial shops typically offer smaller prints (3.5 by 5 inches) is to keep costs down. The cost of raw photographic paper is approximately proportional to the area of the paper. Based on the cost of paper only, what would you expect a 4 by 6–inch print to cost, if a 3.5 by 5–inch print is 17 cents? How about an 8 by 10–inch enlargement?
   c. Check the prices at a local store of 3.5 by 5–inch, 4 by 6–inch, and 8 by 10–inch prints. Compare their prices per unit area. Do you think some sizes have a higher profit margin?
   d. You are probably aware that enlargements sometimes look "grainy" compared to smaller prints. Suppose that a negative has 100 grains (e.g., 100 distinct points of color and brightness) per square millimeter. When it is enlarged to a 4 by 6–inch size, how many grains will there be per square millimeter? What about at an 8 by 10–inch size? How about at a "poster" size of 20 by 24 inches? At what point do you think an enlargement from this negative would begin to look "grainy"? Explain.

49. **Movie Monsters.** Many movies make the impossible seem possible. Consider King Kong, who easily held Fay Ray or Jessica Lange in his hand. In contrast, a real gorilla has a 20-cm hand and body length of 2 meters.
   a. Based on an adult human fitting his hand, estimate the size of King Kong's hand. If King Kong is just a scaled-up gorilla, how tall is he?
   b. How much more pressure would there be on King Kong's joints compared to a real gorilla? Based on this measure, is the story of King Kong realistic? Explain.

50. **Can You Use the 1 to 10 Billion Scale for the Milky Way?** The center of the Milky Way galaxy is about 25,000 light-years from our solar system.
   a. How far is the center of the Milky Way, in kilometers?
   b. Suppose that you tried to make a scale model of the Milky Way galaxy on the 1 to 10 billion scale used for the solar system in subsection 7.4.2. On this scale, how far would the solar system be from the center of the Milky Way? Would this be a practical scale for making a model of the galaxy?
   c. Suppose that you made a 1 to 10 billion model of the solar system with the model Sun located at your own home. Describe where the center of the Milky Way galaxy would be located. For example, would it be somewhere on the Earth, or as far away as the Moon, or as far away as Mars? (The data in Problem 20 will be useful.)
   d. In words, describe the difficulty of visualizing the solar system and the Milky Way galaxy on the same scale. Suggest a more appropriate scale for a model of the Milky Way.

51. **A Dot in the Milky Way.** Draw a small dot on the diagram representing the Milky Way galaxy (Figure 7-3). Next, draw a square around the picture of the Milky Way and a very tiny square around your dot.
   a. Measure the area of the square that contains the Milky Way and the area of the square that contains your dot. Give both areas in square millimeters.
   b. How many of the tiny squares with the dot would fit into the larger square that represents the Milky Way?
   c. Assume that about 400 billion stars exist in the Milky Way galaxy. If the big square represents the entire galaxy of 400 billion stars, how many stars would fit in the small square with your dot? How would this number change if only 100 billion stars were in the galaxy? How would it change if it contained a trillion stars?
   d. You have found the number of stars represented by a dot on the Milky Way galaxy. By combining this information with other ideas about scaling, write a short essay in which you put the size of the galaxy in perspective.

52. **Scaling the Local Group of Galaxies.** The Milky Way galaxy has a diameter of about 100,000 light-years. Another member of the Local Group of galaxies, called the Great Galaxy in Andromeda, also has a diameter of about 100,000 light-years. It is located about 2 million light-years

from the Milky Way. Use a scale of 1 centimeter to 100,000 light-years and draw a sketch showing both galaxies and the distance between them to scale.

53. Scaling the Atom. Pictorial representations of an atom often show a nucleus surrounded by a swarming electron cloud. The actual diameter of an atom is approximately 10^{-10} meters.

a. If a picture shows an atom with a diameter of 5 centimeters, what is the scale of the picture?

b. The nucleus of an atom is about 10^{-15} meters in diameter. How big would it be in the picture from part (a)? Could you see the nucleus in the picture?

c. Suppose that you built a scale model atom in which the nucleus is the size of a tennis ball. About how far would the cloud of electrons extend?

54. The Earth's Timeline. Imagine making a timeline for the Earth's 5-billion-year history. You have a sheet of paper 1 meter long. One edge of the paper will represent when the earth formed; the other will represent the present. Some important moments you might want on your timeline are

Years Ago	Event	Years Ago	Event
4.6 billion	Formation of the Earth	65 million	Extinction of dinosaurs
3.5 billion	Life established	3 million	Earliest humans
600 million	First life on land	30,000	Cave paintings
200 million	Earliest dinosaurs	10,000	Agriculture, villages

a. What is the scale for this timeline?

b. How far is each listed event from the present, in cm?

c. How far from the edge of the paper representing the present would you locate the birth of the United States in 1776? How far from the edge is your own birth? Compare these answers to the size of a typical bacterium (about 10^{-6} meter in diameter) or a typical atom (about 10^{-10} meter in diameter). Discuss your results briefly.

55. A Cosmic Wink. Human history sometimes is characterized as a "blink of the eye" in the 7–20-billion-year history of the universe. Is this characterization accurate? Is it close? Explain.

56. Until the Sun Dies.

a. Suppose that future evolution proceeds at a rate where new intelligent species evolve every 25 million years. In that case, how many new intelligent species could evolve, one after the other, over the next 5 billion years?

b. Suppose that humans survive and that our descendants are still living in 5 billion years. How closely do you think they will resemble us? Write a short essay describing how and why you think humans are likely to change.

57. A Single Century. Return to the much shorter time of a century. Write a short essay that puts one century into perspective by comparing it to things that take place in your daily and yearly lives.

58. Most Populous Countries. The following table lists the 10 most populous nations in the world in 1993.

Nation	Population	Nation	Population
Bangladesh	114 million	Japan	125 million
Brazil	152 million	Nigeria	95 million
China	1.18 billion	Pakistan	122 million
India	897 million	Russia	149 million
Indonesia	188 million	United States	258 million

a. What percentage of the world's 5.5 billion people live in one of the 10 most populous nations?

b. Make a new table showing the 10 countries in order from most to least populous. Then add a column indicating the percentage of the world's population that lives each of the 10 countries.

c. Study a globe to get a feel for the area of each of these countries. What conclusions can you draw about their population densities?

59. The World's Species. Life abounds on planet Earth. More than a million living species of plants and animals are known and classified, although many times more may still be undiscovered. The following table lists the numbers of identified species according to major group:

Group	Number of Species	Group	Number of Species
Insects	750,000	Fungi	69,000
Protozoa	31,000	Monera (Bacteria)	4800
Higher Plants	248,000	All Other Animals	281,000
Algae	27,000		

a. What is the total number of identified species?

b. Make a new table showing the groups in order from most species to fewest. Then add a column indicating the percentage that each group represents of the total number of known species.

c. Recent studies in tropical rain forests suggest that the total number of insect species, including both identified and unidentified species, is approximately 3×10^7. If the percentages of species according to group in the preceding table hold for unidentified species, how many total species (including those unidentified) are there on Earth? How many species in each group?

60. People, Land, and Deforestation. Approximately 0.5 hectare of cropland is needed to support a human with a nutritious, diverse diet (1 hectare = 10,000 square meters). In 1990, the world's population was 5.3 billion people, and it was growing at approximately 100 million people per year.

a. How many hectares of cropland were needed to nourish the 1990 human population?

b. Express your result from part (a) as a percentage of the Earth's total land area of 1.48×10^{10} hectares.

c. How much new cropland needs to be created to support the growing population each year? Do you think the amount of cropland can continue to increase at this rate? Explain.

d. The global rate of tropical deforestation for the early 1990s has been estimated at 140,000 square kilometers per year. How does this rate compare to your answer in part (c)? What are the implications of this result?

61. **Global Deforestation.** Bioclimatic data suggest that tropical rain forests once covered 1.45×10^7 km^2 of the Earth's total land area of 1.48×10^8 km^2. In 1989, the total extent of rain forests was about 5.7×10^6 km^2. The reduction in rain forests is believed to be entirely the result of human activity. What percentage of the Earth's land mass was covered by tropical rain forests before human activity began reducing their size? By how much have the world's rain forests been reduced by human activity, as a percentage? Discuss the implications of this change.

62. **Water Use.** In the United States about 340 billion gallons of water *per day* are used for human activities, of which approximately 10% is used for public tap water; 11% is used by industry; 38% is used to cool electric power plants; and 41% is used for agricultural irrigation.
   a. Calculate the total amount of water used in the United States each year; then calculate the total amounts used for tap water, industry, power plants, and agriculture.
   b. What is the overall per capita use of water in the United States per day? per year?
   c. What is the per capita water use per day of tap water?
   d. You may be surprised at the large per capita tap water use obtained in part (c). Explain how you think that water is used. Do you think your personal water use is more or less than average?

63. **Snow Making.** Pushing to open their slopes early in the season, ski resort operators often rely on making their own snow. The use of 150,000 gallons of water will cover 1 acre of slope 1 foot deep with human-made snow.
   a. Convert the 150,000 gallons into units of acre-feet. How many acre-feet of water are needed to make an acre-foot of snow?
   b. In 1993, Colorado ski resorts used 2000 acre-feet of water to make snow. How many acre-feet of snow did these resorts make? If they try to cover their slopes with 6 inches of human-made snow, how many acres of ski slopes can be opened with this snow?
   c. In contrast to snow making, agricultural irrigation in Colorado used 10 million acre-feet of water in 1993. How does the amount of water used for snow making compare to that for agriculture? Express your answer in both absolute and relative terms.

64. **Space Waste Disposal.** To get rid of radioactive waste, some people have suggested rocketing it into space. Considering only spent fuel rods from nuclear reactors (i.e., not including material from nuclear bombs, radioactive medical waste, etc.), the world has accumulated about 84,000 metric tons of nuclear waste as of 1993.
   a. The cost of launching material on the space shuttle is approximately $7000 per pound. How much would transporting all the world's spent fuel rods into space cost?
   b. A more economical method would involve simpler rockets; the cost could drop to as low as $300 per pound. How much would all these space shots for nuclear waste removal cost?
   c. Do these costs seem justified? Explain. What dangers would such a project pose?

65. **Cars, Bikes, Humans, and Weight.** Getting from place to place in a car or on a bike involves moving different amounts of material.
   a. Estimate the weight of yourself, a car, and a bicycle. Use order of magnitude estimation to provide justification for your numbers.
   b. How much weight is moved between two points when you ride in a car? What percent of the total weight do you represent?
   c. How much weight is moved between two points when you ride a bicycle? What percent of the total weight do you represent?
   d. Based on your findings, explain why car pooling is more efficient, in terms of energy usage, than driving personal cars.
   e. If you were trying to design a more efficient car, what steps would you take? Why?

66. **Automobile Travel.**
   a. Estimate the total number of miles that you traveled in a car (yours or someone else's) during the past year.
   b. Estimate the cost of gasoline for all your traveling by car during the past year.
   c. Estimate the total cost of operating an automobile for all your traveling by car in the past year. Be sure to include all costs (e.g., maintenance, insurance, repairs.)
   d. Estimate the total amount of money spent on gasoline each year in the United States. Compare this amount to the federal deficit.
   e. For every liter of gasoline burned by an automobile, approximately 2.8 kilograms of carbon dioxide are emitted into the atmosphere. Estimate the total amount of carbon dioxide added to the atmosphere by your automobile travel over the past year.
   f. Estimate the total amount of carbon dioxide emitted into the atmosphere each year by all the cars in the United States.

67. **The Corn Car.** Because the supply of oil in the world is limited, alternative fuels for automobiles are being sought. One alternative fuel is ethanol, which can be produced from corn.
   a. Assuming typical driving mileage and fuel efficiency, approximately 6 hectares (1 hectare = 10,000 square meters) of corn-producing land are needed to make enough ethanol for 1 car for 1 year. About how much land, in hectares, must be devoted to corn production in order to supply ethanol for all automobiles in the United States?
   b. About 2 million square kilometers of land currently are devoted to agriculture in the United States. Compare this area to the land needed for ethanol production under the assumptions of part (a). Based on your results, do you think that ethanol is a feasible replacement fuel for gasoline? Explain.

68. **Thunder and Lightning.** When you see a flash of lightning, the roar of thunder usually comes several seconds later. This delay reflects the difference in the speed of light and the speed of sound. Light travels at the extremely high rate of 3×10^8 m/s, but sound travels only 3.3×10^2 m/s in air.
    a. Compare the speed of light to the speed of sound. Express your answer both in absolute terms (the difference in speeds), and in relative terms (the speed of sound as a fraction, or percentage, of the speed of light).
    b. Suppose that you see a bolt of lightning strike a mountain top 1 mile away. How long would the light take to reach you? How long would the sound take to reach you?
    c. Use your results from part (b) to express a general rule relating the time required for thunder to reach you and your distance from a lightning flash. That is, if you see a flash and then count the number of seconds until you hear the thunder, how can you calculate the distance to the lightning strike?
    d. Convert the speed of sound into units of miles per second. Does your result agree with the rule you stated in part (c)? Explain. (It should!)
    e. If you see lightning flash and then hear thunder four seconds later, how far away was the lightning?
    f. Because lightning often strikes water (a relatively good conductor of electricity), you shouldn't be swimming if lightning is striking within 2 miles. Suppose that you are in a pool and see a flash from lightning in the distance. You begin to count the seconds until the sound of thunder. Describe how you will decide whether remaining in the water is safe.
    g. Lightning tends to strike tall, solitary objects such as trees or lightning rods. Imagine that you are on a mountain top above tree-line, where you are taller than most other things. Lightning is flashing around you and you hear thunder from many of the flashes in less than a second. Are you in trouble? What should you do?
    h. During a lightning storm you are safer if you stay away from a lone tree in an open field. However, being in a forest is safer still. Explain why you should stay away from lone trees in open fields, yet run for a nearby forest.

Projects

69. **Scaling Your Bedroom.** Make a scale drawing of your bedroom that fits on a standard sheet of paper. Show objects in it (e.g., bed, furniture, and piles of dirty clothes) at their correct positions and sizes. Label the scale on your drawing verbally and graphically. What is the scale factor?

70. **Collecting Rain.** Colorado has approximately 35 centimeters of rainfall per year (i.e., if all the rain fell at once, it would cover the ground to a depth of 35 cm); this total includes the precipitation in snow.
    a. With an area of 100,000 square miles, what is the total volume of rainwater that falls on Colorado in an average year? Give your answer in cubic meters and gallons.
    b. Much of this water is collected or diverted in some way for human use. If Colorado's population in 1990 was 3.7 million people, how many gallons of rainfall per person does Colorado receive?
    c. Total per capita water use in the United States is about 1400 gallons per day, which includes not only personal water use, but also water use by industry and agriculture (see Problem 62). How many gallons does this quantity represent per year?
    d. Comparing your answers from parts (b) and (c), about what fraction of Colorado's annual rainfall is diverted to human use? Much of the water for the states of Nevada, New Mexico, Arizona, and California comes from the diversion of water from rivers flowing out of Colorado, and these states are even drier than Colorado. What can you say about the overall availability of fresh water in the western states?
    e. Diverting water from its natural flow has consequences for ecosystems. Many ecologists believe that a 10% reduction in stream or river flows will cause long-term degradation of stream and river ecosystems. Do the western states face a water crisis in the near future? Defend your opinion.
    f. Find the average rainfall and population for two other states and repeat this problem. Choose one arid state (e.g., Nevada) and one wetter state (e.g., Louisiana). Discuss water availability in these states, as well as the issue of whether water use is causing ecological damage.

71. **Cold-Blooded Reptiles, Cold-Blooded Dinosaurs?** A warm body is needed for basic metabolic processes, so cold-blooded reptiles bask in the sun to raise their body temperatures. The Sun's heat must be used to raise the temperature of the entire volume of the body, but it is absorbed only by the reptile's surface. Thus the rate at which the Sun warms the reptile is proportional to its surface area to volume ratio.
    a. Explain why the total amount of time that a reptile must bask in the sun is proportional to the *reciprocal* of the surface area to volume ratio (i.e., to the volume to surface area ratio). Explain further why the time a particular cold-blooded reptile must bask in the sun is roughly proportional to its linear size.
    b. Consider a Fence Swift lizard 10 cm in length. It lounges in the morning sun for 10 minutes to raise its body temperature. How long would you expect a 1-meter Monitor lizard to bask to raise its body temperature?
    c. If a dinosaur measuring 5 meters in length was cold-blooded, how much time would you expect it to have needed in the sun to raise its body temperature to a level comparable to the lizards in part (b)?
    d. Does you answer to part (c) make you question whether large dinosaurs were cold-blooded? Explain.
    e. Suggest several problems that might bring the analysis in part (c) into question. Can you think of any ways of learning whether dinosaurs were warm- or cold-blooded? Find and discuss the current beliefs among dinosaur researchers regarding the question of whether dinosaurs were warm- or cold-blooded.

72. **Counting More Stars and More Civilizations.**
    a. Suppose that, rather than 100 billion, there are 1 trillion stars in the Milky Way galaxy. How long would you need to count to 1 trillion?

b. If 1 trillion stars exist in the Milky Way galaxy and 1 in 10 million have supported a civilization at some time, how many civilizations have existed in the galaxy?

c. Astronomers estimate that between 10 and 100 billion galaxies are in the universe. If there are 10 billion galaxies and each has supported (at some time) the number of civilizations you calculated in part (b), how many civilizations have existed in the universe?

d. Compare the number of civilizations in the universe, under the assumptions of parts (a)–(c), to the number of people on Earth. Compare it to the number of stars in the Milky Way galaxy.

e. Suppose that your assumptions are correct and that as many civilizations have existed in the universe as you have calculated. Further, imagine that someday someone can prove that to be the case. Write a short essay about how such knowledge would alter people's "cosmic perspective." Be sure to include any effects you think it should have on human behavior.

f. The galaxy is believed to be about 10 billion years old. Based on the rate of evolution on Earth, an intelligent, space-traveling civilization could evolve in less than 5 billion years. Based on the number of civilizations estimated in part (b) approximately how much time, on average, has elapsed between the development of civilizations in the galaxy? Suppose that humans someday encounter another civilization, either through radio or direct contact. Would you expect them to have a similar level of technological development? Explain.

g. Despite the pseudoscientific claims of UFO believers, no hard evidence exists that Earth has ever been visited by an alien species. This lack of evidence poses a bit of a paradox. According to the assumptions in this problem, many previous civilizations should have been present in the galaxy, with plenty of time to develop the technology and means of colonizing star systems throughout the galaxy. Thus Earth should have been visited, or even colonized, long ago! Form a hypothesis to explain why Earth has not, to anyone's knowledge, been visited. Write a short description of your hypothesis.

8 NUMBERS IN THE REAL WORLD

The numbers encountered in the real world nearly always are approximate and therefore involve uncertainties. Unfortunately, these uncertainties are rarely acknowledged or properly addressed. As a result, many important decisions are made on the basis of mistaken assumptions about the validity of reported numbers. In this chapter we examine uncertainty, emphasizing how it arises and how it can be handled honestly. We call this chapter "Numbers in the Real World" because we discuss numbers as they are used daily, rather than as they are used abstractly in formal mathematics.

CHAPTER 8 PREVIEW:

8.1 ***NUMBERS — NOT ALWAYS WHAT THEY SEEM.*** This introduction describes how numbers as used in daily life often are far less certain than they may at first seem.

8.2 ***SOURCES OF UNCERTAINTY.*** Most numbers in the real world are either measured or estimated. In this section we discuss the various sources of uncertainty in numbers and methods for recognizing these sources.

8.3 ***EXPRESSING UNCERTAINTY.*** In this section we show how to express uncertainty clearly. We contrast this approach with reality, in which numbers usually are reported with little or no indication of uncertainty.

8.4 ***CASE STUDIES IN UNCERTAINTY.*** We conclude the chapter with a set of case studies about uncertainty as it relates to several current issues.

And now for some temperatures around the nation: 58, 72, 85, 49, 77.

— *George Carlin,* comedian

Does anybody really know what time it is?

— *Chicago,* music group

None of us really understands what's going on with all these numbers.

— *David Stockman,* budget director for President Reagan, 1981

8.1 NUMBERS — NOT ALWAYS WHAT THEY SEEM

How much do you weigh? How many people live in the United States? How much has the temperature of the Earth changed because of human activities? How many species will suffer extinction during the next 50 years?

Each of these questions can be answered only approximately. Indeed, the degree of uncertainty increases with each question. You probably know your weight within a pound or two. But the population of the United States may not be accurately known within 10 million or more people. On the question of temperature changes, many scientists disagree about whether a temperature change has even occurred; ascribing the changes to human activities is even more difficult. And the species extinction question is the most difficult of all because it involves predictions.

Unfortunately, numbers reported to the public often disregard uncertainties. The 1990 Census of the United States reported a count of 248,709,873 people. Yet the census undoubtedly missed many people, and double counted others. Further, during the time the census was taken many people were born and many others died. Even if the reported count is supposed to represent the number of responses received by the census, it cannot be exact: The process of collecting surveys, recording responses, and adding the results inevitably involves errors. Moreover, reporting the population of the United States as 248,709,873 is fundamentally dishonest: It implies that the population is known with far more precision than is possible.

The census illustration isn't unusual. Many, perhaps most, of the numbers reported by the media are stated with far more certainty than they deserve. As a result, one of the fundamental aspects of quantitative reasoning is learning how to interpret and deal with uncertainty in a world where it is rarely acknowledged.

> **Time-Out to Think:** The census not only attempts to count the population, but also to determine *where* people live. As this information is used to determine the boundaries of congressional districts and allocations of federal money (e.g., funds for education programs), comment on the consequences of neglecting uncertainty in the census figures.

8.2 SOURCES OF UNCERTAINTY

In Chapter 6 we presented several different ways of classifying numbers. In practical situations, as we showed in Section 6.5, it is useful to think of numbers in terms of just two categories, exact or approximate. Although most of your prior courses in mathematics probably dealt exclusively with exact numbers, approximate numbers are far more common. In fact, exact numbers generally arise in only two situations.

- Numbers from mathematical theory, such as numbers in formulas, are exact. For example, the formula for the circumference of a circle is 2π times the radius, or $C = 2\pi r$. Both 2 and π are exact numbers in this formula. (Although π is an exact number, it cannot be written exactly in decimal form because it is irrational.)
- Numbers obtained by simple counting may be exact, if no counting errors are made. For example, counting the exact number of students in your class (at any particular time) is possible.

In all other cases, numbers are approximate and therefore have an associated uncertainty. More specifically, approximate numbers are obtained from measurement or estimation.

8.2.1 Uncertainty in Measurement

Suppose that you measure your weight by standing on a scale and reading your weight from it. You hope that the weight you measure, which we call the **approximate** (or **measured**) **value** of your weight, is close to its actual or **true value.** However, numerous potential sources of uncertainty exist in your measurement.

Precision

No measurement device is perfect. Different scales offer different levels of **precision**. A scale that allows you to measure to the nearest *tenth* of a kilogram is more precise than one that allows you to measure only to the nearest whole kilogram. The precision of the scale introduces uncertainty because it allows the measured value to differ from the true value.

For example, suppose that you weigh yourself on a scale that can be read to a precision of a tenth of a kilogram, and that your measured weight is 52.3 kilograms. Because you could read the scale only to the *nearest* tenth of a kilogram, this measurement implies that your true weight lies somewhere between 52.25 and 52.35 kilograms.

> **Time-Out to Think:** Explain why a measurement of 52.3 kilograms implies a true value between 52.25 and 52.35 kilograms. (Hint: remember that this scale can be read to a precision of the *nearest* tenth of a kilogram.)

Random Errors

In reality, the true value of your weight may lie *outside* the range of 52.25 to 52.35 kilograms because sources of **error** may exist in your measurement. For example, some scales require balancing small weights, a difficult and tedious process; indeed, obtaining a *perfect* balance is impossible. Furthermore, even if you balance the scale well, reading it properly can be difficult (Figure 8-1).

At some level, errors such as those from balancing and reading a scale are unavoidable when using and reading *any* measurement device. Sometimes such errors cause the measured value to be greater than the true value, and sometimes they cause it to be less than the true value. Hence these types of errors are called **random errors** because their effect on the measured value is unpredictable.

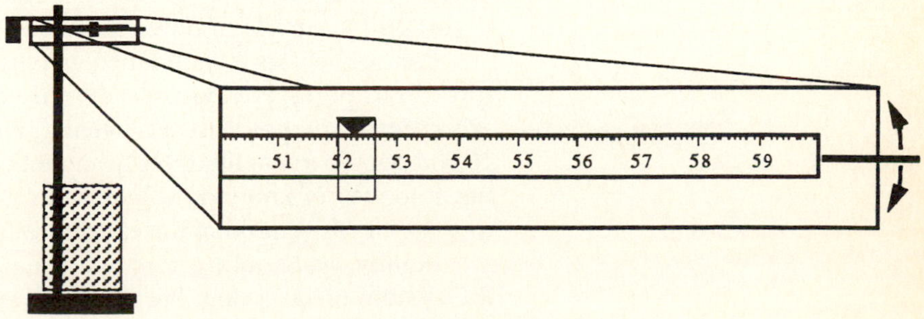

Figure 8-1. Reading a scale introduces random errors both because it can be difficult to read precisely and because achieving a perfect balance is impossible.

How are measuring devices calibrated? The calibration of a scale, for example, can be checked by weighing several objects whose weights have been well established by other measurements. If the scale's weights agree with the expected weights, it is well calibrated. Of course, this result begs the question of how we know the "expected weights." Ultimately, this is the role of organizations such as the International Bureau of Weights and Measures and the National Institute of Standards and Technology (see subsection 6.4.2), whose scientists *define* measurement standards.

Systematic Errors

Imagine that you took great precautions to minimize random errors when you measured your weight. For example, you might have practiced using the scale with known weights to be sure that you could balance and read it well. Can you *then* conclude that your true weight lies within the range of 52.25 to 52.35 kilograms implied by your measured weight of 52.3 kilograms?

The answer still is no. Despite your care in eliminating sources of random error, problems still might exist with the process, or **system**, used to measure your weight. For example, the scale must be *calibrated* properly; that is, it must read 0 when nothing is on the scale, and it must correctly balance at higher weights. If the calibration is off, your measurement will be off. Similarly, shoes or clothing increase your weight; even long hair can affect your weight; and unseen dirt on the scale's balancing system might change the measurements given by the scale.

Errors such as those caused by calibration, clothing, or dirt are called **systematic errors** because they affect the measurement *system*. In general, systematic errors affect all measurements similarly. Clothing, for example, always increases your measured weight, and dirt on the scale's balancing system might make all measured weights lower. Thus if you discover a systematic error, you can adjust earlier measurements. Unfortunately, you can never be sure that you have discovered *all* possible systematic errors.

Time-Out to Think: Call the local phone number that gives the current time. How far off is your clock or watch? Describe possible sources of random error in your personal time keeping. Do you introduce any sources of systematic error, such as deliberately setting your watch a few minutes ahead?

8.2.2 Uncertainty in Estimation

The concepts of precision, random error, and systematic error have analogs in estimation. Ultimately, any process of estimation involves some type of **sampling**. For example, in estimating the popularity of a politician, an opinion poll involves only a small *sample* of the entire population. In estimating the number of apples in a shipment of crates, an inspector might sample just a few crates.

The analog to precision in estimation is the *size* of the sample. In general, the larger the sample, the closer will be an estimate to the true value — as long as no other sources of error are introduced. For example, suppose that you conduct a poll to determine the number of voters expected to vote for Ms. Franklin in the next presidential election. If you poll a sample of 10,000 voters, you are likely to get a better estimate of the actual election results than if you poll a sample of only 500 voters.

Random errors in estimation essentially are caused by uncontrollable factors in sampling. For example, even if you select a 10,000-person sample according to proper statistical techniques, the opinions of this *particular* set of 10,000 people may still differ from those of the general population. Because you have no way of knowing whether this group is more or less likely than another to vote for Ms. Franklin, the effect of this error on our estimate is random.

Finally, systematic errors in estimation occur when a problem exists with the system of sampling. For example, suppose that you conduct the opinion poll by calling telephone numbers selected from a list of contributors to the Republican National Committee. This poll isn't likely to reflect the opinions of voters at large!

8.2.3 Accuracy: True Versus Measured Values

Some statistics texts define accuracy as the *relative* uncertainty of a measured or estimated number. For example, suppose that after accounting for all *known* sources of uncertainty you determine your weight to be 50 ± 2 kilograms. The absolute uncertainty is 2 kilograms, and the relative uncertainty is 2/50, or 4%. By the alternative definition, your measurement is *accurate* to within 4%. Of course, if you are unaware of other sources of uncertainty, you may have underestimated both the absolute and relative uncertainties.

The ultimate goal, in either measurement or estimation, is to obtain an approximate value as close as possible to the true value. The closer that the approximate value lies to the true value, the greater is the **accuracy** of the measurement or estimate.

However, in reality, you can *never* know the true value independent of measurements or estimates. For example, the only way to learn your weight is by weighing yourself, and the only way to learn the population is through some type of census. Thus, because you can know only the approximate values, you have no way of knowing the accuracy of measurements or estimates. Nevertheless, by taking precautions against both systematic and random errors, you can gain confidence that your approximate values accurately reflect true values.

A further problem may arise if a true value changes with time. Your weight, for example, fluctuates even over the course of a single day and can change dramatically if you are dieting either to lose or gain weight. Similarly, the population of the United States is constantly changing with births, deaths, immigration, and emigration. Thus, even if an approximate value accurately reflects a true value at the time of measurement or estimation, it might be less accurate hours, days, or weeks later.

Time-Out to Think: If you have a driver's license, when was it issued? Do you think the values of your height and weight shown on the license accurately reflected true values at the time it was issued? Do they now? Why or why not?

8.2.4 Identifying Sources of Uncertainty

A valid measurement or estimate made in accordance with the scientific method must be repeatable and clearly understandable to others. Thus each step in the measurement or estimation process must be described clearly; we call such a description a **process definition**.

If *you* make a measurement or estimate, you should describe your process definition in detail. To evaluate someone else's reported number, you need to understand the process definition that he or she used. Unfortunately, process definitions aren't always spelled out, so you often must make educated guesses as to how a reported number was obtained.

Example 8-1 *Process Definition for Measuring Heights.* You are asked to measure the height of each member of your class. Describe a process definition for the measurements and ways of minimizing uncertainty.

Solution: A basic way of measuring heights is to place a vertical tape measure on a wall, have students stand against the wall, and measure their heights. To minimize errors of precision, use a tape measure marked in small units — say, millimeters rather than inches.

The difficulty in reading the heights from the tape measure introduces sources of random error. For example, students might not stand straight, hair varies in thickness, and you might press down on different student's hair with different pressure. You can minimize these problems by carefully specifying posture for the measurements and a system for not counting hair thickness. Another problem is that you might simply misread the tape measure. This error can be minimized by asking several people to read the height of each student. If disagreement occurs in the several readings, you can recheck them; if disagreement persists, you could choose the average reading as the best value of the height.

One source of systematic error is introduced by the possibility that the tape is not truly vertical — in that case, all height measurements would be systematically too large. This possibility can be minimized by aligning the tape with the aid of a carpenter's plumb line or level. Another systematic error might be introduced, for example, if you allow students to wear shoes and socks; minimize this possibility by asking everyone to be barefoot.

❖

8.3 EXPRESSING UNCERTAINTY

Properly interpreting approximate numbers requires some idea of the involved uncertainties. Unfortunately, the nature of uncertainty means that you can never know whether you have accounted for all possible sources of error. In other words, you will always be uncertain about the degree of uncertainty! Nevertheless, it's usually possible to identify numerous potential sources of uncertainty. How, then, do you express what you *believe* to be the uncertainty of measured or estimated numbers?

The two common ways to express uncertainty are: (1) *explicitly*, by stating a range of possible values; or (2) *implicitly*, by being careful in rounding.

8.3.1 Explicit Uncertainty Ranges

The best way to express uncertainty is to state a **range** of possible values for a number. In the earlier weight example you were able to read the scale to the nearest tenth of a kilogram. Thus, assuming no other sources of error, a reading of 52.3 kilograms meant that your weight was somewhere within the range of 52.25 to 52.35 kilograms. Stating your weight explicitly as "between 52.25 and 52.35 kilograms" expresses the uncertainty caused by the limited precision of the scale.

Of course, if you believe that other sources of uncertainty may exist, such as problems in the scale's calibration or changes in your weight with time, you should state a larger range of uncertainty. Suppose that, a few days later, someone asks how much you weigh. You might believe that your weight could have increased or decreased by as much as two kilograms. In that case, your weight would lie in the range of 50.3 to 54.3 kilograms. However, because this range itself is uncertain, you should be a bit more conservative and express your weight as being "between about 50 and 55 kilograms."

Time-Out to Think: Suppose that you had measured your weight as 52.3 kilograms but that you were fully clothed, including shoes. In that case, what is the range of your possible weights? How would you express this range explicitly?

The level of confidence essentially is a statement of probability. For example, if you state that you are 90% confident that your weight lies between 50 and 55 kilograms, it means that there is a 90% chance that your weight actually lies in that range and a 10% chance that it does not (i.e., that your weight is less than 50 kilograms or greater than 55 kilograms). Statistics provides methods for quantifying the probabilities.

Even if you state your weight as being "between about 50 and 55 kilograms," you could still be wrong. For example, you might have gained or lost more weight than you thought possible, or the original measurement might have been affected by an undetected systematic error. Therefore it can be helpful if you also express your **level of confidence** in the stated uncertainty range. The science of statistics (see Chapter 14) provides quantitative methods for estimating levels of confidence. Without statistics, however, you can express levels of confidence *subjectively*. For example, you might say "I am very sure that I weigh between 50 and 55 kilograms" or "I'm 90% sure that I weigh between 50 and 55 kilograms." The more confident you are, the greater the percentage you would guess.

A stated range of values can be abbreviated using a plus or minus symbol (±). For example, stating your weight as 52.5 ± 2.5 kilograms is equivalent to stating the range as (52.5 – 2.5) to (52.5 + 2.5), or 50 to 55 kilograms. Again, you should also express your level of confidence in the stated uncertainty range.

Note that, in statistical and scientific work, stated uncertainty ranges generally represent a 95% level of confidence. For example, if a scientist reports that she has measured an increase of 0.5 ± 0.15°C in the global average temperature during the past 50 years, it means that she believes there is a 95% probability that she is correct (and a 5% probability that the actual temperature change lies outside the range of 0.5 ± 0.15°C).

Opinion polls generally express a 95% level of confidence by stating a *margin of error*. For example, suppose that a poll shows that 46% of first graders believe in the tooth fairy, with a margin of error of 3%. This means that the pollsters interviewed a sample of first graders, among whom 46% believed in the tooth fairy. Then, using statistical techniques, the pollsters concluded that there is a 95% probability that between 43% and 49% of *all* first graders believe in the tooth fairy. We discuss opinion polls and the margin of error in detail in Chapter 14.

Example 8-2 *Plus/Minus Notation.* Suppose a census finds that the population of a town lies between 22,000 and 25,000 residents. Express this range using plus or minus notation and interpret the range.

Solution: First find the midpoint of the range in the number of residents by adding the upper and lower limits and dividing by 2:

$$\frac{25,000 + 22,000}{2} = 23,500.$$

Next determine how far the upper and lower limits of this range lie from the midpoint by taking half of the total difference:

$$\frac{25,000 - 22,000}{2} = \frac{3000}{2} = 1500.$$

You now can express the range as 23,500 ± 1500 residents. Note that you can check your work by confirming that subtraction gives the lower limit of the range (23,500 – 1500 = 22,000) and that addition gives the upper limit (23,500 + 1500 = 25,000).

Interpreting the range requires knowing its level of confidence. Because census takers generally employ statistical techniques, you can reasonably assume that they gave a range with a 95% level of confidence. Thus, the census takers believe that there is a 95% chance that the number of residents lies within 1500 of 23,500, and a 5% chance that the number of residents differs from 23,500 by more than 1500.

❖

Example 8-3 *Global Warming.* A scientist reports that she has measured an increase of 0.5 ± 0.15°C in the global average temperature during the past 50 years. What does this statement mean?

Solution: The lower limit of the range is 0.5° – 0.15° = 0.35° and the upper limit is 0.5° + 0.15° = 0.65°. Thus the scientist is claiming that the global average temperature has increased by some amount between 0.35°C and 0.65°C during the past 50 years. Because this result comes from scientific work, you can reasonably assume that she reported the range with a 95% level of confidence.

❖

Order of Magnitude Uncertainty

How do you express uncertainty when dealing with order of magnitude estimates? Again, it's best to express the range explicitly. For example, the age of the universe is *on the order of* 10 billion years, but the range of uncertainty puts the age between about 7 and 20 billion years (see subsection 7.3.2). Note

The uncertainty range for order of magnitude estimates can be expressed with a plus/minus range in the *exponent*. For example, note that $10^{(10-0.3)} = 10^{9.7} \approx 5$ billion and $10^{(10+0.3)} = 10^{10.3} \approx 20$ billion. Thus, saying that the age of the universe is $10^{(10\pm0.3)}$ years is equivalent to saying that it is within a factor of 2 of 10 billion years.

that you cannot express this uncertainty range as a plus or minus range centered on 10 billion years. Instead, you can describe it in terms of a multiplicative factor: The lower limit of the uncertainty range, 7 billion years, is close to $\frac{1}{2} \times 10$ billion yr = 5 billion years, and the upper limit of the uncertainty range is 2×10 billion yr = 20 billion years. Thus you might say that "the age of the universe is within a factor of 2 of 10 billion years."

Similarly, we estimated (also in subsection 7.3.2) that total annual spending on ice cream by all U.S. residents is on the order of \$2.5 billion within a factor of 10, or between about $\frac{1}{10} \times$ \$2.5 billion = \$250 million and $10 \times$ \$2.5 billion = \$25 billion.

8.3.2 Significant Digits and Implied Uncertainty Ranges

In real life, uncertainties rarely are stated explicitly. We must therefore look for other clues in order to make an *educated guess* concerning the range of uncertainty in reported numbers. First we hope that numbers are rounded to reflect only as much precision as they deserve.

Note that we hope that the implied uncertainty range includes the reporter's best estimate of *all* sources of error. Because all sources of error are never known and the implied uncertainty gives no indication of a level of confidence, this method of gauging uncertainty is less satisfactory than explicitly stated ranges.

In dealing with exact numbers, we assume that *infinitely* many digits are known with certainty. For example, we don't distinguish between 5 and 5.00000000 when dealing with exact numbers. In contrast, with measured or estimated numbers we hope that digits are listed only if they have meaning. For example, we hope that a weight reported as 5.00 kilograms means that it was measured to the nearest *hundredth* of a kilogram; a weight of 5 kilograms should mean that it was measured only to the nearest whole kilogram.

Any digit in a number that reflects something that was actually measured, or carefully estimated, is called a **significant digit**. If you assume that only significant digits are reported, you can use them to determine an **implied uncertainty range**. For example, suppose that you are told that the height of a classmate is 167.1 centimeters. This statement *implies* that the height was measured to the nearest tenth of a centimeter or, equivalently, to the nearest millimeter. In that case, the actual height should lie between 167.05 and 167.15 centimeters.

Example 8-4 What is the implied uncertainty range for a measurement reported as 12.92 centimeters?

Solution: The measurement is expressed to the nearest hundredth of a centimeter, so you assume that it was measured to that precision. Thus the actual measurement lies in the range of 12.915 to 12.925 centimeters.

If you are having any difficulty with this solution, you can use a simple trick to help find the implied uncertainty range. First, pretend that the decimal point is *not* there; then the reported number reads 1292. Because the implied uncertainty range for any whole number is ± 1/2, the implied range of uncertainty for 1292 is from 1291.5 to 1292.5. Finally, put the decimal point back where it belongs. Then the implied range of uncertainty of 12.92 is from 12.915 to 12.925.

❖

When Are Zeros Significant?

When zeros are involved, the implied range of uncertainty is ambiguous. Imagine that you are told that the distance between two towns is 210 kilometers. The 2 and the 1 clearly are significant, representing 200 kilometers and 10 kilometers, respectively. But the 0 might simply be a placeholder for the other columns. If the 0 actually was measured and therefore is significant, the range of uncertainty is from 209.5 to 210.5 kilometers. If it is merely a placeholder, the range of uncertainty is from 205 to 215 kilometers. Unfortunately, without further information, you have no way of knowing whether 210 kilometers is precise to the nearest kilometer or only to the nearest 10 kilometers.

Because of this ambiguity with zeros, you generally should assume that they are significant only if they are *necessary* in reporting a measurement. Zeros are not significant if they serve simply as placeholders. Table 8-1 summarizes the circumstances under which digits generally are considered significant.

Table 8-1. When Are Digits Significant?

Nonzero digits	Significant
Zeros that occur *between* nonzero digits (as in 4002 or 3.06)	Significant
Zeros that follow a nonzero digit *and* lie to the right of the decimal point (as in 4.20 or 30.00)	Significant
Zeros to the *right* of the decimal point that serve only to place the location of the decimal point (as in 0.006 or 0.00052)	Not significant
Zeros to the *left* of the decimal point that follow the last nonzero digit (as in 40,000 or 210)	Not significant *unless* stated otherwise

Example 8-5 What is the implied uncertainty range in a volume of 6.05 liters?

Solution: The 0 in 6.05 liters represents a real measurement: Six liters plus 0 tenths of a liter, plus 5 hundredths of a liter actually were measured. Thus it is significant. The number is measured to the nearest hundredth of a liter, so the implied range of uncertainty is 6.045 to 6.055 liters.

❖

Example 8-6 What is the implied uncertainty range for a time reported as 11.90 seconds?

Solution: Assume that this measurement was made to the nearest hundredth of a second; if it had been measured only to the nearest tenth of a second, it should have been reported simply as 11.9 seconds. Thus the zero is significant, and the implied range of uncertainty is 11.895 to 11.905 seconds.

❖

Example 8-7 What is the implied uncertainty range for a weight reported as 0.0067 kilograms?

Solution: In this case, the zeros serve only as placeholders and are not significant. You can see this by rewriting the number in scientific notation as 6.7×10^{-3} kilograms. You can also see it by recognizing that 0.0067 kilograms can just as easily be called 6.7 grams. Thus the implied range of uncertainty is 6.65 to 6.75 grams, $(6.65 \text{ to } 6.75) \times 10^{-3}$ kilograms, or 0.00665 to 0.00675 kilograms.

❖

Example 8-8 What is the implied uncertainty range for a population reported as 240,000?

Solution: Unless told otherwise, you can assume only that the 2 and the 4 are significant. The zeros serve merely as placeholders, which becomes clear if you rewrite the number in scientific notation as 2.4×10^5. By using scientific notation, you can also easily determine the implied range of uncertainty. It is $(2.35 \text{ to } 2.45) \times 10^5$ or 235,000 to 245,000.

❖

Example 8-9 What is the implied uncertainty range for a population reported as 240,000 which, you are told, contains three significant digits?

Solution: If there are three significant digits, the 0 in the thousands column actually was measured, so the population is known to the nearest 1000. Thus the implied uncertainty range is 239,500 to 240,500.

❖

Further Ambiguity

The preceding examples demonstrate that you can remove the ambiguity of zeros by reporting numbers in scientific notation and discarding any zeros that simply are placeholders. In real life, however, numbers usually aren't written in scientific notation and you aren't always told whether zeros are significant. For example, suppose that your professor states that 200 students are enrolled in your class. Following the rules in Table 8-1, you would *assume* that the zeros are not significant so that this is an estimate only to the nearest 100. However, your professor may have estimated to the nearest 10 students or counted exactly 200 students. You have no way of knowing whether the zeros are significant without further information. In the end, the best guidance for determining the significance of these zeros probably is your intuition and any other information you can garner about why the professor stated the number this way.

Time-Out to Think: We stated earlier that the world population is about 6,000,000,000 (6 billion) people. Do you think that any of these zeros are significant? Why or why not?

Another case of ambiguity arises when people round to digits besides 0 or 1. For example, many people commonly round to 5, so a test score of 23 or 24 becomes "about 25." Imagine that you are told that a population is "about 250,000." Following the general rules for determining significance, you would assume that this number was measured to the nearest 10,000 so that the actual population lies between 245,000 and 255,000. However, the number may be far less certain.

In summary, an implied range of uncertainty is a good first step toward making such a guess, but it isn't foolproof. Indeed, it depends on your level of confidence that the reporter has reported only significant figures.

8.3.3 Combining Approximate Numbers

Suppose that a small corporation decides to relocate in a city having a reported population of 460,000. Including both employees and family members, about 100 people make the move. How does this increase affect the reported population?

Clearly, the population of the city will rise by about 100. We might therefore think that we could simply add this to the reported original population: 460,000 + 100 = 460,100. However, such a statement would be wrong. The original reported population of 460,000 has only two significant digits and is therefore known only to about the nearest 10,000. Reporting the new population as 460,100 implies that we know the population to the nearest 100 people, instead of the nearest 10,000. Because the uncertainty in the original population is so much larger than the number of new residents, the *reported* population should remain unchanged at 460,000.

This example illustrates that properly accounting for uncertainty requires special care when combining approximate numbers. Otherwise, you might inadvertently state an answer with more certainty than it deserves. Ideally, you should consider the range of uncertainty on each of the numbers you are combining and then carefully analyze the resulting range of uncertainty after your calculations. In most cases, however, the uncertainties themselves are so vague that following some simpler rules is safe.

Addition and Subtraction

Suppose that you are told the distance between City A and City B is 37 kilometers and that the distance between City B and City C is 14.7 kilometers. How far would you drive in going from City A to City C, via City B?

Note that this assumption — that the stated precision reflects accuracy — is precarious. If the distance to either of the cities is simply wrong because of some unrecognized source of error, all subsequent results will be invalid.

If you assume that the precision with which each number is stated reflects its accuracy — that is, that all sources of error have been accounted for — you can base your calculations on the ranges of uncertainty for the distances between each pair of cities, or

distance between City A and City B:	36.5 to 37.5 km;
distance between City B and City C:	14.65 to 14.75 km.

To find the range of uncertainty when you add these distances, consider the minimum and maximum possible sums, or

$$\text{minimum possible sum} = 36.5\text{ km} + 14.65\text{ km} = 51.15\text{ km};$$

$$\text{maximum possible sum} = 37.5\text{ km} + 14.75\text{ km} = 52.25\text{ km}.$$

Statistically speaking, the uncertainty in the sum should be slightly *less* than the sum of the individual uncertainties. Uncertainty ranges always reflect some level of confidence. For example, suppose that the uncertainty on each distance reflected 95% confidence that the true distances lay within the stated ranges. In that case, it is highly unlikely that the sum would be either the minimum or maximum distances calculated. Thus you could have 95% confidence that the combined uncertainty is less than the sum of the individual uncertainties (but still greater than either of the individual uncertainties).

The range of possible values for the total distance is 51.15 to 52.25 kilometers or 51.7 ± 0.55 km. Note that this range of uncertainty simply is the sum of the uncertainties in each of the added distances: 0.5 km + 0.05 km = 0.55 km. However, because the first distance was known only to the nearest kilometer, reporting the final range to the nearest hundredth of a kilometer seems unreasonable. Stating the range for the final answer as "about 51 to 53 kilometers" is safer.

Time-Out to Think: Adding the numbers in the two ranges assumed a roughly equal confidence in both measurements. Suppose, instead, that you were 99% confident in the range for the distance from City A to City B but only 70% confident in the range given for City B to City C. How would that affect the way you should report your final answer?

Can you get a similar answer without so much work? If you simply add the distances given, without accounting for the uncertainties, you get 37 km + 14.7 km = 51.7 km. As expected, the answer of 51.7 kilometers is the halfway point of the range 51.15 to 52.25 kilometers found earlier. If you round to the nearest kilometer, it becomes 52 kilometers: right in the middle of the range 51 to 53 kilometers stated for the final answer earlier. Generalizing, the rule for adding (or subtracting) approximate numbers is: Do the addition or subtraction and then round the final answer to the same precision as the *least precise* number in the problem.

An answer obtained by adding or subtracting approximate numbers should be rounded to the same precision as the *least precise* number in the problem. (This rule is based on the assumption that the stated precisions account for all sources of error.)

Note one difficulty with this rule: When you report the sum as 52 kilometers, you imply that you know the distance to the nearest kilometer, or in the range of 51.5 to 52.5 kilometers. In reality, however, the range was slightly different and slightly wider. In most situations, the uncertainties are already vague; therefore most people usually just ignore this problem. Keep this practice in mind whenever you read numbers that come from combining other approximate numbers.

Time-Out to Think: Find the minimum and maximum differences when you subtract the distance between Cities B and C from the distance between Cities A and B. Then, convince yourself that the preceding rule applies both to subtraction and addition.

Example 8-10 Suppose that a swimmer in a 100-meter race is timed by a hand-held timer in 58.7 seconds. A few weeks later, a more sophisticated system times the same swimmer in 57.34 seconds. How much has she improved?

Solution: If you assume that all errors have been accounted for in the stated precisions of the measurements, the first time is known to the nearest tenth of a second and the second time to the nearest hundredth of a second. Subtract the two numbers to find their difference: 58.7 s – 57.34 s = 1.36 s. Because the *least precise* number in the problem was known only to the nearest *tenth* of a second, round your answer to this same precision. Thus the swimmer improved her time by about 1.4 seconds between the two races. Of course, if either measurement contained an error (perhaps a systematic error in the second measurement because the pool was short), this improvement wasn't real. ❖

Example 8-11 A book written in 1957 states that, to the nearest 1000 years, the pyramids in Egypt are 5000 years old. How old are they now?

Solution: To the nearest 10 years, 1957 was 40 years ago. Thus the solution is to add 40 years to 5000 years giving an age of 5040, right? Wrong! You must round the final answer to the precision of the least precise number in the problem. The 5000 years was stated only to the nearest 1000 years, so the answer is that the pyramids still are 5000 years old. The 40 years that have passed since the book was written do not affect the *approximate* age of the pyramids.

This example shows that you can inadvertently lie by failing to account for uncertainty. Stating the answer as 5040 years would have implied that you knew the age of the pyramids to the nearest 10 years. In reality, the oldest pyramids were built about 4700 years ago, to the nearest 100 years. The author of the 1957 book had rounded to the nearest 1000 years. If you had reported your result as 5040 years, you would have been wrong by about 300 years. ❖

Multiplication and Division

Suppose that you want to estimate the total amount of money spent on gasoline by the residents of a city. To the nearest 10,000, the population of the city is 80,000; that is, it is between 75,000 and 85,000 people. Based on a survey, to the nearest $10 the average amount of money each resident spends on gasoline is $760 per year; that is, spending is between $755 and $765 per person per year.

Multiplying the numbers at the lower limits of each range yields the minimum possible product and multiplying the numbers at the upper limits of each range yields the maximum possible product:

$$\text{minimum possible product} = 75{,}000 \text{ people} \times \frac{\$755}{1 \text{ person} \times \text{yr}} = \frac{\$5.6625 \times 10^7}{1 \text{ yr}};$$

$$\text{maximum possible product} = 85{,}000 \text{ people} \times \frac{\$765}{1 \text{ person} \times \text{yr}} = \frac{\$6.5025 \times 10^7}{1 \text{ yr}}.$$

Thus the uncertainty in the product extends over a range of almost \$10 million, from about \$56.6 million to \$65.0 million. Let's compare this outcome to what happens by multiplying the given numbers without accounting for the uncertainty:

$$80{,}000 \text{ people} \times \frac{\$760}{1 \text{ person} \times \text{yr}} = \frac{\$6.08 \times 10^7}{1 \text{ yr}} = \$60.8 \text{ million per year.}$$

As expected, the answer of \$60.8 million lies near the middle of the uncertainty range formed by the minimum and maximum products. However, reporting the result in this way implies that you know the amount of money spent per year to the nearest tenth of \$1 million; that is, to the nearest \$100,000. In fact, you know the answer only to within a few million dollars. Similarly, rounding the answer to \$61 million implies that you know the number to the nearest \$1 million. By rounding to \$60 million, you are being much more truthful, because that implies that you know the answer only to the nearest \$10 million.

An answer obtained by multiplying or dividing approximate numbers should be rounded to the same number of significant digits as the number in the problem with the *fewest* significant digits. (This rule is based on the assumption that all sources of error are accounted for in the stated significant digits.)

Can a rule be stated for the uncertainty of multiplied or divided numbers? Clearly, the rule is *not* related to precision. The final answer of \$60 million is precise to the nearest \$10 million, which has nothing to do with the precision of either of the numbers multiplied. Instead, think about significant digits. The population was given with just one significant digit, the amount of money spent per person on gasoline was given with two significant digits, and the answer had only one significant digit. Thus the rule for multiplying or dividing (or powers and roots) is: Do the multiplication or division and then round the final answer to the same number of significant digits as the number in the problem with the *fewest* significant digits.

As with the rule for addition and subtraction, this rule isn't perfect. Reporting the number as \$60 million implies a range of uncertainty of \$55 million to \$65 million, which isn't quite the same as the minimum to maximum range obtained earlier. However, that is simply the reality of how numbers generally are reported.

Example 8-12 Suppose that you measure the side length of a square room to be 3.3 meters. What is the area of the room?

Solution: Square the side length to get the area: $(3.3 \text{ m})^2 = 10.89 \text{ m}^2$. Because the original number had only two significant digits, you should round the final answer to two significant digits. Thus the area of the room is about 11 square meters.

❖

Time-Out to Think: The 3.3-meter side length implies an uncertainty range of 3.25 to 3.35 meters. Use these values to calculate the minimum and maximum possible areas, and compare this range to the number calculated in Example 8-12.

Combining Order of Magnitude Estimates

When you multiply or divide order of magnitude estimates, you can find the resulting uncertainty by multiplying the individual uncertainties.

When you multiply or divide order of magnitude estimates, you can find the resulting uncertainty by multiplying the individual uncertainties. For example, suppose that the population of some city is within a factor of 2 of 100,000 (i.e., between 50,000 and 200,000), and each resident spends \$100 per year on gasoline, within a factor of 10 (i.e., between \$10 and \$1000).

Multiplying these order of magnitude estimates yields an order of magnitude estimate for the total amount of money spent on gasoline by city residents:

$$\frac{\$10^2}{1 \text{ person} \times \text{yr}} \times 10^5 \text{ people} = \frac{\$10^7}{1 \text{ yr}}, \text{ or } \frac{\$10^7}{\text{yr}}.$$

Because the population was known within a factor of 2 and the per capita spending was known within a factor of 10, the uncertainty in their product is $2 \times 10 = 20$. That is, the total amount of money spent on gasoline by city residents is \$10 million per year, within a factor of 20 (i.e., between $\frac{1}{20} \times$ \$10 million = \$500,000 and $20 \times$ \$10 million = \$200 million).

Example 8-13 *A Jar Full of Pennies.* Suppose that you estimate the volume of the jar to be within a factor of 2 of 5 liters and that, on average, 1 penny occupies 1 cubic centimeter, within a factor of 2. How many pennies are in the jar?

Solution: Five liters is the same as 5000 cubic centimeters, so the best estimate of the number of pennies in the jar is $5000 \text{ cm}^3 \times 1 \text{ penny/cm}^3 =$ 5000 pennies. The uncertainty is the product of the individual uncertainties, a factor of 2×2, or 4. Thus the number of pennies in the jar is within a factor of 4 of 5000, or somewhere between about $1/4 \times 5000 = 1250$ and $4 \times 5000 = 20{,}000$. No wonder winning penny-guessing contests is so difficult!

❖

Exact Numbers and Formulas

Whenever a problem involves multiplication or division of approximate numbers, you must keep track of significant digits. Sometimes, a number may have *more* significant digits than appear. For example, suppose that you are asked the number of students assigned to your dorm room. If you have one roommate, the answer is 2. Because this answer is *exact*, it has an infinite number of significant digits — despite that fact that only one digit is written. As usual, determining the significance of the number 2 requires knowing something about where it came from.

Another example of infinitely many significant digits occurs in formulas. For example, the formula for the circumference of a circle is $C = 2\pi r$. Suppose that you measure the radius of a circle to be 4.52 cm. How should you report the circumference? Using your calculator, you will obtain a circumference of 28.39999759 cm. However, you should report the answer as 28.4 cm, because the radius measurement had only three significant digits.

Perhaps you thought that the answer should have only one significant digit because the circumference formula contained the number 2. However, the formula is *exact*: you should consider the 2 in the formula to have infinitely many significant digits. Similarly, the number π is exact; however, because π is irrational, only a finite number of significant digits can be written in decimal form.

To deal with π in problems, you should keep as many significant digits as practical. If you use a calculator, use the π key, rather than using less accurate approximations because most calculators can store at least nine significant digits.

Warning: Round Only in the Final Step

In Example 8-12, you found the area of a square with a side length measured as 3.8 meters to be about 14 square meters, rounded from 14.44 square meters. Suppose that you had 16 such squares. What is the total area? Multiplying the rounded answer yields $16 \times 14 \text{ m}^2 = 224 \text{ m}^2$ which, rounded to two significant digits, is 220 square meters. Working instead with the unrounded figure yields an area of $16 \times 14.44 \text{ m}^2 = 231.04 \text{ m}^2$, which rounded to two significant digits is 230 square meters. Which answer is better?

Note that the uncertainty range on the side length was from 3.75 to 3.85 meters. Thus the minimum and maximum possible areas for sixteen squares are

$$\text{minimum possible area} = 16 \times (3.75 \text{ m})^2 = 225 \text{ m}^2;$$

$$\text{maximum possible area} = 16 \times (3.85 \text{ m})^2 = 237 \text{ m}^2.$$

Always keep track of as many digits as possible in your calculations, rounding only when reporting the final answer.

The answer of 230 square meters, calculated by keeping all the digits until the final step, falls roughly in the middle of the uncertainty range. The answer of 220 square meters, calculated by using rounded numbers for intermediate steps, lies outside the uncertainty range.

The lesson from this example is: Always keep track of as many digits as possible in your calculations, rounding only when reporting the final answer.

8.3.4 The Sad Reality of Uncertainty

Even in an ideal world, the best that you could hope for would be for everyone reporting measured or estimated numbers to state always

- the range of uncertainty within which the reporter believes the true value lies, and
- the reporter's level of confidence in the stated uncertainty range.

Even then you would be left with a feeling of vagueness. After all, stated ranges of uncertainty and levels of confidence at best are educated guesses based on statistical techniques.

In the real world, you rarely get any information with which to evaluate uncertainty. Pick up a recent newspaper and look at some of the numbers given for projections of next year's federal deficit, the projected population in 2050, or the percentage of the population who prefer Cloudy Cola to Clear Cola. All these numbers are likely to be stated as though they are certain. In reality they may be nothing more than pure guesses!

In fact, because calculation of the deficit involves many potential sources of error, even "final" calculations of the deficit are likely to differ from the true deficit by more than \$0.1 billion. For example, government accounting probably cannot keep track of total spending and revenues to within \$0.1 billion, as this amount is less than 0.01% of the total federal budget.

For example, in the 1991 *Budget of the United States Government*, the Bush administration projected a \$6.0 billion surplus in fiscal year 1993. Because the number was stated to a precision of a tenth of \$1 billion, it implied that the number was accurate to within \$0.1 billion, or \$100 million. By the time the 1992 version of the *Budget of the United States Government* was printed, the projection for 1993 had been revised to a deficit of \$201.5 billion; again, note the implied accuracy to within \$0.1 billion! The 1993 budget deficit, when finally calculated in 1994, turned out to be \$254.7 billion. The point should be clear: The original number in 1991 was offered with confidence that it was within \$0.1 billion of the correct value when, in fact, it was off by more than \$250 billion; indeed, not even the sign of the number was right, as the supposed surplus turned out to be a huge deficit!

The sad reality of uncertainty is that nearly everyone pretends to know things with greater certainty than they actually do. The best guidance we can provide for dealing with uncertainty is to view measured or estimated numbers with skepticism. Evaluate both the source of the information and the method of measurement or estimation. Explore whether the people involved in reporting a number have any biases that might cause them to neglect or improperly account for uncertainty. Look for sources of error in their methods, or factors that they might not have considered. Make it the responsibility of those who are providing you with data to give you good reason to believe in their validity.

CALVIN AND HOBBES © 1991 Watterson. Distributed by *Universal Press Syndicate*. Reprinted with permission. All rights reserved.

8.4 CASE STUDIES IN UNCERTAINTY

Human use of many essential resources and generation of many kinds of pollutants have already surpassed rates that are physically sustainable. Without significant reductions in material and energy flows, there will be in the coming decades an uncontrolled decline in per capita food output, energy use, and industrial production. — from *Beyond The Limits* by Meadows, Meadows, and Randers (Chelsea Green, Post Mills, Vermont, 1992)

If the environmental movement has any motto, it's "To hell with you all." — an official of the Center for the Defense of Free Enterprise

The two quotes in the margin present a striking contrast. The authors of the first quote believe that the human race is facing imminent catastrophe unless societies make significant changes in the use of resources. The speaker of the second quote doesn't buy it at all, believing that environmentalists are trying to push a social agenda that is counter to free-enterprise principles. Both would claim to have scientific evidence on their side. Indeed, scientists line up on both sides of this issue — and everywhere in between.

How can there be such controversy? Hasn't science advanced to the point where scientists can determine whether humanity faces an imminent global crisis? Perhaps by now, after studying sources of uncertainty in even relatively simple circumstances (such as measuring your weight), you can recognize how difficult quantitative questions can be. When the questions are complex, answers are subject to great debate. In this section we examine several current issues and the uncertainties involved in making quantitative statements about them.

8.4.1 Population

When the 1990 census was completed, the U.S. Census Bureau reported a U.S. population of 248,709,873 — implying that *every* individual in the United States had been counted. Is that really possible?

About 5 million births take place in the United States each year, which converts to

$$5 \times 10^6 \frac{\text{births}}{\text{yr}} \times \frac{1 \text{ yr}}{365 \text{ day}} \times \frac{1 \text{ day}}{24 \text{ hr}} \times \frac{1 \text{ hr}}{60 \text{ min}} = 10 \frac{\text{births}}{\text{min}} \; .$$

With an average of 10 births per minute, an exact count could be accurate only if it were made perfectly and instantaneously — and it would no longer be correct just 1 minute later!

Of course, a census can be neither perfect nor instantaneous. The following are just a few of the sources of uncertainty in the census, which primarily relies on a survey that is supposed to be completed by every household in the United States.

- Months are required to collect census data, during which time many people move. For example, college students might be counted as household residents by their parents, then counted again in the household where they attend school.
- The homeless have no address, so counting them is difficult.
- Questionnaires may not be filled out correctly. Some people might accidentally report fewer or more household residents than they really have; others might deliberately misreport data.

- Thousands of individuals collect data for the census. Some of these
 individuals might make mistakes, either accidentally or deliberately.
- Because a region's population count determines its apportionment (number of representatives) in state legislatures and Congress, political pressure may influence how census officials account for potential sources of error and report their final results.

Indeed, after investigating complaints about sources of error in their count, census officials eventually concluded that several groups of people (including the homeless) had been systematically undercounted. Using statistical techniques, they then suggested that the census had, in fact, counted only 98.4% of the population of the United States. Hence the real population, rather than being 248.7 million, was reported to be 252.7 million:

$$\frac{2.487 \times 10^8 \text{ people}}{98.4\%} = \frac{2.487 \times 10^8 \text{ people}}{.984} = 2.527 \times 10^8 \text{ people}.$$

Clearly, the U.S. population in 1990 was *not* known to nine significant digits as originally reported. Even the four significant digits in the revised census result look suspicious: How can census officials be sure that they had counted 98.4% of the population? In light of the many uncertainties, the census probably was accurate to no more than two significant digits; that is, the 1990 U.S. population was about 250 million, to the nearest 10 million.

In less affluent countries, population estimates are even more difficult. For example, Nigeria's 1992 census found a total population of 88.5 million — yet projections based on earlier census data had predicted a population of 110 million. Demographers don't know why the census found 25% fewer people than expected, nor do they know which number is more accurate.

Time-Out to Think: Suggest a few reasons why taking a census is more difficult in less affluent countries.

8.4.2 The Global Average Temperature

In Chapter 3 we showed that, because the Earth's climate is a chaotic system, the environmental consequences of adding greenhouse gases to the atmosphere are unpredictable. Ultimately, the only way to determine the environmental impact of the greenhouse effect is to measure it. One of the most important measures is the average temperature of the entire Earth, or the global average temperature. A key question is how the global average temperature today compares to that in the recent past.

Most attempts to measure the global average temperature rely on analyzing data from weather stations. However, weather stations are not evenly distributed around the world. Highly populated areas may have many weather stations, whereas stations are few and far between in unpopulated areas. In particular, the distribution of weather stations is such that

- most weather stations are located in the northern hemisphere (which has most of the world's land mass); and
- most weather stations are on land, yet two-thirds of the Earth's surface is covered by water.

Another complicating factor is that cities, where many weather stations are located, tend to be warmer than their surrounding areas. The causes of this **urban heat island** effect include the heat generated by burning fuel in automobiles,

Because it requires a great deal of effort and expense, most countries attempt a census only occasionally (e.g., every 10 years in the United States). In the intervening years, demographers (people who study population and population trends) use data on birth rates, death rates, and net immigration to estimate the size of a population. Although small-scale surveys, like minicensuses, are used to test the validity of the projections, significant uncertainties remain.

Recall that greenhouse gases, such as carbon dioxide, methane, and water vapor, trap heat through the *greenhouse effect* (see subsection 3.5.3). People add greenhouse gases to the Earth's atmosphere through the burning of fossil fuels (oil, coal, and natural gas), deforestation, and other activities (e.g., raising cattle, rice farming).

Today, the global temperature can be measured using observations from Earth-orbiting satellites. However, because these satellites are recent inventions, past temperatures can be estimated only from historical weather records or other less accurate data.

heat escaping from homes and offices, and the absorption of heat from the Sun by asphalt.

Weather stations themselves can create uncertainty, as temperature is measured essentially by a thermometer inside a small box that usually stands 1–2 meters above the ground. Boxes standing over concrete tend to measure higher temperatures than those standing over grass; in effect, these boxes suffer a miniversion of the urban heat island effect.

To compensate for problems such as those caused by the uneven distribution of weather stations and the urban heat island effect, researchers must use statistical techniques. For example, in regions with many weather stations the many temperature measurements can be averaged. In summary, determining the global average temperature is a very difficult task, subject to numerous uncertainties.

Even greater difficulty arises in determining whether the global average temperature has *changed* during the past 100 years. For example, because fewer weather stations existed in the past (and many existing stations have been moved), data sets from different periods are not entirely consistent.

Despite the many uncertainties, a 1995 report endorsed by more than 2000 scientists internationally concluded that the global average temperature has increased by about 0.5°C over the past 100 years. Nevertheless, because of the many complicating factors, a few scientists still doubt that any warming has occurred.

Time-Out to Think: Trillions of dollars are at stake in policy decisions regarding climate change. For example, reducing the use of fossil fuels probably would harm many industries (and perhaps help others). Do you think that these high stakes influence the opinions of policy makers? Can the stakes affect the way scientists analyze and interpret data? Explain.

8.4.3 Species Extinction Rate

Another important question in assessing the health of the global environment is the rate at which species are driven to extinction by human activity. The loss of species, which reduces the **biodiversity**, or the overall diversity of life on Earth, might have many detrimental consequences. For example, many species may hold genetic secrets that could lead to new drugs, more efficient and safer methods of growing food, or other unguessed benefits; species loss reduces the pool of these potential resources. Of potentially greater consequence is the fact that living organisms help regulate the global environment — the loss of many species therefore might precipitate damaging or even catastrophic environmental changes.

Living organisms affect the global environment in countless ways. For example, virtually all the oxygen in the atmosphere comes from plant respiration. Thus a significant change in the distribution of life on Earth, which we call the **biodistribution** (representing the number of species, the number of individuals within each species, and their geographic distribution), might change even the composition of the atmosphere. Support for this fact comes from the Earth's past: Although life has thrived for most of the Earth's 4.6 billion year history, the atmosphere would have been breathable by humans only for a very small fraction of this time (probably 200 million years or less). Furthermore, because the global environment is a chaotic system (see Section 3.5), the consequences of changing the biodistribution are unpredictable and the safest route for the future is to minimize such changes. The species extinction rate is but one necessary component in determining the extent to which humans are changing the biodistribution.

You might, at first, think that determining the species extinction rate should be relatively easy; after all, many governments and scientific groups make lists of threatened or endangered species. However, such lists generally tabulate only well-known species. In determining the overall species extinction rate, major sources of uncertainty include the following.

- The total number of species living on the Earth is itself highly uncertain. Only about 1.4 million species have been classified and named, but estimates of the total number of species range from a few million to more than 100 million.
- Because the loss of undiscovered species cannot be measured directly, estimates of the species extinction rate generally are based on the amount of habitat loss, such as the area of wetlands paved over or the area of forest cleared for ranching. Habitat loss usually is estimated from satellite photos, but uncertainties arise because photos aren't available for the entire world. Most estimates of habitat loss in tropical forests are on the order of 100,000 square kilometers per year, but uncertainty in this number is at least 50%.

- The number of species that lived in destroyed habitat, and hence were driven to extinction, usually is extrapolated from an estimate of the total number of species living on Earth. The uncertainty in this estimate (described earlier) therefore leads to uncertainty in the extinction rate.

Extinction rates involve assumptions about the impact of habitat loss. For example, suppose that 2 million species live in tropical forests and that 0.5% of the forest is destroyed. Then the *habitat* of roughly 0.5% × 2 million = 10,000 species is lost. The higher estimates for extinction rates come from assuming that half of these 10,000 species, or 5,000 species, become extinct. Note that, if the number of species in the forests is 20 million rather than 2 million, the estimated number of extinctions jumps from 5,000 per year to 50,000 per year.

Because of the many uncertainties, estimates of the species extinction rate are highly controversial. Although most biologists estimate that somewhere between a few hundred and a few thousand species are lost each year, some estimates are dramatically different. On one extreme, estimates range as high as 40,000 species per year, and on the other extreme the estimates are as small as a dozen or fewer species per year.

Time-Out to Think: Some people argue that species extinctions are simply part of the natural process of competition between human beings and other species. Do you agree? Should human activity be considered natural or something separate from the rest of nature? Defend your view.

THINKING ABOUT ... PASSENGER PIGEONS

In 1800, an estimated 5 billion passenger pigeons lived in the United States. Today, passenger pigeons are extinct. The tragedy of their extinction provides an illustration of the difficulties in understanding the forces at work in nature. Because passenger pigeons had been so numerous, no one believed that their existence as a species was threatened until it was too late. The following piece about passenger pigeons is excerpted from an article by Larry Bleiberg in the *Dallas Morning News* (January 1994).

"Imagine billions of birds in flocks so large they darken the sky for hours. Imagine them all disappearing in a human lifetime. That, in brief, is the story of the passenger pigeon — in its heyday, believed to be the most numerous bird on the planet.

"The bird was hunted relentlessly and its habitat destroyed. It never recovered. The bird's history is told in a quiet pavilion at the Cincinnati Zoo and Botanical Garden. This is where Martha, the last passenger pigeon on Earth, died in 1914.

"The Passenger Pigeon Memorial is both a museum piece and a reminder. It was erected as a birdhouse in 1875, two years after the zoo was founded. Ninety-nine years later, it was dedicated as a memorial to endangered species. Along with the passenger pigeon, the pavilion was home to the world's last Carolina parakeet, the only native parrot in the continental United States. It too was hunted to extinction, with the last one, a male named Incas, dying in 1918.

"Scientists tried to breed Martha, offering rewards of up to $5,000 for a mate. Near the end, they also tried in vain to breed her with other pigeon species.

"In earlier years, no one had perceived the need for such desperate measures. Who could imagine a species so numerous could go extinct? Naturalist John James Audubon once recalled seeing a flock that took days to pass. In the evening, the birds would crowd trees, sometimes breaking branches with their weight. The nesting colonies would cover hundreds of square miles. The spectacle, almost impossible to imagine now, flabbergasted those who saw it. The birds awed James Fenimore Cooper, author of *The Last of the Mohicans*, who wrote of being reduced to silence by the work of the Creator.

"The bird was tall and graceful, bearing little resemblance to today's urban pigeons. The male had streaks of blue and the female had a cinnamon-rose-colored breast. The birds were up to 16 inches in length and could fly faster than a mile a minute. The species got its name because of its mass migration — it was a bird of passage. Its habitat ranged from Nova Scotia in Canada in the north, south to Florida and as far west as Texas. Flocks were reported near Lampasas as late as 1883, a few were seen in Corpus Christi six years later.

"But it's all history now. The curious can examine a bronze of Martha near the memorial. They also can look at a few stuffed specimens of passenger pigeons and Carolina parakeets. Both birds were hunted to death.

"Carolina parakeets were shot for sport and captured as pets. The passenger pigeon was considered a delicacy, and hunters would gather squabs by the barrelful. Sportsmen also used the bird for trap shooting.

"In the pigeon's final years, hunters were alerted to flocks by telegraph. Trains were chartered to take them to the site. The birds would be smoked out of the trees with fires.

"Some farmers caught the birds by blinding a pigeon and tying it to a stool. Its erratic motion attracted other birds, who were netted en masse. Thus the term, stool pigeon.

"By the late 1890s, the birds were almost gone. The last one seen in the wild was shot by a boy near Sargents, Ohio. The specimen is on display at the memorial.

"Martha herself was born in captivity at the zoo. In her final years, she attracted tourists and ornithologists from around the world. Finally, on Sept. 1, 1914, at the age of 29, she was found dead at the bottom of her cage. Her body was packed in ice and sent to the Smithsonian Institution. She remains on display on the first floor of the National Museum of Natural History in Washington, D.C.

"'It's so hard to think that there were billions of them,' said Jason Abell, a fifth grader, on a field trip to the zoo from Fort Thomas, Ky. 'Then it got down to the last one.'"

8.4.4 Inflation and the Consumer Price Index

Recall that inflation means higher prices for the same goods, and hence *reduces* the purchasing power of the dollar. For example, an inflation rate of 3% per year means that the cost of living is rising by 3% per year. Thus, unless your earnings also increase by 3% per year, your standard of living will fall. Further discussion of inflation and its effects is presented in subsection 15.4.2.

Let's turn to an issue with less global significance but important effects in the United States. Many financial decisions are tied to inflation. For example, government employee raises and government payments, such as benefits paid to Social Security recipients, are based on the rate of inflation. Income tax rates are **indexed** to inflation: If the inflation is running at 3%, the income levels of different tax brackets also increase by 3%. This adjustment keeps people from moving into higher tax brackets simply because of inflation.

The rate of inflation is measured by the U.S. Bureau of Labor Statistics, which calculates the rate of inflation from changes in the **Consumer Price Index (CPI)**. The CPI is calculated each month by comparing prices of goods from more than 60,000 sources. To provide a fair comparison, the particular goods and the quantities compared must remain constant over time. The CPI from 1980 to 1993 is shown in Table 8-2. The numerical value of the index is arbitrary; currently, it is reported relative to a value of 100 representing average prices during the period 1982–1984.

The government actually measures two consumer price indices: CPI–U is based on products thought to reflect the purchasing habits of *all* urban consumers, whereas CPI–W is based on the purchasing habits of wage earners only (that is, it does not include spending by homemakers, the unemployed, or children). Table 8-2 shows the CPI-U.

Table 8-2. Consumer Price Index 1980–1993

Year	CPI	Year	CPI
1980	82.4	1987	113.6
1981	90.0	1988	118.3
1982	96.5	1989	124.0
1983	99.6	1990	130.7
1984	103.9	1991	136.2
1985	107.6	1992	140.3
1986	109.6	1993	144.5

The annual rate of inflation usually is calculated as the percentage change in the CPI from one year to the next. For example,

$$\text{inflation rate from 1981 to 1982} = \frac{CPI_{1982} - CPI_{1981}}{CPI_{1981}} \times 100\% = \frac{96.5 - 90.0}{90.0} = 7.2\%.$$

The large number of sources used for price comparisons in computing the CPI *should* indicate the rate at which prices are rising. However, sources of uncertainty are numerous. For example, prices of *all* goods are lumped together in calculating the overall CPI. This procedure might not be valid if prices in one sector of the economy are moving at a different rate (or even in the opposite direction) than prices in other sectors. If the price of recreation and entertainment is rising while food is holding steady, the impact of inflation may mean fewer entertainment options for many families. That might be undesirable, but it is certainly less significant than a corresponding increase in food prices.

Moreover, many economists question whether the CPI accurately measures the true rate of inflation. One concern is that the CPI may neglect **substitution effects**, in which consumers simply stop buying an item which increases in price and substitute a similar, lower-priced item. Changes in consumer habits also may affect the validity of using the CPI to measure inflation. For example, because many people purchase personal computers today but few did so in 1980, the CPI did not measure the same set of goods in 1980 as it does today. Also, although typical prices of personal computers have stayed relatively stable for the past 10 years, today's computers are far more powerful: In terms of computing power, at least, the same expenditure today provides a higher standard of living than it did in the past.

Time-Out to Think: Do you think that changes in prices affect the cost of living for all people similarly? For example, do increases in the price of gasoline affect all consumers similarly? How about increases in food prices? Explain.

In summary, measuring changes in consumer prices — and hence the rate of inflation — is far more difficult and controversial than most people would guess. Because raises, benefits, and tax rates depend on the rate of inflation, methods for calculating it tend also to be highly politicized. As a result, different economists and different government groups often disagree about the true rate of inflation.

8.4.5 Who Benefits from a Tax Cut?

In early 1995, the newly elected Republican majority in the House of Representatives proposed a tax cut. Both the Republican Congress and the Democratic administration analyzed the impact of the proposed tax cut. Their results are shown in Figure 8-2. How did they reach such remarkably different conclusions about the same tax cut?

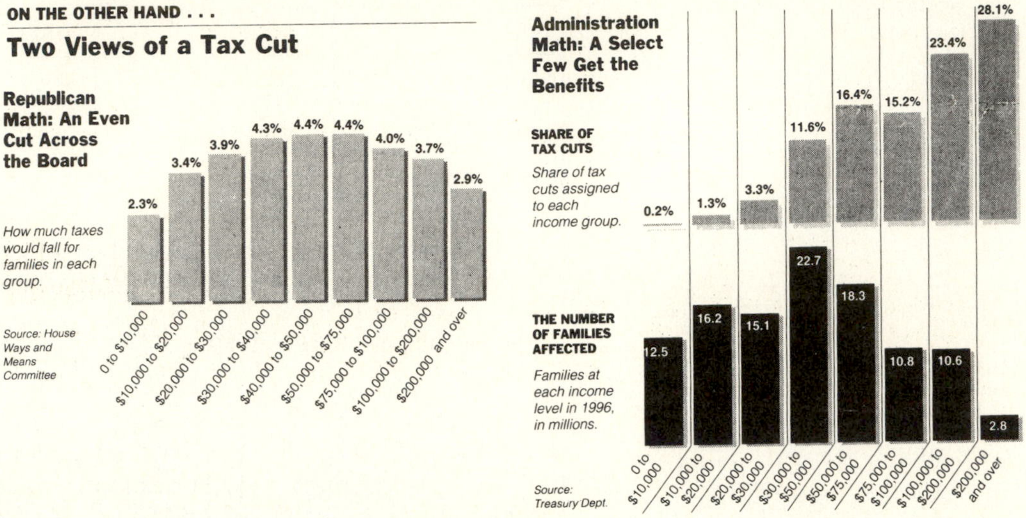

Figure 8-2. These graphs, reproduced from an article in the *New York Times* (4/7/95) show how different methods of calculation lead to drastically different conclusions about the impact of a tax cut. Two Views of a Tax Cut," April 7, 1995. Copyright © 1995 by The New York Times Company. Reprinted with permission.

Each side had a point to make: The Republicans wanted to show that their tax cut would be fair to everyone, whereas the Democrats claimed that it would benefit a few rich people at the expense of most others. However, claiming that the different calculations reflect partisan bias is difficult in this case. Surprisingly, the Democratic calculations were made according to the same basic assumptions used by the previous Republican administrations (Reagan and Bush), and the Republicans used essentially the same methods as the previous Democratic Congresses.

The differences in the analyses result from differing methods of calculating income and total tax paid. To determine income, both groups included not only taxable income (as reported on income tax forms), but also untaxed income such as contributions made to retirement plans and the value of benefits provided by employers. However, the administration's calculations included some sources of "income" that the congressional calculations did not include. For example, the administration calculations estimated how much stock people hold at different income levels and then allocated corporate profits reflected in the value of the stock to individual income. The effect of these different income calculations is dramatic. By the administration's calculations, 2.5% of Americans had incomes of more than $200,000 and 12.5% had incomes of more than $100,000. By the congressional calculations, only 1% had incomes of more than $200,000 and only 5.5% had incomes of more than $100,000.

Politicians and government officials usually abuse numbers and logic in the most elementary ways. They simply cook figures to suit their purpose, use obscure measures of economic performance, and indulge in horrendous examples of chart abuse, all in the name of disgusting unpalatable truths. — from *200% of Nothing* by A. K. Dewdney (Wiley, 1993, p. 95)

The tax burden is assessed from income levels. Because the administration's calculations include the value of corporate profits to individuals, they also include the tax benefits of corporate tax deductions. The combined effects of the different calculations of income and taxation lead to the two very different conclusions about the distribution of benefits from the tax cut.

Finally, the two graphs in Figure 8-2 do not show the same calculations! The "Republican Math" graph shows the tax cut received by individuals in each income group. The "Administration Math" graph shows the *total* share of tax cut benefits going to everyone in each income group, which tends to magnify the effects at upper income levels: Wealthier individuals pay a larger share of total taxes and therefore also receive a larger share of total benefits, even if everyone gets the same percentage tax break.

In conclusion, huge uncertainties are generated simply by attempts to determine income and measure tax benefits — so much so that two different branches of government can arrive at dramatically different conclusions. For the public, further uncertainty is introduced by each side's decision to portray the data in a way that best illustrates the point it wants to make.

Time-Out to Think: The "Administration Math" graph claims that almost half the total benefits of the tax cut would go to the 12% of taxpayers who earn more than $100,000. It fails to show another fact: that these same taxpayers pay about half the total taxes collected by the U.S. government (as of 1994). Does this fact alter your perspective on the administration conclusions? How does it affect your view of the "Republican Math"?

8.5 CONCLUSION

In Chapter 1 we stated that one of the most common misconceptions about mathematics is that it "always requires great precision and exact numbers." In this chapter, we demonstrated that most numbers in the real world are not precise or well known. Key lessons from this chapter include the following.

- The sad reality is that most numbers stated in the media and otherwise in everyday life carry significant uncertainties and that these uncertainties are rarely stated explicitly. As a result, you must make educated guesses about the processes that might have led to the number, as well as any biases that the individual (or group) reporting the number might hold. You must then decide for yourself the uncertainty of the number.
- Uncertainties themselves are uncertain, whether they are expressed explicitly or implicitly. Any time that you deal with a measured or estimated number, the best you can do is to ascribe a subjective level of confidence to the number and its uncertainty.
- *Never* accept a number without question. Even if a number appears to have been measured or estimated well, and to have a small degree of uncertainty, you should always ask yourself: "Does it make sense?"

PROBLEMS

Reflection and Review

Section 8.1

1. **Numbers in the News.** Look through the newspapers from the past week for articles that mention numbers in any context. Get at least two numbers from each of the following sections: national/international news, local news, sports, and business. In each case, describe the number and its context and explain whether you believe the number is exact or approximate. If the number is approximate, does the article state its uncertainty?
2. **Failures to Report Uncertainty in the News.** Search through the past week's newspapers. Determine at least 3 instances in which reported numbers clearly were stated with more precision or accuracy than justified and without any indication of the uncertainty in the numbers. Describe each instance and, for each, briefly explain what might have been a more realistic way to report the numbers.

Section 8.2

3. **Exact or Approximate?** In each case, state whether the indicated number is exact or approximate, and briefly explain why.
   a. The number of letters in the word *mathematics*.
   b. The number 2 in the formula for the area of a square: $A = (\text{side length})^2$.
   c. The number 2 in the statement, "I ate 2 lbs of chocolate."

d. The number 100,000 in the statement, "Ten to the fifth power is 100,000."
e. The number 100,000 in the statement, "The population of this city is 100,000."
f. The number 24,000 in the statement, "The number of students at the university is 24,000."

4. **Statistical Claims.** Each of the following exercises makes a statistical claim. Briefly identify and discuss possible sources of error in each claim.
   a. The life expectancy in Japan is 78.7 years.
   b. Twenty-three and four-tenths percent of Americans are Roman Catholic.
   c. A violent crime occurs in the United States once every 16 seconds.
   d. The population of the United States has been increasing at a rate of approximately 5 million people per year.
   e. The price of memory for microcomputers is decreasing at a rate of approximately $10 per megabyte per year.
   f. Six hundred twenty-three students in my high school class graduated.
   g. Two-thirds of the town voted for the Republican mayoral candidate.
5. **Scale Calibration.** Suppose that, after measuring your weight to be 52.3 kilograms, you check the calibration of the scale. To your surprise, you find that the scale reads 2.4 kilograms even when nothing is on it! How would this systematic error have affected your measured weight?
6. **Accuracy and Precision.** Suppose that you purchase a standard weight at a store, which is accurately calibrated to 1.5 kilograms. You use the standard weight to check the calibration of two different scales. Using the first scale, you are able to measure to the nearest hundredth of a kilogram; you measure the weight to be 1.51 kilograms. With the second scale, you can measure to the nearest ten thousandth of a kilogram, and the measured value is 1.5340 kilograms. If the true value of the standard weight is 1.5 kilograms, which scale gives the more accurate measurement? Which gives the more precise measurement?
7. **Minimizing Random Errors.** Suppose that 25 people, including yourself, are asked to measure the length of a room to the nearest tenth of a millimeter. Assume that everyone uses the same measuring device, such as a tape measure.
   a. All 25 measurements are not likely to be exactly the same; thus the measurements will contain some sources of error. Are these errors systematic or random? Explain.
   b. If you want to minimize the effect of random errors in determining the length of the room, which is the better choice: to report your own, personal measurement as the length of the room, or to report the average of all 25 measurements? Explain.
   c. Describe any possible sources of systematic errors in the measurement of the room length.
   d. Can the process of averaging all 25 measurements help reduce any systematic errors? Why or why not?
8. **Process Definition.** Suggest a process definition for measuring each of the following quantities. Describe possible sources of uncertainty and ways of minimizing the uncertainty.
   a. The amount of weight you lose (or gain) during a month-long special diet.
   b. The amount of time you spend studying for this class.
   c. The total amount of money you pay in taxes each year, including income taxes, Social Security taxes, sales tax, property tax, and other taxes.
   d. The number of students who receive college degrees each year.
   e. Times for the 100-meter dash in the upcoming Olympics. Remember that your measurements must be accurate enough to determine whether a world record has been set.

Section 8.3

9. **Explicit Uncertainty Ranges.** Restate each of the following measurements by using the plus or minus (±) notation to express the range of possible values.

Example: The length of the yard is between 25 and 29 feet.
Solution: The length of the yard is 27 ± 2 feet.

   a. The average score on the exam was between 75 and 80 points.
   b. In city driving this car gets between 31 and 35 miles per gallon.
   c. The average person requires between 1/4 and 1/2 pound of chocolate each week in order to prevent scurvy.
   d. Most people sleep between 7.4 and 8.3 hours per night.
   e. The age of the universe is between 7 and 20 billion years.
   f. The federal deficit this year will be between 150 and 300 billion dollars.
10. **Interpreting Plus–Minus Ranges.** Interpret each of the following statements.
   a. The population of China is 1.2 ± 0.2 billion people.
   b. The age of the Earth, based on radioactive dating techniques, is 4.6 ± 0.1 billion years.
   c. The population of this town is 85,000 ± 6000 people.
11. **Your Own Level of Confidence.** Answer each of the following questions with your own subjective 95% level of confidence. Explain why you are 95% confident in your answer.

Example: How much do you weigh? *Solution:* With 95% confidence, I believe that my weight lies between 154 and 158 pounds. I base this confidence on the fact that I have been measuring my weight at least once a week for the past several months, and my measured weight has never yet fallen outside this range. Thus I think that the probability of its being outside this range today is less than 5%.

   a. How much do you weigh?
   b. How tall are you?
   c. How long do you take to read a page in a paperback novel?
   d. How many hours of TV do you watch each week?

e. How many more semesters (or quarters) will you be in school before you graduate?
f. How much money do you spend on food each day?
g. How many hours do you spend studying for this class each week?

12. Counting Significant Digits. State the number of significant digits in each of the following numbers.
a. 96.2 km/hr **b.** 100.020 seconds
c. 0.00098 mm **d.** 0.0002020 meters
e. 401 people **f.** 200.0 liters
g. 1.00098 mm **h.** 0.000202 meters

13. Implied Range of Uncertainty. State the implied range of uncertainty for each of the following measurements.
a. 241 kg **b.** 0.007 kg
c. 20.0 mm **d.** 420 minutes
e. 27,000 students **f.** 4 kg
g. 8.3050 kg **h.** 0.00000101 mm
i. −27.0° Celsius **j.** $300,003

14. Ambiguity in Zeros. For the number 2000 state the implied range of uncertainty if it was measured to the nearest
a. 1000.
b. 100.
c. 10.
d. 1.
e. tenth.
f. hundredth.

15. More Ambiguity in Zeros. For the number 300,000 state its number of significant digits if the number is correct to
a. within a single unit.
b. the nearest 10.
c. the nearest 100.
d. the nearest 1000.
e. the nearest 10,000.
f. the nearest 100,000.
g. the nearest tenth.
h. the nearest hundredth.

16. Removing Ambiguity with Scientific Notation. Write 300,000 in scientific notation if it has
a. only one significant digit.
b. two significant digits.
c. five significant digits.
d. eight significant digits.

17. Confidence in Implied Uncertainty. For each of the following situations (i) state the implied range of uncertainty; (ii) briefly discuss whether you believe that the implied range of uncertainty reflects the true uncertainty in the number; (iii) give a range of uncertainty for the measured or estimated number with 95% confidence. Discuss the types of further information you would need to determine the range with such confidence.
a. The U.S. Census Bureau reports that the population of a city is 1,452,332.
b. A real estate agent, trying to convince you to list your home with him, tells you that he has sold 500 homes.
c. A stockbroker tells you that she has run some calculations and that by investing with her you will earn $1100 more over the next year than if you put the money in the bank.
d. In 1995, the president states that by 2000 the budget deficit will drop to $23.2 billion.
e. A group of population experts states that world population will level off at 11 billion people in 2050.
f. The owners of your favorite, but last-place, football team state that their rebuilding program will bring a Super Bowl victory in 5 years.

18. More on Confidence. Consider the following statements. For each, state whether you believe either the explicit or implied uncertainty range, and why.
a. My little brother is 4 feet, 2.9051 inches tall.
b. In the 1988 election, President Bush received 1,459,231 votes more than Governor Dukakis.
d. The average amount of money spent on textbooks by students is $120 ± $50 per semester.
e. There are 260 million people in the United States.

19. Combining Uncertainty: Addition and Subtraction. In each case, assume that the measurements are properly stated with the correct number of significant figures. Obtain the range of possible answers by calculating minimum and maximum values. *Example:* Add the volumes 2000 liters and 322 liters. *Solution:* The first number has only one significant figure, implying that its range of uncertainty is from 1500 to 2500 liters. The second number has three significant figures, implying that its range is from 321.5 to 322.5 liters. The minimum possible sum is 1500 + 321.5 = 1821.5 liters; the maximum possible sum is 2500 + 322.5 = 2822.5 liters. Finally, we round the numbers in this range to the precision of the least precise number in the problem; that is, to the nearest 1000 liters. Thus the range of uncertainty for the sum is from 2000 to 3000 liters.
a. Add the distances 36 centimeters and 8.22 centimeters.
b. Add the weights 260 kilograms and 17 kilograms.
c. Subtract the volumes 140 liters and 1.09 liters.
d. Subtract the times 2 hours, 37 minutes and 1 hour, 22 minutes, 15 seconds.
e. Add the distances 4.093×10^{10} kilometers and 6.1×10^8 kilometers.
f. Subtract the weights 72 kilograms and 3.5 kilograms.
g. Add the volumes 1.2 million liters and 5255 liters.
h. Add the times 1.25 years and 245 hours.
i. Add the areas 4.3×10^5 square kilometers and 3.5×10^4 square kilometers.

20. Combining Uncertainty: Multiplication and Division. In each case, assume that the measurements are stated with the correct number of significant figures. Find the range of possible answers by calculating minimum and maximum values.
a. Multiply the lengths 105 meters and 26 meters.
b. Divide the distance 110 kilometers by the time 55 minutes.
c. Multiply the weights 9.7 kilograms and 165 kilograms.
d. Divide the weight 5 grams by the volume 1.3 cubic centimeters.
e. Multiply the times 23 hours, 56 minutes and 366 days, 6 hours.

f. Divide the volume 3600 liters by the time 4 hours, 30 seconds.
g. Multiply the distances 0.23 kilometers by the time 56.4 seconds.
h. Divide 142,000 square kilometers by 1.5 years.

21. **Addition Rule for Approximate Numbers.** Assume that all the following numbers are measured to the indicated precision. Compute all the sums or differences, following the rule that your answer should be given to the precision of the least precise number in the problem.
 a. $48.49 + 4.237 + 12.1$
 b. $(5 \times 10^3) + (2 \times 10^2)$
 c. $(4.326 \times 10^{-6}) + (9.36478 \times 10^{-9})$
 d. $65.7832 + 7.112 + 51{,}009$
 e. $(8.599 \times 10^9) + (7.62 \times 10^7)$
 f. $(6.5 \times 10^2) - (4.2 \times 10^2)$
 g. $8.409 + 1.227 + 13.19$
 h. $(8 \times 10^{80}) + (6 \times 10^{80}) + (2.2 \times 10^{12})$
 i. 4.5 trillion − 235.6 billion
 j. $113.75 - 7.6 + 0.34$
 k. $(7.37 \times 10^{34}) - (4.5 \times 10^{33})$
 l. 37 billionths + 4 thousandths

22. **A Multiplication Rule for Approximate Numbers.** Assume that the following numbers are stated with the proper number of significant digits. Do the indicated operations, following the rule that your answer should be given to the same number of significant digits as the number in the problem with the fewest significant digits.
 a. $(1.3 \times 10^{21}) \times (4.1 \times 10^{-12})$
 b. $(2.871 \times 10^{35}) \times (3 \times 10^{-33}) \times (5.78 \times 10^7)$
 c. $(3.43 \times 10^{-7}) \times (5.661 \times 10^{-5})$
 d. (3 million) ÷ (1.56 thousand)
 e. $(4.448921 \times 10^{13}) \times (1 \times 10^1)$
 f. $(2.4 \times 10^3) \times (5.67 \times 10^3) \div (6.498 \times 10^4)$
 g. $(3.78 \times 10^{30}) \div (2.641 \times 10^{45})$
 h. $(9.4287644 \times 10^{89}) \div (1.2354789 \times 10^{-91})$

23. **Combining Uncertainty: Orders of Magnitude.**
 a. The weight of a cow, within a factor of 2, is 1000 pounds. Within a factor of 4, an average of 4 hamburgers can be made from a pound of cow. How many hamburgers can be made from one cow?
 b. Within a factor of 10, the average American spends \$10 per year on ice cream. Within a factor of 2, the population of a town is 80,000. How much money is spent on ice cream by the residents of this town?
 c. Within a factor of 100, the number of atoms in the universe is 10^{79}. Within a factor of 2, the average atom is made up of eight subatomic particles. How many subatomic particles make up the atoms in the universe?

24. **Range of Uncertainty in Circumference.** The radius of a circle is measured to be 3.6 meters.
 a. Use the formula $C = 2\pi r$ to calculate the circumference of the circle. State your answer with two significant digits. What is the implied range of uncertainty?
 b. Assume the implied range of uncertainty for the radius is 3.55 to 3.65 meters. Suppose that the true radius is at the lower end of this range, or 3.55 m. What is the true circumference? Give your answer with 3 significant digits.
 c. Suppose that the true radius is at the upper end of the implied range, or 3.65 m. What is the true circumference of the circle? Give your answer with 3 significant digits.
 d. The implied range of uncertainty for the radius is 3.55 to 3.65 m, so the range of uncertainty on the circumference should lie between the values you obtained in parts (b) and (c). What is this range? How does it compare with the implied range in part (a)?
 e. Briefly explain the reason for the apparent discrepancy between the actual range of uncertainty of the circumference from parts (b) and (c), and your answer from part (a)? Is the two-significant-digit answer for the circumference valid? Can you think of a better way to state the circumference?

Section 8.4

25. **Home Town Population.** According to local groups (such as the city council or chamber of commerce), what is the current population of your home town? How was this estimate was made? Discuss the uncertainty in this value.

26. **Nigeria's "Missing" Population.** In Subsection 8.4.1, we stated that the 1992 census in Nigeria found only 88.5 million people. Using previous demographic data, demographers had expected a population count of 110 million. Suggest several possible reasons for the discrepancy between the projection and the census data.

27. **Global Average Temperature.** Suppose that a researcher describes the following system for measuring the global average temperature of the world: "My system involves averaging daily temperature readings at every major airport in the world. I graph the resulting daily average temperatures against time. I have averaged the temperatures at all existing major airports from 1945 to the present."

 List possible sources of both systematic and random errors in this measurement process. If this researcher reports that the global average temperature has increased since 1945, how confident would you be of her result?

28. **Global Deforestation Uncertainty.** Ecologist Norman Myers estimated that 142,200 square kilometers of tropical rain forest were cleared globally in 1989, of which 50,000 square kilometers were in Brazil. Later information, based on satellite data, showed that only 26,600 square kilometers were cleared in Brazil in 1989.
 a. What is the implied range of uncertainty in Myers's global deforestation estimate?
 b. How large was Myers's error in estimating the deforestation in Brazil in absolute and relative terms?
 c. Assuming that Myers's relative error for Brazil held globally, how large would actual global deforestation have been in 1989?
 d. Despite the error in his estimate for Brazil, Myers claimed that his global estimate was valid. In part, he argued that the deforestation rate in other countries was higher than he had estimated, thereby offsetting the error caused by his overestimate in Brazil. How could you evaluate his argument? Did he give his original estimate with too many significant digits? Why or why not?

29. Tropical Deforestation. Although the precise rate is uncertain (see problem 28), on the order of 100,000 square kilometers of tropical forest are lost each year due to human activity.

a. Using this order of magnitude estimate, calculate how many acres of forest are lost each day, minute, and second.

b. An estimated (8 to 9) $\times 10^6$ square kilometers of tropical forest remain on Earth. If the rate of deforestation remains constant, by what year will all the forest be destroyed? By what year will 50% of the forest be destroyed?

c. Do you believe that the rate of deforestation is likely to remain constant, increase, or decrease? Why? If you are correct, how will the future of tropical forests differ from your calculation in part (b)?

30. The Passenger Pigeon. In 1800 an estimated 5 billion passenger pigeons lived in the United States. Put this number into perspective. Then, comment on the extinction of the passenger pigeon in 1914.

31. Inflation with the Consumer Price Index.

a. Use the CPI data in Table 8-2 to calculate the rate of inflation for each year in the table. When was the inflation rate highest? lowest?

b. Look up CPI data for years since 1993 and compute rates of inflation for these years.

c. What, if any, is the long-term trend in inflation since 1980? If you identify a trend, explain why you think it exists. If you don't, suggest a reason for the lack of a trend.

Further Topics and Applications

32. This Year's Federal Deficit. Check newspapers or newsmagazines for the latest estimates of this year's federal deficit. What is the projected deficit and what organization is the source of this projection? What is the implied range of uncertainty on the projection? Do you believe that the actual deficit for this year will be within the implied range of uncertainty? Why or why not?

33. Who Do You Believe? Each of the following situations gives two different reported values for a number and the source of each value. In each case, explain whether you believe that one value is more or less certain than the other. What is your level of confidence in the reliability of each value?

a. Gay rights groups claim that 10% of the population is gay, but a telephone survey done by the United States National Opinion Research Center claims that only 2% of the population says that it is gay.

b. Planned Parenthood claims that thousands of women died each year because of illegal abortions before abortion was legalized in the *Roe v. Wade* decision of the Supreme Court. The anti-abortion group Operation Rescue claims that only 39 women died as a result of illegal abortions in the year before *Roe v. Wade.*

c. The U.S. military's Southern Command Post argues that at most 345 soldiers and civilians were killed during the 1989 invasion of Panama, but Alexander Cockburn, writer for *The Nation*, argues that between 1000 and 2500 were killed.

d. In 1992, President Clinton claimed that every $1 billion spent on infrastructure creates 50,000 jobs, but the Republican party claimed that it creates only 12,000 jobs.

34. Should You Get Off the Elevator? Suppose that you step onto an elevator that already has nine people in it. You look at the safety panel and see a sign that says "Maximum Capacity 1300 Pounds." Should you leave the elevator? Explain. If all nine people are members of a girls' high school gymnastics team, how will that affect your decision? What if the nine people are linemen on your school's football team?

35. Comparing Against Standards. Two scientists are testing two new methods of measuring sound waves. They record the sound wave pattern when a certain bell is struck. Past measurements of the sound from this particular bell, using older, accepted methods for measuring sound waves, give the scientists a good idea how the sound wave pattern should look. The following graph below displays the expected data, based on previous measurements, and the data collected by both scientists.

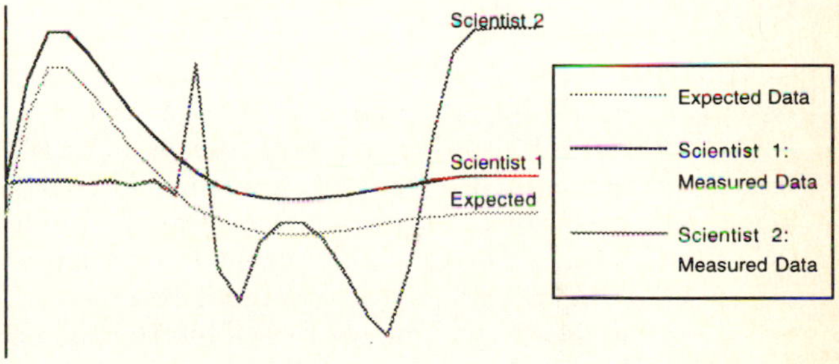

a. Look at each scientist's data. Do you think their measurements contain errors? If so, are their errors more likely to be systematic or random? Explain.

b. Which new method seems to be better? Why?

36. Race Timing. Modern timing devices can easily measure time to the nearest thousandth of a second, yet international sports federations recognize world records (in events such as running or swimming) only to the nearest hundredth of a second. Why do you think that this is so?

37. Absolute and Relative Uncertainty. The absolute uncertainty in a measurement specifies how far the true value lies from the measured value. The relative uncertainty in a measurement specifies how far the true value lies from the measured value in relative or percentage terms. Each of the following involves calculating absolute and relative uncertainties. *Example:* Find the absolute and relative uncertainties for a measurement of 10 ± 1 meters. *Solution:* The absolute uncertainty is 1 meter; that is, if all possible sources of uncertainty have been included, the true value should lie within 1 meter of the measured value of 10 meters. The relative uncertainty is 1/10, or 10%. That is, the true value should lie within 10% of the measured value of 10 meters.

a. What are the absolute and relative uncertainties for a measurement of 7.0 ± 0.3 kg?

b. Consider the following three measurements: 28.7 ± 0.2 m; $(6 \pm 2) \times 10^{-9}$ m; and 4.589 ± 0.001 m. Which has the smallest absolute uncertainty? Which has the smallest relative uncertainty? Explain.
c. Repeat part (b) for the following three measurements: 1.02 ± 0.001 kg; 102 ± 0.1 kg; and $10{,}200 \pm 10$ kg.
d. Repeat part (b) for the following three measurements: 23 ± 0.1 hr; 3.0 ± 0.5 hr; and $(2.1 \pm 0.1) \times 10^4$ hr.

38. **Propagation of Error.** Suppose that you want to cut 10 boards each 1 meter in length. You already have a "master board" measured to be 1.000 meter long. You know that each time you use the master board to mark and cut another board you make an error of ± 0.05 meter due to the thickness of the pencil lead and the saw. Plan 1 is to use the master board to mark and cut each of the other 10 boards. Plan 2 is to use the master board to mark and cut the first board, use the first board to mark and cut the second board, use the second board to mark and cut the third board, and so on, until the ninth board is used to mark and cut the tenth board. Using Plan 1, what is the maximum error expected in cutting the tenth board? Using Plan 2, what is the maximum error expected in cutting the tenth board?

39. **CEO Salaries and Layoffs.** A few years ago, the Chief Executive Officer (CEO) of a major television and publishing corporation earned 1-year compensation of about $80 million (including salary, stock options, and other compensation for his work). During the same year, the corporation laid off 600 employees. Suppose that, instead of paying the CEO the $80 million, the corporation had used the money to continue to pay the salaries of the 600 laid-off employees. Estimate how long the 600 people could have remained employed with the $80 million. Discuss the uncertainties in your estimate.

40. **Overflowing Trash.** With landfills filling up across the United States, concern is growing about the amount of trash produced. Be sure to include an uncertainty with each of your estimates.
a. Estimate the amount of trash you produce on an average day, in units of both volume and weight. Do you think that your personal trash production is average, low, or high for people in the United States? Explain.
b. Estimate the total amount of trash that individuals produce each day in the United States. How much trash is produced each year?
c. Estimate the amount of trash produced by business and industry. Explain your estimates and add them to your estimates from part (b) to estimate the total amount of trash produced in the United States each day and each year.
d. Estimate the volume of a football stadium. How many football stadiums could you fill with the trash produced in the United States every year?
e. How is recycling affecting the need for new landfills? How much more effective could recycling become? What other suggestions can you make for reducing trash production? Do you think your suggestions have a realistic chance of being implemented? Why or why not?

41. **Recycling Newspapers.** Newsprint is made from trees, so recycling it means that fewer trees must be cut down to produce newspapers. In the following problems, be sure to include an uncertainty with each of your estimates.
a. Estimate the weight of one of your local newspapers and the total daily circulation of this newspaper. Use these numbers to estimate the total weight of this newspaper printed each day. Explain your estimates.
b. Estimate the percentage of newspapers in your area that are recycled. Explain your estimate.
c. For every 50 kg of newspaper that is recycled, rather than made from virgin wood, approximately one tree is saved. Approximately how many trees are saved each year from the recycling of your local newspaper?
d. How many more trees could be saved if everyone in your community recycled their newspapers?

42. **Car Cost Per Mile.** The Saturn car company advertises that its 1995 Saturn SW2 costs $0.24 per mile to drive. This figure is based on adding the purchase price of the car to the cost of gasoline, maintenance, insurance, and registration, and then subtracting the estimated resale value of the car at the end of 5 years. To get the units of dollars per mile, the company divides the cost by 60,000 miles; this figure is based on the assumption that the owner will drive the car 60,000 miles in 5 years, then sell or trade it in for a new car.
a. Based on the assumptions stated, what is the total cost of owning a Saturn SW2 for 5 years?
b. Analyze each of the assumptions made. Are they realistic? Overall, is the advertisement reasonable? Explain.

43. **The Amazing Amazon.** The February 1995 issue of *National Geographic* contains the following statement: "Dropping less than two inches per mile after emerging from the Andes, the Amazon drains a sixth of the world's runoff into the ocean. One day's discharge at its mouth — 4.5 trillion gallons — could supply all US households for five months." Answer the following questions and be sure to state the uncertainties in your calculations.
a. Calculate the daily discharge of water from the Amazon in units of gallons per second and units of cubic feet per second.
b. The river with the next greatest daily discharge is the Mississippi River, with an outflow rate of 1,600,000 cubic feet per second. Compare the discharge rate of the two rivers with percentages.
c. Estimate the number of households in the United States and the daily domestic water consumption of one average household.

44. **Hamburger Meat.** Make an order of magnitude estimate of the number of cattle slaughtered each year to provide hamburgers for U.S. fast food. Be sure to explain clearly your method and uncertainties. State your final answer with an explicit uncertainty.

45. **Fast Food and the Central American Rain Forests.** Because of lower prices for land and labor, beef imported from less developed nations generally is cheaper than beef raised in the United States, even after accounting for shipping costs. As a result, a significant amount of beef is imported into the United States each year, much of it for

use in fast-food hamburgers. The following facts are from the article "The Hamburger Connection: How Central America's Forests Become North America's Hamburgers," *Ambio,* January 1981, pp. 3–8. (Note: There is considerable controversy over which fast-food chains use imported beef and in what quantities. Nevertheless, Central American beef clearly is being imported into the United States. However, most supermarket beef is raised in the United States.)

- About 500,000 cattle are slaughtered each year in Central America to provide beef for the United States.
- About 4 years are needed for cattle to grow to maturity for slaughter, so about 2 million cattle are alive and grazing at any one time in Central America.
- Most of the land used for cattle grazing was cleared from tropical rain forests (Central America was almost entirely covered by tropical rain forest before the arrival of human beings).
- In Central America, about 20 cattle are raised per square kilometer of land (considerably less than the number raised per square kilometer in the United States). The reason for this low cattle stocking rate is that land cleared from tropical rain forests is poor in nutrients and supports less pasture.

a. Estimate, with uncertainty, the amount of land devoted to pasture for cattle in Central America.

b. The total amount of forest remaining in Central America is about 150,000 km^2. The pasture land on which cattle are raised usually is completely depleted of nutrients within about 10 years; at that time new pasture land must be obtained by clearing new areas of forest. If the United States continues to import the beef from 500,000 cattle each year, how long will it be before the remaining forest in Central America is completely destroyed in order to provide meat for fast food in the United States? Discuss uncertainties in your result.

46. Bugs Galore! Despite their small individual size, bugs represent a large biomass. For every human, there are approximately 100 million bugs. Assume that the average bug weighs 1 milligram.

a. If the global human population is 6 billion people, how many bugs are alive and crawling (or flying) today? Include an uncertainty in your answer.

b. What is the total weight of all these bugs? Include an uncertainty in your answer.

c. Estimate the total weight of all the human beings alive today. Include an uncertainty in your final answer.

d. How do parts (b) and (c) compare? Discuss the result, as well as possible uncertainties in the assumptions stated for this problem.

47. Tropical Insects. Recent studies suggest that tropical rain forests contain more than 30 million insect species. This estimate is based on a study in which 163 beetle species were found in one species of tree.

a. If there are 50,000 tropical rain forest tree species and each has an equal number of different beetle species how many beetle species exist in rain forest trees?

b. Beetles are estimated to represent 40% of all arthropods (insects and spiders). Estimate the total number of rain forest arthropods.

c. If twice as many arthropods live in the rain forest canopy (above the ground) than on the ground, estimate the total number of rain forest arthropods.

d. How does your answer in part (c) compare to the estimate of 30 million?

e. Discuss the validity and uncertainty in the method used in this problem to estimate the number of insect species.

48. Classification and Number of Species. Biologists classify plants and animals according to the following hierarchical categories: species, genus, family, order, class, phylum, kingdom. The higher the category in the hierarchy, the more and more living forms it contains. There are presently 5 known kingdoms and 89 known phyla.

a. Convert the known number of kingdoms and phyla into order of magnitude estimates.

b. In terms of a power of 10, how many more phyla than kingdoms are there?

c. Assume that the ratio of the number of kingdoms to the number of phyla is the same as the ratio between every pair of categories in the biological classification system. Use this assumption and the known number of kingdoms and phyla to estimate the total number of species on Earth.

d. Another way of estimating the total number of species on Earth is by extrapolating from the total number that have been identified. As of 1993, approximately 1.4 million distinct species of plants and animals have been identified. However, some recent studies (see Problem 47) suggest that the actual number of insect species, alone, may be as high as 30 million. In light of the present state of knowledge, give an order of magnitude estimate for the total number of living plant and animal species. Explain your method. Compare this estimate to the estimate in part (c). If the two estimates differ, suggest possible reasons.

49. How Many Books? Libraries contain a wealth of information in book form. But books take up space which limits the amount of storable information potential.

a. Estimate the total amount of shelf space in the library you use most often. Describe your estimation procedure in detail and include uncertainties.

b. Books come in many shapes and sizes, so estimate an average book width. Explain your reasoning.

c. How many books could your library carry on its shelves? Explain your estimate.

50. The Global Encyclopedia of Earth's Species. Imagine creating an encyclopedia containing one page describing each of Earth's species. Because many species are still unknown, an order of magnitude estimate of 10^8 species is reasonable within a factor of 10.

a. If each volume contains 1000 pages, how many volumes would be needed for this encyclopedia?

b. If each page is 0.2 millimeters thick, how wide is a single volume?

c. How much shelf space would be needed for the complete encyclopedia?

d. Would a typical university library have enough shelf space to hold this encyclopedia?
e. Suppose that the encyclopedia were recorded on CD-ROMs instead of printed pages. Recall that 1 character (letter or number) requires 1 byte of memory, and that a CD-ROM holds approximately 650 megabytes of data. Estimate the total number of CD-ROMs needed for this encyclopedia.

51. **The Length of a Light-Year.** In Chapter 7 we stated that a light-year is approximately 9.5 trillion kilometers. Calculate the length of a light-year more precisely using the following information: The speed of light is 2.99792458×10^5 kilometers per second (defined to be an exact value), and 1 year is 365.2422 days. What was the error (absolute and relative) in our earlier approximation?

52. **Urban Population Data.** Study the table below, which lists 1990 and 1992 population estimates compiled by the United Nations Population Division.
a. Based on the data shown, describe the changes in the populations of Tokyo, Mexico City, Calcutta, and Bombay between 1990 and 1992.

The 10 Most Populous Urban Areas

Rank	City	Millions of People
1990 Estimate		
1	Mexico City	20.2
2	Tokyo	18.1
3	Sao Paulo	17.4
4	New York	16.2
5	Shanghai	13.4
6	Los Angeles	11.9
7	Calcutta	11.8
8	Buenos Aires	11.5
9	Bombay	11.2
10	Seoul	11.0
1992 Estimate		
1	Tokyo	25.8
2	Sao Paulo	19.2
3	New York	16.2
4	Mexico City	15.3
5	Shanghai	14.1
6	Bombay	13.1
7	Los Angeles	11.9
8	Buenos Aires	11.8
9	Seoul	11.6
10	Beijing	11.4

b. Do you think that the populations of these cities actually changed by as much as you found in part (a)? Why or why not? If not, why do you think the estimates are so different for 1990 and 1992?
c. The data shown suggest that the populations of New York and Los Angeles remained quite stable between 1990 and 1992. In light of your answer from part (b), are you confident that the New York and Los Angeles data are accurate? Explain.

Projects

53. **Inflation and the CPI.** Because of substitution effects and other factors, most economists believe that the consumer price index overstates the actual rate of inflation (although a few dissenters claim that it understates the rate!). However, their disagreement is widespread over how much the CPI overstates inflation. In 1995, estimates suggested that the CPI may be overstating real inflation by anywhere from 0.2% to 2%. Although this disagreement may sound insignificant, it has major implications for the battle to balance the federal budget. Because cost-of-living adjustments (COLAs) in Social Security and other federal programs are tied to inflation, the government can save considerable money by assuming that inflation is lower. In addition, because tax brackets are indexed to inflation, a lower inflation rate means that more people remain in higher tax brackets. For example, a February 1995 study by the Congressional Budget Office estimated that, by assuming an annual inflation rate 0.5% lower than the change in the CPI, federal expenditures would decline by $36 billion over a 5-year period while tax revenues would increase by $22 billion.
a. Briefly explain, clearly and in your own words, why assuming a lower inflation rate would mean lower government expenditures and higher government revenues.
b. Investigate the current status of the debate over whether the government should use the CPI to measure inflation. Has the CPI been redefined since 1995?
c. Some people argue that, because it would lead to increased tax revenues, changing the way the government measures the inflation rate is nothing more than a sneaky way to raise taxes without passing any laws. At the same time, groups representing people who live on Social Security argue that it is a way of trying to balance the budget "on their backs." What do you think?
d. Any measure of inflation is apt to be uncertain, so how should the government determine changes in COLAs and in the indexing of tax brackets? After stating your opinion, imagine that you are a member of Congress. Discuss the political prospects for implementation of your suggestion.

54. **The Global Average Temperature.** Research the latest methods for determining whether the global average temperature is changing because of human activity and the latest results of these measurements. Also research potential impacts of changes in the global average temperature. Based on your findings, debate with a classmate whether emissions of greenhouse gases should be more tightly regulated.

55. Endangered Species. Determine how species are classified as either *endangered* or *threatened*. Then do any or all of the following.

a. Choose a favorite species from the endangered species list and research the causes and history of the threat to this species. Discuss the prospects for saving this species from extinction.

b. Research present laws regarding protection of endangered species. Do you believe that the laws are adequate, too weak, or too strong? How would you rewrite these laws?

c. Research the species that have become extinct during the time that human beings have been dominant on the Earth. Choose one species of interest and discuss whether human activity was responsible for the extinction. Discuss the likely impact that the extinction has had on the environment.

d. Consider the argument advanced in subsection 8.4.3 that the biodistribution, interacting with the physical components of the Earth, makes the Earth habitable for human beings. In light of this argument, discuss whether saving endangered species is a necessary or sufficient condition for preserving the long-term habitability of the Earth. Debate with a classmate whether the current "hot" environmental issues have the necessary focus for ensuring the long-term sustainability of civilization.

56. Real Problems in the Real World. Each of the following complex problems has no single, straightforward solution. For each, describe how you would apply the four-step problem solving process described in Section 4.3 (without actually carrying out the process to obtain a solution). You may choose either to enumerate the four steps or to describe your process in essay form. Either way, be sure that you list as many relevant factors as possible and discuss sources of uncertainty associated with each. Also, describe how you would work from these factors to find a solution. Conclude by describing your overall impression of whether the problem *can* be solved and whether any solution would be likely to generate controversy.

a. You are asked to calculate the cost of installing enough bike racks on campus to solve a bicycle parking problem.

b. You want to know the number of new faculty members that would be needed, and the total cost to the university, of making sure that all classes have 50 or fewer students (in order to replace large lecture classes with smaller classes).

c. Some people claim that by spending the money to support a top-quality football program, universities end up with more money for academics. Make a case for or against this claim.

d. You are asked to calculate the cost of sending U.S. troops to intervene in a war in Africa.

e. You want to figure out how much taxes would have to be increased to provide public school teachers with twice their present salaries.

f. You decide that, in the interest of the environment, you will convert your home heating and hot water system to solar power. How much will this conversion cost or save over the next 10 years?

g. How much energy would the United States, as a whole, save by replacing all present lighting sources with the most efficient commercially available lighting technology (for example, replacing ordinary light bulbs with new and more efficient fluorescent light bulbs)?

h. Suppose that China and India decide to use their extensive coal reserves to supply energy to their populations at the same per capita level as in the United States. How much carbon dioxide would be added to the atmosphere?

i. Are automobile insurance companies gouging drivers? Suppose that you want to figure out whether they are justified in raising insurance rates as rapidly as they have during the past few years.

j. A large city of the American Southwest claims that it soon will be facing a severe shortage of water. Would that still be the case if people replace their green lawns with other grasses or xeriscaping (landscaping adapted to a dry climate)?

k. How many jobs will be lost or saved if further cutting of old growth forests in the Pacific Northwest is stopped?

l. Suppose that the city added new bus routes and handed out free bus passes. How many people would give up driving in favor of the bus? How much money, overall, would this cost or save the city?

9 THE POWER OF NUMBERS — MORE PRACTICAL APPLICATIONS

Chapters 6–8 covered the interpretation of numbers: the concept of number, putting numbers into perspective, and dealing with uncertainty. Together, these chapters form the backbone of Part 2 of the text, *Thinking Quantitatively*. In combination with the skills of logic and problem solving developed in Part 1, *Thinking Logically*, you now have all the tools necessary to work with any situation involving numbers. In this concluding chapter of Part 2, we discuss several important and practical subjects.

CHAPTER 9 PREVIEW:

9.1 ***THE UNIFYING POWER OF NUMBERS.*** The title of this section emphasizes that the basic concepts developed to this point in the book allow you to study many diverse topics.

9.2 ***BALANCING THE FEDERAL BUDGET.*** In this section we provide a detailed overview of the difficulties involved in balancing the federal budget.

9.3 ***APPLICATIONS OF DENSITY AND CONCENTRATION.*** In this section we explain how the closely related concepts of density and concentration may be used in many applications, including the study of drunk driving and air pollution.

9.4 ***NUMBERS IN MOTION.*** The everyday concepts of position, velocity, and acceleration have many important practical applications in the natural sciences, which we discuss in this section.

9.5 ***ENERGY: THE FUTURE DEPENDS ON IT.*** In this section we show how the concepts of energy and power enable you to interpret utility bills and interpret the important issues of energy that face the human race.

9.6 ***MATHEMATICS AND MUSIC.*** In this section we explore the relationships between mathematics and music.

*9.7** ***DIRECTIONAL QUANTITIES: THE VELOCITY VECTOR.*** This concluding section introduces the concept of directional quantities, or vectors, which are useful in both the natural and social sciences.

In a sense, we are all mathematicians — and superb ones. It makes no difference what you do. Your real forte lies in navigating the complexities of social networks, weighing passions against histories, calculating reactions, and generally managing a system of information that, when all laid out, would boggle a computer. But if all this is true, why haven't you noticed this ability by now?

— A. K. Dewdney, in *200% of Nothing* (p. 147)

9.1 THE UNIFYING POWER OF NUMBERS

Why does the federal budget show a deficit? Is Hong Kong more crowded than Manhattan? How is air pollution measured? Why does the electric company charge for "kilowatt-hours"?

These diverse questions have one thing in common: Their answers involve numbers. In fact, numbers are involved in almost every facet of life. That's why we devoted the first eight chapters to developing what often is called **number sense:** the analog of common sense when it comes to numbers. In particular, Chapters 6–8 focused on three important aspects of number sense: understanding the concept of numbers (Chapter 6), giving meaning to numbers by putting them in perspective (Chapter 7), and learning to acknowledge and deal with the uncertainty surrounding numbers in the real world (Chapter 8).

Now, if you learn the terminology and context of a particular issue, you can understand it by applying number sense. That's what we mean by the unifying power of numbers: A few *unifying* ideas allow you to address many seemingly different questions by using the same methods. Indeed, the greatest benefit of mathematics may be its power to clarify and illuminate many of the world's mysteries. By tapping the unifying power of mathematics, you can cross boundaries between subject areas and develop a truly interdisciplinary view of the real world.

9.2 BALANCING THE FEDERAL BUDGET

One of the most troublesome quantitative issues facing this country is the federal budget and, specifically, the debt associated with it. It is crucial because budget decisions made today will have implications for decades to come. In Chapter 7, we put the size of the national debt in perspective. Here, we discuss the federal budget in more detail to help you gain an understanding of what is required to balance the federal budget.

There can be no freedom or beauty about a home life that depends on borrowing and debt. — Henrik Ibsen, 1879

In theory, the federal government works like a small business or an individual's personal finances. All have **receipts** (or **income**), and **outlays** (or **expenses**). When receipts exceed outlays, a **surplus** is recorded. When outlays exceed receipts, a **deficit** is recorded. A deficit must be covered either with reserves (accumulated from prior surpluses) or by taking out a loan. If a loan is used to finance the deficit, the borrower must pay interest to the lender. Continuing deficits cause **debt** to accumulate.

A national debt, if it is not excessive, will be to us a national blessing. — Alexander Hamilton, 1781

Since 1934, the U.S. government has had only eight surplus years — and not one since 1970. That is, the U.S. government has been a **debtor** for a long time. The national debt grew from $16 billion in 1930 to nearly $5 trillion in 1995. The debt is financed through loans in the form of bonds, Treasury bills, and other **government securities** on which the government must pay interest. For the past 20 years, the federal government has spent from 10% to 20% of its *entire* budget for interest on the debt (called **debt service**). Currently, annual interest on the debt exceeds $200 billion!

9.2.1 A Small Business

Before studying the federal budget, let's first investigate the simpler books of an imaginary company with not-so-imaginary problems. Table 9-1 summarizes four years of budgets for the Wonderful Widgit Company, which started with a clean slate at the beginning of 1992. The first column shows that during 1992 the company had total receipts, or earnings, of $854. However, because total outlays of $1000 exceeded the receipts, the company incurred a

The net deficit or surplus = receipts – outlays.
If receipts exceed outlays, there is a net surplus. If outlays exceed receipts, there is a net deficit.

first-year deficit of \$854 – \$1000 = –\$146 (we express a deficit with a negative number and a surplus with a positive number). To cover this deficit (so that, for example, checks written for outlays don't bounce), the company borrowed money from a bank, ending the year \$146 in debt.

In 1993, receipts increased to \$908; however, outlays also increased, to \$1082. Among the 1993 outlays was a \$12 interest payment; it was the interest paid to the bank on the first year debt of \$146, at an interest rate of 8.2% (\$146 × 8.2% = \$12). The deficit for 1993 was \$908 – \$1082 = –\$174, which the company had to borrow. Further, it had no money with which to pay off the debt from 1992. As a result, the total debt at the end of 1993 was \$146 + \$174 = \$320. Here is the key point: Because the company failed to balance its budget in 1993, its total debt continued to grow. As a result, its interest payment in 1994 increased to \$26.

Table 9-1. Budget Summary for the Wonderful Widgit Company

	1992	1993	1994	1995
Total Receipts	\$854	\$908	\$950	\$990
Outlays				
Operating	\$525	\$550	\$600	\$600
Employee benefits	200	220	250	250
Social Security	275	300	320	300
Interest on debt	0	12	26	47
Total Outlays	1000	1082	1196	1197
Surplus/Deficit	–146	–174	–246	–207
Debt (accumulated)	–146	–320	–566	–773

After another deficit year in 1994, the company's owners decided on a change of strategy in 1995. They held outlays nearly constant at 1994 levels by freezing operating and employee benefits expenses and cutting security expenses. Note, however, that the interest payment rose substantially because the total debt had increased in 1994. Despite the attempts to curtail outlays and despite another increase in receipts in 1995, the company still ran a deficit and the total debt continued to grow.

Time-Out to Think: Suppose that you are a loan officer for a bank in 1996, when the Wonderful Widget Company comes asking for further loans to cover its increasing debt. Would you agree to lend it the money? What if it offered to pay a higher than normal interest rate? (Note: "Junk bonds" are loans to companies that are considered relatively poor credit risks, but which promise higher interest rates.)

9.2.2 The Federal Budget

The Widget company example shows that a succession of annual deficits leads to a rising total debt. The increasing interest payments on that debt, in turn, make avoiding future deficits difficult. Reversing the explosive growth of the increasing debt is like trying to stop a runaway train. The Widgit

Company story is a mild version of what has happened to the federal budget. By imagining that the units of the Widgit budget are *billions* of dollars, that Widgit security represents the U.S. Defense Department, and that Widgit employee benefits represent the Social Security program, you can begin to get a feel for the scale of the federal budget. Unlike the Widgit company, which started out with no debt in 1992, the U.S. government went into 1992 with an accumulated debt of more than $4 trillion.

Data in Table 9–2 are from the *Economic Report of the President,* February 1994; 1994 and 1995 amounts are estimates.

Table 9-2. U.S. Federal Budget Summary (in $billions)

	1993	1994	1995
Receipts	1,153	1,249	1,354
Outlays			
Defense	291	279	270
Education	50	51	53
Health	99	112	123
Medicare	131	144	156
Social Security	305	320	337
Interest	199	203	213
All others	333	375	367
Total Outlays	1,408	1,484	1,519
Deficit	–255	–235	–165
Federal Debt	–4,351	–4,676	–4,960

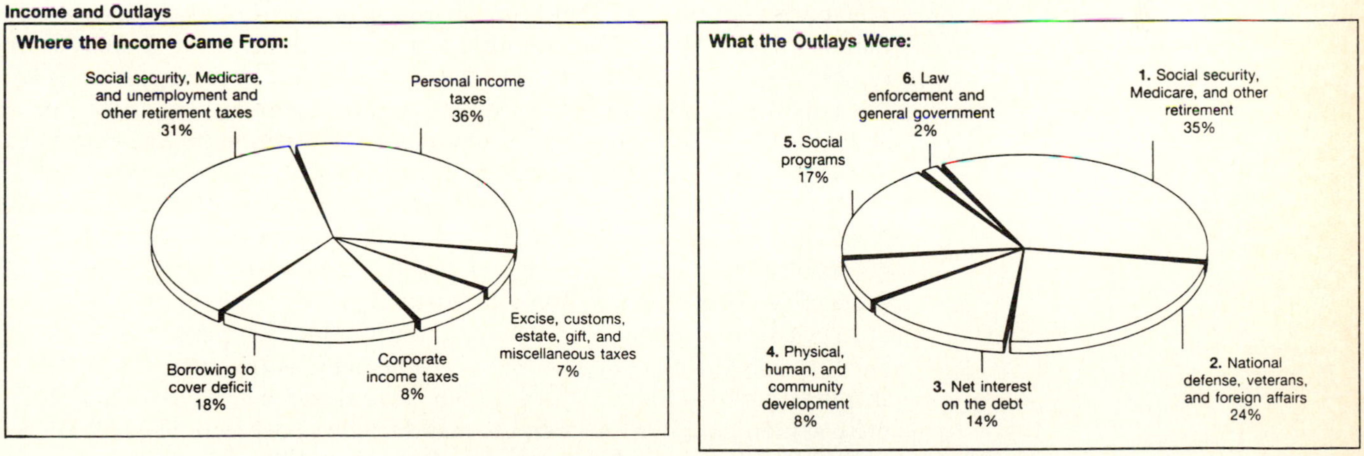

Figure 9-1. (a) This pie chart shows the approximate breakdown of receipts in the federal budget. (b) General distribution of government outlays is shown in this pie chart. Source: Internal Revenue Service, Form 1040,1994.

Table 9-2 provides a three-year summary of the federal budget, and Figure 9-1 shows relative shares of receipts and outlays. Note that receipts for the federal budget come primarily from taxes (individual income and corporate taxes) and from contributions to the Social Security program. Outlays are dominated by Social Security and defense, but note the large and increasing amount going to interest on the debt.

Certain outlays, primarily interest on the debt, are mandatory expenses. If they aren't paid, the government would default on its obligations and the entire economy might be thrown into turmoil.

Other outlays, such as Social Security, are direct payments to individuals. Such payments often are called **entitlements** because the individuals receiving the money have been told that they are entitled to it. Many politicians argue that entitlements should not be cut for budget-balancing purposes. The remaining budget categories are considered to be **discretionary** because, in principle, the government can change this spending at any time.

With receipts increasing more than outlays from 1993 to 1995, the *deficit* dropped during these three years. However, because all three years had a net deficit, the total *debt* continued to increase.

Time-Out to Think: Each security issued by the government to cover the debt promises a particular interest rate, which reflects prevailing interest rates in the economy at large. As old securities become due, new ones are issued to replace them. Suppose that interest rates suddenly rose: What would happen to total interest payments? What would happen to the deficit and debt?

9.2.3 Balancing the Budget

Depending on whose opinion you hear, the federal budget is anything from a minor problem to a major crisis. In any case, balancing the budget involves two related problems. First is the deficit, which arises from the annual failure to balance receipts and outlays. Second is the debt. The government cannot begin to pay off the debt until it eliminates the deficit and achieves a surplus.

What would be required to eliminate the *deficit* in a single year? Suppose that the federal government had tried to balance the budget in 1995 when receipts were $1.354 trillion. If it had frozen total spending at the 1994 level of $1.484 trillion, the result would have been a deficit of *only* $0.13 trillion, or $130 billion — an improvement, but hardly a balanced budget. To wipe out the 1995 deficit, total outlays would need to have been cut by almost 9% from the 1994 level. Decreasing outlays by 9% in a single year is highly unlikely: The cuts would affect so many people that they would be politically unthinkable. Further, most economists believe that such a drastic cut in a single year would greatly damage the economy.

Even if the budget were balanced (zero deficit) in a particular year, that still would not reduce the *debt*. Ultimately, the only reasonable solution to the budget problem is to reduce spending gradually or to increase receipts. Let's make some projections, much as federal budget officials do. Suppose that, beginning in 1995, government outlays held steady. Furthermore, suppose that government receipts rise at a rate of 4% each year. Finally, assume that any budget surpluses are used to retire the debt. Stepping ahead one year at a time with these assumptions, we can compute that the federal debt would be eliminated in 2013. That's not bad, but the assumptions are hardly realistic! The increasing receipts must be generated by some combination of overall growth in the economy and higher tax rates. Because substantial tax increases aren't on any politician's agenda and the economy has never sustained a 4% growth rate for long, the idea of wiping out the debt by 2013 is nothing less than fantasy.

Table 9-3 shows the outcome of similar calculations with different annual increases in receipts and outlays. For example, if receipts increase at an annual rate of 1% and outlays *decrease* at an annual rate of 1%, this simple model projects that the debt will vanish in about 2034.

Time-Out to Think: Explain why three of the projections in Table 9-3 show that the debt would *never* be eradicated.

Table 9-3. Year of Zero Debt

	Receipts +1%	Receipts +2%	Receipts +3%	Receipts +4%
Outlays –2%	2023	2016	2012	2010
Outlays –1%	2034	2020	2014	2011
Outlays 0%	2093	2029	2018	2013
Outlays +1%	Never	2058	2026	2017
Outlays +2%	Never	Never	2046	2023

9.2.4 Further Complications

Now consider the difficulty of even holding outlays steady in a year like 1995. Unchanged spending represents a *decrease* government services because of the effects of inflation. In addition, because the interest payment increased from 1994 to 1995 as shown in Table 9-2, other programs must be cut further to compensate for this mandatory spending. Entitlements such as Social Security pose further problems because more people become eligible for benefits each year. Thus, to hold total Social Security spending steady, payments *per individual* must be cut.

Time-Out to Think: Projections in 1995 indicated that Medicare spending would rise to some $200 billion by 2000. Some politicians proposed changing the program "to reduce its rate of growth." Others claimed that these proposals were "cuts" that would severely hurt ordinary citizens. Can you see both points of view? Do you think one is more compelling than the other? Explain.

Further, as we just showed, even a freeze on spending in 1995 would result in a deficit and therefore in increased debt. And increased debt means higher interest payments the next year, which necessitates deeper spending cuts in programs in order to hold outlays steady. It is an insidious cycle.

Finally, projecting future outlays and receipts involves great uncertainties. A small change in interest rates can dramatically change the amount of money that goes to servicing the debt. If the economy grows more slowly than expected, or goes into a **recession** where the overall economy shrinks, receipts will be lower than projected. Any model of the budget can be useful in helping you understand the budget — whether it is a simple one such as that in Table 9-3 or the far more complex models used by economists in the Congressional Budget Office and the president's Office of Management and Budget. But as predictive tools, models of the budget simply are shots in the dark. Why? The economy, and hence the budget, is a *chaotic* system (see Chapter 3) — it therefore exhibits extreme sensitivity to initial conditions (the butterfly effect) and predictions are necessarily precarious.

THINKING ABOUT ... FEDERAL BUDGET REALITIES

Budget realities make balancing the budget far more difficult than you might imagine. Taking political realities into account, here is how Jessica Matthews, senior fellow of the Council on Foreign Relations, analyzed the problem (*Washington Post*, February 19, 1995).

"Take a box of kitchen matches, count out 160 and spread them on a table. Assign each a value of $10 billion, and you've got the federal budget. Now separate a pile of 35 (Social Security) and another of 26 (interest on the debt), and push them off to the side, out of reach [because interest payments are mandatory and Social Security is politically untouchable].

"Next take 27 (Medicare and Medicaid) and 18 (pensions, unemployment compensation, retirement and disability benefits, food stamps, welfare, student loans, agricultural subsidies, the earned-income tax credit and other so-called mandatory programs), and put these two piles within reach, but just barely.

"Now count out 26 for defense. Count two — yes two — for international spending on everything from controlling nuclear proliferation and drugs to foreign aid and Middle East diplomacy. Set aside 26 for domestic discretionary programs — everything else the federal government does from running the FBI to building highways. Put these three piles front and center. Remember that without any tax cuts balancing the budget requires eliminating 20 of these matchsticks, and you've got a pretty good idea of the problem.

"There are 99 matchsticks at least theoretically within political reach. Assume, boldly, that 10 percent of the activities they represent is "waste, fraud and abuse," and you still must find 10 matchsticks to eliminate.

"The first thing you'll notice is that [most] proposed cuts will do you almost no good.... Eliminating the Arts and Humanities Endowments and the Corporation for Public Broadcasting, for example, save one-fifteenth of a matchstick. Zeroing out international peacekeeping would save a bit more — one fourteenth of a stick. Ending all aid to poor countries would save one-fifth of a matchstick.... Shutting down the State Department, closing every US embassy abroad, ending all contributions to international agencies and turning away from every American interest and responsibility in the world would save two of the needed 10 matchsticks.

"The bottom line is that if one could root out $100 billion of fat there would remain the job of cutting an equal amount of muscle and bone. Remember, too, that the longer balancing is postponed, the faster the 26-matchstick interest pile will grow.

"What's to be done? Bringing Social Security back within reach is part of the answer.... Truth-telling would help a lot more. Balancing the budget will require sacrifices — higher taxes or fewer services. Amending the Constitution may or may not make it happen, but it won't make it any the less painful."

9.3 APPLICATIONS OF DENSITY AND CONCENTRATION

The ancient town of Syracuse is located in Sicily on the coast of the Ionian Sea.

Eureka! This word, used to express triumph upon a discovery, is a Greek word meaning "I have found it." Archimedes (c. 287–212 B.C.), considered by many the greatest scientist of the ancient world, is said to have exclaimed it as he ran naked through the streets of Syracuse.

What inspired this display by Archimedes? (Actually, running naked in ancient Greece wasn't as unusual as it is today.) His king, Hieron, had recently received a new crown from a goldsmith. Hieron asked Archimedes to determine whether the crown was pure gold, as it was supposed to be, or whether the goldsmith had mixed in some silver.

Archimedes was at a loss until, one day, he noticed the water overflowing as he stepped into his bath. In a flash of insight ("aha!"), he realized that the amount of water overflowing was equal to the volume of his body immersed in the bath. Herein lay the solution: Silver is bulkier than gold (a chunk of silver is larger than a chunk of gold of the same weight), so he merely needed to immerse the crown and an equal weight of pure gold into water. If the water level rose the same amount in both cases, the crown was made of pure gold; if the crown caused the water to rise more than the pure gold, it contained silver.

His excitement at this discovery of the *principle of buoyancy* prompted Archimedes' shouting, naked run. For the goldsmith, alas, the news wasn't so great. The crown was partly silver, and the king had the goldsmith executed.

9.3.1 The Concept of Density

At the heart of Archimedes' discovery lies the concept of **density,** which describes how tightly something is packed. In the case of materials, we measure density as mass per unit volume. By immersing the crown and an equal mass of pure gold in water, Archimedes compared their volumes. Because their masses were the same, the one with greater volume was less dense.

The distinction between *mass* and *weight* is discussed in subsection 9.4.4. For the moment, because we are dealing with masses on Earth, you may disregard their distinction.

In the metric system, the most common units of density are kilograms per cubic meter and grams per cubic centimeter. Hence

$$\text{density} = \frac{\text{mass}}{\text{volume}} \qquad \left(\frac{\text{kg}}{\text{m}^3} \text{ or } \frac{\text{g}}{\text{cm}^3}\right).$$

To calculate density, simply divide an object's mass by its volume. For example, if a 40-gram rock has a volume of 10 cubic centimeters, its density is

$$\frac{40 \text{ g}}{10 \text{ cm}^3} = 4\frac{\text{g}}{\text{cm}^3}.$$

A useful guide for putting densities in perspective is the density of water, which is approximately 1 gram per cubic centimeter, or 1000 kilograms per cubic meter:

Recall that

$$\left(\frac{100 \text{ cm}}{1 \text{ m}}\right)^3 = \frac{10^6 \text{ cm}^3}{1 \text{ m}^3}.$$

$$\text{Density of water} = 1\frac{\text{g}}{\text{cm}^3} = 1\frac{\text{g}}{\text{cm}^3} \times \left(100\frac{\text{cm}}{\text{m}}\right)^3 \times \frac{1 \text{ kg}}{1000 \text{ g}} = 1000\frac{\text{kg}}{\text{m}^3}.$$

Rocks, like the one with a density of 4 grams per cubic centimeter, are more dense than water and therefore sink in water. Wood floats because its density is less than that of water.

Time-Out to Think: Explain why you can float better in a swimming pool if your lungs are filled with air than when you fully exhale.

The concept of density may be applied to many other things. For example, **population density** describes the number of people per unit area of a city or country and **information density** can describe the number of megabytes of information stored on each square centimeter of a compact disk.

Example 9-1 *Water mass.* Suppose that you have a 2-gallon jug of water. What is its mass?

Solution: Rearrange the density equation to find the mass by multiplying volume and density:

$$\underbrace{\text{units of volume}}_{\text{m}^3} \times \underbrace{\text{units of density}}_{\frac{\text{kg}}{\text{m}^3}} = \underbrace{\text{units of mass}}_{\text{kg}}.$$

You know the volume of the jug (2 gallons) and the density of water (1 gram per cubic centimeter), so the rest is just unit conversions:

$$\underbrace{2\text{ gal} \times \frac{3.875\ \ell}{\text{gal}} \times \frac{1000\text{ cm}^3}{\ell}}_{\text{volume of jug}} \times \underbrace{1\frac{\text{g}}{\text{cm}^3}}_{\text{density of water}} = 7750\text{ g}.$$

A 2-gallon jug of water has a mass of 7.75 kg.

❖

Example 9-2 *Crowding in Hong Kong and Manhattan.* Compare the population densities of Hong Kong and New York's Manhattan Island.

Solution: You need to know land areas and populations to solve this problem. A standard atlas indicates that Hong Kong has a land area of about 1075 km^2 and a 1990 population of about 6 million. Manhattan has an area of 57 km^2 and a 1990 population of about 1.5 million. Their average population densities are

Note how final answers are rounded, as always, in accord with the rules for combining uncertainty given in Chapter 8.

$$\text{Hong Kong}: \frac{5.8 \times 10^6\text{ people}}{1075\text{ km}^2} = 5400\ \frac{\text{people}}{\text{km}^2};$$

$$\text{Manhattan}: \frac{1.5 \times 10^6\text{ people}}{57\text{ km}^2} = 26{,}000\ \frac{\text{people}}{\text{km}^2}.$$

The average population density of Manhattan is about five times greater than that of Hong Kong. However, most of the population of Hong Kong lives in a relatively small part of the territory. Within that smaller area, the population density probably is much higher than that of Manhattan.

❖

Example 9-3 *Compact Disks.* A standard compact disk stores information with tiny variations on its surface. Each variation represents one **bit** of information, and 8 bits comprise a **byte**. Standard disks can hold 650 megabytes of information on a surface area of about 90 square centimeters. What is the information density on a standard disk? A new technology allows compact disks to store 10 gigabytes of information. How much has the density increased?

Solution: Density on a compact disk is best expressed as information per unit area. Thus a standard compact disk has an information density of:

$$\frac{650 \times 10^6\text{ bytes}}{90\text{ cm}^2} = 7 \times 10^6\ \frac{\text{bytes}}{\text{cm}^2} = 7\frac{\text{megabytes}}{\text{cm}^2}.$$

The newer technology disks have an information density of

$$\frac{10 \times 10^9\text{ bytes}}{90\text{ cm}^2} = 1 \times 10^8\ \frac{\text{bytes}}{\text{cm}^2} = 100\frac{\text{megabytes}}{\text{cm}^2}.$$

The information density on the new disks is an order of magnitude greater than that on the older "standard" compact disks. This single leap in technology happened in a period of just a few years. Imagine what compact disks might hold by 2010!

❖

9.3.2 Blood-Alcohol Concentration

Closely related to the concept of density is the concept of **concentration,** which describes the amount of one substance mixed with another. Various units are used to describe concentration.

Sometimes, concentration is expressed simply as a percentage. For example, alcohol begins to impair brain functions when the **blood-alcohol concentration** reaches 0.05%, or 0.05 liters of alcohol per liter of blood. In most states, a blood-alcohol concentration of 0.1% is enough to warrant a DWI (driving while intoxicated) arrest. A blood-alcohol concentration of 0.5% usually is fatal.

How much beer does an average person need to consume before becoming legally intoxicated? First, we calculate how much alcohol must be in the bloodstream to qualify as legal intoxication. An average person has about 5 liters of blood, so legal intoxication means that 0.1% of this volume is alcohol:

$$0.1\% \times 5 \text{ liters} = 0.001 \times 5 \text{ liters} = 0.005 \text{ liters}.$$

Next, we need to determine how much beer contains 0.005 liters of alcohol. Typical beers are 6% alcohol; that is, alcohol represents 6% of the total volume of the beer. Therefore

$$\text{amount of alcohol} = 0.06 \times \text{amount of beer}.$$

Dividing both sides by 0.06 yields

$$\text{amount of beer} = \frac{\text{amount of alcohol}}{0.06}.$$

Substituting 0.005 liters for the amount of alcohol:

$$\text{amount of beer} = \frac{0.005 \text{ liters}}{0.06} = 0.08 \text{ liters}.$$

That is, for an average person, legal intoxication requires consuming less than 0.1 liters of beer.

Let's think about this surprising result. A liter is slightly more than a quart, so a tenth of a liter is slightly more than a tenth of a quart, or about three ounces. In other words, a mere *three ounces* of beer — less than half a glass — contains enough alcohol to put an average person at the legal intoxication limit!

However, there are some mitigating factors. For example, our calculation was based on the assumption that all the alcohol is absorbed into the bloodstream immediately. Most people don't drink their beer all at once (except when "chugging"). Moreover, food in the stomach slows the absorption of alcohol into the bloodstream. And the alcohol will begin to be metabolized (primarily in the liver) once it enters the bloodstream, removing it at a rate of about 10–15 milliliters per hour. Thus the 5 milliliters of alcohol in 3 ounces of beer will be metabolized within about half an hour. Nevertheless, the basic conclusion holds: If an average-sized person "chugs" 3 ounces of beer on an empty stomach, he or she will be near the legal intoxication limit (for a short time, at least).

Time-Out to Think: Based on your own experiences with alcohol, or observing its effects on others, evaluate the preceding result. Can someone actually become intoxicated on 3 ounces of beer? Suppose that you drink two beers. Given the rate of alcohol metabolism, how long should you wait before attempting to drive?

9.3.3 Air and Water Pollution

Environmental science is one of the most active areas of research today, combining chemistry, physics, engineering, economics, and public policy. Much of environmental science deals with issues of contamination: chemicals that have leaked into water supplies, pollutants spewed into the atmosphere by factories and cars, and radioactive material that must be stored or discarded. Environmental scientists measure the concentration of contaminants such as acid in a mountain lake resulting from acid rain or carbon monoxide in a sample of urban air.

One common measure of contamination is the mass of the contaminant per unit volume of material. For example, suppose that 5 grams of a salt are dissolved in half a liter of pure water. The concentration of the contaminant, salt, is

$$\text{salt concentration} = \frac{\text{mass of dissolved salt}}{\text{volume of water}} = \frac{5\ \text{g}}{0.5\ \ell} = 10\frac{\text{g}}{\ell}.$$

The concentration of the salt in the water is 10 grams per liter or, equivalently, 0.01 gram per cubic centimeter.

Time-Out to Think: Show that 10 grams per liter is the same as 0.01 gram per cubic centimeter.

Example 9-4 ***Impact of Mining.*** A sample of water that flows from a nineteenth century mine is found to be contaminated with lead chloride (a lead salt) at a concentration of 0.0001 gram per milliliter. How much lead chloride is contained in 500 liters of water? Should you be concerned about drinking from this water supply?

Solution: Because each milliliter contains 0.0001 gram of lead chloride, each liter contains 1000 times as much, or 0.1 gram. Therefore the amount of lead chloride contained in 500 liters of water is

$$500\ \ell \times \frac{0.1\ \text{g (lead chloride)}}{\ell} = 50\ \text{g (lead chloride)}.$$

Lead is a toxic substance linked to many health problems, including mental retardation in children, so avoiding this water supply would be a good idea! ❖

Air Pollution

Another common way to describe the concentration of a contaminant is with "parts per" units, such as **parts per million (ppm)** or **parts per billion (ppb)**. For example, an air sample containing carbon monoxide in a concentration of 9 ppm contains 9 carbon monoxide molecules for every 1 million air molecules.

The **U.S. Environmental Protection Agency** (EPA) requires large cities to monitor concentrations of certain gases. Based on health research, the EPA

Units of parts per million or parts per billion simply express a fraction: the number of particles of one substance mixed in with another. Therefore they are equivalent to expressing concentration as a percentage, as in the blood-alcohol concentration. The advantage of "parts per" units is in making numbers easier to read. For example, a concentration of 1 part per million is more difficult to read as a fraction (0.000001) or a percentage (0.0001%).

also sets standards for maximum concentrations. For example, the EPA standard for carbon monoxide is 9 ppm. On a typical winter day in Denver, Colorado, levels of carbon monoxide are about 4 to 5 ppm, or within the limit considered safe. On high-pollution days, however, the carbon monoxide level often reaches 12 ppm.

Another urban pollutant is ozone (O_3). Although ozone in the upper atmosphere protects humans from the Sun's harmful ultraviolet radiation, ozone near the surface is a poisonous gas and one of the primary ingredients in urban smog. The EPA standard for ozone is 120 ppb. Los Angeles and several other cities often exceed this standard.

Besides gaseous pollutants, smog often includes *particulate* pollutants: tiny, solid particles suspended in the air. The concentration of particulate matter usually is measured in units of micrograms (μg) per cubic meter of air. The primary sources of particulate pollution in urban areas are gasoline and diesel engines, wood stoves and fireplaces, and highway gravel (especially in winter when gravel is used on icy roads, then ground into fine particles by passing cars). Particulate matter is a health hazard when it is small enough to be breathed into your lungs. The EPA currently monitors concentrations of particles smaller than about 10 micrometers in diameter; the standard for particulate pollution is 120 micrograms per cubic meter.

Time-Out to Think: If you live in an area with severe pollution, the local media may report concentrations of various pollutants. Check your newspaper and local TV news. Do they ever report on pollution? What units of concentration do they use?

Example 9-5 ***Carbon Dioxide Concentration.*** In 1993, the concentration of carbon dioxide in the Earth's atmosphere was about 360 parts per million. Express this concentration as a percentage and discuss its importance.

Solution: A concentration of 360 parts per million means that for every million molecules of air, there are 360 molecules of carbon dioxide. Thus the fraction of air molecules that are carbon dioxide is 360 millionths, or 0.000360. Converting that to a percentage gives 0.036%. Despite its small percentage, carbon dioxide plays a crucial role in regulating the Earth's temperature through the greenhouse effect (see Subsection 3.5.3). Further, the concentration of carbon dioxide has been rising steadily since the beginning of the industrial age. Annual measurements have been made since 1958, when the concentration was 315 parts per million.

❖

9.4 NUMBERS IN MOTION

Everything moves. Even when an object appears to be stationary, it is moving with the rotation of the Earth, with the Earth's orbit around the Sun, the Sun's orbit around the Milky Way galaxy, and the galaxy's motion through the universe. Because everything is in motion, the study of motion is fundamental in the natural sciences.

The basic goal is to describe how an object's **position** or **distance** from a specified point changes with time. For example, suppose that you are in a car driving at 60 miles per hour along the highway. If you start in Pittsburgh and drive north, you will be a distance of 60 miles north of Pittsburgh after an hour; equivalently, we could say that your position shifts from Pittsburgh to a point 60 miles north of Pittsburgh.

9.4.1 Speed and Velocity

Suppose that you drive at a constant speed of 50 miles per hour for two hours. How far will you travel? The answer, 100 miles, is found by multiplying the speed by the elapsed time, or

$$\text{distance} = \text{speed} \times \text{elapsed time} = 50 \text{ mi/hr} \times 2 \text{ hr} = 100 \text{ mi}.$$

A technical definition of speed is the *rate of change in position with respect to time*. We discuss rates of change generally in Chapter 10. Because the idea of speed is familiar from daily travel, we aren't quite so formal in this section.

Note that speed is distance divided by time (e.g., miles per hour or kilometers per hour). You can use this fact to solve many problems by unit analysis. For example, suppose that you are traveling at a speed of 55 miles per hour. How long does it take to travel 100 miles? The words "how long" indicate that the answer must have units of time, which you can obtain by dividing the distance by the speed:

$$\text{Elapsed time} = \frac{\text{distance traveled}}{\text{speed}} = \frac{100 \text{ mi}}{55 \frac{\text{mi}}{\text{hr}}} = 1.8 \text{ hr}.$$

Because 0.8 hours isn't very meaningful to most people, you should convert it to minutes: 0.8 hr × 60 min/hr = 48 min. Thus 1.8 hours, or 1 hour and 48 minutes, are needed to drive 100 miles at 55 miles per hour.

In most real situations, speed is *not* constant. In driving, for example, your speed usually varies considerably as you respond to traffic conditions, changing terrain, and traffic signs and signals. As a result, thinking in terms of *average speed* is more realistic than thinking of your speed from instant to instant.

Example 9-6 A car goes a distance of 5 miles in 10 minutes. What is its average speed during that period?

Solution: To find the average speed simply divide the distance traveled (5 miles) by the elapsed time (10 minutes):

$$\text{Average speed} = \frac{\text{distance traveled}}{\text{elapsed time}} = \frac{5 \text{ mi}}{10 \text{ min}} = \frac{5 \text{ mi}}{\frac{1}{6} \text{ hr}} = 30 \frac{\text{mi}}{\text{hr}}.$$

The car traveled at an average speed of 30 miles per hour during the 10 minutes. Note that this result does not depend on how the car made this trip — only on the distance traveled and elapsed time.

❖

Example 9-7 ***The Space Shuttle.*** The space shuttle orbits the Earth at an altitude of about 300 *km*. It completes one orbit approximately every 90 minutes (1.5 hours). What is the average speed of the space shuttle as it orbits the Earth?

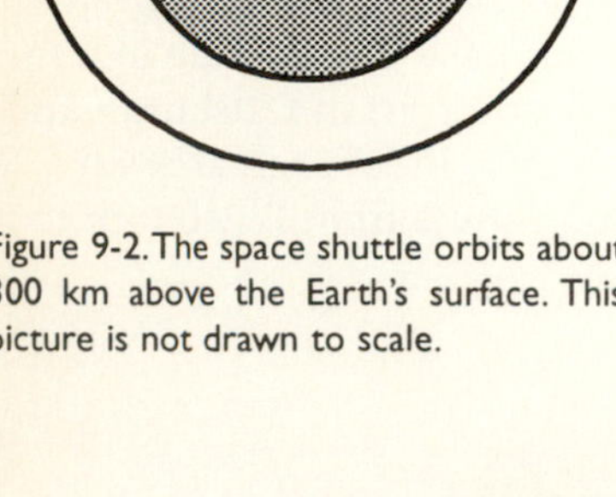

Figure 9-2. The space shuttle orbits about 300 km above the Earth's surface. This picture is not drawn to scale.

Solution: To find the average speed for the space shuttle divide the distance it travels during each orbit by the elapsed time for each orbit, or 1.5 hours. Figure 9-2 shows that the shuttle's orbit has a radius of about 6700 *km* (its altitude of 300 *km* plus the 6400 *km* radius of the Earth). The distance the shuttle travels during each orbit is the circumference of the orbit. Recall that the circumference of a circle is 2π times its radius. Hence

$$\text{average speed} = \frac{\text{distance traveled during one orbit}}{\text{elapsed time for one orbit}}$$
$$= \frac{2 \times \pi \times 6700 \text{ km}}{1.5 \text{ hr}} = 28{,}000 \frac{\text{km}}{\text{hr}}.$$

The average speed of the shuttle is about 28,000 kilometers per hour, or about 17,000 miles per hour.

❖

Example 9-8 *Speeds of the Stars.* Relative to the Earth, nearby stars typically travel at about 20 kilometers per second. How far does a typical nearby star travel (relative to the Earth) every hour?

Because the stars are moving, the shapes of the constellations in the sky change with time. However, despite the high typical speed (70,000 km/hr), the stars are so far away that their motion is undetectable to the naked eye. Indeed, the constellations today have nearly the same shape as those seen by Native Americans, ancient Greeks, and others some 2000 years ago. If you could come back to Earth in 1 million years, however, the constellations would be unrecognizable because of the changes caused by the motion of stars.

Solution: To find the distance traveled, multiply the speed by the elapsed time, and do the necessary unit conversions:

$$\text{Distance traveled in one hour} = \text{average speed} \times 1 \text{ hour}$$
$$= \underbrace{20\frac{\text{km}}{\text{s}}}_{\text{speed}} \times \underbrace{1 \text{ hr} \times 60\frac{\text{min}}{\text{hr}} \times 60\frac{\text{s}}{\text{min}}}_{\text{1 hour, converted to seconds}} = 72{,}000 \text{ km.}$$

Rounding the answer to one significant figure (as speed was given with one significant figure in the problem), a typical nearby star travels about 70,000 kilometers each hour.

❖

Speed Versus Velocity

The terms *speed* and *velocity* often are used synonymously, but there is a subtle distinction between them. **Velocity** is speed with a direction. If you are driving your car due east on Interstate 70 at a *speed* of 55 miles per hour, your *velocity* is "55 miles per hour, due east."

Another example of the distinction between speed and velocity is when you hold your speed steady at 40 miles per hour as you drive around a curve. The direction of your travel is changing as you round the curve; therefore, even though your speed is constant, your velocity is changing.

Because velocity includes information about direction, it is called a **directional quantity** (or **vector**).

Besides velocity, several more examples of directional quantities appear in this chapter. To avoid complications, we keep the directions simple (e.g., due north, or up and down). In Section 9.7, we describe how to handle directional quantities more generally.

Time-Out to Think: Recall a typical trip from home to school or work. In words, contrast your speed and velocity at each point along your route.

Example 9-9 *Average Speed and Velocity.* Jackson, Mississippi, lies approximately 200 kilometers due north of New Orleans, Louisiana. If you fly from Jackson to New Orleans in 1 hour, what is your average speed and velocity for the one-way trip? If you immediately fly back to Jackson, what is your average speed and velocity for the round trip?

Solution: Your average speed to New Orleans is 200 kilometers per hour because you traveled a distance of 200 kilometers in 1 hour. Your average velocity must include direction, so it is 200 kilometers per hour *due south.*

If the return trip back to Jackson also takes 1 hour, the average speed on the return flight also is 200 km/hr. Thus your average speed for the round trip is 200 km/hr. Your return average velocity is 200 km/hr *due north.* Because this velocity is exactly opposite the outbound velocity, your average velocity for the round-trip is zero!

❖

9.4.2 Acceleration

Imagine that you are stopped at a red light. When the light changes you begin driving ahead, going faster and faster. Suppose that after 1 second, you have reached a speed of 5 meters per second (about 11 miles per hour). After 2 seconds, your speed has reached 10 meters per second. After 3 seconds, your speed has reached 15 meters per second. This situation involves **acceleration:** In each second of elapsed time, your speed increased by 5 meters per second. To express the acceleration numerically, we write

$$\text{acceleration} = \frac{\text{change in speed}}{\text{elapsed time}} = \frac{5\ \frac{\text{m}}{\text{s}}}{1\ \text{s}} = 5\ \frac{\text{m}}{\text{s}} \times \frac{1}{1\ \text{s}} = 5\frac{\text{m}}{\text{s}^2}.$$

Note that the units of acceleration are distance divided by time *squared.* The standard units are m/s^2 in SI, or ft/s^2 in the U.S. Customary System. The best way to read these units of acceleration is "meters per second *per* second," or "feet per second *per* second."

Because acceleration is the rate of change in velocity, acceleration also is a directional quantity. In this book, however, we limit mathematical work with acceleration to situations in which only the speed changes.

Acceleration occurs whenever velocity changes; that is, whenever either the speed or direction of motion (or both) changes. For example, you are accelerating in all the following circumstances.

- You gain speed as you drive along an on-ramp to a freeway.
- You slow down for a stoplight. In this case, your acceleration is *negative* because your speed is decreasing.
- You drive around a curve. In this case, you are accelerating because your *direction* is changing, even if you maintain a constant speed.

Example 9-10 Suppose that you get into your Ferrari and accelerate at a rate of 50 ft/s^2. How fast will you be going after 1 second? 2 seconds? 3 seconds? If you continue to accelerate at this rate for a fourth second and are spotted by a police officer, will you get a ticket?

Solution: The acceleration 50 ft/s^2 means that you are increasing your speed by 50 feet per second, each second. Starting from rest, after 1 second your speed is 50 feet per second. During the next second, your speed increases by another 50 feet per second, bringing your speed to 100 feet per second. During the third second, your speed again increases by 50 feet per second, bringing your speed up to 150 feet per second. Similarly, after 4 seconds you would be traveling 200 feet per second. To decide whether you are speeding, convert 200 feet per second to miles per hour:

$$200\frac{\text{ft}}{\text{s}} \times \frac{1\ \text{mi}}{5280\ \text{ft}} \times 60\frac{\text{s}}{\text{min}} \times 60\frac{\text{min}}{\text{hr}} = 136\frac{\text{mi}}{\text{hr}}.$$

The officer definitely will give you a ticket, if she can catch you.

❖

Example 9-11 *Average Acceleration.* Suppose that, on leaving an interstate highway, you decrease your speed from 70 miles per hour to 0 in 10 seconds. What is your *average* acceleration?

Solution: Over the 10 seconds, your speed changes from 70 miles per hour to 0, so your change in speed is 0 – 70 mi/hr = –70 mi/hr. The change is negative because your speed is *decreasing*. Thus

$$\text{average acceleration} = \frac{\text{change in speed}}{\text{elapsed time}} = \frac{-70\frac{\text{mi}}{\text{hr}}}{10\ \text{s}} = -7\ \frac{\text{mi/hr}}{\text{sec}}.$$

Although the units may look strange, they still are units of velocity divided by time. Their meaning should be clear: As you slow down, your speed decreases at an average rate of 7 miles per hour each second.

❖

Example 9-12 *Increase in Speed.* You can use unit analysis to solve problems involving acceleration. Suppose that you accelerate at an average rate of 5 m/s^2 for 10 seconds. How much does your speed increase?

Solution: The question "How much does your speed increase?" indicates that the answer must have units of speed (in this case, units of meters per second). Thus you need to multiply the acceleration by the elapsed time:

$$\text{Change in speed} = \text{average acceleration} \times \text{elapsed time} = 5\frac{\text{m}}{\text{s}^2} \times 10\ \text{s} = 50\frac{\text{m}}{\text{s}}.$$

Your speed increases by 50 meters per second during the 10 seconds.

❖

The Acceleration of Gravity

Why does jumping off a tall building hurt more than jumping off a chair? The answer is that you *accelerate* as you fall. The longer you accelerate, the faster you are going when you hit the ground. And the faster you are going when you hit the ground, the more it hurts!

In a famous (though possibly apocryphal) experiment that involved dropping weights from the *Leaning Tower of Pisa,* Galileo demonstrated that gravity accelerates all objects by the same amount, regardless of their mass. This fact is surprising because it seems to contradict real-life experience: If you drop a piece of paper, it floats gently to the ground, but if you drop a rock, it falls much faster. However, the difference is caused by air resistance. If you dropped a piece of paper and a rock on the Moon, where there is no air, both would fall at exactly the same rate.

The acceleration of gravity sometimes is called "one gee," meaning $1 \times g$, or 9.8 m/s^2. Similarly, 2 gees means $2 \times g$, or 19.6 m/s^2, and so on. Stating gees is especially common when dealing with jet fighters, roller coasters, or rocket launches.

Time-Out to Think: Find a piece of paper and a small rock. Hold both at the same height, one in each hand, and let them go at the same instant. The rock, of course, hits the ground first. Next, crumple the paper into a small ball and repeat the experiment. What happens in this case? Are you now convinced that the *mass* of an object does not affect its rate of fall? Why or why not?

The acceleration of falling objects is called the **acceleration of gravity.** On Earth, the acceleration of gravity is 32 ft/s^2, or 9.8 m/s^2, and it is abbreviated with the letter g. Thus

$$g = \text{acceleration of gravity} = 32\frac{\text{ft}}{\text{s}^2} = 9.8\frac{\text{m}}{\text{s}^2}.$$

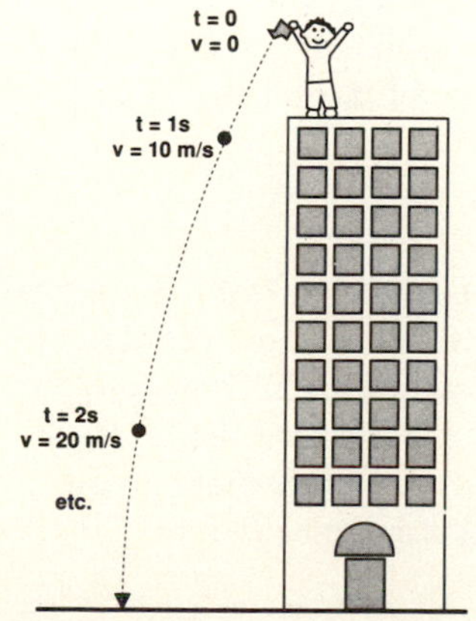

Figure 9-3. A rock dropped from a tall building accelerates at about 10 m/s^2. The downward speed of the rock is 0 when it is released. After 1 second, its speed is about 10 m/s. After 2 seconds, its speed is about 20 m/s.

Suppose that you drop a rock from a tall building (Figure 9-3). At the moment you let it go, its speed is 0 meters per second. After 1 second, the acceleration of gravity causes it to be falling downward at 9.8 meters per second or, rounding, at about 10 meters per second. After 2 seconds, it is falling at a rate of about 20 meters per second. In the absence of air resistance, the speed would continue to increase by about 10 meters per second each second until it hits the ground.

On the Earth, air resistance cannot be avoided. Automobile and airplane manufacturers spend a lot of money researching ways to reduce the effects of air resistance on cars and airplanes. If you ride a bicycle, you probably have experienced another effect: Air resistance increases as your speed increases. As a result, falling objects on Earth do not accelerate indefinitely; eventually, they reach a maximum velocity, or **terminal velocity,** which depends on the object's shape. On the airless Moon, without any terminal velocity, falling objects continue to accelerate until they hit its surface.

Example 9-13 *Tipped Straight Up.* A batter "tips" a fast ball, sending it straight up over home plate at an initial speed of 20 meters per second. Neglecting air resistance, when is the ball caught?

Solution: If you neglect air resistance, the ball's speed is affected only by gravity. Because the ball initially is going *up* at 20 m/s and gravity acts to pull it *down,* the acceleration of gravity is *negative,* or about $-10\ m/s^2$. Thus, after 1 second, the speed of the ball is reduced by 10 meters per second: it is still traveling upward, but at only 10 m/s. After 2 seconds, the speed of the ball is reduced to 0; it is at its highest point. During the third second, the speed of the ball changes from 0 to –10 meters per second; that is, it is traveling *downward* at 10 meters per second. After 4 seconds, the ball is traveling downward at 20 meters per second. Its upward and downward paths should take the same amount of time, so the ball must be back at its starting height after 4 seconds. If the ball is caught at about the same height from which it was hit, it is caught after 4 seconds.

❖

Example 9-14 *Fly Ball.* A batter hits a fly ball to left field where the left fielder catches it. When the ball is hit, it has components of velocity both horizontally (towards left field) and vertically (upward). Suppose that its initial vertical speed is 20 meters per second upward and that its horizontal speed is 15 meters per second toward left field. Neglecting air resistance, when and where is it caught? What *really* happens?

Solution: Gravity affects speed *only* in the vertical direction. The initial vertical speed is the same as in Example 9-13, so the solution is the same. That is, the ball is still caught after 4 seconds. The ball's motion toward left field does not affect its flight time. However, it does affect *where* it lands. Without air resistance, the ball will maintain its horizontal velocity of 15 m/s until it is caught. Thus, in its 4 seconds of travel, the ball travels $15\ m/s \times 4\ s = 60\ m$, or about 200 ft. In reality, air resistance will slow the ball's travel, so it will be caught less than 200 feet from home plate. An exception can occur if the ball is affected by a strong wind, in which case it might travel farther than expected.

❖

Time-Out to Think: An exact calculation of the distance traveled by a baseball, accounting for factors such as air and wind resistance and the rotation of the ball, is extremely complicated. Yet, outfielders catch fly balls. Somehow, the brain computes the ball's flight automatically. Other animals have similar skills: Dogs catch Frisbees, birds land on wobbling wires, and so on. Is there an evolutionary advantage to having a brain that is good at computing motion? Explain.

Example 9-15 *Terminal Velocity.* Suppose that you jump out of an airplane. Neglecting air resistance, how fast would you be falling 15 seconds after your leap? Will you really be falling this fast? Explain.

Solution: Neglecting air resistance, you will simply accelerate with the acceleration of gravity, or 9.8 m/s^2. Thus your change in speed during the 15 seconds would be:

$$\text{change in speed} = \text{acceleration} \times \text{elapsed time} = 9.8\frac{\text{m}}{\text{s}^2} \times 15\ \text{s} = 147\frac{\text{m}}{\text{s}},$$

and converting units gives

$$147\frac{\text{m}}{\text{s}} \times \left(\frac{1\ \text{km}}{1000\ \text{m}} \times \frac{3600\ \text{s}}{1\ \text{hr}}\right) = 530\frac{\text{km}}{\text{hr}}.$$

After 15 seconds, you would be falling toward the ground at 530 km/hr. In reality, because of air resistance, you will not fall nearly that fast. Terminal velocity for a falling person, without a parachute, is between 180 and 250 km/hr, depending on body position (you fall faster if you "streamline" your body, pointing toward the ground, and slower if you "spread eagle" your body). Hitting the ground at this speed would be fatal. Fortunately, the shape of parachutes makes their terminal velocity much lower, allowing you to survive and, with luck, even avoid injury!

❖

The Rotating Earth

As the Earth rotates, its orientation, or direction, in space constantly changes. Thus, even when you are "just sitting here," you are accelerating by virtue of the Earth's rotation. Under most circumstances, the effects of this acceleration are small enough to neglect.

However, the effects of the acceleration due to rotation are noticeable over long distances. Indeed, the effects are crucial to understanding weather. One of these effects, called the **Coriolis effect,** causes storms to circulate counterclockwise in the Northern hemisphere, and clockwise in the southern hemisphere. On a planet that rotates slowly, such as Venus, no similar storms occur, resulting in completely different weather patterns.

A common myth holds that water spirals down drains in opposite directions in the Northern and Southern hemispheres. The Coriolis effect is not noticeable on the size scale of toilets or kitchen sinks, so the difference in drain spirals cannot be observed. The direction in which water spirals down a drain is determined by a complex combination of the shape of the drain and the direction from which the water comes.

Time-Out to Think: Another way to notice the effects of the Earth's rotation is with a **Foucault pendulum,** which swings in such a way as to knock over pegs that indicate the time of the day. Many science museums have Foucault pendulums. Have you ever seen one? If not, try to find one nearby.

9.4.3 The Laws of Motion

Sir Isaac Newton (1642–1727) was one of the greatest mathematicians and scientists of all time. Among his many accomplishments, he consolidated human understanding of the mechanics of motion into three simple laws known as Newton's laws of motion.

Newton's First Law of Motion

When an object is moving with constant velocity, neither its speed nor direction is changing; that is, it is moving at constant speed in a straight line. **Newton's first law of motion** states that

in the absence of external forces, objects move with constant velocity.

Thus objects at rest (velocity = 0) tend to remain at rest, and objects in motion tend to remain in motion (with no change in their speed or direction). On Earth most objects eventually come to rest because of forces of friction; in space, objects keep moving forever.

Time-Out to Think: Have you ever played "air hockey" in which objects glide on a table, supported by air that is forced out of many small holes? If so, explain how this game helps demonstrate Newton's first law of motion. If not, try to find an arcade that has the game, and give it a try!

Newton's Second Law of Motion

When an external force is present, an object will accelerate. **Newton's second law of motion** states that

the acceleration of an object is proportional to the applied force, or

force = mass × acceleration.

For example, suppose that a mass of 1 kilogram accelerates at 10 m/s². The magnitude of the responsible force is

$$\text{force} = \text{mass} \times \text{acceleration} = 1\ \text{kg} \times 10\frac{\text{m}}{\text{s}^2} = 10\frac{\text{kg} \times \text{m}}{\text{s}^2} = 10\ \text{newtons}.$$

The SI unit of force is

$$1\ \text{newton} = 1\frac{\text{kg} \times \text{m}}{\text{s}^2}.$$

Note that the SI unit of force, the **newton,** is equivalent to 1 *kilogram-meter per second squared.*

Dividing both sides of Newton's second law by mass yields

$$\text{acceleration} = \frac{\text{force}}{\text{mass}}.$$

Thus, for a given force, a smaller mass gains a greater acceleration. For example, a force of 10 newtons gives a 0.5 kilogram mass an acceleration

$$\text{acceleration} = \frac{10\ \text{newton}}{0.5\ \text{kg}} = \frac{10\ \dfrac{\text{kg} \times \text{m}}{\text{s}^2}}{0.5\ \text{kg}} = 10\ \frac{\text{kg} \times \text{m}}{\text{s}^2} \times \frac{1}{0.5\ \text{kg}} = 20\frac{\text{m}}{\text{s}^2},$$

which is twice the 10 m/s² acceleration that the same force gives to a 1 kilogram mass.

The only reason you can "feel" your motion in a car or train is because of bumps and other little things that generate forces and cause small accelerations.

Let's think a bit more about Newton's first two laws of motion. For example, according to Newton's first law, traveling at constant velocity should feel no different than being at rest. You may have experienced this sensation during a smooth ride in a car or airplane.

Newton's second law says that you will feel a force anytime you are accelerating. For example, when you drive your car around a curve, you feel yourself lean toward the outside of the curve. What you feel is a force caused by the friction of the wheels against the road.

Finally, the Earth always is accelerating as it orbits the Sun because its *direction* of motion is ever-changing. According to Newton's second law, a force must be causing the acceleration. What is that force? It is the force of gravity between the Sun and the Earth.

Newton's Third Law of Motion

Newton's third law of motion states that

for any force, there is always an equal and opposite reaction force.

This law explains why you don't fall through your chair: Even though gravity is acting in an attempt to accelerate you downward, the chair exerts an equal and opposite upward force that prevents you from falling. It also explains how rockets work: A rocket forces hot gas out behind it, creating an equal and opposite force that propels the rocket forward. And it explains why, when a skater wants to move to the left, she pushes to the right with her right foot.

Time-Out to Think: Inflate a balloon but don't tie the end. Let it go. The balloon will fly all over the room. Explain the balloon's flight as a consequence of Newton's third law of motion. How is the flight of the balloon similar to a rocket launch? Why does a rocket go straight but the balloon goes in many directions?

THINKING ABOUT ... HOW DOES THE TABLE KNOW? (A SHORT PLAY)

Dr. Newton: [standing in front of class with a large rock in his right hand.] We know that gravity acts on this rock because, if I let it go, it will fall immediately. Why doesn't it fall while I'm holding it?

Student: Because your hand provides an equal and opposite force to counteract gravity.

Dr. Newton: Good! As long as the magnitudes of the two forces — that of gravity acting downward and the force from my hand acting upward — are equal, the rock remains stationary. But I have another question for you. We know that the source of the downward force is the Earth's gravity, but what is the source of the upward force from my hand?

Student: Well, it must come from tension in the muscles that hold your hand stationary.

Dr. Newton: Right again! The upward force comes from tension in my muscles, which is in turn produced by my metabolism of energy from food. [Dr. Newton, still holding the rock in his right hand, now points to a paper clip on the table with his left hand.] Suppose that, without changing the upward force from my hand, I quickly replaced the rock with the paper clip. What would happen?

Student: Because the paper clip is much lighter, the force of gravity acting downward on it is much smaller. Therefore, if you continued to apply the same upward force, the paper clip ought to be sent flying upward into the air!

Dr. Newton: That makes sense. Now, let's try it. Here we go! [With his left hand, Dr. Newton quickly replaces the rock (in his right hand) with the paper clip; he sets the rock on the table. The paper clip rests, motionless, in his right hand.] What happened? Why didn't the paper clip go flying toward the ceiling as you predicted?

Student: Because you changed the force you applied with your hand. That wasn't fair!

Dr. Newton: Right. I changed the force so that it is now equal to the force of gravity on the paper clip instead of being equal to the force of gravity on the rock. Fooled you! How did I change the force? Well, my nerves sensed the downward force of the weight in my hand, and my brain helped me adjust the muscle tension in my arm. See, I'm pretty smart!

Dr. Newton: [looking at the rock, which is now on the table.] OK, one more question! Notice that the rock is sitting on the table. Why isn't it falling through?

Student: Because the table exerts an equal and opposite upward force to oppose gravity.

Dr. Newton: Now, imagine that the table continued to exert this same upward force, and I quickly replaced the rock with the paper clip. What would happen?

Student: Uh oh. It would fly up into the air. But that's not going to happen!

Dr. Newton: [quickly replaces the rock on the table with the paper clip.] Right. Why not? Because the table adjusted its upward force to compensate for the switch from the rock to the paper clip. Now, you weren't too surprised when I made this adjustment; after all, I have a brain, I'm smart, and I can figure out how to adjust the force so the paper clip doesn't go flying into the air. But how does the table know to adjust the force? Bet you never knew that tables were so smart!

[The End.]

As this play demonstrates, Newton's third law raises some interesting questions. How *does* the table know? The answer lies in finding the source of the upward force. The table is made of atoms, as are the objects placed upon it. From the outside, an atom looks like a ball of negatively charged electrons (the positively charged nucleus is buried deep inside). You may recall that negative charges repel other negative charges. Thus the electrons in atoms on the surface of the table repel the electrons on the surface of the objects on the table. This electromagnetic repulsion provides the upward force.

The closer the negative charges of the table and the object are pushed together, the greater is their repulsive force. The force of gravity causes heavier objects to push harder against the table than light objects; this force brings the negative charges from the table and the object very slightly closer together, causing the electromagnetic repulsion to increase. The balance is natural and requires no brains on the part of the table. It does, however, require tremendous intellect on the part of the human species. Only in the past century, after thousands of years of civilization, was this explanation finally discovered.

9.4.4 The Distinction Between Mass and Weight

Mass and weight are not the same thing. Mass refers to the amount of "stuff" that comprises an object. Weight describes the *force* on the object. If you have a mass of 50 kilograms, it is the same no matter where you go: The mass is the amount of material in your body. But your weight depends on the strength of gravity and other forces acting upon you.

Note that your weight is different from its "normal" value only while the elevator is *accelerating*, not while it is moving up or down at constant speed.

Imagine standing on a scale in an elevator (Figure 9-4). When the elevator is stationary, or moving at constant speed, the scale reads your "normal" weight. When the elevator is accelerating upward, the floor exerts an additional force that makes you feel heavier, and the scale verifies your greater weight. When the elevator accelerates downward, the floor (and scale) are dropping away, so your weight is less. If the cable breaks, so that the elevator is in **freefall,** the floor drops away at the same rate that you fall. You lose contact with the scale, becoming **weightless.**

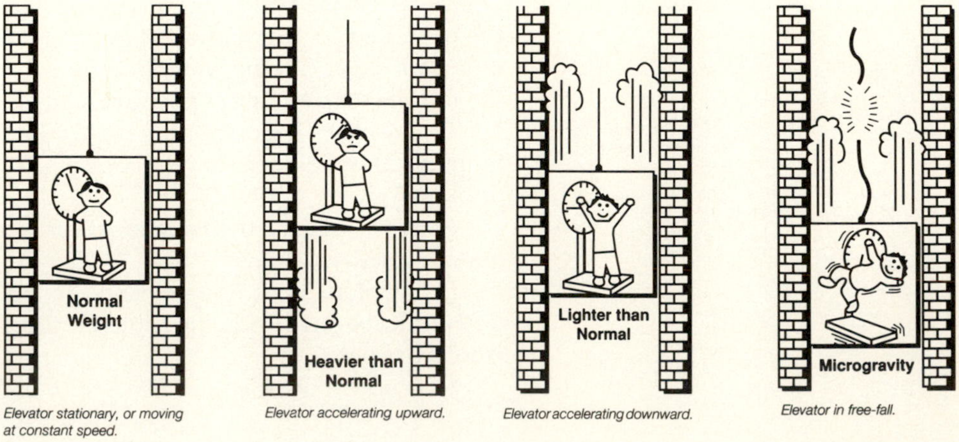

Figure 9-4. Mass is not the same as weight. The boy's mass never changes, but his weight does.

Time-Out to Think: You probably have seen pictures of astronauts floating weightless in space. But you, too, can experience weightlessness simply by being in freefall. For example, you are in freefall, and hence weightless, when you spring from a diving board — at least until you hit the water! Use this fact to explain why divers can perform, albeit for short times, the same tumbles and twists as astronauts in space. Can you think of other circumstances in which you are weightless here on Earth?

As the elevator example shows, weight measures the force that an object "feels"; that is, the force pushing against it. The "normal" weight of an object is the force, due to gravity, that it feels at rest on the Earth. Based on Newton's second law, this force, or weight, is

$$\text{weight} = \text{mass} \times \text{acceleration of gravity.}$$

Like any force, weight has units of mass times acceleration. Thus, although people commonly speak of weights in *kilograms*, that usage isn't technically correct. Kilograms are a unit of mass, whereas the SI units of force are newtons. You may safely ignore this technicality as long as you are dealing with objects on the Earth that are not accelerating. In elevators, or on other planets, the distinction between mass and weight is important and cannot be ignored.

A further subtlety arises in conversions between SI and U.S. customary units because the kilogram is a unit of mass, whereas the pound is a unit of *weight* (Table 9-4). Thus the conversion factor 1 kilogram = 2.2 pounds actually compares two different types of units! However, people usually ignore this subtlety by *pretending* that the kilogram is a unit of weight or that the pound is a unit of mass. Again, as long as you are dealing with masses and weights on Earth that are not accelerating, such pretending poses no problems.

No self-respecting nation would measure mass in slugs. — Megan Donahue, astrophysicist and quantitative reasoning teacher

Table 9-4. Standard Units of Mass and Weight

	U.S. Customary System	SI (metric system)
Mass unit	Slug	Kilogram
Force or weight unit	Pound	Newton

9.5 ENERGY: THE FUTURE DEPENDS ON IT

Cars require the energy supplied by burning gasoline. Light bulbs and appliances require energy, usually produced at power plants from heat (generated by burning oil or coal, or by a nuclear reactor) or by water flowing through a turbine, and carried to homes and industries in the form of electricity. Even living requires energy, which comes from the metabolism of food.

In short, the future of civilization depends on stable and ongoing sources of energy. Yet, energy sources and prices are in constant flux. The Gulf War in 1991 caused energy prices to double in a few weeks and then slowly settle back down. In addition, most energy comes from **nonrenewable sources**, such as coal, oil, and gas, which eventually will be exhausted; even now they are becoming more difficult and expensive to find and extract. As a result, the search for a sensible long-term energy policy is one of the most important issues facing nations today.

Fundamentally, only three basic types of energy exist. **Kinetic energy** is energy of motion. Water flowing through a turbine, the moving blades of an electric mixer, and a car driving down the highway all represent different forms of kinetic energy. Kinetic energy may be calculated from the simple formula

$$\text{Kinetic energy} = \frac{1}{2}mv^2,$$

where m is the mass of the object and v its velocity.

A second fundamental type of energy is **potential energy,** which in effect is stored and available for later use. Potential energy comes in several common forms. For example, if you hold a rock above the ground, it has **gravitational potential energy,** or energy waiting to be released into motion when you drop the rock. Oil contains **chemical potential energy,** or energy that is released when it is burned. Power lines bring **electric potential energy** to your home, which is released when you plug in and turn on an appliance. Finally, as discovered by Einstein, mass is just another form of potential energy that can be released under certain circumstances; the potential energy of mass, or **mass-energy,** is given by the formula $E = mc^2$.

The third fundamental type of energy is **radiative energy,** or energy carried by light (*radiation* is another word for *light*). Different forms of light carry different amounts of energy. For example, X-rays carry more energy than visible light; that's why X-rays can penetrate skin and muscle, allowing doctors to take pictures of your bones. Ultraviolet light carries enough energy to penetrate surface skin and hence can cause cell damage that leads to skin cancer. Microwaves are a lower-energy form of light that happen to be very useful for heating food. The waves received by radio and TV sets also are a form of light, called radio waves.

Microwaves lie between the radio and infrared portions of the light spectrum. The microwaves used in cooking happen to have energies that cause water molecules to absorb them, which gives the water molecules kinetic energy and heats them. Microwave ovens heat food because it contains water; they do not heat plastic dishes, because plastic does not contain water.

A basic understanding of these three types of energy and knowing how to work with them numerically leads to a far better understanding of the energy issues upon which the future depends.

9.5.1 Units of Energy and Power

What are the units of energy? We can obtain them from the formula for kinetic energy. For example, suppose that a one kilogram rock is moving at a speed of 1 meter per second. Its kinetic energy is

$$\frac{1}{2}mv^2 = \frac{1}{2}(1\text{ kg})\left(1\frac{\text{m}}{\text{s}}\right)^2 = 0.5\ \frac{\text{kg}\times\text{m}^2}{\text{s}^2} = 0.5\text{ joule}.$$

Note that the units of energy also can be described as a force multiplied by a distance; that is,

$$1\text{ joule} = 1\text{ newton - meter} = 1\frac{kg \times m^2}{s^2}.$$

Thus the SI unit of energy is the **joule,** which is equivalent to 1 *kilogram-square meter per square second.*

Power is defined as the *rate* of energy use (or production). That is, power is energy use per unit time. The SI unit of power is the familiar **watt,** defined as a power of 1 joule per second, or

$$1\text{ watt} = 1\ \frac{\text{joule}}{\text{s}}.$$

Power is the rate of energy use or production. The unit of power is energy divided by time. In SI, the unit of power is the watt: 1 watt = 1 joule per second.

For example, a 100-watt light bulb consumes 100 joules of energy each second. In 10 seconds, it consumes 1000 joules of energy. A 50-watt light bulb consumes the same 1000 joules of energy in 20 seconds.

Electric Bills: Kilowatt-Hours

Take a look at a home electric bill. It probably lists the amount of electricity consumed in units of **kilowatt-hours.** The hyphenated unit means:

$$1\text{ kilowatt-hour} = 1\text{ kilowatt} \times 1\text{ hour} = 1000\text{ watts} \times 1\text{ hour}.$$

Because this unit is power multiplied by time, which is energy, we can convert it to joules:

$$\begin{aligned}1\text{ kilowatt-hour } &= 10^3 \text{ watts} \times 1\text{ hr}\\ &= 10^3\,\frac{\text{joule}}{\text{s}} \times 1\text{ hr} \times 60\,\frac{\text{min}}{\text{hr}} \times 60\,\frac{\text{s}}{\text{min}}\\ &= 3.6\times10^6 \text{ joule.}\end{aligned}$$

Thus a kilowatt-hour is just another name for 3.6 million joules.

Example 9-16 Suppose that you keep a 100-watt light bulb on all the time for a week. How much energy does it use? Give your answer in both joules and kilowatt-hours.

Solution: A 100-watt light bulb consumes energy at the rate of 100 watts, or 100 joules per second. In a week it uses

$$100\,\frac{\text{joule}}{\text{s}} \times 1\text{ week} \times 7\,\frac{\text{day}}{\text{wk}} \times 24\,\frac{\text{hr}}{\text{day}} \times 60\,\frac{\text{min}}{\text{hr}} \times 60\,\frac{\text{s}}{\text{min}} = 6\times10^7 \text{ joule}.$$

Convert from joules to kilowatt-hours:

$$6\times10^7 \text{ joule} \times \underbrace{\frac{1\text{ kilowatt-hour}}{3.6\times10^6\text{ joule}}}_{\text{conversion factor}} \approx 17\text{ kilowatt-hour}.$$

Left on for a week, a 100-watt light bulb uses 67 million joules, or about 17 kilowatt-hours, of energy.

❖

Example 9-17 *Operating Cost of a Light Bulb.* Suppose that your utility company charges a fairly typical 6¢ per kilowatt-hour of energy. How much does keeping the 100-watt light bulb on for a week cost?

Solution: Example 9-16 demonstrated that the light bulb will use 17 kilowatt-hours of energy in a week. Therefore the cost is

17 kilowatt-hours × 6¢ per kilowatt-hour = \$1.02.

Note that the \$1.02 cost of keeping the light-bulb on for a week is comparable to the purchase price of the light bulb itself.

❖

Food Energy: Calories

A calorie (with a lowercase "c") actually is defined to be 4.184 joules. Because that is such a small amount of energy, "food Calories" (with a capital "C") actually are *kilocalories:* 1 Calorie = 1000 calories = 4184 joules.

The human body requires energy to maintain its body temperature and perform activities such as running, digesting, and thinking. This energy, which comes from food, often is measured in units called **Calories**:

1 Calorie = 4184 joules.

The number of Calories that you need to consume each day depends on many factors. For example, when you exercise you need to consume more Calories to replace the energy that you expended, and you may need more food to maintain your body heat on a cold day than on a warm day.

Time-Out to Think: If you run at the same pace, do you burn more calories on a hot day or a cold day? Explain. Now, extend this idea. You will sweat and lose a lot of fluids in a sauna, but is sitting in a sauna a good way to lose weight for the long term? Why or why not?

Example 9-18 ***Burning Calories Through Running.*** Vigorous running typically consumes about 15 Calories per minute for an average person. How much energy is expended in an hour of running? Compare that amount of energy to a kilowatt-hour.

Solution: An hour of running will use

$$15\frac{\text{Cal}}{\text{min}}\times 60\text{ min} = 900\text{ Cal},\quad \text{or}\quad 900\text{ Cal}\times 4184\frac{\text{joule}}{\text{Cal}}\approx 4\times 10^{6}\text{ joule.}$$

An hour of running consumes roughly 4 million joules, which is slightly more than one kilowatt-hour of energy.

❖

More Energy Units

Many other units are used to measure energy. For example, a commonly used unit for measuring the energy from burning kerosene or natural gas is the **British Thermal Unit (BTU):**

$$1\text{ BTU} = 1055\text{ joules.}$$

Some utility companies bill for gas in units of CCF, or hundreds (centuries) of cubic feet. For example, 5 CCFs means 500 cubic feet of gas.

Also common are units that measure how much gas or oil is used. For example, many utilities charge for natural gas in units of cubic feet, but care is needed in using such units. For example, *liquefied* natural gas is a highly compressed form of natural gas. A cubic foot of liquefied natural gas yields far more energy than a cubic foot of natural gas in its gaseous state. Energies also are expressed in *equivalents*, as in "liters of oil equivalent," which means that the energy is the same as can be generated from 1 liter of oil.

9.5.2 Energy Comparisons

The mass-energy stored in matter can be released only in special circumstances, such as nuclear reactions (i.e., nuclear fission and nuclear fusion). Even then, only a fraction of the mass is converted to energy. For example, nuclear fusion of hydrogen releases only 0.7% of the initial mass-energy. Complete conversion of mass into energy is possible only by combining matter and antimatter. For example, when a proton and an antiproton (which is identical to a proton except in having a negative, rather than positive, charge) meet, their combined mass is completed converted to energy. Thus matter–antimatter reactions are the ultimate means of generating energy. Unfortunately, although tiny amounts of antimatter can be created in laboratories, it is virtually nonexistent in nature.

Despite the many different forms of energy and ways of measuring it, you can always relate energies by comparing them in a consistent set of units (for example, joules, Calories, or kilowatt-hours). The energies of various items and events are listed in Table 9-5.

Note that *mass-energy* is the potential energy of mass calculated from Einstein's equation $E = mc^2$, where E is energy, m is mass, and c is the speed of light. For example, because the mass of an electron is 9×10^{-31} kilogram and the speed of light is 3×10^{8} meters per second, the electron mass-energy is

$$\begin{aligned} E = mc^2 &= \left(9\times10^{-31}\text{ kg}\right)\times\left(3\times10^{8}\ \tfrac{\text{m}}{\text{s}}\right)^2 \\ &= \left(9\times10^{-31}\text{ kg}\right)\times\left(9\times10^{16}\ \tfrac{\text{m}^2}{\text{s}^2}\right) \\ &= 8.1\times10^{-14}\ \tfrac{\text{kg}\times\text{m}^2}{\text{s}^2}, \end{aligned}$$

or 8.1×10^{-14} joules (as shown in Table 9-5). A similar calculation starting with a proton's mass of 1.7×10^{-27} kilograms yields the proton's mass-energy of 1.5×10^{-10} joules. The mass-energy of the universe is an order of magnitude estimate based on the estimated total mass of the universe, 10^{53} kilograms:

$$E = mc^2 = 10^{53}\text{ kg}\times\left(3\times10^{8}\ \tfrac{\text{m}}{\text{s}}\right)^2 = 10^{53}\text{ kg}\times\left(9\times10^{16}\ \tfrac{\text{m}^2}{\text{s}^2}\right)\approx 10^{69}\ \tfrac{\text{kg}\times\text{m}^2}{\text{s}^2} = 10^{69}\text{ joule.}$$

Table 9-5. Energy Comparisons.

Item or Event	Energy (joules)	Item or Event	Energy (joules)
Electron mass-energy	8×10^{-14}	Exploding 1 ton of TNT	5×10^{9}
Fusion of 4 hydrogen atoms into 1 helium atom	4×10^{-12}	One-megaton H-bomb	5×10^{15}
Fission of 1 uranium 235 atom	2.18×10^{-11}	Earthquake, magnitude 8	2.5×10^{16}
Proton mass-energy	1.5×10^{-10}	Hurricane (kinetic energy in winds)	10^{17}
Metabolism of 1 gram of sugar	1.5×10^{4}	Tambora volcano eruption (1815)	8×10^{19}
Sitting for 1 hour (human)	2×10^{5}	U.S. annual energy consumption	10^{20}
Metabolism of 1 candy bar	10^{6}	Estimated world fossil fuel reserves	3×10^{22}
Walking for 1 hour (human)	1.2×10^{6}	Annual energy generation of Sun	10^{34}
Running for one hour (human)	4×10^{6}	Supernova (explosion of a star)	10^{44}
Burning 1 liter oil	1.2×10^{7}	Mass-energy of universe	10^{69}

Example 9-19 ***U.S. Energy Needs.*** Based on Table 9-5, approximately how many liters of oil would be needed to supply all U.S. energy needs for a year?

Solution: The U.S. annual energy consumption is 10^{20} joules. Because 1 liter of oil yields 1.2×10^{7} joules, the amount of oil needed would be

$$\frac{10^{20} \text{ joule}}{1.2 \times 10^{7} \frac{\text{joule}}{\text{liter}}} = 10^{20} \text{ joule} \times \frac{1 \text{ liter}}{1.2 \times 10^{7} \text{ joule}} \approx 8 \times 10^{12} \text{ liters.}$$

The annual energy consumption of the United States is equivalent to the energy of burning about 8 trillion liters of oil, or roughly 2 trillion gallons. ❖

9.5.3 Energy, Temperature, and Heat

The concepts of energy, temperature, and heat are closely related. The material world is made of atoms and molecules. Temperature essentially is a measure of the *average kinetic energy* of these particles. The higher the temperature, the higher is the average energy of the particles.

Strictly speaking, the definition of temperature as average kinetic energy is valid only for **ideal gases** in which *all* the energy comes from molecular *motion*. However, for most practical purposes, this subtlety can be ignored.

Time-Out to Think: A common mistake, even in many textbooks, is to relate temperature to the average *speeds* of particles, rather than to their average energy. Explain why this relation is wrong. (Hint: recall the formula for kinetic energy, $\frac{1}{2} mv^2$.)

Temperature and Heat

Although temperature and heat often are used synonymously, they aren't the same. To understand the difference, imagine the following experiment (but don't try it!). Suppose that you heat your oven to 500°F. Then you open the oven door, quickly thrust your arm inside (without touching anything), and immediately remove it. What will happen to your arm? Not much! Now, suppose that you boil a pot of water. Although the water temperature is only 212 °F, you would be badly burned if you put your arm in the pot, even for an instant. Although the oven is at a higher *temperature,* the pot of water contains more *heat.*

Whereas temperature is related to the *average* energy of molecules, heat is related to their *total* energy. That is, heat involves a combination of temperature and density. When a molecule of air or water strikes your skin, it transfers some of its energy to you. Higher temperatures mean that molecules, on average, strike your skin harder. Higher densities mean that a larger number of molecules strike your skin. Briefly putting your arm in the oven causes little damage because, although the air in the oven is very hot, its density is relatively low. In the pot of boiling water, the density is much higher. Thus, although the temperature of the water is lower than the temperature of the air in the oven, so many more molecules strike your skin that you burn quickly.

Example 9-20 ***Heating Comparisons.*** Use the concepts of temperature and heat to explain why more energy is needed to heat a relatively small swimming pool than to heat an entire house.

Solution: Heating a house means raising the temperature of the air in the house. Because temperature is a measure of average kinetic energy, that means raising the average kinetic energy of the air molecules. Doing so requires energy (usually supplied by burning oil, gas, or wood or by heating baseboards with electricity).

Heating a swimming pool follows the same principle, except that it involves raising the average kinetic energy of the molecules of water. Because water is much denser than air, even a small swimming pool contains more molecules than a house full of air. Thus more energy is needed to heat the larger number of molecules in the pool.

❖

Time-Out to Think: Extend the solution from Example 9-20 to explain why water temperature tends to remain more stable than air temperature, and hence why climates in areas near the ocean tend to be less variable than climates farther inland.

Understanding the Earth's Atmosphere

The two most common elements in the universe are hydrogen and helium—the two lightest gases. Yet these elements are rare in the Earth's atmosphere. The explanation lies with the concept of temperature. At a particular temperature, all molecules in the air have the same average kinetic energy. For example, the average kinetic energy of oxygen molecules and helium molecules must be the same, or

$$\text{Average kinetic energy} = \underbrace{\frac{1}{2} m_{\text{oxygen}} \left(v_{\substack{\text{oxygen} \\ \text{(average)}}} \right)^2}_{\substack{\text{avr. kinetic energy of} \\ \text{oxygen molecules}}} = \underbrace{\frac{1}{2} m_{\text{helium}} \left(v_{\substack{\text{helium} \\ \text{(average)}}} \right)^2}_{\substack{\text{avr. kinetic energy of} \\ \text{helium molecules}}}.$$

Because helium molecules are less massive than oxygen molecules (m_{helium} is smaller than m_{oxygen}), their average speed must be greater to preserve the preceding equality (that is, v_{helium} must be larger than v_{oxygen}). Their higher speeds, in turn, make helium molecules more likely to escape from the atmosphere into space. Thus the hydrogen and helium that was present when the Earth formed was long ago lost to space. In contrast, the strong gravity of stars (e.g., the Sun) and very large planets (e.g., Jupiter and Saturn) prevents hydrogen and helium from escaping; hence these worlds are composed primarily of hydrogen and helium.

The concepts of energy, temperature, and heat also can help you better understand the greenhouse effect (see subsection 3.5.3). Recall that the Earth receives energy from the Sun primarily in the form of visible light, and returns this energy to space primarily in the form of infrared light. The greenhouse effect occurs because molecules of greenhouse gases such as carbon dioxide and methane absorb infrared light, which increases their kinetic energy. Because energy cannot simply disappear, this added energy must have some effects. However, because the Earth's climate is a chaotic system, the effect of the added kinetic energy is *not necessarily* an immediate increase in the global average temperature. Instead, the energy might go to such things as generating storms, melting polar ice, or raising the temperature of the oceans. Measuring any of these potential effects is even more difficult than measuring the global average temperature (subsection 8.4.3).

Time-Out to Think: Many computer climate models suggest that, because of the greenhouse gases human beings have added to the atmosphere, the global average temperature should already have risen substantially. Yet, the measured warming to date is relatively small. Should this fact relieve concerns about the greenhouse effect? Why or why not?

9.6 MATHEMATICS AND MUSIC

The roots of mathematics and music are entwined in antiquity. Pythagoras (c. 500 B.C.) claimed that "all nature consists of harmony arising out of number." He imagined that the planets circled the Earth on invisible heavenly spheres, obeying specific numeric laws and emitting the ethereal sounds known as "music of the spheres."

Evidence abounds of music among all ancient cultures. String and wind instruments were designed at least 5000 years ago in Mesopotamia, and Egyptians played in small ensembles of lyres and flutes by 2700 B.C. Indeed, based on evidence dating back 30,000 years, some archeologists suspect that music may predate both history and speech.

Sound and Music

Sound is produced by any vibrating object. The vibrations produce a **wave** (much like a water wave) that propagates through the surrounding air in all directions. When such a wave impinges on the miraculously designed organ called the ear, it is perceived as sound. Of course, some sounds, such as speech and screeching tires, do not qualify as music. Most musical sounds are made by vibrating strings (violins, cellos, guitars, and pianos), vibrating reeds (clarinets, oboes, and some organ pipes), or vibrating columns of air (other organ pipes, horns, and flutes).

Music is the universal language of mankind. — Henry Wadsworth Longfellow

One of the most basic qualities of sound is **pitch.** For example, a tuba has a "lower" pitch than a flute, and a violin has a "higher" pitch than a bass guitar. To understand the origin of pitch, find a taut string (a guitar or violin string will work best, but a stretched rubber band will do). When you pluck the string, it produces a sound with a certain pitch. Next, use your finger to hold the midpoint of the string in place, and pluck either half of the string. Note that a higher-pitched sound is produced, demonstrating an ancient musical discovery: *The shorter the string, the higher is the pitch.*

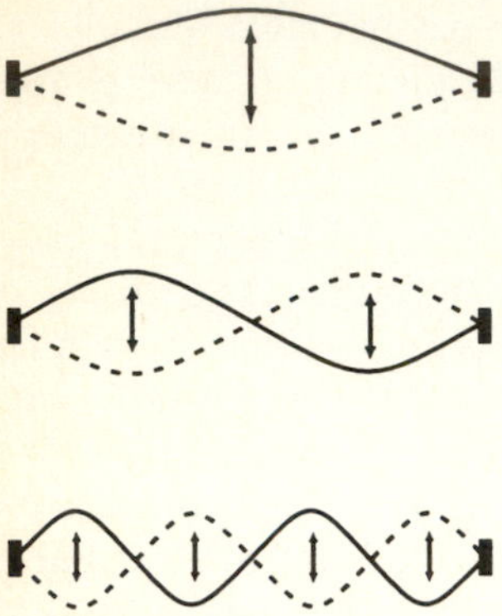

Figure 9-5. (a) Plucking a taut string produces a wave that extends along its full length, vibrating at the fundamental frequency. (b) A wave with half the length of the string has a frequency of twice the fundamental frequency. (c) Each time the frequency is doubled, the pitch of the resulting tone goes up by an octave.

But what *is* pitch? Plucking the fully taught string causes it to vibrate up and down along its full length (Figure 9-5a). For example, if the string vibrates up and down 10 times per second, its **frequency** is said to be 10 **cycles per second (cps)**; each cycle corresponds to one vibration up *and* down. When you pluck only half the string (by holding down its middle), the resulting wave is *half* as long as the first wave (Figure 9-5b), and its frequency therefore is twice as much, or 20 cps. Thus pitch is related to the frequency of the vibrating string; the higher the frequency, the higher is the pitch.

For any particular string, plucking its full length causes it to vibrate at its **fundamental frequency.** Plucking half its length generates a wave with twice the fundamental frequency, called the **first harmonic** (or **overtone**). Halving the length of the string again (Figure 9-5c), doubles the frequency again so that the resulting tone has *four* times the fundamental frequency.

Doubling the frequency of a tone raises the pitch by an **octave.** If you are familiar with musical notes or the piano keyboard (Figure 9-6), you know that an octave is the interval between, say, middle C and the next higher C. For example, middle C on the piano has a frequency of 260 cps, the C above middle C has a frequency of 2×260 cps = 520 cps, and the next higher C has a frequency of 2×520 cps = 1040 cps. Similarly, the C below middle C has a frequency of about $\frac{1}{2} \times 260$ cps = 130 cps.

The frequencies at which a particular string vibrates depend on its length, how tightly it is strung, and several other factors. You might be surprised at the frequencies of common sounds. The human voice consists of sounds with frequencies of 200 to 400 cycles per second. The range of a piano extends from about 27 to 4200 cycles per second. The human ear can detect frequencies between about 20 and 20,000 cycles per second.

Time-Out to Think: The note middle A (above middle C) has a frequency of about 440 cps. What is the frequency of the A an octave higher? What is the frequency of the A an octave below middle A?

Scales

The musical tones that span an octave comprise a **scale.** The Greeks invented the seven-note (or diatonic) scale that corresponds to the white keys on the piano. In the seventeenth century, J. S. Bach adopted a 12-tone scale for his keyboard music, which corresponds to both the white and black keys on a modern piano. Bach's use of the 12-tone scale helped spread it throughout Europe, making it a foundation of Western music. Many other scales are possible: For example, 3-tone scales are common in African music, scales with more than 12 tones occur in Asian music, and 19-tone scales are used in contemporary music.

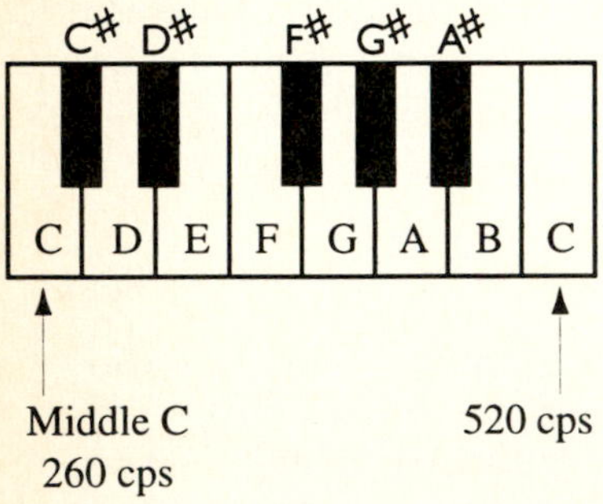

Figure 9-6. The piano keys of a 12-tone scale spanning an octave are shown. Each successive key is one half-step higher than the preceding key. The frequency of the notes changes by a factor of two over an interval of an octave.

On the 12-tone scale, the frequency separating each tone is called a **half-step** (Figure 9-6). For example, E and F are separated by a half-step, as are F# and G. In each half-step, the frequency increases by some *multiplicative* factor; let's call it *f*. Thus the frequency of the note C is the frequency of B times the factor *f*, the frequency of B is the frequency of A# times the factor *f*, and so on. The frequencies of the notes across the entire scale are related as follows:

$$\text{C} \xrightarrow{f} \text{C\#} \xrightarrow{f} \text{D} \xrightarrow{f} \text{D\#} \xrightarrow{f} \text{E} \xrightarrow{f} \text{F} \xrightarrow{f} \text{F\#} \xrightarrow{f} \text{G} \xrightarrow{f}$$
$$\text{G\#} \xrightarrow{f} \text{A} \xrightarrow{f} \text{A\#} \xrightarrow{f} \text{B} \xrightarrow{f} \text{C}$$

Because an octave corresponds to a factor of 2 increase in frequency, the factor *f* must have the property:

$$\underbrace{f \times f \times f \times f \times f \times f \times f \times f \times f \times f \times f \times f}_{12 \text{ times}} = f^{12} = 2.$$

Recall that the twelfth root of 2, or $\sqrt[12]{2}$, is a number that when multiplied by itself 12 times equals 2.

Thus, *f* must be the *twelfth root* of two, or $f = 1.05946$.

Time-Out to Think: Use your calculator to verify that $f^{12} = (1.05946)^{12} = 2$.

We can now calculate the frequency of every note of a 12-tone scale. Starting from middle C, with its frequency of 260 cps, we multiply by $f = 1.05946$ to find that the frequency of C# is 260 cps x 1.05946 = 275 cps. Multiplying again by $f = 1.05946$ gives the frequency of D as 275 cps × 1.05946 = 292 cps. Continuing in this way generates Table 9-6.

All entries in column 3 are the same because the same factor, $f = 1.05946$, separates every pair of notes. The parenthetical terms in column 1 are names used by musicians to describe intervals between the note shown and middle C.

Table 9-6. Frequency of Notes in the Octave Above Middle C

Note	Frequency (cps)	Ratio to Frequency of Preceding Note	Ratio to Frequency of Middle C
C	260		1.00000 = 1
C#	275	1.05946	1.05946
D (second)	292	1.05946	1.12246
D#	309	1.05946	1.18921
E (third)	328	1.05946	1.25992 ≈ 5/4
F (fourth)	347	1.05946	1.33484 ≈ 4/3
F#	368	1.05946	1.41421
G (fifth)	390	1.05946	1.49831 ≈ 3/2
G#	413	1.05946	1.58740
A (sixth)	437	1.05946	1.68179 ≈ 5/3
A#	463	1.05946	1.78180
B	490	1.05946	1.88775
C (octave)	520	1.05946	2.00000 = 2

Human speech is like a cracked kettle on which we tap crude rhythms for bears to dance to, while we long to make music that will melt the stars. — Gustave Flaubert

Column 4 of Table 9-6 shows that a few tones have simple ratios of frequency to middle C. For example, the frequency of G is approximately 3/2 times the frequency of middle C (musicians call this interval a *fifth*), and the frequency of F is approximately 4/3 times the frequency of middle C (musicians call this interval a *fourth*). According to many musicians, the most pleasing combinations of notes, called **consonant tones,** are those whose frequencies have a simple ratio. Referring to consonant tones, the Chinese philosopher Confucius observed that small numbers are the source of perfection in music.

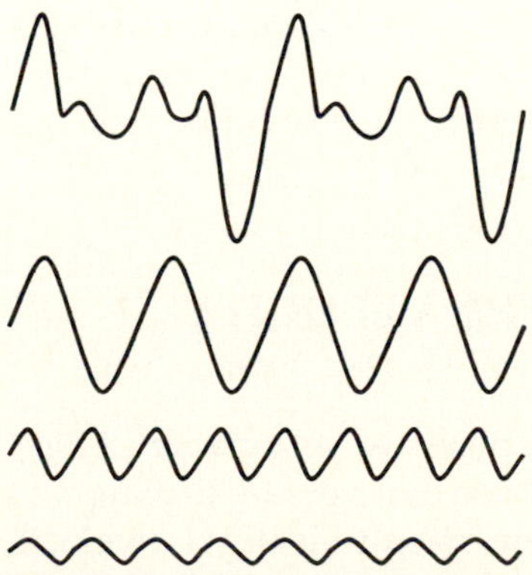

Figure 9-7. The top figure shows a complex sound wave, much like that produced by a musical instrument, consisting not only of the fundamental frequency (second from top), but of higher harmonics as well (lower two figures). A deep mathematical result says that any complex wave can be expressed as the sum of simpler waves.

From Tones to Music

Although the simple frequency ratios of "pure" tones are the building blocks of music, the sounds of music are far richer and more complex. For example, a plucked violin string does much more than produce the simple frequency of its vibration. The string motion is transferred through the bridge of the violin to its top, and the ribs transfer those vibrations to the back of the instrument. With the top and back of the violin in oscillation, the entire instrument acts as a resonating chamber, which excites and amplifies higher harmonics of the original tone.

Similar principles generate rich and complex sounds in all instruments. Figure 9-7 shows a typical sound wave that might be produced by an instrument. Note that it isn't a simple wave like those pictured in Figure 9-5;

instead it consists of a combination of simple waves that are the harmonics of the fundamental. In fact, the complex wave is the *sum* of the other simple waves shown. The fact that a musical sound can be expressed as a sum of simple harmonics surely is the deepest connection between mathematics and music. The French mathematician Jean Baptiste Joseph Fourier first enunciated this principle in about 1810; it was one of the most profound discoveries in mathematics.

Stradivarius was essentially a craftsman of science, one with considerable, demonstrable knowledge of mathematics and acoustical physics.—Thomas Levenson in *Measure for Measure: A Musical History of Science* (pp. 207–208)

Although mathematics helps in understanding music, many mysteries remain. For example, in about 1700, an Italian craftsman named Antonio Stradivarius made what are still considered to be the finest violins and cellos ever produced. Despite years of study by mathematicians and scientists, no one has succeeded in reproducing the unique sounds of a Stradivarius instrument.

The Digital Age

When we talk about sound waves and imagine music to consist of either simple or complex waves, we are working with the **analog** picture of music. Until about 20 years ago, all musical recordings (phonograph cylinders, records, and tape recordings) were based on the analog picture of music. Storing music in the analog mode requires literally storing sound waves; for example, on records, the grooves in the vinyl surface are etched with the shape of the original musical sound wave. If you have listened to analog recordings, you know that this shape easily can become distorted or damaged.

Most recent music recordings use a **digital** picture of music, in which the sound waves are represented by lists of *numbers*. Modern computer technology generally prevents undesired changes to numbers, making digital storage of music much more reliable than analog storage. Today, the most common medium for storing digital music is the **compact disk.** Computers convert music into numbers which are stored on compact disks, and computer chips in your compact disk player convert the numbers back into sound.

Another advantage of digital technology is in providing easy and endless ways to "process" music. Through techniques of **filtering** and **digital signal processing,** the sounds of a musical recording can be modified with computers. For example, extraneous sounds (background noises or "hiss") can be detected and removed. Changing the music by amplifying certain frequencies or attenuating others also is possible. Moreover, once digital music can be modified, it is a short step to creating music. Instruments called **synthesizers** can create and imitate a tremendous variety of sounds without strings, brass tubes, or reeds — they do it digitally! In the digital age, the dividing line between mathematics and music all but vanishes.

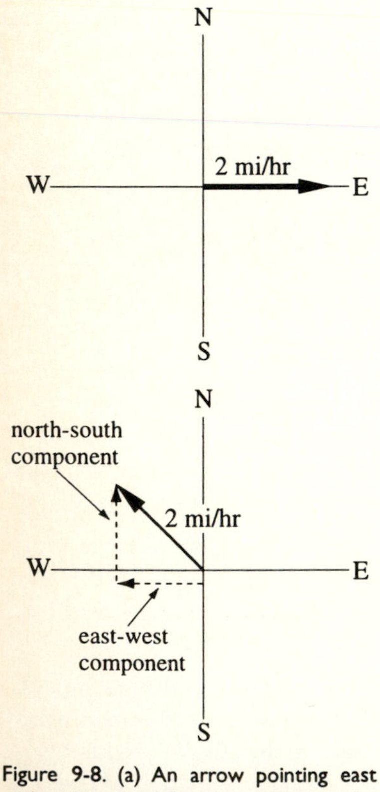

Figure 9-8. (a) An arrow pointing east with a length of 2 units represents the velocity vector for a walk due east at a speed of 2 miles per hour. (b) Representing a walk at the same speed (2 mi/hr) but in a northwest direction simply requires rotating the arrow shown in (a). Note that the vector can be broken down into north–south and east–west components.

9.7* DIRECTIONAL QUANTITIES: THE VELOCITY VECTOR

Any directional quantity, or vector, conveys two pieces of information: a *magnitude*, or size, and a *direction.* For example, suppose that you are driving with a velocity of 55 miles per hour, due east. The magnitude of your velocity vector is your *speed* of 55 miles per hour, and the direction of your velocity vector is due east.

Vectors often are represented as arrows. For example, suppose that you are walking due east with a speed of 2 miles per hour. Figure 9-8(a) represents your velocity vector as an arrow: It points in the direction of travel (east) and its length of 2 units represents your speed of 2 miles per hour.

If you walk at the same speed in some other direction, say, northwest, the arrow simply rotates to point in your new direction of travel (Figure 9-8b).

Note that your velocity vector can be broken down into two components: one representing your speed along the north–south axis and one representing your speed along the east–west axis. Then, from the Pythagorean theorem,

$$(\text{north - south speed})^2 + (\text{east - west speed})^2 = (\text{total speed})^2.$$

Your total speed is 2 miles per hour and, because you are walking northwest, the north–south and east–west speeds are the same. Therefore,

$$2 \times (\text{north - south speed})^2 = (2\ ^{\text{mi}}\!/\!_{\text{hr}})^2 \xrightarrow[\text{take square root}]{\text{divide by 2, then}}$$

$$\text{north - south speed} = \sqrt{\frac{(2\ ^{\text{mi}}\!/\!_{\text{hr}})^2}{2}} = \sqrt{2}\ ^{\text{mi}}\!/\!_{\text{hr}} \approx 1.4\ ^{\text{mi}}\!/\!_{\text{hr}}.$$

That is, your velocity is taking you farther north at a speed of 1.4 miles per hour. Because both speed components are the same, you also are moving westward at a speed of 1.4 miles per hour. However, because the graph shows east as the positive direction and west as the negative direction (along the east–west axis), the east–west component of your velocity is *negative* 1.4 miles per hour (–1.4 mi/hr).

Time-Out to Think: After three hours, how far will you have walked in total? How far west of your starting point will you be? How far north of your starting point?

By convention, symbols for vectors are shown in boldface. Also by convention, when writing the components in parentheses, the component along the horizontal axis is written first and the component along the vertical axis is written second.

Another way to represent your velocity vector is by writing its components in parentheses, separated by a comma. For your northwest walk, the velocity vector, **v**, is written as

$$\mathbf{v} = (\text{east} - \text{west velocity component, north} - \text{south velocity component})$$
$$= (-1.4\ ^{\text{mi}}\!/\!_{\text{hr}},\ 1.4\ ^{\text{mi}}\!/\!_{\text{hr}}).$$

For your walk due east (Figure 9-8a), your velocity had no north–south component, so the east-west component consisted of your entire speed of 2 miles per hour. Thus this velocity vector is written as **v** = (2 mi/hr, 0).

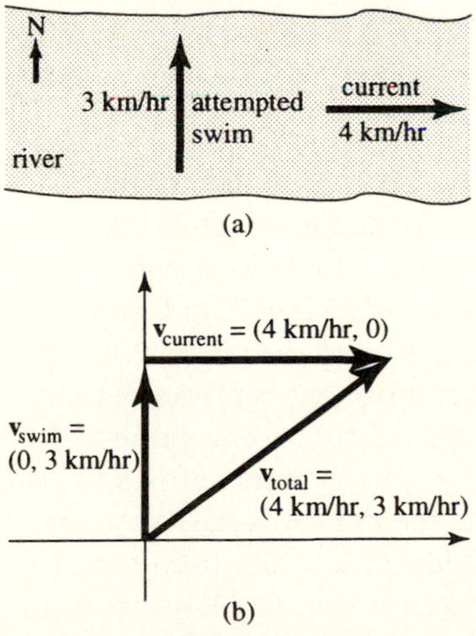

Figure 9-9. (a) You are in a river trying to swim north at 3 km/hr, but a current is moving east at 4 km/hr. (b) To find the sum of the two vectors, place them end-to-end. The vector sum starts from the tail of the first arrow and ends at the tip of the second arrow.

9.7.1 Working with Vector Components

Suppose that you are in a river trying to swim due north at a speed of 3 kilometers per hour, but a current is moving due east at a speed of 4 kilometers per hour. Where will you actually be going?

Your swimming velocity has no east–west component and a north–south component of 3 kilometers per hour; you can write this vector as

$$\mathbf{v}_{\text{swim}} = (0,\ 3\ \text{km/hr}).$$

The current has an east–west component of 4 kilometers per hour and no north–south component, so you can write this vector as

$$\mathbf{v}_{\text{current}} = (4\ \text{km/hr},\ 0).$$

Your total velocity is the *sum* of your swimming velocity and the current velocity. Graphically, you can add the vectors by placing them end-to-end; the vector sum starts from the tail of the first arrow and ends at the tip of the second arrow (Figure 9-9b). Alternatively, you can add the east–west and north–south components separately:

$$\mathbf{v}_{total} = \mathbf{v}_{swim} + \mathbf{v}_{current} = (\underbrace{0 + 3\ \text{km/hr}}_{\text{add east-west components}},\ \underbrace{4\ \text{km/hr} + 0}_{\text{add north-south components}}) = (3\ \text{km/hr},\ 4\ \text{km/hr}).$$

From the Pythagorean theorem, your total speed is

$$\begin{aligned}\text{total speed} &= \sqrt{(\text{east - west speed})^2 + (\text{north - south speed})^2} \\ &= \sqrt{(3\ \text{km/hr})^2 + (4\ \text{km/hr})^2} \\ &= \sqrt{9\ \text{km}^2/\text{hr}^2 + 16\ \text{km}^2/\text{hr}^2} = \sqrt{25\ \text{km}^2/\text{hr}^2} = 5\ \text{km/hr}.\end{aligned}$$

Thus, through the combination of your swimming and the current, you actually will be moving at a speed of 5 kilometers per hour. Your direction will be slightly east of northeast, such that you move eastward at 4 kilometers per hour and northward at 3 kilometers per hour.

Summarizing, this example demonstrated the two basic methods for adding vectors.

To subtract a vector, simply add its opposite. You can draw the opposite vector by reversing the direction of its arrow, and you can find its components by multiplying each of the original vector's components by –1. Multiplying vectors is slightly more complex; consult a textbook on vectors if you want to learn about it.

1. You can add vectors graphically by placing them end-to-end.
2. You can add vector components separately. Because the components have only a magnitude, and not a direction, you can work with them just like you work with other numbers.

9.7.2 Applications of Vectors

The basic techniques we have shown for writing and working with vectors can be extended to vectors with more than two components. For example, an airplane's velocity has *three* components: east–west and north–south components, as before, and an up–down component. You can draw such a vector on a three-dimensional graph, or represent it in the form

$\mathbf{v}_{airplane}$ = (east–west component, north–south component, up–down component).

Although vectors with more than three components are difficult to show graphically (because the graph would need more than three dimensions), working with their components is easy. You may be surprised to learn that such vectors are important in many applications. For example, Einstein proved that a complete analysis of motion requires working with four-component vectors: three of the components are for the three dimensions of space (e.g., length, width, and depth), and the fourth component represents time.

Another typical use of higher dimensional vectors occurs in public health. Imagine that you are an epidemiologist (a person who studies the origin and spread of disease) trying to describe the incidence of AIDS by age group. For example, you might divide the population into the following categories: 0–10 years old, 11–20 years old, 21–30 years old, ... 81–90 years old, and more than 91 years. You will need 11 numbers to represent the incidences of AIDS among these 11 age groups. You can then represent the incidence of AIDS in the entire population with an 11-component vector, **A**:

$$\mathbf{A} = (A_1, A_2, A_3, A_4, A_5, A_6, A_7, A_8, A_9, A_{10}, A_{11}),$$

where each component represents the number of AIDS cases in its age category.

Vectors have countless applications in engineering, architecture, and in all the natural and social sciences, making them one of the most powerful and widely used concepts in mathematics. Indeed, any time that you are dealing with several quantities, or a single quantity that has several categories, you may be sure that vectors aren't far away.

9.8 CONCLUSION

In this chapter we presented various practical applications of the tools developed earlier in this book. Even if you have not studied all of these applications in depth, the chapter should provide a useful reference for future work. Although each section provides important lessons about the application discussed, every section illustrates the one general lesson from this chapter.

- With what you have learned so far, you are capable of handling *any* issue that involves numbers. You need only to learn the definitions and context of the issue and apply your *number sense*. This unifying power of numbers allows you to study seemingly disparate issues with just a few fundamental concepts.

PROBLEMS

Reflection and Review

Section 9.2

1. **The Wonderful Widgit Company Future.** Extending the budget summary of the Widgit Company (Table 9-1), assume that for 1996: Total receipts are $1050, operating expenses are $600, employee benefits are $200, and security costs are $250.
   a. Based on the accumulated debt at the end of 1995, calculate the 1996 interest payment. Assume an interest rate of 8.2%.
   b. Calculate the total outlays for 1996 and the year-end surplus or deficit, and the year-end accumulated debt.
   c. Based on the accumulated debt at the end of 1996, calculate the 1997 interest payment, again assuming an 8.2% interest rate.
   d. Assume that in 1997 the Widgit company has receipts of $1100, holds operating costs and employee benefits to their 1996 levels, and spends no money on security. Calculate the total outlays for 1997, the year-end surplus or deficit, and the year-end accumulated debt.
   e. Imagine that you are the CEO (Chief Operating Officer) of the Wonderful Widgit Company at the end of 1997. Write a three-paragraph statement to shareholders about the company's future prospects.
2. **Analysis of the Federal Budget.** Consider the federal budget summary given in Table 9-2.
   a. In 1995, what percentage of total receipts is devoted to interest on the debt?
   b. In 1995, what percentage of total receipts is devoted to defense spending?
   c. In 1995, what percentage of total receipts is devoted to Social Security?
   d. In 1995, what percentage of total receipts is the deficit?
   e. Comment on these percentages; do they surprise you?
3. **Per Capita Debt.** If half of all Americans work, what is the 1995 federal debt per laborer?
4. **National lottery.** Assume that the federal government could raise $50 million per week in a national lottery and direct of those funds to reducing the federal debt. Then, if the government balances the budget each year (so the debt does not increase), how long would it take to retire the debt using lottery proceeds? (Hint: You need to find the current federal debt to solve this problem.)
5. **Interest on the Federal Debt.**
   a. If the 1994 interest payment is based on the 1993 debt, what interest rate does the government pay?
   b. If the 1995 interest payment is based on the 1994 debt, calculate the interest rate.
   c. Assume that the 1995 interest rate holds for 1996. In that case, what is the 1996 interest payment on the 1995 debt?
   d. Suppose that the 1996 interest rate is 1% *higher* than the 1995 rate. In that case, what is the 1996 interest payment?
   e. Compare your results from (c) and (d). Write a short statement to put the difference in perspective.
6. **Balancing the Federal Budget.** Imagine that you were proposing to balance the federal budget in 1996. For each of the following questions, assume that receipts are $1400 billion in 1996.
   a. If you hold 1996 spending at 1995 levels, what will the 1996 deficit be?
   b. Suppose, instead, that you cut all 1996 spending to 95% of the 1995 levels. What will the 1996 deficit be?
   c. Assume that the 1996 interest payment is based on the 1995 debt and a rate of 4.5%. What will the interest payment in 1996 be?
   d. If you decide to balance the budget in 1996 by spending only the receipts ($1400 billion), how much is left to

spend after making the interest payment? How much must government spending be cut in 1996 (relative to 1995 levels) to balance the budget?

e. Which budget areas would you reduce to achieve the balanced budget in 1996?

7. **Year of Zero Debt.** Study Table 9-3.
   a. If receipts increase at an annual rate of 2% and outlays are held constant, when (according to this model) will the debt be eliminated? Are these reasonable assumptions? Why or why not?
   b. If receipts increase at an annual rate of 3% and outlays decrease at an annual rate of 1%, when (according to the model) will the debt be eliminated? Comment on these assumptions.
   c. Explain briefly why the debt will never be eliminated if receipts increase by 2% and outlays increase by 3%. What would happen in this case?

Section 9.3

8. **Calculating Densities.** Find the average density of the following objects in grams per cubic centimeter.
   a. A rock with a volume of 15 cm^3 and a mass of 0.25 kg.
   b. A box with a volume of 7 liters and a mass of 200 grams.
   c. A sphere with a radius of 10 cm and a mass of 2 kg. Recall the formula for the volume of a sphere: $(4/3) \times \pi \times (\text{radius})^3$.
   d. A cube with side length of 6 in. and a weight of 10 lb.

9. **Granite and Iron.** The density of granite is about 2.7 grams per cm^3. The density of iron is about 7.9 grams per cm^3.
   a. What is the weight, in kilograms, of a granite slab that measures 1 m × 1 m on its face and is 2 cm thick? What is its weight in pounds?
   b. How much would the slab in part (a) weigh if it were made of iron instead of granite?
   c. Compare the volumes of two rocks, one made of granite and one made of iron, that both weigh 1 kg.
   d. When iron is molten (liquid), its density is only 7.0 grams per cm^3. If you start with 1 m^3 of solid iron, what is its volume when it is molten?

10. **Population Density.**
    a. The land area of the United States is about 3.5 million square miles, and the population is about 250 million people. What is the average population density?
    b. Find the land areas and populations for the states of Wyoming and New Jersey. Calculate and compare their population densities.
    c. Find the population and the area of your home town. Calculate its population density.

11. **Compact Disk Density.** Suppose that a future compact disk holds 250 gigabytes of information on a surface area of 90 cm^2.
    a. Calculate the information density on the disk.
    b. The text in a typical 500-page book contains about 1 megabyte of information. How many 500-page books could be stored on 1 cm^2 of the disk? How many such books could be stored on the entire disk?

12. **Blood-Alcohol Concentration.** Humans generally have 70 cm^3 of blood *per kilogram* of weight.
    a. About how many liters of blood are in *your* body?
    b. Suppose that you drink beer containing 7% alcohol by volume. How much would you have to drink to reach a blood-alcohol concentration of 0.1%?
    c. How much beer would you have to drink to reach a near fatal limit of 0.5% blood-alcohol concentration?
    d. Drinking whiskey with 40% alcohol by volume, how much would you have to drink to reach a blood-alcohol concentration of 0.1%?
    e. How much hard alcohol would you have to drink to reach a near fatal limit of 0.5% blood-alcohol concentration?
    f. Suppose that you drank enough to raise your blood-alcohol concentration to 0.2%, but then stopped. If your body metabolizes alcohol at the rate of 10 milliliters per hour, how long would it take to reduce your blood alcohol concentration to below the legal limit of 0.1%?

13. **Lead Chloride Contamination.** Water flowing from an abandoned mine is contaminated with lead chloride (a lead salt) at a concentration of 0.0002 grams per milliliter. How much lead chloride is contained in 1500 liters of water? Should you be concerned about drinking this water? Why or why not?

14. **Particulate Pollution.** A sample of air taken during rush hour in downtown Denver on a December day exceeds the EPA standard for particulate pollution of 120 micrograms per cubic meter by 20%. What is the particulate concentration in micrograms per cubic meter? At this concentration, how many grams of particulate matter would a room that measures 5 meters in length, 3 meters in width, and 2 meters in height contain?

15. **Gaseous Pollution.** The EPA standard for carbon monoxide is 9 ppm (parts per million), and its standard for ozone is 120 ppb (parts per billion). On a day in which both gases are at their maximum EPA levels, what is the ratio of carbon monoxide molecules to ozone molecules? Does this ratio depend on the sample size? Explain.

Section 9.4

16. **Practice with Distance and Speed.**
    a. How far will you travel in a car at a constant speed of 90 kilometers per hour for 3 1/2 hours?
    b. How long would you need to go 85 kilometers at a walking speed of 5 kilometers per hour?
    c. In the Ironman Triathlon, top athletes ride bicycles 112 miles in about 4 hours and 25 minutes. What is their average speed?
    d. A car goes around a circular track with a radius of 1 kilometer in 65 seconds. What is its average speed in miles per hour?
    e. A distance runner is holding a steady pace in which she circles a 400-meter track every 1 minute, 22 seconds. What is her time for a 10-kilometer run?
    f. The world record for the marathon (26 miles, 385 yards) is about 2 hours and 6 minutes. What is the world record pace in miles per hour? in *minutes per mile*?

g. The world record for the 100-meter freestyle (swimming) is about 48.2 seconds (1994). What is the average swimming speed for this race in miles per hour?

h. On the first day of a three day, 160-mile bike trip, Abe rides 55 miles. On the second day he averages 12 miles per hour for six hours. How far must he ride on the third day to reach his destination?

i. Elle can reach her destination on time if she averages 60 miles per hour. Halfway to the destination (in distance) she realizes that she has averaged only 30 miles per hour. How fast must she travel to arrive on time? Explain.

j. Imagine that you ride a ski lift up a mountain at 5 miles per hour and then ski down the same distance as the lift itself at 15 miles per hour. What is your average speed for the round trip? (Be careful on this one!)

17. Practice with Acceleration.

a. As you enter an interstate highway, you drive along a quarter-mile on-ramp and increase your speed from 0 to 55 miles per hour in 12 seconds. What is your average acceleration along the on-ramp in miles per hour per second?

b. As you sled down a steep slick street, your velocity increases from 0 to 23 meters per second in 5 seconds. What is your average acceleration?

c. As you hike up an increasingly steep hillside, your pace decreases from 3 miles per hour to 0.5 mile per hour over a period of 20 minutes. What is your average deceleration in miles per hour per minute? Convert this quantity to miles per hour squared and meters per second squared.

d. If you decelerate at an average rate of –20 miles per hour per second, how long will it take to come to a stop from a speed of 70 miles per hour?

18. Nomo Comes to Class. A baseball player (let's call him Nomo) can throw a fastball well over 90 miles per hour, which is about 40 meters per second. Suppose that, as a contribution to your studies, Nomo agrees to come to class and throw a baseball straight up at 90 miles per hour (class will meet outside that day). Neglect air resistance.

a. How fast, and in which direction (i.e., up or down), will the ball be going 1 second after Nomo throws it?

b. How fast, and in which direction, will the ball be going 2 seconds after it is thrown?

c. When (that is, how many seconds after it was thrown) will the ball reach its highest point, at which it momentarily comes to a stop before it starts falling?

d. How fast, and in which direction, will the ball be going 5 seconds after it is thrown?

e. When will the ball reach the ground?

19. Home Run? A batter hits a ball toward center field. When it is hit, its vertical (upward) speed is 30 meters per second and its horizontal speed (toward center field) is 45 meters per second. The center-field wall is 130 meters (about 427 feet) from home plate, and 6 meters (about 20 feet) tall.

a. Neglecting air resistance and assuming that no walls get in the way, calculate when the ball hits the ground.

b. How far (toward center field) would the ball travel during this time? Does it reach the wall?

c. By estimating, decide whether the ball will clear the wall for a home run. Explain.

20. Can a Penny Kill? Suppose that you drop a penny from the top of a tall building (don't try this experiment!). You find that it hits the ground after 6 seconds (which would happen if you drop a penny from about a 50-story building). Neglect air resistance.

a. How fast is the penny falling 1 second after you drop it?

b. How fast is it falling 2 seconds after you drop it?

c. How fast is it going after 6 seconds (just before it hits the ground)?

d. Convert its speed just before hitting the ground from meters per second to kilometers per hour.

e. Convert the speed from kilometers per hour to miles per hour. If the penny hit someone on the head, would it hurt that person? Could it be fatal? Explain.

21. The Thrill of Speed.

a. Gravity accelerates a skydiver toward the ground at 9.8 m/s^2. Ignoring air resistance and assuming an initial speed of 0, what is the skydiver's speed after 5.0 seconds? Give your answer in km/hr.

b. Imagine getting behind the wheel of a race car that accelerates from 0 to 90 mi/hr in 5.0 seconds. What is the average acceleration rate during this 5-second period? Give your answer in m/s^2.

c. In terms of speed and acceleration, which of these two activities is likely to provide a greater thrill? Give quantitative reasons in your explanation.

22. Storms on a Rotating World. Find a weather map from a recent newspaper on which you can see a storm. Photocopy the map and attach an explanation of how and why the storm is circulating.

23. Newton's Second Law. A typical drive by a professional golfer imparts a velocity of about 120 miles per hour to the golf ball in an interval of about 0.5 second.

a. Express the acceleration of the golf ball in meters per second squared.

b. Assume that the mass of the golf ball is 65 grams. How much force (in newtons) is required to impart the acceleration you found in part (a)?

24. Mass Versus Weight.

a. How much do you weigh, in pounds?

b. Use the conversion that 1 kg weighs 2.2 pounds on Earth to calculate your mass in kilograms.

c. Your weight is your mass times the acceleration of gravity. Express your weight on Earth in newtons.

d. The acceleration of gravity on the Moon is about one-sixth that on Earth. What is your mass on the Moon in kilograms?

e. What is your weight on the Moon? Answer in both pounds and kilograms

f. Explain why measurements made on Earth in pounds are different on the Moon but measurements made in kilograms are the same.

Section 9.5

25. Energy Practice.
a. Calculate the kinetic energy of a 2-kg rock traveling at 5 meters per second.
b. Calculate the kinetic energy of a 1000-kg automobile traveling at 100 km per hour.
c. Your utility bill states that you used 300 kilowatt-hours of electricity last month. Convert this quantity to joules.
d. Suppose that you keep a 60-watt light bulb on continuously for 3 days. How much energy does it use? Give your answer in both joules and kilowatt-hours.

26. Electric Bill. Your electric bill states that you used 1250 kilowatt-hours of energy in the past month.
a. How many joules of energy did you use?
b. The combustion of one liter of oil yields about 1.2×10^7 joules of energy. How many liters of oil would be needed to provide 1250 kilowatt-hours of energy? how many gallons? how many barrels? (1 barrel of oil = 42 gallons of oil.)
c. What is your average *power* use, in watts, if you use 1250 kilowatt-hours in a month?
d. Suggest a possible breakdown in your energy usage. For example, how much energy did you use for lighting, for the refrigerator, and for other appliances? Explain your assumptions and estimates.

27. Human Wattage. The average person requires 2500 food Calories per day.
a. Convert this quantity to joules and kilowatt-hours.
b. What *power* (in watts) does the average human body run at? Compare your answer to the wattage used by some familiar appliance.
c. Estimate the total amount of energy needed from food by an average person over a lifetime.
d. Suppose that, instead of getting energy from food, you were able to live by burning oil. How much oil would you use each year?
e. U.S. annual energy consumption is about 10^{20} joules. What is the *per capita* energy consumption? Compare this value to the energy needed from food alone.

28. Energy Comparisons. Use the data in Table 9-5 to answer each of the following questions.
a. Compare the energy of a 1-megaton hydrogen bomb to the energy of a hurricane.
b. How much sugar do you need to metabolize to fuel a 1-hour run?
c. Compare the Sun's annual energy output to the energy of a supernova (the explosion of a star).

29. Hot Soup and Energy. Suppose that you have hot soup for lunch and want to save your leftover soup in the refrigerator. Briefly explain why letting the soup cool to room temperature before placing it in the refrigerator is more energy efficient than putting it into the refrigerator while it is still hot.

30. Heat, Exercise, and Clothing. People feel cold when their bodies are losing more heat than they are generating.
a. Use this idea to explain why you feel cold when you first jump into a cool swimming pool but feel warmer after a few minutes of vigorous swimming.
b. Explain why clothing (or blankets) keep you warm. Then explain why the most effective strategy for staying warm on a cold day is to dress in layers.

Section 9.6

31. Octaves. Starting with a tone having a frequency of 110 cycles per second, find the frequencies of the tones that are one, two, three, and four octaves higher.

32. Notes of a Scale. Find the frequencies of the 12 notes of the scale that start at the A above middle C, which has a frequency of 440 cycles per second.

33. Experimenting with Waves. Have a friend hold both ends of a cut rubber band so that it is stretched tight. Pluck the entire rubber band, look for the waves on the rubber band, and listen to the pitch of the tone produced; you are hearing the fundamental frequency.
a. Have your friend maintain the same tension in the rubber band. Pinch the rubber band at its midpoint and pluck either half of the rubber band. How does the pitch of the resulting tone change? Explain why the pitch changes, in terms of frequency.
b. Experiment further by pinching the rubber band at various points (keeping the length of the entire rubber band fixed) and plucking on either side of the pinch point. Describe your experiments and the resulting sounds.

Section 9.7

34. Picturing Vectors. On a piece of paper draw axes representing north–south and east–west directions. Let 1 cm on the page represent 1 meter per second. Draw the velocity vectors representing each of the following motions as arrows on your diagram. Indicate clearly both the length and direction of the vector, and find the east–west and the north–south components of the velocity in each case.
a. You walk east at 1 meter per second.
b. A dog runs northeast at 3 meters per second.
c. A car driving southwest at 8 meters per second.

35. Vector Components. An airplane is traveling with a northward component of velocity of 250 miles per hour and a westward component of velocity of 550 miles per hour. Use the Pythagorean theorem to find the overall velocity. In what direction, roughly, is the airplane going (that is, north–northwest, southeast, etc.)?

Further Topics and Applications

36. Air Pollution Reduction. Through a variety of conservation, emissions, and public transportation initiatives, a city is able to reduce its average winter carbon monoxide levels by 0.3 ppm per year. If the worst year was 1985 with average winter levels of 6 ppm, what do you project the levels to be in 2000?

37. Plutonium Release. Fires occurred at the Rocky Flats nuclear weapons plant (west of Denver, Colorado) in 1957 and 1969, releasing as much as 100 pounds of plutonium (in the form of plutonium dioxide) into the atmosphere. Assume that at Rocky Flats the plutonium concentration in the air was 1.5 micrograms per cubic liter (μg/ℓ³), and that as the plutonium was carried downwind its concentration decreased by 0.5 μg/ℓ³ with each mile from the plant. How far downwind were concentrations within than the EPA's "safe" level of 0.5 μg/ℓ³? Did the plutonium pose any danger to workers in downtown Denver, about 5 miles downwind of Rocky Flats? Explain.

38. Mining Gold Ore. The *grade*, or *purity*, of ore (a mixture of precious metal and surrounding rock found in a mine) describes the concentration of the precious metal in the surrounding rock. For example, if gold ore has a grade of 0.3 ounces per ton, 0.3 ounces of gold are found in each ton of the ore. Suppose that a mine is producing gold ore with an average grade of 0.3 ounces per ton, and that pure gold can be sold for $350 per ounce.

a. How many tons of ore must be processed to produce $1000 worth of gold?

b. Suppose that the cost of processing the ore is $20 per ton. What is the net profit on each ton of ore?

c. Suppose that, after the highest grade ore has been removed, the remaining ore in the mine has an average grade of 0.1 ounces per ton. Further, because this ore is more difficult to extract, the cost of processing it is $40 per ton. Can this ore be mined at a profit? Explain.

d. How high would the price of gold need to rise for the ore described in part (c) to be mined at a profit?

39. The Fate of the Sun. A few billion years from now, after exhausting its nuclear engines, the Sun will become a type of remnant star called a **white dwarf.** It will still have nearly the same mass (about 2×10^{30} kg) as the Sun today, but its radius will be only about that of the Earth (about 6400 km). *Hint:* The formula for the volume of a sphere is:

$$\text{volume} = (4/3) \times \pi \times (\text{radius})^3.$$

a. Calculate the average density of the white dwarf in units of kilograms per cubic centimeter.

b. What is the mass of a teaspoon of material from the white dwarf? (Hint: A teaspoon is about 4 cubic centimeters.) Compare this mass to the mass of something familiar (e.g., a person, a car, a tank).

c. A neutron star is a type of stellar remnant compressed to even greater densities than a white dwarf. Suppose that a neutron star has a mass that is 1.4 times the mass of the Sun but a radius of only 10 kilometers. What is its density? Compare the mass of 1 cubic centimeter of neutron star material to the total mass of Mt. Everest (about 5×10^{10} kg).

40. Racing Through Space. In everyday life, people often have the impression that they are just "sitting here." But, in fact, they are attached to a planet that is rotating, revolving around a star (the Sun), moving through a galaxy with that star, and moving with its galaxy through the universe. Relative to some very distant galaxies, the Earth is racing away at almost the speed of light! (Remember that the circumference of a circle is $2\pi r$, where r is the radius.)

a. A dedicated tanner realizes that the best time to "catch rays" is at noon. However, noon doesn't last long anywhere on Earth because of the Earth's rotation. Realizing that the Earth rotates from west to east, the tanner decides to cruise westward in his convertible. By doing so, he hopes to travel against the rotation of the Earth in such a way that he will follow the Sun (so that it remains noon for him). Assume that he is driving along the equator and calculate how fast he must drive to "keep up" with the Earth's rotation. Can he do it? (The radius of the Earth is about 6400 kilometers.)

b. What is the average speed of the Earth in its orbit around the Sun? Assume that the orbit is a circle with a radius of 150 million kilometers.

c. The Sun is located approximately 25,000 light-years from the center of the Milky Way galaxy and orbits the galaxy once in about 225 million years. Assume that the orbit is circular and calculate the average speed of the solar system as it orbits the center of the galaxy.

41. Gees. Acceleration is sometimes measured in *gees*, or multiples of the acceleration of gravity: 1 gee means 1 × g, or 9.8 m/s²; 2 gees means 2 × g, or 2 × 9.8 m/s² = 19.6 m/s²; and so on. Suppose that you experience 6 gees of acceleration in a rocket.

a. What is your acceleration in meters per second squared?

b. You will feel a compression force from the acceleration. How does this force compare to your normal weight?

c. Do you think that you could survive this acceleration for long? Explain.

42. A Power Plant. A new power plant can generate a gigawatt (a billion watts) of power.

a. How much energy, in kilowatt-hours, can it generate each month?

b. If the average home uses 1000 kilowatt-hours per month, how many homes can this power plant supply with energy?

c. If the power plant generates its energy from oil, how many *barrels* of oil (1 barrel = 42 gallons) does it require each month? each day?

43. Energy and Transportation. Different modes of transportation require different amounts of energy. The following data are based on estimates from *The Bicycle: Vehicle for a Small Planet*, by M. Lowe (World Watch Institute, Washington, D.C., 1989).

Mode	Calories per passenger mile
Automobile, one occupant	1860
Bus transit	920
Rail transit	885
Walking	100
Bicycling	35

a. Discuss the assumptions that must have been used to come up with the data in the table. Do the numbers seem reasonable? Explain.

b. Suppose that you want to going shopping at the local mall. It is 2.25 miles away. Using percentages, compare the amounts of energy needed to get you there by car, bus (if available), bicycle, and walking.

c. Imagine that you were appointed transportation director for your state. Based on the information in the table, describe the transportation system you would propose.

d. Suppose that your recommendation in part (c) is to be placed on the ballot. Would voters be likely to support your proposal? Why or why not?

44. **The Energetic Caesar Salad.** You are having a dinner party for five guests and yourself, and want to use a recipe for Caesar salad dressing from a European cookbook.

Caesar Salad Dressing (4 servings, 75 Calories each)

230 grams oil	5 grams garlic
115 grams Parmesan cheese	5 grams salt
15 grams Worcestershire sauce	5 grams pepper
60 grams lemon juice	1 egg (about 60 grams)

a. Rewrite the recipe to make 6 servings that you need.

b. Suppose that you don't have a metric scale. Instead, you have measuring spoons and cups with U.S. customary units of teaspoons, tablespoons, and cups; you also have a scale that measures weights in ounces and pounds. Explain how you will measure each ingredient properly.

c. Each serving supplies 75 Calories. How many joules of energy are generated in metabolizing one serving of salad dressing?

d. What is the *mass* of one serving of salad dressing? Suppose that instead of being metabolized chemically in your body, the mass was completely converted to energy (perhaps by an unfortunate encounter with some *anti*-salad dressing!). Use $E = mc^2$ to calculate the energy that would be released by one serving of salad dressing.

e. Compare the efficiency of the energy generation from metabolizing the salad dressing in your body with complete conversion of its mass into energy.

45. **Nuclear Fission Bomb.** A 235-gram sample of uranium-235 (U-235) contains Avogadro's number (6×10^{23}) of atoms.

a. If all these atoms undergo fission, how much energy is produced? (Consult Table 9-5.)

b. How many tons of TNT would be needed to create an explosion of equivalent energy?

c. How does the answer to part (b) compare to the energy of a 1-megaton hydrogen bomb?

d. How does the answer to part (b) compare to the energy of the roughly 20-kiloton bomb that destroyed Hiroshima in 1945?

e. Suppose that a terrorist got hold of 235 grams of U-235, and used it to make an atomic bomb. Assume that the rest of the bomb weighs about 500 pounds and hence is transportable by car, boat, or airplane. Comment on the damage that such a bomb could do.

46. **Einstein and the Sun.** The Sun generates its energy through nuclear fusion: Every second approximately 600 million tons of hydrogen are fused into 596 million tons of helium. The "missing" 4 million tons of matter is transformed into energy according to $E = mc^2$. How much energy does the Sun produce each second?

47. **Nuclear Power Plants.** Operating at full capacity, the Fort St. Vrain Nuclear Power Station in Platteville, Colorado, can generate 330 megawatts of power.

a. If the Fort St. Vrain station operated at full capacity for a month, how much energy would it produce? Give your answer in both kilowatt-hours and joules.

b. If a typical household uses 1000 kilowatt-hours of electricity each month, how many households could have their energy needs met by the Fort St. Vrain station?

c. Nuclear power plants generate energy through fission of U-235 (see Table 9-5). How many atoms must be fissioned to supply 1 month's worth of power for the number of households in part (b)?

d. Calculate the total weight, in kilograms, of U-235 needed to provide 1 month's worth of power for the number of households in parts (b) and (c). (The mass of 6×10^{23} U-235 atoms is 235 grams.)

e. Find out whether a nuclear power plant is serving your own region. If so, how much power does it generate at full capacity? Does it typically run at full capacity or at some lesser power? What fraction of the electricity in your region is supplied by the power plant?

48. **Spontaneous Human Combustion.** Suppose that, through a horrific act of an angry god (or very powerful alien, if you prefer), all the mass in your body suddenly was converted into energy according to $E = mc^2$. How much energy would be produced? Compare to the energy of a nuclear bomb (see Table 9-5). What effect would your disappearance have on the surrounding region?

49. **Energy from Junk Mail.** The flow of junk mail through the average person's mailbox seems endless. Most of it goes directly into the trash; a small percentage is recycled. Suppose, instead, that junk mail were burned to make energy. Burning 1 gram of paper releases 2×10^4 joules of energy.

a. Estimate the total mass of junk mail, in grams, received in U.S. mailboxes each year. Clearly explain your assumptions and associated uncertainties.

b. Estimate the total amount of energy, in joules, that could be produced from America's annual supply of junk mail.

c. Use your answer from part (b) to estimate the average power, in watts, that could be supplied by burning junk mail. Compare this quantity to the power output of a 1 gigawatt power station.

d. Total U.S. electrical power production is about 400 gigawatts. What fraction of U.S. electricity needs could be produced by junk mail? Is junk mail a realistic source of energy? Explain.

50. **The Power of Photovoltaics.** Photovoltaic cells convert sunlight directly into electricity. In direct sunlight, each square meter receives about 1400 watts of solar power. That is, if a panel of photovoltaic cells were 100% efficient and had an area of 1 square meter, it would generate about 1400 joules of energy each second.

a. Suppose that a 1 m^2 panel of photovoltaic cells has an efficiency of 12%. Further, suppose that it receives an average of 6 hours of sunlight per day. How much energy would it produce during those 6 hours? Give your answer in both kilowatt-hours and joules.
b. Suppose that the energy produced by panel in part (a) can be stored in batteries so that it can be released at a uniform rate day and night (i.e., 24 hours a day). What is the *average power* that can be supplied by the solar panel?
c. A typical U.S. household requires 1 kilowatt of power. Under the assumptions of parts (a) and (b), how large an array of photovoltaic cells would be needed to meet this demand? Would it fit on the roof of a typical single-family home?
d. Under the assumptions of parts (a) and (b), how large an array of photovoltaic cells would be needed to supply the total U.S. electricity usage of 400 gigawatts of power?
e. The land area of the continental United States is approximately 10 million km^2. If all U.S. electricity were supplied by photovoltaic cells, what fraction of the land area would need to be covered by photovoltaic arrays? Comment on the environmental impact of such a system, and contrast with the environmental impact of current electrical energy sources (e.g., fossil fuels, nuclear energy, and hydroelectric energy).

51. **Could Wood Be Burned for Energy?** A total of about 180,000 terawatt-years of solar energy reaches the Earth's surface each year, of which 0.06% is used by plants in photosynthesis. Of the energy used in photosynthesis, 1% is stored in plant matter (e.g., wood). (1 terawatt-year = 1 terawatt $\times$ 1 year; 1 terawatt = 10^{12} watts.)
a. Calculate the total amount of energy stored in plants each year over the entire Earth. Give your answer first in terawatt-years, then convert to joules.
b. Suppose that power stations generated electricity by burning plant matter. If *all* the energy stored in plants each year could be converted to electricity, what average power would be possible?
c. Total world power use for electricity is on the order of 10 terawatts. What fraction of the total energy stored in plants each year would have to be burned to supply this much power?
d. Considering your results from parts (b) and (c), can you draw any conclusions about why humans depend on *fossil* fuels, such as oil and coal, which are the remains of plants that died long ago? Explain.

52. **Nuclear Materials.** Natural uranium ore consists primarily of uranium-238 (U-238) with only small amounts of uranium-235 (U-235). Mined uranium can be *enriched* by discarding much of the U-238 so that the remaining material has a larger fraction of U-235; producing 1 ton of enriched uranium requires about 8 tons of natural uranium. A typical 1 gigawatt nuclear power plant uses about 30 metric tons of enriched uranium each year.
a. The world's existing nuclear power plants, combined, generate about 400 gigawatts of power (1990 estimate). How much natural uranium (in tons) is needed to supply the world's nuclear reactors each year?
b. World uranium reserves (i.e., the amount of uranium available for future mining) are estimated to be about 6 million tons. Estimate how long these reserves will last. Explain your assumptions and uncertainties.
c. **Project.** Uranium reserves would last longer if nuclear power plants convert to *breeder reactors*, in which plutonium is created as a by-product of the reactions. Because this plutonium also can be used as nuclear fuel, a breeder reactor can generate up to 100 times as much energy as an ordinary reactor from the same amount of natural uranium. Research the pros and cons of breeder reactors and discuss the prospects that breeding will be used in future nuclear power plants.

53. **Plutonium in Nuclear Weapons.** Nuclear weapons use the fission of plutonium to generate an even more powerful fusion reaction with hydrogen. A typical nuclear weapon uses about 5 kilograms of plutonium.
a. The density of plutonium is about 16 grams per cubic centimeter. What is the volume of a typical nuclear weapon's plutonium? Put that volume in perspective.
b. The efficiency of nuclear fission of plutonium is approximately 0.0006; that is, about 6 of every 10,000 grams of mass is converted into energy during plutonium fission. How much mass is converted to energy by the fission of 5 kg of plutonium? Using $E = mc^2$, calculate the energy yield, in joules, from this fission.
c. Compare the energy yield from plutonium fission to the overall energy yield of a 1-megaton H-bomb (see Table 9-5). What fraction of the overall energy comes from plutonium fission? What is the remaining fraction that comes from hydrogen fusion?
d. In the 1980s the Soviet Union had at least 10,000 nuclear weapons. Today, many defense experts are concerned that some of the plutonium from these weapons may find its way into the hands of terrorists. Estimate the amount of plutonium in the old Soviet weapons. Based on your results, discuss the need for security against nuclear terrorism.

54. **The Hibernating Bear.** Mammals such as the Alaskan brown bear slow their metabolic rate during hibernation. During a 6-month hibernation, a brown bear metabolizes about 100 kilograms of fat from a typical body mass of 450 kg. Metabolizing animal fat yields about 3.8×10^7 joules per kilogram of mass.
a. Calculate the total amount of energy used by a brown bear during hibernation. Then calculate the average metabolic rate of the bear, in watts, during hibernation.
b. Biologists have found that the normal (non-hibernating) resting metabolic rate of mammals is roughly proportional to the 3/4 power of body weight. For example, if one mammal weighs 4 times as much as another, its metabolic rate is $4^{3/4} = 2.8$ times higher. Given that the resting metabolic rate of a 75 kg human is about 75 watts, estimate the resting metabolic rate of the brown bear.

c. Based on your results from parts (a) and (b), compare the metabolic rate of the brown bear during hibernation and non-hibernation.
d. In animals that enter "true" hibernation (e.g., many rodents), their hibernating metabolic rate is less than half their normal metabolic rate. In light of this fact, are brown bears true hibernators? Is it safe to walk into the den of a hibernating brown bear? Explain.

55. **Circle of Fifths.** The circle of fifths is generated by starting at a particular musical note and stepping upward by intervals of a fifth (seven half-steps). For example, starting at middle C, a circle of fifths includes the notes C → G → D′ → A′ → E″ → B″ → F‴ →..., where each (′) denotes a higher octave. Eventually the circle comes back to C several octaves higher.
a. Show that the frequency of a tone increases by a factor of $2^{7/12} = 1.498$ if it is raised by a fifth. (Hint: Recall that each half-step corresponds to an increase in frequency by a factor of $f = 1.05946$.)
b. By what factor does the frequency of a tone increase if it is raised by two fifths?
c. Starting with middle C at a frequency of 260 cycles per second, find the frequencies of the other notes in the circle of fifths.
d. How many notes are required for the circle of fifths to return to a C? How many octaves are covered by a complete circle of fifths?
e. What is the ratio of frequencies of the C at the beginning of the circle and the C at the end of the circle?
f. A circle of *fourths* is generated by starting at any note and stepping upward by intervals of a fourth (five half-steps). By what factor is the frequency of a tone increased if it is raised by a fourth? How many steps are needed to complete the entire circle of fourths? How many octaves are covered in a complete circle of fourths?

56. **Rhythm and Mathematics.** Musical rhythm and mathematics are closely related. For example, in "4/4 time," there are four **quarter notes** in a **measure**. If two quarter notes have the duration of a **half note**, how many half notes are in one measure? If two **eighth notes** have the duration of a quarter note, how many eighth notes are in one measure? If two **sixteenth notes** have the duration of an eighth note, how many sixteenth notes are in one measure?

Projects

57. **Balanced Budget?** In 1995, Congress and the president agreed that the federal budget should be balanced by 2002. Research current progress toward a balanced budget. Is the goal of a balanced budget in 2002 likely to be achieved? If so, explain how it is happening. If not, explain why not.

58. **Entitlement Spending.** The following table shows the distribution of entitlement payments (see subsection 9.2.2) to U.S. families, broken down by family income and family type (data from the Congressional Budget Office, early 1995).

Family Description	% of Families Receiving Benefits	Average Benefit
All families	49%	$10,320
By income:		
$0–$29,999	58%	$9,950
$30,000–$99,000	37%	$11,710
$100,000 or more	31%	$15,220
By type:		
With children	39%	$8,200
Elderly	98%	$13,970
All other	32%	$6,930

a. Note that 58% of low-income families ($0–$29,999 income) receive an average benefit of $9950. If the total amount of these benefits were distributed equally among all low-income families (rather than 58%), what would the average benefit be?
b. Do a calculation similar to that in part (a) to find the average benefit among *all* high-income families (more than $100,000). Compare to your result in part (a).
c. Suppose that, based on your results in parts (a) and (b), someone claims that the government spends almost as much money on entitlements to high-income families as to low-income families. Is this statement supported by the data? Why or why not? (Hint: Does the table give any information about the *number* of families in each income category?)
d. In 1995, entitlements comprised approximately 55% of the total federal budget of $1.4 trillion. How much money was spent on entitlements?
e. In this table, "elderly" means people over age 65. Estimate the fraction of the U.S. population in this category. Then estimate the total amount of money spent on benefits for the elderly. Explain your estimates and uncertainties. Based on your estimate, what fraction of 1995 entitlement spending went to the elderly? What fraction of the overall 1995 federal budget?
f. Note that nearly half of all U.S. families receive entitlement benefits. Identify any benefits that *you* personally receive (don't forget to include any federal education grants or student loans). If you receive any benefits, compare the total amount you receive each year to the total amount you pay in taxes. Overall, are you supporting the government or is the government supporting you? For additional research, survey your class or a group of friends and identify all of the entitlement benefits received.
g. Based on the information in this problem, and other information you have gathered about entitlements received by yourself and friends, discuss the political difficulties of cutting entitlement spending.

59. **The Mortgage Interest Deduction.** Interest on home mortgages in the United States is tax deductible. For example, suppose that a person has an income of $80,000 after accounting for all other deductions. If the person paid

mortgage interest of $10,000 during the year, she will have to pay tax on only $70,000 instead of $80,000. If the tax rate on this income is 33%, the $10,000 deduction results in a tax savings of 0.33 × $10,000 = $3300.

a. Find or estimate the number of people who own homes in the United States and the total amount of money saved by taxpayers through the mortgage interest deduction.

b. Compare the amount of money saved by taxpayers through the mortgage interest deduction to the amount that the federal government spends on various other items such as national defense, interest on the debt, programs for feeding and housing the poor, fighting crime, and education. Also compare to the federal deficit. Discuss your findings.

c. Proponents of the mortgage interest deduction cite benefits such as its encouragement of home ownership, which can help people break the cycle of poverty, and its promotion of jobs and the construction, home maintenance, and real estate industries. Opponents argue that it essentially is a government handout to the wealthy: Higher-income taxpayers tend to pay larger amounts of mortgage interest (because they tend to own more expensive homes) and to be in higher tax brackets, so the mortgage interest deduction saves them much larger amounts of money than it does for low-income homeowners. Opponents also note that the deduction provides no tax break for people who are too poor to own a home. Considering these arguments and other relevant factors, write a short essay expressing your opinion about the mortgage interest deduction. Should it be eliminated, modified, left alone, or expanded? Explain any estimates you make to support your opinion.

d. Suppose that the U.S. government eliminated the mortgage interest deduction. Consider all the effects of this action on such things as home sales, employment, tax fairness, the health of the real estate industry, and more. Hold a group debate concerning how the elimination of the mortgage interest deduction would affect (i) the federal deficit, and (ii) the price of homes.

60. Densities of Storage Media. Find out how the storage densities of various electronic media (e.g., diskettes, RAM, CD-ROMs) are measured. Trace the history of the storage densities over the past 10 years. How quickly is the density of commercially available storage media increasing? What do you expect to happen in the future? Is the density of information limited? Why or why not? Summarize your findings in a 1-page essay or 10-minute oral presentation.

61. Compact Fluorescent Light Bulbs. Compact fluorescent light bulbs fit most existing light sockets and produce high-quality light with less power than ordinary (incandescent) light bulbs. The average life of a compact fluorescent bulb is estimated to be about 10,000 hours, whereas the average life of an incandescent bulb is about 1000 hours.

a. Suppose that you replace a 100-watt incandescent light bulb with a 27-watt compact fluorescent bulb that supplies the same lighting intensity. Over its 10,000 hour life, how much energy (kilowatt-hours) is saved with the new bulb?

b. If the energy is generated by the combustion of oil, how much oil will be saved? (The combustion of 1 liter of oil yields about 1.2×10^7 joules of energy.)

c. Determine the cost of electricity per kilowatt-hour in your area. Based on this cost, how much will you save on your electric bill over 10,000 hours by switching from a 100-watt incandescent bulb to a 27-watt compact fluorescent?

d. Find the purchase prices of both 100-watt incandescent bulbs and 27-watt compact fluorescent bulbs. By accounting for the purchase price, the lifetimes of the bulbs, and their energy costs, calculate the *total* cost of each type of bulb for 10,000 hours of lighting. Which is more cost effective?

e. Making reasonable assumptions about how long lights are on in a commercial establishment, estimate how many times each incandescent bulb must be replaced each year. Would a switch to compact fluorescents generate any additional savings besides those calculated in part (d)? Explain.

f. Based on your findings, suggest several reasons why compact fluorescent bulbs have become very popular among businesses, but have yet to become popular for home use. Do you have any suggestions for changing this situation?

g. To promote the use of compact fluorescent bulbs, many utility companies make special offers to help customers purchase them. Investigate any such programs by your local utility company and comment on its effectiveness. Also explain why utility companies, which sell electricity for money, are trying to help people use less electricity.

62. Energy Audit. Do a thorough electrical energy audit of your home, apartment, or dormitory. Although there are several ways to approach this problem, you might try the following.

- Determine the power requirement, in watts, of all electrical appliances.
- Estimate the number of hours that each appliance is used during a typical month.
- Calculate the total energy, in kilowatt-hours, used by each appliance for a month.
- Sum all these amounts to get your total monthly electricity use, in kilowatt-hours.

a. What is your estimated daily, monthly, and annual use of energy for electricity? If you live in a home or apartment where you have an electric meter, compare your daily estimate to a direct daily meter reading. Compare your monthly estimate to the amount shown on your monthly electricity bill. Discuss any discrepancies.

b. Find the cost of electricity (per kilowatt-hour) in your area. Based on your findings in part (a), calculate your average daily, monthly, and annual costs for electricity.

c. Burning 1 kilogram of coal yields about 1.6×10^9 joules of energy. Assume that your local power plant generates energy from coal and that it delivers electricity with an efficiency of about 30% (that is, 30% of the energy released from the coal ends up as electricity at your residence). How much coal is needed to supply your daily, monthly, and annual electrical needs?

d. Find out how electricity actually is generated in your area. Discuss the use of resources and environmental impacts associated with your personal electricity consumption.

e. Based on your personal audit, estimate total energy use by your dormitory, apartment building, or neighborhood.

f. Prepare a plan by which you (and your dorm, apartment building, or neighborhood) could save both energy and money. This plan might include changing to more energy-efficient appliances, conserving electricity through less use, or making use of solar or wind energy (among many other possibilities). Write a two-page (maximum) summary of your plan and its benefits.

g. If you are free to implement your plan, will you do it? When would you expect to see results? If you live in an apartment building or dormitory, present your plan to a manager. Do you think it will be implemented? Why or why not?

63. **Solar Power Plants.** Because not all areas receive the same amount of sunshine, solar energy might be "harvested" at giant power stations in areas with a lot of sunshine. For example, a test facility in the southern California desert uses parabolic mirrors to reflect sunlight onto a tube of liquid, which heats up, forms steam, and turns a turbine to produce electricity. In 1993, this facility had a peak power production of 354 megawatts. Investigate various proposals for harvesting solar energy at large power stations. Discuss the prospects that large-scale solar power plants will become common in the future.

64. **It's All Solar.** Except for geothermal and nuclear energy, virtually all other sources of energy come from sunlight. For example, if you burn wood, you are releasing energy which the tree received from sunlight and used through the process of photosynthesis. Similarly, because oil and coal come from the decomposition of ancient dead plants (and, perhaps, animals that ate the plants), they also store energy that traces ultimately to sunlight.

    The *conversion efficiency* of sunlight into energy describes the fraction of the original solar energy that is available for human use. For example, wood has a conversion efficiency of 1/1000, meaning that for every 1000 units of solar energy striking a tree, enough wood is produced to generate 1 unit of energy when the wood is burned.

    a. Burning coal formed from the decomposition of ancient wood generates only half as much energy as burning wood itself (for the same mass of coal or wood). A modern power plant converts coal to electricity with an efficiency of about 25% (i.e., every 4 units of energy generated by burning coal produces 1 unit of electrical energy; the rest is wasted). What is the *overall* conversion efficiency from *sunlight* to electricity when the source of the electricity is a coal-burning power plant?

    b. The conversion efficiency of modern solar photovoltaic cells is about 10%. Compare this efficiency to that found in part (a) and comment on the differences. Given the much higher efficiency of photovoltaic cells, why aren't they in common use?

    c. Research the current state of technology for solar photovoltaic cells and their current costs. How does the cost of solar energy compare to the cost of generating energy from oil or coal? How does it compare to the cost of generating nuclear energy or hydroelectric energy?

    d. The costs cited for energy usually are production and delivery costs, but do not include the costs of any impact on the environment or public health. Discuss possible methods of accounting for the costs of environmental impacts. If these costs were included, is solar energy cost effective today in your opinion? Write a short essay, or hold a debate on this question. Be sure to consider whether photovoltaic cells themselves have an environmental impact.

65. **Playing Scales and Intervals.** If you have access to a musical instrument (for example, piano, guitar, or clarinet), play a 12-tone scale. Then play two notes at intervals of an octave, a fifth, a fourth, and a third. Experiment with other intervals. Can you hear which intervals are consonant? Bring your instrument and give a class demonstration of the relationship between music and mathematics.

66. **Bach and Mathematics.** Three books that explore the relationships between mathematics and music are *Godel, Escher and Bach: An Eternal Golden Braid,* by Douglas Hofstadter; *Measure for Measure,* by Thomas Levenson; and *Emblems of Mind,* by Edward Rothstein. Use these or other sources and write a short essay on the following passage from Levenson:

    > *And because of that extraordinary ability, Bach's music retains in the rigor of its mathematical structures a sense of the absolute, of impartial, impersonal accuracy—of the truth that we discern when we can identify the patterns a phenomenon creates for us.*

67. **Stradivarius and the Scientific Method.** The seventeenth century instrument maker Stradivarius experimented extensively with materials, designs, and processes to create violins and cellos that have never been reproduced. Research his instruments further and write an essay on the claim that Stradivarius's success was the result of his careful application of the scientific method.

68. **Well-Tempered Scales.** Refer to Table 9-6, and note that two tones a fifth apart have a frequency ratio of 1.49831, which is very close to, *but not exactly* 1.5 or 3/2. Similar discrepancies exist between the actual ratios and the ideal ratios for some of the other intervals in this table. In tuning a musical instrument, these discrepancies are "smoothed out" in a variety of ways called *tempering*. Through reading or observing and talking to instrument tuners, describe the different methods of tempering that are used today.

PART 3

THINKING MATHEMATICALLY

10 THE LANGUAGE OF CHANGE

In Part 1 of the text you developed logical thinking skills and in Part 2 you added quantitative thinking skills. Now in Part 3 we take you deeper into the language of mathematics and show you how it can help you model and understand the real world. In this chapter we explore a universal feature of the world around us: *change*. We demonstrate how change can be described with mathematical models, both graphically and with equations. You will find that mathematics can help you understand and explore the past, present, and future. By the end of the chapter you should appreciate the utility of mathematics for making sense of a rapidly changing world.

CHAPTER 10 PREVIEW:

10.1 ***MODELS FOR A CHANGING WORLD.*** We begin the chapter by discussing the role of mathematical models in understanding change.

10.2 ***WHAT IS A RELATION?*** The mathematical way to describe change is through the use of relations. In this section we introduce relations visually, with graphs, and investigate their properties and applications.

10.3 ***RATES OF CHANGE.*** In this section we focus on graphs of linear relations, showing how the slope of the graph is interpreted as the rate of change of the relation.

10.4 ***LINEAR EQUATIONS.*** We develop the general form of a linear equation in this section, and show how to solve and interpret linear equations with specific examples.

10.5 ***CREATING LINEAR MODELS FROM DATA.*** In this section we discuss methods for creating a linear models from two data points and a best fit line from more than two data points.

10.6 ***FURTHER APPLICATIONS OF LINEAR MODELS.*** We conclude our discussion of linear modeling with additional applications from business and finance and a peak at the principles involved in more complex models.

10.7* ***CHANGING RATES OF CHANGE — VISUAL CALCULUS.*** Conceptually, extending our work to relations in which the rate of change is not constant is straightforward. In this section we explore such relations graphically and point the way to calculus.

Nothing endures but change.

— Heraclitus, c. 500 B.C.

Nothing in the world lasts save eternal change.

— Honorat de Bueil, Marquis de Racan, c. 1600

There is nothing in this world constant, but inconstancy.

— Jonathan Swift, c. 1707

Man's yesterday may ne'er be like his morrow; Nought may endure but Mutability.

— Percy Bysshe Shelley, 1816

The only thing that one knows about human nature is that it changes.

— Oscar Wilde, 1895

10.1 MODELS FOR A CHANGING WORLD

Change is ever-present in the physical, social, psychological, and economic environments. Quantities such as the prices of stocks and bonds, your blood sugar level, or the position of the Earth as it orbits the Sun all change with time. Other quantities exhibit change in different ways. For example, pressure and temperature change as you move up or down in altitude, and the width of the Mississippi River changes as you drift southward with its flow.

If you hope to understand the world, you must consider how change affects it. Fortunately, mathematics offers many methods for describing change, from the arithmetic of calculating absolute and relative change (see Chapter 3) to the advanced techniques of calculus. Indeed, it is accurate to say that mathematics is the language of change.

As we first discussed in Chapter 1, and again in several subsequent chapters, mathematics is more than just a language: It also provides a way of modeling the real world. We now begin a deeper investigation into the creation and use of mathematical models.

Many problems, such as the effects of carbon dioxide emissions on the global climate or the economic impact of a tax cut, are too complex to describe exactly. Yet gaining some understanding of such problems may be crucial to a business decision, a political position, or the survival of civilization. The goal of mathematical modeling is to find a mathematical representation of an actual problem or situation, and then to use mathematics to analyze and understand it.

Mathematical models take many different forms. Sometimes they are visual representations such as pictures, graphs, or charts (Figure 10-1). Other times they are sets of equations that must be solved or mathematical simulations of real phenomena. Once a mathematical model is formed, it can be used for discovery, understanding, or prediction. Of course, as the Roman philosopher Seneca (c. 4 B.C. – A.D. 65) said, "All art is but imitation of nature." Mathematical modeling is an art and its imitation of nature, while useful, may not be exact.

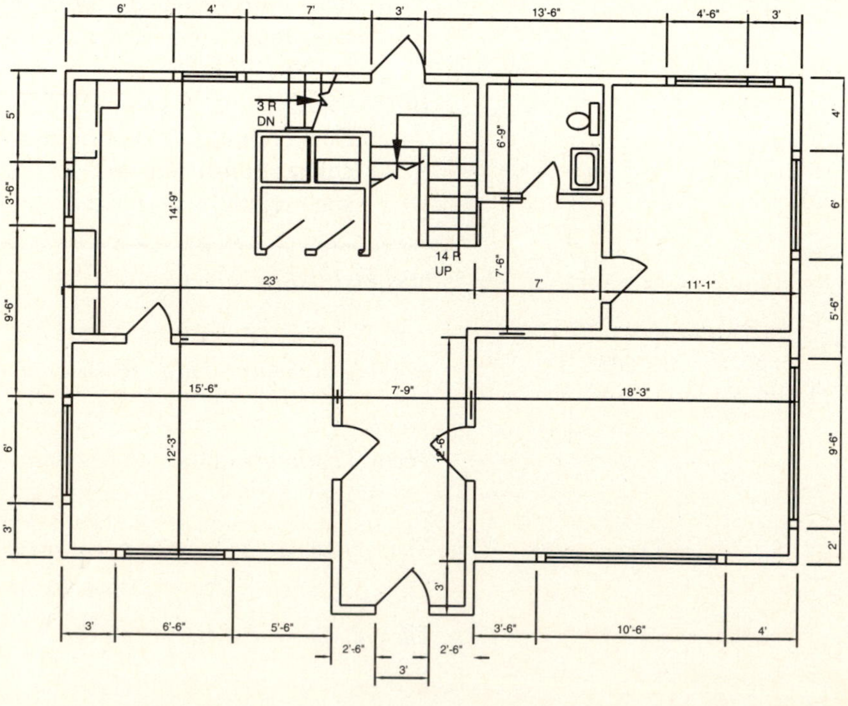

Figure 10-1. Familiar everyday models include maps and scale models or scale drawings, such as floor plans. Mathematical models, such as graphs, charts, and equations, also provide representations of reality.

In this chapter, we present mathematical relations that describe how quantities change. Relations, whether expressed visually or in the form of equations, represent some of the most powerful and prevalent forms of mathematical models. We focus on linear relations because, in addition to being the simplest relations, they capture the essence of modeling and provide a prototype for many other kinds of models. Linear models often give very good descriptions of particular problems. In other cases, linear models provide merely a starting point for more complex models.

Time-Out to Think: A road map is a good example of a model. Find a road map for an area with which you are familiar. Discuss some of the differences between the map and the reality it is modeling. Even with these differences, why is the map still useful?

10.2 WHAT IS A RELATION?

Imagine that you have a thermometer outside a window and that you monitor temperature changes hourly throughout the day. As Table 10–1 shows, the temperature rises from early morning until about 2:00 P.M., and then steadily drops as evening approaches. Can you describe the change in the temperature more precisely?

Table 10-1. Temperature Data for One Day

Time	Temperature	Time	Temperature
6:00 A.M.	50°F	1:00 P.M.	73°F
7:00 A.M.	52°F	2:00 P.M.	73°F
8:00 A.M.	55°F	3:00 P.M.	70°F
9:00 A.M.	58°F	4:00 P.M.	68°F
10:00 A.M.	61°F	5:00 P.M.	65°F
11:00 A.M.	65°F	6:00 P.M.	61°F
12:00 noon	70°F		

A relation involves one quantity changing with respect to another quantity.

What we call a relation is called a **function** in most mathematics books. We use the term *relation* because it makes it clear that we are talking about a *relationship* between two quantities.

The first step in describing how the temperature changes is to recognize that *two* quantities are involved: the temperature and the time. There is a **relation** between the time of day and the temperature, which you can show by writing them together in the form (*time, temperature*). By convention, you should write *time* first because the temperature depends on the time of day — it wouldn't make sense to say that the time of day depends on the temperature. That is, the temperature changes *with respect to* the time.

Quantities that change with respect to time are common. Children grow, planets revolve, glaciers crawl, and the national debt swells: in each case, something is changing with respect to time. However, other types of change don't involve time.

Imagine a ride in a hot-air balloon. As the balloon rises, the surrounding atmospheric pressure decreases (which can cause your ears to "pop"). The change in pressure depends primarily on the altitude, not on the time of day or year; that is, pressure changes *with respect to* altitude. Therefore you should write *altitude* first in showing this relation as (*altitude, pressure*).

The Mississippi River itself runs 2340 miles from Lake Itasca, Minnesota, to the Gulf of Mexico. The Mississippi River system, however, which includes the Red Rock River in Montana and the Missouri River, has a length of 3710 miles, or almost 6000 kilometers.

Now imagine a trip down the Mississippi River. The width of the river changes as you travel southward with the current; that is, the river width depends on your distance from the river's source. Therefore the river width changes *with respect to* your distance from the source and you should write *distance from source* first in showing the relation as (*distance from source, river width*).

Time-Out to Think: Think of a relation in which a quantity changes with respect to time. Think of another relation in which a quantity changes with respect to a quantity other than time.

Once you've identify the quantities involved in a relation, the next step is to describe the relation itself. You can do so in at least four different ways.

1. You can describe a relation in *words*. Returning to the first example, you might say that "the temperature increased from 50°F at 6 A.M. to a 2 P.M. high of 73°F, and then steadily decreased to 61°F at 6 P.M."
2. You can describe a relation by using *data*, such as those shown in Table 10–1. A table provides detailed information but can become unwieldy with great quantities of data.
3. You can draw a *picture* or graph of a relation. A graph is visual and easy to interpret and consolidates a great deal of information that might be difficult to state either in words or in a table.
4. You can write a compact mathematical description of a relation in the form of an *equation*.

The first two methods (words and tables) are straightforward and probably familiar to you already. So let's move directly to pictures of relations in this section and to equations in the following sections.

10.2.1 Pictures of Relations

A picture that shows a relation between two quantities is called a **graph**. The most common and important way to draw a graph is to use a **coordinate plane**.

In some books, the origin is defined as the point (0, 0). We define it as the intersection of the two axes, which may or may not occur at (0, 0). (Note, for example, that the origin is at (0, 0) in Figure 10-4(a) but is at (0 hr, 50°F in Figure 10-4(b).)

Recall that a number line extends in both the positive and negative directions from the zero point (Figure 10–2a). Every point on the number line is associated with a real number, and every real number has a unique point on the number line. To create a coordinate plane, draw two perpendicular number lines. Each of the number lines is called an **axis** (plural, **axes**), and the intersection point of the axes is called the **origin**. Normally, numbers increase to the right on the horizontal axis and upward on the vertical axis (Figure 10–2b).

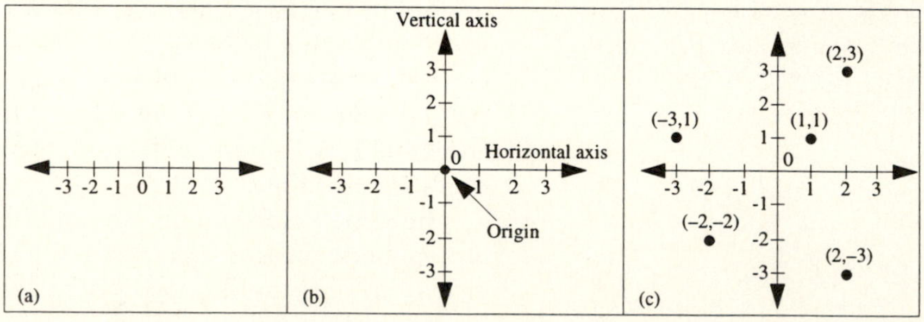

Figure 10-2. (a) The construction of a coordinate plane begins with a number line. (b) Two perpendicular number lines make the axes of a coordinate plane. (c) Several points are plotted and labeled with their coordinates in the form (*horizontal coordinate, vertical coordinate*).

Points in the coordinate plane are described by two numbers, called the **coordinates**, which give the horizontal and vertical distances between the point and (0, 0). Coordinates are written in parentheses in the form (*horizontal coordinate, vertical coordinate*). Note that the horizontal coordinate is always specified first. For example, the point (2, 3) is 2 units to the right and 3 units up from the point (0, 0). The point (–3, 1) is 3 units to the left and 1 unit up from (0, 0). These and several other points are shown in Figure 10–2(c).

The process of drawing a graph requires three basic steps, which we now describe in detail.

Step 1: Identify the Variables

A relation involves two variables and is represented by pairs of coordinates in the form (*first variable, second variable*).

The first (or independent) variable is plotted on the horizontal axis of a graph and the second (or dependent) variable is plotted on the vertical axis.

The two quantities in a relation are called **variables** because they change or vary. For example, in the (*time, temperature*) relation shown in Table 10–1, both *temperature* and *time* vary: The time steadily advances moment by moment, while the temperature rises, then falls.

We wrote *time* first in the (*time, temperature*) relation because it makes more sense to say that the temperature depends on the time than vice versa; after all, time advances no matter what else occurs. That is, *time* has a certain *independence* and therefore is called the **first**, or **independent**, **variable** in the (*time, temperature*) relation. *Temperature*, because it *depends* on time, is called the **second**, or **dependent**, **variable**.

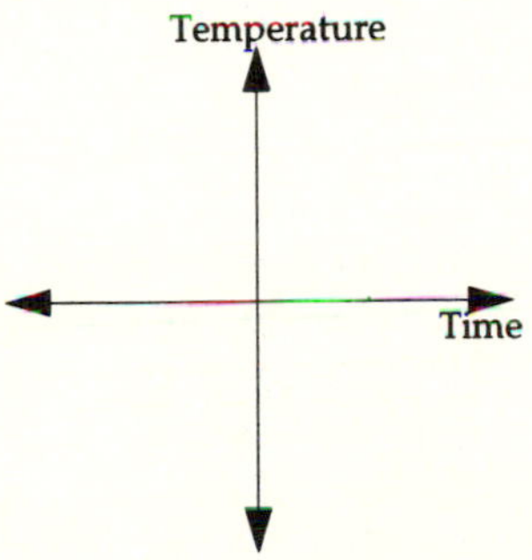

Figure 10-3. The first step in constructing a graph of the (*time, temperature*) data from Table 10-1 is to identify the axes: The first variable, *time*, should be plotted on the horizontal axis and the second variable, *temperature*, should be plotted on the vertical axis.

Because dependence often is not clear-cut, we prefer to use the terms *first variable* and *second variable* when discussing relations.

When making a graph, values of the first variable are plotted on the horizontal axis and values of the second variable on the vertical axis. Thus, in the (*time, temperature*) relation, *time* is plotted on the horizontal axis and *temperature* on the vertical axis (Figure 10–3).

Dependence — and hence the choice of first and second variables — isn't always clear-cut. For example, consider a simplistic case of the *law of supply and demand*. From one viewpoint, supply depends on demand: When demand increases, supply decreases because consumers are buying the available product; when demand decreases, supplies increase as the unsold product collects dust on store shelves. From another point of view, however, demand depends on supply: An increase in supply often is followed by a decrease in demand — especially if a product has saturated the market and is no longer appealing. Similarly, if supplies of a commodity decrease to the point that it is rare, demand often increases. Because both views of dependence are reasonable, there is no obvious way to decide which quantity is the first variable and which quantity is second variable.

Dependence is even more difficult to assess in complex systems. For example, the U.S. economy has many variables: the prime lending rate, the rate of inflation, the unemployment rate, the national debt, and many more. All these variables may be closely related and continually changing, and showing that one particular variable depends on another probably is impossible. As a result, in graphing a relation between, say, inflation and the unemployment rate, the choice of first variable and second variable is somewhat arbitrary. Sometimes the relation may be shown as (*inflation, unemployment rate*) with *inflation* as the first variable on the horizontal axis and *unemployment rate* as the second variable on the vertical axis, and sometimes it may be shown with the first and second variables switched.

Time-Out to Think: Does dependence imply causality? That is, if the second variable depends on the first variable, do changes in the first variable *cause* changes in the second variable? Think of some examples to make your case.

Step 2: Identify the Domain and Range

The domain of a relation consists of all values of the first variable that make sense and are of interest.

The second step in preparing to graph a relation involves deciding which values of the quantities should be shown. For example, in graphing the (*time, temperature*) data from Table 10–1, it makes no sense to show times over a period of many years. In fact, the only times that are of interest in this case are those from 6 A.M. to 6 P.M., during which measurements were made. The set of all values of the *first* variable that both make sense and are of interest is called the **domain** of the relation. In the (*time, temperature*) relation the domain consists of times between 6:00 A.M. and 6:00 P.M.

The range of a relation consists of the values of the second variable that correspond to the domain.

Note that only temperatures between 50°F and 73°F occur between 6:00 A.M. and 6:00 P.M (the domain of the (*time, temperature*) relation). Thus we say that 50°F to 73°F is the **range** of the (*time, temperature*) relation. In general, the range of a relation consists of the values of the second variable that correspond to the domain.

Because the first variable is plotted on the horizontal axis, the domain tells you which values must be shown on the horizontal axis. Similarly, the range tells you which values of the second variable must be shown on the vertical axis.

Step 3: Draw the Graph

Having identified the first and second variables, you know what the horizontal and vertical axes represent, respectively. Having identified the domain and range, you also know what values must be shown on the horizontal and vertical axes, respectively. All that remains is to draw the graph.

Figure 10–4 shows two possible graphs of the (*time, temperature*) relation. As required, in both graphs the horizontal axis shows all times in the domain of 6 A.M. to 6 P.M., or, equivalently, times between 0 and 12 hours *after* 6 A.M. Similarly, in both graphs the vertical axis shows all temperatures in the range of 50°F to 73°F.

The axes in Figure 10–4(a) include time and temperature values outside the domain and range, giving the impression that the temperature variations were relatively small. Alternatively, if you want to emphasize details of how the temperature changed with respect to time, you would zoom in on the domain and range as shown in Figure 10–4(b).

Choices concerning tickmarks on the axes are a matter of personal taste. However, you should seek an appropriate balance between showing too many tickmarks, which can crowd an axis and make it difficult to read, and too few tickmarks, which can make it difficult to read values from the graph. Typically, 3 to 10 tickmarks should be shown along each axis. In addition, you should choose tickmarks that show round numbers. For example, it is easier to read the temperature axis if its tickmarks are labeled in 5°F increments rather than 3.7° increments.

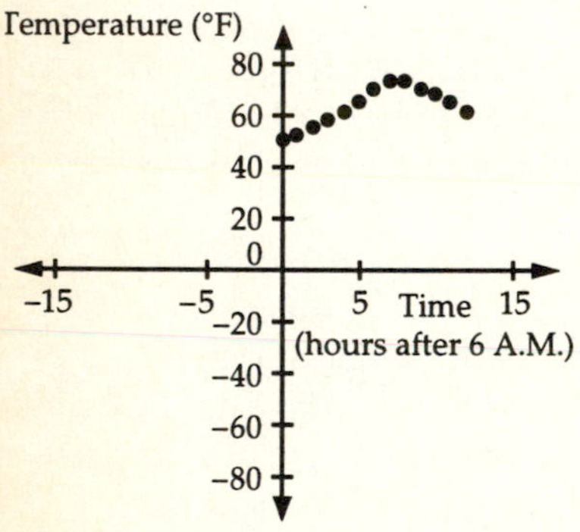

(a)

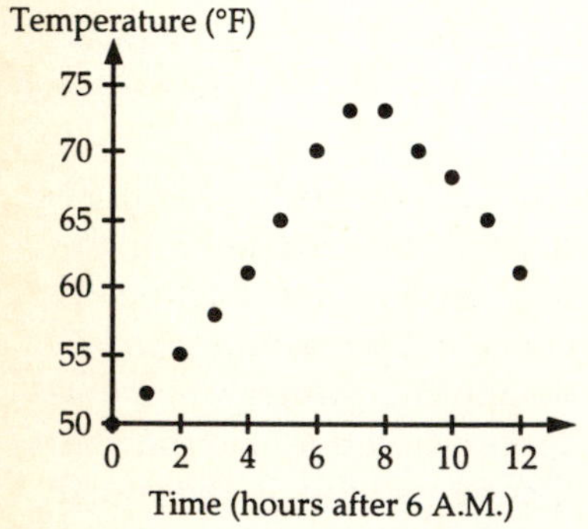

(b)

Figure 10-4. (a) This graph of the (*time, temperature*) relation shows a complete coordinate plane. (b) By zooming in on the domain and range, this graph makes it easier to see details of how the temperature changed with respect to time.

Summary Box 10-1. Graphing Relations.

Step 1: **Identify the two variables in the relation. If dependence is clear, choose the first (independent) variable for the horizontal axis and the second (dependent) variable for the vertical axis; otherwise, the choice is arbitrary.**

Step 2: **Identify the domain and the corresponding range. The domain tells you which values of the first variable must be shown on the horizontal axis, and the range tells you which values of the second variable must be shown on the vertical axis.**

Step 3: **Label your axes and draw the graph. The choice of whether to zoom in solely on the domain and range, as opposed to including additional values on the axes, depends on what you want to show. Similarly, the choices of how many tickmarks to show and how to label them is a matter of personal taste but should be made so that the graph is easy to read.**

Finally, the method you choose for drawing the actual graph depends on what you know about the relation. If you have a set of data points, simply plot them as we did in Figure 10–4. We discuss other methods of graphing later in the chapter.

10.2.2 Relations as Models

The (*time*, *temperature*) graphs of Figure 10–4 show only 13 data points, corresponding to the hourly measurements made from 6 A.M. to 6 P.M. However, the air temperature had some value at *every* instant of the day. Because the measured data appear to change smoothly with time, connecting the data points with a smooth continuous curve seems *reasonable* (Figure 10–5a).

Filling in the gaps between data points in the (*time*, *temperature*) graph creates a *mathematical model* of the change in the temperature at *every* moment throughout the day. Using this model, you can estimate the temperature at times other than those at which it was measured. For example, at 11:30 A.M., or 5 1/2 hours after 6 A.M., the model predicts that the temperature was about 67°F to 68°F (Figure 10–5b).

Because the model was based on assuming smooth temperature variations, you should not be surprised that it predicts an 11:30 A.M. temperature that is about halfway between the temperatures at 11 A.M. and noon. Nevertheless, because the temperature was not measured at 11:30 A.M., the model prediction may be incorrect. For example, if the temperature remained at 65°F until 11:45 A.M., then quickly rose to 70°F by noon, the model is incorrect.

This example contains an important lesson about relations. Nearly all relations are constructed from a limited set of data or assumptions and hence represent mathematical models. A model's predictions can be only as good as the data and the assumptions from which the model is built. Therefore it is crucial to evaluate the validity of the model represented by a relation before accepting any of its predictions or implications.

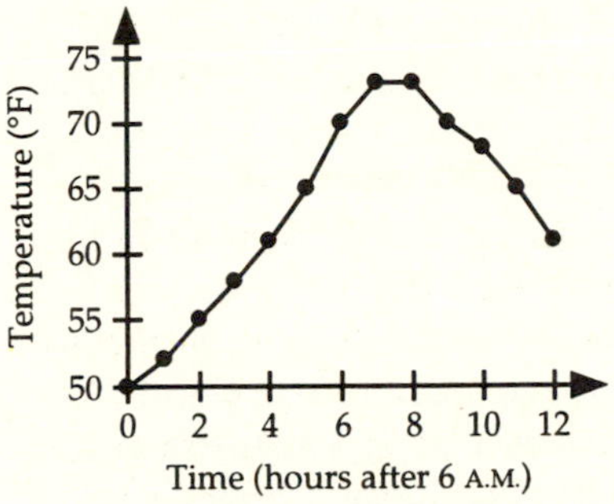

(a)

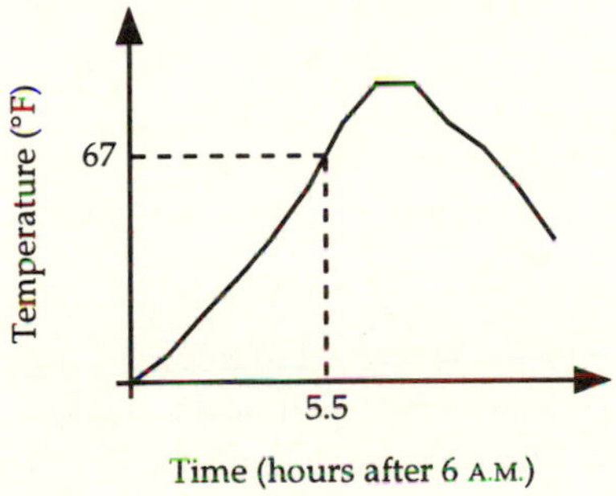

(b)

Figure 10-5. (a) Filling in the gaps between data points creates a mathematical model that you can use to estimate the temperature at times other than those at which it was measured. (b) At 11:30 A.M. (or 5 1/2 hours after 6 A.M.) on the *time* axis, the model predicts that the temperature was about 67°F or 68°F.

Example 10–1 *A Balloon Ride.* Consider a ride in a hot-air balloon, in which the atmospheric pressure changes with respect to altitude. Make a *rough* sketch of the (*altitude*, *pressure*) relation and discuss its validity.

Solution: Step 1: As discussed previously, pressure depends on altitude, so identify the horizontal axis with the first variable *altitude* and the vertical axis with the second variable *pressure*.

Step 2: Assume that the hot-air balloon can rise as high as a typical airplane, or about 30,000 feet. Then the domain extends from 0 feet (sea level) to 30,000 feet. A common unit for atmospheric pressure is *inches of mercury*, or the height of a column of mercury in a pressure gauge. By consulting an encyclopedia or a book on meteorology, you will find that atmospheric pressure is about 30 inches of mercury at sea level and that the pressure falls to one-third of its value with every 30,000 feet of altitude. Thus the pressure at 30,000 feet is about $1/3 \times 30$ inches = 10 inches of mercury, so the range of the pressure is from 10 to 30 inches of mercury.

Step 3: Draw the graph. One approach, illustrated in Figure 10–6 on the next page, is to show the altitude from 0 to 30,000 feet and the pressure from 0 to 30 inches of mercury. As required, this approach shows all values in the domain and range. From Step 2, you have two data points for the (*altitude*, *pressure*) relation: the sea level pressure of 30 inches of mercury, or the point (0, 30 inches), and the pressure of 10 inches of mercury at an altitude of 30,000 feet, or the point (30,000 feet, 10 inches).

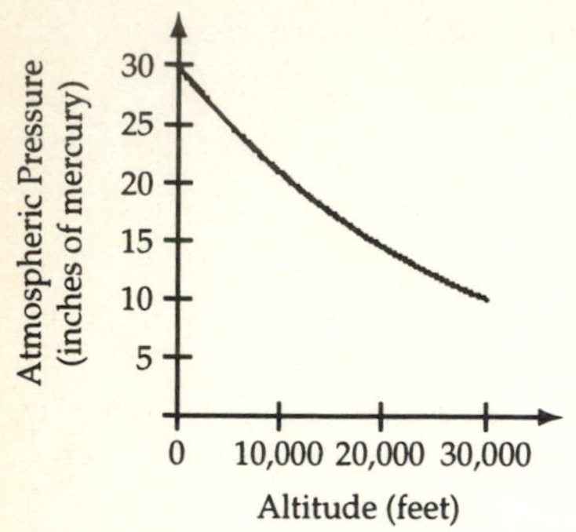

Figure 10-6. A rough sketch of the (*altitude, pressure*) relation shows that pressure decreases smoothly with increasing altitude.

Between the two data points, you have no data, so you must guess the shape of the graph. You can reasonably assume that pressure decreases smoothly with increasing altitude. That is, the surrounding pressure doesn't suddenly jump or change as your balloon rises. Furthermore, because the Earth's atmosphere doesn't end abruptly (it becomes gradually thinner with increasing altitude), the pressure must fall more gradually at higher altitudes. The smooth, increasingly gradual fall in pressure is shown by the curve in Figure 10–6.

This graph represents a *model* of the actual relation between altitude and pressure, so the validity of its predictions depends on the validity of the assumptions on which it was based. In particular, because atmospheric pressure is affected by wind and other weather conditions, this model can yield rough estimates at best. Nevertheless, because this graph is based on just a few simple facts about the (*altitude, pressure*) relation, you can be reasonably confident that it is a good model, at least for average weather conditions. For example, the graph shows a value of about 22 inches of mercury for the pressure at an altitude 10,000 feet, which is a good estimate of the actual pressure at 10,000 feet under average weather conditions.

❖

Example 10–2 ***A River Trip.*** Make a rough sketch of the (*distance from source, river width*) relation for the Mississippi River trip. Discuss the validity of the sketch.

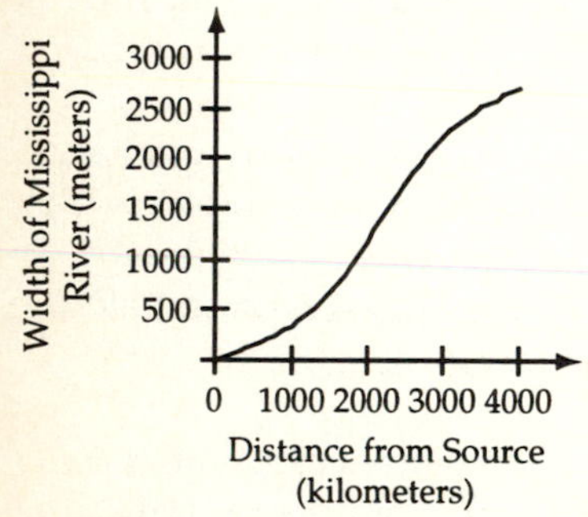

Figure 10-7. A rough sketch of the relation (*distance from source, river width*) for the Mississippi River suggests that the river width increases smoothly with distance from the source.

Solution: Step 1: Identify the horizontal axis with the first variable, *distance from the source,* and the vertical axis with the second variable, *river width.*

Step 2: The Mississippi River runs about 3800 kilometers from Lake Itasca, Minnesota, to the Gulf of Mexico, so the domain is 0 to 3800 kilometers from the source. At its delta (entrance to the Gulf of Mexico), the river width is about 3 kilometers, or 3000 meters. Thus, assuming that the river starts with a width of 0, the range is from 0 to about 3000 meters.

Step 3: The river width of 0 at the river's source represents a data point (0, 0) and its 3000 meter width at its delta represents a data point (3800 km, 3000 m). Between these two data points you have no data, so you must guess the shape of the curve. The curve shown in Figure 10–7 represents *one reasonable model* for the (*distance from source, river width*) relation, based on the two data points and the assumption that the river becomes gradually wider with distance from the source.

Is the model valid? Generally speaking, it seems reasonable that the river should gradually widen with distance from the source because more water enters the river from tributaries along its route. However, this model neglects any deviations from a gradual widening. For example, the river likely flows through occasional narrow, deep channels where the width decreases temporarily. Similarly, the width probably changes dramatically where major tributaries (such as the Missouri River) enter. Thus, although the simple model in Figure 10–7 probably captures the general features of the true (*distance from source, river width*) relation, the real relation probably is far more complex than that shown.

❖

Time-Out to Think: Suppose that the model from the preceding example suggested that the Mississippi River width is exactly 1575 meters at a distance of 2135 kilometers from the source. How reliable would you consider such a prediction? Why?

10.2.3 The Many Faces of Relations

Classifying relations according to *how* one variable changes with respect to the other can be useful. Look back at the graph of the (*altitude, pressure*) relation (Figure 10–6). It is an example of a **decreasing relation** because the graph falls from left to right along the horizontal axis. That is, pressure decreases as altitude increases.

In contrast, the graph of the (*distance from source, river width*) relation (Figure 10–7) rises from left to right along the horizontal axis; it is an example of an **increasing relation**. Other relations, such as the (*time, temperature*) relation (Figure 10–5), *both* increase and decrease over their domains.

Periodic Relations

Graphs of relations can take an endless variety of forms besides strict increases or decreases. One important form repeats a pattern over and over, and is called a **periodic relation**. An example of a periodic relation is provided by the Earth's orbit about the Sun. Because the shape of the orbit is elliptical (Figure 10–8a), the distance between the Earth and the Sun is constantly changing. In particular, the Earth is farthest from the Sun on about July 3, when the distance is about 152,000,000 kilometers. Over the next six months, the Earth gets closer to the Sun until it reaches a minimum distance of about 147,000,000 kilometers on about January 3. The distance then increases until it returns to its maximum on the next July 3. This pattern repeats over and over.

Because the Earth's distance from the Sun changes with respect to time, we can graph the relation *(time, distance from Sun)*. And, because the relation has a pattern that repeats, we can show its essential features by graphing only a few cycles. We therefore choose (somewhat arbitrarily) a domain that covers 3 years, beginning on July 3 (of any year). The range is between 147,000,000 and 152,000,000 kilometers, which covers the values of the second variable, *distance from Sun*, during each year. The graph of the relation shows its repeating pattern (Figure 10–8b).

Many other phenomena, including heartbeats, musical sounds, and the cycling of an automobile engine, exhibit periodic relations. Although their variables and axes differ from the relation graphed for the Earth's orbit about the Sun, their basic repeating patterns are very similar.

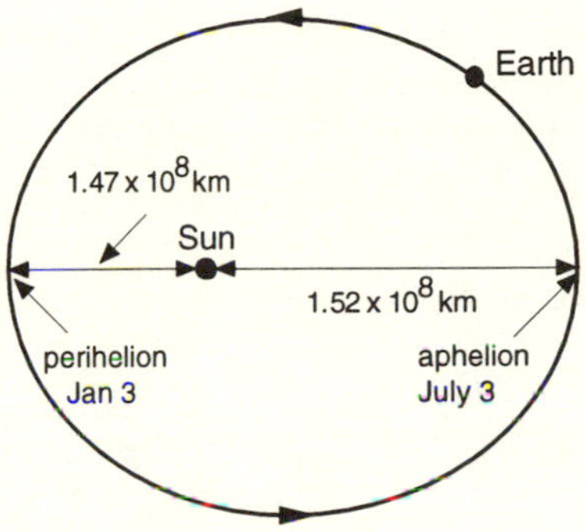

(a)

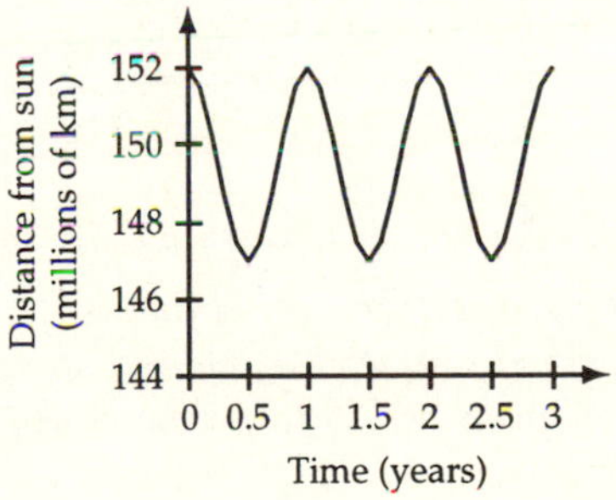

(b)

Figure 10-8. (a) The shape of the Earth's orbit around the Sun is an ellipse (not drawn to scale); the point nearest the Sun is *perihelion*, and farthest from the Sun is *aphelion*. (b) The relation between the Earth's distance from the Sun and time is periodic.

A Domain of Individual Points

Suppose that you are a traffic engineer who must ensure that a road connects every pair of towns in a certain county. Figure 10–9 shows the number of roads needed to connect as many as five towns. If the county has only one town, no road is needed. If the county has two towns, one road is needed to connect them. For three towns, three roads are needed. Four and five towns require six and ten roads, respectively.

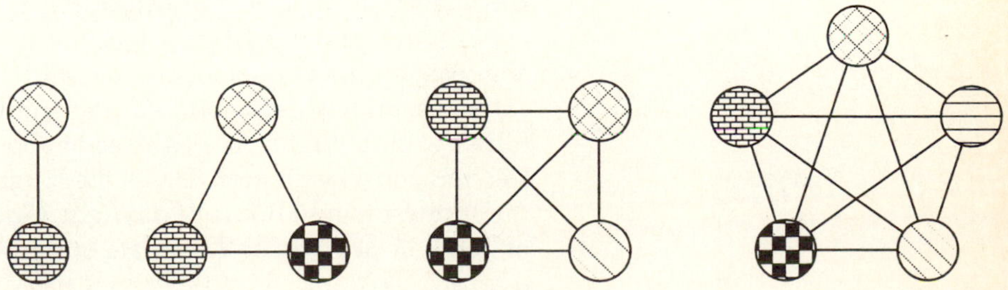

(a) (b) (c) (d)

Figure 10-9. The number of roads needed to connect all pairs of towns in a county may be shown using models in which towns are represented by circles and roads are represented by lines.

The number of towns in the county and the number of roads needed to connect every pair of towns are related. To investigate the (*number of towns, number of roads*) relation, let's look for a pattern in Figure 10–9:

- Two towns require just 1 road (Figure 10–9a).
- Three towns require 3 roads (Figure 10–9b), and 3 = 1 + 2.
- Four towns require 6 roads (Figure 10–9c), and 6 = 1 + 2 + 3.
- Five towns require 10 roads (Figure 10–9d), and 10 = 1 + 2 + 3 + 4.

Generalizing, the total number of roads needed to connect n towns is

$$1 + 2 + 3 + \ldots + (n - 1) \text{ roads.}$$

Time-Out to Think: By adding a sixth town (circle) to the diagram in Figure 10-9(d), confirm visually that 5 more roads are needed to make all the connections; thus 1 + 2 + 3 + 4 + 5 = 15 roads are required for the six towns.

Table 10–2 shows this relation for several more values of n.

Table 10-2. Number of Roads Needed to Connect Towns

Towns	Roads	Towns	Roads	Towns	Roads
1	0	6	15	11	55
2	1	7	21	12	66
3	3	8	28	13	78
4	6	9	36	14	91
5	10	10	45	15	105

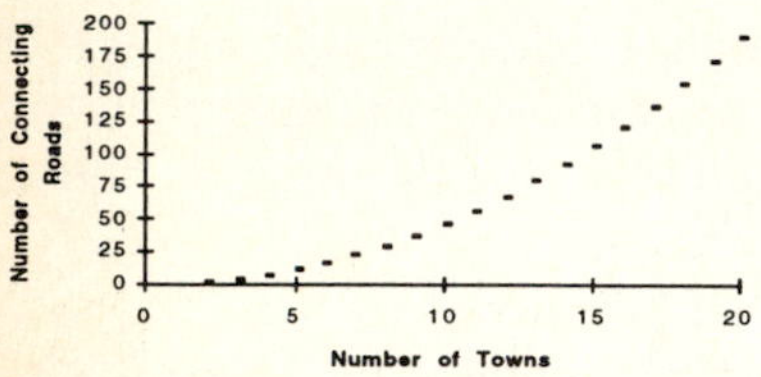

Figure 10-10. A graph of the relation between the number of towns and the number of roads needed to connect each pair of towns consists of individual points rather than a continuous curve.

Following the basic graphing procedure, we identify the first variable, *number of towns,* with the horizontal axis and the second variable, *number of roads,* with the vertical axis. The data points from Table 10–2, and a few more, are plotted in Figure 10–10.

The graph shows only individual data points. Should we fill in the gaps between the data points as we did with the (*time, temperature*) relation in Figure 10–5? The answer is *no*: Thinking about the number of roads needed for 2 1/2 towns or 5.33 towns makes no sense. That is, the number of towns must be a natural number (1, 2, 3, 4...). As a result, the graph consists only of *individual points* instead of a continuous curve. Cases like this one, in which the domain consists only of specific numbers (rather than *all* numbers) within an interval, represent another common and important type of relation.

Example 10–3 *Hours of Daylight.* In the Northern Hemisphere, the number of hours of daylight is greatest on the summer solstice (about June 21) and least on the winter solstice (about December 21). Halfway between these extremes, on the spring and autumn equinoxes, there are 12 hours (half of a full day) of daylight. At a latitude of about 40°N (for example, San Francisco, Denver, and Washington, D.C.), the longest day has 14 hours of daylight and the shortest has 10 hours of daylight. Graph the relation between the number of hours of daylight and the date of the year for 40°N latitude.

Solution: The number of hours of daylight changes with respect to the day of the year. Identify the horizontal axis with the first variable, *time,* and the vertical axis with the second variable, *hours of daylight*. The number of hours

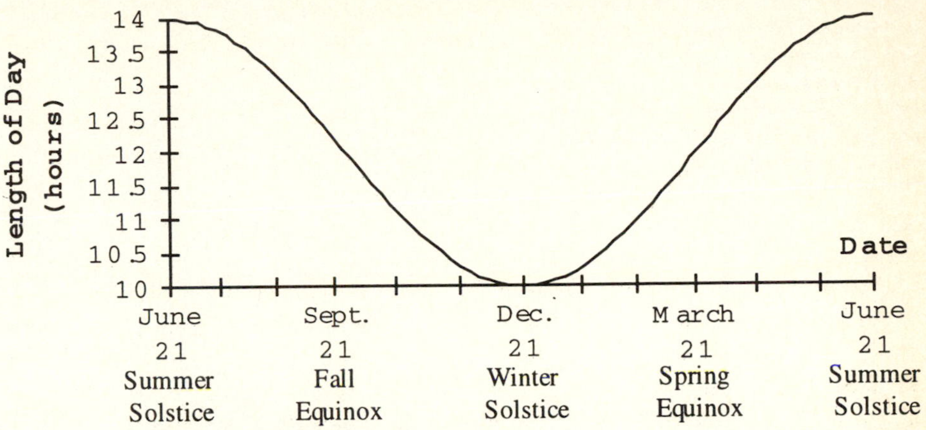

Figure 10-11. The relation that gives the number of hours of daylight for each day of the year (at latitude 40°N) is periodic. Although the graph looks like a continuous curve, it actually consists of 366 individual points: one for each date.

of daylight decreases from 14 hours at the summer solstice to 10 hours at the winter solstice, passing through 12 hours at the fall equinox. It then increases back to 14 hours at the next summer solstice, passing through 12 hours at the spring equinox. The cycle repeats every year, so it is a periodic relation.

The number of hours of daylight has a single value for each individual date, so considering fractions of a date (such as May 4.5 or June 12.7) makes no sense. Thus, like the (*number of towns, number of roads*) relation, the domain consists only of isolated numbers and its graph consists of individual points rather than a continuous curve.

To show the complete pattern of the relation, you must show at least one complete cycle on the graph; that is, choose a domain extending over a full year, beginning on the summer solstice (Figure 10–11). Note that, although the graph consists of 366 individual data points (for the dates of the year), the points are so close together that they *appear* continuous.

❖

10.3 RATE OF CHANGE

Having shown several different kinds of relations, we now focus on one special but important kind of relation: those with straight line graphs. Let's begin with an example.

10.3.1 Slope and Rate of Change

Imagine that you are measuring the rainfall in a storm by monitoring the depth of rain accumulating in a rain gauge. After collecting rainfall data for 6 hours, you plot the graph shown in Figure 10–12. Note that your graph shows a (*time, rain depth*) relation in which *time* is the first variable (on the horizontal axis) and *rain depth* is the second variable (on the vertical axis). Because the graph is a straight line, or **linear graph**, the (*time, rain depth*) relation that it represents is called a **linear relation**.

During the first hour, the rain depth increased from 0 to 1.5 inches. During the second hour, it increased from 1.5 to 3.0 inches. During the third hour, it increased from 3.0 to 4.5 inches. Continuing on, the graph shows that the rain depth increased by the same amount, 1.5 inches, during each hour of the storm. We say that the **rate of change** of rain depth with respect to time is 1.5 inches per hour, or 1.5 in/hr. Note that the units for this rate of change are the units of the second variable divided by the units of the first variable (in this case, inches divided by hours).

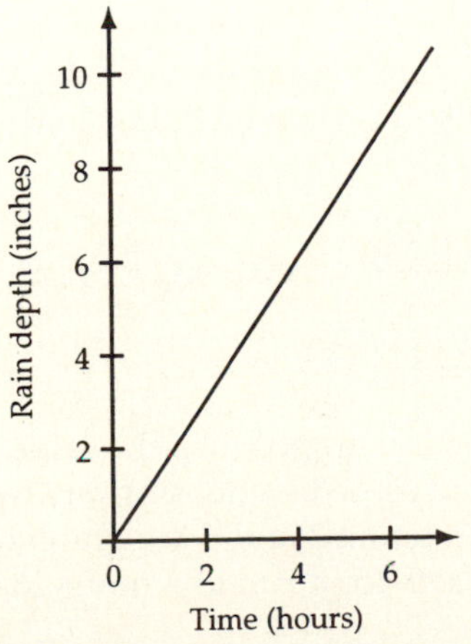

Figure 10-12. The graph of the (*time, rain depth*) relation shows the water depth in a rain gauge during the first 6 hours of a storm. The rain depth rises by 1.5 inches with each passing hour.

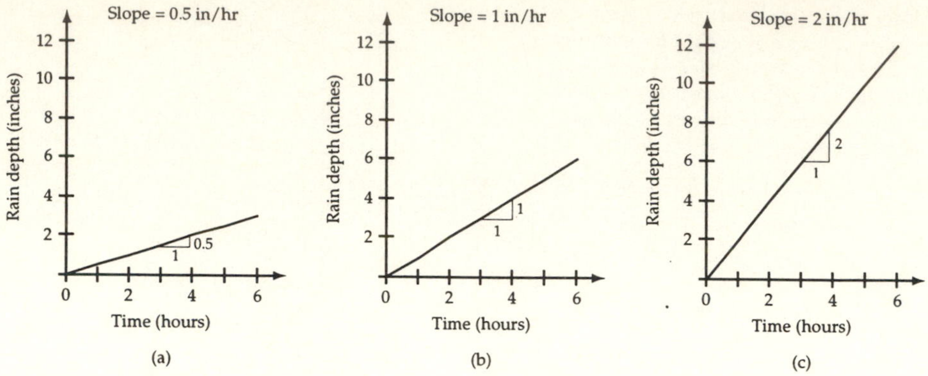

Figure 10-13. The (*time, rain depth*) relation is shown for three storms, along with the slope of each graph. (a) The rain depth rises at a rate of 0.5 inch per hour. (b) The rain depth rises at a rate of 1 inch per hour. (c) The rain depth rises at a rate of 2 inches per hour.

The rate of change measures how quickly the second variable changes with respect to the first variable.

Figure 10–13 shows linear graphs for three other rain storms. In Figure 10–13(a), the rain depth increased from 0 to 0.5 inches during the first hour, from 0.5 inch to 1.0 inch during the second hour, from 1.0 inch to 1.5 inches during the third hour, and so on. That is, the rain depth increased by 0.5 inch during each hour of the storm, so the *rate of change* of rain depth with respect to time was 0.5 $^{in}/_{hr}$ for this storm. Similarly, Figure 10–13(b) shows a storm in which the rate of change of rain depth with respect to time was 1 $^{in}/_{hr}$. For the storm graphed in Figure 10–13(c), the rate of change of rain depth with respect to time was 2 $^{in}/_{hr}$. Comparing the three graphs leads to a crucial observation:

The greater the rate of change, the steeper is the graph.

The rate of change of a relation is the slope of its graph.

To be more precise, the rate of change of a relation is the **slope** of its graph. The term *slope* is used as in ordinary language, when we talk about the slope of a hill or a mountain trail. That is, the slope of a graph describes its vertical *rise* for each unit that it *runs* horizontally. Because the rise of the graph represents the change in the second variable and the run represents the change in the first variable,

$$\text{slope of graph} = \frac{\text{rise}}{\text{run}} = \frac{\text{change in second variable}}{\text{change in first variable}} = \text{rate of change of relation}\,.$$

For example, between hours 3 and 4 the graph in Figure 10–13(a) shows a rise of 0.5 inches in the second variable, *rain depth*, for a run of 1 hour in the first variable, *time*. Therefore its slope is

$$\text{slope} = \frac{\text{change in second variable}}{\text{change in first variable}} = \frac{0.5\text{ in}}{\underbrace{1\text{ hr}}_{\substack{\text{changes between}\\ \text{hours 3 and 4}}}} = 0.5\ ^{in}/_{hr}\,.$$

Linear relations have a constant rate of change and their graphs are straight lines with constant slope.

Moreover, because the slope of a straight line is constant (i.e., the same everywhere on the line), you would find the same slope by choosing *any* two points on the linear graph. For example, between hours 2 and 6 the graph in Figure 10–13(a) shows a rise of 2 inches in *rain depth* for a run of 4 hours in *time*, yielding the same slope found previously:

$$\text{slope} = \frac{\text{change in second variable}}{\text{change in first variable}} = \frac{2\text{ in}}{\underbrace{4\text{ hr}}_{\substack{\text{change between}\\ \text{hours 2 and 6}}}} = 0.5\ ^{in}/_{hr}\,.$$

In Figure 10–13(b), the rain depth rises 1 inch for each 1 hour that time runs along the horizontal axis, so its slope is

$$\text{slope} = \frac{\text{change in second variable}}{\text{change in first variable}} = \frac{1 \text{ in}}{1 \text{ hr}} = 1 \ {}^{\text{in}}\!/_{\text{hr}}.$$

Similarly, in Figure 10–13(c), the rain depth rises 2 inches for each 1 hour that time runs along the horizontal axis, so its slope is

$$\text{slope} = \frac{\text{change in second variable}}{\text{change in first variable}} = \frac{2 \text{ in}}{1 \text{ hr}} = 2 \ {}^{\text{in}}\!/_{\text{hr}}.$$

Time-Out to Think: Choose *any* two points on the straight-line graph of Figure 10-12. By measuring the change in the first and second variables between the two points, verify that the slope of the line is the rate of change found earlier, or 1.5 inches per hour. Also verify that you obtain the same slope using *any* two points on the line, and explain why we therefore say that the slope is constant.

10.3.2 Interpreting a Linear Graph

Let's use what you've learned about slope and rate of change to interpret linear graphs. Figure 10–14 shows the graph of a linear relation between price and demand for pineapples (the number of pineapples sold on a particular day) at a small store. Note that the graph shows *price* as the first variable and *demand* (or *number of pineapples sold*) as the second variable; that is, the *demand* depends on the *price* in the (*price, demand*) relation. As shown, the domain covers prices between \$0 and \$10, and the range extends from 0 to 100 pineapples sold per day.

The graph shows that demand decreases as price increases. That is, fewer people buy pineapples when the price is high. The slope, which is the rate of change in the (*price, demand*) relation, can be calculated by choosing *any* two points on the graph. For example, demand falls from 80 to 70 pineapples as the price increases from \$2 to \$3, so

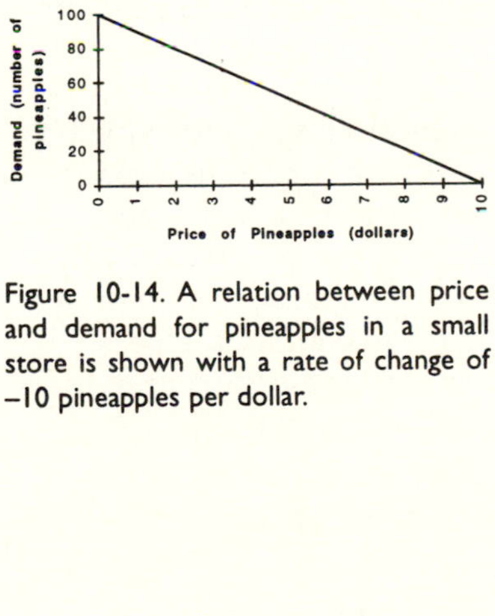

Figure 10-14. A relation between price and demand for pineapples in a small store is shown with a rate of change of –10 pineapples per dollar.

$$\text{slope} = \frac{\text{change in second variable}}{\text{change in first variable}} = \frac{80 - 70 \text{ pineapples}}{\$2 - \$3} = -10 \frac{\text{pineapples}}{\$}.$$

The slope of the graph, or rate of change of the relation, is –10 pineapples per dollar. That is, for every \$1 increase in the price, 10 fewer pineapples are sold. The slope is negative because (*price, demand*) is a decreasing relation; that is, its graph falls from left to right.

Example 10–4 Use the graph shown in Figure 10–14 to predict how demand would change if the store owners increased the price of pineapples from \$3 to \$5. Then use the graph to predict the change in demand if the price falls from \$2 to \$0. Comment on the validity of the graph.

Solution: According to the graph, a change in the price from \$3 to \$5 would decrease demand from 70 to 50 pineapples. A decrease in price from \$2 to \$0 would increase demand from 80 to 100 pineapples.

Remember that the graph represents a *model* of the true (*price, demand*) relation. The store cannot have sales data for every possible price of pineapple, so the model probably is based on data collected on a few selected days. The model therefore may be reasonably accurate for typical prices of pineapple, so the predicted change in demand with a price change from \$3 to \$5 may be reasonably accurate. However, the model appears to fail for the price change from \$2 to \$0: If the store can *sell* 80 pineapples at a price of \$2, it probably could give away many more than 100 pineapples if they were free!

❖

10.3.3 Drawing a Linear Graph

Interpreting graphs is important, but you may also need to create them. Imagine that you set out hiking from a trailhead at an elevation of 8000 feet and that the trail gains elevation at a constant rate of 650 feet per mile for its 8-mile length. Let's draw a graph that shows how the elevation changes with respect to distance walked.

Because elevation varies with respect to distance, *distance* is the first variable and *elevation* is the second variable. The domain is all distances between 0 and 8 miles because that is the trail's length. The key words *constant rate* indicate that the relation is linear, so the constant rate of change 650 feet per mile is the slope of the graph.

There are two basic ways to construct the linear graph. The first way is to find and connect two points that satisfy the relation. For example, when the distance is 0 (at the trailhead), the elevation is 8000 feet; therefore one point that satisfies the relation is (0 mi, 8000 ft). After a mile, the elevation has increased by 650 feet to 8650 feet; therefore a second point that satisfies the relation is (1 mi, 8650 ft). Plotting these two points and connecting them with a straight line generates the graph (Figure 10–15a).

The second basic way to construct a linear graph is to start with just one point and draw a line with the required slope. In Figure 10–15(b), we begin with the point (0 mi, 8000 ft) and then draw a line that rises 650 feet in the vertical direction for every mile that it runs in the horizontal direction. Although the two graphing procedures are slightly different, both yield the same linear graph.

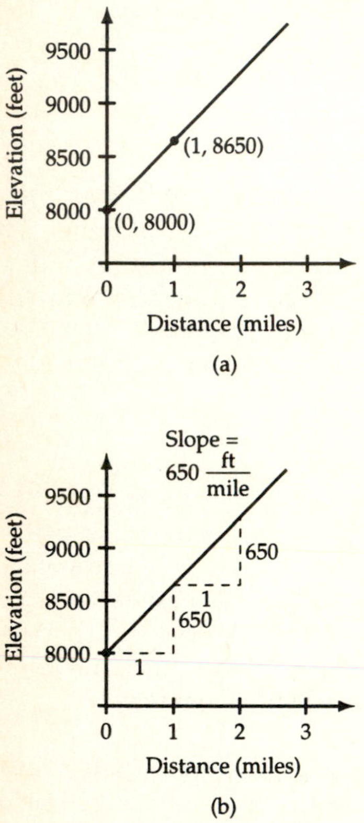

Figure 10-15. The two basic ways to draw the graph of a linear relation are shown for the hiking example. (a) Identify two points that satisfy the relation and connect them with straight line. (b) Start at one point and draw a line with the required slope.

Example 10–5 ***Heavy Loads.*** Suppose that you drive a freight truck and that one of your standard routes has a long uphill section. Over the years you notice that, with no freight on board, you can drive up the hill at a maximum speed of 45 miles per hour (mph). If the truck is loaded with 5 tons of cargo, you can drive up the hill at 30 miles per hour. Based on these two observations, graph a linear relation that describes how the maximum truck speed varies with the cargo weight. Use this relation to predict the maximum truck speed with a 10-ton cargo and discuss the validity of this prediction.

Solution: Choose *weight* (measured in tons) as the first variable and *maximum speed* (measured in mph) as the second variable in the (*weight, maximum speed*) relation.

Method 1: Your observations give you two points: (0 tons, 45 mph) and (5 tons, 30 mph). Plot these two points and connect them with a straight line (Figure 10–16a).

Method 2: The required slope can be determined using the two points that satisfy the relation.

$$\text{Slope} = \frac{\text{change in second variable}}{\text{change in first variable}} = \frac{(45-30)\text{ mph}}{(0-5)\text{ tons}} = -\frac{15\text{ mph}}{5\text{ tons}} = -3\frac{\text{mph}}{\text{ton}}.$$

As a rate of change, this means that for every ton of freight, the maximum speed *decreases* by 3 mph. As shown in Figure 10–16(b), the graph of the relation can be drawn by choosing one point that satisfies the relation, say (0 tons, 45 mph), and drawing through it a line with slope –3 mph/ton.

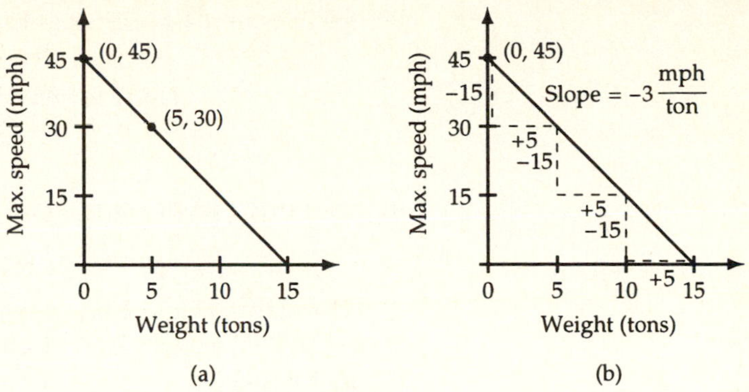

Figure 10-16. The relation between freight weight and maximum speed can be graphed by (a) plotting two points and connecting them with a straight line, or (b) plotting one point and drawing a line through it with the correct slope.

Reading from the graph, the model predicts that with a load of 10 tons, the maximum truck speed is 15 mph. As always, the graph represents a model of the true (*weight, maximum speed*) relation. Because of the many factors that play a role in determining the truck's maximum speed (e.g., friction with the road, air resistance, engine power), the true true relation is unlikely to be linear. Nevertheless, the linear model probably provides a reasonable *estimate* of the top speed with a 10-ton cargo.

❖

Summary Box 10-2. Graphs, Relations, and Rates of Change

- **A relation between two quantities may be represented by a graph in a coordinate plane. The first variable is plotted on the horizontal axis and the second variable on the vertical axis.**
- **The *slope* of the graph (at any point) is the *rate of change* of the second variable with respect to the first variable.**
- **The units of the rate of change (or slope) are the units of the second variable divided by the units of the first variable.**
- **If the relation is linear, the rate of change is *constant* and the graph is a straight line.**
- **The graph of a linear relation may be drawn by either**
  (i) drawing a line through any two points that satisfy the relation, or
  (ii) drawing a line with the required slope through any point that satisfies the relation.

10.3.4 The Rate of Change Rule

Consider again the (*time, rain depth*) relation shown in Figure 10–12, in which the rate of change in the rain depth with respect to time is 1.5 in/hr. We therefore could say that the *change in rain depth* over a 1-hour period is 1.5 inches and the change in rain depth over a 2-hour period is 3.0 inches. Note that the latter result requires a simple calculation:

$$\text{change in rain depth} = \underbrace{1.5\frac{\text{in}}{\text{hr}}}_{\text{rate of change}} \times \underbrace{2\text{ hr}}_{\substack{\text{elapsed}\\ \text{time}}} = 3\text{ in}.$$

Similarly, over a 4-hour period

$$\text{change in rain depth} = \underbrace{1.5\frac{\text{in}}{\text{hr}}}_{\text{rate of change}} \times \underbrace{4\text{ hr}}_{\substack{\text{elapsed}\\ \text{time}}} = 6\text{ in.}$$

In fact, during any time interval of the storm,

$$\text{change in rain depth} = \text{rate of change} \times \text{elapsed time.}$$

The elapsed time is the change in the first variable, *time*. Thus we have a general rule that connects the change in the second variable to the change in the first variable.

Rate of Change Rule for Linear Relations

Change in second variable = rate of change × change in first variable

Example 10–6 Look back at Figure 10–14, which shows the graph of the relation between price and demand for pineapples at a small store. Use the rate of change rule to predict the change in demand for pineapples if the price increases by $3.

Solution: The first variable is the *price* of the pineapples, and the second variable is the *demand* (number of pineapples sold). As found earlier, the rate of change in demand with respect to price is –10 pineapples per dollar. The rate of change rule for a price increase of $3 gives

$$\begin{aligned}\text{change in demand} &= \text{rate of change} \times \text{change in price}\\ &= -10\frac{\text{pineapples}}{\$}\times \$3\\ &= -30\text{ pineapples.}\end{aligned}$$

If the price increases by $3, the demand changes by –30 pineapples; that is, it decreases by 30 pineapples. For example, increasing the price from $4 to $7 reduces demand from 60 to 30 pineapples. Of course, the validity of this result depends on the validity of the linear model, discussed in Example 10–4.

❖

Example 10–7 ***Below Sea Level.*** Consider a submarine that descends from sea level at a constant rate of 60 meters per minute. What is the elevation of the submarine after 12 minutes? Assume that the elevation is 0 at sea level, positive above sea level, and negative below sea level.

Solution: *Time* (in minutes) is the first variable and *elevation* (in meters) is the second variable. The submarine is *descending*, so the rate of change of elevation with respect to time is –60 meters per minute. Applying the rate of change rule with an elapsed time of 12 minutes yields

$$\begin{aligned}\text{change in elevation} &= \text{rate of change} \times \text{elapsed time}\\ &= -60\frac{\text{m}}{\text{min}}\times 12\text{ min}\\ &= -720\text{ m.}\end{aligned}$$

The answer, –720 meters, indicates that the submarine is 720 meters *below* the surface after 12 minutes.

❖

10.4 LINEAR EQUATIONS

We have discussed three of the four ways to represent a relation between two variables: in words, with a data table, and with pictures or graphs. Although these representations are useful and informative, the fourth method, the use of equations, is even more general and powerful. We focus here on writing equations for linear relations, which we then use to demonstrate important principles of modeling with *any* kind of relation.

10.4.1 General Form of a Linear Equation

Suppose that your job is to oversee the operation of an automated assembly line that manufactures computer chips. You arrive at work one day to find a stock of 25 chips that were produced during the night. If you work an 8-hour shift and chips are produced at the constant rate of 4 chips per hour, how large is the stock of chips at any particular time during your shift?

Answering this question requires finding a relation that describes how the number of chips depends on the time of day. Our goal in this section is to describe this relation with an equation. First, however, let's create a table and a graph. We begin by identifying *time*, measured in hours, as the first variable and *number of chips* as the second variable in the (*time, number of chips*) relation. At the start of your shift *time* = 0, and your initial stock is the 25 chips produced during the night. One hour into your shift you will have those 25 chips plus the four produced during the first hour, or 25 + 4 = 29 chips. After two hours, you will have the initial stock of 25 chips plus the 8 chips produced during the first 2 hours, or 25 + 8 = 33 chips. Continuing to calculate the number of chips at the end of each hour gives the entries listed in Table 10–3.

Table 10-3. Stock of Chips Over the Course of the Workday

Time (since arrival at work)	Stock of Chips	Time (since arrival at work)	Stock of Chips
0	25	5	45
1	29	6	49
2	33	7	53
3	37	8	57
4	41		

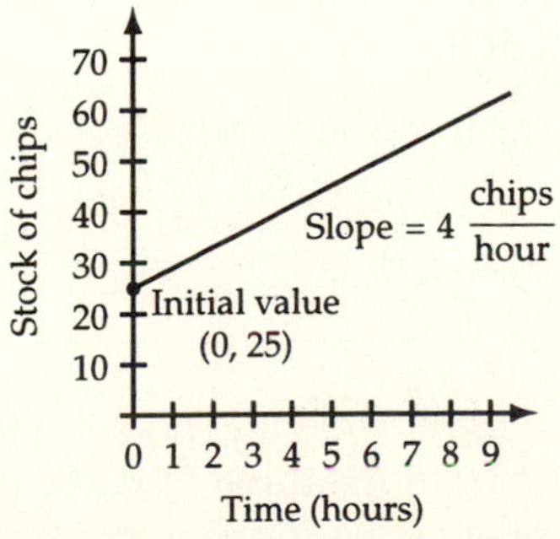

Figure 10-17. The graph shows the (*time, stock of chips*) relation over the course of an 8-hour day. The initial stock was 25 chips and chips are produced at a rate of 4 chips per hour.

Because the stock grows by 4 chips every hour, the rate of change in the (*time, number of chips*) relation is 4 chips per hour. Because this rate of change is constant, the relation is linear. To construct a graph, we start at the point (0 hr, 25 chips), representing the number of chips at the start of your shift, and draw a straight line with a slope of 4 chips per hour (Figure 10–17).

We are now ready to write an equation describing the relation. The 25 chips produced before your shift represent your initial number of chips. The number of chips produced since the start of your shift represents a *change* in the number of chips. Thus the stock of chips at any particular time during your shift is

$$\text{number of chips} = \text{initial number of chips} + \text{change in number of chips}.$$

From the rate of change rule, the change in the number of chips is

$$\underbrace{\text{change in number of chips}}_{\text{change in second variable}} = \text{rate of change} \times \underbrace{\text{elapsed time}}_{\text{change in first variable}}.$$

Substituting the rate of change of 4 chips per hour, we get

change in number of chips = $4 \frac{\text{chips}}{\text{hr}} \times \text{elapsed time}$.

Using this result and 25 for the initial number of chips, the equation for the number of chips becomes

number of chips = 25 chips + $(4 \frac{\text{chips}}{\text{hr}} \times \text{elapsed time})$.

We may write this equation more compactly by using letters to stand for the variables. Choosing t for the time and N for the number of chips, the equation becomes

$$N = 25 \text{ chips} + 4 \frac{\text{chips}}{\text{hr}} \times t.$$

This equation provides a compact **formula** that gives the number of chips, N, at any time, t. For example, after 3.5 hours, the stock of chips is

$$N = 25 \text{ chips} + 4 \frac{\text{chips}}{\text{hr}} \times 3.5 \text{ hr} = 39 \text{ chips}.$$

Note that this quantity doesn't appear in the table and would be difficult to read from the graph. Yet calculating it from the equation is easy.

Time-Out to Think: Verify that the equation *does* reproduce the numbers of Table 10-3 by substituting a few values of t into the equation and calculating the corresponding values of N. What is the stock of chips after 5.5 hours?

To generalize from this example to any linear relation, note that:

- The number of chips, N, is the second variable;
- the time, t, is the first variable;
- your initial stock of 25 chips represents the initial value of the second variable, N; and
- the term $4\frac{\text{chips}}{\text{hr}}$ is the **rate of change** in the second variable, N, with respect to the first variable, t, so $4\frac{\text{chips}}{\text{hr}} \times t$ represents the *change* in the value of the second variable, N.

Making these substitutions in the formula for the number of chips yields a general way to write equations for linear relations.

General Linear Equation

second variable = initial value + change in second variable

or

second variable = initial value + (rate of change \times first variable).

Example 10–8 A loaf of bread is taken out of a 350°F oven and allowed to cool at room temperature. Suppose that the interior temperature of the bread decreases at a constant rate of 4°F per minute. Find an equation that gives the bread temperature after the bread leaves the oven. Graph and interpret the equation.

Solution: The temperature changes with respect to time, so the first variable is *time*, t, and the second variable is *temperature*, T. The constant rate of change of the bread temperature is -4° F per minute; it is negative because the temperature

In other books or courses you may have seen equations written without units. For example, the linear equation for computer chips might appear as N = 25 + 4t. We use units in equations because they help to remind you of the meaning of numbers and symbols. In addition, as we have shown in previous chapters, they also can help you to prevent errors.

The initial value of the second variable is always the value of the second variable when the first variable equals 0.

is decreasing. The initial temperature of the bread is the oven temperature, 350°F. The linear equation that describes the temperature at any time is

$$\text{temperature} = \text{initial value} + (\text{rate of change} \times \text{time}),$$

or

$$T = \underbrace{350°\text{F}}_{\text{Initial value}} + \underbrace{\left(-4\frac{°\text{F}}{\text{min}}\right)}_{\text{rate of change}} \times t.$$

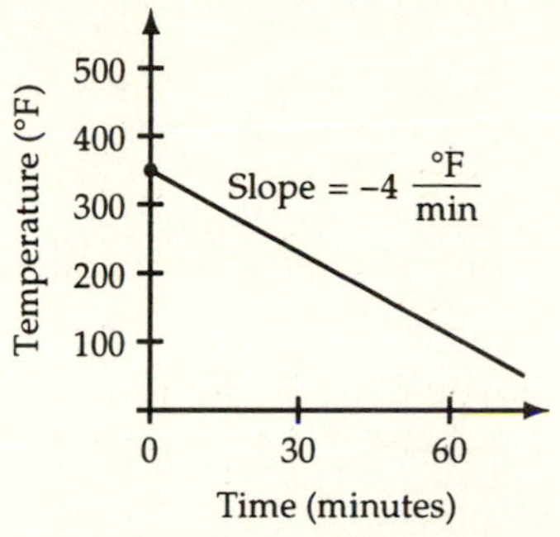

Figure 10-18. The graph of the linear relation for bread temperature begins at the initial temperature of 350°F and decreases at a rate of 4°F per minute.

Make a graph by starting at the point when the bread leaves the oven (0 minutes, 350°F) and drawing a line with a slope of –4°F per minute (Figure 10–18).

The equation allows you to find the bread temperature at various times. For example, 20 minutes after the bread leaves the oven (t = 20 minutes), its temperature is

$$T = 350°\text{F} + \left(-4\frac{°\text{F}}{\text{min}}\right) \times (20 \text{ min}) = 350°\text{F} - 80°\text{F} = 270°\text{F}.$$

Similarly, after 1 hour (t = 60 minutes), the bread temperature is 110°F — just about cool enough to eat! As always, this equation represents a model of how bread really cools. This model seems reasonable for the first hour or so, as the bread cools enough to eat, but it can't be valid for all times. For example, the equation suggests that the bread temperature continues to decrease indefinitely. In fact, the bread temperature will level off as it approaches room temperature. ❖

10.4.2 Solving Linear Equations

Let's continue working with the (*time, number of chips*) equation, or

$$N = 25 \text{ chips} + 4\,\frac{\text{chips}}{\text{hr}} \times t.$$

Suppose that you want to know when your stock will reach 43 chips. That is, you want to find the value of the first variable, *time*, when the second variable, *number of chips*, is 43. Setting $N = 43$, the equation becomes

$$43 \text{ chips} = 25 \text{ chips} + 4\frac{\text{chips}}{\text{hr}} \times t.$$

Finding the time at which you will have 43 chips requires *solving* this equation for the time, t. You can do so as follows.

Solving linear equations may appear simpler without using units. In this case, the solution would be written as

$25 + 4t = 43$
$4t = 18$ (subtracting 25)
$t = 4.5$ (dividing by 4).

However, as noted previously, using units can help you to avoid mistakes.

starting equation: $$43 \text{ chips} = 25 \text{ chips} + 4\frac{\text{chips}}{\text{hr}} \times t$$

subtract 25 chips from both sides: $$43 \text{ chips} - 25 \text{ chips} = 25 \text{ chips} + 4\frac{\text{chips}}{\text{hr}} \times t - 25 \text{ chips}$$

simplify equation: $$18 \text{ chips} = 4\frac{\text{chips}}{\text{hr}} \times t$$

divide both sides by $4\,\frac{\text{chips}}{\text{hr}}$: $$\frac{18 \text{ chips}}{4\,\frac{\text{chips}}{\text{hr}}} = \frac{4\,\frac{\text{chips}}{\text{hr}} \times t}{4\,\frac{\text{chips}}{\text{hr}}}$$

simplify equation: $$\frac{18}{4} \text{ hr} = t, \quad \text{or} \quad t = 4.5 \text{ hr}.$$

Thus your stock will be 43 chips after 4.5 hours.

Time-Out to Think: Check the preceding answer by substituting $t = 4.5$ hr into the starting equation. Verify that the stock is, indeed, 43 chips after 4.5 hours.

Following a similar procedure, you may solve *any* linear equation for the first variable in terms of the second by using just two essential rules from algebra.

A more detailed review of these and other algebraic rules is provided in Appendix B.

- You can add or subtract the same quantity on both sides of an equation without changing the meaning of the equation.
- You can multiply or divide both sides of an equation by the same quantity (provided that it is not zero) without changing the meaning of the equation.

Example 10–9 Use the relation for the cooling time of bread (Example 10–8) to determine when (after it leaves the oven) the bread temperature is 120°F.

Solution: When you set the temperature to $T = 120°\text{F}$ in the relation from Example 10–8, the equation becomes

$$120°\text{F} = 350°\text{F} + \left(-4\frac{°\text{F}}{\text{min}}\right) \times t.$$

Now solve for the time, t.

subtract 350°F from both sides:
$$120°\text{F} - 350°\text{F} = 350°\text{F} + \left(-4\frac{°\text{F}}{\text{min}} \times t\right) - 350°\text{F}$$

simplify equation:
$$-230°\text{F} = -4\frac{°\text{F}}{\text{min}} \times t$$

divide both sides by $-4\frac{°\text{F}}{\text{min}}$:
$$\frac{-230°\text{F}}{-4\frac{°\text{F}}{\text{min}}} = \frac{-4\frac{°\text{F}}{\text{min}}}{-4\frac{°\text{F}}{\text{min}}} \times t$$

simplify equation:
$$\frac{230}{4}\ \text{min} = t, \quad \text{or} \quad t = 57.5\,\text{min}.$$

According to the model represented by this equation, the bread will reach a temperature of 120°F when it has been out of the oven 57.5 minutes.

❖

10.4.3 The Equation of a Line (x's and y's)

The general linear equation is also called the **equation of a line**. In many mathematics books, the first variable is denoted x and the second variable is denoted y. In that case, the general linear equation takes the form:

$$y = \text{initial value of } y + (\text{slope} \times x).$$

In addition, it is customary to let b stand for the initial value and m stand for the rate of change (or slope). With these substitutions, the equation of the line becomes

$$y = mx + b.$$

The symbol b also is called the ***y*-intercept**, because the straight-line graph passes through the y-axis at a value of b; that is, the point $(0, b)$ lies on the

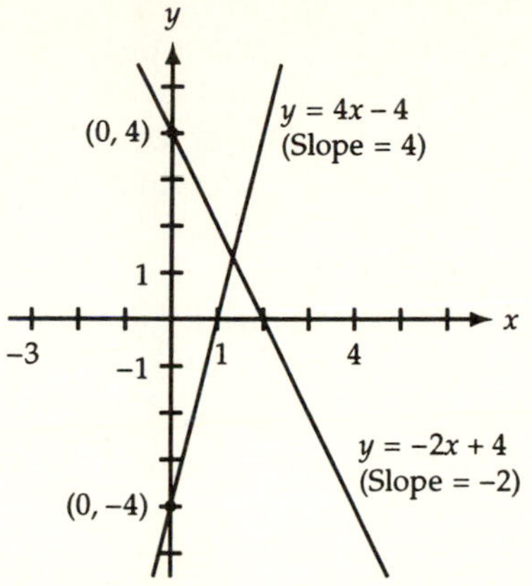

Figure 10-19. The line $y = 4x - 4$ has a slope of 4 and a y-intercept at -4; the line $y = -2x + 4$ has a slope of -2 and a y-intercept at 4.

graph. For example, the equation $y = 4x - 4$ represents a straight line with a slope of 4 and a y-intercept at the point $(0, -4)$. Similarly, $y = -2x + 4$ is a straight line with a slope of -2 that passes through the point $(0, 4)$. Both examples are graphed in Figure 10–19.

Writing $y = mx + b$ provides a compact version of the general linear equation for use in studying properties of linear equations. However, remember that only x and y represent variables; m and b represent constant values (called **coefficients**) in any relation.

Figure 10–20(a) shows several lines, all passing through the same point, with different slopes. Note that when the slope is positive ($m > 0$), the line rises to the right. When the slope is negative ($m < 0$), the line falls to the right. If the slope m is zero ($m = 0$), the line is horizontal. That is, changing the slope of a line changes its orientation on a graph.

Figure 10–20(b) shows the effect of changing the y-intercept on a set of lines that have the same slope. All the lines rise at the same rate, but cut the *y-axis* at different points. That is, changing the y-intercept shifts a line vertically.

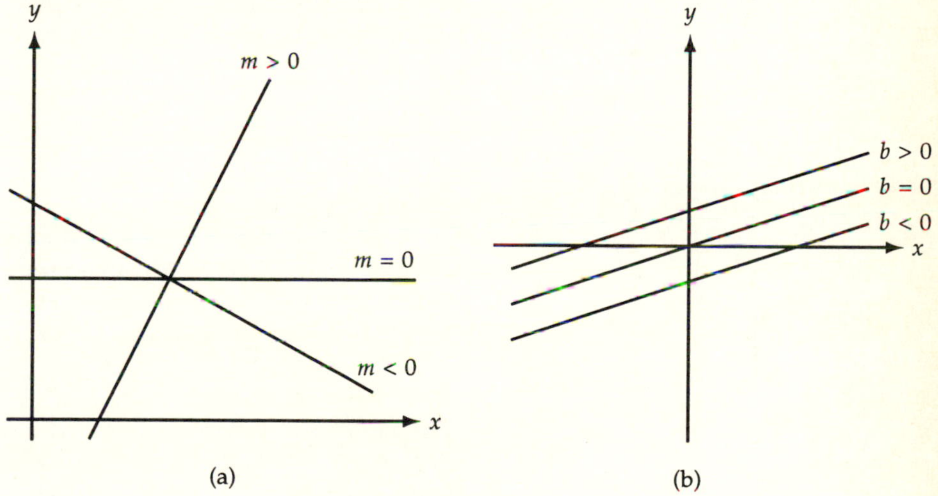

Figure 10-20. (a) When several lines pass through the same point, changing the slope changes the orientation of the line: A positive slope rises to the right, a negative slope falls to the right, and a zero slope is flat. (b) When several lines have the same slope, changing the y-intercept shifts the line vertically.

THINKING ABOUT ... *THE BAGHDAD CONNECTION*

Algebra, which we use to solve equations in this chapter, was born in the city of Baghdad (located in modern-day Iraq). After the fall of Rome in the fifth century, European civilization entered the period known as the Dark Ages. Much of the great scholarly work of Greece and Rome already was lost. But much more would have been lost, if not for the emergence of a new center of intellectual achievement in Baghdad. During the eighth and ninth centuries, in particular, the Muslim empire centered in Baghdad translated and thereby saved many of the ancient Greek works.

Jews, Christians, and Muslims in Baghdad worked together in scholarly pursuits during this period. An Islamic leader, or *caliph*, named Al-Mamun established a "House of Wisdom" comparable in scope to the ancient Library of Alexandria. Many of the methods of algebra were developed here.

The word *algebra* itself comes from the title of a book written by the mathematician and astronomer Muhammad ibn Musa al-Khwarizmi, a faculty member in Al-Manun's House of Wisdom who died sometime before A.D. 850. Al-Khwarizmi wrote several books on astronomy and mathematics including one titled *Hisab al-jabr wal-muqabala*, which means "the science of reduction and confrontation" or "the science of equations." Within the Arabic title, you can see the origin of the word *algebra* (*al-jabr*).

Interestingly, Al-Khwarizmi's name also became a common English word. Another of his achievements was a treatise on arithmetic, in which he described the number system developed by Hindu mathematicians. Although he did not claim credit for the Hindu work, later writers often attributed such credit to him, which probably explains why modern numerals are known as *Hindu-Arabic* rather than solely as Hindu. Some later authors even attributed the numerals to Al-Khwarizmi personally and, in a sloppy writing of his name, the use of Hindu numerals became known as *algorismi*, and later as the words *algorism* or *algorithm*.

10.5 CREATING LINEAR MODELS FROM DATA

In this section we discuss methods for creating linear models designed to fit a set of data. First we discuss the simple case of finding a linear equation that fits two data points, which always is possible. Then we introduce the concept of finding a line that best fits a set of more than two data points, which is a crucial concept in statistics.

10.5.1 A Linear Equation from Two Points

When you are given a set of data, the process of creating a linear model involves finding a straight line that "fits" the data. In the simplest case, in which you have only two data points, you always can find a line that passes through both of them. Graphically, you can do so simply by plotting the two points and connecting them with a straight line (with the aid of a ruler). Alternatively, you can find a linear equation that fits the two data points in the form

$$y = \text{initial value of } y + \text{slope} \times x.$$

For example, suppose that you want to find the equation for a linear relation that includes the two data points (1, 4) and (3, 8). A simple way to find this equation entails finding the slope and the initial value of y (or the y-intercept) of the line that passes through the two points (Figure 10–21a). You can find the slope by fnding the change in the variables between the two data points and then dividing:

$$\text{slope} = \frac{\text{change in second variable}}{\text{change in first variable}} = \frac{8-4}{3-1} = \frac{4}{2} = 2.$$

Substituting this value of the slope, the general equation of the line becomes

$$y = \text{initial value of } y + 2x.$$

To complete the equation, you need to determine the initial value of y. You can do so by substituting either of the two data points into the preceding equation; let's use the first point (1, 4).

set $x = 1$ and $y = 4$:	$4 = \text{initial value of } y + 2 \times 1$
subtract $2 \times 1 = 2$ from both sides:	$4 - 2 = \text{initial value of } y + 2 - 2$
simplify equation:	$2 = \text{initial value of } y.$

Knowing both the slope and the initial value of y, you can write the complete equation for a line passing through the points (1, 4) and (3, 8) as

$$y = 2x + 2.$$

The graph of the equation is shown in Figure 10–21.

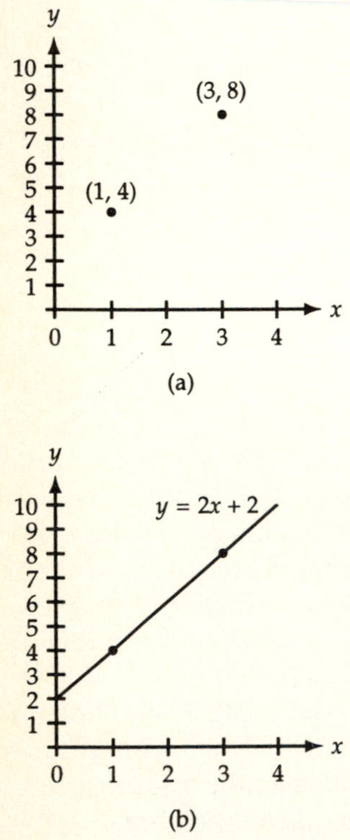

Figure 10-21. (a) You are looking for a line that passes through the points (1, 4) and (3, 8). (b) The linear relation that fits the two points has a slope of 2 and an initial value (y-intercept) of 2; its equation is $y = 2x + 2$.

Time-Out to Think: Substitute the second point (3, 8) for the first. Do you get the same value for the initial value of y? Do both data points satisfy the equation? Explain.

Oil facts (as of 1994): Of the total U.S. energy needs (about 80 quadrillion BTUs per year), 40% are met by oil, 22% by coal, 22% by natural gas, and 7% by nuclear reactors. The United States imports 40% of the oil it uses. The total world production of oil increased from 21 million barrels per day in 1960 to 60 million barrels per day in 1990. The estimated world reserve of crude oil is 1 trillion barrels.

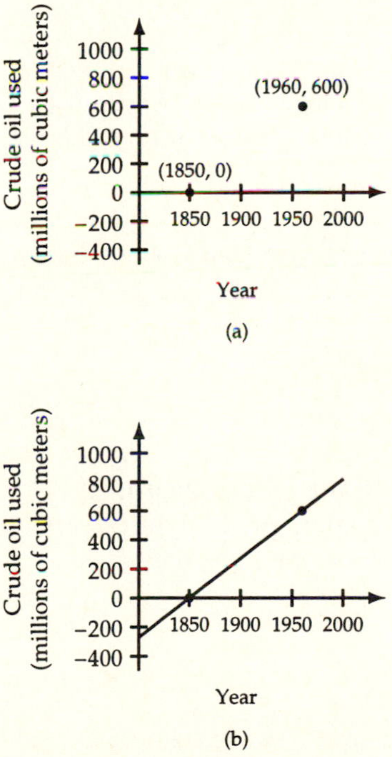

Figure 10-22. (a) Two data points are shown for the (*time, total crude oil used*) relation. (b) The graph shows a linear model that fits the data.

Example 10–10 *Oil Reserves.* Suppose that in 1850 the world had used none of its crude oil reserves but that by 1960 it had used a total of 6×10^{11} cubic meters of oil. Create a linear model tracing world oil use over time based on these two data points. Discuss the validity of the model.

Solution: The problem involves a relation (*time, total crude oil used*). Two data points are provided. The first is the point (1850, 0 m^3) indicating that no oil was used by 1850; the second is the point (1960, 6×10^{11} m^3), which represents the amount of oil used by 1960. To create a linear model, you must find a straight line that passes through these two data points (Figure 10–22a). Letting T stand for the second variable, *total crude oil used*, and t for the first variable, *time*, the linear equation should have the general form

$$T = \text{initial value of } T + \text{slope} \times t.$$

Find the slope by dividing the change in the variables between the two data points:

$$\text{slope} = \frac{\text{change in second variable}}{\text{change in first variable}} = \frac{(6\times10^{11} - 0)\ \text{m}^3}{(1960-1850)\ \text{yr}} = \frac{6\times10^{11}\ \text{m}^3}{110\ \text{yr}} = 5.45\times10^9\ \frac{\text{m}^3}{\text{yr}}.$$

Substituting this value of the slope, the equation of the line becomes

$$T = \text{initial value of } T + \left(5.45\times10^9\ \frac{\text{m}^3}{\text{yr}}\right)\times t.$$

You may substitute either data point to find the initial value of T. Using the first point, $t = 1850$ and $T = 0$, yields

$$0 = \text{initial value of } T + \left(5.45\times10^9\ \frac{\text{m}^3}{\text{yr}}\right)\times 1850\ \text{yr},$$

or

$$0 = \text{initial value of } T + 1.00825\times10^{13}\ \text{m}^3.$$

Solving for the initial value of T gives

$$\text{initial value of } T = -1.00825\times10^{13}\ \text{m}^3.$$

Now the linear equation becomes

$$T = -1.00825\times10^{13}\ \text{m}^3 + \left(5.45\times10^9\ \frac{\text{m}^3}{\text{yr}}\right)\times t.$$

However, because the data given for the oil use had only one significant digit, you should include only one significant digit in the final linear model, so it becomes

$$T = -1\times10^{13}\ \text{m}^3 + \left(5\times10^9\ \frac{\text{m}^3}{\text{yr}}\right)\times t.$$

This linear model is graphed in Figure 10–22(b); note that it does, indeed, fit the two given data points. To interpret the model, first note that it gives negative values for the total amount of oil used, T, prior to 1850. This result clearly is nonsense; because the problem assumed zero oil consumption prior to 1850, the domain should include only the years *after* 1850. Next, consider the validity of the linear model within this domain. For any other data points, such as the amount of oil used by 1900 or 1925, would the line still provide a good fit to the data? In fact, the rate of oil consumption has increased over time due to increasing population and increasing per capita use of oil. This fact implies that the rate of change in the amount of oil used over time has not remained constant, but rather has increased; therefore the true (*time, total crude oil used*) relation isn't linear. Nevertheless, the linear model provides a useful first approximation to the actual relation for oil use.

❖

Example 10–11 Use the linear model in the preceding example to predict the total amount of oil that will have been used by 2050. Do you think the prediction of this model will be accurate?

Solution: Substitute 2050 into the equation as the value for the time, t, and compute the total amount of oil used, T. To deal properly with uncertainty, work with the *unrounded* numbers first, then round the answer after you have finished:

$$T = -1.00825 \times 10^{13}\ \text{m}^3 + \left(5.45 \times 10^9\ \frac{\text{m}^3}{\text{yr}}\right) \times 2050 = 1.09 \times 10^{12}\ \text{m}^3 \approx 1 \times 10^{12}\ \text{m}^3.$$

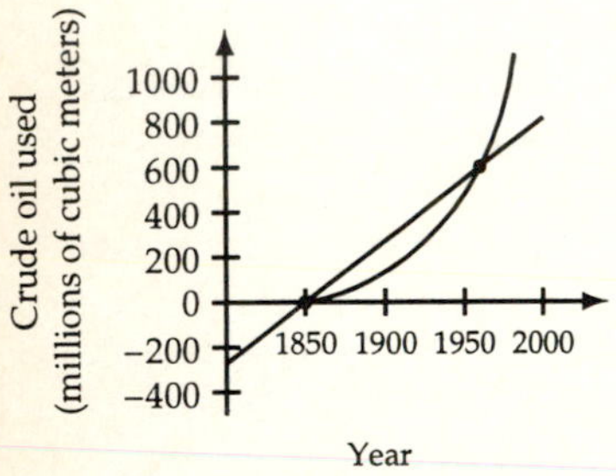

Figure 10-23. The rate of change in the amount of oil used over time is not constant; instead, it has increased with time as shown by the curve. Thus because the linear model was "forced" to fit the data points for 1850 and 1960, it overestimates oil use before 1960 and underestimates it after 1960.

According to the linear model, about 1 trillion cubic meters of oil will have been used by 2050. As described in Example 10–10, however, the rate of change in the amount of oil use has increased over time. In that case, the real shape of the relation probably looks more like the curve shown in Figure 10–23. Note that the linear model overestimates oil use before 1960 and underestimates it thereafter. Thus 1 trillion cubic meters probably is an underestimate of the true amount of oil that will have been used by 2050.

❖

10.5.2 Linear Regression Models

If more than two data points are given, finding a single straight line that passes through all of them usually is impossible. However, you can still create a linear model that represents a "best fit" to the data.

Table 10–4 shows changes over several decades in the men's and women's world records for the mile run. The two data sets (men and women) are plotted in Figure 10–24. In addition, we have drawn a straight line to represent the **best linear fit**, or **linear regression** line, for each data set. If you take a course in statistics, you will learn specific methods for finding linear regression lines; here, we recommend drawing them "by eye." We also discuss linear regression lines further in Chapter 14.

As expected, the data points do not fall exactly on the linear regression lines for the two sets of data. However, they are close, so the linear regression lines represent reasonable models for the changes in the world records. Because the linear models fit the data so well, you might use them to make predictions about world records in the future. The men's and women's lines intersect in the year 2038. That is, according to these models, the women's

Table 10-4. Women's and Men's World Records in the Mile Run: Selected Records (Minutes:Seconds)
Data courtesy of Hal Bateman, USA Track and Field

Date	Women's Record	Date	Men's Record
1967	4:37	1942	4:06
1969	4:36	1945	4:01
1971	4:35	1954	3:59
1973	4:29	1958	3:55
1979	4:22	1964	3:54
1981	4:21	1975	3:51
1985	4:17	1981	3:48
1989	4:15	1993	3:44

and men's mile records will be equal in 2038, and the women's record will be faster in the years beyond. Is this a plausible conclusion?

Clearly, the models must eventually break down: If you extend both lines even farther into the future, they make the implausible prediction that the mile records eventually become zero! As with many linear models, these have only a limited range of validity. Where do they stop being valid? That's difficult to say; close examination of the data suggests that the linear model for the men's record may already be breaking down in the 1990s. However, the women's record may continue to decrease linearly for several years to come. Will it continue to decrease beyond the men's record? Only time will tell.

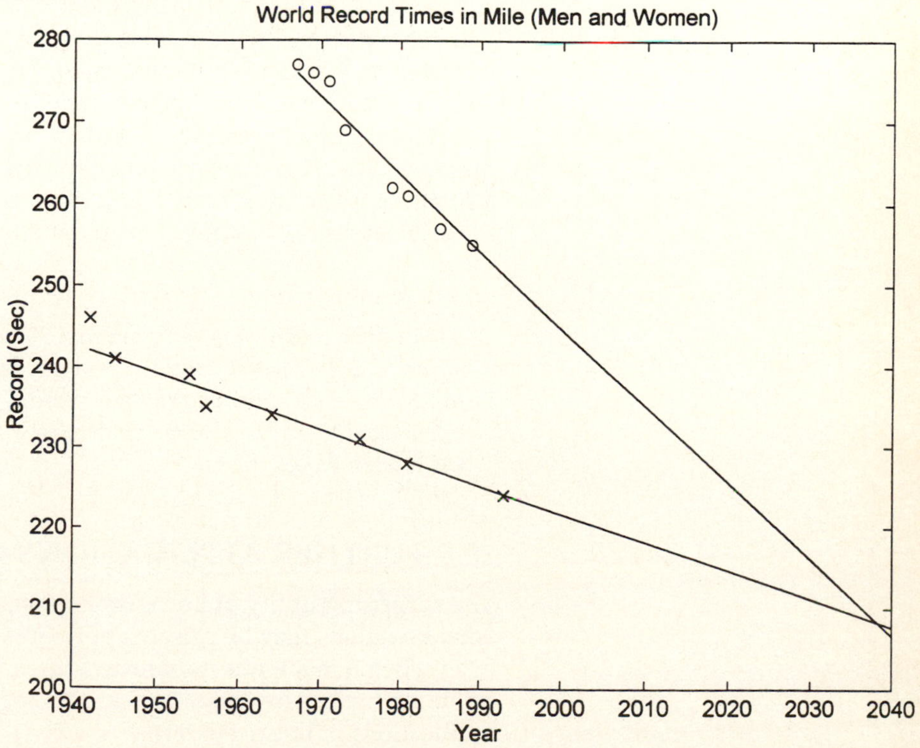

Figure 10-24. Data points show the women's and men's world records in the mile run as it has changed over several decades. The best linear fit to each data set also is shown.

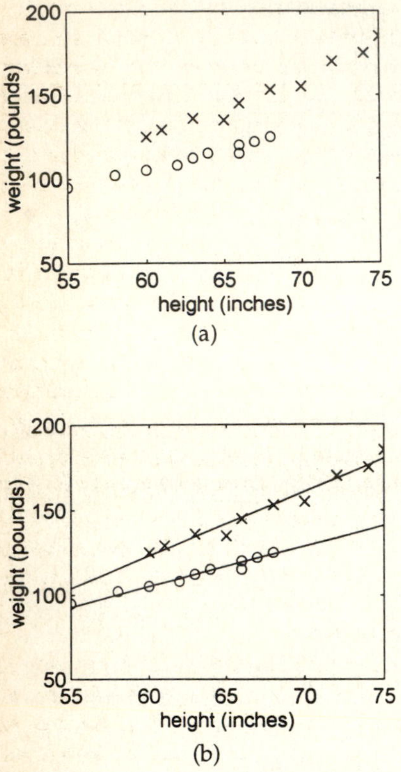

Figure 10-25. (a) Each data point represents the height and weight for 1 of 20 people. (b) The points seem to form two groups, presumably men and women. Best fit lines were drawn "by eye" to model the data.

Example 10–12 *Visual Linear Regression.* Figure 10–25a shows height and weight data for 20 people. Discuss the data, find the best linear fit "by eye," and find the equation for the best linear fit.

Solution: The data seem to cluster in two sets of points. If you were to draw one line that best fits *all* the data, it would fall cleanly between the two sets of points. As this approach doesn't seem very reasonable, your intuition might suggest that the upper set of points represents men in the data set and that the lower set represents women. Therefore draw two lines "by eye," one through each of the two sets of points. Do you agree that the lines shown in Figure 10–25(b) look like best fits to the data?

The lines represent linear models for the (*height*, *weight*) relations for both men and women. To find equations for these models, use one data point and the slope. By measuring the slope with the aid of a ruler, you can find that the slope of the men's line is about 3.8 pounds per inch. Let h stand for height, w for weight, and write the linear equation:

$$w = (3.8\ \tfrac{\text{lb}}{\text{in}} \times h\ ) + \text{initial value of } h.$$

The men's best fit line passes through the point (60 in, 123 lb). Now use these values for (h, w) to find the initial value of h:

$$123\text{ lb} = (3.8\ \tfrac{\text{lb}}{\text{in}} \times\ 60\text{ in.}) + \text{initial value of } h$$

Solving for the initial value of h gives

$$\text{initial value of } h = -105\text{ lb.}$$

Thus the overall equation for the men's best fit line is

$$w = (3.8\ \tfrac{\text{lb}}{\text{in}} \times h) - 105\text{ lb.}$$

For the women's data, the best fit line passes through the point (55 in, 93 lb) and has a slope of approximately 2.4 pounds per inch. Following the same procedure as for the men's line, the equation of this line is

$$w = (2.4\text{ lb/in} \times\ h) - 37\text{ lb.}$$

You can use these linear models to make predictions about people not included in the data set. For example, the upper line predicts that a 6′10″ man (h = 82 in) would weigh about 206 pounds. Are such predictions reliable? The lines fit the data closely, so the linear models are good approximations to the *given* data. However, the data represent only 20 people. Because it is such a small sample of the entire population, you should be cautious about using these linear regression lines for predictions about the population at large. ❖

Time-Out to Think: We stated that 20 people is too small a sample from which to draw general conclusions about heights and weights. How large a sample do you think is needed to draw reasonable conclusions about the entire population? If we had a larger sample, say 100 or 1000 people, do you think linear models would still fit?

10.6 FURTHER APPLICATIONS OF LINEAR MODELS

The surprising utility of linear models stems from two key facts. First, many actual relations may be closely approximated with linear relations. Second, even when a relation is not linear, a linear model may be a good starting point for further investigation. In this section we present several more applications of linear models.

10.6.1 Linear Models in Financial Applications

As shown earlier by the (*price, demand*) relation in Figure 10–14, linear relations often model financial situations well. Indeed, some of the most common and important applications of linear models occur in the worlds of business and finance. The following three examples demonstrate three important financial applications of linear modeling.

Example 10–13 ***Depreciation of Equipment.*** The owner of a printing company depreciates the value of his printing press by 10% of its initial value each year for tax purposes. If the press was purchased for $5000, find the relation that gives the depreciated value of the press each year after its purchase. Do you think that this model reflects the true value?

Solution: The value of the printing press is depreciated annually by 10% of its $5000 purchase price, or by $500 per year. That is, for tax purposes its value is $5000 when purchased, $4500 after 1 year, $4000 after 2 years, and so on. The depreciation therefore is a rate of change: It is constant with a value of –$500 per year (it is negative because the value of the press decreases with time).

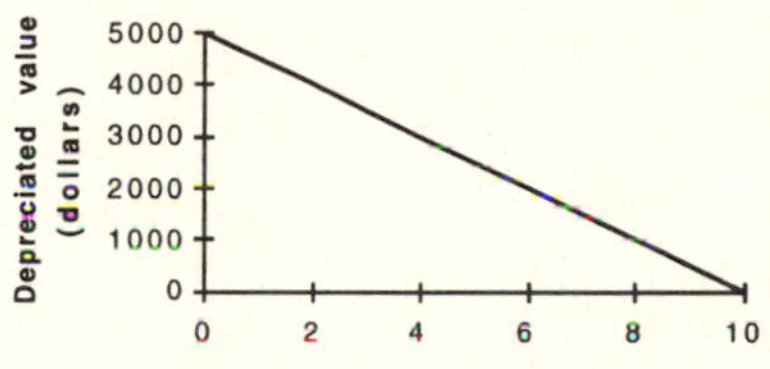

Figure 10-26. The graph shows the depreciation line for a printing press with a purchase price of $5000 that decreases in value at a rate of $500 per year.

Identify the first variable as *time*, denoted t, and the second variable as *depreciated value*, denoted V. Calling the time at which the press was purchased $t = 0$ years, the initial value is $V = \$5000$. The rate of change is –$500 per year, so the equation for this relation takes the form:

$$V = \$5000 - (\$500/\text{yr}) \times t.$$

The relation may be graphed by starting at the initial point (0 years, $5000) and drawing a line with a slope of –$500 per year (Figure 10–26).

The depreciation line is a model of the printing press value over time. Does it reflect the *true* value at all times? Not likely! According to this model, the value of the equipment reaches 0 after 10 years. If the printer cares for the press well, it may remain in excellent condition and still be quite valuable after 10 years. But, if it breaks down and can't be fixed, it may be worthless much sooner. Nevertheless, for tax purposes the Internal Revenue Service (IRS) requires business owners to depreciate capital equipment (such as the printing press). With no practical way to determine the actual value of every piece of equipment used by every business, the IRS uses linear models of depreciated value. Thus, for tax purposes, this model provides an *exact* depreciated value of the equipment each year (assuming that it is depreciated on a 10-year schedule).

❖

Time-Out to Think: In using a 10-year depreciation schedule, the IRS assumes that the value of capital equipment like the printing press typically reaches $0 after 10 years. In contrast, the IRS requires owners of rental property to depreciate its value on a 27-year schedule. Why do you think the IRS assumes different depreciation times for capital equipment and rental property? Do you think these depreciation schedules are reasonable?

Example 10–14 ***Fund Raising.*** A common form of fund raising is to invite a well-known speaker and charge an entrance fee for the event. Consider an event in which a prestigious speaker charges $500 for an appearance. Tickets will be sold for $10 per person. How many tickets must be sold for the event to clear $1000?

Solution: This rate of change problem is in a slightly disguised form: You are interested in how the *profit* from the event changes with respect to the *number of tickets* sold. Note that, for every ticket sold, the profit increases by \$10; the rate of change in the profit therefore is \$10 per ticket. Because this rate of change is constant, the relation is linear. Let P stand for profit and n stand for the *number of tickets* sold; then the linear equation will have the form:

$$P = (\text{rate of change} \times n) + \text{initial value of } P.$$

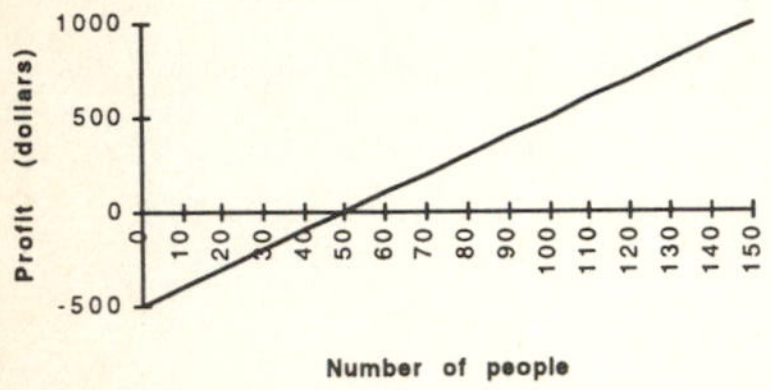

Figure 10-27. The profit from a fund-raiser that charges \$10 per ticket and must pay a speaker \$500 is represented by a linear relation.

You already know the rate of change (\$10 per ticket). The initial value is the profit earned if *no* tickets are sold. In that case, the organization is \$500 in the hole because the speaker must be paid \$500. In other words, the initial value of the profit is –\$500 if no tickets are sold. The complete linear equation is

$$P = (\$10/\text{ticket} \times n) - \$500.$$

The linear graph is shown in Figure 10–27. To find the number of tickets sold (n) that produces a profit of \$1000, substitute P = \$1000 into the equation, and solve for n:

$$\$1000 = \left(\frac{\$10}{\text{ticket}} \times n\right) - \$500.$$

Adding \$500 to both sides gives

$$\$1500 = \frac{\$10}{\text{ticket}} \times n.$$

Dividing both sides by \$10/ticket yields

$$n = \frac{\$1500}{10\frac{\$}{\text{ticket}}} = 150 \text{ tickets}.$$

Selling 150 tickets will generate a profit of \$1000. Always check your result to be sure that it fits the original question: 150 tickets at \$10 each generate \$1500 in revenue; subtracting the \$500 speaker cost yields a profit of \$1000. Thus, the answer *does* make sense. Besides answering the specific question of how many tickets must be sold to generate a \$1000 profit, the linear relation simplifies predicting profits at other sales levels. In fact, the graph shows directly that the break-even point for the event is 50 tickets; if 50 tickets are sold the speaker can be paid but the profit would be \$0.

❖

Example 10–15 *Quantity Discounts.* It is not uncommon for a retailer who sells large-volume items such as computer disks to quote unit prices that vary with the number of units purchased. Let's call p the unit price for disks and n the number of disks purchased. Suppose that the price of disks is listed as

$$p = \text{unit price of disks} = \begin{cases} \$2.00 \text{ if } 0 \le n \le 9 \\ \$1.50 \text{ if } 10 \le n \le 19 \\ \$1.25 \text{ if } n \ge 20. \end{cases}$$

For example, if you buy five disks, the price is \$2.00 per disk; if you buy 19 disks, the price is \$1.50 per disk; and if you buy 21 disks, the price is \$1.25 per disk. Find a relation that gives the total cost of purchasing n disks. How much would 18 disks cost? Does this pricing scheme represent a good strategy by the seller?

Solution: The relation sought is (*number of disks, total cost*), abbreviated (n, C). The basic fact underlying the relation is that the total cost is the number of disks purchased, n, times the unit price, p:

$$\text{Total cost} = \text{price per disk} \times \text{number of disks}, \quad \text{or} \quad C = p \times n.$$

Note that the unit price plays the role of a rate of change with units of dollars per disk. Because the price per disk varies with the quantity, however, the form of the relation also varies with the quantity. This dependence is similar to the pricing brackets:

$$C = \text{total cost of disks} = \begin{cases} \$2.00 \times n \text{ if } 0 \le n \le 9 \\ \$1.50 \times n \text{ if } 10 \le n \le 19 \\ \$1.25 \times n \text{ if } n \ge 20. \end{cases}$$

A relation of this type is said to have a **conditional definition** because its form depends on certain conditions — in this case, the number of disks purchased. According to the relation, the price is \$1.50 per disk if you buy 18 disks, or a total price of 18 × \$1.50 = \$27.00.

Is this type of quantity discount pricing a good strategy for the seller? If it encourages greater sales, and thereby higher profits, the seller is likely to be satisfied. Nevertheless, the strategy produces some results that may not be optimal from a business point of view. For example, the cost of 8 disks is \$16, but the cost of 10 disks is only \$15. Thus for a buyer who wants only 8 disks, the seller effectively is giving away two additional disks (plus a \$1 rebate!). ❖

Time-Out to Think: Can you think of an alternative pricing strategy for the previous example that would preserve the incentive for buyers to buy in bulk, without the disadvantage of effectively giving away disks?

10.6.2 Models with Two Linear Relations

All the models described so far involved using only one relation at a time. Many mathematical models in business, management, science, and engineering consist of more than one relation; in fact, they may involve hundreds or even thousands of relations. Although we won't create such complex models in this text, you should understand the principles behind them. Each day, decisions that affect your life are based on the results of complex mathematical models.

The following two examples offer a glimpse of the principles at work in complex models, but each example involves only two linear relations. In Example 10–16, we show how to create a model for profits in book publishing in which two independent equations can be combined into one. Example 10–17 is a mixing problem that involves solving two equations simultaneously; that is, finding the point at which two straight-line graphs intersect.

Example 10–16 ***Combining Two Equations.*** Suppose that the cost of publishing a particular book is \$1000 up-front (for example, setup costs for printing equipment) plus a unit cost of \$5 per book. You plan to sell the book for \$10 per copy. Find a relation that describes the profit from the book. How many books must you sell to break even?

Solution: Your goal is to determine the profit from the book, so you begin by recognizing that profit is revenue minus cost:

$$\text{Profit} = \text{revenue from sales} - \text{publishing costs}.$$

The information given states that both revenue and cost depend on the number of books sold. That is, they are *each* described by a separate relation. Let R stand for *revenue*, C stand for *cost*, and n stand for the *number of books sold*.

To find the relation (n, R) between the number of books sold and the revenue, begin with the initial value of the revenue, which is \$0: if no books are sold, there is no revenue. The rate of change in revenue is \$10 per book. Therefore the linear relation that gives the revenue as it varies with number of books sold is

$$R = (\$10/\text{book}) \times n.$$

For the relation (n, C) between the number of books sold and the cost, the initial value of the cost is \$1000, which is the up-front cost that must be paid before any books are printed. The cost then rises by \$5 per book printed, and the relation is

$$C = (\$5/\text{book} \times n) + \$1000.$$

Note that simplifying this equation makes use of the distributive property, which is reviewed in Appendix B.

Combining the relations for the revenue and cost gives the profit (denoted P):

$$P = R - C = \left[\frac{\$10}{\text{book}} \times n\right] - \left[\left(\frac{\$5}{\text{book}} \times n\right) + \$1000\right] = \left(\frac{\$10}{\text{book}} - \frac{\$5}{\text{book}}\right) \times n - \$1000$$

$$\text{or:} \quad P = \frac{\$5}{\text{book}} \times n - \$1000$$

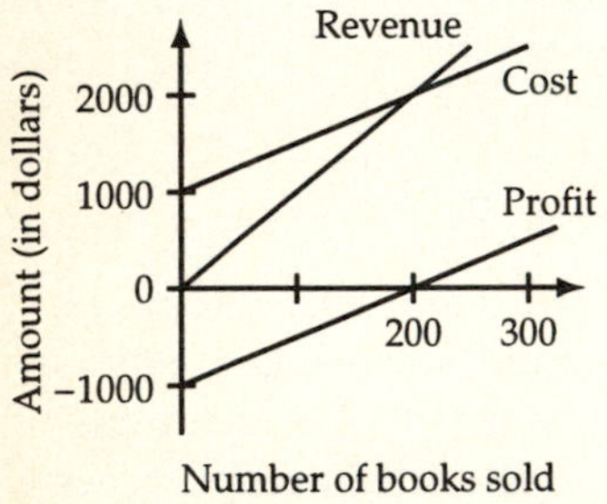

Figure 10-28. The revenue and cost associated with a publishing project are shown as straight-line graphs. The profit, which is the difference between revenue and cost, also is shown.

All three relations — revenue, cost, and profit each with respect to the number of books sold — are graphed in Figure 10–28. You may use these relations to answer many questions about the publishing project. For example, the break-even point for the project is a profit of \$0. Setting $P = 0$ in the profit relation and solving for n yields

$$P = 0 = \left(\frac{\$5}{\text{book}} \times n\right) - \$1000, \quad \text{or} \quad n = 200 \text{ books}.$$

The break-even point for the project is 200 books. The rate of change in the profit relation is \$5 per book, so each additional book sold will generate \$5 of profit. For example, selling 300 books (100 books more than the break-even point) generates a profit of \$500.

As always, you must also consider the validity of the model. It involves many simplifications from reality. For example, the model is based on the assumption that all the books printed are sold. In reality, books are printed in large lots and then put in stores for sale; if not all the books are sold, the model will have overestimated the profit. Nevertheless, this model is a good starting point for estimating profits. Book publishers often start with models like this one and then refine them to account for additional factors such as unsold books, marketing costs, and shipping costs.

❖

Example 10–17 ***Simultaneous Equations.*** Whole wheat flour contains 16 grams of protein per cup and white flour contains 12 grams of protein per cup. Suppose that you want a 4-cup mixture with 60 grams of protein. How much whole wheat and white flour should be mixed together?

Solution: The two variables in this problem are the amount of white flour (denoted w) and the amount of whole wheat flour (denoted h). The problem requires satisfying two conditions at the same time:

1. The total amount of flour in the final mixture must be 4 cups.
2. The total amount of protein in the final mixture must be 60 grams.

The first condition simply means that $w + h = 4$ cups. Subtracting w from both sides gives the linear relation (w, h):

$$h = 4 \text{ cups} - w.$$

To express the second condition as an equation use the information given that each cup of white flour contains 12 grams of protein and that each cup of wheat flour contains 16 grams of protein. The total amount of protein from each is its protein content multiplied by its amount of flour in the final mixture:

$$\text{Amount of protein in final mixture from white flour} = 12\,\tfrac{\text{g}}{\text{cup}} \times w, \text{ and}$$

$$\text{amount of protein in final mixture from wheat flour} = 16\,\tfrac{\text{g}}{\text{cup}} \times h.$$

The total amount of protein in the mixture, which must be 60 g according to condition (2), is the sum of these amounts:

$$\left(\frac{12\text{ g}}{\text{cup}} \times w\right) + \left(\frac{16\text{ g}}{\text{cup}} \times h\right) = 60\text{ g}.$$

Solving for h yields a linear relation that describes condition (2):

subtract ($12\,\tfrac{\text{g}}{\text{cup}} \times w$) from both sides: $$\left(\frac{16\text{ g}}{\text{cup}} \times h\right) = 60\text{ g} - \left(\frac{12\text{ g}}{\text{cup}} \times w\right).$$

divide both sides by $16\,\tfrac{\text{g}}{\text{cup}}$: $$h = \frac{15}{4}\text{cup} - \frac{3}{4}w.$$

equate to first relation $h = 4$ cups $- w$ $$\underbrace{4\text{ cup} - w}_{\text{first relation for } h} = \underbrace{\frac{15}{4}\text{ cup} - \frac{3}{4}w}_{\text{second relation for } h}.$$

rearrange equation: $$w - \frac{3}{4}w = 4\text{ cup} - \frac{15}{4}\text{ cup}$$

simplify equation: $$\frac{1}{4}w = \frac{1}{4}\text{ cup}$$

multiply both sides by 4: $$w = 1\text{ cup}$$

The two conditions are met if the amount of white flour is 1 cup. Because the total amount of flour in the final mixture is 4 cups, the amount of wheat flour must be 3 cups. Check your work by noting that 1 cup of white flour has 12 grams of protein and that 3 cups of wheat flour has $3 \times 16 = 48$ grams of protein, for a total of $12 + 48 = 60$ grams of protein — as required. Thus a mixture of 1 cup white flour and 3 cups wheat flour yields 4 cups with a total protein content of 60 grams.

You may also solve this problem graphically. Figure 10–29 shows the two relations obtained for (w, h). The solution is the point (1 cup, 3 cups) at which the two lines intersect. In general, the solution to two simultaneous equations is the *intersection* point of their graphs.

❖

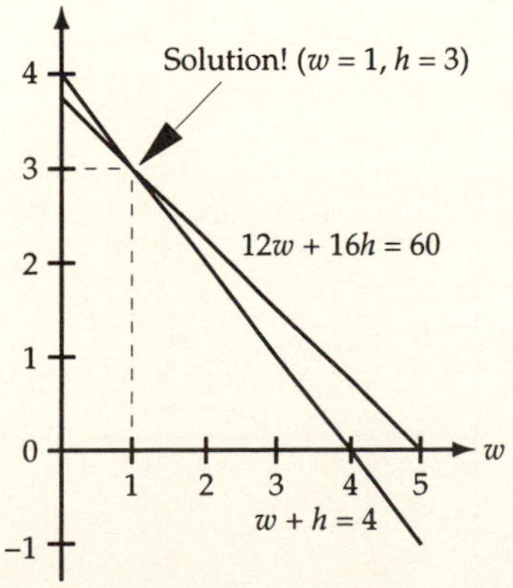

Figure 10-29. The mixing problem involves a model of two relations having the same variables w and h. The solution to the problem is the intersection point of the two lines.

Time-Out to Think: Unless you are a baker, the previous example may seem removed from everyday life. But mixing problems in general are very common. Think of a few analogous problems that involve mixing things other than two types of flour, such as two types of investments, two prices of tickets, and so on.

10.7* CHANGING RATES OF CHANGE — VISUAL CALCULUS

Suppose that you drive 400 miles in 8 hours. Your average speed for the trip is 400 miles ÷ 8 hours = 50 miles per hour. Speed is a rate of change — the rate of change of *distance* traveled with respect to *time*. If you hold a steady speed of 50 miles per hour throughout the trip, the rate of change is constant. A linear relation, with a slope of 50 miles per hour, would describe the distance traveled.

However, you probably don't travel *exactly* 50 miles per hour for much of the trip. Instead, you may drive faster or slower at times or stop for a break. At any particular instant your speed might be anywhere from 0 (a stop) to 70 miles per hour or more. Because your speed is changing, and speed is the rate of change in distance traveled, the rate of change *itself* changes! In the car, you can determine your speed at each instant by watching the speedometer. But how do you deal mathematically with a changing rate of change?

10.7.1 Rates of Change from Graphs

The easiest way to deal with changing rates of change is visually, using graphs. Consider the graphs of two possible 400-mile trips shown in Figure 10–30. The first variable, on the horizontal axis, is *time* since the start of your trip. The second variable, on the vertical axis, is your *distance* from home. Figure 10–30(a) represents a trip during which you hold a steady speed of 50 miles per hour. The dotted lines show that after one hour ($t = 1$ hour) you have traveled 50 miles, as expected. After 4 hours you have traveled 200 miles, and the entire 400-mile trip takes eight hours.

Time-Out to Think: Suppose that you traveled at a faster but still constant speed of 70 miles per hour. What would the graph look like then? Would the slope of the line be steeper, the same, or less steep? Why?

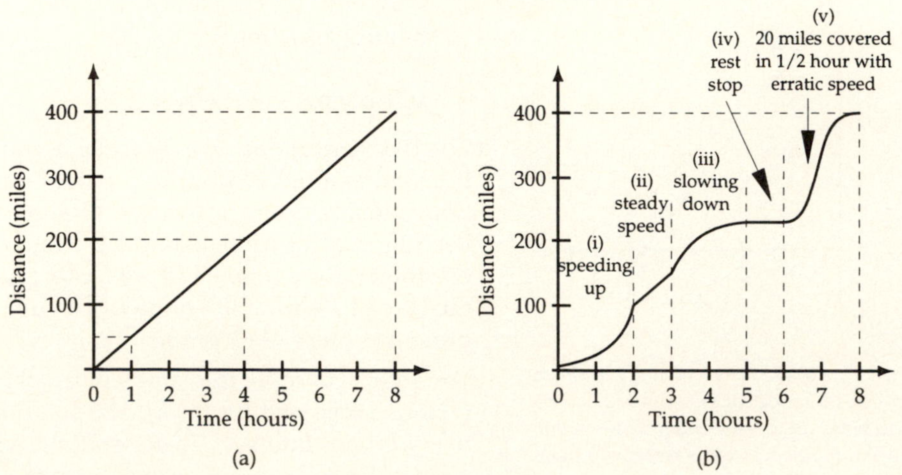

Figure 10-30. The graphs show two possible trips of 400 miles in 8 hours: (a) at a constant speed of 50 miles per hour; and (b) at varying speeds.

Figure 10–30(b) also shows a 400-mile trip that takes 8 hours, but it is far more realistic. To interpret this trip, let's look at the labeled regions.

- In *region i* the graph curves upward; that is, the slope of the curve is *increasing*. The slope is your speed, so it is increasing during this part of the trip.
- In *region ii* the graph is linear; the slope, or your speed, is constant.
- In *region iii* the graph curves downward; the slope decreases with time, so your speed is decreasing.
- In *region iv* the graph is linear, but flat; your distance from home is not changing, so you must be at a rest stop.
- In *region v*, your speed varies quite a bit; at first it increases, then holds steady, and finally decreases as you come to a stop at the end of your trip.

In summary, the speed is constant when the graph is linear, it is increasing when the graph curves upward, and it is decreasing when the graph curves downward. We can generalize these three principles for any relation.

1. If the rate of change is constant, the graph is linear.
2. If the rate of change is increasing, the graph curves upward.
3. If the rate of change is decreasing, the graph curves downward.

10.7.2 Measuring Slope with Tangent Lines

Another example of a changing rate of change is the graph of rain depth measured in a gauge during a 10-hour storm shown in Figure 10–31(a). Note that this graph is *not* linear: Its slope changes throughout the 10-hour period. The slope is the rate of change in the rain depth, so this graph represents a storm in which the rate of change in the rain depth is not constant. Can you use this graph to estimate the rate of change in the rain depth at various times during the storm?

Figure 10–31(b) shows **tangent lines**, or lines that just touch the curve, drawn at points corresponding to three particular times during the storm: $t = 1$ hr, $t = 3$ hr, and $t = 8$ hr. Each tangent line indicates the slope of the graph at a *specific point*; that is, at a specific instant in time. Figure 10–31(c) shows the measured slopes of the three tangent lines. The slope of the tangent line at $t = 1$ hour is 1 in/hr; thus, 1 hour into the storm, the rate of change in the rain depth was 1 in/hr. The slope of the tangent line at $t = 3$ hours is 0, which means that 3 hours into the storm the rate of change in the rain depth was 0

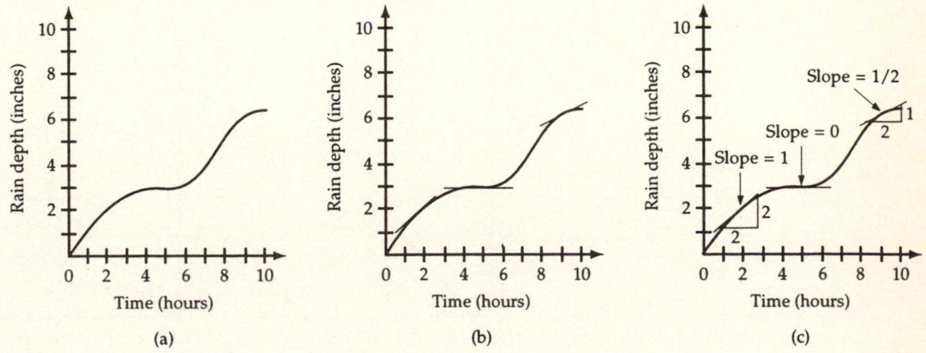

Figure 10-31. (a) The graph shows rain depth accumulation during a 10-hour storm. (b) Tangent lines just touch the curve at a single point of interest. Tangent lines are shown for points 1 hour, 3 hours, and 8 hours after the storm begins. (c) The slope of each tangent line gives the rate of change in rain depth at that specific time.

— the rain must have stopped at that time. The slope of the tangent line at $t = 8$ hours is 0.5 in/hr, so the rate of change in the rain depth 8 hours into the storm was 0.5 in/hr.

The rate of change at any particular point on a graph is the slope of a tangent line to the graph at that point.

Generalizing for any relation, the rate of change at any particular point on a graph is the slope of a line tangent to the graph at that point.

10.7.3 Finding Tangent Lines

One way to draw tangent lines on a graph is simply "by eye." However, this method introduces some uncertainty. For example, Figure 10–31 shows a tangent line with a slope of 0.5 in/hr at $t = 8$ hours. However, it is difficult to say whether the tangent line is precise; perhaps the slope of the true tangent line is 0.55 in/hr or 0.48 in/hr or some other value close to 0.5 in/hr.

Fortunately, a mathematical technique allows much more precision in drawing tangent lines. Recall that the slope of a *straight line* is simply the change in the second variable divided by the change in the first variable over some interval on the graph:

$$\text{Slope} = \text{rate of change} = \frac{\text{change in second var iable}}{\text{change in first variable}}.$$

We can extend this definition to find slopes on nonlinear graphs. Figure 10–32(a) shows the graph of a relation (x, y) in which the slope changes from point to point. How can you measure the slope at the point where $x = 3$?

You might begin by finding the **average slope** in some interval around the point corresponding to $x = 3$. For example, you can connect the points on the graph where $x = 2$ and $x = 4$ with a straight line (Figure 10–32b). The slope of this line is the average slope of the graph in the interval from $x = 2$ to $x = 4$.

You can make a better approximation to the slope at $x = 3$ by measuring the average slope in a smaller interval, such as from $x = 2.5$ and $x = 3.5$ (Figure 10–32c). Even better approximations would come from the average slopes of lines connecting the points where $x = 2.9$ and $x = 3.1$, or $x = 2.99$ and $x = 3.01$. If you continued the process of connecting points that are closer and closer to $x = 3$, eventually you would end up with a line that is **tangent** to the graph (Figure 10–32d). The slope of this tangent line is exactly the slope of the graph at $x = 3$. This example illustrates the general principle of finding a tangent line at any particular point on any graph: Simply find the average slope of the graph over smaller and smaller intervals around the point of interest.

Time-Out to Think: Visually estimate the slopes of each of the lines in Figure 10-32. What pattern to you see? In your own words, explain why the tangent line in Figure 10-32(d) is the actual slope at $x = 3$.

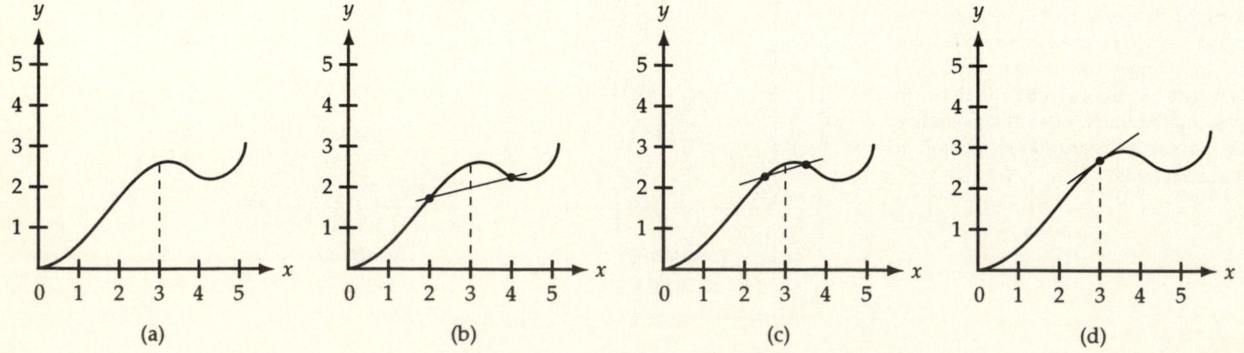

Figure 10-32. (a) The problem is to measure the slope at the point where $x = 3$ on this nonlinear graph. (b) A first approximation to the exact slope is the average slope between $x = 2$ and $x = 4$. (c) A better approximation comes from the average slope between $x = 2.5$ and $x = 3.5$. (d) Continuing this approximation process eventually leads to the exact slope of a tangent line at $x = 3$.

10.7.4 Limits

Returning to the example of driving along a highway, how can you calculate your speed at a specific *instant*, as shown by the speedometer in your car? Although observing your speed at any instant is difficult, your average speed over a short period of time is a good approximation.

For example, suppose that you want to know your **instantaneous speed** at $t = 4{:}21.25$ (4 minutes, 21.25 seconds) into the trip. Your average speed between $t = 4{:}00$ (4 minutes) and $t = 5{:}00$ (5 minutes) is probably a good approximation to this instantaneous speed, unless you suddenly stepped on the accelerator or hit the brakes during this 1-minute period.

An even better approximation to the instantaneous speed at $t = 4{:}21.25$ comes from averaging over a period of 1 second: from $t = 4{:}21$ to $t = 4{:}22$. The average speed over a time interval of 0.01 second will be closer still to your instantaneous speed. If you continue this process to its **limit**, in which the average speed is measured over the smallest possible amount of time, you will have your precise instantaneous speed.

Let's briefly examine how the concept of a limit is handled mathematically. The Greek letter capital delta, Δ, is commonly used to denote change. If we let x denote your distance from home, a change in distance is written Δx. If we use t to represent time, the elapsed time (or change in time) is Δt. With this notation, the definition of average speed becomes:

$$\text{average speed during any period of time} = \frac{\text{change in distance}}{\text{change in time}} = \frac{\Delta x}{\Delta t}.$$

Your instantaneous speed can be found by computing the average speed over smaller and smaller time intervals. That is, it is the limit of the average speed as the change in time (Δt) gets smaller and smaller. Mathematically, we express the idea that the change in time (Δt) gets smaller and smaller by saying that it approaches zero, which we write as $\Delta t \to 0$. Using this notation, and abbreviating limit as "lim," we can write your instantaneous speed as

$$\text{instantaneous speed} = \lim_{\Delta t \to 0} \frac{\Delta x}{\Delta t}.$$

Because $\Delta x/\Delta t$ is the average speed during the time interval Δt, the entire statement reads: "Instantaneous speed is the limit of the average speed as the interval of time shrinks to zero."

Note that the process of finding the instantaneous speed is just like that used earlier to find the tangent line. Thus the slope of a tangent line is the limit of the average slope on the graph near the point of interest. Using the new notation and the definition of slope, we can say that the slope of the tangent line on any graph, whether or not it is linear, is

$$\text{slope of tangent line} = \lim_{\Delta(\text{first variable}) \to 0} \frac{\Delta(\text{second variable})}{\Delta(\text{first variable})}.$$

Called one of greatest intellectual achievements of all times, calculus was invented (or discovered) in the late 1600s by the Englishman Isaac Newton and the German Gottfried Wilhelm Leibniz. These men were working independently and were unaware of the other's work.

The concept of a limit is the cornerstone of the branch of mathematics called **calculus**. Indeed, we may reasonably describe calculus as nothing more than a set of mathematical methods for calculating limits and thereby measuring change. Slopes of tangent lines and rates of change are associated with **derivatives**, which comprise one half of calculus. The other half of calculus deals with questions such as how to find the distance traveled if you know the rate of change (the speed). This process, which is simply the reverse of the process used to find slopes or rates of change, is called **integration**.

Although we don't discuss calculus further in this text, we have already illustrated its essential principles by analyzing changing rates of change on graphs. That is why "visual calculus" is included in the title of this section. If you choose to study calculus in the future, you will find that much of your course simply formalizes the visual description presented here.

10.8 CONCLUSION

In this chapter we demonstrated the use of relations, which allow you to study change mathematically. Key lessons that you should remember from this chapter include the following.

- Change is ever-present. Mathematically, we deal with change through relations, which describe how one variable changes with respect to another. Relations may be described (1) in words; (2) with a data table; (3) with a graph; or (4) in the form of an equation.
- A linear relation has a straight-line graph and a constant rate of change. Although they are only one special type of relation, linear relations illustrate many basic mathematical principles. They also are exceedingly useful because many real phenomena can be closely approximated by linear models.
- Relations (linear or otherwise) can be used to construct models of real situations. These models may be based on measured or estimated data, and certain assumptions. A model is only as good as the data and assumptions on which it is built.

PROBLEMS

Reflection and Review

Section 10.1

1. **Everyday Models.** Describe three different models that you use or encounter frequently in everyday life. What is the underlying "reality" that those models represent? What simplifications are made in constructing those models?

Section 10.2

2. **Related Quantities.** For each of the following relations, write a short statement that expresses a possible relation between the quantities. *Example: (interest rate, monthly payment). Solution:* The monthly payment on a loan may change with respect to the interest rate. If the interest rate increases, the monthly payment also increases. If the interest rate decreases, the monthly payment decreases.
   a. (*time, price of movies*)
   b. (*price of a product, demand for a product*)
   c. (*distance from Earth, strength of gravity*)
   d. (*slope of hill, speed of skateboard*)
   e. (*rate of peddling, speed of bicycle*)
   f. (*tax rate, revenue collected*)
   g. (*number of cars on road, air quality*)
   h. (*price of oil, oil consumption*)

3. **Relations in the News.** In today's newspaper, identify at least three different quantities (variables) that change. For each quantity (a) state what other quantity it changes with respect to; at least one of your examples should involve a variable other than time; and (b) write a paragraph or two that explains the relation in which the quantity is involved.

4. **Points in the Coordinate Plane.** Draw a set of axes in the coordinate plane. Plot and label the points: (0, 1), (–2, 0), (1, 5), (–3, 4), (5, –2), (–6, –3).

5. **Relations from Data Tables.** Each of the following data tables describes a relation. In each case
   (i) Identify the first and second variables.
   (ii) Describe the domain and range.
   (iii) Plot the data in a coordinate plane. Be sure to label all axes clearly and choose units appropriately. To highlight details of the relation "zoom in" as appropriate.
   (iv) State whether filling in the gaps between the data points would make sense. Explain.

a.

Date	Average High Temperature	Date	Average High Temperature
Jan 1	42°F	Jul 1	85°F
Feb 1	38°F	Aug 1	83°F
Mar 1	48°F	Sep 1	80°F
Apr 1	58°F	Oct 1	69°F
May 1	69°F	Nov 1	55°F
Jun 1	76°F	Dec 1	48°F

b.

Year	Tobacco produced (billions of lbs)	Year	Tobacco produced (billions of lbs)
1975	2.2	1986	1.2
1980	1.8	1987	1.2
1982	2.0	1988	1.4
1984	1.7	1989	1.4
1985	1.5	1990	1.6

c.

Year	Projected U.S. Population (millions)	Year	Projected U.S. Population (millions)
1995	258.3	2015	289.0
2000	268.3	2020	294.4
2005	275.6	2025	298.2
2010	282.6	2030	302.0

6. **(*Altitude, Pressure*) Graph.** Study Figure 10–6, which shows a graph of (*altitude, pressure*).
   - **a.** Use the graph to estimate the pressure at altitudes of 5000, 10,000, 15,000, 20,000, and 25,000 feet.
   - **b.** Use the graph to estimate the altitude at which the pressure is 25, 20, and 15 inches of mercury.
   - **c.** Estimating beyond the boundaries of the graph, at what altitude do you think the atmospheric pressure reaches 5 inches of mercury? At what altitude is the pressure zero? Explain your reasoning.
   - **d.** Suppose that the graph shown was based on measured data on a particular day at a particular location. Should you use the same relation for other locations and other days? Why or why not?

7. **(*Distance from Source, River Width*) Graph.** Study Figure 10–7, which shows a graph of the relation (*distance from source, river width*).
   - **a.** Use the graph to estimate the river width at distances of 1000, 2000, and 3000 kilometers from the source.
   - **b.** Use the graph to estimate the distances from the source at which the river width is: 500, 1500, and 1750 meters.
   - **c.** The model used to sketch this graph is based on the assumption that the river width increases smoothly along the river path. How valid is this assumption? Suppose that you actually measured the river width every kilometer along its path and graphed the resulting data. How different would this graph look from the graph shown? Explain.

8. **Types of Relations.** Imagine graphing (but you don't need to actually do it) each of the following relations. For each, do the following.
   - **(i)** Describe the form you think the graph will take. For example, will it be increasing, decreasing, periodic, or have some other form?
   - **(ii)** State whether the domain consists of only individual numbers, or of all numbers within some interval. Explain.
   - **a.** (*time of day, your heart rate*)
   - **b.** (*time of day, elevation of tide*)
   - **c.** (*tax rate, revenue collected by government*)
   - **d.** (*college grade point average, salary of college graduates*)
   - **e.** (*number of hours of study per week, grade point average*)
   - **f.** (*time, storage capacity of computer memory chips*) where time covers all years between 1970 and 2000
   - **g.** (*day of year, average high temperature*)
   - **h.** (*mortgage interest rate, number of homes sold*)
   - **i.** (*population of deer, population of mountain lions*)

9. **Rough Sketches of Relations.** For each of the following relations use your intuition to do the following.
   - **(i)** Describe any restrictions on the domain and choose an appropriate domain for graphing the relation.
   - **(ii)** Describe the range.
   - **(iii)** Make a rough sketch of the relation and explain the assumptions that go into your graph.
   - **(iv)** Briefly discuss the validity of your graph as a model of the true relation between the variables.
   - **a.** (*altitude, temperature*) when climbing a mountain.
   - **b.** (*day of year, average high temperature*) for the town in which you are living over 2-year period
   - **c.** (*blood alcohol level, coordination*)
   - **d.** (*days since last haircut, hair length*)
   - **e.** (*minutes after lighting, length of candle*)
   - **f.** (*level of effort, grade in course*) for your own course work and grades
   - **g.** (*time, per capita consumption of cigarettes*) for the United States over the past 50 years
   - **h.** (*time, world population*) since A.D. 1000
   - **i.** (*time, population of China*) where time is measured in years

10. **Connecting Towns.** Use the formula

$$\text{number of roads} = 1 + 2 + \ldots + (n - 1),$$

where n is the number of towns, to verify the data in Table 10–2. Show your work.

Section 10.3

11. **Analyzing Linear Graphs.** For each graph shown, do the following.
    - **(i)** Explain the relation that is represented by the graph.
    - **(ii)** Find the slope of the graph and express it as a rate of change (be sure to include units).
    - **(iii)** Explain whether you think a linear model is realistic for the particular relation.

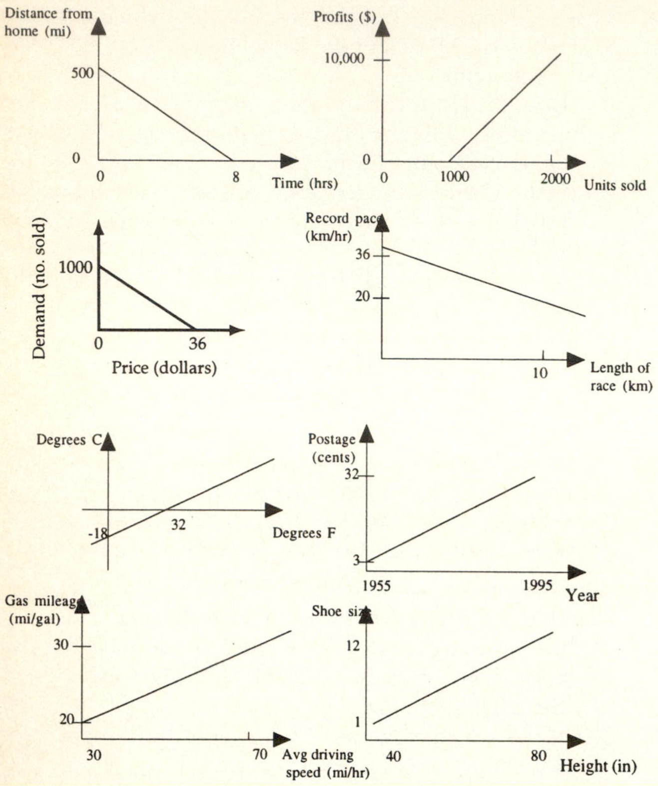

12. Drawing Linear Graphs. For each situation described:

(i) identify the variables involved in the relation;

(ii) identify the domain and range of the relation;

(iii) identify two points that satisfy the relation or one point and the rate of change;

(iv) draw an accurate linear graph of the relation; and

(v) based on your graph, answer any additional questions asked.

a. You drive along a highway at exactly 40 miles per hour for 4 hours. How does your distance traveled vary in time?

b. A birdbath contains 3 inches of water at dawn. As the sun shines, the water evaporates at a rate of 0.25 inch per hour. How does the depth of the water in the birdbath vary in time? Does the water disappear in one day?

c. A certain tree that you are studying increases in diameter by 0.2 inches per year by adding annual rings. When you started observing the tree, its diameter was 4 inches. How does the diameter of the tree vary in time? What is the diameter of the tree after 10 years of observation? Estimate the time at which the tree started growing. Estimate the time at which the tree reaches a diameter of 10 inches.

d. A new candle is 20 centimeters long. Once lit it burns at a rate of 2 centimeters per hour. Find the length of the candle as it varies in time. How long does the candle last?

e. A snowplow has a maximum speed of 30 miles per hour on a dry highway. Its maximum speed decreases by 0.5 mile per hour for every inch of snow depth on the highway. Find the maximum speed of the plow as it varies with snow depth. According to this model, at what snow depth will the plow be unable to move?

f. Beginning in 1960, the population of BoomTown began increasing at a rate of 200 people per year. If the 1960 population was 2000 and the growth rate remains constant, find the population at all times after 1960. What is your projection for the population in the year 2000?

g. The cost of publishing a book has two components: fixed cost (one-time costs for setting up equipment, publicity, and overhead) and the unit cost for each book (paper, ink, and binding). The fixed cost for one printing of a new computer manual is $2000, and the unit cost is $5 per manual. How does the production cost vary with the number of manuals printed? How many manuals must be sold to make a profit?

h. Postage for international airmail is 50 cents per half ounce. How does the postage of a letter vary with its weight? How many ounces can be mailed for $3?

i. In the fermentation process that produces beer, sugar is consumed by a yeast culture to produce carbon dioxide. Assume that a batch of beer originally has 5 grams of sugar that is accessible to the yeast and that the sugar is assimilated at a rate of 0.1 gram per day. Find the amount of sugar in the brew over time. When is the sugar gone?

j. Assume that your heart beats 60 times each minute. How does your total number of heartbeats vary with your age? When did (will) you have your one millionth heartbeat?

k. A semitrailer truck with no cargo can maintain a maximum speed of 50 miles per hour up a steep hill. With 20 tons of cargo, its maximum speed drops to 40 miles per hour. Assume that maximum speed is related linearly to load. How does the maximum speed vary with the load? At what load does the model predict a maximum speed of 0 miles per hour?

l. According to the U.S. Department of Justice, in 1965, 90 percent of all murder cases in America ultimately were "cleared" or solved. In 1990, only 70 percent of all cases were solved. Find a linear relation that relates the percentage of solved cases to the year. According to this model, what percentage of cases will be solved in the year 2010?

13. Rate of Change Rule. Each of the following situations describes a rate of change, which you may assume to be constant. State the rate of change numerically, with units, and answer any questions. Show your work. *Example:* Every week your fingernails grow 5 millimeters. How much will your fingernails grow in 2.5 weeks? *Solution:* At 5 millimeters per week, your fingernails will grow 12.5 millimeters in 2.5 weeks (5 mm/wk × 2.5 wk = 12.5 mm).

a. A gas station owner finds that for every penny increase in the price of gasoline, he sells 150 fewer gallons of gas per week. How much more or less gas will he sell if he raises the price by 8 cents per gallon? If he decreases the price by 2 cents per gallon?

b. Your maximum heart rate, in beats per minute, is 220 minus your age (in years). How much does your maximum heart rate change from age 35 to age 50? What is your maximum heart rate at age 65?

c. After a steady rain depth, water is 50 centimeters deep in a rain barrel. As you watch the water depth over the next few days you notice that it decreases by 5 centimeters per day. When will the water depth be 30 centimeters? Assuming that no more rain falls, when will the rain barrel be empty?

Section 10.4

14. Linear Equations. For each of the situations described:

(i) identify the first and second variables;

(ii) identify the initial value of the second variable and the rate of change;

(iii) write a linear equation to represent the value of the second variable in the relation; and

(iv) answer any questions posed.

a. Suppose that the average price of a new car increases at a constant rate of $1200 per year. If the average price today is $12,000, how much will a new car cost in 2.5 years?

b. The price of a megabyte of computer memory is decreasing linearly at a rate of $5 per year. In January 1995, a megabyte of memory cost $40. What price is predicted for July 1998?

c. Assume that the world record time in the 100-meter butterfly (swimming) has been decreasing at a constant average rate of 0.05 second per year. If the record in 1988 was 53.0 seconds, what can you expect it to be in 2004? What does the model predict for the record in 1972?

15. Solving Linear Equations. Answer the following questions about each relation.

a. The depth of water in a swimming pool as it is being filled is given by the relation

$$d = 2.5\ \frac{\text{in}}{\text{min}} \times t.$$

What is the depth after 8 minutes? When is the water 15 inches deep? When is the water 75 inches deep?

b. The population of PleasantVille is given by the relation

$$P = 550 \text{ people} + \left(150\ \frac{\text{people}}{\text{yr}} \times t\right).$$

What will the population be in 2001? When will the population reach 1000?

c. The profit for a fund-raiser is given by the relation

$$P = \left(8\ \frac{\$}{\text{ticket}} \times n\right) - \$200,$$

where n is the number of tickets sold. How much money will be raised (or lost) if 20 tickets are sold? How much money will be raised (or lost) if 40 tickets are sold? How many tickets must be sold for the organizers to make $400?

16. Creating and Solving Linear Equations. Each of the following problems involves a linear relation. Write an equation for the linear relation and answer the questions that follow.

a. A YMCA fund-raiser offers raffle tickets for $5 each. How much does the revenue increase with each ticket sold? The prize for the raffle is a $350 television set. If the prize must be bought from the proceeds of the ticket sales, how many tickets must be sold before the raffle can begin to make a profit?

b. As you drive along the interstate highway, your cruise control holds your car speed at 63 miles per hour. At what rate is your distance traveled changing with respect to time? After how many hours will you have traveled 150 miles?

c. Births and immigration are adding 120 people per year to a small town. At the same time, 85 people "leave" the town each year through death or emigration. At what net rate is the population of the town changing? After how many years will the population have increased by 140 people?

17. Graphing General Linear Equations. For each of the following equations, find the slope of the graph and the *y-intercept*; then sketch the graph of the relation for values of x between –10 and 10.

a. $y = 2x \times 6$ b. $y = -3x + 3$

c. $y = -5x - 5$ d. $y = 4x + 1$

Section 10.5

18. Creating Linear Equations. Following the procedure given in the text determine the linear equation of the line that passes through each pair of data points and draw a graph of the linear relation. Be sure to confirm that the line passes through the data points.

a. (2, 7) and (4, 13) b. (–1, 3) and (0, 7)

c. (14, 2) and (–4, –20) d. (17, –9) and (4, 8)

e. (–13, –5) and (–10, 3) f. (1, 1) and (6, 6)

g. (7, –11) and (–2, –7) h. (87, 30) and (45, 21)

19. Linear Models. Consider the situations described in Problem 12 parts (a) – (l). For each part, do the following.

(i) Create a linear equation that represents the model you have already graphed.

(ii) Use your equation to confirm the answers to any questions asked about each situation. Show all your work.

(iii) Discuss the validity of the model and the answers you have found for each situation.

20. Linear Regression. Consider the data sets given in Problem 5 parts (a) – (c). For each part, do the following.

(i) Plot the data with appropriate axes and labels.

(ii) By eye, draw the straight line through the data points that appears to give the best fit to the data.

(iii) Pick two points on the line of best fit, find the coordinates of those points and write the equation for the line, which becomes your linear regression line.

(iv) Assess whether this linear model is a valid representation of the data.

Section 10.6

21. Depreciation of Equipment. A $1000 washing machine in a laundromat is depreciated for tax purposes at a rate of $50 per year. Find the depreciated value of the washing machine as it varies in time. When does the depreciated value reach $0?

22. **Fund-Raising Strategies.** The Psychology Club plans to pay a visitor $75 to speak at a fund-raiser. If tickets are sold for $2 apiece, find the profit/loss for the event as it varies with the number of tickets sold. How many people must attend the event for it to break even? How would the break-even attendance change if tickets were sold for $3 apiece?

23. **Quantity Discounts.** A record and tape store is having a sale on blank cassettes. For quantities between 1 and 10 cassettes, the unit price is $3.50; for quantities between 11 and 19, the unit price is $3.00; and for quantities of more than 20, the unit price is $2.60. As a consumer, does it make sense to buy 10 cassettes? Does it make sense to buy 20 cassettes? Explain.

24. **Book Publishing.** Suppose that the initial cost of publishing a book is $2000 and that the unit cost is $15 per book. Once published, the book can sell for $20 per copy.
   a. Find the relation (n, C) that gives the cost C of printing n books.
   b. Find the relation (n, R) that gives the revenue R that results from selling n books.
   c. Find the relation (n, P) that gives the profit P that results from printing and selling n books.
   d. Sketch the graph of the profit relation (n, P).
   e. What is the profit (or loss) when 100 books are sold?
   f. What is the profit (or loss) when 500 books are sold?
   g. What is the break-even point (the number of books that must be sold to cover the initial costs)?

25. **Mixing Flours.** Assume that whole wheat flour contains 16 grams of protein per cup and that white flour contains 12 grams of protein per cup. You want to produce 10 cups of a flour mixture that contains 100 grams of protein.
   a. Following the procedure in Example 10–17, let w and h represent the unknown amounts of white and wheat flour, respectively. Find a linear relation between w and h that expresses the total amount of flour needed.
   b. Find a linear relation between w and h that expresses the total amount of protein needed.
   c. Make a graph of the two relations from parts (a) and (b). Use your graph to determine how many cups of each kind of flour are needed to make the required mixture. Explain.
   d. Solve the two simultaneous equations from parts (a) and (b) to confirm the graphical solution in part (c).

26. **Box Office Dilemma.** Imagine that you are selling tickets at a box office. At the end of the evening you know that you have sold a total of 300 tickets and collected $2000. Unfortunately, you do not know how many child (at $5 each) and how many adult tickets (at $10 each) were sold.
   a. Let a and c represent the unknown numbers of adult and child tickets sold, respectively. Find a linear relation between a and c that gives the total number of tickets sold.
   b. Find a linear relation between a and c that expresses the total amount of money collected.
   c. Make a graph of the two relations from parts (a) and (b) and use your graph to determine how many adult and child tickets were sold.
   d. Solve the equations from parts (a) and (b) to confirm your solution in part (c).

Section 10.7

27. **A Variable Rain Storm.** The following graph shows how the depth of rain in a rain gauge changed during a 6-hour storm. Use the graph to answer the following questions.
   a. How much rain had fallen after 1 hour? after 2 hours? after 6 hours?
   b. During what period of time did the rain stop falling?
   c. At what time was the rain depth the heaviest?
   d. Draw a tangent line to the graph (by eye) at a time 2 hours after the storm began. Use your tangent line to estimate the rain depth rate 2 hours after the storm began.
   e. Draw a tangent line to the graph (by eye) at a time 4 hours after the storm began. Use your tangent line to estimate the rain depth rate 4 hours after the storm began.

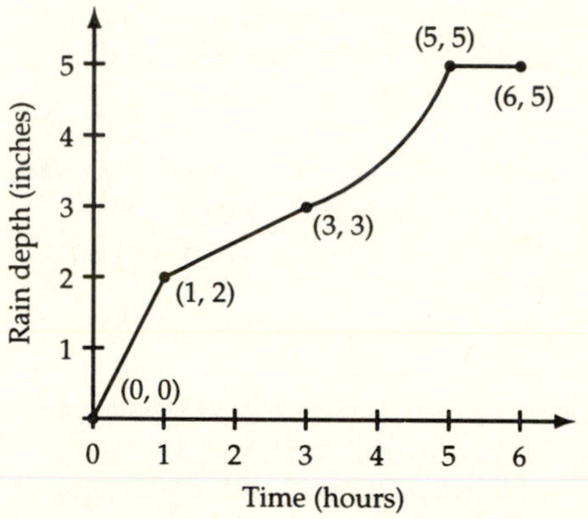

28. **Bike Trip Graph.** Suppose that you set out on a bicycle trip from home. For the next 8 hours, your distance from home, d, increases with time, t, as shown in the following graph. Use the graph to answer the following questions.
   a. How far had you traveled after 1 hour? after 2 hours? after 5 hours?
   b. According to the graph, when did you stop for lunch?
   c. At what time were you traveling the fastest?
   d. Draw a tangent line to the graph (by eye) and use it to estimate your speed 3 hours into the journey.
   e. What was your speed 5 hours into the journey?

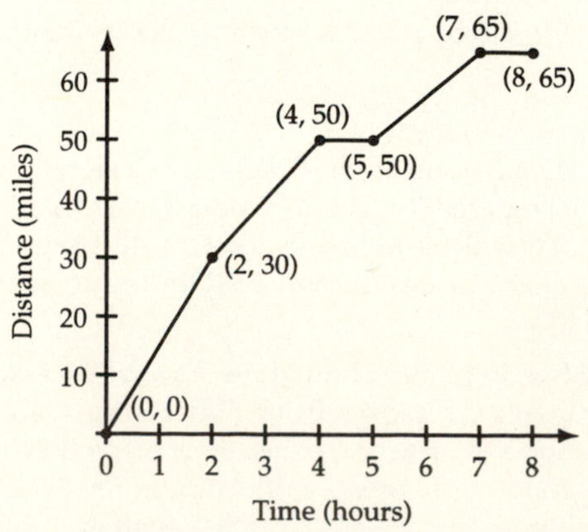

Further Topics and Applications

29. **Sketches of Relations.** For each of the following relations use your intuition or additional research, if necessary, to do the following.
    (i) Describe any restrictions on the domain, and choose an appropriate domain for graphing the relation.
    (ii) Describe the range.
    (iii) Make a rough sketch of the relation, and explain the assumptions that go into your graph.
    (iv) Briefly discuss the validity of your graph as a model of the true relation between the variables.
    a. (*time of day, traffic flow at intersection of two interstate highways*) over a two-day period
    b. (*price of gasoline, number of tourists in Yellowstone*)
    c. (*time, world record in the 100-meter dash*) over the last 30 years
    d. (*time, postage for a first-class letter*) over the last 30 years
    e. (*temperature of stove, time to boil a gallon of water*); recall that water boils at 212°F (or 100°C)
    f. (*driving speed, gas mileage*)
    g. (*day of the year, high temperature in Nome, Alaska*)
    h. (*radius of snowball, time required for snowball to melt*)
    i. (*angle of cannon, horizontal distance traveled by cannonball*)
    j. (*time of year, water temperature in Lake Michigan*)
    k. (*shoe size, person's height*)
30. **A Linear Banking Model.** Suppose that you put \$1000 in the bank and one year later you have \$1065.
    a. Create a linear equation that models how much money you have in the bank over time.
    b. Draw a graph of your equation. Label the graph clearly.
    c. Use your equation to predict how much money you will have in the bank after 5 years.
    d. Do you think that this model is valid for a bank account? Explain.
31. **Linear Growth.** Suppose that you were 20 inches long at birth and 4 feet tall on your tenth birthday.
    a. Create a linear equation that models your growth based on the two data points (birth and age 10).
    b. Draw a graph of your equation. Label the graph clearly.
    c. Use your equation to predict your height at ages 2 and 6.
    d. Use your equation to predict how tall you would be at ages 20 and 50.
    e. Comment on the validity of this linear model.
32. **Rental Cars.** Assume that you are renting a car that costs \$40 per day plus an additional 10 cents per mile. Find a relation that describes how the rental cost varies with the number of miles driven *for a single day*. How far can you drive on a one-day trip with \$90? How far could you drive on a two-day trip with \$150?
33. **Computer Rentals.** Suppose that you can rent time on computers at the local copy center for a \$5 setup charge and an additional \$3 for every 5 minutes. Find a relation that describes how the cost of renting the computer depends on time. How long do you have to finish your term paper if you have \$15 to spend for computer time?
34. **Bracket Pricing.** Suppose that daffodil bulbs are priced at \$3 each if they are bought in quantities of fewer than 10; at \$2 each in quantities between 10 and 30; and at \$1 each in quantities greater than 30. Let T be the total cost of a purchase of n bulbs. Write the relation between T and n. Are 11 bulbs or 9 bulbs less expensive? Are 29 bulbs or 31 bulbs less expensive? Find the intervals in which the total cost is less for buying more bulbs.
35. **Pricing Strategies.** The owner of a hubcap store devises the following pricing strategy for hubcaps in order to promote sales. He lets h be the number of hubcaps a customer buys and p be the corresponding price per hubcap. He then sets the prices as follows.
    If $1 \le h < 10$, then $p = \$30$;
    if $10 \le h < 20$, then $p = \$25$; and
    if $h \ge 20$, then $p = \$20$.
    Does this pricing strategy seem sensible? Why or why not?
36. **Lease vs. Purchase.** Suppose that you can lease a motorcycle for \$100 down and \$120 per month. Find a relation between time (in months) and the amount spent on leasing. How long can you lease the motorcycle before you pay more than the \$4500 purchase price of a new motorcycle?
37. **Wildlife Management.** A common technique for estimating populations of birds or fish is to tag and release individual animals in two different outings. If the wildlife remain in the sampling area and are randomly caught, a fraction of the animals tagged during the first outing are likely to be caught again during the second outing. Based on the number caught and the fraction caught twice, the total number of animals in the area can be estimated.
    Consider a case in which 200 fish are tagged and released during the first outing. During a later outing in the same area, 200 fish are again caught and released, of which a fraction p are already tagged.
    a. Suppose that half the fish caught during the second outing are tagged ($p = 0.5$). Estimate N, the *total* number of fish in the entire sampling area. Explain your reasoning.
    b. Suppose that one-fourth of the fish caught during the second outing are tagged ($p = 0.25$). Estimate the total number of fish in the entire sampling area N.
    c. Generalize your results from parts (a) and (b) to find a formula for the relation (p, N) between the total number of fish, N, and the fraction tagged during the second outing, p.
    d. Graph the relation obtained in part (c). What is the domain? Explain.
    e. Suppose that 15% of the fish in the second sample are tagged. Use the formula from part (c) to estimate the total number of fish in the sampling area. Confirm your result on your graph.
38. **Nutrition.** A growing boy needs an average of 1.5 grams of protein in his diet each day for each kilogram of body weight, whereas a growing girl should have 1.0 gram of protein per day per kilogram of body weight. Graph the relations that give the recommended daily protein levels for boys and girls with body weights of between 10 and 50 kilograms. Why are the recommended amounts of protein different for boys and girls? Do you think that these recommendations are worth following?

39. **Physical Linear Relations.** Each of the following linear relations describes a physical phenomenon. Graph each relation and briefly describe its meaning.
   a. The relation (F, C) that converts temperatures from a Celsius temperature scale, C, to a Fahrenheit scale, F, is given by $C = (5/9) \times (F - 32)$.
   b. The relation (h, p) that relates the water pressure p to the depth h below the surface of a lake is $p = \rho gh$, where $\rho = 1\ \text{g/cm}^3$ is the density of water and $g = 9.8\ \text{m/s}^2$ is the acceleration of gravity.
   c. Let L represent latitude north of the equator, in degrees, and let d be the distance north of the equator, in miles. The relation (L, d) is $d = 69 \times L$.
   d. Imagine that you are on the outer edge of a merry-go-round with a radius of $R = 10$ meters. The number of revolutions that the merry-go-round makes each second is called its **angular frequency**. It is measured in revolutions per second and is denoted ω. The relation (ω, v) that gives your actual velocity v as the merry-go-round spins, measured in meters per second, is $v = R\omega$.
   e. Einstein's famous equation $E = mc^2$ is a relation (m, E) that expresses the energy content E in an amount of mass m, where $c = 3 \times 10^8$ m/s is the speed of light.
   f. If a length of electrical wire has a resistance of $R = 20$ ohms and carries a current i, the voltage drop V across the length of the wire is the relation (i, V) given by $V = i \times R = 20i$.

40. **Hiking on a Loaf of Bread.** One kilogram of bread, when metabolized, yields about 1.2×10^7 joules of energy. The human body can convert approximately 15% of the energy in the food it digests into muscular work.
   a. If you eat 0.5 kg of bread, how much energy is available to your body for muscular work?
   b. After eating, you decide to hike up a mountain. The amount of energy (or work) required to lift your mass m to a height h against the force of gravity is $W = mgh$, where g is the acceleration of gravity (9.8 m/s^2). Using the energy you found in part (a) and your own body mass, determine how high you could climb after eating your bread. What assumptions are involved in this calculation?
   c. Generalize your results from parts (a) and (b). Assume that you want to take a hike in which you expect to climb h meters, where h could vary between 0 and 3000 meters. Find a formula for the relation (h, B) that describes how much bread, B, you would need to eat to fuel the hike. Graph the relation and discuss any assumptions that you made.

41. **Sales Taxes.** In the town of Paradise Valley a 3% local sales tax and a 2% state sales tax are charged on all retail sales. Let p be the before-tax amount of a purchase in dollars. Let T be the after-tax amount of the purchase. Find a linear relation that describes how T varies with p.

42. **Women in the Senate.** In 1991, two women were U.S. Senators. In 1993, six women were U.S. Senators. Assume that this trend is linear and will continue.
   a. Graph the projected growth in women senators over the 20 years between 1991 and 2011. What is the rate of change of this relation?
   b. According to this model, when will women comprise half the Senate?
   c. How many women are in the Senate at present? Is the model prediction accurate? Discuss the validity of the model.

43. **Boiling Point of Water.** Water boils at 100°C at sea level and at 90°C at an elevation of 3000 meters. If boiling temperature varies linearly with elevation, find the relation that fits the data given. At what elevation is the boiling point predicted to be 0°C? Discuss the validity of a linear model for this problem.

44. **Heat Transfer.** A thin copper bar has a length of 100 centimeters. The left end of the bar is held at a constant temperature of 10°C, and the right end of the bar is held at a constant temperature of 40°C. Once the bar has reached thermal equilibrium, the temperature across the length of the bar varies linearly between the two end temperatures. Find the temperature as it varies with position along the bar. At what point is the temperature 25°C?

45. **Salesperson Strategies.** Imagine that you sell greeting cards and have a choice of (A) earning a salary of $800 per month plus a 10% commission on all sales, or (B) earning a 20% commission on all sales.
   a. Let s represent your monthly sales (in dollars) and E represent your monthly earnings. Find the relation (s, E) that describes your monthly earnings under option (A). Graph the relation. Does it give the earnings that you expect if $s = 0$? Explain.
   b. Find the relation (s, E) that describes your monthly earnings for a given amount of sales under option (B). Graph the relation on the same set of axes used in part (a). Does it give the earnings that you expect if $s = 0$? Explain.
   c. From your graphs, determine which option is preferable if your monthly sales are $2000. Explain.
   d. Which option is preferable if your monthly sales are $4000?
   e. Where is the "trade-off" point at which the two options give you the same earnings?

46. **Antifreeze Solutions.** You have two antifreeze solutions that you need to mix in the right proportions so that your car will survive the winter. Solution A contains 10% pure antifreeze by volume, and solution B contains 20% pure antifreeze by volume. Your car needs 6 liters of a solution that contains 1 liter of pure antifreeze.
   a. Following the procedure of Example 10–17, let a and b represent the unknown amounts of antifreeze A and B, respectively. Find a linear relation between a and b that expresses the total amount of antifreeze needed.
   b. Find a linear relation between a and b that expresses the total amount of pure antifreeze needed.
   c. Make a graph of the two relations, and use your graph to determine how many liters of each kind of antifreeze are needed to make the required mixture.
   d. Solve the equations from parts (a) and (b) to confirm your answer to part (c).

47. Minimum Wage. The following table gives a history of the minimum wage (in dollars) in the United States from 1950 to 1991, showing only the years in which the minimum wage changed. Look carefully at the data. Suppose that someone makes the following claim: "Minimum wages in the United States increased linearly between 1950 and 1991."

Year	Minimum Wage ($)	Year	Minimum Wage ($)
1950	0.75	1976	2.30
1956	1.00	1978	2.65
1961	1.15	1979	2.90
1963	1.25	1980	3.10
1967	1.40	1981	3.35
1968	1.60	1990	3.80
1974	2.00	1991	4.25
1975	2.10		

a. Plot the data points on a graph.
b. Draw a straight line that you think best fits the data.
c. Do the data support the claim of a linear relation? Explain.
d. What is the slope of your best fit line? Use your line to predict the minimum wage this year and in the year 2000. Do you think that the linear model is reasonable for predicting the minimum wage after 1991? Explain.

48. Median Income Data. The following graph shows the median income for men and women in the United States for selected years between 1960 and 1992. Draw the lines of best fit for each data set by eye. For each line, use a point on the line and an estimated slope to find an equation for the line. Does a linear model describe the data well? According to this model, when (if ever) will parity be achieved in men's and women's income?

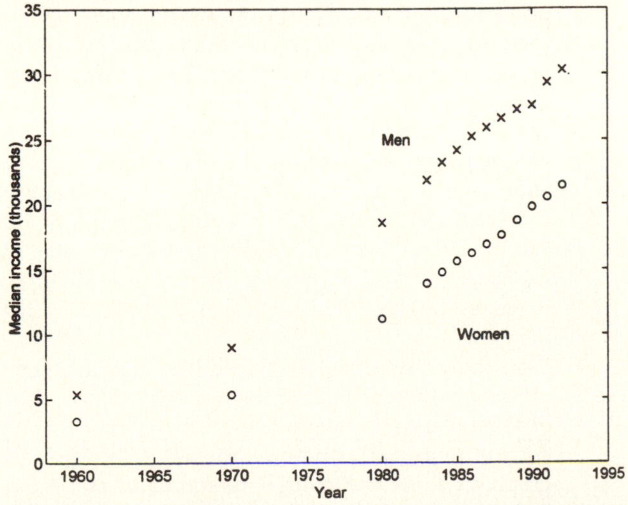

49. Elevation/Latitude Relation. Climate is a primary factor that determines the plant and animal species living in a particular region. Because climate tends to change both with latitude and elevation, species living at low elevations in northerly latitudes often are similar to species living at higher elevations farther south. Roughly speaking, the climate change is about the same in moving 100 miles north (maintaining the same elevation) as in gaining 300 feet in elevation (maintaining the same latitude). Thus, in terms of climate, you gain the equivalent of 300 feet in elevation per 100 miles that you move northward in latitude. (This is a very rough relationship, as many factors besides elevation and latitude affect climate.)

a. In the central Rocky Mountains, around 40° north latitude, tree line (the elevation above which trees are not found) occurs at approximately 11,500 feet. If you travel 500 miles further north, at what elevation might you expect to find tree line?
b. Alpine tundra is a common ecosystem in Alaska and northern Canada. Would you be surprised to find alpine tundra at mid-latitudes in the United States? Explain.

50. Rate of Change Puzzles. Answer each of the following "puzzle-like" questions, which involve rates of change. Be sure to explain your reasoning. Some questions are challenging!

a. Twenty people can produce 4 necklaces in 2 hours. How long will 15 people take to make 30 necklaces? How many people are needed to make 40 necklaces in 2 hours? How many necklaces will 5 people make in 12 hours?
b. Ralph can reach his destination on time if he averages 60 miles per hour. Halfway to the destination (in distance) he realizes that he has averaged only 30 miles per hour. How fast must he travel to arrive on time?
c. If 5 girls pack 5 boxes in 5 minutes, how many girls are needed to pack 50 boxes in 50 minutes?
d. Pipes A and B can fill a tank in 2 hours and 3 hours, respectively. Pipe C can empty the full tank in 5 hours. If all the pipes are opened at the same time when the tank is empty, how many hours are needed to fill the tank?
e. Two candles of equal length are lighted at the same time. One candle takes 6 hours and the other takes 3 hours to burn completely. After how much time will one candle be exactly twice as long as the other?
f. If a woman walks to work and rides home, her commute takes one and a half hours. When she rides both ways, her commute takes half an hour. How long would it take her to walk the round trip?
g. A woman usually takes the 5:30 train home from work, arriving at the station at 6:00 where her husband meets her to drive her home. One day she left work early and took the 5:00 train, arrived at the station at 5:30, and began walking home. Her husband, leaving home at the usual time, met his wife along the way and brought her home 10 minutes earlier than usual. How many minutes did the woman walk?
h. Two boats leave opposite shores of a river at the same time and travel at constant but different speeds. They pass each other 700 yards from one shore. After they reach the opposite shore, they turn around. On their return trip the boats pass 400 yards from the opposite shore. How wide is the river?

Projects

51. The Colorado River. The Colorado River flows from Rocky Mountain National Park in Colorado to the Gulf of California. As with the Mississippi, there is a relation (*distance from the source, river width*) for the Colorado River. In the past, the graph of this relation for the Colorado would have looked quite similar to that of the Mississippi. Today, however, much of the water in the Colorado is diverted (to cities, agriculture, and industry) before it reaches the Gulf of California. In addition, major dams have created reservoirs along its path. Make a sketch of the relation (*distance from the source, river width*) for the Colorado River today. You will need to consult a map that shows the path of the Colorado River with the locations of dams and their resulting reservoirs.

52. Daylight Hours. Make a graph, like Figure 10–11, of the relation between the number of hours of daylight and the date for the town in which you live. To make the graph, you will need to know the number of hours of daylight on the summer and winter solstices and on the equinoxes. You can find this information in a variety of sources. An easy method is to calculate the number of hours of daylight by comparing the times of sunrise and sunset on any particular day. Most local newspapers publish the time of sunrise and sunset each day on their weather page, and most libraries carry old newspapers either in print or on microfilm.

53. Bread Cooling. Bake a loaf of bread. After removing it from the oven, use a cooking thermometer to monitor its interior temperature. You should make measurements at least every five minutes until the temperature holds steady (because it has reached room temperature). Plot your points on a graph. Draw, by eye, a best fit linear model to fit the data. Is the linear model a good fit during the time that the bread is cooling? How do your results compare to the cooling relation described in Example 10–8?

54. Collecting Data for Linear Regression. Collect data for two variables that are plausibly related in a linear way. For example, you might collect data for the relations (*weight, shoe size*) or (*height, weight*) from your classmates. Or within your community, you might collect data for the variables (*time in years, population*) or (*time in years, sales tax revenue*). Then carry out the following analysis on your data set.

a. Plot the data with appropriate axes and labels.

b. By eye, draw a straight line through the data points that appears to give the best fit to the data.

c. Pick two points on the line of best fit, find the coordinates of those points, and determine the equation for the line. This becomes your linear regression line.

d. Assess whether this linear model gives a valid representation for the data set.

55. Federal Taxes. Federal taxes in the United States consist of income taxes and a set of special taxes collectively known as FICA.

a. Like many countries, the United States uses a bracket system of income taxes. Consider a simplified system with only two brackets in which individuals earning less than $20,000 per year pay 20% of their earnings in taxes, and individuals earning more than $20,000 pay 30% of their income in taxes. Find and graph a relation that gives net earnings (take-home pay) as it varies with gross (pre-tax) earnings. How much total tax does a person earning $20,000 pay? a person earning $21,000? Does your graph show that some individuals in the upper tax bracket actually have less take-home pay than individuals in the lower bracket? From your graph, find this interval of "negative earnings."

b. Next, consider a modified system with two brackets in which individuals earning less than $20,000 per year pay 20% of their earnings. Individuals earning more than $20,000 pay the same 20% on their first $20,000 of income, and 30% on all additional income. How much total tax does a person earning $20,000 pay? a person earning $21,000? Does this system correct the unfairness of the system in part (a)? Why or why not? Find and graph a relation that gives net earnings (take-home pay) as it varies with gross (pretax) earnings.

c. After the 1986 tax reform in the United States, the tax bracketing was as follows: For a single individual, the tax rate was 0% on the first $6,000 of income; 15% on the remaining income up to $30,000; and 28% on all income above $30,000. Find and draw a graph of the relation that gives net earnings (take-home pay) as it varies with gross (pretax) earnings under this system. What is the overall tax rate paid by an individual earning $100,000?

d. In addition to the income tax structure described in part (c), wage earners paid 7.5% of their income for FICA (primarily social security and Medicare); employers also paid 7.5% to FICA. Thus the total FICA tax amounted to 15% (self-employed individuals pay the entire 15% themselves). However, most of FICA (assume *all* for the purposes of this problem) was paid only on the first $53,000 of wage income. Modify your graph from part (c) to include the 15% FICA taxes in addition to the income tax structure.

e. Non-wage income, such as income from capital gains (stock sales, sales of real estate, etc.) is exempt from FICA. Based on the results from parts (c) and (d), calculate the overall tax rates paid by (i) an individual earning $50,000 in wages; (ii) an individual earning $100,000 in wages; (iii) an individual earning $100,000 from capital gains.

f. Alas, the tax structure described has been changed significantly since 1986. Find the current bracket structure of federal income taxes in the United States, and the current rates and structure for FICA taxes. Draw a graph to show the overall relation (*income, tax rate*), including both federal income taxes and FICA.

g. Now that you have the overall tax structure for the United States from part (f), you can calculate anyone's federal taxes, right? Wrong! We have not yet included the effects of tax deductions. Briefly discuss the effects of tax deductions on actual tax rates paid by individuals. What are the major deductions claimed in the United States?

h. Do you believe that the current tax system is fair? Why or why not? If not, how would you recommend changing it? Is anyone in Congress currently advocating the same changes that you favor? If so, why haven't they yet been implemented?

56. Islamic Scholars of the Middle Ages. Do additional research on the achievements of Islamic scholars during the Middle Ages. Write a short paper (3–5 pages) summarizing your findings. You may focus on a particular individual, a group of individuals who worked in related fields, or an overall picture of Islamic scholarship during the period.

11 THE LANGUAGE OF SIZE AND SHAPE

Things are, and things change. That is, most things have a present size and shape that is likely to change in the future. In Chapter 10, we concentrated on mathematics as the language of change. In this chapter, we focus on geometry, the language of size and shape. Of course, we already have used geometric ideas in presenting such topics areas, volumes, and scale models. Here, we examine size and shape in more depth. This process will expand the range of quantitative issues that you can analyze and understand.

CHAPTER 11 PREVIEW:

11.1 ***A BRIEF HISTORY OF GEOMETRY.*** In this section we trace the origins of geometry, concentrating on the development of Greek geometry through its culmination in Euclid's work.

11.2 ***FUNDAMENTAL CONCEPTS OF GEOMETRY.*** In this section we review the essential geometric elements: points, lines, angles, planes, and dimension.

11.3 ***PROBLEMS IN TWO DIMENSIONS.*** In this section we show how "flat" figures such as triangles, rectangles, and circles can be used for applications such as surveying, design, and much more.

11.4 ***PROBLEMS IN THREE DIMENSIONS.*** In this section we consider problems that can be solved using surface area and volume formulas for three-dimensional solids. Applications include solar energy and the threat posed by melting the polar ice caps.

11.5 ***FURTHER APPLICATIONS OF TRIANGLES.*** In this section we look at the properties and applications of triangles in more depth.

*11.6** ***NON-EUCLIDEAN GEOMETRY.*** In this section we introduce basic concepts of non-Euclidean geometry, with its important applications to black holes and to the fate of the universe.

*11.7** ***FRACTAL GEOMETRY.*** In this section we introduce fractal geometry, a recent mathematical invention, with applications in everything from science to special effects in movies.

[The] knowledge at which geometry aims is of the eternal, and not of the perishing and transient.

— Plato

It is the glory of geometry that from so few principles … it is able to accomplish so much.

— Isaac Newton

Equations are just the boring part of mathematics. I attempt to see things in terms of geometry.

— Stephen Hawking,
Everybody Counts: A Report to the Nation on the Future of Mathematics Education, p. 35.

Geometry seems to stand for all that is practical, poetry for all that is visionary, but in the kingdom of the imagination you will find them close akin, and they should go together as a precious heritage to every youth.

— Florence Milner, School Review, 1898

11.1 A BRIEF HISTORY OF GEOMETRY

The origins of **geometry** extend so far into the past that we cannot determine exactly how or when it first arose. Nevertheless, a clue is found in the literal translation of *geometry* as "Earth measure." Ancient peoples developed geometric ideas to survey the boundaries of flood basins around agricultural fields and to establish patterns of planetary and star motion in the sky. However, from the earliest times geometry was more than just a practical science, as demonstrated by the artistic use of geometric shapes and patterns in cave paintings, pottery decorations, and ancient architecture.

Legend has it that Thales used his knowledge of geometry to predict the year in which a total eclipse of the Sun would be visible in the area where he lived, which is now part of Turkey. The eclipse occurred as two opposing armies (the Medes and the Lydians) were massing for battle. The eclipse so frightened the armies that they put down their weapons, signed a treaty, and returned home. Modern research shows that the only eclipse visible in that part of the world at about that time occurred on May 28, 585 B.C., making this the first historical event that can be dated precisely.

Formal geometry usually is traced back to the Greek philosopher Thales (624–546 B.C.), who is credited with being the first person to introduce ideas of abstraction into geometry, envisioning lines of zero thickness and perfect straightness. He also developed deductive methods of proof. Although some historians claim that he merely restated or extended ideas discovered by earlier civilizations, Thales was the first to put these ideas into the form that has passed down to the present.

Following the work of Thales, geometry became so important to the ancient Greeks that many considered it the ultimate human endeavor. Mystic societies, like that of the Pythagoreans (see Chapter 6), devoted themselves to its study. More rationally, Plato emphasized the study of geometry at the Academy that he founded in Athens in 387 B.C. Above the doorway to the Academy, which was in effect the world's first university, Plato had the following words inscribed: "Let no one ignorant of mathematics enter here." Plato's Academy remained a center of learning for more than 900 years, until it was ordered closed by the Eastern Roman Emperor Justinian in A.D. 529.

The culmination of Greek geometry came with the mathematician Euclid (c. 325–270 B.C.). Although he may have studied at Plato's Academy in Athens, Euclid did most of his work at a university called the Museum (because it honored the Muses, the patron goddesses of the sciences and the fine arts) in Alexandria. Euclid's greatest fame comes not from his own research, but from a textbook he wrote. In 13 volumes, Euclid's *Elements* consolidated all of Greek mathematics into a clear and logical treatise. For almost 2000 years, Euclid's *Elements* was the primary textbook for geometry throughout the Western world. Until recently, it was the second-most reproduced book (after the Bible) and, by almost any measure, the most successful textbook in history.

Time-Out to Think: The Muses were the patrons of both the sciences and the fine arts. Comment on the connection between science and art in the ancient world and contrast it with the view held today. Do you believe that the sciences and the arts should be considered distinct, or as two sides of the same coin?

The geometry described in Euclid's work, now called **Euclidean geometry**, is the familiar geometry of lines, angles, and planes. We devote most of this chapter to Euclidean geometry and its wide range of applications. Euclidean geometry was the *only* geometry for more than 2000 years. Recently, however, mathematicians have discovered new types of geometry that have important applications. We briefly discuss *non-Euclidean geometry* and *fractal geometry* in the final two sections of the chapter.

THINKING ABOUT ... PLATO, KEPLER, AND ATLANTIS

Plato, like many of the ancient Greek philosophers, wasn't interested in knowledge for its practical use. Rather, he believed that knowledge should be used to elevate the soul. He believed that the abstraction used in mathematics represented pure thought, which he considered greatly preferable to the gross and imperfect everyday world. Although he was uninterested in earthly applications of mathematics, he believed that mathematics could be applied to develop an understanding of the heavens.

Plato believed that the heavens exhibit perfect geometric form. He especially valued the "perfection" found in symmetry. For example, he believed that all heavenly bodies — the Sun, Moon, planets, and stars — must move in perfect circles (the most symmetric shape). Similarly, he believed that the heavenly bodies were held in place by crystalline spheres (the most symmetric solid). Thus the Platonic view of the universe consisted of a spherical Earth at the center, surrounded by a set of concentric crystalline spheres. Even today, the phrase "the music of the spheres," is used to describe heavenly music or other things of perfection.

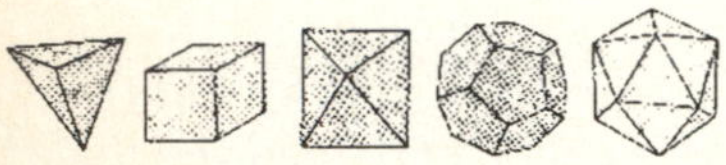

Figure 11–1. The five Platonic solids have 4, 6, 8, 12, and 20 identical faces. Only these five solids have faces that are regular polygons.

Another symmetry in which Plato placed great faith was that of the five perfect solids. As first discovered by the Pythagoreans, only five solids exist in which all the faces are regular polygons (closed shapes in which all sides have the same length, such as a square). These five solids, called the **Platonic solids** (Figure 11–1) because of their importance to Plato, are: the **tetrahedron**, with 4 triangular faces; the **cube**, with 6 square faces; the **octahedron**, with 8 triangular faces; the **dodecahedron**, with 12 pentagonal faces; and the **icosahedron**, with 20 triangular faces. Plato believed that four of the perfect solids represented the four elements of earth, water, fire, and air. The dodecahedron, he believed, represented the universe as a whole.

Some 2000 years later, when Johannes Kepler finally solved the mystery of planetary motion, Plato's ideas still had profound influence. Until he was forced to abandon the idea (see "*Thinking About* ... Eight Minutes of Arc," in Section 11.2), Kepler sought a model in which planets moved in perfect circles around the Sun. In addition, Kepler recognized six planets (Mercury, Venus, Earth, Mars, Jupiter, and Saturn). Because this array left five gaps between the planets, he imagined that planetary distances from the Sun could be explained by nesting the five Platonic solids within one another.

A disciple of Socrates, Plato presented his ideas in a series of dialogues between Socrates and others. In the same dialogue (*Timaeus*) in which Plato discussed the role of the perfect solids in the universe, he also invented a moralistic tale about a fictitious land he called *Atlantis*. Interestingly, although Plato's cosmology was abandoned long ago, millions of people today believe that Atlantis really existed. Plato's fiction, in the end, has more adherents than the ideas in which he firmly believed. Commenting on this irony, the prolific popular writer Isaac Asimov wrote (in Asimov's *Biographical Encyclopedia of Science and Technology*, Doubleday, 1982):

> *If there is a Valhalla for philosophers, Plato must be sitting there in endless chagrin, thinking of how many foolish thousands, in all the centuries since his time ... who have never read his dialogues or absorbed a sentence of his serious teachings ... believed with all their hearts in the reality of Atlantis.*

11.2 FUNDAMENTAL CONCEPTS OF GEOMETRY

Geometric objects, such as points and lines, represent *idealizations* that do not exist in the real world. As a result, defining them precisely can be surprisingly difficult. For example, we might say that a line is a collection of points that all lie in the same direction and continue forever, but this description depends on the concept of a point. Or, we might say that a point is the intersection of two lines, but this definition brings us circularly back to where we began! Rather than seeking precise definitions, we rely on intuition. Recognizing that geometric objects are abstractions, we can define them by analogy to physical objects.

A mathematical point is the most indivisible and unique thing that art can present.
— poet John Donne (1573–1631)

11.2.1 Points, Lines, and Planes

A geometric **point** is imagined to have zero size. Of course, no physical object has zero size, but many real phenomena approximate geometric points (Figure 11–2a). Stars, for example, appear as points of light in the night sky. The intersection of two chalk lines on a playing field, or of two roads, also are approximations to a point.

A geometric **line** is formed by connecting two points along the shortest possible path. A line is imagined to extend infinitely in length and to have no thickness (Figure 11–2b). Because no physical object is infinite in length, we work with **line segments**, or pieces of a line. Long straight sections of highway make good approximations to line segments, as do laser beams in a light show.

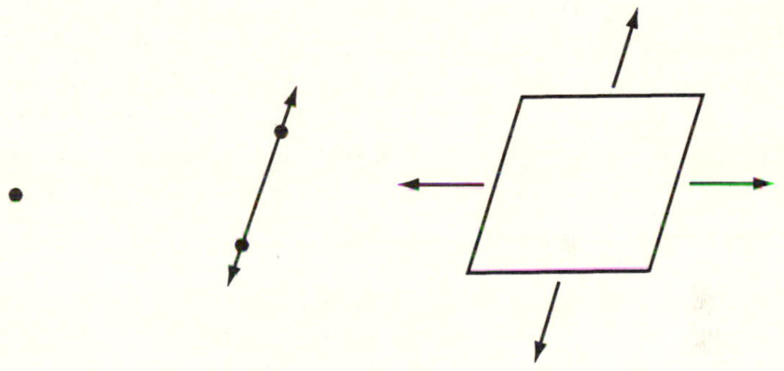

Figure 11–2. (a) A geometric point has zero size. Drawing something with zero size is impossible, so a point is represented by a dot. (b) A line can be drawn between any pair of points. Unlike the drawing, a geometric line has no thickness and extends infinitely in length. (c) A geometric plane is a perfectly flat surface that extends infinitely in length and width but has no thickness.

A geometric **plane** is a perfectly flat surface that extends infinitely in length and width but has no thickness (Figure 11–2c). A sheet of paper, a smooth table top, and the surface of a chalkboard are everyday approximations to segments of planes.

Any two lines in a plane eventually intersect unless they happen to be **parallel**. The shortest (perpendicular) distance between two parallel lines is the same at all points (Figure 11–3).

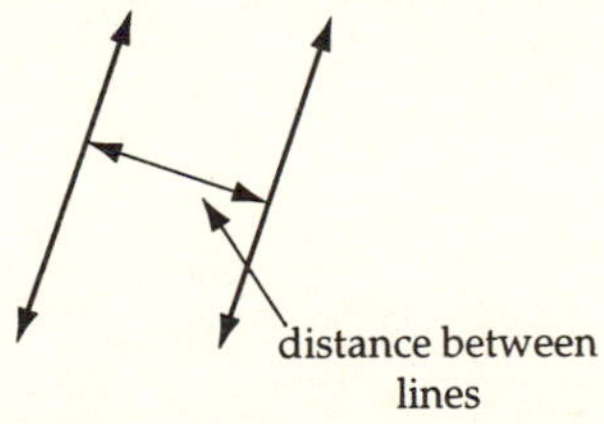

Figure 11–3. Parallel lines are lines in the same plane that never meet. The distance between parallel lines is the same at any point along their lengths.

> **Time-Out to Think:** List at least three additional everyday realizations of points, line segments, and segments of planes. How does each physical realization compare to the geometric concept that it represents?

11.2.2 Dimension

The concept of **dimension** describes the progression from a point to a line to a plane. A point is said to have **zero dimension**. If we sweep a *point* back and forth in one direction, we generate a line (Figure 11–4a on the next page). The line is said to have **one dimension**. If we sweep a *line* back and forth in one direction we generate a **two-dimensional** plane (Figure 11–4b). If we sweep a *plane* back and forth in one direction, we generate a volume or **three-dimensional space** (Figure 11–4c).

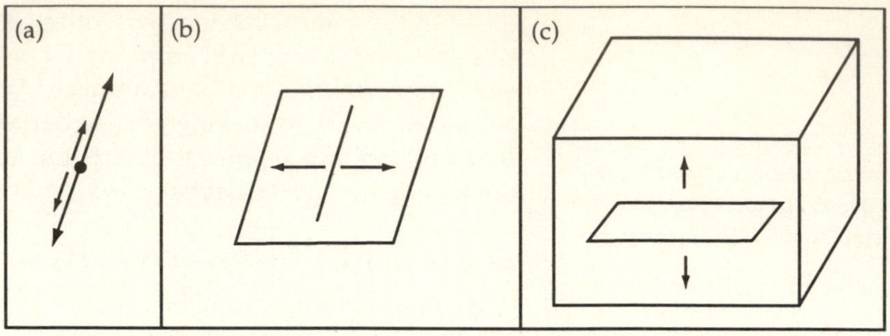

Figure 11–4. (a) A line is generated by sweeping a point back and forth in one direction. (b) Sweeping a line back and forth in one direction generates a plane. (c) Sweeping a plane back and forth in one direction generates three-dimensional space.

Note that *dimension* describes the number of independent directions in which movement is possible. If you were a geometric prisoner confined to a point, which has zero dimension, you would have no place to go. If you lived on a line, you could move back and forth along its one dimension. In a plane, you could move in two directions, such as a north–south and east–west (or any combination of those two directions). In addition to the two directions of a plane, a space allows you to move up and down.

Another way to think about dimension is by the number of **coordinates** required to locate a point. A one-dimensional line requires only one coordinate; for example, the coordinate x specifies a precise location along the number line (Figure 11–5a). Two coordinates, such as x and y, are required to locate a point in a two-dimensional plane (Figure 11–5b). Three coordinates, such as x, y, and z, are required to locate a point in three-dimensional space (Figure 11–5c).

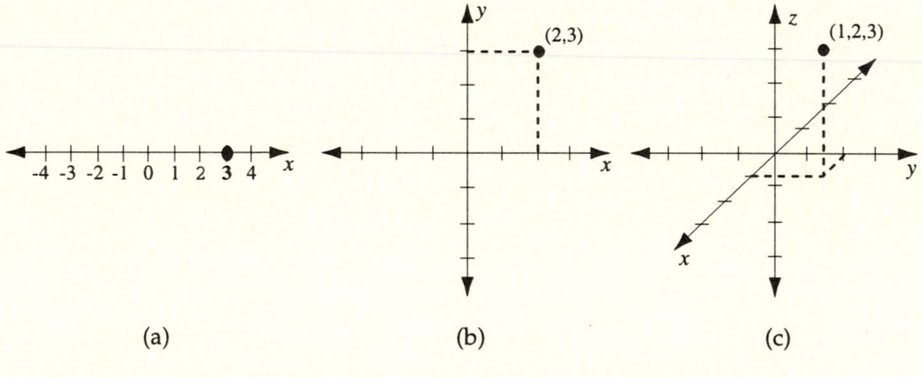

Figure 11–5. (a) One coordinate locates a point on a line. The point 3 is highlighted. (b) Two coordinates locate a point in a plane. The point (2, 3) is shown. (c) Three coordinates locate a point in three-dimensional space. The point (1, 2, 3) is shown.

Time-Out to Think: Anticipating the discussion of non-Euclidean geometry in Section 11.6, let's consider the surface of the Earth. Although the Earth isn't flat like a plane, we can specify any point on the Earth's surface with two coordinates: latitude and longitude. Should the Earth's surface be considered two-dimensional? Explain.

Beyond Three Dimensions

We created a line by sweeping a point back and forth in one direction, a plane by sweeping a line back and forth, and a three-dimensional space by sweeping a plane back and forth. Can we sweep a three-dimensional space back and forth to create a *fourth* dimension? Physically, the answer is *no*: Because we live in a three-dimensional space, we cannot find a fourth direction through which to sweep a space. Mathematically, however, a fourth dimension is easy to describe, simply by creating a fourth coordinate to specify a point.

Although visualizing a space with more than three dimensions is impossible, such spaces can be very useful. For example, Einstein showed that *time*, together with the three dimensions of space, makes a four-dimensional space known as **space-time**. To specify the location of a point (or **event**) in space-time requires four coordinates: three coordinates for its location in space and one coordinate to specify *when* the event occurs.

Imagine a data table that describes people by six characteristics: height, weight, age, eye color, hair color, and skin color. If we tried to draw a graph, each point (one point for each person) would have these six characteristics as its coordinates. Thus the graph would require six dimensions! Although we cannot draw such a graph, we can easily work with six dimensions of physical characteristics. Thus spaces with more than three dimensions are common in many fields of research.

Time-Out to Think: Think of three more examples in which working with more than three dimensions might be useful.

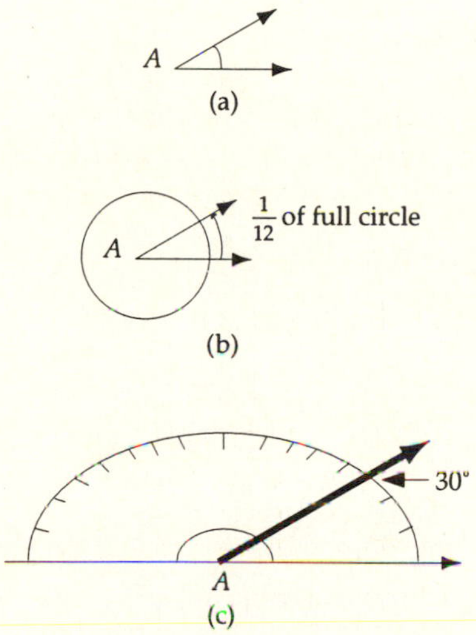

Figure 11–6. (a) This angle, with its vertex at point A, is called *angle A* or $\angle A$. (b) $\angle A$ subtends 1/12 of a full circle, so $\angle A$ measures $1/12 \times 360° = 30°$. (c) The protractor also measures $\angle A = 30°$.

11.2.3 Angles

The intersection of two lines or line segments forms an **angle**. The point of intersection is called the **vertex**. Figure 11–6(a) shows an arbitrary angle with its vertex at point *A*, so we call it *angle A*, denoted as $\angle A$.

The most common way to measure angles is in **degrees** (°), derived from the ancient base-60 numeral system of the Babylonians. By definition, a full circle encompasses an angle of 360°, so an angle of 1° represents 1/360 of a circle. To measure an angle, imagine its vertex as the center of a circle. Figure 11–6(b) shows that $\angle A$ **subtends**, or cuts, one-twelfth of a circle; thus $\angle A$ measures 1/12 of 360°, or 30°. Angles may be measured with a simple device called a **protractor**, in which the vertex is placed at the center of a semicircle (Figure 11–6c). Other devices for measuring angles, such as theodolites, transits, and sextants, operate on the same basic principle.

Figure 11–7. The first three angles shown are acute. The 90° angle is a right angle. The 135° angle is obtuse. The 180° angle is a straight angle.

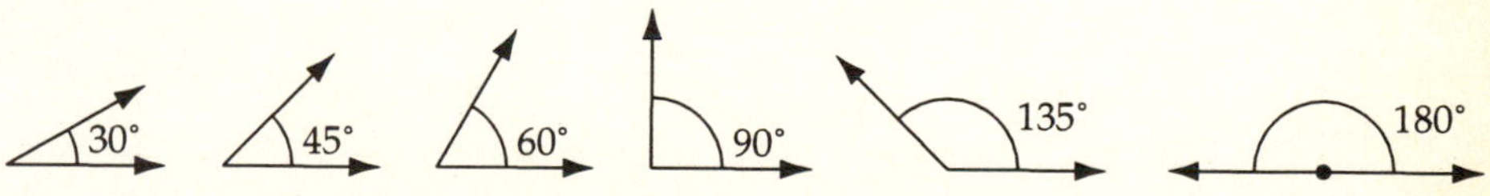

Some angles have special names. A 90° angle is a **right angle**, and lines or line segments forming a right angle are **perpendicular** (or **orthogonal**). A 180° angle is a **straight angle** because it is formed by a straight line. Any angle of less than 90° is called an **acute angle**, and an angle between 90° and 180° is called an **obtuse angle**. Several common angles are shown in Figure 11–7.

Time-Out to Think: The terminology of angles is common in everyday usage. For example, what does "doing a 360" mean? If two people are "180° opposed," what does that mean? How is the term *acute* in an acute illness related to its meaning in an acute angle? If someone is being obtuse, does the term *obtuse* bear any relation to its meaning in an obtuse angle? Explain.

Example 11–1 Find the angles that subtend a semicircle, a quarter circle, an eighth of a circle, and a hundredth of a circle.

Solution: A semicircle is half a circle, so an angle subtending it is $360° \div 2 = 180°$. An angle subtending a quarter circle is $360° \div 4 = 90°$. An angle subtending an eighth of a circle is $360° \div 8 = 45°$, and an angle subtending a hundredth of a circle is $360° \div 100 = 3.6°$.

❖

Subdividing Degrees

Although fractions of a degree can be specified with decimals (e.g., 3.6°), subdividing each degree into 60 **minutes** and each minute into 60 **seconds** is more common (Figure 11–8). Symbols for minutes and seconds are (') and ("), respectively. (That is, 1° = 60' and 1' = 60".) For example, 29°33'21" is read as 29 degrees, 33 minutes, and 21 seconds.

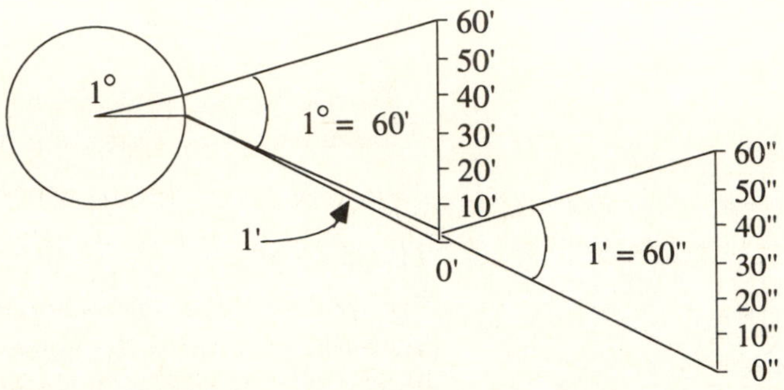

Figure 11–8. Each degree is subdivided into 60 equal parts called minutes, and each minute is subdivided into 60 equal parts called seconds. The angles are exaggerated in size for clarity.

Because the symbols °, ', and " can be confusing, we use the abbreviations *deg* (degrees), *min* (minutes), and *sec* (seconds) when working with these units.

Example 11–2 (a) Convert 3.6° into degrees, minutes, and seconds. (b) Convert 29°33'21" into a decimal fraction of degrees.

Solution: a. Note that 3.6° = 3° + 0.6°, so first convert 0.6° to minutes:

$$0.6 \text{ deg} \times \frac{60 \text{ min}}{1 \text{ deg}} = 36 \text{ min}$$

Thus, 3.6° = 3° + 0.6° = 3° + 36' = 3°36'.

b. To convert 29°33'21" to a decimal, first convert 21" to minutes:

$$21 \text{ sec} \times \frac{1 \text{ min}}{60 \text{ sec}} = 0.35 \text{ min.}$$

Thus 33'21" = 33.35', so convert 33.35' to degrees:

$$33.35 \text{ min} \times \frac{1 \text{ deg}}{60 \text{ min}} = 0.55583 \text{ deg.}$$

Thus 29°33'21" = 29.55583°.

❖

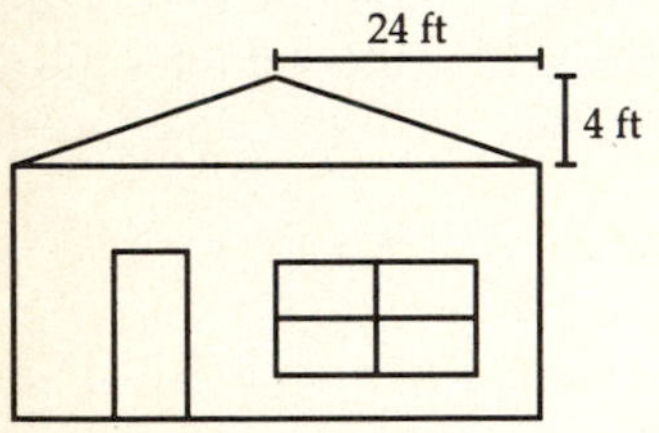

slope = $\frac{1}{6}$

(a)

slope = $\frac{1}{10}$ = 10%

(b)

Figure 11–9. (a) A roof with a pitch of "2 in 12" rises 2 feet vertically for every 12 feet horizontally, which is the same as 4 feet vertically for every 24 feet horizontally. (b) A road with a grade of 10% rises 10 feet vertically for every 100 feet horizontally, or 2 feet vertically for every 20 feet horizontally.

Angles and Slope

The pitch of a roof often is described by a phrase such as "2 in 12," and the steepness of a road may be described by a term such as "10% grade." These descriptions refer to the *slope* of an angle relative to a horizontal line. The phrase "2 in 12" means that the roof rises 2 feet vertically for every 12 feet horizontally. Thus the slope of a "2 in 12" roof is $2 \div 12 = 1 \div 6$, or 1/6 (Figure 11–9a). Similarly, a road with a grade of 10% rises 10 feet in elevation for every 100 feet of horizontal distance, or 1 foot vertically for every 10 feet in the horizontal; its slope is 10%, or 0.1 (Figure 11–9b).

Example 11–3 ***Comparing Roads and Roof Lines.*** Which is steeper, a road with an 8% grade or a road that rises "5 in 45"? Which has a greater pitch, a "2 in 12" roof or a "3 in 15" roof?

Solution: To compare the roads, determine their slopes. The road with a "5 in 45" grade has a slope of 5 ÷ 45 = 0.11, or 11%. Therefore it is steeper than the road with an 8% grade. Similarly the first roof has a slope of 2 ÷ 12 = 1/6 or 16.67%, whereas the second roof has a steeper pitch of 3 ÷ 15 = 1/5 or 20%.

❖

THINKING ABOUT *... EIGHT MINUTES OF ARC*

Plato's ideas about geometric perfection in the heavens (see "*Thinking About* ... Plato, Kepler, and Atlantis" in Section 11.1) profoundly influenced astronomy for 2000 years. Every attempt to understand planetary motion began with Plato's assumption of circular orbits. Because this assumption was incorrect and because it generally was combined with the further incorrect assumption of an Earth-centered universe, creating a model to accurately predict planetary positions was very difficult.

Nevertheless, Greek astronomer Ptolemy (A.D. 100–170) succeeded in creating a model that was sufficiently accurate to remain in use for almost 1500 years. It was enormously complex, however, requiring planets to move on small circles, which in turn moved on larger circles. When Copernicus finally overturned the idea of an Earth-centered universe (in a book published in 1543), his model still imagined circular orbits. As a result, to make reasonably accurate predictions of planetary positions, he was forced to continue the Ptolemaic idea of smaller circles turning on larger ones. Thus, even though many scientists quickly accepted the Copernican premise of a Sun-centered solar system, the details of his model remained unsatisfyingly complex.

Danish nobleman Tycho Brahe (1546–1601) compiled a set of greatly improved observations of planetary positions over a thirty-five year period toward the end of the fifteenth century. In about 1600, Johannes Kepler (1571–1630) set out to find a model that would match Tycho's observations while maintaining his great faith in the perfection of geometry. Kepler's early models sought to explain planetary distances by nesting the five Platonic solids and to match Tycho's observations with perfect circles for planetary orbits.

Kepler worked with particular intensity to find an orbit for Mars because Tycho had told him that the Mars observations would be the most difficult to reconcile with a circular orbit. After years of calculation, Kepler found a circular orbit that matched nearly all of Tycho's observations of Mars to within two minutes of arc. In two cases, however, this orbit predicted a position for Mars that differed from Tycho's observations by the slightly greater margin of eight minutes of arc.

Kepler surely was tempted to ignore these two observations and attribute them to an error by Tycho. After all, eight minutes of arc is barely one-fourth the angular diameter of the full Moon. But Kepler trusted Tycho's careful work, and the missed eight minutes of arc led him finally to abandon the idea of circular orbits. A few months later, he found the correct solution: Planets orbit the Sun in elliptical orbits with the Sun at one focus, and the orbit of a planet sweeps out equal areas in equal times as it travels about the Sun.

Astronomer Carl Sagan asserts that Kepler's discovery represented the true birth of modern science. For the first time, a scientist was willing to cast off long-held beliefs in a quest to match theory to observation. As a reward, Kepler became the first person in history to understand the motions of the planets through the heavens. About this event, Kepler wrote:

If I had believed that we could ignore these eight minutes [of arc], I would have patched up my hypothesis accordingly. But, since it was not permissible to ignore, those eight minutes pointed the road to a complete reformation in astronomy.

11.3 PROBLEMS IN TWO DIMENSIONS

Just as a symphony can be created from a few fundamental components, all of geometry can be constructed from the basic ideas presented in Section 11.2. We now turn to consideration of problems that can be modeled using these ideas of geometry. Here, we focus on problems in two dimensions and then move on to problems in three dimensions in Section 11.4.

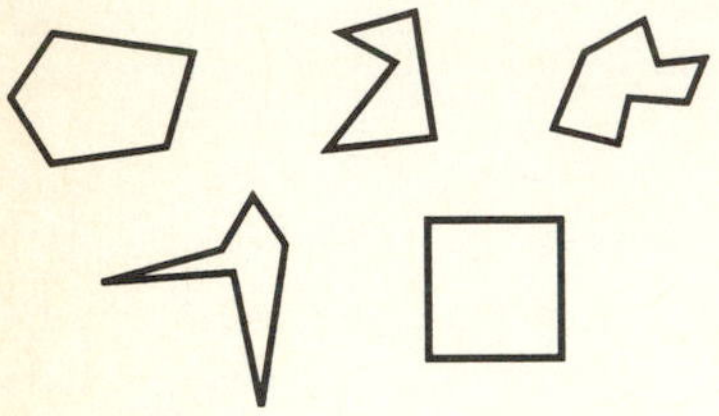

Figure 11–10. All of these figures are polygons: closed, straight-sided figures in the plane. Only the last polygon is regular.

11.3.1 Polygons

Polygons (literally, "many sided" figures) consist of line segments in a plane that are pieced together to form closed shapes (Figure 11–10). A polygon is classified by its number of sides and is **regular** if all its sides have the same length (Table 11–1).

Table 11–1. A Few Regular Polygons

Sides	Name	Picture	Sides	Name	Picture
3	Equilateral triangle		6	Regular hexagon	
4	Square		8	Regular octagon	
5	Regular pentagon	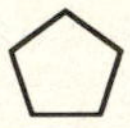	10	Regular decagon	

> **Time-Out to Think:** Regular polygons are used for many purposes. For example, the U.S. Department of Defense is housed at the Pentagon, which has the shape of a regular pentagon. Think of several more uses of regular polygons.

Triangles

With three sides, triangles are the simplest polygons. The sum of the three angles of *any* triangle measures 180°. The three general types of triangles are: **equilateral triangles**, in which all three sides and angles are equal (Figure 11–11a); **isosceles triangles** (pronounced eye-sos-o-lees), in which two sides and two angles are equal (Figure 11–11); and **scalene triangles**, in which all three sides and three angles are different (Figure 11–11c).

If one of the angles is 90°, the triangle is a **right triangle** (Figure 11–11d). Recall that, if we label the sides of a right triangle a, b, and c, with c being the **hypotenuse**, or longest side, the Pythagorean theorem states that $c^2 = a^2 + b^2$.

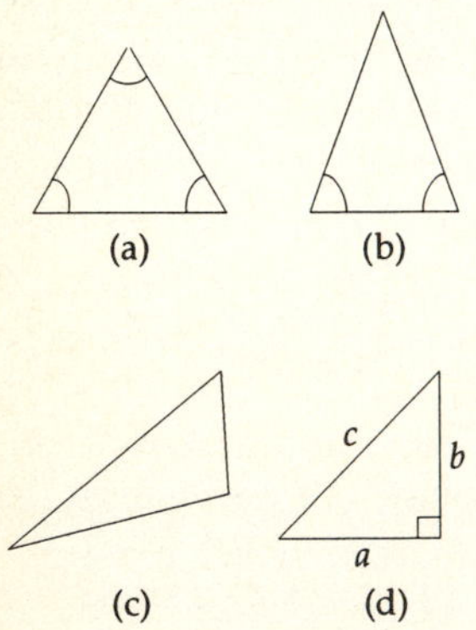

Figure 11–11. (a) An equilateral triangle has three sides of equal length and three 60° angles. (b) An isosceles triangle has two equal sides and two equal angles. (c) All sides and angles are different in a scalene triangle. (d) A right triangle includes one 90° angle. The longest side, opposite the right angle, is the hypotenuse.

> **Time-Out to Think:** With the three angles of a triangle having a total measure of 180°, explain why each angle in an equilateral triangle is a 60° angle. Explain why a right triangle may be either isosceles or scalene, but not equilateral.

Consider the rectangle in Figure 11–12a, with base b and height h. The area of this rectangle is the product of its base and height, or $b \times h$. The diagonal through the rectangle divides its area into two equal triangles, so the area of each triangle must be $(1/2) \times b \times h$. Although slicing a rectangle produces only right triangles, the same formula applies to all triangles. Once a side is

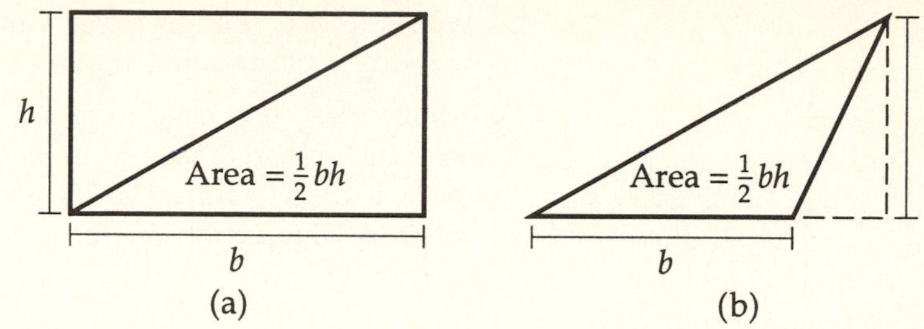

Figure 11–12. (a) Slicing a rectangle along a diagonal creates two right triangles. The area of the rectangle is $b \times h$, so the area of each triangle is $A = (1/2)b \times h$. (b) Once a side is chosen as the base, the height is the *perpendicular* distance to the top of the triangle.

chosen as the base, the height is the perpendicular distance between the base and the top of the triangle (Figure 11–12b). Hence the area of any triangle is

$$\text{Area} = \frac{1}{2} \times \text{base} \times \text{height}, \qquad \text{or} \qquad A = \frac{1}{2}\,bh.$$

Look around you and note how often right angles appear: walls and floors intersect at right angles, roads meet at right angles, the cardinal points of the compass are separated by right angles, and so on.

Time-Out to Think: Identify several more everyday examples of right angles.

Because they are so commonplace, right triangles arise in countless problems. The following examples illustrate their tremendous utility.

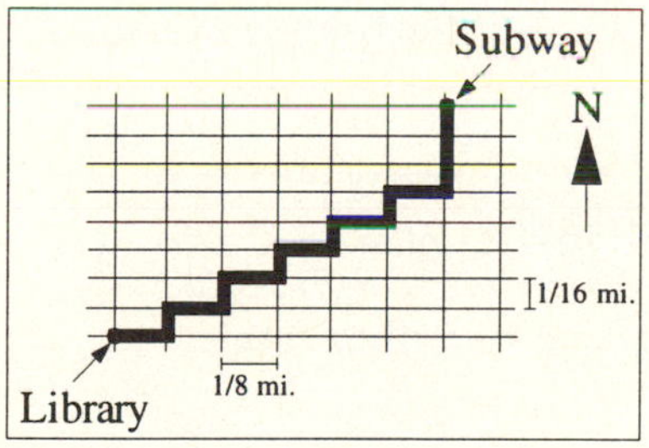

Figure 11–13. A map of several city blocks shows the locations of the library and the subway station.

Example 11–4 ***What Do Distances Mean?*** Consider the map in Figure 11–13 showing several city blocks. Assume that these blocks are 1/8 of a mile long east–west, and 1/16 of a mile long north–south. How far is the library from the subway?

Solution: The first issue is what is meant by "how far." If *how far* means the shortest distance between the library and the subway (as a bird might fly), you will get one answer. But, if it refers to the distance traveled by conventional means (e.g., walking, taxi, or bicycle), your answer will be different because the route cannot pass through buildings. By conventional means, *any* route (e.g., the one shown in a heavy solid line) requires that you travel six blocks east and eight blocks north. Because each block east is 1/8 mile and each block north is 1/16 mile, the total distance is

$$6\times\left(\frac{1}{8}\text{ mi}\right)+8\times\left(\frac{1}{16}\text{ mi}\right)=\frac{3}{4}\text{ mi}+\frac{1}{2}\text{ mi}= 1.25\text{ mi}.$$

The direct path (shown by a dashed line) is the hypotenuse of a right triangle, with the other two sides having lengths of 3/4 mile (0.75 mi) and 1/2 mile (0.5 mi). Find the hypotenuse, c, by applying the Pythagorean theorem with $a = 0.75$ mi and $b = 0.5$ mi:

$$c^2 = a^2 + b^2 = (0.75\text{ mi})^2 + (0.5\text{ mi})^2 = 1.25\text{ mi}^2$$

$$\xrightarrow{\text{square root}} \quad c = \sqrt{1.25\text{ mi}^2} \approx 1.12\text{ mi}.$$

Not surprisingly, the direct path is shorter. As this problem demonstrates, distance may be interpreted in different ways.

❖

Example 11–5 ***Building Stairs.*** You have built a stairway in a new house and want to enclose the space beneath the stairs. The stairway is 15 feet long and rises at an angle of 45° (Figure 11–14 on the next page). How much plywood do you need to cover the triangular surface below the stairway? How many rectangular 4 foot by 8 foot sheets of plywood should you buy?

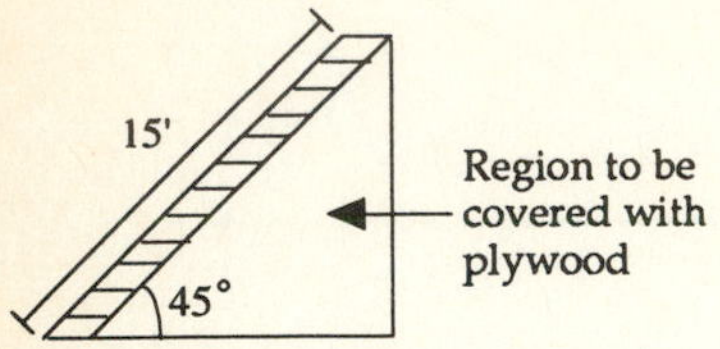

Figure 11–14. The triangular region below a 15-foot-long stairway is to be covered with plywood.

Solution: First, you need to find the area of the triangular region beneath the stairs, which means that you must find both the height, h, and the length of the base, b, of the stairs. The 15-foot length of the stairs represents the hypotenuse of a right triangle, so the Pythagorean theorem states that

$$b^2 + h^2 = (15 \text{ ft})^2 = 225 \text{ ft}^2.$$

You already know that two angles of this triangle are 45° and 90°. From the property that the angles of a triangle sum to 180°, you conclude that the third angle is 45°. Because two of the angles are the same (45°) the triangle is isosceles and the base and height must be the same. You may therefore replace b with h in the Pythagorean theorem, and solve:

Recall that if you know the square of a number, you can find the number itself by taking the square root. For example, since $4^2 = 16$, you know that $\sqrt{16} = 4$.

$$b^2 + h^2 = 225 \text{ ft}^2 \xrightarrow[b=h]{\text{substitute}} h^2 + h^2 = 225 \text{ ft}^2 \xrightarrow{\text{simplify}} 2h^2 = 225 \text{ ft}^2$$

$$\xrightarrow{\text{divide by 2}} h^2 = \frac{225 \text{ ft}^2}{2} = 112.5 \text{ ft}^2 \xrightarrow{\text{square root}} h = \sqrt{112.5 \text{ ft}^2} = 10.7 \text{ ft.}$$

The height of the stairway is 10.7 feet and, because the base and height are the same, the base also is 10.7 feet. The area of the triangular region is

$$A = \frac{1}{2} b \times h = \frac{1}{2}(10.7 \text{ ft}) \times (10.7 \text{ ft}) = 57.25 \text{ ft}^2.$$

Slightly less than 60 square feet of plywood are needed to cover the triangular region. A rectangular 4 foot by 8 foot plywood sheet has an area of $4 \times 8 = 32$ square feet, so two sheets contain enough area to cover the triangular region. ❖

Time-Out to Think: How would you cut the two rectangular (4 foot by 8 foot) plywood sheets to cover the space beneath the stairs in Example 11–5? How many cuts would be necessary? Would working from three sheets instead of two be easier?

Example 11–6 ***Area of Property.*** You are considering the purchase of a mountain lot that borders a stream and an adjacent piece of private property. The stream frontage is 250 feet, the distance along the adjacent lot is 1200 feet, and the shape of the lot is a right triangle (Figure 11–15). What is the acreage of the lot?

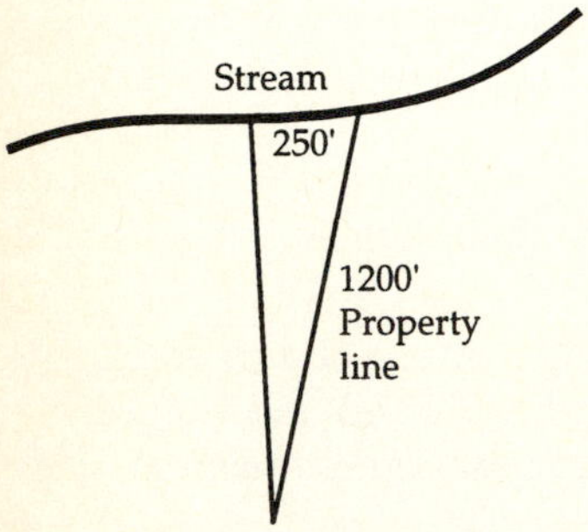

Figure 11–15. You are considering the purchase of this mountain lot which has the shape of a right triangle.

Solution: If you consider the stream frontage as the base of the triangle, the "height" (h) of the triangular lot is the side perpendicular to the stream. The 1200-foot property line is the hypotenuse and the 250-foot stream frontage is the third side. Thus apply the Pythagorean theorem and solve for h:

$$h^2 + (250 \text{ ft})^2 = (1200 \text{ ft})^2 \xrightarrow[(250 \text{ ft}^2)]{\text{subtract}} h^2 = (1200 \text{ ft})^2 - (250 \text{ ft})^2 \xrightarrow{\text{square root}}$$

$$h = \sqrt{(1200 \text{ ft})^2 - (250 \text{ ft})^2} \approx 1174 \text{ ft}.$$

The height is 1174 ft and the area of the lot is

$$A = \frac{1}{2} bh = \frac{1}{2} \times (1174 \text{ ft}) \times (250 \text{ ft}) = 146{,}750 \text{ ft}^2, \text{ or } 146{,}750 \text{ ft}^2 \times \frac{1 \text{ acre}}{43{,}250 \text{ ft}^2} \approx 3.4 \text{ acre}.$$

The lot has an area of about 3.4 acres. Note the importance of using the *height* of the triangle, rather than the hypotenuse, in the area formula. ❖

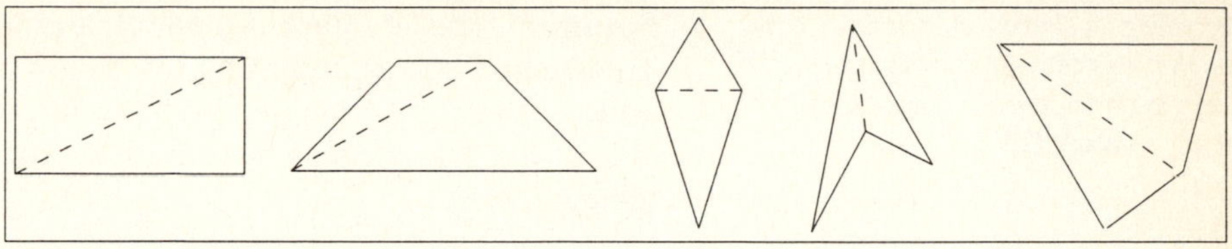

Figure 11–16. The area of any quadrilateral, or four-sided polygon, can be found by dividing it into two triangles (dotted lines).

Quadrilaterals

After triangles, the next simplest polygons are quadrilaterals. The word *quadrilateral* literally means "four-sided"; several types of quadrilaterals are shown in Figure 11–16. The area of a quadrilateral can always be found by dividing it into two triangles and adding the areas of each triangle. However, the areas of some special quadrilaterals can be found by using a simpler formula. For example, a rectangle is a quadrilateral in which all the angles are right angles, and the area is simply the *base* times the *height*. A square is a rectangle in which all sides have the same length.

Another special quadrilateral is a **parallelogram**, in which the opposite sides are parallel and have the same length. As shown in Figure 11–17, the area of a parallelogram also is its base times its height, or $b \times h$. Summarizing, squares, rectangles and parallelograms, all have the same simple area formula:

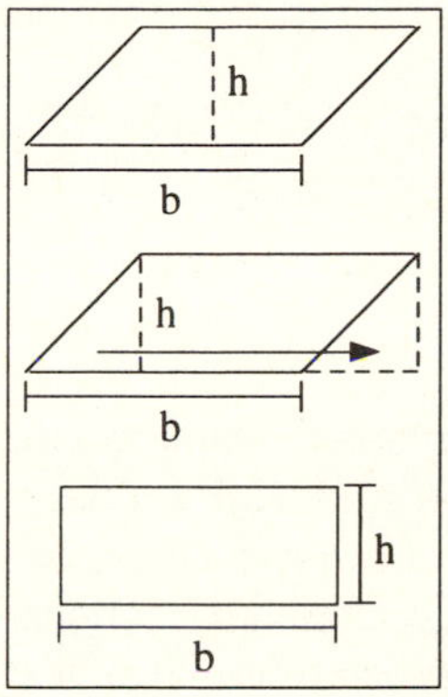

Figure 11–17. To find the area of a parallelogram, cut and rearrange the triangular portions to make a rectangle. Thus, as with a rectangle, the area of the parallelogram is $b \times h$. Note that the height is *not* the length of the slanted sides.

$$\text{Area of squares, rectangles, and parallelograms} = \text{base} \times \text{height}, \quad \text{or} \quad A = b \times h.$$

Example 11–7 *Laying Carpet.* The three rooms and hallway shown in the floor plan of Figure 11–18 are to be carpeted wall to wall. The carpet for the bedrooms costs \$15 per square yard, and the carpet for the bathroom and hall costs \$12 per square yard. What is the cost of the entire project?

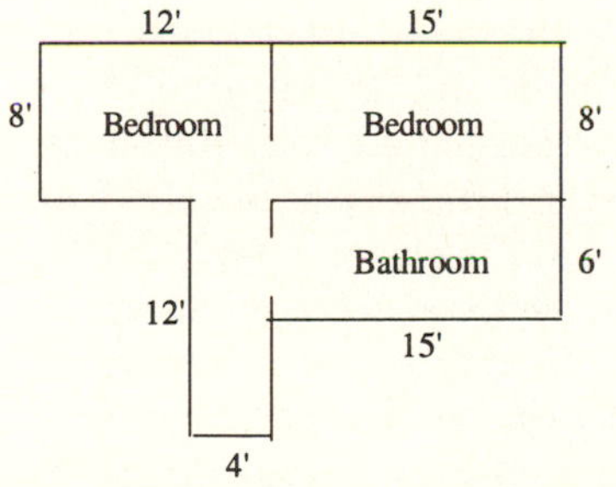

Figure 11–18. Carpet is to be installed in the three rooms and hallway represented by this floor plan.

Solution: The area of the two bedrooms is $(12 \times 8) + (15 \times 8) = 216$ square feet, or

$$216\ \text{ft}^2 \times \left(\frac{1\ \text{yd}}{3\ \text{ft}}\right)^2 = 24\ \text{sq yd}.$$

At \$15/yd², 24 square yards of carpet will cost 24 yd² × \$15/yd² = \$360.
For the hall and bathroom the total area is

$$\underbrace{12\ \text{ft} \times 4\ \text{ft}}_{\text{area of hall}} + \underbrace{15\ \text{ft} \times 6\ \text{ft}}_{\text{area of bathroom}} = 48\ \text{ft}^2 + 90\ \text{ft}^2 = 138\ \text{ft}^2, \text{ or } 138\ \text{ft}^2 \times \frac{1\ \text{yd}^2}{9\ \text{ft}^2} \approx 15.3\ \text{yd}^2.$$

At \$12/yd², 15.3 square yards of carpet will cost 15.3 yd² × \$12/yd² ≈ \$184.
The total cost for the carpet is \$360 + \$184 = \$544. However, this cost assumes that no carpet is wasted; in fact, because carpet is sold in rolls of fixed length, some waste almost always occurs when carpet is cut to fit. As a rule of thumb, you should expect waste of 10% to 20%; thus the actual cost of the project is likely to be 10% to 20% more than \$544, or about \$600 to \$650. ❖

> **Time-Out to Think:** Suppose that the carpet in Example 11–7 comes in 12-foot-wide rolls. How should it be cut to fit the area? How much carpet would be wasted?

Example 11–8 *Landscaping.* A city park is to be developed on an open block bounded by two sets of parallel streets (Figure 11–19). The streets along the block are each 55 yards long and intersect at a 45° angle. How much sod should be purchased to cover the entire area in grass?

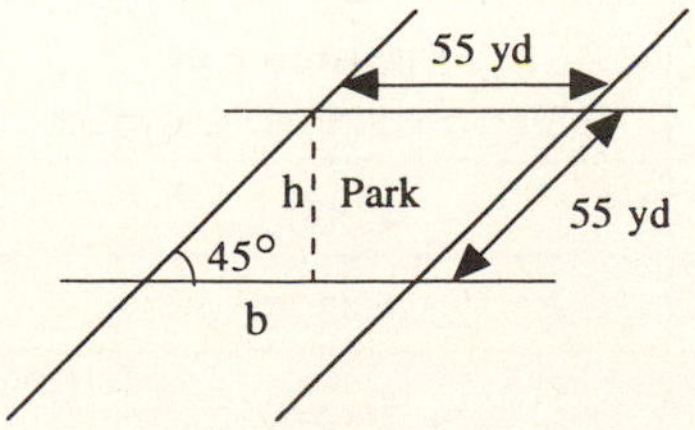

Figure 11–19. A city park is bounded by streets that are 55 yards long and that intersect at a 45° angle.

Everything an Indian does is in a circle, and that is because the power of the world works in circles, and everything tries to be round. —Black Elk

Solution: The city park is a parallelogram with a base length of 55 yards. The dashed line representing the height h of the parallelogram is one side of an isosceles right triangle that has a hypotenuse of 55 yards; thus, the third side of the triangle also has length h. Applying the Pythagorean theorem yields:

$$h^2 + h^2 = (55\text{ yd})^2 \xrightarrow{\text{simplify}} h^2 = \frac{(55\text{ yd})^2}{2} \xrightarrow{\text{square root}} h = \sqrt{\frac{(55\text{ yd})^2}{2}} \approx 38.9\text{ yd}.$$

The area of the parallelogram is its base times its height: $A = 55\text{ yd} \times 38.9\text{ yd} \approx 2140\text{ yd}^2$. Approximately 2140 square yards of sod are needed. ❖

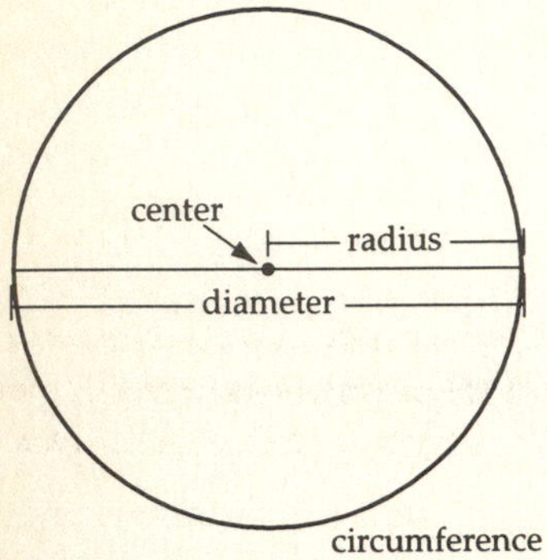

Figure 11-20. A radius and diameter are shown for a circle. The distance around the circle is its circumference.

11.3.2 Circles

With their symmetry and simplicity, circles always have fascinated mystics and mathematicians alike. Geometrically, a circle is defined as the set of points whose distance from a fixed point, the **center**, is constant. The constant distance between the center and the circle is the **radius**, denoted r. The distance across the circle, passing through the center, is twice the radius or the **diameter**, denoted d ($d = 2r$). The distance around the circle is its **circumference** (Figure 11–20).

> **Time-Out to Think:** Describe how you can draw a circle by using a piece of string, a thumb tack, and a pencil. Describe how to determine the circumference of your circle if you have a longer piece of string and a ruler.

You have already worked with formulas for both the circumference ($C = 2\pi r$) and area ($A = \pi r^2$) of a circle. Where do these formulas come from? Archimedes (287–212 B.C.; see also Section 9.3) made the first recorded attempt to find an exact formula for the circumference of a circle. His strategy begins with two squares: one *inscribed* in a circle and the other *circumscribed* around the circle (Figure 11–21a). He reasoned that the circumference of the circle must be greater than the perimeter of the inscribed square and less than the perimeter of the circumscribed square. Next, he doubled the number of sides of his figures, so that each square became an octagon (Figure 11–21b). He repeated the process again, producing inscribed and circumscribed 16-gons (16-sided polygons as shown in Figure 11–21c), 32-gons, and so on.

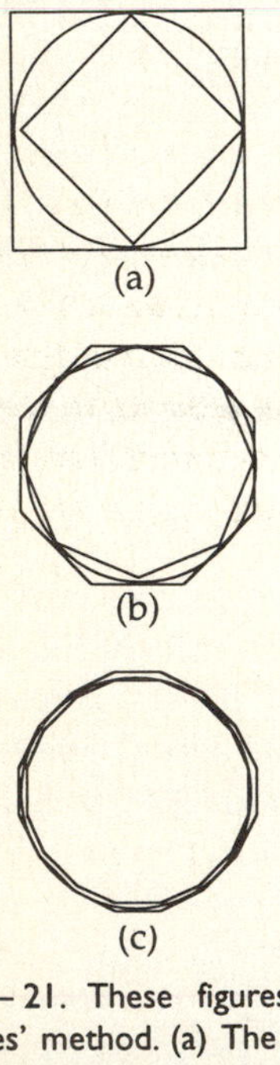

Figure 11–21. These figures illustrate Archimedes' method. (a) The circumference of the circle lies between the perimeters of the inscribed and circumscribed squares. (b) Replacing the squares with octagons yields a better approximation to the circumference. (c) As the number of sides in the polygons increases (16-gons shown), their perimeters approach the circumference of the circle.

Table 11–2 shows the perimeters for inscribed and circumscribed polygons around a circle with a diameter of 1 unit. The circle's circumference is $\pi \times$ diameter, or π units. Thus as the number of sides of the inscribed and circumscribed polygons increases, their perimeters converge toward π. In the

Table 11–2. Archimedes Method for Approximating π

Number of sides of polygons	Perimeter of inscribed polygon	Perimeter of circumscribed polygon
4	2.8284	4.0000
8	3.0615	3.3137
16	3.1214	3.1826
32	3.1365	3.1517
64	3.1403	3.1441

language of calculus, π is the limit of the series of numbers obtained with Archimedes' strategy. Thus Archimedes almost invented calculus nearly 2000 years before Newton and Leibniz finally developed it.

Time-Out to Think: Almost every aspect of modern science and technology makes use of calculus. How might the world be different if Archimedes had succeeded in inventing calculus 2000 years before it actually was?

Archimedes' approximation to π was about 3.14. Because π is irrational, it cannot be written exactly. Today, supercomputers can compute π to billions of digits; the first few digits are 3.141592653589793....

The problem of finding the *area* of a circle was no less vexing to ancient mathematicians. They knew that the ratio of the area of a circle to its radius squared is the same constant for all circles, but they didn't realize that the constant is π. The formula for the area of a circle can also be found using inscribed and circumscribed polygons, much as Archimedes did for the circumference.

Time-Out to Think: In an act of stunning innumeracy, the Indiana legislature in 1897 decided that π should have an exact value, so they legislated that π should be exactly 3! What is wrong with this approach?

Optimization problems ask that some quantity (like area, volume, or cost) be *optimized* under specified conditions.

The following three examples show how the circle can be used to model practical situations. Note that Example 11–11 demonstrates an **optimization problem**, which asks: What is the *greatest* area that can be enclosed by a fixed perimeter? In general, optimization problems deal with the greatest or least value of some quantity under specified conditions.

Example 11–9 ***Backyard Geometry.*** In designing your backyard you decide to build a circular patio and connect it to the house by a walkway (Figure 11–22). The patio is to have a diameter of 30 feet, the walkway is to be 20 feet long and 4 feet wide, and the entire yard measures 75 feet by 60 feet. How much material do you need for the patio and walkway, and what is the area of the remaining yard?

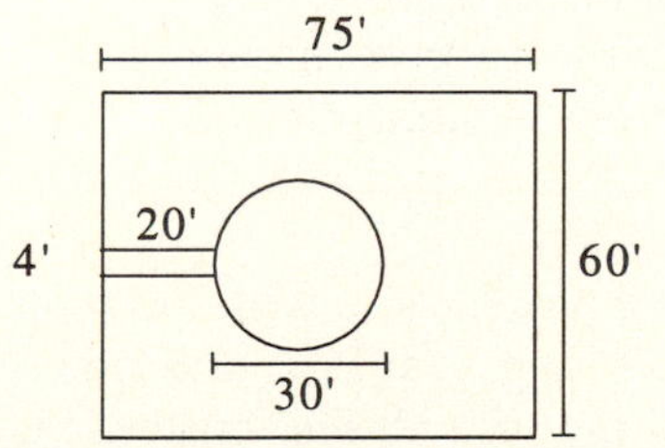

Figure 11–22. A patio and a walkway are to be built in a rectangular yard.

Solution: The walkway is a rectangle whose area is 20 ft × 4 ft = 80 ft^2. The patio is a circle with a radius of 15 feet and an area of $\pi \times (15\text{ ft})^2 = 707\text{ ft}^2$. A total of 707 ft^2 + 80 ft^2 = 787 ft^2 of material is needed for the patio and walkway. The entire yard is a rectangle whose area is 75 ft × 60 ft = 4500 ft^2. This leaves an area of 4500 ft^2 – 787 ft^2 = 3713 ft^2 for the remaining yard. ❖

Example 11–10 ***Interior Design.*** A window consists of a 4-foot by 6-foot rectangle capped by a semicircle (Figure 11–23). How much glass is needed for the window? How much trim is needed for its perimeter?

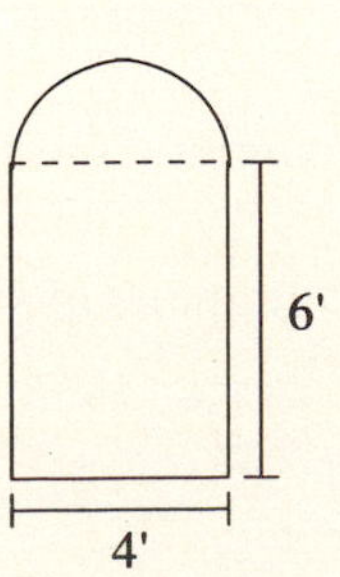

Figure 11–23. A window is to made in the shape of a rectangle capped by a semicircle.

Solution: The total area of the window is the area of the 4 × 6-foot rectangle plus the area of a semicircle with a diameter of 4 feet, or a radius of 2 feet. (The area of a semicircle is half of the area of a full circle, or $\frac{1}{2}\pi r^2$.)

$$A = \underbrace{4\text{ ft} \times 6\text{ ft}}_{\text{rectangle area}} + \underbrace{0.5 \times \pi \times (2\text{ ft})^2}_{\text{semicircle area}} \approx 24\text{ ft}^2 + 6.3\text{ ft}^2 = 30.3\text{ ft}^2.$$

The window requires a little more than 30 square feet of glass. The trim around the window must cover the 4-foot base of the rectangle, two 6-foot sides, and the semicircular cap. The straight edges have a total length of 6 ft + 6 ft + 4 ft = 16 ft. The perimeter of the semicircular cap is half the circumference of a circle with a diameter of 4 feet, or $0.5 \times \pi \times 4$ ft = 6.3 ft. Therefore a total of 16 ft + 6.3 ft = 22.3 ft of trim is needed. As always in construction, some waste will occur when the glass and trim are cut. Thus at least 32 square feet of glass and 24 feet of trim should be purchased.

❖

Example 11–11 *Optimizing Area.* You have 132 meters of fence that you want to use to enclose a corral on a ranch. What shape would you choose for the corral if you want it to have the *maximum* area?

The circle has the greatest *area* of all two-dimensional figures with a fixed perimeter.

Solution: Before answering this question, take a few moments to experiment. Find a piece of string, tie the ends together, and make it into various shapes. If you stretch the string into a long narrow shape, it doesn't enclose much area. What shape enclosed by the string has the maximum area? After some experimentation, you should be able to conclude that two likely candidates for the best shape are a square and a circle. Which has the greater area, a square with a perimeter of 132 meters or a circle with a circumference of 132 meters?

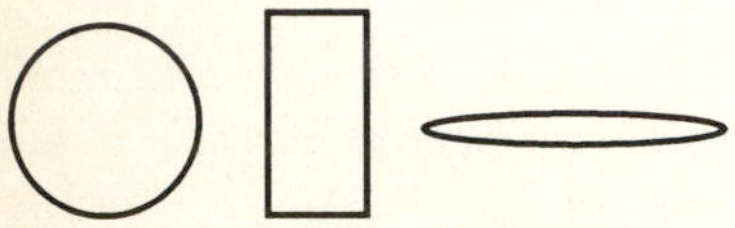

Figure 11–24. All the figures shown have the *same perimeter*, but the circle has the greatest area.

A square with a perimeter of 132 meters has sides with length 132 m ÷ 4 = 33 m. Therefore the area of the square is $A = (33 \text{ m})^2 = 1089 \text{ m}^2$. A circle with a circumference of 132 m has a diameter of 132 m ÷ π = 42 m, or a radius of 21 m; its area is $A = \pi \times (21 \text{ m})^2 = 1385 \text{ m}^2$. Clearly, the circle has the greater area. In fact, among all figures having the same perimeter, a circle has the maximum area (Figure 11–24).

❖

Similar triangles have the same set of three angles, but may have different side lengths. Corresponding pairs of sides have the same length ratio.

11.3.3 Similar Triangles

Two triangles are **similar** if they have the same shape (but not necessarily the same size). Two similar triangles are shown in Figure 11–25; angles (vertices) are labeled with capital letters and sides with lower case letters. Note that side *a* is *opposite* ∠*A*, side *b* is *opposite* ∠*B*, and side *c* is opposite ∠*C*.

Imagine picking up triangle *ABC* and laying it on top of triangle *A′B′C′* so that the vertices *A* and *A′* coincide. You can then see that ∠*A* and ∠*A′* are equal. A similar procedure applied to the pairs ∠*B*, ∠*B′* and ∠*C*, ∠*C′* shows that all three pairs of angles are equal in measure.

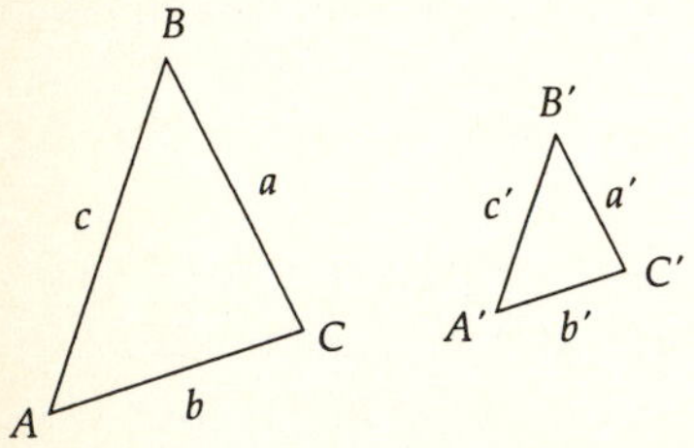

Figure 11–25. Triangles *ABC* and *A′B′C′* are similar.

Next, consider the lengths of the sides. Using a ruler, you will find that side *a* is twice as long as side *a′*, side *b* is twice as long as side *b′*, and side *c* is twice as long as side *c′*. That is, the ratio of each pair of sides is constant in similar triangles, or

$$\frac{a}{a'} = \frac{b}{b'} = \frac{c}{c'}.$$

Example 11–12 Figure 11–26 shows two similar triangles. Find the length of the unknown side labeled *a*.

Solution: Because the triangles are similar, corresponding pairs of sides have the same ratio. That is,

$$\frac{a}{7} = \frac{4}{6} = \frac{6}{9}.$$

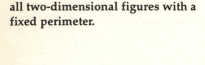

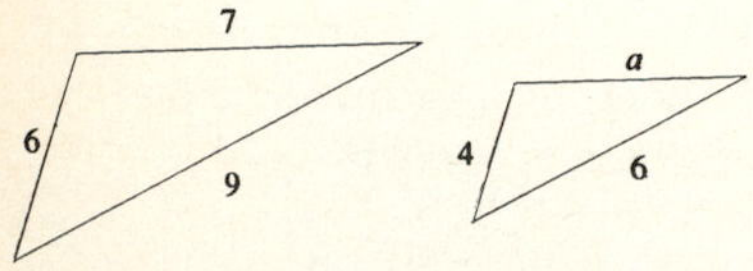

Figure 11–26. Two similar triangles are shown, with one side unknown.

Note that the ratios 4/6 and 6/9 both are equivalent to 2/3, which must also be the ratio of $a/7$. That is,

$$\frac{a}{7} = \frac{2}{3} \xrightarrow{\text{multiply by 7}} a = \frac{2}{3} \times 7 = \frac{14}{3}.$$

Thus the length of side a is 14/3.

❖

Example 11–13 ***Eratosthenes and the Circumference of the Earth.*** In about 240 B.C., Greek astronomer Eratosthenes (c. 276–196 B.C.) measured the circumference of the Earth. He learned that on noon of the summer solstice, the Sun was directly overhead in the city of Syene (present day Aswan) in southern Egypt. At the same time, it was 7° from directly overhead in Alexandria. He estimated the north–south distance between Syene and Alexandria to be 5000 stadia (plural of stadium), a Greek unit of distance thought to be about 1/6 kilometer. How did Eratosthenes use properties of similar triangles to find the circumference of the Earth? What value did he calculate?

Solution: Figure 11–27 shows the geometry of the problem. Two radii are drawn from the center of the Earth, one passing through Syene and one through Alexandria. The radii point to the zenith (the point directly overhead) for each location. The radius passing through Syene represents the direction to the Sun because the Sun is directly overhead in Syene. The radius passing through Alexandria makes a 7° angle with the direction of the Sun, forming the pair of similar angles shown. By including the dashed lines in the figure, we get similar triangles; thus the angle formed at the center of the Earth also must be 7°. Further, because 7° represents 7/360 of a full circle, the distance from Syene to Alexandria must be 7/360 of the circumference of the Earth:

Eratosthenes' value of about 43,000 km is only about 3000 km ÷ 40,000 km = 7.5% more than the actual circumference of 40,030 km (equatorial). Because the exact length of Greek stadia is not known, his actual measurement may have been somewhat closer to or farther from the actual circumference of the Earth. As Eratosthenes claimed only to be making a reasonable estimate, he was extraordinarily successful.

$$5000 \text{ stadia} = \frac{7}{360} \times \text{Earth circumference} \xrightarrow{\text{multiply by } \frac{360}{7}}$$

$$\text{Earth circumference} = \frac{360}{7} \times 5000 \text{ stadia} \approx 257{,}000 \text{ stadia}.$$

Assuming that stadia are 1/6 kilometer, Eratosthenes found the circumference of the Earth to be about

$$257{,}000 \text{ stadia} \times \frac{1 \text{ km}}{6 \text{ stadia}} = 42{,}800 \text{ km}.$$

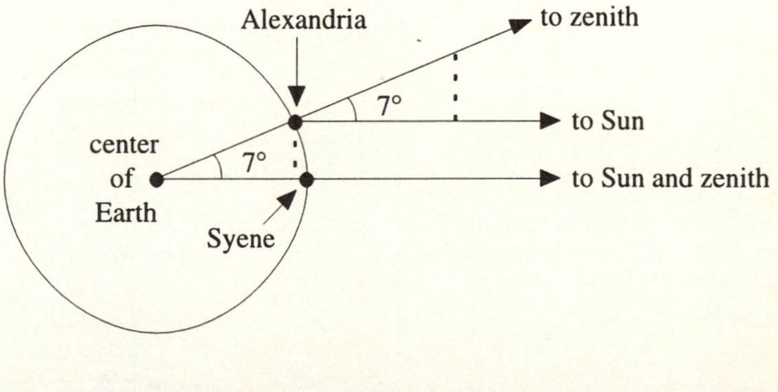

Figure 11–27. Eratosthenes used properties of similar triangles to calculate the circumference of the Earth in about 240 B.C.

❖

Time-Out to Think: Eratosthenes, who did his work more than 1700 years before Columbus, was well aware that the Earth is round! In fact, the Greeks had accepted the idea of a round Earth at least the time of Pythagoras (c. 500 B.C.). Can you think of any evidence that the Greeks might have used to support their belief that the Earth is round?

Shadows are longest on the shortest day of the year. In the Northern Hemisphere, that day is the winter solstice (about December 22). The shadow lengths given in this problem correspond to those at a latitude of about 35°, approximately the latitude of Santa Barbara, CA, Albuquerque, NM, or Charlotte, NC.

Example 11–14 *Solar Access.* Many cities have policies that prevent new houses and additions from casting shadows on neighboring houses. The intent of these policies is to allow everyone access to the Sun for the use of solar energy devices. Suppose that a city's solar access policy reads as follows:

On the shortest day of the year, a house cannot cast a noontime shadow longer than the shadow that would be cast by a 12-foot high fence on the property line.

Suppose that your house is set back 30 feet from the north property line. Suppose further that, on the shortest day of the year, a 12-foot fence casts a noontime shadow 20 feet in length. If you make an addition to your house, how tall can your house be (after the addition) under the solar access policy?

Solution: Figure 11–28 shows the geometry of the problem. As you are concerned with "how tall" the house can be, we drew it at its maximum allowed height: Its shadow therefore extends to the same place as the shadow of the 12-foot-high fence on the property line. If the house were any taller than the height h shown, its shadow would extend past the fence shadow and violate the solar access policy.

Note the two similar triangles: One is formed by the ground, the line of sunlight, and the *fence;* the other by the ground, the line of sunlight, and the *house*. The triangles are similar, so

$$\frac{\text{max. house height } (h)}{\text{house shadow length}} = \frac{\text{fence height}}{\text{fence shadow length}}.$$

The problem specifies that the fence height and the fence shadow length are 12 feet and 20 feet, respectively. The house shadow length is the length of the fence shadow plus the setback from the property line, or 20 ft + 30 ft = 50 ft. Solve for h by multiplying both sides by the *house shadow length*:

$$h = \left(\frac{\text{fence height}}{\text{fence shadow length}}\right) \times (\text{house shadow length}) = \left(\frac{12 \text{ ft}}{20 \text{ ft}}\right) \times 50 \text{ ft} = 30 \text{ ft}.$$

The new addition must be built so that the total height of the house does not exceed 30 feet. However, the calculation may contain several sources of uncertainty. For example, one assumption was that the ground is flat; if the house is on a sloping lot, this solution won't be valid. Note also that the answer is based on the shadow cast by the *north* wall of the house. Figure 11–28 indicates that a sloping roof, or an addition on the south side of the house, might be able to extend higher than 30 feet without making the shadow any longer.

Figure 11–28. The figure represents a 12-foot high fence on a property line and the 20-foot-long shadow that it casts on the shortest day of the year. Set back 30 feet from the property line, the house is drawn at its maximum allowed height so that its shadow falls to the same place as the fence shadow.

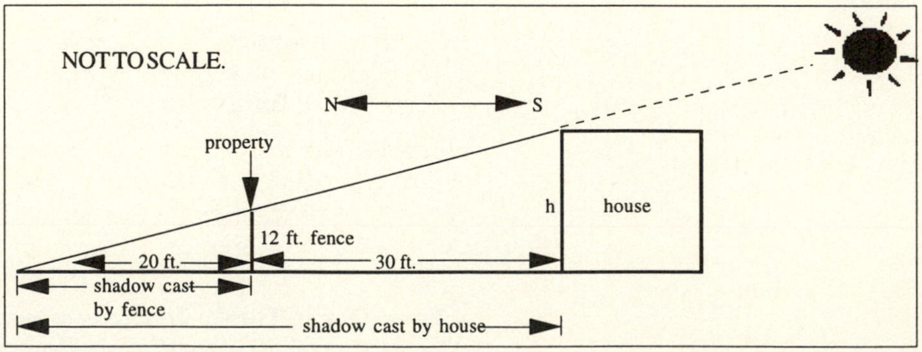

❖

Time-Out to Think: Suppose that you were the architect for the new addition in Example 11–14. On Figure 11–28, show how you could build an addition to exceed 30 feet in total height without violating the solar access policy.

11.4 PROBLEMS IN THREE DIMENSIONS

Because the world is three-dimensional, many problems require working with three-dimensional objects, or **solids**. Just as perimeter and area are fundamental properties of plane (two-dimensional) shapes, the fundamental properties of solids include surface area and volume. In this section, we explore problems involving boxes, cylinders, and spheres, which illustrate basic principles that can be extended to other solid shapes.

11.4.1 Problems with Boxes (Rectangular Prisms)

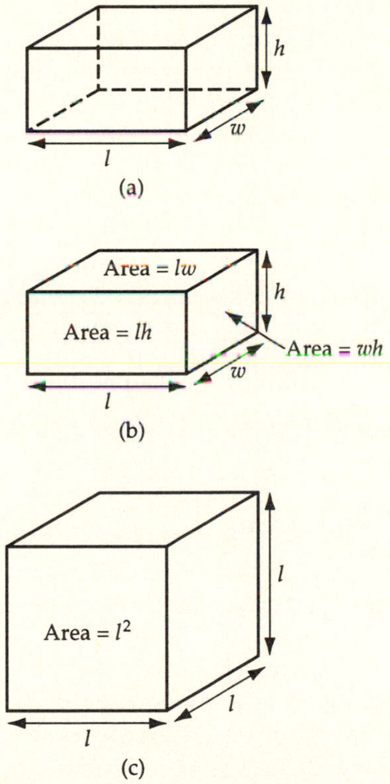

Figure 11–29. (a) A 6-sided box, or rectangular prism, has a length, *l*, width, *w*, and height, *h*. (b) The top and bottom (base) have area *lw*; the front and back sides have area *lh*, and the remaining two sides have area *wh*. (c) A cube is a box in which all side lengths are the same.

A typical 6-sided box (Figure 11–29a) — technically known as a **rectangular prism** — has a length, l, width, w, and height, h. Its total surface area is the sum of the areas of its six sides (Figure 11–29b), or

$$\text{Surface area} = S = 2(lw + lh + wh).$$

The volume of the box is the area of its base times its height or, equivalently, the product of its length, width, and height:

$$\text{Volume} = \text{area of base} \times \text{height}, \quad \text{or} \quad \text{volume} = \text{length} \times \text{width} \times \text{height}.$$

A **cube** (Figure 11–29c) is a rectangular prism in which the length, width, and height all are equal, so that its volume simply is its side length cubed (l^3), and its surface area is six times the area of one side ($6l^2$).

Example 11–15 *Water Reservoir.* A water reservoir has a rectangular base that measures 30 meters by 40 meters, and vertical walls 15 meters high. At the beginning of the summer, the reservoir was filled to capacity. At the end of the summer, the water depth is 4 meters. How much water was used?

Solution: The reservoir has the shape of a rectangular prism. When filled, the volume of water was 30 m × 40 m × 15 m = 18,000 m^3. At the end of the summer, the amount of water remaining is 30 m × 40 m × 4 m = 4800 m^3. Therefore the amount of water used was 18,000 m^3 – 4800 m^3 = 13,200 m^3. The *fraction* of the original water remaining is 4800 m^3 ÷ 18,000 m^3 = 0.267. That is, about 27% of the water remains and about 73% was used.

❖

Example 11–16 *Molecules in a Room.* Estimate of the amount of air in an average room in a house, in liters. How many molecules does this volume contain?

Solution: Assume that a typical room has a floor measuring 10 feet by 10 feet and a ceiling height of 10 feet. The volume of the room is 10 ft × 10 ft × 10 ft = 1000 ft^3. Converting to liters yields

$$1000\ \text{ft}^3 \times \left(\frac{1\ \text{m}}{3.28\ \text{ft}}\right)^3 \times \left(\frac{100\ \text{cm}}{1\ \text{m}}\right)^3 \times \left(\frac{1\ \text{liter}}{1000\ \text{cm}^3}\right) \approx 30{,}000\ \text{liter}.$$

The room holds about 30,000 liters of air. Obtaining the number of molecules requires the **Ideal Gas Law** from chemistry: 22 liters of gas (such as air) at room temperature and pressure contains 6×10^{23} (Avogadro's number) molecules. That is, air contains 6×10^{23} molecules per 22 liters. The total number of molecules, then, is

$$30{,}000 \text{ liter} \times \frac{6 \times 10^{23} \text{ molecules}}{22 \text{ liter}} = 8 \times 10^{27} \text{ molecules}.$$

To an order of magnitude, the typical number of molecules of air in a room is about 10^{27}. Note that this quantity is comparable to the number of *atoms* in the human body (see Section 7.3).

❖

Example 11–17 *Optimal Container Design.* You are designing wooden crates (rectangular prisms) that must have a volume of 2 cubic meters. The wood used to make the crates costs \$12 per square meter. What is the most economical design for the crates, and how much will each crate cost to manufacture?

Solution: This problem involves optimization because the design is to have the *smallest* surface area (and hence the lowest manufacturing cost) for a volume of 2 cubic meters. You might begin by trial and error. For example, if the base is a square with sides 1.2 meters long, its area is $1.2 \text{ m} \times 1.2 \text{ m} = 1.44 \text{ m}^2$. In this case, the height of the crate must be

$$\text{height} = \frac{\text{volume}}{\text{area of base}} = \frac{2 \text{ m}^3}{1.44 \text{ m}^2} = 1.39 \text{ m}.$$

The total surface area of this box is the area of its top and bottom (each 1.44 m^2) plus the area of its four side walls that have dimensions of 1.2 meters by 1.39 meters, or

$$\text{Surface area} = (2 \times 1.44 \text{ m}^2) + [4 \times (1.2 \text{ m} \times 1.39 \text{ m})] = 9.55 \text{ m}^2.$$

As your next guess, you might try a square base with sides of 1.1 meters. In that case, the height and surface area would be

$$\text{height} = \frac{\text{volume}}{\text{area of base}} = \frac{2 \text{ m}^3}{(1.1 \text{ m})^2} = 1.65 \text{ m};$$

$$\text{Surface area} = [2 \times (1.1 \text{ m} \times 1.1 \text{ m})] + [4 \times (1.1 \text{ m} \times 1.65 \text{ m})] = 9.68 \text{ m}^2.$$

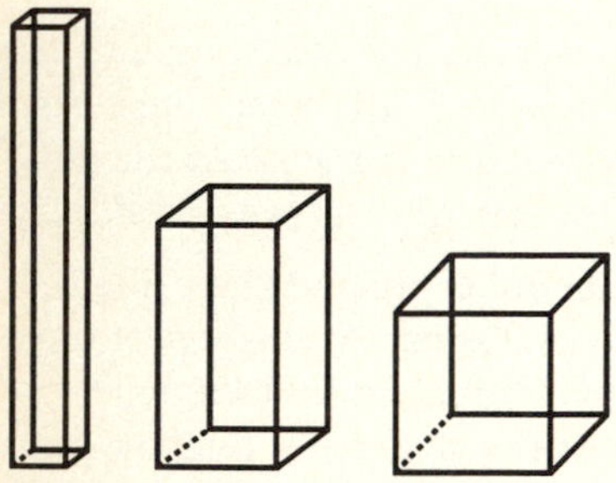

Figure 11–30. All three boxes shown have the same volume, but the cube has the *smallest* total *surface area*.

Note that this box is more "elongated" (taller, with a smaller base) than the first box, and that its surface area is larger. Thus you might guess that the optimal design has the minimum elongation: a *cube* (Figure 11–30). Because all sides of a cube have the same length, and the volume must be 2 cubic meters, the side length is

$$\text{side length} = \sqrt[3]{\text{volume}} = \sqrt[3]{2 \text{ m}^3} = 1.26 \text{ m}.$$

The six sides of this cube have a total surface area of $6 \times (1.26 \text{ m})^2 = 9.53 \text{ m}^2$. At a cost of \$12 per square meter, the wood for each crate made with this optimal design will cost $\$12/\text{m}^2 \times 9.53 \text{ m}^2 = \$114.31$, assuming that no material is wasted.

❖

11.4.2 Cylinders

An aluminum can has the shape of a **cylinder** (technically, a **right circular cylinder**). The two key dimensions of a cylinder are its height, h, and the radius, r, of its circular base (Figure 11–31a). The volume of a cylinder is the area of its circular base, πr^2, multiplied by its height h:

$$\text{Volume} = \text{area of base} \times \text{height} = \pi r^2 h.$$

The surface area of a cylinder is the sum of the areas of its top and bottom, each of which is πr^2, and the area of its curved surface. As shown in Figure 11–31(b), the area of the curved surface is $2\pi rh$, making the total surface area of the cylinder

$$\text{Surface area} = \underbrace{2\pi\ rh}_{\substack{\text{area of} \\ \text{curved surface}}} + \underbrace{2\pi r^2}_{\text{area of ends}}.$$

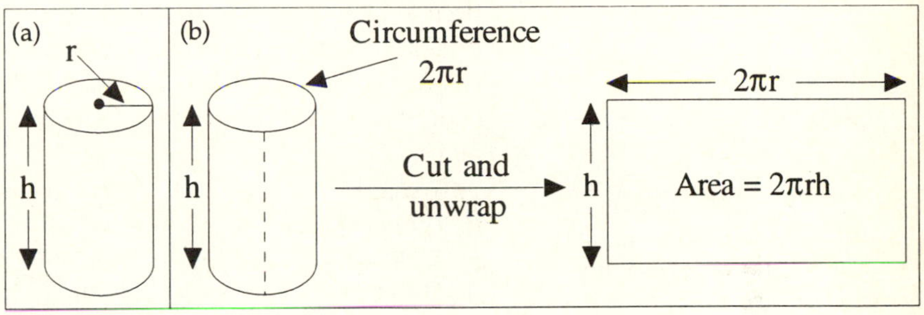

Figure 11–31. (a) A (right circular) cylinder has a circular base with radius *r* and height *h*. (b) To find the area of the curved surface imagine cutting it lengthwise and unfolding it to form a rectangle. The areas of the rectangle and the curved surface are the same.

Example 11–18 ***Comparing Volumes.*** Which holds more soup: a can with a radius of 3 inches and a height of 4 inches or a can with a radius of 4 inches and a height of 3 inches?

Solution: Find the volumes of two cylinders.

$$\text{Can 1: } V = \pi \times r^2 \times h = \pi \times (3 \text{ in})^2 \times 4 \text{ in} = 113.1 \text{ in}^3$$

$$\text{Can 2: } V = \pi \times r^2 \times h = \pi \times (4 \text{ in})^2 \times 3 \text{ in} = 150.8 \text{ in}^3$$

The second can, with the larger radius but shorter height, has the larger capacity.

❖

Example 11–19 ***Container Design.*** Your company manufactures oil drums that have a radius of 0.75 meter and a height of 1.2 meters (Figure 11–32). The cost of the material for the tops and bottoms of the drums is \$5.25 per square meter; the cost of the material for the curved wall is \$4.25 per square meter. What is the capacity of a single drum? What is the cost of materials for a single drum?

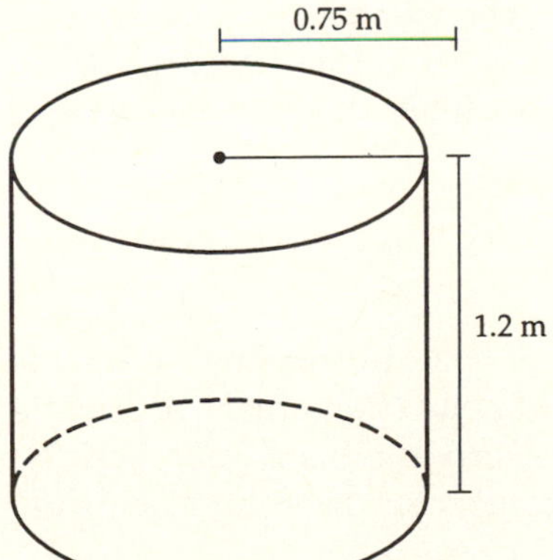

Figure 11–32. Each oil drum is a cylinder with radius 0.75 m and height of 1.2 m.

Solution: Each drum has the shape of a cylinder, so its volume is

$$V = \pi r^2 h = \pi \times (0.75\ m)^2 \times (1.25\ m) = 2.21\ m^3.$$

The capacity of each drum is about 2.2 cubic meters. The cost of materials for a drum is the sum of the costs for the top and bottom and the curved wall, or

$$\text{Total cost} = \left[\underbrace{2 \times \pi (0.75 \text{ m})^2}_{\substack{\text{area of top} \\ \text{and bottom}}} \times \underbrace{\frac{\$5.25}{1 \text{ m}^2}}_{\substack{\text{material cost for} \\ \text{top and bottom}}} \right] + \left[\underbrace{2\pi (0.75 \text{ m})(1.2 \text{ m})}_{\text{area of curved wall}} \times \underbrace{\frac{\$4.25}{1 \text{ m}^2}}_{\substack{\text{material cost for} \\ \text{curved wall}}} \right] = \$43.59.$$

The manufacturing cost for each drum is \$43.59, assuming no wastage.

❖

Example 11–20 ***Optimal Can Design (Challenge Problem).*** Standard soft drink cans hold 12 ounces, or 355 milliliters. What are the optimal dimensions of a 12-ounce can, in terms of minimizing the surface area?

Solution: Begin with the formula for the total surface area, A, of the can:

$$A = \text{area of top and bottom} + \text{area of curved side} = 2\pi^2 + 2\pi rh.$$

The total volume of the can must be 355 milliliters, which is the same as 355 cm^3. From the cylinder volume formula, you can now solve for the can height, h, in terms of r:

$$V = \pi r^2 h = 355 \text{ cm}^3 \xrightarrow[\text{sides by } \pi r^2]{\text{divide both}} h = \frac{355 \text{ cm}^3}{\pi r^2}.$$

Substituting this expression for h into the area formula gives a relation (*radius, area*) that describes how the total surface area of a 355 cm^3 can depends on the radius of the can:

$$A = 2\pi r^2 + 2\pi r \times \frac{355 \text{ cm}^3}{\pi r^2} \quad \Rightarrow \quad A = 2\pi r^2 + \frac{710 \text{ cm}^3}{r}.$$

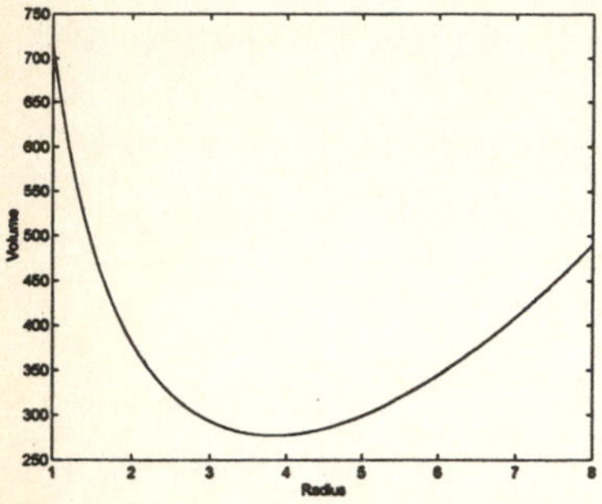

Figure 11–33. The relation between the surface area and the radius of a 355 cm^3 can is given by $A = 2\pi r^2 + 710/r$. The graph of the relation indicates that a radius of about 3.8 cm yields the minimum surface area.

Now, find the radius that gives the minimum value of the surface area. Although the (*radius, area*) relation isn't linear, you can graph it by using the point-by-point method or a graphing calculator or computer (see Appendix B); the graph is shown in Figure 11–33. The graph indicates that the relation yields a minimum area at a radius of about $r = 3.8$ cm. The height of the can, then, is

$$h = \frac{355 \text{ cm}^3}{\pi r^2} = \frac{355 \text{ cm}^3}{\pi (3.8 \text{ cm})^2} = 7.8 \text{ cm}^3.$$

The optimal design (minimum surface area) for a cylindrical can with a capacity of 355 cubic centimeters has a radius of about 3.8 cm and a height of about 7.8 cm. You can check these results to be sure that they yield the correct volume: $V = \pi r^2 h = \pi(3.8 \text{ cm})^2 \times (7.8 \text{ cm})^2 = 354 \text{ cm}^3$. The slight discrepancy from 355 cm^3 is the result of rounding.

❖

Time-Out to Think: Measure the dimensions of a soda can. How do its dimensions compare to those that minimize the amount of aluminum calculated in Example 11–20? Why do you think that manufacturers chose a design that does not have the minimum area?

11.4.3 Spheres

A **sphere** (Figure 11–34) is the set of all points in *three*-dimensional space that are a constant distance (the radius) from a fixed point (the center). A geometric sphere is an idealization of common objects such as billiard balls, ball bearings, marbles, atoms, and planets. The formulas for the volume and surface area of a sphere are

$$\text{volume} = \frac{4}{3}\pi r^3 \text{ and surface area} = 4\pi r^2.$$

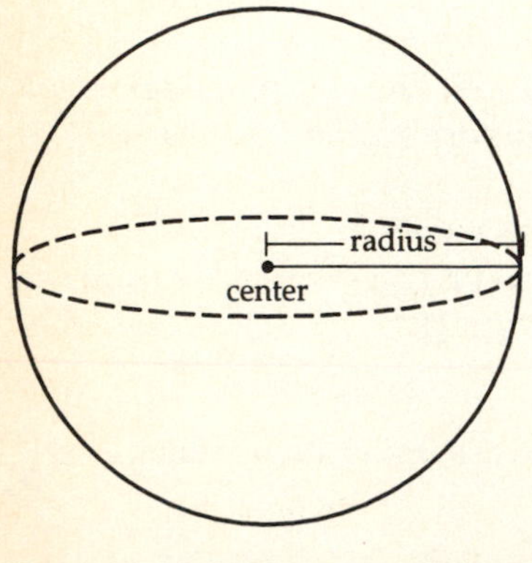

Figure 11–34. A sphere is an idealization of many common objects.

You can confirm these formulas through experimentation. For example, you can measure the volume of water needed to fill a hollow sphere of some particular radius and then verify that the volume matches the prediction of the volume formula. Similarly, you can measure the surface area of a sphere by covering it with graph paper and summing the areas of the squares needed to cover the entire surface.

Why Is the Moon Round?

Of all solid shapes with a fixed volume, a sphere has the minimum surface area.

The same principle explains the spherical shape of the Earth, the Sun, and the planets. Only small bodies in space, like some asteroids and comets, are distinctly nonspherical; that is because their gravity is too weak to "squeeze" them into spheres.

If you are a parent, at some point your children are likely to ask, "Why is the Moon round?" The answer relates to an important *optimization* property of spheres: For a particular volume, a sphere has the minimum surface area of any shape. That is, spheres have a smaller *surface area to volume ratio* (see Chapter 7) than any other solid shape.

This optimization property also means that spheres are the most tightly packed solids. If you have a lump of clay and squeeze it as tightly as possible, you will end up with a sphere. Because the Moon formed through the action of gravity, which effectively "squeezes" the Moon as you might squeeze a lump of clay, the Moon is a sphere.

Example 11–21 *Density of the Earth.* The radius of the Earth is about 6400 kilometers and its mass is about 6.0×10^{24} kg. What is the Earth's average density? Given that the density of rocks found on the Earth's surface averages about 3 grams per cubic centimeter, what can you say about the density of the Earth's core? Explain.

Solution: Recall from Chapter 9 that density is mass per unit volume. To make comparison easy, use units of grams per cubic centimeter. Note that 6400 km = 6.4×10^8 cm and that 6.0×10^{24} kg = 6.0×10^{27} grams. Thus the average density of the Earth is

$$\text{average density} = \frac{\text{mass of Earth}}{\text{volume of Earth}} = \frac{6.0 \times 10^{27}\,\text{g}}{\frac{4}{3}\pi\left(6.4 \times 10^{8}\,\text{cm}\right)^{3}} = 5.5\ \frac{\text{g}}{\text{cm}^{3}}.$$

The interior of the Earth is denser because denser material sank during the time when the Earth was molten, early in its history. Today, only the Earth's *outer core,* buried deep in the interior, remains molten.

The average density of the Earth is about 5.5 grams per cubic centimeter. As that is considerably more than the average density of surface rocks (3 grams per cubic centimeter), the interior of the Earth must be much denser than the surface. More detailed analyses of the Earth's density, coupled with measurements from seismic waves, reveal that the Earth's core is made primarily of nickel and iron and has an average density of about 15 grams per cubic centimeter.

❖

Packing Spheres

Whereas a single sphere is the most tightly packed solid of a given volume, packing many spheres is a much different problem. Imagine packing a crate of oranges. If you arrange them one on top of the other, a lot of empty space remains between them (Figure 11–35a). You can reduce the amount of empty space in the crate by allowing the spheres to roll into the gaps (Figure 11–35b). How should you pack them to fit the maximum number of oranges into the crate? This question illustrates a general type of problem, called a **packing problem,** in which the goal is to figure out how to pack small objects into a larger one.

Obtaining exact solutions to packing problems may be very difficult; in fact, many packing problems remain unsolved. Nevertheless, as Example 11–22 shows, you can always set limits on the number of small objects that can be packed into a larger object.

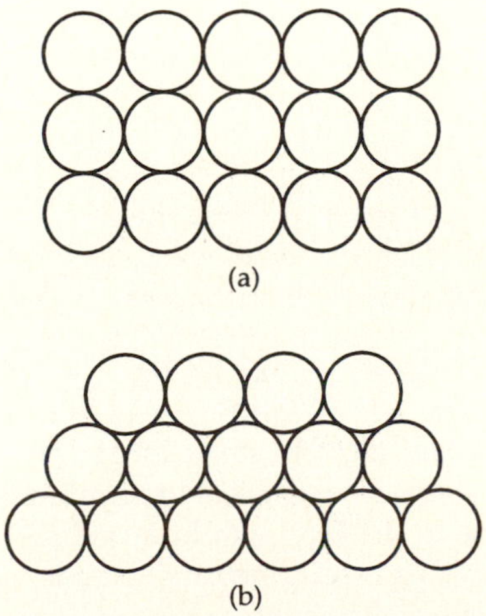

Figure 11–35. (a) Packing spheres always leaves some empty space. (b) These spheres are packed more efficiently, but some empty space remains.

Example 11–22 *Packing Tumor Cells.* An oncologist (cancer specialist) measures the diameter of a solid tumor in a patient to be 2 centimeters and notes that it is nearly spherical in shape. She also notes that a single cancer cell is roughly spherical and has a diameter of 2 micrometers (radius of 1 μm). Approximately how many cells are in the tumor?

Recall that 1 micrometer, or micron, denoted 1μm, is 10^{-6} m (or 10^{-4} cm).

Solution: Make a first estimate by finding the ratio of the tumor volume to the volume of a single cell:

$$\frac{V_{\text{tumor}}}{V_{\text{cell}}} = \frac{(4/3)\pi(2\text{ cm})^3}{(4/3)\pi\left(10^{-4}\text{ cm}\right)^3} = 8\times10^{12}.$$

The volume of the tumor is 8 trillion times the volume of a single cell. However, the actual number of cells in the tumor must be *fewer* than 8 trillion because the volume ratio does not account for gaps between the cells (which allow water and nutrients to flow among the cells).

Because cubes can be packed without leaving any gaps, next suppose that instead of being spherical with a diameter of 2 μm, each cell is a cube with a side length of 2 μm. Because each of these cubes is larger than a cancer cell, the number of these cubes that would fit in the tumor is an *underestimate* of the actual number of cells. The volume of a cube 2 μm on a side is

$$V = (2\ \mu\text{m})^3 = (2\times10^{-4}\text{ cm})^3 = 8\times10^{-12}\text{ cm}^3.$$

The number of cubes that can be packed into the tumor is the ratio of the tumor volume to the volume of a single cube:

$$\frac{V_{\text{tumor}}}{V_{\text{cube}}} = \frac{33.5\text{ cm}^3}{8\times10^{-12}\text{cm}^3} = 4.2\times10^{12}$$

The 4.2 trillion cubes that would fit in the tumor is a lower limit on the actual number of cells. Combining this fact with the previous upper limit of 8 trillion cells, the actual number of cancer cells in the tumor is between about 4 and 8 trillion.

❖

11.4.4 Melting Ice Caps and Other Catastrophes

Mathematical models used to analyze the greenhouse effect (see subsection 3.5.3) suggest that any global warming will be magnified near the poles. That is, a small increase in the *average* temperature of the Earth would mean a much greater increase in temperatures near the poles, which could begin to melt the polar ice caps. Because ice near the north pole is floating in the Arctic Ocean, melting it would not affect sea level. However, the ice sheets on Greenland and Antarctica would melt into the oceans and thereby raise sea level.

Time-Out to Think: Place an ice cube in a glass of water and note the water level. As the ice melts, does the water level change? Explain how this simple experiment confirms that melting the *floating* ice at the north pole would not affect sea level.

If some volume of material is spread over an area bounded by vertical walls, the depth of material *is* the volume divided by the area. For example, if you pour a volume of water, V, into a box with base area A, the depth of the water will be $V \div A$. If the area is bounded by nonvertical walls (e.g., a cone), the depth is proportional to, but not equal to, $V \div A$.

How much would sea level rise? Because oceans already cover most of the Earth's surface, and because most continental areas have coastal mountain ranges or inland plateaus, even a substantial rise in sea level would inundate only the low-lying coastal regions of continents. As a result, the total *surface area* of the Earth's oceans isn't likely to change much. Thus the *volume* of water from melted ice would be distributed over the *area* of the oceans.

To calculate the rise in sea level, divide the volume of water added to the oceans by the surface area of the oceans. The result of this calculation, shown in Example 11–23, indicates potentially devastating effects: most of humanity lives in densely populated coastal regions that would be inundated with water.

The principle of dividing a volume by an area to find a depth is useful in many other problems. For example, this principle can be used to estimate how lands along rivers might be affected by floods or how much trash can fit in a landfill.

Example 11–23 ***Melting Ice Caps.*** Estimate the rise in sea level if the entire Antarctic ice cap melts.

Solution: Finding the rise in sea level requires dividing the volume of water contained in the Antarctic ice cap by the surface area of the oceans. Begin by estimating the volume of water in the Antarctic ice cap. By studying a globe you will see that the shape of Antarctica is *roughly* circular with a radius of about 2000 kilometers; the area of the continent therefore is about $\pi \times (2000 \text{ km})^2 = 1.3 \times 10^7 \text{ km}^2$. By consulting an atlas or encyclopedia, you will find that the average depth of ice on Antarctica is about two kilometers. Multiplying this depth by the surface area of the continent gives the total volume of Antarctic ice:

$$\text{area} \times \text{average depth} = (1.3 \times 10^7 \text{ km}^2) \times 2 \text{ km} = 2.6 \times 10^7 \text{ km}^3.$$

Because ice floats in water, its density is *less* than that of water. By looking it up or noting how much of an ice cube floats above the water level, you can find that ice is roughly 5/6 as dense as water. Thus, for example, if you melt 1 liter of ice you will obtain about 5/6 liter of water. Similarly, if the Antarctic ice cap melts, its water volume will be about

$$\text{water volume} = \frac{5}{6} \times \text{ ice volume} = \frac{5}{6} \times \left(2.6 \times 10^7 \text{ km}^3\right) = 2.2 \times 10^7 \text{ km}^3.$$

Even with severe global warming, complete melting of the ice caps probably would take a hundred years or more. A more immediate threat is that the warming will weaken the ice, leading to the occasional breakoff of large chunks that would crash into the sea. These huge icebergs could raise sea level a little bit, and might also generate tidal waves.

That is, the melting of the Antarctic ice cap would add about 22 million cubic kilometers of water to the oceans. Next, find the total surface area of the oceans. As the Earth's radius is about 6400 kilometers, its surface area is about $4\pi \times (6400 \text{ km})^2 = 5.1 \times 10^8 \text{ km}^2$. The oceans cover about 2/3 of this area, or $2/3 \times 5.1 \times 10^8 \text{ km}^2 = 3.4 \times 10^8 \text{ km}^2$. Therefore the rise in sea level would be

$$\text{rise in sea level} = \frac{\text{volume of water from melted ice}}{\text{total surface area of oceans}} = \frac{2.2 \times 10^7 \text{ km}^3}{3.4 \times 10^8 \text{ km}^2} = 0.06 \text{ km}.$$

Converting the rise in sea level to more meaningful units yields

$$0.06 \text{ km} \times \frac{1000 \text{ m}}{1 \text{ km}} = 60 \text{ m}, \quad \text{and} \quad 60 \text{ m} \times \frac{3.38 \text{ ft}}{1 \text{ m}} \approx 200 \text{ ft}.$$

If the entire Antarctic ice cap melted, sea level would rise by some 60 meters, or about 200 feet! Any coastal area that currently is less than about 200 feet above sea level would be submerged following melting of the Antarctic ice cap.

❖

Example 11–24 ***A Neutron Star Comes to Town.*** Some massive stars end their lives as **neutron stars**, which have about the same mass as the Sun but are compressed to radii of about 10 kilometers. Suppose that, through horrifically bad luck, a neutron star magically appeared in your home town. What would happen? The density of a neutron star is about 7×10^{11} kilograms per cm^3. (Fortunately, although astronomers have discovered hundreds of neutron stars, there is no danger that any of them will visit the Earth.)

Solution: Although the radius of a neutron star is small enough (10 kilometers) to "fit" in many cities, its mass is about that of the Sun — about 300,000 times the Earth's mass. Hence the gravitational pull of the neutron star would be far greater than that of the Earth. The Earth would be crushed onto the surface of the neutron star, where it would spread out and form a thin, spherical shell on the neutron star's surface.

Assuming that the Earth would be crushed to the average density of the neutron star (7×10^{11} kg per cm^3), the *volume* of the crushed Earth would be about

$$\text{volume of crushed Earth} = \frac{\text{mass of Earth}}{\text{density of crushed Earth}} = \frac{6.0 \times 10^{24}\ \text{kg}}{7 \times 10^{11}\ \frac{\text{kg}}{\text{cm}^3}} = 9 \times 10^{12}\ \text{cm}^3.$$

Because the Earth's mass is so small compared to that of the neutron star, adding the Earth's mass would not appreciably change the neutron star's size. Therefore the *volume* of the crushed Earth would be spread in a thin layer over the surface area of the neutron star:

$$\text{thickness of layer} = \frac{\text{volume of crushed Earth}}{\text{surface area of neutron star}}$$
$$= \frac{9 \times 10^{12}\ \text{cm}^3}{4\pi \times \left(10\ \text{km} \times \frac{10^5\ \text{cm}}{1\ \text{km}}\right)^2} = \frac{9 \times 10^{12}\ \text{cm}^3}{4\pi \times 10^{12}\ \text{cm}^2} \approx 0.7\ \text{cm}.$$

Thus, if a neutron star unfortunately appeared in your home town, a short while later the remains of the Earth would be nothing more than a thin layer, about 7 millimeters (0.7 cm) thick, covering the surface of the neutron star. ❖

Summary Box 11–1. Volume and Surface Area Formulas

Solid Figure	Picture	Volume	Surface Area
Rectangular prism	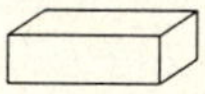	$V = l \times w \times h$	$A = 2(lw + lh + wh)$.
Cube	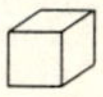	$V = l^3$	$A = 6l^2$
Right circular cylinder	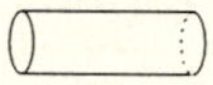	$V = \pi r^2 h$	$A = 2\pi r^2 + 2prh$
Sphere	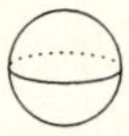	$V = (4/3)\pi r^3$	$A = 4\pi r^2$

11.5 FURTHER APPLICATIONS OF TRIANGLES

The importance of triangles in both the history and applications of mathematics cannot be overestimated. Indeed, triangles are so important that an entire branch of mathematics, **trigonometry** (literally "triangle measure"), is devoted to their study. In this section we briefly explore some of the many applications of trigonometry.

11.5.1 Angles in Radians

Besides using *degrees*, another common way to measure angles is in **radians**. The radian is defined so that an angle corresponding to a full circle (360°) is 2π radians (approximately $2 \times \pi \approx 6.28$ radians). Thus, for example,

- an angle subtending a semicircle (half circle) is one-half of 2π, or π radians;

Figure 11–36. Angles are measured counterclockwise on a unit circle, beginning at the radius extending to the right. The circle is labeled with angles in both degrees and radians. Note that the point representing 0° or 0 radians also represents 360° or 2π radians because going around the full circle returns to this point.

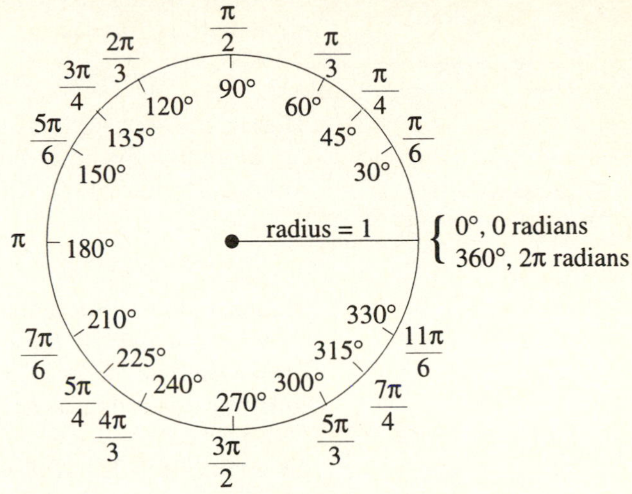

- an angle subtending a quarter circle is one-fourth of 2π, or $\pi/2$ radians;
- an angle subtending an eighth of a circle is one-eighth of 2π, or $\pi/4$ radians; and
- an angle subtending one-hundredth of a circle is $2\pi/100$, or $\pi/50$ radians.

Drawing a **unit circle** — a circle with a radius of one unit — provides a useful way to contrast angles in degrees and radians (Figure 11–36).

Because 2π radians and 360° are alternative ways of describing a full circle, the conversion factor between radians and degrees is

$$2\pi \text{ radians} = 360^\circ \quad \text{or} \quad \frac{2\pi \text{ radians}}{360^\circ} = 1 \quad \text{or} \quad \frac{360^\circ}{2\pi \text{ radians}} = 1.$$

For example, to convert an angle of 15° into radians, you apply the form of the conversion factor with degrees in the denominator:

$$15^\circ \times \frac{2\pi \text{ radians}}{360^\circ} = \frac{\pi}{12} \text{ radians} \approx 0.26 \text{ radians}.$$

Similarly, to convert an angle of 1.2 radians into degrees, you apply the form of the conversion factor with radians in the denominator:

$$1.2 \text{ radians} \times \frac{360^\circ}{2\pi \text{ radians}} = \frac{216^\circ}{\pi} \approx 68.75^\circ.$$

Radians and Arc Length

Because they are directly related to a length of **arc** (a piece of a circle), radians are a more "natural" angle measure than degrees. For example, suppose that a circle has a radius of *one* kilometer. If you walk all the way around this circle, you will have walked a distance of 2π kilometers (the circumference of the circle) and through an angle of 2π radians. If you walk halfway around the circle, you will have covered an angle of π radians and a distance of π kilometers. If you walk along an arc of 1 radian (about 57°), you will walk one kilometer — which is the same length as the radius of the circle.

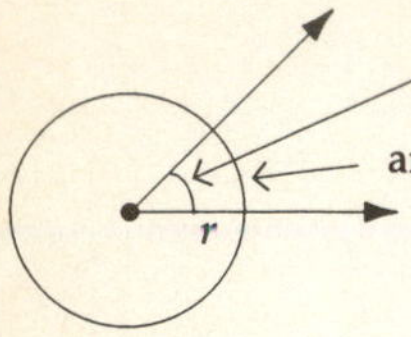

Figure 11–37. The arc length along a circle subtended by an angle of a radians is a radii, or $a \times r$, where r is the radius of the circle.

Generalizing, the length of arc subtended by an angle of a radians in a circle with radius r is

$$\text{length of arc} = (\text{angle in radians}) \times (\text{radius of circle}) = a \times r.$$

That is, the arc length is a radii (Figure 11–37).

Note that the preceding formula provides an alternative way to define radians. Dividing both sides of the formula by the radius of the circle yields

$$\text{angle in radians} = \frac{\text{length of arc}}{\text{radius of circle}}.$$

That is, radians represent a *ratio* (arc length divided by the radius of the circle); they are not a *physical* unit like length, mass, or time. Thus radians are **dimensionless**, and you do not need to keep track of them as you do with units representing physical quantities.

Example 11–25 ***Walking in Circles.*** Suppose that a circle has a radius of 15 meters. If you walk along the edge, how far will you walk in sweeping an angle of 1 radian? 2.5 radians? What fraction of the circle will you have walked in each case?

Solution: An angle of 1 radian subtends a length of arc equal to the radius of the circle, or 15 meters. Thus, if your walk sweeps an angle of 1 radian, you walk a distance of 15 meters. With 2π radians in a full circle, a walk of 1 radian represents $1/2\pi$, or about 16% of a full circle. For 2.5 radians, the length of arc is 2.5 radii, or

$$\text{length of arc} = \underbrace{2.5 \text{ radians}}_{\text{angle in radians}} \times \underbrace{15 \text{ meters}}_{\text{radius}} = 37.5 \text{ meters}.$$

If your walk sweeps an angle of 2.5 radians, you walk a distance of 37.5 meters. This distance represents a fraction $2.5/2\pi$, or about 40%, of the full circle.

❖

Example 11–26 Suppose that a circle has a radius of 7 centimeters. What is the length of an arc subtended by an angle of 10°?

Solution: Whenever you deal with arc length, you must use radians. Begin by converting the 10° angle into radians:

$$10° \times \frac{2\pi \text{ radians}}{360°} = \frac{\pi}{18} \text{ radians} \approx 0.17 \text{ radians}.$$

Now find the length of arc:

$$\text{length of arc} = \underbrace{0.17 \text{ radians}}_{\text{angle in radians}} \times \underbrace{7}_{\text{radius}} \text{cm} = 1.2 \text{ cm}.$$

An angle of 10° subtends an arc length of 1.2 cm on a circle having a radius of 7 cm.

❖

Angular Size, Physical Size, and Distance

If you look at an object in the distance, its **angular size** is the angle that it subtends from your vantage point. Thus the angular size of an object depends *both* on its physical size and its distance from you. To illustrate, close one eye and hold a book in front of your open eye. Your normal left-to-right field of view (Figure 11–38a) covers an angle of 180°. Thus, when the book is directly in front of your eye (Figure 11–38b), its angular width is 180° (it blocks your entire view). Next, extend your arm and hold the book farther in front of you. Now its angular size is smaller, and it blocks only part of your field of view (Figure 11–38c).

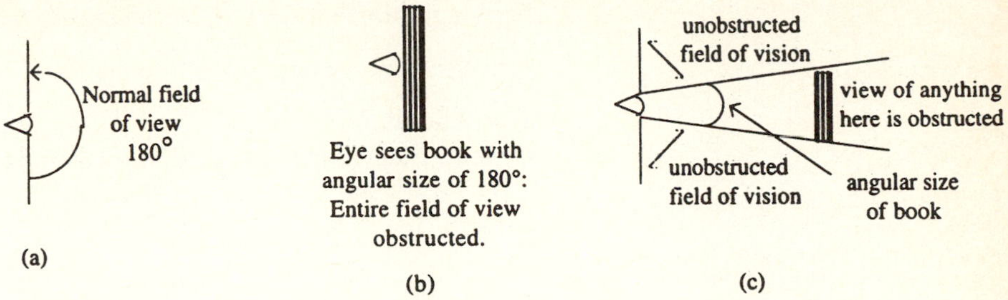

Figure 11–38. (a) A schematic illustration shows that your normal field of vision extends over an angle of 180°. (b) If you place a book directly in front of your eye, it completely blocks your field of view; the angular size of the book is 180°. (c) If you hold the book farther from your eye, its angular size is smaller.

Time-Out to Think: Note that the book can block your view of objects that are physically larger than it. For example, hold the book so that it blocks your view of a nearby building. Although the Sun has a diameter of more than 1 million kilometers, can you block the Sun with the book? Explain.

Note that we are using the term *size* to represent any linear dimension, such as a width or a diameter. The formulas presented in this section do not hold for areas.

Figure 11–39(a) shows a schematic diagram of an object located at a distance r from your eye. In Figure 11–39b, a piece of a circle with radius r also is shown. Note that, as long as the angular size of the object is relatively small (less than about 10° or 20°, as shown), the arc length subtended by the object is approximately the same as the physical length of the object. Earlier, we showed that the relation between angle and arc length is

$$\text{length of arc} = (\text{angle in radians}) \times (\text{radius of circle}).$$

From Figure 11–39, we can make three substitutions: The arc length is approximately the physical size of the object, the radius of the circle is the distance to the object from your eye, and the angle is the angular size of the object. Thus,

$$\text{physical size of object} \approx (\text{angular size of object (in radians)}) \times (\text{distance to object}).$$

This simple formula has many practical applications. For example, a surveyor might measure the angular size and distance of a distant mountain, then use the preceding formula to calculate the height of the mountain.

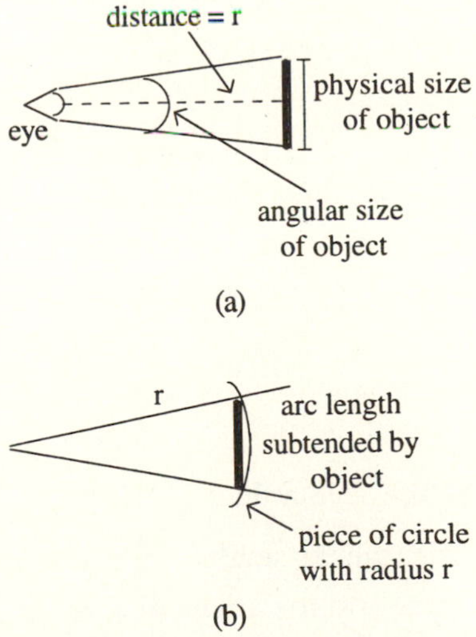

Figure 11–39. (a) An object is shown at a distance r from your eye. Its angular size depends on both its physical size and its distance r from you. (b) Part of a circle with radius r is added to the diagram. As long as the angular size is relatively small, the physical size of the object is approximately the same as the arc length it subtends.

Example 11–27 ***The Size of a Dime.*** Suppose that you hold a dime at a distance of 1 meter from your eye (try it!). What is its angular size?

Solution: The diameter of a dime is about 1.8 centimeters. Solve the preceding formula for the angular size, then substitute the dime's diameter of 1.8 cm for the *physical size* and the 1-m distance from your eye as the *distance* to object:

$$\text{physical size} = \text{angular size} \times \text{distance} \xrightarrow{\text{divide by distance}}$$

$$\text{angular size} = \frac{\text{physical size}}{\text{distance}} = \frac{1.8\text{ cm}}{100\text{ cm}} = 0.018\text{ radians.}$$

At a distance of 1 meter, the angular size (diameter) of a dime is about 0.018 radians, or $0.018 \times 360°/2\pi \approx 1°$.

❖

Example 11–28 ***Size of the Moon.*** The angular diameter of the Moon (as seen from the Earth) is about 0.5° and the Moon is approximately 400,000 kilometers from the Earth. What is the actual diameter of the Moon?

Solution: First convert the Moon's angular diameter into radians:

$$0.5° \times \frac{2\pi\text{ radians}}{360°} \approx 0.0087\text{ radians.}$$

Now substitute into the physical size formula (here, *size* refers to diameters):

physical size ≈ angular size × distance = 0.0087 radians × 400,000 km = 3500 km.

This calculation yields a diameter of about 3500 kilometers, which is very close to the Moon's actual diameter of 3480 kilometers.

❖

Example 11–29 ***The Hubble Space Telescope.*** The angular resolution of the Hubble Space Telescope is about 0.05"; that is, it can distinguish two objects if they are separated by an angle of at least 0.05 seconds of arc. How far away would you have to place a dime to make its angular diameter 0.05"?

Solution: First, convert from seconds of arc to radians:

$$0.05'' \times \frac{1'}{60''} \times \frac{1^\circ}{60'} \times \frac{2\pi \text{ radians}}{360^\circ} = 2.4 \times 10^{-7} \text{ radians.}$$

Now, rearrange the physical size formula to find the distance that makes a dime (diameter = 1.8 cm) have this angular size:

$$\text{physical size} = \text{angular size} \times \text{distance} \xrightarrow{\text{divide by angular size}}$$

$$\text{distance} = \frac{\text{physical size}}{\text{angular size}} = \frac{1.8 \text{ cm}}{2.4 \times 10^{-7}} = 7.5 \times 10^{6} \text{ cm;}$$

$$7.5 \times 10^{6} \text{ cm} \times \frac{1 \text{ km}}{1000 \text{ m}} \times \frac{1 \text{ m}}{100 \text{ cm}} = 75 \text{ km}, \quad \text{and} \quad 75 \text{ km} \times \frac{1 \text{ mi}}{1.6 \text{ km}} \approx 47 \text{ mi.}$$

The 0.05" resolution of the Hubble Space Telescope corresponds to the angular size of a dime at a distance of 75 kilometers, or almost 47 miles. In other words, the telescope could distinguish two dimes placed side by side at a distance of 47 miles. (It couldn't resolve words on the dime because they are smaller than the dime itself.)

❖

Time-Out to Think: Spy satellites look down on the Earth from a typical altitude of about 300 kilometers. If a spy satellite has a telescope of the same resolution as the Hubble Space Telescope, can it tell when you leave your house? Explain.

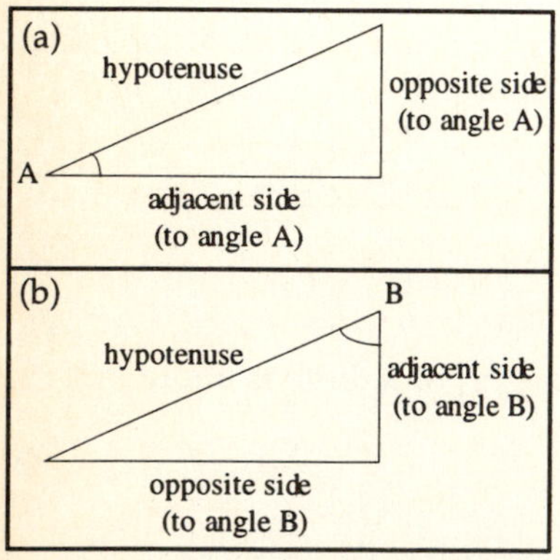

Figure 11– 40. The hypotenuse of a right triangle is the longest side and is opposite the right (90°) angle. (a) The adjacent and opposite sides are shown for ∠*A*. (b) The adjacent and opposite sides are shown for ∠*B*.

11.5.2 Sine, Cosine, and Tangent

Consider the right triangle shown in Figure 11–40(a). Note that:

1. The *hypotenuse*, or longest side, is opposite the 90° angle.
2. Angle *A* (∠*A*) is formed by the hypotenuse and the **adjacent side**.
3. The **opposite side** is opposite to ∠*A*.

Identification of the hypotenuse is unambiguous, because a right triangle has only one 90° angle. In contrast, identification of the adjacent and opposite sides depends on which of the remaining angles you are studying. Figure 11–40(b) shows the same right triangle, but this time the sides adjacent and opposite to ∠B are labeled.

Six different ratios may be formed from pairs of sides in a right triangle, but we work with only three of them in this book. They are called the **sine** ratio (abbreviated **sin**), the **cosine** ratio (abbreviated **cos**), and the **tangent** ratio (abbreviated **tan**). Using A to represent $\angle A$, their definitions are

$$\sin A = \frac{\text{opposite side}}{\text{hypotenuse}}, \quad \cos A = \frac{\text{adjacent side}}{\text{hypotenuse}}, \quad \text{and} \quad \tan A = \frac{\text{opposite side}}{\text{adjacent side}}.$$

Sine, Cosine, and Tangent Ratios on a Calculator

Calculating sine, cosine, and tangent ratios is easy for only a few special angles. In general, obtaining these ratios requires a table of trigonometric values or a calculator.

Appendix D describes how to set your calculator to the correct mode of degrees or radians, and offers further examples of calculating the trigonometric ratios.

Using the *sin, cos,* and *tan* buttons on your calculator involves a slight complication: To obtain correct answers, you first must set your calculator according to whether you are measuring angles in degrees or radians. Before you continue, set your calculator to *degree* mode, and confirm that

$$\sin 45°=0.707, \quad \cos 15° = 0.966, \quad \sin 15° = 0.259,$$
$$\tan 15° = 0.268, \quad \text{and} \quad \sin 1° = 0.017.$$

Then you set your calculator to *radian* mode, and confirm that

$$\sin \pi/4 = 0.707, \quad \cos 15 = 0.650, \quad \tan \pi/6 = 0.577,$$
$$\tan 15 = -0.86, \quad \text{and} \quad \sin 1 = 0.841.$$

When working with trigonometric ratios, be sure that your calculator is set in the correct mode: either degrees or radians.

> **Time-Out to Think:** Note the importance of selecting the correct mode on your calculator. For example, cos 15° is much different from cos 15 *radians*. Why is sin 45° the same as the sin $\pi/4$?

Calculating Unknown Side Lengths

If you know the length of one side and one angle (not the right angle) in a right triangle, you can use the trigonometric ratios to calculate the unknown side lengths. For example, look back at Figure 11–40(a) and suppose that angle $\angle A$ is 37° and the hypotenuse is 5 centimeters long. By rearranging the sine and cosine formulas you can find formulas for the lengths of the opposite and adjacent sides:

$$\sin A = \frac{\text{opposite side}}{\text{hypotenuse}} \quad \Rightarrow \quad \text{opposite side} = \sin A \times \text{hypotenuse};$$

$$\cos A = \frac{\text{adjacent side}}{\text{hypotenuse}} \quad \Rightarrow \quad \text{adjacent side} = \cos A \times \text{hypotenuse}.$$

Now, substitute 5 cm for the hypotenuse and 37° for $\angle A$ to obtain the lengths of the opposite and adjacent sides (your calculator must be in *degree* mode to confirm these results):

$$\text{opposite side} = \sin A \times \text{hypotenuse} = \sin 37° \times 5\text{ cm} = 0.602 \times 5\text{ cm} = 3.009\text{ cm};$$

$$\text{adjacent side} = \cos A \times \text{hypotenuse} = \cos 37° \times 5\text{ cm} = 0.799 \times 5\text{ cm} = 3.993\text{ cm}.$$

> **Time-Out to Think:** A good way to check the preceding results is by using the Pythagorean theorem, which can be written as
>
> $$(\text{adjacent side})^2 + (\text{opposite side})^2 = (\text{hypotenuse})^2.$$
>
> Verify that the adjacent and opposite sides just calculated satisfy the Pythagorean theorem for a right triangle with hypotenuse 5 cm.

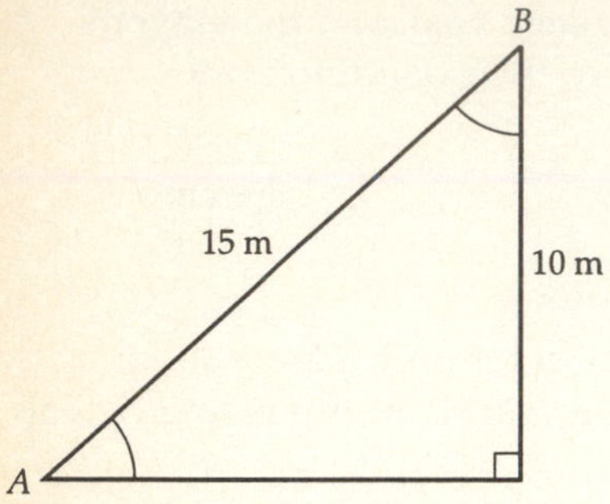

Figure 11–41. In this triangle $\angle A$ and $\angle B$ initially are unknown, but can be calculated using the inverse sine and inverse cosine operations, respectively.

You will get $\sin^{-1} 0.667 = 41.8°$ only if your calculator is set to degree mode. If your calculator is set to radian mode, you will get $\sin^{-1} (0.667) = 0.730$ radians. These answers are equivalent because 0.730 radians = 41.8°.

Calculating Unknown Angles

You also can use the trigonometric ratios to find unknown angles in a right triangle if you know two side lengths. For example, suppose that one side of a right triangle is 10 meters long and that the hypotenuse is 15 meters long (Figure 11–41). Because you know the lengths of the hypotenuse and the side opposite to $\angle A$, you can calculate

$$\sin A = \frac{\text{opposite side}}{\text{hypotenuse}} = \frac{10 \text{ m}}{15 \text{ m}} = 0.667.$$

You now know that $\sin A = 0.667$, but how do you find $\angle A$? You could proceed by trial and error, guessing various values of $\angle A$, calculating the sine of your guessed values, and continuing until you find the value for which $\sin A = 0.667$. But an easier way is to use an operation called the **inverse sine**, abbreviated as either **arcsin** or **sin^{-1}** (note that the "–1" does *not* represent a power). On your calculator, you should find the button for the inverse sine and use it to confirm that

$$A = \sin^{-1} 0.667 = 41.8°.$$

Similarly, because you know the lengths of the hypotenuse and the side adjacent to $\angle B$, you can calculate

$$\cos B = \frac{\text{adjacent side}}{\text{hypotenuse}} = \frac{10 \text{ m}}{15 \text{ m}} = 0.667.$$

Using the **inverse cosine (cos^{-1})** operation, you then can find that

$$B = \cos^{-1} 0.667 = 48.2°.$$

Time-Out to Think: As a check on the preceding results, use the calculated values of $\angle A$ and $\angle B$ to confirm that the sum of the angles in Figure 11–41 is 180°.

The **inverse tangent (tan^{-1})** operation allows you to calculate an unknown angle when you know both the opposite and adjacent sides, as shown in Example 11–30.

Example 11–30 ***Solar Access Revisited.*** Look back at the geometry of the solar access example (Example 11–14) shown in Figure 11–28. What is the angle of elevation of the Sun?

Solution: The angle to the Sun is the angle between the ground and the hypotenuse (line of sunlight) in Figure 11–28; let's call it ∠A. In the smaller triangle, the opposite side to ∠A is the 12-foot height of the fence and the adjacent side is the 20-foot shadow cast by the fence. You have all the information needed to find the tangent of ∠A:

$$\tan A = \frac{\text{opposite side}}{\text{adjacent side}} = \frac{12 \text{ ft}}{20 \text{ ft}} = 0.6.$$

Now, use the *inverse tangent* operation to find ∠A.

$$A = \tan^{-1} 0.6 = 31°.$$

Thus the angle of elevation of the Sun is 31°, which is equivalent to 0.54 radians.

❖

11.5.3 Triangulation

Although many types of problems can be solved using the sine, cosine, and tangent ratios, one of the most practical is **triangulation**. It involves forming a triangle from three points of interest and then determining angles or distances. Triangulation is essential to navigation and surveying.

Example 11–31 *Ocean Rescue.* Imagine that you are in a helicopter as part of a search party looking for a small, lost sailboat. You spot a flare in the distance, presumably fired by the lost crew. Focusing on the spot where you saw the flare, you estimate the angle to the boat to be 75° (Figure 11–42). According to your altimeter, you are hovering at an altitude of 1000 feet. If you drop a buoy to mark your current location, how far away is the boat along the surface?

Solution: The altitude of the helicopter represents the side *adjacent* to the 75° angle. The unknown distance to the boat is the side *opposite* the 75° angle. Thus you can find the distance using the tangent ratio:

$$\tan 75° = \frac{\text{opposite side}}{\text{adjacent side}} = \frac{\text{distance to boat}}{1000 \text{ ft}} \xrightarrow{\text{solve for distance}}$$

$$\text{distance to boat} = 1000 \text{ ft} \times \tan 75° = 1000 \text{ ft} \times 3.73 = 3730 \text{ ft.}$$

If you drop a buoy to mark your location, it will be about 3700 feet from the missing boat. Be sure to note the direction to the boat as well!

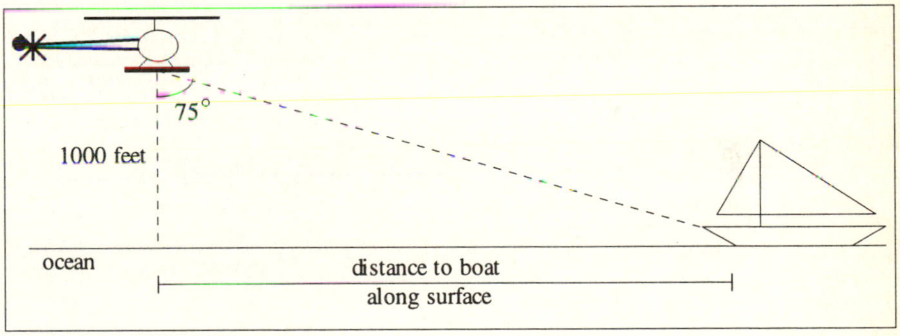

Figure 11–42. A helicopter spots a lost boat when it sees a flare sent up by the crew. The angle to the boat is estimated as 75°, and the altimeter shows that the helicopter is at an altitude of 1000 feet. Not drawn to scale.

❖

Example 11–32 *Surveying.* A surveyor, equipped with instruments for measuring angles precisely, needs to determine the distance to a nearby peak. As shown in Figure 11–43, she first uses a transit to spot the peak from point 1. She marks point 1 with a rod and walks 25 meters to point 2 (walking in a direction perpendicular to the line of sight to the peak from point 1). At point 2, she sets up the transit and takes another sighting of the peak. She then swings the transit to take a sighting of the rod at point 1 and reads an angle of 87.6°. How far (along the ground) is the peak from point 2? From point 1?

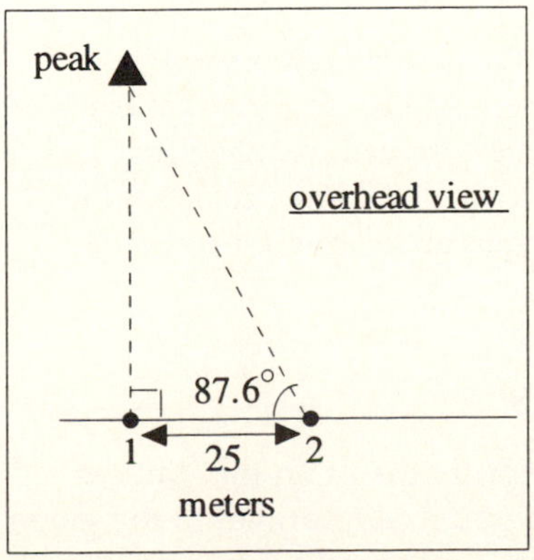

Figure 11–43. A surveyor sights the peak of a hilltop from two different points, separated by 25 meters, measuring angles as shown. Note that this is an overhead view, not drawn to scale.

Solution: The 25-meter distance is the side of a right triangle *adjacent* to the 87.6° angle, and the distance from point 2 to the peak is the *hypotenuse* of the triangle. The surveyor therefore uses a cosine ratio:

$$\cos 87.6° = \frac{\text{adjacent side}}{\text{hypotenuse}} \Rightarrow \text{hypotenuse} = \frac{\text{adjacent side}}{\cos 87.6°} = \frac{25 \text{ m}}{0.0419} = 597 \text{ m.}$$

The hypotenuse, or the distance to the peak from point 2, is 597 meters. The surveyor could find the distance from point 1 to the peak (the *opposite* side) by using the tangent ratio. Alternatively, with two sides of the right triangle known, she can use the Pythagorean theorem. Writing *opp* for the opposite side, *adj* for the adjacent side, and *hyp* for the hypotenuse, the Pythagorean theorem becomes

$$opp^2 + adj^2 = hyp^2.$$

Solving for the opposite side yields

$$opp = \sqrt{hyp^2 - adj^2} = \sqrt{(597\text{ m})^2 - (25\text{ m})^2} = \sqrt{355{,}784\text{ m}^2} \approx 596.5\text{ m}.$$

The distance from point 1, 596.5 meters, is within half a meter of the distance from point 2 (597 meters). What would Figure 11–43 look like if it were drawn to scale?

❖

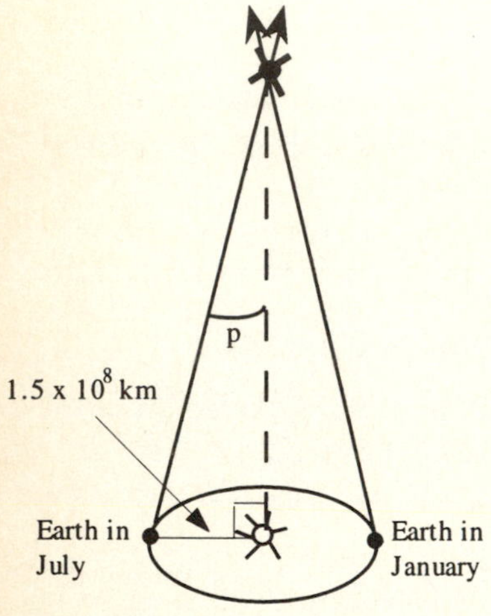

Figure 11–44. The orientation of an arrow from the Earth to a star shifts over the course of a year. Here, arrows are shown for January and July, which, because they are six months apart, represent opposite sides of the Earth's orbit. The side opposite to angle *p* is the Earth–Sun distance (about 150 million kilometers).

Example 11–33 *Stellar Distances.* Imagine drawing an arrow from the Earth to a star. As the Earth orbits the Sun over the course of a year (and the star essentially remains fixed in space), the orientation of this arrow shifts. Figure 11–44 shows the angle *p* created by this shift as the star is viewed from opposite sides of the Earth's orbit. Suppose that, for a particular star, the angle *p* is measured to be 0.46"; how far away is the star? (This technique for measuring stellar distances is called *stellar parallax*, and the angle *p* is called the *parallax angle*.)

Solution: The distance to the star is the *hypotenuse* of a right triangle and the Earth–Sun distance is the side *opposite* to angle *p*. Therefore use the sine ratio to find the distance. First, however, you must convert the angle *p* from arc seconds into degrees (calculators generally do not have a mode for arc seconds):

$$0.46'' \times \frac{1'}{60''} \times \frac{1^\circ}{60'} = (1.28 \times 10^{-4})^\circ .$$

Now find the distance to the star:

$$\sin\left(1.28 \times 10^{-4}\right)^\circ = \frac{\text{opposite side}}{\text{hypotenuse}} = \frac{\text{Earth - Sun distance}}{\text{star distance}} = \frac{1.5 \times 10^8\text{ km}}{\text{star distance}} \xrightarrow{\text{solve}}$$

$$\text{star distance} = \frac{1.5 \times 10^8\text{ km}}{\sin\left(1.28 \times 10^{-4}\right)^\circ} = \frac{1.5 \times 10^8\text{ km}}{2.2 \times 10^{-6}} = 6.8 \times 10^{13}\text{ km}.$$

The distance to the star is about 68 trillion kilometers. As 1 light-year is about 10 trillion (10^{13}) kilometers (see Section 7.4), this distance is about 6.8 light-years.

Stellar parallax allows astronomers to determine distances to nearby stars. It actually is measured by comparing photographs taken at different times of year. Relative to very distant stars, nearby stars are found in different positions on the photograph at different times of year.

Time-Out to Think: The measurement of stellar parallax represents *direct proof* that the Earth orbits the Sun because there would be no parallax if the Earth were stationary in space. Many ancient Greeks argued that the Earth must be stationary because they could *not* detect parallax. Why couldn't they detect stellar parallax? (Hint: Can you resolve angles as small as 1 arc second with your naked eye?)

11.6* NON-EUCLIDEAN GEOMETRY

I attach special importance to [non-Euclidean geometry] because without it I should have been unable to formulate the theory of relativity. — Albert Einstein

Suppose that you have a straight line and a point that isn't on that line, as shown in Figure 11–45(a). How many lines can you draw parallel to the given line through that point? A moment or two of thought should convince you that only one parallel line can pass through the point (Figure 11–45b); any other line passing through the point (in the same plane) eventually would cross the first line. This apparently simple conclusion opens the door to some amazing results.

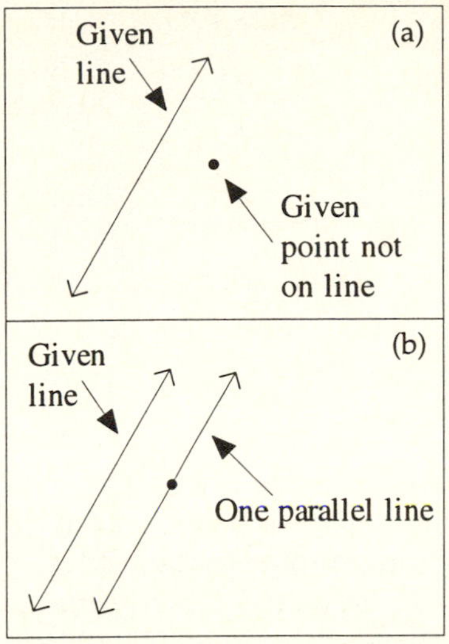

Figure 11–45. (a) Given a line and a point that is not on that line, how many lines can be drawn through the point that are parallel to the given line? (b) Euclid's parallel postulate asserts that through any given point exactly one line can be drawn parallel to a given line.

Janos Bolyai's work was certainly influenced by his father, a mathematician who spent much of his life attempting to prove the parallel postulate. In a letter to Janos in 1820, the father wrote: "You should detest [the parallel postulate], it can deprive you of all of your desire, your health, your rest, and the whole happiness of your life. This abysmal darkness might perhaps devour a thousand towering Newtons, it will never be light on earth."

Euclid (see Section 11.1) built his geometry by proving theorems. The starting point for his proofs were five basic postulates, which he assumed to be self-evident and not in need of proof. Paraphrased, the five postulates state that

1. a line can be drawn through any two points;
2. a line segment can be extended indefinitely along the same line;
3. given any point as a center, and any radius, a circle can be drawn;
4. all right angles are equivalent (90°); and
5. given a line and a point not on the line, only one parallel line can be drawn through the point.

The fifth postulate, illustrated by Figure 11–45, is called the **parallel postulate**. Euclid found that many of his theorems could be proved without using the parallel postulate. Perhaps as a result, he and many later mathematicians guessed that the parallel postulate might be less fundamental than the other four postulates.

For the next 2000 years, mathematicians offered hundreds of "proofs" claiming the parallel postulate to be a consequence of Euclid's other four postulates. Unfortunately, all these "proofs" were found to be invalid under close scrutiny. Finally, in about 1830, a Russian mathematician named Nikolai Ivanovich Lobachevsky and a Hungarian mathematician named Janos Bolyai independently took the radical view that the parallel postulate is not provable. Instead, they proceeded to create systems of geometry in which the parallel postulate is *not* true!

This step gave birth to the field of **non-Euclidean geometry**, in which Euclid's parallel postulate is replaced by some other postulate. During the remainder of the nineteenth century numerous mathematicians helped advance the understanding of non-Euclidean geometries. At the time, this work generally was considered an abstract exercise; however, non-Euclidean geometries are now known to have direct relevance to the Earth and the universe.

11.6.1 Elliptical and Hyperbolic Geometries

The first of the two general types of non-Euclidean geometry, for which the surface of the Earth makes a good model (Figure 11–46a on the next page), is called **elliptical geometry**. Note that, as we are considering only the *surface* of the Earth, we are dealing with a *two-dimensional* geometry. Therefore only two coordinates are needed to specify a location: latitude and longitude.

In elliptical geometry, the statement that *no* lines parallel to a given line can be drawn through a given point replaces Euclid's parallel postulate. Understanding this statement requires defining a "line" on the surface of the Earth. Because the surface itself is curved, any line drawn on the Earth's surface necessarily is curved. In other words, no truly straight lines can be drawn on the surface of the Earth. We therefore need another way to define lines. In Euclidean geometry, lines not only are straight, but they also represent the *shortest distance* between two points. What is the shortest distance between two points on the Earth's surface?

Time-Out to Think: To help you answer this last question, find a globe (not a flat map!) and determine the shortest distance between Los Angeles and Tokyo. Note that both cities are at approximately the same latitude (close to 35° N). Are you surprised by how far north the shortest path goes? Is this the route that airplanes travel? Explain.

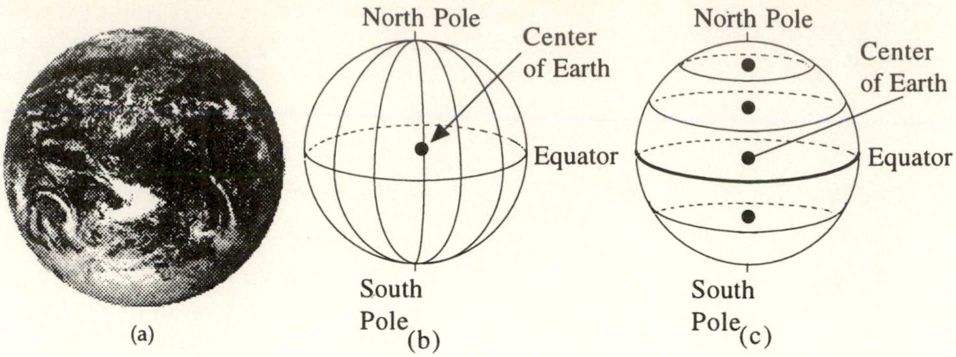

Figure 11–46. (a) The surface of the Earth is a good model of a non-Euclidean, elliptical geometry. (b) Great circles — circles with their centers at the center of the Earth — represent *lines* in elliptical geometry. The equator and all circles of longitude are great circles. (c) Except for the equator, circles of latitude do not have their centers at the center of the Earth and therefore are not great circles.

Paths that represent the shortest possible distance between points on the surface of the Earth always are segments of **great circles** — circles that have their centers at the center of the Earth. Circles of longitude are great circles (Figure 11–46b), as is the equator. In contrast, circles of latitude (except the equator) are not great circles (Figure 11–46c). Because great circles help determine the shortest distance between two points, great circles represent "lines" in elliptical geometry.

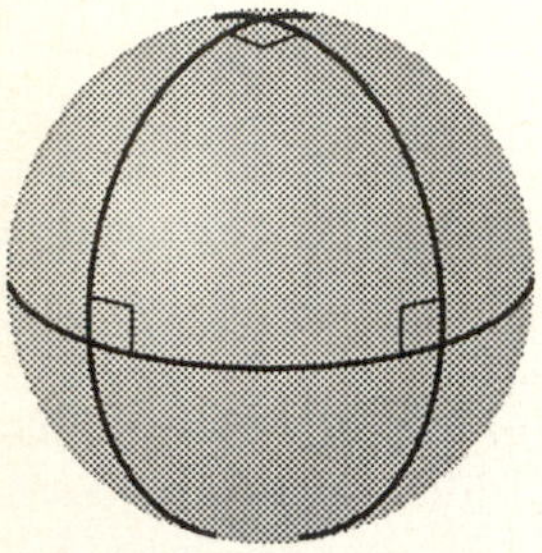

Figure 11–47. The sum of the angles in a triangle in elliptical geometry always is *greater* than 180°. In the triangle shown, all three angles are 90°, so the sum of the angles is 270°.

Because great circles all have the same center (the center of the Earth), they invariably intersect other great circles; drawing two great circles that never intersect is impossible. Thus, because great circles represent lines, no parallel lines can be drawn in elliptical geometry. For example, although lines of longitude *appear* to be parallel at the Earth's equator, they all intersect at the north and south poles.

Another consequence of elliptical geometry concerns triangles. Recall that, in Euclidean geometry, the total measure of the three angles of a triangle always is 180°. In contrast, the total measure of the angles of a triangle in elliptical geometry (comprised of segments of three great circles) always is *greater* than 180° (Figure 11–47).

Hyperbolic geometry is the second general type of non-Euclidean geometry. A good model of a two-dimensional hyperbolic geometry is the surface of a saddle (Figure 11–48a). Note that "lines" on the hyperbolic surface that appear parallel at first tend to diverge from one another with distance. As a result, an *infinite* number of parallel lines may be drawn parallel to a given line through a given point. This property, which replaces the parallel postulate of Euclidean geometry, again leads to many other consequences. For example, the total measure of the angles in a triangle always is *less* than 180° in hyperbolic geometry (Figure 11–48b).

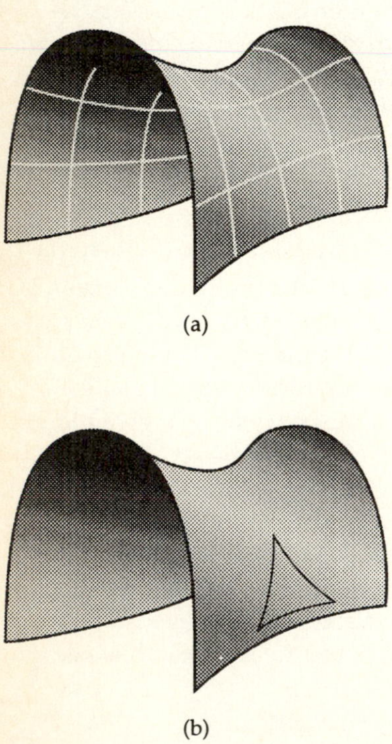

Figure 11–48. (a) A saddle-shaped surface makes a good model of a two-dimensional hyperbolic geometry. Lines that appear parallel near the center of the region shown diverge with distance. (b) The sum of the angles in a triangle in hyperbolic geometry always is *less* than 180°.

Note that two-dimensional non-Euclidean geometries may be visualized only by looking in *three* dimensions. For example, a three-dimensional globe is required to show the two-dimensional surface of the Earth. Extending this idea, visualizing a *three-dimensional* non-Euclidean geometry would require looking in four or more dimensions — which isn't possible. Mathematically, however, working with non-Euclidean geometries in three or more dimensions is straightforward.

11.6.2 The Geometry of the Universe

Imagine two smart ants trying to determine whether or not the surface of the Earth is Euclidean. Over small areas, the surface of the Earth *appears* to be flat. For example, roads can be made parallel within the confines of a town. Thus, to the ants, the Earth would look flat, and they could draw line segments that would appear to obey the parallel postulate. Imagine, however, that the ants decide to perform a major experiment. They send an ant expedition on a journey to unexplored territory. To their great surprise, although the expedition

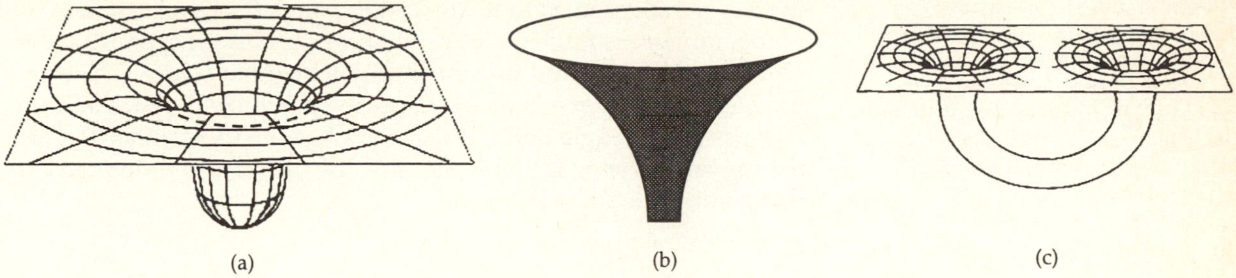

Figure 11–49. Two-dimensional representations illustrate basic ideas of the non-Euclidean universe. (a) The Earth travels around the Sun along the straightest possible path. Because the Sun curves the fabric of space, this path is an ellipse. (b) Black holes are regions of the universe where space is curved so much that it becomes a sort of bottomless pit. (c) Some scientists speculate that the bottomless pits of black holes might connect to form tunnels, or worm holes. The worm hole itself is *not* part of the universe, so it might allow travel from one point in the universe to another without passing through the space in between.

never turns around, it eventually returns home! Why? Because the expedition eventually circles the Earth. The ants will have proved that the surface of the Earth is non-Euclidean, despite its flat appearance on small scales.

What happens if we try a similar experiment with the universe? That is, does the three-dimensional space of the universe obey the parallel postulate, or is it non-Euclidean? Because the universe is so large, we cannot send out an expedition and wait for its return. However, we can look for curvature of space by using the fact that rays of light travel the shortest distance between two points. For example, light from distant stars has been observed to deviate from a straight path when it passes near the Sun. Similarly, light from very distant objects called *quasars* bends when it passes by large, distant galaxies; in some cases, the result is multiple images of a single quasar, as light from the quasar is bent along several different paths towards Earth. From these observations, we conclude that our universe is non-Euclidean!

More specifically, three-dimensional space itself is curved in the vicinity of massive objects such as stars and galaxies. Einstein first predicted this property of the universe in his **general theory of relativity**, published in 1915, which shows that gravity can be understood as the curvature of space. In this view, the orbit of a planet takes on a new and different meaning. Instead of being the consequence of a complex mathematical law, a planet simply travels along the straightest possible path. However, because the Sun curves space, this path is an ellipse (Figure 11–49a). (*Caution*: the representations in Figure 11–49 are highly schematic, compressing the curvature of space, which can be "seen" only in four (or more) dimensions, into a mere two dimensions.)

With this view of gravity, **black holes** literally are holes in the fabric of the universe (Figure 11–49b). If such holes connect with other holes in the universe, they might form a sort of tunnel, or **worm hole** (Figure 11–49c), through which it might be possible to travel from one point in the universe to another without passing through the space in between! Because of their exotic properties, worm holes have become a staple of science fiction writers.

In fact, the true geometry of the universe is even more interesting. Einstein found that time must be considered a dimension along with the three dimensions of space; hence the four dimensions of the universe sometimes are referred to as **space-time**. According to the general theory of relativity, space-time is non-Euclidean. Further, the curvature may vary from point to point; in some places the geometry of the universe may be elliptical, in other places hyperbolic, and in still other places flat (Euclidean).

Although its geometry may vary from point to point, the universe must have some *overall* geometry. By analogy, close study of the Earth's surface reveals that its geometry varies from point to point: saddle-shaped valleys have hyperbolic geometry, hills have elliptical geometry, and plateaus are flat. Nevertheless, the overall geometry of the Earth's surface is elliptical.

The overall geometry of the universe is one of the outstanding questions in science. If the overall geometry is elliptical, the universe is *closed*; that is, all

The entire universe is expanding; that is, the distances between groups of galaxies is growing. Edwin Hubble first discovered the expansion in 1927. However, it can also be predicted from Einstein's general theory of relativity (published in 1915).

lines eventually return to their starting points (like great circles), and the size of the universe is finite. In a closed universe, current expansion eventually will come to a halt, and the universe will begin to collapse. The end of the universe would come in a "Big Crunch," in which all space and time would collapse into a single point. If the overall geometry is either flat or hyperbolic, the universe is *open;* if that is the case, the universe is infinite in extent and will continue to expand forever.

11.7* FRACTAL GEOMETRY

It is remarkable that **classical geometry**, developed more than 2000 years ago, remains so relevant and powerful today. Yet, despite its utility, classical geometry has clear limitations. For example, although lines, angles, and planes effectively describe many human creations (e.g., buildings, city streets, bridges, and towers) they are far less satisfactory for most natural objects. From the complicated skyline of a mountain landscape to the patterns of a head of broccoli, from the branching of river deltas to the labyrinths of the mammalian lung, natural forms often involve a complexity that defies complete description by classical geometry.

To describe natural objects better, an entirely new field of mathematics, called **fractal geometry**, has been developed during the past 30 years. Fractal geometry already has found widespread applications in mathematics and science. In addition, fractal geometry allows computers to generate realistic looking objects and landscapes. Indeed, many of the "landscapes" (of other worlds) in recent science fiction movies are, in fact, images created through fractal geometry.

11.7.1 What Are Fractals?

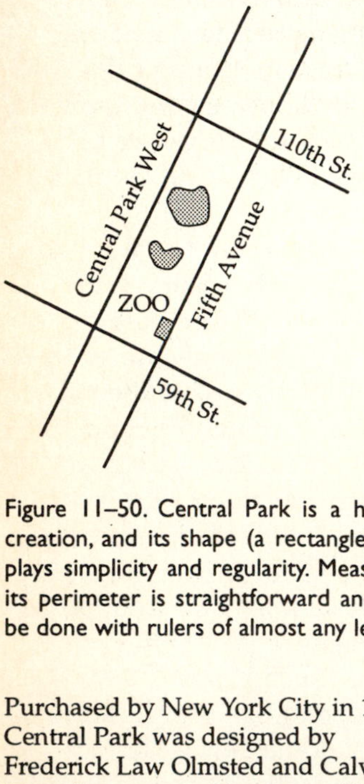

Figure 11–50. Central Park is a human creation, and its shape (a rectangle) displays simplicity and regularity. Measuring its perimeter is straightforward and can be done with rulers of almost any length.

Purchased by New York City in 1856, Central Park was designed by Frederick Law Olmsted and Calvert Vaux. It was one of the first public parks laid out by landscape architects. Olmsted also worked to preserve such natural areas as Yosemite National Park in California.

Imagine that you want to measure the perimeter of Central Park in New York City. Central Park was deliberately laid out in the shape of a rectangle, conforming to the regular Manhattan street plan. Let's assume that you have several rigid "rulers" for measurement, ranging in length from 100 meters to 1 micrometer (perhaps unrealistic, but useful for a *thought* experiment). You can use the various rulers to measure length; however, you may count only whole numbers of ruler lengths (no fractions) (Figure 11–50).

Suppose that you first measure the park perimeter with the 100-meter ruler. You begin at one corner of the park, laying the 100-meter ruler along the edge. You then walk to the end of the ruler, mark the location, and slide it down so that it measures the next 100 meters along the edge of the park. You complete the measurement by counting the number of times that the 100-meter ruler fits along the perimeter of the park. Each length laid out by the ruler is called an **element** (of length). With this terminology, the perimeter of the park is

$$\text{perimeter} = \text{number of elements} \times \text{length of each element}.$$

In this first case, the length of each element is the 100-meter ruler length. For example, if you find 96 ruler lengths around the park, the park perimeter is about 9.6 kilometers (96×100 meters = 9.6 km).

Note that the 100-meter ruler may not fit the park boundaries perfectly. For example, suppose that one side of the park actually measures 15.6 of the 100-meter ruler lengths. Because you are permitted to count only whole numbers of elements in this thought experiment, your measurement for this side of the park would be 15 ruler lengths and your measurement of the park perimeter would be somewhat short of the actual perimeter. You can make a better measurement of the park perimeter by using a 1-meter ruler as your element. Again, count the number of elements, or whole ruler lengths, that fit around the park boundary and multiply by the one-meter length of each element.

Imagine that you continue to repeat the measuring process, each time using a shorter ruler as your element: a 1-centimeter ruler, then a 1-millimeter ruler, and finally a 1-micrometer ruler. As you use smaller elements, you will be able to make more accurate measurements of the park perimeter. The discrepancies between subsequent measurements also will become smaller. In fact, rulers with a length of less than about 1 meter will yield nearly identical measurements.

This result shouldn't be surprising. After all, measuring the perimeter of a rectangle isn't difficult! However, consider measuring the perimeter (coastline) of an *island*. So that you don't have to worry about tides or waves, imagine that it is winter and the water around the island is frozen. Your task is to measure the coastline, as defined by the ice–land boundary, by using a variety of rulers with successively shorter lengths.

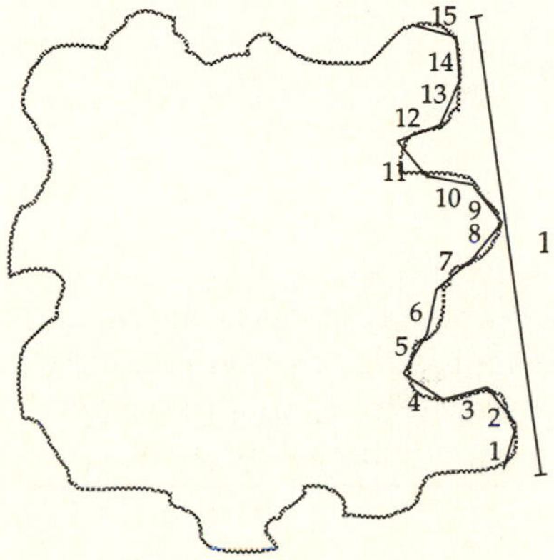

Figure 11–51. When a straight ruler is used to measure a natural object like a coastline, details that are smaller than the ruler are not measured. Note that, along the measured coast, we can fit only one 100-meter ruler, but we can fit 15 10-meter rulers. As a result, using shorter rulers yields longer measurements.

As before, you begin by using a 100-meter ruler. Because you are allowed to count only whole numbers of ruler lengths, the 100-meter ruler will adequately measure large-scale features such as bays and estuaries, but will miss smaller features such as promontories and inlets (Figure 11–51). Using a 10-meter ruler, you can follow smaller features that were missed by the 100-meter ruler and therefore you will measure a *longer* perimeter. By switching to a 1-meter ruler, you will be able to follow the contours of even smaller details of the coastline and obtain an even greater perimeter.

Figure 11–52 shows graphs representing measurements of the Central Park perimeter and the length of the island coastline obtained by using rulers of varying length. As expected, the curve for Central Park shows that different rulers yield essentially identical measurements of the perimeter (unless the ruler is very long). In contrast, the curve for the island shows that you obtain larger measurements for the length of the island coastline by using smaller rulers. Whereas the perimeter of Central Park is clearly defined simply by using a small enough ruler, agreeing on the "true" length of the coastline is difficult because it depends on the length of ruler used for measurement.

Another way to think about the difference between the results for Central Park and the island is in terms of magnification. Central Park's perimeter comprises the straight line segments of a rectangle. If you view a line segment under a magnifying glass, no new details are revealed: it still is just a line segment. Such objects are said to have a *single* **characteristic scale**, below which are no new geometric features.

In contrast, if you view a piece of the coastline under a magnifying glass, you will see details that were not visible without magnification. Thus the

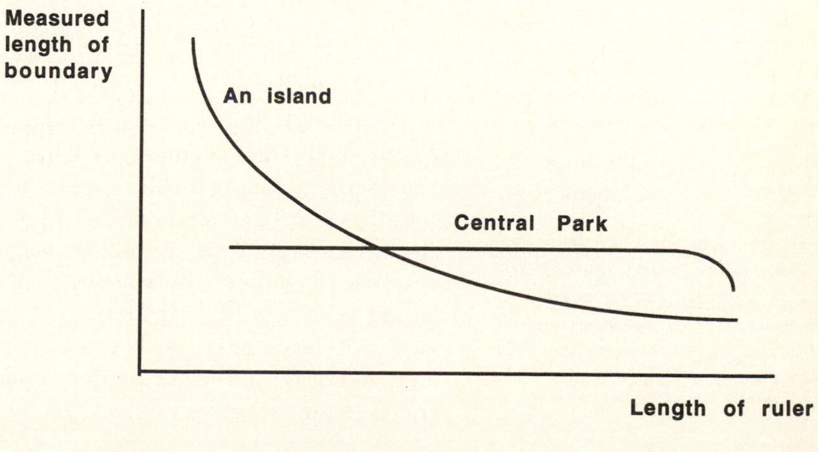

Figure 11–52. Shown are the outcomes of two thought experiments in which the boundaries of Central Park and an island are measured by using rulers of varying lengths. The measured perimeter of Central Park essentially is independent of the ruler length, but the measured length of the island coastline becomes larger as smaller rulers are used.

coral mountains

Figure 11–53. These natural objects have a complexity that cannot be described with classical geometry; instead, their properties are best described with fractal geometry.

coastline does *not* have a single characteristic scale; instead, viewing the coastline at ever smaller scales continues to reveal new features and details that cannot be seen at larger scales. Such objects, for which further magnification reveals further detail, are **fractals**. Classical geometry cannot properly deal with such objects. Instead, they lead to the strange new world of fractal geometry (Figure 11–53).

11.7.2 Fractal Dimension

The boundary of Central Park is one-dimensional because you can locate any point with one coordinate. For example, you might choose the northwest corner of the park as a starting point and measure distances clockwise around the boundary; then you can locate precisely the point with a coordinate of 375 meters.

In contrast, if you define a particular starting point, the location of a point 375 meters farther along the coastline depends on the length of ruler you use to measure the distance. Thus, because you *cannot* specify a point on the coastline with a single coordinate, the coastline is not an ordinary one-dimensional object. But neither is it two-dimensional; the coastline is a boundary around the area of the island. Instead, the coastline has a **fractal dimension** that falls "in-between" the ordinary dimensions. The fractal dimension of the coastline lies *between* 1 and 2, indicating that it has some properties that are best thought of in terms of length (one dimension) and others that are more like area (two dimensions). Note that, unlike ordinary dimensions, fractal dimensions need not be integers.

A Definition of Dimension

Suppose that we measure the length of a (straight) line segment with an element, or ruler, of a particular length. We simply count the number of elements (or ruler lengths) that fit along the line segment, and multiply by the length of each element:

$$\text{length of segment} = \underbrace{(\text{number of elements})}_{\text{number of ruler lengths that fit along line segment}} \times \underbrace{(\text{element length})}_{\text{length of ruler}}.$$

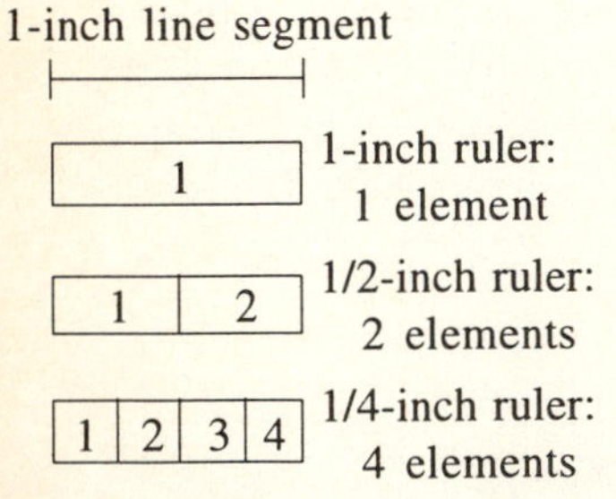

Figure 11–54. Measuring a 1-inch line segment with a 1-inch ruler yields only one element. A 1/2-inch ruler yields two elements, and a 1/4-inch ruler yields four elements. In general, reducing the ruler length by a factor of R increases the number of elements by a factor of $N = R$.

Figure 11–54 shows a 1-inch line segment. Using a ruler that is the same length as the line segment, or 1-inch in length, yields only one element along the segment. Using a ruler that is smaller by a *factor* of 2, or 1/2 inch in length, yields two elements along the 1-inch line segment. That is, decreasing the ruler length by a "reduction factor" of $R = 2$ leads to an increase in the number of elements by a factor of $N = 2$. Similarly, we would find 10 elements with a 1/10-inch ruler. That is, reducing the element length by a factor of $R = 10$ leads to an increase in the number of elements by a factor of $N = 10$. In general, for a line segment, reducing the element length by a factor of R increases the number of elements by a factor of $N = R$.

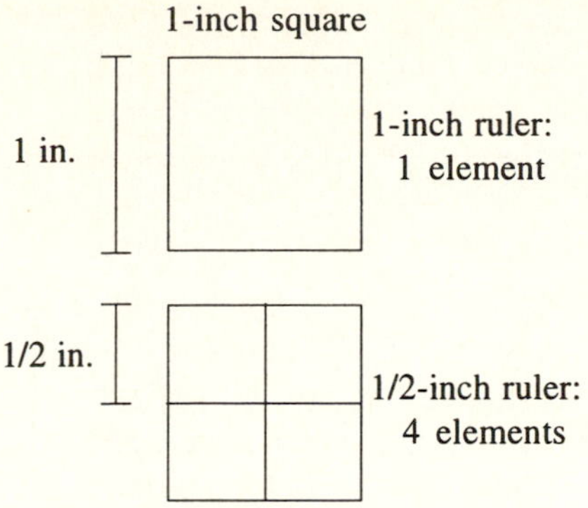

Figure 11–55. A 1-inch ruler yields only one *area element* in measuring a 1-inch square. Reducing the ruler length by a factor $R = 2$, to 1/2 inch, increases the number of elements by a factor of $N = 4$. In general, reducing the ruler by a factor of R increases the number of area elements by a factor of $N = R^2$.

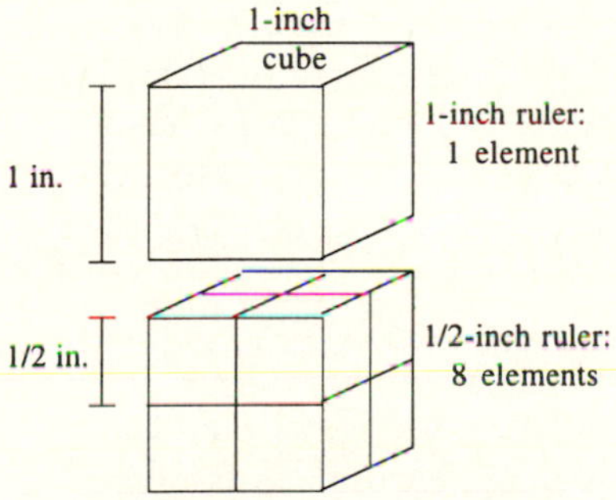

Figure 11–56. A 1-inch ruler yields one *volume element* in a 1-inch cube. Reducing the ruler length by a factor $R = 2$, to 1/2 inch, increases the number of elements by a factor of $N = 8$. In general, reducing the ruler length by a factor R increases the number of volume elements by a factor of $N = R^3$.

We can use a similar process to determine the area of a square by counting the number of *area elements* that fit within it (Figure 11–55):

Area of square = (number of area elements) × (area of each element).

Using a ruler that has the same 1-inch length as a side of the square, we can make an area element of 1 square inch. Only one such element fits in the square. With a 1/2-inch ruler, we make an area element of $(1/2 \text{ in})^2 = 1/4 \text{ in}^2$; four of these elements will fit in the square. That is, decreasing the ruler length by a factor of $R = 2$ leads to an increase in the number of area elements by a factor of $N = 4$. With a 1/10-inch ruler, the area element is $(0.1 \text{ in})^2 = 0.01 \text{ in}^2$ and 100 elements are needed to cover the square: reducing the ruler length by a factor of $R = 10$ increases the number of elements by a factor of $N = 100$. In general, reducing the ruler length by a factor of R increases the number of area elements by a factor of $N = R^2$.

Finally, we can measure the volume of a cube by counting the number of *volume elements* that fit within it (Figure 11–56):

Volume of cube = (number of volume elements) × (volume of each element).

Again, we begin with a ruler of the same 1-inch length as a side of the cube, creating a 1-cubic-inch volume element. Only one such element fits in the cube. Reducing the ruler length by a factor of $R = 2$, to 1/2 inch, creates a volume element of $(1/2 \text{ in})^3 = 1/8 \text{ in}^3$; therefore the number of elements increases by a factor of $N = 8$. If we reduce the ruler length by a factor $R = 10$ to 1/10 inch, the volume element is $(0.1 \text{ in})^3 = 0.001 \text{ in}^3$ and $N = 1000$ times as many elements fit in the cube. In general, reducing the ruler length by a factor of R increases the number of volume elements by a factor of $N = R^3$.

If we define a number D so that $N = R^D$, we can summarize our findings as follows:

- For a one-dimensional object (e.g., a line segment), we found $N = R$; as $N = R^D$, we conclude that $D = 1$.
- For a two-dimensional object (e.g., a square), we found $N = R^2$; as $N = R^D$, we conclude that $D = 2$.
- For a three-dimensional object (e.g., a cube), we found $N = R^3$; as $N = R^D$, we conclude that $D = 3$.

The number D is called the **fractal dimension** of the object. For ordinary objects such as line segments, squares, and cubes, the fractal dimension is an integer and is equal to the ordinary dimension.

The Snowflake Curve

The snowflake curve sometimes is called a **Koch curve**, after Helga von Koch who first described it in 1906.

We can investigate the concept of fractal dimension further with a special object called a **snowflake curve**. The snowflake curve is an abstraction that is created, in principle, by a drawing *process*. The process begins with a straight line segment, which we designate L_0 as shown at the bottom of Figure 11–57 (on the next page), and proceeds in three steps.

1. Divide the line segment L_0 into three equal pieces.
2. Remove the middle piece.
3. Replace the middle piece with two segments of the same length arranged as two sides of an equilateral triangle.

The result, designated L_1, is shown above L_0. Note that L_1 consists of four line segments and that each is 1/3 the length of L_0 (because L_0 was divided into three equal pieces).

Next, we repeat the three steps on *each* of the four segments of L_1. The result is L_2, which comprises 16 line segments, each 1/9 the length of L_0.

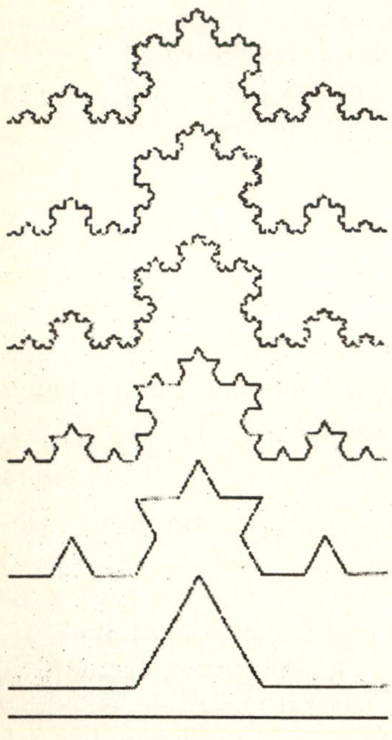

Figure 11–57. Starting with the line segment L_0, the first six approximations to the snowflake curve are shown. From *The Science of Fractal Images*, Heinz-Otto Peitgen & Dietmar, Saupe, Eds. © 1988 Springer-Verlag, NY. Reprinted with permission.

Consult Appendix C for a review of logarithms.

Time-Out to Think: Count the segments shown in Figure 11–57 to confirm that L_2 comprises 16 line segments. Measure to confirm that each is 1/9 the length of L_0. *Why* are the segments 1/9 the length of L_0? How long are the segments of L_3?

Continuing to repeat the three-step process generates L_3, L_4, L_5, and so forth. If we could repeat this process an *infinite* number of times, the ultimate result, denoted L_∞, would be the snowflake curve. Of course, because we can't repeat the process infinitely, we can never draw the snowflake curve, L_∞, exactly — only finite approximations to it, such as L_6, can be drawn.

Now, imagine measuring the complete snowflake curve, L_∞, with a ruler the length of L_0. Such a ruler would simply lay across the base of the curve, missing all the curve's fine detail and measuring only the straight-line distance between the endpoints. Thus this ruler would yield only one element along the snowflake curve.

Next, suppose that we reduce the ruler length by a factor of $R = 3$, so that it is 1/3 the length of L_0. This ruler is the same length as each of the four segments of L_1, so it yields four elements along the snowflake curve. That is, reducing the ruler length by a factor of $R = 3$ increases the number of elements by a factor of $N = 4$. Reducing the ruler length by a factor of $R = 9$ makes it 1/9 the length of L_0, or the length of the 16 segments of L_2. The number of elements found is now increased by a factor of $N = 16$. Note the pattern: Every time we reduce the ruler in length by a factor of 3, the number of elements detected increases by a factor of 4. That is,

when $R = 3$, $N = 4$; when $R = 9$, $N = 16$; when $R = 27$, $N = 64$; and so on.

What is the fractal dimension of the snowflake curve? Recall that the fractal dimension D is defined by the relation $N = R^D$. For the snowflake curve, the number D is such that $4 = 3^D$, $16 = 9^D$, $64 = 27^D$, and so on. Choosing the first pair of numbers, we solve for D using logarithms:

$$4 = 3^D \xrightarrow{\text{logs of both sides}} \log_{10} 4 = \underbrace{\log_{10} 3^D = D \log_{10} 3}_{\text{recall } \log x^n = n \log x} \Rightarrow D = \frac{\log_{10} 4}{\log_{10} 3} = 1.2619.$$

Time-Out to Think: Use your calculator to confirm that, if $D = 1.2619$, then $16 = 9^D$ and $64 = 27^D$.

The fractal dimension of the snowflake curve is approximately 1.2619. What does that mean? The fact that the fractal dimension is greater than 1 means that the snowflake curve has more "substance" than an ordinary one-dimensional object. In a sense, the snowflake curve begins to fill the part of the plane in which it lies. The closer the fractal dimension of an object is to 1, the more closely it resembles a collection of line segments. The closer the fractal dimension is to 2, the more closely it comes to filling a part of a plane.

In general, finding the fractal dimension of any object requires finding the factor increase in the number of elements, N, that corresponds to a ruler reduction factor, R. The fractal dimension is the number D that satisfies the equation $N = R^D$, or solving for D,

$$D = \text{fractal dmension} = \frac{\log_{10} N}{\log_{10} R}.$$

Example 11–34 ***Computing a Fractal Dimension.*** In measuring the length of a certain strange jagged curve, you find that every time you decrease the length of your ruler by a factor of 2, the number of elements increases by a factor of 3. What is the fractal dimension of the curve?

Solution: The fact that a decrease in ruler length by a factor of 2 leads to a factor of 3 increase in the number of elements implies that, when $R = 2$, $N = 3$, and that when $R = 4$, $N = 9$, and so on. With the fractal dimension, D, defined by the relation $N = R^D$,

$$3 = 2^D \xrightarrow{\text{logs of both sides}} \log_{10} 3 = \underbrace{\log_{10} 2^D = D \log_{10} 2}_{\text{recall } \log x^n = n \log x} \Rightarrow D = \frac{\log_{10} 3}{\log_{10} 2} = 1.585.$$

The fractal dimension of the curve is about 1.585. Because this dimension lies between 1 and 2, the curve fills the segment of the plane in which it resides to some extent. The fact that its fractal dimension is somewhat greater than the fractal dimension of the snowflake curve indicates that it fills its plane segment to a greater extent than does the snowflake curve. In a sense, this curve is "denser" than the snowflake curve.

❖

The Snowflake Island

The **snowflake island** is an *area* (island) bounded by three snowflake curves. To draw the snowflake island, we begin with an equilateral triangle (Figure 11–58). Then, we convert *each* of the three sides of the triangle into a snowflake curve, L_∞, following the steps previously described. As with the snowflake curve, we cannot draw the snowflake island exactly because it would require an infinite number of steps. Therefore only a few stages in the generation of the snowflake island are shown. Because its border consists of snowflake curves, the fractal dimension of the border (or coastline) of the snowflake island is the same as that of the snowflake curve, or about 1.2619.

How long is the coastline of a snowflake island? Because it comprises three snowflake curves, we first need to know the length of a single snowflake curve. Look back at the sequence of curves L_0, L_1, L_2,... in Figure 11–57. Note that the length of L_1 is 4/3 times the length of L_0. The length of L_2 is 16/9 times the length of L_0, or 4/3 times the length of L_1. Continuing, the length of L_3 is 4/3 times the length of L_2, and so on. If we set the length of L_0 to be 1 unit, the lengths of L_1, L_2, L_3, L_4, L_5, L_6, ... respectively, are

$$\frac{4}{3} = 1.333,\ \left(\frac{4}{3}\right)^2 = 1.777,\ \left(\frac{4}{3}\right)^3 = 2.370,\ \left(\frac{4}{3}\right)^4 = 3.160,\ \left(\frac{4}{3}\right)^5 = 4.214,\ \left(\frac{4}{3}\right)^6 = 5.619.$$

Each successive approximation to the snowflake curve is longer than the previous one by a factor of 4/3. The length of L_n therefore is $(4/3)^n$ times the length of L_0. Because $(4/3)^n$ approaches infinity as n continues to increase, we

Figure 11–58. The first, second, third, and sixth stages in the construction of the snowflake island are shown. If the process is repeated endlessly, the ultimate object has a finite area, but a coastline with an infinite length. From *Fractals: Form, Chance, and Dimension* by Mendelbrot. Copyright © 1977 by Benoit B. Mendelbrot. Used with permission of W.H. Freeman and Company.

conclude that the complete snowflake curve, L_∞, must be infinitely long! Being made up of three snowflake curves, the coastline of the snowflake island also must be infinitely long.

What is the area of the snowflake island? As Figure 11–58 shows, the island is contained within the bounds of the page. Because the area of the page is finite, the snowflake island also has a finite area. Thus we have the intriguing result that a snowflake island is an object with a finite area and an infinitely long coastline!

11.7.3 The Dimension of Coastlines

Note that each of the four pieces of L_2 in Figure 11–57 looks exactly like L_1, except they are smaller, and that L_3 consists of four pieces, each of which looks like L_2. In fact, if you examine any piece of the snowflake curve, L_∞, under magnification, it will look exactly like one of the earlier curves, L_0, L_1, L_2,..., used in its generation. Because the snowflake curve looks similar to itself when examined at different scales, this property is called **self-similarity**. It is a consequence of repeated application of a simple set of rules.

Natural objects, such as real coastlines, also possess details at many different spatial scales, so it makes sense to assign them a fractal dimension. However, unlike the case with self-similar fractals, a magnified view of a natural object isn't likely to look *exactly* like the original object. We must find an alternative approach to finding the fractal dimension.

Lewis Fry Richardson was an eclectic and eccentric English scientist who proposed computer methods for predicting the weather in 1920 — before computers were invented!

Figure 11–59 shows a graph of experimental data collected by L. F. Richardson in about 1960. Richardson's data represent measurements and estimates of the lengths of various coastlines and international borders (called *frontiers* on the graph) measured by "rulers" of varying sizes. Along the horizontal axis, the graph shows the logarithm of the "length of side in kilometers," which essentially is the length of the ruler used for the measurement. Along the vertical axis, the graph shows the logarithm of the total length of the coastline or border measured. Thus, except for the use of logarithms, it has the same axes as the graphs for Central Park and an island (Figure 11–52).

Richardson's graph allows a simple determination of the fractal dimension D: the *slope* of each line is $1 - D$ or, equivalently, $D = 1 -$ slope. For example, Richardson includes a horizontal line, which has a slope of 0, representing measurements of a circle. Thus the fractal dimension of the circle is 1 ($D = 1 -$ slope $= 1 - 0 = 1$), as expected: The fractal dimension of a circle is the same as its ordinary dimension of 1.

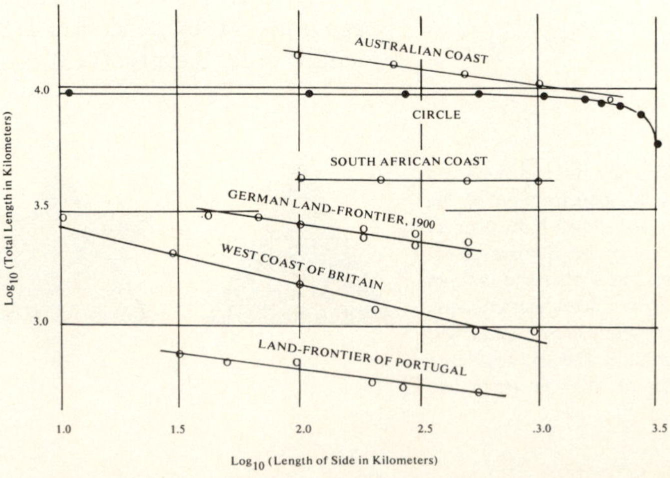

Figure 11–59. The experimental data of L. F. Richardson show the measured lengths of various coastlines and borders as they vary with the length of the measuring device. The slope of each line is $1 - D$, where D is the fractal dimension of the corresponding boundary. From *Fractals: Form, Chance, and Dimension* by Mendelbrot. Copyright © 1977 by Benoit B. Mendelbrot. Used with permission of W.H. Freeman and Company.

The graphs for the coastlines of Portugal, Great Britain, Germany, and Australia are straight lines with a slope of approximately –0.25, suggesting a fractal dimension of $D = 1.25$. Note how close that is to the fractal dimension of the coastline of the snowflake island ($D = 1.26$). We conclude that natural coastlines must be have properties similar to those of the snowflake curve. (The coastline of South Africa appears to be an exception, at least according to these data.)

Time-Out to Think: Lines for borders (frontiers) in Richardson's data also have slopes close to –0.25 and therefore fractal dimensions of about 1.25. Explain how this result accounts for the fact that, although the location of the border is not disputed, the length of the border between Portugal and Spain ("land frontier of Portugal") is claimed to be 987 kilometers by Portugal and 1214 kilometers by Spain.

11.7.4 Fractals by Iteration

Self-similar fractals, like the snowflake curve and snowflake island, are generated by repetition of simple rules. The process of repeating a rule over and over is called **iteration**. All fractals can be generated by iteration; different rules produce fractals that differ in wild and wonderful ways.

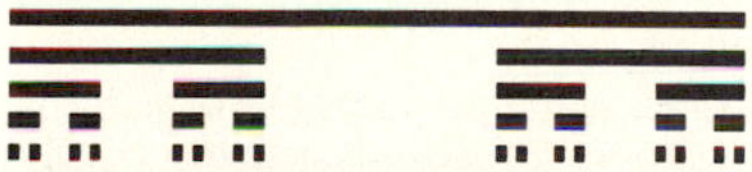

Figure 11–60. The dust-like Cantor set is generated by starting with a single line segment at the top. With each iteration, moving downward, the middle third of each line segment is removed. The Cantor set has a fractal dimension of less than 1.

For example, consider a fractal generated from a line segment to which the following rule is applied repeatedly: *Delete the middle third of each line segment of the current figure* (Figure 11–60). With each iteration, the line segments becomes shorter until eventually the line turns to dust. The limit (after infinitely many iterations) is a fractal called the **Cantor set**. Because this ephemeral structure results from diminishing a one-dimensional line segment, its fractal dimension is *less* than 1.

Two general approaches are used to create fractals by iteration. The first, used in generating both the snowflake curve and the Cantor set, is called **deterministic iteration**: the *exact* same set of rules is repeated in each iteration. Another fractal produced by deterministic iteration is the **Sierpinski triangle**. It is produced by starting with a solid black equilateral triangle and iterating with the following rule: *For each black triangle in the current figure, connect the midpoints of the sides and remove the resulting inner triangle* (Figure 11–61). The complete Sierpinski triangle, which would require *infinite* iterations to produce, is a strange fractal beast. For example, the total area of its black regions is zero! The fractal dimension of the Sierpinski triangle is between 1 and 2; it is less than 2 because "material" has been removed from the initial two-dimensional triangle.

Figure 11–61. The first three stages, and the eighth stage, in the development of the Sierpinski triangle are shown. Each step involves removing equilateral triangles on smaller and smaller scales. From *The Science of Fractal Images*, Heinz-Otto Peitgen & Dietmar, Saupe, Eds. © 1988 Springer-Verlag, NY. Reprinted with permission.

Figure 11–62. A Sierpinski sponge is generated by successively removing cubes from an initial solid cube. Shown is the result of several applications of the iteration rule. From *Fractals: Form, Chance, and Dimension* by Mendelbrot. Copyright © 1977 by Benoit B. Mendelbrot. Used with permission of W.H. Freeman and Company.

Figure 11–63. Two views of Barnsley's fern, one more magnified than the other, reveal that this mathematically generated object closely resembles a real fern. From Michael Barnsley, *Fractals Everywhere*, © 1988 Academic Press, Inc. Reprinted with permission.

A closely related object is the **Sierpinski sponge** (Figure 11–62). It is generated by starting with a solid cube and iterating with the rule: *Divide each cube of the current object into 27 subcubes, and remove the central subcube and the center cube of each face*. The resulting object has a fractal dimension between 2 and 3; it is less than a full three dimensions because material has been removed from the space occupied by the sponge.

The second approach to creating fractals, called **random iteration**, allows very slight, random variations at each step of the iteration process. Often, the slightest variability in the rules can result in fractals that resemble natural forms. Barnsley's fern (Figure 11–63) is an example of a fractal produced by random iteration. Note how closely this mathematically generated object resembles a real fern.

The fact that fractals so successfully replicate natural forms suggests an intriguing possibility: Perhaps *nature* produces the many diverse forms that we see around us through simple rules that are applied repeatedly and with a hint of randomness. Because of this observation and because modern computers allow iterations to be carried out thousands or millions of times, fractal geometry surely will remain an active field of research for decades to come.

11.8 CONCLUSION

In this chapter, we explored mathematics as the language of size and shape, which the study of geometry embodies. We covered the fundamental concepts of geometry, emphasizing how geometric objects in two and three dimensions can be used to solve real problems. Key ideas to take away from this chapter include the following.

- A wealth of problems can be solved with Euclidean geometry, which dates back more than 2000 years. In particular, two- and three-dimensional shapes (polygons, circles, boxes, cylinders, and spheres) can be used to model many real objects.
- Many other types of problems can be solved through application of the properties of triangles and similar triangles. Right triangles, which obey the Pythagorean theorem and allow calculation of sine, cosine, and tangent ratios, are particularly important.
- Extensions of traditional Greek geometry include non-Euclidean geometry, which has helped explain the geometry of the universe, and fractal geometry, which offers new ways of understanding nature.

PROBLEMS

Reflection and Review

Section 11.2

1. **Geometry Around You.** Look around your room. Briefly describe at least three realizations each of (a) points, (b) lines, and (c) planes.

2. **Dimension.** Examine a closed book.
   a. How many dimensions are needed to describe the book? Explain.
   b. How many dimensions describe the surface (cover) of the book? Explain.
   c. How many dimensions describe an edge of the book? Explain.
   d. Describe some aspect of the book that represents zero dimensions.

3. **Angles and Circles.** Find the angles that subtend:
   a. 1/3 circle **b.** 1/12 circle **c.** 1/20 circle
   d. 1/30 circle **e.** 1/90 circle **f.** 3/4 circle

4. **Fractions of Circles.** Find the fraction of a circle subtended by an angle of:
   a. 1° **b.** 4° **c.** 15° **d.** 30°
   e. 60° **f.** 90° **g.** 180° **h.** 235°

5. **Minutes and Seconds in the Circle.**
   a. Calculate the number of minutes of arc in a full circle.
   b. Calculate the number of seconds of arc in a full circle.

6. **Converting to Minutes and Seconds.** Convert each of the following into degrees, minutes, and seconds.
   a. 26.5° **b.** 41.333° **c.** 15.0161°
   d. 0.002° **e.** 126.9971° **f.** 0.0001°

7. **Converting to Decimal Degrees.** Convert each of the following into a decimal fraction of degrees:
   a. 10°15' **b.** 10°15'15" **c.** 36°45'8"
   d. 90°35'27" **e.** 1°1'1" **f.** 0°22'58"

8. **Angle Practice.** Find each of the unknown angles. Explain your reasoning.

a.

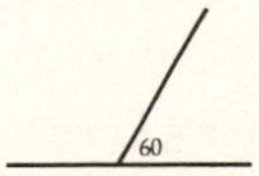

b.

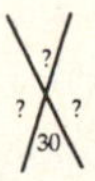

c.

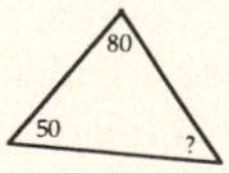

d.

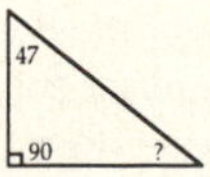

9. **Pitch and Grade.**
   a. What is the slope of a 5 in 12 roof? If the roof rises for 10 feet along the horizontal, how high is it vertically?
   b. How much does a road with a 3% grade rise for each horizontal foot? If you drive for 8 miles along this road, how much elevation will you gain?
   c. What is the angle (relative to the horizontal) of a 6 in 6 roof? Is it possible to have a 7 in 6 roof? Explain.
   d. What is the grade (in percentage) of a path that rises 1000 feet every mile?
   e. What is the grade of a road that rises 10 feet for every 120 horizontal feet?
   f. How much does a 15% grade trail rise for each 100 horizontal yards?

Section 11.3

10. **Properties of Right Triangles.** Find the length of the unknown sides or angles in each right triangle.

a.

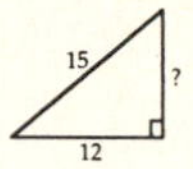

b.

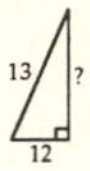

c.

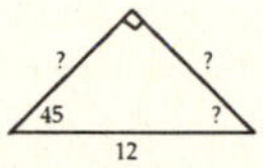

d.

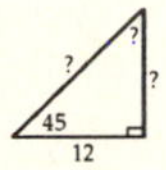

11. **Distance Measurement.** The following city map shows the bus stop nearest to a movie theater. The city blocks are square and are 40 meters long.

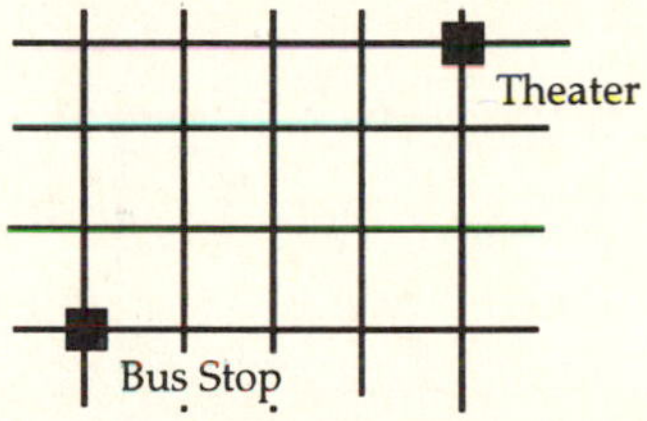

   a. What is the shortest *direct* path (as the crow flies) from the bus stop to the theater?
   b. What is the shortest path along the sidewalk?
   c. Draw two possible routes on the figure that both have the shortest length along sidewalks.

12. **Building Stairs.** Refer to Figure 11–14 showing the area to be covered with plywood under a set of stairs. Suppose that instead of being 15 feet long, the stairs will be 20 feet long (rising at a 45° angle).
   a. How much plywood would you need to buy? How much more is that amount, by percentage, than calculated in the example for a 15-foot staircase?
   b. Explain how you would cut 4-foot by 8-foot plywood sheets to cover the area; how many would you need?

13. **More Right Triangle Problems.**
   a. Refer to Figure 11–15, but suppose that the stream frontage is only 125 feet and the property line is 600 feet. How many acres does the property contain?
   b. The distance between bases along the base paths on a baseball diamond is 90 feet. Because the baseball diamond actually is a square, how far does the catcher throw the ball from home plate to second base?

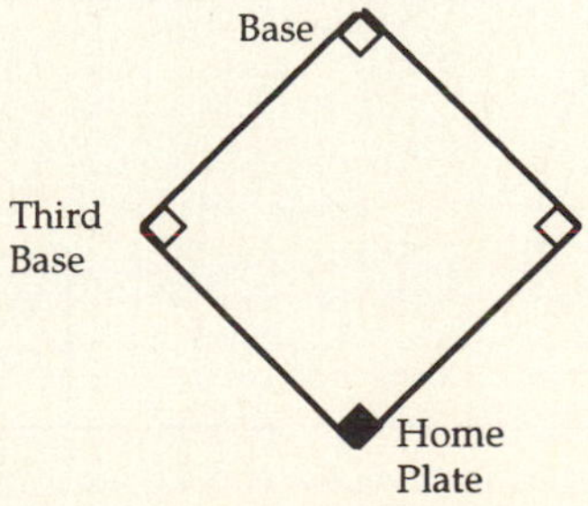

c. A set of phone lines runs east along the edge of a field for 0.5 mile and then north along the edge of the same field for 0.75 mile. How much phone cable would be needed to run the line diagonally across the field?

d. You have a choice of either walking west from a cabin along the shore of a lake for 1.5 miles and then walking south along the shore of the lake for 0.8 mile to a second cabin or swimming directly to the second cabin. How far is the swim? If you can walk twice as fast as you can swim, which is the fastest route?

14. Plane Area Problems.

a. Refer to the floor plan in Figure 11–17. Suppose that the bedroom floors are to be covered with carpet costing $15 per square yard, the bathroom floor is to be covered with tile costing $2 per square foot, and the hallway is to be covered with tile costing $5.50 per square foot. What is the total cost of materials for this job?

b. The end views of two different barns are shown in the following figure. Which end will require the most paint? Explain.

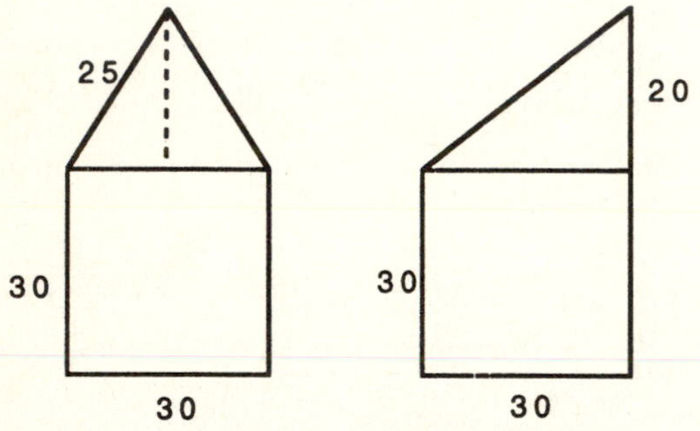

c. The following figure shows a city park in the shape of a parallelogram with a rectangular playground in its center. If all but the playground is covered with grass, what area is covered by grass?

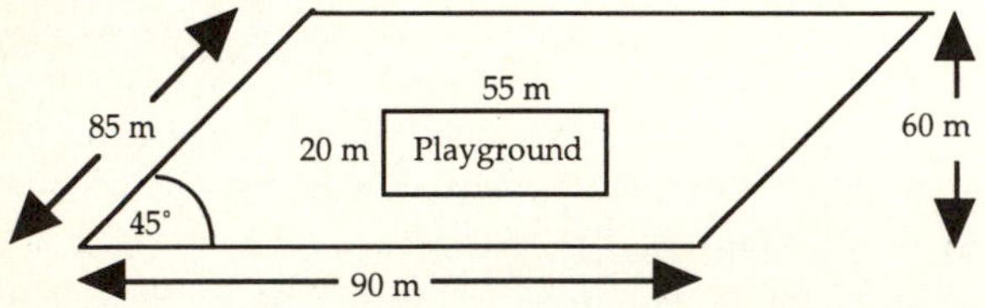

d. The following figure shows the layout of a back yard that is to be seeded with grass with the exception of the patio and flower bed. What area is to be seeded with grass?

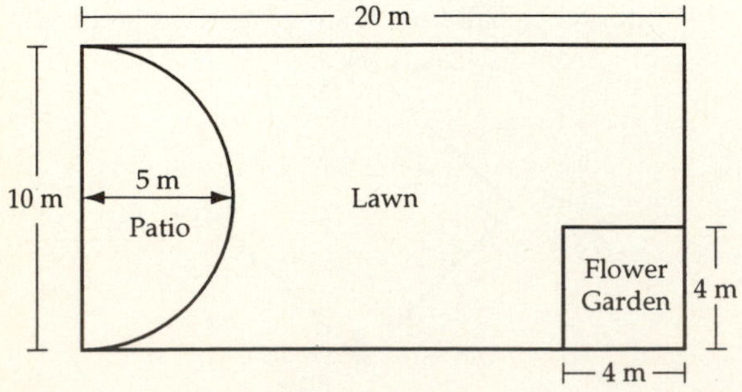

15. Similar Triangles. Look at each of the following pairs of triangles. Which are pairs of similar triangles? Explain.

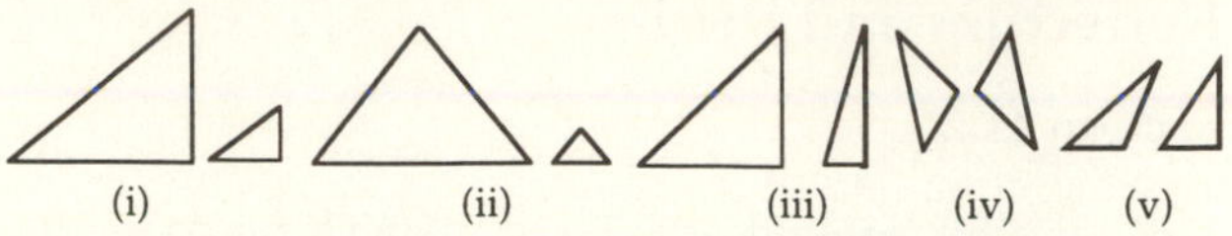

16. Properties of Similar Triangles. Find the lengths of the unknown sides in each of the following pairs of similar triangles.

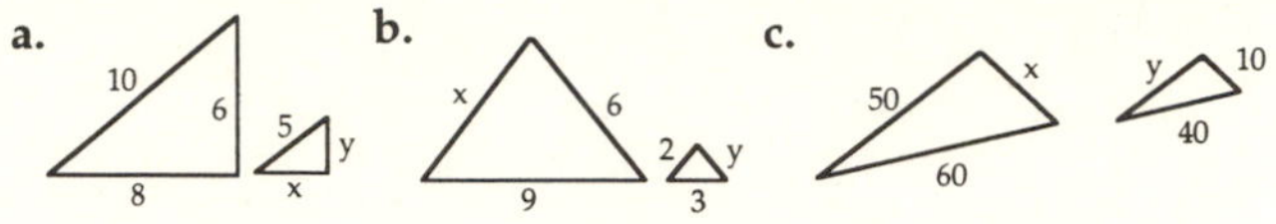

17. Planet Nearth. You are a geographer on the planet Nearth that orbits a distant star (Nearth's sun). Nearth is spherical in shape, but no one knows its size. One day, you learn that on the equinox your sun is directly overhead at noon in the city of Nyene, which lies due north of Alectown. On the day of the equinox, you observe that the altitude of the Sun is 80° (10° from the zenith) at noon in Alectown. If the distance between Nyene and Alectown is 1000 kilometers, what is the circumference of Nearth?

18. Solar Access Laws. Suppose that you live at a latitude where the shadow cast by a 12-foot high fence is 25 feet long at noon on the winter solstice. Assume that your town enforces the solar access ordinance stated in Example 11–14. If your house is set back 30 feet from the north property line, what is the maximum house height allowed?

Section 11.4

19. Box Problems.

a. A competition swimming pool is 50 meters long, 25 yards wide, and has an average depth of 6 feet. How much water does the pool hold?

b. A large convention center measures 40 meters by 60 meters in floor area, with a 10-meter-high ceiling. How much air does it hold? How many molecules of air?

c. How many cubical dice, 3 cm on a side, can be packed into a box that measures 30 cm by 60 cm by 90 cm?

20. Optimizing Boxes. You design boxes for a moving company that must have square bottoms and a volume of 1 cubic meter. Material for the boxes costs 10¢/m². The boxes currently in use measure 0.5 m by 0.5 m by 4 m.

a. Confirm that the volume of the current boxes is 1 cubic meter. How much do the materials for producing one of the current boxes cost?

b. Design a box that minimizes the production cost, while maintaining a volume of 1 cubic meter.

c. Given that your new boxes are cheaper to produce, should the moving company still offer the older box size to customers? Why or why not?

21. Cylinder Problems.

a. A heat duct in the college library has a circular cross section with a radius of 10 inches and a length of 25 feet. What is the volume of the duct and how much paint (in square feet) is needed to paint the duct?

b. Your company manufactures oil drums that have a radius 0.25 meter and a height of 1 meter. The material used for the tops and bottoms of the drums costs \$7.25 per square meter and the material used for the sides costs \$4.50 per square meter. What is the capacity of a single drum? What is the cost of materials for a single drum?

c. Three tennis balls fit exactly (no room to spare) in a cylindrical can. Which is greater, the circumference of the can or the height of the can? Explain your reasoning.

22. Sphere Problems.

a. A lacrosse ball has a diameter of 4.5 inches. What is its volume and surface area?

b. Calculate the radius of a spherical water tank that holds enough water to fill a swimming pool that measures 50 meters by 25 yards and has an average depth of 6 feet.

23. Density of the Sun. The radius of the Sun is about 696,000 km and its mass is about 2.0×10^{30} kg. Calculate its average density. Compare it to the average density of the Earth. Do you think that the center of the Sun is more or less dense than the center of the Earth? Explain.

24. Packing Spheres. You have been asked to serve as a consultant for a children's museum that is building a "wading pool" filled with plastic balls. The pool is to be 30 feet long, 15 feet wide, and 3 feet deep. The plastic balls have a diameter of 4 inches. Approximately how many balls are needed to fill the pool?

25. Melting Ice. A glacier's surface is approximately rectangular with a length of about 100 meters and a width of about 20 meters. The ice in the glacier averages about 3 meters in depth. Suppose that the glacier melts into a lake that is roughly circular with a radius of 1 kilometer. Assuming that the area of the lake does not expand significantly, about how much would the water level rise if the entire glacier melts into the lake?

26. White Dwarf Catastrophe. A white dwarf star has a density of about 1800 kg per cubic centimeter and a radius about the same as the Earth's (6400 kilometers). Suppose that, instead of becoming wrapped around a neutron star, as in Example 11–24, the Earth became wrapped around a white dwarf star. How thick a layer would the Earth make on the white dwarf?

Section 11.5

27. Radians and Circles. Find the angles, in radians, that subtend:

a. 1/3 circle **b.** 1/12 circle **c.** 1/20 circle
d. 1/30 circle **e.** 1/90 circle **f.** 3/4 circle

28. Degree–Radian Conversions. Convert each of the following angles from degrees into radians.

a. 1° **b.** 2° **c.** 10° **d.** 30°
e. 45° **f.** 135° **g.** 180° **h.** 270°
i. 25° **j.** 37° **k.** 14.6° **l.** 22°30'15"

29. Radian–Degree Conversions. Convert each of the following angles from radians into degrees.

a. π **b.** $\pi/6$ **c.** $\pi/15$ **d.** $\pi/30$
e. $\pi/45$ **f.** $\pi/90$ **g.** $\pi/180$ **h.** 1
i. 0.25 **j.** 1.35 **k.** 2.7 **l.** 0.866

30. Angles and Arc Length.

a. A circle has a radius of 10 meters. If you walk along the edge, how far will you walk in sweeping an angle of (i) 1 radian? (ii) 2.5 radians? (iii) π radians? (iv) $\pi/10$ radians? In each case, state what fraction of the circumference of the circle you will have walked.

b. A circle has a radius of 2 kilometers. What is the length of an arc subtended by an angle of (i) 1 radian? (ii) 1°? (iii) $\pi/4$ radians? (iv) 15°? In each case, what fraction of the circle is represented by these angles?

c. A circle has a circumference of 10 km. How long is an arc that subtends (i) 1°; (ii) 1'; (iii) 1"?

31. Angular Size, Physical Size, and Distance.

a. The angular size of the Sun is about the same as that of the Moon: one-half degree (that's why the Moon can eclipse the Sun). The distance to the Sun is approximately 150 million kilometers. Calculate the actual diameter of the Sun.

b. What is its angular diameter of a dime if it is held 1 centimeter from your eye? What is its angular diameter if you observe the dime from a distance of 100 meters?

c. By measuring your arm length and the width of your index finger, calculate the angular width of your finger when held up with your arm extended. Compare your result to the 8 minutes of arc that led Kepler to his discovery of the laws of planetary motion (see "*Thinking About* ... Eight Minutes of Arc" in Section 11.2).

32. Calculator Practice. Use your calculator to evaluate each of the following expressions. Be sure that your calculator is in degree or radian mode, as needed. (If no degree symbol is indicated, then the angle is in radians.)

a. sin 27° **b.** cos 80° **c.** cos 245° **d.** sin 180°
e. tan 0.7 **f.** sin 2.2 **g.** tan 50° **h.** sin 125°
i. tan 88° **j.** cos 10° **k.** cos $\pi/6$ **l.** cos 1.5

33. More Calculator Practice. Use your calculator to evaluate each of the following inverse relations. Give all answers in degrees.

a. $\sin^{-1} 0$ **c.** $\cos^{-1} 0$ **e.** $\cos^{-1} 0.87$ **g.** $\sin^{-1} 0.3$
b. $\tan^{-1} 1$ **d.** $\sin^{-1} 0.55$ **f.** $\tan^{-1} 2.6$ **h.** $\cos^{-1} 0.9$

34. Practice Using Sine, Cosine, and Tangent.

a. Suppose that angle A in the right triangle shown is 30° and that the hypotenuse represents a length of 7 cm. Find the lengths of the sides adjacent and opposite to angle A. Check your answers by confirming that they satisfy the Pythagorean theorem.

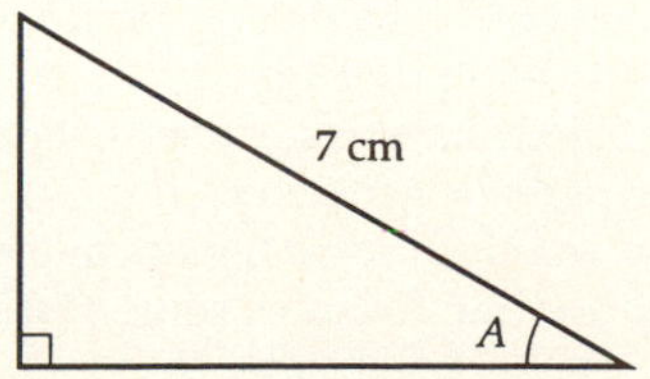

b. Draw a right triangle in which one of the angles is 25° (use a protractor) and the hypotenuse is 2.5 cm. Calculate the lengths of the other sides by using sine and cosine ratios and confirm your results by measuring the sides.

c. Suppose that one side of a right triangle measures 8 meters and that the hypotenuse measures 11 meters. How large is the angle opposite the 8-meter side? How large is the adjacent angle?

35. **Applications of Sine, Cosine, and Tangent.**
a. Suppose that you live at a latitude where the shadow cast by a 12-foot-high fence is 25 feet long at noon on the winter solstice. In that case, what is the angle to the Sun at noon on the winter solstice?
b. Suppose that you own a triangular piece of property similar to the one shown Figure 11–15 with 250 feet of stream frontage. However, your property differs in that the angle between the stream and the property line is only 65°. Calculate the lengths of the other two sides of your property and the total area of your property in acres.
c. From your boat you can see the lighthouses of both Port Windsong and Port Landfall. From your current position Port Landfall is due east, and Port Windsong is northeast (45° angle to due east). On a map, you see that Port Windsong is 30 kilometers due north of Port Landfall. How far are you from each Port? (Hint: Draw a picture!)

36. **Triangulation.**
a. Study Figure 11–42. Suppose you find the angle to the boat is 80°, rather than 75°. In that case, how far away is the boat along the surface?
b. Refer to Figure 11–43. Suppose that the distance between the two surveying points 1 and 2 is 50 meters, rather than 25 meters. In that case, how far away is the peak?

37. **Stellar Distance.** How far away is a star with a parallax angle $p = 0.02''$?

Section 11.6

38. **Distances on the Earth.** (This problem requires use of a globe.) The cities of Los Angeles and Kabul (Afghanistan) are both at latitudes of approximately 35° N. By following the circle of 35° N latitude on a globe, about how far would you travel in going from Los Angeles to Kabul? Compare this distance to the *shortest* distance between the two cities, which follows a great circle. Describe the great circle route (i.e., what other places does it pass over?).

39. **Models for Non-Euclidean Geometry.** In each of the following explain why your objects are non-Euclidean and how you determined whether they are elliptical or hyperbolic.
a. Describe two common objects having surfaces that provide a good model for elliptical geometry.
b. Describe two other objects having surfaces that provide a good model for hyperbolic geometry.
c. Can an object be hyperbolic in some regions and elliptical in others? Explain. (Hint: Examine some vases.)

Section 11.7

40. **Ordinary Dimensions for Ordinary Objects.**
a. Suppose that you want to measure the length of the sidewalk in front of your house. Describe a thought process by which you can conclude that $N = R$ for the sidewalk and hence that its fractal dimension is the same as its ordinary dimension of 1.
b. Suppose that you want to measure the area of your living room floor, which is square shaped. Describe a thought process by which you can conclude that $N = R^2$ for the living room and hence that its fractal dimension is the same as its ordinary dimension of 2.
c. Suppose that you want to measure the volume of a cubical swimming pool. Describe a thought process by which you can conclude that $N = R^3$ for the pool and hence that its fractal dimension is the same as its ordinary dimension of 3.

41. **Fractal Dimensions for Fractal Objects.**
a. Suppose that you are measuring the length of the stream frontage along a piece of mountain property. You begin with a 15-meter ruler and find just one element along the length of the stream frontage. When you switch to a 1.5-meter ruler, you are able to trace finer details of the stream edge and you find 20 elements along its length. Switching to a 15-centimeter ruler, you find 400 elements along the stream frontage. Based on these measurements, what is the fractal dimension of the stream frontage?
b. Suppose that you are measuring the area of a very unusual square leaf with many holes, perhaps from hungry insects, in a fractal pattern (e.g., similar to the Sierpinski triangle, Figure 11–61). You begin with a 10-cm ruler, and find that it lies over the entire square, making just one element. When you switch to a 5-cm ruler, you are better able to cover areas of leaf while skipping areas of "holes" and you find 3 area elements. You switch to a 2.5-cm ruler and find 9 area elements. Based on these measurements, what is the fractal dimension of the leaf? Explain *why* the fractal dimension is less than 2.
c. Suppose that you are measuring the volume of a cube cut from a large rock that contains many cavities forming a fractal pattern. Beginning with a 10-m ruler, you find just one volume element. Smaller rulers allow you to ignore cavities, gauging only the volume of rock material. With a 5-m ruler you find 6 volume elements. With a 2.5-m ruler you find 36 volume elements. Based on these measurements, what is the fractal dimension of the rock? Explain why a fractal dimension between 2 and 3 is reasonable. (Ignore the practical difficulties of making this measurement caused by the fact that you cannot see through a rock to find all its holes!)

42. **Fractal Patterns in Nature.** Describe at least five natural objects that exhibit fractal patterns. In each case, explain the structure that makes the pattern a fractal, and estimate its fractal dimension.

Further Topics and Applications

43. **Keep Off the Grass.** An old principle of public landscaping says that "sidewalks come last." In other words, let the people find their chosen paths and then build the sidewalks. The figure shows a campus quadrangle measuring 40 meters by 30 meters. The location of the doors of the

library, the chemistry building, and the humanities building are shown. How long are the new sidewalks planned to connect the library to the other two buildings?

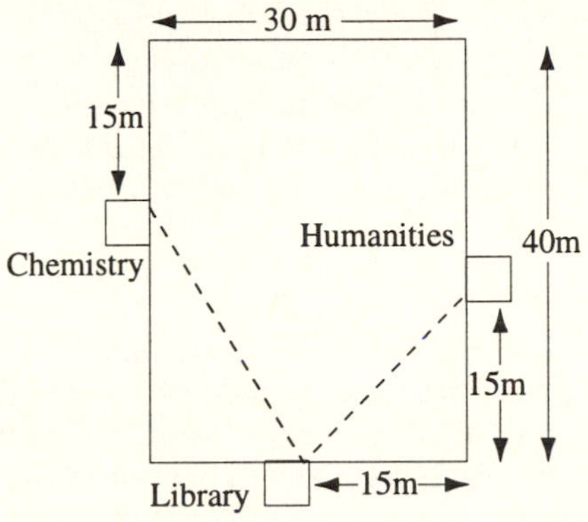

44. Optimization Problems. The following problems all ask for optimal solutions. Draw a picture in each case to help you solve the problem.

a. You have 100 yards of fence and want to enclose a rectangular garden. You decide to make a garden that measures 20 yards by 30 yards, with a total perimeter of 100 yards. What is the area of this garden? Can you find another rectangular design that requires 100 yards of fence but yields a garden with a larger area?

b. A rancher must design a rectangular corral with an area of 400 square meters. She decides to make a corral that measures 10 meters by 40 meters (which has the correct total area). How much fence is needed for this corral? Has the rancher found the most economical solution? Can you find another design for the corral that requires less fence but still provides 400 square meters of corral?

c. You are making boxes and begin with a rectangular piece of cardboard that measures 1.75 meters by 1.25 meters. From each corner of that rectangular piece you cut out square pieces that are 0.25 meter on a side, as shown in the following figure. You fold up the "flaps" to form a box without a lid. What is the surface area of the box? What is its volume? If the corner cuts were squares 0.3 meter on a side, would the resulting box have a larger or smaller volume than the first box? What is your estimate of the size of the square corner cuts that will give a box with the largest volume?

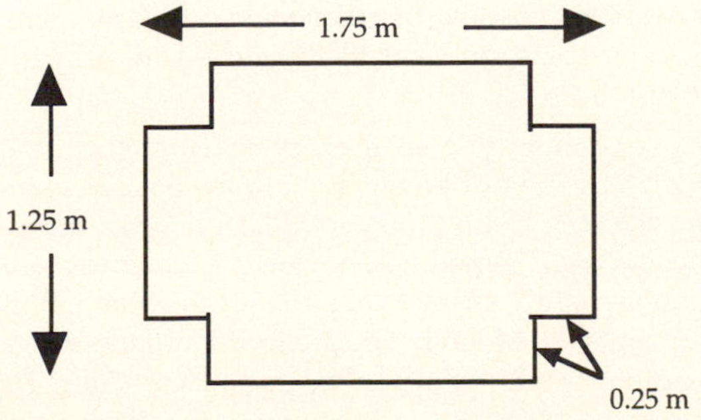

d. You are designing a cylindrical can to hold a volume of 236 cubic centimeters (8 fluid ounces). What should you make the dimensions so that each can requires the minimum possible amount of material? Compare them to the actual dimensions of an 8-ounce soup can.

e. Telephone cable must be laid from a terminal box on the shore of a large lake to an island. The cable costs $500 per mile to lay underground and $1000 per mile to lay underwater. The locations of the terminal box, island, and the shore are shown in the following figure. As an engineer on the project you decide to lay 3 miles of cable along the shore underground and then lay the remainder of the cable along a straight line underwater to the island. How much will this project cost? Your boss examines your proposal and asks whether laying 4 miles of cable underground before starting the underwater cable would be more economical. How much would your boss's proposal cost? Will you still have a job?

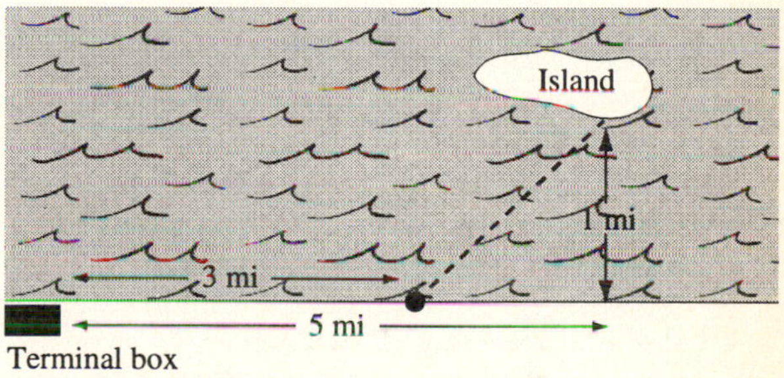

45. Surveying Problems.

a. You are at sea and have a sighting of a lighthouse to the northeast. You radio to a ship directly east of you that has a sighting directly north to the same lighthouse. If you are 4.3 miles from the other ship, how far are you from the lighthouse?

b. In trying to estimate the height of a nearby building you make the following observations. If you stand 15 feet away from a light post, you can line up the top of the building with the top of the light post. Furthermore you can easily determine that the top of the light post is 10 feet above your head and that the building is 50 feet from your sighting position. What is the height of the building?

46. Movie Theater Seat Selection. According to many movie connoisseurs, the optimal way to view a movie is so that the screen fills approximately 35° of your field of view (left to right). Suppose that a movie theater has a screen that is 15 meters wide. Further, suppose that each row of seats is separated from the preceding row by 1 meter and that the front row starts 5 meters from the screen. How many rows back should you sit for an optimal movie experience?

47. Design of the Human Lung. The human lung has approximately 300 million nearly spherical air sacs (alveoli) each with a diameter of about 1/3 millimeter. The key feature of the air sacs is their surface area because on their surfaces gas is exchanged between the bloodstream and the air.

a. What is the total surface area of the air sacs? What is the total volume of the air sacs?
b. Suppose that a single sphere were made that had the same volume as the total volume of the air sacs. What is the radius and surface area of such sphere? How does this surface area compare to that of the air sacs?
c. If a single sphere had the same surface area as the total surface area of the air sacs, what would be its radius?
d. Based on your results, comment on the design of the human lung.

48. **Water Bed Safety.** Consider a king-size water bed that is 7 feet long, 7.5 feet wide, and 9 inches deep when filled with water.
a. What is the volume of water in the bed in units of cubic feet, cubic meters, liters, and gallons?
b. Recall that water has a density of 1 kilogram per liter. How much does the water bed (neglecting the frame) weigh? Give your answer in both kilograms and pounds.
c. If an average person weighs 140 pounds, how many people standing would weigh the same as the water bed? Would you say that a water bed on the second floor of an apartment is safe? Explain.
d. If the water bed burst and drained into the 8 foot by 12 foot room on the floor below, how deep would the water be?

49. **How Much Air?** The Earth's atmosphere gets thinner (less dense) with increasing altitude. However, you can make a rough estimate of the total amount of air in the atmosphere by assuming that its density is about the same as that at sea level and that the height of the atmosphere is about 10 kilometers.
a. Assume that the Earth's atmosphere consists of a layer 10 kilometers thick that surrounds the Earth. What is the total volume of the atmosphere? (Hint: The atmosphere is very thin compared to the radius of the Earth, so it can be thought of as a 10-km-thick layer spread over the surface area of the Earth.)
b. Assume that each 22 liters of air contains 6.02×10^{23} (Avogadro's number) molecules of gas. What is the total number of molecules in the Earth's atmosphere?
c. A typical deep breath draws in about 4 liters of air. How many molecules are in such a breath?
d. Assume that a gaseous pollutant (such as carbon monoxide or ozone) is in the air in a concentration of 400 parts per billion. How many molecules of the pollutant would be inhaled with a deep breath?

50. **Automobile Engine Capacity.** The size of car engines generally is stated as the total volume of their pistons. Pistons have the shape of circular cylinders.
a. American cars often state their engine sizes in units of cubic inches. Suppose that a six-cylinder car (a car with six pistons) has pistons that have a radius of 2.22 inches and a height of 3.25 inches. What is the engine size?
b. Foreign cars generally state engine sizes in liters. Compare the engine size of the car in part (a) to a foreign car with a 2.2-liter engine.
c. Look up the number of pistons and the engine size for your car (if you don't own a car, choose a car that you'd like to own). Estimate the dimensions (radius and height) of the pistons in your car.

51. **Sand Cones.** Think back to your days in the sandbox and imagine pouring sand through a funnel. The stream of sand forms a cone on the ground that becomes higher and wider as sand is added. The proportions of the sand cone remain the same as the cone grows; with sand, the height of the cone is roughly one-third the radius of the circular base (see figure). The formula for the volume of a cone with height h and a base with radius r is $V = (1/3)\pi r^2 h$.
a. If you build a sand cone 2 feet high, how many cubic feet of sand does it hold?
b. Suppose that you excavate a basement and extract 1000 cubic feet of dirt. If the dirt is stored in a conical pile, how high will the cone be? Assume that a dirt cone has the same proportions as a sand cone.
c. Estimate the total number of grains of sand in the cone in part (a). Explain your assumptions and uncertainties.

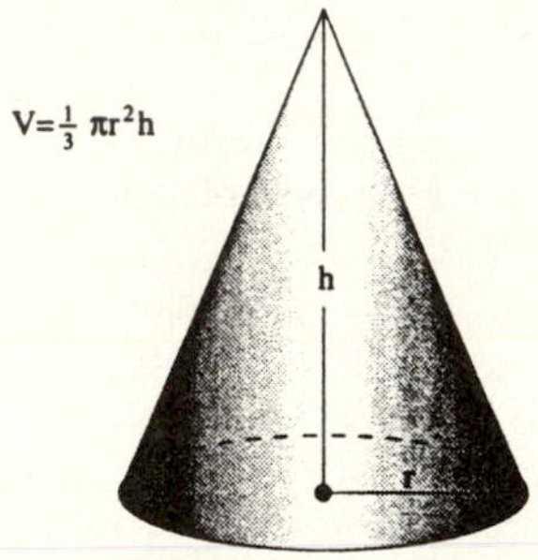

52. **The Chunnel.** The world's longest tunnel is the English Channel Tunnel, or "Chunnel," connecting Dover, England, and Calais, France. The Chunnel consists of three separate, adjacent tunnels. Its length is approximately 50 km. Each of the three tunnels is shaped like a half-cylinder 4 meters high. How much earth (volume) was removed to build the Chunnel?

53. **The Trucker's Dilemma.** It is a dark and stormy night. Through the beat of the wipers on your truck you see a rickety country bridge marked "Load Limit 40 Tons." You know your truck, *White Lightning*, like the back of your hand: it weighs 16.3 tons. Trouble is, you are carrying a cylindrical steel water tank that is full of water. The empty weight of the tank is printed on its side: 1750 pounds. But what about the water? Fortunately, the Massey-Fergusson trucker's almanac in your glove box lists the density of water as 0.03613 pounds per cubic inch. So you dash out into the rain with a tape measure to find the dimensions of the tank: length = 22 ft, diameter = 6 ft, 6 in. Back in the truck, dripping and calculating, do you risk crossing? Explain.

54. **Eclipse Geometry.** A solar eclipse occurs when the Moon lies directly in your line of sight to the Sun, blocking light from the Sun. If the Moon completely blocks the Sun, the eclipse is total. Sometimes, however, the angular size of the Moon is slightly smaller than that of the Sun; in that case, even when the Moon is directly in front of the Sun, a small ring of the Sun is visible, called an *annular eclipse*. The variation in eclipse types occurs because the distances to both

the Sun and Moon vary (the orbits of the Earth around the Sun and of the Moon around the Earth are elliptical). Because the physical sizes of the Moon and Sun do not change, the varying distances lead to varying angular sizes. Refer to the following data in this problem.

	Moon	**Sun**
Physical diameter	3480 km	1,392,000 km
Average distance from Earth	384,400 km	1.496×10^8 km
Minimum distance from Earth	356,400 km	1.471×10^8 km
Maximum distance from Earth	406,700 km	1.521×10^8 km

a. Under what circumstances is the angular size of the Moon the *greatest* in comparison to the angular size of the Sun? In these circumstances, calculate the angular sizes of the Moon and Sun. Explain why solar eclipses that occur under these circumstances have the longest possible duration of totality (the time during which the Moon completely blocks the Sun).

b. Under what circumstances is the angular size of the Moon the *smallest* in comparison to the angular size of the Sun? Calculate the angular sizes of the Moon and Sun under these circumstances. Explain why this circumstance creates an annular, rather than total, eclipse.

c. Suppose that a solar eclipse occurs when both the Moon and Sun are at their average distances. Is the eclipse annular or total? Explain.

d. Think about the geometry of a solar eclipse (draw a picture if necessary) and explain why a solar eclipse can be seen only from a small area of the Earth at one time. (As the Earth rotates and the Moon moves in its orbit, the shadow from the eclipse moves across the surface of the Earth forming a narrow eclipse track; if you want to view an eclipse, you must be somewhere along the eclipse track during the time of the eclipse.)

e. A lunar eclipse occurs when the Earth comes between the Moon and the Sun. In that case, the Earth's shadow can be seen on the full Moon. Explain why a lunar eclipse can be seen from anywhere on the night side of the Earth. Further, explain how observation of lunar eclipses provides evidence that the Earth is round.

55. Reading with the Hubble Space Telescope. Imagine that you had access to the Hubble Space Telescope (angular resolution of about 0.05") for the purpose of reading this page. Approximately how far away could you place this page and still read it clearly? Explain any assumptions. (Hint: To read the page you need to be able to distinguish between different letters, not just to see that letters are present.)

56. The Angular Size of Stars. Assume that a star the same size as the Sun is 10 light-years away. Calculate its angular diameter as seen from the Earth. Would the Hubble Space Telescope be able to see any features on the surface of this star, or would it appear only as a geometric *point* of light? Explain.

57. Stellar Parallax. The nearest star, *Proxima Centauri,* is approximately 4.2 light-years away (it is the nearest star of the three-star system called Alpha Centauri).

a. Calculate the parallax angle of *Proxima Centauri.* (Hint: refer to Figure 11–44; you will need to use an inverse sine). Use your result to explain why stellar parallax is undetectable to the naked eye and hence why it wasn't observed until the advent of powerful telescopes in the nineteenth century.

b. Explain why stellar parallax constitutes *direct* proof that the Earth orbits the Sun. Why did their inability to detect stellar parallax lead most ancient Greeks to conclude that the Earth does *not* go around the Sun?

58. The Parsec. A unit of distance commonly used by astronomers, as well as by the *Star Trek* crew of the starship Enterprise, is the parsec. A parsec is defined as the distance to an object that has a parallax angle of 1". Refer to Figure 11–44 and calculate how far a parsec is in kilometers and in light-years. Given that the nearest star is 4.2 light-years distant, how many stars are located within one parsec of our solar system?

59. Sphere Geometry. Explain why the ordinary Euclidean formula for volume can be used to measure the volume of any chunk of material cut from a sphere, but the Euclidean formulas for area do not work for shapes drawn on the *surface* of the sphere. Then explain why a complete sphere is a Euclidean object in three dimensions, but its surface represents a *non*-Euclidean, two-dimensional object.

60. The Quadric Koch Curve and Quadric Koch Island. The *quadric Koch curve,* many variations of the snowflake (or Koch) curve, one of begins with a horizontal line segment to which the following rule is applied: Divide each line segment into four equal pieces; replace the second piece with three line segments of equal length that make the shape of a square above the original piece; and replace the third piece with three line segments making a square below the original piece. The quadric Koch curve would result from infinite applications of the rule; the first three stages of the construction are shown in the following figure.

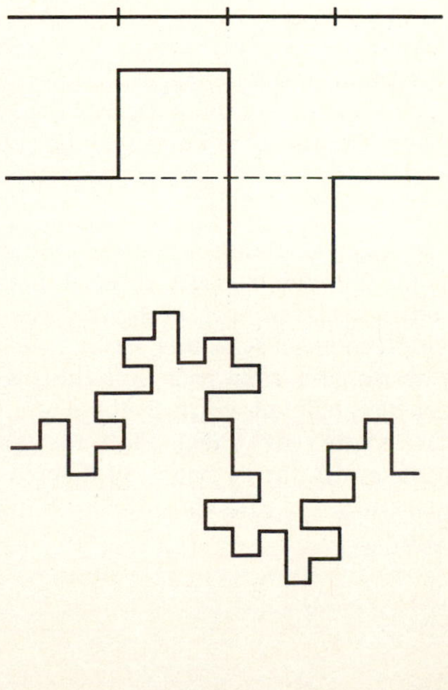

a. Determine the relation between N and R for the quadric Koch curve.
b. What is the fractal dimension of the quadric Koch curve? Can you draw any conclusions about the length of the quadric Koch curve? Explain.
c. The *quadric Koch island* is constructed by beginning with a square. Then each of the four sides of the square is replaced with a quadric Koch curve. Explain why the total area of the quadric Koch island is the same as the area of the original square. How long is the coastline of the quadric Koch island?

61. **The Cantor Set.** Recall that the Cantor set is formed by starting with a line segment and then successively removing the middle one-third of each segment in the current figure (Figure 11–60). If a ruler the length of the original line segment is used, it detects one element in the Cantor set because it can't "see" details smaller than itself. If the ruler is reduced in size by a factor of $R = 3$, it will find two elements (only solid pieces of line, not holes, are measured). If the ruler is reduced in size by a factor of $R = 9$, how many elements does it find? Based on these results, what is the fractal dimension of the Cantor set? Explain why this number is less than 1.

62. **Fractal Dimension from Measurements.** An ambitious and patient crew of surveyors has used various rulers to measure the length of the coastline of Dragon Island. The following table gives the measured length of the island, L, and the length of the ruler used, r. Graph these data on the same axes used in Figure 11–59. Use the slope of your graph to estimate the fractal dimension of the coastline. (Hint: You will need to make a logarithmic graph by plotting pairs ($\log_{10} r$, $\log_{10} L$)).

The Coastline of Dragon Island

r meters	100	10	1	0.1	0.01	0.001
L meters	315	1256	5000	19905	79244	315,479

63. **Natural Fractals Through Branching.** One way that natural objects reveal fractal patterns is by branching. For example, the intricate structure of the human lung, the web of capillaries in a muscle, the branches or roots of a tree, or the successive division of streams in a river delta all involve branching at different spatial scales. Explain why structures formed by branching resemble self-similar fractals. Further, explain why fractal geometry rather than ordinary geometry leads to greater understanding of such structures.

64. **Sine, Cosine, and Tangent for Special Angles.** This problem leads you through the calculation of sine, cosine, and tangent ratios for a few special angles. Refer to the ratio definitions given in subsection 11.5.2.
a. Draw an isosceles right triangle (with two 45° angles and one 90° angle) in which the base and height are 1 unit in length (any units). Find the length of the hypotenuse with the Pythagorean theorem. Using this triangle, find sin 45°, cos 45°, and tan 45° (without your calculator!).
b. Draw an equilateral triangle with sides 2 units in length. Then draw a line representing the height of this triangle, and find its length with the Pythagorean theorem. Use this triangle to find all of the following without your calculator: sin 60°, sin 30°, cos 60°, cos 30°, tan 60°, tan 30°.
c. Draw a right triangle in which one of the angles is very small, say, 1°. Note that the side adjacent to the 1° angle is now almost identical in length to the hypotenuse and that the side opposite the 1° angle is very short. Imagine shrinking the 1° angle all the way to 0°; in that case, the adjacent side would be equal in length to the hypotenuse and the opposite side would have 0 length. Use these facts to find sin 0°, cos 0°, and tan 0°. Explain.
d. In the triangle you drew for part (c), note that the third angle is about 89°. By imagining what would happen if you "opened up" this angle until it reached 90°, find sin 90° and cos 90°; be sure to explain your reasoning. Based on this reasoning, what can you say about tan 90°?

65. **Sum of the Angles in a Triangle.** In this problem you prove that the total measure of the angles in a triangle is 180°. Consider triangle ABC in the following figure. A straight line L is drawn parallel to side AC.
a. Note that the two dashed lines form right angles with both line L and side AC. The dashed lines form a rectangle. Side AB becomes a diagonal in this rectangle. Explain why the two triangles in this rectangle must be similar triangles. Use this fact to prove that the angle A′ must be the same as angle A.

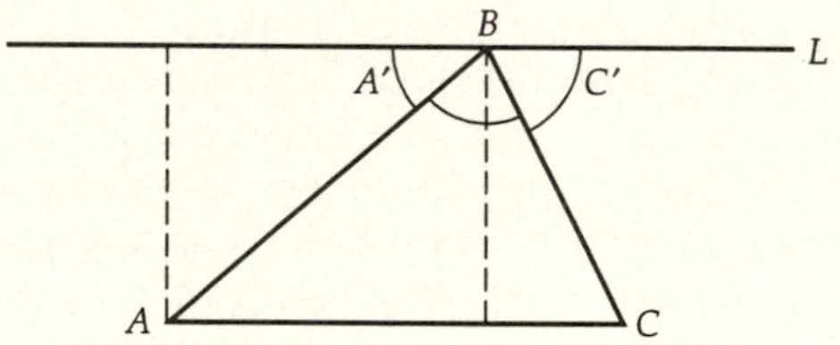

b. Use a similar argument to prove that angle C' must be the same as angle C.
c. Explain why the sum of the angles A', B, and C' must be 180°. Complete the proof that the sum of the angles in the triangle is 180°.

Projects

66. **Impact of Global Warming.** Study a globe or map that gives topographic (elevation) information.
a. Discuss the impact of a 200-foot rise in sea level. Estimate the fraction of the world's population now living in regions that would be underwater.
b. Even with extreme global warming, the Antarctic ice cap would melt slowly. Without calculating, explain how you could estimate the time required for it to do so.
c. Of more immediate concern than the melting of all polar ice is the possibility that the Ross Ice Shelf of Antarctica will break off and spill into the ocean. What is the Ross Ice Shelf? How big is it? Through research and estimation, as needed, discuss what would happen if the Ross Ice Shelf were to break off. Discuss any evidence regarding whether that might happen any time soon.

67. Road Grade.
  a. Estimate the grade of various roads in your state. Explain your estimates.
  b. How steep do you think a road can be before cars would be unable to climb it? Why?
  c. What is a "runaway truck ramp"? What types of roads need them? Suppose that you are driving a truck that becomes a runaway two miles from the nearest runaway truck ramp. Describe the rest of your trip.
  d. Research how climbs are categorized in the *Tour de France* bicycle race. How steep are these climbs?

68. The Largest Living Creature. The largest individual living creature on Earth is the General Sherman Tree in Sequoia National Park, California. It is approximately 83 meters tall, 11 meters in diameter (at its base), and estimated to be 2000 to 3000 years old. It is still growing. According to a sign at the tree, it adds as much wood each year as a normal 30-foot-tall tree. Is this claim reasonable? Explain the assumptions and calculations you use in evaluating the claim.

69. The Great Pyramids of Egypt. Egypt's Old Kingdom began in about 2700 B.C. and lasted for 550 years. During that time at least six pyramids were built as monuments to both the life and afterlife of the pharaohs. These pyramids remain among the largest and most impressive structures constructed by any civilization. The building of the pyramids required a mastery of art, architecture, engineering, and social organization at a level unknown before that time. The collective effort required to complete the pyramids transformed Egypt into the first nation-state in the world. Of the six pyramids, the best known are those on the Giza plateau outside Cairo, and the largest of those is the Great Pyramid built by Pharaoh Khufu (or Cheops to the Greeks) in about 2550 B.C. With a square base 756 feet on a side and a height of 481 feet, the pyramid is laced with tunnels, shafts, corridors, and chambers, all leading to and from the deeply concealed king's burial chamber. The stones used to build the pyramids were transported, often hundreds of miles, with sand sledges and river barges, by a labor force of 100,000, as estimated by the Greek historian Herodotus. Historical records suggest that the Great Pyramid was completed in approximately 25 years.
  a. To appreciate the size of the Great Pyramid of Khufu, compare its height to the length of a football field (which is 100 yards long).
  b. The volume of a pyramid is given by the formula $V = (1/3) \times (\text{area of base}) \times (\text{height})$. Use this formula to estimate the volume of the Great Pyramid. State your answer in both cubic feet and cubic yards.
  c. The average size of a limestone block in the Great Pyramid is 1.5 cubic yards. How many blocks were used to construct this pyramid?
  d. A modern research team, led by Mark Lehner of the University of Chicago, estimated that the use of winding ramps to lift the stones and desert clay and water for lubrication would allow placing one stone every 2.5 minutes. If the pyramid workers labored 12 hours a day, 365 days a year, how long would building the Great Pyramid have taken? How does this estimate compare with historical records? Why did Lehner's research team conclude that the Great Pyramid could have been completed with only 10,000 laborers, rather than the 100,000 laborers estimated by Herodotus?
  e. Constructed in 1889 for the Paris Exposition, the Eiffel Tower is a 980-foot-high iron lattice structure supported on four arching legs. The legs of the tower are at the corners of a square with sides of length 120 feet. If the Eiffel Tower were a solid pyramid, how would its volume compare to the volume of the Great Pyramid?

70. More on the Pyramids of Egypt. The pyramids of ancient Egypt are a marvel of ancient technology. Do some detailed research and report on one or more of the following topics.
  a. Investigate further geometric aspects of the pyramids. For example, investigate why the pyramid shape was chosen for these tombs and how the dimensions of the pyramids were chosen. How are the pyramids oriented (e.g., east–west or other directions), and why?
  b. Investigate the construction of the pyramids. For example, research evidence that suggests precise quarry locations for the stones used in the pyramids and methods of transporting the stones to the construction sites and of raising them onto the pyramids.
  c. Pyramids are found in many other places. For example, the pyramids of central America bear a striking resemblance to the Egyptian pyramids. Research pyramids at other sites and compare them to the pyramids constructed in ancient Egypt.
  d. Many pseudoscientists claim that the construction of the Pyramids by the ancient Egyptians was "impossible." Some have gone so far as to claim that aliens aided in the construction! Investigate the technology and know-how of the ancient Egyptians. Show that, although the pyramids were certainly a technological marvel of their time, their construction was by no means beyond the abilities of the Egyptians. Discuss why so many people hold the erroneous belief that the Egyptians could not have built the pyramids.

71. Moon Illusions. The full Moon appears larger when seen near the horizon than when it is higher overhead, but this appearance is an illusion. Find a way to measure the angular diameter of the Moon with reasonable accuracy (say, within about 10 to 20%). Describe your measurement technique. Then, on a night of a full Moon, measure the angular diameter as the Moon rises. You should find that the angular diameter remains constant. Given this constancy, hypothesize about why the Moon seems to look larger near the horizon.

72. Greek Mathematics and Science. The mathematics and science of the ancient Greeks had a profound and direct influence on the development of modern mathematics and science. Do a research project, culminating in a term paper or class presentation, on ancient Greek mathematics and science. Some possible areas of focus: (1) describe the Greeks, where they lived, and how their civilization rose and fell; (2) make a time-line of important Greek philosophers whose ideas have been transmitted through the ages to the present day and explain how each built upon work of previous philosophers; (3) research the history of Plato's academy, and discuss its relationship to modern universities; or (4) research the publication history of Euclid's *Elements*.

73. **The Geometry of Ancient Cultures.** Although the work of the ancient Greeks became the basis of modern geometry, many ancient cultures made extensive use of geometry. Research the use of geometry in an ancient culture of your choice. Some possible areas of focus: (1) study the use of geometry in ancient Chinese art and architecture; (2) investigate the geometry and purpose of Stonehenge; (3) compare and contrast the geometry of the Egyptian pyramids to those of Central America; (4) study the geometry and possible astronomical orientations of Anasazi buildings and communities; or (5) research the use of geometry in the ancient African empire of Aksum (in modern-day Ethiopia).

74. **Why 360?** Do research to determine why the ancient Babylonians chose to divide the circle into 360°. Be sure to include some discussion of Babylon itself, such as: Where was it? when did its civilization flourish? what happened to it? Comment on the possibility that 360 represented either a mismeasurement or an approximation to the number of days in a year. What other possible explanations are there for the choice of 360?

75. **Fractal Nature.** The applicability of fractal geometry to nature is a topic of current debate. Consider the hypothesis that a few simple processes (rules) can lead to the vast variety of forms found in living plants (or animals). This hypothesis is based on the assumption that natural forms are created by processes similar to those that generate fractals by iteration. Working alone or in groups, compile evidence for and against this hypothesis. Report your conclusions.

12 THE DISCRETE SIDE OF MATHEMATICS

In Chapters 10 and 11 we explored two major areas of mathematics: change, which might be called the road to calculus, and geometry, the study of size and shape. In this chapter, we survey a much different part of the mathematical landscape called discrete mathematics. Taken altogether, these three chapters comprise three of the great pillars of mathematics. Discrete mathematics is a highly visual subject with applications ranging from simple probability to the design of computer networks and the analysis of business decisions.

CHAPTER 12 PREVIEW:

12.1 ***DISCRETE THINKING.*** Broadly speaking, the real world may be modeled in two ways: with continuous models, in which every point is considered, or with discrete models, which consider only distinct points. We begin the chapter by contrasting these two ways of thinking.

12.2 ***PRINCIPLES OF COUNTING.*** An important part of discrete mathematics involves learning to count numbers of possibilities — such as the number of possible telephone numbers or lottery ticket numbers. In this section we examine these principles, which we apply later in probability and statistics.

12.3 ***NETWORK ANALYSIS.*** In this section we survey network analysis (or graph theory), a subject that allows us to describe how people or things are connected.

12.4 ***PROJECT DESIGN.*** How can large projects with many parts be organized? In this section we describe the critical path method, a technique used for analyzing schedules that arise in business applications and your own life.

*12.5** ***INSIDE NETWORK ANALYSIS.*** In this section we reprise Section 12.3, taking a deeper look into two mathematical methods used in network analysis: Euler circuits and minimum spanning trees.

The wrinkles progress among themselves in a phalanx — beautiful under networks of foam, and fade breathlessly while the sea rustles in and out of the seaweed.

— Marianne Moore (1887–1972), "A Grave"

The union of the mathematician with the poet, fervor with measure, passion with correctness, this surely is the ideal.

— William James (1842–1910)

You could not step twice into the same river; for other waters are ever flowing on to you.

— Heraclitus (c. 540–480 B.C.)

12.1 DISCRETE THINKING

By convention there is color, by convention sweetness, by convention bitterness, but in reality there are atoms and space.
— Democritus, c. 460–400 B.C.

The surface of a table feels smooth and solid. But, in reality, the surface consists of distinct atoms, which in turn consist of protons, neutrons, and electrons. The table surface illustrates a fundamental distinction between two types of thinking. Most of the time, we think of the table surface as a smooth, or **continuous**, surface. However, when we consider its atoms, we are thinking of the table as a **discrete** collection of distinct particles.

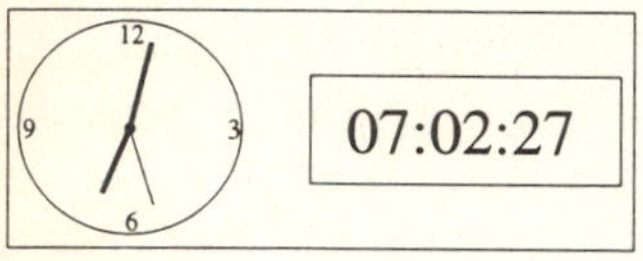

Figure 12-1. The clock with hands seems to sweep continuously through all instants of time, while the digital clock steps ahead in (discrete) one-second intervals.

Clocks provide another illustration of the distinction between the discrete and the continuous (Figure 12–1). On continuous (or analog) clocks, the hands (hours, minutes, and seconds) sweep smooth circles as they move relentlessly around the clock face. This continuous motion gives the impression that the clock hands pass through *every instant* of time. In contrast, a discrete (or digital) clock shows the time with digits that click through the day one second at a time, giving the impression that time moves ahead in short steps.

The fundamental distinction between continuous and discrete thinking also is reflected in mathematics. Up to this point in the book, we primarily have covered continuous mathematics, in which we model quantities that undergo continuous or uninterrupted change. For example, the flight of a baseball is continuous because the ball passes through all points along its flight, as shown in Figure 12–2(a). At any particular moment, the height of the baseball might have *any* (positive) numerical value. That is, quantities in continuous mathematics can be any of the *real* numbers.

In contrast, discrete mathematical problems usually can be modeled only with the *integers* because they focus only on distinct objects or points as shown in Figure 12–2(b). In this chapter, we discuss two important types of discrete modeling. In section 12.2 we study principles of counting, or **combinatorics**. For example, on a team of 10 runners, how many different 4×100-meter relays can be put together? Such counting principles have many important applications and are indispensable in probability and statistics. The remaining sections focus on an area of mathematics called **network analysis** (or **graph theory**). This rapidly growing field is of particular importance in the analysis of computer and telecommunication networks and in the operation of large businesses.

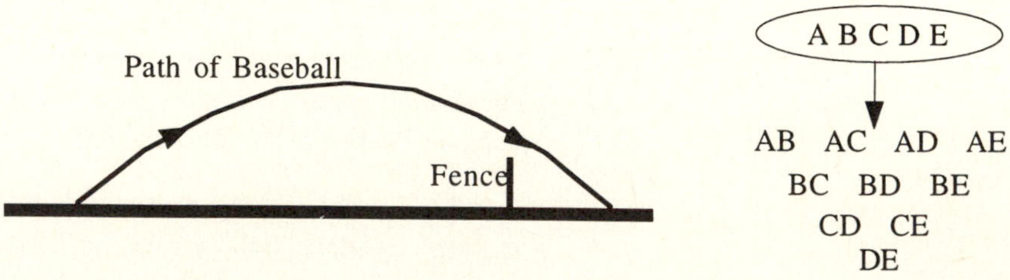

(a) (b)

Figure 12-2. (a) Continuous mathematics describes quantities that undergo continuous or uninterrupted change, as illustrated by the flight of a baseball. (b) The problem of finding all the two-letter combinations that can be made from five letters is a simple illustration of a problem in discrete mathematics.

12.2 PRINCIPLES OF COUNTING

Counting is one of the most fundamental and primitive quantitative skills of human beings. Indeed, one of the first mathematical accomplishments of young children is learning to count to a new and higher number. Early exercises in counting usually involve enumerating the objects in a collection; for example, finding the number of grapefruits in a basket or tallying a classroom vote. In this section, we build on that fundamental counting skill, exploring principles that help enumerate and arrange sets of objects.

12.2.1 The Multiplication Principle

How many outfits can be formed from three pairs of shorts and four shirts? One way to approach the problem is to enumerate the possibilities in an **array** as shown in the following table. The four rows correspond to the four choices of shirts, and the three columns correspond to the three choices of shorts. Each of the 3 × 4 = 12 squares in the table represents a different outfit.

	Shorts #1	Shorts #2	Shorts #3
Shirt #1	Outfit #1	Outfit #2	Outfit #3
Shirt #2	Outfit #4	Outfit #5	Outfit #6
Shirt #3	Outfit #7	Outfit #8	Outfit #9
Shirt #4	Outfit #10	Outfit #11	Outfit #12

To illustrate an alternative way of listing arrangements, imagine that a restaurant menu offers a choice of soup or salad for the first course and a choice of a vegetarian, fish, or meat dish for the main course. How many different meals can you create? The possibilities can be enumerated with a **tree** (Figure 12–3). Note that that each of the three second-course choices is connected with two first-course choices. You can follow the tree branches to see the six possible meals along the bottom row.

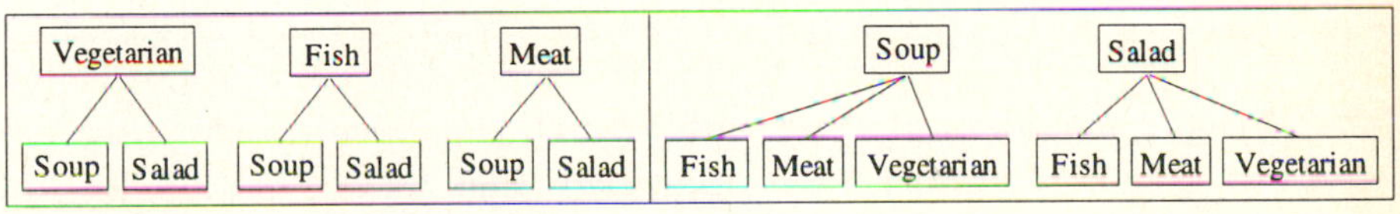

Figure 12-3. A tree structure identifies six possible meals created from two first courses and three main courses. Regardless of which course is in the first row the six choices are the same: vegetarian + soup; vegetarian + salad; fish + soup; fish + salad; meat + soup; and meat + salad.

> **Time-Out to Think:** Explain how we could have used an array, instead of a tree, to find the number of possible meals.

The multiplication principle: If you choose one item from a group of *M* items, and another from a group of *N* items, the total number of two-item choices is $M \times N$.

The **multiplication principle** captures the thinking behind these examples. In its simplest form it may be stated as:

> If you choose one item from a group of *M* items, and another item from a group of *N* items, the total number of two-item choices is $M \times N$.

Applying the multiplication principle to the menu, $M = 2$ is the number of first courses and $N = 3$ is the number of main courses, making $M \times N = 2 \times 3 = 6$ possible meals. Now, suppose that you manage an apartment complex and label the units with a letter (one of the 26 letters of the alphabet A–Z) followed by a number (0–9): The number of possible apartment labels is 26 × 10 = 260. And, with eight boys and nine girls in a dancing class, 8 × 9 = 72 different boy–girl combinations are possible.

What happens when we have combinations of more than two items? For example, how many meals would be possible if a restaurant menu offered four desserts in addition to two first courses and three main courses? The method of making a table becomes awkward when there are more than two groups — we need a three-dimensional table to list all of the choices from three groups. The tree method can be extended if we are willing to draw a lot of lines and boxes (Figure.12–4 on the next page). Perhaps the easiest way to solve the problem is by extending the multiplication principle. With three choices for the first course, two for the main course, and four for dessert, the total number of possible meals is $3 \times 2 \times 4 = 24$.

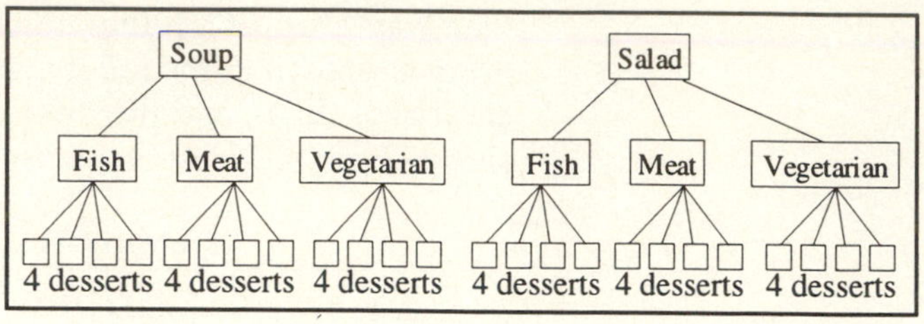

Figure 12-4. Each of the two first courses (top row) is connected to each of the three main courses (middle row), which in turn are connected to each of the four desserts (bottom row). A total of 24 meals is possible, found by tracing the branches from each of the 24 items in the bottom row.

The multiplication principle may be extended to any number of groups of items. For example, suppose that, to meet your core requirements you have a choice of 12 classes in natural science, 15 classes in social science, 10 classes in English, and 8 classes in the fine arts. The total number of possible ways of meeting the requirements is $12 \times 15 \times 10 \times 8 = 14{,}400$. Note that the multiplication principle makes huge numbers of arrangements for relatively few choices. This incredible growth in the number of arrangements that can be formed from a few choices sometimes is called a "combinatorial explosion."

Example 12–1 *Choosing Computer Systems.* A computer catalog offers the following choices. Four different companies are listed that make both color and monochrome (black-and-white) monitors. Five different companies offer the basic processor, each offering a choice of three different speeds. With any of the computer disk drives six different capacities are available, and a choice of seven different printers is offered. How many different systems can you create?

Solution: Begin by splitting this problem into smaller pieces. When you apply the multiplication principle, it gives $4 \times 2 = 8$ different monitors (four companies each with a choice of either color or monochrome). Similarly, there are $5 \times 3 = 15$ choices for the basic processor (five companies, three speeds). Combining the 8 monitors, the 15 processors, the 6 disk drives, and the 7 printers, you find that a total of $8 \times 15 \times 6 \times 7 = 5040$ different computer systems may be created.

❖

12.2.2 Arrangements with Repetition

Imagine that license plates in your county display seven numerals. How many different license plates are possible? On a license plate, numerals may be used more than once. For example, both 3333333 and 1234567 are acceptable plates. Thus, there are 10 choices (the numerals 0 through 9) for each of the seven numerals on the plate. Following the multiplication principle, we can make $10 \times 10 = 100$ different arrangements for the first two numerals. For the first three numerals, the total is $10 \times 10 \times 10 = 1000$ arrangements. Continuing in this manner gives $10 \times 10 \times 10 \times 10 \times 10 \times 10 \times 10 = 10^7$, or 10 million, different license plates.

The license plate problem involves selecting items from only a *single* group — the numerals 0–9. However, the items in the group may be chosen over and over again. We can generalize the license plate problem to state a rule for counting **arrangements with repetition** of items. If we make r selections (the seven digits of the license plates) from a group of n different items (the numerals 0–9 for the license plates), a total of n^r different arrangements is possible.

Example 12–2 *License Plates.* Assume that your state uses letters of the alphabet and numerals in any order to make seven-symbol plates. How many plates are possible?

Solution: With 26 letters in the alphabet and 10 numerals, 36 choices are possible for each of the seven symbols on the plate. Thus $36^7 \approx 7.84 \times 10^{10}$, or about 78.4 billion, different license plates may be formed.

❖

Example 12–3 *Passwords.* How many six-letter passwords can be made from consonants of the alphabet and the characters @, $, and &?

Solution: If you count "y" as a consonant, the alphabet contains 21 consonants (and 5 vowels for a total of 26 letters). Combined with the characters @, $, and &, we have 24 characters to use in the six-letter passwords. Repetition is allowed in passwords, so $24^6 = 191{,}102{,}976$ different passwords are possible.

❖

12.2.3 Permutations

Imagine that you coach a team of four swimmers. How many different ways can you put together a four-person relay team? In this example, repetition is *not* allowed because a swimmer can swim only once in the relay; that is, you aren't allowed to have the same person swim all four relay legs. You may put any of the *four* swimmers in the first position. Once the first swimmer has been chosen any of *three* swimmers can fill the second position. Once the first and second swimmers have been chosen, either of *two* swimmers can fill the third position. That leaves just *one* swimmer for the last (anchor) position. The total number of possible arrangements is $4 \times 3 \times 2 \times 1 = 24$. If we call the four swimmers A, B, C, and D, the 24 different relay orders are

ABCD	ABDC	ACBD	ACDB	ADBC	ADCB
BACD	BADC	BCAD	BCDA	BDAC	BDCA
CABD	CADB	CBAD	CBDA	CDAB	CDBA
DABC	DACB	DBAC	DBCA	DCAB	DCBA

The relay example illustrates arrangements of items known as **permutations**. The crucial properties of permutations are that

(a) all items are selected from the same collection,

(b) no item may be used more than once, and

(c) *order matters* (ABCD is a different relay team than DCBA).

Example 12–4 *Class Schedules.* A middle school principal needs to schedule the eighth grade algebra, English, history, Spanish, science, and gym classes. Excluding the lunch hour, the school day has six periods. How many different arrangements of classes are possible?

Solution: In this example of permutations, the "items" are the classes selected from the collection of six classes. No class may be scheduled more than once during the day. And the order of the classes matters because a schedule that begins with Spanish is different from one that begins with gym. The principal may choose any of the six classes for the first period. After that class has been chosen five choices are left for the second period. Continuing in this manner, four choices will remain for the third period, three choices for the fourth period, two choices for the fifth period, and only one choice will be left for the sixth period. Altogether, there are $6 \times 5 \times 4 \times 3 \times 2 \times 1 = 720$ different arrangements for the classes.

❖

Factorial Notation

Products of the form $6 \times 5 \times 4 \times 3 \times 2 \times 1$ come up so frequently in counting problems that they have a special name. Whenever a positive integer n is multiplied by all the preceding positive integers, the result is called ***n* factorial** and is denoted $n!$ (the exclamation mark is read as "factorial"). Thus

$$n! = n \times (n-1) \times (n-2) \times (n-3) \times \cdots \times 3 \times 2 \times 1.$$

By convention, 0! is defined to be 1.

The following examples illustrate calculations with factorials.

$$1! = 1$$

$$2! = 2 \times 1 = 2$$

$$3! = 3 \times 2 \times 1 = 6$$

$$4! = 4 \times 3 \times 2 \times 1 = 24$$

$$5! = 5 \times 4 \times 3 \times 2 \times 1 = 120$$

$$9! = 9 \times 8 \times 7 \times 6 \times 5 \times 4 \times 3 \times 2 \times 1 = 362{,}880$$

$$\frac{6!}{4!} = \frac{6 \times 5 \times 4 \times 3 \times 2 \times 1}{4 \times 3 \times 2 \times 1} = \frac{6 \times 5 \times 4!}{4!} = 6 \times 5 = 30$$

$$\frac{25!}{22!} = \frac{25 \times 24 \times 23 \times 22!}{22!} = 25 \times 24 \times 23 = 13{,}800$$

Table 12-1. Values of *n*!

n	*n*!
0	1
1	1
2	2
3	6
4	24
5	120
6	720
7	5040
8	40,320
9	362,880
10	3,628,800
11	39,916,800
12	479,001,600

Table 12–1 shows the values of $n!$ for the first 13 integers. Note that the value of $n!$ grows very rapidly as n increases.

Time-Out to Think: Use the factorial key on your calculator to verify a few of the values in Table 12–1. Then compute $n!$ for $n = 13, 14, \ldots$ For what value of n does n! exceed the capacity of your calculator?

The Permutations Formula

Suppose that the swimming team coach needs 4 women for a relay event and has 10 women from which to choose. How many different ways can the relay team can be selected?

The coach has 10 choices for the first position, 9 choices for the second position, 8 choices for the third position, and 7 choices for the fourth (anchor) position. Thus $10 \times 9 \times 8 \times 7 = 5040$ different arrangements are possible for the four-woman relay team. We say that there are 5040 *permutations of 10 swimmers selected four at a time.*

We could write the result ($10 \times 9 \times 8 \times 7 = 5040$) in factorial notation as

$$10 \times 9 \times 8 \times 7 = \frac{10 \times 9 \times 8 \times 7 \times 6 \times 5 \times 4 \times 3 \times 2 \times 1}{6 \times 5 \times 4 \times 3 \times 2 \times 1} = \frac{10!}{6!}.$$

What do the numbers 10 and 6 have to do with the relay problem? Ten is the number of swimmers from which the coach may choose the relay participants. Four swimmers are chosen for the relay team, so 6 is the number of swimmers who do *not* swim. Thus, the number of permutations of 10 swimmers selected 4 at a time is

$${}_{10}P_4 = \frac{10!}{(10-4)!} = 5040.$$

We can now generalize the formula for permutations. If we make r selections from a group of n items, the number of permutations (orders of arrangement) is

$$_nP_r = \frac{n!}{(n-r)!},$$

where $_nP_r$ is read as "the number of permutations of n items taken r at a time."

Example 12–5 *Computer Passwords.* If no letter may be used more than once, how many six-letter passwords can be made from the letters of the alphabet?

Solution: This is a permutation problem because it involves selecting from one group (letters of the alphabet), selecting each letter only once (according to the rules stated in the problem), and recognizing that order matters (the password *donkey* is different from the password *knodey*). In this case, you must select 6 letters from the 26 letters of the alphabet. Using the permutations formula, the number of permutations of 26 letters taken 6 at a time is:

$$_{26}P_6 = \frac{26!}{(26-6)!} = \frac{26!}{20!} = 26 \times 25 \times 24 \times 23 \times 22 \times 21 = 165,765,600.$$

Even with the rule that each letter may be used only once, more than 165 million passwords can be formed from the letters of the alphabet.

❖

Example 12–6 *Batting Order.* How many ways can the manager of a baseball team form a (9-player) batting order from a roster of 15 players? Assume that any player may bat in any position.

Solution: Again, the order matters and each player may be used only once in the batting order. The number of different batting orders is the number of permutations of 15 items (players) taken 9 at a time:

$$_{15}P_9 = \frac{15!}{(15-9)!} = \frac{15!}{6!} = 1,816,214,400.$$

Nearly 2 *billion* batting orders are possible for a 15-player baseball team!

❖

12.2.4 Combinations

Suppose that the five members of a city council — let's call them Zeke, Yolanda, Wendy, Vern and Ursula — decide that a three-person subcommittee is needed to study the impact of a new shopping center. How many subcommittees could be formed from the five members of the council?

Unlike permutation problems, this problem isn't concerned with order. For example, only *one* committee with members Zeke, Wendy, and Ursula may be formed, regardless of the order in which we name the members. Thus we are interested only in the number of **combinations** of members on committees, rather than permutations. Note that a combinations problem is just like a permutations problem except that order does *not* matter.

One way of solving this problem is simply to list all the possible committees. Using the letters Z, Y, W, V, and U for Zeke, Yolanda, Wendy, Vern, and Ursula, respectively, we reason as follows. Six subcommittees can be formed in which Zeke is a member: ZYW, ZYV, ZYU, ZWV, ZWU, and ZVU. Next, consider subcommittees without Zeke, consisting of the four other members of the council. Three such subcommittees can be formed in which Yolanda is a member: YWV, YWU, and YVU. Finally, only one subcommittee with neither Zeke nor Yolanda, WVU, is possible. The total number of possible subcommittees is 6 + 3 + 1 = 10.

There is an alternative approach to this problem that can be easily extended to larger groups. The number of possible ways of writing three names of city council members *in order* is the number of permutations of five names ($n = 5$) selected three at a time ($r = 3$):

$$_nP_r = {_5P_3} = \frac{5!}{(5-3)!} = \frac{5!}{2!} = \frac{120}{2} = 60 \text{ arrangements.}$$

However, because permutations include order, they include many duplicate committees. For example, the six arrangements, ZYW, ZWY, YZW, YWZ, WZY, and WYZ, all represent the same subcommittee with members Z, Y, and W. Because *each* three-member subcommittee has 3! = 6 permutations, the permutations formula *overcounts* the actual number of subcommittees by a factor of 3! = 6. Therefore the actual number of subcommittees is $_5P_3 \div 6 = 60 \div 6 = 10$ — the same result we obtained by listing all the three-person subcommittees.

We can generalize this result by recognizing that, for this example, $3! = r!$ Thus we can always find the number of combinations by dividing the number of permutations, $_nP_r$, by $r!$ That is,

$$_nC_r = \frac{_nP_r}{r!} = \frac{n!}{(n-r)!r!}$$

The notation $_nC_r$ is read "the number of combinations of n items taken r at a time." We can confirm that this formula yields the result of 10 subcommittees when we select $r = 3$ members from the $n = 5$ city council members:

$$_5C_3 = \frac{5!}{(5-3)!\times 3!} = \frac{5!}{2!\times 3!} = \frac{120}{2\times 6} = \frac{120}{12} = 10.$$

Time-Out to Think: Write down all the two-letter permutations of the letters a, b, c, d, and e. Write down all the two-letter combinations of the letters a, b, c, d, and e. Verify that the permutation and combination formulas work in this case.

Example 12–7 ***Relay Combinations.*** A coach has 10 swimmers from which to choose a 4-woman relay team. Suppose that the coach wants to notify the 4 relay participants the night before the race but won't decide the actual relay order until the next day. How many combinations of 4 relay swimmers are possible from 10 swimmers?

Solution: This is a combination problem, not a permutation problem, because the order doesn't matter the night before the race. The number of permutations, or ordered relays, for this team, found earlier, is

$$_{10}P_4 = \frac{10!}{(10-4)!} = 5040.$$

However, there are 4!, or 24, different ways to arrange any group of 4 women in a relay. That is, the permutations formula counts each distinct group of 4 women as 24 different ordered relays, so the number of possible groups of 4 women is 5040/24 = 210. You may also get this result from the combinations formula with $r = 4$ and $n = 10$:

$${}_{10}C_4 = \frac{10!}{(10-4)!\times 4!} = \frac{10!}{6!\times 4!} = \frac{10\times 9\times 8\times 7\times 6}{4\times 3\times 2\times 1} = \frac{5040}{24} = 210.$$

❖

Example 12–8 *Counting Aces.* A standard deck of cards contains four aces: the aces of hearts, diamonds, spades, and clubs. How many different sets of three aces can be selected from a standard deck of cards? How many different sets of two aces can be selected?

Solution: Let ♥♦♠♣ represent the suits hearts, diamonds, spades, and clubs, respectively. There are four possible ways to choose three aces: ♥♦♣, ♥♦♠, ♥♠♣, and ♦♠♣. The combination formula also gives the same result: taking $n = 4$ aces $r = 3$ at a time yields 4 different groupings:

$${}_4C_3 = \frac{4!}{(4-3)!\times 3!} = \frac{4!}{1!\times 3!} = \frac{4!}{3!} = \frac{4\times 3\times 2\times 1}{3\times 2\times 1} = 4.$$

Similarly, two aces may be arranged

$${}_4C_2 = \frac{4!}{(4-2)!\times 2!} = \frac{4!}{2!\times 2!} = \frac{4\times 3\times 2\times 1}{2\times 2} = 6$$

different ways. The 6 different arrangements of aces are: ♥♦, ♥♠, ♥♣, ♦♣, ♦♠, and ♠♣.

❖

Example 12–9 *Poker Hands.* How many different 5-card poker hands can be dealt from a deck of 52 cards?

Solution: Order does *not* matter in card hands. The number of combinations of 52 cards drawn 5 at a time is

$${}_{52}C_5 = \frac{52!}{(52-5)!5!} = \frac{52!}{47!5!} = 2,598,960.$$

Fifty-two cards may be arranged in more than *two and one-half million* 5-card hands.

❖

12.2.5 Counting and Probability

The use of combinations to count card hands suggests that counting techniques can be used to calculate the odds or chances of events occurring. That actually is a bit of an understatement! The subject of combinatorics (counting techniques) and the subject of probability (determining odds) go hand-in-hand both historically and in problem solving. We devote Chapter 13 entirely to probability. For the moment, a few examples will offer a tantalizing preview of this subject.

Example 12–10 *(Not) Winning the Lottery.* Suppose that a lottery involves drawing 6 balls at random from a collection of 45 numbered balls. How many different drawings are possible? What are your chances of choosing the winning 6 balls?

Solution: Winning depends only on choosing the correct 6 balls; order doesn't matter. Thus in this combinations problem, $r = 6$ balls are selected from a total of $n = 45$ balls. The number of combinations is

$$_{45}C_6 = \frac{45!}{(45-6)!6!} = \frac{45!}{39!6!} = 8{,}145{,}060.$$

With more than 8 million possible draws of the 6 balls, your chances of winning this lottery are *less* than 1 in 8 million.

❖

Example 12–11 ***Spade Flushes.*** A flush is any set of five cards of the same suit. How many different spade flushes are there? What are the chances of being dealt a spade flush from a standard deck of cards?

Solution: A standard deck contains 13 spades. Because order doesn't matter in a card hand, the number of different 5-card spade flushes is

$$_{13}C_5 = \frac{13!}{(13-5)!\times 5!} = \frac{13!}{8!\times 3!} = 1287.$$

From Example 12–9, you know that 2,598,960 different 5-card hands can be dealt from a standard deck of cards. Of these 5-card hands, 1287 are spade flushes. Therefore the chance of being dealt a spade flush (or a flush of any suit for that matter) is $1287 \div 2{,}598{,}960 \approx 0.0005$ or about 1 chance in 2000.

❖

When order of selection matters, ABC and CBA count as different arrangements. When order of selection does not matter, ABC and CBA count only as a single arrangement because they both contain the same three letters.

Table 12-2. Summary of Counting Arrangements

Multiplication Principle	Arrangements with Repetition	Permutations	Combinations
One choice from a group of M items; one choice from a group of N items	r items selected from a group of n items, where the same item may be selected over and over again. Order of selection matters.	r items selected from a group of n items, where any item may be chosen only once. Order of selection matters.	r items selected from a group of n items, where any item may be chosen only once. Order of selection does *not* matter.
$M \times N$ arrangements	n^r arrangements	$_nP_r = \frac{n!}{(n-r)!}$ arrangements	$_nC_r = \frac{n!}{(n-r)!\times r!}$ arrangements

12.3 NETWORK ANALYSIS

Networks (or **graphs**) are among the most visual of all mathematical models. Networks are used to model connections, or possible connections, between discrete items: telephones, computers, cities, and many more. The story of network analysis begins in the early eighteenth century in the Prussian town of Königsberg (now Kaliningrad, Russia) on the Pregel River.

With over 700 books and publications to his name, Leonhard Euler is one of the most prolific mathematician that ever lived. He made fundamental contributions to many areas of mathematics, as well as to astronomy, music, hydraulics, optics, and mechanics.

12.3.1 The Bridges of Königsberg

Königsberg had seven bridges that straddled the Pregel River (Figure 12–5a). A popular pastime of the day was to try to find a path, beginning and ending at the same point, that crossed each bridge *exactly once*. In other words, could someone start at one bridge and take a tour that crossed every bridge once and returned to the starting point? Most citizens of Königsberg believed that such a tour was impossible, but they couldn't be sure. The problem came to the attention of the eighteenth-century Swiss mathematician Leonhard Euler (pronounced *oiler),* whose solution to the Königsberg bridge problem gave birth to the subject of **network theory**.

Euler used a schematic drawing to capture the essential aspects of the bridge problem. As shown in Figure 12–5(b), the dots in the drawing represent points of land and lines or curves represent bridges. This array of dots and curves is called a **network**, or **graph**. The dots in the network are called **vertices** (plural of **vertex**), or **nodes**; the lines and curves that connect vertices are called **edges**, or **arcs**. We label vertices with uppercase letters (*A*, *B*, *C*, ...) and edges with lowercase letters (*a*, *b*, *c*, ...).

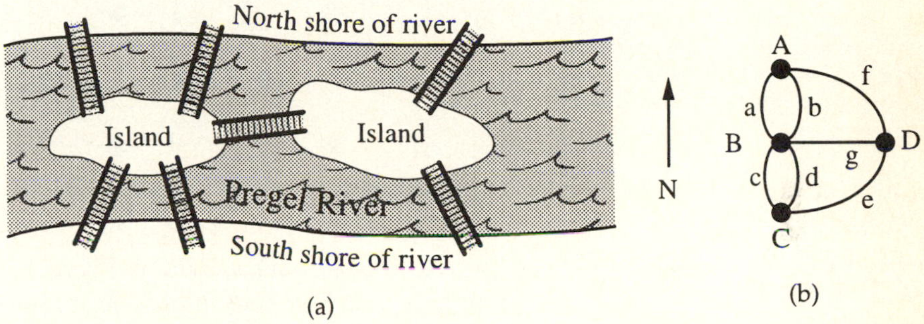

Figure 12-5. (a) The seven bridges of Königsberg spanned the Pregel River. (b) In the network associated with the bridges, the edges (curves) represent bridges and the vertices (dots) represent the land between bridges.

The network representation of the bridges contains two particularly important abstractions.

- Although an edge represents each of the seven bridges, vertices represent only four "pieces" of land: the north side of the river (vertex *A*), the south side of the river (vertex *C*), the west island (vertex *B*), and the east island (vertex *D*). For example, even though the three bridges on the south side of the river connect to land in different places, in the network these three bridges all connect to vertex *C*, which represents *all* land on the south side of the river.
- The lengths and the shapes of the bridges is unimportant to this problem; what matters is which pieces of land they connect. Thus the length or shape of the edges in the network don't matter. In this example, all seven edges are drawn as simple curves.

Euler Circuits

With these abstractions in mind, the problem of finding a path that crosses all the real bridges once, starting and ending at the same point, becomes equivalent to that of finding a path through the network (Figure 12–5b) that traverses each of the edges once, while starting and ending at the same vertex. Such a path, if one exists, is called an **Euler circuit** in honor of Euler's work on the Königsberg bridge problem. The advantage of using the network, rather than the real bridges, is that it captures the essential elements of the problem while disposing of unimportant details (such as the exact lengths and shapes of the bridges).

Two basic questions must be asked ask about the Königsberg bridge network: Does it have an Euler circuit and, if it does, how do we find it? First, however, we can practice with some simpler networks.

Figure 12–6(a) on the next page shows a simple network that forms a closed loop; we can think of each vertex as an island and each edge as a bridge between islands. An Euler circuit can be found by beginning at any vertex and simply "walking" ahead. The resulting path is an Euler circuit because it crosses all the bridges exactly once and starts and ends at the same place. In this circuit, we never have to make any choices: Each vertex has only one path into it and one path out of it.

Next, consider the network in Figure 12–6(b). If we start at vertex A and follow the edges in the order $a \rightarrow b \rightarrow c \rightarrow d \rightarrow e$, our path traverses every edge once. However, this path is *not* an Euler circuit because it ends at vertex C, rather than returning to vertex A. An analogous situation arises if we start at vertex C. Moreover, if we start at vertex B (or D), traversing all the edges requires traversing some edges more than once (such as the path $b \rightarrow c \rightarrow a \rightarrow b \rightarrow e \rightarrow d$). Thus no Euler circuits exist for this network.

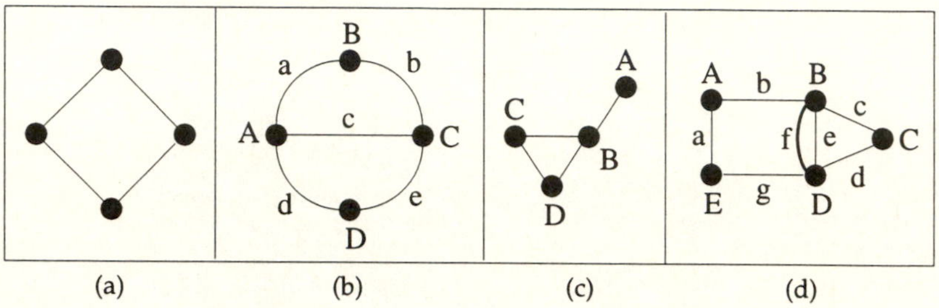

Figure 12-6. Networks (a) and (d) have Euler circuits, but networks (b) and (c) do not. The crucial property is that networks (a) and (d) have *even* number of edges connected to each vertex. Networks (b) and (d) do not have this property.

In the network shown in Figure 12–6(c), the presence of vertex A with only one edge connected to it precludes an Euler circuit because any circuit must use the edge to vertex A more than once. The network of Figure 12–6(d) is more complex, but a little trial and error should convince you that, starting at any vertex, you *can* create a circuit by using each edge exactly once; for example, the path $a \rightarrow b \rightarrow c \rightarrow d \rightarrow e \rightarrow f \rightarrow g$ is an Euler circuit. What property makes an Euler circuit possible in this case? First, note that all the vertices have two or four edges. When a path enters a vertex with two edges, only one edge remains for leaving that vertex. When a path enters and leaves a vertex with four edges, two edges can still be used to enter and leave that vertex on a subsequent visit to that vertex. We can now generalize:

> *An Euler circuit exists for a network only if all of the vertices have an even number of edges.*

Test for an Euler circuit: all vertices must have an even number of edges connected to them.

Now we can apply this test to the Königsberg bridge problem. Returning to the bridge network (Figure 12–5), we see that vertices A, C, and D each has 3 edges and vertex B has 5 edges. Thus, because at least one vertex has an odd number of edges (in fact, *all* the vertices have an odd number of edges), this network has no Euler circuit. The people of Königsberg were searching in vain for a path that crossed each bridge once and returned to its starting point. Although this result may seem anticlimactic, the Euler circuits have many practical uses beyond analyzing Sunday afternoon walks in the eighteenth century.

Example 12–12 ***The Optimal Mail Route.*** An ingenious mail carrier is loath to work any harder than necessary. Her assignment is to deliver mail in the several city blocks shown in Figure 12–7a. Can she park her truck in one place and do the job without walking any of the sidewalks more than once?

Solution: The mail carrier seeks an Euler circuit for her route. First, she transforms the city map into a network (Figure 12–7b) in which edges represent sidewalks and vertices represent intersections. All the shaded sidewalks along which mail must be delivered are included; thus, when both sides of a street (such as a and j) must be covered, two edges connect two vertices in the network. A quick inspection shows her that an Euler circuit exists because all vertices of the network have an *even* number of edges. One of several possible Euler circuits is the path $a \rightarrow b \rightarrow c \rightarrow d \rightarrow \ldots \rightarrow i \rightarrow j$. That is, by parking her truck at the leftmost point in the network and delivering mail along the path of the Euler circuit, the mail carrier completes her route with the minimum possible effort.

Figure 12-7. (a) A section of city blocks shows the (shaded) sidewalks along which mail must be delivered by a single carrier. (b) The route is modeled by a network in which each sidewalk is an edge and each intersection is a vertex.

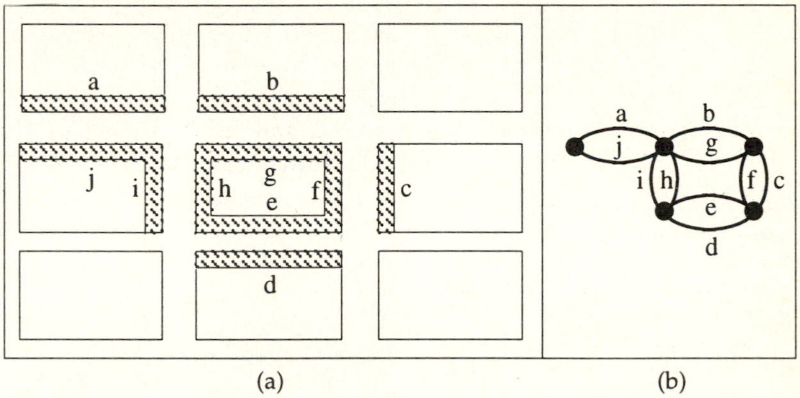

❖

12.3.2 Applications of Networks

Networks may be used as mathematical models for a variety of purposes. For example, the vertices might represent power stations, railway terminals, jobs on an assembly line, or computers connected through the Internet. In each case, the process of representing the real system as a network is straightforward: Vertices represent a physical location (a power station, a railway terminal, a job station, or a computer), and edges represent physical connections between them (power lines, railroad tracks, links between job stations, or cables between computers).

However, networks may be used to model many other types of relationships. For example, vertices might represent countries and edges the relationship "share a common border" (Figure 12–8). We could also make a network in which the vertices represent countries and the edges the relationship "is a

Figure 12-8. Several countries of Europe are represented by a network. Each edge represents the relationship "share a common border."

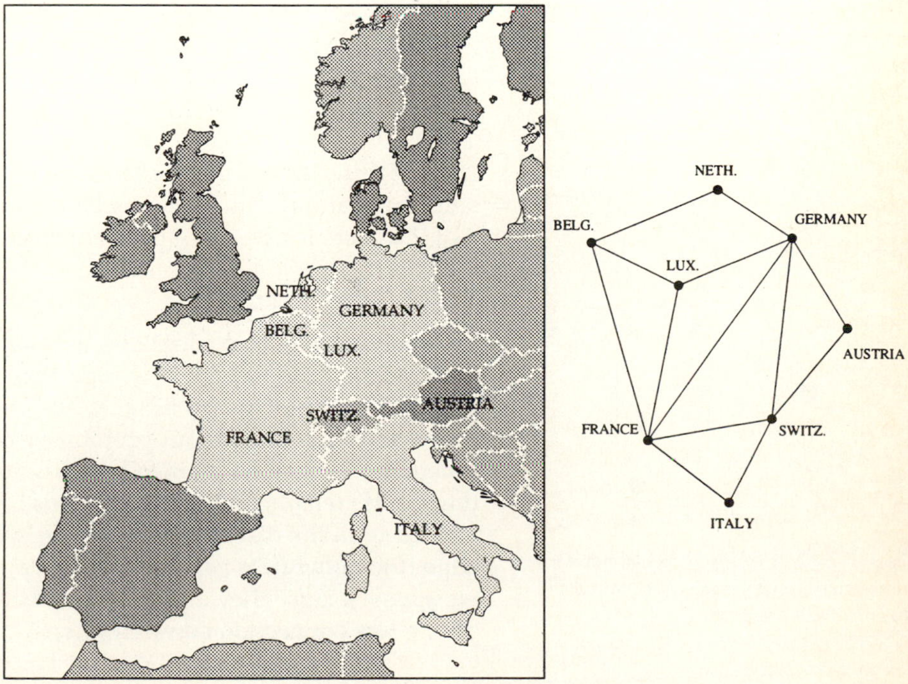

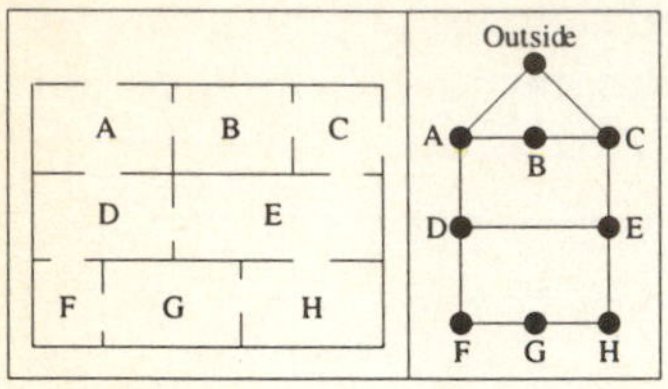

Figure 12-9. The floor plan of a house is drawn as a network showing connections between rooms. Each vertex represents a room (or the outside), and the edges represent the relationship "has a door to."

trading partner of." Another example of networks in everyday use is the representation of a floor plan of a house, as shown in Figure 12–9. In this case, an edge stands for the relationship "has a door to."

Before we go on to specific problems, a bit of terminology will be helpful. In the network in Figure 12–10(a), each vertex is connected to *exactly two* other vertices, forming a closed ring; this type of network is called a **cycle**. Closely related to a cycle is a **circuit**, which is any path within a network that begins and ends at the same vertex. For example, the path that connects the vertices in the order $A \rightarrow B \rightarrow C \rightarrow A$ in Figure 12–10(b) is one of several circuits in this network; another is the path connecting the vertices $A \rightarrow G \rightarrow F \rightarrow D \rightarrow B \rightarrow A$. An Euler circuit is a special circuit that uses every edge of the network once.

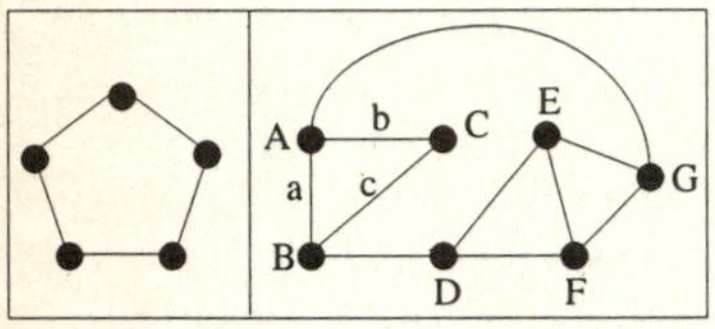

(a) (b)

Figure 12-10. (a) This network forms a simple closed ring, or cycle. (b) This network contains several circuits — closed paths that begin and end at the same vertex (e.g., $A \rightarrow B \rightarrow C \rightarrow A$).

Time-Out to Think: Can you find at least four other circuits in Figure 12–10b? How many can you find?

In the two networks of Figure 12–11, each vertex is connected to *every* remaining vertex. Such networks are said to be **complete**. A complete network always includes at least one circuit; in each of the two complete networks shown, a path around the outer edges forms a circuit.

A network in which all the vertices are connected — but which has no circuits — is called a **tree** (Figure 12–12). Among their many uses, tree networks can be used to represent family trees.

The **order** of any network is its number of vertices. For example, the two networks in Figure 12–11 have order 4 and order 5, respectively; the tree network in Figure 12–12 has order 9. Each vertex in a network may be described further by its **degree**, which is the number of edges connected to it. For example, each of the vertices in a cycle (Figure 12–10) has degree 2; all the vertices in the network of Figure 12–11(b) have degree 4.

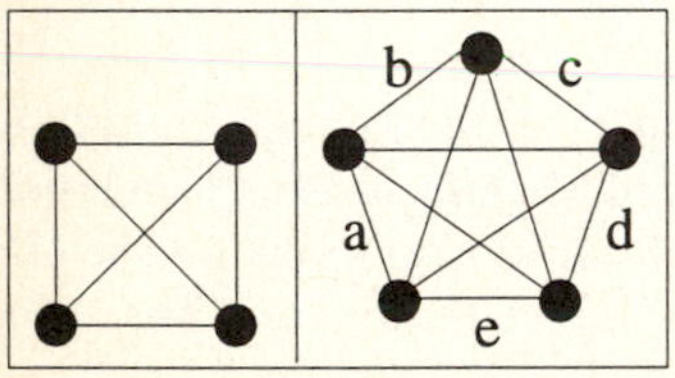

Figure 12-11. Both networks shown are complete, meaning that each vertex in the network is connected to the others. The order of the networks is 4 and 5, respectively. (Note that some of the edges cross each other in these networks, but that does *not* mean that a vertex exists at the intersection points. Vertices exist only where large dots appear.)

Time-Out to Think: Identify the order of each of the networks shown in Figure 12–10 through Figure 12–12. Within each network, identify the degree of each vertex.

With this terminology in mind, we are ready to explore network problems in greater detail. In the remainder of this section, we present examples that illustrate the range of problems that can be modeled with networks.

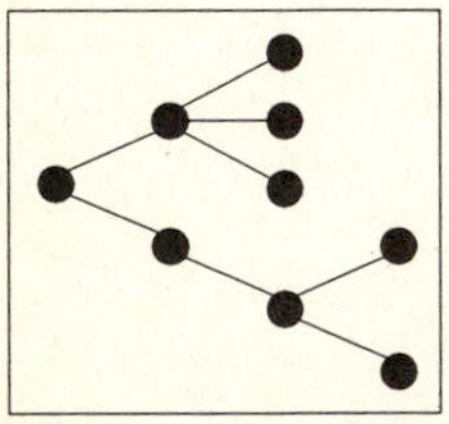

Figure 12-12. A tree is a special network in which all the vertices are connected but no circuits appear.

12.3.3 Minimum Cost Networks: Connecting All the Towns

Imagine that the local telephone company needs to connect the seven towns of Sunshine County with new high-speed telephone lines. Because of geographic constraints (such as lakes and mountains), some towns cannot be connected directly without prohibitive cost. Figure 12–13(a) shows a schematic map of Sunshine County with the seven towns as vertices and the *possible* telephone line connections between towns as edges. Each edge is labeled with the distance, in kilometers, between the towns connected by that edge (the network is *not* drawn to scale). These numbers are called **costs** (or **weights**) because they are related to the cost of installing a particular telephone line connection between towns.

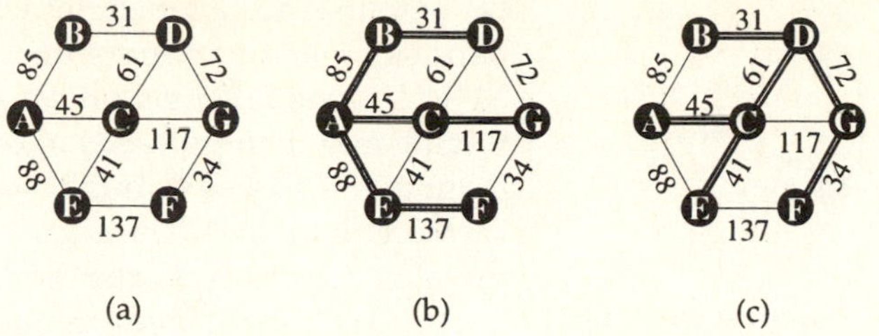

Figure 12-13. (a) The seven towns of Sunshine County are represented by the vertices of a network. The feasible links for new telephone cables are shown as edges, with the corresponding distances (in km) indicated. (b) One of several spanning trees for the Sunshine County network is shown with bold edges. Its total length is 503 kilometers. (c) The minimum cost spanning tree (bold) has a length of 284 kilometers.

In order to solve this problem, the telephone company must find a set of edges in the original network that

- either directly or indirectly links every town to every other town, and
- has the minimum possible cost (or length).

A set of edges that links all the towns is called a **spanning network**. Many possible spanning networks exist for the Sunshine County problem; two are shown in Figure 12–13(b) and (c). Adding the costs, or distances, along the spanning network shown in Figure 12–13(b) gives a total of 503 kilometers. The total length of cable needed for the spanning network in Figure 12–13(c) is only 284 kilometers; it has the minimum cost (or distance) of all possible spanning networks for Sunshine County; hence it is called the **minimum cost** spanning network. Finding the minimum cost spanning network can be immensely practical. If the cost of installing cable is, say, several thousand dollars per kilometer, the telephone company's profits depend heavily on finding the correct solution.

So far, we have calculated only the "costs" (lengths of cable in this case) of spanning networks that are shown. The problem of *finding* the minimum cost spanning network is slightly more complex and is addressed in Section 12.5.

Time-Out to Think: Find two other spanning networks for Sunshine County and calculate their "costs" by adding the distances along the edges. Can you convince yourself that Figure 12–13(c) represents the *minimum* cost spanning network?

12.3.4 The Traveling Salesman Problem: Visiting All the Towns

Imagine that the edges of the Sunshine County network represent *roads* between the towns (rather than possible telephone lines). Suppose that you work as a salesperson, living in one of the seven towns, and that your job requires you to service customer accounts in each of the other six towns of Sunshine County. Can you find a route that allows you to visit all the towns with the minimum amount of travel *and* to return home at the end of the trip? This is an example of a famous problem in mathematics called the **traveling salesman problem** which requires finding the shortest circuit (closed path) that visits all of the vertices exactly once.

As is always the case in mathematics, the traveling salesman problem has applications far beyond traveling salespeople! However, before we investigate the problem and its applications, we need to explore some more aspects of the mathematics of networks.

Hamiltonian Circuits

Earlier, we defined an Euler circuit as a path through a network that traverses *every edge* once and returns to the starting point. An Euler circuit also must pass through every vertex. Relating this fact to the traveling salesman problem,

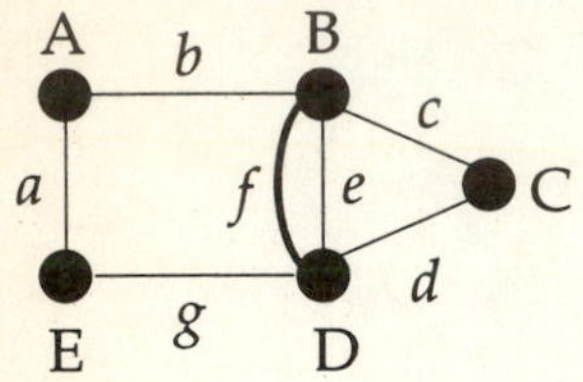

Figure 12-14. This network has the Euler circuit $a\to b\to c\to d\to e\to f\to g$, which visits every vertex and traverses every edge. However the circuit $a\to b\to c\to d\to g$ also visits every vertex and is shorter. The latter circuit would be preferable for a traveling salesperson.

we note that an Euler circuit will take the salesman to every town, but perhaps to some towns more than once. For example, the network in Figure 12–14 has the Euler circuit $a\to b\to c\to d\to e\to f\to g$. This circuit begins at town E and then proceeds in order to towns: A, B, C, D, B, D, and E. Although it passes through every town, it passes through some towns (B and D) more than once. The circuit $a\to b\to c\to d\to g$ also passes through every town, but is much shorter. For a traveling salesperson, it is a better solution.

A circuit that passes through every vertex of a network exactly once (but doesn't necessarily traverse every edge) is called a **Hamiltonian circuit**, after the nineteenth-century Irish mathematician William Rowan Hamilton. As shown in Figure 12–15, many networks do not have Hamiltonian circuits. The arrows in Figure 12–15(a) and (b) trace Hamiltonian circuits through those networks; as required, these circuits pass through every vertex once and form a closed loop. The networks in Figure 12–15(c) and (d) have no Hamiltonian circuits: it is impossible to find a path through either of these networks that passes through all vertices and returns to its starting point without passing through some vertices more than once. (Remember that the *possible* paths through the networks must use only the edges that are shown; don't try to make a path by adding edges that don't exist!)

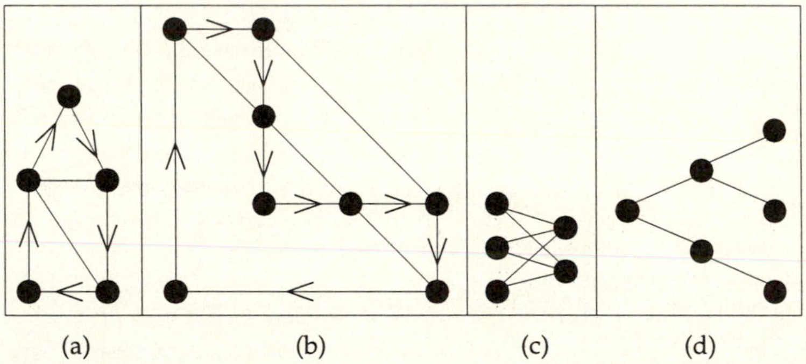

Figure 12-15. (a) and (b) These networks have a Hamiltonian circuit (following the arrows). (c) and (d) These networks have no Hamiltonian circuit.

Time-Out to Think: The Sunshine County network shown in Figure 12–13(a) has two Hamiltonian circuits. Can you find them? Mark them on the figure with arrows.

Two basic questions arise about Hamiltonian circuits: First, under what circumstances does a network have a Hamiltonian circuit? Second, how do we find them when they exist? Unfortunately, neither question has a simple answer. In fact, there is no known rule for *efficiently* determining whether a Hamiltonian circuit exists. Except in special cases — such as a *cycle* (e.g., Figure 12–10a), which always has itself as a Hamiltonian circuit, and a *tree* (e.g., Figure 12–12), which can never have a Hamiltonian circuit — the only way to search for Hamiltonian circuits is by trial and error.

Unfortunately, because large networks may have large numbers of Hamiltonian circuits, the process of finding them all by trial and error can be worse than just tedious. In the case of *complete* networks, in which every vertex is connected to every other vertex, the number of Hamiltonian circuits is given by the formula

$$\text{number of Hamiltonian circuits in a complete network (order } n) = \frac{(n-1)!}{2},$$

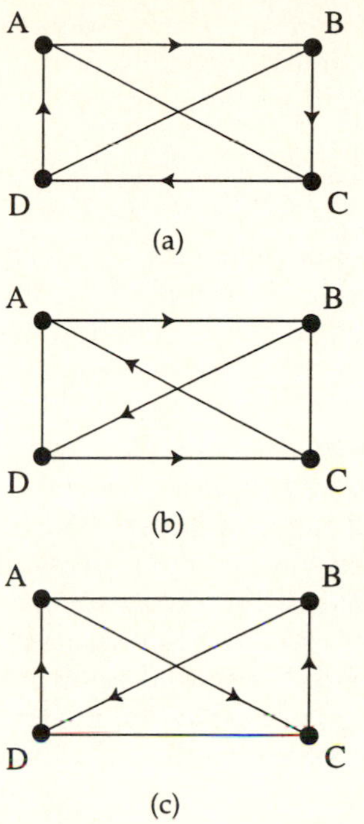

Figure 12-16. A complete network of order 4 has the 3 Hamiltonian circuits shown.

where n is the order of the network. For example, the number of Hamiltonian circuits in a complete network of order 4 is

$$\frac{(4-1)!}{2} = \frac{3!}{2} = \frac{6}{2} = 3,$$

as shown in Figure 12–16.

Solving Traveling Salesman Problems

The traveling salesman problem requires not only finding a Hamiltonian circuit (which visits all the towns and returns home), but finding the one that has the *minimum* total length of all Hamiltonian circuits in the network. For example, suppose that you are planning a summer vacation during which you want to visit five national parks. Let's make the problem slightly more exotic (and more mathematically interesting) by assuming that you will rent a private airplane at one park and do all your traveling between parks by air. What is the most efficient path by which you can visit all five parks? Although sales are not involved in this problem, it is mathematically equivalent to the traveling salesman problem: You must begin and end at the same place (you must return the rented airplane), and you seek the shortest path that visits all five parks.

Table 12-3. Air Miles Between National Parks

—	Bryce	Canyonlands	Capitol Reef	Grand Canyon	Zion
Bryce	—	136	69	108	52
Canyonlands	136	—	75	202	188
Capitol Reef	69	75	—	151	123
Grand Canyon	108	202	151	—	101
Zion	52	188	123	101	—

Table 12–3 lists the five national parks and the distances (by air) between them. Figure 12–17(a) shows a network representation in which vertices represent parks and edges represent the possible air routes between them. The edges are labeled with their "costs"; in this case, the "costs" are the flight distances in miles. Your task is to find the Hamiltonian circuit through this network having the minimum total flight distance.

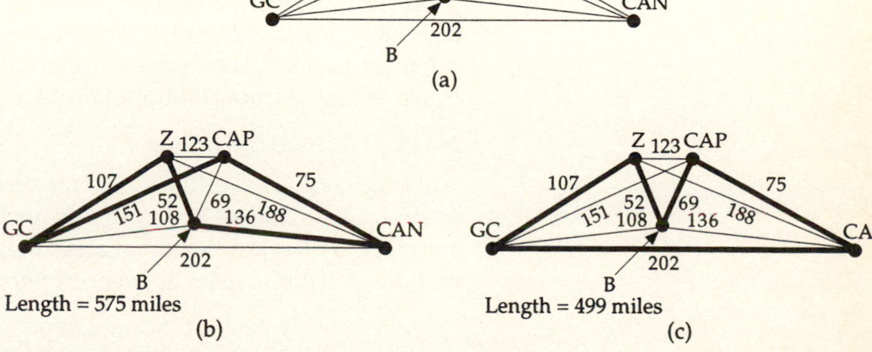

Figure 12-17. (a) The vertices represent five national parks, and edges represent air routes between them. Each edge is labeled with a "cost" representing the distance between parks. (b) A Hamiltonian circuit obtained by the nearest neighbor method, originating at Bryce, has a total length of 515 miles. (c) The optimal circuit, with a length of 499 miles, is shown.

The only guaranteed way to find a solution is by trial and error. Because the park network is a *complete* network (any park can be reached from any other park) of order $n = 5$ (for the 5 parks), it has 12 Hamiltonian circuits:

$$\text{number of Hamiltonian circuits} = \frac{(n-1)!}{2} = \frac{(5-1)!}{2} = \frac{4!}{2} = 12.$$

Figure 12–17(b) and (c) show 2 of the 12 Hamiltonian circuits among the five national parks. In principle, all 12 circuits could be listed and the total length of each circuit could be calculated. If you did so, you would find that the circuit in Figure 12–17(c) is the shortest and it therefore is the solution to this problem.

In practice, applications of the traveling salesman problem often involve hundreds or thousands of vertices, making any kind of trial and error approach prohibitive even with the most powerful supercomputers. The first two columns of Table 12–4 list the number of vertices in a complete network and the corresponding number of Hamiltonian circuits. The fastest supercomputers today can calculate the lengths of about 1 million Hamiltonian circuits each second, or about 1 circuit per microsecond. The third column of Table 12–4 shows the time needed by such a supercomputer to find the traveling salesman (shortest) circuit by checking *all* possible circuits.

Because Hamiltonian circuit problems present such a challenge, Professor Leonard Adelman, a computer scientist at the University of California, chose one for the first-ever test of a computer using DNA in late 1994. He synthesized DNA pieces representing seven cities and the possible connections between them. When mixed together, the DNA pieces chemically joined to form longer chains. By finding chains that included DNA segments from every city, Dr. Adelman identified the Hamiltonian circuits. Theoretically, DNA computers could be a billion times faster than current supercomputers. Although such computers still could not test every possible circuit for large networks, finding approximate solutions would be far faster.

Table 12-4. Finding Traveling Salesman Problem Circuits

Number of Vertices, n	Number of Hamiltonian Circuits = $(n-1)!/2$	Computer Time Required
5	12	12 microseconds
10	181,440	1.8 milliseconds
15	4.4×10^{10}	7.3 minutes
20	6.1×10^{16}	19.7 years
25	3.1×10^{23}	3.1×10^{8} years
30	4.4×10^{30}	1.4×10^{15} years
50	3.0×10^{62}	1.0×10^{55} years
70	8.5×10^{97}	2.7×10^{82} years
100	4.7×10^{155}	1.5×10^{140} years

The combinatorial explosion associated with the traveling salesman problem is vividly illustrated. Even for moderate values of n (such as $n = 30$) the computation time greatly exceeds the present age of the universe! Even the most powerful computers *imaginable* would scarcely have an impact on this prodigious computational problem.

Nearly Optimal Solutions

Although only trial and error *guarantees* a solution to the traveling salesman problem, some faster procedures have been developed to find *nearly* optimal solutions. For example, the **nearest neighbor method** begins at any vertex and, at each step, visits the nearest vertex that has not yet been visited.

Let's apply the nearest neighbor method to the National Parks problem beginning at Bryce. The *nearest neighbor* to Bryce is Zion, 52 miles away. From Zion, the nearest neighbor not already visited is Grand Canyon, 101 miles away. From Grand Canyon, the nearest unvisited neighbor is Capitol Reef, 151 miles away. Finally, from Capitol Reef we have only one place left to visit: Canyonlands, 75 miles away. This is the circuit shown in Figure 12–17(b). Although it is *not* the optimal circuit (which is shown Figure 12–17c), it is longer only by a factor of about

$$\frac{575 \text{ mi}}{499 \text{ mi}} = 1.15,$$

or 15%. Thus it represents a *near-optimal* solution and is one that can be found much more quickly than proceeding by trial and error. Although the nearest neighbor method *usually* finds near-optimal solutions, it occasionally produces very expensive circuits when a long trip across the entire network is needed to close the loop.

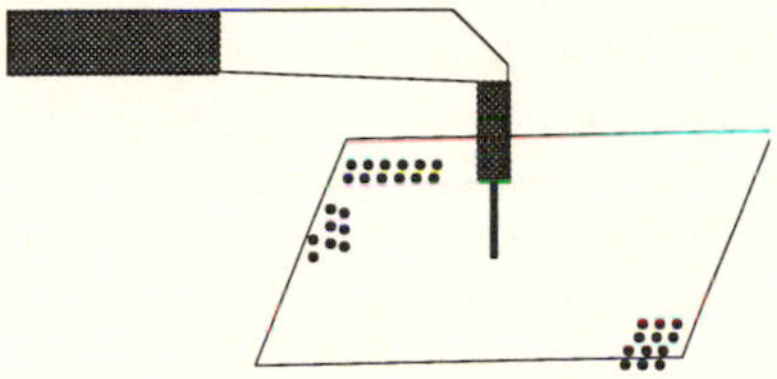

(a)

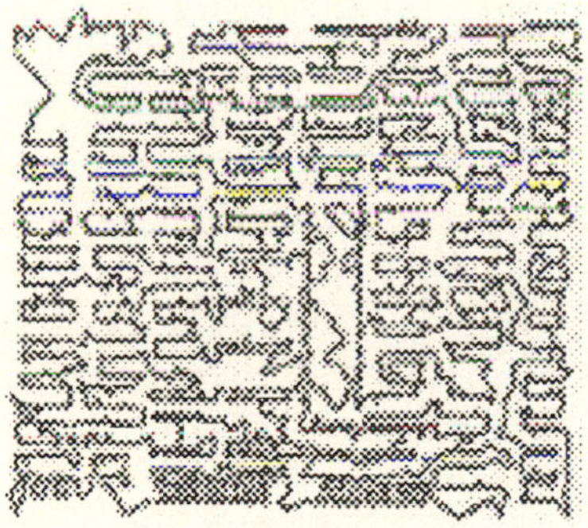

(b)

Figure 12-18. (a) Holes must be drilled, by a robot, at precise locations on circuit boards for computers. Finding the most efficient drilling scheme is equivalent to a traveling salesperson problem. (b) The path shows a near-optimal solution for a circuit board with 3038 holes. (b) from David Applegate, AT&T, Robert Bixby, Rice University, Vasek Chvatal, Rutgers University, and William Cook, Bellcore. Reprinted with permission of William Cook.

Further Applications of the Traveling Salesman Problem

Applications of the traveling salesperson problem are endless: determining optimal itineraries for air, sea, or land trips; scheduling doctors' rounds or salespersons' visits; routing deliveries of mail, supplies, or merchandise; and planning inspection and maintenance routes. In each application, the edges and the vertices of the network may have different meanings: The weights on the edges may represent distance, cost, time, or any other quantity that must be minimized.

One of the more vivid and unexpected applications of the traveling salesman problem is in the design of circuit boards for computers. Imagine a small thin board in which holes must be drilled at precise locations so that computer components can be inserted. A drilling machine (a robot that works day and night) can start at any location, but it must drill each hole on the board and finish at its starting point so that it is ready for another board. The holes form the vertices of a network; edges in this network represent *possible* moves between pairs of holes. The cost or weight of each edge is the physical distance between the holes. Finding the most efficient way of drilling the holes involves finding a traveling salesman circuit through this network. The number of vertices on a circuit board may easily run into the hundreds or thousands and hundreds or thousands of boards usually need to be drilled. Further, because the drilling machine is expensive to set up and operate, efficiency is essential. A near-optimal solution for a circuit board with 3038 holes is shown in Figure 12–18.

THINKING ABOUT ... OPERATIONS RESEARCH

Every day people are confronted with the need to find efficient solutions to practical problems. Often it is done deliberately with considerable forethought, but at times they optimize their lives in almost unconscious ways. The problem may be to minimize the travel time to several stores on a shopping trip, maximize the revenue from a garage sale, find the best schedule for building a house, or choose a wise investment strategy for some extra money.

Decisions concerning strategies that minimize or maximize an important commodity are the subject of a branch of applied mathematics called **operations research,** or **management science.** It is a subject that seeks strategies for carrying out complex tasks in organized and efficient ways. Questions in operations research often take a "how to ...?" form. The answers to these questions must describe a *process*, and preferably an efficient one.

Developed in the years immediately following World War II, operations research was born in part to solve business and management problems.

Today, a large percentage of operations research applications still originate in these fields. However, the subject has grown far beyond its original purposes. It is now applied to problems in city planning (scheduling of maintenance vehicles, such as snowplows and street cleaners, and designing sewer, power, and water lines), communications (designing telephone switching devices for routing long distance calls and selecting optimal locations for transmitters and receivers), and transportation (laying out highways and railroads, designing distribution strategies for merchandise, and devising efficient airline routes and schedules). The following passages, all taken from recent publications, indicate the scope and the successes of operations research.

"Delta Airlines flies over 2500 domestic flight legs every day, using about 450 aircraft from 10 different fleets. The fleet assignment problem is to match aircraft to flight legs so that seats are filled with paying passengers. Recent advances in mathematical programming algorithms [a powerful operations research technique] and computer hardware make it possible to solve optimization problems of this scope for the first time. Delta is the first airline to solve to completion one of the largest and most difficult problems in the industry. Use of the Coldstart model is expected to save Delta Air Lines $300 million over the next three years." — From "Coldstart: Fleet Assignment at Delta Airlines" by R. Subramanian et al., *Interfaces* 24:1, Jan–Feb 1994, pp. 104–120.

"Using CAPS (call processing simulator), AT&T can model a network of call centers utilizing advanced 800 network features before its customers make capital investments to start or change their call centers. In 1992, AT&T completed about 2000 CAPS studies for its business customers, helping increase, protect and regain more than one billion dollars in an eight billion dollar 800-network. While this is impressive alone, the CAPS tool is the turnkey for more than $750 million in annual profit for AT&T's business customers." — From "Improving Pupil Transportation" by T. Sexton et al., *Interfaces* 24:1, Jan–Feb 1994, pp. 87–103.

"North Carolina uses data envelopment analysis (DEA) to produce a pupil transportation funding process that encourages operational efficiency and reduces expenditures. The new process has led to changes in bus routes and schedules, adjustments in school start and stop times, and reductions in the inventory of buses. Between 1990 and 1993, the state saved $25.2 million in capital costs and $27.9 million in operating costs, and it expects savings to increase." —From "AT&T's Call Processing Simulator (CAPS)" by A. Brigandi et al., *Interfaces* 24:1, Jan–Feb 1994, pp. 6–28.

Clearly, operations research affects everyone's daily life, often in hidden and taken-for-granted ways. However it also has a greater influence. In previous chapters we discussed many crucial issues of looming proportions such as overpopulation, disease, environmental devastation, and energy conservation. Identifying and analyzing these crises is one thing. Solving them is quite another matter. If human beings are to make progress in solving today's pressing issues, massive organizational and unprecedented policy changes will be required. Operations research is sure to play a part in meeting these complex challenges.

12.3.5 Planar Networks and Coloring Networks

In the street in front of three adjacent houses are outlets for gas, water, and electricity (Figure 12–19). Can each house be connected to each utility without any of the service lines crossing? We can approach this problem with a network in which vertices represent the houses and utilities and edges represent the lines that connect the utilities to the houses. Figure 12–19(b) shows one attempt to solve the problem. Unfortunately, it fails: We cannot connect the electricity to house 1 without crossing either the gas line or the water line.

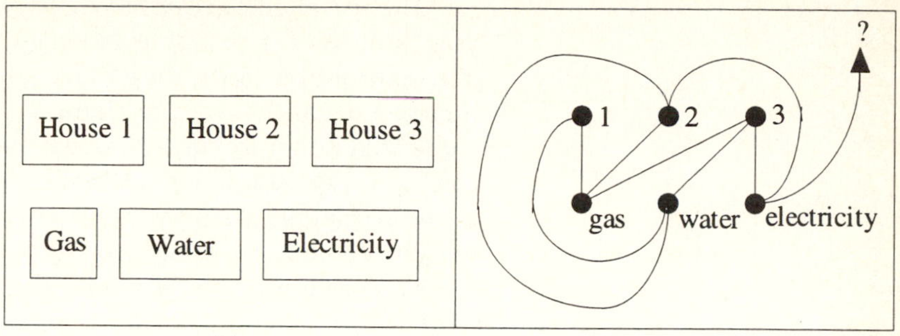

(a) (b)

Figure 12-19. (a) The houses and utilities problem involves connecting each of three houses to each of three utilities without any crossing of lines. (b) The problem may be modeled by a network in which vertices represent houses and utilities and edges represent connecting lines. The network shown doesn't solve this problem, because the electrical line can't be connected to house 1 without crossing another line.

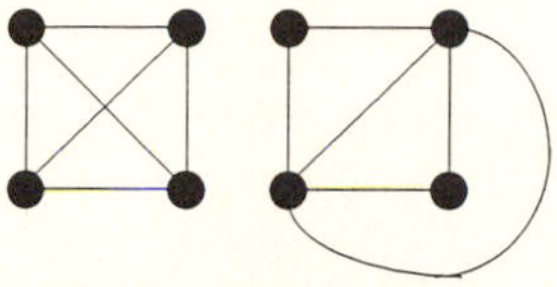

(a) (b)

Figure 12-20. (a) This network is planar because it can be redrawn so that (b) the edges intersect only at vertices. Note that all network connections are preserved when it is redrawn, even though their shapes change.

> **Time-Out to Think:** Look for a solution (besides the failed one in Figure 12–19b) to the three-house three-utility problem. Convince yourself that making the connections is impossible without at least a pair of utility lines (edges) crossing.

If a network can be drawn in such a way that its edges intersect *only* at vertices, it is called a **planar network**. The three-house three-utility problem is **nonplanar**, because finding a solution in which no lines cross is impossible. In determining whether a network is planar, remember that the length and shape of edges is unimportant because they are only abstractions of real connections. In other words, you can always redraw a network in any way you like so long as you preserve all the *connections* between pairs of vertices. For example, the network in Figure 12–20(a) at first appears to be nonplanar because two edges cross. However, redrawing it as shown in Figure 12–20(b) shows that it actually *is* planar. This redrawing is "legal" because it retains all the connections between vertices.

Table 12-5

	A	B	C	D	E
Judo	x	x			
Yoga	x		x		
CPR			x		x
Aerobics				x	
Pottery			x		x
WW		x		x	

Another type of network problem, closely related to finding planar networks, uses an idea called **coloring** networks. Imagine that you are the director of a recreation center and need to schedule classes in judo, yoga, CPR, aerobics, pottery, and water-walking. In addition, you have the requests, shown by Xs in Table 12–5, of five people (represented by the letters A through E) who want to take some of these classes. Each person can take at most one class per day. Clearly, you could accommodate all the requests by scheduling each of the six classes on a different day of the week. To minimize staff time, however, you prefer to accommodate all the class requests in the fewest number of days. How can you?

You can use a network to solve this problem. Begin with six vertices, one for each class. Connect two vertices with an edge *only* if a person has requested both classes represented by the vertices. For example, person A wants both yoga and judo, so you draw an edge from the yoga vertex to the judo vertex. Similarly, the judo and water-walking vertices should be connected with an edge because person B wants both of those classes. When all the edges have been drawn, the network is as shown in Figure 12–21.

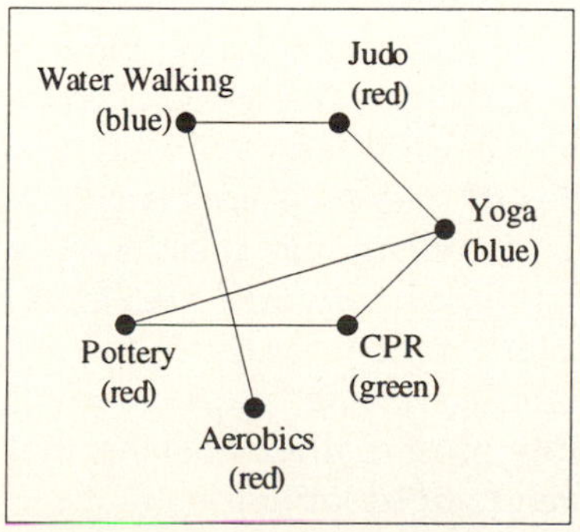

Figure 12-21. A network can be used for scheduling five people for six recreation classes. Vertices represent the classes, and edges show the classes that *cannot* meet on the same day because of participants' requests. Colors are assigned to vertices, making sure that an edge never connects two vertices of the same.

Because CPR and pottery are connected by an edge, someone wants both classes; thus, you can't schedule these two classes on the same day. This constraint suggests a simple rule for solving the problem of optimum scheduling: Two classes may be scheduled on the same day provided that they are *not* connected by an edge.

The use of colors can help solve this problem. Begin at the vertex for yoga and color it, say, blue. Now, go to any other class not connected to yoga and color it blue, indicating that it can be held on the same day as yoga; therefore color the water-walking vertex blue. All the remaining classes are connected either to yoga or water-walking, so they must be scheduled on different days and colored different colors. Use red to represent another day of the week and use it to color the vertex for judo. Aerobics isn't connected to judo, so you can color it red also. You can also color the vertex for pottery red because it is connected neither to judo nor aerobics. Only one vertex, CPR, is left which you can color, say, green. The resulting colors are labeled in Figure 12–21.

Finally, the solution may be completed by assigning days of the week to the colors; for example, you might assign blue to Monday, red to Tuesday, and green to Wednesday. This means that yoga and water-walking would meet on Monday; judo, aerobics, and pottery on Tuesday; and CPR on Wednesday. Thus all class requests can be accommodated on just three days.

Time-Out to Think: Verify that the preceding solution leads to no scheduling conflicts: Each person can attend at most one class per day and also take the classes of his or her choice. Find another schedule (coloring scheme) by starting with the CPR class (instead of yoga). Is this schedule any more efficient (that is, does it require fewer days of the week)? How can you know whether you have found the optimal schedule?

THINKING ABOUT ... THE FOUR-COLOR PROBLEM

A famous problem in mathematics, the **Four-Color Problem**, deals with both planar networks and colorings. It seems to have originated in about 1850 and has inspired a tremendous amount of mathematical investigation since then. Look back at Figure 12–8, which shows a partial map of Europe and the associated network. Recall that each vertex in the network represents a country and that two vertices are connected by an edge if they share a boundary on the map. A likely question (at least if you are a mapmaker) is: How many colors are needed to paint the map of Europe so that adjacent countries never have the same color? This network coloring problem requires coloring the vertices of the network so that pairs of connected vertices have different colors.

Because the edges represent common boundaries between countries, two edges can never intersect (except at vertices); thus the network associated with a map is always *planar*. For the map coloring problem, the question becomes: How many colors are needed to color a planar network? This tantalizing question was given the name the Four-Color Problem for the following reason. If you spend some time drawing and coloring various maps, you will probably come to the conclusion, as have generations of mathematicians, that four colors suffice to color any map. But a few hours of scribbling on maps doesn't constitute a mathematical proof. The claim that *any* map can be colored with no more than four colors is called the Four-Color Conjecture.

For more than 100 years a rigorous proof of the Four-Color Conjecture eluded mathematicians. Then, in 1976, University of Illinois mathematicians Kenneth Appel and Wolfgang Haken announced a proof of the conjecture. But the drama didn't end there. The proof offered by Appel and Haken was different from other proofs of that time: It relied on a computer to routinely check nearly 2000 special networks. The mathematics community was divided over the validity of such a proof. Some argued that as long as a human could *in principle* carry out all the checking, the use of a computer was legitimate; others argued that direct human involvement was necessary at every step of the way, if for no other reason than to avoid computer errors! Today, most mathematicians accept the computer proof of the Four-Color Conjecture, and the conjecture is now a theorem. As postage stamps in 1976 proclaimed: Four colors suffice.

12.4 PROJECT DESIGN

We now turn to another type of problem that relies on networks for its formulation, although the use of networks is rather different. It is the problem of planning a project that has many individual stages or tasks. Some of the tasks may be performed simultaneously, whereas others have a definite precedence (that is, one must precede another). The scheduling techniques in this section have been applied to projects from house building to space trips.

12.4.1 Finding the Critical Path: A House-Building Project

A house-building project may be broken into many smaller jobs. Figure 12–22 shows a simplified **flowchart** for a complete project. We can regard the flowchart as a network in which the edges represent tasks and the vertices represent completion points for the tasks. Note that time flows from left to right through this network, specifying the precedence of the various tasks. To interpret this network note, for example, that

- each edge, or task, is labeled with a weight that represents the number of months needed to complete the task (for example, the label on edge c indicates that 2 months are required to obtain financing for the house);
- in some phases of the project, such as between points A and B, only one task can be undertaken at a time (design, in this case), so the network contains one path between A and B;
- during other phases, such as between points D and G, two or more tasks can be carried out concurrently (for example, construction can take place at the same time that paint, appliances, and carpet are ordered); and
- the edge connecting vertices C and D serves only to connect the finance task (edge c) with the rest of the network, so its weight is zero, because no time is needed for its completion.

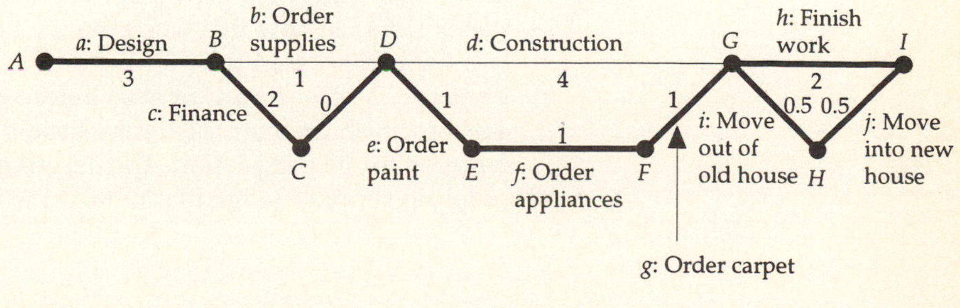

Figure 12-22. The schedule for a house-building project is represented with a network. Edges represent tasks, vertices represent completion stages, and the weights on the edges give the completion time (in months) for the tasks. The critical path, shown with bold edges, requires 11 months.

The most important question both for minimizing labor costs and minimizing stress on the homeowners is: What is the minimum time required for the entire job? Let's use the network to explore this question. In the parts of the network where only one task takes place at a time, the completion time is simply the weight (time) on its edge of the network. For example, going from point *A* to point *B* offers no choice but to allow three months for design.

In the parts of the network where two or more tasks take place simultaneously, the project cannot continue until *all* the simultaneous tasks have been completed. For example, the project cannot reach point *G* until all the tasks represented by edges *d*, *e*, *f*, and *g* are complete. When more than one task takes place simultaneously, the **limiting task** is the one that requires the most time. For example, between stages *D* and *G*, task *d* (construction) is the limiting task. As a result, we arrive at the following rule:

> *the minimum completion time for the project is the path through the network with the largest total weight.*

For this house-building project, we start at point *A*. The design phase takes us to *B* in 3 months. Financing and ordering supplies can take place concurrently, but both must be completed to reach point *D*. Financing takes longer, so it is the limiting task here; thus 2 months must be allowed between points *B* and *D*. In going from *D* to *G*, construction is the limiting task, requiring 4 months, whereas the sequential tasks of ordering appliances, paint, and carpet require a total of only 3 months. Finally, moving out of the old home and into the new home requires 1 month, whereas the concurrent task of doing the finishing work on the house requires 2 months. The minimum time required for the entire project is the total weight (time) on edges *a*, *c*, *d*, and *h*, which is 11 months. This path through the network is called the **critical path**.

Time-Out to Think: Find the total time along the path formed by edges *a*, *b*, *d*, and *h*. Explain why, although this path requires a shorter time than the critical path, it is *not* the minimum completion time for the project.

12.4.2 Critical Path Analysis

We can also ask when particular tasks can be started and completed during the project. This information is crucial for ensuring that the project is completed in the minimum possible time. To answer this question, we determine the **earliest start time** (**EST**) and **earliest finish time** (**EFT**) for each edge (task). Let's see how this works on the project shown in Figure 12–22.

The earliest start time for the design task (edge *a*), which "leaves" from vertex *A*, is zero: it cannot start before the beginning of the project! The earliest finish time for the design task is 3 months because that is the time required for its completion. The design must be completed before the task of ordering supplies (edge *b*) can begin, so the earliest start time for ordering supplies is 3 months into the project; that is, it is the earliest *finish* time of the preceding task (design). Because ordering supplies will take 1 month, and its earliest start time is 3 months, its earliest finish time is 4 months. Generalizing, two rules determine the earliest start and finish times for any task:

- the earliest start time of any task leaving a particular vertex is the *largest* of the earliest finish times of the tasks entering that vertex; and
- the earliest finish time of any task is the earliest start time of that task *plus* the time required for the task, or EFT = EST + time for task.

Task c (finance) leaves vertex B, so the earliest *start* time for task c is the earliest finish time of the tasks entering vertex B (only the design task in this case), which is 3 months. The earliest finish time for task c is 5 months because it will take 2 months to complete after its earliest start time of 3 months. The results of a complete analysis of the network are shown in Table 12–6, where the earliest start and finish times for each task are shown.

Closely related to the earliest start and finish times are the **latest start time** (**LST**) and **latest finish time** (**LFT**). The latest start time of any task is the latest time at which it can be started *without delaying* the overall project. Similarly, the latest finish time of a task is the time by which it must be completed if the entire project is to be completed in the minimum possible time. Violations of latest start or finish times represent *overruns* that mean higher cost, later finish, or crisis. Following logic similar to that used for the earliest start and finish times, we use two rules to find the *latest* start and finish times for each task:

- the latest finish time of any task *entering* a particular vertex is the *smallest* of the latest start times of the tasks *leaving* that vertex.
- the latest start time of any task is the latest finish time of that task *minus* the time required for the task, or LST = LFT – time for task.

To find the latest start and finish times, we begin at the final vertex. The latest finish time for all tasks entering the final vertex always is the critical path completion time (the minimum time) for the project. For the house-building project the critical path completion time is 11 months. Thus the latest finish times for both tasks h (finishing work) and j (moving in) are 11 months; if they finish any later, the project will fall behind schedule. We continue to apply the two rules.

- The latest start time for task h is 9 months because it requires 2 months and its latest finish time is 11 months.
- The latest start time for task j is 10.5 months, 0.5 month before its latest finish time of 11 months.
- Task i (moving out) enters vertex H, so its latest finish time is the smallest of the latest start times of the tasks *leaving* vertex H. In this case, the only task leaving vertex H is task j, which has a latest start time of 10.5 months; thus 10.5 months is the latest finish time for task i.
- The latest start time for task i is 10 months, or its latest finish time of 10.5 months minus the 0.5 month it requires for completion.

The earliest and latest finish times for the whole project also are shown in Table 12–6 on the next page.

One more quantity of interest for each task is the *difference* between its earliest and latest start times (or, equivalently, its earliest and latest finish times), which is called the **slack time**. For example, the earliest start time for task *b* (ordering supplies) is 3 months, and its latest start time is 4 months. Thus its slack time is 1 month, meaning that a delay of up to 1 month in starting this task will not affect the total time required for the project. In general, if the slack time for a task is positive, its start may be delayed from its earliest start time without harming the overall schedule. If the slack time for a task is 0, then it is *on* the critical path for the project and it must be started at its earliest start time. Table 12–6 summarizes the entire critical path analysis for this project.

Table 12-6. Critical Path Analysis for House-Building Project

Task	EST	LST	EFT	LFT	Slack (LST–EST)	Critical Path
a	0	0	3	0	0	Yes
b	3	4	4	5	1	No
c	3	3	5	5	0	Yes
d	5	5	9	9	0	Yes
e	5	6	6	7	1	No
f	6	7	7	8	1	No
g	7	8	8	9	1	No
h	9	9	11	11	0	Yes
i	9	10	9.5	10.5	1	No
j	9.5	10.5	10	11	1	No

Time-Out to Think: Study Table 12–6. What would happen if task e started 6 months into the project, instead of at its earliest start time of 5 months? Would the overall schedule be affected? What would happen if task *h* is not started until 10 months into the project? What would happen if task e took longer than expected and wasn't completed until 8 months into the project? Overall, explain how the critical path analysis can be used to ensure on-time completion of a project.

The house-building example illustrates a large class of important problems in operations research known as **scheduling problems**. The method of analysis we used here is called the **critical path method (CPM)**. In practice, scheduling problems can become both large and complicated if many levels of detail are included in the project. Consider how much more complex the critical path analysis of the house-building project would become if we broke down the tasks more realistically by separating the construction task, for example, into subtasks like laying the foundation, framing, and drywall. As with the networks we discussed earlier, even the most powerful computers cannot analyze many real scheduling problems completely. As a result, mathematicians are actively seeking better algorithms for their analysis.

12.4.3 Applications of Scheduling Problems

Scheduling problems appear in almost every application imaginable. Anytime a project must be completed, a scheduling problem is involved. Your instructor faces a scheduling problem in determining how all the necessary material can be covered in the allotted class time. You face a scheduling problem in determining how you can do your laundry, eat dinner, complete your homework, and get dressed in time for a date tonight. When more than one person is involved in a project, the scheduling problems become far more complex. For example, consider the difficulties a software company faces in producing the next generation of word processor (that might involve hundreds of systems analysts and programmers) on time or a movie studio faces in releasing its hoped-for blockbuster in time for the holiday season.

Although CPM analysis is useful, real-world problems often present additional complexities. For example, the completion times given for the tasks of the house-building network were taken to be certain. In reality, completion times rarely are certain; instead, project managers work with *estimated* completion times or maximum and minimum estimated times for completion of a task. Uncertainty is particularly evident when a project is of a type that has never before been attempted. In movie production, past experience with other movies might yield fairly good estimates of the time required for tasks such as filming, editing, adding a soundtrack, and marketing. In that case CPM analysis is extremely useful. In contrast, estimating completion times and costs for producing a new generation of computers, or for a mission to Mars, may be little better than guesswork.

The Apollo project was born in 1961, when President Kennedy called for the United States to "commit itself to achieving the goal, before this decade is out, of landing a man on the Moon and returning him safely to Earth." Before the decade was out, in July 1969, Apollo 11 carried Neil Armstrong and Buzz Aldrin to the Moon and back (along with a third astronaut, Michael Collins, who orbited the Moon but did not land).

During the 1950s, when the space program was beginning and the Cold War was intense, ambitious and uncertain projects such as developing missiles and launching rockets presented unprecedented scheduling challenges. To address these problems, mathematicians, engineers, and scientists developed a new process called the **project evaluation and review technique (PERT)**. Among its many applications, PERT often is credited for the success of the Apollo project Moon landings, the first of which occurred in July 1969.

In addition to uncertain task completion times, project management can be extremely complicated. For instance, what happens to the schedule if a particular task falls behind? How does the additional labor that must be recruited, or overtime wages that must be paid, affect the project budget? Today, large computer programs are used to answer these and similar questions through PERT-based analyses of projects ranging from the design of a space station to preparing a military operation to building a shopping center. Indeed, it would be difficult to overestimate the importance of CPM and PERT to the success of complex projects undertaken today by businesses and governments.

Although we have only touched on CPM analysis — haven't done any problems using PERT in this book — remember that both are extensions of the network ideas studied in this chapter. The long road from the Bridges of Königsberg to the Moon is paved with mathematical models of networks.

12.5* INSIDE NETWORK ANALYSIS

In this section we explore in greater detail the methods used to find Euler circuits and minimum cost spanning trees. Answers to "how to" questions must give a procedure, and the more efficient the procedure, the better. Let's look at systematic ways to find Euler circuits and minimum cost spanning trees.

12.5.1 Euler Circuits

Earlier in the chapter we stated that a network has an Euler circuit only if every vertex has *even degree* (has an even number of connecting edges). We now address a related question: Once we know that a network has an Euler circuit, how can we find it? One approach is by trial and error. A more systematic algorithm, called the **burning bridges rule**, is as follows:

> *To find an Euler circuit begin at any vertex and follow the edges in such a way that you never use an edge that is the only connection to a part of the network that has not already been covered.*

In other words, never use an edge that is the last bridge to territory that hasn't been visited. The burning bridges rule always will lead to an Euler circuit in networks that have one.

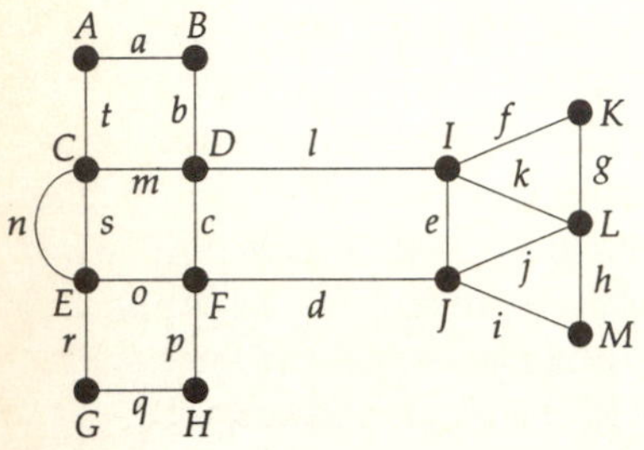

Figure 12-23. This network has an Euler circuit because all vertices have even degree. We can find an Euler circuit using the burning bridges rule.

Let's apply the burning bridges rule to the network shown in Figure 12–23. This network is of order 13. Because all the vertices have even degree, we know that an Euler circuit exists. Starting arbitrarily at vertex *A*, we begin tracing the edges in the order $a \rightarrow b \rightarrow c \rightarrow d \rightarrow e$. At most vertices we have several choices for a path, but usually the choice is unimportant. However, occasionally the choice is critical! Having arrived at vertex *I* by following edges *a* through *e*, we have a choice of turning left and using edge *l* or going down edge *f* or edge *k*. Using edge *l* violates the burning bridges rule because that edge is the only remaining link to edges *f* through *k* — edges that have not yet been covered. Therefore we must proceed along either edge *f* or *k*. Choosing edge *f* leads to the complete Euler circuit $a \rightarrow b \rightarrow c \rightarrow d \rightarrow e \rightarrow \ldots \rightarrow s \rightarrow t$ (going in alphabetical order).

> **Time-Out to Think:** Figure 12–23 contains several other Euler circuits. Can you find one?

12.5.2 Minimum Cost Spanning Trees

Let's now develop an algorithm to find the *minimum cost spanning network* for a network. Recall that the minimum cost spanning network is the collection of edges that ensures that every vertex is connected to every other vertex *at the minimum cost*. First, we need to make one preliminary observation: The minimum cost spanning network cannot have any circuits (closed loops) because that would mean that some vertices are "overly connected." If a possible spanning network has a circuit, at least one of the edges can be eliminated and the remaining network would still connect all the vertices. Since the minimum cost spanning network can have no circuits, it must be a tree network; for this reason we call it the **minimum cost spanning tree**.

The best method for finding minimum cost spanning trees, **Kruskal's algorithm**, turns out to be unexpectedly simple. Like the nearest neighbor method used for the traveling salesperson problem (subsection 12.3.4), this method is a "greedy algorithm" — a procedures that does what seems best at the moment. Although *not* usually a good strategy (not just in network theory), occasionally it does work!

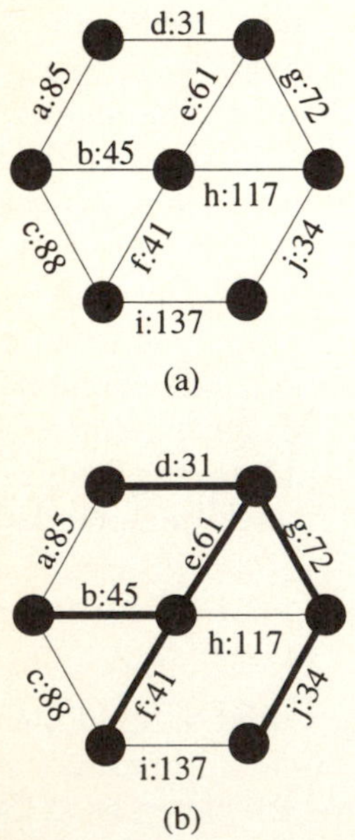

Figure 12-24. (a) The network of Sunshine County shows the distances between seven towns. Kruskal's algorithm can be used to find the minimum cost spanning tree for this network (shown in (b)).

Let's return to the Sunshine County network shown in Figure 12–24(a) and use it to describe Kruskal's algorithm. The first step is to make a list of the edges from the least expensive to the most expensive. Here is the list for the Sunshine County network.

edge *d*: weight = 31	edge *g*: weight = 72
edge *j*: weight = 34	edge *a*: weight = 85
edge *f*: weight = 41	edge *c*: weight = 88
edge *b*: weight = 45	edge *h*: weight = 117
edge *e*: weight = 61	edge *i*: weight = 137

Now, beginning with the edge with the smallest weight, we assemble the edges in order of weight. When we have assembled enough edges so that every vertex is connected, either directly or indirectly, to every other vertex, we have created the minimum cost spanning tree.

In this example, the least expensive edge (edge *d*) does not by itself connect every vertex to every other vertex. The *two* least expensive edges (edges *d* and *j*) do not by themselves connect every vertex to every other vertex. The *three* least expensive edges (edges *d*, *j*, and *f*) do not by themselves connect every vertex to every other vertex. Continuing in this manner, we will discover that edges *d*, *j*, *f*, *b*, *e*, and *g* are needed to connect every vertex, either directly or indirectly, to every other vertex. Thus, edges *d*, *j*, *f*, *b*, *e*, and *g* form the minimum cost spanning tree for the Sunshine County network as shown in Figure 12–24(b).

The only other rule that must be observed is that the final collection of edges cannot include a circuit. If it does, the most expensive edge of the circuit must be removed. In the Sunshine County example, no circuits were created by edges *d*, *j*, *f*, *b*, *e*, and *g*, so these edges do form the minimum cost spanning tree. Its length is the sum of the individual costs, or 284 miles.

Example 12–3 ***Minimum Cost Power Lines.*** Figure 12–25a shows a network of towns (vertices) and possible power-line connections (edges). The cost of building a power line (in $ millions) along each edge also is shown. How should the power lines be built to minimize the total cost?

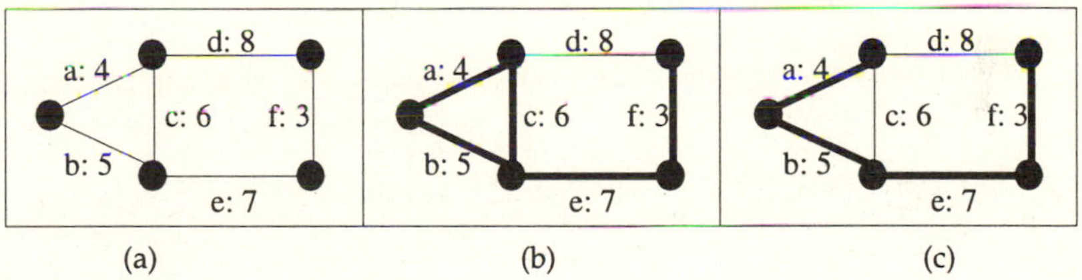

Figure 12-25. (a) The task is to obtain the minimum cost spanning tree for this network. (b) The process of adding edges in order of increasing cost until all of the vertices are connected creates a circuit. (c) Removing the edge of the circuit with the highest cost leaves the minimum cost spanning tree.

Solution: First list the edges in order of increasing cost. Then assemble the edges in order of increasing cost until all the towns are connected (that is, until every town can be reached from every other town). The ordering of edges is

> Edge *f*: weight = 3 → edge *a*: weight = 4 → edge *b*: weight = 5 → edge *c*: weight = 6 → edge *e*: weight = 7.

This step produces the network shown in (see Figure 12–25b) in which edges *abc* form a circuit that must be removed. The most expensive edge of this circuit (edge *c* with a cost of $6 million) is easily removed. Removing it doesn't alter the connections of the towns: Each town remains connected to the other towns via the spanning tree. The resulting network, shown in Figure 12–25(c), is the minimum cost spanning tree and has a length of 20, or a cost of $20 million.

❖

12.6 CONCLUSION

We devoted this chapter to the broad subject of discrete mathematics. Key lessons to be remembered from this chapter include the following:

- Amazingly, the mathematics in this chapter involves nothing more than arithmetic, yet discrete mathematics can be used to address some of the most complex problems imaginable.
- Arrangements of people and items can take many different forms, but in every case precise methods are available for counting them. These principles of counting are crucial to probability and statistics.
- Networks are powerful modeling tools that allow visualization of both everyday situations and complex problems with great ease. Even if you are unlikely to design networks in the future, those designed and used by others are certain to have a major impact on your life.

PROBLEMS

Reflection and Review

Section 12.1

1. **Discrete or Continuous?**
   a. As you wait at a bus stop, is the arrival of buses discrete or continuous?
   b. Suppose that you are skydiving. Does your altitude as you fall vary discretely or continuously?
   c. Consider the possible results of a census of the number of people in your home town. Are the possible results discrete or continuous?
   d. Identify three activities in your daily life that represent discrete processes.
   e. Identify three activities in your daily life that represent continuous processes.

Section 12.2

2. **Factorial Practice.** Use your calculator (but preferably not the factorial key) to find the value of the following quantities.

   a. $7!$ b. $\frac{7!}{3!}$ c. $\frac{7!}{4!3!}$ d. $\frac{12!}{5!}$ e. $\frac{12!}{8!(12-8)!}$ f. $\frac{21!}{20!}$

3. **Permutation and Combination Practice.** Rewrite each of the following permutation and combination expressions in terms of factorials; then find the numerical value of each expression. In each case make up a simple problem that is answered by the expression. For example, part (a) could answer the question "How many three-person committees can be formed from a group of nine people"?
   a. ${}_9C_3$ b. ${}_9P_3$ c. ${}_4P_2$ d. ${}_4C_2$ e. ${}_{12}C_8$ f. ${}_{12}P_8$
4. **Choosing from Two Groups.** Answer each of the following questions by (i) making an array; (ii) drawing a tree; and (iii) using the multiplication principle. Be sure that you get the same answer by all three methods!
   a. How many different choices of cars do you have if a particular model comes in eight different colors and three different styles (sedan, station wagon, or hatchback)?
   b. Assume that your car radio has five preset buttons, each of which can be tuned to an AM or FM station. How many different stations can you preset?
   c. The local paint store offers wallpaper in eight colors, each of which comes in four different patterns. How many different styles of wallpaper are available?
5. **Choosing from More Than Two Groups.** For each of the following questions (i) answer the question using the multiplication principle; (ii) explain how you could obtain the same answer by drawing a tree (drawing the tree isn't necessary though you may if you want).
   a. A local ski sale features eight different kinds of skis, on each of which six different bindings can be mounted. In addition, seven different brands of boots are on sale. How many different ski/boot packages could be bought on sale?
   b. Of the nine members of a university's governing board, three members are up for reelection. The first member up for reelection is running against only one other candidate, the second member is running against two other candidates, and the third is running against three other candidates. After the election, how many different governing boards are possible?
   c. Suppose that you need to take five courses next semester, one each in humanities, sociology, science, math, and music. You have a choice of four humanities courses, three sociology courses, five science courses, two math courses, and three music courses. If scheduling conflicts can be avoided, how many different five-course schedules are possible?

6. **Arrangements with Repetition.** Answer each of the following questions about arrangements of items with repetition. In each case, be sure that you understand why it is a case of arrangements with repetition.
   a. How many different four-digit house addresses can be formed with the numerals 0–9?
   b. How many different 10-note "tunes" can be created from the notes C, D, E, F, G, A, and B?
   c. Supposedly, if a monkey types randomly on a typewriter long enough, it eventually will type the works of Shakespeare. How many different 5-letter words could a monkey type from the 26 letters of the alphabet?
   d. Imagine a lock that consists of three dials, each of which can be set at the letters A, B, C, D, E, F, G, and H. How many locks of this type could be made, each with a different three-dial combination?
7. **Permutations.** Answer each of the following questions about permutations. In each case, be sure that you understand why it is a case of permutations.
   a. Only 6 of your 13 cookbooks will fit on a small shelf above your stove. How many different ways can you arrange the books if the order of the books makes a difference to you?
   b. How many different four-letter passwords can be formed from the five vowels if no repetition of letters is allowed?
   c. Ten finalists in a talent show must give their final performance. How many different ways can their appearances be scheduled?
   d. Ten finalists in a two-day talent show must give their final performance. Five contestants will perform on the first night of the show, and five will perform on the second night. How many different ways can the schedule for the first night be made?
8. **Combinations.** Answer each of the following questions about combinations. In each case, be sure that you understand why it is a case of combinations.
   a. How many different 4-person subcommittees can be formed from a group of 12 people?
   b. How many different 5-card hands can be dealt from a deck that has only face cards and aces (16 cards altogether)?
   c. Your compact disk player can hold 5 disks at one time. If you own 9 compact disks, how many different ways can you place them in the compact disk player? The order of the 5 disks in the player is not important.
   d. The coach of the debate club must choose a team of 4 people to travel to the next meet. How many different teams can be chosen from the 12 members of the club?
9. **Lessons in the Ice Cream Shop.** Josh and John's Ice Cream shop offers 12 different flavors of ice cream and 6 different toppings. Answer the following questions by *first* determining the type of arrangement required: selections from more than one collection, arrangements with repetitions, permutations, or combinations.
   a. How many different sundaes can you create using one ice cream flavor and one topping?
   b. How many different triple cones can you create from the 12 flavors if the same flavor may be used more than once?
   c. Using the 12 flavors, how many different triple cones can you create with 3 *different* flavors if you specify which flavor goes on the bottom, middle, and top?
   d. Using the 12 flavors, how many different triple cones can you create with 3 *different* flavors if you don't care about the order of the flavors on the cone?
   e. How do the answers for parts (a)–(d) change if Josh and John's carries 15 flavors?
10. **Manager's Job.** On a baseball team with 25 players, a manager must choose a batting order having only 9 players. How many different 9-player batting orders are possible? Suppose that the manger tried a different batting order every day (with no breaks for rest days or the off-season). How long (in years) would it take to try every possible batting order?
11. **License Plates.** Suppose that license plates are made with three letters followed by three numbers. How many different license plates are possible? How many license plates are possible if they just consist of any six characters (either letters or numbers)? How many license plates are possible if they consist of six characters, none of which can be used twice on a plate?
12. **Making a Toast.** If eight people make a complete toast (each pair of people touches glasses), how many different chimes of glass should you hear? Are you counting permutations or combinations?
13. **Choosing Finalists.** Ten students are deadlocked after the regular competition of the South Dakota Super Spelling Bee. The winners will be determined by a drawing. How many different ways can the judges award the top three places?
14. **Standing Room Only.** Sixty people crowd onto a bus that has seats for only 48 people. How many different groups of people could end up with a seat?
15. **House Numbers.** The houses on the west side of the 800 block of Sierra Drive can have any odd three-digit house number that begins with 8. How many different house numbers are possible?
16. **Seating Arrangements.** Five women and four men must be seated on a platform at a graduation ceremony. How many different seating arrangements are possible? How many different seating arrangements are possible if the men and women must sit in alternate seats?
17. **Passwords of Symmetric Letters.** Each of the 11 letters A, H, I, M, O, T, U, V, W, X, Y appears the same when it is flipped right to left (or looked at in a mirror). They are the symmetric letters. How many six-letter computer passwords can be formed using only the symmetric letters of the alphabet?
18. **Committees.** How many different ways are there to choose the 18 members of the Senate Foreign Relations Committee from the 100 members of the U.S. Senate? (If you don't want to calculate the actual number of committees; at least give a formula for obtaining the number.)

Section 12.3

19. Practical Uses of Euler Circuits. The mail carrier example in subsection 12.3.1 illustrates a practical use of Euler circuits: Mail carriers *do* park their trucks in one place and seek efficient delivery routes that return them to their trucks. Describe, in words, at least three other practical situations in which Euler circuits are useful.

20. Practical Applications of the Traveling Salesman Problem. Subsection 12.3.4 gives several practical applications of the traveling salesman problem. Describe, in words, at least three other practical situations in which the traveling salesman problem might arise.

21. Euler Circuits. The town of Sleepy Waters is entwined around the Tranquility River with eight bridges and three islands as shown.

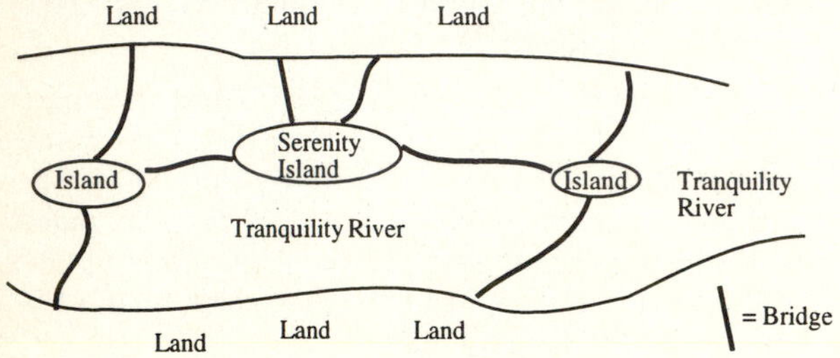

a. How many vertices are needed to draw a network representing the town? How many edges are needed?
b. Draw a network that represents the town.
c. Can someone begin at the pub on Serenity Island and walk in a closed loop that crosses each bridge exactly once and returns to the pub? In other words, does the network have an Euler circuit?

22. Network Models. Describe at least three examples from your own experience that could be modeled with networks. In each case, explain what the vertices and edges in the network would represent.

23. Drawing Networks. Draw networks having each of the following properties.
a. A network of order 6 whose vertices all have an odd degree
b. A tree of order 7 with at least one vertex with an even degree
c. A network of order 5 with no vertices having an odd degree
d. A network of order 6 in which all vertices have degree 2
e. A network of order 8 in which all vertices have degree 3

24. Neighboring States. Which states have the most neighbors? One is Tennessee. Find a map of Tennessee and its immediate neighbors. Letting vertices represent states and edges represent the relationship "is a neighbor of," draw a network showing all connections among Tennessee and its neighbors. What is the order of the network? What is the degree of the vertex corresponding to Tennessee? What other state has as many neighbors as Tennessee?

25. Layout of an Art Gallery. The floor plan of a small art gallery is shown. Draw a network that has each of the gallery rooms as vertices and the connections between each room and its neighboring rooms.

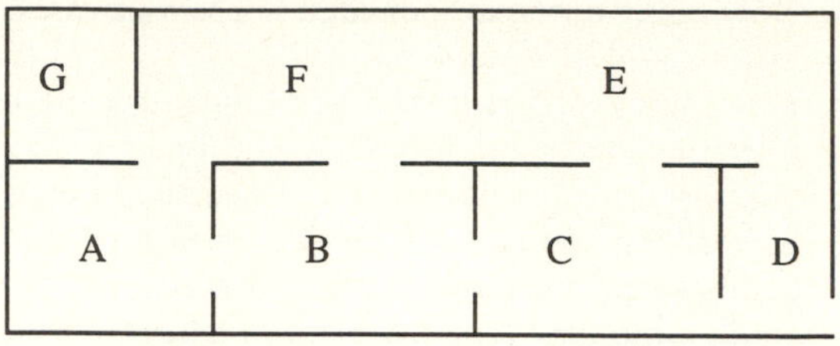

26. Friendships and International Trade. Among the members of a fourth grade class, Amy trades baseball cards with Beth, Cate, and Daniel; and Beth trades with Daniel and Cate. Draw a network that represents these trading relations. Edges in the network represent the relationship "trades with." Based on this example, explain how networks might also be used to describe international trade relations.

27. Networks Terminology. For each of the six networks shown in the accompanying figure, answer the following questions.
a. What is the order of the network?
b. What is the degree of each vertex of the network?
c. Does the network have a special form (for example, a tree, a cycle, or a complete network)?
d. Does the network have an Euler circuit?

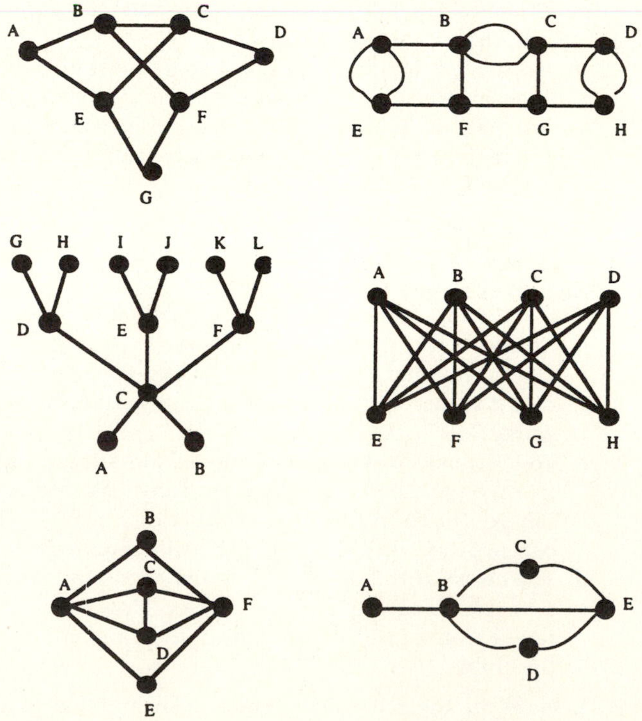

28. Number of Odd Vertices. Draw several different networks to convince yourself that every network has an even number of vertices with an odd degree. Can you use this fact to show that the number of people on Earth who have shaken an odd number of hands must be even?

29. Cycle Networks. Recall that a cycle is a network in which every vertex is connected to two other vertices to form a closed ring. Draw a cycle of order 5. How many edges does it have? How many edges does a cycle of order 6 have? Does a cycle with n vertices have $n - 1$ edges where n is any positive integer?

30. Complete Networks. Draw complete networks of order 5 and order 6. How many edges do they have? Does a complete network with n vertices have $n \times (n - 1)/2$ edges, where n is any positive integer? Explain.

31. Telephone Calls in Sunshine County. The telephone company's cable installation problem for Sunshine County (subsection 12.3.3) requires a minimum cost spanning tree solution, which is shown in Figure 12–13. In actually routing a call, this network does *not* necessarily provide the most efficient path between towns. For example, Figure 12–13(c) shows that a call from town A to town B would be routed through towns C and D, rather than taking the direct path from A to B. In words, explain why providing the most efficient routes for calls between towns is *not* important to the telephone company in this particular problem.

32. Practice with Spanning Networks. Consider network (a) in the following figure and the three spanning trees shown in (b), (c), and (d) of the figure.

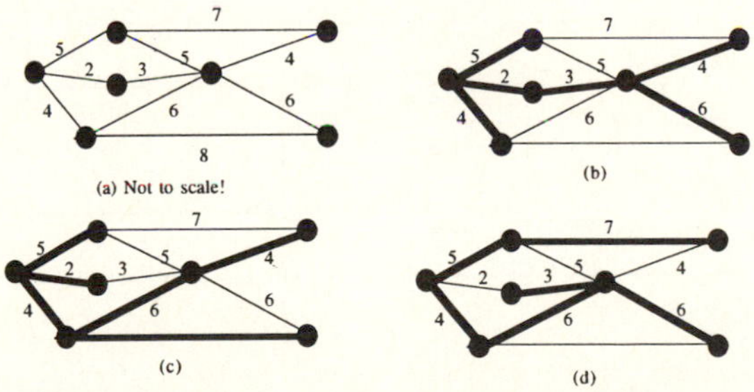

a. Find the cost of each of the three spanning trees.
b. Which of the three spanning trees has the minimum cost?

33. Practice with Hamiltonian circuits. Consider the networks shown. In each case, follow the edges according to the arrows, and determine whether that path forms a Hamiltonian circuit. Justify your reasoning and explain your answer carefully. If a path does not form a Hamiltonian circuit, try to find a different path that does.

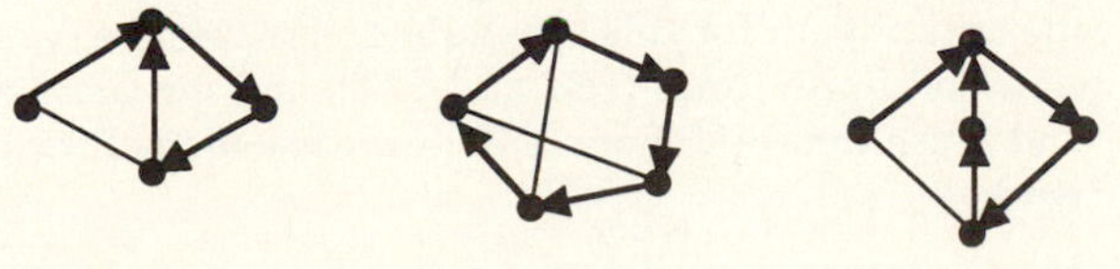

(a) (b) (c)

34. Traveling the National Parks. Suppose that you are taking a trip to Bryce, Canyonlands, Capitol Reef, and Grand Canyon National Parks. Use the data Table 12–3 to answer the following questions.

a. Draw a network that shows the four national parks and the distances between them.
b. What is the total length of the trip that visits the parks in the order of Bryce, Canyonlands, Capitol Reef, and Grand Canyon?
c. Use the nearest neighbor method and start at Grand Canyon to find a circuit that passes through all the parks and returns to Grand Canyon. Describe your work.
d. Use the nearest neighbor method to find a circuit that passes through all the parks, starting and ending at Bryce. Describe your work.
e. Is either of the circuits in parts (c) or (d) the shortest possible circuit? Explain.

35. A Traveling Salesperson Problem. The network in the following figure shows the flight times (in hours) between cities that a salesperson must visit. The goal is to find the shortest circuit that allows a visit to each city exactly once, beginning and ending in the same city. Start at various cities and use the nearest neighbor method to find a possible shortest circuit (at each step go to the nearest city that hasn't already been visited). Do you get different circuits for different starting cities? Can you conclude which is *the* shortest circuit that passes through each city exactly once? If so, name it.

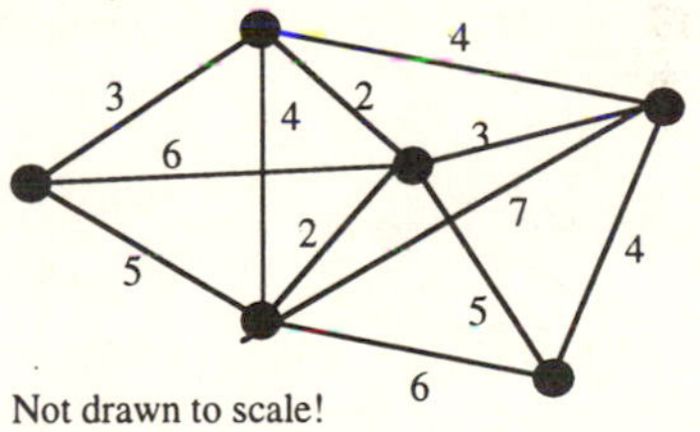

36. Five-City Vacation. A section of a road atlas looks like the following figure showing distances between various pairs of the cities St. Louis, Chicago, Kansas City, Memphis, and Louisville. For the connections shown find the shortest circuit that passes through each city exactly once and starts and finishes in the same city. Some connections between cities are missing. Is a shorter circuit possible if all connections between the cities were available?

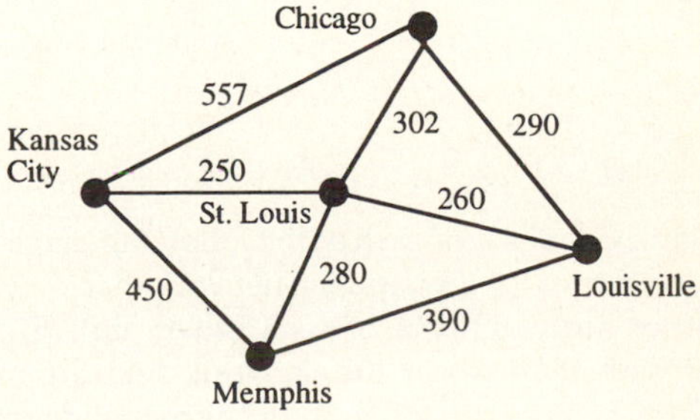

37. Car Shuttles. Each week Abe, Barbara, Carl, or Dolores hosts a bridge game at his or her house. Everyone is eager to save gasoline, so the host always picks up and drops off the other three players. The location of each player's house and the distances between them are shown. What path should Abe use to pick up his guests to minimize the distance traveled? Would Barbara, Carl, or Delores use a different route? If so, name it.

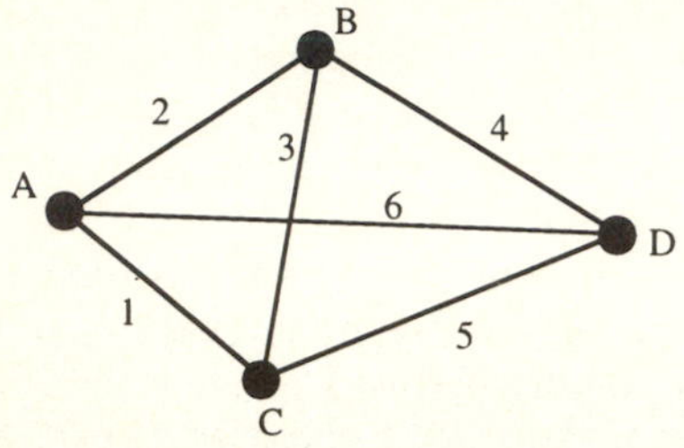

38. Taxi Routing. A taxi driver based at a train station (the black square) must deliver five riders to five different hotels. The locations of the hotels are shown and the travel times (in minutes) are indicated on the edges. If no time is spent waiting at the hotels and the taxi must return to the train station, what route minimizes the total travel time?

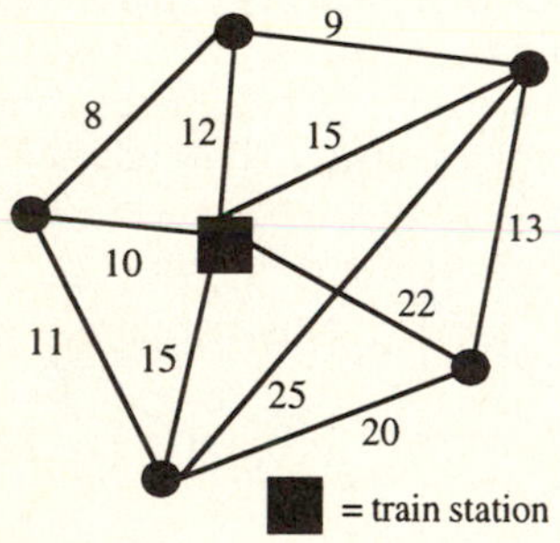

39. A Utilities Problem. Three houses must be connected to a single water main and a single gas main as shown in part (a) of the following figure. Can the remaining gas connections be made without any connecting lines crossing? If *three* houses must be connected to water, gas, and electric mains, as shown in part (b), can the connections be made without any lines crossing? Explain.

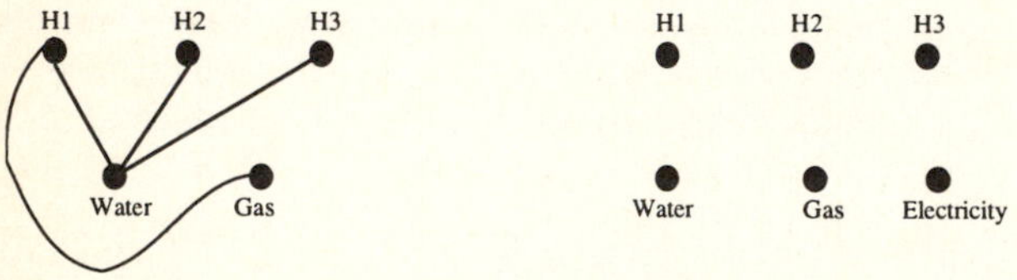

(a) (b)

40. Coloring Graphs. For each of the following maps of imaginary countries: (i) First draw a network that shows which countries are neighbors of each other; and (ii) working either with the map or the network, find the minimum number of colors needed to color the map (adjacent countries cannot have the same color).

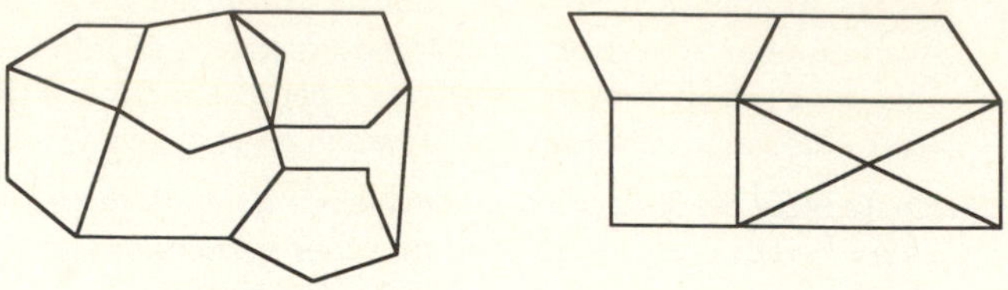

(i) (ii)

41. The Amiable Ambassadors. After signing a peace treaty, five ambassadors decide to go out to dinner. Unfortunately certain pairs of ambassador still cannot bring themselves to ride together in the same car. The pairs of ambassadors that can't ride together are (a, b), (a, c), (b, d), (c, e) and (d, e). First draw the network that shows the excluded pairs of ambassadors (letting each vertex of the network represent an ambassador). Then color the vertices of the network to find the fewest cars needed to take the ambassadors to dinner. Give one possible assignment of ambassadors to cars.

42. A Coloring Network for Kids and Clubs. The following table shows the activities of five kids. Assume that each activity meets once a week and no child can attend more than one activity in a day. Imagine also that you are a devoted parent of these kids. What is the least number of days (if you are lucky) on which you will be taking kids to activities? Recall that a network can be made in which each vertex represents a different activity.

	Ann	Bob	Cid	Don	Ed
G Scouts	x		x		
B Scouts		x		x	x
Choir	x				x
Soccer		x	x		
Dance			x	x	
B ball					

Section 12.4

43. Critical Path Analysis. Continue the analysis of the network for the house-building project, as described in the text, to confirm all of the EST, EFT, LST, and LFT values shown in Figure 12–23. Explain your work clearly.

44. Scheduling a Paint Job. A large room is scheduled for painting according to the schedule shown in the figure on p. 427. The estimated time for the completion of each task is shown on the corresponding edge of the network. Find the critical path for this scheduling network, the earliest start and finish times (EST and EFT), and the latest start and finish times (LST and LFT) for each edge of the network.

45. Mars Mission. The idea of sending humans to Mars has been discussed seriously ever since the Apollo Moon landings. At present, the most likely scenario for a Mars mission involves a multinational project with a journey of

Figure for Problem 44

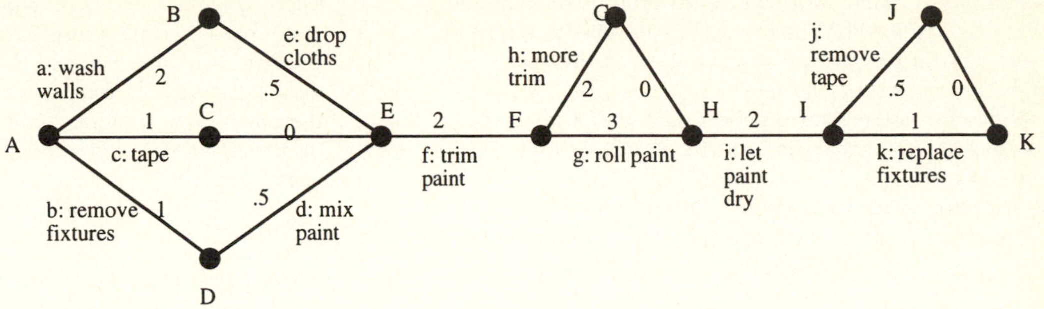

several months in each direction and a stay of a year or more on the surface by 8 to 12 astronauts. Discuss some of the tasks that should appear in a CPM or PERT analysis of such a mission, as well as the uncertainties in both the completion times and budgets for those tasks.

Section 12.5

46. Burning Bridges Rule. For each of the networks shown in the figure for problem 27, determine whether an Euler circuit exists. In those cases in which an Euler circuit does exist, apply the burning bridges rule to find an Euler circuit. In each case explain the steps of the method and note the vertices at which critical decisions must be made.

47. Practice with Kruskal's Algorithm. Use Kruskal's algorithm to find the minimum cost spanning tree of the networks shown.

a.

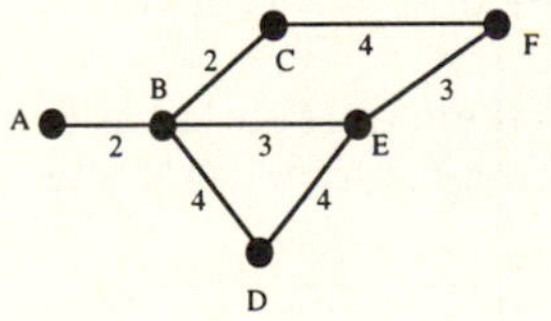

b.

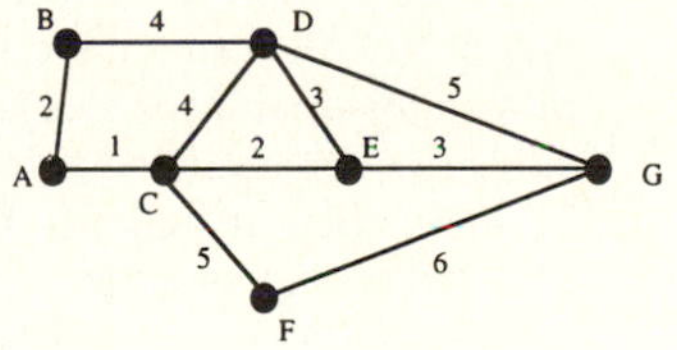

Further Topics and Applications

Combinatorics Problems

48. Lotteries. Imagine a game of Lotto in which you choose six numbers between 1 and 42. Each week lottery officials announce the six winning numbers. If all six of your numbers match the six winning numbers (in any order), you win big bucks.

a. How many different sets of six numbers are possible?

b. What are your chances of choosing the six winning numbers?

c. Are your chances of winning any better with the numbers 16, 22, 3, 40, 33, 11 than with the numbers 1, 2, 3, 4, 5, 6? Is any set of six numbers more likely than any other? Why or why not?

49. Telephone Numbers. A typical phone number in the United States consists of a three-digit area code followed by a three-digit exchange followed by a four-digit number. How many different phone numbers can be formed within a single area code, if all numerals 0–9 can be used? (Be sure to take note of whether repetitions are allowed and whether order matters.) Can a city of 2 million people be served by a single area code? How many exchanges are needed to serve a city of 80,000 people? Comment on the reality of these estimates, in light of the fact that certain numerals are not used as the first digit of an area code or an exchange.

50. Pizza Hype. Luigi's Pizza Parlor advertises 84 different three-topping pizzas. How many individual toppings does Luigi actually use? Ramona's Pizzeria advertises 45 different two-topping pizzas. How many toppings does Ramona actually use? (Hint: in these problems, you are given the total number of combinations, and you must find the number of toppings that are used.)

51. ZIP Codes. The U.S. Postal Service uses both five-digit and nine-digit ZIP codes.

a. How many five-digit ZIP codes are available to the U.S. Postal Service?

b. For a U.S. population of 250 million people, what is the average number of people per ZIP code, if all possible ZIP codes are used? Are all possible ZIP codes being used? Explain.

c. In some parts of the country nine-digit ZIP codes are used. By what factor is the number of possible ZIP codes increased in changing from five-digit to nine-digit ZIP codes?

52. Radio Call Letters. Call letters for American radio stations generally begin with W east of the Mississippi River and with K west of the Mississippi River. How many four-letter call letters can be formed west of the Mississippi River if letters can be repeated?

53. Pascal's Triangle. For each $n = 1, 2, 3, \ldots,$ make a row with the numbers ${}_nC_r$ for $r = 0, 1, 2, \ldots, n$. The following result is an array of numbers known as Pascal's triangle.

$${}_0C_0 = 1$$

$${}_1C_0 = 1 \quad {}_1C_1 = 1$$

$${}_2C_0 = 1 \quad {}_2C_1 = 2 \quad {}_2C_2 = 1$$

$${}_3C_0 = 1 \quad {}_3C_1 = 3 \quad {}_3C_2 = 3 \quad {}_3C_3 = 1$$

$${}_4C_0 = 1 \quad {}_4C_1 = 4 \quad {}_4C_2 = 6 \quad {}_4C_3 = 4 \quad {}_4C_4 = 1$$

$${}_5C_0 = 1 \quad {}_5C_1 = 5 \quad {}_5C_2 = 10 \quad {}_5C_3 = 10 \quad {}_5C_4 = 5 \quad {}_5C_5 = 1$$

The numbers in the nth row give the number of combinations of n items taken r at a time. Any number in the table is the sum of the two nearest numbers in the preceding row.

a. From the table read off the number of three-person subcommittees that can be formed from a group of five people.

b. Write the next row of the table (corresponding to $n = 6$).

c. How many three-person committees can be formed from a group of six people?

d. How many committees with one, two, three, or four members can be formed from four people?

e. How many committees with one, two, three, four, or five members can be formed from five people?

f. Can you generalize? Explain why $2^n - 1$ committees of *any* size can be formed from a group of n people.

54. More Card Hands. From a standard deck of 52 cards,

a. how many ways are there to be dealt five hearts?

b. how many ways are there to be dealt three jacks?

c. how many ways are there to be dealt two kings?

d. how many ways are there to be dealt *any* three like cards (for example, three aces, or three 2s)?

e. how many ways are there to be dealt three 10s and two aces?

55. Tournaments. At the state basketball tournament 16 teams are vying for the championship. The next round will consist of eight games to eliminate half of the teams.

a. How many different pairs of teams are possible?

b. How many different ways are there to schedule the next eight games of the tournament?

c. Consider the next round of the tournament when only eight teams (and four games) remain. How many different pairs of teams are possible? How many different ways are there to schedule these four games of the tournament?

d. Consider the next round of the tournament when only four teams (and two games) remain. How many different pairs of teams are possible? How many different ways are there to schedule these two games of the tournament?

56. Forming a Soccer Team. Assume that a soccer team consists of three front-line positions, four midfield positions, three defensive positions, and a goalie. On the entire team, a coach has five players that can play on the front line, seven that play at midfield, five that play at the defensive positions, and two goalies. How many different lineups can be formed? (Hint: work this problem in stages. First, determine how many ways there are to choose players for each of the four positions.)

57. Shuffling Cards. Each of the 52 cards in a deck is distinct (different face value or suit).

a. How many arrangements of the 52 cards in a deck are possible? That is, if you shuffle the cards, what is the probability of getting a *specific* arrangement? (State whether order matters.)

b. Suppose that, through the magic of modern medicine, you become immortal. Wondering what to do with your time, you decide to arrange a deck of cards in as many different ways as possible. You get pretty good at it, so that you can try one arrangement every minute. Further, you take miracle drugs that alleviate your need for sleep and food. If you continue to rearrange the cards every minute, could you try all of the possible arrangements before the Sun dies in about 5 billion years? Explain.

Network Problems

58. Business Applications. Consult the business section of a newspaper and identify at least three stories in which network analysis or the critical path method has been used *or could be used*.

59. Family Trees. Draw a network of your family tree that includes all your siblings, your parents and their siblings, and your parents' parents. What kind of network have you created?

60. DNA Computers. Suppose that, in the future, computers based on DNA can check 1 trillion circuits of a network per second (about a billion times faster than today's computers). How long would such a computer need to find a traveling salesperson circuit (by checking all possible circuits) in a network with 100 vertices? Is this approach practical? Explain.

61. Soccer Tournament. In the three weeks that remain in the season, the five teams of the Skyline Soccer League (A, B, C, D, and E) must play the games indicated in the following table. Draw a network that shows the teams and their opponents for the remaining games of the season. What is the order of the network you have drawn? What is the degree of each vertex of the network? Is the network complete? Not all teams have the same number of games to play in the final three weeks. If there were six teams in the league with three weeks remaining, would all the teams be able to play three games?

—	A	B	C	D	E
A	—	X	X		X
B	X	—		X	
C	X		—	X	X
D		X	X	—	X
E	X		X	X	—

62. Reading Meters. A gas meter reader must visit all the houses on the streets of the map shown. Letting edges represent the streets and vertices represent the intersections, sketch a network for the street plan. Can you find a circuit that traverses each street exactly once and returns to the same starting point? If so, show the circuit on your network.

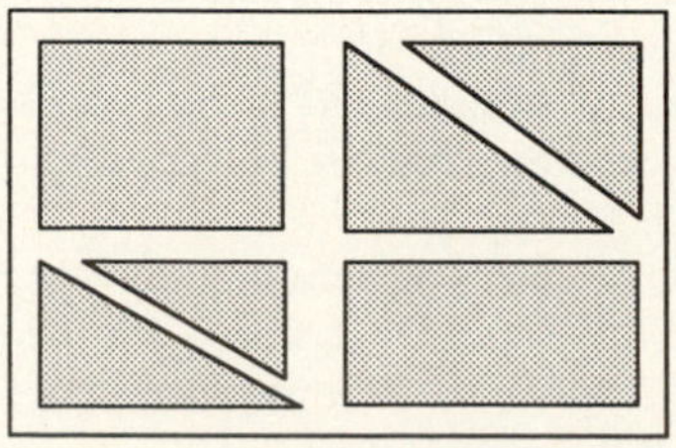

63. Another Soccer League. As a wise coach who knows some network theory, you have drawn the following network to show the games remaining in the final four weeks of the soccer league. Some of the players on your team don't understand how to read the network. Convert the network to a table to show all of the games.

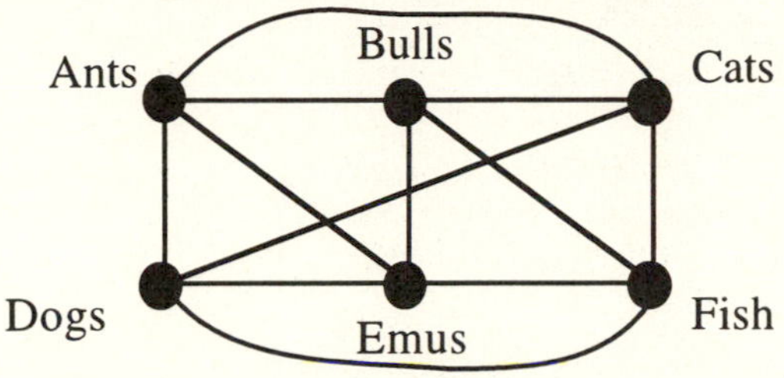

64. City Streets. Consider the map of streets and intersections shown.

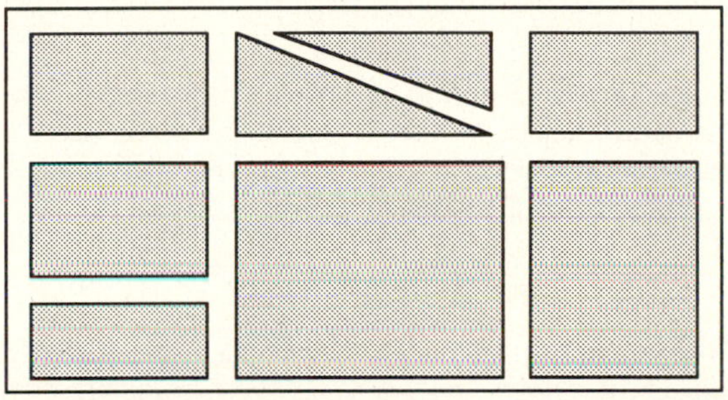

a. Let edges represent the streets and let vertices represent the intersections and draw the network that results.

b. Now use the same city map and assume that each street has a sidewalk on both sides of it. Draw the network that results when edges are used to represent sidewalks.

65. Plowing Streets. Can a snowplow clear the streets of the town shown, starting at Garage 1 and finishing at Garage 2, without driving over a lane that has not already been plowed? Assume that the plow can drive in either direction on any street but that each street has two lanes that must be plowed. Can you find a circuit that begins and finishes at Garage 1? Explain your answers.

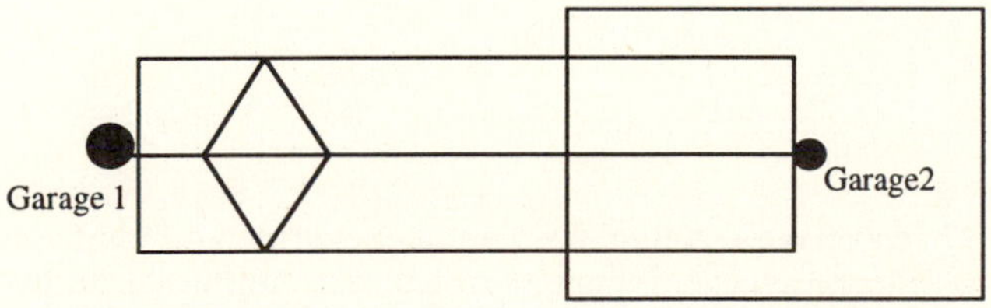

66. Checking Parking Meters. An attendant must make hourly inspections of parking meters along the sidewalks (shaded strips) on the following map. Convert the map to a network in which edges represent sidewalks and vertices represent intersections. Can you find a circuit beginning and ending at City Hall that covers each sidewalk exactly once? If so, show it on your network.

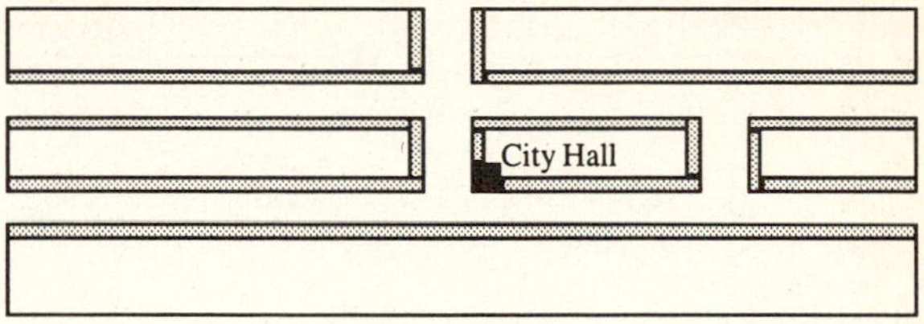

67. Community Planning. A mountain community is being planned and provision must be made for water. The following map shows the building sites (the vertices of the network) and the feasible routes for water pipelines (the edges of the network). The weights on the vertices give the distances between sites in kilometers. A single well can supply all the houses and be drilled on any of the housing sites. Find the least amount of pipe that must be used to connect all the housing sites.

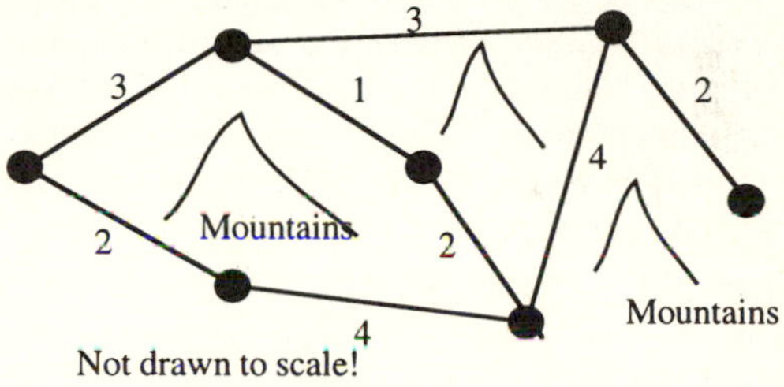

68. Local Area Networks. If every workstation in a computer network is connected to at least one other workstation, every user can communicate with every other user. Consider the layout shown, in which vertices represent workstations and edges represent feasible connections. The weight on each edge gives the cost of the wire needed for that link. What is the minimum cost need to achieve full connectivity in the network?

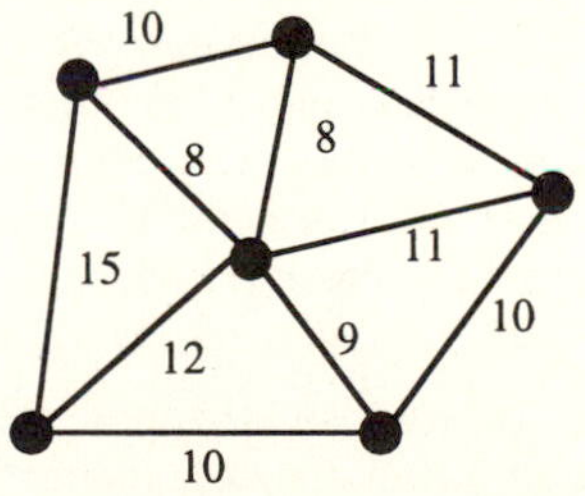

69. Building a Hotel. The owners of a proposed new hotel have developed the scheduling network (shown in the figure on p. 430) for the project. Completion times, in months, for each task are shown on the edges. What is the total time for completion of the project? What are the earliest/latest start/finish times for each stage of the project? Which stages are critical and must be completed as scheduled to avoid delays in the project? Which stages can be delayed if necessary and for how long?

Figure for Problem 69

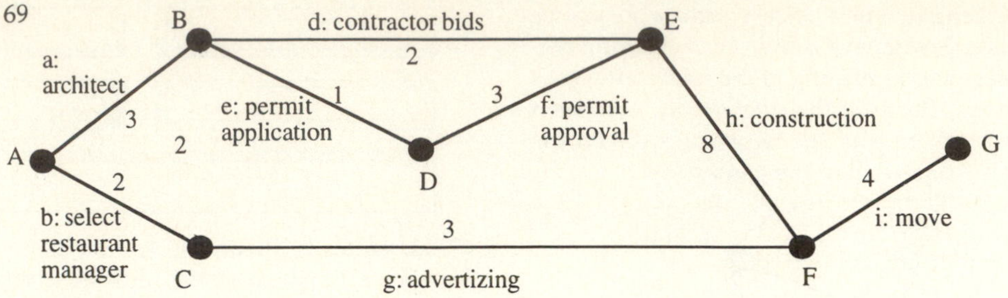

70. **Designing Circuit Boards.** The design of a circuit board calls for each of components *a, b,* and *c* to be connected to each of components A and B with wires. How many wires are needed? Can the circuit board be made in such a way that no wires cross?

71. **Equal Playing Time.** Imagine that you are the coach of a kids soccer team with 12 players. At any particular time 8 players are on the field and a regulation game consists of two 20-minute halves. Can the players be scheduled so that they all have exactly equal playing time? If not, make a schedule that gives everyone nearly equal playing time.

72. **Two Procedure Problems.** The following two problems involve the idea of finding the best way to carry out a task. Explain your thinking and solutions carefully.
    a. **The fox, goose, and mouse problem.** A traveler arrives at a river with his companions: a fox, a goose, and a mouse. The only way to cross the river is a small boat that holds at most the traveler and one animal. At no time can the fox and the goose or the goose and the mouse be left alone on either shore. How can the traveler get himself and all three animals across the river (alive). What is the minimum number of trips needed?
    b. **The cannibals and missionaries problem.** Three cannibals and three missionaries arrive at the bank of a river and want to cross to the other side. They have a boat that holds at most two people. At no time can the cannibals outnumber the missionaries on either bank. How can all six people get to the other side? What is the minimum number of trips needed?

73. **Making Shelves.** You need to make 10 book shelves with lengths of 3, 3, 3, 4, 4, 6, 6, 6, 8, and 8 feet. The lumber yard has a sale on 12-foot boards that cost \$5 dollars each. Assuming that you can cut a new board anywhere along its length, what is the least amount that you need to spend for the shelves? Is there a less expensive solution if you can buy only 10-foot boards for \$4 each?

74. **Seating Guests.** You are planning a large reception and want to seat guests at separated rectangular tables that hold eight people each. If the guests come in groups of 8, 7, 6, 6, 5, 5, 5, 4, 3, 3, 2, 2, and 1, and want to be seated with their groups, how many tables (at a minimum) are needed?

75. **Optimal Location.** A company has three branch offices and needs to build a new office for its headquarters. The locations of the branch offices are shown with sets of coordinates (*a, b*) that give the distance east and north of the crossroads.

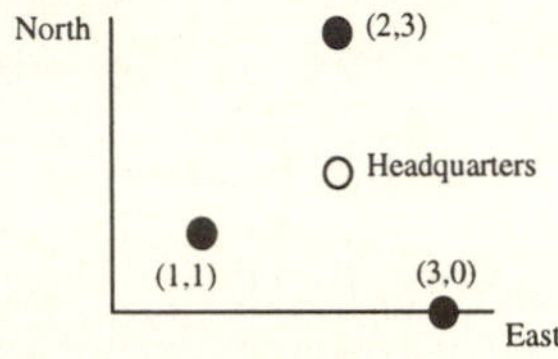

Neglecting the location of roads, where would you locate the headquarters so that it is as close to the three branch offices as possible? Recall that the distance between two points with coordinates (*a, b*) and (*c, d*) is

$$d = \sqrt{(a-c)^2 + (b-d)^2} \; .$$

Proceed by trial and error to find the approximate coordinates of the headquarters.

76. **Distribution of Rental Cars.** The following figure shows the four distribution sites for a rental car company. At one particular time two sites have an excess of cars and two sites have a shortage of cars, as shown. Each of the sites with an oversupply is connected to each of the sites with an undersupply by a route that has a certain cost per car (shown as a weight on the edge). Find the best (least expensive) scheme for redistributing the rental cars.

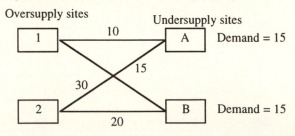

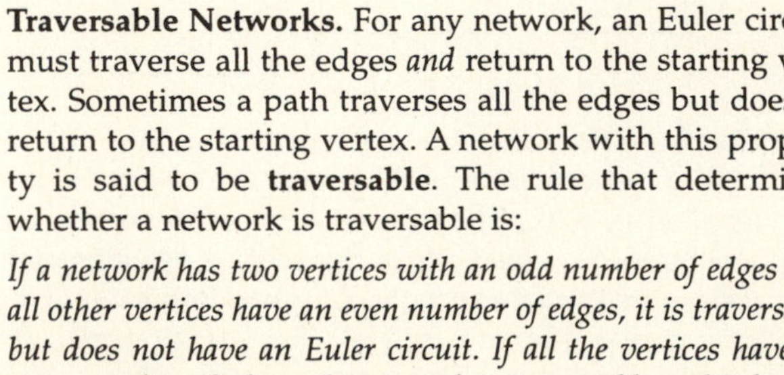

77. **Traversable Networks.** For any network, an Euler circuit must traverse all the edges *and* return to the starting vertex. Sometimes a path traverses all the edges but doesn't return to the starting vertex. A network with this property is said to be **traversable**. The rule that determines whether a network is traversable is:

    If a network has two vertices with an odd number of edges and all other vertices have an even number of edges, it is traversable but does not have an Euler circuit. If all the vertices have an even number of edges, the network is traversable and it has an Euler circuit.

    a. Explain why a traversable network can have two vertices with an odd degree.

b. Consider the networks shown in Figure 12–26 (Problem **27**). Determine whether the networks are traversable and show a traversing path.

Projects

78. Analyzing the Lottery. In an example in Section 12.2, you discovered that the chances of buying *one* ticket and winning a lottery with six lucky numbers is approximately 1 in 8 million. How does the lottery in your state (or a neighboring state) work? First, if the lottery is not a six-number lottery, compute the theoretical chances of winning the lottery. Then call or visit the state lottery office and collect some data (this information should be part of the public record). Determine the approximate number of grand prize winners and the total number of lottery tickets that are sold for each grand prize winner. Is the ratio of winners to tickets less than, greater than, or approximately equal to the theoretical chances of winning? Can you explain the differences? How are prizes paid out (for example, by a single check or over an extended period of time)? After the grand prize is paid to the winner, what is the typical amount of money that the state retains for its own uses?

79. Network Theory and Public Works. Call or visit your local public works department, which provides services such as street sweeping and snow removal. Determine whether the department has staff people who do network analysis to optimize the department's operations. If so, report on some of the network analysis techniques that are being used. If not, are there any vehicle routing problems that you could solve using the techniques of this chapter? If so, describe them.

80. Scheduling and Construction. Locate the site of a residential or commercial construction project in your area and try to meet a contractor or supervisor. Is there a master plan for the construction project that has deadlines for individual stages of the project? Has the construction company used CPM (critical path analysis) to develop a schedule for the project? Report on the techniques that are being used and comment on how you could improve the scheduling of the project with what you have learned in this chapter.

81. Searching a List. Every time you call directory assistance or a customer service department or use a dictionary on a word processor, a computer must search huge lists of names or numbers.

For example, assume that you have an *alphabetized* list of 32 names. To find a specific name you could start at the top of the list and search. If the name is "Addison" it will be near the top, and you should find it quickly. But, if the name is "Wesley," it is near the bottom of the list and it will take a while to find. This strategy, called a **sequential search**, is inefficient. The sequential search can be improved by a "divide and conquer" strategy. If you go to the middle of the list, checking one name will determine whether the name you're looking for is in the first or last half of the list. This step reduces the number of names to search from 32 to 16. You can now use the half of the list that contains your name and repeat the process. Go to the middle of the list of 16 names and determine whether the name is in the first or last half of that list. This step will reduce the number of names to be checked to 8. Repeating this process successively halves the length of the list to be checked ($32 \to 16 \to 8 \to 4 \to 2 \to 1$) until only one name remains — the one you seek. In this case, a total of five checks is required. This strategy is called a **binary search**. In general, if you have a list with 2^n ordered items, you can find any item in n searches. The savings in using a binary search instead of a sequential search can be astonishing. For example, at least $2^{14} = 16{,}384$ words begin with the letter *a* in a large dictionary. A sequential search could take as many as 16,384 checks to find a particular word (if your word happened to be *azygous*). A binary search is guaranteed to find the word in a mere 14 searches!

There about 300 words beginning with the letter *y* in an average dictionary. How many words must be checked (at worst) to find a particular word with a sequential search? Can you describe a method for finding one word in the list that requires checking at most nine words? How many words need to be checked (at most) to find one of the approximately 3900 words that begin with the letter *r* in an average dictionary?

PART 4

EXPLORING MORE APPLICATIONS

13 LIVING WITH THE ODDS

Each year the number of people who win the lottery is roughly the same as the number of people struck by lightning. Yet the average American spends several hundred dollars per year on state lotteries. Even if you don't gamble on lotteries or hike in thunderstorms, probability is a principal factor in your life. Whether considering which college is most likely to suit your needs or how much money society should spend to avoid environmental damage from the greenhouse effect, probability is involved. We dealt informally with probability earlier in the contexts of inductive reasoning and estimating uncertainties, but in this chapter we take a more formal approach to probability.

CHAPTER 13 PREVIEW

13.1 ***PROBABILITY IN LIFE.*** We begin the chapter with a brief demonstration that everyone is affected by probability.

13.2 ***FUNDAMENTALS OF PROBABILITY.*** In this section we begin the formal study of probability with a bit of history by which we explain the fundamental techniques for determining probabilities.

13.3 ***COMBINING PROBABILITIES.*** This section shows you how to calculate the combined probability of two or more events for which you know individual probabilities.

13.4 ***THE LAW OF AVERAGES.*** If the odds of something are 1 in a million, we expect it to happen to 200–300 people in the United States. This example demonstrates the law of averages, which we discuss in this section.

13.5 ***PROBABILITY AND COINCIDENCE.*** Chances are better than 50–50 that 2 people in a group of 25 have the same birthday. This surprising fact shows that coincidences are more likely than we usually guess. In this section we examine probability and coincidence and their associated psychology.

13.6 ***RISK ANALYSIS AND DECISION MAKING.*** In this section we consider the role of probability in analyzing risks and making relevant decisions.

13.7 ***THE MATHEMATICS OF SOCIAL CHOICE.*** Once you know the odds, how do you go about making social choices? Surprisingly, mathematics offers some answers. In this section we briefly discuss voting systems and the prisoner's dilemma.

*13.8** ***COMPETITIVE OR ZERO-SUM GAMES.*** In this section we demonstrate the mathematics of zero-sum games, a further application of game theory.

The principal means for ascertaining truth — induction and analogy — are based on probabilities; so that the entire system of human knowledge is connected with the theory of probability.

— Pierre Simon, Marquis de Laplace (1819)

Anybody can win unless there happens to be a second entry.

— George Ade (1844–1866)

13.1 PROBABILITY IN LIFE

Probability is involved in virtually every decision that anyone makes, as well as in the processes of learning and reasoning. Sometimes, the role of probability is clear. For example, in deciding whether to plan a picnic you might want to know the probability of rain. Or, in deciding whether to draw a card in a game of blackjack, you might want to know the probability that it will improve your hand.

In other cases probability guides decisions on a far deeper level. For example, presumably you enrolled in the college that you think is most likely to meet your personal needs. How did you (possibly unconsciously) evaluate the probabilities involved in making this decision?

Another example concerns business decisions. In deciding whether to increase inventory for the coming summer, a store owner must evaluate the probability that more customers will visit the store. Probability *is* the business of insurance companies, which sell policies designed to protect people from events that have a low probability of happening to them, but which are bound to happen to someone.

Indeed, probability is at the heart of many important current issues. For example, what is the probability that the greenhouse effect will cause the destruction of civilization? If this probability is low, there is no need to change lifestyles. But, if it is high, spending trillions of dollars to prevent the disaster might be worthwhile.

Because it is so prevalent, probability is crucial to quantitative reasoning and making intelligent decisions. Unfortunately, research has shown that most people, sometimes even mathematicians, have a poor intuition for probability. This inability may explain why history is littered with those who have gambled and lost — individuals who lost their fortunes at casino games and lotteries, financial "experts" who gambled on stock prices or interest rates and lost their fortunes, or national leaders who gambled on a new social policy and found unintended consequences.

Fortunately, *calculating* many probabilities is easy. Mathematically, a probability can be assigned to any individual **event** or group of events. A probability is expressed as a fraction between 0 and 1. If an event is certain to occur, it has a probability of 1; equivalently, the chance of the event occurring is 100%. If an event is impossible, its probability is 0. A fraction indicates the likelihood of an event. For example, if we flip a fair coin the probability of heads is 1/2; that is, there is a 1/2, or 50%, chance that the coin will land heads.

In this chapter, we show how to work with and calculate probabilities and how to investigate some of the many ways that probability affects everyday life. With practice, you can improve your intuition about probability and thereby become better at making all kinds of decisions.

> **Time-Out to Think:** Many idioms simply are expressions of probability, such as "a snowball's chance in hell" (a probability of nearly 0), or "is the Pope Catholic?" (a probability of nearly 1). What are some others?

13.2 FUNDAMENTALS OF PROBABILITY

Chevalier was a title of low-ranking nobility in France. The Chevalier de Méré was a man by the name of Antoine Gombaud.

Modern mathematical study of probability may be traced to the gambling of a French con man known as the Chevalier de Méré. In 1654, the Chevalier bet other gamblers even money that he could roll at least one 6 in four rolls of a standard six-sided (cube-shaped) die (Figure 13–1). The Chevalier made quite a profit on this bet, despite having calculated his odds of winning incorrectly!

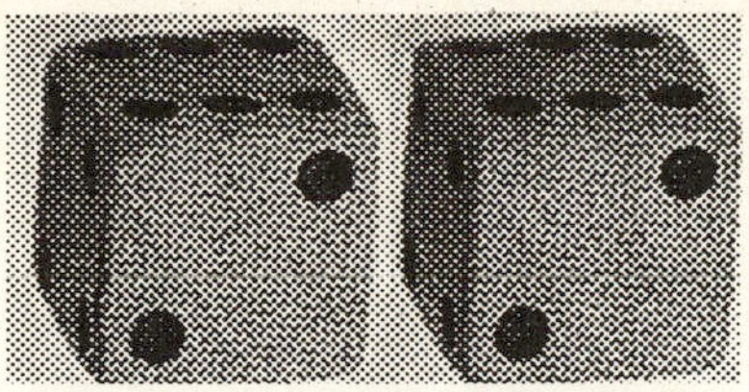

Figure 13-1. Standard six-faced dice are shown. In his first game, the Chevalier rolled one die, betting that he could roll at least one 6 in 4 tries. In his second game, the Chevalier rolled two dice, betting that he could roll a double 6 within 24 tries.

His faulty reasoning went as follows: The probability of rolling a 6 on one roll of a die is 1/6, so he assumed that his chances would be four times as great in four rolls. That is, he thought that the probability of getting at least one 6 in four rolls would be 4/6, or 2/3. In fact, the odds of rolling a 6 within four tries are only about 0.52. Nevertheless, because odds of 0.52 are favorable, the Chevalier profited over the long run.

Time-Out to Think: According to the Chevalier's reasoning, the odds of rolling a 6 on one roll is 1/6, so it must be 2/6 within two rolls, 3/6 within three rolls, and 4/6 within four rolls (his bet). What happens if you extend this reasoning to six rolls? Are you guaranteed to roll a 6 if you roll the die six times? Could you have convinced the Chevalier that his reasoning was faulty? Explain.

As gamblers caught on to this "con game," they were no longer willing to play. The Chevalier therefore proposed a new game to entice more players. This time he rolled two dice and bet that he could throw a *double 6* within 24 rolls (Figure 13–1b). Again, he expected to win big. He began with the probability of rolling a double 6 on a single try, or 1/36, which is an application of the multiplication principle (subsection 12.2.1). Each die can land in six different ways, so two dice can land in $6 \times 6 = 36$ ways, of which only one way is a double 6. Alternatively, we can make a table of all the possible ways the two dice can land (Figure 13–2), observing that only one of the 36 outcomes is a double 6.

Figure 13-2. The 6 possible outcomes from tossing each of two dice are shown across the top and down the left. Combined they show the 36 possible outcomes from tossing two dice. Only 1 of the 36 is a double 6; hence the probability of a double 6 is 1/36.

	1	2	3	4	5	6
1	1, 1	1, 2	1, 3	1, 4	1, 5	1, 6
2	2, 1	2, 2	2, 3	2, 4	2, 5	2, 6
3	3, 1	3, 2	3, 3	3, 4	3, 5	3, 6
4	4, 1	4, 2	4, 3	4, 4	4, 5	4, 6
5	5, 1	5, 2	5, 3	5, 4	5, 5	5, 6
6	6, 1	6, 2	6, 3	6, 4	6, 5	6, 6

The Chevalier then used his faulty reasoning to conclude that, if he rolled the pair of dice 24 times, his odds of rolling at least one double 6 would be 24/36, or 2/3. Alas, the actual probability is about 0.49, which is *less* than 50%. The Chevalier began to lose a lot of money. For help, he turned to his friend, mathematician Blaise Pascal, who, in turn, began a correspondence about the mathematics of probability with Pierre de Fermat (whose fame includes Fermat's last theorem, discussed in Chapter 2). Thus modern probability was born.

The Chevalier had correctly determined the probabilities of *individual* events; that is, he realized that the probability of a 6 on a *single* roll of one die is 1/6, and that the probability of a double 6 on a *single* roll of two dice is 1/36. His failure was in combining the individual probabilities to determine the odds of winning on multiple tries. We follow a similar strategy in explaining the fundamentals of probability. First, we explore the three general methods of determining probabilities for individual events in this section. Then, in the next section, we combine probabilities from individual events to calculate probabilities for more complex situations.

Time-Out to Think: In the Chevalier's first game, his actual odds of winning were 0.52, or 52%. In the second game, his odds of winning were 0.49, or 49%. Comment on how the difference between 49% and 52% can be the difference between rags and riches in gambling.

13.2.1 A Priori Techniques

The statement that the probability of heads on a single coin toss is 1/2 is based on preconceived notions about the coin and the toss. That is, we assume a *model* in which the coin is equally likely to land heads or tails. Of course, real coins may not necessarily behave this way: If a coin is dented or dirty, it may land heads more often than tails, or vice versa.

The Latin words *a priori* mean "before the fact." A priori probability implies a probability determined "before" any experimentation, which means that it is based on preconceived ideas about possible outcomes.

Any probability determined from preconceived notions or models is called an **a priori** probability. The basic method of calculating an a priori probability, assuming that all possible outcomes are equally probable, is as follows.

Step 1: Count all the possible outcomes of an event.
Step 2: Count the number of outcomes that represent *success* — that is, the number of outcomes that represent the sought after result.
Step 3: Determine the probability of success by dividing the number of successes by the total number of possible outcomes.

Looking back at Examples 12–10 and 12–11 (subsection 12.2.5), you will see that they gave a preview of a priori probabilities. Indeed, the counting techniques in Section 12.2 are fundamental to finding a priori probabilities.

Let's apply these ideas to a simple coin toss in which success means that the coin lands heads. The first step is to recognize that there are only two possible outcomes: heads and tails. The second step is to recognize that only one of these outcomes, heads, meets the criterion for success. The third step is to find the probability by dividing the one success (heads) by the two possible outcomes (heads or tails) to find a probability of 1/2. Note that this calculation depends on the assumption that all possible outcomes — heads and tails in this case — are equally probable; that is, the coin is "fair."

Example 13–1 *A Priori Die Probability.* Use the three-step process to find the probability of a 4 on a single roll of a die. Discuss why it is an a priori probability.

Solution: Step 1: The faces numbered 1 through 6 give a total of six possible outcomes. **Step 2:** A success is defined in the problem statement as a roll of 4; as 4 shows on only one face of the die, only one outcome represents success. **Step 3:** Success (rolling a 4) happens on one of six possible outcomes, so the probability of rolling a 4 is 1/6. This is an a priori probability because it is based on preconceived notions about the die, such as that the die has six faces numbered 1 through 6 and that the die is "fair." If the die is loaded (weighted on any particular side) or has more or less than six faces or has two faces with the same number, the a priori probability would be incorrect. ❖

Example 13–2 *Drawing an Ace.* Find the probability of drawing an ace from a deck of 52 cards. Discuss why doing so is an a priori probability.

Solution: Step 1: The 52 distinct cards offer 52 possible outcomes when drawing a particular card. **Step 2:** Success is defined in the problem statement as drawing an ace; with 4 aces in a standard deck, 4 of the possible outcomes represent success. **Step 3:** Divide the 4 successful outcomes (drawing any one of the four aces) by the total of 52 possible outcomes to find that the probability of drawing an ace is 4/52, or 1/13. This is an a priori probability because it is based on assumptions about the deck of cards such as that the deck is standard with 52 distinct cards including 4 aces and that the deck is well shuffled so that the likelihood of drawing any particular card is the same as that of drawing any other card.

❖

Example 13–3 ***Monkeys at the Typewriter.*** What is the probability that a monkey at a typewriter will type the word "cool" if it hits four keys?

Solution: Step 1: First, count the total number of possible outcomes. A standard typewriter has 26 letters, 10 numbers, and 10 punctuation keys for a total of 46 keys. According to the multiplication principle (Section 12.2), $46^4 = 46 \times 46 \times 46 \times 46 = 4{,}477{,}456$ different four-key sequences can be typed. **Step 2**: Success is defined as the word "cool," and only one way exists to type this word: specifically hitting the keys in the sequence c-o-o-l. **Step 3:** Only one successful outcome (the word "cool") can be obtained from the 4,477,456 possible outcomes. Thus the probability that the monkey will type the word "cool" is 1 in 4,477,456, or slightly better than 1 in 5 million. This is an a priori probability because it is based on the assumption that the monkey chooses keys at random, with an equal probability of striking any particular key.

❖

Time-Out to Think: Can you extend Example 13–3 to the probability that monkeys at typewriters eventually would write the play *Macbeth*? Without attempting to calculate the probability, discuss the idea.

Probability Distributions

Generally, you can count the total number of possible outcomes by using the counting techniques of Section 12.2. However, a table can be very helpful in counting the number of successes. For example, suppose that you toss two coins simultaneously. What are the probabilities of the possible outcomes?

Because each coin can land in two possible ways (heads or tails), the multiplication principle indicates that $2 \times 2 = 4$ outcomes are possible, as shown in the following table. Note that the table has a row for each of the four ways the coins can fall.

Coin 1	Coin 2	Outcome	Probability
H	H	HH	1/4
H	T	HT	1/4
T	H	HT	1/4
T	T	TT	1/4

If you don't care which coin is which, two of the possible outcomes in the table are equivalent: the second and third rows both show a result of one head and one tail. A **probability distribution** summarizes the probabilities of the distinct results (Table 13–1).

Thus when you toss two coins the probability of getting two heads is 1/4 or 25%; the probability of two tails also is 1/4 or 25%; and the probability of one head and one tail is 1/2 or 50%. *Note that the sum of the probabilities in a probability distribution must always be 1 (or 100%)!*

Table 13-1. Probability Distribution for Two Tossed Coins.

Result	Probability
HH	¼
HT	½
TT	¼
Total	1

Example 13–4 *Three Coins.* Make a probability distribution for three coins tossed simultaneously.

Solution: By the multiplication principle, the total number of outcomes from three tossed coins is $2 \times 2 \times 2 = 8$. You can show all eight outcomes by making a table with a column for each of the three coins. The pattern of entries ensures that each row represents a different outcome.

Coin 1	Coin 2	Coin 3	Outcome	Probability
H	H	H	HHH	1/8
H	H	T	HHT	1/8
H	T	H	HTH	1/8
H	T	T	HTT	1/8
T	H	H	THH	1/8
T	H	T	THT	1/8
T	T	H	TTH	1/8
T	T	T	TTT	1/8

Table 13-2. Probability Distribution for Three Tossed Coins

Result	Probability
3 H	1/8
2 H, 1 T	3/8
1 H, 2 T	3/8
3 T	1/8
Total	1

Several of the outcomes represent the same result because no distinction among the three coins was made. Rows 2, 3, and 5 all represent two heads and one tail, and rows 4, 6, and 7 represent one head and two tails. The results can therefore be consolidated into a concise probability distribution, as shown in Table 13–2.

The distribution shows that the probability of all three coins landing heads is 1/8, as is the probability of all three landing tails. The probability of two heads and one tail is 3/8, which also is the probability of one head and two tails. As always, the sum of the probabilities in the distribution is 1.

❖

13.2.2 Empirical Techniques

A second method for determining probabilities is through **empirical techniques**; that is, through experiments and observations. For example, if you suspect that a particular coin is unfair, you might perform an experiment in which you toss the coin 100 or 1000 times. Suppose that you find that the coin lands heads about 75% of the time. Because this rate is significantly different from the 50% a priori probability of heads, you would conclude that the coin is unfair. The empirical result for this particular coin suggests that its probability of landing heads is 0.75, or 75%.

Empirical techniques are particularly useful when you have no means of establishing an a priori probability. For example, suppose that you want to know the probability that a heavy smoker will die of lung cancer. Unlike the flipping of coins, there is no basis for an a priori estimate of this probability. Instead, you must rely on data for the causes of death of thousands of heavy smokers. If you find, say, that 30% of heavy smokers died of lung cancer, you might conclude that the probability that a particular smoker will die of lung cancer is 0.3.

The severity of a flood generally is measured in terms of the volume of water involved, or by how much a river rises above normal. For example, records might suggest that a river rises by 10 feet only about once every 100 years (a 100-year flood). The same river might rise by 20 feet about once every 500 years (a 500-year flood). Thus an area might be prone to both 100-year floods and 500-year floods, which could occur in any particular year.

Suppose that you want to know the probability that a particular region will experience a major flood. Again, you must rely on empirical techniques because no means of knowing an a priori probability exists. By studying historical and geologic records, you learn that the region has experienced a

Time-Out to Think: Consider the connection between empirical techniques and statistical research (which we cover in depth in Chapter 14). Because it is impractical to find the cause of death for *all* heavy smokers, researchers would study a relatively small number of heavy smokers and propose that causes of death for this small group are representative of causes of death for all heavy smokers. Discuss several ways that probability comes into play in this process.

major flood nine times in the past 1000 years. You might therefore conclude that the probability of a major flood in any year is about 9 in 1000, or 0.009. Because this probability is close to 0.01, or 1%, such an event is sometimes called a "100-year flood" to indicate that a flood of this severity happens *on average* about once every 100 years.

Sometimes, empirical probabilities are called **relative frequency probabilities**. The frequency of an event is the number of times that it happens; the relative **frequency** is the fraction of times that it happens. In the case of the unfair coin, imagine that you toss it 1000 times and that it lands heads 750 times. Then the frequency of heads in the experiment is 750 and the relative frequency is 750/1000, or 0.75. Similarly, in the case of heavy smokers, the relative frequency of death from lung cancer was 0.3; in the case of the floods, the relative frequency of floods was 0.009.

Example 13–5 *Empirical Coin Testing.* A single round of an experiment in which two coins are tossed has three possible outcomes: 0 heads, 1 head, or 2 heads. Suppose that you repeat the two-coin toss 200 times and that the following table presents the results. Compare the empirical probabilities from the experiment to the a priori probabilities for two tossed coins. Are the coins unfair? (Note that, as they must, the relative frequencies sum to 1.)

Number of Heads	Frequency	Relative Frequency
0	43	43/200 = 0.215
1	104	104/200 = 0.520
2	53	53/200 = 0.265

Solution: Based on the experimental data, the empirical probability of no heads is 0.215, which is slightly less than the expected a priori probability of 0.25 (see Table 13–1). Similarly, the empirical probability of one head, 0.52, is slightly greater than the a priori probability of 0.5; and the empirical probability of two heads, 0.265, is slightly greater than the a priori probability of 0.25. Despite the differences between the a priori and empirical probabilities, you cannot conclude that the coins are unfair. Empirical probabilities rarely match a priori probabilities *exactly*. Because the a priori and empirical probabilities are close, and the coins were tossed only 200 times, concluding that the coins are unfair seems premature.

❖

Time-Out to Think: Suppose that a slot machine advertises that the probability of hitting the jackpot as 1%. You play the machine 200 times without winning. Can you conclude that the advertisement was false and the machine unfair? Why or why not?

13.2.3 Subjective Techniques

The third general method for estimating probabilities is through personal judgment or intuition. Such techniques are subjective because they depend on individual beliefs and biases; nevertheless, they are the most commonly used probabilities. The validity of a subjective probability depends on the experience and ability of the individual making the estimate.

Consider your personal estimate of the probability that going to college today will make you more successful tomorrow. Unless you believe that this probability is relatively high, you probably wouldn't be in college. Yet this probability depends on so many unpredictable events that justifying any particular subjective value for it is very difficult.

Now imagine that a TV meteorologist states: "There is a 20% chance of rain for tomorrow." Presumably, her estimate is based on experience at reading weather maps and charting local weather. Thus, although her estimate is subjective, it carries a fair amount of validity because of her expertise.

Time-Out to Think: Describe a method by which you could determine empirically the accuracy of predictions by a particular TV meteorologist.

13.3 COMBINING PROBABILITIES

The Chevalier de Méré incorrectly combined the probabilities from a single roll of the dice into probabilities for multiple rolls. To avoid his errors, you must learn methods of combining probabilities for individual events into an aggregate probability for multiple events. You might wish to combine probabilities in two basic types of situation.

Joint probabilities represent the probability of event A *and* event B. Either/or probabilities represent the probability of event A *or* event B. See Chapter 2 for further discussion of the logical connectors *and* and *or*.

1. You might want to know the probability that two or more events will occur together (jointly). Examples of situations in which you must calculate **joint probabilities** include finding the probability of getting heads twice in a row on a coin toss or of having a baby *and* getting a pay raise in the same year. Joint probabilities always involve *multiplication* of individual probabilities.
2. You might want to know the probability that *either* of two or more events will occur. Examples of situations in which you must calculate **either/or probabilities** include finding the probability of drawing either an ace or a king from a deck of cards or of losing your home either to a fire *or* to a hurricane. Either/or probabilities involve *addition* of individual probabilities.

13.3.1 Joint Probabilities

We first consider joint probabilities. We use the letter P to represent probability; next to the P, in parentheses, we describe the event for which we are seeking the probability. For example, to express the probability of obtaining a 4 on a tossed die we write P(4 on tossed die) or, if it is clear that we are talking about a tossed die, P(4). With this notation, we express the probability as

$$P(4) = P(4 \text{ on tossed die}) = \frac{\text{number of ways to get a 4}}{\text{total number of possible outcomes}} = \frac{1}{6}.$$

Now, suppose that we toss two dice. What is the probability that *both* will come up 4? One way to answer this question is by thinking of it as an *individual* event — a single toss of two dice — and then follow the three-step procedure for finding a priori probabilities.

Step 1: Each of the two dice can fall in 6 ways (the faces 1 through 6), so the total number of possible outcomes is $6^2 = 36$.

Step 2: A success in this case, rolling a double 4, can happen in only one way — both dice showing a 4.

Step 3: With 36 possible outcomes, of which only 1 is a success, the probability that both dice will show 4 is 1/36.

However, we obtain the same result more easily by considering the toss of the two dice as two distinct events — one for each of the die — and multiplying the two probabilities:

$$\text{P(4 on first die } \textit{and} \text{ 4 on second die)} = \text{P(4 on first die)} \times \text{P(4 on second die)}$$

$$= \frac{1}{6} \times \frac{1}{6} = \frac{1}{36} .$$

Now suppose that we toss two coins. What is the probability that both land heads? Again, we can consider this action as two separate events — the tossing of two individual coins. In each case, the probability of heads is 1/2. Thus the joint probability is

$$\text{P(heads on coin 1 } \textit{and} \text{ heads on coin 2)} = \text{P(heads on coin 1)} \times \text{P(heads on coin 2)}$$

$$= \frac{1}{2} \times \frac{1}{2} = \frac{1}{4} .$$

Independent Events

When we toss one coin, the result doesn't affect what happens when we toss a second coin. Similarly, when we roll two dice, the result on one die doesn't affect the result on the other. Tossing coins and dice are examples of **independent events**; that is, the outcome of one event doesn't affect the outcome of other events in the experiment. As the preceding examples show, calculating joint probabilities for independent events is simply a matter of multiplying the individual probabilities:

$$\text{P(A } \textit{and} \text{ B)} = \text{P(A)} \times \text{P(B)} .$$

Note that this rule is related to the *multiplication principle* described in subsection 12.2.1 and may be easily extended to additional events. For example, with a third event C, the joint probability of A and B and C all occurring is

$$\text{P(A } \textit{and} \text{ B } \textit{and} \text{ C)} = \text{P(A)} \times \text{P(B)} \times \text{P(C)} .$$

Example 13–6 ***Three Coins.*** Suppose that you toss three coins. What is the probability of getting three tails?

Solution: Because coin tosses are independent events, simply multiply the probability of tails on each individual coin:

$$\text{P(tails on coin 1 } \textit{and} \text{ tails on coin 2 } \textit{and} \text{ tails on coin 3)}$$

$$= \underbrace{\text{P(tails)}}_{\text{coin 1}} \times \underbrace{\text{P(tails)}}_{\text{coin 2}} \times \underbrace{\text{P(tails)}}_{\text{coin 3}} = \frac{1}{2} \times \frac{1}{2} \times \frac{1}{2} = \frac{1}{8} .$$

The probability that three tossed coins all land tails is 1/8. This a priori probability is based on the assumption that the coins are fair.

❖

Example 13–7 ***Drawing Cards with Replacement.*** Suppose that you draw a card from a deck, replace the card, reshuffle the deck, and draw another card. What is the probability that both draws will be clubs? If you continue to replace the card and shuffle the deck after each draw, what is the probability of drawing a club three times time in a row?

Solution: Because you put the first card back and reshuffled the deck, the selection of the second card doesn't depend on the first selection. Thus each event of drawing a card is independent of other events. You may therefore use the multiplication principle to find the probabilities. With four suits in a deck of cards (hearts, spades, clubs, and diamonds), the probability of a club on any single draw is 1 in 4. The probability of drawing two clubs on two draws, assuming replacement and reshuffling, is

$$\text{P(club } \textit{and} \text{ club)} = \text{P(club)} \times \text{P(club)} = \frac{1}{4} \times \frac{1}{4} = \frac{1}{16} = 0.0625 = 6.25\%.$$

The probability of a club on three straight draws (replacing the card each time) is

We are using the word *odds* as a synonym for *probability*. In gambling, odds sometimes are stated in a different way. For example, if the probability of success is 1 in 16, the probability of failure is 15 in 16. Gamblers might say that the odds are "15 to 1" against success.

$$\text{P(3 clubs)} = \text{P(club)} \times \text{P(club)} \times \text{P(club)} = [\text{P(club)}]^3 = \left(\frac{1}{4}\right)^3 = \frac{1}{64} = 0.015625.$$

The odds of getting two clubs in a row are 1 in 16, or about 6%. The odds of getting three clubs in a row are 1 in 64, or just over 1.5%.

❖

Example 13–8 ***Consecutive 100-Year Floods.*** What is the probability that a "100-year flood" will strike a city in two consecutive years? If that actually happens, what conclusions might you draw?

Solution: Begin by recalling that a "100-year flood" describes an event whose empirical probability in any year is 1 in 100. Because the weather in any year doesn't affect the weather the next year, weather-related events are independent. Thus the probability of a 100-year flood in two consecutive years is the product of the probabilities in each individual year, or

$$\text{P(100-year flood in 2 consecutive years)}$$
$$= \underbrace{\text{P(100-year flood)}}_{\text{year 1}} \times \underbrace{\text{P(100-year flood)}}_{\text{year 2}} = 0.01 \times 0.01 = 10^{-4}.$$

That is, the probability of a 100-year flood occurring in two consecutive years is about 1 in 10,000. Of course, the assumption is that the probability of a 100-year flood actually is 0.01 in any year. If a 100-year flood were to occur in two consecutive years, you might wonder whether the probability of a single flood was correct. Two primary sources of uncertainty might exist. First, the probability of a 100-year flood is determined empirically from study of historical or geological records, so the probability may be understated. For example, historical records might have failed to record some floods, leading to an undercount in the number of past floods. Alternatively, conditions may have changed, making floods more likely than they were in the past. For example, global climate change might affect local weather conditions and make flooding more common, or building in natural flood plains might exacerbate an otherwise minor flood. Because of the inherent uncertainty of the empirical probability for 100-year floods, two consecutive years of such flooding suggest that the empirical result *probably* isn't reliable.

❖

Dependent Events

Suppose that you draw candy from a bag that initially contains 10 pieces: 5 chocolate and 5 caramel. Clearly, the probability of getting a piece of chocolate on your first try is 5 in 10, or 1/2. Now, suppose that you get a chocolate on the first try and quickly eat it. What is the probability of getting another piece of chocolate on your second try? Nine pieces remain in the bag, of which only 4 are chocolate. Thus the probability of getting a piece of chocolate on the second try is 4 out of 9, or 4/9. In this case, the probability of the second event *is* affected by the outcome of the first event. We therefore say that these are **dependent events.**

Calculating the joint probability of dependent events still involves multiplying the individual probabilities. However, with dependent events we must take into account how prior events affect subsequent ones. Continuing our example, what is the probability of getting two pieces of chocolate in a row? Simply multiply the probabilities of getting chocolate on the first and second tries:

$$\text{P(chocolate first try and chocolate second try)}$$
$$= \text{P(chocolate first try)} \times \text{P}\begin{pmatrix}\text{chocolate second try} \\ \text{given chocolate on first try}\end{pmatrix} = \frac{1}{2} \times \frac{4}{9} = \frac{2}{9}.$$

Thus the probability of drawing two pieces of chocolate in a row is 2/9.

We can generalize this example to any set of events, A and B. To show that event B is dependent on the outcome of event A, we write its probability as P(B given A), which we read as: "the probability of event B given the prior occurrence of event A." Hence the joint probability for dependent events is

$$\text{P(A } \textit{and} \text{ B)} = \text{P(A)} \times \text{P(B given A)}.$$

Example 13–9 ***Drawing Cards Without Replacement.*** Suppose that you draw a card from a standard deck and then draw a second card *without* replacing the first. What is the probability that both draws will be clubs? What is the probability of drawing three clubs in a row?

Solution: These events are *dependent* since your selection of the first card prevents that card from being chosen again on the second draw. The probability of a club on the first draw is 1 in 4 because a deck of 52 cards contains 13 clubs. However, after the first draw only 51 cards remain. Given that the first draw was a club, only 12 clubs are left. Thus the probability of a club on the second draw, given a club on the first draw, is 12 in 51. The combined probability is

$$\text{P(club } \textit{and} \text{ club)} = \underbrace{\text{P(club)}}_{\text{first draw}} \times \underbrace{\text{P(club given club on first draw)}}_{\text{second draw}}$$
$$= \frac{13}{52} \times \frac{12}{51} = \frac{156}{2652} = \frac{12}{204} = 0.0588.$$

The probability of getting a club on three straight draws is

$$\text{P(3 clubs)} = \underbrace{\text{P(club)}}_{\text{first draw}} \times \underbrace{\text{P(club given club on first draw)}}_{\text{second draw}} \times \underbrace{\text{P(club given clubs on first 2 draws)}}_{\text{third draw}}$$
$$= \frac{13}{52} \times \frac{12}{51} \times \frac{11}{50} = 0.013.$$

Thus the probability of drawing two clubs in a row without replacement is a little less than 6% and the probability of drawing three clubs in a row is barely more than 1%.

❖

Time-Out to Think: Compare the probabilities for getting three clubs in a row when the cards are replaced and reshuffled (Example 13–7) to when the cards are not replaced in the deck (Example 13–9).

Example 13–10 *Dependent Surprise.* Suppose that you meet a girl named Ali, who tells you that she has only one sibling. (a) What is the probability that the sibling is a brother? (b) Suppose that she also tells you that she is the oldest of the two children in her family; does this condition change the probability that her sibling is a brother? Explain.

Solution: A family with two children could have, in order of birth: boy–boy, boy–girl, girl–boy, or girl–girl. (a) If you know only that one child is a girl, you can rule out only the boy–boy possibility. That leaves the three other possibilities, two of which have a boy as the other child. Hence the probability that Ali's sibling is a brother is 2/3. (b) If you also know that Ali is the oldest child, then you can rule out both the boy–boy and boy–girl possibilities. That leaves two possibilities (girl–boy and girl–girl), of which one includes a boy. Thus, if you know that Ali is the oldest child, the probability that her sibling is a brother is 1/2.

Surprised? Why does the additional information that Ali is the oldest child change the probability? In part (b), you know that Ali is the oldest child, so you are concerned only with the gender of *one* child (the second one). Because the gender of one child cannot affect the gender of the other, this case yields the *independent* probability of a boy, which is 1/2. However, in part (a) you must consider *both* children at the same time because you don't know whether Ali is older or younger. Therefore the probability in this case is *dependent* because you are seeking the probability that the *other* child is a boy, *given* that one of the two children is a girl.

❖

13.3.2 Either/Or Probabilities

What is the probability that a tossed coin will land *either* heads *or* tails? These are the only two possible outcomes, so the probability that one or the other occurs must be 1. Thus either/or probabilities involve addition; in this case,

$$\text{P(heads } or \text{ tails)} = \text{P(heads)} + \text{P(tails)} = \frac{1}{2} + \frac{1}{2} = 1.$$

Mutually Exclusive Events

A coin can land either heads *or* tails, but it can't land both heads *and* tails at the same time. When one event (heads) precludes another (tails) from occurring at the same time, they are said to be **mutually exclusive events**. In general, the rule for finding the either/or probability of two mutually exclusive events A and B is

$$\text{P(A } or \text{ B)} = \text{P(A)} + \text{P(B)}.$$

Example 13–11 *Either/Or Dice Probability.* Suppose that you roll a single, six-faced die. What is the probability of getting either a 2 or a 3?

Solution: The outcomes of 2 or 3 are mutually exclusive because a single die can yield only one result. The probability of a 2 is 1/6, and the probability of a 3 also is 1/6. Therefore the combined probability is

$$P(2 \text{ } or \text{ } 3) = P(2) + P(3) = \frac{1}{6} + \frac{1}{6} = \frac{2}{6} = \frac{1}{3}.$$

The probability of rolling either a 2 or a 3 is 1 in 3.

❖

Example 13–12 ***Both Heads or Both Tails.*** Suppose that you toss two coins. What is the probability that you get *either* both heads or both tails?

Solution: As Example 13–5 demonstrated, the probability that both coins land heads is 1/4, which also is the probability of two tails. To find the probability of *either* both heads or both tails, add:

$$P(2 \text{ heads } or \text{ 2 tails}) = P(2 \text{ heads}) + P(2 \text{ tails}) = \frac{1}{4} + \frac{1}{4} = \frac{1}{2}.$$

The probability of *either* both heads or both tails is 1 in 2.

❖

Example 13–13 ***Either/Or Card Probability.*** What is the probability of drawing a *face* card (jack, queen, or king) from a standard deck of cards?

Solution: A standard deck of 52 cards contains 4 jacks, 4 queens, and 4 kings. Thus the probability of drawing a jack is 4/52, or 1/13. The probability of drawing a queen also is 1/13, as is the probability of drawing a king. These three events — drawing a jack, queen, or king — are mutually exclusive: Any face card drawn can be only one of these. Therefore add to find the combined probability:

$$\begin{aligned} P(\text{face card}) &= P(\text{jack } or \text{ queen } or \text{ king}) \\ &= P(\text{jack}) + P(\text{queen}) + P(\text{king}) = \frac{1}{13} + \frac{1}{13} + \frac{1}{13} = \frac{3}{13} \approx 0.23. \end{aligned}$$

The probability of drawing a face card is 3/13, or about 0.23.

❖

Non–Mutually Exclusive Events

When two events *can* happen simultaneously, they are *not* mutually exclusive. For example, suppose that you want to know the probability that a card drawn from a deck will be either a queen or a club. These events — drawing a queen or drawing a club — are not mutually exclusive because drawing both a queen and a club at the same time *is* possible: the queen of clubs.

To illustrate the calculation of either/or probabilities with non–mutually exclusive events, imagine that 8 people are in a room: 2 Democratic men, 2 Republican men, 2 Democratic women, and 2 Republican women. Using the letter M for each man and W for each woman, and subscripts of D for Democrat and R for Republican, we can represent these 8 people as M_D, M_D, M_R, M_R, W_D, W_D, W_R, and W_R.

If you select 1 person at random, what is the probability that the person will be *either* a woman or a Democrat? These events are not mutually exclusive, for 1 person may be both a woman and a Democrat. With four women and two Democrats who are men, the probability of choosing either a woman or a Democrat is 6 in 8, or 3/4.

Suppose that we had tried to add the individual probabilities. The probability of choosing a woman is 1/2 because of the 8 people, 4 are women. The probability of choosing a Democrat also is 1/2 because of the 8 people, 4 are Democrats. Adding these probabilities gives 1, which is not the correct answer! Why not? Two of the people meet *both* criteria: the women Democrats. We can adjust for this double counting by subtracting the probability of choosing a female Democrat:

$$P(\text{woman } or \text{ Democrat}) = \underbrace{P(\text{woman})}_{\text{4 of 8 people}} + \underbrace{P(\text{Democrat})}_{\text{4 of 8 people}} - \underbrace{P(\text{woman } and \text{ Democrat})}_{\text{2 of 8 people are counted twice}}$$

$$= \frac{1}{2} + \frac{1}{2} - \frac{1}{4} = \frac{3}{4}.$$

Generalizing, the either/or probability of any non–mutually exclusive events A and B is

$$P(A \ or \ B) = P(A) + P(B) - P(A \ and \ B).$$

Example 13–14 *Queen or Club.* What is the probability of drawing either a queen or a club from a deck of 52 cards?

Solution: The probability of a queen is P(queen) = 4/52 because a deck of cards contains 4 queens. The probability of a club is P(club) = 13/52 because the deck contains 13 clubs. However, because one card, the *queen of clubs*, is both a queen and a club, it will be double counted if you simply add the probabilities of a queen and a club. You must therefore subtract the double-counted probability. With only one queen of clubs in the deck, the probability of selecting it is 1/52, so

$$P(\text{queen } or \text{ club}) = P(\text{queen}) + P(\text{club}) - P(\text{queen and club})$$

$$= \frac{4}{52} + \frac{13}{52} - \frac{1}{52} = \frac{16}{52} = \frac{4}{13} = 0.3077.$$

The probability of drawing either a queen or a club is 4/13, or about 30%. ❖

Table 13–3. Summary of Combining Probabilities.

Joint Probability: Independent Events	Joint Probability: Dependent Events	Either/Or Probability: Mutually Exclusive Events	Either/or Probability: Non–Mutually Exclusive Events
$P(A \ and \ B)$ $= P(A) \times P(B)$	$P(A \ and \ B)$ $= P(A) \times P(B \text{ given } A)$	$P(A \ or \ B)$ $= P(A) + P(B)$	$P(A \ or \ B)$ $= P(A) + P(B) - P(A \ and \ B)$

13.3.3 Return to the Chevalier de Méré

We are now ready to obtain the correct probabilities for the Chevalier de Méré's two games and thereby see why he won in the first case and lost in the second. Let's begin with the first game, in which the Chevalier bet that he could roll at least one 6 in four rolls of a die.

The Chevalier could have won this bet in any of four ways: besides winning if he rolled a 6 once in the four tries, he also would have won if he rolled a 6 twice, three times, or four times. We can therefore express the probability of winning the bet as a *sum* of four individual (mutually exclusive, either/or) probabilities:

$$P(\text{at least one 6 in four rolls}) = P(\text{one 6}) + P(\text{two 6s}) + P(\text{three 6s}) + P(\text{four 6s}).$$

Note that finding this combined probability requires a fair amount of work because each of the four individual probabilities must be found. Fortunately, there is a much easier way!

Instead of looking at the circumstances in which the Chevalier could have won the bet, let's examine the circumstances in which he could have lost. He lost only if *no* 6 showed up in the four rolls. This probability is quite easy to calculate. As only one of the six faces on a die is a 6, the probability of *not* getting a 6 on a single roll is 5/6. From the joint probability rule, the probability of getting no 6 in four rolls is

$$\begin{aligned}
\text{P(no 6 in four rolls)} &= \underbrace{\text{P(not 6)}}_{\text{roll 1}} \textit{ and } \underbrace{\text{P(not 6)}}_{\text{roll 2}} \textit{ and } \underbrace{\text{P(not 6)}}_{\text{roll 3}} \textit{ and } \underbrace{\text{P(not 6)}}_{\text{roll 4}} \\
&= \underbrace{\text{P(not 6)}}_{\text{roll 1}} \times \underbrace{\text{P(not 6)}}_{\text{roll 2}} \times \underbrace{\text{P(not 6)}}_{\text{roll 3}} \times \underbrace{\text{P(not 6)}}_{\text{roll 4}} \\
&= \frac{5}{6} \times \frac{5}{6} \times \frac{5}{6} \times \frac{5}{6} = \left(\frac{5}{6}\right)^4 = 0.4823.
\end{aligned}$$

Now that we have found the probability that the Chevalier *lost* the bet, we can find the probability that he won by subtracting from 1:

$$\text{P(winning the bet)} = 1 - \text{P(losing the bet)} = 1 - 0.4823 = 0.5177.$$

Thus, as stated when we first discussed the Chevalier's bet, his probability of winning at his first game was about 0.52, or 52%.

Time-Out to Think: Explain why the probability of winning the bet is simply 1 minus the probability of losing.

Next, consider the Chevalier's second game at which he sought to roll a double 6 within 24 tries. Again, we can easily find the probability of losing at this game, for that happens only if *no* double 6 is rolled in the 24 tries. The probability of rolling a double 6 on a single roll is 1/36, so the probability of *not* rolling a double 6 is 35/36. Using the multiplication principle, the probability of not rolling a double 6 in any of the 24 rolls is

$$\begin{aligned}
\text{P(no double 6 in 24 rolls)} &= \underbrace{\text{P(not double 6)}}_{\text{roll 1}} \times \underbrace{\text{P(not double 6)}}_{\text{roll 2}} \times \cdots \times \underbrace{\text{P(not double 6)}}_{\text{roll 24}} \\
&= [\text{P(not double 6)}]^{24} = \left(\frac{35}{36}\right)^{24} = 0.5086.
\end{aligned}$$

This is the probability that the Chevalier lost the bet, so the probability that he won is

$$\text{P(winning the bet)} = 1 - \text{P(losing the bet)} = 1 - 0.5086 = 0.4914.$$

The Chevalier's probability of winning on this latter bet was about 0.49, which is less than 50%. Although this game proved to be disastrous for the Chevalier, it proved to be lucky for us: His consultations with Pascal and Fermat led to the discovery of important and practical methods of calculating probabilities.

The *at least once* rule:
To find the probability that an event occurs at least once, first find the probability that it does not occur. The probability that it occurs at least once is simply: 1 – probability it does not occur.

13.3.4 The *At Least Once* Rule

The method used to analyze the Chevalier's games invokes what we call the *at least once* rule: To find the probability that an event occurs at least once, first find the probability that it does not occur. The probability that it occurs at least once is simply: 1 – probability it does not occur.

The *at least once* rule applies to any question of the form: "What is the probability of *at least one* particular outcome in a series of events?" The following examples show the tremendous utility of the technique in applications to state lotteries, flooding, and AIDS.

Example 13–15 ***Lottery Odds.*** Many State lotteries include "scratch and win" games, in which you purchase a ticket for \$1 and scratch its surface to reveal any winnings. A "winning ticket" typically means winning \$2 or more. A common quote from lottery officials is that the "odds of winning are 1 in 10." That is, approximately 1 in 10 tickets is a winner of some type. If you study in detail the odds quoted by the lottery, you will find that almost all winning tickets are winners of the minimum (typically, \$2) value. Suppose that you purchase 10 lottery tickets. What is the probability that you will have at least 1 winner among these 10 tickets? Discuss your findings.

Solution: Because this is an "at least one" type of problem, the easy way to solve it is by first finding the probability that *none* of the 10 tickets is a winner. The odds of winning on a single ticket are 1/10, so the odds of losing on that ticket are 9/10, or 0.9. From the joint probability rule, the odds of all 10 tickets being losers are

$$\text{P(no winners in 10 tickets)} = \left(\text{P(not winning on one ticket)}\right)^{10} = (0.9)^{10} = 0.349.$$

Because this is the probability of losing on all ten tickets, the probability of getting at least 1 winner is

$$\text{P(at least one winner)} = 1 - \text{P(no winners)} = 1 - 0.349 = 0.651.$$

Lottery: a tax on people who are bad at math. — message circulated on the Internet.

The probability of at least one winning ticket when you buy ten lottery tickets is only about 65%. Further, any winnings are unlikely to be more than \$2. In other words, if you spend \$10 on lottery tickets, you should expect to lose most of it, and you stand a better than 1 in 3 (35%) chance that you will lose all of it.

❖

Example 13–16 ***Floods.*** Suppose that historically a particular area is expected to get one extreme flood every 100 years. What are the chances that a 100-year flood will occur at least once during the twenty-first century?

Solution: This is an "at least once" question, so begin by calculating the odds that a 100-year flood does *not* occur during the twenty-first century. A 100-year flood has a 0.01 chance of occurring in any particular year; thus there is a 0.99 chance that a 100-year flood will not occur. The probability of no flood in a 100-year period is

$$\text{P(no flood in 100 years)} = \left(\text{P(no flood in one year)}\right)^{100} = (0.99)^{100} = 0.366.$$

The probability that a flood *will* occur, then, is 1 – 0.366 = 0.634, or about 64%. That is, there is only about a 64% chance that a 100-year flood will occur in the 100 years of the twenty-first century.

❖

Time-Out to Think: Does the result in Example 13–16 surprise you? Suppose that the area in question did not flood any time during the twentieth century. Does that make a flood during the twenty-first century more likely? Why or why not? Does it suggest that the data on which the probability of flooding was based were flawed or misinterpreted? Explain.

Example 13–17 *AIDS Among College Students.* The overall incidence of AIDS in the population is not well known. Estimates of the fraction of the general population infected with the virus that causes AIDS, called HIV, run from well under 1% of the population to 3% or more. Of course, the incidence of HIV infection varies greatly among different segments of the population; it may be lower than 0.1% among monogamous heterosexuals and as high as 20% among the highest risk groups (intravenous drug users or promiscuous homosexuals). Suppose that, among students at a particular college, the incidence of HIV is 3%. Further, suppose that an unusually promiscuous (and foolish, as this example will show) student has sex with an average of 6 partners in each of the student's 4 years of college. What is the probability that at least 1 of those partners is infected with HIV?

Solution: Because this is an "at least one" question, begin by calculating the odds that none of the partners is infected. The probability that a particular partner is infected with HIV is given in the problem as 0.03; therefore the probability that the partner is *not* infected with HIV is 0.97, or 97%. With 6 partners a year for 4 years, the student has a total of 24 partners. The probability that none of the 24 is infected with HIV is

$$\text{P(no infection in 24 partners)} = \left(\text{P(no infection in one partner)}\right)^{24} = (0.97)^{24} = 0.48.$$

Because the probability that none of the 24 partners is infected is 0.48, the probability that at least 1 partner is infected with HIV is 1 – 0.48 = 0.52. That is, there is a *better than 50% chance* that at least one of the student's partners is infected with HIV! Contact with an HIV-infected partner doesn't necessarily mean that the disease will be transmitted, and the 3% infection rate given in this problem is uncertain. Nevertheless, the risk should not be ignored. ❖

THINKING ABOUT ... DOES GOD PLAY DICE?

Is the future predestined by the past, or does it have many possible forms? Although this question may seem tied to superstitious notions about fate, it has been a topic of serious study for centuries. Amazingly, science now has an answer to this question. First, let's pose the question in a slightly more scientific way: *Suppose that we knew all of the laws of nature and, at some particular moment in time, we knew what every single particle in the universe was doing. Could we then predict the future of the universe for all time?*

Until the twentieth century, nearly every philosopher would have answered *yes*. The idea was so pervasive that many philosophers concluded that God was like a watchmaker: God simply started the universe up, and the future was forever after determined. This idea that everything in the universe is predictable from its initial state is called **determinism**; a universe that runs predictably like a watch is called a **deterministic universe**.

Twentieth-century physics shattered the idea of a deterministic universe. In 1926, German physicist Werner Heisenberg discovered the **uncertainty principle**, which essentially states that knowing *both* exactly where a particle is currently located *and* where it is going is impossible. As a result, the precise future location of the particle cannot be predicted; the best we can do is make statements about the *probability* that the particle will be found in certain areas. On the macroscopic scale of everyday life, this uncertainty is so small as to be unnoticeable. On the subatomic level, however, uncertainty clouds all attempts at prediction. Because everything, ultimately, is made of atoms, the uncertainty principle therefore implies a built-in randomness to the universe.

The idea that nature is governed by probability rather than certainty unsettled many people. Even Einstein, who helped lay foundations for the

Heisenberg discovered the uncertainty principle by thinking about trying to observe an electron. To see the electron, we must shine light on it. Suppose that we try to see the electron with visible light, which has a wavelength of about 500 nanometers. Then we will know the electron's position only to within 500 nanometers, which is about *5000 times* the size of a typical atom! To locate the electron more precisely, we could use light with a shorter wavelength; say, an X-ray with a wavelength of 1 nanometer. However, because X-rays carry a lot of energy, the electron's motion will be changed when the X-ray strikes it. Thus, even though we will know the electron's location quite well, we will have destroyed any information about where it was going. Heisenberg recognized that this is not a mere experimental problem but rather a fundamental limitation on the predictability of nature.

quantum physics upon which Heisenberg's discovery was based, was deeply disturbed by the uncertainty principle. Although he was well aware that the theories of quantum physics had survived many experimental tests, Einstein maintained a belief that the theories were incomplete. He believed that scientists one day would discover a deeper level of nature at which uncertainty would be removed. To summarize his philosophical objections to uncertainty Einstein said, "God does not play dice."

Einstein did more than simply object on philosophical grounds. He also proposed a number of thought experiments in which he showed that the uncertainty principle implied paradoxical results. Claiming that such paradoxes made no sense, he argued that the uncertainty principle must not be correct. In the years since Einstein's death in 1955, advances in technology have made performing some of Einstein's thought experiments possible. The results have come out essentially as Einstein had guessed: They appear to be paradoxical. That is, the experiments have confirmed the reality of the uncertainty principle — at the same time that they have posed apparent logical paradoxes. The lesson of quantum physics is that nature, at its most fundamental level, is guided by probability.

What can we make of an idea, such as the uncertainty principle, that seems to violate common sense at the same time that it survives every experimental test? Under the tenets of science, experiment is the ultimate judge of theory, and we must accept the validity of quantum theory despite philosophical objections. After all, why should we expect common sense developed on the macroscopic scale to hold on the scale of the subatomic? Perhaps the best answer to Einstein's objection that "God does not play dice" comes from Niels Bohr, another pioneer of quantum physics, who replied, "Stop telling God what to do."

13.4 THE LAW OF AVERAGES

Suppose that you toss a coin just once. What will happen? If the coin is fair, the probability of getting either heads or tails is the same. That is all you can say; you cannot predict the outcome of the coin toss.

Next, suppose that you toss a coin 10 times. How many times will you get heads? If the coin is fair, you will likely get heads *close to* half of the time; for example, 5 heads and 5 tails, or 6 heads and 4 tails. You aren't likely to get, say, all heads. If you toss the coin 100 times, you probably will not get heads *exactly* 50 times; nevertheless, you can reasonably expect to get *close to* 50 heads. If you toss the coin 1000 times, you can expect to get heads close to 500 times.

A more formal definition of the law of averages states that, if an experiment (a coin toss or roll of dice) is conducted repeatedly, the fraction of successes eventually approaches the theoretical probability of success. Each experimental trial must be independent; that is, the outcome of one experiment must have no effect on any others.

The coin tosses illustrate a principle often called the law of averages, which states that the larger the number of trials, the more likely it is that the overall outcome will be close to that predicted by the probability. The Chevalier de Méré's "luck" was a clear demonstration of the law of averages. With a 52% chance of winning at his first game, he made a profit if he found enough people willing to bet; with a 49% chance of winning at the second game, he lost money.

The law of averages, as its name implies, concerns only *average* results from many experiments. It does *not* affect the outcome of individual experiments. Unfortunately, many people mistakenly believe in what is sometimes called the **gambler's fallacy**, which suggests that the law of averages can help a person recover losses.

As a simple illustration of the gambler's fallacy, suppose that you toss a (fair) coin 100 times, getting 56 heads and 44 tails. What will happen if you continue to toss the coin another 100 times? The gambler's fallacy holds that, because you have had more heads than tails for the first 100 tosses, you should expect more tails than heads for the next 100 tosses. However, the outcomes of the first 100 tosses cannot affect subsequent tosses. The law of averages suggests only that, as always, the next 100 tosses are most likely to yield equal numbers of heads and tails. If you are betting on tails, you will be losing after the first 100 tosses. Once you are on the losing side, the law of averages tells you that you are likely to stay there.

Another way to contrast the law of averages and the gambler's fallacy is to realize that the former makes a statement about *ratios*, whereas the latter wrongly introduces the notion of *differences*. The law of averages indicates that, if you toss 1000 coins, the fraction of heads to tails will likely be closer to 0.5 than if you toss only 100 coins. The gambler's fallacy takes this outcome to mean that the total number of heads and tails will merge as more coin tosses are made. In fact, however, the difference in the number of heads and tails is likely to be larger with more coin tosses, as Example 13–18 shows.

Example 13–18 *The Gambler's Fallacy.* Suppose that you perform two coin tossing experiments. In the first experiment, you toss 100 coins and 56 land heads. In the second, you toss 1000 coins and 485 land heads. Find the fraction of heads in each experiment and the difference between the number of heads and tails. Explain how this latter outcome illustrates the difference between the law of averages and the gambler's fallacy.

Solution: In the first experiment you got 56 heads and 100 – 56 = 44 tails. The fraction of heads is 56/100, or 0.56, and you got 56 – 44 = 12 more heads than tails.

In the second experiment you got 485 heads and 1000 – 485 = 515 tails. The fraction of heads is 485/1000, or 0.485, and you got 515 – 485 = 30 more tails than heads.

Note that the second experiment yields a *fraction* of heads closer to 0.5. Yet the difference between the *number* of heads and tails is larger in this experiment (30 instead of 12), which is what you should expect from the law of averages.

A person suffering from the gambler's fallacy would expect the number of heads and tails to be closer in the larger experiment. That is why gamblers tend to think that, if they have been on a losing streak, they are "due" for a winning streak. Alas, it simply isn't true.

❖

13.4.1 Mean and Expected Value

If an experiment of tossing 100 coins is repeated many times, the **mean** (or average) outcome would be 50 heads. The mean also is called the **expected value** because we expect 50 heads to be the average of many experiments.

> **Time-Out to Think:** Explain why you should *not* expect to get exactly 50 heads when you toss 100 coins. Then explain why, if you repeat the experiment of tossing 100 coins many times, you *should* expect 50 heads to be the *mean* result. Is "expected value" a good name for the mean? Explain.

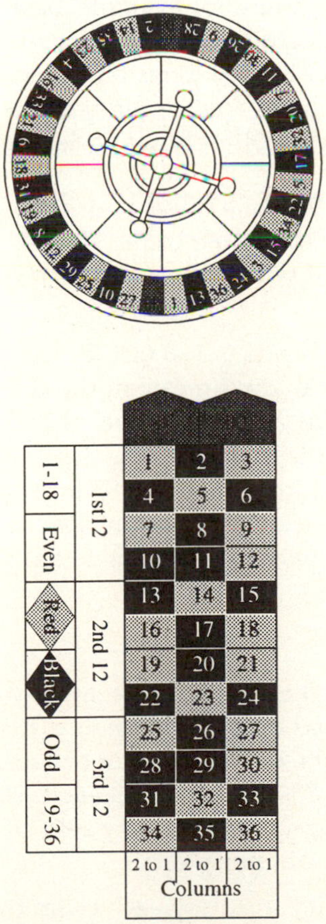

Figure 13–3. (a) A roulette wheel has the numbers 1 through 36, with half colored red and half colored black, and the green numbers 0 and 00. (b) Bets can be placed on individual numbers or on any of the combinations shown.

Consider the game of roulette (Figure 13–3). A roulette wheel has 38 numbered slots: 18 are red, 18 are black, and 2 are green (the green numbers are 0 and 00). Suppose that you place a \$1 bet on red. With 18 red slots out of the 38 total, your odds of winning are 18/38; hence your odds of losing are 20/38. A win means that you gain \$1 and a loss means that you lose \$1.

Imagine that you play the game many times. Of each 38 games, you expect to win 18 times on average, earning a total of \$18. Similarly, you expect to lose 20 times on average, losing a total of \$20. Thus, on average your expected losses are \$20 – \$18 = \$2 in every 38 games, or \$2/38 = \$0.53 = 5.3¢ per dollar bet. You can obtain the same result by multiplying each gain or loss by its probability:

$$\underbrace{\$1}_{\text{gain}} \times \underbrace{\frac{18}{38}}_{\substack{\text{probability}\\ \text{of gain}}} + \underbrace{-\$1}_{\text{loss}} \times \underbrace{\frac{20}{38}}_{\substack{\text{probability}\\ \text{of loss}}} = -\$0.053.$$

The result is negative because you will lose money, on average, if you place this bet many times. We say that your **expected earnings** are a **loss** of 5.3¢ for every dollar bet. If you place larger bets, your losses will be proportionally higher. For example, if you place \$10 bets, you should expect to lose an average of 53¢ per bet.

Because you are losing an average of 5.3¢ per dollar bet, the house (casino) that owns the roulette wheel is winning this amount. That is, the casino has a house edge of 5.3%. If many people place this bet many times, the casino will keep 5.3% of the money gambled.

The idea of expected earnings is important not only in gambling, but also in many business decisions. Auto insurance companies, for example, base rates on the number of accidents and the accident payoffs they expect with a large population of insured individuals. Banks set interest rates on loans by considering an expected default rate. Businesses often analyze probabilities in setting prices, choosing locations, and making many other types of decisions.

Example 13–19 ***10 to 1 Odds.*** Suppose that someone offers you 10 to 1 odds that you will not get three heads when you toss three coins. That is, you lose \$1 if you fail to get three heads but win \$10 if you succeed. Is this a good bet for you?

Solution: The probability that you *will* get three heads is 1/8 (see Table 13–2). By the law of averages, if you play this bet many times you should expect to win about 1/8 of the time and lose about 7/8 of the time. Thus, on average, you will lose \$7 on the 7 losses in each 8 tries. But you will win \$10 on the one win. Your expected winnings therefore are \$10 – \$7 = \$3 for every 8 times you play the game. Or, equivalently, your expected earnings are \$3/8 = 37.5¢ per game. It's a good bet for you; if your partner is willing to keep playing, you will win a lot of money!

❖

Example 13–20 ***Insurance Rates.*** Suppose that an insurance company sells a catastrophic medical insurance policy. Based on past experience, the company expects that 1 of every 500 policyholders will file a claim of about \$100,000, 1 of every 200 will file a claim of about \$30,000, and 1 of every 50 will file a claim of \$10,000. What is the expected payout per policy sold? If the policies sell for \$1000, what is the expected profit per policy?

Solution: To obtain the expected payout per policy multiply each claim by its probability:

$$\underbrace{\$100,000 \times \frac{1}{500}}_{\substack{\text{1/500 policies file}\\ \text{\$100,000 claim}}} + \underbrace{\$30,000 \times \frac{1}{200}}_{\substack{\text{1/200 policies file}\\ \text{\$30,000 claim}}} + \underbrace{\$10,000 \times \frac{1}{50}}_{\substack{\text{1/50 policies file}\\ \text{\$10,000 claim}}} = \$550.$$

If policies are sold for \$1000, the expected profit per policy is \$1000 – \$550 = \$450. Of course, this expected profit is realized only when a large number of policies are sold (so that the law of averages comes into play) and if the company correctly estimated the number and amount of claims.

❖

Example 13–21 *The House Edge.* If you play roulette (Figure 13–3) and bet on a single number, say, 3 or 15, the payoff odds are 35 to 1. That is, your $1 bet pays $35 if you win. Calculate your expected earnings if you bet on single numbers. Compare that to your expected earnings on a red bet, as discussed previously. Suppose that, on a given day at a particular casino, $1,000,000 is bet at the roulette table. How much does the casino expect to earn?

Solution: With 38 numbers on the roulette wheel, your odds of winning when you bet on a single number are 1/38. Your odds of losing are 37/38. Thus your expected earnings on a $1 bet are

$$\underbrace{\$35}_{\text{gain}} \times \underbrace{\frac{1}{38}}_{\substack{\text{probability}\\\text{of gain}}} + \underbrace{-\$1}_{\text{loss}} \times \underbrace{\frac{37}{38}}_{\substack{\text{probability}\\\text{of loss}}} = -\$0.053.$$

The negative sign indicates that you should expect an average *loss* of 5.3¢ per dollar bet. Note that this amount is the same as the loss calculated for a red bet earlier. In fact, payoff odds in roulette generally are set so that your expected earnings are –5.3¢ per dollar bet *at best*. That is, the casino has a house edge of at least 5.3% in roulette. If $1 million is bet on a particular day, the casino should expect to earn a profit of at least 5.3% × $1 million = $53,000. A lot of people will go home a lot poorer.

❖

Example 13–22 *Lottery Expectations.* Suppose that you purchase "scratch and win" lottery tickets for $1 each and that the odds of winning are 1 in 10 for a $2 winner, 1 in 50 for a $5 winner, 1 in 500 for a $10 dollar winner, and 1 in 10 million for a $1 million winner. What are your expected earnings per lottery ticket? The average lottery player in the United States spends about $500 per year on lottery tickets. For the odds given here, how much does the average person win (or lose)? Discuss.

Solution: The expected winnings on each lottery ticket are

$$\underbrace{\$2 \times \frac{1}{10}}_{\substack{\text{earnings from}\\\text{\$2 winners}}} + \underbrace{\$5 \times \frac{1}{50}}_{\substack{\text{earnings from}\\\text{\$5 winners}}} + \underbrace{\$10 \times \frac{1}{500}}_{\substack{\text{earnings from}\\\text{\$10 winners}}} + \underbrace{\$10^6 \times \frac{1}{10^7}}_{\substack{\text{earnings from}\\\text{\$1 million winners}}} = \$0.42.$$

State lotteries are the only game in some towns and not the only game in other towns. They are the worst game in all towns. — From *Can You Win?* by Mike Orkin. Copyright © 1991 by M. Orkin. Used with permission of W. H. Freeman and Company.

The expected winnings per ticket are $0.42. However, because each ticket costs $1, the net expected earnings are a *loss* of $0.58 ($0.48 – $1.00 = –$0.58). That is, you should expect to lose an average of 58¢ for every $1 that you spend. A person spending $500 per year on lottery tickets should expect to lose an average of 0.58 × $500 = $290. Aside from a few big winners, most lottery players are losing hundreds of dollars every year.

❖

13.4.2 The Binomial Probability Formula

We have stated that the expected value should not be expected in any single trial; rather it is the aggregate average expected from many trials. How can we calculate the probability of a particular result, such as getting 50 heads in 100 coin tosses? Mathematicians have developed techniques for determining probabilities in many complex situations. Here, we consider only the common case in which a particular experiment has just two possible outcomes.

For example, in an experiment of tossing a coin the two possible outcomes are heads or tails; in an experiment to test the effectiveness of a new drug for treating cancer, the two possible outcomes might be *cured* or *not cured*.

We use the letter p to represent the probability of one outcome (success) and q to represent the probability of the other outcome (failure). Note that, as only two outcomes are possible, the sum of the probabilities must be 1, or

$$p+q=1 \quad or \quad q=1-p.$$

The word *binomial* literally means "two names"; the names are the possible outcomes of the experiment.

If the experiment is repeated n times, the probability of r successes is given by the **binomial probability formula**

$$P(r) = {}_nC_r \times p^r \times q^{n-r} = \frac{n!}{r!(n-r)!} p^r \times q^{n-r},$$

where ${}_nC_r$ is the number of possible combinations of n objects taken r at a time (see Section 12.2).

Although the binomial formula looks complex, it is relatively easy to use. For example, suppose that you want to know the probability that you will get 50 heads if you toss a coin 100 times. The probability of success (heads) is $p = 0.5$, and the probability of failure (tails) is $q = 0.5$. The number of coin tosses is $n = 100$. The probability of $r = 50$ successes (heads) is:

$$\begin{aligned} P(\text{50 heads in 100 tosses}) &= {}_{100}C_{50} \times 0.5^{50} \times 0.5^{100-50} \\ &= \frac{100!}{50!(100-50)!} 0.5^{50} \times 0.5^{100-50} = 0.0796. \end{aligned}$$

The probability of obtaining 50 heads in 100 tosses is about 8%.

Time-Out to Think: Check the preceding calculation on your calculator. Some calculators have a key (usually labeled ${}_nC_r$) for entering combinations; it will save you the work of entering the factorials. A more sophisticated calculator, or a mathematical computer program, may even have a short-cut for the entire binomial formula.

Example 13–23 ***Lottery Tickets.*** Suppose that you purchase 20 lottery tickets in which the chance of a winner on any single ticket is 1 in 10. What is the mean (expected value) for the number of winning tickets you will receive? What are the odds that you will actually receive the expected number of winners in your particular set of 20 tickets?

Solution: Your expected value is 2 winning tickets (20 tickets × 0.1 = 2 tickets). That is, according to the law of averages, if you buy many sets of 20 tickets, the average set will have 2 winners. For any particular set, however, you need the binomial formula to calculate the odds of 2 winners. The probability of success (winning) is $p = 0.1$, and the probability of failure (losing) is $q = 0.9$. The number of trials is the $n = 20$ tickets purchased. Thus the probability of $r = 2$ successes is

$$\begin{aligned} P(\text{2 winners in 10 tickets}) &= {}_{10}C_2 \times 0.1^2 \times 0.9^{10-2} \\ &= \frac{10!}{2!(10-2)!} \times 0.1^2 \times 0.9^8 = 0.19. \end{aligned}$$

The odds are only about 20% that you will get the expected value of 2 winning tickets from the 20 you purchased.

❖

Example 13–24 ***Family Ratio.*** What is the probability that a family with 6 children will have 3 boys and 3 girls?

Solution: Assume that the probability of a boy is $p = 0.5$ and that the probability of a girl is $q = 0.5$. With $n = 6$ children, the probability of $r = 3$ boys is

$$\begin{aligned} \text{P(3 boys and 3 girls)} &= {}_6C_3 \times 0.5^3 \times 0.5^{6-3} \\ &= \frac{6!}{3!(6-3)!} \times 0.5^3 \times 0.5^3 = 0.31. \end{aligned}$$

The odds are about 30% that a family with 6 children will have 3 boys and 3 girls.

❖

Example 13–25 ***Unexpected Values.*** Suppose that you toss a coin 6 times. What is the expected number of heads? What is the probability that you get heads once? What is the expected number of heads if you toss a coin 60 times? What is the probability that you get 10 heads with 60 tosses? Based on these results, what can you say about probabilities of results that are far from the expected value as the number of trials increases?

Solution: With 6 tosses, the expected number of heads is 3 (as heads occur on average half the time). The probability of getting just 1 head may be found with the binomial formula. The probability of heads is $p = 0.5$, and the probability of tails is $q = 0.5$. With $n = 6$ tosses, the probability of $r = 1$ head is

$$\text{P(1 heads in 6 tosses)} = {}_6C_1 \times 0.5^1 \times 0.5^{6-1} = \frac{6!}{1!(6-1)!} \times 0.5^6 = 0.094.$$

With 60 tosses, the mean number of heads is 30. To find the probability of getting 10 heads use $n = 60$ coin tosses and look for $r = 10$ heads:

$$\text{P(10 heads in 60 tosses)} = {}_{60}C_{10} \times 0.5^{10} \times 0.5^{60-10} = \frac{60!}{10!(60-10)!} \times 0.5^{60} = 6.5 \times 10^{-8}.$$

The odds of getting 1 head in 6 coin tosses are close to 10%, but the odds of getting 10 heads in 60 coin tosses are 6.5 in 100 million, or less than 1 in 10 million! In both cases, the result is 1/3 of the expected number (1 compared to 3, or 10 compared to 30). The latter case is simply "scaled up" by a factor of 10, yet it reduces the odds enormously. This outcome illustrates one of the principal lessons of the law of averages: As the number of trials increases, obtaining a result that is proportionally far from the mean becomes *much* more difficult.

DILBERT® by Scott Adams

13.4.3 Probable Outcomes

If you use the binomial probability formula to calculate the probabilities of every possible outcome in a given number of trials, then you will have a complete probability distribution for the result. For example, Table 13–4 shows the probability distributions for tossing 5 coins, 10 coins, and 20 coins; all of these values can be calculated with the binomial formula.

Table 13–4. Probability Distributions for 5, 10, and 20 tossed coins.

5 Coins

# heads	prob.
0	0.031
1	0.156
2	0.313
3	0.313
4	0.156
5	0.031

10 Coins

# heads	prob.
0	0.001
1	0.010
2	0.044
3	0.117
4	0.205
5	0.246
6	0.205
7	0.117
8	0.044
9	0.010
10	0.001

20 Coins

# heads	prob.	# heads	prob.
0	10^{-6}	11	0.160
1	2×10^{-5}	12	0.120
2	2×10^{-4}	13	0.074
3	0.001	14	0.037
4	0.005	15	0.015
5	0.015	16	0.005
6	0.037	17	0.001
7	0.074	18	2×10^{-4}
8	0.120	19	2×10^{-5}
9	0.160	20	10^{-6}
10	0.176		

> **Time-Out to Think:** To practice using the binomial formula and convince yourself that the tables are correct, verify a few of the entries in Table 13–4.

Figure 13–4 shows the graphs, called **histograms**, of each probability distribution shown in Table 13–4. In each case, the smooth curve superimposed on the histogram approximates the actual distribution. The smooth curves show clearly the mean, or expected values: 2.5 heads for 5 coin tosses, 5 heads for 10 coin tosses, and 10 heads for 20 coin tosses.

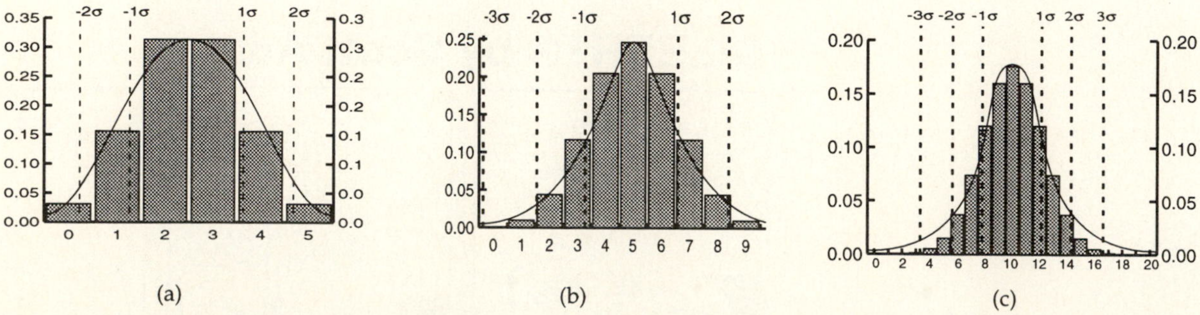

Figure 13–4. The graphs of the probability distributions from Table 13–4 are shown. Note that, as the number of coins increases, the probability distribution becomes relatively more peaked and narrow. The dotted lines on each graph show distances of 1, 2, and 3 standard deviations on either side of the mean.

The graphs differ not only in their means but also in their spread, or **dispersion**, about the means. That is, the graph for 5 coin tosses is rather flat and broad, whereas the graph for 20 coin tosses is relatively more peaked and narrow. One way to describe the dispersion is to state the fraction of possible outcomes that lie within a certain range. For example, adding the probabilities near the center of the distribution for 20 coins (Table 13–4) shows that the outcome will be between 7 and 13 heads about 90% of the time.

Another way to quantify the dispersion is with the **standard deviation**. The standard deviation is an important statistical measure of dispersion, and we discuss it further in Chapter 14. For the moment, however, you can use the formula for the standard deviation of a binomial distribution easily:

$$\text{standard deviation} = \sigma = \sqrt{npq} \qquad (\sigma \text{ is the Greek letter } sigma).$$

As before, p is the probability of success, q is the probability of failure, and n is the number of times the experiment is repeated. The formula for the mean (expected value) is

$$\text{mean} = \mu = np \qquad (\mu \text{ is the Greek letter } mu).$$

Before interpreting the standard deviation, let's calculate the mean and standard deviation for the 5, 10, and 20 coin tosses. In each case, $p = 0.5$ is the probability of heads and $q = 0.5$ is the probability of tails; only n varies.

For 5 coins: $\mu = np = 5 \times 0.5 = 2.5;$ $\sigma = \sqrt{npq} = \sqrt{5 \times 0.5 \times 0.5} = 1.1.$

For 10 coins: $\mu = np = 10 \times 0.5 = 5;$ $\sigma = \sqrt{npq} = \sqrt{10 \times 0.5 \times 0.5} = 1.6.$

For 20 coins: $\mu = np = 20 \times 0.5 = 10;$ $\sigma = \sqrt{npq} = \sqrt{20 \times 0.5 \times 0.5} = 2.2.$

The means come out as shown earlier on the graphs — as they should! Going from 5 coins to 20 increases the standard deviation, but only slightly compared to the increase in the number of coins. That's why the graphs become relatively *narrower* as the number of trials increases.

The dotted lines in Figure 13–4 show distances of 1, 2, and 3 standard deviations on either side of the mean. Note that the number of successes (heads) usually is within 1 standard deviation of the mean and nearly always is within 3 standard deviations. In fact, for any number of trials the chances are

These rules for interpreting the standard deviation are called *Chebyshev's Rules*. In fact, they apply to almost all distributions, not just binomial distributions.

- *at least* 75% that the number of successes will be within 2 standard deviations of the mean, and
- *at least* 89% that the number of successes will be within 3 standard deviations of the mean.

For large numbers of trials (large values of n) and cases where both p and q are *near* 1/2, the binomial distribution approximates a **normal distribution**. In that case

The normal distribution is extremely important in statistics, and we discuss it further in Chapter 14.

- the number of successes will be within 1 standard deviation (1σ) of the mean about 2/3 (68.3%) of the time,
- the number of successes will be within 2 standard deviations (2σ) of the mean about 95% (95.4%) of the time, and
- the number of successes will be within 3 standard deviations (3σ) of the mean over 99% (99.7%) of the time.

Example 13–26 ***Dispersion for 100 Coin Tosses.*** If a coin is tossed 100 times, what is the mean number of heads? What is the standard deviation of the possible number of heads? Interpret the results.

Solution: In this problem, $n = 100$, $p = 0.5$ is the probability of heads, and $q = 0.5$ is the probability of tails. The mean and standard deviation are

$$\mu = np = 100 \times \frac{1}{2} = 50, \text{ and } \sigma = \sqrt{npq} = \sqrt{100 \times \frac{1}{2} \times \frac{1}{2}} = 5.$$

The mean of 50 heads is the most likely number of heads obtained from tossing 100 coins. Recall, however, that the probability of getting 50 heads in 100 tosses is only about 8%.

The standard deviation is 5. Thus, according to the general rules for interpreting standard deviation, there is *at least* a 75% chance that the number of heads will be within 2 standard deviations of the mean, or between 40 and 60 heads. There is *at least* an 89% chance that the number of heads will be within 3 standard deviations of the mean, or between 35 and 65 heads.

❖

Example 13–27 ***Dispersion for 1 Million Coin Tosses.*** A coin is tossed 1 million times. Find and interpret the mean and standard deviation for the possible number of heads.

Solution: Again, $p = q = 0.5$. This time, however, $n = 1{,}000{,}000$. The mean (expected value) is $\mu = np = 106 = 500{,}000$, which is the most likely number of heads, even though the chance of getting *exactly* 500,000 heads is quite small. The standard deviation is

$$\sigma = \sqrt{npq} = \sqrt{10^6 \times \frac{1}{2} \times \frac{1}{2}} = \sqrt{250000} = 500.$$

Because the number of trials, 1 million, is very large, this distribution approximates a normal distribution. Thus the probability is 95% that the number of heads will be within 2 standard deviations ($2 \times 500 = 1000$) of the mean, or between 499,000 and 501,000 heads. There is a better than 99% chance that the number of heads will be within 3 standard deviations of the mean: $500{,}000 \pm 1500$ or between 498,500 and 501,500 heads.

Note that $1500 \div 500{,}000 = 0.003 = 0.3\%$. Thus, although you shouldn't expect *exactly* 500,000 heads, the outcome almost always will be within about 0.3% of 500,000. This result is another illustration of the law of averages: With a very large number of trials, the outcome almost always is very close to the expected value.

❖

Example 13–28 ***Why the House Always Wins.*** A patron's odds of winning at a particular game in a gambling casino are 49.7% (about the best odds that a patron can get on any game). Suppose that this game is played 1 million times each week in the casino. Calculate the mean and standard deviation for the number of times that patrons will win and interpret the results.

Solution: In this game the probability that the patron wins is $p = 0.497$, or a bit less than 50–50. The probability that the casino wins is $q = 1 - p = 0.503$. With the game played 1 million times ($n = 10^6$), the mean and standard deviation are

$$\text{mean} = \mu = np = 10^6 \times 0.497 = 497{,}000, \text{ and}$$

$$\text{standard deviation} = \sigma = \sqrt{npq} = \sqrt{10^6 \times 0.497 \times 0.503} = \sqrt{249991} = 499.99 \approx 500.$$

The expected, or most likely, outcome is that patrons will win 497,000 of the 1 million games. The odds are 99.7% that the number of games won by patrons will be within 3 standard deviations, or 3 (500 = 1500 games, of the mean; that is, 497,000 ± 1500, or between 495,500 and 498,500 games. That also means that the *casino* is virtually certain (99.7%) to will win *at least* 501,500 games, or *at least* 3000 more games than the patrons. Ever wonder how casinos can afford large hotels, fancy tables, and free drinks?

❖

13.5 PROBABILITY AND COINCIDENCE

What is the probability that *you* will win a big prize ($1 million or more) in the lottery this year? Typically, several million lottery entries are required for each $1 million prize. Your odds of winning are exceedingly small: much less than 1 in a million on a particular play!

What is the probability that *someone* will win a big prize in the lottery this year? This probability is a virtual certainty: Nationwide, several dozen people are such winners each year.

People who win the lottery often attribute their winning to "good luck" or to some special "system" they used in choosing a lottery number. Yet, from the standpoint of probability, someone was bound to win — no luck needed!

Although a *particular* coincidence may be highly unlikely, *some* such coincidence may be extremely likely (or even certain) to occur.

This example illustrates a surprising relationship between probability and coincidence. Although a *particular* coincidence may be highly unlikely, *some* such coincidence may be extremely likely (or even certain) to occur.

If you are in a class with 24 other students, what is the probability that someone in the class has the same birthday as yours? This is an "at least once" question, so begin by recognizing that, with 365 days in a year, the probability that any particular student has *your* birthday is 1/365 and the probability that a particular student does *not* have your birthday is 364/365. Hence the probability that *none* of the 24 other students has your birthday is $(364/365)^{24} \approx 0.94$, or 94%. Therefore the chance that someone in the class will have your birthday is only about 6%.

The small probability that someone in your class of 25 has your birthday probably isn't surprising. Next, however, consider the probability that *any* two people in your class have the *same* birthday. As usual, calculating this "at least one" probability first requires finding the probability that *no* two people have the same birthday. Suppose that you randomly select two people from the class. The first person has *some* birthday (a probability of 365/365), so the probability that the second person has a different birthday is 364/365.

Now, imagine selecting a third person. If the first two people have two different birthdays, two days of the year are "taken." Hence the probability that the third person has a different birthday from either of the other two is 363/365. Similarly, if the first three people all have different birthdays, the probability that a fourth person has a different birthday from any of the first three is 362/365. And so on. The probability that all 25 students have different birthdays is

$$\frac{365}{365} \times \frac{364}{365} \times \frac{363}{365} \times \cdots \times \frac{365-22}{365} \times \frac{365-23}{365} \times \frac{365-24}{365} = \frac{4.92 \times 10^{63}}{365^{25}} = 0.43 .$$

The complementary probability that at least two of the 25 students have the same birthday is 1 – 0.43 = 0.57, or 57%! Thus, although a *particular* birthday coincidence is fairly unlikely in a class of 25, chances are better than 50% that there will be *some* birthday coincidence in the class. Surprised?

13.5.1 Outrageous Coincidences?

The difference between the probability of some coincidence and a particular coincidence is a simple extension of the idea of expected values. With lottery odds of 1 in a million, we *expect* several winners if several million people play.

This idea may be extended to seemingly amazing coincidences. Imagine that you meet a person who lives in a different state but has the same birthday, drives the same kind of car, and has a mother who attended the same elementary school as your mother. What is the probability of this amazing coincidence?

The probability that you both have the same birthday is 1/365. The probability that you both drive the same kind of car depends on the popularity of the model; let's say that this probability is 1/100. For the probability that your mothers attended the same elementary school, we can make an order of magnitude estimate of 10,000 elementary schools in the United States. Then the probability that your mothers attended the same school is about 1/10,000. Overall, the probability of this particular coincidence is

$$\underbrace{\frac{1}{365}}_{\text{same birthday}} \times \underbrace{\frac{1}{100}}_{\text{same car}} \times \underbrace{\frac{1}{10{,}000}}_{\text{same school}} = \frac{1}{3.65\times 10^{8}} \approx 3\times 10^{-9}.$$

To an order of magnitude, the probability is about 1 in a billion!

To be more exact, the number of *pairs* of people that can be formed from a set of 100 million people is the number of combinations of 100 million people taken 2 at a time. Note that, for any n,

$$_{n}C_{2} = \frac{n(n-1)}{2}.$$

For $n = 10^8$, we find about $10^{16}/2$, or 5×10^{15} pairs. To an order of magnitude, that still is 10^{16} pairs.

Naturally, you would be quite amazed if this 1 in a billion coincidence happened to you. But this is a *particular* coincidence. How often should you expect 1 in a billion coincidences like this one to occur? To an order of magnitude, 100 million (10^8) people drive cars and have mothers who went to school in this country. The number of *pairs* of such people is on the order of $10^8 \times 10^8 = 10^{16}$. Therefore the 1 in a billion event has 10^{16} opportunities to occur. As $10^{16} \div 10^9 = 10^7$, or 10 million, you should expect on the order of 10 million such "1 in a billion" coincidences among pairs of people in the United States. Seemingly amazing coincidences will be discovered only if the pair of people meet and ask the right questions of each other; nevertheless, you shouldn't be surprised that such coincidences crop up now and then. Indeed, if you look hard enough, you probably can find a 1 in a billion coincidence between you and your next door neighbor.

A card example offers another illustration of why you shouldn't be surprised by amazing coincidences. In five-card poker, there are 2,598,960 possible hands (see Example 12–9). Thus your chances of being dealt any *particular* hand are less than 1 in 2 million! Yet it is inevitable that you will get *some* set of 5 cards, despite the exceedingly low probability of being dealt that particular hand.

This example, and several others in this section, are based on examples described in *Innumeracy* by John Allen Paulos (Hill and Wang, 1988).

Proper interpretation of coincidences can be especially important in legal proceedings. For example, suppose that an assault takes place in a city of 2 million people. Eyewitnesses agree on several characteristics of the assailant such as height, skin color, weight, hair color and length, beard, and glasses; they also agree on the getaway car's model, year, and color. A few days later, police pick up a man matching the eyewitness description. Although he has no prior arrest record, and proclaims his innocence, he is charged with the crime. Prosecutors show that the probability of any particular man matching the eyewitness description is only about 1 in 2 million (e.g., by estimating the fraction of men in the city that have a beard, the fraction of cars that match the description of the getaway car, and so on). On this basis, the prosecution case looks quite strong.

We lead you through the steps of the probability calculation for this legal example in Problem 66 at the end of the chapter.

However, the more relevant question is whether there might be *another* person in the city matching the eyewitness description, given that one such person has already been found. Surprisingly, a calculation shows this probability to be more than 20%! Given a 20% chance that the police picked up the wrong man, the prosecution case is extremely weak without additional evidence.

13.5.2 Streaks

Although we discuss streak probabilities in general terms, no simple formula for calculating them exists.

Another type of perfectly natural coincidence that may seem remarkable is in the phenomenon of streaks. For example, you might be very surprised if you toss a coin and get heads six times in a row, but you shouldn't be.

The probability of heads is 1/2, so the probability of getting a *particular* streak of 6 heads in a row is $(1/2)^6 = 1/64$. However, in 100 coin tosses there are 95 sequences of 6 tosses: 1 through 6, 2 through 7, 3 through 8, and so on. With 95 opportunities and odds of 1/64, the odds are quite good for getting at least one streak of 6 heads in 100 tosses.

Similarly, there are 91 opportunities for a streak of 10 heads in 100 coin tosses. Although the odds for a particular streak of 10 heads are $(1/2)^{10} = 1/1024$, such a streak is likely if you repeat the game of 100 coin tosses just a few times. The lesson: Surprising streaks should not be surprising!

Example 13–29 ***Hot Hand at the Craps Table.*** Suppose that you are playing the game of craps at a casino, rolling the dice and making your bets. Suppose that you are making bets (e.g., those called pass line, come, don't pass, and don't come) in which your odds of winning are 48.6%. Suddenly, you find yourself with a "hot hand": You roll winners on 8 consecutive bets. Is your hand really "hot"? Should you increase your bet because you are on a hot streak?

Solution: With odds of winning of 0.486, your odds of winning 8 games in a row are $0.486^8 = 0.003$, or about 1 in 330. At first, such small odds might make you think that you really are "hot." However, look around the casino: If it is a large casino, several hundred people may be playing at the craps tables. Thus at any particular moment, *someone* probably is having a "hot streak" of 8 straight wins; in fact, even longer hot streaks are likely during any given night. Your apparent "hot hand" is a mere coincidence, and your odds of winning on your next bet are still only 0.486. To increase your bet based on a mere coincidence would be foolish, to say the least!

❖

13.5.3 Probability, Psychology, and Pseudosciences

Despite the fact that coincidences are bound to happen, not to be impressed when a *particular* coincidence occurs is difficult. Psychologists have learned that people therefore tend to seek some *special* explanation for a coincidence, even when no explanation besides probability is needed. A few key ideas summarize the psychological effects of coincidences.

- Many seemingly astounding coincidences are to be *expected* under the laws of probability. Although a *particular* coincidence may be highly unlikely, it may be extremely likely (or even certain) that *some* coincidence will occur.
- The *psychological perception* of coincidences, when they happen, tends not to reflect their probabilistic nature. For example, if you personally experience a seemingly remarkable coincidence, you are likely to feel

that it is something special because amazing coincidences so rarely happen to you *in particular.*

- As time passes, people tend to remember the rare times that amazing coincidences happen but ignore the many more times that they don't. After all, if it doesn't happen, there's nothing to remember.
- Finally, people's minds tend to embellish their perception of coincidences over time. A coincidence with a 1 in 100 probability, for example, might later be remembered as something far more improbable. The result is that many people's claims of amazing coincidences are exaggerations of actual events.

Many superstitions and pseudoscientific beliefs stem from these psychological effects of coincidences. For example, a "psychic" who makes a guess that has a 1 in 50 chance of being correct has an expected number of 2 correct guesses in each 100. Thus, if 100 people pay to see this psychic, 1 to 3 of them probably will leave believing that the psychic read their minds. They may tell their friends of the wonderful mental powers of this person. The rest of the people, whom the psychic cannot "read," will have nothing to be impressed about; they may even keep quiet about their visits because they are embarrassed at spending money on a psychic.

Most professional "seers," such as psychics and astrologers, make claims with much better odds than 1 in 50. For example, a common prediction involves a guess at who will win the next election; if only 3 candidates are running, the odds are 1 in 3. Predicting whether the stock market will rise or fall has a 1 in 2 chance of being right. Moreover, if a seer is evaluating you, the seer may have additional evidence on which to base predictions, such as your approximate age and your wedding ring (or lack of), and they may be very good at reading your facial signals and body language.

Scientific studies of large numbers of these predictions indicate that no psychic or astrologer has ever done substantially better than should be expected based on probability. The only reasonable conclusion is that the claims of special powers by these seers are bunk.

Oct. 18–24 [1987]: New Moon on the 22nd helps to stimulate the positive aspects of the sky right now. Look for a strong rally [in the stock market] that could take us into December.... — prediction from the 1987 edition of a popular astrology guide. The biggest stock market crash in history occurred on October 19, 1987.

Time-Out to Think: Americans spend billions of dollars each year on psychics, astrologers, faith healers, and other pseudoscientific practitioners who routinely fail all scientific tests of their validity. If Americans, as a whole, had a better understanding of probability, do you think that so much money would be spent in this way? Based on what you know, is spending more money to study pseudoscientific claims worthwhile?

Example 13–30 *Predictive Dreams.* Perhaps you or someone you know has had a dream in which you see some event, and the event later happens. Such predictive dreams often are cited as evidence for mysterious mental powers. Rather than being supernatural, could such predictive dreams merely be coincidence? Explain.

Solution: One way to approach this problem is to estimate the probability that a dream is predictive. Suppose that a dream simply is a sequence of images that your brain selects from your experiences and concerns. For example, you might see your aunt, you might see your dog, and then you might see your dog biting your aunt. Let's suppose that the probability that this sequence of images, or some similar one, actually occurs is a relatively low 1 in 50,000. Psychologists say that people dream an average of five times per night, so the odds of a predictive dream on any particular night would be 1 in 10,000. Of the approximately 250 million people in the United States, then, some 250 million ÷ 10,000 = 25,000 people have predictive dreams on any particular night!

What about the probability of having predictive dreams two nights in a row? Because the probability of such a dream on one night is 1 in 10,000, the probability of having a predictive dream on two consecutive nights is 1 in $10{,}000^2$, or 1 in 100 million. Again, somewhere in the United States, a couple of people probably have had dreams come true both last night and the night before. Because of the very low probability of two consecutive nights of predictive dreams by *any* particular person, those who experience them are likely to think that they have some special mental powers. Yet, their occurrence is easily explained as coincidence: With 250 million dreamers, some will have dreams come true on consecutive nights, or even several times in a year. ❖

> **Time-Out to Think:** Do you think that the 1 in 50,000 estimate of the probability of a predictive dream is too high, too low, or about right? Based on this estimate and five dreams per night, the average person should expect a predictive dream about every 10,000 nights, or every 27 years. Have you ever had one?

13.6 RISK ANALYSIS AND DECISION MAKING

Approximately 100 to 200 people are killed by lightning each year in the United States, and several times this number are injured.

Probability also is involved in the analysis of risks and hence in the decisions that are based on risk assessment. Again, the probabilities often are surprising. For example, in the United States, several hundred people win prizes of a million dollars or more in state lotteries each year. Not bad — until you realize that roughly the same number are struck by lightning!

13.6.1 Will You Buy This Product?

Imagine a smooth-talking salesman whose advice you've followed many times in the past. He has been reliable, and his products always have performed as advertised. One day, he comes to you with a new product:

> "I can't reveal the details but, trust me, you will love this product! There is just one problem — it will kill everyone who uses it. Will you buy one?"

Not likely! After all, could any product be so great that you would die for it? A few weeks later, the salesman shows up again:

> "No one was buying, but we've made some wonderful improvements. Your odds of being killed by the product are now only 1 in 10. Interested?"

Despite the improvement, most people still would send the salesman home, and wait for his inevitable return:

> "Okay, this time the product is really ready. You'll love it. We've made it so safe, that it would take some 18 years for it to kill as many people as live in San Francisco. It can be yours, for a mere $10,000!"

From the early 1970s through the early 1990s, annual deaths from automobile accidents declined from a peak of more than 50,000 to less than 40,000. Most researchers trace the decline to two factors: improvements in automobile safety features and a reduction in the speed limit. In the mid-1990s, improvements in safety features continue to be made, but speed limits are being raised. Predicting the overall impact on the fatality rate is a matter of great debate.

You may be surprised to realize that, if you are like most people, you will jump at this offer. The product is, after all, the automobile. Approximately 40,000 Americans are killed each year in auto accidents; thus the equivalent of the population of San Francisco (1990 census about 725,000) is killed in auto accidents in about 18 years. The majority of adult Americans own an automobile, and new car prices today average well over $10,000.

Time-Out to Think: Because of the risk of death in an automobile and the price of a typical car, how much would you be willing to spend to buy a car with better safety features? How much time would you sacrifice to take a safer form of transportation (e.g., a train, if it were available for your commute)?

13.6.2 Dealing with Risk

Assessing the risk of accident or death in an automobile is relatively easy, for detailed records are kept by auto insurance companies and used to set rates. In general, however, risk assessment is far more complex and fraught with uncertainties.

For example, consider the case of asbestos, which is a known risk factor for lung cancer (see Chapter 3). However, by the time this risk was recognized, asbestos was already present in many homes, schools, and hospitals. During the past two decades, billions of dollars have been spent removing asbestos. Is the cost worth the benefit in reduced risk?

Answering this question is very difficult. The lung cancer risk from asbestos was assessed by studying people who worked with the material. The actual risk for people with very limited exposure to asbestos — which is nearly everyone — is much more difficult to quantify. Further, before spending money to remove asbestos you might wonder how this risk compares to other risks. For example, radon gas in homes is another risk factor for lung cancer. Suppose that you have a house with asbestos insulation and radon buildup. The cost of properly removing and disposing of the asbestos is likely to be at least $5000. The cost of mitigating radon buildup (by improving circulation of air between indoors and outdoors) typically is less than $1000. Are either or both of these expenditures sensible?

Another complication arises from the fact that the risk from asbestos is relatively well known, compared to many other potential risks. For example, more than 50,000 synthetic chemicals are commonly used throughout the world. In the vast majority of cases, these chemicals have never been tested rigorously for toxicity.

Time-Out to Think: When a new medical drug is discovered, the law requires that its safety and effectiveness be shown before it is brought to market. When a new industrial chemical is discovered, it can be used immediately; the law forces its withdrawal only if it proves to be dangerous. Any thoughts on the legal distinction between chemicals and drugs?

Imagine that a chemical spill releases a gas into a neighborhood and that the risk of lung damage from this gas is 1 in 1000. The neighborhood probably would be evacuated. Yet recent evidence suggests that the risk of lung damage from the urban air that most Americans breathe every day may be much higher than 1 in 1000. If anyone suffers from the chemical spill, the chemical company probably will be sued. Who gets sued for urban smog?

The most difficult risks to deal with are those that vary over wide extremes. An example is provided by the greenhouse effect (see Chapters 3 and 8). Some scientists believe that the greenhouse effect will cause only minor problems to which humanity can easily adapt. Others believe the consequences could be so severe that they might lead to the destruction of civilization, or even the extinction of the human species.

What can be done in the face of such uncertainty? Preventing further damage from the greenhouse effect requires severe curtailment of the use of fossil fuels, which might cost trillions of dollars to the global economy. If the greenhouse effect would cause only minor problems, spending trillions of dollars is foolish. However, doing so certainly is worthwhile if the consequences of doing nothing would be the destruction of civilization.

Time-Out to Think: Deciding what to do about the greenhouse effect, because it has the potential to destroy civilization, may be the most important decision made during your lifetime. What do *you* think should be done about it?

13.7 THE MATHEMATICS OF SOCIAL CHOICE

Once we have analyzed probabilities, how do we actually go about *making* a decision? In a personal decision, such as choosing a college, a logical choice is the school that you believe offers the highest probability of helping you attain your goals. In group or social situations, however, decision making is complicated because it involves the choices of many people simultaneously.

13.7.1 Voting Systems

The most direct application of mathematics to social choice is in the analysis of voting. Such analysis can be surprisingly complex because a common misconception about democracy is that it involves only majority rule. However, that often is not the case. For example, a two-thirds vote of Congress is required to override a presidential veto of legislation. In the Senate, a filibuster allows two-fifths of the members to prevent a vote on legislation. The Supreme Court can nullify a proposition passed by a majority vote if it is deemed to violate the Constitution.

In 1960, President Kennedy won the electoral vote over Richard Nixon by 303 to 219. However, in the popular vote, the result was 34,226,731 for Kennedy and 34,108,157 for Nixon.

For that matter, although it has never yet happened, a candidate for president could win without receiving the most votes. The reason is that presidential elections are decided by the **electoral college**, which is a "winner take all" system within each state. For example, suppose that candidate A gets 50.01%

of the vote in California, while candidate B gets 49.99%. Although this outcome is a virtual dead heat, candidate A would receive *all* of California's 54 electoral votes — which represents one-fifth of the total of 270 electoral votes required to win the presidency. Imagine that candidate A similarly wins several other large states by tiny margins and that candidate B wins the rest of the states by large margins. Candidate B might have more total votes, yet lose the election by receiving fewer *electoral* votes.

Even simple elections can have complications. For example, consider a three-way race for mayor. In some cities, the winner of the election is the person who receives the most votes, called a **plurality**. However, in other cities the two leading vote-getters are then matched one-on-one in a runoff election to decide the winner. Which system do you think is fairer?

Time-Out to Think: In the 1992 presidential election, Bill Clinton won with 43% of the popular vote. George Bush received 38% and independent Ross Perot received 19%. If national law required a runoff between Clinton and Bush, could Bush have been reelected over Clinton? Would that have been likely? Explain.

Further complications arise in **weighted voting systems**. For example, in one type of weighted voting system with four candidates for office, the voter selects first, second, and third choices. Three points are assigned for a first place vote, two for a second place vote, and one for a first place vote. The candidate with the largest number of total points is the winner. This type of voting system is used by sportswriters voting in the weekly college football poll.

The analysis of the subtleties of voting systems can become mathematically complex and can lead to major political battles. For example, every decade states redraw the maps of congressional districts in a process called **apportionment**. Researchers working for the political parties are constantly looking for ways to redraw districts so that their party is likely to gain more representatives in Congress. Clearly, the selection of a voting system can have a major impact on social decision making.

13.7.2 The Prisoner's Dilemma

Another aspect of social choice involves situations in which one person's success depends on choices made by another person. The mathematical subject of **game theory** involves playing out hypothetical situations of this type in order to explore and analyze various possible outcomes. For our purposes, game theory can be divided into two areas: **cooperative games** and **zero-sum**, or **competitive**, games. Cooperative games involve situations in which the players have their own individual interests but also can benefit by cooperating. In contrast, in zero-sum games one player's loss is the other's gain.

Joint undertakings stand a better chance when they benefit both sides.
— Euripides, c. 450 B.C.

The prototype cooperative game is called the **prisoner's dilemma**. Although it may seem a bit unrealistic on its own, this game is representative of many common situations. In its original form, the game features two people suspected of committing a crime who are taken into custody by the police. The two prisoners are isolated so that they cannot coordinate their statements to the police. Depending on their statements, three outcomes are possible.

- If both prisoners confess, both will be convicted with moderate sentences (some leniency is given because they confessed).
- If both prisoners maintain their innocence, both will receive light sentences (lighter than with confession because the prosecutor has no hard proof that they committed the crime).

- If one prisoner confesses but the other maintains innocence, the prisoner who confesses will be set free (for turning state's evidence) but the other will receive the maximum sentence.

The dilemma is clear. The first prisoner reasons that by maintaining his innocence he will either get a light sentence (if his partner also maintains her innocence) or the maximum sentence (if his partner confesses). If he confesses, the worst that can happen is a moderate sentence (if his partner also confesses); he will be set free if his partner maintains her innocence. The second prisoner reasons the same way. Overall, the *confess* strategy seems the safest choice for both prisoners because it eliminates the prospect of the maximum sentence and holds the possibility of going free. Yet, if they choose to confess, both get moderate sentences — a less desirable outcome than if they both proclaim their innocence.

Time-Out to Think: Put yourself in the position of one of the prisoners in the prisoner's dilemma game. If you knew that your partner was rational and trustworthy, which option would you choose? If you knew that your partner would try to take advantage of you, which option would you choose?

The prisoner's dilemma captures the familiar predicament of deciding when cooperation is advisable and when it might lead to being exploited. To illustrate how the prisoner's dilemma can be quantified, let's consider a more practical example. Suppose that two superpowers have just signed a disarmament agreement that requires some measure of trust on both sides. In the months following the signing, each country has two options. The *cooperative option* is to abide by the agreement and move toward disarmament, which we call the *dove* strategy. The *uncooperative option* is to cheat on the agreement and secretly maintain armaments in preparation for a possible conflict, which we call the *hawk* strategy

Table 13–5. Payoff Table (A, B)

	B: Dove	B: Hawk
A: Dove	(6, 6)	(0, 10)
A: Hawk	(10, 0)	(3, 3)

Table 13–5 shows the dove and hawk choices for power A in column 1 and for power B in row 1. For each of the four possible outcomes (dove–dove, dove–hawk, hawk–dove, and hawk–hawk) a number representing the benefit to each country on a scale of 0 (minimum benefit) to 10 (maximum benefit) is assigned. For example, if power A chooses the *dove* option and power B chooses the *hawk* option, power A has put itself at risk, leaving power B with an advantage. We show the relative **payoffs** for this situation with the notation (0, 10), which means no benefit for power A and maximum benefit for power B. If both powers choose the dove option, the mutual benefit is relatively high (6 on the scale of 10) for both powers, whereas if both powers choose the *hawk* option, the mutual benefit is relatively small (3 on the scale of 10) for both powers.

Time-Out to Think: Thinking in economic, political, and nationalistic terms, what might the numbers in Table 13–5 represent?

Now the thinking of the prisoner's dilemma applies. If each power is unaware of the other's decision, how will each power act? Consider the options of power A: If it chooses the *dove* option, its possible benefits are 6 (if B also chooses *dove*) and 0 (if B chooses hawk); if it chooses the *hawk* option, its possible benefits are 10 (if B chooses *dove*) and 3 (if B chooses hawk). Thus power A does at least as well as power B if it chooses the *hawk* option, *regardless* of how power B acts. A similar analysis holds for power B: It also will gain greater benefits with the *hawk* option. Thus the likely outcome will be that both countries choose the hawk option, even though both would be better off by trusting each other to choose the *dove* option.

The situation in which both powers choose the *hawk* position is said to be *stable*: If one power changes from the hawk option (while the other maintains that position), it will have a smaller payoff, so little incentive exists to switch to the *dove* choice. If both powers choose the *dove* option, they realize the greatest *mutual* benefit. However, this situation is *unstable* because, if one power switches to *hawk* before the other, it can realize even greater benefits.

To the extent that the prisoner's dilemma applies to arms races and social issues, it seems to carry a pessimistic message. The safest course of action is for countries or individuals to act in their own interests (the *hawk* option) because it removes the possibility of being exploited. However, this strategy does not yield the greatest overall benefits. Furthermore, even if cooperation is chosen, yielding the highest mutual benefit, the temptation remains to act selfishly for even greater benefits. A vivid example of the prisoner's dilemma is played out in many environmental issues. In many instances individuals or societies realize short-term gains by polluting, overpopulating, or depleting natural resources. But the long-term global health of the planet does not benefit from this kind of behavior.

Fortunately, the prisoner's dilemma is unrealistic in several ways. First, more than two options and two players generally are involved in complex political or social situations, and the choices are rarely so well defined. More important, in real situations, the prisoner's dilemma may be played out more than once. Surprisingly, mathematical study of repeated play of the prisoner's dilemma shows that cooperation *can* be the best choice.

13.7.3 The Mathematics of Cooperation

University of Michigan political scientist Robert Axelrod studied repeated play of the prisoner's dilemma in the early 1980s. He created a computer model of the game, and programmed the computer to take the role of the two players. His model included various strategies for the players in the repeated "rounds" of the prisoner's dilemma so that he could compare the outcomes of the different strategies over time. The six overall strategies that he tested can be summarized as follows.

1. Be cooperative in every round: Always act in the greatest interest of all players.
2. Be uncooperative in every round: Always act for immediate personal gain.
3. Flip a coin and make a random response in each round: Keep your opponent guessing with unpredictable responses.
4. Alternate between cooperative and uncooperative each round: Change your response, but in a predictable way.
5. Play *tit-for-tat*: Be cooperative during the first round and then do what your opponent did in the previous round.
6. Be cooperative as long as your opponent cooperates, but once the opponent fails to cooperate, be uncooperative for the rest of the game.

His results were quite surprising and probably would not have been discovered without the aid of the computer model. First, he found that the results depend on whether the players know how many rounds will be played. If the players have that information, as in the original prisoner's dilemma, being uncooperative every round was the best overall strategy.

However, things become very interesting if neither player knows how long the game will last — which generally is the case in most real social and political situations. For example, if both players always are uncooperative, both players always have a *small* mutual benefit. If player A chooses strategy 1, of always being cooperative, player B can exploit the situation by always being uncooperative; player B always would get the maximum benefit and player A always would get the minimum. But what happens if the players choose an *adaptive* strategy (each response depending on the opponent's previous response) such as that of strategies 5 and 6?

Axelrod's computer tournaments showed that the superior strategy is *tit-for-tat*. With a motto of "do unto others as they last did to you," tit-for-tat is cooperative in spirit, but cannot be systematically exploited. In his book *The Evolution of Cooperation*, Axelrod describes this effect:

> *[The tit-for-tat strategy has the] combination of being nice, retaliatory, forgiving, and clear. Its niceness prevents it from getting into unnecessary trouble. Its retaliation discourages the other side from persisting whenever defection is tried. Its forgiveness helps restore mutual cooperation. And its clarity makes it intelligible to the other player, thereby eliciting long-term cooperation.*

More recent work with computer-assisted prisoner's dilemma games involving more than two players has reaffirmed the long-term benefits, even in a competitive environment, of cooperative strategies. Although these games may still be idealizations of real situations, they offer hope that the best way to respond to each other and to our surroundings is with some measure of cooperation.

13.8* COMPETITIVE OR ZERO-SUM GAMES

Game theory is a relatively recent mathematical topic. John von Neumann, often called the "father" of modern computing, worked on two-person, zero-sum games as early as 1928. In 1944, he joined forces with economist Oskar Morgenstern to publish the book *Theory of Games and Economic Behavior*, regarded as the first testament of modern game theory.

Soon after the book's appearance, hopes ran high that at last a firm mathematical framework had been developed for resolving social and political conflicts (compare these hopes to Leibniz's dream of Chapter 5). Researchers discovered many applications of game theory, but its limitations also became apparent. Real situations often involve many players (governments, businesses, or individuals) with many possible strategies and motivations. Furthermore, the payoffs (often called *utility functions*) for the players in real situations cannot always be measured in dollars. Important, but intangible, factors such as national pride, revenge, generosity, and public image often enter into decisions but are difficult to quantify. For these reasons, game theory lost some of its promise and following during the 1960s and 1970s.

However, interest in game theory has been resurgent in recent years. The availability of powerful computers has made possible devising and analyzing game theory models that are far more realistic, leading to surprising discoveries about complex social and political systems. Indeed, the 1994 Nobel Prize in Economics was awarded to mathematician John Nash for work in game theory that he began in the early 1950s.

We described game theory briefly with our discussion of the prisoner's dilemma and cooperative games in the Section 13.7. Here, as a further example of how game theory works, we examine competitive games, which differ from cooperative games in some fundamental ways.

13.8.1 Pure Strategies and Saddle Points

Imagine that Ann and Bob are two high-powered real estate developers, both of whom are planning to build new shopping centers. Ann has three possible locations for her new shopping center: locations (or strategies) A_1, A_2, and A_3. Bob also has three possible locations (or strategies): B_1, B_2, and B_3. The two shopping centers will compete for customers, and the relative value of one shopping center depends both on its location and the location of the other shopping center. Thus for each pair of locations the developers have a projected gain or loss in terms of future profits, which are shown (in millions of dollars) in a Table 13–6.

Table 13–6. Payoff Table for Shopping Center

		Bob			
		Location B_1	**Location B_2**	**Location B_3**	**Row minimum**
	Location A_1	0	–2	–4	–4
Ann	**Location A_2**	6	5	8	5
	Location A_3	–5	3	–1	–5
	Column maximum	6	5	8	

Ann's three strategies (A_1, A_2, and A_3) label the rows, and Bob's three options (B_1, B_2, and B_3) label the columns. The standard convention in payoff tables is that the numbers in the table are *the payoff from the column player* (Bob) *to the row player* (Ann). For example, if Ann chooses location A_2 and Bob chooses location B_2, Ann will gain \$5 million in future sales at Bob's expense. If Ann chooses strategy A_1 and Bob chooses strategy B_3, because the payoff is negative Bob will gain \$4 million at Ann's expense.

Time–Out to Think: What pair of locations will give Ann the greatest payoff? What pair of locations will give Bob the greatest payoff? Remember that negative numbers in the table represent the amount that Ann loses to Bob.

Let's assume that neither Bob nor Ann knows which location the other will choose. Because the numbers in the table give the amount that Ann gains at Bob's expense, she hopes to pick the location that will give her the greatest payoff regardless of where Bob builds. The *worst* she can do at each location is represented by the minimum numbers in each row; these minimum payoffs are listed in the column labeled *row minimum* (in this case, –4, 5, –5). For example, if she chooses location A_1, the worst that can happen is that Bob chooses location B_3 and that she *loses* \$4 million. Her best overall choice, then, is the location that has the maximum of these minimum payoffs, which is location A_2 with a minimum payoff of \$5 million.

Ann is using what often is called a **maximin strategy**: For her three choices of location, she finds the maximum of the minimum payoffs.

Bob's thinking is complementary to Ann's. The payoffs in the table represent how much he loses, so he seeks to minimize the payoff number. He first looks down the columns associated with his three locations and finds the *maximum* payoff in each column, which represents the worst possible outcome at each location; these payoffs are listed in the column labeled *column maximum* (in this case, 6, 5, 8). His best choice is the location that minimizes these worst possible outcomes, which is location B_2. This choice assures Bob that regardless of the location that Ann chooses, he will lose *at most* \$5 million.

Bob has used what often is called a **minimax strategy**: For his three choices of location, he has found the minimum of the maximum payoffs.

The net result is that Ann chooses location A_2 and Bob chooses location B_2, with Ann gaining \$5 million at Bob's expense. Although both Ann and Bob probably hoped to do better, this choice ensures that both are protected against even worse outcomes. That is, the choice of locations A_2B_2 is stable: If Ann changes her strategy (with Bob's location B_2 fixed), she would do worse; if Bob changes his choice of location (with A_2 fixed), he would do worse.

This problem is an example of a competitive game with a special kind of solution called a **saddle point.** The point in the table that represents the joint locations A_2B_2 (the "solution" to the game) is like a saddle or a mountain pass: It lies at the *low point of a row* and at the *high point of a column* in the payoff table. The players are said to have a **pure strategy** because one choice of location works best for each player all the time. The payoff associated with the saddle point (in this case, \$5 million) is called the *value* of the game. By the way, the conclusion of this analysis is that this real estate "game" is biased towards Ann perhaps because she has a better choice of locations to begin with. Bob might be well advised to devise some different strategies, or perhaps take up a different occupation

Time–Out to Think: If Ann were to change her mind and choose location A_1 for her shopping center, could she get a higher payoff (assuming that Bob stays with B_2)? If Bob were to change his mind and choose location B_3 for his shopping center, could he get a better payoff (assuming that Ann stays with A_2)?

Any competitive game can be formulated in terms of a payoff table in which the options and rewards for each player are specified. Competitive games also are called **zero–sum** games because one player's loss is the other player's gain.

A Note on Dominated Strategies

The payoff table for a competitive game often shows that one player has a strategy that *always* does worse than another strategy. For example, in Table 13–7, strategy A_1 always gives player A a better result than strategy A_3. The poorer strategy (A_3 in this case) is said to be a *dominated* by the better strategy (A_1 in this case), and a rational player would never choose it. Assuming that the players are rational, we can therefore eliminate the dominated strategy (row A_3 in this case) without changing the nature of the game. In the rest of our work, we assume that any dominated strategies have already been eliminated from payoff tables.

Table 13–7. Payoff Table With a Dominated Strategy.

	B_1	B_2
A_1	1	–2
A_2	4	1
A_3	0	–3

13.8.2 Mixed Strategies

Both players used pure strategies in the preceding real estate game, leading to a saddle point solution. However, this strategy isn't always possible. Consider Table 13–8, which shows the strategies of a swimsuit manufacturer (player A). The numbers in this payoff table represent thousands of dollars.

Table 13–8. Payoff Table for Swimsuits.

		Player B	
		B_1 (Cool)	B_2 (Warm)
Player A	A_1 (Small)	3	–2
	A_2 (Large)	–1	4

In this case, the swimsuit manufacturer (player A) has two strategies: A_1 is to make a small quantity of swimsuits for the coming summer, and A_2 is to make a large quantity of swimsuits for the coming summer. Here, the opponent (player B) is not a human competitor, but rather Mother Nature, who has two "strategies": B_1 is that the coming summer is relatively cool, and B_2 is that the coming summer is relatively warm. If the manufacturer correctly guesses the weather conditions for the coming summer, the payoffs will be positive: \$3000 if the manufacturer makes a small number of swimsuits for a cool summer and \$4000 if the manufacturer makes a large number of swimsuits for a hot summer. If the manufacturer miscalculates and produces too many swimsuits for a cool summer or too few swimsuits for a hot summer, the company will take a loss.

The swimsuit manufacturer is the only player in this game who is able to choose a strategy. Player A might reason that she could lose as much as \$2000 if she chooses A_1; to be safe, she might choose A_2, which guarantees that her worst outcome is a \$1000 loss. In thinking this way, player A has found her optimal strategy.

The next question is: If the swimsuit manufacturer must play this game for several summers in a row, how often should she use strategy A_1? How often should she use strategy A_2?

In this case a single pure strategy will *not* give the swimsuit manufacturer the best payoff in the long run. She must therefore use a **mixed strategy.** Let's say that, over the long run (several summers), she will use option A_1 a fraction p of the time, where p is a number between 0 and 1 (we could also think of p as the probability of using option A_1). With only two options, she will use option A_2 a fraction $1 - p$ of the time. Now, suppose that the summers always are cool; that is, strategy B_1 always applies. Then player A will have a payoff of 3 (remember that numbers represent thousands of dollars in this case) with a probability of p and a payoff of –1 with a probability of $1 - p$. Therefore the expected value of player A's payoff against strategy B_1 is

$$(3 \times p) + [(-1) \times (1 - p)] = 4p - 1.$$

For example, if the swimsuit manufacturer uses strategy A_1 one-fourth of the time ($p = 1/4$) and A_2 three-fourths of the time, she would expect a payoff of $4 \times (1/4) - 1 = 0$ over the long run for cool summers.

Similarly, if strategy B_2 (warm summers) always applies, player A's expected payoff is

$$[(-2) \times p] + [4 \times (1 - p)] = -6p + 4.$$

For example, if the swimsuit manufacturer uses strategy A_1 with a probability of 1/3 and A_2 with a probability of 2/3, she would expect a payoff of $-6 \times (1/3) + 4 = 2$ over the long run for warm summers.

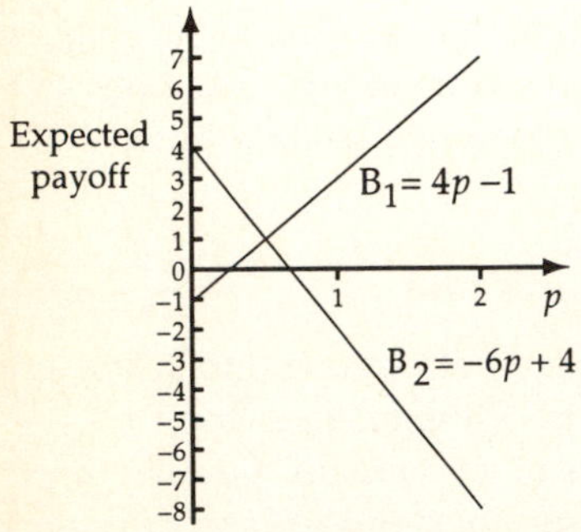

Figure 13-5. The two payoffs for Player A are based on conditions B_1 (rising line) and B_2 (falling line). The variable p is the fraction of the time that Player A uses strategy A_1. The best value of p for Player A corresponds to the intersection of the two lines.

To determine player A's best overall strategy, we can graph the two payoff relations (the expected payoff against B_1 and the expected payoff against B_2). Note that both relations are linear with respect to the variable p. Using the methods described in Chapter 10, we plotted the graphs shown in Figure 13–5, with the expected payoff on the vertical axis and the probability p (that player A will use strategy A_1) on the horizontal axis. The domains of the relations are the allowed values for p, which are between 0 and 1.

If condition B_1 applies all the time, player A (look at the downward sloping line) does best choosing $p = 0$. This choice corresponds to using strategy A_2 always with its payoff of 4. Alternatively, if condition B_2 always applies, player A (look at the upward sloping line) should choose $p = 1$, which means always using strategy A_1 for the payoff of 3. However, Mother Nature is *not* likely to use the same "strategy" all the time, so player A must choose a value of p that gives the best payoff *on average* over many repetitions of the game. This best choice is the value of p that corresponds to the intersection point of the two expected payoff lines. She can find this value of p by setting the two payoff values equal to each other:

$$\underbrace{4p-1}_{\substack{\text{Expected payoff} \\ \text{against } B_1}} = \underbrace{-6p+4}_{\substack{\text{Expected payoff} \\ \text{against } B_2}} \xrightarrow{\text{solve}} p = \frac{1}{2}.$$

In other words, if player A uses strategy A_1 half the time ($p = 1/2$), in the long run she can expect the best payoff. Evaluating either of the expected payoff relations with $p = 1/2$, she finds that the expected payoff is 1. Thus the value of the game is 1, or $1000, as the payoffs are thousands of dollars.

This method has another useful outcome. Suppose the swimsuit manufacturer knows (perhaps from records of past summers) that the coming summer is more likely to be warm than cool; for example, suppose that the chances of a warm summer are 2 in 3. That is, the probability that Mother Nature applies strategy B_1 is 1/3 and the probability that Mother Nature plays strategy B_2 is 2/3. In this case, the expected payoff for player A is

$$\begin{aligned}\text{Expected payoff} &= \frac{1}{3}\times\text{Expected payoff against } B_1 + \frac{2}{3}\times\text{Expected payoff against } B_2 \\ &= \frac{1}{3}\times(4p-1)+\frac{2}{3}\times(-6p+4) = -\frac{8}{3}p+\frac{7}{3}.\end{aligned}$$

In this case, how should player A choose p to obtain the best possible payoff? Note that when $p = 0$ (player A always uses strategy A_2), the expected payoff is $7/3 = 2.33$. But when $p = 1$ (player A always uses strategy A_1), the expected payoff is $-1/3$ (a loss). So in this case, when the summer is more likely to be warm, the best strategy is A_2 — making a lot of swimsuits.

Because of the structure of the payoff table, some of the strategies that arise are surprising. You should confirm that, if there is a 50–50 chance for a warm summer, the swimsuit manufacturer still is better off if she makes a large quantity of swimsuits. If the chances of a cool summer are 3 in 5, the swimsuit manufacturer can expect a payoff of 1 from *either* strategy!

The preceding analysis is typical of the approach used for all two-person competitive games. The graphic analysis can also be carried out if the two players have more than two strategies. A fundamental result of game theory, proved by John von Neumann, is that all games of this nature (two-person, zero-sum) have a unique value and an optimal mixed strategy for each player.

CONCLUSION

In this chapter we explored the mathematics of probability and how it influences our lives. Key lessons to remember from this chapter include the following.

- Probability plays a role in virtually all decisions. Understanding probability therefore is necessary for making wise decisions.
- Because most people's natural intuition for probability is poor, calculating probabilities whenever possible is preferable to simply guessing at them. With practice, intuition for probabilities can improve.
- Whether gambling at a casino, guessing the best business strategy, or deciding how to evaluate the risk of an environmental problem, you should always evaluate the probabilities involved. The results often are surprising, and can help you to live your life more sensibly.

PROBLEMS

Reflection and Review

Section 13.1

1. **Probability in Your Life.** If you look deeply, you will find that probability is involved in *every* decision you make. For each part of this problem trace your decision and identify and briefly discuss the places where probability entered into your decision.
   a. Why did you decide to enroll in college?
   b. If you exercise regularly, explain why; if you don't, explain why not.
   c. If you are like most drivers, you probably exceed the posted speed limit occasionally. Explain how you decide when to drive at or below the speed limit and when to exceed it.
   d. Explain your choice for president in the last election. If you did not vote, explain why not.
   e. Most grocery stores offer a choice between paper and plastic bags; in addition, you can always bring your own reusable bags for groceries. Explain your choice.
   f. Automobiles cause significant damage to the Earth's environment. For example, they require many raw materials for construction, they use some of the world's limited remaining supply of oil to operate, and they emit pollutants into the air. Nevertheless, most people drive cars, or at least travel in cars driven by others. Explain why *you* drive a car (or travel in cars driven by others).
2. **Certainty and Impossibility.** We often hear "nothing is impossible," or "anything is possible if you try hard enough." Do such sayings mean that if you try hard enough, you can live without food for 10 years? Can you jump out of an airplane and fall up? Briefly discuss the concepts of *impossible* and *certain*. How are they related to probability?
3. **Induction and Analogy.** Study the quotation from Laplace on the title page of this chapter. Briefly explain, with examples, how induction and analogy are based on probability. Do you agree with Laplace's conclusion? Why or why not?
4. **Probability and Learning.** Write two paragraphs about the role of probability in learning. For example, how was probability involved in the development of your study habits? How does probability help explain why "practice makes perfect"? Taking this idea a step further, can you use probability to explain how a child learns to speak? Explain.

Section 13.2

5. **A Priori Card Probabilities.** A deck of 52 cards has four suits (13 cards each of diamonds, hearts, clubs, and spades); each suit has cards numbered 2–10, jack, queen, king, and ace. Follow the three-step process in the text to find a priori probabilities for each of the following.
   a. If you draw 1 card from a well-shuffled deck, what is the probability that it will be a king?
   b. What is the probability that it will be a card *other than* a king?
   c. What is the probability of drawing a spade?
   d. What is the probability of drawing a card with a number on it (i.e., *not* an ace, jack, queen, or king)?
6. **Bag of Marbles.** Suppose that you have a bag of colored marbles containing 10 white marbles, 5 blue marbles, and 2 red marbles. You reach into the bag and draw out 1 marble. Follow the three-step process to find the a priori probability that the marble is red.
7. **Four-Coin Probability Distribution.**
   a. Make a table showing all the possible outcomes from tossing four coins at once.
   b. Consolidate duplicate results to show the *probability distribution* when you toss four coins.
   c. What is the probability of getting three heads and one tail when you toss four coins at once?
   d. What is the probability of getting *anything but* all tails?
8. **Dice Probability Distribution.** Consider two dice (each die has six faces, showing 1 through 6 dots).
   a. As shown in Figure 13–2, the two dice may land in 36 possible ways. We can modify that table to show the sum of the two dice by adding each pair of numbers. Complete the following table to show these sums.

+	1	2	3	4	5	6
1	2					
2						
3		5				
4						
5						11
6						

   b. Note that the sum of the two dice can range from 2 through 12. Make a table showing the *probability distribution* for all possible sums when tossing two 6-faced dice.
   c. What is the probability of getting a total of 5 on a toss of two dice?
   d. What is the probability of getting *anything but* a 5?
   e. What is the most common result from tossing two dice? What is the probability of this result?
   f. Suppose that someone offers you 10 to 1 odds in betting that you won't roll a 7. That is, if you don't roll a 7 you lose $1, but if you do roll a 7 you win $10. In the long run, would you expect to win or lose at this game? Why?
9. **Models of Probability.** Although a priori probabilities depend on some kind of model, empirical probabilities do not. Briefly explain why empirical probabilities do not depend on a model.

10. Coin Expectations. Suppose that you toss a fair coin 1000 times. Should you *expect* it to come up heads exactly 500 times? Why or why not? Explain without making any calculations.

11. California Floods. Some areas of California have already experienced more than one "100-year flood" during the 1990s. Can you conclude that the historical or geological records, used to determine the probability of a major flood, were interpreted incorrectly? Why or why not? Discuss some of the assumptions that underlie using a historical or geologic record to estimate a present-day probability.

12. Coin Experiment. Take three coins and toss them simultaneously. Record your results as either zero heads (0H), one head (1H), two heads (2H), or three heads (3H). Repeat this experiment 40 times.

a. Make a table of your results, similar to that in Example 13–5, showing both frequency and relative frequency.
b. State your empirical probabilities for each of the four possible results: P(0H), P(1H), P(2H), and P(3H).
c. Compare your empirical probabilities to the a priori probabilities obtained in Example 13–4. Explain any differences. If your experiment produced the a priori probabilities exactly, should you be surprised?

13. Fair Coins? Imagine that you do an experiment in which you toss three coins 1000 times and that the following table shows the results.

Number of Heads	Frequency
0	260
1	495
2	245
3	0

a. Compute the empirical probability for each outcome.
b. Compare the empirical probabilities from the experiment to the a priori probabilities for three tossed coins. Are the coins unfair?
c. Someone suggests that the reason for your strange result is that one of the three coins has tails on both sides. Do you believe her? Why or why not?

14. Subjective Probabilities. State your own subjective probabilities for each of the following and justify each with a brief explanation.

a. The probability that your school's football team will rank in the top 10 at the end of the next football season.
b. The probability that, 20 years from now, you will feel that your collegiate experience was of great value to your life and career.
c. The probability that you will be married in 2010.
d. The probability that a Democrat will be elected president in the next presidential election.
e. The probability that terrorists will obtain and use a nuclear weapon within the next 10 years.
f. The probability that your state legislature's latest "get tough on crime" legislation will significantly reduce the crime rate.
g. The probability that the United States will suffer a major catastrophe (e.g., a world war, anarchy, or environmental devastation) during the next 75 years.

Section 13.3

15. Card Probabilities with Replacement. Suppose that you draw cards from a deck, replacing each card after it is drawn and shuffling the deck.

a. What is the probability of drawing an ace on the first try?
b. What is the probability that both of your first two draws will be aces?
c. What is the probability that your first four draws all will be aces?
d. What is the probability that your first five draws all will be aces?

16. Joint Probabilities. Use the rules for joint probabilities to answer each of the following questions. Be sure to decide whether the events are dependent or independent.

a. If you throw two dice, what is the probability of getting a 6 on both?
b. If you toss a (fair) coin six times, what is the probability of getting all heads?
c. Assume that the odds of getting a winning lottery ticket ($2 or more) are 1 in 10. What is the probability of purchasing three winning tickets in a row?
d. Suppose that you draw cards from a deck, but after each draw you replace the card and shuffle the deck. What is the probability of drawing an ace on the first try and a king on the second?

17. Consecutive 500-Year Floods. Ordinarily, you might feel relatively safe living in an area that, according to historical and geologic records, floods only once every 500 years. Thus, if a flood destroys your home you might well rebuild, imagining that another flood during your lifetime is unlikely. Suppose, however, that your area floods in *two* consecutive years. Should you just chalk it up to bad luck, or should you look for another explanation? Explain your reasoning. Based on your answer, would you rebuild your house a second time? Why or why not?

18. Drawing Candy. Suppose that you draw candy from a bag that initially contains 10 pieces of candy: 5 chocolate and 5 caramel. Suppose that after drawing the first piece of candy, you put it back in the bag. In that case, what is the probability of drawing two pieces of chocolate in a row? Compare this result to the case where you eat the first-drawn candy, as discussed in subsection 13.2.2.

19. Card Probabilities Without Replacement. Suppose that you draw cards from a deck, without replacement.

a. What is the probability of drawing an ace on the first try?
b. What is the probability that both of your first two draws will be aces?
c. What is the probability that your first four draws all will be aces?
d. Suppose that you draw five cards from a deck. What is the probability that they will be, *in order*, the ace of

spades, king of spades, queen of spades, jack of spades, and 10 of spades?

20. **Child Probabilities.** Suppose that you meet a boy named Samuel, who tells you that he has two siblings.
    a. What is the probability that both siblings are brothers?
    b. Samuel then tells you that he is the oldest of the three children in his family. In that case, what is the probability that both siblings are brothers? Explain.

21. **Either/Or Probabilities.** Use the rules for either/or probabilities to answer each question. Be sure to decide whether the events are mutually exclusive.
    a. What is the probability, on a single draw, of drawing either a 2 or a 4 from a deck of 52 cards?
    b. A group of people comprises of 25 Democratic men, 25 Republican men, 25 Democratic women, and 25 Republican women. If you assign each person in the group a number and then randomly select one of these numbers, what is the probability that the person selected will be a woman *or* a democrat?
    c. When drawing from a deck of cards, what is the probability that you will draw either a king, an ace, or a diamond?
    d. When rolling a pair of dice, what is the probability that you will roll a total of 6, 7, or 8?

22. **Better Bet for the Chevalier.** Suppose that the Chevalier de Méré had bet that he could roll a double 6 within 25 rolls, rather than 24.
    a. In that case, what would have been his odds of winning?
    b. Had he made this bet, would he still have lost over time? Do you think that he would still have called the mathematician Pascal for help? Explain.
    c. Comment on how the Chevalier's influence on mathematics illustrates how seemingly minor events can lead to major changes in human history. How is this idea related to probability? How is it related to the butterfly effect in chaos, as discussed in Chapter 3?

23. **Lottery Odds.** The probability of a $2 winner in a particular state lottery is 1 in 10, the probability of a $5 winner is 1 in 50, and the probability of a $10 dollar winner is 1 in 500.
    a. What is the probability of getting a $2, $5, or $10 dollar winner? Compare to the probability of getting only a $2 winner.
    b. If you buy 50 lottery tickets, what is the probability that you will get at least one $5 winner?
    c. If you buy 500 lottery tickets, what is the probability that you will get at least one $10 winner?

24. **Miami Hurricanes.** Studies of the Florida everglades show that, historically, the Miami region is hit by a hurricane about every 40 years.
    a. Based on the historical record, what is the empirical probability that Miami will be hit by a hurricane next year?
    b. What is the probability that Miami will be hit by hurricanes in 2 consecutive years?
    c. What is the probability that Miami will avoid a hurricane for 10 consecutive years?
    d. Suppose that the greenhouse effect causes hurricanes to become twice as common in the future as they have been in the recent past. How would that alter your answers to parts (b), (c), and (d)?
    e. What kinds of building codes would you recommend for Miami? Explain.

25. **AIDS Among College Students.** Suppose that 2% of the students at a particular college are infected with HIV (rather than the 3% assumed in Example 13–17).
    a. If a student has 6 sexual partners in each of 4 years of college, what are the odds that at least 1 of these partners is infected with HIV?
    b. If a student has sex, on average, with a new partner each month, what is the probability that at least 1 partner encountered during a 4-year period is infected with HIV?

Section 13.4

26. **Gambler's Fallacy.** You have lost several games in row at a casino, and are now deep in a hole.
    a. Does the law of averages suggest that you will now start winning if you continue to bet? Why or why not?
    b. What would the gambler's fallacy "predict" in this situation?
    c. Explain how the gambler's fallacy can therefore lead a gambler to ruin.

27. **Behind in Coin Tossing: Can You Catch Up?** Say that you have tossed a fair coin 100 times, getting 38 heads and 62 tails, which is 24 more tails than heads.
    a. Explain why, on your next toss, the difference in the number of heads and tails is as likely to grow to 25 as it is to shrink to 23.
    b. Extend your explanation from part (a) to explain why, if you toss the coin 1000 more times, the final difference in the number of heads and tails is as likely to be larger than 24 as it is to be smaller than 24.
    c. Suppose that you continue to toss the coin forever. Explain why the following statement is true: If you stop at any random time, you always are more likely to have fewer heads than tails, in total.
    d. Suppose that you are betting on heads with each coin toss. After the first 100 tosses, you are well on the losing side (having lost the bet 62 times while winning only 38 times). Explain why, if you continue to bet, the odds are that you will remain on the losing side.
    e. Explain how your answer in part (d) illustrates the gambler's fallacy.

28. **Number of Children.** Assume that boys and girls are equally probable (each 50%) at birth and that a family has 4 children.
    a. What is the mean (expected value) number of boys in a family of 4 children?
    b. What is the probability that all 4 children are girls?
    c. What is the probability that the family will have 2 boys and 2 girls?
    d. What is the probability that the family will have 1 boy and 3 girls?
    e. If a family has 4 children, should you expect that they have 2 boys and 2 girls? Explain.

29. Expectations in Coin Tossing. Suppose that you toss a (fair) coin 50 times.
- a. What is the probability of getting heads 25 times?
- b. What is the probability that you will *not* get 25 heads?
- c. If you toss a coin 50 times, should you *expect* to get 25 heads? Explain.

30. Where Do Lottery Winnings Go? If you spend $10 on lottery tickets, you likely will lose most or all your money. In fact, *most* people lose *most* of the money that they spend on the lottery. Nevertheless, lottery officials typically advertise that they return half of the money received on tickets to the public in the form of cash prizes. If they are telling the truth, how can most people lose most of the money they spend on the lottery?

31. Defective Products. You randomly select 5 samples of a product that has a 10% defect rate (1 of 10 are defective on average).
- a. With the aid of the binomial probability formula, calculate the probabilities for getting 0 defective products, 1 defective product, 2 defective products, and so on, in your sample of 5. Use your results to complete the following probability distribution. We have calculated the probability of 2 defective products for you. (Hint: Be sure that the total of the probabilities is 1.)

Number of Defectives	Probability
0	
1	
2	0.07290
3	
4	
5	
Total	**1.0000**

- b. What is the probability of getting 2 *or more* defective products in the sample of 5?

32. One Hundred Million Coin Tosses. Suppose that 100 million coins are tossed simultaneously. Find and interpret the mean and standard deviation in the possible number of heads.

33. Casino Profitability. Example 13–28 showed that the most likely number of games patrons will win is 497,000, with a 99.7% chance that the patrons will win between 495,500 and 498,500 games.
- a. What is the most likely number of games that the casino will win?
- b. What is the 99.7% confidence range in the number of games that the casino will win?
- c. What is the likelihood that patrons will win *more* than 498,500 games? (Hint: Don't forget that the remaining 0.3% is the total probability that patrons will win *either* fewer than 495,500 games or more than 498,500 games.)
- d. Based on your results so far, briefly discuss the likelihood that the casino will lose more games than it wins.
- e. Under the most likely results, how many more games will the casino win than the patrons?
- f. If the average bet on this game is $5, how much profit can the casino expect on the 1 million times this game is played each week?

34. Casino Estimation. Imagine a casino that only has *craps* tables where, if players make the best possible bets, the house (casino) wins 51% of the time.
- a. Estimate the number of people that might come to a given casino on any day and the average amount of cash that people bring. Casinos come in many sizes, so explain your estimate.
- b. Based on your estimates from part (a), how much money would people bring into the casino on an average day?
- c. Assume that the casino can match the amount of money estimated in part (b) and that the total money in play is thus twice this amount. At the end of the day, what is the casino's expected profit? What is the patrons' expected net loss?
- d. Is the casino ever likely to have a losing day? Explain.
- e. How much profit do you expect this casino to make in a year?

Section 13.5

35. Chance Encounter. While on vacation hundreds of miles from home, you meet someone and strike up a conversation. To your surprise, you find that you both have a common acquaintance.
- a. For this problem, assume that the average American has 1000 acquaintances (whose names would be recognized in a conversation). Do you think this estimate is reasonable? Explain.
- b. If each person's 1000 acquaintances are spread throughout the country, how many *linkages* exist between the acquaintances of two people? (Hint: Suppose that one person knows only John and Bob and that the other knows Sue and Tracy. That gives four possible linkages: John knows Sue, John knows Tracy, Bob knows Sue, and Bob knows Tracy.)
- c. For the 250 million people in the United States, what is the probability that you will have a common acquaintance with a person that you meet at random?
- d. Next, consider a linkage through two intermediates; that is, you know someone who knows someone who knows your new friend. How many linkages are there among three groups of 1000 people (assuming random distribution, with no overlap in the groups)? Compare this number to the population of the United States.
- e. The result in part (d) suggests that you are likely to be linked at random to any person by just two intermediate acquaintances (someone who knows someone). Discuss any complicating factors in this result.

36. Streaks. Toss a coin 100 times and record your results (heads or tails) in order.
- a. What were your overall results (numbers of heads and tails)? Are these overall results in any way surprising? Explain.
- b. Describe each streak of three or more consecutive heads or tails. Where does it appear? How many consecutive heads or tails did you get?

c. Calculate the odds of each particular streak you had.
d. Were any of your streaks unlikely? Explain.

37. **Birthday Coincidences.** Suppose that 30 students are in your class.
a. Calculate the probability that someone in the class has the same birthday as yours.
b. Calculate the probability that at least two people in the class have the same birthday, though not necessarily *your* birthday. (Hint: If you have set up the problem correctly, the following fact will save you a lot of time in punching buttons on your calculator:
$$365 \times 364 \times \ldots \times 337 \times 336 = 2.2 \times 10^{76}.)$$

38. **Birthday Intuition.** Ask a group of your friends the following two questions, and keep track of their answers.

Question 1: How many people must be in a group to be sure that at least two people have the same birthday?

Question 2: How many people must be in a group to have a 50–50 chance that two people have the same birthday?
a. What is the correct answer to question 1? Explain.
b. What fraction of your friends answered question 1 correctly? Are you surprised? Why or why not?
c. Show that the correct answer to question 2 is 23. (Hint: Follow the technique used for birthdays in a group of 25 people in Section 13.5.)
d. Did any of your friends answer question 2 correctly? How do the answers given by most of your friends compare to the correct answer?
e. Based on your results, can you make any general statements about "natural" intuition for probability?

39. **Predictive Dreams.** Assume that the probability of a predictive dream during any night is 1 in 10,000.
a. What is the probability that you will *not* have a predictive dream during any given night?
b. What is the probability that you will not have a predictive dream over a period of 1 year? What is the complementary probability that you will have at least one predictive dream during a 1-year period?
c. Based on your result in part (b), about how many people in the United States have predictive dreams each year? About how many people in the world have predictive dreams each year?
d. How old are you? Based on your age, what is the probability that you would not have had a predictive dream in your life? What is the probability that you have had a predictive dream?
e. Have you ever had a predictive dream? If so, describe it.
f. We have assumed a 1 in 50,000 probability for a dream being predictive (with five dreams per night). Does this seem reasonable? Explain.

40. **The Predictive Power of Eclipses.** Some astrologers claim that major events often are preceded by eclipses. The truth of this claim, of course, depends on what is considered a "major" event.
a. The number of eclipses (both lunar and solar) in any year varies. As many as seven eclipses are possible in any single year, but the average number is about four per year. On average, what is the probability that a major event will occur within 3 weeks after an eclipse?
b. Based on the probability you calculated in part (a), should you be surprised if eclipses often precede major events? Explain.
c. Astrologer Grace Morris claimed (Associated Press, November 3, 1990) that the Iraqi invasion of Kuwait was precipitated by a lunar eclipse. Evaluate her claim. The following facts may be useful: (i) the invasion of Kuwait occurred on August 2, 1990; (ii) the lunar eclipse occurred on August 6, 1990; (iii) the timing of eclipses is easily predicted; both astronomers and astrologers knew *years* in advance that there would be an eclipse on August 6, 1990; and (iv) note that her statement about the prediction was made on November 3, 1990.

41. **Joe DiMaggio's Record.** One of the longest standing records in sports is the 56-game hitting streak (in baseball) of Joe DiMaggio. In this problem, you are to estimate the odds of such a record. To do so assume that a player on a long "hot streak" is batting about .400, which is about the best that anyone ever hits over a period of 50 or more games. (A batting average of .400 means the batter gets a hit 40% of the time. Typically, only a handful of players each year hit that well for a period as long as 56 games.)
a. Suppose that a player gets to bat about four times per game. What is the probability that a player batting .400 will *not* get a hit (in four tries) in a game? (Hint: If the player has a probability of 0.4 of getting a hit in any particular at-bat, what is the probability that he will *not* get a hit in that at-bat? that he will not get a hit in four at-bats?)
b. What is the probability that he will get at least one hit in four at-bats in a game?
c. Use the result in part (b) to calculate the probability of a .400 hitter getting a hit in 56 consecutive games.
d. Suppose that, instead of batting .400, a player has a more ordinary average of .300. In that case, what is the probability of the player's getting a hit in 56 consecutive games? Again, assume four at-bats per game.
e. Considering the results in parts (c) and (d) and that baseball has been played for about 100 years, are you surprised that *someone* set a record in which he got a hit in 56 consecutive games? Explain clearly. Do you think that DiMaggio's record ever will be broken? Why or why not?

42. **Tabloid Predictions.** At the beginning of the year (or end of the previous year) many tabloids and magazines publish a set of predictions for the coming year made by psychics, astrologers, and other "seers." Find a set of these predictions from last year.
a. State whether each prediction came true.
b. Does the overall rate at which the predictions came true seem consistent with probability? Explain.
c. Look for predictions by the same person for next year. Are the person's successes from the previous year cited? the person's failures? Discuss.

Section 13.6

43. Risks from Automobiles. Approximately 40,000 Americans are killed in automobile accidents each year.

a. Assume that the distribution of automobile accidents is random with equal probabilities for all Americans. What is the probability that you will be killed in an automobile accident this year? (In reality, the probability will vary depending on factors such as whether you are intoxicated, whether you drive in poor weather, etc.)

b. What is the probability that you will *not* be killed in an automobile accident this year?

c. What is the probability that you will not be killed in an automobile accident during the next 50 years (assuming that the accident rate remains unchanged in the future)?

d. What is the probability that you will die in an automobile accident sometime during the next 50 years? Are you surprised that auto accidents are the leading cause of accidental death in this country? Why or why not?

44. Automobiles and Vietnam. Advances in automobile safety (e.g., seat belts and better bumpers) have lowered mortality in auto accidents significantly. During the 1960s and 1970s, approximately 50,000 people were killed in traffic accidents each year.

a. The Vietnam War saw Americans being killed in combat over a period of about a decade spanning most of the 60's and the early 70's. Approximately 50,000 Americans were killed in the Vietnam War, and their names are listed on the beautiful and moving Vietnam Memorial in Washington, D.C. Compare the number of Americans killed in the Vietnam War with the number killed at home in auto accidents during the same period of time.

b. The Vietnam War spawned a vast protest movement intended to get the United States out of Vietnam. No similar protest movement has ever been launched against the automobile. Write a short essay on why the public judges some risks to be "acceptable," even when they may be far more dangerous than other risks deemed "unacceptable." Support your position with further examples of relatively low-risk activities that have generated organized protest in contrast to high-risk activities that go unchallenged.

45. Airline Safety. According to the *1992 World Almanac and Book of Fact*, there were 6.9 million commercial airplane flights during 1990. Of these, six fatal flights resulted in a total of 39 deaths.

a. Estimate the total number of passengers on all commercial flights in 1990. Explain your estimate. What was the probability that a particular passenger on a particular flight would be killed in 1990?

b. Assume that the probability you found in part (a) is still the same (i.e., that 1990 was a typical year for airline safety). What is the probability that a person who takes 10 plane trips will be killed?

c. Another way of considering airplane safety is in terms of passenger miles, or the number of miles flown by all passengers. Estimate the total number of passenger miles for all of the flights in 1990. Explain your estimate.

d. Use the result in part (c) to calculate the probability of death in a plane crash per passenger mile. What is the probability of death if you fly 1000 miles? What is the probability of death if you fly 20,000 miles?

e. Estimate how far the average person travels by automobile each year. Explain your estimate.

f. Suppose that you drive and fly the same number of miles. Which is safer?

g. Compare the relative safety of driving and flying. Defend your conclusion. For example, are comparisons based on passenger miles (which you calculated in parts d through f) a good way to evaluate safety? Explain.

h. Many people are afraid to fly, whereas few are afraid to travel in an automobile. Is this fear justified? Explain. How would you coax a person who is afraid of flying to take a trip by air rather than by car?

Section 13.7

46. Parliamentary Power. To show how strict majority voting can cause surprising consequences, evaluate each of the following.

a. Suppose that the Senate has the following breakdown in party representation: 49 Democrats, 49 Republicans, and 2 Independents. Further, suppose that all Democrats and Republicans vote along party lines. Assuming that a majority is required to pass a bill, explain why the 2 Independents, despite holding only 2% of the Senate seats, effectively hold power equal to that of either of the large parties.

b. Imagine that a small company has four shareholders who hold 26%, 26%, 25%, and 23% of the company's stock. Assume that votes are assigned in proportion to share holding (e.g., if there are a total of 100 votes, the four people get 26, 26, 25, and 23 votes, respectively). Also assume that decisions are made by strict majority vote. Explain why, although each individual holds roughly one-fourth of the company's stock, the individual with 23% holds *no* effective power in voting.

47. Three-Way Race. Imagine a three-way race for governor among Smith, Jones, and Johnson. About 65% of the population are strongly opposed to Johnson. The remaining 35% are avid Johnson supporters.

a. Assume that the person with the most votes wins the election. Describe a scenario in which Johnson ends up being governor, despite being actively disliked by the vast majority (65%) of voters.

b. If the state has a runoff in which the top two vote-getters in the primary face each other in a second election, can Johnson still win? Explain.

c. Which of the two systems would you consider to be more democratic? Why?

48. A Prisoner's Dilemma. Consider a prisoner's dilemma in which the possible sentences for two suspects are given in the following table. The first number in each payoff pair is the sentence for suspect A (the row player), and the second number is the sentence for suspect B. Subscript 1 indicates confession, and subscript 2 indicates a claim of innocence. For example, if both suspects confess (A_1, B_1), both get 6 years; if suspect A confesses and suspect B maintains innocence (A_1, B_2), suspect A gets 10 years and suspect B gets 1 year.

	B_1	B_2
A_1	(6, 6)	(10, 1)
A_2	(1, 10)	(3, 3)

a. If suspect A confesses (A_1), what is the minimum sentence that he can expect? If suspect A maintains his innocence (A_2), what is the minimum sentence that he can expect? Which strategy seems safest for suspect A?
b. If suspect B confesses (B_1), what is the minimum sentence that she can expect? If suspect B maintains her innocence (B_2), what is the minimum sentence that she can expect? Which strategy seems safest for suspect B?
c. If both suspects maintain their innocence (A_2, B_2), could either suspect get a shorter sentence by switching from either A_2 or B_2? Is the choice (A_2, B_2) stable or unstable?
d. If both suspects confess (A_1, B_1), could either suspect benefit by switching from either A_1 or B_1? Is the choice (A_1, B_1) stable or unstable?
e. Which of the strategies, A_1 or A_2, is cooperative and which is uncooperative?
f. Which of the strategies, B_1 or B_2, is cooperative and which is uncooperative?

49. **International Diplomacy.** Recall the international diplomacy game discussed in subsection 13.7.2 with the payoff table shown below. The first number in the payoff pairs is the relative benefit for power A, and the second number is the relative benefit for power B.

		Power B	
		Dove	Hawk
Power A	Dove	(7, 7)	(1, 10)
	Hawk	(10, 1)	(3, 3)

a. If you were power A and believed that relations with power B had improved during the past year, what strategy would you choose? Why?
b. If you were power A and believed that relations with power B had deteriorated significantly during the past year what strategy would you choose? Why?

Section 13.8

50. **Interpreting a Competitive Game.** Consider the following payoff table for a competitive (zero-sum) game. Recall the convention that the positive numbers in the table are the amounts that player B pays player A, and that negative numbers in the table are the amounts that player A pays player B.

a. If player A chooses strategy A_2 and player B chooses strategy B_3, who pays whom how much?

	B_1	B_2	B_3
A_1	0	−3	4
A_2	1	0	2
A_3	2	−2	0

b. If player A chooses strategy A_3 and player B chooses strategy B_1, who pays whom how much?
c. For each of player A's rows, find the minimum value in each row. What is the maximum value of these row minima? Call this number the maximin.
d. For each of player B's columns, find the maximum value in each column. What is the minimum value of these column maxima? Call this number the minimax.
e. Does the minimax equal the maximin? Explain.

51. **Dominated Strategies.** A payoff table is labeled *Original Game* (remember the convention that the numbers in the table are the amounts that player B pays player A).
a. For player A, is one strategy *always* less preferable than another strategy regardless of how player B plays? Show that if this less desirable (or dominated) strategy is removed the payoff table becomes the one labeled *Reduced No. 1*.
b. Player B in looking for the *smallest* numbers in *Reduced No. 1* table. Is one strategy for player B *always* less preferable than another strategy regardless of how player A plays? Show that by removing this dominated strategy, the payoff table becomes the one labeled *Reduced No. 2*.
c. Now, player B has only one choice. What is the best choice for player A?

Original Game

	B_1	B_2
A_1	1	−2
A_2	4	1
A_3	0	−3

Reduced No. 1

	B_1	B_2
A_1	1	−2
A_2	4	1

Reduced No. 2

	B_2
A_1	−2
A_2	1

52. **Finding Saddle Points.** Which of the following payoff matrices has a saddle point solution? Find the optimal strategies for each player and the value of the game for the games with saddle points.

	B_1	B_2
A_1	1	−2
A_2	4	−3

	B_1	B_2
A_1	1	2
A_2	−1	−5
A_3	0	4

	B_1	B_2	B_3
A_1	1	3	4
A_2	4	1	2
A_3	0	2	0

53. **Dueling Pennies.** Both you (player A) and your opponent (player B) have a penny that you can lay on the table (at the same time) to show either heads or tails. The payoff table (in dollars) for this game is shown (as always the numbers give the amount that player B pays player A).

	B plays H	B plays T
A plays H	1	1
A plays T	−1	1

a. Verify that this game has no saddle point solution.
b. Assume that the game is to be played repeatedly and that you (player A) want to know in what fraction, p, of the games you should play heads and in what fraction of the games $1 - p$ you should play tails. In terms of p, what is your expected payoff if player B plays heads? In terms of p, what is your expected payoff if player B plays tails?
c. Use the straight-line graphing method shown in the text to determine your best value of p. With this value of p, what is the value of the game?
d. Repeat the analysis for player B to determine her best value of p.
e. Is the value of the game the same for both you and your opponent?

Further Topics and Applications

54. The Morality of Gambling. In a 1995 speech, Republican Senator and presidential candidate Richard Lugar attacked state supported lotteries and the trend toward increasing legalization of gambling. Specifically, he said: "The spread of gambling is a measure of the moral erosion taking place in our country.... It says that if you play enough, you can hit the jackpot and be freed from the discipline of self-support through a job or the long commitment to ongoing education."

Write a short essay explaining what you think Senator Lugar meant and describing whether you agree with his assessment of gambling.

55. Counting Cards. In blackjack, the object of the game is to receive cards whose value totals as close to 21 as possible, without exceeding it. The probability that you are dealt, say, an ace depends on which cards have already been dealt.

a. Suppose that you receive the first card dealt from a deck of 52 cards. What is the probability that it is an ace?
b. Suppose that you can see 10 of the cards that have been dealt (either because they are showing face up on the table or because they are in your own hand) and that 2 of them are aces. What is the probability that the next card dealt will be an ace?
c. Based on a comparison of the results in parts (a) and (b), explain why card counting (counting the number of cards of each type that have appeared on the table) can improve your odds of winning at blackjack.
d. If you have an excellent memory and a good system for computing the probabilities in blackjack, you can tilt the odds of winning in your favor (odds of winning more than 50%) by card counting. To combat such card counters, many casinos deal blackjack by shuffling more than one deck of cards together. Explain why this strategy helps the casinos combat card counters.
e. Suppose that you are an excellent card counter and that you tilt the odds of winning in your favor by 52% (your odds) to 48% (casino odds). Suppose that each game of blackjack takes 3 minutes. If your average bet is $10 per game, what are your expected earnings per hour of play? Is this a lucrative job?

56. Mail Sweepstakes. You receive a notice in the mail saying that you are eligible to win $1 million in a sweepstakes, simply by filling out and returning the enclosed card. As it turns out, you are one of 10 million people who return the card, which makes your odds of winning 1 in 10 million.

a. Assume that you spend 32¢ to mail the card back to the sweepstakes. Calculate your expected earnings (or losses) in sending back the card.
b. Suppose that you spend 10 minutes filling out and returning the card and that your time is worth $6 per hour. Including the monetary value of the time you devote to the sweepstakes card, what are your expected earnings?
c. Suppose that you decide to purchase three magazine subscriptions, which also are offered to you as an option when you return the card. Later, you learn that you could have purchased the same subscriptions through a student discount service for $7 less than you paid. What are your expected earnings now?
d. Suppose that, on average, each of the 10 million entrants spends an extra $7 on subscriptions. After paying out the $1 million prize, how much profit will the sweepstakes company have earned?

57. Lottery Psychology. Imagine a lottery that will offer a prize of $1 million dollars to just one person. In order to offer the prize, the lottery must collect $2 million in revenue.

a. Suppose that you are part of a crowd of 50,000 attending a college football game. To generate $2 million in revenue, each person in the stadium would have to contribute $40 to the lottery. Imagine looking around the stadium at all the other people. Would *you* hand over $40 for the chance at the $1 million prize? Do you think that other people would pay $40 to enter this lottery? Why or why not?
b. Per capita lottery spending in the United States is several hundred dollars per person, for odds about the same as described in part (a). Presumably few, if any, of the people in the stadium would enter the lottery described. Why then do you think so many people buy lottery tickets at a store?

58. Poker Hands. In a deck of 52 cards, each card is unique (e.g., king of spades, 5 of diamonds, etc.). Suppose that you are playing 5-card poker, in which you receive 5 cards from the initial deal.

a. How many different hands are possible?
b. A "royal flush" means that your 5 cards are the ace, king, queen, jack, and 10 — all of the same suit. What is the probability of getting a royal flush of spades?
c. What is the probability of getting a royal flush of any suit?
d. (*Challenge Problem*) What is the probability of getting four aces in a five-card hand? What is the probability of getting four of a kind? (Hint: You need to know only two facts: the total number of possible five-card hands, and the number of different hands that have four aces, or four of a kind.)

59. **Roulette.** In the game of roulette there are 18 black numbers, 18 red numbers, and two green numbers (0 and 00). Spinning the wheel randomly selects any one of the 38 numbers. The highest probability bet that you can make in a game of roulette is to bet on a color; for example, you could bet that the result will be a red number.
   a. If you bet on red (or black), what is your probability of winning on a single spin of the wheel? Explain why it is the highest probability bet that you can make in roulette.
   b. What is the probability that the casino will win (and you will lose) when you bet on a color?
   c. Consider a single table in a casino that is open 24 hours a day. Suppose that, on average, the wheel spins once each minute. How many times does it spin per day? How many times does it spin per year?
   d. Write a paragraph discussing the implications of the probabilities you obtained to a casino in which the roulette wheel spins tens of million of times each year.

60. **Probable Weather.** Weather forecasts often state probabilities. Consider a forecast calling for a 20% chance of snow on a given day.
   a. What is the probability of *no* snow on that day?
   b. Suppose that the same forecast is made for 5 different days. What is the probability of snow on at least 1 of those days? (Hint: Consider the probability that no snow falls on any of those days).
   c. What is the probability of snow on 2 of the 5 days?
   d. (*Challenge Problem*) Write a general formula for the probability of snow on at least 1 day during a period of n days if each day has a 20% chance of snow.
   e. Based on the formula obtained in part (d), calculate the number of "20% snow" days must be forecast to be 90% sure of snow on at least 1 of those days. (Hint: You will need to take the logarithm of both sides.)
   f. How many days would need to be forecast to be 99% sure of snow on at least one "20% snow" day?
   g. Based on the result in part (e), if no snow fell on any of those days, what would you conclude about the weather forecaster's ability?
   h. How do the results obtained in parts (a)–(g) change if the chance of snow is 50% rather than 20%?

61. **Finding Mutant Cells.** Biologists often are interested in studying mutant cells. To do so, they often subject a group of cells to radiation which causes many of the cells to mutate "artificially." If a biologist irradiates 1000 cells and, after exposure, the frequency of mutant cells is 0.005, how many cells must be examined to give a 95% chance of finding at least one mutant cell?

62. **Termite Genetics.** Geneticists sometimes study relatively simple organisms, such as termites, to make their work easier. A particular gene in termites has two variants, designated A and a. The A variant is much more common, occurring in 90% of termites, whereas the a variant occurs in only 10% of termites. Each termite carries a pair of these genes; that is, each one has AA, Aa, or aa genes. The termite receives one gene from its mother and one gene from its father. Because of the frequency of the A variant, most termites carry AA, some carry Aa, and only a few carry aa.

   A typical termite colony has one queen, one king, and many workers, all of which are the offspring of the single queen and king; thus all the workers are brothers and sisters. Suppose that a particular genetic test examines the genes of 24 termites at the same time. Furthermore, suppose that you can't find the queens and kings and can test only once. To maximize the probability of finding a termite with the a gene variant, should you test 24 termites from a single colony, test 12 termites each from two colonies, or use some other strategy? Explain.

63. **Major Bash.** The Earth is bombarded continually by interplanetary debris, ranging from dustlike particles to large meteorites. When large meteorites hit the earth, noticeable impact craters are created. The probability of a meteorite 100 meters in diameter hitting the Earth is approximately 1 in 10,000 per year.
   a. What is the probability that a 100-meter meteorite will *not* hit the Earth in a given year?
   b. Give an order of magnitude estimate of the number of years left in your life. Explain.
   c. What is the probability that at least one 100-meter meteorite will hit the Earth during the remainder of your lifetime?
   d. After how many years does the probability that at least one 100-meter meteorite hits the Earth reach 90%?
   e. Does your answer in part (c) suggest that you should worry about such an event during your lifetime? Explain.
   f. Research the expected effects of the impact of a 100-meter meteorite. How large a crater would it make? How might it affect human civilization? Are similar impacts from the recent past known?

64. **Fate and the Uncertainty Principle.** If the universe were deterministic, what role would probability play in nature? Do you think that the uncertainty introduced by quantum physics has any bearing on the idea of fate?

65. **A Legal Case.** Consider the legal example presented in subsection 13.5.1 in which an assault takes place in a city of 2 million people. Eyewitnesses agree on several important characteristics of the assailant such as height, skin color, weight, hair color and length, beard, and glasses; they also agree on the getaway car's model, year, and color. A few days later, police pick up a man matching the physical description and driving the right type of car. Although he has no prior arrest record, and proclaims his innocence, he is charged with the crime.

   Prosecutors show that the probability of any particular man matching all of the described is only about 1 in 2 million (e.g., by estimating the fraction of men in the city that have a beard, the fraction of cars that match the description of the getaway car, and so on). However, the more relevant question is whether there might be *another* person in the city matching the eyewitness description, given that one such person has already been found. This problem leads you through the calculation of this probability.

   a. The probability of that a particular man matches the eyewitness description is $p = 1/2{,}000{,}000$ and the probability of a non-match is $q = 1{,}999{,}999/2{,}000{,}000$. The number of males in the city is $n = 1{,}000{,}000$ (half of the 2 million population). Calculate the probability that *no* male in the city matches the description.

b. Use the binomial probability formula to calculate the probability that exactly one male matches the eyewitness description.

c. Use your result from part (a) to calculate the probability that *at least one* male in the city matches the eyewitness description.

d. Combine your results from parts (a) and (b) to find the probability that *more than one* male (i.e., at least two) in the city matches the eyewitness description.

e. Let event A represent at least one male matching the description, and let event B represent more than one male matching the description. Recall the formula for the joint dependent probability: P(A *and* B) = P(A) × P(B given A).

Note that P(A and B) represents the probability that *both* at least one *and* more than one male match the description, which is the same as the probability from part (d) that more than one male matches the description. P(A) is the probability from part (c) that at least one male matches the description. P(A given B) is the probability that more than one male matches the description, given that one such person has already been found; use the conditional probability formula to calculate this probability.

f. Briefly discuss the ramifications of probability to the legal case described in this problem.

66. School Voting System. Find out and describe how representatives are selected for the student government at your school. Is the system an example of strict majority rule? Do you think that the system is fair? Why or why not?

67. Prisoner's Dilemma in a Restaurant. Two people dining together have already agreed to split the check. To keep things simple, assume that there are two menu items, one that costs $10 and one that costs $20. As each person silently peruses the menu, the choice is between selecting the $10 item (and possibly contributing to the cost of the other person's $20 dinner) and selecting the $20 item (and possibly getting a similar benefit). A possible payoff table for this situation is shown. Each payoff is the *difference* between the actual price of the meal and what the player ends up paying.

	B chooses $10 meal	**B chooses $20 meal**
A chooses $10 meal	(0, 0)	(5, –5)
A chooses $20 meal	(–5, 5)	(0, 0)

a. How were the numbers in the table determined? First, if both diners choose the $10 meal, how much does each pay after splitting the check? Did either diner pay extra?

b. If both diners choose the $20 meal, how much does each pay after splitting the check? Did either diner pay extra?

c. If diner A chooses the $10 meal and diner B chooses the $20 meal, what is the cost to each diner after splitting the check? How much more does diner A pay than the cost of the meal he selected?

d. If diner A chooses the $20 meal and diner B chooses the $10 meal, what is the cost to each diner after splitting the check? How much more does diner B pay than the cost of the meal she selected?

e. If you were either diner, which dinner would you order? Would your choice depend on whether you were dining with a friend? Would your choice depend on whether you dined with this person once a week? Explain.

68. Vacation Planning. Imagine that Adele and Bruce are planning a vacation, and though they love each other very much, they cannot agree on where and when to travel. The following table shows the options that Bruce and Adele are considering. They can travel in one of three months (April, August, or January), and they have three resorts on their list (A, B, and C). The table shows the expected temperature (°F) at each of the three resorts in each month; and here is where the problem arises. Bruce wants to vacation in the coolest possible weather, but Adele is looking for heat. Knowing a little game theory, Adele and Bruce agree that Adele should choose the month for the vacation and Bruce should choose the resort.

		Bruce			
		Res A	**Res B**	**Res C**	**Row min.**
Adele	**Apr**	50	60	55	
	Aug	90	80	100	
	Jan	40	50	45	
	Col. max.				

a. Think along with Adele first. She is looking for the maximum temperature in the table, so she starts by finding the minimum temperature in each row. Fill in these values in the column marked *row minimum*.

b. What is the maximum of Adele's row minima? Which month should she choose? With this choice, what is the lowest temperature she can expect?

c. Now think along with Bruce. He is looking for the minimum temperature in the table, so he starts by finding the maximum temperature in each column. Fill in these values in the row marked *column maximum*.

d. What is the minimum of Bruce's column maxima? Which resort should he choose? With this choice, what is the maximum temperature he can expect?

e. Is Bruce's optimal solution equal to Adele's optimal solution? Argue that this game has a saddle point.

f. Once Adele and Bruce have chosen their saddle point choices, can Bruce improve his situation by changing (assuming that Adele remains with her choice)? Can Adele improve her situation by changing (assuming that Bruce remains with his choice)?

69. **Soccer Strategy.** The leading scorer in the league has a penalty kick against the best goalie. The kicker (player A) has two strategies: A_1 is to kick the ball to the center of the goal, and A_2 is to kick to the side. The goalie (player B) also has two strategies: B_1 is to defend the center of the goal, and B_2 is to defend the side. The probability of scoring a goal in each of the four cases is shown in the following payoff table.

	B plays center	B plays side
A kicks center	0.2	0.9
A kicks side	0.7	0.3

a. What is the best strategy for both the kicker and the goalie?
b. What is the value of this game?
c. Generalize your results to formulate a slightly more realistic soccer game in which the kicker has three strategies: kick center, kick right, and kick left. The goalie also has three strategies: defend center, defend right, and defend left. Assign payoffs (probabilities of a goal) to each of the nine situations. Then find the optimal strategies for the kicker and goalie. What is the value of your game?

70. **Parking Meter Wars.** You (player A) have a choice of paying a machine $10 at a self-pay parking lot or trying to get away without paying. The parking attendant has a choice of patrolling the lot or patrolling elsewhere. The cost to you of each of these strategies is shown in the following payoff table. Note that the payoffs are zero or negative because, in most cases, you will pay either a fine or the parking fee. The payoff for the no pay/no patrol choices reflects the additional frustration of paying the fee ($10) and not being checked.

	Patrol	No patrol
Pay	–$10	–$15
No pay	–$25	0

a. If you regularly pay the parking machine half the times you park and the attendant patrols three-fourths of the time you park, what is your long-term cost of parking in the lot?
b. What is your optimal strategy in this game? What is the value of the game?

Projects

71. **State Lotteries.** Research the history and status of lotteries in the United States.
a. How many states now have lotteries? Is there a lottery in your state?
b. What is the per capita spending on lotteries in the United States? Can you find a state-by-state breakdown in per capita spending?
c. Describe at least three different lottery games. Choose games from your state, if it has a lottery. Otherwise, choose games from a state of your choice. For each game, find or calculate the odds of wining prizes of various amounts.
d. Research any general trends in who spends money on lotteries. For example, is per capita lottery spending different among people in different income groups?
e. Some people believe that lotteries encourage compulsive gambling, but others disagree. Research and describe the evidence on both sides. What do *you* think?

72. **AIDS Probability.** Find out how many people are currently infected with the virus that causes AIDS in the United States and worldwide. Then do any or all of the following.
a. Obtain a breakdown in the number and fraction of HIV-infected individuals by gender and behavior. For example, find the number of infected heterosexual women and the fraction of the population of heterosexual women infected. Do the same for other categories of people, such as heterosexual men, homosexual men, intravenous drug users, and so on. Based on your findings, discuss the probability that a person from each group met through a random encounter will be HIV infected.
b. Obtain a breakdown in the number and fraction of HIV-infected individuals by nation. Discuss your findings. For example, is the infection rate growing at a particularly high rate in some nations? If so, why?
c. Research the probability that the AIDS virus is transmitted in a single sexual encounter. How is the probability affected by the use of condoms? by other forms of contraception?
d. Sexual relations are not the only way to contract the AIDS virus. Research and discuss other ways that the virus can be contracted, along with estimated probabilities for contracting the disease by these methods.
e. Research one or more aspects of AIDS itself. For example: What are its symptoms? How does it kill people? What are the origins of the disease? Why is it proving so hard to cure? What progress has been made toward AIDS prevention or cure?

73. **Floods.** Research a recent episode of extreme flooding somewhere in the world. Describe the flooding and its effects on both people and the environment. For whatever flooding you choose to study, be sure to answer the following.
a. What is the historical probability of a flood of this magnitude in the region where the flood occurred?
b. Is there any evidence that human activity has increased the natural probability of flooding? Explain.

74. **Greenhouse Effect Policy Debate.** Research the current state of predictions about the greenhouse effect. For example, by how much do different models predict that sea level will rise? Do researchers assign probabilities to any of these possible results? Based on your research, hold a policy debate in class concerning what should be done to reduce or eliminate greenhouse emissions.

75. **Random Numbers.** The probabilities calculated for many games and many problems depend on the idea that num-

bers are truly random. However, generating truly random numbers is a difficult task.

a. **Roulette:** In the early 1980s a pair of graduate students in physics studied the game of roulette in detail. They discovered that, for any particular roulette wheel, the outcome of spins was *not* truly random. For example, a particular roulette wheel might tend to have more balls fall into one particular octant (a "slice" of the wheel containing one-eighth of the numbered slots) more often than others. By observing this pattern, and then making bets on the actual probabilities for the particular roulette wheel, the students found that the odds of winning could be raised to slightly over 50%. Suggest some reasons why a particular roulette wheel is unlikely to generate truly random results.

b. **Random Number Generator:** Many calculators have a key that is designed to generate a random number when pressed. Find the random number key on a calculator. Either through experimentation, or consulting the calculator manual, determine what kind of number is generated (e.g., is it always between 0 and 1?). Press the random number button at least 100 times and record your results. What is the average value of these random numbers? Do any apparent patterns emerge in these numbers? Explain why *some* patterns (but not *particular* patterns) are to be expected, even if the numbers are truly random. How could you determine whether the random number button is generating truly random numbers?

c. **Code Breaking:** Many coding schemes rely on random numbers to generate codes that are supposed to be hard to break. Research one of the following three important episodes of code breaking: (1) the British breaking of German codes in World War II; (2) the American breaking of Japanese codes in World War II; or (3) the American breaking of Soviet Union codes in the early 1960s. For whichever code you choose, discuss the role (if any) played by random numbers in the code and in breaking it.

76. Voting Systems. Choose one or more of the following voting systems to research.

a. **College Football Poll.** Describe the voting for one of college football's polls. Do you think that the voting *system* is fair? Do you think that the outcome is generally reasonable? Discuss why some people believe that a national playoff should decide the number 1 team. Do you agree? Why or why not?

b. **Quota Queen?** In 1994, President Clinton nominated Lani Guinier for a position in his administration. Facing Senate opposition, the nomination eventually was withdrawn. Guinier was labeled a "quota queen" by some members of the Senate because she had written articles advocating what is called a cumulative voting procedure. Research Guinier's suggested voting system. Explain how it differs from current voting systems. Explain why she believed that this system would be more democratic than current systems. Do you agree? Do you think that the label of "quota queen" was justified? Explain.

c. **Shareholder Voting.** Choose a large public corporation and find out how shareholders vote on corporate decisions. Which decisions are made by shareholders? How are votes assigned (e.g., one person–one vote, one share–one vote, etc.)? Do you think that the system is fair? Explain.

77. Let's Make a Deal. In September 1990, controversy about a probability question erupted over an item in the "Ask Marilyn" column in *Parade* magazine (carried by many newspapers across the country). Her column answered a question loosely based on the once-popular TV game show called "Let's Make a Deal." The question was:

> *Suppose that you're on a game show and you're given the choice of three doors: Behind one door is a car; behind the others, goats. You pick a door, say No. 1, and the host, who knows what's behind the doors, opens another door, say No. 3, which has a goat. He then says to you, "Do you want to pick door No. 2?" Is it to your advantage to switch your choice?*

Marilyn answered that the probability of winning was higher if the contestant switched. This answer generated a huge number of letters, including a few from mathematicians, claiming that she was wrong.

a. Marilyn answered with the following logic. When you first pick door No. 1, the chance that you picked the one with the car is 1/3. The probability that you chose a door with a goat is 2/3. When the host opens door No. 3 to reveal a goat, it does not change the 1/3 probability that you picked the right door in the first place. Thus, as only one other door remains, the probability that it contains the car is 2/3. Briefly discuss this logic. Do you agree with it?

b. Another way to evaluate the problem is by analyzing all the possibilities. The prizes could be arranged behind the three doors in three possible ways:

	Door 1	**Door 2**	**Door 3**
Case 1:	Car	Goat	Goat
Case 2:	Goat	Car	Goat
Case 3:	Goat	Goat	Car

Assume that you choose door 1. Clearly, your initial probability of winning the car is 1/3. Now, suppose that the host opens *one of the two* remaining doors to reveal a goat: In case 1, which of the two doors he opens doesn't matter because both have goats behind them; in cases 2 or 3, he must use his knowledge of where the car is located in deciding which of the two doors to open. Analyze your probability of winning if you hold to door 1 versus switching to the one the host did not open. Can you show that you have a 2/3 chance of winning by switching?

c. If you still are not convinced, here is a third way to analyze the problem. Suppose that there are 100 doors instead of 3 but that only one has a car behind it. After you pick a door, the host opens 98 of the remaining 99 doors to reveal goats. Explain why, at this point, you'd be a fool not to switch to the one remaining door. Explain why this similar problem shows that switching in the original problem also is best.

d. Find back issues of *Parade* magazine and read the columns about this controversy. Discuss what happened and any lessons that were (or should have been) learned.

e. This problem differs from the real "Let's Make a Deal" show in one crucial respect: We evaluated only a *single* instance of the game. In the real game, the host did not always open a second door and give the contestant the opportunity to switch. Martin Gardner, a popular writer on mathematical subjects, wrote: "If the host is malevolent, he may open another door only when it's to his advantage to let the player switch, and the probability of being right by switching could be as low as zero." Explain Gardner's statement.

78. Repeated Play of Prisoner's Dilemma. Consider the international diplomacy game discussed in subsection 13.7.2 with the payoff table shown. Remember that the first number in the payoff pair is the relative benefit for power A, and that the second number is the relative benefit for power B.

		Power B	
		Dove	**Hawk**
Power A	**Dove**	(7, 7)	(1, 10)
	Hawk	(10, 1)	(3, 3)

a. Choose a friend or classmate as an opponent and play a round of the diplomacy game. Each player should make his or her choice by secretly writing *Dove* or *Hawk* on a slip of paper; then both slips of paper are revealed at the same time and the payoffs are recorded.

b. Now experiment with repeated rounds of the diplomacy game. Without telling your opponent, start playing one of the first four strategies listed in subsection 13.7.3. Comment on the responses of your opponent to your strategy. What can you say about your cumulative payoffs as the game progresses?

c. Next, start the repeated play again. This time (without telling your opponent) choose the *tit-for-tat* strategy. Comment on the responses of your opponent to your strategy. What can you say about your cumulative payoffs as the game progresses? Do you agree that *tit-for-tat* is the best strategy for repeated play?

79. Game Theory in Your Life. Describe a conflict or a decision in your own life that might be modeled as a two-person game in game theory. Is your situation modeled best by a competitive game or a cooperative game? Can you make a payoff table that accurately gives the benefits of each of your options for each of your opponent's options? Based on the model that you created, comment on the effectiveness of game theory. What are its limitations? Do you think that game theory models provide useful results and advice in certain situations? Explain.

14 SIGNIFICANT STATISTICS

Does living near a toxic waste dump increase the risk of cancer? What fraction of the public believes that the president is doing a good job? How much is the cost of health care rising? All such questions can be answered through statistical research. Indeed, you can hardly pick up a newspaper without immediately seeing the results of statistical studies and surveys. In this chapter, we present the basic principles of statistics, emphasizing how you can interpret properly the statistical data that inundates you every day.

CHAPTER 14 PREVIEW

14.1 *THE SCIENCE OF STATISTICS.* We begin the chapter by defining statistics and discussing why it is so important.

14.2 *FUNDAMENTALS OF STATISTICS.* In this section we outline the basic principles of statistical research, including the relationship between samples, populations, and methods of sampling.

14.3 *VISUAL DISPLAYS OF DATA.* We briefly discuss some of the most common ways of displaying statistical data with tables and graphics in this section.

14.4 *CHARACTERIZING DATA DISTRIBUTIONS.* In this section we introduce mean, median, mode, the five-number summary, and the standard deviation — all used to characterize data distributions.

14.5 *STATISTICAL INFERENCE.* The process of inference from samples to populations is the key to most statistical research. In this section we cover the prevalence and interpretation of the normal distribution and surveys and opinion polls.

14.6 *SAMPLE ISSUES IN STATISTICAL RESEARCH.* We conclude the chapter with several case studies that illustrate important issues in statistical research.

Statistical methods … are necessary in reporting the mass data of social and economic trends.… But without writers who use [statistics] with honesty and understanding, and readers who know what they mean, the result can only be semantic nonsense.

— Darrell Huff, from *How to Lie with Statistics*

There are three kinds of lies: lies, damned lies, and statistics.

— Benjamin Disraeli

Statistical thinking will one day be as necessary for efficient citizenship as the ability to read and write.

— H. G. Wells

14.1 THE SCIENCE OF STATISTICS

In 1990, medical doctors revised the standard tables used to determine a healthy weight for adults. According to the 1990 tables, adults face no medical danger in adding 20 to 30 pounds as they approach middle age. For the many Americans whose weight increases in middle age, these new guidelines offered a great sense of relief.

Only 5 years later, in 1995, the results of a large study that had followed the health of 115,000 nurses for 16 years contradicted the 1990 findings. This study found that women weighing just 10 to 20 pounds more than the healthiest women faced a 20% greater risk of early death from cancer or heart disease. Further, the study found that, for a given height, the healthiest women weighed at least 15% *less* than the average American woman. Instead of relief for people who gain weight in middle age, this new study suggests that any weight gain is unhealthy.

The word *statistics* can have two meanings: First, statistics are the *numbers* that describe data (e.g., an average or a percentage increase in risk of death); second, the word statistics sometimes is used to mean the *science* of statistics.

How could two sets of findings about weight, published only 5 years apart, directly contradict each other? Each finding was based on **statistics** — numbers, such as the 20% increase in risk of early death — that are used to describe large sets of data. Understanding the contradiction in the reports and forming an opinion about the validity of either study requires investigating the *science* of statistics — the science of collecting, organizing, and interpreting data.

If you follow the news, you will be hard-pressed to find a lead story that is *not* based on an analysis or interpretation of statistical data, and you often will find studies that contradict previous ones. Whether in opinion polls, marketing research, tracing the spread of AIDS, determining the cost of health care, or deciding whether a particle accelerator has produced a new type of fundamental particle, the science of statistics plays a vital role.

We have already covered many of the concepts needed to understand statistics. For example, we discussed bias, correlations, and causality, in Chapter 3. We presented methods of estimation in Chapter 7. We examined ideas of sampling, error, and uncertainty in Chapter 8. We described linear regression and the best fit line to a data set in Chapter 10. Most important, the ideas of probability covered in Chapter 13 underlie much of modern statistics. In this chapter, we build upon that previous work to help you understand the many uses and misuses of statistics in our society.

Time-Out to Think: Find as many examples as possible of statistics or of news based on statistical reports in the front section of today's newspaper.

14.2 FUNDAMENTALS OF STATISTICS

- The U.S. Labor Department publishes a monthly unemployment rate, determined by surveying 60,000 households out of the entire U.S. population.
- Working for the government, meat inspectors randomly select samples of meat from markets to be tested for contamination.
- To determine the general properties of stars, astronomers study a relatively small number of stars in great detail.

One common characteristic of all these statistical studies is the drawing of conclusions about some large **population** by studying a much smaller **sample**. In the unemployment survey, the population is the entire United

States, and the sample is the 60,000 households surveyed. In the inspection of meat, the population is all meat products sold in the United States, and the sample is those few chosen for testing. For astronomers, the population is all the stars in the universe, and the sample comprises the relatively few stars studied in detail.

Technically, a *census* involves studying every member of the population, whereas a *survey* studies only a small sample of the population. Because of the difficulties in conducting an accurate census, the U.S. Census Bureau now includes statistical methods of surveying to adjust its results and estimate errors.

A second common characteristic of these examples is that studying the entire population would be impractical. In Chapter 8, we discussed the difficulties that the U.S. Census Bureau has in conducting a complete census once every decade; imagine the difficulties that would be involved in surveying all households every month! For the meat inspectors, testing a sample may mean culturing it for bacteria; if all meat were tested, nothing would be left to sell. The astronomers face the most daunting problem: As shown in Chapter 7, more than 3000 years would be needed just to count the stars in the Milky Way galaxy (at a rate of one per second); the entire universe contains more stars than grains of sand on all the beaches of the world. Clearly, astronomers will never be able to study more than a tiny fraction of all stars.

These examples illustrate that a primary purpose of statistics is *generalization*. As with all processes of generalization, uncertainties are necessarily involved at almost every step of statistical generalization.

Get the facts first and then you can distort them as much as you please.
— Mark Twain

Because statistical generalization involves so many uncertainties, people sometimes say that statistics can be used to support *any* conclusion. Although this claim has a degree of truth, many statistical studies are carefully researched and provide valuable insights into important issues. The key to dealing with statistics in daily life is evaluating whether a statistical analysis is reasonable and drawing your own conclusions by assessing the results of the analysis.

14.2.1 Descriptive and Inferential Statistics

Consider the study mentioned earlier that tracked 115,000 nurses over a 16-year period. To draw conclusions about weight and health, the researchers collected an enormous amount of information: the height and weight changes of each nurse during the 16 years, the illnesses suffered by the nurses, and much more. This huge base of information is called the **raw data** of the study. Clearly, before any general conclusions could be drawn, the researchers needed to consolidate and summarize these data in a meaningful way.

The task of finding clear and concise ways to summarize raw data is the essence of the branch of statistics called **descriptive statistics** (because it deals with *describing* the raw data), which deals only with the sample. Descriptive statistics includes methods for choosing a sample to study, for collecting the raw data from the sample, and for reducing the raw data to a few pertinent statistics.

Once the sample is adequately described, the next task is to generalize the findings to the larger population. In the study of the nurses, the larger population is *all* women (perhaps even all adults). Methods for generalizing from the sample to the population are the task of the second major branch of statistics, called **inferential statistics** (because it deals with *inferring* characteristics of the population from the sample).

Inferential statistics involves probability, because statisticians can never be certain that a small sample will reflect accurately the characteristics of a large population. The methods of inferential statistics allow estimates of the probability that the sample, if chosen in an unbiased manner, will reflect the characteristics of the population.

The field of inferential statistics has advanced considerably during the past century. For example, pre-election polls typically interview only 1000 to 2000 people across the entire United States. Yet, if a poll is conducted according to the scientific principles of inferential statistics, it can accurately predict the winner of an election in which tens of millions of people cast ballots.

Time-Out to Think: Listen to the news or check the paper for the results of an opinion poll. Does the report state how many people were surveyed? Does it state whether the poll was "scientific" (did it follow the principles of inferential statistics)?

14.2.2 Choosing a Sample

The success of a statistical study depends on selection of the sample. One of the more famous sampling errors in history occurred in 1936, involving a magazine called the *Literary Digest*, which had successfully predicted the outcomes of several previous elections through large polls. After compiling data from an unusually large sample of 2.4 million voters, the *Literary Digest* predicted that Alf Landon would win the presidency by a large margin over Franklin Roosevelt. Instead, Roosevelt won in a landslide.

What went wrong? The *Literary Digest* chose its sample by taking names from telephone books, rosters of clubs and associations, voter registration lists, mail-order lists, and similar sources. The *Literary Digest* then sent "ballot" postcards, on which recipients could indicate their choice for president, to 10 million people drawn from these lists. The final sample on which the prediction was based was the 2.4 million people who returned the postcards.

Two forms of bias introduced by the sampling methods probably caused the error in the *Literary Digest* prediction:

- The initial selection of the 10 million names was not representative of voters as a whole. Although more than 95% of voters have telephones today, that wasn't the case in 1936. By selecting names from telephone books and club membership lists, the sample was biased toward the more affluent voters who could afford telephones and club memberships. This method is an example of **selection bias** because it reflects bias in the selection of the initial sample by the people who conducted the survey.
- Of the 10 million postcards, only about 25% were returned. Because President Roosevelt was the incumbent, people who were satisfied with his performance were less likely to return the cards than those who objected to his vast programs for combating the Great Depression of the 1930s. This outcome is an example of **participant bias** because participants voluntarily choose whether to answer survey questions. This method introduces personal bias because those who feel strongly about an issue are more likely to respond.

Participant bias sometimes is called **self-selection bias.** We don't use this term in order to avoid confusion with *selection bias* introduced by researchers or poll-takers.

As the *Literary Digest* example suggests, any form of bias in sample selection means that the sample is highly unlikely to reflect accurately the population. Today, proper statistical research demands that samples be selected without any human judgment, so as to avoid human biases. Such selection is accomplished through **random sampling** (or **probability sampling**), in which members of the sample are chosen from the population randomly with the aid of a random number table or a random number generator on a computer. If done properly, random sampling ensures that every member of the population has an equal probability of being selected for the sample.

Of course, even a randomly chosen sample doesn't guarantee that representation of the larger population; by sheer bad luck, a well-chosen sample might still have characteristics very different from those of the population. For example, let's say that we survey 50 randomly selected students at some school about the importance of a class in economics. The selection is random, so it is unbiased. Yet, by accident, the 50 students selected may be mostly business students, who are more likely than the student body at large to consider an economics class important.

In summary, a definition of an unbiased sample selection process is:

> The sample selection process is unbiased if nothing in the sampling process *preordains* the sample to be unrepresentative of the population under study.

Time-Out to Think: So far we have discussed bias in dealing with people, but bias also can affect physical studies. Recall the problem of trying to measure the global average temperature (see subsection 8.4.2). Explain how the *urban heat island effect* introduces a form of *selection bias* if temperatures are recorded only at weather stations in major cities.

Finally, one more general form of bias can affect sampling: bias in the selection of the population. In any statistical study, the first step is to identify the population to be studied. For example, suppose that you wanted to conduct a study to find out whether regular exercise can increase longevity. At first, you might think that the population under study is all people. However, to include smokers in the population might not be wise because exercise might affect smokers and non-smokers differently. Similarly, including people who already have heart disease and children might be pointless. In fact, you likely would choose to study the population of healthy adults; but who defines *healthy?* As you must choose the population before you can choose the sample, you may introduce bias even before beginning the sampling process.

Example 14–1 ***Pre-Election Survey.*** Suppose that you are conducting a poll of 1,500 people to predict the outcome of the next presidential election. What is the population under study? Suggest an unbiased method for choosing your sample. Also, discuss potential problems with your sampling method and potential solutions.

Solution: Because you are trying to predict the outcome of an election, the population consists of all people *who will vote* in the election; people who will *not* vote in the election are not part of this population, as they will not be able to affect the outcome of the election. How can you sample the population of people who will vote? One approach might be to select from all adult citizens. However, because citizens must register to vote, a better method for choosing a sample is to select individuals randomly from a list of all registered voters in the United States. Because this sample is random, it is unbiased. However, this method presents at least two problems.

1. Only a fraction of registered voters actually will vote on election day, so the sample isn't truly representative of the population under study. One way of alleviating this problem might be to also ask whether respondents plan to vote and to include this answer in the survey results.

2. Because presidential elections are decided by *electoral* votes (state by state), the popular vote doesn't necessarily reflect the outcome of the election (see subsection 13.7.1). Thus, even if the sample accurately reflects the popular vote, it might not predict the winner of the election. Alleviating this problem might require sampling voters within each state, rather than nationally. Then, based on which candidate is predicted to win each state, predict the overall election results.

❖

14.2.3 The Process of Statistical Study

A statistical study is an interplay between describing the sample and inferring characteristics of the population. That is, it is an interplay between the methods of descriptive and inferential statistics. Five basic steps are involved in a statistical study.

1. Clearly express the goal of the study (e.g., predicting the outcome of an election, or evaluating the health risk from weight gain) and identify the *population* to which the study applies.
2. Choose an unbiased sample from the population and collect data from the sample.
3. Consolidate the data collected from the sample into a set of **sample statistics**, such as the average for the sample and other numbers that summarize the data.
4. Infer the characteristics of the population, called the **population parameters**, based on the sample results. Note that the population parameters aren't actually measured; rather they are inferred by generalization from the sample. Techniques of inferential statistics also provide methods for assessing the uncertainty in the population parameter estimates.
5. The final step is to return to the goal of the study. Analyze logically the inferred population parameters and draw appropriate conclusions about the population.

Figure 14–1 shows the interrelationships among the population, the sample, the sample statistics, and the population parameters.

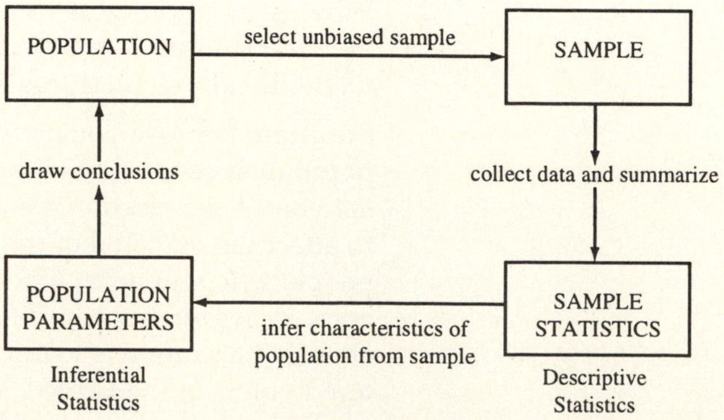

Figure 14-1. The diagram illustrates the inter-relationships among the concepts of sample, sample statistics, population, and population parameters in a statistical study.

Example 14–2 ***Sample and Population.*** Suppose that you want to learn about the physical characteristics of students at your school (height, weight, resting heart rate, percentage of body fat, etc.). You decide to select 100 students randomly and measure these characteristics. Describe the population, sample, sample statistics, and population parameters for your study.

Solution: The *population* is the entire student body at your school, as that is the group you want to learn about. The *sample* is the 100 students selected for measurement. The *sample statistics* describe the results of your measurements on the sample. For example, they might include the average heights and weights of the 100 students in the sample. Note that, because they are calculated from actual measurements, the sample statistics will reflect accurately the characteristics of the sample (as long as the measurements and calculations are done properly). The *population parameters* include the average heights and weights of *all* students at your school. You won't actually make measurements for the entire population, so you must infer the population parameters from your analysis of the sample statistics.

❖

14.3 VISUAL DISPLAYS OF DATA

Visual depictions, like tables and graphs, are often the most meaningful way to summarize large amounts of raw data. While statisticians have always found innovative ways to produce striking tables and graphs, the computer age has made advanced graphical techniques available to anyone using a spreadsheet or statistical software package. The result is that modern media are filled with imaginative methods for displaying statistical data. Fortunately, although we cannot possibly summarize all of the ways of visualizing data in a few pages, there are just a few basic principles.

14.3.1 Frequency Tables

The most basic data table organizes data into a set of categories and indicates the **frequency** (number of times) with which data appear in each category. For example, imagine that a company offers a taste test of a new cola to 15 people. Each individual is asked to rate the taste of the cola on a 5-point scale:

(bad taste) 1 2 3 4 5 (excellent taste).

Now, imagine that the 15 ratings (the raw data) are

1 3 3 2 3 4 3 4 2 3 5 3 3 4 3.

To construct a **frequency table**, count the number of responses in each category. Only one person gave the new cola the worst rating of 1; two people gave it a rating of 2; eight gave it a rating of 3; four gave it a rating of 4; and only one person gave it the highest rating of 5. The frequency table is shown in Table 14–1.

Table 14-1. The responses of 15 people to a taste test of a cola are tabulated.

Taste Scale	Frequency	Relative Frequency	Cumulative Frequency
1	1	1/15 = 6.7%	1
2	2	2/15 = 13.3%	1 + 2 = 3
3	8	8/15 = 53.3%	1 + 2 + 8 = 11
4	3	3/15 = 20%	1 + 2 + 8 + 3 = 14
5	1	1/15 = 6.7%	1 + 2 + 8 + 3 + 1 = 15
Total	**15**	**1 = 100%**	**15**

Occasionally, the **cumulative relative frequency** also is tabulated. It simply is the fraction (or percentage) of scores in a particular category and all preceding categories. For example, the cumulative frequency at row 3 of Table 14-1 indicates that 11 of the 15 people gave the cola a taste rating of 3 or worse. The cumulative relative frequency for this row is 11/15 = 73.3%; that is, 73.3% of the respondents gave a rating of 3 or worse.

Two other useful ways of describing data also are tabulated. The **relative frequency** is the fraction or percentage of each response category. For example, since 2 of the 15 people gave the cola a taste rating of 2, the relative frequency for this response is 2/15, or 13.3%. The **cumulative frequency** is the number of responses in a particular category *and all preceding* categories. For example, in row 3 of Table 14–1 the cumulative frequency is 11 because 8 people gave the cola a rating of 3, 2 people gave it a rating of 2, and 1 person gave it a rating of 1: 1 + 2 + 8 = 11.

One of the reasons that statistical data can be easily misinterpreted is because it can be presented in so many different ways. Note, for example, that the cumulative frequency of 11 for row 3 of Table 14–1 means that 11 people rated the cola as 3 *or worse* on the taste scale. The makers of the new cola probably wouldn't want to advertise this result. Instead, they probably would turn the table upside-down, as shown in Table 14–2. Note that, with the inverted table, the third row now shows that 12 people rated the cola as 3 *or better* on the taste scale.

Table 14-2. The data of Table 14–1 are inverted to show the cumulative frequency in a more favorable light.

Taste Scale	Frequency	Relative Frequency	Cumulative Frequency
5	1	1/15 = 6.7%	1
4	3	3/15 = 20%	1 + 3 = 4
3	8	8/15 = 53.3%	1 + 3 + 8 = 12
2	2	2/15 = 13.3%	1 + 3 + 8 + 2 = 14
1	1	1/15 = 6.7%	1 + 3 + 8 + 2 + 1 = 15
Total	**15**	**1 = 100%**	**15**

Time-Out to Think: Suppose that you are the product development manager for the new cola and want to convince company executives that it will be successful in the market. Which of the two tables (Table 14–1 or 14–2) would you present? Why?

Binning of Data

In the taste test example, the tables were fairly short because only 15 tasters and 5 ratings were used. Often, however, many more measurements are obtained. For example, suppose the times (in seconds) of the 20 finishers of the 100-meter dash at a particular track meet were

9.92, 9.97, 9.99, 10.01, 10.06, 10.07, 10.08, 10.10, 10.13, 10.13,
10.14, 10.15, 10.17, 10.17, 10.18, 10.21, 10.24, 10.26, 10.31, 10.38.

This process of grouping data is called **binning** because it is similar to creating several different *bins* into which we place the measurements.

Note that each time is given to a hundredth of a second and that the times extend from 9.92 to 10.38 seconds — a range of 46 hundredths of a second. A frequency table listing times to a hundredth of a second would therefore require 47 lines (one line for the fastest time, plus 46 additional lines to span the range of 46 hundredths of a second). Rather than make such an unwieldy table, a better approach is to **group**, or **bin**, the data. In this case, we can bin the data by rounding each time to the nearest tenth of a second. The resulting frequency table, with the binned data, is shown in Table 14–3.

Table 14-3. A set of 20 times, measured to a hundredth of a second, is consolidated into a manageable frequency table by binning the times to the nearest tenth of a second.

Time (to nearest 0.1 second)	Frequency	Relative Frequency	Cumulative Frequency
9.9	1	1/20 = 5%	1
10.0	3	3/20 = 15%	1 + 3 = 4
10.1	7	7/20 = 35%	1 + 3 + 7 = 11
10.2	6	6/20 = 30%	1 + 3 + 7 + 6 = 17
10.3	2	2/20 = 10%	1 + 3 + 7 + 6 + 2 = 19
10.4	1	1/20 = 5%	1 + 3 + 7 + 6 + 2 + 1 = 20
Total	**20**	**1 = 100%**	**20**

Example 14–3 ***Binned Exam Scores.*** Consider the following set of 20 scores from a 100-point exam. Determine appropriate bins, and make a frequency table.

76, 80, 78, 76, 94, 75, 98, 77, 84, 88, 81, 72, 91, 72, 74, 86, 79, 88, 72, 75

Solution: The scores range from 72 to 98. Of the many ways to bin the data, one is to group data in 5-point bins. For example, the first bin would represent scores from 95 through 99, the second bin would represent scores from 90 through 94, and so on. After you have identified the bins, count the frequency, or number of scores, in each bin. For example, one score is in bin 95–99 (the high score of 98) and two scores are in bin 90–94 (the scores of 91 and 94). These values are shown in the frequency column of Table 14–4.

To find the relative frequencies, divide each frequency by the total number of scores. With 20 exam scores, the relative frequency of the 95–99 bin is $1 \div 20 = 0.05$. To find cumulative frequencies, add all the scores at or above a certain level. For example, the cumulative frequency for bin 90–94 is 3 for the total of 3 scores of 90 or above.

Table 14–4. The exam scores from Example 14–3 are binned and displayed in a frequency table.

Score	Frequency	Relative Frequency	Cumulative Frequency
95–99	1	0.05	1
90–94	2	0.10	3
85–89	3	0.15	6
80–84	3	0.15	9
75–79	7	0.35	16
70–74	4	0.20	20
Total	**20**	**1.0**	**20**

Another common practice with exams is binning the scores in 10-point ranges: 90–99, 80–89, and so on.

❖

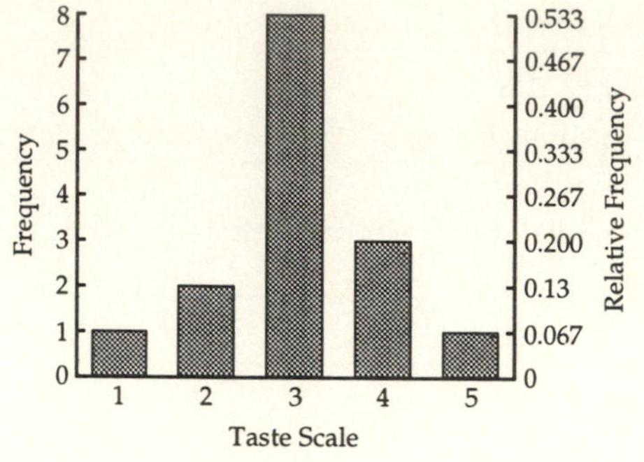

Figure 14-2. The data from Table 14–1 are depicted with a bar graph.

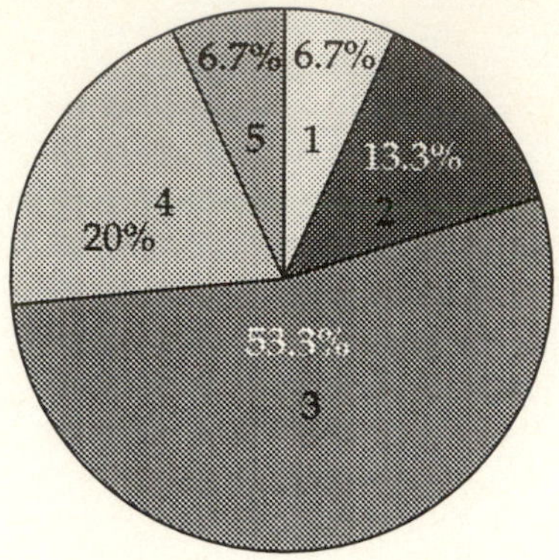

Figure 14-3. The data from Table 14–1 are depicted with a pie chart that shows the relative frequency of each category.

14.3.2 Pie Charts and Bar Graphs

Once data are summarized in a table, a graph is a good way to represent the summary visually. For example, suppose that you want to graph the data from the cola taste test shown in Table 14–1. One simple way of doing so is with a **bar graph**, in which the height of each bar corresponds to the frequency of each rating in the taste test (Figure 14–2). Note that the bars do not touch one another. The graph shows both the frequency (left-hand vertical scale) and the relative frequency (right-hand vertical scale) for the data.

The same data also can be represented with a **pie chart** (Figure 14–3). Each slice of the pie proportionally represents the relative frequency of one of the five taste ratings. The entire pie must always represent 100%.

Bar graphs and pie charts are used primarily when data represent either a set of categories or a subjective numerical rating scale. For example, the cola taste test ratings all are opinions of the individual tasters. We therefore cannot make a quantitative statement about *how much* better tasting a rating of 3 is compared to a rating of 2. Similarly, look back at Figure 9–1, which shows pie charts for government receipts and outlays. Each slice of those pie charts describes the amount of money in a particular category; the order in which categories are listed is arbitrary.

Pie charts and bar graphs are most useful for data that represent either a list of categories or a subjective numerical rating scale.

Although bar graphs and pie charts are the most common ways of illustrating categorical or subjective data, many variations exist. For example, Figure 14–4 represents the various categories under which legal immigrants were admitted to the United States in 1994. Each category of data is represented by a slice of a cylinder, which makes the graph conceptually equivalent to a pie chart. Presumably, the designers of this graph chose the cylinder because showing so many categories in a standard, circular pie chart would have been difficult.

Time-Out to Think: Check a few newspapers or magazines for bar graphs and pie charts. Can you find other examples, similar to Figure 14–4, in which the basic principles of a bar graph or pie chart were used to make an innovative graph?

Example 14–4 *Interpreting the Immigration Chart.* Study Figure 14–4. Suppose that the Smith immigration bill had been in place in 1994. How many legal immigrants admitted in 1994 would have been barred from immigration?

Solution: According to the key at the bottom of the chart, categories labeled with a square would be eliminated by the Smith bill. Thus the following categories and numbers of 1994 immigrants would have been barred from immigration:

First preference	13,181
Second preference	26,327
Third preference	22,191
Fourth preference	61,589
Total:	**123,288**

A total of 123,288 of the 1994 immigrants would have been barred from immigration under the Smith bill. According to the key, a circle indicates categories that would be restricted. The only such category shown is parents of U.S. citizens, under which 56,370 legal immigrants were admitted in 1994; however, the graph does not indicate by how much, so you can't draw any conclusions about the number of people in this category who would have been affected by the Smith bill.

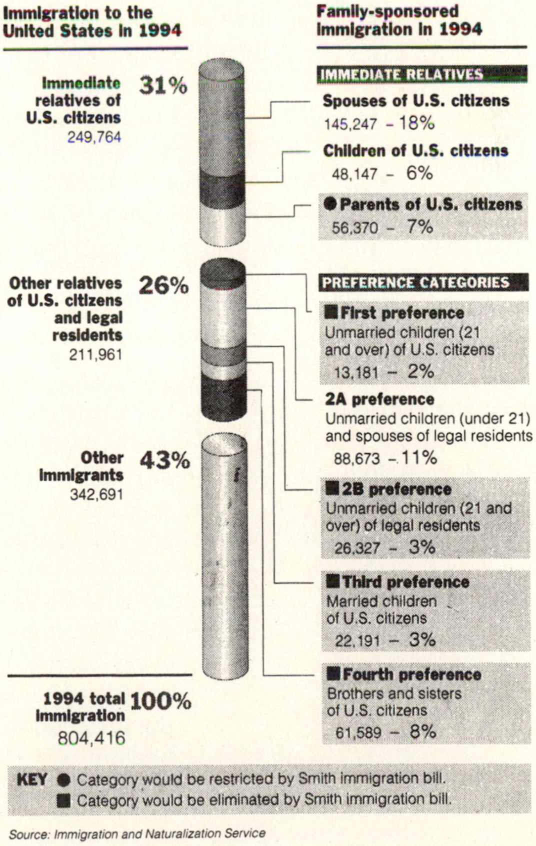

Figure 14-4. Just as the slices of a pie chart must total 100%, the slices of the cylinder add up to 100%. Three main slices represent the three general categories of immigrants. The two main categories that represent family-sponsored immigration are further subdivided. The dots and squares (see the key at the bottom) represent how different categories would be affected by a particular proposal for limiting legal immigration. *The New York Times*, 9/25/95. Copyright © 1995 by The New York Times Company. Reprinted with permission.

❖

14.3.3 Histograms and Line Charts

Suppose that you want to graph the data on the 100-meter times from Table 14–3. You could make a bar graph or pie chart, but the difference between 9.9 seconds and 10.0 seconds is a specific and meaningful amount of time in this case. Further, 9.95 seconds forms a clear dividing line between these categories: If a time is less than 9.95 seconds, it falls into the bin for 9.9 seconds; if it is equal to or more than 9.95 seconds, it falls into the bin for 10.0 seconds. You can therefore depict these data with a variation on the bar graph, called a **histogram** (Figure 14–5a). Several features of the histogram are important:

- Each rectangle in the histogram is *centered* on the corresponding data value. For example, the rectangle for 9.9 seconds covers the region from 9.85 to 9.95 seconds on the graph.
- Both frequency and relative frequency are plotted on the same graph, simply by labeling the vertical axis in two different ways.
- The two small slanted lines on the horizontal axis (between the labels for 0 and 9.9 seconds) indicate a gap in the axis. That is, this graph is focused on the region from 9.9 to 10.5 seconds and is not concerned with times between 0 and 9.9 seconds.

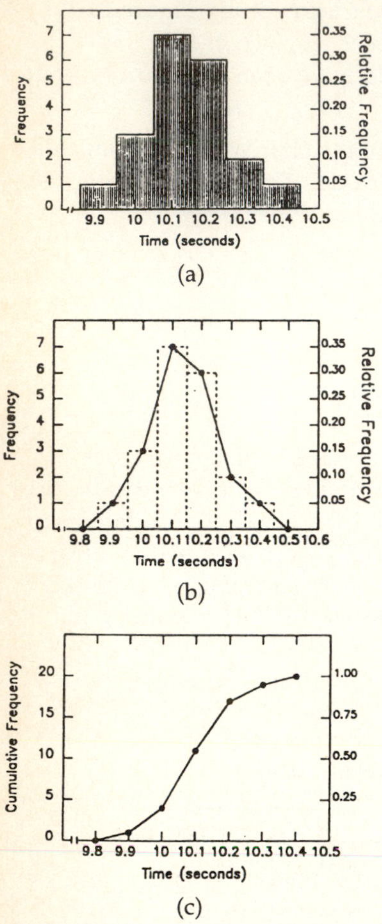

Figure 14-5. (a) This histogram shows the data for times in a 100-meter race from Table 14–3. (b) To construct a line chart, a dot is placed at the midpoint of each time bin and the dots are connected. For comparison, the histogram is superimposed (dashed lines) on the line chart. (c) A line chart for the data also can be constructed to show the *cumulative* frequency.

An alternative way of graphing these data is with a **line chart**, which is constructed by placing dots at the *midpoint* of each bin and connecting the dots with lines (Figure 14–5b). For example, as all times between 9.95 and 10.05 seconds were binned at 10.0 seconds, the dot is placed directly over 10.0 seconds on the horizontal axis. For aesthetic purposes, the line chart includes dots on the horizontal axis showing that no runners had times of 9.8 or 10.5 seconds.

Finally, a variation on the line chart shows the *cumulative* frequency, rather than the frequency (Figure 14.5c). For example, the dot at 10.1 seconds in Figure 14–5c shows a cumulative frequency of 11, indicating that 11 runners had times of 10.1 seconds or faster (to the nearest tenth of a second).

Graphic Distortions

Each of the two basic types of frequency graphs — histograms and line charts — distorts the data in some way.

- The histogram seems to imply that all values in a particular bin are the same. For example, the histogram in Figure 14–5a shows that three runners had times in the bin for 10.0 seconds. In fact, the original data from which Table 14–3 was constructed shows that the three times in this bin were all different: 9.97 seconds, 9.99 seconds, and 10.01 seconds.

- Because the line chart connects the dots representing the frequency of each bin, it gives the impression that the data are *continuous* (that every time is represented); in fact, the data comprise a set of individual (discrete) times. For example, the line chart in Figure 14–5b gives the impression that a time of 10.05 seconds occurred with a frequency of about 4.5; clearly, that makes no sense and is a misinterpretation of the chart.

Example 14–5 ***Drawing Histograms and Line Charts.*** Create a histogram, a line chart, and a cumulative frequency line chart from the data in Example 14–3.

Solution: To draw the histogram, label the horizontal axis with the exam scores and the vertical axis with the frequency (left-hand scale) and relative

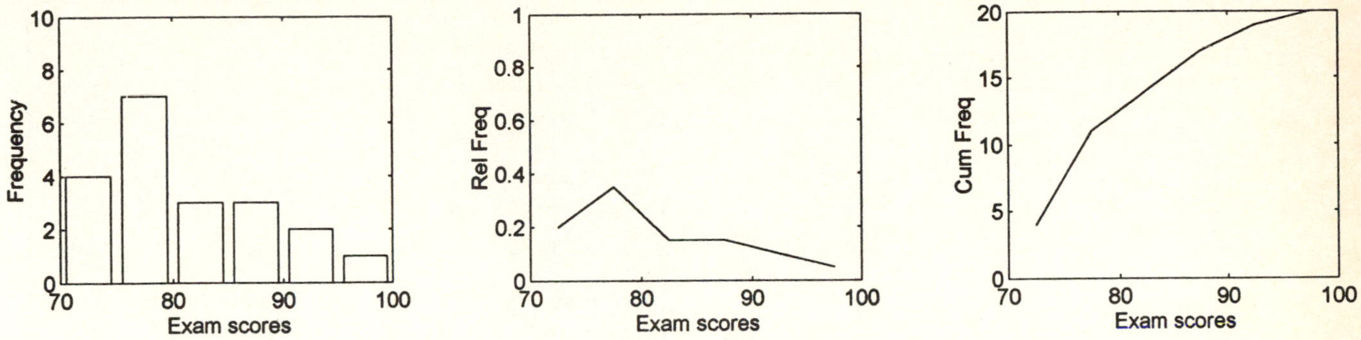

Figure 14-6. Three graphs of the data from Example 14–3 are shown.

frequency (right-hand scale). With no scores below 75, you don't need to show the interval from 0 to 74. Each rectangle in the histogram spans the scores for its bin, and its height equals the relative frequency for the bin.

To make a line chart, simply place a dot at the center of each bin, corresponding to the frequency, and connect the dots with lines. For example, scores in the bin for 90–94 points are shown with a dot at a score of 92.5 points; the height of the dot is 2 units above the horizontal axis because that is the frequency of scores in this bin. For aesthetic purposes, also show the scores in the bins for 100–104 points and 70–74 points with dots on the horizontal axis to indicate zero frequency.

The cumulative frequency line chart also has dots in the center of bins; in this case, however, the height of each dot is the cumulative frequency of the bin, rather than the frequency. The resulting graphs are shown in Figure 14–6.

❖

14.3.4 Scatter Plots

Suppose that you have annual rainfall and snowfall data for 30 cities and wonder whether the amount of rain a city receives in a year can be used to predict the amount of snow that will fall. Because you hope that the amount of snowfall depends on the amount of rainfall, you should graph the rainfall data on the horizontal axis and the snowfall data on the vertical axis, as shown in Figure 14–7(a). Each dot represents the amount of rain and snow received in one of the 30 cities. The resulting graph is called a **scatter plot**.

What does the scatter plot tell you? In the case of the rainfall and snowfall data, the 30 points truly look scattered. Some cities have high annual rainfall and low annual snowfall; others have both low rainfall and low snowfall. No pattern emerges that might allow you to predict annual snowfall from annual rainfall.

Figure 14–7(b) shows a scatter plot for another set of data. In this case, data on life expectancy and infant deaths were collected in 21 countries. Each dot on the scatter plot represents one of the 21 countries. The horizontal axis shows the average life expectancy of citizens in each country and the vertical axis shows the infant death rate (measured as the number of infant deaths per year per 1000 live births). The data points in this scatter plot exhibit a clear pattern. In general, countries with lower rates of infant death have higher life expectancies and vice versa. In this case we can conclude that a strong correlation exists between these two variables and that life expectancy is a good predictor of the infant death rate.

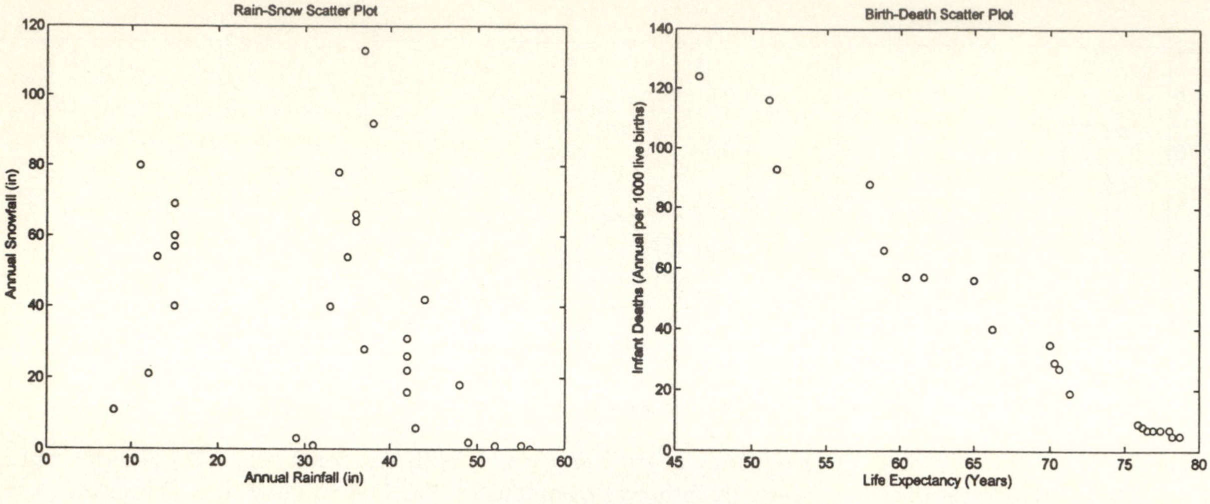

(a) (b)

Figure 14-7. (a) Based on this scatter plot of rainfall-snowfall data for 30 cities, there is no obvious relation between the variables. (b) By contrast, the life expectancy and infant death data of this scatter plot, for 21 countries, suggests a strong relation between these variables.

Time-Out to Think: Recall from Chapter 3 that correlations may occur either by coincidence, because of a common underlying cause, or because one variable is a causal factor for the other. Which do you think is the case in Figure 14–7(b)? Explain.

14.3.5 Time-Series Diagrams

When one of the two variables in a statistical table is time, the resulting graph is called a **time-series** diagram. Time-series diagrams are particularly useful because they can reveal trends. In a time-series diagram, a relation of the form *(time, quantity)* is plotted to show how the quantity changes with time.

For example, Figure 14–8 shows relative values of stock, bond, and gold prices over a 12-week period. The stock prices shown are an average for the 500 stocks tracked by the Standard and Poor's 500 stock list (S&P 500). If you wanted to invest money that would reflect this average, you could purchase shares in a **stock index fund** that spreads its investments over the S&P 500 stock list. Similarly, the bond prices shown are an average for many bonds, as tracked by the Lehman Treasury Bond Index. Gold prices are from New York traders.

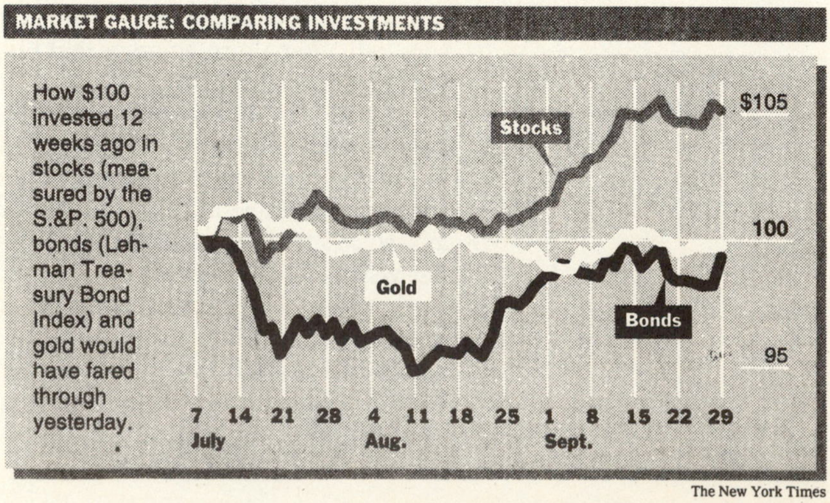

Figure 14-8. This time-series diagram shows three quantities — relative bond prices, gold prices, and stock prices — over a 12-week period in 1995. Time is plotted on the horizontal axis allowing depiction of trends or changes. *The New York Times*, 9/30/95. Copyright © 1995 by The New York Times Company. Reprinted with permission.

Example 14–6 *Investment Markets.* Study Figure 14–8. Suppose that on July 7, 1995, you had invested $100 each in a stock index fund that tracks the S&P 500, a bond fund that follows the Lehman Index, and gold. How much would your investment portfolio have been worth on August 4? On September 29? What was the overall rate of return for the 12-week period?

Solution: According to the graph, the $100 in the stock fund would have been worth about the same $101 on August 4. The $100 bond investment would have declined in value to about $96. The gold investment would have held its initial value of $100. Thus on August 4 your complete portfolio would have been worth $101 + $96 + 100 = $297, or slightly less than your initial investment of $300.

By September 29, the stock fund had increased in value to $105, the bond fund had regained much of its earlier losses, ending at about $99, and the gold fund still would have been worth $100. Thus the total value of your portfolio on September 29 would have been $105 + $99 + $100 = $304. Based on your initial investment of $300, your overall rate of return for the 12 weeks was $304 ÷ $300 = 1.0133; that is, your investment increased in value by 1.33% during the 12 weeks.

❖

14.3.6 Graphics Galore

We have covered the basic types of graphs of statistical data when only two variables are involved, such as 100-meter race time and frequency, infant deaths and life expectancy, and stock prices and time. Figure 14–8 also shows one way to depict more than two variables at once, in this case simply by plotting three time-series graphs (stocks, bonds, and gold) on the same set of axes. A survey of newspapers, television news shows, magazines, or corporate reports will reveal countless variations of these basic methods. Because graphs may be created in many different ways, we can't survey all of them. In this final subsection on visual displays of data, we offer a small sampling of some common graphic techniques.

Stack Plots

See Section 9.2 for a review of the components of the federal budget.

Figure 14–9 shows the changes in major spending categories of the federal budget from 1960 through 1999; the chart was created in 1995, and the figures for 1996 through 1999 are projections. Like a pie chart, this graph shows only percentages of the total budget; it gives no indication of the actual amount of money spent by the government. Each category has its own wedge in this chart, and its percentage of the budget in any given year is the height of the wedge. This type of graph, in which the wedges are stacked, sometimes is called a **stack plot**; it can be difficult to read because determining the height of a wedge requires finding the difference in its height at its lower and upper edges.

For example, in 1960 the national defense wedge represents about 50% of the budget (it begins at about 30% and extends to about 80%, making its height 80% – 30% = 50%). By 1980, this wedge had declined to about 20% of the budget (extending from about 55% to about 75% on the chart). By 1999, the national defense wedge is projected to decline to only about 15% of the budget.

Time-Out to Think: The category of "all other" includes everything except payments to individuals (e.g., social security and Medicare), national defense, and interest on the federal debt. That is, it includes spending for education, environmental cleanup, scientific research, and the space program. Comment on the change in the proportion of the federal budget available for this category over time.

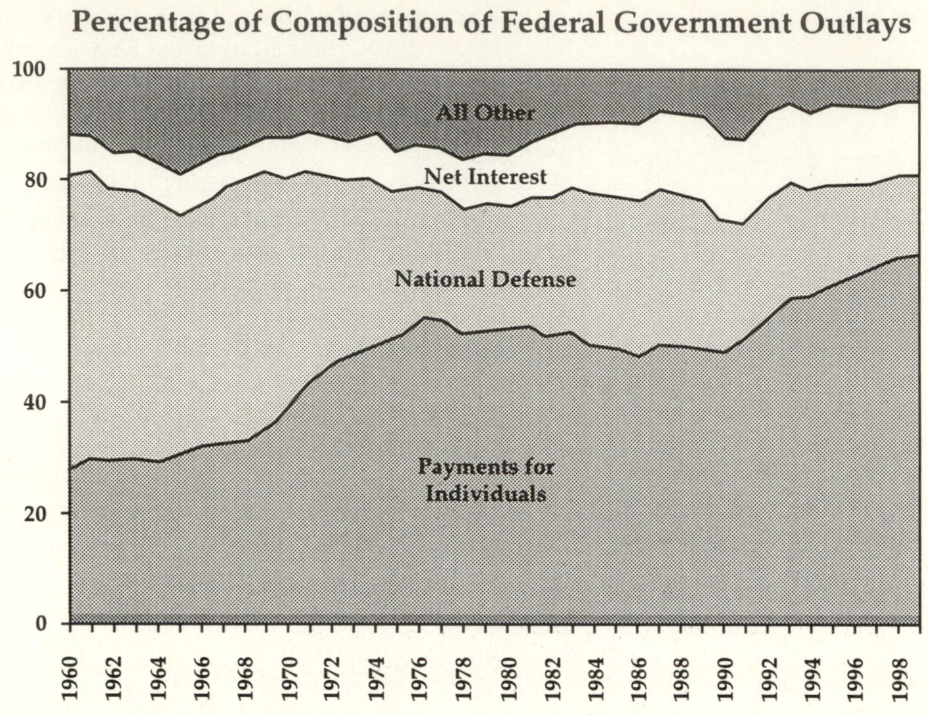

Figure 14-9. This stack plot shows the changes in major spending categories of the federal budget from 1960 through 1999.

Rate of Change Graph

See subsection 8.4.4 for a review of the consumer price index and its interpretation.

Figure 14–10 shows a bar graph on which three separate variables are plotted simultaneously: annual increase in tuition at private colleges, annual increase in tuition at public colleges, and annual increase in the consumer price index. Because each bar represents a *rate of change* (annual increase) special care is required in interpreting this graph. For example, because the bars for public colleges shorten from 1991 to 1994, tuition appears to have gone down. However, the shrinking bars mean only that the *rate of increase* in tuition has declined; the absolute cost of tuition continues to rise.

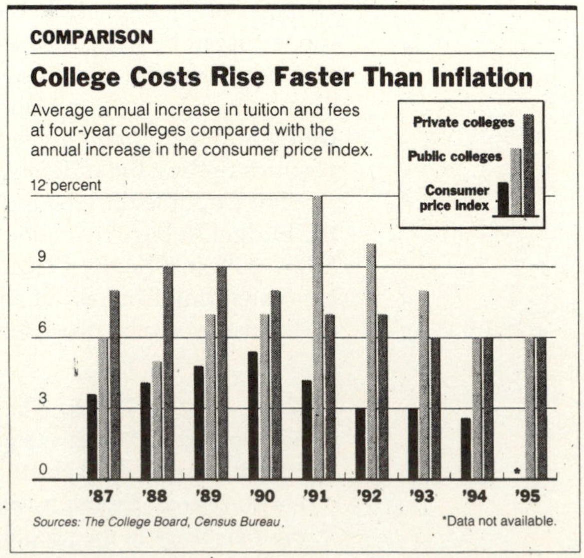

Figure 14-10. Shown are average annual percentage increases from 1987 to 1995 for three sets of data: private college tuition, public college tuition, and the consumer price index. The vertical scale is percentage *increase*, not absolute cost. *The New York Times*, 9/29/95. Copyright © 1995 by The New York Times Company. Reprinted with permission.

The primary message of this graph is that, in every year shown, the rate of increase for both public and private college tuition exceeded the increase in the consumer price index. Because the consumer price index is a measure of inflation, college tuition costs have risen faster than inflation.

Example 14–7 *Tuition Increases.* Use Figure 14–10 to answer each of the following questions. (a) In which year did average public college tuition rise the most compared to the consumer price index? (b) In which year did average public college tuition rise the least compared to the consumer price index? (c) Suppose that tuition at an average public college was $5000 in 1990. Estimate the tuition in 1991, 1992, and 1993.

Solution:(a) In 1991 tuition at public colleges rose by about 12%, some three times the approximately 4% increase in the consumer price index. (b) In 1988, average public college tuition rose by about 5%, or slightly more than the approximately 4% rise in the consumer price index. (c) In 1991, average public college tuition rose by about 12%. Thus, if tuition was $5000 in 1990, it rose by 12% × $5000 = $600 in 1991 to a total of $5600. In 1992, public college tuition rose by about 10%. Thus the 1991 tuition of $5600 rose by 10% × $5600 = $560 in 1992, to a total of $5600 + $560 = $6160. In 1993, public college tuition rose by about 8%. Thus, the 1992 tuition of $6160 rose by 8% × $6160 ≈ $490 in 1993, to a total of about $6650.

❖

Three-Dimensional Graphics

In ecological studies, researchers often seek to learn the migration patterns of animals, especially of birds. Figure 14–11 shows data about the migration of the bobolink in New York State. This example of a three-dimensional graph shows three different variables plotted on three different axes. The vertical axis represents the number of birds flying over a particular city at a particular

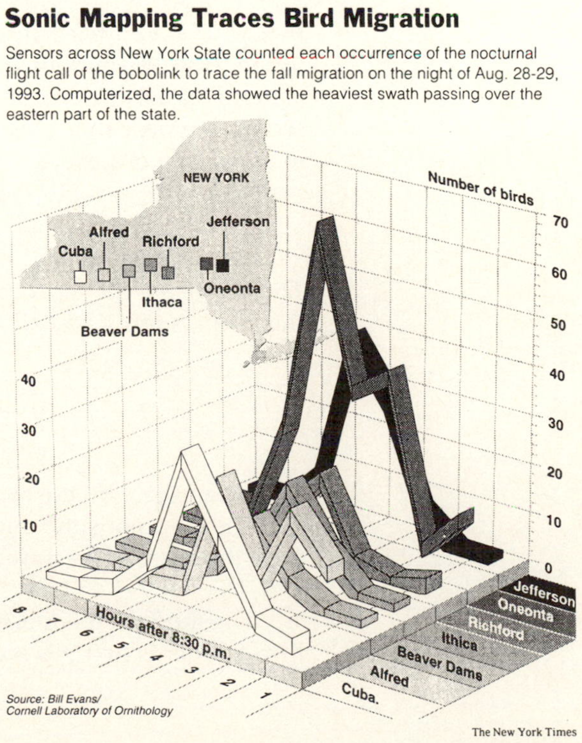

Figure 14-11. In this example of a three-dimensional graph, three variables are plotted on three different axes. The three axes are (1) location moving west to east in New York State, plotted by identifying cities; (2) time of day, plotted as hours after 8:30 P.M.; and (3) number of birds detected at each location and time. *The New York Times*, 10/3/95. Copyright © 1995 by The New York Times Company. Reprinted with permission.

time. The number of birds is statistical data, based on counting the occurrences of the bobolink's nocturnal flight call (sound). The other two axes plot the time of night and the city for which the data were recorded.

Note that the cities were selected to form an east–west line across New York State, representing a *sample* of all locations along that line. All the data shown were collected on a single night.

Example 14–8 ***Interpreting the Three-Dimensional Graph.*** Answer each of the following questions by studying Figure 14–11. (a) At about what time did the sensors detect the most bird calls? (b) Approximately how many birds passed over Oneonta at about 12:30 A.M.? (c) Over what part of the state did most of the birds fly?

Solution: (a) The number of birds detected in all the cities peaked between 4 and 6 hours after 8:30 P.M., or between about 12:30 A.M. and 2:30 P.M. (b) On the time axis, 12:30 A.M. is 4 hours after 8:30 P.M. At that time the data appears to align with the lower peak on the line at Oneonta. Reading the exact height of this peak is somewhat difficult, but an estimated 30 to 40 birds were flying over the city at that time. (c) Clearly, more birds flew over the two easternmost cities of Oneonta and Jefferson than cities farther west. Thus most of the birds were flying over the eastern part of the state.

❖

14.4 CHARACTERIZING THE DATA DISTRIBUTION

Aside from the use of tables and graphs, data sets often are described by numerical statistics. You've probably heard of some of them before — mean, median, and standard deviation. In this section we briefly describe these and several other important statistical measures.

14.4.1 Mean, Median, and Mode

Statisticians generally use the term *average* only for the *mean*, but in ordinary English *average* can be the mean, median, or mode.

Everyone is familiar with the term *average*. However, instead of a single definition, three definitions of average — the mean, median, and mode — are commonly used. To illustrate their differences, let's consider a set of exam scores. Suppose that 27 students take an exam and that the resulting scores, in ascending order, are

47, 52, 56, 57, 61, 65, 65, 69, 70, 71, 71, 72, 73, 75,
77, 77, 77, 78, 81, 83, 85, 87, 87, 91, 93, 96, 97.

The **mean** is what most people usually think of as an average. To calculate the mean, simply add all 27 scores and then divide by 27, or

$$\text{mean} = \frac{\begin{pmatrix} 47+52+56+57+61+65+65+69+70+71+71+72+73+75 \\ +77+77+77+78+81+83+85+87+87+91+93+96+97 \end{pmatrix}}{27} = \frac{2013}{27} = 74.6\,.$$

The mean score on the exam is 74.6.

The **median** is the middle score in the distribution. In this case, the median is the score 75 because 13 scores were below 75 and 13 scores were above 75.

The **mode** is the most common score. In this case, the mode is 77 because three students had this score and no other score occurred more than twice.

Example 14–9 *Averages.* Find the mean, median, and mode of the following 20 times for a 100-meter race (these are the times binned in Table 14–3).

9.92, 9.97, 9.99, 10.01, 10.06, 10.07, 10.08, 10.10, 10.13, 10.13,
10.14, 10.15, 10.17, 10.17, 10.18, 10.21, 10.24, 10.26, 10.31, 10.38

Solution: The mean is the sum of all the times divided by 20:

$$\text{mean} = \frac{\begin{pmatrix} 9.92+9.97+9.99+10.01+10.06+10.07+10.08+10.10+10.13+10.13, \\ 10.14+10.15+10.17+10.17+10.18+10.21+10.24+10.26+10.31+10.38 \end{pmatrix}}{20}$$

$$= \frac{202.67}{20} = 10.13$$

The mean time is 10.13 seconds. For the 20 times (an even number), there are two "middle" times: the tenth and eleventh fastest, or 10.13 and 10.14 seconds, respectively. The median time therefore is between 10.13 and 10.14 seconds. Times of 10.13 seconds and 10.17 seconds both occurred twice in the distribution, so both are modes and we say that this distribution is **bimodal.**

For an even number of data points, there are two "middle" values. Technically, the median is halfway between these two values. In Example 14–9, the median is 10.135 seconds because that is half-way between the two middle times of 10.13 and 10.14 seconds. We ignore this technicality and simply say that the median time is between 10.13 and 10.14 seconds.

Average Confusion

Suppose that a city newspaper surveys wages at local supermarkets, and reports that the average employee makes $10 per hour. The 15 employees of Stella's Grocery Store immediately request a pay raise, claiming that they work as hard as employees at other stores but that their average wage is only $8.10. The store manager quickly rejects their request, telling them that they are *overpaid* because their average wage, in fact, is $11.29. How can the manager and the employees calculate the average differently?

In practice, both the mean and median are commonly used as the average. The mode is used as the average much more rarely.

Suppose that the hourly wages of the 15 employees, in ascending order, are

$7.00, $7.00, $7.25, $7,25, $7.50, $7.80, $7.80, $8.10,
$8.10, $8.45, $8.45, $8.45, $12.15, $16.80, $22.50.

The employees used the median of this set of wages, or $8.10 (7 employees earn less than $8.10 and 7 earn more). However, the manager calculated the mean:

$$\text{mean} = \frac{\begin{pmatrix} \$7.00+\$7.00+\$7.25+\$7,25+\$7.50+\$7.80+\$7.80+\$8.10 \\ +\$8.10+\$8.45+\$8.45+\$8.45+\$12.15+\$16.80+\$22.50 \end{pmatrix}}{15} = \$11.29.$$

This example illustrates the confusion that can arise because the mean, median, and mode all legitimately can be called the *average*. It also shows that the choice of which quantity to call the average often is based on expediency, to prove some point.

> **Time-Out to Think:** Why does the manager prefer to regard the mean as the average? Why do the employees prefer to regard the median as the average?

14.4.2 The Five-Number Summary

Consider again the 27 exam scores for which we found the mean, median, and mode, with three special scores highlighted:

47, 52, 56, 57, 61, 65, <u>**65**</u>, 69, 70, 71, 71, 72, 73, <u>**75**</u>,
77, 77, 77, 78, 81, 83, <u>**85**</u>, 87, 87, 91, 93, 96, 97.

The highlighted score of 75 is the median, as we found earlier. Note that the highlighted 65 is the median of the scores *below* 75; this score of 65 is called the **lower quartile**. Similarly, the score of 85 is called the **upper quartile** because it is the median of the scores *above* 75. In general, one quarter of the data lies *at or above* the upper quartile, and one-quarter lies *at or below* the lower quartile.

The median describes the middle of the overall distribution, and the quartiles tell us something about the spread or **dispersion** of the data. Another simple measure of the dispersion is the **range** of the data; in the case of the exam scores, the range is from 47 to 97 (the lowest and highest scores, respectively).

A five-number summary consists of the median, the upper and lower quartiles, and the high and low extremes.

A common way to provide a quick characterization of a distribution is with a **five-number summary**: the median, the upper and lower quartiles, and the high and low extremes. For the exam scores, the five-number summary is

- median = 75,
- upper quartile = 85,
- lower quartile = 65,
- high score = 97, and
- low score = 47.

Example 14–10 ***Five-Number Summary.*** Write a five-number summary for the twenty 100-meter times from Example 14–9:

9.92, 9.97, 9.99, 10.01, 10.06, 10.07, 10.08, 10.10, 10.13, 10.13,
10.14, 10.15, 10.17, 10.17, 10.18, 10.21, 10.24, 10.26, 10.31, 10.38.

Solution: In this case, the times divide neatly into quarters:

$$\underbrace{9.92, 9.97, 9.99, 10.01, 10.06,}_{\text{fastest 25\% of times}}\ \underbrace{10.07, 10.08, 10.10, 10.13, 10.13,}_{\text{second fastest 25\% of times}}$$

$$\underbrace{10.14, 10.15, 10.17, 10.17, 10.18,}_{\text{third fastest 25\% of times}}\ \underbrace{10.21, 10.24, 10.26, 10.31, 10.38}_{\text{slowest 25\% of times}}$$

The median is between 10.13 and 10.14 seconds, as found in Example 14–9. The upper quartile is between 10.06 and 10.07 seconds, which are the middle scores of the 10 fastest times. The lower quartile is between 10.18 and 10.21 seconds. The extremes are the fastest and slowest times. Thus the five-number summary is

- the median is between 10.13 and 10.14 seconds,
- the upper quartile is between 10.06 and 10.07 seconds,
- the lower quartile is between 10.18 and 10.21 seconds,
- the fastest time of 9.92 seconds, and
- the slowest time of 10.38 seconds.

❖

14.4.3 Variance and Standard Deviation

Besides specifying the range and the quartiles, another method of describing the dispersion of a data set is by focusing on how far each number lies from the mean. Consider another set of exam scores, this time with only 10 students:

45, 55, 63, 72, 77, 79, 81, 84, 88, 97

As always, we calculate the mean of this set of scores by adding them up and dividing by the number of scores, in this case 10:

$$\text{mean} = (45+55+63+72+77+79+81+84+88+97) \div 10 = \frac{741}{10} = 74.1.$$

To find the **deviation** of a particular score from the mean, we simply take the score and subtract the mean. For example, the high score is 97, so its deviation from the mean is

$$\text{deviation} = \text{score} - \text{mean} = 97 - 74.1 = 22.9.$$

For scores below the mean, the deviation is negative. For example, the second lowest score is 55, and its deviation from the mean is

$$\text{deviation} = \text{score} - \text{mean} = 55 - 74.1 = -19.1.$$

In Table 14–5 the 10 scores are listed in ascending order in column 1 and the deviation for each score is calculated in column 2. Because the deviations describe how far each score lies from the mean, you might at first think that the average of the deviations would be a good measure of the dispersion. However, the total of the deviations always is *zero* (because they are deviations from the mean). By convention, we therefore square each of the deviations, which yields the positive numbers shown in column 3.

Table 14-5. The 10 exam scores listed have a mean of 74.1. Also shown for each score is the deviation of the score from the mean and the square of the deviation.

	score	deviation from mean	deviation^2
	45	45 – 74.1 = –29.1	$(-29.1)^2 = 846.81$
	55	55 – 74.1 = –19.1	$(-19.1)^2 = 364.81$
	63	63 – 74.1 = –11.1	$(-11.1)^2 = 123.21$
	72	72 – 74.1 = –2.1	$(-2.1)^2 = 4.41$
	77	77 – 74.1 = 2.9	$(2.9)^2 = 8.41$
	79	79 – 74.1 = 5.9	$(4.9)^2 = 24.01$
	81	81 – 74.1 = 6.9	$(6.9)^2 = 47.61$
	84	84 – 74.1 = 9.9	$(9.9)^2 = 98.01$
	88	88 – 74.1 = 13.9	$(13.9)^2 = 193.21$
	97	97 – 74.1 = 22.9	$(22.9)^2 = 524.41$
Total	**741**	**0**	**2234.9**

If we divide the sum of the squared deviations, 2234.9 by *1 less* than the total number of scores, or 9, we get the **variance**, or

$$\frac{2234.9}{10-1} = \frac{2234.9}{9} = 248.3.$$

The variance is approximately the average of the squared deviations. However, interpreting it is difficult because the number may be very large. An easier number to interpret is the **standard deviation**, which is the square root of the variance. For the 10 exam scores,

$$\text{standard deviation} = \sqrt{\text{variance}} = \sqrt{248.3} = 15.8.$$

Strictly speaking, the mean of the squared deviations is their total divided by the number of data points, which we can call n. The variance is obtained by dividing by $n - 1$ because statisticians have found that to be a better measure of the dispersion. Note that, if n is large, $n \approx n - 1$.

Time-Out to Think: Compare the standard deviation of 15.8 to the individual deviations shown in Table 14–5. Does the standard deviation represent a "typical" deviation from the mean? Explain.

This procedure for calculating the variance and standard deviation is cumbersome for large data sets. An alternative formula exists for such cases but, because it is less intuitive, we do not cover it in this book. In a few special cases other formulas exist. For example, in subsection 13.4.3 we showed that the standard deviation of a binomial distribution is $\sqrt{npq}$, where p and q are the two probabilities involved and n is the number of trials.

Summary Box 14–1. Steps to the Variance and Standard Deviation

Step 1: Find the mean of the data set.

Step 2: For each value in the data set, find the deviation from the mean: Deviation = value – mean.

Step 3: Square each of the deviations and find the total of these squared deviations.

Step 4: Divide the sum from step 3 by *1 less* than the total number of data values to obtain the variance.

Step 5: Take the square root of the variance to obtain the standard deviation.

Example 14–11 ***Calculating Standard Deviation.*** Two sets of five students take a quiz, with the following results.

Group 1 scores: 19, 20, 21, 22, 23.

Group 2 scores: 12, 16, 19, 28, 30.

Find the mean and standard deviation for each group and compare these statistics for the two groups.

Solution: First, find the mean for each group.

$$\text{Group 1:} \quad \text{mean} = \frac{19+20+21+22+23}{5} = \frac{105}{5} = 21.$$

$$\text{Group 2:} \quad \text{mean} = \frac{12+16+19+28+30}{5} = \frac{105}{5} = 21.$$

Note that the mean is the same for both groups, even though the distributions are very different. To find the standard deviations, calculate the deviations and squared deviations for each data point (steps 2 and 3 in Summary Box 14–1). The easiest way to do so is by making a simple table:

Group 1			Group 2		
Score	**Deviation**	**(Deviation)^2**	**Score**	**Deviation**	**(Deviation)^2**
19	19 – 21 = –2	$(-2)^2 = 4$	12	12 – 21 = –9	$(-9)^2 = 81$
20	20 – 21 = –1	$(-1)^2 = 1$	16	16 – 21 = –5	$(-5)^2 = 25$
21	21 – 21 = 0	$0^2 = 0$	19	19 – 21 = –2	$(-2)^2 = 4$
22	22 – 21 = 1	$1^2 = 1$	28	28 – 21 = 7	$7^2 = 49$
23	23 – 21 = 2	$2^2 = 4$	30	30 – 21 = 9	$9^2 = 81$
		Sum = 10			**Sum = 240**

Following Step 4 of Summary Box 14–1, find the variance by dividing the sum of the squared deviations by 1 less than the number of scores, or 5 – 1 = 4. Then find the standard deviation (Step 5) by taking the square root of the variance.

Group 1: variance = 10 ÷ 4 = 2.5, and standard deviation = $\sqrt{2.5}$ = 1.6.

Group 2: variance = 240 ÷ 4 = 60, and standard deviation = $\sqrt{60}$ = 7.7.

Although both groups of scores have the same mean, the standard deviation of group 2 is much larger because its scores are spread out to a greater degree. ❖

14.4.4 Shape of the Distribution

Thus far we have discussed several ways of characterizing the distribution of a set of data: the mean, median, and mode; the five-number summary; and the variance and standard deviation. When data is represented with a histogram or a line chart, characterizing the general shape of the distribution also is useful.

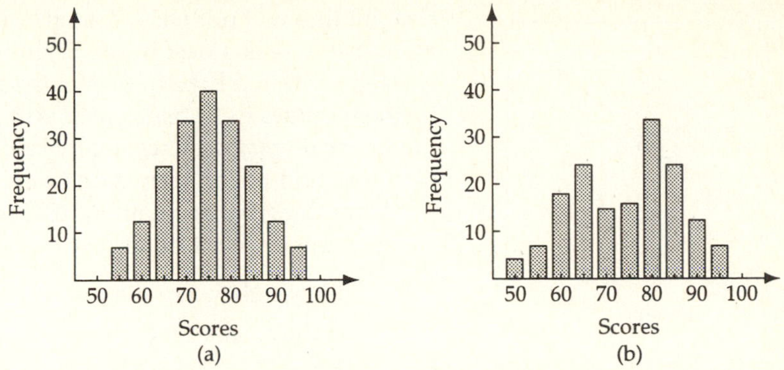

Figure 14-12. (a) The exam scores shown in this histogram form a single-peaked distribution. (b) The histogram for another exam shows two peaks: a bimodal distribution.

A particularly important type of symmetric, single-peaked distribution is the *bell-shaped,* or *normal distribution*. Because it is so important in statistical inference, we defer discussion of it to Section 14.5.

One of the many ways to characterize the shape is by the number of peaks in the distribution. Figure 14–12 shows the distribution of exam scores, in 5-point bins, for two exams. In Figure 14–12(a), the histogram has a peak for scores of 70–74, and tails off to either side. It is called a **single-peaked** distribution and indicates that many students scored near the mode of 70–74 with fewer at extreme scores. In contrast, the exam scores shown in Figure 14–12(b) show two peaks, indicating that one group of students did very well and the other did very poorly. It is called a **bimodal** distribution because it has modes at both 60–64 and 80–84 points. Many other arrangements of peaks are possible, but hereafter we consider only single-peaked distributions.

Three more exam distributions are shown in Figure 14–13, this time with line charts on which a smooth curve has been drawn, rather than with histograms. In Figure 14–13(a), the distribution is **symmetric**, and the mean, median, and mode all are the same. In Figure 14–13(b), most of the scores are relatively low, with a mode (most common score) at about 65. Although most students scored near this mode, a few received much higher scores. These high scores pull the mean to a higher value than the median and the median to a higher value than the mode. Because of the few very high scores, this distribution is said to be **positively skewed**. The exam distribution shown in Figure 14–13(c) is similar, except that most students did well, with a mode at about 85. Here, in this case, the few very low scores pull the mean to a lower value than the median and mode, and the distribution is said to be **negatively skewed**.

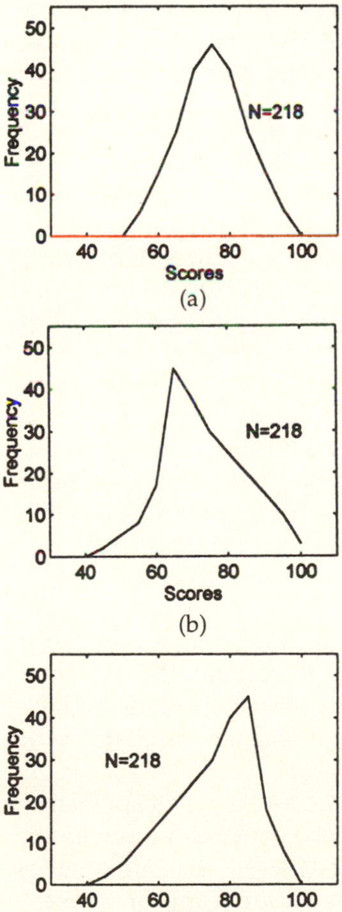

Figure 14-13. (a) A line chart shows exam scores distributed symmetrically about the mean. The mean, median, and mode all are the same. (b) On another exam, most scores were low, but a few high scores pull the mean to a higher value than the median. This distribution is positively skewed. (c) A third exam is negatively skewed, with most students doing well but a few low scores pulling down the mean.

Example 14–12 ***Distribution of Family Income.*** Consider family income for all families in the United States. Is this distribution symmetric, positively skewed, or negatively skewed? Explain.

Solution: To determine the symmetry of the distribution consider how the mean and median are related. Median family income is relatively low (about $35,000). Half of all families earn less than the median but cannot earn less than the minimum income of $0. For the half of all families who earn more than the median, income has virtually no upper limit. These relatively few wealthy families pull up the mean so that the mean income is higher than the median income. Thus the distribution includes an excess of large income values, making it positively skewed.

❖

14.4.5 Scatter Plots and Correlation

So far in this section we have focused on single-variable distributions such as test scores and family income. When two variables are involved, we often are interested in whether a correlation exists between them. In subsection 14.3.4 we showed that data for two variables may be displayed in a scatter plot.

Recall from Chapter 10 that only two data points are needed to determine a unique straight line. In general, for *more* than two data points, a single

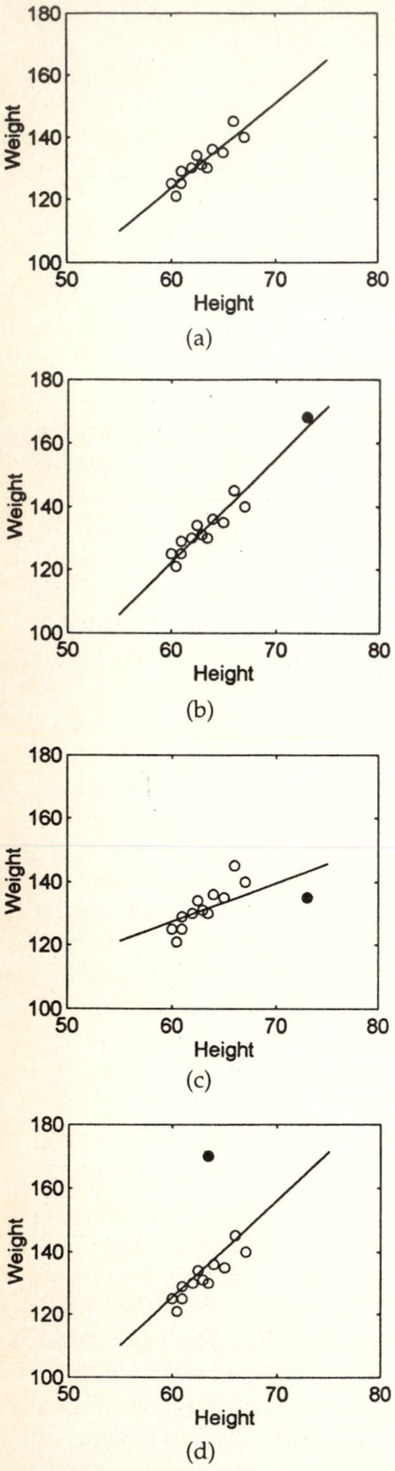

Figure 14-14. Scatter plots of height (in inches) on the horizontal axis and weight (in pounds) on the vertical axis are shown for a girls' high school basketball team. (a) These data are for the 11 original team members, along with the best-fit line. (b), (c), and (d) One point is added to the original 11, and its effect on the best-fit line is shown.

straight line will not pass through all of them. Nevertheless, we can use *linear regression* to seek a best linear fit through a set of data. In subsection 10.5.3, we demonstrated how to draw linear regression lines "by eye." If you take further courses in statistics, you will learn more systematic methods for finding linear regression lines; many calculators and computer programs also can help you find linear regression lines.

Describing the correlation between two variables on a scatter plot involves answering two important questions.

1. How tightly are the points clustered near the best fit line?
2. If the correlation is strong, what is the equation of the correlation line?

If the data points generally are close to the best-fit line, the correlation is strong. For example, Figure 14–7(b) shows a strong correlation between the infant death rate and life expectancy within a country. If the data points are scattered far from the best-fit line, the correlation is weak or nonexistent. The rainfall–snowfall data of Figure 14–7(a) shows no correlation. Once a regression line is drawn (by eye), its equation can be obtained with the methods described in subsection 10.5.3.

Effects of Outliers

Perhaps the greatest difficulty in searching for correlations is caused by **outliers**, or certain points that are separated conspicuously from the other points. For example, Figure 14–14(a) shows a set of height and weight data for a girls' high school basketball team. Each data point represents one girl; height in inches is plotted on the horizontal axis and weight in pounds on the vertical axis. Note that the best fit line shown indicates a good correlation between height and weight.

Suppose that the team adds one new member who is 6 feet (72 inches) tall and weighs 170 pounds. The height–weight data for the team is shown in Figure 14–14(b). The point representing the new player is an outlier because it lies far above and to the right of the other points. This outlier happens to fall very close to the best-fit line of the original data set, so there is little change in the line. Indeed, this added point makes the correlation between the team members' heights and weights look even stronger.

Figure 14–14(c) shows data for the original team and one new member, but this member is 72 inches tall and weighs only 130 pounds. The outlier representing this new team member pulls the entire best-fit line down. The slope is much different from that of the original line, and the correlation now appears to be much weaker.

Figure 14–14(d) shows data for the original team and a new team member who is 64 inches tall and weighs 170 pounds. The data point representing this new player is an outlier far above the original data set. This outlier gives the best-fit line a greater slope, again making the correlation appear weaker.

Thus even one outlier can have a significant effect on the correlation, which in turn affects the predictive power of the data. The treatment of outliers is both important and subtle. If an outlier is a genuine data point, it should be included and allowed to alter the best-fit line accordingly. However, if an outlier is spurious, it should be removed so that it won't improperly skew the data.

THINKING ABOUT ... MEANINGLESS CORRELATIONS

In 1988, Oxford physician Richard Peto submitted a paper to the British medical journal *Lancet* showing that heart attack victims had a better chance of survival if they were given aspirin within a few hours after their heart attacks. The editors of *Lancet* asked Peto to break down the data into subsets, to see whether the benefits of the aspirin were different for different groups of patients. For example, was aspirin more effective for patients of a certain age or for patients with certain dietary habits?

Peto objected to the request, claiming that breaking the sample under study into too many subgroups would result in some correlations found by chance alone. Writing about this story in the *Washington Post*, journalist Rick Weiss said, "When the editors insisted, Peto capitulated, but among other things he divided his patients by zodiac birth signs and demanded that his findings be included in the published paper. Today, like a warning sign to the statistically uninitiated, the wacky numbers are there for all to see: Aspirin is useless for Gemini and Libra heart-attack victims but is a lifesaver for people born under any other sign."

In another case, St. Louis surgeon Sherman Silber decided to highlight statistical errors in infertility research by studying 28 patients receiving infertility treatments, 7 of whom successfully had children. Taking the last names of each of the patients he programmed a computer to search for any statistically significant correlations. He found that patients with last names beginning with G, Y, or N were most likely to have success with infertility treatment.

The moral of these stories is that a "fishing expedition" for correlations usually produces some. That doesn't make the correlations meaningful, even though they may appear significant by standard statistical measures.

14.5 STATISTICAL INFERENCE

In Sections 14.3 and 14.4, we explored methods of describing data visually and with numbers that characterize a data distribution. We have been dealing with methods of descriptive statistics that describe the data collected from a sample. Because the goal of a statistical study usually is to learn something about a population, let's now investigate the process of inferring population parameters from sample statistics.

14.5.1 The Normal Distribution

Recall that the symmetric distribution shown in Figure 14–13(a) has a shape that looks like a bell. This special type of symmetric distribution is called the **bell-shaped curve**, or the **normal distribution**. The normal distribution is extremely important because it occurs so often. We can describe the prevalence of the normal distribution by considering a human attribute such as the height of adult women. Most women tend to be fairly average in height, which creates the peak of the distribution. Moving away from the average of the distribution, we find fewer and fewer women near the extremes of very short or very tall, which produces the *tailing off* of the normal curve on either side of the mean.

The fact that a quantity that is itself the product of many individual, random "trials" (or factors) tends to be normally distributed is known in mathematics as the **central limit theorem**. We demonstrated this theorem in subsection 13.4.3 by showing that the binomial distribution approximates a normal distribution when the number of trials is large.

In fact, any quantity that is the result of many factors is likely to follow a normal distribution. For example, the height of an adult is the product of many genetic and environmental (e.g., nutrition) factors. Similarly, scores on SAT tests or IQ tests tend to be normally distributed because each test score is the sum of the results from many individual test questions.

All normal distributions have the same characteristic bell shape; they differ only in their values for the mean and standard deviation. As with any distribution, the mean indicates the center of the normal distribution. However, the standard deviation has a special interpretation with the normal distribution — an interpretation that we presented in subsection 13.4.3 — involving three rules. Because they are so important, we repeat those rules here. Moreover,

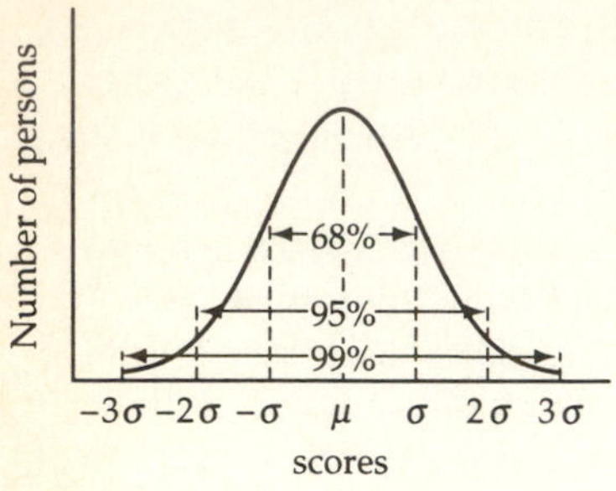

Figure 14-15. The percentages of scores falling within 1, 2, and 3 standard deviations of the mean in a normal distribution are shown.

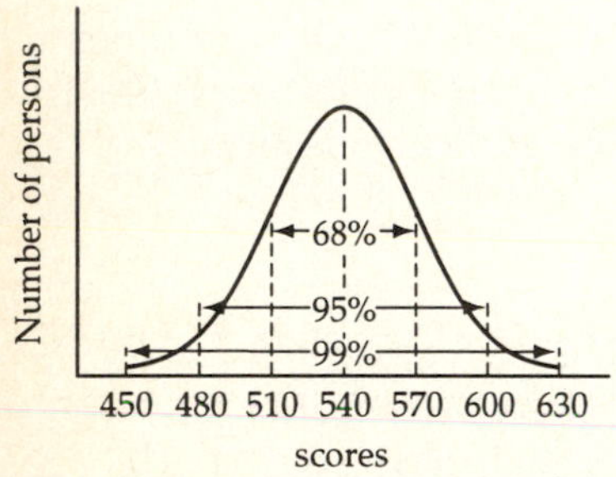

Figure 14-16. These SAT scores are normally distributed with a mean of 540 and a standard deviation of 30.

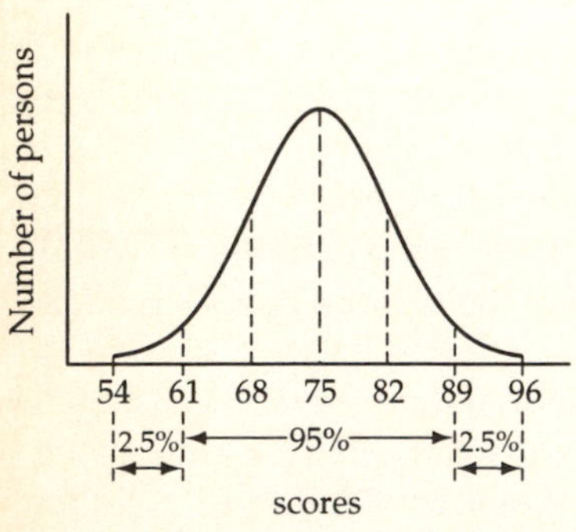

Figure 14-17. These exam scores are normally distributed with a mean of 75 and a standard deviation of 7.

because the standard deviation is such a common measure of dispersion, *memorizing* the following rules concerning normal distributions will be helpful (shown graphically in Figure 14–15).

- About two thirds (actually 68.3%) of the scores in a population will fall within 1 standard deviation of the mean.
- About 95% (actually 95.4%) of the scores in a population will fall within 2 standard deviations of the mean.
- Over 99% (actually 99.7%) of the scores in a population will fall within 3 standard deviations of the mean.

Example 14–13 *SAT Score Distribution.* Suppose that a college reports that the mean score of its entering freshmen on the mathematics SAT is 540, with a standard deviation of 30. Interpret this statement.

Solution: As an exam built from many questions, SAT scores should follow a normal distribution. Because the normal distribution is symmetric, the mean, median, and mode of the distribution are the same score: 540. That is, most students scored near the mean of 540.

Because the distribution is normal, about 68% of the students scored within 1 standard deviation of the mean. For the standard deviation of 30, "within 1 standard deviation" means anywhere from 30 points *below* the mean, or 540 − 30 = 510, to 30 points *above* the mean, or 540 + 30 = 570. That is, about 68% of the students scored between 510 and 570.

Similarly, about 95% of the students scored within 2 standard deviations (60 points) of the mean, or between 480 and 600 (60 points below the mean is 540 − 60 = 480, and 60 points above the mean is 540 + 60 = 600). Finally, more than 99% of the students scored within 3 standard deviations of the mean; as 3 standard deviations is 90 points for this set of SAT scores, more than 99% of the students scored between 450 and 630. The overall distribution is shown in Figure 14–16.

❖

Example 14–14 *One Side of the Curve.* Suppose that 1000 students take an exam. If the mean score is 75, with a standard deviation of 7, about how many students scored above 89?

Solution: Again, because you are dealing with an exam, assuming that the scores are normally distributed is reasonable; the distribution is shown in Figure 14–17. A score of 89 is 14 points, or 2 standard deviations, above the mean of 75. About 95% of the scores are *within* 2 standard deviations of the mean, so about 5% of the scores are *farther* than 2 standard deviations from the mean. Half of this 5%, or 2.5%, are scores 2 standard deviations *below* the mean; the other half, or 2.5%, of the scores are 2 standard deviations *above* the mean. Thus 2.5% of 1000 students, or about 25 students, scored higher than 89.

❖

Example 14–15 *Normal Auto Prices.* Because the price of an automobile involves many variables (e.g., options and dealer preparation charges), and often bargaining at the dealership as well, different people tend to pay different prices for the same model of car. Suppose that a survey shows that the mean price paid for a particular new car model is $16,400, with a standard deviation of $400. Assume that 100,000 cars of this model are sold. How many people paid between $16,000 and $16,800? How many paid less than $16,000? How many paid more than $17,200?

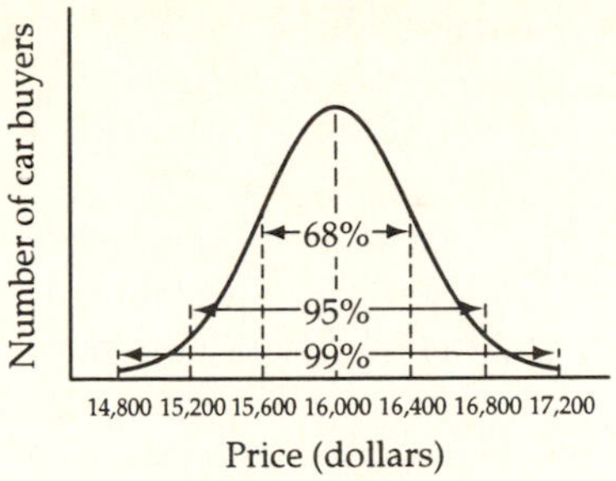

Figure 14-18. This graph shows normally distributed auto prices with a mean of $16,400 and a standard deviation of $400.

Solution: Because there are many variables involved in determining the price paid by any single consumer, the prices paid by many consumers should follow a normal distribution (Figure 14–18). Hence about 68% of the 100,000 car buyers, or about 68,000 consumers, paid a price within 1 standard deviation of the mean. For the standard deviation of $400 and the mean of $16,400, about 68,000 consumers paid between $16,000 and $16,800. The remaining 32,000 car buyers paid *either* less than $16,000 or more than $16,800. Because a normal distribution is symmetric, half of these people, or about 16,000 consumers, paid less than $16,000.

Finally, to determine how many paid more than $17,200, you should recognize that this price is 2 standard deviations (2 × $400 = $800) above the mean price of $16,400. About 95% of consumers paid within 2 standard deviations of the mean, or prices between $15,600 and $17,200. Therefore about 5%, or 5000 people, paid *either* less than $15,600 or more than $17,200. About half of these 5000 people, or roughly 2500 people, paid more than $17,200.

❖

14.5.2 Surveys and Opinion Polls

Suppose that a random sample of 1000 television viewers reveals that 320, or 32%, of the viewers in the sample were watching *Friends* during its time period. The survey takers then report that 32% of *all* television viewers were watching *Friends* at that time, with a **margin of error** of 3%. What do they mean?

In the jargon of opinion polls, adding and subtracting the margin of error from the sample result gives a *95% confidence interval*. In this case, the 95% confidence interval is from 32% – 3% = 29% to 32% + 3% = 35%. Thus the survey takers are claiming to be 95% confident that the true percentage of *all* television viewers were watching *Friends* at that time was between 29% and 35%. In other words, they believe that there is a 95% probability that the true percentage was within the 29% to 35% range, and only a 5% probability that the true percentage was outside this range.

In general, the margin of error for any unbiased survey is easy to estimate. If the number of people surveyed is n,

$$\text{margin of error} \approx \frac{1}{\sqrt{n}}.$$

For the survey of 1000 television viewers, the margin of error is $1/\sqrt{1000} \approx$ 3%, as claimed by the survey takers.

Based on taking many samples and finding the percentage of viewers watching in each sample (also called the *mean*), the resulting distribution of the means from every sample is called the **sampling distribution of means**.

The margin of error is related to the standard deviation as follows. Imagine that the survey takers repeat the survey of 1000 viewers many times, with a different random sample each time, and then tabulate (a frequency table) how many of the surveys find that 32% of the sample are watching *Friends*, how many find that 33% are watching, and so on. The results of these many surveys should follow a normal distribution, and the *mean* of these surveys will be the best estimate possible of the true percentage of all viewers watching *Friends*. The *standard deviation* indicates the dispersion in the results from the surveys. The margin of error is what statisticians calculate to be 2 standard deviations in the distribution of surveys.

Example 14–16 ***Interpreting the Margin of Error.*** Suppose that an opinion poll shows that 65% of consumers in a sample prefer the taste of Clear Cola to that of Cloudy Cola, with a margin of error of 5 percentage points. What is the probability that *less* than 60% of the population prefers Clear Cola?

Solution: From the definition of the margin of error, there is about a 95% chance that the true population percentage is within 5 percentage points of the 65% sample percentage; that is, there is a 95% chance that between 60% and 70% of the population prefers Clear Cola. Thus there is about a 5% chance that Clear Cola is preferred *either* by less than 60% of the population or more than 70% of the population. The probabilities should be symmetric, so there is a 2.5% (half of the 5%) probability that less than 60% of the population prefers Clear Cola.

Example 14–17 ***Estimating the Margin of Error.*** On the same day, three polling organizations survey voters on their preferences for the next election. Company A surveys 500 people and finds that 52% plan to vote for Smith. Company B surveys 1500 people and finds that 49% plan to vote for Smith. Company C surveys 3000 people and finds that 47% plan to vote for Smith. Find the margin of error for each survey. If 50% of the vote is needed to win the election, is Smith likely to win? Are the three surveys consistent?

Solution: The margins of error for surveys of 500, 1500, and 3000 people are

$$\text{500 people: Margin of error} \approx \frac{1}{\sqrt{500}} = 0.045 = 4.5\%;$$

$$\text{1500 people: Margin of error} \approx \frac{1}{\sqrt{1500}} = 0.026 = 2.6\%;$$

$$\text{3000 people: Margin of error} \approx \frac{1}{\sqrt{3000}} = 0.018 = 1.8\%.$$

By adding and subtracting the margin of error in each survey, the 95% confidence intervals are as follows.

Company A: The percentage of voters supporting Smith is 52% ± 4.5%, or between 47.5% and 56.5%.

Company B: The percentage of voters supporting Smith is 49% ± 2.6%, or between 46.4% and 51.6%.

Company C: The percentage of voters supporting Smith is 47% ± 1.8%, or between 45.2% and 48.8%.

Neither Company A's survey nor Company B's survey can be used to predict the outcome of the election because both include values on either side of 50%. However, Company C's survey represents a 95% probability that Smith will receive between 45.2% and 48.8% of the vote. Based on this result, Smith apparently will lose. Note also that an outcome showing Smith receiving less than 48.8% of the vote falls within the 95% confidence limits of both Company A's and Company B's surveys. Thus all three surveys are consistent with the prediction that Smith will receive less than 48.8% of the vote.

❖

Bias in Surveys

Surveys and opinion polls are notorious for introducing bias. The *Literary Digest* case (discussed in Section 14.2) illustrated two common forms of bias: selection bias and participant bias. A third common bias in surveys comes from poorly worded questions.

All three forms of bias are all too common. The best way to guard against bias is through common sense. Think about how the survey was conducted and decide whether the conclusions are justified. If they aren't, the context for interpreting the survey results may be drastically different.

Example 14–18 *Call-In Survey.* A TV talk show is featuring a discussion of immigration. The host asks viewers to call in with an answer to the question: Should immigration be further restricted? Two phone numbers are given, one for *yes* answers and one for *no* answers. Each phone number is a 900 number, for which callers are billed \$1 by their telephone company. At the end of the show, the host announces that 65% of the callers thought immigration should be further restricted and that 35% didn't. What can you conclude from these results?

Solution: The only thing that you can conclude is that 65% of the people who bothered to make the \$1 call thought that immigration should be further restricted. You cannot draw any conclusions about any population larger than this self-selected sample of callers.

The survey suffers badly from at least three forms of participant bias. First, only viewers interested in immigration are likely to be watching the show. Second, viewers who feel strongly about the issue are more likely to want to call. Third, of those who want to call, only a small number will spend the dollar to make the phone call.

❖

Example 14–19 *Choice in Questions.* Following carefully constructed methods for random sampling, a survey polls 1000 Americans to find out which issues are most important in the upcoming election. The survey asks participants to: "Rank each of the following issues in order of their importance to you."

Inflation	Pollution	Other
Crime	Immigration	
Unemployment	Abortion	

Will the responses on this survey be meaningful?

Solution: Although the selection of participants is unbiased because it is random, the question suffers from a great deal of bias. Most important, the survey limits the choices of respondents. Even though there is an *Other* category, most people will be thinking along the lines of the listed categories (the *availability error* discussed in Section 3.3). The majority of Americans may well believe that education is the most important issue; yet, because it isn't on the list, only a small number of people are likely to select *Other* and fill in education.

❖

14.6 SAMPLE ISSUES IN STATISTICAL RESEARCH

Now that we have covered the general principles of statistical research, the best way to understand statistics in everyday life is by examining case studies. In this section, we discuss a few interesting statistical studies, chosen both because they show how statistics are used, and because they illustrate the care that must be taken in interpreting statistical research.

14.6.1 Weight and Health

We first return to the study described at the beginning of the chapter, which concluded that either excess weight or a midlife weight gain increases the risk of early death. This conclusion is one of many drawn from a sample of 115,000 female nurses whose health has been monitored since 1976. All the women in the group were healthy and between 30 and 55 years old when the study began.

Before going any further, we first must consider whether the sample is unbiased. After all, the sample includes only women of a single profession. However, because we have no reason to think that nurses are any different from other people *in terms of health*, the extrapolation from this sample to the general population probably is reasonable.

To emphasize the effects of weight, the study excluded women who smoked or who were lean because of other health problems. The researchers used a measure of weight in proportion to height called the body mass index:

$$\text{body mass index} = \frac{\text{weight in kilograms}}{(\text{height in meters})^2} = \frac{0.45 \times (\text{weight in pounds})}{(0.0254 \times \text{height in inches})^2}.$$

Between 1976 and 1995, 4,726 of the nurses died. To determine death rates by weight, the researchers binned the nurses according to body mass index (those with stable weights were considered separately from those with weight gains during the study). For example, suppose that 10,000 nurses in the study had a body mass index of less than 19 and that 100 of these nurses died between 1976 and 1995. Then the death rate for these women was 100 in 10,000, or 1%.

The study reported that the leanest women, defined as those with a body mass index of less than 19, had the lowest death rate. It also reported that, as body mass index rose, death rates rose dramatically. For example, obese women, defined as those with a body mass index of more than 32, suffered four times as many deaths from heart disease and twice as many deaths from cancer as women whose weights were below average for their height and age.

Because of the huge amount of data available, the researchers were able to draw numerous conclusions about the health of the nurses. Figure 14–19 shows a line chart of the increased risk of death for women having different body mass indexes; for example, an increased risk of 100% means double the death rate of the leanest women.

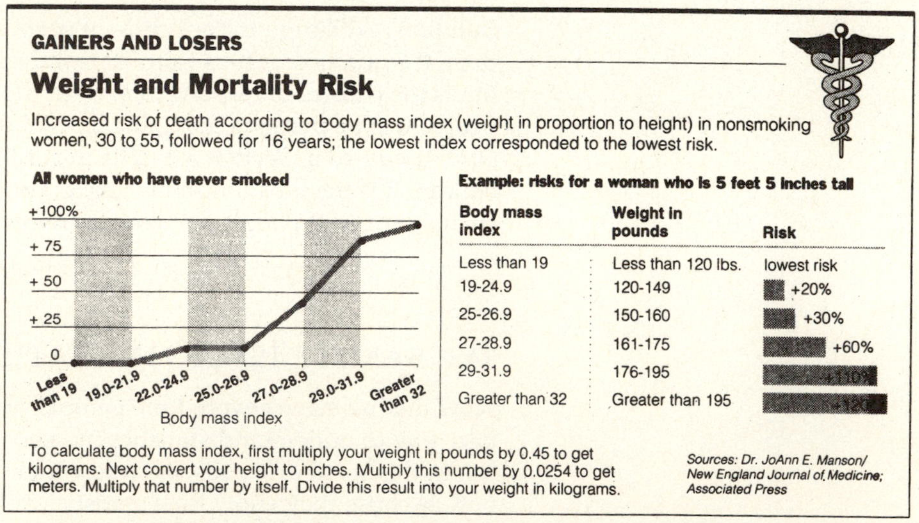

Figure 14-19. The graph shows the increased risk of death, in percent (vertical scale), for women of different body mass indexes in the study of the 115,000 nurses. The example shows the risk for a woman 5 feet 5 inches tall, with weights in pounds that correspond to body mass indexes for this height. *The New York Times*, 9/14/95. Copyright © 1995 by The New York Times Company. Reprinted with permission.

Time-Out to Think: Compare the graph shown in Figure 14–19 to the risks shown in the accompanying example. Note that the numbers are not completely consistent, pointing out the difficulty in a making sense of statistical studies even when they are carefully reported in newspapers. Can you explain the discrepancies or do you think that the figure is in error?

Another study finding was that a moderate weight gain in middle age resulted in an increased risk of premature death. This finding directly contradicted the doctors who, only 5 years earlier, had stated that a midlife

The researchers *did* suggest a potential physical mechanism for the increased mortality with higher body mass index: that bile acids formed from fats in the diet and estrogens (hormones) produced in body fat raise the risk of certain cancers. At present, this assertion is only a hypothesis and certainly hasn't been confirmed.

weight gain was harmless. Why did this study differ from earlier ones on which the 1990 conclusions were based? The earlier studies apparently failed to take into account the fact that many lean people are smokers or have other diseases. As a result, these studies mixed the low death rates of healthy lean people with the far higher death rates of smokers and others with life-threatening diseases.

The flaws in the earlier studies beg the next question: Have the researchers in the new study accounted for all potentially confounding factors? For example, could it be that weight isn't really the main factor in the low death rates, but rather some aspect of diet or exercise? Ultimately, the complexity of human beings makes knowing whether all potentially confounding factors have been taken into account very difficult. Thus, although this research involves a large sample and appears to include a careful and reasonable analysis, it still demands critical evaluation. Until a physical mechanism to explain how weight affects death rates is demonstrated, the argument is far from over.

14.6.2 Dan Quayle and the Missing American Children

During the 1992 presidential campaign, Vice-President Dan Quayle started a major public debate on the issue of out-of-wedlock births. In particular, he singled out a fictional television character for criticism. The character, named Murphy Brown and played by actress Candice Bergen, was a single woman who decided to have a baby. Quayle suggested that this decision made her a poor role model for other women.

Meanwhile, the U.S. Census Bureau was conducting a biennial (every two years) door-to-door survey of births. To check the validity of the survey, the Bureau compares its numbers to data collected from hospital birth records. Although agreement had been found in the past, the 1992 survey reported 300,000 fewer children of unmarried mothers than expected! It also reported a smaller number of unmarried mothers than expected and fewer teenage mothers than expected (Figure 14–20).

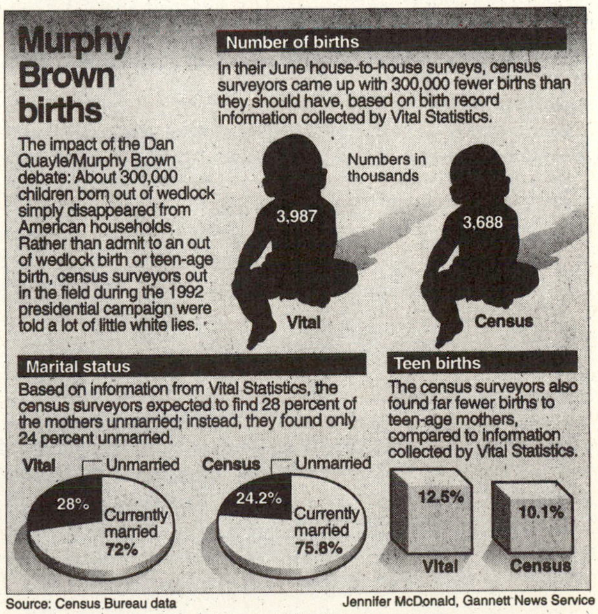

Figure 14-20. This graphic, from a newspaper, summarizes how the results of the U.S. Census Bureau survey compared to the expected results from birth records. *The Denver Post*, 12/30/93. Reprinted with permission.

Where are the 300,000 missing children? The researchers concluded that the children weren't missing but rather that their unmarried mothers had told the door-to-door surveyors either that they were married or did not have children. That is, lying skewed the survey results.

Why was lying a problem for the 1992 survey but not for earlier surveys? Presumably, Dan Quayle's inspired debate on single motherhood made some single mothers less than candid about their marital or parental status. This example illustrates how survey results, even when conducted in an unbiased manner, can be affected by current social conditions. Just as literature is best understood in the context of the times in which it was written, so is a social context required in interpreting statistical surveys.

14.6.3 Trends in the SAT

Scholastic Aptitude Test (SAT) scores have been widely used to assess the general state of American education. School districts, states, and the nation routinely compare SAT scores from year to year. From the 1960s to the 1990s, mean SAT scores generally declined. In 1995, SAT scores rose dramatically.

Common sense raises two immediate questions that must be answered to assess the validity of SAT scores as a measure of trends in American education.

1. Are trends among the *sample* of high school students who take the SAT representative of trends for the *population* of all high school students?
2. As the test changes every year, can scores on the test from one year be legitimately compared to scores in other years?

Unfortunately, neither question can be answered with a clear *yes*. For example, many people argue that much of the long-term decline in SAT scores is the result of changes in the makeup of students who take the test. In the 1960s, only about the top third of high school students took the SAT. Today, more than half of all high school students take the SAT. Thus the decline in score averages might simply reflect changes in the sample: *Average* students now take the exam, whereas only *above average* students took it in the past. Because of the change in the sample, inference from sample trends to population trends is extremely difficult.

Comparison of SAT scores becomes even more difficult beginning with the 1996 test. In 1996, the College Board "recentered" SAT scores so that the average score on both the verbal and math sections is 500. Clearly, this makes year-to-year comparisons far more difficult.

The second question is even more difficult to answer. The College Board, which administers the SAT, asserts that the exams are comparable in difficulty from one year to the next, but that claim is difficult to quantify. As a result, most statements about difficulty are subjective. Further complications arise when the test undergoes major changes. For example, the 1995 test was the first to allow the use of calculators on the mathematics section, additional time was allotted for the entire test, and an entire section of prior verbal SATs (the section on antonyms) was replaced with a new section (on vocabulary words in context). With these changes, can it really be claimed that the 1995 test was equal in difficulty to prior tests and that the jump in average scores in 1995 reflects improvements in education?

Time-Out to Think: Some people have proposed to improve the quality of U.S. education by creating national achievement tests to be taken by *all* students in grades 4, 8, and 12. Do you think such tests would help or hurt education? Would they make assessment of the quality of education easier? Explain.

14.6.4 Medical Research and the Placebo Effect

Most studies of the effectiveness of a new drug or treatment are conducted by comparing results of the treatment with results from a **placebo** — something that looks and feels like the drug being tested but lacks its active ingredients. Neither patients nor researchers are supposed to know who received the placebo and who received the real drug until the end of the study.

Surprisingly, the placebo often proves to be a highly effective treatment itself! Indeed, some studies have shown that as many as two thirds of patients receiving a placebo improve, at least temporarily. The placebo effect is enhanced when patients trust their physician and the physician is enthusiastic about the new "treatment."

The mechanism of the placebo effect is unknown, and an important question is whether the placebo effect can produce a real cure or only a perception of feeling better. With mild medical problems such as colds, distinguishing a true cure from a psychological perception would be difficult. However, with serious diseases such as AIDS or cancer, understanding the placebo effect could have important implications. For example, if the placebo effect actually can cure such diseases, an important aspect of treatment should be psychological, urging patients to believe that they will be cured.

The placebo effect also causes problems in research. After all, if two thirds of patients receiving the placebo will report improvement, how can the researcher determine whether the real treatment is making a difference? This problem is most serious in the evaluation of treatments used primarily for mild medical conditions (e.g., vitamins, herbs, and acupuncture). Because most people trying such treatments *believe* that they will be beneficial, as do the practitioners who prescribe them, the placebo effect is likely to be especially strong. That makes evaluating the true effectiveness of such treatments very difficult.

Time-Out to Think: Suppose that the effectiveness of some new treatment turns out to be entirely the result of the placebo effect. Should the treatment still be used? Should it be used only under certain circumstances? Defend your opinion.

14.6.5 Are Unbiased Surveys a Thing of the Past?

In early 1990, the Nielsen television rating system reported a large drop, 8%, in the number of people watching television. Such a drop would make television advertising time less valuable and might have a severe impact on the economics of the television industry. The industry claimed that the Nielsen survey results were misleading and that viewership had not declined. The industry argued that only 47% of the people that Nielsen contacted agreed to use the Nielsen meter, which monitors when the television is on and what is being watched. A decade earlier, 68% of the people contacted agreed to use the meter. Robert Niles, head of research for NBC at the time, said "We don't know whether the people who turned down the meter are like those who agreed to take it."

Market research is a multibillion dollar industry in which researchers survey consumers to determine everything from whether car buyers are willing to pay for more safety features to whether a market exists for a new sugar substitute. Because surveys ask questions of only a small sample of the population, the sample must be unbiased. Unfortunately, as the Nielsen issue demonstrates, unbiased sampling is increasingly difficult to achieve.

Probability techniques allow researchers to select an unbiased sample of people to interview relatively easily. However, if some of the selected people refuse to participate, *participant bias* is introduced into the survey. Unfortunately for market researchers, more than a third of all Americans routinely shut the door or hang up the phone when contacted for a survey.

Lack of cooperation on surveys also affects political and opinion polling. However, large polling organizations claim to be able to correct for the biases introduced by their analyses of who refuses to participate and by comparing polls to election results over long periods of time. For most market researchers, who conduct smaller scale surveys, such corrections are too difficult and expensive to undertake.

Why? Although surveying those who refuse to participate is difficult, researchers suspect several factors. First, the proliferation of surveys means that many people are contacted repeatedly for different surveys and simply tire of the inconvenience. Americans generally are working more hours and may resent the intrusion of a survey during their few hours of relaxation at home. Second, people concerned about privacy are fearful that information they provide to a survey may put them on yet another mailing list or find its way into a credit report. Perhaps most important, researchers have found that many people are upset by a practice known in the marketing industry as "selling under the guise" of market research, or by its acronym of "sugging." This practice has become so common that even people who are willing to participate in a legitimate survey may be uncooperative because they suspect sugging is taking place. Figure 14–21 summarizes cooperation rates found by, of course, another survey.

Time-Out to Think: Have you been contacted to participate in any kind of survey recently? Was it a real survey or sugging? Did you agree to participate in the survey? Why or why not?

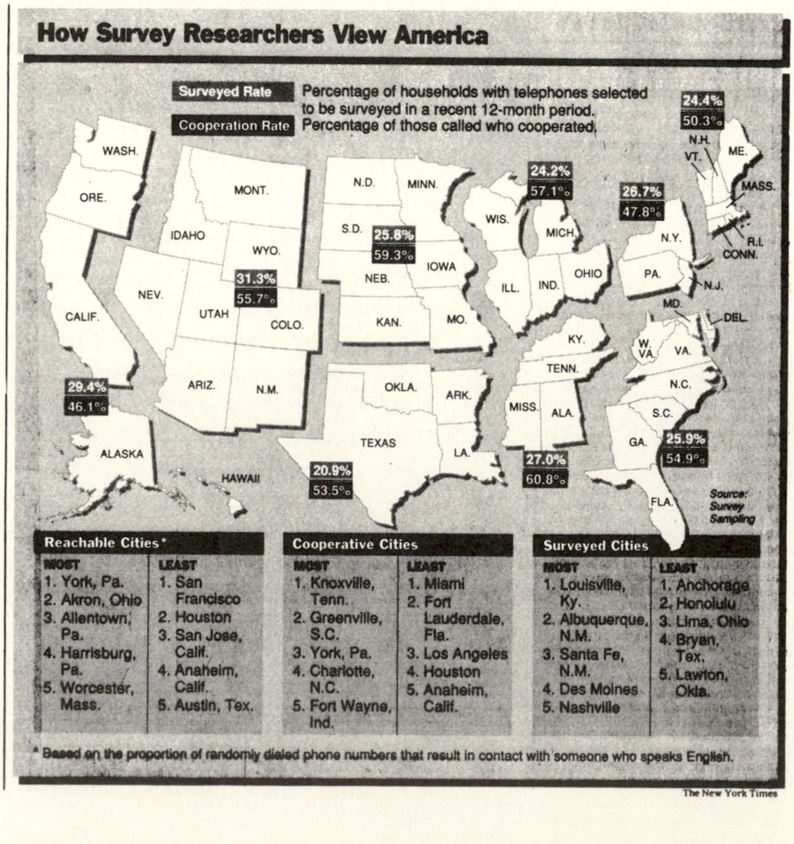

Figure 14-21. The results of a survey, designed to find how cooperative people are when contacted for surveys, are shown. Cooperation means agreeing to participate in the survey, as opposed to hanging up the phone or walking away. *The New York Times*, 10/5/90. Copyright © 1990 by The New York Times Company. Reprinted with permission.

14.6.6 The Genetics of Intelligence

In the 1910s, psychologist Henry Goddard did intelligence testing on immigrants arriving at Ellis Island and concluded that 80% of Jews, Italians, and Hungarians and 90% of Russians were "feeble-minded." As a result, hundreds were deported each year.

In Germany, the Nazis latched onto similar ideas in proclaiming northern Europeans to be superior to other people. Jews, blacks, gypsies, homosexuals, and other groups were labeled "subhuman" and chosen for extermination. The Nazis deliberately killed more than 13 million civilians, including 6 million Jews — more than two thirds of Europe's prewar Jewish population.

The murders of millions of people, justified on the basis of so-called intelligence testing, demonstrates the troubled history of attempts to link heredity and intelligence. Perhaps no other subject in statistics is more widely debated and more widely abused.

The idea that brain size is correlated with intelligence might have been rejected immediately by noting that women have, on average, considerably smaller brain volumes than men; even the most sexist intelligence testers do not claim comparable differences in intelligence.

Psychologists long have sought to measure intelligence through a variety of techniques. During the 1800s, measurements of brain size became popular, based on the assumption that larger brains mean greater intelligence. Later, much of the research was shown to be in error, with brain volumes routinely measured incorrectly. Subsequent attempts to measure intelligence involved various types of tests. The best known measure of intelligence, the *intelligence quotient* or IQ, was first developed by French psychologist Alfred Binet (1857–1911). The original definition of the IQ was a child's "mental age," defined by how well the child did on Binet's tests, divided by the child's actual age and converted to a percentage. For example, if a 5-year-old child performed at the mental level of a typical 6-year-old child, the IQ of the tested child was $6/5 \times 100\%$ or an IQ of 120. Today, the IQ test is defined so that the average person has an IQ of 100, and scores are normally distributed with a standard deviation of 15 points.

Binet believed that intelligence could be increased through properly directed schooling and warned against taking his tests as a measure of innate intelligence. Indeed, his primary objective in developing IQ tests was to identify children who needed special help in school and then to provide them with that help.

Two primary questions lie at the heart of any statistical attempt to demonstrate a hereditary link to intelligence. First, does the IQ, or any other similar number, by itself, truly measure intelligence? Second, is intelligence an innate quality or one that can be manipulated through education?

Neither question has a clear answer. Many psychologists argue that IQ is a valid measure of intelligence, but others believe that a single number can never adequately describe something as complex as human intelligence.

Even if IQ is a valid measure of intelligence, determining whether it is inherited (nature) is extremely difficult because separating the effects of upbringing (nurture) is nearly impossible. For example, research has shown that *physical* connections between neurons in the brains of young children develop in response to external stimuli almost from the moment of birth and perhaps even before. Because a child's environment and surroundings can affect his or her physical brain development, separating genetic and environmental effects may well be impossible even when testing the intelligence of very young children.

An excellent summary of how Cyril Burt's fraud was detected, along with a summary of the sad history of attempts to link intelligence with race, is provided in *The Mismeasure of Man*, by Stephen Jay Gould (W.W. Norton and Company, 1981).

Perhaps the only clear way to distinguish the effects of heredity and environment is through the study of identical twins separated at birth. Such twins would have the same genetic makeup but would be raised in different environments. Unfortunately for scientists, identical twins separated at birth are extremely rare. Further, the history of this research is badly tainted by a famous case of scientific fraud. In the 1950s, British psychologist Cyril Burt said that he had studied more than 50 pairs of identical twins separated at birth. He claimed that the twins had nearly identical IQ scores, despite the different environments in which they had been raised. Burt concluded that IQ is hereditary and further claimed that different races had different innate levels of intelligence. However, Burt's work was later shown to be fraudulent; indeed, Burt's pairs of twins probably never even existed; in all likelihood he simply invented the data reported in his studies!

The debate over the genetics of intelligence isn't likely to be resolved soon. Moreover, in light of the long and troubled history of this debate, you probably should regard new claims about nature versus nurture in intelligence skeptically.

Time-Out to Think: Do you think anyone will ever be able to determine whether or how much of intelligence is inherited? Why or why not?

14.7 CONCLUSION

Data and statistics are everywhere, so an understanding of statistical methods is crucial to understanding society. Despite the complexity of many statistical techniques and problems, evaluating statistical research ultimately comes down to answering just a few key questions.

- Is the sample representative of the population under study and, if so, are the inferred conclusions about the population reasonable?
- Has the study carefully and properly described the *sample* chosen for the study?
- Is the process of inference from the sample to the population reasonable? Do any hidden variables or hidden sources of bias make the results subject to misinterpretation?
- Does the research and its conclusions have real value to society, or is it merely being used to support some preconceived belief?

PROBLEMS

Reflection and Review

Section 14.1

1. **The Word Statistics.** The word statistics has the same root as the word *status*. Briefly explain how the two words are related.
2. **Statistics in the News.** Identify at least three stories reported in today's news that involve statistics in some way. In each case, briefly describe (one to two paragraphs) the role of statistics in the story.

Section 14.2

3. **Descriptive Statistics in Sports.** Most sports keep track of many descriptive statistics (as opposed to inferential), such as won–loss percentage. Choose a sport and describe at least three different statistics commonly tracked by participants or spectators of the sport. In each case, briefly describe the importance of the statistic to the sport.
4. **Samples and Populations.** For each of the following, state the population under study and suggest a method for choosing an unbiased sample from this population.
   a. You want to determine the average weight of collegiate football players.
   b. You want to determine the number of smokers who die from lung cancer each year.
   c. You want to determine the average mercury content of the tuna fish consumed by U.S. residents.
5. **Drinking Survey.** Suppose that you want to determine the (alcoholic) drinking habits of students by asking 100 students to fill out a survey about their alcohol consumption.
   a. What is the population?
   b. Suggest an unbiased method for choosing your sample of students to be surveyed.
   c. How would you administer the survey to ensure honest responses from all students selected?
6. **Pre-Election Survey.** Suppose that you are conducting a poll to determine who is most likely to win an upcoming student government election at your school.
   a. What is the population?

b. About how many students do you think should be included in your sample? Why?
c. Suggest an unbiased method for choosing your sample.
d. Discuss any practical problems with your proposed sampling procedure and possible solutions.

7. **Football Player Body Fat.** You decide to make a study of the average percentage body fat for collegiate football players.
   a. What is the population?
   b. Of the many ways to choose a sample from the population, which of the following would be the best sample for your study, and why? Also, briefly explain why each of the other choices would not make a good sample for your study.
   - The linebackers at your school.
   - The freshmen football recruits at your school.
   - The quarterbacks for each of the top 20 teams in the nation.
   - The entire team at your school.
   - The entire team at the school with the longest losing streak in the nation.

   Which of the above would be the best sample for your study, and why? Also, briefly explain why each of the other choices would not make a good measurement sample for your study.
   c. Describe a method for obtaining the sample statistics from the sample.
   d. Once you have your sample statistics, what can you conclude about the population? Explain.

8. **Poor Sampling.** Find an article in a newspaper or magazine of the past two weeks that attempts to describe some characteristic of a population, but which you believe involved poor sampling (e.g., a sample that was too small, or unrepresentative of the population under study).
   a. Attach a photocopy of the article.
   b. What is the population reported on in the article?
   c. Describe the sample chosen for study.
   d. What conclusion was drawn about the population in the article?
   e. Explain why you believe that the sample was improper.
   f. Explain whether you believe that the conclusion of the article is valid.

9. **Statistics in the News.** Find an article in a newspaper or magazine from the last two weeks that describes a statistical study that was well done.
   a. Attach a photocopy of the article.
   b. Briefly explain the purpose of the study.
   c. Describe the population and the sample for the study. How was the sample chosen?
   d. Describe the sample statistics found in the study.
   e. What conclusions did the study draw about the population parameters?
   f. Briefly explain why the results of this study are important if they are valid.

Section 14.3

10. **Term Paper Results.** The following student grades were given on a term paper in a small class:

    A+, A, A–, B+, B+, B, B, B, B–, B–, C+,
    C, C, C, C–, C–, C–, D+, D, F, F, F.

    a. Sort the data into five bins: A, B, C, D, F. Then make a frequency table for the binned data. Include columns for relative frequency and cumulative frequency.
    b. Make a pie chart of the binned data. On each slice of the pie indicate its percentage of the total.
    c. Make a bar graph of the binned data.

11. **Exam Scores.** The results of an exam given to 25 students are as follows:

    67, 54, 89, 67, 98, 68, 88, 78, 87, 82, 91, 80, 73, 59, 85, 86,
    76, 99, 38, 75, 84, 87, 77, 69, 90.

    a. Sort the scores into bins of 10 (e.g., 90–99, 80–89, etc.). Then make a frequency table for the binned data. Include columns for relative frequency and cumulative frequency.
    b. Make a pie chart of the binned data. On each slice of the pie indicate its percentage of the total.
    c. Make a bar graph of the binned data.

12. **Exploring Tables.** The following table shows the times for the 10 fastest runners in the semifinals of a men's 100-meter dash; times were rounded to the nearest tenth of a second. Only a few of the blanks are filled in. Based on your understanding of frequency tables, complete the remaining entries in the table.

Time	Frequency	Relative Frequency	Cumulative Frequency
10.0			
10.1	3		4
10.2		0.2	
10.3			10
Total			**10**

13. **Scatter Plot.** The following table lists the heights and weights of players on a college basketball team. Make a scatter plot with height (inches) on the horizontal axis and weight (pounds) on the vertical axis. Does there appear to be a strong correlation? Explain.

Height	Weight	Height	Weight
6 ft 0 in.	**175**	**6 ft 2 in.**	**202**
6 ft 7 in.	**255**	**5 ft 11 in.**	**170**
5 ft 9 in.	**155**	**6 ft 1 in.**	**181**
6 ft 4 in.	**188**	**6 ft 6 in.**	**215**
6 ft 1 in.	**190**	**6 ft 7 in.**	**230**

14. **Time-Series Business Data.** Find an example of a time-series diagram in the business section of a newspaper from this week.
    a. Attach a copy of the graph and cite its source.
    b. Briefly explain the quantity shown in the time-series graph.
    c. Briefly interpret any trends shown by the graph.

15. **Government Outlays.** Analyze the stack chart of government outlays shown in Figure 14–9.
    a. Use a ruler to determine a scale for the graph. State the vertical height on the graph that represents 100%, 10%, and 1%. (Hint: For example, if the vertical height of the graph were 10 cm, the scale would be 10 cm = 100%, 1 cm = 10%, or 1 mm = 1%.)
    b. Use your ruler and the scale you found in part (a) to determine the proportion of the budget that went to net interest on the debt in 1960, 1970, 1980, 1990, and 1999 (projected).
    c. Write one to two paragraphs summarizing the data shown in the graph.

16. **Tuition Increases.** Study Figure 14–10 and answer each of the following questions.
    a. In which year did average *private* college tuition rise the most compared to the consumer price index?
    b. In which year did average *private* college tuition rise the least compared to the consumer price index?
    c. Suppose that tuition at an average *private* college was $10,000 in 1990. Estimate the tuition in 1991, 1992, and 1993.

17. **Find a Rate of Change Graph.** Look through current newspapers and magazines to find an example of a graph that shows rates of change rather than absolute numbers (e.g., Figure 14–10).
    a. Attach a copy of the graph and cite its source.
    b. Briefly explain the quantity or quantities shown in the graph. Why do you think the graph plotted rates of change rather than absolute numbers?
    c. If someone mistakenly believed that the graph showed absolute numbers rather than rates of change, what incorrect conclusions might that person draw?
    d. Briefly summarize the point being made by the graph.

18. **Three-Dimensional Graph.** Answer each of the following by studying Figure 14–11.
    a. Approximately how many birds passed over Richford at about 9:30 P.M.?
    b. Over which town did sensors detect the largest number of birds? Second largest? Fewest?

Section 14.4

19. **Mean, Median, and Mode.** The scores on a homework assignment for 10 people are

    5, 5, 6, 7, 7, 9, 9, 9, 9, 10.

    a. What is the *mean* of the set?
    b. What is the *median* of the set?
    c. What is the *mode* of the set?

20. **Race Statistics.** Consider the following set of 100-meter times.

    10.5 s, 10.3 s, 11.1 s, 10.8 s, 10.9 s,
    11.1 s, 11.1 s, 11.3 s, 11.0 s, 11.1 s.

    a. Calculate the mean for this set of times.
    b. What is the median for this set of times?
    c. Give a five-number summary for this set of data.
    d. Calculate the standard deviation for this set of times.
    e. Look again at the data. Does the standard deviation from part (d) make sense? Explain.

21. **Deviations in Height.** You measure the heights, to the nearest centimeter, of seven students. Your results are

    172 cm, 145 cm, 166 cm, 178 cm,
    190 cm, 188 cm, 169 cm.

    a. Give a five-number summary for this set of data.
    b. What is the mean of the data? Give your answer with four significant digits.
    c. What is the standard deviation for this set of data.
    d. Briefly explain why the standard deviation you found makes sense for this set of data.

22. **Comparing Standard Deviations.** Two heats of a 100-meter dash are run with six runners in each heat, with the following results.

    Heat 1: 9.98 s, 10.01 s, 10.05 s, 10.09 s, 10.10 s, 10.15 s
    Heat 2: 9.91 s, 9.95 s, 10.14 s, 10.23 s, 10.30 s, 10.35 s

    Find the mean and standard deviation for each heat, and interpret the differences. Show all work, carefully organized as in the examples in the text.

23. **Air Force Cadets.** Suppose that you are evaluating a new group of 50 pilot candidates for the Air Force. You are performing a hand–eye reflex test, in which you time how fast (to a precision of 0.1 s) each candidate can react by pressing a buzzer in response to a flashing light. The following table lists the results of your test in terms of frequency.

Reflex Time	Frequency
1.1	5
1.2	10
1.3	20
1.4	8
1.5	4
1.6	2
1.7	1
Total	**50**

    a. Add columns to this table for the relative frequency and cumulative frequency.
    b. Construct a histogram for this data set.
    c. Construct a line chart of the frequency data.
    d. Construct a line chart for the cumulative frequency data.
    e. Calculate the mean of the data (to three significant digits).
    f. What is the median for the data?
    g. What is the mode for the data?
    h. Would you say that this distribution is symmetric, positively skewed, or negatively skewed? Why?

24. Kids and TV. You've studied the habits of a sample of 25 fourth-grade children, finding the number of hours of television, to the nearest hour, watched by these children each weekday:

5, 8, 4, 1, 2, 3, 0, 2, 6, 4, 4, 4, 5,
1, 6, 6, 2, 4, 3, 4, 0, 2, 4, 4, 3.

a. Construct a frequency table with columns for the frequency, relative frequency, and cumulative frequency.
b. Construct a histogram for this data set.
c. Construct a line chart for this data set.
d. Calculate the mean for the data (to two significant digits).
e. What is the median for the data?
f. What is the mode for the data?
g. Would you say that this distribution is symmetric, positively skewed, or negatively skewed? Why?
h. Give a five-number summary for this set of data.

25. Cumulative Frequency Line Chart. The following cumulative frequency line chart shows the results of a hypothetical 100-meter race. Study the line chart and answer each question. Explain your answers.

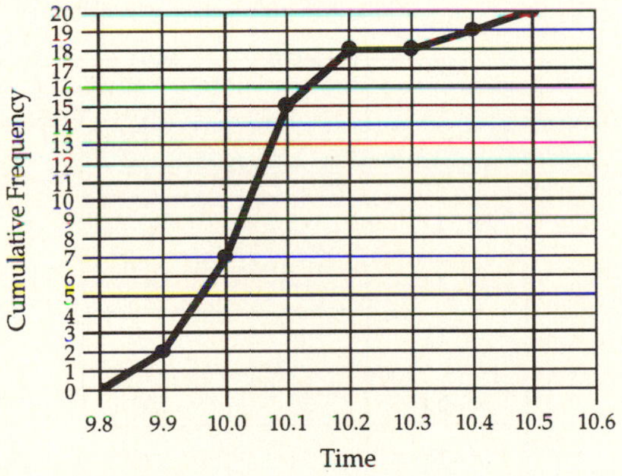

a. How many runners were timed in the race?
b. How many runners ran 10.0 s or *faster*?
c. How many runners ran 10.4 s?
d. What was the median time for this set of data?
e. What was the mode for this set of data?

26. Skewness. For each of the following distributions, state whether you expect it to be symmetric, positively skewed, or negatively skewed and briefly explain why.

a. An exam (with a maximum possible score of 100) in which most students do very well, say, in the 80s and 90s, but a few students do very poorly.
b. An exam (with a maximum possible score of 100) in which most students do very poorly, say, in the 30s and 40s, but a few students do very well.
c. The speeds of cars on a freeway.
d. The weights of newborn babies.
e. Amount of alcohol, per week, consumed by college students.

27. U.S. Family Income. To study family incomes in the United States you choose an unbiased sample of 200 families. At the end of the study, you summarize the results as follows:

- The mean family income in your sample was $20,000.
- The median family income in your sample was $18,000.
- The standard deviation in your sample was $5000.
- The extremes of family incomes in your sample were a low of $2300 and a high of $862,000.

a. In your sample, about how many families earned less than $18,000? How do you know? (Hint: Think about the meaning of the median.)
b. Would you characterize this distribution as symmetric, positively skewed, or negatively skewed? Why?
c. Based on what you know about U.S. family incomes, how well do you think your sample statistics compare to the true population parameters? Explain.

28. Speed Limits and Death Rates. The following table (D. J. Rivkin, *New York Times*, November 25, 1986) gives the death rates and speed limits in 10 countries.

Country	Death Rate (per 100 million vehicle miles)	Speed Limit (miles per hour)
Norway	3.0	55
United States	3.3	55
Finland	3.4	55
Britain	3.5	70
Denmark	4.1	55
Canada	4.3	60
Japan	4.7	55
Australia	4.9	6
Netherlands	5.1	60
Italy	6.1	75

a. In what order are the countries in this table listed?
b. Make a scatter plot of these data. Based on the 10 data points, are speed limits and highway death rates strongly correlated?
c. Are there any outliers in this plot? If the outlier(s) is (are) removed does that change the strength of the correlation?
d. The source article title was "fifty-five mph speed limit is no safety guarantee." Based on the data, do you agree? Explain.

Section 14.5

29. Unemployment Survey. The U.S. Census Bureau estimates the unemployment rate in the United States monthly by surveying 60,000 households.

a. What is the margin of error in a survey of 60,000 potential workers?
b. The unemployment rate generally is stated each month to the nearest tenth of a percent. For the margin of error, is this precision reasonable? Explain.

30. Textbook Survey. Suppose that the amount of money spent by students at a large university on books is normally distributed. If the mean amount spent on books per semester per student is $150, with a standard deviation of $25, about how many of the 25,000 students at a large university spend *less* than $100 on books each semester?

31. **Close to Normal Distributions.** The three rules for interpreting the standard deviation in normal distributions provide reasonable estimates whenever a distribution is *close* to being normal. Using the nearly-normal data for Problem 27, estimate the number of families in the sample that earned between $20,000 and $25,000.

32. **Opinion Poll.** A poll finds that 54% of the population approves of the job that the president is doing; the poll has a margin of error of 4%.
    a. What is the 95% confidence range on the true population percentage that approves of the president's performance?
    b. According to this poll, what is the probability that *less* than half the population approves of the president's performance?

Section 14.6

33. **Personal Health.** The following questions are based on the study of weight and health described in subsection 14.6.1.
    a. Calculate your own body mass index. (Note: If this question is too personal for your taste, you may "invent" a hypothetical person for answering the question. State the person's height in feet and inches and weight in pounds, then calculate the person's body mass index from these values.)
    b. Based on your body mass index and the information contained in the article, do you think that you are at higher risk than necessary? If yes, explain why, and if you think that you should try to do anything about it. If no, explain why not, and describe whether you have any other significant risk factors (such as smoking). (Again, if you prefer, you may answer this question for the hypothetical person you described in part a.)
    c. According to the study, being very thin is not unhealthy. Does this conclusion hold for a person who is thin because of an eating disorder (e.g., anorexia or bulimia)? How do you think an eating disorder affects mortality risk? Why?

34. **Lying to an Election Poll.** In February 1990, Nicaragua held democratic elections for the first time in many years. The leader of the ruling Sandinista party, Daniel Ortega, ran against the publisher of an opposition newspaper, Violeta Chamorro. Pre-election polls produced conflicting results, but most suggested that Ortega would win the election. However, Chamorro won easily. Apparently, many Nicaraguans had lied to pollsters, saying they would vote for Ortega but then, in the privacy of the polling booth, voting for Chamorro.
    a. Suggest a reason why Nicaraguans might have lied to pollsters about their voting plans.
    b. Briefly explain why secret ballots are so important to democratic elections.

35. **Herbal Cure for the Common Cold.** A local clinic urges people suffering from a common cold to purchase their new herbal remedy. The clinic claims that three fourths of the patients who take the herbal remedy report feeling better within 2 days.
    a. Is the clinic's claim believable? Why or why not?
    b. Assuming the claim is true, do you think this clinic has found a "cure" for the common cold? Explain.

36. **IQ Superstitions?** Steven Jay Gould, author of *The Mismeasure of Man*, offers the following suggestion about the interpretation of IQ scores:

    A certain skepticism about what there is to IQ besides being good at certain sorts of tests may make us less superstitious about its importance.

    Explain what you think Gould means. Do you agree with him? Why or why not?

Further Topics and Applications

37. **U.S. Suicide Rates.** The following table gives suicide rates in the United States (measured in deaths per 100,000) for eight different age groups.

Age	1970	1991
5–14	1	2
15–24	8	13
25–34	14	16
35–44	18	15
45–54	20	16
55–64	22	16
65–74	21	17
75–84	21	24

    a. How do you suppose these data were collected?
    b. Describe any patterns or trends that you see in these data. Be sure to note that the suicide rate varies both by age and year.
    c. How would you display these data in the most informative way? Experiment with pie charts, histograms, and line charts. Show all the data on a single visual display.

38. **Income and Degrees.** A 1993 survey collected the following income data for U.S. residents by educational level.

Educational Level	Median Income
No high school diploma	$9,000
High school graduate	$11,000
Bachelor's degree	$27,000
Master's degree	$35,000
Professional degree	$56,000
Doctoral degree	$48,000

a. Speculate on how these data were gathered and identify some potential sampling errors.
b. Why do you think that *median* salary is displayed? How might the figures change if *mean* annual salary were displayed?
c. Make a bar graph to display the data.
d. From these data can the median income of all U.S. residents be estimated? Explain.

39. Television Time. The following table shows the average daily TV viewing time per U.S. household over the past 40 years.

Year	1954	1964	1974	1984	1994
Viewing hours	4.6	5.3	6.1	7.0	7.2

a. Comment on how these figures might have been determined and how reliable they might be.
b. Make a histogram and a line chart of the data.
c. Draw a straight line (linear regression) on the histogram of part (b) that appears to fit the data. Does the line provide a good fit to these data?
d. Based on the linear fit of part (c), what do you project the average TV viewing time will be in 2004?

40. Analyzing the Reporting of a Survey. The relevant parts of an article entitled "Number of Poor Declines" by Associated Press writer Kim I. Mills (October 1995) reads:

"The segment of Americans living in poverty dropped under 15% last year to 38.1 million, the first year in a decade that both the rate and the total declined. There were no signs the gap between the rich and the poor was closing....

"However, the total was still 5.6 million above 1989, when the poverty rate was 13.1%, Census Bureau statistician Daniel Weinberg said Thursday in releasing the report....

"Poverty in America was defined as being below an income of $15,141 for a family of four.

"The poverty rate went from 15.1% of the population in 1993 to 14.5% in 1994, according to the Commerce Department report which was based on a Census survey of about 60,000 households." From Kim I. Mills, "Number of Poor Declines," Associated Press, October 1995. Reprinted with permission.

a. How were the data for this survey collected? The article specifies the defined poverty level for a family of four. Do you think that *only* families of four were surveyed? Explain.
b. Do you think that the definition of poverty (incomes below $15,141) has been the same since 1989? If so, what effect might this have on the conclusions of the survey?
c. Use the data in the article and draw a histogram to display the poverty rate for each year between 1989 and 1994 (you will need to estimate it for three of the years).
d. The article notes that *both* the "rate and total" declined in 1994. Is it possible for the poverty rate to decline while the total number of families living in poverty rises (or vice versa)? Explain.
e. Use the data in the article to estimate the U.S. population in 1989 and 1994.
f. Give your overall assessment of this survey. Be sure to note whether you are commenting on the survey itself or the reporting of the survey. Based on the article and your assessment, do you believe that progress has been made in the war against poverty?

41. Biased Questioning About the Holocaust. Nazi Germany killed 6 million Jews — more than two thirds of Europe's prewar Jewish population — in the horror known as the Holocaust. However, some anti-Semites and racists claim that the Holocaust never occurred. A 1992 Roper poll asked the following question about the Holocaust.

"Does it seem possible or does it seem impossible to you that the Nazi extermination of the Jews never happened?"

Only 65% of those surveyed answered "impossible it never happened." About a year later, a Gallup poll asked the following rephrased question about the Holocaust.

"The term Holocaust usually refers to the killing of millions of Jews in Nazi death camps during World War II. In your opinion, did the Holocaust: definitely happen, probably happen, probably not happen, or definitely not happen?"

The results of the Gallup poll: 83% said the Holocaust definitely happened, 13% said it probably happened, 2% said it probably did not happen, 1% said it definitely did not happen, and 1% had no opinion.

a. The Roper poll suggested that a significant fraction of Americans doubt the reality of the Holocaust. Do the results of the Gallup poll support this conclusion? Explain.
b. Identify a double negative in the 1992 question. Could this double negative have affected responses to the question? Explain.
c. Another source of potential bias in the 1992 question lies in respondents' interpretation of the words *impossible* and *extermination*. Could someone who accepts the reality of the Holocaust have chosen the answer "possible" in the 1992 survey? Explain.
d. Why do you think that some people try to deny the reality of the Holocaust?
e. Some college newspapers have printed advertisements from groups selling literature denying the reality of the Holocaust. Is it ethical to accept such ads? Defend your opinion.

42. Baseball Standings. The 1994 baseball season ended on August 11 because of a player's strike. At the end of the season, the American League East standings appeared as follows.

Team	**Games Won**	**Games Lost**
New York	70	43
Baltimore	63	49
Toronto	55	60
Boston	54	61
Detroit	53	62

How would you present these data graphically and completely and as informatively as possible? Consider using bar graphs, stack plots (with wins and losses stacked in a single bar), and including the won–lost percentage. Draw the graph that you think best represents these data.

43. **The Shape of Data.** The following three histograms show the distribution of data for (i) time between eruptions of the Old Faithful Geyser in Yellowstone National Park (source: *Handbook of Small Data Sets* by Hand, et. al., 1994), (ii) failure time of computer chips, (iii) and weights of rugby players. The number of data points is given as N in each case. Answer the following questions about each data set, estimating the data values from the histograms.
   - **a.** What is the mean of the data?
   - **b.** Give the five-number summary of the data.
   - **c.** Describe the shape of the distribution and explain why each distribution has that shape.

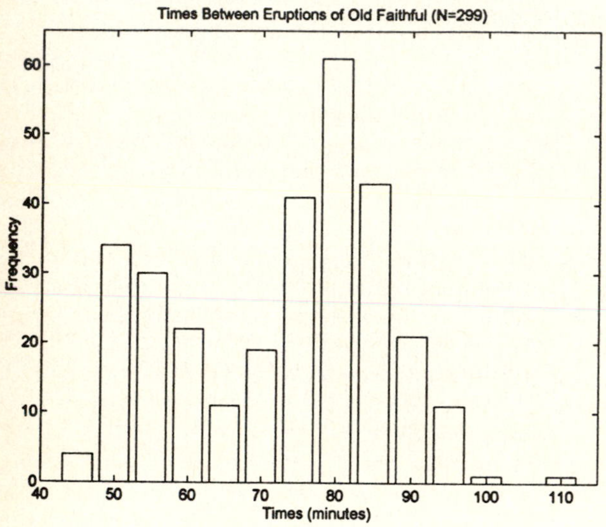

(i)

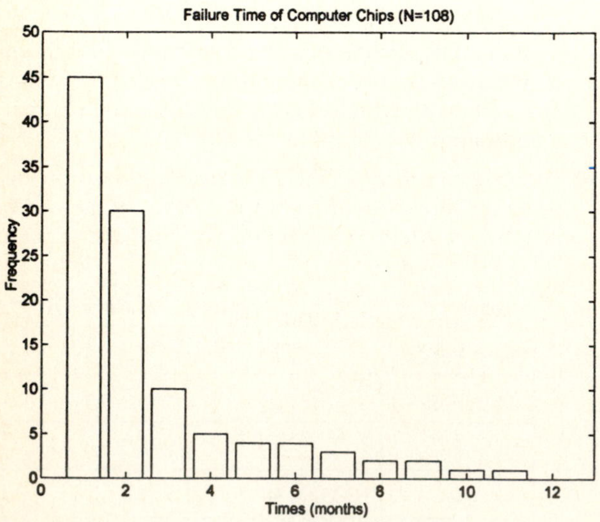

(ii)

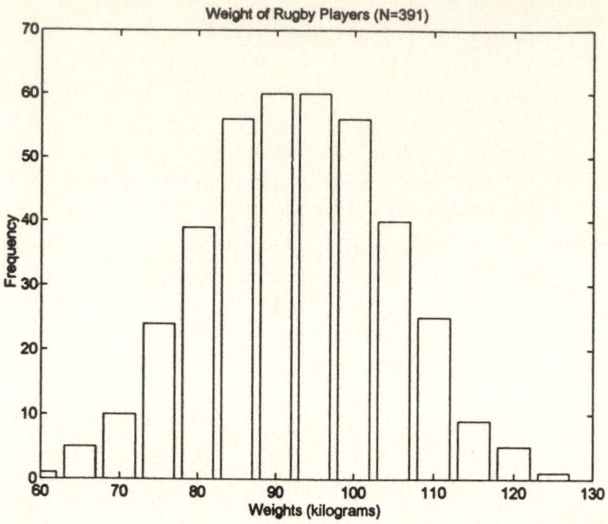

(iii)

44. **Displaying the World Population.** Choose a graphic method to display the following data on the regional distribution of the world population in 1994. Data are in millions of people.

North America	290
Latin America	470
Europe (with Russia)	728
Asia	3392
Africa	700
Oceania	28
Total	**5607**

45. **Women in the Labor Force.** The following data (source: *1995 Information Please Almanac*) show how the number of women in the U.S. labor force has changed since 1900. (Data for 1950–1990 includes women over age 16.)

Year	Number of Female Workers (thousands)	% of Female Population Working	% of Total Labor Force
1900	5319	18.8	18.3
1910	7445	21.5	19.9
1920	8637	21.4	20.4
1930	10,752	22.0	22.0
1940	12,845	25.4	24.3
1950	18,389	33.9	29.6
1960	23,240	37.7	33.4
1970	31,543	43.3	38.1
1980	45,487	51.5	42.5
1990	56,554	57.5	45.3

a. Use a time-series diagram to show how each variable (total number, percentage of female population, and percentage of total labor force) changed over time.
b. Imagine drawing a best-fit line (linear regression) for each of the variables shown. State whether the line fits the data well. Explain.
c. Assume that the numbers of men and women in the population are equal. What can you conclude about the percentage of men that work?
d. Draw a pie chart for the 1990 data that has four regions: working and nonworking women, and working and non-working men.

46. Fruit Fly Genetics. Genetic theory predicts that the four eye colors — red (R); pink (P); brown (B); white (W) — in mated fruit flies will occur in the ratio 9R:3P:3B:1W. You conduct an experiment in which you check the eye colors of 32 fruit fly offspring and observe:

R, R, P, R, B, W, R, R, R, R, R, P, P, R, R, P,
P, B, B, B, R, R, W, R, B, P, B, P, R, R, P, R.

a. Make a frequency table from these data for the four different eye colors. Include the relative frequency, but not the cumulative frequency, in your table.
b. Construct a pie chart for the data in your frequency table. Label each slice with the appropriate percentage.
c. Construct a pie chart that shows the expected ratio of eye colors (9R:3P:3B:1W). (Hint: From the expected ratio calculate the expected percentage of fruit flies that have each eye color, then use these percentages to make the slices in your pie chart.)
d. Compare the results from the experimental data with the results predicted from theory. For example, relative to white eyes, did red occur more often, less often, or exactly as often as predicted by theory? What about brown and pink?
e. Does the fact that your results do not agree perfectly with the theory mean that the theory is invalid? Why or why not?

47. Meditation and Crime. Robert Parks of the American Physical Society writes an electronic newsletter called *What's New*. On October 7, 1994, he offered the following report on the 1994 "Ig Nobel" prizes, selected for especially poor research:

The Physics prize went to the Japanese Meteorological Agency for its seven-year study of a possible link between catfish wiggling their tails and earthquakes. The Peace prize went to physicist John Hagelin for his experiment to reduce crime in Washington, DC by the coherent meditation of 4,000 [meditation] experts. By coincidence, Hagelin was holding a press conference to announce his final results. It was a data analysis clinic; violent crime, he proudly declared, decreased 18%! Relative to what? To the predictions of "time-series analysis" involving variables such as temperature and the economy. So although the weekly murder count hit the highest level ever recorded, it was less than predicted. From Robert Parks, "What's New," 1974, American Physical Society. Reprinted with permission.

a. Based on the information in this report, were the number of murders above or below average during the week of meditation?
b. Explain how Hagelin's belief that meditation can reduce crime has the characteristics of an irrefutable hypothesis (as discussed in Chapter 3).

Projects

48. Height–Weight Survey in Class. Carry out and write a report on a statistical study of your class to determine how well height and weight are correlated. If you believe that there might be issues of confidentiality regarding weight data (or the possibility of false reporting), you could choose height and shoe size as the variables to study. Be sure to include the following considerations in your study and report.

a. After collecting the weight and height data, present each data set individually using the most appropriate display methods discussed in the chapter.
b. Find the sample statistics for each variable.
c. Draw a scatter plot for the two variables.
d. Determine whether the two variables are strongly correlated; that is, is height a good predictor of weight (or shoe size)?
e. Do you think that your class is a good sample for the population of all students at your school? Is your class a good sample for the population of all people in this country? Why or why not?

49. Journals and Statistical Studies. One of the best sources of statistical studies are medical journals. Visit your library and find the area where recent journals are kept. Two medical journals with many general interest articles are *The New England Journal of Medicine* and the *Journal of the American Medical Association*. Find a statistical study reported in one of these journals (or any other journal) and write a summary of the study and its findings. Identify key aspects of the study, as discussed in this chapter (e.g., identify the population and the sample, comment on the sampling methods and whether the sample is well-chosen, and identify sample statistics that are presented). Conclude your report with your own personal reaction to the study. Do you believe the findings or did you find reasons to be skeptical of the results?

50. Study Habits and Grades. As a student, you might occasionally wonder why some students receive better grades than others. Making the reasonable assumption that grades are not determined solely by hereditary factors, you decide to determine to what extent study habits affect grades. Specifically, you choose to focus on two possible predictors of grades: hours spent outside of class doing homework (a quantitative or numerical variable) and the type of music listened to while studying (a categorical variable). Design, carry out, and write a report on a statistical study designed to link grades to these two variables. Imagine that the survey will be published by the campus newspaper to be of use to all students. Here are some ideas, tasks and questions that you should consider for inclusion in the study and report.

a. What is the measure of grades? You might choose overall grade point average, grades received by students in a single class, or some other measure.

b. What is the population and how will you choose your sample?
c. How will you collect data? Be sure to include a copy of any written surveys that you use or a list of questions that you use for an interview or phone survey.
d. How will you measure the type of music variable? Be sure to design the categories carefully (including the category *no music*).
e. Be sure to anticipate and avoid sources of bias in your sampling. Describe how you have done so.
f. Use the most appropriate visual displays to present the raw data in your study.
g. Summarize the statistics for the sample using the methods described in the chapter.
h. To the extent reasonable, try to draw general conclusions about study habits and grades.
i. Do you think that different explanatory variables (besides study time and music) should have been used in this study? Why or why not?

51. Sampling Problems. Sampling techniques can be used to estimate physical quantities. To estimate a large quantity, you might measure a representative small sample and find the total quantity by "scaling up." To estimate a small quantity, you might measure several of the small quantities together and "scale down." Use sampling techniques to answer the following questions. Explain your sampling technique in each case.

Example: How thick is a sheet of a paper. *Solution:* One way to estimate the thickness of a sheet of paper is to measure the thickness of a ream (500 sheets) of paper. A particular ream was 7.5 centimeters thick. Thus a sheet of paper from this ream was 7.5 cm ÷ 500 = 0.015 centimeters thick, or 0.15 millimeters.

a. **Paper Weight.** How much does a sheet of paper weigh?
b. **Coin Counting.** How thick is a penny? a nickel? a dime? a quarter? Would you rather have your height stacked in pennies, nickels, dimes or quarters? Explain.
c. **Tree Rings.** Find a slice from an old tree (e.g., in a museum or Botany department) and use its tree rings to estimate the age of the tree when it was cut.
d. **Grains of Sand.** How much does a grain of sand weigh? How many grains of sand are in a typical playground sand box?
e. **Star Counting.** How many stars are visible in the sky on the clearest, darkest nights? How could astronomers estimate the total number of stars in the universe?

15 UNDERSTANDING EXPONENTIAL GROWTH → KEY TO HUMAN SURVIVAL?

During the 1990s, the population of the Earth is increasing by about 90 million people per year. Human consumption of vital resources also is increasing every year, but supplies of those same resources are decreasing. Radioactive wastes from nuclear weapons and power plants will remain toxic for millennia. Although these facts may seem alarming, the response should not be panic but a quest for understanding and solutions. In this chapter we investigate the mathematical laws of exponential growth and decay that underlie population growth, resource depletion, radioactive decay, and much more. As you will see, an understanding of these mathematical laws may be the key to finding solutions to many of the most severe challenges facing the human race in the twenty-first century.

CHAPTER 15 PREVIEW:

15.1 ***WHAT IS EXPONENTIAL GROWTH?*** We begin the chapter by describing exponential growth and contrasting it with linear growth.

15.2 ***EXPONENTIAL ASTONISHMENT.*** In this section we present four parables to illustrate the astonishing nature of exponential growth.

15.3 ***DOUBLING TIME AND HALF-LIFE.*** Any exponential process has a fixed doubling time (for growth) or half-life (for decay). In this section, we use approximate formulas to calculate doubling times and half-lives.

15.4 ***EXPONENTIAL MODELS AND APPLICATIONS.*** In this section we derive and use the general exponential growth (or decay) law as a model for processes including population growth, monetary inflation, drug assimilation, radioactive decay, and resource depletion.

15.5 ***REAL POPULATION GROWTH.*** Although population growth is often described as an exponential process, the reality is far more complex. In this section we examine *real* population growth and the related concepts of carrying capacity, logistic growth, overshoot and collapse, and the balance of nature.

Everyone born before 1950 has witnessed a doubling of world population, the first generation ever to do so. This unprecedented growth is destroying agriculture's environmental support systems at record rates, reducing the living standards of hundreds of millions. Mounting population pressures are driving growing numbers of people from the countryside into the city and across national boundaries, exacerbating ethnic, religious, and tribal conflicts within societies.

— Lester R. Brown and Hal Kane, *Full House*

The greatest shortcoming of the human race is [its] inability to understand the exponential relation.

— Albert A. Bartlett, Professor of Physics, University of Colorado

Table 15-1. World Population

Year	Population (billions)	Increase (millions)
1950	2.555	—
1960	3.038	483
1970	3.704	666
1980	4.457	753
1990	5.295	838
1991	5.381	86
1992	5.469	88
1993	5.557	88
1994	5.644	87
2020	7.9 (estimate)	

Preliminary data (reported early in 1996) suggests that world population grew in 1995 by 100 million — by far the largest increase in a single year in human history.

15.1 WHAT IS EXPONENTIAL GROWTH?

As you look at more than 45 years of world population data in Table 15-1, what do you see? Continuing a trend that began several centuries ago, world population is undergoing explosive growth: Each decade the number of people *added* to the world population has itself increased. During the four years 1990–1994, despite famines and raging epidemics of AIDS and other diseases, world population rose by about 88 million people each year.

To put this number in perspective, every 3 years the world adds as many people as live in the entire United States. Each *month* the population increases by the equivalent of the population of Switzerland (or New York City). In 1992, 300,000 people died of starvation in Somalia; worldwide, those numbers were replaced by new births in just *29 hours*. During the next hour, while you read this chapter, the world population will increase by about 10,000 people.

Although projections of future population growth have large uncertainties, most projections based on current trends suggest that the world population will reach 8 billion people by 2020 — more than *four times* the population just a century earlier and nearly double the population at the time that most of today's college students were born. Most of that growth (9 out of 10 people) will take place in the developing regions of the world. More than 1 billion people will be added to Asia alone, where India is expected to overtake China as the most populous nation. Significant growth also is expected in Latin America (where Mexico City is expected to overtake Tokyo as the world's largest city) and Africa (where Ethiopia, Egypt, and Nigeria will double their populations). Even in the United States, where population is growing at a slower rate, the population is expected to increase by *100 million* people within the next 50 years.

Clearly, these population facts pose significant challenges. Where will the resources come from to house and feed all of these people? How can anarchy and chaos be prevented in nations where population growth outstrips the government's ability to provide services and prevent crime? What steps must be taken to alleviate environmental damage as more people use more resources and release more pollution?

The first step in finding solutions is coming to grips with these large numbers and growth rates. The mathematics of the world's population, as well as the growth and decline of resources to sustain that growth, involve exponential growth and decay. In this chapter we explore the essential properties of exponential processes and use exponential models of populations to make projections about what might lie ahead. Moreover, we show that exponential growth and decay processes may be applied to many other problems besides population growth.

Exponential Versus Linear Growth

Exponential growth (sometimes called geometric growth) is one of two fundamental growth mechanisms. The other is **linear growth** (or straight-line growth), which we discussed in Chapter 10.

We can contrast linear and exponential growth with a simple population problem. Imagine that a community of 10,000 people is flourishing and growing at a rate of 500 people per year. This growth is *linear* because the population increases by the same *absolute* number of people (500) each year. After 1 year the population will reach 10,500; after 2 years it will reach 11,000; after 3 years it will reach 11,500; and so on.

By contrast, suppose that the community grows by 5% per year. In the first year, a 5% increase raises the population from 10,000 to 10,500 — the

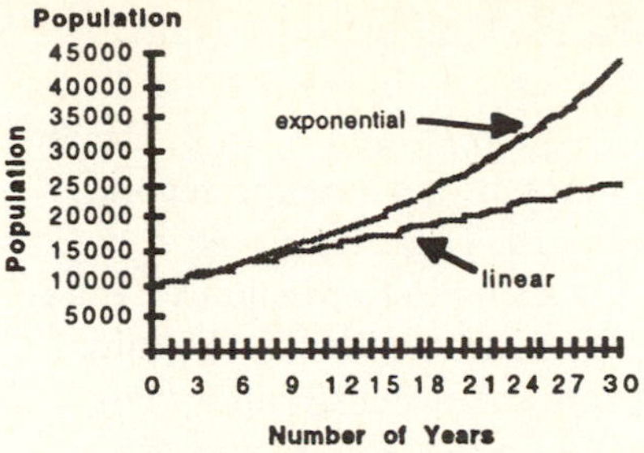

Figure 15-1. The two curves contrast linear and exponential growth over 30 years for a community with an initial population of 10,000. In the linear case, the rate of growth is 500 people per year, regardless of the current size of the population. In the exponential case, the rate of growth is 5% per year; as the population grows larger, the yearly increase also grows larger.

same as with linear growth. However, in the second year the population increases by 5% of 10,500, or by 525 people, to 11,025. In the third year, it increases by 5% of 11,025, or by 551 people. This growth is *exponential*: The amount of increase in the population itself grows larger each year because it is a fixed fraction (or percentage) of the growing population.

Figure 15-1 shows the community's population with both linear and exponential growth over a period of 30 years. Although the two growth patterns yield similar results for the first few years, the exponential growth soon "takes off." By the end of 30 years, the population with exponential growth is nearly twice what it would be with linear growth.

Because exponential growth leads to vastly different results from linear growth, distinguishing properly between these two fundamental types of growth is essential. In most cases, exponential processes may be identified by key words or phrases such as "grew by 3% last year" or "decreased by a factor of 2 last year."

These key words help to identify many common quantities besides populations that grow or decay exponentially. The growth of the U.S. economy, as measured by the gross domestic product (GDP) usually is reported in terms of a percentage increase for a month or year, indicating that the economy tends to grow exponentially. Bank accounts and annuities, with their fractional interest rates, also exhibit exponential growth. Examples of exponential decay (which simply is negative growth) include the decay of radioactive substances, such as uranium or plutonium, and the decline in the real value of the dollar through inflation.

Time-Out to Think: Watch tonight's news or check today's newspaper. Identify at least three quantities that are described by exponential growth. Remember, if the change in quantity is described by a percentage of its previous value, it is growing (or decaying) exponentially.

15.2 EXPONENTIAL ASTONISHMENT

The tendency of exponentially growing quantities to take-off, as shown in Figure 15-1, leads to what we call *exponential astonishment*: the intuition-defying reality of exponential growth. The following four parables illustrate the surprising nature of exponential growth.

15.2.1 From Hero to Headless in 64 Easy Steps

An entertaining, and almost certainly *untrue*, legend arose concerning the invention of the game of chess. According to this legend, chess was invented in ancient times by a man who presented the game to his king. The king was so enchanted by the new game that he said to the inventor, "Name your reward." The king was willing to give the inventor gold, jewels, or any number of other riches, and was surprised when the man asked only for some wheat.

"If you please, king, I would like you to put one grain of wheat on the first square of my chess board," said the inventor. "Then, place two grains on the second square, four grains on the third square, eight grains on the fourth square, and so on." In other words, the king was to double the number of grains on each succeeding square. The king gladly agreed to the reward, thinking that this man was a fool for asking only for a few grains of wheat when he could have had any riches of his choice.

Let's see how it adds up. On the first square, the king placed 1 grain of wheat. On the second square he placed twice as many as on the first, or 2. Thus, the *total* number of grains on the first two squares was 1 + 2 = 3. On the third square he had to place 4 grains (twice as many as the 2 on the second square), making a total of 7 grains on the chessboard. Then, he placed 8 grains on the fourth square (twice as many as the 4 on the third square), making a total of 15 grains on the board. On the fifth square he placed 16 grains; 32 on the sixth square; and so on. Table 15-2 summarizes the situation: For each square, it shows the number of grains placed on that square and the total number of grains on the board up to that point. Note that each quantity can be described in terms of powers of 2 and that a chess board has a total of 64 squares.

Table 15-2. Placing Grains of Wheat on a Chess Board

Square	Grains on This Square	Total Grains Thus Far
1	1 ($=2^0$)	1 ($=2^1-1$)
2	2 ($=2^1$)	1+2=3 ($=2^2-1$)
3	4 ($=2^2$)	3+4=7 ($=2^3-1$)
4	8 ($=2^3$)	7+8=15 ($=2^4-1$)
5	16 ($=2^4$)	15+16=31 ($=2^5-1$)
6	32 ($=2^5$)	31+32=63 ($=2^6-1$)
7	64 ($=2^6$)	63+64=127 ($=2^7-1$)
⋮	⋮	⋮
64	2^{63}	$2^{64}-1$

The table immediately reveals one of the "facts" of exponential growth. The 8 grains on the fourth square, for example, are slightly more than the total of 7 grains on the first three squares combined! Similarly, the 64 grains on the seventh square are slightly more than the total of 63 grains on the first six squares combined. Thus, our first exponential growth fact is: The increase with a single doubling is approximately equal to the sum of all preceding doublings.

Exponential Growth Fact 1:

With a single doubling the amount of increase is approximately equal to the sum of *all* preceding doublings.

Incidentally, the king never finished paying the man as agreed. Instead, the king became quite irritated, and had the man beheaded. Why? Well, aside from the fact that the squares soon were too small for the required number of grains, imagine the king's predicament in simply trying to find enough wheat in his kingdom. In the last column of Table 15-2, the total number of grains on the board after n squares are filled is $2^n - 1$. When n becomes large, subtracting 1 hardly affects the total value. That is:

$$\text{for large } n: \qquad 2^n - 1 \approx 2^n.$$

If the king had been able to continue, he would have needed 2^{64} grains to fill the board. With a calculator, you can check that $2^{64} = 1.8 \times 10^{19}$; the man was asking for 18 *billion billion* grains of wheat!

Recall from Chapter 6 that a "grain" is an ancient measure of weight, based on the weight of a typical grain of wheat, and that 1 pound is defined as 7000 grains. Thus the weight of these 18 billion billion grains would be about

$$\left(1.8\times10^{19}\text{ grains}\right)\times\left(\frac{1\text{ pound}}{7000\text{ grains}}\right)\times\left(\frac{1\text{ ton}}{2000\text{ pounds}}\right)=1.3\times10^{12}\text{ tons.}$$

The requested grain would weigh about a *trillion* tons. According to the U.S. Department of Agriculture, the current world harvest of all grain (wheat, rice, and corn) is somewhat less than 2 billion tons per year — less than one fifth of 1% of the amount needed to fill the chess board. Indeed, a trillion tons is much more than the total amount of wheat harvested in all of human history!

15.2.2 The Magic Penny

One lucky day you meet a leprechaun. After much fanfare and discussion about mythology, the leprechaun promises to give you fantastic wealth — and then proceeds to hand you a penny before disappearing. Although disappointed with this gift, you head home excited about meeting a leprechaun and place the penny under your pillow. The next morning, to your surprise, you find two pennies under your pillow. The following morning you find four pennies and the fourth morning eight pennies. Apparently, the penny given to you by the leprechaun was a *magic* penny: Each night while you sleep, each magic penny turns into *two* magic pennies.

Table 15-3. Wealth from the Magic Penny

Number of Days	Wealth
1	\$0.01
2	\$0.02
3	\$0.04
4	\$0.08
5	\$0.16
6	\$0.32
7	\$0.64
8	\$1.28
9	\$2.56
10	\$5.12
t	$\$0.01 \times 2^{t-1}$

We can write a simple formula for your wealth. The first day you have 1 cent, which is 2^0 cents. On the second day you have 2 cents, or 2^1 cents. On the third day you have $2^2 = 4$ cents, and so on. Thus the formula for your wealth is

$$\text{wealth after } t \text{ days} = 2^{t-1} \text{ cents} = \$0.01 \times 2^{t-1}.$$

(Note that t, chosen to remind you of time, is the number of days.) The last step is simply a unit conversion from cents to dollars. Table 15-3 shows your wealth, in dollars, over the first 10 days.

Although the magic penny certainly is a neat trick, you might be wondering when you will get the fantastic wealth that was promised. After all, by the tenth day, you still have barely enough money to buy lunch!

How much money will you have after 20 days? Perhaps, as you have barely \$5 after 10 days, you might expect to have something in the neighborhood of \$10 after 20 days. But that would not be exponential growth! Instead,

$$\text{wealth after 20 days} = 2^{20-1} \text{ cents} = \$0.01 \times 2^{19} = \$5242.88.$$

Although 10 days brought you only \$5, 20 days brings you more than *\$5000*. Clearly, you're going to need a much bigger pillow! After 30 days,

$$\text{wealth after 30 days} = 2^{30-1} \text{ cents} = \$0.01 \times 2^{29} = \$5{,}368{,}709.12.$$

At the end of a month, you are a multimillionaire. Now, in an act of great patriotism, you decide to keep the money under your pillow and let it continue to grow until you are able to pay off the about \$5 trillion national debt of the United States. To find out when your wealth will reach this level, you need to solve for the number of days t required to reach that amount; that is, you must find that value of t such that

$$\$0.01 \times 2^{t-1} = \$5 \times 10^{12} \qquad \text{or} \qquad 2^{t-1} = \frac{\$5 \times 10^{12}}{\$0.01} = 5 \times 10^{14}.$$

There are two ways to find the value of t that satisfies this equation. One way is to use logarithms (see Appendix C). The other way is to proceed by trial and error, noting that

$$2^{49-1} = 2.8 \times 10^{14} \qquad \text{and} \qquad 2^{50-1} = 5.6 \times 10^{14}$$

Thus you will have the \$5 trillion after 50 days. Think about it: In the first 5 days the magic penny yields only 16¢, in ten days enough to buy lunch, and in 20 days enough to buy a used car. But in less than 2 months, every person in the United States will thank you for paying off the national debt on their behalf!

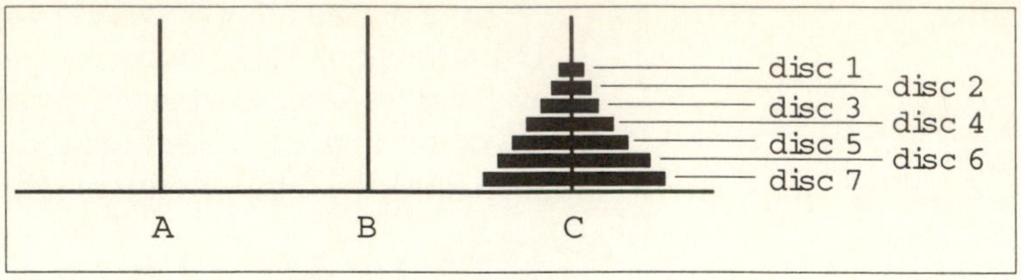

Figure 15-2. The towers of Hanoi puzzle consists of three wooden pegs, labeled A, B, and C, and a set of wooden disks of varying sizes. The puzzle is shown here with seven disks, numbered in order of increasing size.

15.2.3 The Towers of Brahma: Is the End Nigh?

Perhaps you have tried the simple and popular puzzle called the Towers of Hanoi. It consists of three wooden pegs and a set of wooden disks, of varying sizes with holes in their centers, that can be moved from peg to peg. The puzzle, in a version with seven disks, is shown schematically in Figure 15-2; the pegs are labeled A, B, and C, and the seven disks are numbered in order of increasing size. The object of the puzzle is to move the entire stack of disks from one peg to another peg. Two rules must be followed:

- only one disk can be moved at a time, and
- the disk that is moved may be placed only on an empty peg or on top of a larger disk.

Time-Out to Think: Before reading on, try to solve a simpler version of this puzzle with a quarter, a nickel, a penny, and a dime to represent four different-sized disks. Draw three dots on a sheet of paper, and label them A, B, and C. Place the stack of four disks (coins) on dot C. Following the rules of the game, try to move the stack to dot A or dot B.

Let's solve the puzzle assuming that all disks initially are on peg C. We proceed with a set of minigoals, first moving one disk to another peg, then forming a stack of two disks on another peg, then stacking three disks on another peg, and so on. Achieving the first goal is easy: Simply move disk 1 to empty peg A (B would work just as well), as shown in Figure 15-3(a). The next goal is to stack two disks on another peg. Starting from the configuration in Figure 15-3(a), achieving this goal requires two moves (Figure 15-3b):

1. move disk 2 to peg B, and
2. move disk 1 from peg A to peg B.

The moves needed to meet the third goal, getting a stack of three disks on another peg, are shown in Figure 15-3(c):

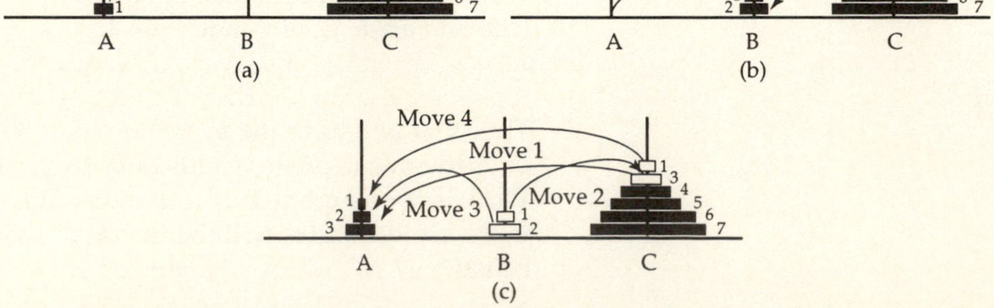

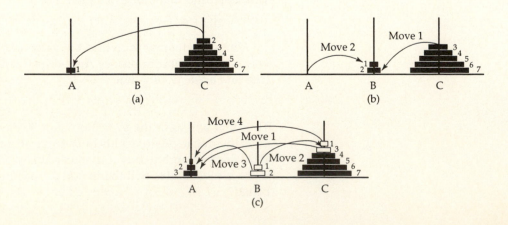

Figure 15-3. The solution to the towers of Hanoi puzzle is shown for the first three disks. (a) Only one move is required to put one disk on another peg. (b) Two more moves are needed to stack two disks on another peg. (c) Four more moves are needed to stack three disks on another peg.

1. move disk 3 to the empty peg A;
2. move disk 1 from peg B back to peg C (this move is legal because disk 1 then sits on the larger disk 4);
3. move disk 2 from peg B to peg A, where it is now on the larger disk 3; and
4. move disk 1 from peg C to peg A, completing the stack of three disks on peg A.

If you continue working on the puzzle, you will see that achieving each successive goal (a stack of 4 disks on another peg, followed by a stack of 5 disks, and so on) requires twice as many moves as the previous goal. Table 15-4 shows the computation of the number of moves required to complete the puzzle.

Table 15-4. Towers of Hanoi with 7 Disks

Goal	Moves Required for This Step	Total Moves in Game So Far
1 disk on another peg	1	1
2 disks on another peg	2	1 + 2 = 3
3 disks on another peg	4	3 + 4 = 7
4 disks on another peg	8	7 + 8 = 15
5 disks on another peg	16	15 + 16 = 31
6 disks on another peg	32	31 + 32 = 63
7 disks on another peg	64	63 + 64 = 127

Note the similarity of Table 15-4 to Table 15-2. Again, our first exponential growth fact is in action: Each successive goal requires one more move than the *total* number of moves in the entire game to that point. Note also that achieving the goal of n disks on another peg requires 2^{n-1} moves in itself, and a total of $2^n - 1$ moves to that point in the game (be careful not to confuse 2^{n-1} and $2^n - 1$).

How long would completing the puzzle with seven disks take? If you get very good at the puzzle, making the moves without making a mistake or having to backtrack, you might be able to move one disk each second. In that case, you could complete the puzzle in 127 seconds, or barely over 2 minutes. That may seem like no big deal, but...

According to an ancient Hindu legend, at the beginning of the world the Brahma placed three large diamond needles on a brass slab in the Great Temple. On one of the needles the Brahma placed 64 disks of solid gold, with varying sizes and holes in their centers, arranged with the largest disk on the bottom and the smallest on the top. Working in shifts, day and night, the temple priests moved the golden disks from their original diamond needle to another, following the two laws laid down by Brahma: Only one disk may be moved at a time, and a larger disk may never be placed on a smaller disk. Upon completion of their task, the temple will crumble and the world will come to an end.

In effect, the legend simply has the priests working a 64-disk version of the towers of Hanoi puzzle. Assuming that they move disks at the rate of one per second, they moved the first seven disks — more than 10% of the stack — in a mere two minutes. How much time did they spend moving half the stack, or 32 disks, to another needle? A total of $2^n - 1$ moves is needed to get a stack of n disks on another needle, so $n = 32$ disks requires

$$2^{32} - 1 \approx 4.3 \times 10^9 \text{ moves.}$$

Because the priests move one disk each second, moving a stack of 32 disks to another needle required 4.3 billion seconds, or about

$$\left(4.3 \times 10^9 \text{ s}\right) \times \frac{1 \text{ min}}{60 \text{ s}} \times \frac{1 \text{ hr}}{60 \text{ min}} \times \frac{1 \text{ day}}{24 \text{ hr}} \times \frac{1 \text{ yr}}{365 \text{ day}} = 136 \text{ years.}$$

What if this story is more than a legend? What if the priests are working, at this very moment, transferring disks at a rate of one per second without ever missing a beat? If they succeeded in moving half the stack in a mere 136 years, should we start printing signs that read "the end is nigh"?

Let's calculate: Moving the entire stack of 64 disks will require $2^{64} - 1$ moves, or about 2^{64} seconds — which is about 585 *billion* years! As the present age of the universe (since the Big Bang) is only about 10 billion years, we still have some 575 billion years to spare. We hope that you are relieved!

Time-Out to Think: Use your calculator to confirm, with a unit conversion, that 2^{64} seconds is about 585 billion years.

15.2.4 Bacteria in a Bottle

The bacteria in a bottle parable is by Professor Albert A. Bartlett of the University of Colorado.

For our final example of exponential astonishment, we turn to a topic of greater import: population growth. In Section 15.5, we consider how populations *really* grow. Here, we consider a highly idealized thought experiment concerning the growth of a colony of bacteria in a bottle.

To begin, imagine placing a single bacterium in a bottle filled with nutrients. After a certain time, the one bacterium divides into two smaller bacteria, thus doubling the population of the bottle. Each of these progeny then grow in size until they, in turn, are ready to divide into two (again doubling the population of the bottle). This process continues as long as adequate resources (space and nutrients) are available and as long as waste products (from cellular metabolism) do not interfere.

To keep the experiment simple, let's suppose that the bacteria grow quickly, dividing every minute. If you place the first bacterium into the bottle at, say, 11 o'clock, then it will divide after 1 minute so that there are 2 bacteria in the bottle at 11:01. These *each* divide to make 4 bacteria at 11:02, which in turn divide into 8 at 11:03, and so on. Finally, let's imagine that the bacteria fill the bottle after 60 minutes, at 12 o'clock.

Table 15-5 shows the progress of the bacterial growth. Like the three preceding parables, the growth is a simple case of repeated doublings, where the population of bacteria after n minutes is given by 2^n. However, analyzing this problem leads to some interesting insights. Let's consider a few basic questions about the bacterial growth.

- *Question 1*: Given that it took sixty minutes to fill the bottle; when was the bottle half full?

Time-Out to Think: Before looking ahead, try to answer question 1 for yourself. Do the same with each of the questions that follow.

Table 15-5. Bacteria in a Bottle (dividing every minute)

Time	Minutes Since Start of Colony	Number of Bacteria	Fraction of Bottle Filled
11:00	1	1 (= 2^0)	$1/2^{60}$
11:01	2	2 (= 2^1)	$2/2^{60} = 1/2^{59}$
11:02	3	4 (= 2^2)	$4/2^{60} = 1/2^{58}$
11:03	4	8 (= 2^3)	$8/2^{60} = 1/2^{57}$
11:04	5	16 (= 2^4)	$16/2^{60} = 1/2^{56}$
⋮	⋮	⋮	⋮
11:56	56	2^{56}	$2^{56}/2^{60} = 1/16$
11:57	57	2^{57}	$2^{57}/2^{60} = 1/8$
11:58	58	2^{58}	$2^{58}/2^{60} = 1/4$
11:59	59	2^{59}	$2^{59}/2^{60} = 1/2$
12:00	60	2^{60}	$2^{60}/2^{60} = 1$ (Full)

Many people, when asked this question, offer 30 minutes as the answer. After all, if the bottle was full after 60 minutes, shouldn't it have been half full after 30 minutes? Not with *exponential* growth. In fact, the population of the bottle *doubled* during the last minute, so it was half full just before this final doubling took place. That is, the bottle was half full at 11:59, one minute before the end!

- *Question 2*: When was the bottle one-quarter full?

Following the same logic, the bottle was one-quarter full a minute before it was one-half full, or at 11:58. That is, because the population doubles each minute, the bottle was one-quarter full at 11:58; one-half full at 11:59, and full at 12:00. Continuing backward, it was one-eighth full at 11:57, one-sixteenth full at 11:56, and so on. The last column of Table 15-5 shows the fraction of the bottle that is filled at each time.

- *Question 3*: When do the bacteria begin to feel overcrowded?

Overcrowding is a relative term and it is difficult to say that the bacteria begin to feel overcrowded at a particular moment. Nevertheless, let's consider the situation at 11:56, a mere 4 minutes before the end. If the bacteria were to look around the bottle at that time, they would see that it is only 1/16 full. Thus 15/16 of the bottle is still unoccupied by bacteria (and filled with pristine nutrients waiting to be eaten!). At 11:56, then, the amount of unused space is *15 times* that of the used space; the bacteria probably would still feel relatively comfortable. The intriguing lesson is the rapidity with which the bottle fills in the final minutes. For the first 56 minutes of the colony's existence, the growth can proceed with little concern about available space in the bottle. Then, in the final four minutes, the bottle goes from being barely 6% full (1/16 = 6.25%) to full. Figure 15-4 shows the percentage of the bottle filled at each minute after the colony began.

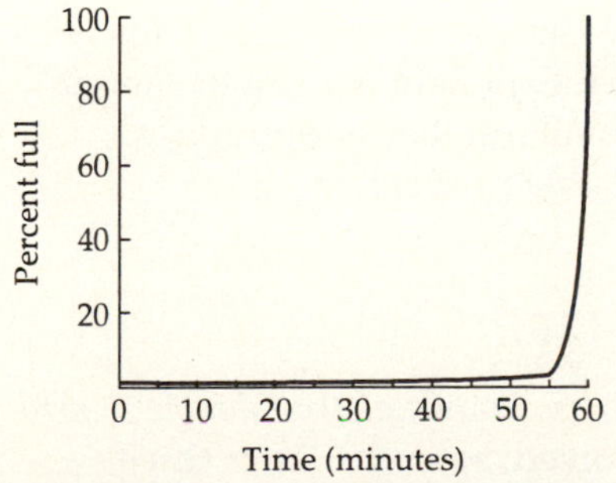

Figure 15-4. The percentage of the bottle filled with bacteria is shown over the 60 minutes during which the bacterial population grows exponentially. For most of the 60 minutes, the percentage of the bottle that is full is very close to zero. In the final few minutes, the continued doublings fill the bottle rapidly.

What happens when the bottle fills? At the very least, the colony's growth ends and many, if not all, bacteria will die. Taking this example to an extreme, let's imagine that the bacteria foresee this problem and decide to take steps to avoid it. Through some last minute technological triumph, the bacteria discover three new, nutrient-filled bottles in the laboratory (making a total of four bottles, including the original). They quickly build little bacterial spaceships and set out to colonize the other three bottles.

- *Question 4*: Suppose that the bacteria distribute their population evenly among the four bottles at 12:00. When will they have filled *all four* bottles?

Again, the answer at first may seem deceptive. After all, if it took 1 hour to fill 1 bottle, it might seem reasonable that it would take 4 hours to fill 4 bottles. However, the population grows by doubling. If one bottle is full at 12:00, two bottles fill by 12:01, and four bottles fill by 12:02! The discovery of three new bottles buys the colony only 2 additional minutes of growth.

In fact, the colony can do nothing to accommodate its unbounded growth. To see why, let's do some calculations. Suppose that, somehow, the bacterial population continues to multiply. After a second hour, the population of the colony would reach 2^{120} bacteria. The *smallest* bacteria measure approximately 0.1 micrometer across; if the bacteria are cube-shaped, the volume of a single bacterium is approximately $(0.1 \times 10^{-7}\ \text{m})^3 = 10^{-21}\ \text{m}^3$. Therefore, the colony of 2^{120} bacteria would occupy

$$2^{120}\ \text{bacteria} \times \left(10^{-21}\ \frac{\text{m}^3}{\text{bacteria}}\right) = 1.3 \times 10^{36}\ \text{bacteria} \times \left(10^{-21}\ \frac{\text{m}^3}{\text{bacteria}}\right) = 1.3 \times 10^{15}\,\text{m}^3.$$

Because the total surface area of the Earth is about 5×10^{14} square meters, we can calculate the thickness of the layer of bacteria on the Earth simply by dividing (see subsection 11.4.4):

$$\frac{1.3 \times 10^{15}\,\text{m}^3}{5.1 \times 10^{14}\,\text{m}^2} = 2.5\ \text{m}.$$

That is, after a mere two hours of exponential growth, the bacteria would be so numerous that they could cover the entire Earth in a layer 2 ½ meters, or about 8 feet, thick!

- *Question 5*: If the bacterial population continues to grow exponentially, when will its volume exceed *the total volume of the universe* (roughly $10^{79}\ \text{m}^3$)?

We know that, after n minutes, 2^n bacteria occupy a volume of $2^n \times 10^{-21}$ cubic meters. Thus, we simply set this expression for the bacterial volume equal to the volume of the universe, and find the value of n that makes

$$2^n \times 10^{-21}\ \text{m}^3 = 10^{79}\ \text{m}^3.$$

Exponential Growth Fact 2:

Exponential growth cannot continue indefinitely. After only a relatively small number of doublings, exponentially growing quantities reach impossible proportions.

Again, we may use logarithms or trial and error. We find that after 333 minutes — just over 5 ½ hours — the volume of bacteria would exceed the volume of the entire universe. Needless to say, that cannot happen. Thus we arrive at our second fact of exponential growth: Exponential growth cannot continue indefinitely. After only a relatively small number of doublings, exponentially growing quantities reach impossible proportions.

15.3 DOUBLING TIME AND HALF-LIFE

Each of the four parables in Section 15.2 involves quantities that *double* in size at every step of the process. When time is involved, as in the latter three parables, each doubling occurs during a fixed period of time, called the **doubling time**. As you will see shortly, in the case of exponential decay, a quantity decreases to half its original size in a characteristic time called the **half-life**.

The doubling time for a particular rate of exponential growth is a *constant* that is independent of the current size of the quantity. In the case of the

Exponential Growth Fact 3:

The doubling time for an exponential rate of growth is a constant, regardless of the present size of the quantity.

Exponential Growth Fact 4:

Exponential growth occurs whenever a quantity increases in size by a fixed percentage during a fixed period of time.

community that grows exponentially at an annual rate of 5% per year (Figure 15-1), the population doubles from 10,000 to 20,000 in about 15 years. The next doubling, from 20,000 to 40,000, takes the *same* amount of time — another 15 years! Thus, after about 30 years, the population has doubled twice, for a total increase by a factor of 4. The population would double again, to 80,000 in the next 15 years.

The four parables also show that, at each step, a quantity increases by the same *percentage*. The number of grains of wheat increases by 100% on each square of the checkerboard, and the number of bacteria in the bottle increases by 100% each minute. Exponential growth occurs whenever a quantity increases in size by *any* fixed percentage during a fixed period of time. A population that increases in size by 10% per year grows exponentially. Prices grow exponentially in an economy that has an inflation rate of 4%. In this section, we explain how to find and work with the doubling time or half-life for any exponentially growing or decaying quantity.

15.3.1 Number of Doublings

Recall that, after one doubling time, a quantity has increased in value by a factor of 2 (it doubles). After two doubling times, it has increased from its original value by a factor of $2^2 = 4$. After three doubling times, it has increased by a factor of $2^3 = 8$. In general, after n doubling times, it has increased by a factor of 2^n. The relation between the number of doublings, n, and the total amount of elapsed time is

$$n = \text{number of doublings} = \frac{\text{total elapsed time}}{\text{doubling time}} = \frac{t}{T_2}, \text{ where } \begin{cases} t = \text{total elapsed time} \\ T_2 = \text{doubling time.} \end{cases}$$

For example, if a population has a doubling time $T_2 = 40$ years, in $t =$ 200 years, it doubles five times (200 yr ÷ 40 yr = 5).

Because a quantity increases by a *factor* of 2^n after n doublings, we may also write

$$\text{increase factor} = 2^n = 2^{t/T_2}.$$

For example, if the doubling time of a population is $T_2 = 40$ years, in 200 years it increases by a factor of

$$2^{t/T_2} = 2^{200\text{ yr}/40\text{ yr}} = 2^5 = 32.$$

If the population is 1 million initially, it would be 32 million after 200 years.

Example 15-1 If the doubling time of a bank account balance is 13 years, by what factor does it increase in 50 years?

Solution: Over a 50-year period, the bank balance increases by a factor of

$$2^{t/T_2} = 2^{50\text{ yr}/13\text{ yr}} = 2^{3.85} = 14.4.$$

For example, if the account has $1 initially, it grows to $14.40 after 50 years; if it has $10,000 initially, it grows to $144,000 in 50 years. Note that the doubling time is independent of the size of the bank balance.

❖

15.3.2 The Doubling Time Formula

We are now ready to make the connection between an exponential rate of growth and the doubling time. Suppose that an ecological study of a prairie dog community shows that, at the beginning of the year, it contains 100 prairie dogs. Taking weekly measurements, the researchers are able to

r = fractional growth rate

determine that the prairie dog population increases at a rate of 5% per week. In this case the **percentage growth rate** is 5% per week, or, equivalently, the **fractional growth rate** is 0.05 per week. We refer to the fractional growth rate as r. Table 15-6 shows the growth of the prairie dog population over a period of 30 weeks.

Table 15-6. Prairie Dog Population with a Growth Rate of 5% per Week

Number of Weeks	Prairie Dog Population	Number of Weeks	Prairie Dog Population
0	100	15	208
1	105	16	218
2	110	17	229
3	116	18	240
4	121	19	253
5	128	20	265
6	134	21	279
7	142	22	293
8	148	23	307
9	155	24	323
10	163	25	339
11	171	26	356
12	180	27	373
13	189	28	392
14	198	29	412

Exponential Fact 5:

If a quantity is growing at a fractional rate, r, per unit of time, its doubling time is

$$\textbf{Doubling Time} = T_2 \approx \frac{0.7}{r}.$$

This approximation is valid *only* for small rates of growth.

The first doubling (from 100 to 200 prairie dogs) occurs over approximately 14 weeks. The second doubling (from 200 to 400 prairie dogs) also occurs over 14 weeks. If we continued this table, we would see that the population doubles *every* 14 weeks. This example illustrates a general rule: The doubling time is *approximately* 0.7 divided by the fractional rate of growth. In this case, where $r = 0.05$ per week, the doubling time is about

$$\frac{0.7}{0.05/\text{week}} = 14 \text{ weeks}.$$

Generalizing, if a quantity is growing at a fractional rate, r, per unit of time, its doubling time is

$$\text{Doubling Time} = T_2 \approx \frac{0.7}{r}, \text{ where } r \text{ is the fractional growth rate.}$$

Many books use the *percentage* rate of growth p instead of the *fractional* rate of growth r. Because $p = 100 \times r$, the doubling time approximation sometimes is called the **Rule of 70** and written as:

$$T_2 \approx \frac{70}{p}.$$

This approximation is so important and so useful that you should commit it to memory. Keep in mind, however, that it works well only when the growth is relatively small, say, less than about 15% ($r = 0.15$).

Example 15-2 ***Inflation.*** Suppose that prices rise with inflation at a rate of 1.2% per month. Over what period of time will prices double?

Solution: A growth rate of 1.2% per month means that $r = 0.012$ per month. Use the doubling time formula:

$$T_2 \approx \frac{0.7}{0.012 \text{ per month}} = 57.8 \text{ months} \approx 4 \text{ years and } 10 \text{ months.}$$

At an inflation rate of 1.2% per month, prices double in a little under 5 years.

❖

Example 15-3 *Bank Accounts.* Suppose that a bank account increases in value by 9% per year. If you invest $100 today, how much will be in the account when your great-great-grandchildren inherit it in 100 years?

Solution: With growth at 9% per year, the fractional growth rate is $r = 0.09$ per year. Thus the doubling time for the account is

$$T_2 \approx \frac{0.7}{0.09 \text{ per year}} = 7.7 \text{ years.}$$

Substitute this doubling time into the factor increase formula to find that after $t = 100$ years,

$$\text{increase factor} = 2^{t/T_2} = 2^{100 \text{ yr}/7.7 \text{ yr}} = 2^{12.98} = 8079.$$

That is, using the approximate doubling time, the amount of money in the account would increase by a factor of about 8000. Thus, the original $100 would grow to about $800,000 in 100 years. However, a more precise answer, found in Example 15-9 with the exact doubling time formula, is a doubling time of 8.04 years and a balance of $555,000 in 100 years. Note that the approximate formula yields a doubling time fairly close to the actual doubling time: 7.7 years is only about 5% less than 8.04 years. In contrast, because the increase factor formula compounds this small error over a long period of time, the approximate value of $800,000 for balance after 100 years is about 45% larger than the actual value of $555,000. Nevertheless, the approximate result still represents a good order of magnitude estimate.

❖

Example 15-4 *Termites!* Unbeknownst to you, a family of 100 termites invades your house. Suppose that the termite population grows at a rate of 15% per week. How many termites would be in your house after a year?

Solution: Use the doubling time formula (with $r = 0.15$ per week) to find that

$$T_2 \approx \frac{0.7}{0.15 \text{ per week}} = 4.67 \text{ weeks.}$$

Substitute this doubling time into the factor increase formula to find that after 1 year, or about 52 weeks,

$$\text{increase factor} = 2^{t/T_2} = 2^{52 \text{ weeks}/4.67 \text{ weeks}} = 2^{11.13} = 2248.$$

That is, using the approximate doubling time, the number of termites would increase by a factor of about 2250 in a year, and the original 100 termites would increase to about 225,000 termites. A more precise answer, calculated with the exact doubling time formula in subsection 15.3.5, is a doubling time of 4.96 weeks, and an actual termite population of about 143,000 at the end of a year. As in the previous example, the approximate formula yields a doubling time within a few percent of the actual value, but the error in the increase factor is much larger and represents only an order of magnitude estimate.

❖

Example 15-5 ***Breakdown in the Approximation!*** Suppose that prices are increasing at a rate of 100% per year. How long does it take for prices to double?

Solution: The rate of growth is 100% per year, so prices double in 1 year. Note that the approximate doubling formula does *not* apply in this case because a growth rate of 100% per year, or $r = 1.0$ per year, is *not* small. If you mistakenly used the approximate doubling time formula you would find $T_2 \approx 0.7 \div 1$ yr = 0.7 year — much shorter than the actual doubling time of 1 year.

❖

15.3.3 Exponential Decay and Half-Life

Rather than *growing*, some quantities decrease or *decay* exponentially. Decay may be viewed as negative growth, so the rate of decay should be a negative number. Therefore, you can deal with exponential decay simply by using a negative value for the rate of growth, r, in the exponential growth formulas. For example, if a population *decreases* at a rate of 5% per year, $r = -0.05$ per year. The doubling time of this population also is negative, or

$$T_2 \approx \frac{0.7}{r} = \frac{0.7}{-0.05/\text{year}} = -14 \text{ years}.$$

Exponential Fact 6:

When a quantity decays exponentially, its half-life is constant and independent of quantity at any given time. The fractional rate of decay, r, is negative, and the half-life is

$$\textbf{Half-Life} = T_{1/2} \approx -\frac{0.7}{r}.$$

This formula works well only when r is relatively small.

What does this *negative* doubling time mean? It means that the population will decrease to *half* its original size after 14 years. The time required for any exponentially decaying quantity to fall to half its present size is its **half-life**, given by

$$\text{Half-Life} = T_{1/2} \approx -\frac{0.7}{r}, \text{ where } r \text{ is the decay rate (negative).}$$

Because the rate of decay, r, is a negative number, the minus sign in the half-life formula ensures that the half-life always is a positive number. Apart from the minus sign, it is the same formula used for the doubling time. As with the doubling time formula, this formula is an approximation that is accurate only for small decay rates.

Suppose that a particular quantity decays with a half-life of 1 year. Then, after 1 year, it decreases to one half its original value. After 2 years, it decreases to one fourth of its original value; after 3 years to one eighth of its original value; and so on. Generalizing, after n half-lives, the value falls to a fraction $(1/2)^n$ of its original value. The relation between the number of half-lives, n, and the total amount of elapsed time, t, is

$$n = \text{number of half-lives} = \frac{\text{total elapsed time}}{\text{half-life}} = \frac{t}{T_{1/2}}, \text{ where} \begin{cases} t = \text{total elapsed time} \\ T_{1/2} = \text{half-life.} \end{cases}$$

For example, if a radioactive substance has a half-life of $T_{1/2} = 40$ years, its quantity is cut in half five times in $t = 200$ years (200 yr ÷ 40 yr = 5).

Because a quantity decreases by a *factor* of $(1/2)^n$ after n half-lives, we also can write

$$\text{fraction remaining} = \left(\frac{1}{2}\right)^n = \left(\frac{1}{2}\right)^{t/T_2}.$$

Example 15-6 *Decreasing Population.* Suppose that development of a mountainous area causes the population of mountain lions to decay at a rate of 4% per year. When will the number of mountain lions be half its current value?

Solution: The 4% annual decay rate means that $r = -0.04$ per year. The half-life is

$$T_{1/2} \approx -\frac{0.7}{r\,/\text{year}} = -\frac{0.7}{-0.04\,/\text{year}} = 17.5 \text{ years.}$$

The mountain lion population will fall to half its present size in about 17 to 18 years.

❖

Example 15-7 *Devaluation of Currency.* Suppose that inflation is causing the value of the ruble to fall against the value of the dollar at a rate of 12% per month. How long does it take for the ruble to lose half its value?

Solution: The rate of decay is $r = -0.12$ per month. Therefore

$$T_{1/2} \approx -\frac{0.7}{r\,/\text{month}} = -\frac{0.7}{-0.12\,/\text{month}} = 5.8 \text{ months.}$$

The half-life is about 6 months, meaning that the ruble loses half its value (against the dollar) during that time.

❖

Example 15-8 *Radioactive Decay.* If the half-life of a radioactive substance is 40 years, what fraction of it remains after 200 years?

Solution: For a half-life of $T_{1/2} = 40$ years, the fraction remaining after $t = 200$ years is

$$\text{fraction remaining} = \left(\frac{1}{2}\right)^{t/T_2} = \left(\frac{1}{2}\right)^{200\text{ yr}/40\text{ yr}} = \left(\frac{1}{2}\right)^{5} = \frac{1}{32}.$$

For example, if the radioactive material originally weighed 32 grams, only 1 gram remains after 200 years. You can check whether this answer is reasonable by dividing the original amount in half every half-life, or 40 years. After 40 years, the original 32 grams will decrease by half to 16 grams. After 80 years, it will have decreased again by half, to 8 grams; after 120 years, to 4 grams; after 160 years, to 2 grams; and after 200 years, to 1 gram.

❖

Time-Out to Think: For each of Examples 15–6 to 15–8, suggest *why* the decrease should be exponential; that is, by a fixed *percentage* each year, rather than by a fixed amount each year.

15.3.4 Rate Confusion in Financial Applications

We pause here for a cautionary note about exponential growth rates. Most applications are unambiguous about what an annual growth rate means. For example, a 5% annual growth rate means that the size of the quantity grows by 5% from one year to the next. However, genuine confusion can arise in banking and investment applications. Most banks advertise an interest rate called the **annual percentage rate (APR)**. However, because most banks add interest to their accounts more than once a year, the APR usually is *not* the actual annual growth rate. For example, if interest is *compounded* (see subsection 3.6.2) every month, interest is computed and added to the account *twelve* times a year. The effect of compounding is that the actual growth in the account is slightly *more* than the APR. The actual growth in the account is called the **effective yield**. Banks generally quote both the effective yield and the APR of their accounts. We explore compounding in bank accounts in greater detail in Chapter 16.

15.3.5 Exact Formula for Doubling Time

The doubling time and half-life formulas given earlier are approximate formulas that are valid for small growth rates. The *exact* formula for the doubling time that applies to any growth rate, ***r***, is

Tracking units in the exact formulas is a subtle process that we don't explain here; as stated, the formulas work as long as the doubling time or half-life are given with the same time units as the fractional growth rate, r. For example, if r is 0.5 per day, then the calculated doubling time has units of days.

$$T_2 = \frac{\log_e 2}{\log_e(1+r)} \approx \frac{0.693}{\log_e(1+r)},$$

where the units of T_2 are the same units of time used in the growth rate ***r***. In this formula $\log_e$ represents the **natural logarithm**. Most calculators have a button for the natural logarithm (usually labeled *ln*), so you can use this formula easily. When the growth rate, r, is small (less than about 0.15), $\log_e(1 + r) \approx r$; furthermore, $\log_e 2 \approx 0.693 \approx 0.7$, which explains the approximate formula, $T_2 \approx 0.7/r$.

Time-Out to Think: Use the natural logarithm button on your calculator to confirm that $\log_e 2 \approx 0.693$. Then calculate the value of $\log_e(1 + r)$ for $r = 0.01$, $r = 0.05$, $r = 0.1$, $r = 0.15$, and several larger values of r. Do your results confirm the approximation $\log_e(1 + r) \approx r$ for small values of r? Explain.

Similarly, for exponential decay with a negative rate of growth (r is negative), the exact formula for the half-life is

$$T_{1/2} = \frac{\log_e 2}{\log_e(1-r)} \approx \frac{0.693}{\log_e(1-r)}.$$

Again, the units of T_2 are the same units of time used in the decay rate, ***r***.

Example 15-9 ***Bank Account Revisited.*** Suppose that a bank account increases its value by 9% per year. What is the exact doubling time of the account? If you invest \$100 today, how much will be in the account when your great-great-grandchildren inherit it in 100 years? Compare this result to the answers obtained with the approximate doubling time formula in Example 15-3.

Solution: The fractional growth rate is $r = 0.09$ per year. Thus the doubling time for the account is:

Note that the approximation $\log_e 2 \approx 0.693$ has 3 significant digits. Thus, we state our answers with 3 significant digits.

$$T_2 = \frac{\log_e 2}{\log_e(1+0.09)} = \frac{0.693}{\log_e(1+0.09)} = 8.04 \text{ years}.$$

After $t = 100$ years, the account balance will have increased in value by a factor of

$$\text{increase factor} = 2^{t/T_2} = 2^{100 \text{ yr}/8.04 \text{ yr}} = 2^{12.44} = 5550.$$

Thus the \$100 will grow to about \$100 × 5550 = \$555,000 in 100 years. Although \$555,000 is considerably less than the \$800,000 obtained with the approximate formula, the approximate formula still gave a reasonable order of magnitude estimate.

❖

Example 15-10 *Large Growth Rates.* Suppose that prices are increasing at a rate of 100% per year. Show that the exact formula yields the correct doubling time of 1 year.

Solution: Example 15-5 showed that the approximate formula with $r = 1.0$ per year gives an incorrect doubling time of 0.7 ÷ 1 yr = 0.7 years. The exact formula yields

$$T_2 = \frac{\log_e 2}{\log_e(1+1)} = \frac{\log_e 2}{\log_e 2} = 1.$$

The doubling time of 1 year is exact, as you would expect if prices were increasing 100% per year.

❖

Example 15-11 *Ruble Revisited.* Suppose that inflation is causing the value of the ruble to fall against the value of the dollar at a rate of 12% per month. How long does it take for the ruble to lose half its value? Compare this answer to the one obtained with the approximate formula in Example 15-7.

Solution: The fractional decay rate is $r = -0.12$ per year. Thus the half-life is

$$T_{1/2} = \frac{\log_e 2}{\log_e(1-r)} = \frac{0.693}{\log_e(1-(-0.12))} = \frac{0.693}{\log_e(1.12)} = 6.11 \text{ months.}$$

The ruble loses half its value against the dollar in 6.11 months, or slightly longer than the 5.8 months obtained with the approximate formula.

❖

15.4 EXPONENTIAL MODELS AND APPLICATIONS

Like linear relations, exponential relations can be used to model real-world situations. As with a linear model (see Chapter 10), an exponential model can describe how a particular quantity changes over time and allow predictions about its future state. Of course, you must always beware of the model's limitations.

15.4.1 The General Exponential Growth Law

Suppose that a town starts with a population of 1000 that then grows at a rate of 5% per year, or $r = 0.05$ per year. How does the population change with time?

The growth rate, $r = 0.05$ per year, means that each year the population is 5% larger than in the previous year. Thus after any one year period,

new population = previous population + increase in population
= previous population + 0.05 × previous population
= 1.05 × previous population.

That is, the population each year is 1.05 times the population from the previous year. Let's use Q to represent the population (we use Q to remind us of *quantity*) and t to represent the time, in years. If we call the starting time

$t = 0$ years, then the **initial value** of Q is $Q = 1000$ people. Stepping forward 1 year at a time:

$$\text{Initially, at } t = 0 \text{ years, } Q = 1000.$$

$$\text{After } t = 1 \text{ year, } Q = 1000 \times 1.05 = 1050.$$

$$\text{After } t = 2 \text{ years, } Q = \underbrace{1050}_{\substack{\text{population}\\ \text{after 1 year}}} \times 1.05 = (\underbrace{1000 \times 1.05}_{1050}) \times 1.05 = 1000 \times (1.05)^2 = 1102.$$

$$\text{After } t = 3 \text{ years, } Q = \underbrace{1102}_{\substack{\text{population}\\ \text{after 2 years}}} \times 1.05 = (\underbrace{1000 \times 1.05^2}_{1102}) \times 1.05 = 1000 \times (1.05)^3 = 1158.$$

The pattern should be clear; after t years the population has increased by a factor of $(1.05)^t$. That is, after a time t, the population is

$$Q = 1000 \times (1.05)^t.$$

This formula describes an **exponential relation** between the variables t and Q. Using the terminology of Chapter 10, the formula defines a relation (t, Q). In principle, this relation is valid for *any* time; t need not be an integer. The data in Table 15-7 were generated with this relation and are graphed in Figure 15-5. As with any exponential growth, the number of new individuals added each year increases every year. That's why the graph of the exponential relation gets steeper and steeper as time goes on.

Table 15-7. Population Data Generated by the Exponential Relation $Q = 1000 \times (1.05)^t$

Year	Population	Year	Population
0	1000	4.0	1216
1.0	1050	5.0	1276
1.5	1076	6.0	1340
2.0	1102	8.0	1477
2.5	1130	10.0	1629
3.0	1158	12.0	1796
3.5	1186	14.0	1979

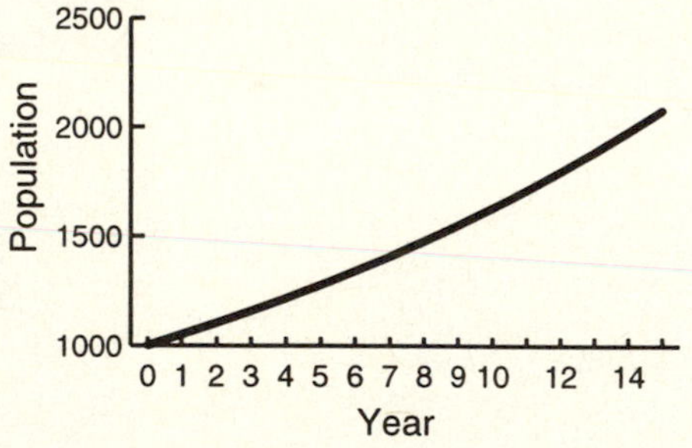

Figure 15-5. The graph of the exponential relation $Q = 1000 \times (1.05)^t$ gets steeper with time, meaning that the number of people *added* to the population increases each year.

The general exponential growth law is

$$Q = Q_0 \times (1 + r)^t,$$

where the units of time used for r and t must be the same.

If you try to insert the units of r and t in this formula, the units of the answer will be very strange! The reason is the way we use r and t, which is designed to make the formula easy to apply. Technically, r should be the dimensionless *fraction* by which the quantity increases in a period of time. Similarly, t should be the dimensionless *number* of these time periods. As long as you follow the rule that the units of time used in r and t are the same, you may safely ignore this technicality.

> **Time-Out to Think:** At a growth rate of 10% per year, by how much does a population of 2000 people increase in a year? in two years?

We can now generalize to *any* quantity Q. If the initial value of the quantity is Q_0 ($Q_0 = 1000$ in the preceding example) and Q grows exponentially at a rate r per unit time ($r = 0.05$ in the preceding example), for any time t after the initial time:

$$Q = Q_0 \times (1 + r)^t. \qquad \text{(The units of time for } r \text{ and } t \text{ must be the same!)}$$

This formula is the **general exponential growth law**. Note that, in the form stated, the law works only if the units of time for r and t are the same. For example, if $r = 0.1$ per year, t must be in years; if $r = 0.25$ per week, t must be in weeks; and if $r = -0.07$ per hour, t must be in hours.

Time-Out to Think: Confirm that substituting an initial population of 1000 and a growth rate of 5% per year in the general exponential growth law yields the relation found previously: $Q = 1000 \times (1.05)^t$.

Example 15-12 ***Bank Accounts Again.*** Suppose that a bank account increases its value by 9% per year. If you invest $100 today, how much will be in the account in 9 months? 2 years? 30 months? 10 years? (Assume that interest is paid at least every month; in Chapter 16, we show how the annual increase in account value is related to the monthly interest.)

One further subtlety is involved in using the general exponential growth formula. We stated that the law is valid *in principle* for any time *t*. Strictly speaking, that is true only if the growth is *continuous*. If the growth is *discrete*, the formula is valid only for integer multiples of the discrete time step. For example, if bank interest is paid monthly, you can use the formula to calculate the account value after 9 months, but not after 9 1/2 months.

Solution: In this problem the quantity Q is the bank balance. Its initial value is $Q_0 = \$100$; the growth rate is $r = 0.09/\text{yr}$. Using the general exponential growth law, the bank balance Q after t years is

$$Q = Q_0 \times (1 + r)^t = \$100 \times (1.09)^t.$$

Now you can answer the specific questions, but remember that, because r is a rate per *year*, t must have units of years.

After 9 months, or $t = 0.75$ year, the balance is $Q = \$100 \times (1.09)^{0.75} = \$106.67$.

After $t = 2$ years, the balance is $Q = \$100 \times (1.09)^2 = \$118.81$.

After 30 months, or $t = 2.5$ years, the balance is $Q = \$100 \times (1.09)^{2.5} = \$124.04$.

After $t = 10$ years, the balance is $Q = \$100 \times (1.09)^{10} = \$236.73$.

❖

Example 15-13 ***U.S. Population Growth.*** The population of the United States is growing at a rate of about 0.7% per year. Beginning with the 1990 population of about 250 million, what will the population be in 2045 (when today's high school students are reaching retirement age)? Comment on the validity of this projection by comparing it to the population if the growth rate falls to 0.2% per year or rises to 1.2% per year.

Solution: In this case the quantity Q is the U.S. population. Its initial value is $Q_0 = 250$ million, and $t = 0$ corresponds to the year 1990. The growth rate is $r = 0.007/\text{yr}$. At any time after 1990 the general exponential growth law gives the population:

$$Q = Q_0 \times (1 + r)^t = 250 \text{ million} \times (1.007)^t.$$

The year 2045 corresponds to $t = 55$ years (from 1990 to 2045). Substituting $t = 55$ years into the general exponential growth law gives the population in 2045.

At 0.7% per year, $Q = 250 \text{ million} \times (1.007)^{55} = 370$ million people.

If the U.S. population continues to grow at a rate of 0.7% per year, the population will reach about 370 million — about 50% greater than its current size — by the time today's high school students retire. Growth rates of 0.2% and 1.2% per year give the following populations in 2045.

At 0.2% per year, $Q = 250 \text{ million} \times (1.002)^{55} = 279$ million people.

At 1.2% per year, $Q = 250 \text{ million} \times (1.012)^{55} = 481$ million people.

Within a one percent range in the growth rate (0.2% to 1.2%), the projected population for 2045 varies by more than 200 million people (from about 280 million to 480 million). This *difference* of 200 million people is nearly as much as the entire current population of the United States!

Clearly, population projections are very sensitive to changes in the growth rate. In the United States, population grows not only because of differences between the birth and death rates but also because of immigration. Therefore the validity of an exponential growth model for the U.S. population depends on at least three factors: personal choices about family size; medical care and its influence on death rates; and political decisions about immigration policy. Changes in any of these may change the overall growth rate, which can translate to a difference of hundreds of millions of people in just 50 years!

❖

Example 15-14 ***Metropolitan Population Growth.*** A small metropolitan area had a population of 85,000 in 1990. Because of its pleasant environment, many people want to move there. Concerned about rapid growth, the residents passed a growth control ordinance limiting population growth to 3% each year. If the population grows at this 3% annual rate, what will the population be in 2010? in 2100?

Solution: In this case, the initial value of the population is $Q_0 = 85{,}000$ and the rate of growth is $r = 0.03$ per year. For 2010, the elapsed time is $t = 20$ years and for 2100 it is $t = 110$ years. Substitute into the general exponential growth law.

In 2010, $Q = Q_0 \times (1 + r)^t = 85{,}000 \times (1.03)^{20} = 153{,}519$.

In 2100, $Q = Q_0 \times (1 + r)^t = 85{,}000 \times (1.03)^{110} = 2.2 \times 10^6 = 2.2$ million.

At a 3% rate of growth per year, the community of 85,000 people in 1990 will grow to a moderately-sized city of about 150,000 by the year 2010 and to a metropolis of more than 2 million people by 2100.

❖

15.4.2 Inflation

In economics, the term **inflation** refers to the growth in prices over time. For example, suppose groceries that cost \$100 at the beginning of 1995 cost \$103 (for the same items) at the beginning of 1996. The cost of the groceries had risen, or had been *inflated* by, 3% during that year. Because the prices of different items change differently over time, the federal government estimates the rate of inflation by averaging the prices of many items. One common measure of the rate of inflation is the consumer price index (see subsection 8.4.4). The media report the monthly price change, along with the annual rate of inflation that the monthly change implies.

Let's examine how inflation works. If prices rise by 0.8% during the month of July, the rate of inflation for July is 0.8% per month. What annual rate of inflation does this imply? To find out, we can use the general exponential growth law with $t = 12$ months and $r = 0.008$ per month. Letting Q be the price 12 months later and Q_0 be the current price,

$$Q = Q_0(1+r)^t = Q_0(1+0.008)^{12} \xrightarrow{\text{divide by } Q_0} \frac{Q}{Q_0} = (1+0.008)^{12} = 1.100.$$

The new price Q is 1.1 times the old price Q_0, so the fractional *increase* in the price is 0.1, or 10%.

That is, the new price Q after 1 year is 1.1 times the old price Q_0, which means the price rose by 10%. Thus the *annual* rate of increase is 0.1 or 10%, based on the July increase of 0.8%. Note that the annual rate of inflation is *not* 12 times the monthly rate ($12 \times 0.8\% = 9.6\%$). The news report should state that "prices rose 0.8% in July, corresponding to an annual inflation rate of 10%."

Comparing prices today with prices in the past requires accounting for inflation. For example, the average rate of inflation from 1990 to 1995 was about 3% per year ($r = 0.03$ per year). In other words, over that 5-year period ($t = 5$ years) typical prices rose by a factor of

$$\frac{Q}{Q_0} = (1+r)^t = (1+0.03)^5 = 1.16.$$

That is, a group of items costing $100 in 1990 would be expected to cost $116 in 1995.

Because items that cost $100 in 1990 would cost $116 in 1995, economists say that $116 in "1995 dollars" is equivalent to $100 in "1990 dollars." Let's use the symbol $\$_{1990}$ to represent 1990 dollars and the symbol $\$_{1995}$ to represent 1995 dollars. Then we can express that 100 of the 1990 dollars have the same value as 116 of the 1995 dollars by writing

$$\underbrace{100\ \$_{1990}}_{\text{value of items in "1990 dollars"}} = \underbrace{116\ \$_{1995}}_{\text{value of items in "1995 dollars"}}.$$

Rearranging this equation yields

$$\frac{\$_{1990}}{\$_{1995}} = \frac{116}{100} = 1.16\text{, or } \$_{1990} = 1.16 \times \$_{1995}.$$

Although this terminology for describing how prices change with inflation is quite common, it can be a bit confusing. Note that because a 1990 dollar is *more* valuable than a 1995 dollar, the *price* of items in 1990 dollars is *lower* than the price of the same items in 1995 dollars.

That is, the value of the dollar in 1990 was 1.16 times its value in 1995. This result should be expected: Because inflation reduces the value of the dollar, the equation shows that 1990 dollars represent more value than 1995 dollars. Similarly, dividing both sides of the preceding equation by 1.16 yields

$$\$_{1995} = \frac{\$_{1990}}{1.16} = 0.86 \times \$_{1990}.$$

That is, the value of a 1995 dollar is 0.86 times that of a 1990 dollar. Equivalently, we could say that one 1995 dollar is equivalent to only 86¢ in 1990 dollars.

The preceding equations represent conversions between 1995 dollars and 1990 dollars. Following the methods introduced in Chapter 4, we can write the *conversion factor* in three equivalent forms:

$$1\ \$_{1990} = 1.16\ \$_{1995} \quad \text{or} \quad \frac{1.16\ \$_{1995}}{1\ \$_{1990}} = 1 \quad \text{or} \quad \frac{1\ \$_{1990}}{1.16\ \$_{1995}} = 1.$$

With a conversion factor, you can compare prices of *particular* items at different times. For example, suppose that a gallon of gasoline cost $1.20 in 1990 and $1.35 in 1995. What is its price change in *constant dollars*? Converting the 1995 price to 199 dollars yields

$$\underbrace{1.35\ \$_{1995}}_{\text{price in 1995 dollars}} \times \underbrace{\frac{1\ \$_{1990}}{1.16\ \$_{1995}}}_{\text{conversion factor}} = \underbrace{1.16\ \$_{1990}}_{\text{value in 1990 dollars}}.$$

The $1.35 price in 1995 dollars is equivalent to a value of $1.16 in 1990 dollars. Because the *actual* price in 1990 was $1.20, the "real" price of gasoline fell 4¢ (in 1990 dollars) over the 5-year period.

Example 15-15 *Computer Price Drop.* Suppose that a particular computer system cost $2500 in 1990. An identical system sold for only $900 in 1995. What is the real decrease in the price of the computer over the 5-year period?

Solution: First, convert the 1995 price to 1990 dollars:

$$\underbrace{900\ \$_{1995}}_{\substack{\text{price in} \\ \text{1995 dollars}}} \times \underbrace{\frac{1\ \$_{1990}}{1.16\ \$_{1995}}}_{\text{conversion factor}} = \underbrace{776\ \$_{1990}}_{\substack{\text{value in} \\ \text{1990 dollars}}}.$$

The 1995 price of $900 is equivalent in value to only $776 in 1990 dollars. Because the actual price of the computer was $2500 in 1990, its real price decrease over the 5-year period was $2500 – $776 = $1724 in 1990 dollars. That is, its real value dropped by a factor of $1724 ÷ $2500 = 0.69, or 69%, in just 5 years. This tremendous drop in computer prices reflects rapidly advancing technology, which enables computer manufacturers to produce the same or more powerful machines at lower prices.

❖

Example 15-16 ***Mattress Investments.*** Suppose that your grandfather put $100 under his mattress 50 years ago. If he had invested it instead in a bank account paying interest of 3.5% per year (roughly the average U.S. rate of inflation during that period), how much would it be worth now?

Solution: To find the value of the $100 today use the general exponential growth law with $r = 0.35$ per year, $t = 50$ years, and $Q_0 = \$100$:

$$Q = Q_0 \times (1 + r)^t = \$100 \times (1.035)^{50} = \$558.$$

Invested at the rate of inflation, the $100 would be worth well over $500 today. Unfortunately, the $100 was put under a mattress, so it still has a face value of only $100. That's why saving money under a mattress generally is a very poor investment decision.

❖

Example 15-17 ***Hyperinflation.*** During the worst periods of hyperinflation in Brazil during the past two decades, prices rose as rapidly as 80% per month. How much do prices rise at this rate in 1 year? in 1 day?

Solution: Use the general exponential growth law, with $r = 0.8$ per month, to compare prices by dividing current prices Q by original prices Q_0. Over a $t = 12$ month period, prices change by a factor of

$$\frac{Q}{Q_0} = (1+r)^t = (1+0.8)^{12} = 1157.$$

That is, prices increase more than a thousand fold over a year! During a period of t = 1 day = 1/30 month, prices increase by a factor of

$$\frac{Q}{Q_0} = (1+r)^t = (1+0.8)^{1/30} = 1.02.$$

Again, the new price Q is 1.02 times the old price Q_0, so the fractional *increase* in the price is 0.02, or 2%.

Thus prices rise by about 2% in a single day! The difficulties caused by hyperinflation are immense. Store owners, for example, typically raise prices every day; sometimes they even close for a couple of hours at midday to raise prices. At the same time, wage earners must demand similar pay increases to keep pace with prices. And, with the value of the currency dropping by a factor of more than 1000 in a year, the government has to keep printing bills with more and more zeros. Eventually, governments usually *devalue* their currency by simply chopping off three zeros, or six zeros, or whatever is needed. (Over the past two decades Brazil's currency has been devalued by a factor of more than 1 trillion.)

❖

Time-Out to Think: Discuss some of the social upheaval likely under hyperinflation. Are any of the world's countries currently experiencing this phenomenon? If so, what are the consequences? How can hyperinflation be stopped?

15.4.3 Radioactive Decay and Dating

The term *radioactive* came about for historical reasons and has little to do with radio waves.

Everything is made of atoms, which in turn consist of two major components: the **nucleus and electrons**. Most atomic nuclei are *stable;* that is, under normal circumstances they hold together forever. However, some nuclei, are inherently *unstable;* that is, they tend to break apart, or decay, over time. A substance with unstable nuclei is called **radioactive**.

Why are some nuclei radioactive? All nuclei are in constant internal motion. If observing an atomic nucleus with extreme magnification were possible, it would appear to be constantly and rapidly expanding, contracting, and distorting its shape. A nucleus undergoes **radioactive decay** — that is, it breaks apart — if it expands so much that it can no longer hold itself together.

Because the internal motion of a nucleus is random, radioactive decay is a probabilistic process. For example, just as the probability of obtaining heads in coin toss is 50%, nuclei of a particular radioactive substance might have a 50% probability of decay during a 1-year period. If you had a million atoms of this substance, you would expect about 50%, or 500,000, of their nuclei to undergo radioactive decay in 1 year. During the following year, you would expect about 50% of the remaining 500,000 nuclei to undergo radioactive decay, leaving 250,000 atoms of the substance. During the third year, half the nuclei of these 250,000 atoms would decay, leaving 125,000 atoms, and so on.

Thus radioactive decay is an exponential process, characterized by the *half-life* of the radioactive substance. For example, tritium (a radioactive form of hydrogen used in nuclear weapons) has a half-life of about 12 years. If you start with 100 grams of tritium, half of it will decay in 12 years, leaving 50 grams. By the end of 24 years, or two half-lifes, only 25 grams of tritium remain. After 36 years, or three half-lifes, only 12.5 grams of tritium remain, and so on.

An atomic nucleus consists of protons and neutrons. The number of protons in a particular atom is called its **atomic number**; each of the elements in the periodic table has a different atomic number. The *sum* of the number of protons and the number or neutrons is called the **atomic weight**. If a single element (i.e., a substance with a particular number of protons) comes in varieties with different atomic weights, each variety is called an **isotope** (which literally means the same number of protons). Plutonium, for example, has an atomic number of 94. The isotope plutonium–239 has an atomic weight of 239; hence it has 239 – 94 = 145 neutrons.

Radioactivity occurs naturally. Although human-made sources of radioactivity, such as plutonium-239, are highly publicized, the greatest danger from radioactivity to the average person comes from naturally occurring radon gas, which can accumulate in homes. Indeed, some researchers estimate that radon gas causes as many lung cancer deaths each year as tobacco smoking.

Although many people associate radioactivity with danger, it can be very useful. For example, radioactive materials are used in modern medicine to make images (X-rays and CAT scans) of internal organs and to treat cancer. Radioactive materials also are used by archaeologists, paleontologists, and geologists to estimate the age of relics and other materials.

Example 15-18 ***Plutonium Decay.*** Plutonium-239 has a half-life of about 24,000 years. Write an exponential law that describes its decay. If 100 grams of plutonium-239 is buried, how much will remain after 1000 years?

Solution: First determine the growth rate from the substance's half-life. Recall the half-life approximation formula from Section 15.3. Substituting $T_{1/2}$ = 24,000 years into that formula yields

$$T_{1/2} \approx -\frac{0.7}{r} \xrightarrow{\text{solve for } r} r = -\frac{0.7}{T_{1/2}} = -\frac{0.7}{24{,}000 \text{ yr}} = -2.92 \times 10^{-5} \text{ /yr}.$$

The growth rate is *negative* because it represents *decay*. Now use the general exponential growth law with Q_0 = 100 grams (the initial amount of plutonium–239); Q is the amount of plutonium–239 remaining after time t = 1000 years, or

Note that $r = -2.92 \times 10^{-5}$,
$1 + r = 1 + (-2.92 \times 10^{-5})$
$= 1 - 2.92 \times 10^{-5}$.

$$Q = Q_0 \times (1+r)^t = (100 \text{ g}) \times (1 - 2.92 \times 10^{-5})^{1000} = 97 \text{ g}.$$

Approximately 97 of the original 100 grams remain after 1000 years. The long half-life means that the decay rate is extremely small, so very little of the plutonium–239 decays over a 1000-year period.

❖

Time-Out to Think: Plutonium-239 is one of the most toxic substances known. Given its half-life and the large amount of it produced during the Cold War, discuss some ideas about how to protect future generations from its dangers.

American scientist Willard Libby won the Nobel Prize in Chemistry in 1960 for inventing the method of radioactive dating.

Example 15-19 ***Radiocarbon Dating.*** Ordinary carbon, called carbon–12, is stable and does not decay. However, carbon–14 is a relatively rare form (isotope) of carbon that decays radioactively with a half-life of about 5700 years. Carbon–14 is constantly produced in the Earth's atmosphere through the absorption of radiation from the Sun. When living organisms breathe or eat, they ingest some carbon–14 along with ordinary carbon–12. After an organism dies, no more carbon–14 is ingested, so the age of its remains can be calculated by determining how much carbon–14 has decayed. Suppose that you find a human bone at an archeological site. You analyze the bone and discover that it contains only one tenth of the carbon–14 that it contained when the person died (various methods may be used to determine how much carbon–14 the bone originally contained). How long ago did the person die?

Solution: The easiest way to solve this problem is to use the formula for the fraction remaining after time t with a half-life $T_{1/2}$. In this case, the fraction remaining is 0.1, so the formula becomes

$$\text{fraction remaining} = \left(\frac{1}{2}\right)^{t/T_2}, \quad \text{or} \quad 0.1 = \left(\frac{1}{2}\right)^{t/5700 \text{ yr}}$$

As discussed in Appendix C, $\log_{10} x$ is the power to which 10 must be raised to give x. The important properties of the logarithm are as follows:
1. $\log (x \times y) = \log x + \log y$
2. $\log (x/y) = \log x - \log y$
3. $\log x^y = y \log x$

The goal is to find the time, t, that makes this equation a true statement. As before, you may proceed by trial and error, testing various values of t, or you may use logarithms as follows.

Take base-10 log of both sides: $$\log_{10}(0.1) = \log_{10}\left(\frac{1}{2}\right)^{t/5700}$$

Apply property that $\log x^y = y \log x$: $$\log_{10}(0.1) = \frac{t}{5700 \text{ yr}} \log_{10}\left(\frac{1}{2}\right)$$

Substitute $\log_{10}(0.1) = -1$ and $\log_{10}(1/2) = -0.301$ (find these values with your calculator): $$-1 = \frac{t}{5700 \text{ yr}} \times (-0.301)$$

Multiply both sides by 5700 yr and divide both sides by −0.301: $$t = 18{,}036 \text{ yr}.$$

Thus the bone comes from a person who died about 19,000 years ago.

❖

Time-Out to Think: Would carbon–14 be useful for dating a fossil 100 million years old? Why or why not? (Hint: Consider its half-life.)

Example 15-20 ***Uranium Decay.*** Uranium-238 has a half-life of 4.5 billion years. When it decays, it ultimately turns into lead, which therefore is called a **decay product** of uranium-238. Suppose that you find a rock containing a mixture of uranium-238 and lead. The distribution of lead atoms among the uranium atoms appears random, suggesting that the rock originally contained *only* uranium and that the lead was formed by uranium decay. By measuring the relative quantities of lead and uranium and taking into account their different atomic weights, you determine that 35% of the original uranium-238 has decayed. When did the rock form?

Solution: Again, use the formula for the fraction remaining, this time with a half-life of $T_{1/2} = 4.5$ billion years and a fraction remaining of 100% – 35% = 65% = 0.65. Substituting into the formula for the fraction remaining yields

$$\text{fraction remaining} = \left(\frac{1}{2}\right)^{t/(4.5\times10^9\ \text{yr})}, \quad \text{or} \quad 0.65 = \left(\frac{1}{2}\right)^{t/(4.5\times10^9\ \text{yr})}.$$

Solve for t using logarithms, as follows.

Take base-10 log of both sides:

$$\log_{10}(0.65) = \log_{10}\left(\frac{1}{2}\right)^{t/4.5\times10^9\ \text{yr}}$$

Apply property that $\log x^y = y \log x$:

$$\log_{10}(0.65) = \frac{t}{4.5\times10^9\ \text{yr}}\log_{10}\left(\frac{1}{2}\right)$$

Substitute $\log_{10}(0.65) = -0.187$ and $\log_{10}(0.5) = -0.301$ (find these values with your calculator):

$$-0.187 = \frac{t}{4.5\times10^9\ \text{yr}}\times(-0.301)$$

Multiply both sides by 4.5×10^9 yr and divide both sides by –0.301:

$$t = 2.8 \times 10^9\ \text{yr}.$$

The rock formed (solidified from a prior molten state) about 2.8 billion years ago.

❖

Time-Out to Think: Rocks are dated by comparing the amount of the radioactive substance present in a mixture to the amount of the substance's decay products. Thus rocks can be dated only since the time they last solidified; if a rock becomes molten (liquid), the mixture separates and information about the original content of the rock is lost. Although the solar system is known to be about 4.6 billion years old, the oldest rocks found on Earth are only about 3.5 billion years old. Why?

15.4.4 Physiological Processes

Many physiological processes are exponential. For example, a cancer tumor grows exponentially, at least in its early stages, and most drugs in the bloodstream break down exponentially.

The metabolism of alcohol is a notable exception to the exponential breakdown of drugs in the bloodstream: Its metabolism is linear, at a typical rate of 10 to 15 milliliters per hour (see subsection 9.3.2).

Example 15-21 ***Drug Metabolism.*** Most drugs (e.g., aspirin or antibiotics) have a characteristic half-life, or the time required for the drug concentration in the blood to be reduced by a factor of one half. Consider an antibiotic that has a half-life in the bloodstream of 12 hours. A 10-milligram injection of the antibiotic is given at 12:00 P.M. What is the amount of antibiotic in the blood at 8:00 P.M.? When does the antibiotic reach 1% of its original concentration?

Solution: Use the formula for the fraction remaining at any time t. The half-life is $T_{1/2} = 12$ hours; 8:00 P.M. is 8 hours after the injection, so $t = 8$ hours. Hence

$$\text{fraction remaining} = \left(\frac{1}{2}\right)^{t/T_{1/2}} = \left(\frac{1}{2}\right)^{8\text{ hr}/12\text{ hr}} = \left(\frac{1}{2}\right)^{2/3} = 0.63.$$

After 8 hours, 63% of the 10 milligrams injected initially, or about 6.3 milligrams, of the antibiotic remain. To find the time when the antibiotic reaches 1% of its original concentration set the fraction remaining to 0.01:

$$\text{fraction remaining} = \left(\frac{1}{2}\right)^{t/T_{1/2}}, \quad \text{or} \quad 0.01 = \left(\frac{1}{2}\right)^{t/12\text{ hr}}.$$

Again, solve for t using logarithms.

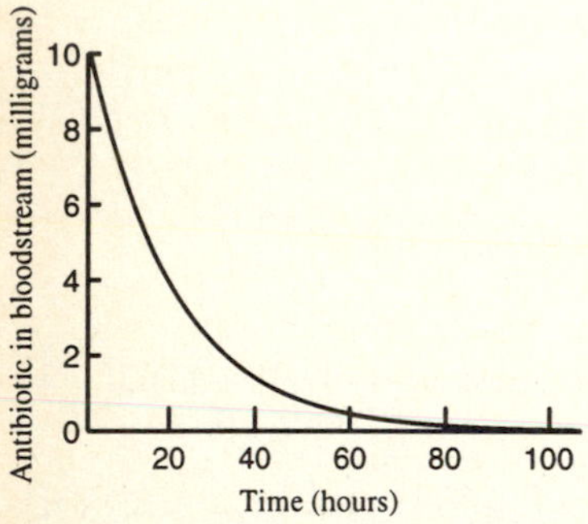

Figure 15-6. When an initial dose of 10 milligrams of an antibiotic with a half-life of 12 hours is injected into the blood, its concentration drops exponentially.

Take base-10 log of both sides: $$\log_{10}(0.01) = \log_{10}\left(\frac{1}{2}\right)^{t/12\text{ hr}}$$

Apply property that $\log x^y = y \log x$: $$\log_{10}(0.01) = \frac{t}{12\text{ hr}}\log_{10}\left(\frac{1}{2}\right)$$

Substitute $\log_{10}(0.01) = -2$ and $\log_{10}(0.5) = -0.301$ (find these values with your calculator): $$-2 = \frac{t}{12\text{ hr}} \times (-0.301)$$

Multiply both sides by 12 hr and divide both sides by -0.301: $$t = 79.7\text{ hr}.$$

The concentration falls to 1% of its initial value in about 80 hours. Figure 15-6 shows the decrease in the antibiotic concentration with time.

❖

15.4.5 Environment and Resources

Among the most important applications of exponential models are those involving the environmental and resource depletion. Concentrations of many pollutants in the water and atmosphere have been increasing exponentially, as has consumption of nonrenewable resources such as oil and natural gas. Two basic factors can be responsible for such exponential growth.

1. The *per capita* demand for a resource, or creation of waste products (pollution), often increases exponentially. For example, per capita energy consumption in the United States increased exponentially during most of the twentieth century.
2. An exponentially increasing population can lead to an exponentially increasing demand for resources (or creation of waste products) even if per capita demand remains constant.

In most cases, the growth or decay rate is determined by a combination of both factors.

Example 15-22 ***World Oil Production.*** In 1950, world oil production was 518 million tons. From then until the late 1970s production increased at a rate of 7% per year. Had growth continued at this rate, how much oil would have been produced by 1993? Comment on the fact that world oil production in 1993 was approximately 3000 million tons.

Solution. Let $t = 0$ represent 1950 and Q represent world oil production in millions of tons. The initial oil production in 1950 was $Q_0 = 518$ million tons. The general exponential growth law, with $r = 0.07$ per year, gives the amount of oil production t years after 1950:

$$Q = Q_0 \times (1 + r)^t = 518 \text{ million tons} \times (1.07)^t.$$

Had this rate of growth continued until 1993, which corresponds to $t = 43$ years, world oil production would have reached

$$Q = Q_0 \times (1 + r)^t = 518 \text{ million tons} \times (1.07)^{32} = 9500 \text{ million tons}.$$

This model prediction of 9500 million tons overestimated the actual figure by a factor of more than 3. During the early 1980s the rate of oil production fell, and between 1988 and 1993, world oil production nearly was constant. The lesson here is that exponential models can be extremely useful when used with care. However, in complex systems such as the world oil economy, too many factors (e.g., wars, embargoes, collapsing economies, decreasing reserves) are involved for them to be fully modeled with a simple exponential growth law.

❖

Example 15-23 ***Extinction By Poaching.*** Suppose that poaching reduces the population of an endangered animal by 10% per year. Further, suppose that when the population of this animal falls below 30, its extinction is inevitable (owing to the lack of reproductive options without severe inbreeding). If the current population of the animal is 1000, when will it face extinction? Comment on the validity of the exponential model.

Solution: In this case Q is the population of the endangered animal. The current size of the population is $Q_0 = 1000$, and the decay rate is $r = -0.1$ per year (negative because the population is decreasing). Thus the decrease in the population can be modeled by the general exponential growth equation:

$$Q = 1000 \times (1 - 0.1)^t = 1000 \times 0.9^t.$$

To find when the population faces extinction set $Q = 30$, the minimum population size that can survive, and solve for the time, t.

Set $Q = 30$:	$30 = 1000 \times (0.9)^t$
Switch left and right sides and divide both sides by 1000:	$(0.9)^t = 30 \div 1000 = 0.3$
Take base-10 log of both sides:	$\log_{10}(0.9)^t = \log_{10}0.3$
Apply property that $\log x^y = y \log x$:	$t \times \log_{10}(0.9) = \log_{10}0.3$
Divide both sides by $\log_{10}(0.9)$ and evaluate logarithms with your calculator:	$t = \dfrac{\log_{10} 0.03}{\log_{10} 0.9} = \dfrac{-1.522}{-0.0458} = 33.22$ years.

The animal would face extinction in about 33 years. Does an exponential model of poaching make sense? At first, it does: As the animals become rarer, they become harder for poachers to find. That a certain fraction of the population is "found" and killed each year seems reasonable. However, the exponential model is likely to have several shortcomings. If the animal congregates in herds, finding one means finding many and the rate of killing may be higher than predicted by the exponential model. If the poaching is driven by black market trading of the animal (as with ivory tusks from elephants or rhinoceros horns), the black market value of a dead animal may rise as the animal becomes rarer, thereby increasing the incentive to poachers and increasing the rate of killing. Conversely, as the animal becomes rarer, protection efforts may be increased, thereby promoting survival of the species. ❖

15.4.6 Creating Models by Finding Growth Rates

So far, the examples in this section have used the general exponential growth law in situations where the growth rate or doubling time was given. Another important situation occurs when you have two (or more) values of a quantity that grows exponentially and are asked to find the growth law. In this case you must determine the growth rate from the information given and then create a model with the form of the general exponential growth law. The process is similar to that used in Section 10.5 to create a linear model.

In 1850, world population was about 1 billion. By 1950, it had reached 2.5 billion. From these two facts let's create an exponential model with the form

$$Q = Q_0 \times (1 + r)^t.$$

To complete the model for this *particular* problem, we need to find the growth rate, r. We begin by letting $t = 0$ correspond to 1850, which means that $t = 100$ years corresponds to 1950. The initial value of the population in 1850 is $Q_0 = 1$ billion. In 1950, the population is $Q = 2.5$ billion and $t = 100$ years. We now can solve for r as follows.

Recall that for any number a and any power $n > 1$,

$$(a^n)^{1/n} = a.$$

That is, taking the nth root of the nth power returns the original number.

Set $Q_0 = 1$ billion, $Q = 2.5$ billion, and $t = 100$ yr in the general exponential growth law:	$2.5 \text{ billion} = (1 \text{ billion}) \times (1 + r)^{100}$
Divide both sides by 1 billion:	$2.5 = (1 + r)^{100}$
Raise both sides to the power of 1/100 (i.e., take the 100th root of both sides):	$2.5^{1/100} = \left((1+r)^{100}\right)^{1/100} = 1 + r$
Substitute $2.5^{1/100} = 1.00920$ (find this value with your calculator):	$1.00920 = 1 + r$
Subtract 1 from both sides:	$0.00920 = r.$

Substituting this rate of growth into the general exponential growth law yields the specific growth law for this model:

$$Q = Q_0 \times (1+r)^t = 1 \text{ billion} \times 1.00920^t.$$

Now, let's use this model to predict the 1995 population. The elapsed time from 1850 to 1995 is $t = 145$ years, so the model predicts

$$Q = 1 \text{ billion} \times 1.00920^{145} = 3.77 \text{ billion}.$$

The model underestimates the actual 1995 population of about 6 billion by about 2 billion people! What went wrong? We created the model by finding an *average* rate of growth for the years 1850 to 1950. Apparently, the rate of world population growth since 1950 has been higher than it was (on average) from 1850 to 1950.

Example 15-24 ***New Population Growth Rate.*** Find the average growth rate of world population between 1950 and 1995. Compare the result to the average growth rate between 1850 and 1950.

Solution: Begin by writing the general form of the exponential growth law:

$$Q = Q_0 \times (1 + r)^t.$$

To complete the model for this particular problem, find the growth rate, r, by letting $t = 0$ correspond to 1950, when the population was the initial value $Q_0 = 2.5$ billion. The elapsed time between 1950 and 1995 is 45 years, so the 1995 population is $Q = 6$ billion when $t = 45$ years. Substituting these values into the general exponential equation yields

$$6 \text{ billion} = (2.5 \text{ billion}) \times (1 + r)^{45}.$$

Now solve for r.

Divide both sides by 1 billion:	$2.4 = (1 + r)^{45}$
Raise both sides to the power of 1/45 (i.e., take the 45th root of both sides):	$2.4^{1/45} = \left((1+r)^{45}\right)^{1/45} = 1+r$
Switch left and right sides, subtract 1 from both sides, and use your calculator:	$r = 2.4^{1/45} - 1 = 1.0206 - 1 = 0.0206.$

The average rate of world population growth since 1950 has been about 0.02, or 2%, per year — considerably larger than the average growth rate of 0.92% between 1850 and 1950. Indeed, a growth rate of 2% per year corresponds to a doubling time of only 70 ÷ 2 = 35 years!

❖

Example 15-25 ***Natural Gas Consumption.*** World natural gas consumption in 1960, in units of "oil equivalents" (i.e., the amount of oil that would generate the same amount of energy), was about 400 million tons. In 1994, it reached about 2 billion tons. Create an exponential model of the increase in the use of natural gas. If oil production remains constant at about 3 billion tons per year (see Example 15-22), when will natural gas overtake oil as the world's primary source of energy?

Solution: To create an exponential model of the form $Q = Q_0 \times (1 + r)^t$, let the 1960 gas consumption be the initial value of $Q_0 = 400$ million tons. The elapsed time from 1960 to 1994 is 34 years, so the 1994 gas consumption is $Q = 2$ billion tons when $t = 34$ years. Substituting these values into the exponential growth law yields

$$2 \times 10^9 \text{ tons} = 4 \times 10^8 \text{ tons} \times (1 + r)^{34}.$$

Now solve for r.

Divide both sides by 4×10^8 tons:	$\dfrac{2 \times 10^9}{4 \times 10^8} = (1+r)^{34}$
Raise both sides to the power of 1/34 (i.e., take the 34th root of both sides):	$5^{1/34} = \left((1+r)^{34}\right)^{1/34} = 1+r$
Subtract 1 from both sides and use your calculator:	$r = 5^{1/34} - 1 = 1.048 - 1 = 0.048.$

Thus natural gas consumption increased at an average annual rate of about 4.8% between 1960 and 1994. The exponential model for natural gas consumption, where t represents the number of years after 1960, is

$$Q = 4 \times 10^8 \text{ tons} \times (1.048)^t,$$

To find out when the consumption will reach 3 billion tons, substitute $Q = 3 \times 10^9$ tons into this growth law:

$$3 \times 10^9 = 4 \times 10^9 \text{ tons} \times (1.048)^t.$$

Now solve for t.

Switch left and right sides and divide both sides by 4×10^9 tons:	$(1.048)^t = \dfrac{3 \times 10^9 \text{ tons}}{4 \times 10^8 \text{ tons}} = 7.5$
Take base-10 log of both sides:	$\log_{10}(1.048)^t = \log_{10}(7.5)$
Apply property that $\log x^y = y \log x$:	$t \log_{10}(1.048) = \log_{10}(7.5)$
Substitute $\log_{10}(7.5) = 0.875$ and $\log_{10}(1.048) = 0.0204$ (from calculator):	$t \times 0.0204 = 0.875$
Divide both sides by 0.0204:	$t = 42.9$ yr.

The model predicts that natural gas consumption will reach 3 billion tons per year about 43 years after 1960; that is, in about 2003. If the growth in gas consumption continues at a rate of 4.8% per year, and oil production remains constant (or decreases), natural gas will become the world's primary source of energy shortly after the turn of the century.

❖

15.5 REAL POPULATION GROWTH

Perhaps the most important application of exponential modeling is in population growth. For most of human history, from the earliest humans more than 2 million years ago to the advent of agriculture less than 10,000 years ago, human population probably never exceeded about 10 million. The advent of agriculture brought about the development of cities and more rapid population growth. By A.D. 1, human population had reached 250 million. The increase continued slowly, to about half a billion by 1650.

The beginning of the industrial age brought about huge increases in human ability to grow food and use natural resources. Improvements in medical and health science lowered death rates dramatically. As a result, human population began to grow exponentially. Although it has varied somewhat over time, the population growth rate currently is about 1.6% per year. With a current population of 6 billion, this rate means that the population is rising by about 0.016×6 billion = 96 million people each year.

Clearly, the number of people on the Earth cannot grow without bound. The Earth has only a finite capacity to support human beings, and perhaps the key factor in limiting human growth is the food supply. From the beginning of the industrial revolution until 1992, food production grew at an even faster rate than population. Hence the per capita food supply was greater in 1992 than ever before in history. However, since 1992 this trend has reversed, leading many experts to suggest that the three major food production systems — fishing, livestock, and agriculture — have reached or surpassed their natural limits. If so, the world may be headed toward a global food crisis in the relatively near future.

> **Time-Out to Think:** If per capita food production is higher than at most other times in history, why do some people still starve?

The rate of population growth is simply the difference between the birth rate and the death rate. For example, suppose that there are 85 births per 1000 people and 65 deaths per 1000 people per year. Then the population growth rate is

$$\text{growth rate (per year)} = \text{birth rate} - \text{death rate} = \frac{85}{1000} - \frac{65}{1000} = \frac{20}{1000} = 0.02 = 2\%.$$

The tremendous growth in human population during the past two centuries is directly attributable to a rapid decrease in the death rate because birth rates have plummeted. In nearly every country of the world, birth rates are at historic lows. However, death rates have fallen even faster (particularly among children), resulting in exponential population growth.

15.5.1 Carrying Capacity and Logistic Growth

Exponential growth cannot continue indefinitely; indeed, at its current rate, the human population cannot continue to grow much longer. As a result, most projections assume that human population will level off at somewhere between about 7 and 15 billion during the next century. How do models account for the *end* of exponential growth?

Instead of simply continuing with unchecked exponential growth, a model can take into account the **carrying capacity** of the Earth — the number of people that the Earth can support for long periods of time. For example, suppose that the carrying capacity of the Earth is 10 billion people. Then a reasonable assumption might be that the rate of population growth will slow as the carrying capacity is approached. One simple way to account for this possibility in an exponential model is to modify the rate of growth as follows:

$$\text{growth rate} = r \times \left(1 - \frac{\text{population}}{\text{carrying capacity}}\right).$$

This formula yields a growth rate very close to the exponential growth rate r when the population is small. However, the growth rate decreases as the population approaches the carrying capacity. Note that, when the population *equals* the carrying capacity, this formula yields a growth rate of $r \times (1 - 1) = 0$. A model based on the assumption that the rate of growth decreases smoothly, and becomes zero when the carrying capacity is reached, is called a **logistic growth model**.

Figure 15-7 contrasts a logistic growth model with an exponential growth model. Note the following features of the logistic model.

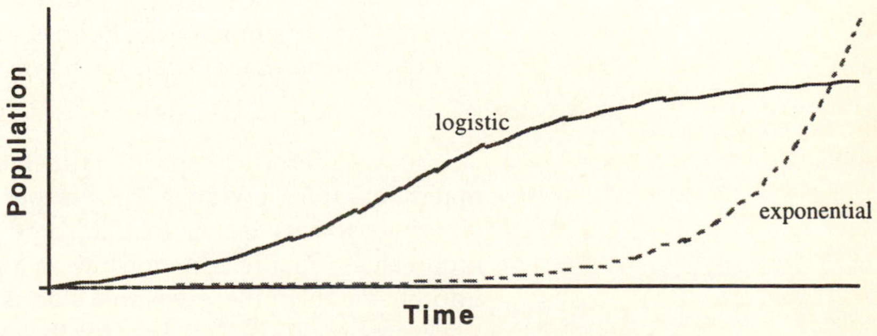

Figure 15-7. The general shapes of logistic growth curves and exponential growth curves are contrasted (different scales are used for the two curves to emphasize the different shapes). The exponential model increases without bound. The logistic model grows exponentially at first but then levels off at the carrying capacity.

- The population eventually reaches a maximum, after which it remains constant; this stable population represents the carrying capacity.
- While the population is small (compared to the carrying capacity), the shape of the logistic curve is the same as the shape of an exponential curve; the carrying capacity essentially has no effect on the rate of growth of a small population.
- As the population approaches the carrying capacity, the rate of growth decreases; thus the population smoothly levels out at the carrying capacity and remains constant after that.

What is the Earth's carrying capacity? Unfortunately, three primary factors make answering this question extremely difficult.

1. The carrying capacity depends on the standard of living assumed and the quantity of various resources required to maintain that standard of living. For example, one factor in the carrying capacity is the availability of energy; therefore the Earth might support a larger population if per capita energy consumption is reduced.
2. The carrying capacity can change over time, depending both on human technology and the environment. For example, many people base estimates of the carrying capacity today on the availability of fresh water; however, if new sources of energy (e.g., fusion) are developed, nearly unlimited amounts of fresh water may be obtained through the desalinization of sea water. Conversely, climate change induced by human activity might alter the environment and reduce the ability to grow food, thereby lowering the carrying capacity.
3. Even if anyone understood exactly how the first two factors affect the carrying capacity, no one knows how to calculate it.

The history of attempts to guess the carrying capacity of the Earth is fraught with missed predictions. Among the most famous was that made by English economist Thomas Malthus (1766–1834). In a 1798 paper entitled *An Essay on the Principle of Population as It Affects the Future Improvement of Society*, Malthus argued that mass deaths through starvation and disease would soon hit Europe and America. He based his argument on the fact that the populations of Great Britain, France, and America were doubling every 25 years, and he didn't believe that food production could keep up. Malthus's prediction did not come true, primarily because advances in technology *did* allow food production to keep pace with population growth.

Time-Out to Think: Malthus wrote that the greatest problem facing humanity is "that the power of population is indefinitely greater than the power in the Earth to produce subsistence for man." Although his immediate predictions didn't come true, some people argue that his fundamental idea was correct. According to this view, Malthus was wrong only in his prediction of *when* a catastrophe would occur, not *if*. What do *you* think?

Because carrying capacity is such a difficult quantity to determine, estimates vary over a wide range. Further complicating attempts to understand carrying capacity is the fact that some people have clear biases toward low or high values. People who depend on a growing population of customers for improvements in their personal standard of living are likely to be biased toward believing high values for the carrying capacity. People who believe that consumption and materialism are vices are likely to be biased toward believing low values for the carrying capacity. Typical estimates of carrying capacity range from less than three billion people to more than 20 billion.

15.5.2 Overshoot

How can anyone estimate a carrying capacity lower than the current world population? The answer is that carrying capacity represents the population that can be sustained for *long* periods of time. Any population of animals, plants, or bacteria can exceed the carrying capacity of its environment *temporarily*. This phenomenon is called **overshoot**. If a population overshoots the carrying capacity, then a decrease in the population is inevitable.

Overshoot is a fundamentally different concept from that of the logistic model. The logistic model is based on the assumption that the growth rate *automatically* adjusts as the population approaches the carrying capacity. However, because of the astonishing rate of exponential growth, real populations often increase beyond the carrying capacity in a relatively short period of time. If the overshoot is substantial, the decrease can be rapid and severe and is known as **collapse**. Figure 15-8 contrasts a logistic growth model with overshoot and collapse.

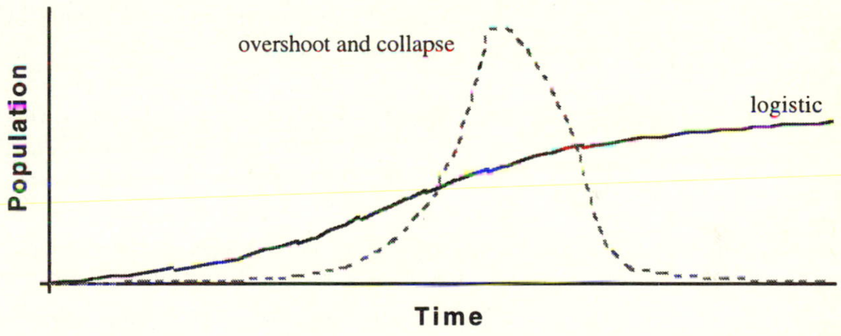

Figure 15-8. The general shape of an overshoot and collapse model is contrasted with that of a logistic model. The logistic model offers a smooth transition to population stability but makes the often unrealistic assumption that growth rates automatically adjust as the carrying capacity is approached. If, instead, the population overshoots the carrying capacity, a decrease or collapse is inevitable.

It is conceivable that human population has already exceeded the Earth's carrying capacity and is now in an overshoot phase. If so, the future holds only two possibilities: Either the population faces collapse, or a way must be found to increase the carrying capacity. However, few believe that the carrying capacity already has been reached, in which case time remains to find a way to follow a logistic model and to avert overshoot and collapse.

Time-Out to Think: The concept of carrying capacity can be applied to a localized environment as well as to the entire Earth. Consider the decline of past civilizations such as the ancient Greeks, Romans, Mayans, Anasazi, and others. Do overshoot and collapse models describe the fall of any of these or other civilizations? Explain.

15.5.3 Is There a Balance of Nature?

In examining the Earth's human population carrying capacity, an implicit question always is: How does human activity affect the natural environment? For the past several centuries, most people have assumed that an underlying **balance of nature** restores natural conditions when they have been upset. Unfortunately, recent discoveries show that the reality is far more complex.

For example, consider a world with only two species of animal, one of which (the predator) eats the other (the prey). Under the concept of a simple

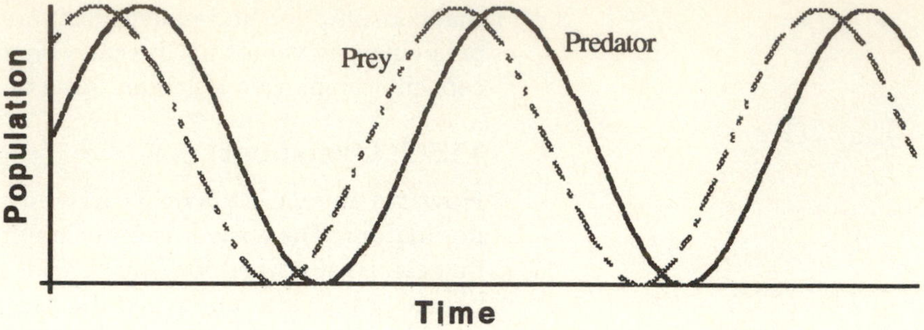

Figure 15-9. In this simple model one species of predator feeds on one species of prey. Rather than achieving stability or balance, both populations oscillate in cycles of overshoot and collapse. Note also that the predator population lags behind the prey population in time.

balance of nature, prey and predator populations eventually would become stable at some optimum level. In reality, even a simple model shows otherwise. Both prey and predator tend to have *oscillating* populations (Figure 15-9). To understand why, imagine that, when the predator population is low, the prey population can increase rapidly. The increase in the prey population means more food for predators, so the predators breed and their population grows. But the predator population soon overshoots the level (the carrying capacity) that can be supported by the prey population. The prey population collapses because more are being eaten than are being born. When the prey population collapses, the collapse of the predator population becomes inevitable. Eventually the predator population becomes small enough that the prey can again begin to multiply and the cycle repeats.

Thus, rather than achieving a *balance*, the predator–prey cycle is one of constant overshoot and collapse. And this model includes only *two* species. In nature, with millions of species competing for resources and eating one another, changes in populations are infinitely more complex. Indeed, biologists today have largely abandoned the notion of a balance of nature and instead regard nature as an ever-changing system that often shifts in sudden and dramatic ways. Not surprisingly, the ideas of *chaos* theory (see Chapter 3) can be applied to models of real populations.

Further, these observations apply even *before* human activity is figured into the mix. The ways that humans affect the environment and the carrying capacity of the Earth are enormously more complex than anyone would have guessed even a few decades ago.

Time-Out to Think: Discuss factors that may increase or decrease the carrying capacity of the Earth during the next 50 to 100 years. For example, fusion energy, resources from space, and conservation might increase carrying capacity. Climatic change, species extinction, and pollution might decrease it. Overall, do you think that human activity will tend to increase or decrease the carrying capacity during the next century? Is the growth of human population likely to follow a logistic model or an overshoot and collapse model? Explain your reasoning.

THINKING ABOUT ... THE DIFFERENCE BETWEEN PEOPLE AND BACTERIA

Look back at the bacteria in the bottle parable in Section 15.2. Imagine that, at 11:56, a prescient bacterium had climbed on a bacterial soap box and proclaimed: "We must stop our growth for, if we continue, we will fill our bottle and face ruin in a mere four minutes!"

Looking around at that time, other bacteria would have seen 15 times as much unused space in the bottle as their "bacterial civilization" had used up in its entire 56-minute existence. This apparent abundance might have made them scoff at their soapbox friend, yet his prediction was correct. The lesson herein is so powerful that Professor Al Bartlett, inventor of the parable of the bacteria in the bottle, refers to it as a *law*: Toward the end of the period of exponential growth, the population is growing so rapidly that detecting the impending crisis is extremely difficult.

Facts do not cease to exist because they are ignored. — Aldous Huxley

Because human population has grown exponentially for the past several centuries, it is tempting to conclude that humans face the same fate as the bacteria in the bottle. After all, if the current population of about 6 billion continued growing exponentially with a doubling time of about 40 years, population would reach 12 billion in 40 years, 24 billion in 80 years, and 48 billion in 120 years. In only about 650 years, humanity would be forced to stand elbow-to-elbow on every piece of land on Earth. Thus, like the bacteria in a bottle, the exponential growth of human population *will* stop; the only questions are *when* and *how*.

Unlike the bacteria, humans have the ability to foresee a potential crisis and make choices that can avert it. Let's first consider the question of *when* the exponential growth will stop. The time at which a crisis would appear depends on the Earth's carrying capacity. Although some people believe that human population already has exceeded the carrying capacity, others argue that the carrying capacity is 10 billion, 20 billion, or even 40 billion. Yet, at the current rate of exponential growth, the difference between today's population and a population of 40 billion is only about 100 years — a very short time on the scale of human history.

Because the answer to *when* the exponential growth will stop is *soon*, the more important question becomes *how*. There are only two basic ways to slow the growth of a population:

- a decrease in the birth rate, or
- an increase in the death rate.

Most people already are choosing the first option, as birth rates now are lower than at almost any other time in history. Indeed, population actually is decreasing in a few European nations. Nevertheless, world-wide birth rates still are much higher than death rates, and exponential growth therefore continues.

If a decrease in birth rates doesn't halt the exponential growth, an increase in the death rate will. If population significantly exceeds the carrying capacity by the time this process begins, the increase in the death rate will be dramatic — probably on a scale never before seen. This forecast is not a threat, a warning, or a prophecy of doom. It is simply a law of nature: *Exponential growth always stops*. As human beings, we can choose to halt our population growth through intelligent and careful decisions. Or, we can do nothing, leaving ourselves to the mercy of natural forces over which we can have no more control than we do over hurricanes, tornadoes, earthquakes, or the explosion of a distant star.

15.6 CONCLUSION

We devoted this entire chapter to a single mathematical idea: exponential growth and decay. Exponential change occurs in a tremendous variety of everyday situations, both locally and globally. Understanding exponential change and what it implies is crucial. The key ideas to remember from this chapter include the following.

- Any exponential change is characterized by a constant doubling time (for exponential growth) or half-life (for exponential decay).
- Exponential growth (and decay) are characterized by a fixed fractional or percentage growth rate. Phrases such as "increases by 12% per year" imply exponential growth.
- With each successive doubling, the increase in the size of a quantity is approximately equal to the sum of *all* preceding doublings. As a result, exponentially growing quantities quickly reach incredible proportions, at which point something will force them to stop growing. Exponential growth *cannot* continue indefinitely.
- Because exponential growth results in such astonishing change, understanding it can be extremely difficult and foreseeing its consequences even more difficult. Because so many important phenomena in society involve exponential growth — including financial situations, radioactive processes, and population growth — an understanding of exponential change may be the single most important mathematical concept for everyone to grasp.

PROBLEMS

Reflection and Review

Section 15.1

1. **Identifying Exponential Growth and Decay.** The signal that exponential growth is present is a phrase such as "*Q* has increased by 45% per year" or "*P* has been increasing by one third each year." Identify three news stories (radio, TV, newspapers, or magazines) that describe a quantity that is increasing exponentially. Does the story actually mention exponential growth? Identify three news stories that describe a quantity that is increasing linearly. Does the story actually mention linear growth?

2. **Linear Versus Exponential.** Do the following statements describe linear or exponential relationships? Why?
   a. The population of Reno is increasing at a rate of 5050 people per year.
   b. The price of food in Brazil is increasing at a rate of 30% per month.
   c. The birth rate in Austria is declining at a rate of 1.5% per year.
   d. You can depreciate the value of office equipment by $200 per year.

Section 15.2

3. **Chessboard Wheat.** According to the parable described in subsection 15.2.1, how many grains of wheat should be placed on square 30 of the chessboard? How many total grains would there be at that point? How much would the wheat on the chessboard weigh, in pounds, at that point?

4. **Magic Money.** Imagine that you receive a magic penny as described in subsection 15.2.2.
   a. How much money would you have after 15 days?
   b. Remember that all your money is in pennies. Suppose that you stacked the pennies after 15 days. How high would the stack rise? (Hint: Find a few pennies and a ruler!)
   c. How many days would elapse before you have your first billion dollars?
   d. Suppose that you could keep making a single stack of the pennies. After how many days would the stack be long enough to reach the nearest star (besides the Sun), which is about 4.3 light-years away? (See subsection 7.4.2 to review light-years.)

5. **The Towers of Hanoi.**
   a. In the Towers of Hanoi puzzle in subsection 15.2.3, the moves required for achieving a stack of three disks on another peg are listed. Make a list of the eight moves required to achieve a stack of four disks on another peg.
   b. According to the rules of the Hindu legend, how long would the priests need to move 63 of the 64 disks onto another peg in a stack at 1 second per move? How long do they need to make the final set of moves that gets all 64 disks on another peg? Comment on how this problem illustrates the first "fact" of exponential growth.

6. **Bacteria in a Bottle.** Consider the bacteria in a bottle described in subsection 15.2.4.
   a. How many bacteria are in the bottle at 11:50? What fraction of the bottle is full at that time?
   b. If the bacteria continued to grow, they could cover the Earth in a layer 2.5 meters thick after 2 hours. When would they cover the Earth in a layer 5 meters thick? 10 meters thick? 1 kilometer thick?

Section 15.3

7. **Change from the Doubling Time.** Answer each of the following questions.
   a. If the doubling time of a population of bacteria is 3 hours, by what factor does the population increase after 24 hours? after 1 week?
   b. If the doubling time of a bank account balance is 10 years, by what factor does it grow in 30 years? in 50 years?
   c. If the doubling time of a city's population is 22 years, how long does it take for the population to triple?
   d. If prices are increasing with a doubling time of 3 weeks, by what factor do prices increase in a year?

8. **Doubling Time Practice.** Answer each of the following questions; as needed, use the approximate formula for the doubling time.
   a. Suppose that the consumer price index of a country is increasing at a rate of 7% per year. What is its doubling time? By what factor will prices increase in 3 years?
   b. If a certificate of deposit increases its value by 5.5% per year, what is its doubling time? If you deposit $100 today, how much will you have after 5 years? after 50 years?
   c. Suppose that a population is growing at a rate of 1.6% per year. What is its doubling time? By what factor will the population increase in 75 years?
   d. If prices are rising at a rate of 0.7% per month, what is their doubling time? By what factor do prices increase in 1 year? in 8 years?
   e. Suppose that world population is rising at a rate of 1.6% per year. What is its doubling time? If population is 6 billion now, what will it be in 10 years? in 100 years? in 1000 years? Can such growth continue for 1000 years? Why or why not?

9. **Half-Life Practice.** Answer each of the following questions; as needed, use the approximate formula for the half-life.
   a. Suppose that predation is causing a population of rabbits to decline at a rate of 6% per year. What is the half-life for the rabbit population? Approximately when will it fall to 50% of its present size?
   b. Suppose that you have a radioactive substance that decays at a rate of 0.0005% per year. What is its half-life? If you start with 100 kilograms, when will only 1 gram remain?
   c. Suppose that an extremely low birth rate is causing the population of a country to decline at a rate of 0.3% per year. When will the population fall to half its current size?
   d. If the value of the dollar is falling against the value of the yen at a rate of 6% per year, how long will it take before the dollar loses half its value against the yen?
   e. Suppose that deforestation is causing the area of a particular forest to decline at a rate of 15% per year. When will half the forest be gone?
   f. A clean-up project is causing the concentration of a particular pollutant in the water supply to decline a rate of 8% per week. What is the half-life of the concentration of the pollutant? When will it fall to 1% of its original concentration?

10. **Using the Formulas.** Answer each of the following questions; as needed, use the approximate doubling time and half-life formulas (remember that these formulas work best for small growth and decay rates). In each case state whether your answer is exact, a good approximation, or a poor approximation.
    a. If a quantity doubles every 4 days, what is its doubling time?
    b. If a quantity increases four-fold every 4 days, what is its doubling time?
    c. If a quantity is halved every 4 days, what is its half-life?
    d. If a quantity increases by 5% per year, what is its doubling time?
    e. If a quantity decreases by 5% per year, what is its half-life?
    f. If a quantity declines by 5% per hour, what is its half-life?
    g. If a quantity falls by 3% per week, what is its half-life?
    h. If a quantity has a doubling time of 300 days, what is its growth rate?
    i. If a quantity has a doubling time of 20 years, what is its growth rate?
    j. If a quantity has a half-life of 200 years, what is its decay rate?
    k. If a quantity rises by 25% per year, what is its doubling time?
    l. If a quantity drops by 18% per hour, what is its half-life?

11. **Working with Growth Rates.** Answer each of the following statements true or false. Explain your reasoning.
    a. If inflation is causing prices to rise at a rate of 10% per year, items that cost $1000 last year now cost $1100.
    b. If a country with a population of 100 million people is growing at a rate of 3% per year, its population grew by 3 million people during the past year.
    c. A growth rate of 12% per year is equivalent to a growth rate of 1% per month.
    d. If hyperinflation is driving up prices at a rate of 100% per month, items that cost $100 two months ago now cost $400.
    e. If a bank account increases its value at a rate of 1% per month, an initial deposit of $100 will increase to $112 in a year.

12. **Exponential Decay.**
    a. Poaching is causing a population of elephants to decline by 7% per year. When will the population be half its present size?
    b. In each of the past several years, the production of a particular gold mine has declined by about 10%. When will the mine be producing only half as much gold ore as it is at present?

13. **Exact Formula for the Doubling Time.** Rework each of the parts of Problem 8 by using the exact formula for the doubling time. In each case, compare the answers obtained from the exact and approximate formulas.

14. **Exact Formula for the Half-Life.** Rework each of the parts of Problem 9 by using the exact formula for the half-life. In each case, compare the answers obtained from the exact and approximate formulas.

Section 15.4

15. **Linear Versus Exponential Monetary Growth.**
    a. Suppose that you decide to save money by putting it under your mattress and that each year you save \$100. Make a table that shows how much money you have saved in each of the first 20 years and graph the linear relation (*time, money saved*). How much will you have at the end of 20 years?
    b. Suppose that you place the \$100 in an account that increases its value by 8% per year. Make a table that shows how much money you have saved in each of the first 20 years and graph the exponential relation (*time, money saved*). How much will you have at the end of 20 years?
    c. In words, contrast the results of the linear growth of your savings under the mattress and the exponential growth involved in the savings account. How much of your own money (not counting investment interest) did you contribute in each case?

16. **Using the General Exponential Growth Law.** In each of the following problems, apply the general exponential growth law to answer each of the questions. Make rough sketches of the exponential relations.
    a. The population of a town with a current population of 85,000 is growing at a rate of 2.4% per year. How much does the population increase in 1 year? What is the population after 25 years of this growth?
    b. A mutual fund increases in value at a rate of 7.25% per year. If you deposit \$10,000 initially, how much will you have after 2 years? after 15 years?
    c. The crime rate in a town is rising at a rate of 3% per year. If 1500 crimes were committed in 1995, how many will be committed in 2000? 2010?
    d. A forest is being clear cut at a rate of 7% per year. If the forest currently occupies an area of 1 million acres, what area will remain in 3 years?
    e. A lack of employment is causing people to leave a town, reducing its population at a rate of 0.3% per month. If the population of the town is initially 10,000, what will be its population after 1 year? When will its population reach 6000?
    f. Inflation is causing prices to rise by 0.7% per month. If a basket of groceries costs \$125 now, how much would you expect it to cost in 1 year? in 10 years?
    g. Suppose that the value of the peso is falling against the value of the dollar at a rate of 10% per week. If a peso is currently worth 25¢, what will it be worth in 4 weeks? When will its value fall to 1¢?
    h. A particular drug breaks down in the human body at a rate of 15% per hour. If the initial concentration of the drug in the bloodstream is 8 milligrams per liter, what is the concentration after 8 hours? When does the concentration fall below 1 microgram per liter?
    i. Suppose that your starting salary at a new job is \$2000 per month. Each year, your employer gives you a raise that averages 5%. How much will you be earning (monthly salary) at the end of 7 years?
    j. Suppose that you hid 100,000 rubles in a mattress at the end of 1991. Suppose further that inflation causes the value of rubles to decline by 15% per month (against the dollar). If the rubles were worth \$10,000 when you hid them under the mattress, how much are they worth now?

17. **Valium Metabolism.** The drug valium is eliminated from the bloodstream exponentially with a half-life of 36 hours. Suppose that a patient receives an initial dose of 20 milligrams of valium at midnight.
    a. Determine the decay rate, *r*.
    b. What is the valium concentration at noon the next day?
    c. When will the valium concentration reach 10% of its initial level?

18. **Radioactive Decay and Dating.** The half-life of carbon–14 is about 5700 years, and the half-life of potassium–40 (a radioactive form of potassium) is about 1.3 billion years. Answer each of the following questions.
    a. Suppose that you find a piece of cloth painted with organic dyes. By analyzing the dye in the cloth, you find that only 77% of the carbon–14 originally in the dye remains. When was the cloth painted?
    b. A well-preserved piece of firewood has 6.2% of the carbon–14 of a live tree. Estimate when the wood was cut.
    c. When potassium–40 decays, it turns into argon–40, which is stable. Suppose that you find a meteorite that contains a mixture of potassium–40 and argon–40: 20% of the mix is potassium–40, and 80% is argon–40. If all the argon–40 originally was potassium–40, how long ago did this meteorite form?
    d. Is carbon–14 very useful for establishing the age of the Earth? Why or why not? Would potassium–40 dating work better? Explain.

19. **Practice Creating Exponential Models.** Create an exponential model for each of the following situations. Then discuss whether the exponential model provides a good description of the growth process.
    a. Suppose that the value of your stock market portfolio was \$300 in 1990 and \$450 in 1995. Create an exponential model for the growth in value of your portfolio, and use it to predict your portfolio's value in 2000.
    b. The number of murders in a large city was 160 in 1970 and 475 in 1990. Create an exponential model and use it to predict the number of murders in 1995.
    c. The cost of a particular computer was \$8000 when it was purchased in January 1990 and it is worth \$600 today. Create an exponential model for its decline in value and use the model to predict the computer's value in June 1995. What will be its value 1 year from now? When will its value fall to \$100?
    d. Suppose that you have 100 kg of a radioactive substance. You observe that after 6 months 88 kg of the substance remain (12 kg have decayed). Create an exponential model for the decay of the substance. How much will remain after 18 months?

e. Suppose that a basket of groceries that cost \$100 in 1985 costs \$135 in 1993. Create an exponential model for the increase in price and use it to predict the current price of these groceries.

Further Topics and Applications

20. **What Can Your Descendants Do with Your Dollar?** Suppose that you invest \$1 in a bank account that increases its value annually by 7%. Amazingly, the bank is able to hold its interest rate constant for a very long time. In your will, you leave this bank account to your great-great-great-great-great-great-great-great-great-great-great-great-great-great grandchildren, who ought to be your present age in about 400 years.
    a. How much money will be in the account for them 400 years from now?
    b. Unfortunately, inflation has depreciated the value of the dollar by a constant rate of 5% per year. In 400 years, how much money or "buying power" will the \$1 have?

21. **Modeling World Population Growth.** The human population was 3.9 billion in 1970. In 1991, it was 5.5 billion. Create an exponential model for the growth of the human population over this period. Does the model accurately predict today's population? If not, does it underestimate or overestimate the population? Has the rate of population growth increased or decreased? Explain.

22. **Postage Inflation.** In 1958, the cost of a first-class stamp was 4¢. Using the current cost of a first-class stamp, create an exponential model of the increase in first class postage prices. Use the model to "predict" the price of a stamp in 1992; does the prediction match the actual 1992 postage rate of 29¢? Use the model to estimate the cost of a first-class stamp 5 years and 10 years from now.

23. **Movie Prices.** In 1960 the typical price of admission at a movie theater was 75¢. Use the current prices for first-run theaters in your area to create an exponential model of rising movie prices. According to your model, how much will a movie ticket cost in 2000? When will the price of theater admission reach \$15? Do you believe that this model is accurate? Explain.

24. **Compounding in Bank Accounts.** Most banks advertise an annual percentage rate (APR) but actually add interest to accounts more frequently than once a year. Suppose that a bank offers an APR of 12% but actually compounds interest monthly.
    a. For an APR of 12%, you may assume that the monthly rate is $12\% \div 12 = 1\%$. How much will an initial deposit be worth a year later if interest is compounded monthly at a rate of 1%? (Hint: Use the general exponential growth law with $r = 1\%$ per month.)
    b. Is the increase in the account balance less than, equal to, or more than 12%? Why? (The amount the account actually increased is called the **annual yield**. See Chapter 16 for further details.)

25. **Growth Control Mediation.** A city with a 1990 population of 100,000 has a growth control policy that limits the increase in residents to 2% per year. Naturally, this policy causes a great deal of dispute. On one side, some people argue that *growth* costs the city its small-town charm and clean environment. On the other side, some people argue that *growth control* costs the city jobs and drives up housing prices. Finding their work limited by the policy, developers suggest a compromise of raising the allowed growth rate to 5% per year. Contrast the population of this city in 2000, 2010, and 2050 for 2% annual growth and 5% annual growth. If you were asked to mediate the dispute between growth control advocates and opponents, explain the strategy you would use.

26. **Discovering Radioactive Waste.** A toxic radioactive substance with a density of 2 milligrams per square centimeter is detected in the ventilating ducts of a nuclear processing building that was used 45 years ago. If the half-life of the substance is 20 years, what was the density of the substance when it was deposited 45 years ago?

27. **Population Growth in Your Lifetime.** The world population in 1990 was about 5.5 billion. Assume that population currently is growing, and will continue to grow, by about 1.6% per year. What is the doubling time? What will be the world population when you are 25 years old? 50 years old? 80 years old? 100 years old? Discuss the challenges or benefits of population growth that might be expected over your lifetime.

28. **Pesticide Decay and Your Cat.** The concentration of a particular pesticide that is applied to lawns declines exponentially with a half-life of 1 week. Suppose that the pesticide is applied with an initial concentration of 10 grams per square meter of lawn.
    a. When does the concentration reach 1 gram per square meter?
    b. Suppose that your cat eats about 5 square centimeters of grass each day. On the day the pesticide is applied, how much pesticide, in grams, would the cat consume?
    c. Suppose that the pesticide is toxic to cats if consumed in amounts greater than 100 milligrams per day. For how many days should you keep your cat in the house after the pesticide is applied to ensure the cat's safety?

29. **Dollars and Yen.** At the beginning of 1989, the dollar was worth about 135 Japanese yen. At the beginning of 1995, the dollar was worth about 97 yen. Create both a linear and an exponential model for the decline in value of the dollar against the yen. What does the linear model predict for the present value of the dollar in yen? What does the exponential model predict? Check the newspaper to find the actual exchange rate at present. Which model yields a better prediction? Based on your findings, discuss the wisdom of using some of your money to purchase Japanese yen and then investing it in Japanese bank accounts or stocks.

30. **The Future of Mexico City.** According to one recent estimate, the population of Mexico City in 1990 was 20 million. In 1960 the population was about 5 million.
    a. Create an exponential growth model for the population of Mexico City, and use it to predict the population in 2000.

b. The 1990 population of all of Mexico was about 90 million, with a growth rate of about 2% per year. Suppose that Mexico's population continues to grow at 2% per year, and that Mexico City continues to grow according to the model you developed in part (a). According to these simplistic assumptions, when would the entire population of Mexico live in Mexico City?

c. Explain what is wrong with the assumptions in part (b). Suggest a more realistic scenario for the growth of the population of Mexico and Mexico City. Be sure to discuss reasons for Mexico City's rapid growth and factors that might cause that growth to slow (or even reverse).

31. **Consumer Price Index.** The consumer price index (CPI) is an aggregate measure of prices in the United States (see subsection 8.4.4). Use the CPI data in the following table to make a case for or against the claim that the cost of living (as measured by the CPI) increased exponentially between 1970 and 1990.

If you argue in favor of this claim, find the growth rate for the CPI (which must be constant if the growth is exponential) and find a relation that fits the CPI data as closely as possible. What do you predict the CPI will be in 2000? If you argue against the claim, give your reasons.

U.S. Consumer Price Index (CPI)

Year	CPI
1970	38.3
1975	53.8
1980	82.4
1985	107.6
1987	113.6
1988	118.3
1989	124.0
1990	130.7

32. **Rethinking Sedentary Human History.** Digging at an archaeological site near the Tigris river, archaeologists recently discovered a wooden implement whose appearance suggested that it was used like a farming hoe. After dating the tool by its carbon–14 content, the archaeologists were shocked. The implement appeared to be 20,000 years old! Prior to this discovery, archaeologists had thought that agriculture developed only in the past 10,000 years.

a. The half-life of carbon–14 is 5700 years. What is its nominal decay rate per year?

b. Suppose that the tool originally contained 5.0 milligrams of carbon–14 and it actually is 20,000 years old, how much carbon–14 should it contain now?

c. You learn that the sample now contains 1.7 milligrams of carbon–14. What age does this suggest for the tool?

d. Based on this discovery, should history books now be revised to say that agriculture developed 20,000 years ago? Why or why not? Explain.

33. **The Meaning of Exponential Decay.** Suppose that the population of mountain lions near an urban area declines exponentially at a rate of –4% per year. The exponential decay of the mountain lion population implies that the total number of mountain lions lost each year gets smaller over time. For example, for 1000 mountain lions initially, a 4% loss in the first year drops the population by 40 to 960. Some time later, when the population has fallen to 500, a 4% loss drops the population by 4% of 500, or only about 20 lions. Why is the decline of a population of mountain lions, caused by development, exponential, rather than, say, a fixed number of lions per year? (Hint: development is an indirect cause of the population decline — the lions aren't killed by the construction; consider the direct causes of the decline in the lion population spawned by the development.)

34. **Increasing Atmospheric Carbon Dioxide.** Between 1860 and 1990, carbon dioxide (CO_2) concentration in the atmosphere rose from roughly 290 parts per million to 350 parts per million.

a. Use the 1860 and 1990 values to create a linear model of the change in CO_2 concentration over the 130 year period. Use your model to predict when the concentration will be double its 1990 level.

b. Use the 1860 and 1990 values to create an exponential model of the change in CO_2 concentration over the 130 year period. Use this exponential model to predict when the concentration will be double its 1990 level, and compare this prediction to that in part (a).

c. Some scientists believe, based on climate research, that doubling the atmospheric CO_2 concentration would increase the global average temperature by 3°C. If the average global temperature was 15.5°C in 1990, what would the projected temperature be if the CO_2 concentration doubled?

Projects

35. **Carrying Capacity Estimates.** Do some research to obtain several different opinions concerning the Earth's human population carrying capacity. Based on your research, draw some conclusions about whether overpopulation presents an immediate threat. Write a short essay detailing the results of your research and clearly explaining your conclusions.

36. **U.S. Population Growth.** Research population growth in the United States to determine the relative proportions of the growth resulting from birth rates and from immigration. Then research both the problems and benefits of the growing U.S. population. Form your own opinions about whether the United States has a population problem. Write an essay covering the results of your research and stating and defending your opinions. If you believe that the United States has a population problem, discuss some of the implications for society and suggest policies that might be implemented to alleviate the situation. If you believe that the United States doesn't have a population problem, state how large a population the country could sustain and explain your reasoning.

37. **Resource Depletion.** Choose one important natural resource (e.g., oil, wood, fish, or aluminum). Do some research to determine the depletion of this natural resource over the past 100 years. In words, describe a

relation that expresses how the resource has been depleted over time. For example, is the depletion linear, exponential with a constant rate of decay, or exponential with a changing rate of decay? If the resource is nonrenewable, estimate when it will be completely exhausted and discuss alternative resources that might be used in its place. If the resource is renewable, discuss whether the resource currently is being used in a sustainable manner. If it isn't, discuss whether the depletion of the resource poses any problems for society and, if so, how further depletion might be prevented.

38. Radon in the Home. One of the leading causes of lung cancer in the United States is radon gas that accumulates in well-sealed homes. Radon-222 is a gas created as a natural decay product of uranium-238, which is present in the ground in much of the United States. Outside, even in areas with the highest uranium content, the concentration of the radon gas released into the air generally is too low to pose a health risk. In well-sealed homes, however, radon gas can leach into the house through the foundation, and then remain trapped indoors. As a result, the concentration of radon gas can rise to levels that pose a health risk. The U.S. Environmental Protection Agency (EPA) measures the concentration of radon gas in units of picocuries per cubic meter. (The curie is a unit used in measuring radioactivity that corresponds to the decay of 37 billion atoms per second; a picocurie is 10^{-12} curie.) According to the EPA, concentrations of radon gas above 4 picocuries per cubic meter are unsafe and steps should be taken to mitigate against the radon buildup.

a. The half-life of radon-222 is 3.8 days. Suppose that a sealed box has a radon gas concentration of 100 picocuries per cubic meter. If no gas is allowed in or out of the box, when will the concentration fall below the safe level of 4 picocuries per cubic meter?

b. For its half-life of 3.8 days, what is the decay rate of radon-222 per second?

c. The easiest way to mitigate radon buildup is to open doors and windows so that ventilation allows the radon to escape. Unfortunately, this solution can be very energy inefficient (if the house is being heated or air conditioned); it also may not help in basements that have little ventilation. Suggest several other methods of mitigating radon buildup. You might contact local builders, environmental consultants, or the EPA to find out about effective means of radon mitigation.

d. Investigate whether radon gas is a problem in your community. If so, determine whether most homes and apartments are adequately mitigating radon buildup. Research the most cost-effective methods of mitigation in your community.

e. (Challenge.) To determine the concentration of radon gas in a home you need to analyze three factors: the rate of inflow of radon gas to the home (through the foundation), the rate of outflow through ventilation, and the rate at which the radon decays radioactively. Discuss how a mathematical model could be created to account for all three factors. Why would such a model be useful?

39. Return from Near Extinction. After facing near extinction in the mid 1980s, the Kemp Ridley turtle has made a comeback. The following table shows the number of turtle nests at their primary nesting site in Mexico.

Year	Turtle Nests	Year	Turtle Nests	Year	Turtle Nests
1978	924	1984	798	1990	992
1979	954	1985	702	1991	1155
1980	868	1986	744	1992	1275
1981	897	1987	737	1993	1184
1982	750	1988	842	1994	1568
1983	746	1989	878		

a. On average, each female has 2.7 nests per year. How many females had nests in 1978, the first year of data; 1985, the low point; and in 1994, the high point?

b. Write a linear equation to model the decline in the number of nesting females between 1978 and 1985.

c. When does your linear model based on 1978 and 1985 predict that the turtles would have gone extinct?

d. Fortunately, the Ridley turtle made a dramatic comeback. Write an exponential equation to model the increase in number of nesting turtles between 1985 and 1994.

e. Researchers hope that the number of nesting turtles will reach 10,000 by 2010. What does your model predict for the year 2010? How does this number compare to the prediction of 10,000 nesting turtles?

f. If researchers had based their 2010 hope on an exponential model, what exponential rate would they have needed to use to get from the actual 1985 figure to their 2010 figure? How does this rate compare to the exponential rate you calculated based on the 1985 and the 1994 data? Why do you think they chose 10,000 as a target population?

40. Group Activity: Disease and Exponential Growth. The following group activity demonstrates the exponential spread of a disease transmitted through personal contact. A group of 30–50 people is ideal. Each person in the group has a die, a tally sheet, and a personal ID number (which could be determined by rolling dice). The activity consists of five 3-minute stages. During each stage, everyone mingles and interacts. Each time a pair of people interact they each record the other's ID number on the tally sheet and then roll their dice: If the total on the dice is 5 or less, they have been exposed and the ID number of the other person is circled; otherwise, the ID number is left uncircled. The action breaks at the end of each stage and then resumes.

At the end of five stages, one ID number is selected to represent an infected person. Everyone looks at his or her interactions from the first stage, and then anyone who interacted with the infected person *and* was exposed (total of 5 or less on the dice) becomes infected. The total number of infected people at the end of stage 1 is recorded.

The number of new infected people is recorded after the interactions of stages 2 through 5. Once infected a person remains infected.

a. Having carried out this activity, graph the six data points (initial number of infected people is one, and the five stages are the other data points).

b. Determine the applicable exponential growth law.

c. Repeat the activity with different conditions for exposure (e.g., the total on the dice is 6 or 4). How is this change likely to affect the outcome?

d. Can you incorporate the effect of public education programs into the activity? Can you incorporate the effect of a cure into the activity? Explain.

16 FAITH IN FORMULAS

A formula is a mathematical recipe for computing a particular quantity. You already have worked with many formulas — for linear relations, statistics, permutations and combinations, to mention a few. In this final chapter, we present a few useful formulas that you are likely to encounter in other classes and in daily life. They range from the complex formulas used in finance to the exotic formulas of Einstein's special theory of relativity. Although some of these formulas may look intimidating at first, you now have all the skills needed to use and interpret them. Your success in dealing with the formulas in this final chapter should give you the confidence you need for future mathematical work.

CHAPTER 16 PREVIEW:

16.1 ***FORMULA POWER.*** The introduction describes the power of mathematical formulas and emphasizes that you have already developed all the skills needed to work with *any* formula.

16.2 ***WORKING WITH FORMULAS.*** In this section we discuss a few general guidelines to help you in working with formulas and present several examples.

16.3 ***LOGARITHMIC SCALES.*** Among the most commonly used formulas are those that define logarithmic scales. In this section we describe the logarithmic scales used to measure sound (in decibels), the magnitude of earthquakes, and acidity (or pH) of solutions.

16.4 ***FINANCIAL FORMULAS.*** In this section we deal with compound interest, investment plans, and loan payments. If you are concerned about your financial future, this section is indispensable.

16.5 ***GRAVITY: FROM APPLES TO THE MOON AND BEYOND.*** For students interested in the physical sciences, this section offers a general introduction to gravity and formulas for calculating gravitational effects.

16.6* ***THE SPECIAL THEORY OF RELATIVITY.*** Einstein's special theory of relativity reveals that "common sense" ideas about space and time are wrong. Find out why in this section; remarkably, the formulas of special relativity involve nothing more than elementary algebra.

In science, the chosen way to paint a picture of reality is to build a model, often expressed in the compact language of mathematics. We try to encode our experiences of the real world into the symbols and rules of a mathematical formalism, and then use this formalism to generate predictions of what will transpire in the future.

— John L. Casti

It is one thing for the human mind to extract from the phenomena of nature the laws which it has itself put into them; it may be a far harder thing to extract laws over which it has no control. It is even possible that laws which have not their origin in the mind may be irrational, and we can never succeed in formulating them.

— Sir Arthur Stanley Eddington, 1920

16.1 FORMULA POWER

Formulas arise in countless situations and take endless forms. Many calculations on an income tax form (e.g., figuring depreciation of property or deductions) are simple formulas. If you hear a temperature given on the Celsius scale and want to know what it is on the Fahrenheit scale, you can use a conversion formula. A formula specifies the maximum size of parcels that the U.S. Postal Service will handle. Check your utility bill; it probably states a formula explaining how your charges were calculated. The same is true of most credit card statements.

Indeed, the importance of formulas in everyday life can't be overstated. As you continue your studies, you will undoubtedly encounter many more formulas. Even after you finish school, you are bound to come across many more formulas in this complex and highly quantitative society.

Fortunately, at this point in the book you have developed essentially all the tools needed for working with formulas of any kind. This concluding chapter provides an opportunity for you to hone your skills and build confidence that you now have the ability to make formulas work for you.

The specific examples offered in this chapter should be quite valuable. After a few general observations about working with formulas, we focus on the formulas that define the logarithmic scales commonly used to measure the loudness of sounds, the strength of earthquakes, and the acidity of chemicals. Then we introduce the practical formulas of finance, which you can use to calculate bank interest, loan payments, and investment values. Finally, we cover some important physical formulas that you are sure to encounter if you take any further natural sciences courses.

Time-Out to Think: Check a tax form and identify several examples of formulas that are designed to help you put the correct numbers on the appropriate lines. Also identify some formulas on utility bills or credit card statements.

16.2 WORKING WITH FORMULAS

Before we consider formulas for specific applications, a few general observations about working with formulas will be helpful. First, a formula is simply a mathematical recipe. Or, using the terminology introduced in Chapter 10, a formula expresses a *relation* between two (or more) quantities. Let's briefly review a few important characteristics of relations.

- When the graph of a relation is a straight line, the relation is said to be *linear*. Although linear relations are extremely useful and apply to many realistic situations, they are a special case (Figure 16–1).
- Any relation that is *not* linear, and therefore doesn't have a straight-line graph, is said to be *nonlinear*.
- A linear relation is easy to interpret because it involves a constant rate of change. The behavior of nonlinear relations can be much more complex. In some situations, nonlinear relations can lead to the phenomenon of *chaos*, described in Chapter 3.
- Any relation may be described in words, with tables, with graphs, and with equations. Always choose the method that is most useful in a particular situation.

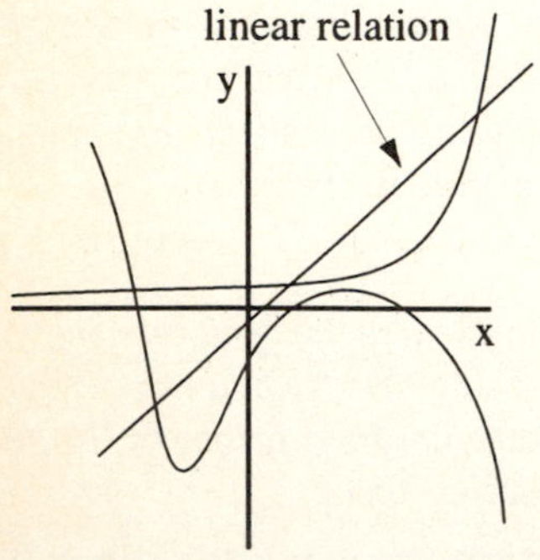

Figure 16-1. The graph of a linear relation is a straight line. Nonlinear relations have graphs that are curves and can be much more complex than linear relations.

In Chapter 10, we described *all* the important properties of linear relations. In subsequent chapters, we discussed properties of a few nonlinear relations, such as formulas for area, volume, and exponential growth. However, because the variety of nonlinear relations is infinite, we can't summarize the properties of all of them. Indeed, entire mathematics books are devoted to classifying and analyzing nonlinear relations.

Fortunately, you can make *practical* use of many nonlinear relations or formulas without years of mathematical study. Just as you can use a cookbook without creating the recipes yourself, you can use mathematical formulas developed by others. The only requirements are that you understand the meaning, purpose, and limitations of the formula and that you have the mathematical skills necessary to work with it.

16.2.1 Variables and Constants

Formulas often appear to be a jumble of different symbols. An important first step in putting a formula to use is making sense of these symbols. Keep in mind the following four general guidelines when you first look at any formula.

- Some symbols in formulas represent quantities that can change, or vary; they are called **variables**. Other symbols represent quantities that have specific numerical values that are fixed; they are called **constants**. For example, the formula for the circumference of a circle is $C = 2\pi r$. Because a circle can have any radius and the circumference varies with the radius, this formula has the two variables r and C. Both the 2 and the π in this formula are constants because they are fixed numbers.
- Many formulas contain more than two variables. For example, the formula for the area of a rectangle is $A = l \times w$, where l is the length and w is the width; the symbols A, l, and w all are variables because their values vary depending on the dimensions of the rectangle.
- When you graph or use a formula, usually you are interested in how *one* of the variables changes with respect to only one other variable. For example, you might want to know how the area of a rectangle varies with width, say, for a fixed length of 1 meter. In that case, the length l is a constant ($l = 1$ meter), and the width w and area A are the variables. Thus depending on your purpose, you may regard some variables as constants. Deciding which symbols in a formula will be variables and which will be constants is crucial.
- Unlike abstract mathematical formulas, the variables in most formulas rarely are written as x and y; other symbols, or even complete words, may be used. Choosing symbols that will help remind you of their meaning usually is a good idea.

The following two examples illustrate these four ideas. In addition, they offer a brief review of algebra and graphing.

Example 16–1 ***Time, Speed, and Distance.*** You are planning a 500-mile trip. Describe how the time required for the trip depends on your average speed.

Solution: Recall the formula relating distance, speed, and time:

$$\text{distance traveled} = (\text{average speed}) \times (\text{time elapsed}), \text{ or } d = v \times t.$$

You should first confirm that this formula makes sense by testing it with a simple example. At an average speed of $v = 50$ miles per hour for $t = 10$ hours, the formula indicates that you would travel, as you should expect,

$$d = 50\,\tfrac{\text{mi}}{\text{hr}} \times 10\text{ hr} = 500\text{ mi}.$$

Note that this formula has three variables: *distance, speed,* and *time* (or d, v, and t). In this problem, you are told to regard *distance* as a constant (500 miles); thus *speed* and *time* are the variables. Because you are asked how the time required depends on speed, you are looking for a relation of the form (*speed, time*), or (v, t), where speed is the first (independent) variable and time is the second (dependent) variable. Put the relation into this form by solving it for the time:

$$d = v \times t \xrightarrow{\text{divide both sides by v}} t = \frac{d}{v}.$$

As d is a constant in this problem, you may replace it by its numerical value, $d = 500$ miles:

$$t = \frac{500 \text{ miles}}{v}.$$

This equation describes how the time, t, varies with respect to the average speed, v, for the 500-mile trip. However, because the variable v appears in the denominator, this equation is nonlinear and its interpretation is not immediately obvious. A more meaningful way to describe this relation is with a graph, which you can sketch by using the point-by-point graphing method (see Appendix B). Begin by making a short table of points that gives values of t for various speeds v (you should use your calculator to confirm that these points satisfy the relation).

v (mi/hr)	5	10	20	30	40	50	60	70	80
t (hr)	100	50.00	25.00	16.67	12.50	10.00	8.33	7.14	6.25

A graph of the relation is shown in Figure 16–2. It shows that at faster speeds (larger values of v), the trip requires less time. At very slow speeds, the time required for the trip becomes quite large. For example, if you decided to make the trip on foot, with an average speed of 1 mile per hour, it would take t = (500 miles) ÷ (1 mi/hr) = 500 hours for the trip. If you traveled at only 0.1 mile per hour, it would take 5000 hours. If you traveled at a speed of 0 — that is, if you never left — you would never get there! Hence the time required for the trip approaches infinity as the speed approaches zero, which explains why the graph gets closer and closer to the horizontal axis (as the speed increases) without ever touching it. Similarly, as the speed approaches infinity, the time required to get to the destination approaches zero; thus, as the speed decreases, the graph gets closer and closer to the vertical axis without ever reaching it.

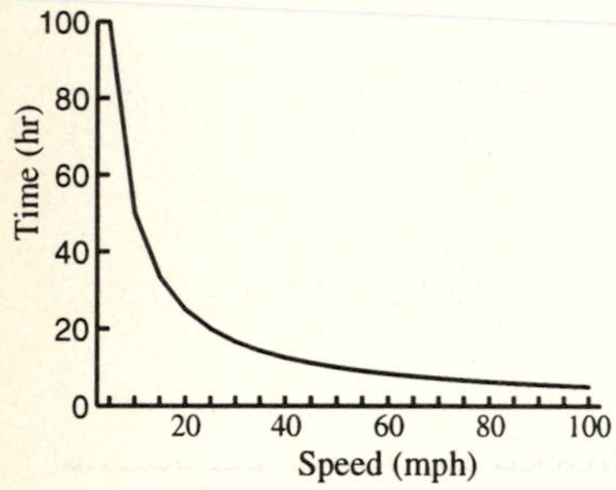

Figure 16-2. The graph shows how the time required for a 500-mile trip varies with respect to the average speed. That is, it is a graph of the relation t = (500 mi)/v.

❖

Time-Out to Think: Why didn't we consider negative values for either variable in the (*speed, time*) relation?

Example 16–2 ***The Speed of Light.*** Light is a wave (it is also a particle — see Chapter 5). The **frequency** of a particular light wave is defined as the number of waves that pass a fixed position during a 1-second interval, and the **wavelength** is the distance between two adjacent crests or troughs of the wave (Figure 16–3a). Together, all forms of light comprise the **electromagnetic spectrum** (Figure 16–3b). Although different forms of light have different frequencies and wavelengths, all light travels at the same speed. The letter c represents the speed of light, and $c = 3 \times 10^8$ m/s. A simple formula relates the wavelength, L, and the frequency, f, of a light wave to the speed of light, c:

$$\text{wavelength} \times \text{frequency} = \text{speed of light, or } L \times f = c.$$

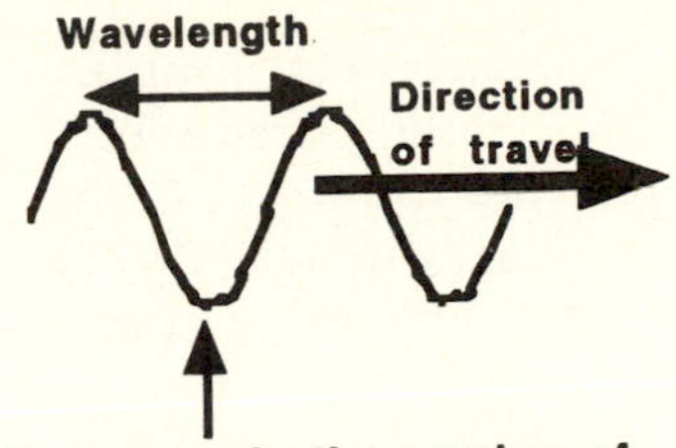

(a)

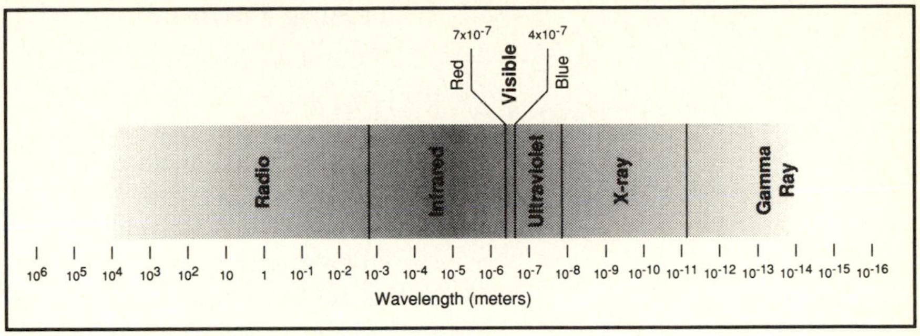

(b)

Figure 16-3. (a) The speed of light is the product of the wavelength and frequency of a light wave. (b) The complete electromagnetic spectrum, or spectrum of light, extends over all wavelengths. The commonly identified categories of light are shown with their wavelengths.

Find and interpret a relation that shows how the frequency of a light wave depends on its wavelength.

Solution: Because $c = 3 \times 10^8$ m/s is a constant, the formula $L \times f = c$ has only two variables. You are asked how the frequency depends on wavelength, so solve the formula for the frequency by dividing both sides by L:

$$f = \frac{c}{L} = \frac{3 \times 10^8 \ \text{m/s}}{L}.$$

The formula now has the form of a relation (L, f) that describes how the frequency varies with respect to wavelength. Note that this relation is similar in form to the one in Example 16–1. Hence the interpretation is similar: As the wavelength gets larger, the frequency gets smaller and vice versa. That is, light with long wavelengths has relatively low frequency, and light with short wavelength has relatively high frequency.

Note that the units of frequency are 1/s, or "per second." For example, the wavelength of light that looks *green* is approximately 5×10^{-7} meter, or 500 nanometers; therefore its frequency is

$$\text{the frequency of green light:} \quad f = \frac{3 \times 10^8 \ \text{m/s}}{L} = \frac{3 \times 10^8 \ \text{m/s}}{5 \times 10^{-7} \ \text{m}} = 6 \times 10^{14} \ \frac{1}{\text{s}}.$$

Because saying "per second" without saying something else first sounds a bit strange, the units of frequency usually are expressed as "waves per second" or "cycles per second"; alternatively, these units are called **hertz**. Thus the frequency of green light is about 600 trillion cycles per second, or 600 trillion hertz. Figure 16–4 shows the graph for wavelengths of light up to 1 millimeter (0.001 meter). For clarity, the vertical axis shows frequencies only up to 3 trillion (3000 billion) cycles per second. Note that the data point for green light, with a wavelength of 500 nanometers and frequency of 600 trillion hertz, would lie close to the vertical axis and far above the top of the graph. ❖

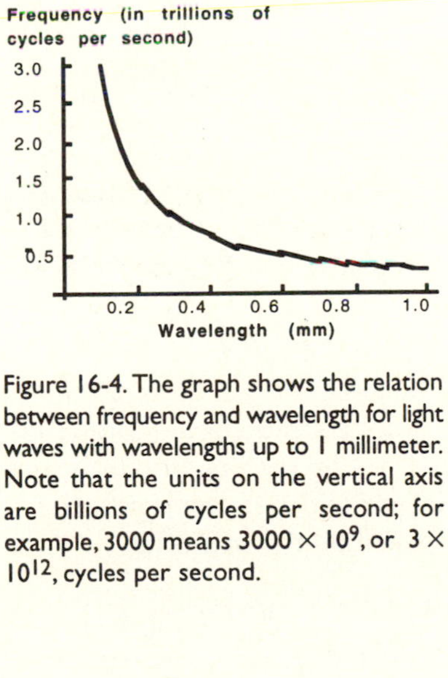

Figure 16-4. The graph shows the relation between frequency and wavelength for light waves with wavelengths up to 1 millimeter. Note that the units on the vertical axis are billions of cycles per second; for example, 3000 means 3000×10^9, or 3×10^{12}, cycles per second.

Time-Out to Think: The frequency of your favorite FM radio station is given in megahertz (MHz), or millions of cycles per second. For example, a radio station found at "97.3 on your dial" broadcasts radio waves with a frequency of 97.3 MHz, or 97.3 million cycles per second. Calculate the wavelength of the radio waves from your favorite radio station. Where would this wavelength appear on the graph in Figure 16–3(b)?

16.2.2 Creating Formulas

Where do formulas come from? In Example 16–1, the relatively simple formula relating distance, speed, and time probably is familiar from everyday life, as well as from your work earlier in this book. In Example 16–2, the formula that relates frequency and wavelength to the speed of light was developed through careful scientific work. Thus you work with many formulas that either are familiar or simply given to you.

Alternatively, with the skills and experience that you have acquired through this book, you can devise formulas for many situations that you encounter. The next two examples illustrate this process.

Example 16–3 *Landscaping.* A swimming pool is to be built in a yard that measures 10 meters by 15 meters. The landscape architect has determined that, for aesthetic reasons, the space between the pool and the edge of the yard should form a border of uniform width (Figure 16–5a). Describe how the area of the pool depends on the width chosen for the border.

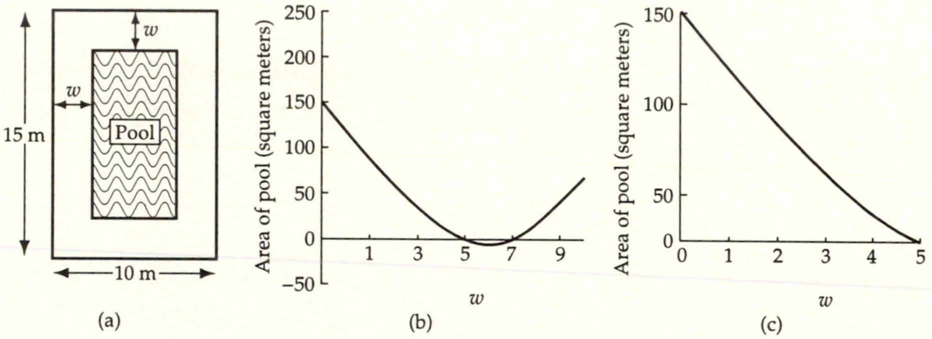

Figure 16-5. (a) A landscape architect has requested a border of uniform width, w, around a swimming pool to be built in a yard measuring 10 meters by 15 meters. (b) The graph shows the relation between the area and the border width, $A = 4w^2 - 50w + 150$. (c) Redrawn, the graph shows only the sensible values of w.

Solution: The area of the rectangular pool is its length multiplied by its width. Figure 16–5(a) shows that the length of the pool is the length of the yard (15 meters) *minus* the width of the border on either side. If w represents the width of the border, the length of the pool is $15 - 2w$ meters. Similarly, the width of the pool is the width of the yard minus the width of the border, or $10 - 2w$ meters. Thus the area, A, of the pool is given by the formula:

$$\text{area of pool}: \quad A = (15-2w)(10-2w) = 4w^2 - 50w + 150.$$

This formula is a relation describing how the area, A, depends on the border width, w. Figure 16–5b shows a graph of this relation. Note that only nonnegative values of w are shown, because a border with negative width makes no sense. Note, however, that $A = 0$ when $w = 5$ meters, and then the graph shows negative areas before turning upward. Why? If w is 5 meters, the total amount of border on either side of the pool is 10 meters; because the yard is only 10 meters wide this leaves no room for the pool! Thus values of w greater than 5 meters make no sense either and shouldn't be shown. Figure 16–5c shows the final graph, with only sensible values of w. Not surprisingly, the graph indicates that the pool has a maximum area when the yard is all pool ($w = 0$) and a minimum area of 0 when the yard is all border ($w = 5$).

> **Time-Out to Think:** Confirm the algebraic expansion of the expression $(15 - 2w)(10 - 2w)$ to $4w^2 - 50w + 150$. What is the area of the pool if the border is 2 meters wide? What are the dimensions of the pool in that case?

❖

Example 16–4 *Pricing Merchandise.* A marketing survey shows that an auto supply store can sell 25 sets of windshield wipers a day if they are priced at \$3 apiece and 20 sets a day if they are priced at \$4 apiece. For a linear price–demand relation, create a formula that gives total sales revenue as it varies with the price of the wipers. What is the optimal price for maximizing revenue?

Solution: Begin by identifying the variables in the problem. One variable is the price because it varied in the marketing survey; call it p. Price determines the number of wipers that can be sold in a day (in economics that is called *demand*), so the number of wipers sold also is a variable; call it n. Finally, the total revenue from sales depends on both the price and the number of wipers sold, so it is a third variable; call it R.

The total sales revenue is the number of wipers sold multiplied by their price, or

$$R = n \times p.$$

To emphasize the work with variables, we do not show the units in this example. Keep in mind that the constants –5 and 40 *do* have units: –5 *sets of wipers per dollar* and 40 *sets of wipers*. Note also that the units work out properly: Because p has units of dollars, $-5p$ has units of sets of wipers, as does n.

The problem also states that the number of wipers sold depends linearly on the price: At lower prices larger numbers of wipers are sold. Find this price–demand relation by creating a linear model (see subsection 10.5.1), based on the two data points from the marketing survey. The formula for the relation (*price, number of wipers*), or (p, n), has the general form:

$$n = \text{rate of change} \times p + \text{initial value}.$$

The data points given in the problem, in the form (p, n), are (\$3, 25 wipers) and (\$4, 20 wipers). Use the methods described in subsection 10.5.1 to find the rate of change (or slope of graph) of the linear relation (–5 sets per dollar) and the initial value of n (40 sets of wipers). Thus the price–demand relation is:

$$n = -5p + 40.$$

This linear relation is graphed in Figure 16–6(a). Note that prices above \$8 per set of wipers result in a demand less than zero; that is, no sales occur because prices are too high. The price–demand relation gives n in terms of p, so you can substitute this expression for n into the revenue formula:

$$R = np = (-5p + 40)p = -5p^2 + 40p.$$

This formula allows you to calculate the revenue for *any* price. A graph of this formula, showing only the possible prices between \$0 and \$8, is shown in Figure 16–6(b). Note that, as expected, a price of \$0 leads to no revenue — the store would be giving the wipers away! But a price of \$8 (or more) also yields no revenue because at that price no one buys any wipers. Between those extremes the graph reveals that a price of \$4 gives the maximum revenue. Thus the optimal price for the wipers, in terms of maximizing revenue, is \$4 per set of wipers.

❖

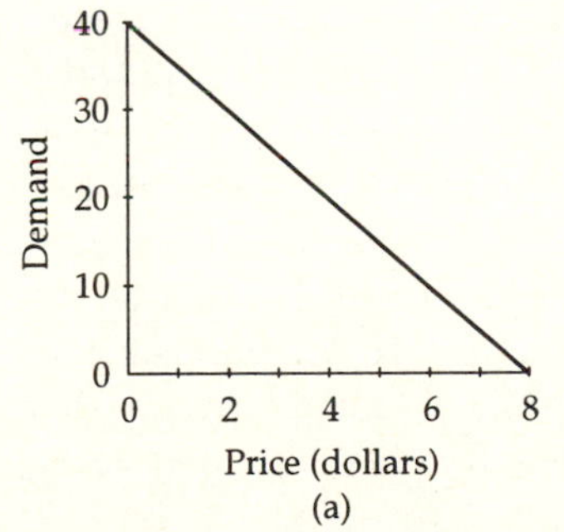

(a)

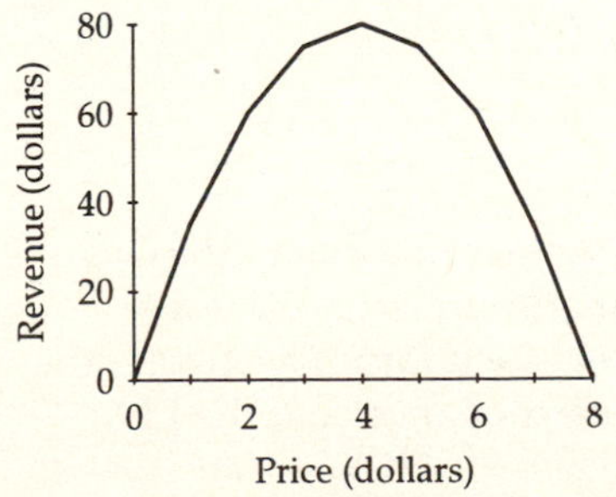

(b)

Figure 16-6. (a) The price–demand relation in Example 16–3 is linear, with the form $n = -5p + 40$. (b) The price–revenue relation is nonlinear. A price of \$4 maximizes revenue.

> **Time-Out to Think:** As discussed in Chapter 10, a linear model for the price–demand relation may not be completely realistic; for example, if the wipers are given away, demand probably would be far more than just 40 sets per day. In words, describe a more realistic model for the price–demand relation. Then describe how this model would affect the form of the price–revenue relation.

16.2.3 Faith in Formulas

Like using a recipe from a cookbook, you often will rely on formulas without really knowing much about how they were created or why they work. Nevertheless, you shouldn't simply accept formulas with blind faith. Just as you can test a recipe by tasting the result, you can test a formula by trying it out. In particular, before using a formula more generally, you should always check the following.

- Does the relation described by the formula make sense if it is described in words or with a picture?
- Does the formula yield the proper units?
- Does testing a few sample values in the formula, especially at extreme values, give reasonable results?

If the formula fails any of these tests, something probably is wrong with it! If it passes these tests and comes from a reputable source, you probably can feel comfortable using the formula even without knowing the details of its origin or derivation.

Finally, remember that formulas are mathematical *models* of something real. The validity of a model is only as good as the data and assumptions from which it is built. Thus the most important step in using a formula is to understand what is being modeled, and how, so that you can properly interpret our results.

Example 16–5 *A Misprint?* Suppose that your marketing textbook tells you that the total sales revenue, R, for automobiles depends on the price, p, according to the formula $R = 2000p + 300p^2$. Do you trust this formula?

Solution: The formula comes from a marketing textbook, which should be a reputable source. However, because both terms in the formula are positive, as the price per car rises, revenue rises — no matter how high the price! Clearly, this result makes no sense; if it were true, automobile dealers could increase their revenue simply by raising prices, without regard to customer demand. Something must be wrong with this formula.

If you compare this formula to the price–revenue relation in Example 16–4, you might guess that the book contains a simple misprint: Perhaps the formula should be $R = 2000p - 300p^2$. Even if you later learn that this revised formula is correct, be careful how you use it. As Example 16–4 demonstrated, this model is based on the assumption of a linear price–demand relation; that might be reasonable for some values of the price, but it breaks down at very low and very high prices.

❖

Table 16–1.

y	$\log_{10} y$
0.000001	–6
0.00001	–5
0.0001	–4
0.001	–3
0.01	–2
0.1	–1
1	0
10	1
100	2
1000	3
10,000	4
100,000	5
1,000,000	6

16.3 LOGARITHMIC SCALES

When a quantity varies over a wide range of values, working with the *logarithm* of the quantity often is easier. As a result, many common formulas involve logarithms; in this section we present three such formulas: define the decibel scale for sound, the magnitude scale for earthquakes, and the pH scale for acidity.

Before we consider these formulas, let's briefly review logarithms; for a more detailed review, consult Appendix C. The expression $\log_{10} y$ can be thought of as "the power to which 10 must be raised to obtain y." For example, $\log_{10} 1000 = 3$ because $10^3 = 1000$. Similarly, $\log_{10} 10{,}000{,}000 = 7$ because

$10^7 = 10{,}000{,}000$, and $\log_{10} 0.1 = -1$ because $10^{-1} = 1/10$. Table 16–1 shows $\log_{10} y$ as y ranges from 1 millionth to 1 million. Note that, in general,

$$\log_{10} 10^x = x.$$

This expression is valid because $\log_{10} 10^x$ means "the power to which 10 must be raised to obtain 10^x"; clearly, this power simply is x. Finding the logarithm of any power of 10 therefore is easy. However, a calculator is indispensable for finding logarithms of other numbers.

> **Time-Out to Think:** Use your calculator to verify each of the following: $\log_{10} 3 = 0.477$, $\log_{10} 700 = 2.845$, $\log_{10} 1245 = 3.095$, and $\log_{10} 0.06 = -1.222$. Then verify that $10^{0.477} = 3$, $10^{2.845} = 700$, $10^{3.095} = 1245$, and $10^{-1.222} = 0.06$. Do you see how logarithms and exponents are related? Explain.

Working with logarithms in formulas can *look* quite complex. However, only three simple rules about logarithms and exponents (again, see Appendix C for a more detailed review) are used in this chapter.

1. You can take the logarithm of both sides of an equation (as long as both sides are positive). This rule is useful for "bringing down" exponents. For example, suppose that you have the equation $10^x = 50$ and want to solve for x. "Bring down" the x by taking the logarithm of both sides:

Starting equation:	$10^x = 50$
Take base-10 log of both sides:	$\log_{10} 10^x = \log_{10} 50$
Apply property that $\log_{10} 10^x = x$:	$x = \log_{10} 50 = 1.7.$

2. You can make both sides of an equation an exponent of 10. This rule is useful when you need to remove a logarithm. For example, suppose that you have the equation $\log_{10} x = 2.7$ and want to solve for x. Make both sides powers of 10 and solve as follows:

Starting equation:	$\log_{10} x = 2.7$
Make both sides powers of 10:	$10^{\log_{10} x} = 10^{2.7}$
Apply property that $10^{\log_{10} x} = x$:	$x = 10^{2.7} = 501.2.$

3. To "bring down" an exponent within a logarithm, you can apply the power rule for logarithms:

   $$\log_{10} a^x = x \log_{10} a.$$

   For example, $\log_{10} 5^2 = 2 \log_{10} 5$.

Alexander Graham Bell was born in Edinburgh, Scotland, in 1847. His grandfather had studied the mechanics of sound, his mother was deaf, and his father was a pioneer in teaching speech to the deaf. Alexander, too, became interested in speech, hearing, music, and public speaking. His family moved to Ontario in 1870 and, shortly thereafter, Bell moved to the United States where he became a professor of vocal physiology at Boston University. There, he married a deaf pupil and continued his work on speech and hearing. His most famous invention, the telephone, was patented in 1876. The first message ever transmitted by telephone was Bell calling his assistant to say, "Mr. Watson, come here, I want you."

16.3.1 The Decibel Scale for Sound

The loudness (or relative intensity) of sound usually is measured in units called **decibels**, abbreviated dB. As its name implies, a decibel, is one tenth of a **bel**, a unit named for telephone inventor Alexander Graham Bell. The decibel scale is defined so that a sound of 0 dB represents the softest audible sound (to the human ear). The loudness of any other sound in decibels can then be found from the following formula:

$$\text{loudness in dB} = 10 \log_{10}\left(\frac{\text{intensity of the sound}}{\text{intensity of softest audible sound}}\right).$$

What does this formula mean? The term in parentheses compares the intensity of the sound you want to measure to the intensity of the softest audible sound; that is, it indicates *how many times* louder the sound is than the softest audible sound. Thus, the formula yields the loudness in decibels if you know the *ratio* of the intensity of the sound to the intensity of the softest audible sound.

Recall that $\log_{10} 1 = 0$ because $10^0 = 1$ (see Table 16–1).

One simple check to be sure that the formula makes sense is to confirm that it gives 0 dB for the softest audible sound. If you substitute "intensity of softest audible sound" for the term "intensity of sound" in the numerator, the formula yields

$$\text{loudness in dB} = 10\log_{10}\left(\frac{\text{intensity of the softest audible sound}}{\text{intensity of the softest audible sound}}\right)$$
$$= 10\log_{10}(1) = 10 \times 0 = 0 \text{ dB}.$$

This result should give you at least some confidence in the formula.

Suppose that you know the loudness of a particular sound in decibels. How many times louder is it than the softest audible sound? Answering this question requires solving the decibel formula for the intensity ratio, which requires applying the rules of logarithms as follows.

Start with decibel formula: $\text{loudness in dB} = 10\log_{10}\left(\frac{\text{intensity of the sound}}{\text{intensity of softest audible sound}}\right)$

Divide both sides by 10: $\frac{\text{loudness in dB}}{10} = \log_{10}\left(\frac{\text{intensity of the sound}}{\text{intensity of softest audible sound}}\right)$

Make both sides exponents of 10: $10^{\frac{\text{loudness in dB}}{10}} = 10^{\log_{10}\left(\frac{\text{intensity of the sound}}{\text{intensity of softest audible sound}}\right)}$

Apply property that $10^{\log_{10} x} = x$, and switch left and right sides: $\frac{\text{intensity of the sound}}{\text{intensity of softest audible sound}} = 10^{\frac{\text{loudness in dB}}{10}}.$

In this form, the decibel formula allows you to find the intensity ratio when you know the loudness of the sound in decibels.

To summarize, two equivalent ways of writing the decibel formula are

(1) $\text{loudness in dB} = 10\log_{10}\left(\frac{\text{intensity of the sound}}{\text{intensity of softest audible sound}}\right)$, and

(2) $\frac{\text{intensity of the sound}}{\text{intensity of softest audible sound}} = 10^{\frac{\text{loudness in dB}}{10}}.$

Recall that $\log_{10} 100 = 2$ because $10^2 = 100$.

Example 16–6 ***Computing Decibels.*** If a sound is 100 times as intense as the softest audible sound, what is its loudness in decibels?

Solution: The *ratio* of the sound intensity to the intensity of the softest possible sound is 100. Thus from formula (1) for the decibel scale, the loudness in decibels is

$$\text{loudness in dB} = 10\log_{10}\left(\frac{\text{intensity of the sound}}{\text{intensity of the softest audible sound}}\right)$$
$$= 10\log_{10}(100) = 10 \times 2 = 20 \text{ dB}.$$

A sound that is 100 times more intense than the softest possible sound has a loudness of 20 dB.

❖

Example 16–7 *How Much Louder?* How much louder is a sound of 80 decibels than the softest audible sound?

Solution: In this case, you are given the loudness of a sound in decibels and asked to find the ratio of its intensity to that of the softest audible sound. Use decibel formula (2), substituting 80 for "loudness in dB":

$$\frac{\text{intensity of the sound}}{\text{intensity of softest audible sound}} = 10^{\frac{(\text{loudness in dB})}{10}} = 10^{\frac{80}{10}} = 10^{8}.$$

The ratio of sound intensities is 10^8; that is, a sound of 80 dB is 10^8, or 100 million, times louder than the softest audible sound.

❖

Common Sounds

What humans perceive as sound actually is tiny pressure changes in the air; sound travels through the air as a wave of such pressure changes, called a **sound wave**. The pressure changes carry energy; thus, when a sound wave strikes the eardrum, its energy causes the ear drum to move in response. The brain analyzes the motions of the eardrum, perceiving sound.

Why did anyone ever decide to measure sound on a logarithmic scale? Because the human ear has a logarithmic response to sound; that is, the ear is sensitive to *relative* differences in sound intensity rather than absolute differences. As a result, the ear can distinguish relatively small differences in sound at low intensities, such as the small differences in the loudness of two whispers. At high intensities, however, sound perception is much coarser — two nearby jets will sound about the same unless the sound from one is many times the intensity of the sound from the other. Table 16–2 lists the approximate loudness of some common sounds.

Table 16–2: Typical Sounds in Decibels

Decibels	Times Louder than Softest Audible Sound	Example
140	10^{14}	jet at 30 meters
120	10^{12}	strong risk of damage for human ear
100	10^{10}	siren at 30 meters
90	10^{9}	threshold of pain for human ear
80	10^{8}	busy street traffic
60	10^{6}	ordinary conversation
40	10^{4}	background noise in average home
20	10^{2}	whisper
10	10^{1}	rustle of leaves
0	1	threshold of human hearing
–10	0.1	inaudible sound

Example 16–8 *Comparing Sounds.* How much more intense is the sound of a jet at 30 meters than a sound that poses a strong risk of damage to your ear?

Solution: According to Table 16–2, at a distance of 30 meters the sound from a jet is 140 dB, or 20 dB louder than the 120-dB sound that can cause damage to your ear. How much louder is the 140-dB sound than the 120-dB sound? Begin with decibel formula (2):

$$\frac{\text{intensity of the sound}}{\text{intensity of softest audible sound}} = 10^{\frac{(\text{loudness in dB})}{10}}.$$

Multiplying both sides by the intensity of the softest audible sound yields

$$\text{intensity of a sound} = 10^{\frac{(\text{sound in dB})}{10}} \times (\text{intensity of softest audible sound}).$$

To compare the two sounds simply divide their intensities:

$$\frac{\text{intensity of 140 dB sound}}{\text{intensity of 120 dB sound}} = \frac{10^{\frac{140}{10}} \times (\text{intensity of softest audible sound})}{10^{\frac{120}{10}} \times (\text{intensity of softest audible sound})}$$

$$= \frac{10^{14}}{10^{12}} = 10^{14-12} = 10^2 = 100.$$

Thus the 140-dB sound is 100 times as intense as the 120-dB sound.

❖

Example 16–9 ***Another Comparison.*** How much more intense is a 57-dB sound than a 23-dB sound?

Solution: Use the formula developed in the preceding example to compare the intensities of the two sounds:

$$\frac{\text{intensity of 57 dB sound}}{\text{intensity of 23 dB sound}} = \frac{10^{\frac{57}{10}} \times (\text{intensity of softest audible sound})}{10^{\frac{23}{10}} \times (\text{intensity of softest audible sound})}$$

$$= \frac{10^{5.7}}{10^{2.3}} = 10^{3.4} = 2512.$$

A sound of 57 dB is about 2500 times more intense than a sound of 23 dB.

❖

Measuring Sound Intensity

The decibel scale formula allows comparison of the *relative* intensities of two sounds. But how can the *absolute* intensity of a sound be measured? Because a sound wave carries energy, a device is needed that records the amount of energy it receives from the sound. Because the intensity of a sound can vary with time, it should record the amount of *energy received per unit of time,* or *power* (see Section 9.5). Moreover, a larger detector will receive more sound waves than a small one; thus a fair comparison of sound intensity requires measuring the power received *per unit of area.* Recall that the unit of power is watts; then the unit of intensity is:

$$\text{units of intensity: watts per square meter } \left(\frac{\text{watt}}{\text{m}^2}\right).$$

The intensity of the softest audible sound, 0 on the decibel scale, is about 10–12 watts per square meter.

The intensity of a sound decreases with distance from its source; for example, a nearby jet is much louder than one that is far away. In fact, sound intensity and distance are related by an **inverse square law.** That is, doubling the distance from the source causes the intensity to *drop* by a factor of $2^2 = 4$; tripling the distance from the source reduces the intensity by a factor of $3^2 = 9$; and increasing the distance by a factor of 10 reduces the intensity by a factor of $10^2 = 100$.

Time-Out to Think: From what you learned about scaling laws in Chapter 6, explain *why* sound intensity and distance are related by an inverse square law. (Hint: Imagine a short, sharp sound, like a bell ringing. The sound waves spread out in all directions from the bell. For any given distance, the sound waves are dispersed over a larger *area*. Recall how area scales with changes in linear size.)

Example 16–10 *Determining Sound Intensity.* What is the intensity, in watts per square meter, of a sound of 95 decibels?

Solution: First use the formula for the decibel scale to find the ratio of the intensity of the 95-dB sound to the intensity of the softest audible sound:

$$\frac{\text{intensity of the sound}}{\text{intensity of softest audible sound}} = 10^{\frac{\text{(loudness in dB)}}{10}} = 10^{\frac{95}{10}} = 10^{9.5} \approx 3 \times 10^{9}.$$

A sound of 95 dB is about 3 billion times more intense than the softest audible sound. Thus its intensity is

$$\text{intensity of the sound} = \left(3 \times 10^{9}\right) \times \left(\text{intensity of softest audible sound}\right).$$

Substitute the intensity of the softest audible sound, about 10^{-12} watts per square meter, to obtain

$$\text{intensity of the sound} = \left(3 \times 10^{9}\right) \times \left(10^{-12}\,\frac{\text{watt}}{\text{m}^2}\right) = 3 \times 10^{-3}\,\frac{\text{watt}}{\text{m}^2}.$$

A sound of 95 decibels has an intensity of about 0.003 watt per square meter. ❖

Example 16–11 *Sound Advice.* How far should you be from a jet to avoid a strong risk of damage to your ear?

Solution: According to Table 16–2, at a distance of 30 meters the sound from a jet is 20 dB louder, or 100 times more intense, than the 120-dB sound that can cause damage to your ear. Thus, to be safe, you should be far enough from the jet that its sound intensity is 100 times less than it is at 30 meters. As sound intensity and distance are related by an inverse square law, you must be 10 times farther away for the sound intensity to be reduced by 100 times; that is, you must be 300 meters from the jet. ❖

16.3.2 The Magnitude Scale for Earthquakes

The **magnitude scale** for earthquakes is another common example of a logarithmic scale. The modern definition of this scale is closely related to the **Richter scale**, which was originally created by geologist Charles Richter in 1935. At the time, earthquakes usually were recorded by machines called *seismometers* in which a pen was attached to a mechanical arm that would swing up and down during an earthquake. Richter defined his scale so that a magnitude of 0 represented the smallest quakes that could be registered by the seismographs of the time and so that an increase of 1 magnitude on the scale corresponded to an increase by a factor 10 in the distance that the mechanical arm would swing. Thus a *ten*fold increase in the intensity of the earthquake corresponds to an increase of 1 unit on the logarithmic Richter scale.

Geologists have since learned that earthquakes are much more complex than they thought in 1935. The energy released from earthquakes is transmitted through the ground by **seismic waves**, some of which travel along the surface of the Earth and others of which travel through its interior. The surface waves cause damage during earthquakes, making the land move up and down. Depending on how the energy of the earthquake is distributed between surface waves and interior waves — and on the type of bedrock in the vicinity of an earthquake — earthquakes of similar total intensity can cause vastly different amounts of damage and produce varying responses on seismometers.

> **Time-Out to Think:** Mexico City is built in an area where the underlying material is relatively soft (e.g., sand). The ground beneath Los Angeles is much harder rock. Explain why two otherwise identical earthquakes would likely cause more damage in Mexico City than in Los Angeles.

Today, geologists determine the magnitude of an earthquake by comparing measurements made by seismometers all over the world and correcting for factors, such as the depth of the earthquake, based on current knowledge about earthquake energy. The total energy, E, released by an earthquake is related to its magnitude, M, by the formula:

$$\log_{10} E = 4.4 + 15\text{M}, \text{ where } \begin{cases} E \text{ is the total energy released by the earthquake (in } \textit{joules}) \\ M \text{ is the magnitude of the earthquake.} \end{cases}$$

Because this formula gives the logarithm of the energy, you can use the rules of logarithms to find the actual energy, E:

Start with magnitude formula:	$\log_{10} E = 4.4 + 1.5M$
Make both sides exponents of 10:	$10^{\log_{10} E} = 10^{4.4+1.5M}$
Apply property that $10^{\log_{10} x} = x$:	$E = 10^{4.4+1.5M}$
Apply property that $10^{a+b} = 10^a \times 10^b$:	$E = 10^{4.4} \times 10^{1.5M}$
Simplify ($10^{4.4} = 2.5 \times 10^4$):	$E = (2.5 \times 10^4) \times 10^{1.5M}$.

The energy calculated by this formula is in units of joules.

Geologists classify earthquakes according to strength categories as shown in Table 16–3.

Table 16–3. Earthquake Categories and Their Frequency

Category	Magnitude	Approximate Number per Year (worldwide average since 1900)
Great	8 and up	1
Major	7–8	18
Strong	6–7	120
Moderate	5–6	800
Light	4–5	6000
Minor	3–4	50,000
Very minor	less than 3	magnitude 2–3: 1000 per day magnitude 1–2: 8000 per day

Table 16–4 lists some of the notable earthquakes of the twentieth century. Note that the intensity of the earthquake isn't always directly correlated with the number of casualties.

Time-Out to Think: Why do you think that there is so much uncertainty in the number of estimated deaths for major earthquakes such as those in China (1976) or Armenia (1988)? How reliable should you consider the numbers of estimated deaths for other earthquakes?

Table 16–4. Some Notable Earthquakes of the Twentieth Century

Date	Location	Magnitude	Estimated Deaths
April 1906	San Francisco	8.3	700
December 1908	Sicily	7.5	160,000
December 1920	Gansu, China	8.6	100,000
September 1923	Sagami Bay, Japan	8.3	100,000
February 1931	New Zealand	7.9	255
August 1950	Assam, India	8.4	30,000
February 1960	Morocco	5.8	12,000
March 1964	Alaska	8.5	114
February 1971	Los Angeles	6.6	51
December 1972	Nicaragua	6.2	10,000
July 1976	Tangshan, China	7.9	250,000–700,000
September 1985	Mexico City	8.1	7000
December 1988	Armenia	6.9	25,000–45,000
October 1989	San Francisco	7.1	90
June 1990	Iran	7.7	40,000
January 1994	Los Angeles	6.7	61

Example 16–12 ***The Meaning of 1 Magnitude.*** How much more energy is represented by an increase of 1 magnitude on the earthquake scale?

Solution: You are asked to compare energies, so begin with the version of the magnitude formula that gives the energy, E:

$$E = \left(2.5 \times 10^4\right) \times 10^{1.5M} \text{ joules.}$$

Compare the energies of two quakes (1 and 2) by finding the ratio of their energies; that is, divide the energy of one quake, E_1, by the energy of the other, E_2:

$$\frac{E_1}{E_2} = \frac{\left(2.5 \times 10^4\right) \times 10^{1.5M_1} \text{ joules}}{\left(2.5 \times 10^4\right) \times 10^{1.5M_2} \text{ joules}} = \frac{10^{1.5M_1}}{10^{1.5M_2}} = 10^{1.5M_1 - 1.5M_2} = 10^{1.5(M_1 - M_2)}.$$

A magnitude difference of 1 means that $M_1 - M_2 = 1$. Therefore the energy ratio E_1/E_2 is $10^{1.5}$, or about 32. Thus each step on the earthquake scale corresponds to an increase by a factor of about 32 in total energy.

Recall the *division rule* for powers: $10^a \div 10^b = 10^{a-b}$.

Many news reports and books mistakenly say that 1 magnitude represents a factor of 10 in energy; that was the correct definition for the Richter scale, which is no longer used. On the modern magnitude scale, as shown in Example 16–12, 1 magnitude represents a factor of about 32 in energy.

Example 16–13 ***Comparing Disasters.*** Compare the estimated energies released by the 1906 and 1989 San Francisco earthquakes.

Solution: Use the energy ratio formula developed in Example 16–12. Comparing the 1906 quake (magnitude $M^{1906} = 8.3$) with the 1989 quake (magnitude $M^{1989} = 7.1$) gives

$$\frac{E\,(1906\text{ quake})}{E\,(1989\text{ quake})} = 10^{1.5(M_{1906} - M_{1989})} = 10^{1.5(8.3-7.1)} = 10^{1.8} = 63.$$

The 1906 quake released more than 60 times as much energy as the 1989 quake.

❖

The words *one megaton* used to describe a nuclear bomb mean that it releases as much energy as a megaton, or 1 million tons, of TNT.

Example 16–14 ***Bomb–Earthquake Comparison.*** Compare the energy released by a 1 megaton nuclear bomb — about 5×10^{15} joules — to the energy released by a great earthquake of magnitude 8.

Solution: The energy, *E*, released by the earthquake of magnitude 8 is

$$\begin{aligned} E &= \left(2.5\times10^{4}\right)\times10^{1.5M}\text{ joule} = \left(2.5\times10^{4}\right)\times10^{1.5\times8}\text{ joule} \\ &= \left(2.5\times10^{4}\right)\times10^{12}\text{ joule} = 2.5\times10^{16}\text{ joule.} \end{aligned}$$

Thus the energy of the earthquake is about 2.5×10^{16} joules. Now, compare the energy of the earthquake to the energy of the nuclear bomb by dividing:

$$\frac{E\,(\text{magnitude 8 earthquake})}{E\,(\text{1 megaton nuclear bomb})} = \frac{2.5\times10^{16}\text{ joule}}{5\times10^{15}\text{ joule}} = 5.$$

A magnitude-8 earthquake releases about 5 times as much energy as a 1-megaton nuclear bomb. Of course, that doesn't mean that the earthquake will cause more damage: Much of the earthquake's energy will dissipate in the Earth's interior, whereas nearly all the nuclear bomb's energy would be released at the Earth's surface. Further, the earthquake does not leave behind radioactive by-products.

❖

16.3.3 The pH Scale for Acidity

If you check the labels of many household products, including cleansers, drain openers, and shampoo, you will see that they state a quantity called the **pH.** The pH is used by chemists to classify substances as **neutral, acidic**, or **basic** (**alkaline** is a synonym for basic). The classification of acids and bases originally was based on taste; in fact, the word *acid* comes from the Latin *acidus*, which means "sour." Thus, for example, fruits taste sour because they are acidic. Bases were classified as substances that could neutralize the action of acids; to neutralize stomach acid, common antacid tablets (for upset stomachs) are basic.

By definition, pure water is neutral and has a pH of 7. Acids have a pH lower than 7, and bases have a pH higher than 7 (Table 16–5). A few typical pH values are given in Table 16–6.

Table 16–5. pH Classification

pH	Type of Solution
less than 7	acid
7	neutral
more than 7	base

Table 16–6. Typical pH Values

Solution	pH
pure water	7
drinking water	6.5–8
vinegar	3
lemon juice	2
stomach acid	2–3
baking soda	8.4
household ammonia	10
drain opener	10–12

Time-Out to Think: Check around your house or apartment for labels that state a pH. Are the substances acids or bases?

The pH is related to the concentration of free protons (positively charged hydrogen ions) in a solution. The defining formula is

$$pH = -\log_{10}\left[H^+\right],$$

where $\left[H^+\right]$ is the hydrogen ion concentration in units of *moles per liter*. From chemistry, you may recall that 1 mole contains about 6×10^{23} particles (this is Avogadro's number; see subsection 7.3.2).

If you know the pH, and want to find the hydrogen ion concentration, you need to solve the pH formula for $\left[H^+\right]$ as follows.

Start with pH formula:	$pH = -\log_{10}\left[H^+\right]$
Multiply both sides by –1:	$-pH = \log_{10}\left[H^+\right]$
Make both sides exponents of 10:	$10^{-pH} = 10^{\log_{10}\left[H^+\right]}$
Apply property that $10^{\log_{10} x} = x$ and switch left and right sides:	$\left[H^+\right] = 10^{-pH}$

Example 16–15 *Finding pH.* What is the pH of a solution with a hydrogen ion concentration of 0.001 mole per liter?

Solution: The first version of the formula for the pH yields

$$pH = -\log\left[H^+\right] = -\log 0.001 = -\log 10^{-3} = 3.$$

A solution with a hydrogen ion concentration of 0.001 mole per liter has pH 3, which means that it is a strong acid.

❖

Example 16–16 *Concentration from pH.* What is the hydrogen ion concentration of a substance with a pH of 5?

Solution: In this case, the second version of the pH formula is more convenient:

$$\left[H^+\right] = 10^{-pH}.$$

Thus, for pH 5, the hydrogen ion concentration is 10^{-5} moles per liter. Because a mole contains about 6×10^{23} hydrogen ions, each liter of a pH 5 solution contains about $6 \times 10^{23} \times 10^{-5} = 6 \times 10^{18}$ hydrogen ions.

❖

Acid Rain

Normal raindrops are not *pure* water; that's why their pH isn't exactly 7, but instead a slightly acidic pH of 6.

The problem of **acid rain** can be discussed in terms of pH. Normal raindrops usually are slightly acidic, with a pH slightly under 6. In contrast, acid rain often has pH 4 or less. Acid rain in the northeastern United States and acid fog in the Los Angeles area have been observed with pH as low as 2 — the same acidity as pure lemon juice!

The sources of acid in acid rain generally are compounds of sulfur or nitrogen (sulfuric or nitric acids) formed by emissions from burning fossil fuels. For example, the large, coal-burning power plants and industries in the Midwest and along the East Coast are thought to be responsible for acid rain throughout the northeastern United States and Canada. The acid fog of Los Angeles probably comes from automobile emissions.

Surprisingly, the water in acidified lakes is exceptionally clear, which can give observers the mistaken impression that the water is particularly pure. However, the clarity of the water is the result of its *lack* of life — living organisms tend to make the water cloudy.

A liter is the same as a cubic decimeter, or 1/1000 cubic meter. Thus a lake containing 1 million liters of water has a volume of 1000 cubic meters —a *very* small lake. For example, if the lake's average depth is only 1 meter, its surface area would be 1000 square meters, or about one-fourth the size of a football field.

Acid rain is a serious environmental problem because acids can kill trees and other plants. For example, acid rain has been blamed for damaging nearly half the trees in the famous Black Forest in Germany. Similarly, thousands of lakes in the northeastern United States and Canada are "dead," meaning that all the organisms that once lived in the lake have died. Because the cause of death in these lakes has been traced to their acidity, acid rain is assumed to be the culprit.

Example 16–17 ***Reviving a Lake.*** Suppose that a small lake containing about 1 million liters of water has died because its pH has fallen to 4. Imagine that you want to revive the lake by adding lime with a pH of 12. How much lime must be added to raise the pH of the lake to 6?

Note: This problem requires two pieces of additional knowledge from chemistry. First, the concentration of negative ions in an acid or base, denoted $[OH^-]$, is given by $[OH^-] = (10^{-14})/[H^+]$. Second, 1 mole of these negative ions is required to neutralize 1 mole of positive hydrogen ions.

Solution: The lake's pH 4 means that its hydrogen ion, $[H^+]$, concentration is $10^{-pH} = 10^{-4}$ moles per liter. The lake contains 1 million (10^6) liters of water, so the amount of hydrogen ions in the lake is

$$(10^6 \text{ liters}) \times (10^{-4} \text{ moles/liter}) = 10^2 \text{ moles of hydrogen ions.}$$

To have pH 6, which means a hydrogen ion concentration of 10^{-6} moles per liter, the lake should have only

$$(10^6 \text{ liters}) \times (10^{-6} \text{ moles/liter}) = 1 \text{ mole of hydrogen ions.}$$

You want to reduce the currently 100 moles of hydrogen ions to only 1 mole, so 99 moles of hydrogen ions must be neutralized by adding 99 moles of OH^- ions. As the numbers are approximate already, use 100 moles instead of 99 moles.

Because lime has pH 12, it contains $10^{-pH} = 10^{-12}$ moles of hydrogen ions per liter; thus its concentration of negative ions is:

$$[OH^-] = (10^{-14})/(10^{-12}) = 10^{-2} \text{ moles per liter of lime.}$$

Call the number of liters of lime added n. Then the number of moles of OH^- ions added is $n \times 10^{-2}$. Because you need to add 100 moles of the OH^- ions, you must find the value of n such that $n \times 10^{-2} = 100$, or

$$n \times 10^{-2} = 10^2 \xrightarrow{\text{solve for } n} n = 10^2 \div 10^{-2} = 10^4.$$

Thus 10,000 liters of lime must be added to neutralize the excess of 100 moles of hydrogen ions in the lake.

❖

Time-Out to Think: Preliminary investigations of a few small lakes suggest that the method of adding lime can be successful. Comment on the practicality of adding lime to revive the tens of thousands of lakes that have died from acidity. Once revived, do you think that the lakes would remain healthy? Why or why not?

16.4 FINANCIAL FORMULAS

Rare is the individual in today's society who is not either receiving interest on a bank account or paying interest on a loan or credit card. Indeed, many people have multiple accounts or loans that involve interest. If you've ever looked closely at the interest payments you receive from your bank or at the payment schedule for a loan you've taken out, you may have been a bit mystified by the numbers you see. Fortunately, relatively simple formulas govern these calculations.

16.4.1 Compound Interest

Banks pay **interest** on the amount of money, or **principal**, on deposit in a bank account. Suppose that the bank promises to pay interest of 5% at the end of each year. If you put $100 into an account, after one year the bank will pay you 5% of $100, or $5, and your new balance will be $105. Because that bank pays interest only once at the end of the year, it is called **simple interest**, which is interest paid only on the original principal.

We briefly covered compounding in Section 3.6 and we covered the growth of bank balances through compounding as an example of exponential growth in Chapter 15.

What happens if you leave your money in that bank for more than one year? During the second year, the bank again will pay you 5% of your principal. However, because of the interest paid at the end of the first year, your principal when the second year begins is $105. Thus at the end of the second year you would receive 5% or $105, or $5.25, and your new balance would be $110.25. Although the interest rate is 5% per year, you gain more than 10% after 2 years. This phenomenon, called **compounding**, produces more interest than simple interest because you receive money not only on your original principal, but also on any interest already received.

Most banks pay interest more often than once a year, which further magnifies the effects of compounding. For example, suppose that a bank pays interest compounded quarterly at an **annual percentage rate** (*APR*) of 8% per year. Quarterly compounding means that the interest is paid *four* times during the year (every quarter of a year, or every three months). If the annual percentage rate is 8% per year,

$$\text{quarterly percentage rate} = \frac{\text{annual percentage rate}}{4} = \frac{8\%\,/\,\text{yr}}{4\ \text{quarters/yr}} = 2\%\,/\,\text{quarter}.$$

Thus, with an annual percentage rate of 8% compounded quarterly, the bank actually pays 2% each quarter. Let's examine what happens to your balance over the year for an initial deposit of $1000.

- At the end of the first quarter you receive interest of 2% of $1000, or $20. Your new balance is 102% of $1000, or 1.02 × $1000 = $1020.
- At the end of the second quarter you receive 2% or $1020, making your new balance 1.02 × $1020 = $1040.40.
- At the end of the third quarter your interest is 2% of $1040.40, for a new balance 1.02 × $1040.40 = $1061.21.
- At the end of the fourth quarter, you receive 2% of $1061.21, bringing your balance to 1.02 × $1061.21 = $1082.43.

We could continue this process to calculate your balance over periods of longer than a year, or modify the process to calculate your balance if the *APR* is compounded over some other period, such as monthly or daily. However, the easier way is to use the **compound interest formula**:

$$A = P\left(1+\frac{APR}{n}\right)^{nY}, \quad \text{where} \begin{cases} P = \text{principal} \\ APR = \text{annual percentage rate in decimal form} \\ n = \text{number of compounding periods per year} \\ Y = \text{number of years (can be a fraction)} \\ A = \text{compounded amount after } Y \text{ years.} \end{cases}$$

Except for using different variables, the compound interest formula has the form of the general exponential growth law from Chapter 15:

$$Q = Q_0 \times (1+r)^t.$$

To obtain the compound interest formula, make the following substitutions:

- Q_0 becomes the principal P;
- r becomes the interest rate per payment period APR/n;
- t becomes the number of payment periods nY; and
- Q becomes the compounded amount A.

To be sure that this formula makes sense, let's check it against the results already calculated for quarterly compounding. We assumed an $APR = 8\%$, and a starting principal of $P = \$1000$. With quarterly compounding, the number of compounding periods per year is $n = 4$. Using the compound interest formula, we obtain the following.

- For $Y = 1/4$ year (one quarter) the new amount of the balance is

$$A = P\left(1+\frac{APR}{n}\right)^{nY} = \$1000\left(1+\frac{0.08}{4}\right)^{4\times(1/4)} = \$1000(1.02)^1 = \$1020.$$

- For $Y = 2/4$ year (two quarters) the new amount of the balance is

$$A = P\left(1+\frac{APR}{n}\right)^{nY} = \$1000\left(1+\frac{0.08}{4}\right)^{4\times(2/4)} = \$1000(1.02)^2 = \$1040.40.$$

- For $Y = 3/4$ year (three quarters) the new amount of the balance is

$$A = P\left(1+\frac{APR}{n}\right)^{nY} = \$1000\left(1+\frac{0.08}{4}\right)^{4\times(3/4)} = \$1000(1.02)^3 = \$1061.21.$$

- For $Y = 1$ year (four quarters) the new amount of the balance is

$$A = P\left(1+\frac{APR}{n}\right)^{nY} = \$1000\left(1+\frac{0.08}{4}\right)^{4\times(1)} = \$1000(1.02)^4 = \$1082.43.$$

All these results are the same as those obtained earlier.

Because the year-end balance is \$82.43 more than the starting principal, the **effective yield** (or **annual yield**) is

$$\text{annual yield} = \frac{\text{change in balance during 1 year}}{\text{balance at beginning of year}} = \frac{\$82.43}{\$1000} = 0.08243 = 8.243\%.$$

Thus, even though the annual percentage rate is only 8%, the account brought an annual yield of 8.243%.

The difference between the annual percentage rate, *APR*, and the effective yield is the source of the "rate confusion" discussed briefly in subsection 15.3.4. The formulas in this chapter are stated in terms of the *APR*, which is how they usually are stated in financial applications. The formulas in Chapter 15 also can be applied to compound interest but only by using the *effective yield* (rather than the *APR*) for the growth rate *r*.

Time-Out to Think: Find a newspaper and look through it for advertisements by banks. Do interest-bearing accounts advertise the *APR*, effective yield, or both?

Finally, before you use the compound interest formula remember that, if any of the assumptions made in creating the formula is invalid, it won't work. For example, the compound interest formula is based on the assumptions that the annual percentage rate remains constant and that you never withdraw money from the account; unless these assumptions hold, the formula will not accurately predict the future balance of your bank account.

The compound interest formula can fail under many other circumstances. For example, the formula will fail if a bank fails (see subsection 7.5.4). Similarly, suppose that your money is in a bank account in a country whose currency collapses because of hyperinflation; your entire account could become worthless, despite the compound interest formula!

Time-Out to Think: Think of several more assumptions that must hold for the compound interest formula to predict correctly the future value of a bank account.

Calculator Hints

The following hints will help ensure that you properly use your calculator when working with financial formulas.

- The interest rate must be entered as a decimal fraction. For example, an *APR* of 8% is entered as *APR* = 0.08.
- Always compute values inside parentheses first, *then* raise to powers (using the y^x key) or take logs.
- Never round until the calculation is completed; rounding during the calculation may lead to significant errors.
- Most calculators keep track of more digits than they display, so make calculations in a series of steps (rather than stopping, writing a value down, and using it later). If you must do a calculation in more than one series of steps, use your calculator's memory functions (store and recall) to keep track of numbers that you will need again.
- With so much button pushing, mistakes are easily made. Always check your calculations at least twice.

Example 16–18 *Daily Compounding at 8%.* Suppose that you deposit \$1000 in a bank account that pays an *APR* of 8% and compounds interest *daily*. How much will you have after 1 year? Compare this result to that for quarterly compounding.

Solution: In this case, the number of compounding periods per year is $n = 365$. All other variables in the compound interest formula are as they were in the prior calculation for quarterly compounding; that is, $P = \$1000$, $APR = 0.08$, and $Y = 1$. Substituting into the compound interest formula yields

$$A = P\left(1+\frac{APR}{n}\right)^{nY} = \$1000\left(1+\frac{0.08}{365}\right)^{365\times(1)} = \$1000(1.000219178)^{365} = \$1083.28.$$

Your year-end balance would be \$1083.28, or 85¢ higher than the \$1082.43 from quarterly compounding. Thus the increased number of compounding periods per year results in a higher effective yield. In this case, the effective yield rises from 8.243% to 8.328%.

❖

Example 16–19 *Monthly Compounding at 6%.* Suppose that you deposit \$5000 in a bank account that pays an *APR* of 6% and compounds interest monthly. How much money will you have after 1 year? after 5 years? What is the annual yield of the account? What is the total yield over 5 years?

Solution: For the compound interest formula, the principal is $P = \$5000$, the annual percentage rate is $APR = 0.06$, and, because interest is compounded monthly, the number of compounding periods per year is $n = 12$. For a period of 1 year, $Y = 1$, and for a period of 5 years, $Y = 5$. Substituting into the formula gives

$$\text{after 1 year}: \quad A = P\left(1+\frac{APR}{n}\right)^{nY} = \$5000\left(1+\frac{0.06}{12}\right)^{12\times 1} = \$5000(1.005)^{12} = \$5308.39;$$

$$\text{after 5 years}: \quad A = P\left(1+\frac{APR}{n}\right)^{nY} = \$5000\left(1+\frac{0.06}{12}\right)^{12\times 5} = \$5000(1.005)^{60} = \$6744.25.$$

After 1 year you have \$5308.39, and after 5 years your balance is \$6744.25. Because your money increased by \$308.39 over 1 year, the annual yield is

$$\text{annual yield} = \frac{\$308.39}{\$5000} = 0.0617 = 6.17\%.$$

Over 5 years, the total yield is

$$\text{yield} = \frac{\$6744.25 - \$5000}{\$5000} = 0.3489 = 34.89\%.$$

The annual (or effective) yield is 6.17%, which is higher than the *APR* of 6%. The yield after 5 years is 34.89%, which is more than five times the *APR* ($5 \times 6\% = 30\%$) *and* more than five times the annual yield ($5 \times 6.17\% = 30.85\%$)! Thus compounding has more than a simple multiplying effect on the *APR* and the annual yield.

❖

Example 16–20 *Annual Yield from APR.* A bank advertises a certificate of deposit with an *APR* of 7% compounded monthly. What is the annual yield?

Solution: Monthly compounding means that $n = 12$ compounding periods per year; the *APR* is 7% = 0.07. The easiest way to find the annual yield is to take a sample principal, say, $P = \$100$, and compute the new balance amount A after $Y = 1$ year:

$$A = P\left(1+\frac{APR}{n}\right)^{nY} = \$100\left(1+\frac{0.07}{12}\right)^{12\times(1)} = \$100(1.0058333)^{12} = \$107.23.$$

The increase on a \$100 principal is \$7.23, so the annual yield of this account is 7.23%. As always, note that this yield is somewhat more than the *APR* of 7%.

❖

Continuous Compounding

We have shown that, with an *APR* of 8%, quarterly compounding gives an annual yield of 8.243% and daily compounding gives an annual yield of 8.328%. What if interest is compounded even more frequently, like every second, or every trillionth of a second?

Table 16–7 lists the annual yield at 8% *APR* with varying numbers of compounding periods. Each value can be calculated by following the method in Example 16–20. For example, at an $APR = 0.08$ with $n = 12$ compounding periods (monthly), a balance of \$100 would grow in $Y = 1$ year to

$$A = P\left(1+\frac{APR}{n}\right)^{nY} = \$100\left(1+\frac{0.08}{12}\right)^{12\times(1)} = \$100(1.00666667)^{12} = \$108.29995.$$

This annual yield is 8.29995%, as shown; of course, your balance would be rounded to the nearest cent, or \$108.30.

Table 16–7. Annual Yield (*APR* = 8%) for Various Numbers of Compounding Periods, *n*

n	1	4	12	365	500	1000	10,000	1,000,000
Annual yield	8%	8.2432%	8.29995%	8.32776%	8.32801%	8.32836%	8.32867%	8.32870%

Figure 16–7 shows a graph of the data in Table 16–7, illustrating how the annual yield varies with respect to the number of compounding periods for an *APR* of 8%. Note that the annual yield doesn't grow indefinitely with greater numbers of compounding periods; instead, it approaches a limit that must be very close to the 8.3287% found for $n = 1$ million. Although the idea of having infinitely many compounding periods may seem strange, it sometimes is done and is called **continuous compounding**. Mathematically, the annual yield with continuous compounding is given by the formula:

$$\text{annual yield with continuous compounding} = e^{APR} - 1,$$

where e is a special irrational number, with a value of about $e \approx 2.71828$. Thus for continuous compounding at an *APR* of 8%, you can use a calculator to

Named after the great mathematician Leonhard Euler, the number e is one of the fundamental constants of mathematics. It is as important and arises as often as the number π. Most calculators have a key for evaluating e^x.

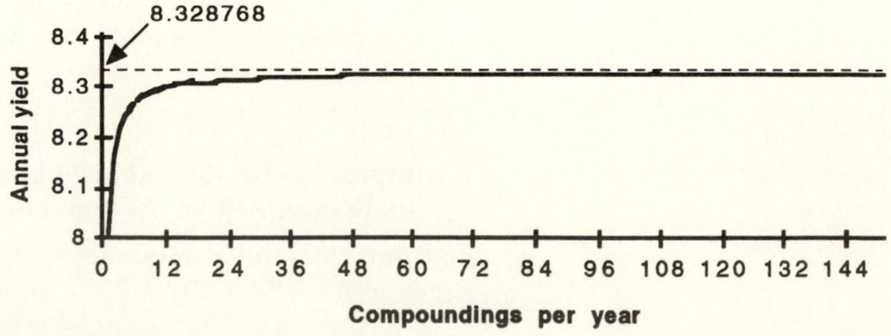

Figure 16-7. A graph of the annual yield at an *APR* of 8% with respect to the number of compounding periods *n*. An increasing *n* approaches *continuous compounding* and the annual yield approaches 8.328768%.

find that the annual yield is $e^{0.08} - 1$, or 8.328768%. The amount of money you would have after Y years is given by the formula:

$$\underbrace{\text{Compounded Amount } A}_{\text{(after } Y \text{ years with } continuous \text{ compounding)}} = Pe^{(APR \times Y)}.$$

Example 16–21 *Yields with Continuous Compounding.* Suppose that you deposit \$200 in an account with an APR of 5.5% and continuous compounding. What is your annual yield? How much will you have in your account after 10 years?

Solution: Use a calculator (either by entering the value of e or using the e^x key) to calculate the annual yield of

$$e^{APR} - 1 = e^{0.055} - 1 = 0.05654 = 5.654\%.$$

After $Y = 10$ years, the balance is

$$\underbrace{\text{Compounded Amount } A}_{\text{(after } Y \text{ years with } continuous \text{ compounding)}} = Pe^{(APR \times Y)} = (\$200)e^{(0.055 \times 10)} = (\$200)e^{0.55} = \$346.65.$$

With continuous compounding at an APR of 5.5%, the annual yield is 5.654%. In 10 years, a \$200 principal will grow to \$346.65.

❖

Solving the Compound Interest Formula

So far, all the compound interest formula problems have involved calculating a compounded amount, A, when all the other variables (principal, APR, number of compounding periods, and number of years) are known. Sometimes, however, you may want to calculate values for other variables.

For example, suppose that you want to know how long it takes for money to grow by some factor, such as growing by 10% or doubling or tripling. In this case, the variable of interest is time, or number of years, Y, so the compound interest formula must be solved for Y. Although the process looks cumbersome, it is straightforward and proceeds as follows.

Start with compound interest formula: $$A = P\left(1 + \frac{APR}{n}\right)^{nY}$$

Divide both sides by P: $$\frac{A}{P} = \left(1 + \frac{APR}{n}\right)^{nY}$$

Take $\log_{10}$ of both sides: $$\log_{10}\left(\frac{A}{P}\right) = \log_{10}\left(1 + \frac{APR}{n}\right)^{nY}$$

Apply property that $\log ax = x \log a$: $$\log_{10}\left(\frac{A}{P}\right) = nY \log_{10}\left(1 + \frac{APR}{n}\right)$$

Divide both sides by $n \log_{10}\left(1 + \frac{APR}{n}\right)$: $$Y = \frac{\log_{10}\left(\frac{A}{P}\right)}{n \log_{10}\left(1 + \frac{APR}{n}\right)}.$$

In Example 16–22 we use this formula for Y to calculate the amount of time required for money to grow by various factors.

Another common question with compound interest is: How much principal must you start with in order to have a certain amount of money later? For example, you might want to know how much you should deposit in an account in order to have \$10,000 (perhaps for a new car) after 5 years. Problems of this type are called **present value problems** because they ask questions like: What is the *present* value that must be deposited to give a *future* value of $10,000? In this case, you need to solve the compound interest formula for the principal, P, by dividing both sides by the term in parentheses:

Recall that a negative power is simply the reciprocal of the corresponding positive power. That is, $1/a^x = a^{-x}$.

$$A = P \times \underbrace{\left(1+\frac{APR}{n}\right)^{nY}}_{\text{divide both sides by this term}} \Rightarrow P = A \div \left(1+\frac{APR}{n}\right)^{nY} = A\left(1+\frac{APR}{n}\right)^{-nY}.$$

Example 16–22 *Growth Times.* With a 6% APR and monthly compounding, how long must you wait for your money to double? triple? grow by 10%?

Solution: The question is "how long?" so you need to calculate the time, or number of years Y, with the given $APR = 0.06$ and number of compounding periods $n = 12$.

Doubling of money means that $A/P = 2$, and tripling means that $A/P = 3$. Substitute these values into the compound interest formula solved for Y to find

You may want to compare this result with the approximate doubling time formula in Chapter 15: Doubling time = $0.7/r$. In this case, as it is an approximation, you can use $r = APR = 0.06$. Then the doubling time is $0.7/0.06 = 11.67$ years. Note that this approximate value is very close to the actual value for the doubling time obtained in this example.

$$\text{doubling time}: \; Y = \frac{\log_{10}(2)}{12 \times \log_{10}\left(1+\frac{0.06}{12}\right)} = \frac{0.301029995}{12 \times \log_{10}(1.005)} = 11.58 \text{ years}.$$

$$\text{tripling time}: \; Y = \frac{\log_{10}(3)}{12 \times \log_{10}\left(1+\frac{0.06}{12}\right)} = \frac{0.477121254}{12 \times \log_{10}(1.005)} = 18.36 \text{ years}.$$

Your money doubles in about 11½ years, and triples in just over 18 years.

For 10% growth A is 10% greater than the original principal P, or $A/P = 110\% = 1.1$. Substitute this value into the formula for Y:

$$Y = \frac{\log_{10}\left(\frac{A}{P}\right)}{n\log_{10}\left(1+\frac{APR}{n}\right)} = \frac{\log_{10}(1.1)}{12 \times \log_{10}\left(1+\frac{0.06}{12}\right)} = \frac{0.041392685}{12 \times \log_{10}(1.005)} = 1.59 \text{ years}.$$

Your money would grow by 10% in slightly over a year and a half, with monthly compounding and an APR of 6%.

❖

Example 16–23 *Planning Ahead.* Suppose that your family is having a baby and that you want to invest money to pay for the baby's college education. You estimate that a college education will cost $100,000 in 18 years. If you can invest in a long-term account that guarantees an interest rate of 9.5%, compounded daily, how much do you need to deposit?

Solution: This is a present value problem because it asks how much you need to deposit now to give a future value (in 18 years) of $A = \$100{,}000$. The interest rate is $APR = 0.095$, the number of compounding periods per year is $n = 365$ (for daily compounding), and the time is $Y = 18$ years. Substituting into the compound interest formula solved for P yields

$$P = A\left(1+\frac{APR}{n}\right)^{-nY} = \$100{,}000\left(1+\frac{0.095}{365}\right)^{-365\times 18} = \$100{,}000(1.00026027)^{-6570} = \$18{,}091.$$

An investment of $18,091 will yield the $100,000 you expect to need for a college education in 18 years. (A good way to check this answer is to put this principal into the original compound interest formula and verify that it yields the amount $A = \$100{,}000$ after 18 years.)

❖

16.4.2 Investment Plans

Suppose that you want to save for retirement, your child's college expenses, or any other reason. Rather than putting a lump sum in the bank today to collect interest, you probably will set up an interest-bearing **investment plan** into which you make regular payments. Popular types of investment plan include Individual Retirement Accounts (IRAs), 401(k) plans, Keogh plans, employee pension plans, and annuities.

Suppose that your investment plan has an *APR* of 12%, compounded monthly, so that the monthly interest rate is 1%. Assume that at the *end* of each month you deposit \$100 into the plan. Then

- at the end of the first month, you make your first deposit, giving you a balance of \$100;
- at the end of the second month, you receive interest of 1% on the \$100, or \$1, and deposit another \$100, bringing your balance to \$201; and
- at the end of the third month, you receive interest of 1% on your prior balance of \$201, or \$2.01, and deposit another \$100, bringing your balance to \$201 + \$2.01 + \$100 = \$303.01.

Note: This formula is based on the assumption that the compounding period is the same as the payment period. That is, if payments are made monthly, interest is compounded and paid monthly as well. If the compounding period is different from the payment period, the term *APR/n* should be replaced by the effective yield for each payment period. Otherwise, the formula will give only an approximation.

In principle, you could continue this procedure to find the balance after 4 months, 5 months, and so on. Fortunately, an easier way is to use the **investment plan formula**:

$$A = PMT \times \left[\frac{\left(1+\frac{APR}{n}\right)^{nY} - 1}{\frac{APR}{n}} \right], \quad \text{where} \begin{cases} A = \text{accumulated investment plan balance} \\ PMT = \text{regular payment amount} \\ APR = \text{annual percentage rate} \\ n = \text{number of payment periods per year} \\ Y = \text{number of years.} \end{cases}$$

Example 16–24 ***Retirement Plan.*** To save money for your retirement, you start an IRA with an *APR* of 8%, compounded and paid monthly, at age 30. At the end of each month, you deposit \$100 to the account. How much will the IRA have when you retire at age 65? Compare its value to the total amount of deposits over this time period.

Solution: The IRA will accumulate deposits and interest for $Y = 35$ years. For monthly payments $n = 12$, the monthly payment amount is $PMT = \$100$, and the $APR = 8\%$, or 0.08. The investment plan formula yields

$$A = PMT \times \left[\frac{\left(1+\frac{APR}{n}\right)^{nY} - 1}{\frac{APR}{n}} \right] = \$100 \times \left[\frac{\left(1+\frac{0.08}{12}\right)^{(12\times 35)} - 1}{\frac{0.08}{12}} \right] = \$229{,}388 .$$

Your actual contributions over 35 years, in regular monthly payments of \$100, amount to 35 yr × 12 months/yr × \$100/month = \$42,000.

However, thanks to compounding, your IRA will have a balance of almost \$230,000 when you retire — *more than five times* the amount of your contributions!

❖

Example 16–25 ***College Fund.*** To save for your baby's college education, you want an investment plan that will yield \$100,000 at the end of 18 years. Suppose that you can invest in a long-term plan that guarantees an interest rate of 9.5%. What regular monthly payments should you make?

Remember that dividing is the same as multiplying by the reciprocal.

Solution: You seek the monthly payment, *PMT*, so you must solve the investment plan formula for *PMT* by dividing both sides by the term in brackets:

$$A = PMT \times \underbrace{\left[\frac{\left(1+\frac{APR}{n}\right)^{nY}-1}{\frac{APR}{n}}\right]}_{\text{divide by this term}} \quad \Rightarrow \quad PMT = A\left[\frac{\frac{APR}{n}}{\left(1+\frac{APR}{n}\right)^{nY}-1}\right].$$

You seek a compounded amount A = \$100,000; the interest rate is $APR = 0.095$ (9.5%), and the number of payment periods per year is $n = 12$ (monthly). Substituting into the formula yields

$$PMT = A\left[\frac{\frac{APR}{n}}{\left(1+\frac{APR}{n}\right)^{nY}-1}\right] = \$100{,}000 \times \left[\frac{\frac{0.095}{12}}{\left(1+\frac{0.095}{12}\right)^{(12\times18)}-1}\right] = \$176.24.$$

Monthly payments of \$176.24 will provide \$100,000 for college after 18 years. (Check this answer by substituting a payment amount of \$176.24 into the original investment plan formula.) Note that over the 18-year period the total amount of your contributions would be

$$(18\text{ yr}) \times (12\text{ months/yr}) \times (\$176.24/\text{month}) = \$38{,}068.$$

Therefore the interest that accrues over 18 years accounts for almost two thirds of the final balance of \$100,000.

❖

Time-Out to Think: Example 16–23 showed that just over \$18,000 invested in an account with a 9.5% *APR* would yield \$100,000 after 18 years. For the investment plan in Example 16–25, you would contribute a total of about \$38,000, in monthly payments of \$100, to reach the same \$100,000 in 18 years. Discuss the pros and cons of investing a lump sum now with those of making regular payments over the 18 years.

Investment Plan Realism

As with any model, the investment plan formula involves many assumptions. In this case, some of the assumptions are especially egregious. For example, finding an investment plan that will guarantee a fixed interest rate for many years is nearly impossible. Instead, most investment plans vary their rates in response to prevailing interest rates; interest rates often are adjusted each month or each year. In addition, most people vary their monthly payments into an investment plan over time; if you get a raise, for example, you might want to increase your monthly contribution; if you get laid off, you might reduce or suspend your monthly contributions, or even withdraw money from your investment plan early.

Further complications arise from the tax treatment of investment plans. Contributions to many popular retirement plans are tax-deferred, meaning that no tax is paid on income placed in the plan until it is withdrawn at retirement. Depending on a person's particular tax situation, the same monthly pension contribution can effectively cost two different people two different amounts. In addition, Congress frequently changes tax laws. Thus the effective value of an investment plan may change simply because of a change in the law.

Finally, none of the investment plan calculations presented reflect the effects of inflation. For example, suppose that you create a college fund designed to have \$100,000 after 18 years. If inflation is higher than you expect, \$100,000 might be well short of the actual cost of college in 18 years.

Your marginal tax rate is the rate paid on the last portion of your income. The federal income tax system is based on different rates for different portions of income. For example, you might pay no tax on your first $500 per month of income, 15% on the next $1000 (income from $501 to $1500), and 31% on the remainder (income above $1501). Thus your marginal tax rate is 31%.

Example 16–26 *Tax-Deferred Contributions.* Sara and Jim each make tax-deferred contributions of $100 per month to their pension plans. Sara, who has a relatively high income, is in a tax bracket with a 31% marginal tax rate. Jim, who has a lower salary, is in the 15% marginal tax rate bracket. How much is each person's take-home (after-tax) pay reduced by the $100 pension contribution?

Solution: At first, you might think that a contribution of $100 per month would reduce take-home pay by $100 per month. However, that wouldn't take into account the tax-deferred status of the contributions. First consider Sara. Her contribution of $100 reduces her gross pay (before taxes and contributions) by $100. If she didn't make the pension contribution, she would have to pay 31% of the $100, or $31, in taxes. Thus, without the pension contribution, her take-home pay would be $100 – $31 = $69 greater. Effectively, she adds $100 to her pension plan for a cost of $69.

Note that things between Jim and Sara are not likely to even out at retirement, either. Suppose that, when they retire, both live off of their pension plan savings. If they both withdraw the same amount per month, they will both be receiving the same annual income and hence will pay taxes at the same rate. Again, Jim will have paid more over his lifetime to receive the same pension as Sara.

The analysis for Jim is similar, except that he would pay only 15% of $100, or $15, in taxes if he didn't put the $100 into his pension plan. His take-home pay would be $100 – $15 = $85 greater without the tax-deferred contribution. Thus the effective cost to Jim for adding $100 to his pension plan is $85.

Even though both Sara and Jim are making the same $100 per month contribution, Sara's take-home pay is reduced by only $69, whereas Jim's is reduced by $85. At retirement, both individuals end up with the same amount in their pension plans (assuming the same interest rates), but it will have cost Jim more than Sara.

❖

Time-Out to Think: The preceding example illustrates a fundamental fact about tax deductions: The higher your tax bracket, the more a particular deduction is worth to you. Does this fact have any bearing on the "fairness" of tax deductions? Why or why not?

16.4.3 Loan Payments

Suppose that you borrow money from the bank and agree to pay back the loan in regular payments over a specified period of time. Such a loan is called an **amortized loan**, and your regular payments are called the **amortization payments**. The amount you borrow is called the **loan principal**, denoted P. The length of time over which you are supposed to repay the loan is called the **loan term**, usually in years, denoted Y. You are expected to make a certain number of payments, n, each year (e.g., for monthly payments, $n = 12$). Finally, when you take out the loan it will be at some APR, which, as always, must be expressed in decimal form (e.g., 0.05 instead of 5%). Your regular amortization payment, PMT, is given by the following formula.

$$PMT = \frac{P \times \left(\frac{APR}{n}\right)}{1 - \left(1 + \frac{APR}{n}\right)^{-nY}}, \quad \text{where} \begin{cases} P = \text{loan principal} \\ PMT = \text{regular payment amount} \\ APR = \text{annual percentage rate} \\ n = \text{number of payment periods per year} \\ Y = \text{loan term in years.} \end{cases}$$

The loan payment formula can be used to calculate payments any time the interest is **fixed** (not allowed to change while the loan is being repaid).

Example 16–27 ***Student Loan.*** Suppose that you have a student loan of \$7500 with a fixed rate of 9% to be paid back over 10 years. What are your monthly payments? How much will you pay over the lifetime of the loan? What is the total interest you will pay on the loan?

Solution: The loan principal is P = \$7500 and the interest rate is $APR = 0.09$. Monthly payments mean that $n = 12$, and the loan term is $Y = 10$ years. Use the loan payment formula:

$$PMT = \frac{P \times \left(\frac{APR}{n}\right)}{1-\left(1+\frac{APR}{n}\right)^{-nY}} = \frac{\$7500 \times \left(\frac{0.09}{12}\right)}{1-\left(1+\frac{0.09}{12}\right)^{-(12\times 10)}} = \frac{\$7500 \times (0.0075)}{1-(1.0075)^{-120}} = \$95.01.$$

Monthly payments on this student loan are \$95.01 (rounded to the nearest cent). Over the lifetime of the loan you will pay

$$(10 \text{ yr}) \times (12 \text{ months/yr}) \times (\$95.01/\text{month}) = \$11{,}401.$$

With a principal of \$7500, your total interest payments will be \$11,401 – \$7500 = \$3901 over the 10-year period.

❖

Example 16–28 ***Home Mortgage.*** Suppose that you buy a home by taking out a \$100,000 mortgage with a fixed term of 30 years. What are your monthly payments if the interest rate is 8%? if the interest rate is 7%? Compare the payments over 30 years at these two rates.

Solution: The loan principal is P = \$100,000. Monthly payments mean that $n = 12$, and the loan term is $Y = 30$ years. For an APR of 8%,

$$PMT = \frac{P \times \left(\frac{APR}{n}\right)}{1-\left(1+\frac{APR}{n}\right)^{-nY}} = \frac{\$100{,}000 \times \left(\frac{0.08}{12}\right)}{1-\left(1+\frac{0.08}{12}\right)^{-(12\times 30)}} = \frac{\$100{,}000 \times (0.0067)}{1-(1.0067)^{-360}} = \$733.76.$$

If the interest rate is only 7%, the monthly payments are

$$PMT = \frac{P \times \left(\frac{APR}{n}\right)}{1-\left(1+\frac{APR}{n}\right)^{-nY}} = \frac{\$100{,}000 \times \left(\frac{0.07}{12}\right)}{1-\left(1+\frac{0.07}{12}\right)^{-(12\times 30)}} = \frac{\$100{,}000 \times (0.0058)}{1-(1.0058)^{-360}} = \$666.43.$$

At 8%, the monthly payments on the \$100,000 loan are \$733.76. At 7%, the payments are \$666.43, or lower by \$733.76 – \$666.43 = \$67.33 per month. Thus each year the lower interest rate would save you 12 × \$67.33 = \$807.96. Over the 30-year life of the loan, the difference in payments is 30 × \$807.96 = \$24,238.80. A difference of only 1% in interest rate yields a difference in total payments of more than \$24,000 over 30 years.

❖

Loan Realism

Unlike investment plans, loans offering fixed interest rates are quite common. Short-term loans, such as 3- to 5-year auto loans, usually offer fixed rates. Student loans, with repayment periods of 10 years or more, also offer fixed rate. Student loans often offer an added benefit: Repayment may not be required to begin until some time after graduation. However, some student loans accrue interest while the student is in school, which is added to the principal; in that case when payment begins, the principal is higher than the amount originally borrowed.

Most ARMs also specify a maximum by which the interest rate can rise or fall in any 1 year, as well as a rate "cap" above which the interest rate is guaranteed never to rise.

Home mortgages (loans secured by the value of property) offer some especially difficult choices. Although fixed rate mortgages (usually with terms of 15 years or 30 years) are common, most banks and mortgage companies offer the option of an **adjustable rate mortgage** (**ARM**). The initial interest rate on an ARM usually is lower than the interest rate for a fixed rate loan, but the interest rate can change. Most ARMS specify how the interest rate is calculated in relation to prevailing rates. For example, the rate may be set each month at 2% above the rate offered by U.S. Treasury bills; if this government rate changes, your mortgage rate also changes.

Buyer beware! Many ARMs offer an initial rate that is below the prevailing rates. This special low rate often ends after just 6 months to a year, at which time the rate jumps. If you are offered an ARM with an enticingly low rate, be sure to check how long the rate will last. As with most financial decisions, *if it sounds too good to be true, it probably is.*

Making a decision between taking a fixed rate loan or an ARM can be one of the most important financial decisions of your life. As demonstrated in Example 16–28, just a 1% difference in rate can mean tens of thousands of dollars saved or lost over the life of the loan. Unfortunately, the decision often comes down to guesswork about which way and how far interest rates will move over the next 15 or 30 years. For example, suppose that you are taking out a 30-year loan and that your choice is between a fixed rate of 8% or an ARM that starts at 7%. If you think that the future rate on the ARM is likely to rise above 8%, you should take the fixed rate loan. But, if you believe that rates are likely to stay the same or fall, you may be better off with the ARM.

Another factor that may come into play is your ability to qualify for the loan. Because ARMs have lower initial payments, you can generally qualify for a larger loan with an ARM than with a fixed rate. However, as your payments may rise in the future if interest rates rise, you must be sure that you will be able to afford the rising payments.

Why are there so many choices and complications in long-term loans? To understand change your perspective to that of the mortgage *holder* (the bank, mortgage company, or individual) to whom you owe your money. Imagine that a bank holds your 30-year mortgage with a fixed *APR* of 8% and a loan principal of $100,000. The bank therefore *receives* your monthly payments of $733.76 (as calculated in Example 16–28).

Suppose that after a few years your outstanding loan balance is $90,000 (that is, you have paid off $10,000 of the loan principal). Further, suppose that for some reason (perhaps to raise cash), your bank decides to sell the mortgage on the open market. Will another bank pay $90,000 for it? If another bank buys your mortgage, it will receive your monthly payments (still $733.76) at an interest rate of 8%. If interest rates have risen above 8%, this bank could make a new loan at a higher rate and hence your loan is a poor value at a price of $90,000. As a result, your bank will be forced to sell your loan at a **discount**; that is, for less than its **face value** of $90,000.

If you own a home, you may have experienced a change in the bank or company to which you make your payments; such changes usually are the result of your loan being sold from one company to another.

But, suppose that interest rates have fallen below 8%. In that case, new loans will carry a lower rate than 8% and therefore lower monthly payments than you are making. In that case, another bank might be tempted to pay more than the face value of $90,000 for your loan; that is, your loan would sell for a **premium**.

This analysis of the market for buying and selling mortgages also applies to buying and selling fixed rate investments such as bonds. For example, the price of existing bonds rises above their face value (they sell at a *premium*) when interest rates fall because these bonds are then paying a rate that is higher than rates on new bonds. Similarly, bond prices fall (sell at a *discount*) when rates rise because the existing bonds are then paying a lower rate of interest than new bonds. Because there are many different types of bonds, with many different rules about interest and different tax treatments, analysis of the bond market is extremely complex.

Note that, if prevailing interest rates rise substantially, the bank holding your loan at 8% effectively takes a large loss because the bank's money could otherwise be invested at higher rates. However, if rates fall substantially, the bank makes a nice profit because it could not invest its money at such a high rate elsewhere. Because predicting future interest rates is difficult, banks try to protect themselves against long-term losses by generally charging higher rates for longer-terms loans. Similarly, banks can offer lower initial interest rates for ARMs because the ability to adjust the interest rate protects the bank against losses if prevailing interest rates rise.

Time-Out to Think: In this week's newspapers find and compare the current rates for fixed 15- and 30-year mortgages and the initial rates for ARMs.

A further complication arises because most loans allow you to accelerate your payments and pay them off early. More damaging from the bank's point of view, if interest rates fall substantially you might simply **refinance** your loan. In other words you could take out a new loan at a lower rate and use this money to pay off your old loan. As a result, some banks charge **prepayment penalties** if you pay off the loan early; in addition, most banks charge **closing costs**, or up-front fees, at the time you borrow the money. Thus, even if interest rates fall, you may not gain by refinancing because of the fees you would have to pay as closing costs or prepayment penalties.

If all that sounds complicated, it is! From your point of view as a borrower, you must deal not only with differences in interest rates among loans of different terms or adjustable loans, but also with differences in closing costs from one bank to another, differences in prepayment penalties, and differences in the levels of service provided. In addition, when deciding whether to prepay a mortgage, you also need to consider the effects of any special tax treatment given to mortgage interest.

From the point of view of a bank, the situation is even more complicated because it must worry about whether the borrower might default (fail to pay) on the loan. Not surprisingly, bankers and others who specialize in investments tend to live highly stressful lives!

Example 16–29 ***Accelerating a Loan.*** Suppose that you want to retire the student loan described in Example 16–27 in less than 10 years. If you make payments of \$195 each month, instead of the required \$95, how soon can you pay off the loan? How much will you save in total interest?

Solution: To find the number of years required to pay off the loan solve the loan payment formula for Y. Again, although the algebra *looks* intimidating, it is straightforward and proceeds as follows.

Start with amortized payment formula:

$$PMT = \frac{P \times \left(\frac{APR}{n}\right)}{1-\left(1+\frac{APR}{n}\right)^{-nY}}$$

Multiply both sides by $1-\left(1+\frac{APR}{n}\right)^{-nY}$ and divide both sides by PMT:

$$1-\left(1+\frac{APR}{n}\right)^{-nY} = \frac{P \times \left(\frac{APR}{n}\right)}{PMT}$$

Subtract 1 from both sides:

$$-\left(1+\frac{APR}{n}\right)^{-nY} = \frac{P \times \left(\frac{APR}{n}\right)}{PMT} - 1$$

Multiply both sides by –1:

$$\left(1+\frac{APR}{n}\right)^{-nY} = -\left(\frac{P \times \left(\frac{APR}{n}\right)}{PMT} - 1\right) = 1 - \frac{P \times \left(\frac{APR}{n}\right)}{PMT}$$

Take $\log_{10}$ of both sides:

$$\log_{10}\left(1+\frac{APR}{n}\right)^{-nY} = \log_{10}\left(1 - \frac{P \times \left(\frac{APR}{n}\right)}{PMT}\right)$$

Apply property that $\log a^x = x \log a$:

$$-nY \log_{10}\left(1+\frac{APR}{n}\right) = \log_{10}\left(1 - \frac{P\left(\frac{APR}{n}\right)}{PMT}\right)$$

Divide both sides by $-n \log_{10}\left(1+\frac{APR}{n}\right)$:

$$Y = -\frac{\log_{10}\left(1 - \frac{P \times \left(\frac{APR}{n}\right)}{PMT}\right)}{n \log_{10}\left(1+\frac{APR}{n}\right)}.$$

Although this formula looks forbidding, you simply need to substitute values of the monthly payments, PMT = \$195, along with APR = 0.09, n = 12 payments per year, and P = \$7500 to get

$$Y = -\frac{\log_{10}\left(1 - \frac{P \times \left(\frac{APR}{n}\right)}{PMT}\right)}{n \log_{10}\left(1 + \frac{APR}{n}\right)} = -\frac{\log_{10}\left(1 - \frac{\$7500 \times \left(\frac{0.09}{12}\right)}{\$195}\right)}{12 \log_{10}\left(1 + \frac{0.09}{12}\right)} = -\frac{\log_{10}(0.711538461)}{12 \log_{10}(1.0075)} = 3.8 .$$

By making the larger payments of \$195, instead of \$95, paying off the loan will take about 3.8 years (approximately 3 years and 10 months) instead of 10 years. The total of your loan payments will be

$$(3.8 \text{ yr}) \times (12 \text{ months/yr}) \times (\$195 \text{ /month}) = \$8892.$$

With the principal of \$7500, your total interest payment is \$8892 – \$7500 = \$1392. Compared to the \$3901 paid in interest if you took 10 years to pay off the loan (see Example 16–27), you would save about \$2500.

❖

16.5 GRAVITY: FROM APPLES TO THE MOON AND BEYOND

If the ancient Greeks knew that the Earth was round, why did Columbus, more than 1500 years later, have so much trouble convincing people of the wisdom of his voyage? In fact, scholars in Columbus's time were well aware that the Earth was round; the uneducated, which at that time was the vast majority of the population, believed the Earth to be flat. Scholars also knew the circumference of the Earth, first determined by Eratosthenes in 240 B.C. Based on that knowledge, and the estimated extent of the Eurasian continent, some scholars concluded that Columbus could not survive the westward voyage to Asia because it was too far. As it turned out, they were right! Columbus's voyage already was in trouble by the time he reached the Americas; if the Americas had not been "in the way," he would not have made it all the way to Asia.

The ancient Greeks imagined the world to be made of four elements: fire, water, earth, and air. Aristotle further imagined that two forces exist in nature: he called them **gravity**, the tendency of heavy objects to fall, and **levity** (which we now know is not a real force), the tendency of light objects to rise. Using these ideas, Aristotle "explained" the Greek belief in an Earth-centered universe by arguing that the heaviest element, earth, would naturally fall toward the center; that also explained why the Earth is round, a fact that the Greeks had known at least since the time of Pythagoras in 500 B.C. Aristotle explained other physical behaviors similarly. Water, because it is lighter than earth, sits on the earth in lakes and oceans; because it is heavier than air, water falls through air as rain. Air, being lighter still, lies above the earth and oceans, and bubbles of air rise up through water. Flames of fire, the lightest of the four elements, always rise from wherever they were released. Aristotle further believed that the "heavens," the realm of the Sun, Moon, planets, and stars, were totally distinct from the Earth. He suggested that the heavens were made of a fifth element, which he called the *ether* (literally, "upper air").

Time-Out to Think: The literal meaning of *quintessence* is "fifth element." The literal meaning of *ethereal* is "made of ether." Look up each of these words in the dictionary. Explain how their modern meanings are related to Aristotle's ancient beliefs.

Aristotle's beliefs held sway for more than a thousand years. During the sixteenth and seventeenth centuries, however, motion was studied in greater detail. Why do objects fall to the ground? Why does a pendulum swing as it does? Are the motions of the heavens distinct from motions on Earth? Galileo and others investigated these questions and began to apply mathematics to their solutions.

Then, in one of the most famous "aha!" insights of all time, Isaac Newton is said to have seen an apple falling to the ground and to have looked up to see the Moon in the daytime sky. He suddenly realized that the force that brought the apple to the ground and the force that held the Moon in orbit were the same!

16.5.1 Newton's Law of Universal Gravitation

Newton's insight helped him to discover the **universal law of gravitation**. In words, this law is summarized by three statements.

1. Every mass attracts every other mass through the force called gravity.
2. The force of attraction between any two objects is *directly proportional* to the product of their masses. For example, doubling the mass of *one* object doubles the force of gravity between the two objects.
3. The force of attraction *decreases* with the *square* of the distance between the centers of the objects; that is, the force follows an **inverse square law** with distance. For example, doubling the distance between two objects weakens the force of gravity by a factor of 2^2, or 4.

Mathematically, Newton's law of universal gravitation is written:

$$F_g = G\frac{M_1 M_2}{d^2}, \quad \text{where} \begin{cases} F_g = \text{graviatational force of attraction} \\ M_1 = \text{mass of object 1} \\ M_2 = \text{mass of object 2} \\ d = \text{distance between } \textit{centers} \text{ of two objects} \\ G = \text{the } \textit{gravitational constant}. \end{cases}$$

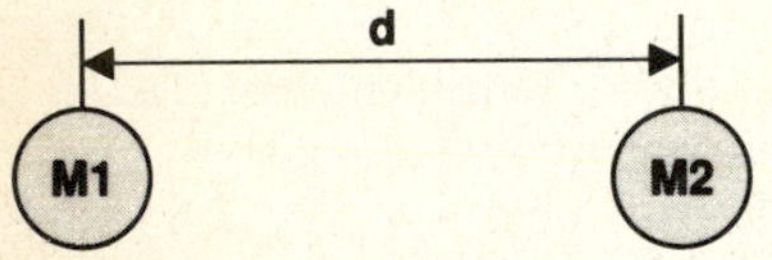

Figure 16-8. Any two objects interact through gravity. The masses of the two objects are denoted M_1 and M_2, respectively. The distance between their centers is denoted *d*. The gravitational force attracting the two objects is expressed by the law of universal gravitation.

Depending on the problem at hand, masses M_1 and M_2, the distance d between the centers of the two masses, and the force of gravity F_g all may be regarded as variables (see Figure 16–8). However, G is the **gravitational constant**, which has an experimentally measured value of

$$G = 6.67 \times 10^{-11} \frac{\text{m}^3}{\text{kg} \times \text{s}^2} = 6.67 \times 10^{-20} \frac{\text{km}^3}{\text{kg} \times \text{s}^2}.$$

Time-Out to Think: Confirm, by doing a unit conversion on the first value given for the gravitational constant, that the two values are the same.

Example 16–30 ***Earth–Moon Versus Earth–Sun.*** . The mass of the Sun is about 25 million times greater than the mass of the Moon, and the Earth–Sun distance is about 400 times the Earth–Moon distance. Compare the gravitational attraction between the Earth and Sun to that between the Earth and Moon.

Solution: Use the gravitational law to write expressions for the force of gravity between the Earth and Sun and between the Earth and Moon:

$$F_{\text{Earth-Sun}} = G\frac{M_{\text{Earth}} M_{\text{Sun}}}{\left(d_{\text{Earth-Sun}}\right)^2}; \quad F_{\text{Earth-Moon}} = G\frac{M_{\text{Earth}} M_{\text{Moon}}}{\left(d_{\text{Earth-Moon}}\right)^2}.$$

To compare the gravitational forces, divide (multiply by the reciprocal):

$$\frac{F_{\text{Earth-Sun}}}{F_{\text{Earth-Moon}}} = \frac{GM_{\text{Earth}} M_{\text{Sun}}}{\left(d_{\text{Earth-Sun}}\right)^2} \div \frac{GM_{\text{Earth}} M_{\text{Moon}}}{\left(d_{\text{Earth-Moon}}\right)^2} = \frac{GM_{\text{Earth}} M_{\text{Sun}}}{\left(d_{\text{Earth-Sun}}\right)^2} \times \frac{\left(d_{\text{Earth-Moon}}\right)^2}{GM_{\text{Earth}} M_{\text{Moon}}}$$

$$= \frac{M_{\text{Sun}}}{M_{\text{Moon}}} \left(\frac{d_{\text{Earth-Moon}}}{d_{\text{Earth-Sun}}}\right)^2.$$

From the problem statement, the mass of the Sun is about 25 million (2.5×10^8) times that of the Moon and the Earth–Sun distance is about 400 times that of the Earth–Moon distance. That is,

$$\frac{M_{\text{Sun}}}{M_{\text{Moon}}} = 2.5 \times 10^8 \quad \text{and} \quad \frac{d_{\text{Earth-Moon}}}{d_{\text{Earth-Sun}}} = 400.$$

Tides also are caused by gravity. However, the tidal force is related to the *difference* between the force of gravity on the part of the Earth nearer the Moon (or Sun) and the force of gravity on the part of the Earth farther from the Moon (or Sun). Thus, because the Moon is much closer to the Earth than the Sun, the Moon has a greater tidal effect on the Earth.

Substituting these values into the relation comparing the gravitational forces yields

$$\frac{F_{\text{Earth-Sun}}}{F_{\text{Earth-Moon}}} = \frac{M_{\text{Sun}}}{M_{\text{Moon}}}\left(\frac{d_{\text{Earth-Moon}}}{d_{\text{Earth-Sun}}}\right)^2 = \left(2.5 \times 10^8\right) \times \left(\frac{1}{400}\right)^2 = 1563.$$

The force of gravity between the Earth and the Sun is more than 1500 times greater than the force of gravity between the Earth and the Moon. That is why the Earth orbits the Sun, and not the Moon!

❖

Finding the Acceleration of Gravity

What can we do with this mathematical statement about the law of gravity? By combining it with Newton's second law of motion (subsection 9.4.3) we can explain the motion of falling objects, baseballs hit into the air, the Moon going around the Earth, and the Sun orbiting in the Milky Way galaxy! Let's consider the case of falling objects.

If you drop a rock, the force acting on the rock is the force of gravity. The two masses involved are the masses of the Earth and the rock, denoted M_{Earth} and M_{rock}, respectively. The distance between their *centers* is the distance from the *center of the Earth* to the center of the rock. If the rock isn't too far above the Earth's surface, this distance is approximately the radius of the Earth, R_{Earth} (about 6400 kilometers); that is, $d = R_{\text{Earth}}$. Thus the force of gravity acting on the rock is

$$F_g = G\frac{M_{\text{Earth}}M_{\text{rock}}}{d^2} = G\frac{M_{\text{Earth}}M_{\text{rock}}}{\left(R_{\text{Earth}}\right)^2}.$$

According to Newton's second law of motion, the force is equal to the product of the mass and acceleration of the rock, or

$$F_g = G\frac{M_{\text{Earth}}M_{\text{rock}}}{\left(R_{\text{Earth}}\right)^2} = M_{\text{rock}}a_{\text{rock}}.$$

Dividing both sides by the mass of the rock yields an expression for the acceleration of the rock:

$$a_{\text{rock}} = G\frac{M_{\text{Earth}}}{\left(R_{\text{Earth}}\right)^2}.$$

Note that the mass of the rock does *not* appear in the expression for acceleration, reflecting the fact that all objects fall at the same rate under gravity (in the absence of friction, wind resistance, or other forces). Thus this acceleration applies not only to the rock, but to *any* object falling near the surface of the Earth: It is the **acceleration of gravity**, g, which has the same effect on all falling objects (see subsection 9.4.2).

Knowing the Earth's mass (6.0×10^{24} kg) and radius (6.4×10^3 km), you can confirm that g is 9.8 meters per second:

$$g = G\frac{M_{\text{Earth}}}{\left(R_{\text{Earth}}\right)^2} = \left(6.67 \times 10^{-20}\,\frac{\text{km}^3}{\text{kg} \times \text{s}^2}\right) \times \frac{6.0 \times 10^{24}\ \text{kg}}{\left(6.4 \times 10^3\ \text{km}\right)^2} = 0.0098\,\frac{\text{km}}{\text{s}^2} = 9.8\,\frac{\text{m}}{\text{s}^2}.$$

With this mathematical description of the acceleration of gravity, you can address many more problems. Suppose, for example, that you want to know the weight of an object on the Moon. Recall that weight is mass times the acceleration of gravity (see Section 9.4), so the difference in weight on the Earth and the Moon is the difference in their accelerations of gravity. If you look at the formula for the acceleration of gravity on the Earth, it should be clear that the acceleration of gravity on the Moon must be

$$g_{\text{Moon}} = G\frac{M_{\text{Moon}}}{(R_{\text{Moon}})^2}.$$

If you substitute all the numbers, you will find that the acceleration of gravity on the Moon is about 1/6 that on the Earth. Thus objects on the Moon weigh about one sixth of what they would weigh on Earth. If you can lift a 50 kilogram bag of rock on the Earth, you'll be able to lift six such bags on the Moon!

Time-Out to Think: Use your calculator to find the acceleration of gravity on the Moon. You will need these facts: The mass of the Moon is about 7.4×10^{22} kg and its radius is about 1.74×10^3 km. Is your result about 1/6 of the acceleration of gravity on the Earth?

Example 16–31 ***Gravity in Earth Orbit.*** What is the gravitational acceleration of the space shuttle, orbiting 300 kilometers above the surface of the Earth? Compare the result to the acceleration of gravity in this low Earth orbit to that at the Earth's surface.

Solution: Set the gravitational force on the shuttle equal to its mass times acceleration, then solve for its acceleration:

$$G\frac{M_{\text{Earth}}M_{\text{shuttle}}}{d^2} = M_{\text{shuttle}}a_{\text{shuttle}} \quad \Rightarrow \quad a_{\text{shuttle}} = G\frac{M_{\text{Earth}}}{d^2}.$$

In this case, the distance d is the radius of the Earth (6400 km) *plus* the 300 km altitude of the shuttle, or $d = 6700$ km. Thus the gravitational acceleration of the shuttle when orbiting the Earth is

$$a_{\text{shuttle}} = G\frac{M_{\text{Earth}}}{d^2} = \left(6.67\times10^{-20}\ \frac{\text{km}^3}{\text{kg}\times\text{s}^2}\right)\times\frac{6.0\times10^{24}\ \text{kg}}{(6,700\ \text{km})^2} = 0.0089\frac{\text{km}}{\text{s}^2} = 8.9\frac{\text{m}}{\text{s}^2}.$$

Recall that a common misconception is that shuttle astronauts are weightless because of an absence of gravity. However, as this example shows, gravity is nearly as strong in Earth orbit as on the ground. The astronauts are weightless because they are in free fall — see Section 9.4 for further discussion.

The acceleration of gravity in low Earth orbit is 8.9 meters per second squared, or only slightly less than the acceleration of gravity at the surface.

❖

16.5.2 Kepler's Third Law

Using the gravitational formula he had discovered, Newton was able to derive a formula relating the period p of an orbiting object to its average orbital distance a. Often called **Kepler's third law**, because it was first stated in 1617 (in a slightly different form) by Johannes Kepler as his third law of planetary motion, the formula is

$$p^2 = \frac{4\pi^2}{G(M_1 + M_2)}a^3, \quad \text{where} \begin{cases} M_1 \text{ and } M_2 \text{ are the masses of the two objects} \\ p = \text{orbital period (time to complete one orbit)} \\ a = \text{average orbital distance (between object centers)} \\ G = \text{the universal constant of gravitation.} \end{cases}$$

Kepler's third law includes the masses of the orbiting objects, which leads to an amazing result. Through observation of a pair of orbiting objects, both the orbital period, p, and the average orbital distance, a, often can be determined. Then, Kepler's third law allows calculation of the *masses* of the orbiting objects. It is through this law that astronomers have determined the mass of the Earth, the mass of the Sun, the masses of all the planets, and the masses of distant stars!

Example 16–32 *The Mass of the Sun.* The Earth orbits the Sun once each year, at an average distance of about 150 million kilometers. Use these facts and Kepler's third law to find the mass of the Sun.

Solution: In this case, the two masses are the Earth and the Sun. The Earth's orbital period is $p_{Earth} = 1$ year and its average orbital distance is $a_{Earth} = 1.5 \times 10^8$ km, so Kepler's third law for this problem becomes

$$(p_{Earth})^2 = \frac{4\pi^2}{G(M_{Sun} + M_{Earth})}(a_{Earth})^3.$$

Note that this formula includes the *combined* mass of the Earth and Sun. However, because the Earth is much less massive than the Sun, the sum of their masses is approximately the mass of the Sun alone, or

$$M_{Sun} + M_{Earth} \approx M_{Sun}.$$

Use this approximation to rewrite Kepler's third law for this problem as

$$(p_{Earth})^2 = \frac{4\pi^2}{GM_{Sun}}(a_{Earth})^3.$$

Now, solve for the mass of the Sun by multiplying both sides by M_{Sun} and dividing both sides by $(p_{Earth})^2$:

$$(p_{Earth})^2 = \frac{4\pi^2}{GM_{Sun}}(a_{Earth})^3 \quad \Rightarrow \quad M_{Sun} = \frac{4\pi^2(a_{Earth})^3}{G(p_{Earth})^2}.$$

Because G, as given earlier, carries units of seconds, you must convert the period, $p_{Earth} = 1$ year, to seconds:

$$p_{Earth} = 1 \text{ yr} \times \frac{365 \text{ day}}{1 \text{ yr}} \times \frac{24 \text{ hr}}{1 \text{ day}} \times \frac{60 \text{ min}}{1 \text{ hr}} \times \frac{60 \text{ s}}{1 \text{ min}} = 3.15 \times 10^7 \text{ s}.$$

Substituting this value and $a_{Earth} = 1.5 \times 10^8$ km into the formula yields

$$M_{Sun} = \frac{4\pi^2(a_{Earth})^3}{G(p_{Earth})^2} = \frac{4\pi^2(1.5 \times 10^8 \text{ km})^3}{\left(6.67 \times 10^{-20} \frac{\text{km}^3}{\text{kg} \times \text{s}^2}\right)(3.15 \times 10^7 \text{ s})^2} = 2 \times 10^{30} \text{ kg}.$$

The mass of the Sun is about 2×10^{30} kg.

❖

Time-Out to Think: Contemplate the momentous nature of the result in the preceding example. Knowing only the length of a year and the distance to the Sun (both easily measured quantities), you can calculate the mass of a star! Explain how a similar procedure could be used to calculate the mass of Jupiter by observing one of its satellites, the masses of stars in binary systems (where two stars orbit one another), or the mass of the Milky Way galaxy from knowing the Sun's location and orbital period.

16.6* THE SPECIAL THEORY OF RELATIVITY

The idea that all motion is relative may seem strange. For example, imagine that you are sitting still at a traffic light and see another car drive by. Isn't it obvious who is moving and who is not? To understand why it is *not* obvious, consider this: The rotation of the Earth every 24 hours means that even when you think you are "sitting still" you are moving around the Earth's center at a speed of more than 1000 kilometers per hour (at the latitudes of the continental United States). You also are traveling around the Sun at a speed of about 100,000 kilometers per hour, carried with the Earth in its orbit. The Sun, in turn, orbits the center of the Milky Way galaxy at nearly 1 million kilometers per hour. And there are many other motions in the cosmos.

By the end of the 1800s, several discoveries had been made that could not fully be explained by Newton's laws. Unaware of these discoveries at the time, a 16-year-old school dropout from Germany began to contemplate an astounding question: What would the world look like if you could travel at the speed of light? The boy, Albert Einstein, found that he inevitably encountered paradoxes when attempting to answer the question logically. As with many paradoxes, the eventual solution led to deep, new insights.

The theory of relativity encompasses the very ideas of space, time, and motion. Einstein developed and published the theory of relativity in two parts. The first part, called the *special theory of relativity*, was published in 1905. It is *special* because it applies only to the special case of motion at constant velocity; that is, it cannot be used for problems that involve acceleration. After 10 years of further work, Einstein generalized the theory to include all motion in his *general theory of relativity*, published in 1915. One of the astonishing results of the general theory of relativity is that gravity can be understood as the curvature of space and time, which we discussed briefly in Section 11.6. In this section, we discuss briefly the special theory of relativity.

The theory of relativity can be derived from just two basic facts. The first is that all motion is relative. To understand this statement, imagine that you are sitting still and you see someone move past you; that is, you claim to be at rest and claim that the other person is in motion. According to this first "principle of relativity," the other person can equally well claim that *you* are moving past him or her, while he or she is at rest.

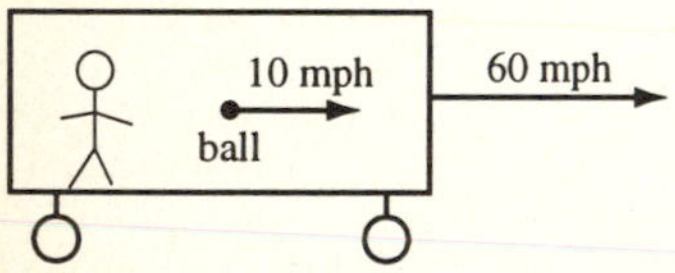

Figure 16-9. Suppose that you toss a ball at a speed of 10 miles per hour inside a train moving at 60 miles per hour. To someone on the ground outside the train, the ball appears to be moving at 70 miles per hour.

The second fact underlying relativity is that the speed of light always is the same, c = 300,000 km/s, regardless of motion. Note that this is not what you would expect in everyday life! For example, imagine that you are on a train traveling at 60 miles per hour from New York to Boston. Inside the train, you toss a ball to a friend sitting a few rows ahead, throwing the ball with a speed of 10 miles per hour. A person outside the train would see the ball traveling at 70 miles per hour: the 60 miles per hour that the train is moving, plus the additional 10 miles per hour at which you throw the ball (Figure 16–9).

Time-Out to Think: Suppose that you throw the ball to someone sitting a few rows behind you, instead of in front of you. In that case, how fast does a person outside the train see the ball moving?

Adding the velocities of the train and the ball is a natural, common-sense idea. *But light is different.* Imagine that you are on a train traveling at half the speed of light (a fast train!). Inside, you turn on a flashlight and shine it toward the front of the train. You, of course, see the light move through the train at the speed of light. But how fast would a person outside the train see the light moving? If light were like baseballs, the answer would be 1½ times the speed of light: the speed of light that the flashlight beam travels plus the speed of the train, which is half the speed of light. But that's wrong! Instead, the answer is that the outside person would see the light traveling at the speed of light (Figure 16–10).

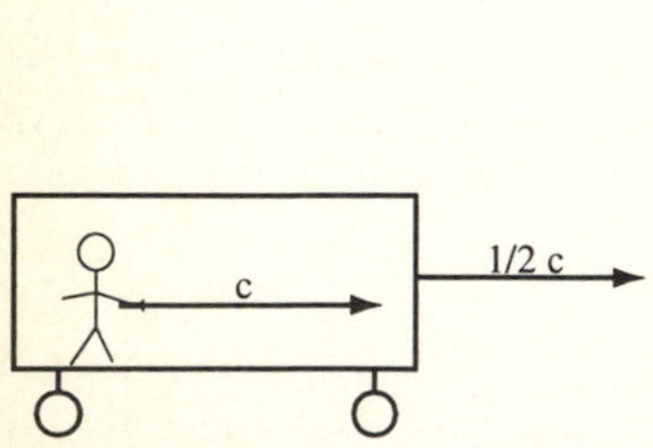

Figure 16-10. If you turn on a flashlight inside a train moving at 1/2 the speed of light, or 1/2 c, you will see the light beam going at the speed of light. "Common sense" suggests that someone outside the train should see the light moving at 1.5 c, but that is not the case. Instead, the person also sees the light traveling at c.

Before continuing, we need one more piece of terminology. Two people (or objects) are said to have the same frame of reference, or **reference frame**, if they have the same motion. For example, two people riding in the same car have the same reference frame. In contrast, if you are standing on the street and another person drives past you in a car, the two of you have different reference frames.

Time-Out to Think: Of the two starting points for relativity, one says that all motion is *relative*, whereas the other states that the speed of light is an *absolute*. Indeed, the fact the speed of light always is the same, regardless of motion, makes it one of the only "absolutes" in nature. Do you think that the theory of relativity is misnamed? If so, what name might you have given it instead?

16.6.1 Time Dilation

Starting from the two basic facts underlying relativity, a few simple *thought experiments* can show that time and space measurements depend on an observer's reference frame. These thought experiments are easiest if you imagine them to involve "space trains" traveling through space.

Suppose that you are in a space train, called train A, which you believe to be at rest. Space train B is moving past you to the right at some speed v, which is very close to the speed of light, as shown in Figure 16–11(a). Because all motion is relative, the passengers on train B could validly claim that *they* are at rest and that you are moving to the *left* at a speed v!

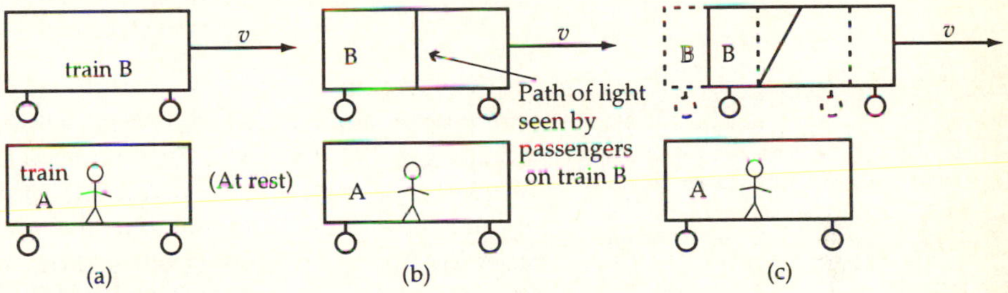

Figure 16-11. (a) As a passenger on train A, you believe that you are at rest while train B moves by at a speed *v*. (b) On train B, a light is shining from the floor to the ceiling. According to passengers on train B, the light traces a vertical path from floor to ceiling. (c) From your point of view, train B is moving forward while the light makes its way from the floor to the ceiling. As a result, you see the light in train B follow the slanted path shown.

Next, imagine that a light shines from the floor to the ceiling in train B. What does the path of the light beam look like? According to the passengers in train B, who believe that they are at rest, the light simply follows a vertical path from the floor to the ceiling (Figure 16–11b). However, from your vantage point in train A, you see something different: While the light is on its way from the floor to the ceiling of train B, the entire train is moving forward. As a result, the path of the light beam that you see is slanted in the direction that the train is moving (Figure 16–11c). The faster the train is moving, the more the path would slant.

Time-Out to Think: The speed of light is extremely fast, so light travels from the floor to the ceiling in a tiny fraction of a second. Thus, in order to see a slanted path for the light, the train would have to be moving very fast — a substantial fraction of the speed of light. Convince yourself that at "normal" speeds (like the speeds of cars, airplanes, or space shuttles) the slant would be scarcely noticeable. Thus, at normal speeds, the effects of relativity on time and space are virtually unnoticeable.

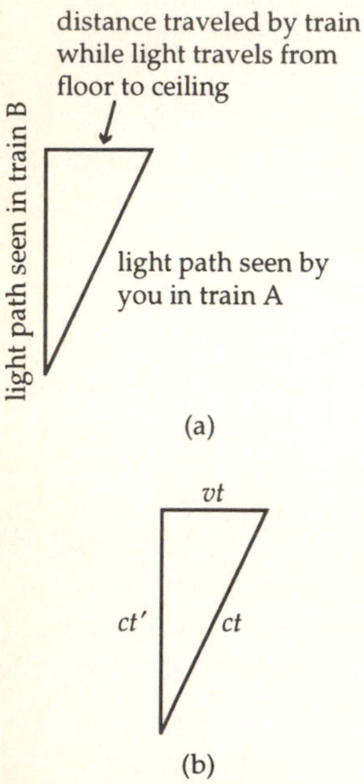

Figure 16-12. (a) For greater clarity, the right triangle from Figure 16–11(c) is shown. Its three sides are formed by: the light path seen by you in train A; the light path seen by passengers in train B; and the motion of train B. (b) The lengths of the triangle sides are relabeled in terms of the speed of light, c, the speed of the train, v, and the elapsed time for the light to travel each path. Note that the time you measure in train A, or t, is *different* from the time measured by passengers in train B, or t'.

The two different views of the light path shown in Figure 16–11(c), along with the moving ceiling of train B, form a right triangle (Figure 16–12a). The height of this triangle is the vertical path of the light seen by passengers in train B; the hypotenuse is the slanted path that *you* see from train A; and the remaining side is the distance that the train travels (as seen by you in train A) while the light is traveling from the floor to the ceiling.

To work with this triangle mathematically, you need to express the lengths of the sides in terms of the speed of light, c, and the speed of the train, v. Recalling that distance = speed × time, you can do so as follows.

- First, measure the time required for the light to travel from the floor of the train to the ceiling, and call the result a time t. Because you observe the light traveling on the slanted path and the light moves at speed c, the length (distance) of the slanted path must be $c \times t$.
- The top of the triangle is the distance the train travels during the time, t, that the light is traveling from the floor to the ceiling. Because the train is traveling at a speed v, the length of this side must be $v \times t$.
- The vertical side of the triangle is the light path observed by *passengers in train B*. Suppose that they measure the time required for the light to travel from the floor of the train to the ceiling and that their result is a time t' (read as "t-prime"). Because the light moves at speed c, the length of the vertical path must be $c \times t'$.

The lengths are shown on the triangle in Figure 16–12(b). Upon close examination, you should notice something astonishing: Because the distance ct is longer than the distance ct' and c is the same constant for everyone, t must be longer than t'! Both you (on train A) and the passengers on train B are measuring the same phenomenon — the time required for light to travel from the floor to the ceiling of the train. Yet you measure a *longer* time than the passengers on train B. Therefore *time itself* must be slower on train B than it is for you on train A. You have uncovered one of the most mind-bending consequences of relativity, called **time dilation**: Time is slower in moving reference frames than in your own reference frame.

How much slower is time on train B? You can find the answer with the help of the Pythagorean theorem, which has the form: $(\text{side})^2 + (\text{side})^2 = (\text{hypotenuse})^2$. Applying it to the triangle in Figure 16–12(b) yields $(ct')^2 + (vt)^2 = (ct)^2$, which can be solved for t' with a bit of algebra.

Start with Pythagorean theorem: $(ct')^2 + (vt)^2 = (ct)^2$

Expand the squares: $c^2t'^2 + v^2t^2 = c^2t^2$

Subtract v^2t^2 from both sides: $c^2t'^2 = c^2t^2 - v^2t^2$

Divide both sides by c^2: $t'^2 = \dfrac{c^2t^2 - v^2t^2}{c^2}$

Factor out t^2 on right side and simplify: $t'^2 = t^2 \times \dfrac{c^2 - v^2}{c^2} = t^2 \times \left(1 - \dfrac{v^2}{c^2}\right)$

Take square root of both sides: $t' = t\sqrt{1 - \dfrac{v^2}{c^2}}$, or $t' = t\sqrt{1 - \left(\dfrac{v}{c}\right)^2}$.

If the results so far are not bizarre enough, consider the following. Suppose that we look at the problem from the perspective of the passengers on train B, who believe that *they* are at rest and that *you* are in motion. The only change in the thought experiment is that, now, *you* see the light on the shorter vertical path and *they* see it on the longer slanted path. The result is that the passengers on train B will say that *your* time is slow, not theirs! As all motion is relative, both points of view are equally valid. As long as you are in motion relative to them, *you* will claim that their time is slow and *they* will claim that your time is slow. Despite the apparent paradox, this argument poses no physical problems. As long as you are moving relative to one another, you can't be together comparing your clocks.

The final result is called the **time dilation formula**. Because no material object can ever reach the speed of light, v always is less than c, so v/c always is less than 1. Therefore the term in the square root of the formula always is less than 1, so t' always is less than t. In other words, time passes more slowly in the moving reference frame of train B.

Example 16–33 *Comparing Times on Two Trains.* Suppose that you are on train A and that train B is moving past you at 70% of the speed of light. While 1 hour passes for you, how much time passes on train B?

Solution: Train B's speed is $v = 0.7c$ (70% of the speed of light), or $v/c = 0.7$. The variable t represents *your* time, so $t = 1$ hour. Substituting into the time dilation formula yields

$$t' = t\sqrt{1-\left(\frac{v}{c}\right)^2} = (1\text{ hr})\sqrt{1-(0.7)^2} = (1\text{ hr})\sqrt{1-0.49} = (1\text{ hr})\sqrt{0.51} = 0.714\text{ hr}.$$

Converting to minutes, this result is about

$$0.714\text{ hr} \times \frac{60\text{ min}}{1\text{ hr}} \approx 43\text{ min}.$$

While one hour passes for you, only 43 minutes passes on train B.

❖

Example 16–34 *A Long Trip.* Suppose that you stay home on Earth while your twin sister takes a trip to a distant star in a spaceship that travels at 99% of the speed of light. If both of you are 25 years old when she leaves and you are 45 years old when she returns, how old is your sister when she gets back?

Solution: Your sister's speed is $v = 0.99c$, or $v/c = 0.99$. The time that passes for you is $t = 20$ years (you age from 25 to 45). The time that passes for your sister is

$$t' = t\sqrt{1-\left(\frac{v}{c}\right)^2} = (20\text{ yr})\sqrt{1-(0.99)^2} = (20\text{ yr})\sqrt{1-0.9801} = (20\text{ yr})\sqrt{0.0199} = 2.8\text{ yr}.$$

This example underlies the *twin paradox* in relativity. A full discussion of the paradox and its resolution is beyond the scope of our work here, but the end result is that you would get the same answer: your sister really will be younger than you when she returns!

While 20 years pass for you, only a little less than 3 years pass for your traveling sister. This is a *real* difference: Although 20 years will have passed on Earth, only 3 years will have passed for your sister — she will return only 3 years older at age 28. Although you were the same age before she left, you now are 17 years older than she is.

❖

Example 16–35 *Not an Everyday Occurrence.* Suppose that a friend takes a trip in an airplane traveling at 500 miles per hour. If the trip takes 10 hours according to *you* (on the ground), how much time passes in the airplane?

Solution: The first step is to express the speed of the airplane as a fraction of the speed of light. Divide the airplane speed of 500 miles per hour by the speed of light and carry out the necessary unit conversions:

$$\frac{v}{c} = \frac{500\ \frac{\text{mi}}{\text{hr}} \times \left(\frac{1}{60}\frac{\text{hr}}{\text{min}}\right) \times \left(\frac{1}{60}\frac{\text{min}}{\text{s}}\right) \times \left(1.61\frac{\text{km}}{\text{mi}}\right)}{3\times10^5\ \frac{\text{km}}{\text{s}}} = \frac{0.22\ \frac{\text{km}}{\text{s}}}{3\times10^5\ \frac{\text{km}}{\text{s}}} = 7.5\times10^{-7}.$$

Substitute this number and the $t = 10$ hours that pass for you into the time dilation formula:

$$t' = t\sqrt{1-\left(\frac{v}{c}\right)^2} = (10\text{ hr})\sqrt{1-\left(7.5\times 10^{-7}\right)^2}$$
$$= (10\text{ hr})\sqrt{1-5.6\times 10^{-13}} = (10\text{ hr})\sqrt{0.99999999999944} = 10\text{ hr}.$$

To the accuracy of most calculators, the final square root is indistinguishable from 1. Thus the time that passes on the airplane is virtually indistinguishable from the time that passes on the ground. Note that the speed of an airplane is less than *one millionth* (10^{-6}) of the speed of light, demonstrating that time dilation is difficult to notice at speeds that are small compared to the speed of light.

❖

16.6.2 Effects on Distance and Mass

Just as time is different in different reference frames, distances and masses also are affected. Although we don't present detailed derivations here, you can easily see why that must be the case. Because time is affected by motion, and distance and time are related (distance = speed × time), distance must also be affected by motion. The resulting **length contraction formula** is

$$\text{length in moving reference frame} = (\text{rest length})\sqrt{1-\left(\frac{v}{c}\right)^2}.$$

The mass increase formula provides one way to see that exceeding the speed of light is impossible. Note that, as v approaches c, the moving mass approaches infinity. Because force = mass × acceleration, more and more force is needed to continue to accelerate an object when its speed nears the speed of light. With mass going to infinity at the speed of light, infinite force would be needed to make the object move any faster. Infinite force is not possible, so no object can exceed (or even reach) the speed of light.

To understand why mass is different in different reference frames, recall that Newton's second law states that mass is related to acceleration (force = mass × acceleration) and that acceleration has units of distance divided by time squared. As both distance and time are affected, acceleration and mass also must be affected. In fact, mass increases as velocity increases. The **mass increase formula** is

$$\text{moving mass} = \frac{(\text{rest mass})}{\sqrt{1-\left(\frac{v}{c}\right)^2}}.$$

Example 16–36 ***Length Contraction.*** Suppose that passengers on train B measure it to be 100 meters long. Because the passengers believe the train to be at rest, their measurement is its rest length. If train B is moving past you at 99% of the speed of light, how long would you measure its length to be?

Solution: Use the length contraction formula to calculate the length of the moving train:

$$\text{length in moving reference frame} = (\text{rest length})\sqrt{1-\left(\frac{v}{c}\right)^2} = (100\text{ m})\sqrt{1-(0.99)^2} = 14\text{ m}.$$

You would measure the train to be only 14 meters long, not 100 meters long.

❖

Example 16–37 ***Mass Increase.*** A small fly has a mass of 1 gram at rest. It is an unusual fly, however, in that it can travel at 99.99% of the speed of light. What is the mass of the fly at that speed?

Solution: Use the mass increase formula to calculate the mass of the moving fly:

$$\text{moving mass} = \frac{(\text{rest mass})}{\sqrt{1-\left(\frac{v}{c}\right)^2}} = \frac{1\text{ g}}{\sqrt{1-(0.9999)^2}} = 70.7\text{ g}.$$

The mass of the fly is almost 71 grams, or more than 70 times its rest mass.

❖

16.6.3 Is It True?

A natural reaction upon first learning about relativity is to think: That can't be right. After all, it seems to violate common sense. However, the effects of relativity are different from common-sense expectations only at speeds thousands of times faster than those we travel at in our daily lives. As a result, we don't have any *common* sense regarding such speeds because we don't commonly travel at them!

One way of convincing yourself of the validity of relativity is through a logical argument, similar to the logic that Einstein followed when he first encountered the paradox of imagining travel at the speed of light. However, the ultimate test of relativity is through experimentation. Like any good theory, relativity makes concrete predictions that can be tested and verified. For example, the prediction of time dilation can be tested by observing subatomic particles at high speeds.

Subatomic particles can be produced in particle accelerators that are used by physicists to study atoms and nuclei. Many of these particles have very short lifetimes; that is, they quickly decay into other particles. For example, a particle called the π^+ meson, when not moving, has a lifetime of about 18 billionths of a second. But, when π^+ mesons are produced at high speed in particle accelerators they last much longer than 18 billionths of a second — as predicted by the time dilation formula.

Such experiments have been carried out thousands of times. In every case, the results have matched the predictions of Einstein's theory of relativity to within the uncertainty range of the experiments. Because modern instruments make such measurements with extreme precision and accuracy, the theory of relativity may well be the best tested theory of all time (no pun intended!).

Example 16–38 ***Time Dilation with Subatomic Particles.*** Suppose that a π^+ meson is produced in a particle accelerator traveling at $0.998c$. According to the time dilation formula, how quickly will the particle decay?

Solution: The π^+ meson will always "think" it is at rest and therefore decay after 18 billionths of a second (1.8×10^{-8} s) in its *own* reference frame. However, from the point of view of the scientists conducting the experiment the π^+ meson represents the *moving* reference frame. Hence the π^+ meson's lifetime of 1.8×10^{-8} s represents t' in the time dilation formula, or $t'=1.8 \times 10^{-8}$ s. Solve for the lifetime observed by the scientists, t, dividing both sides of the time dilation formula by the square root term:

$$t' = t\sqrt{1-\left(\frac{v}{c}\right)^2} \quad \Rightarrow \quad t = \frac{t'}{\sqrt{1-\left(\frac{v}{c}\right)^2}} = \frac{1.8\times10^{-8}\text{ s}}{\sqrt{1-(0.998)^2}} = \frac{1.8\times10^{-8}\text{ s}}{0.0632} = 2.8\times10^{7}\text{ s}.$$

When produced at 0.998*c*, the π^+ meson is expected to last about 280 billionths of a second, rather than its "normal" lifetime of 18 billionths of a second. That is, its moving lifetime is more than 15 times as long as it is at rest. Predictions such as this, based on relativity, have been verified by experiments.

❖

16.6.4 $E = mc^2$

In addition to direct testing of the theory, relativity also predicts numerous consequences that can be tested *indirectly*. The most famous of these consequences is $E = mc^2$. Let's investigate.

We begin with the mass increase formula, calling the moving mass m and the rest mass m_0.

$$\text{moving mass} = \frac{\text{(rest mass)}}{\sqrt{1-\left(\frac{v}{c}\right)^2}}, \quad \text{or} \quad m = m_0\left(1-\frac{v^2}{c^2}\right)^{-\frac{1}{2}}.$$

To further explore this formula, we need a special mathematical approximation. For small values of x (i.e., values much less than 1), the following approximation holds:

$$(1+x)^{-\frac{1}{2}} \approx 1-\frac{1}{2}x, \text{ if } x \text{ is small compared to 1.}$$

Time-Out to Think: Try this approximation with a few small values of x, such as $x = 0.05$ or $x = 0.001$. Can you convince yourself that the approximation works?

Note what happens if we substitute $x = -v^2/c^2$ in this approximation.

Start with approximation: $$(1+x)^{-\frac{1}{2}} \approx 1-\frac{1}{2}x$$

Substitute $x = -v^2/c^2$: $$\left(1+\underbrace{-\frac{v^2}{c^2}}_{x}\right)^{-\frac{1}{2}} \approx 1-\frac{1}{2}\underbrace{\left(-\frac{v^2}{c^2}\right)}_{x}$$

Simplify: $$\left(1-\frac{v^2}{c^2}\right)^{-\frac{1}{2}} \approx 1+\frac{1}{2}\frac{v^2}{c^2}.$$

We can use this new form of the approximation to rewrite the mass increase formula. The condition that x must be small compared to 1 for this approximation to be valid now means that $-v^2/c^2$ must be small; that is, the speed v must be small compared to the speed of light c.

Start with mass increase formula: $$m = m_0\left(1-\frac{v^2}{c^2}\right)^{-\frac{1}{2}}$$

Substitute $\left(1-\frac{v^2}{c^2}\right)^{-\frac{1}{2}} \approx 1+\frac{1}{2}\frac{v^2}{c^2}$: $$m \approx m_0\left(1+\frac{1}{2}\frac{v^2}{c^2}\right)$$

Expand right side: $$m \approx m_0 + \frac{1}{2}\frac{m_0 v^2}{c^2}$$

Multiply both sides by c^2: $$mc^2 \approx m_0 c^2 + \frac{1}{2} m_0 v^2.$$

You may recognize the last term on the right as the *kinetic energy* (see Section 9.5) of an object with mass m_0. Because the other two terms also have units of mass multiplied by speed squared, they also must represent some kind of energy. Einstein recognized that the term on the left represents the *total* energy of a moving object. He then noticed that, even if the speed is zero ($v = 0$) so that there is *no* kinetic energy, the equation states that the total energy is *not* zero; instead, it is m_0c^2. In other words, when an object is not moving at all, it still contains energy by virtue of its mass! The amount of energy, as we have just shown, is $E = m_0c^2$.

Thus $E = mc^2$ is a direct consequence of Einstein's theory of relativity. In that sense, any experiment that confirms the validity of $E = mc^2$ also confirms all the other bizarre consequences of the theory. The successful operation of nuclear reactors, the destructive force of nuclear bombs, and the generation of energy by the Sun all therefore provide concrete evidence that Einstein's theory is valid.

Time-Out to Think: Step back and consider what you have just read. Starting from the fact that the speed of light always is the same, we derived the time dilation formula and obtained the related length contraction and mass increase formulas. Then, by applying a bit of algebra to the mass increase formula, we proved that $E = mc^2$! A common myth holds that the mathematics of relativity is too hard for the average person. Any comments?

16.6.5 Faster Than Light?

Another consequence of the theory of relativity is that nothing can go faster than the speed of light. We can prove this fact with a logical argument, based on the fact that the speed of light always is the same.

Imagine that you have just built the most incredible rocket imaginable. You get in it to go for a test spin and blast off. Soon you are going faster than anyone had ever imagined possible — then you put it into second gear! You keep accelerating, and keep going faster and faster and faster. Note that we haven't set any limits on your speed; you simply are going as fast as you can possibly go. Here is the key question: Are you going faster than the speed of light?

To answer this question, imagine that a friend of yours back on Earth turns on a flashlight and shines the beam in your direction. Based on the fact that everyone always measures the same speed of light, you eventually will observe the flashlight beam passing by you at the speed of light. Now, consider your friend's perspective on Earth. She also must see the light beam going the speed of light. Hence, because the light eventually catches and passes you, she concludes that your speed is *slower* than the speed of light.

As evidence that the inability to exceed the speed of light is well established, consider that not even science fiction writers try to violate this law of the universe; instead, they look for ways to "leave" the universe so that they won't have to obey its laws. In the *Star Trek* series, for example, the *Enterprise* leaves the ordinary universe by going into "warp"; in the *Star Wars* series, spaceships got around the speed of light barrier by entering "hyperspace" (or four-dimensional space).

Time-Out to Think: Take a couple of moments to review the preceding thought experiment. Are you convinced?

The logic of our thought experiment is airtight. As long as everyone always measures the same speed of light (which can be verified by experiment), exceeding the speed of light is impossible. That is, exceeding the speed of light isn't merely a technological challenge; rather, it simply cannot be done.

THINKING ABOUT ... ALBERT EINSTEIN

In Chapter 1, we used Einstein's equation $E = mc^2$ to demonstrate the interdisciplinary thinking necessary in the modern world. In fact, Einstein's entire life dramatizes the linkages among society, politics, individual behavior and choices, and mathematics and science. Einstein was born on March 14, 1879, in Ulm, Germany. As a toddler, he was so slow in learning to speak that relatives feared he might be retarded. Even when he began to show promise, it was in unconventional ways. He dropped out of high school at the suggestion of a teacher who told him that he would "never amount to anything."

An often told story that Einstein "failed" mathematics is not true; rather, his difficulty with teachers probably arose because he asked questions that they couldn't answer, and they therefore considered him disruptive to classroom discipline.

Einstein's ruminations about relativity began during a precollege vacation in northern Italy. He took the vacation deliberately to avoid military service in Germany — Einstein was a pacifist from the beginning. Despite dropping out of high school, Einstein was admitted to college in Switzerland, in large part because of his proficiency with mathematics. However, he didn't really enjoy college, preferring to spend his time reading theoretical physics while cutting most of his classes. He nevertheless managed to graduate, with the help of lecture notes taken by a friend.

After graduation, Einstein sought a teaching position but, because he was Jewish and not a Swiss citizen, no one would hire him. Eventually, the Swiss patent office hired him in 1901. While working there he spent his free time contemplating unsolved questions in physics and working toward his Ph.D. In 1905, sometimes called his "miracle year," Einstein published five papers in the *German Yearbook of Physics* and earned his Ph.D. Three of the papers solved what were, in retrospect, probably the three greatest mysteries in physics at the time.

The first paper dealt with what is called the photoelectric effect; this paper essentially presented the first concrete evidence of the wave–particle duality of light (see Section 5.3). The second paper dealt with the fact that suspended particles in water jiggle about even after the water has been still for a long time. Einstein analyzed this motion, and his work led to the first direct measurements of molecular motion and molecular and atomic sizes. The third paper presented the special theory of relativity.

Despite his tremendous discoveries, four more years passed before Einstein obtained a professorship. His reputation grew, and in 1913 he returned to Germany to work at the Kaiser Wilhelm Physical Institute in Berlin. When World War I broke out, many German scientists signed a prowar proclamation; Einstein (a Swiss citizen at the time) was one of a small number to sign a counter-proclamation calling for peace.

Einstein published the general theory of relativity in 1915. After some of its central predictions were observationally confirmed in 1919, Einstein's fame began to extend beyond the world of science. He received the Nobel Prize for Physics in 1921 and was by this time a household name. He was a visiting professor at the California Institute of Technology when Hitler came to power in Germany. Recognizing Hitler's evil, he decided to remain in the United States. He became an American citizen in 1940.

In 1939, the discovery of uranium fission led to the recognition that bombs could be built that would tap the energy stored in matter according to $E = mc^2$. Despite his horror at the thought of such bombs, Einstein was even more fearful that Hitler's scientists might develop them. He wrote a letter to President Franklin Roosevelt urging a gigantic research program to develop a nuclear bomb. The result was the Manhattan Project which, six years later, successfully developed the first nuclear weapons. To Einstein's further dismay, the bombs were used not against Hitler's Germany but against Japan; Germany had been defeated by the time the bombs were developed.

Einstein spent much of the rest of his life arguing for a worldwide agreement to ban the further manufacture and use of nuclear weapons. He also argued vociferously for democracy and human rights around the world. Scientifically, he continued to contribute to physics through insightful questions and teaching.

Einstein died in Princeton, New Jersey, on April 18, 1955. Because of Einstein, our view of the world is radically different than it was before his birth. We now know that space and time are intertwined in ways that would have been difficult to imagine before Einstein's work. The significance of this discovery was stated eloquently by Einstein himself, shortly before his death, when he said: *"This death signifies nothing ... the distinction between past, present, and future is only an illusion, even if a stubborn one."*

16.7 CONCLUSION

In this final chapter we presented several formulas that are among the most important and commonly used today. Although some of them *look* complex, you have the all the skills needed to work with them — and any other formula you encounter. Further, the concepts covered in this book will enable you to evaluate formulas critically as models and thereby to interpret the results in a proper context. That, in the end, is what we mean by *quantitative reasoning*, the title of this book. We hope that you now have much more confidence in your ability to reason quantitatively than you did a few months ago and wish you luck in your future endeavors.

PROBLEMS

Reflection and Review

Section 16.1

1. **Formulas You Use.** Identify and describe at least three formulas that you have used in the past in your work, your personal life, or in another course.

Section 16.2

2. **Holding Time Constant.** Suppose that you drive for 10 hours at a steady speed.
   a. Describe in words how the distance you travel depends on your speed.
   b. Write a relation that gives the distance traveled (in 10 hours) for any speed.
   c. Draw a graph that shows how the distance traveled (in 10 hours) varies with speed.

3. **Holding Distance Constant.** Suppose that you take a 400-mile trip and drive at a steady speed.
   a. Describe in words how the time for the trip depends on your speed.
   b. Write a relation that gives the travel for any speed.
   c. Draw a graph that shows how the travel time varies with speed.

4. **The Speed of Light.** Use the formula that relates the wavelength and frequency of a light wave to the speed of light to answer the following questions (1 nanometer = 10^{-9} meter, 1 Hertz = 1 cycle per second).
   a. What is the frequency of red light, which has a wavelength of about 700 nanometers?
   b. What is the frequency of an X-ray with a wavelength of 0.1 nanometer?
   c. What is the frequency of a gamma ray with a wavelength of 0.0005 nanometers?

d. What is the wavelength of infrared light with a frequency of 50 terahertz (trillion hertz)?
e. What is the wavelength of an FM radio station that broadcasts at 106.5 megahertz?
f. What is the wavelength of an AM radio station that broadcasts at 700 kilohertz?
g. Redraw the graph of the relation shown in Figure 16–4, but show only wavelengths of visible light (400 to 700 nanometers) on the horizontal axis.

5. **Landscaping.** Refer to Example 16–3. Suppose that the swimming pool is to be built in a yard that measures 12 meters by 20 meters. Find a relation that gives the area of the pool as it depends on the width of the border. Describe the meaning of the relation in words.

6. **Pricing Merchandise.** Study Example 16–4. Suppose that a marketing survey found that the store could sell 20 sets of tires a day at $75 per set and 10 sets of tires per day at $125 per set. What is the optimal price for maximizing revenue?

7. **Valid Formula?** Suppose that you read an article claiming to have found a formula that describes how the number of trees in a forest relates to the area of the forest. With n for the number of trees and A for the area of the forest, the author claims that n = constant ÷ A^2. Is this formula reasonable? Why or why not?

Section 16.3

8. **The Decibel Scale.** Answer each of the following questions. As necessary, refer to Table 16–2.
   a. What is the loudness, in decibels, of a sound 45 million times louder than the softest audible sound?
   b. What is the loudness, in decibels, of a sound 18 trillion times louder than the softest audible sound?
   c. What is the loudness, in decibels, of a sound 1000 times softer (less intense) than the softest audible sound?
   d. How much louder (more intense) is a 35-dB sound than a 10-dB sound?
   e. How much louder (more intense) is an 85-dB sound than a 15-dB sound?
   f. How much louder (more intense) is a 125-dB sound than a 95-dB sound?
   g. Suppose that a sound is 100 times louder (more intense) than a whisper. What is its loudness in decibels?
   h. Suppose that a sound is 2 million times louder (more intense) than the threshold of pain for the human ear. What is its intensity in decibels? How does this sound compare to that of a jet at 30 meters?

9. **Sound Intensity.** Calculate the absolute intensity, in watts per square meter, for each of the following sounds.
   a. A jet at 30 meters.
   b. A whisper.
   c. A sound of 62 decibels.
   d. A sound of 160 decibels.

10. **Variation in Sound with Distance.** Suppose that a siren is placed 0.1 meter from your ear.
    a. How many times louder will the sound you hear be than that of a siren at 30 meters (see Table 16–2)?
    b. How loud will the siren next to your ear sound, in decibels?
    c. How likely is it that this siren would cause damage to your ear drum? Explain.

11. **Earthquake Magnitudes.** Use the formula for the magnitude scale for earthquakes and data from Table 16–4, as needed, to answer each of the following questions.
    a. How much energy, in joules, is released by an earthquake of magnitude 5?
    b. How much more energy is released by an earthquake of magnitude 8 than one of magnitude 6?
    c. How much energy, in joules, was released by the 1985 earthquake in Mexico City?
    d. How much energy, in joules, was released by the 1960 earthquake in Morocco?
    e. Compare the energy of a magnitude 7 earthquake to that released by a 1 megaton nuclear bomb (5×10^{15} joules).
    f. What magnitude of earthquake would release an energy equivalent to that of a 1 megaton nuclear bomb? Which would be more destructive? Why?

12. **LA and China Earthquakes.** Compare the energies of the 1994 Los Angeles earthquake that killed 61 people and the 1976 Tangshan earthquake that killed on the order of half a million. Do you think that an earthquake of magnitude 7.9 could cause a similar number of deaths in Los Angeles? Why or why not?

13. **The pH Scale.** Consider the following questions about the pH scale.
    a. If the pH of a solution increases by 1 (e.g., from 4 to 5 or from 7 to 8), how much of a change does that represent in the hydrogen ion concentration? Does the increase in pH make the solution more acidic or more basic?
    b. If the pH of a solution increases by 1.5 (e.g., from 4 to 5.5 or from 7 to 8.5), how much of a change does that represent in the hydrogen ion concentration?
    c. What is the hydrogen ion concentration of a solution with pH 8.5?
    d. What is the hydrogen ion concentration of a solution with pH 3.5?
    e. What is the pH of a solution with a hydrogen ion concentration of 0.1 moles per liter? Is this solution an acid or base?
    f. What is the pH of a solution with a hydrogen ion concentration of 10^{-9} moles per liter? Is this solution an acid or base?

14. **Reviving a Larger Lake.** Refer to Example 16–17. Suppose that the lake is much larger, containing 10 billion liters of acidified water with pH 4. In this case, how much lime must be added to raise the pH of the lake to 6? Explain.

15. **Neutralizing Acid.** You can answer the following questions by using the method in Example 16–17.
    a. Suppose that you have 10 liters of vinegar with pH 3. How much lime of pH 12 would be needed to neutralize it? Explain.
    b. If you decide to neutralize the 10 liters of vinegar with baking soda (pH 8.4), how much would you need?

Section 16.4

16. **Compound Interest and Annual Yield.** Use the compound interest formula to answer each of the following.
   a. Suppose that you deposit $500 in a bank that offers an *APR* of 6.5% compounded daily. What is your balance after 6 months? after 1 year? What is the annual yield for this account?
   b. Suppose that you deposit $500 in a bank that offers an *APR* of 4.5% compounded monthly. What is your balance after 1 year? after 10 years? What is the annual yield for this account?
   c. Suppose that you deposit $800 in a bank that offers an *APR* of 7.25% compounded quarterly. What is your balance after 1 year? after 5 years? after 20 years? What is the annual yield for this account?

17. **Your Bank Account.** Find the current interest rate (*APR*) for your personal bank savings account (choose just one account if you have more than one, or pick a rate from a nearby bank if you don't have an account).
   a. How often is interest compounded on your account?
   b. Calculate the annual yield on your account by using the compound interest formula. Your bank also should state the annual yield on your account (call the bank and ask if you can't find it elsewhere). Does your calculation agree with what the bank claims?
   c. Suppose that you receive a gift of $10,000 and place it in your account. If you leave the money there, and the interest rate never changes, how much will you have in 10 years?
   d. Pick a compounding period that is different from yours. For example, if your bank compounds daily, you might pick monthly compounding. Calculate how much you would have after 10 years from the $10,000 in part (c) with this different compounding period. Briefly discuss your results.
   e. Suppose that you find another account that offers interest at an *APR* 2% higher than yours, with the same compounding period as your account. Calculate how much you would have after 10 years from the $10,000 in part (c) with this higher *APR*. Briefly discuss your results.

18. **Continuous Compounding.** The following questions allow you to explore continuous compounding further.
   a. Make a table similar to Table 16–7 for an *APR* of 12%.
   b. Use the formula for continuous compounding to find the annual yield at an *APR* of 12%.
   c. Show the results of parts (a) and (b) on a graph similar to that of Figure 16–7.
   d. Explain, in words, what you found about continuous compounding at an *APR* of 12%.
   e. Suppose that you deposit $500 in an account with an *APR* of 12%. With continuous compounding, how much money will you have at the end of 1 year? at the end of 5 years?
   f. Repeat parts (a)–(e) with an *APR* of 6%.

19. **Solving the Compound Interest Formula.** Answering each of the following questions involves solving the compound interest formula for a particular variable.
   a. How long will it take your money to triple at an *APR* of 8% compounded daily?
   b. How long will it take your money to triple at an *APR* of 8% compounded quarterly? Explain why your answer is either longer or shorter than your answer in part (a).
   c. How long will it take your money to grow by 50% at an *APR* of 7% compounded daily?
   d. If you deposit $1000 in an account that pays an *APR* of 7% compounded daily, how long will it take for your balance to reach $100,000?
   e. Suppose that you are giving a gift for a niece. You decide to give her a 10-year certificate of deposit that pays an *APR* of 9% compounded monthly. If you want the value of the account to be $10,000 at the end of the 10 years, how much should you deposit now?

20. **Investment Plans.** Answer each of the following questions with the help of the investment plan formula.
   a. To save money for retirement, you set up an IRA with an *APR* of 8% at age 25. At the end of each month you deposit $50 in the account. How much will the IRA contain when you retire at age 65? Compare that amount to the total amount of deposits made over the time period.
   b. A friend creates an IRA with an *APR* of 8.25%. Like you, she starts the IRA at age 25 and deposits $50 per month. How much will her IRA contain when she retires at age 65? Compare the value of her IRA to yours in part (a). Comment on the effect of a difference of only 0.25% in *APR*.
   c. You put $200 per month in an investment plan that pays an *APR* of 7%. How much money will you have after 18 years? Compare this amount to the total amount of deposits made over the time period.

21. **Investment Planning.** Answer each of the following questions with the help of the investment plan formula.
   a. You intend to create a college fund for your baby. If you can get an *APR* of 7.5% and want the fund to have $150,000 in it after 18 years, how much should you deposit monthly?
   b. At age 35 you start saving for retirement. If your investment plan pays an *APR* of 9% and you want to have $2 million when you retire in 30 years, how much should you deposit monthly?
   c. Suppose that you can afford to put $100 per month in an investment plan that pays an *APR* of 7%. When will the value of your investment reach $50,000? $1 million? (Hint: You will need to solve the investment plan formula for the number of years, *Y*.)

22. **Tax-Deferred Contributions.** Suppose that you are making $300 per month pension plan contributions that are tax deferred. After taxes and your contribution, your take-home pay is $1500 per month. Assume that, if you didn't make the tax-deferred contribution, you would have to pay 21% of the $300 in taxes.
   a. What would be your take-home pay if you didn't make the tax-deferred contributions to your pension plan?
   b. What is the effective cost to you of putting $300 per month into your pension plan?

c. Explain how the tax-deferred nature of the pension plan works to your benefit.

23. **Comparing Tax-Deferred Contributions.** Adrienne and Alphonse each make tax-deferred contributions of $200 per month to their pension plans. Adrienne is in a tax bracket having a marginal tax rate of 28%. Alphonse's marginal tax rate is 15%. By how much do the $200 pension contributions reduce each person's take-home pay? Discuss.

24. **Loan Payments.** Answer each of the following questions with the help of the loan payments formula.
a. Calculate the monthly payments on a student loan of $25,000 at a fixed rate of 10% for 20 years.
b. Calculate the monthly payments on a home mortgage of $150,000 with a fixed rate of 9.5% for 30 years.
c. Calculate the monthly payments on a home mortgage of $150,000 with a fixed rate of 8.75% for 15 years.
d. Suppose that you run up a $2000 credit card bill. You make arrangements with the bank to pay it off in 3 years at an interest rate of 14%. What are your monthly payments?

25. **Accelerated Payment.** Suppose that you have a student loan of $25,000 with a fixed rate of 9% for 20 years.
a. What are your required monthly payments?
b. If you decide to accelerate your payments and pay $350 per month, how long will it take to pay off the loan?

Section 16.5

26. **The Gravitational Law.** Use the gravitational law to answer each of the following questions.
a. How does tripling the distance between two objects affect the gravitational force between them?
b. Compare the gravitational force between the Earth and the Sun to that between Jupiter and the Sun. The mass of Jupiter is about 318 times the mass of the Earth.
c. Suppose that the Sun were magically replaced by a star with twice as much mass. What would happen to the gravitational force between the Earth and the Sun?
d. The mass of the Moon is about 7.4×10^{22} kg and its radius is about 1740 kilometers. What is the acceleration of gravity on the Moon?

27. **Using Kepler's Third Law.** The Moon orbits the Earth in about 27.3 days at an average distance of about 384,000 kilometers, and the Moon is much less massive than the Earth (by a factor of about 80). Use Kepler's third law to determine the mass of the Earth.

Section 16.6

28. **Formulas from Relativity.**
a. Suppose that you stay home on Earth while your friend takes a trip to a distant star in a spaceship that travels at 98% of the speed of light. If you both are 20 years old when she leaves and you are 60 years old when she returns, how old is your friend when she gets back?
b. Train B's passengers measure it to be 100 meters long. If train B is moving past you at 85% of the speed of light, what length would you measure for train B?
c. A person has a mass of 50 kg at rest. What is her mass when she is traveling past you at 98% of the speed of light?
d. A π^+ meson traveling at $0.9997c$ is produced in a particle accelerator. According to the time dilation formula, how long will particle decay take?

Further Topics and Applications

29. **Book Publishing Profits.** A marketing survey estimates that the linear demand relation for a best-selling applied mathematics textbook is $n = 500 - 5p$ where n is the number of books that can be sold in a week at a price of $\$p$.
a. Find the relation that gives total revenue R earned when the price of the book is set at $\$p$. Graph the revenue relation and estimate the optimal price (the price that maximizes revenues).
b. Now include the cost of producing the book. Assume that the printing cost consists of a one-time expense of $2000 for overhead plus a unit cost of $10 per book. Find the linear relation that gives the cost C of producing n books.
c. Now combine the revenue and cost relations. Let the profit P be the difference between revenue and costs (i.e., $P = R - C$), and develop a relation that describes the profit that can be expected if the retail price of the book is set at $\$p$. From your graph of the profit relation, estimate the retail price that will maximize profits.

30. **Sound and Distance.** Use the formula that relates the intensity of a sound to distance (subsection 16.3.1) to answer each of the following questions. Refer to Table 16–2, as necessary.
a. The decibel level for busy street traffic in Table 16–2 is based on the assumption that you stand very close to the noise source — say, 1 meter from the street. If your house is 100 meters from a busy street, how loud will the street noise be, in decibels?
b. At a distance of 10 meters from the speakers at a concert the sound level is 135 dB. How far away should you be sitting to avoid the risk of damage to your hearing?
c. Imagine that you are a spy at a restaurant. The conversation you want to hear is taking place in a booth across the room, about 8 meters away. The people speak in soft voices so that, to each other, they hear voices of about 20 dB (assume that they sit about 1 meter apart). How loud is the sound of their voices when it reaches your table? If you have a miniature amplifier in your ear and want to hear their voices at 60 dB, by what factor must their voices be amplified?

31. **Thinking in Logarithmic Scales.** Briefly describe, in words, the effects you would expect in each of the following situations.
a. You have your ear against a new speaker when it emits a sound with an intensity of 160 dB.

b. Your friend is calling you from across the street in New York City, with a shout that registers 90 dB. Traffic is heavy, and several emergency vehicles are passing by with sirens.
c. A jet plane flies overhead at an altitude of 1 kilometer.
d. An earthquake of magnitude 2.8 strikes the Los Angeles area.
e. An earthquake of magnitude 8.5 strikes the New York City area.
f. A forest situated a few hundred miles from a coal-burning industrial area is subjected regularly to acid rain, with pH 4, for many years.
g. A young child (too young to know better) finds and drinks from an open bottle of drain opener with pH 12.

32. Toxic Dumping in Acidified Lakes. Consider a situation in which acid rain has heavily polluted a lake to a level of pH 4. An unscrupulous chemical company dumps some acid into the lake illegally. Assume that the lake contains 100 million gallons of water and that the company dumps 100,000 gallons of acid with pH 2.
a. What is the hydrogen ion concentration, $[H^+]$, of the lake polluted by acid rain alone?
b. Suppose that the unpolluted lake, without acid rain, would have pH 7. If the lake were then polluted by company acid alone (no acid rain), what hydrogen ion concentration, $[H^+]$, would it have?
c. What is the concentration, $[H^+]$, after the company dumps the acid into the acid rain-polluted lake (pH 4)? What is the new pH of the lake?
d. If the U.S. Environmental Protection Agency can test for changes in pH of only 0.1 or greater, could the company's pollution be detected?

33. Neutralizing Acid Lakes. A "dead" lake has pH 4.8. What is its hydrogen ion concentration? How much lime (pH 12) must be added to neutralize the lake? Discuss the pros and cons of using lime to neutralize acidified lakes.

34. Acid Rain Analysis. The following quotation is from J. MacKenzie, "Breathing Easier: Taking Action on Climate Change, Air Pollution, and Energy in Security," Washington, DC: World Resources Institute, 1989, p. 12. Reprinted with permission. "Acid deposition and ground-level ozone, spread across the countryside and acting separately and together, are believed to be important contributors to the decline of forests in Europe, the United States, and Canada. Trees weakened by air pollution are more susceptible to such natural stresses as drought, insects, and weather extremes. Ozone, in turn, is causing widespread losses in crop productivity in the United States. Levels of these pollutants are high where tree and crop damages are found. Along the Appalachian Mountain chain, the average acidity of cloud moisture is ten times greater than that at nearby lower elevations and about one hundred times greater than that of unpolluted precipitation. The peak cloud acidity (pH of 2.3) at several eastern mountains is 2000 times greater than that of unpolluted rain water (pH of 5.6). In fact it approximates that of lemon juice."
a. Calculate the $[H^+]$ of the peak cloud acidity with pH 2.3.
b. Calculate the $[H^+]$ of unpolluted rain water with pH 5.6.
c. The report claims that the peak cloud acidity is 2000 times greater than the acidity of unpolluted rain water. Is this claim justified? Explain.
d. Pure water has pH 7.0, which is significantly different from the pH 5.6 of unpolluted rain water. Find the percentage difference in pH between pure water and rain water.
e. Calculate the $[H^+]$ of pure water and compute the percentage difference in $[H^+]$ between pure water and rain water.
f. Is the percentage change between the pH of unpolluted rain water and pure water large? What about the percentage change of $[H^+]$? Are these percentages significant? Who (what types of groups) might report pH changes and who might report $[H^+]$ changes?
g. The report claims that the average cloud acidity along the Appalachians is about 100 times greater than unpolluted rain water. How did the writer come to this conclusion? Do you agree with it? Support your answer.
h. What would be the pH of the average cloud if it *is* 100 times more acidic than unpolluted rain water?
i. The opening statement in the quotation claims that a correlation exists between cloud acidity and the decline of forests. What additional information is needed to support this conclusion? Other research has found that a factor of 100 increase in the hydrogen ion concentration, $[H^+]$, of water causes 1% of a forest to become weak and eventually die off. Develop an argument to support the opening statement of the quotation.

35. Mutually Assured Destruction. During the Cold War, the United States and Soviet Union sought to deter possible attacks against each other through a doctrine often called "mutually assured destruction" (MAD). At the height of the Cold War, the two nations had about 25,000 strategic nuclear weapons (weapons capable of reaching the other country) between them, each with a typical yield of about one megaton (5×10^{15} joule). Compare the total energy represented by these weapons to the energy of a great earthquake of magnitude 8. Was "mutually assured destruction" an apt name? Explain.

36. Millions from Millidollars. When interest is paid on bank accounts, the calculated interest often involves fractions of a cent. Bank computer systems usually pay the interest to the nearest cent but keep track of the difference between the theoretical interest (without rounding) and the actual interest paid. In the next period of interest payment, this difference is applied to ensure that the correct amount is paid over long periods of time. However, in one famous case of bank fraud a programmer changed the computer program to deposit any remaining fractional cents into his own account. Suppose that a bank computes interest daily, and handles 100,000 accounts. If the program diverts an average of 0.4¢ from each account each day, how much will the criminal accrue in his account in 1 year?

37. **Biweekly Mortgage Payments.** Some financial analysts recommend making biweekly (every 2 weeks) rather than monthly mortgage payments. Let's examine this strategy. Suppose that you have a home mortgage of $100,000 at a fixed *APR* of 9% for 30 years.
 a. What are your monthly payments? How much, in total, do you pay in a year?
 b. Suppose that you make a payment equal to half your monthly payment every 2 weeks, or 26 times per year. What are your biweekly payments? How much, in total, do you pay in a year? Explain why this strategy accelerates your payoff of the loan.
 c. Solve the amortization payment formula for the number of years, *Y*, required to pay off the loan by making biweekly payments. How long will paying off the loan with biweekly payments take?
 d. Based on your results, discuss the pros and cons of the biweekly payment strategy. For whom is the strategy best suited (e.g., people who receive monthly paychecks or people who receive biweekly paychecks)? Why?

38. **How Much House Can You Afford?** Suppose that you can afford monthly payments of $500. If current mortgage rates are 9% for a 30-year loan, what loan principal can you afford? If you are required to make a 20% down payment and have the cash on hand for it, what price home can you afford?

39. **Choosing an Auto Loan.** Suppose that you can afford monthly car payments of $220 and need to borrow $10,000 to buy the car you want. The bank offers three choices of car loan with the following rates and terms: 7% for a 3-year loan; 7.5% for a 4-year loan; or 8% for a 5-year loan. Which loan best meets your needs? Explain. (Hint: Calculate your amortization payment for each loan.)

40. **Analyzing Formulas.** The following problems describe various relations and a formula for the relation. In each case, answer any questions asked.
 a. **Free fall.** The distance *d* that an object falls in a gravitational field *t* seconds after it is dropped from rest is

$$d = \frac{1}{2}gt^2$$

 where $g = 9.8\ \text{m/s}^2$ is the acceleration due to gravity. If the object is released from the top of a building 100 meters high, when does it hit the ground?
 b. **Probability.** The probability *p* (or chances) of rolling a die *n* times and rolling a 1 each time is $p = 6 - n$. What are the chances of rolling three 1's in a row?
 c. **Postage Inflation.** The following formula is claimed to describe the cost of postage for a first class letter (in cents), where *t* is the number of years since 1955 and *p* is the postage: $p = 3 \times 2^{t/10}$. According to this formula, what is the current cost of a first class stamp? Do you believe the formula? Does it give correct predictions? Explain.
 d. **DNA.** The number of possible three-letter "words" that can be formed from an alphabet with *m* letters (with repeated letters allowed) is $w = m^3$ words. DNA, the material that makes up genes, uses four bases called adenine (A), guanine (G), thymine (T), and cytosine (C) to encode genetic information. These four bases are grouped into three-base "words" and each word codes for a specific amino acid. For example, the "word" CAG codes for the amino acid glutamine, and GAC codes for aspartic acid. How many three-letter words can be formed from this four-base genetic alphabet? As only 20 amino acids are used in the human body, are this many words needed? Explain why biologists say that the genetic code contains redundancy.
 e. **Earth's Orbit.** If an *x*–*y* coordinate system is placed on the solar system, the elliptical orbit of the Earth is described by the equation

$$\frac{x^2}{150^2} + \frac{y^2}{140^2} = 1,$$

 where *x* and *y* are the coordinates of the Earth (see the following figure) measured in millions of kilometers. Solve this equation for *y* in terms of *x* to find a relation that gives the *y* coordinate of the position in terms of the *x* coordinate. What is the smallest distance between the Earth and the Sun?

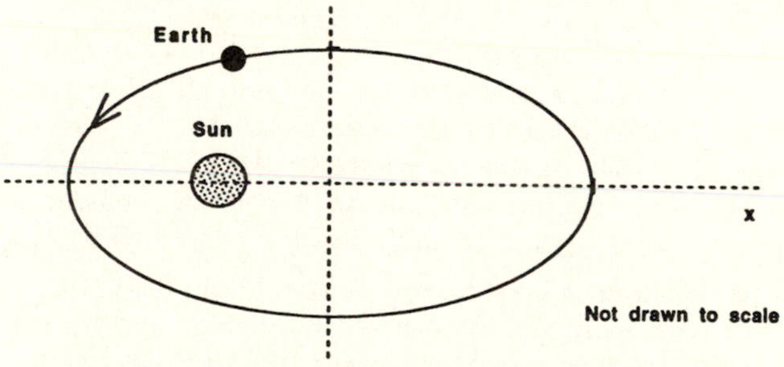

 f. **Draining a Tank.** A large cylindrical water tank has a height of $H = 10$ meters, a radius of $R = 10$ meters, and a hole in the bottom with a radius of $r = 0.1$ meters, as shown in the figure on page 625. Assume that the tank is full of water when the hole in the bottom is unplugged. The depth of the water as it drains from the tank is given by the relation

$$h = \left(\sqrt{H} - \frac{r^2}{R^2}\sqrt{2gt}\right)^2,$$

 where *t* is time and $g = 10\ \text{m/s}^2$ is the acceleration of gravity. Do the following: (i) confirm that at $t = 0$ the height of water in the tank given by the formula is the full tank height of 10 meters; (ii) find the height of the water in the tank after 60 seconds and after 1 hour; (iii) find the length of time required to drain the entire tank (hint: be sure to keep careful track of units!); and (iv, for extra credit) find a cylindrical can, cut a hole in the bottom, fill it with water, and time how long it takes to drain; does the formula yield the correct prediction?

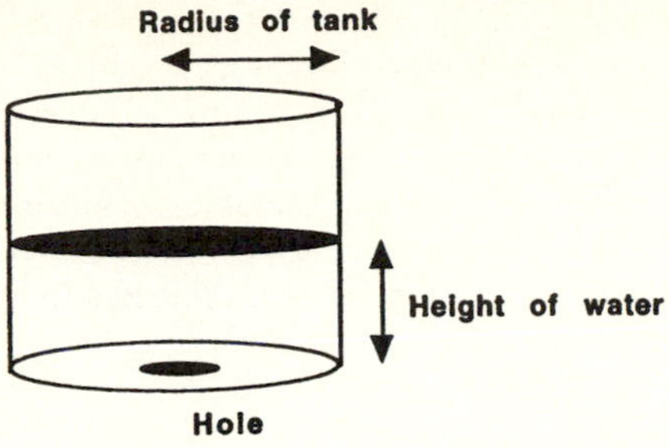

g. **Tape deck dynamics.** Many audio and video tape decks have a counter that records the number of revolutions of the take-up reel. As you may have noticed, this counter does *not* move at a constant rate as a tape plays. The relation that gives the number of revolutions N as it varies with the length of time t that the tape has been playing is

$$N = \sqrt{\frac{vt}{2\pi h}}.$$

This relation is based on the assumption that the tape has a uniform speed of v inches per second through the tape deck, that the thickness of the tape is h inches, and that the take-up reel initially is empty. A typical deck runs at $v = 1.25$ inches per second and $h = 0.001$ inches. Substitute these values of v and h into the relation and find how many revolutions the take-up reel makes in playing a 30-minute tape. What will the counter read after 15 minutes (assuming one count per revolution)?

41. **Geostationary Orbit.** Communications satellites are placed into a special orbit, called geostationary orbit, in which the satellite appears to remain stationary in the sky. In other words, once a satellite-receiving dish is pointed at a particular satellite, it never has to be moved.

To be geostationary, a satellite must have an orbital period that exactly matches the rotation period of the Earth; that way, the Earth and the satellite rotate at the same rate, so the satellite appears to remain in the same position in the sky as seen from the ground (it must also orbit in the plane of the Earth's equator).

The rotation period of the Earth actually is slightly shorter than a day; it is about 23 hours and 56 minutes (called a sidereal day). Use Kepler's third law to find the altitude of a geostationary satellite. (Hint: The mass of the satellite is negligible compared to the mass of the Earth, which is 6.0×10^{24} kg.)

Projects

42. **Noise Ordinances.** Most municipalities have laws (ordinances) regarding noise. Suppose that you are asked to draft an ordinance governing the level of noise, in decibels, allowed in your city. Based on the data in Table 16–2 showing decibel measurements of common sounds, suggest legal limits on noise for each of the following circumstances (be sure to justify your proposed limit in each case):

- residential neighborhoods during the daytime;
- residential neighborhoods at night (e.g., after 10 P.M.);
- commercial districts during the daytime; and
- noise limits inside an airport passenger terminal.

After you have proposed your own noise limits, determine the actual legal limits in your city. How do they compare to the limits you proposed? After studying the real limits, do you think that yours are better? Why or why not? Extra credit: Find the noise limits for several nearby major cities. How do they compare to each other? How do they compare to your proposed limits?

43. **Choosing a Mortgage.** This problem should be analyzed in small groups. Imagine that you work for an accounting firm and a client family has told you that they are buying a house and need a loan of $120,000. Find the current rates available from local banks for both fixed rate mortgages and adjustable rate mortgages (ARMs). Analyze the offerings and summarize orally or in writing the best option for your client. You may need to recommend different options, depending on your client's situation (e.g., number of children, income, prospects for raises or pay cuts in the future, etc.).

44. **Investment Advisor.** Recently, your long-lost aunt passed away. In her will, she bequeathed to you her entire estate, worth $500,000. However, the will also specifies that you cannot spend any of the money for 10 years. Decide how to invest this money in order to maximize the return on it in 10 years. Consider investments in bank accounts, stocks, bonds, and other funds. Explain your investment decisions and estimate how much money you will have at the end of 10 years. Discuss any risks involved in your investment plan.

45. **Owning Versus Renting.** The amortization payment formula gives you the total monthly payment needed to pay off a loan. However, part of the payment goes toward the loan principal and part goes toward interest. Initially, because the loan balance is high, the interest payments are relatively high and the principal payments are relatively low. As time passes, more and more of your monthly payments go toward principal and less and less toward interest. Because the principal and interest portions of the payments vary over time, their formulas are more complex than the basic amortization payment formula. (An easy way to calculate the principal and interest portions of payments is to use the built-in functions in a computer spreadsheet program or a financial calculator.) Nevertheless, you can easily estimate the annual interest payment simply by using the *APR* and the loan balance. For example, if you borrow $100,000 at an *APR* of 10%, you will pay *approximately* $10,000 (10% of $100,000) in interest during the first year. In subsequent years, as the loan balance shrinks, you will pay less interest. However, if the loan term is long, such as a 30-year mortgage, the loan balance shrinks very slowly in the early years; thus the estimate from the first year remains close to the actual interest payments for several years.

a. Suppose that you take out a mortgage of $100,000 at a fixed interest rate of 10% for 30 years. What are your monthly payments?
b. Because mortgage interest is tax deductible, your actual cost is less than you might at first think. Suppose that you are in a tax bracket where you are paying federal and state income tax at a marginal rate of 36%. Because you will not have to pay tax on the interest portion of your mortgage payments, approximately how much will you save on your tax bill because of your mortgage payments? (Use the estimated interest payment during the first year as described in the problem statement.)
c. Based on your yearly tax savings in part (b), how much are you saving in taxes each month? Suppose that, not counting your mortgage, your take-home pay is $3000 per month. Calculate your remaining pay after your mortgage payments by subtracting the mortgage payment *and* adding the monthly tax savings from the mortgage interest tax deduction.
d. Generalize from part (c): What is the effective cost of your monthly mortgage payment after taking into account the tax savings? Note that this effective cost depends only on your 36% tax rate, not on your monthly income.
e. Suppose that you are trying to decide whether to rent or buy. Assume that you have found a house which you can buy with the loan you've been considering ($100,000 at a fixed *APR* of 10% for 30 years). Alternatively, you have found an apartment you like that rents for $650 per month. Which option will cost you less money? Why? (Note that additional costs such as property taxes, insurance, and maintenance are being ignored.)
f. Suppose that, instead of being in a 36% marginal tax bracket, you earn less and your marginal tax bracket is only 24%. In this case, what is the effective cost of your mortgage after accounting for the tax savings? Could your lower marginal tax rate change your decision about renting versus buying (as in part e)? Explain.
g. Suppose that you are retired and are living off an annuity that you wisely accumulated during your career. As a result, you have only a small amount of income from continuing interest and you pay no income taxes. How would this situation affect your decision to rent or buy (as in part e)? Explain.

46. **Distribution of Wealth.** Economists use **Lorenz relations** to describe the distribution of wealth in a culture or society. Imagine that the entire population under consideration is ranked according to wealth. Then the value, w, of the Lorenz relation for a fraction, p, is the fraction of the total wealth that is owned by the poorest pth of the population. For instance, if $w = 1/3$ when $p = 1/2$, one third of the wealth is owned by one half the population. Zero percent of the population owns zero percent of the wealth and 100 percent of the population owns 100 percent of the wealth, so all Lorenz relations must have $w = 0$ when $p = 0$, and $w = 1$ when $p = 1$.

Look up data concerning the distribution of wealth for the United States. Based on these data, draw a sketch of the Lorenz relation. For additional research, do the same for several other countries; try to choose countries with a variety of overall economic conditions. What conclusions can you draw about how the distribution of wealth affects the overall economic health of a nation?

47. **The Bombarded Earth.** The Earth is bombarded with interplanetary debris, ranging from dust-sized particles to large meteorites. Small objects burn up in the atmosphere, but larger objects can reach Earth and create craters. The following table shows how the diameter of an object is related to the frequency with which such objects enter the Earth's atmosphere.

Object Description	Diameter (meters)	Frequency (no./year)
dust	10^{-6}	10^{12}
bright meteor	10^{-3}	10^{6}
meteorite	10^{0}	10^{0}
Arizona Crater meteorite	10^{2}	10^{-4}

From these data you can base predictions about the frequency of impact on the diameter of the object. Of particular interest is the meteorite that may have helped bring about the end of the dinosaurs. Evidence suggests that this meteorite, believed to have left a large crater near the Yucatan peninsula (Mexico) about 65 million years ago, was approximately 10 km in diameter.

a. Make a graph of the data from the table, but use logarithmic scales. That is, plot logarithms (base-10) of the diameters on the x-axis and the logarithms of the frequencies on the y-axis. Briefly explain why graphing these data with such a "log–log" plot is easier than plotting them directly.
b. The data should fall on a straight line on the log–log graph. Extend the line and use the extension to predict the frequency of impacts of 10-km objects (10^4 meters).
c. Based on the result in (b), could a 10-km object have struck the Earth at the time the dinosaurs became extinct? Why or why not?
d. Some researchers argue that any 10-km object hitting the Earth would unleash a chain of events leading to a mass extinction, similar to the mass extinction that killed off the dinosaurs. Based on your results, how often might mass extinctions be expected to occur on the Earth?
e. Research the evidence for past mass extinctions. Were the dinosaurs the only animals that became extinct 65 million years ago? How many other mass extinctions are known?
f. Even the impact of a 10-km object causes relatively few direct deaths; that is, being hit on the head by the meteorite is not the primary cause of death in a mass extinction. Research and describe, briefly, the mechanism by which an impact is thought to lead to mass extinction. Does this mechanism suggest any cause for concern over how human activities, especially in driving many species to extinction, might affect the Earth? Explain.

A ARITHMETIC SKILLS REVIEW

This appendix provides a short review of some basic arithmetic principles and operations. If you need this review, begin by reading the sections and then working on the problems at the end of the appendix.

A.1 BASIC OPERATIONS

The art of performing calculations with numbers is called **arithmetic**. The four basic operations of arithmetic are **addition**, **subtraction**, **multiplication**, and **division**. To review basic terminology, consider operations you can perform on the number 5:

- Addition. Adding 0 doesn't change the original number: $5 + 0 = 5$. Adding a number besides 0 *does* change it; for example, $5 + 1 = 6$. The result of addition is called a **sum**; in this case we say that *the sum of 5 and 1 is 6*.
- Subtraction. Subtracting 0 doesn't change the original number: $5 - 0 = 5$. Subtracting a nonzero number *does* change it: for example, $5 \times 3 = 2$. The result of subtraction is called a **difference**; in this case we say *that the difference between 5 and 3 is 2*.
- Multiplication. Multiplying by 1 leaves the number unchanged: $5 \times 1 = 5$. Multiplying by any other number *does* change it: for example, $5 \times 4 = 20$. The result of multiplication is called a **product**; in this case we say that *the product of 5 and 4 is 20*.
- Division. Dividing by 1 leaves the number unchanged: $5 \div 1 = 5$. Dividing by any other number *does* change it: for example, $5 \div 10 = 1/2$. The result of division is called a **quotient**; in this case we say *the quotient of 5 and 10 is one-half*.

The order of operation isn't important for addition or multiplication. For example, $3 + 4 = 4 + 3$ and $4 \times 6 = 6 \times 4$. This property, called the **commutative property**, holds for any numbers added or multiplied together. In contrast, subtraction and division are *not* commutative. For example, $5 - 3 \neq 3 - 5$, and $25 \div 5 \neq 5 \div 25$.

A.1.1 Exploring Multiplication and Division

Several alternative notations express multiplication and division. If we use the symbols a and b to represent arbitrary numbers, the three common ways of representing multiplication are the use of the "times" symbol (×), use of a dot, or simply placing two symbols next to one another.

Three ways to write multiplication: $a \times b$, $a \cdot b$, ab.

Showing multiplication by writing symbols next to one another should be used only when there is no ambiguity. For example, 54 is the number "fifty-four," and does not mean 5 times 4. You can show 5×4 using parentheses: $(5)(4) = 5 \times 4$.

The three common ways to express division are the division symbol (÷), the long division indicator ($\overline{)\ \ }$), and as a fraction.

Three ways to write division: $a \div b$, $b\overline{)a}$, $\dfrac{a}{b}$.

Multiplication as Addition

Multiplication is the process of *adding* a number multiple times. For example, 4×5 means adding 5 four times, starting from 0:

$$4 \times 5 = 0 + \underbrace{5 + 5 + 5 + 5}_{4 \text{ times}} = 20.$$

Similarly, 6×5 means adding 5 six times:

$$6 \times 5 = 0 + \underbrace{5 + 5 + 5 + 5 + 5 + 5}_{6 \text{ times}} = 30.$$

The symbol "…" is called an **ellipsis** and is used to indicate a continuing pattern.

We can *generalize* by using a symbol, such as n, to represent any natural number:

$$n \times 5 = 0 + \underbrace{5 + 5 + 5 + \cdots + 5 + 5 + 5}_{n \text{ times}}.$$

With this definition of multiplication you can understand why, for example, any number multiplied by 0 is 0: If you begin at 0, and don't add a number, you still have 0.

Division as Subtraction

Division can be interpreted as the process of counting how many times you can *subtract* one number from another until reaching 0. For example, $10 \div 2 = 5$ because 2 can be subtracted from 10 five times:

$$10 - \underbrace{2 - 2 - 2 - 2 - 2}_{5 \text{ times}} = 0 \quad \Rightarrow \quad 10 \div 2 = 5.$$

Sometimes a **remainder** is left after you subtract as many times as possible. For example, $20 \div 3$ asks how many times you can subtract 3 from 20. After subtracting 3 from 20 six times you have 2 remaining, which isn't enough to subtract 3 again. Thus $20 \div 3 = 6$ with remainder 2:

$$20 - \underbrace{3 - 3 - 3 - 3 - 3 - 3}_{6 \text{ times}} = 2 \quad \Rightarrow \quad 20 \div 3 = 6, \text{ remainder } 2.$$

Because the remainder, 2, is two-thirds of 3, you can reach zero by subtracting two-thirds of 3, or 2. Thus instead of showing the remainder explicitly you can use fractions to state the answer:

$$20 - \underbrace{3 - 3 - 3 - 3 - 3 - 3}_{6 \text{ times}} - \left(\frac{2}{3} \times 3\right) = 0 \quad \Rightarrow \quad 20 \div 3 = 6\frac{2}{3}.$$

Dividing by Zero

What if you divide a number by 0? Consider $10 \div 0$, which asks how many times you can subtract 0 from 10. If you subtract 0 once, you still have 10; if you subtract it twice, you still have 10; if you subtract it three times, you still have 10. In fact, the number of times you can subtract 0 is unlimited. Therefore division of any nonzero number by 0 is **undefined**:

The symbol ≠ means "does not equal."

$$n \div 0 = \frac{n}{0} = \text{undefined} \qquad (n \neq 0).$$

What happens if you divide *zero* by another number? If you divide 0 by any nonzero number, you still have 0. For example, $0 \div 5$ asks how many times you can subtract 5 from 0, until you reach 0. Because you are starting from 0, you cannot subtract at all. That is,

$$0 \div n = 0 \qquad (n \neq 0).$$

Suppose that you try to divide $0 \div 0$, which asks how many times you can subtract 0 until you reach 0. This question poses a paradox: Because you are starting from 0, you cannot subtract; but because subtraction of 0 shouldn't change anything, you ought to be able to do it an unlimited number of times. Because of this paradox, the answer to $0 \div 0$ is said to be **indeterminate**:

$$0 \div 0 = \frac{0}{0} = \text{indeterminate}.$$

A.1.2 Working with Negative Numbers

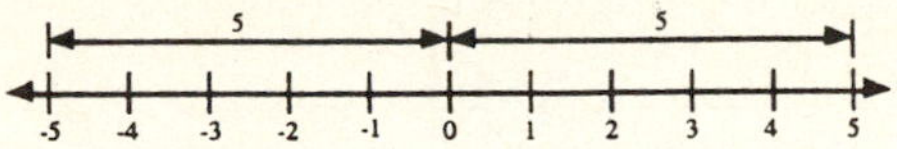

Figure A-1. The numbers 5 and –5 are equidistant from 0 on the number line.

Negative and positive numbers lie on opposite sides of 0 on the number line. For example, the number 5 lies five units to the *right* of 0, while –5 lies five units to the *left* of 0 (Figure A-1).

Because 5 and –5 are opposites, they "cancel" each other when added:

$$5 + (-5) = 0.$$

Addition and Subtraction of a Negative Number

You can always replace subtraction with the addition of a negative number. For example, $10 - 6 = 10 + (-6) = 4$. Generalizing, with a and b representing any numbers,

$$a - b = a + (-b).$$

Note that subtracting a *negative* number is equivalent to adding a positive number. For example, $10 - (-6) = 10 + [-(-6)] = 10 + 6 = 16$. In general,

$$a - (-b) = a + b.$$

Balancing your checkbook provides a practical example that explains *why* the preceding rules are true. The first five columns in the following illustrative checkbook should look similar to those in your own checkbook. The sixth column, *amount added to balance*, shows each step as addition: Making a deposit adds a positive amount to the balance, whereas writing a check or making a withdrawal adds a *negative* amount to the balance.

Check Number	Date	Description	Amount Withdrawn	Amount of Deposit	Amount Added to Balance	Balance
—	4/1	Opening deposit		100.00	100.00	100.00
1	4/2	Groceries	25.00		–25.00	75.00
2	4/4	Bookstore	40.00		–40.00	35.00
—	4/6	Deposit from Mom		5000.00	5000.00	5035.00
—	4/7	Cash withdrawal	150.00		–150.00	4885.00
3	4/7	Tuition	4800.00		–4800.00	85.00
—	4/8	Paycheck		120.00	120.00	205.00

Multiplication and Division with Negative Numbers.

Thinking of negative and positive numbers as opposites leads to three basic rules for multiplication and division (summarized in Table A-1).

Table A-1. Rules for Multiplication or Division

× or ÷	*Positive*	*Negative*
Positive	Positive	Negative
Negative	Negative	Positive

- Multiplying or dividing two positive numbers yields another positive number, as no opposites are involved.

  positive × positive = positive, and positive ÷ positive = positive.

- Multiplying or dividing a positive number by a negative number yields a negative number because the opposite in the question requires an opposite in the answer.

  positive × negative = negative, and positive ÷ negative = negative.

- Multiplying or dividing two negative numbers yields a positive result because such an operation is taking the opposite of an opposite.

  negative × negative = positive, and negative ÷ negative = positive

Real world situations also show *why* the preceding rules hold. For example, suppose that the price of a sandwich is \$3; then the price of five sandwiches is: 5 × \$3 = \$15, showing that the product of two positive numbers is another positive number. Similarly, if a pizza with 8 slices is divided among 4 people, there are 8 slices ÷ 4 people = 2 slices per person, showing that the quotient of two positive numbers is another positive number.

Next, suppose that you are \$100 in debt or, equivalently, that your net worth is –\$100. If your debt doubles, your new debt is \$200 and your new net worth is 2 × (–\$100) = –\$200, showing that the product of a positive number and a negative number is a negative number. To show that the quotient of a positive number and a negative number is a negative number, suppose you scuba dive to a depth of 20 meters, which is equivalent to an *elevation* of –20 meters (20 meters *below* sea level). On your return, you will be half as deep when your elevation is (–20 meters) ÷ 2 = –10 meters.

Describing a situation to obtain the rule for multiplying or dividing two negative numbers is slightly more subtle. If you have done any mountain climbing or taken a ride in a hot air balloon, you know that (on most days) temperature drops as you rise to higher altitudes. Imagine that you are riding in a hot air balloon on a day when the temperature *drops* by 2°C (Celsius) for every 100 meters you rise (Figure A-2). Then the change in temperature with altitude is *negative* 2°C per 100 meters.

$$\text{Change in temperature with altitude} = \frac{-2\,°\text{C}}{100\text{ meters}} = -\frac{1}{50}\frac{°\text{C}}{\text{meter}} = -0.02\frac{°\text{C}}{\text{meter}}.$$

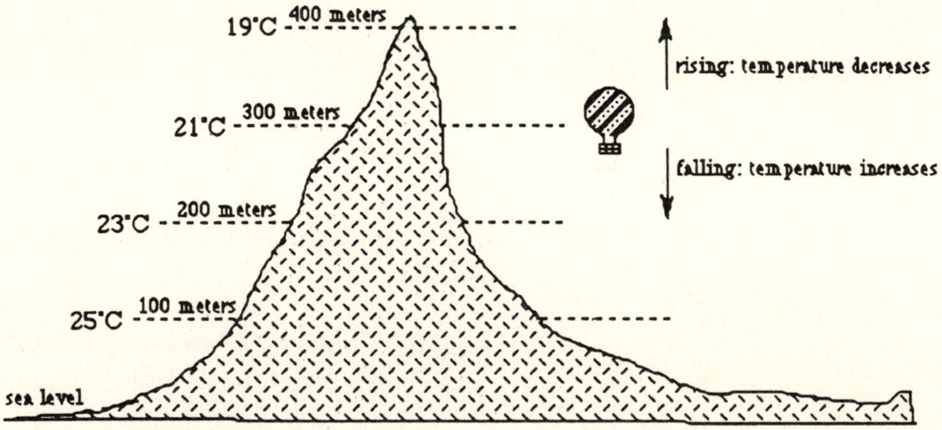

Figure A-2. If the temperature falls by 2°C for every 100 meters of altitude, rising balloonists will experience a temperature drop of 0.02°C per meter, and descending balloonists will experience a temperature rise of 0.02°C per meter.

The negative sign indicates that the temperature *falls* as the altitude *rises*. Suppose that you start from the ground, which has altitude 0, and rise to an altitude of 500 meters. How much does the temperature change? To obtain the answer multiply the rate of temperature change with altitude by the change in altitude:

$$-\frac{1}{50}\frac{°\text{C}}{\text{meter}} \times 500\text{ meters} = -10\,°\text{C}.$$

The temperature *drops* by 10°C, confirming that a negative number multiplied by a positive number yields a negative number.

Now suppose that you *descend* from an altitude of 500 meters to an altitude of 300 meters. How much does the temperature change? You have lost 200 meters of altitude, so the change in your altitude is *negative* 200 meters. Moreover, because you know that the temperature increases as you descend, the answer must be positive. Thus the change in temperature after a 200 meter descent must be

$$-\frac{1}{50}\frac{°\text{C}}{\text{meter}}\times(-200\text{ meters}) = +4\ °\text{C},$$

confirming that the product of two negative numbers is a positive number.

A.1.3 Operation Priority

Whenever calculations require several successive arithmetic operations, the following rules apply to the *order* in which the operations should be done.

1. First complete any operations inside parentheses, starting with the innermost set. Note that brackets, [ ], and braces, { }, are used in mathematics like larger parentheses.
2. Complete any powers or roots, in order, from left to right.
3. Complete any multiplication or division, in order, from left to right.
4. Finally, complete any addition or subtraction, in order, from left to right.

Most calculators also follow these rules.

Example A-1. Compare $4 + 8 \times 3$ to $(4 + 8) \times 3$.

Solution: Although both look similar, the two calculations have different answers because the parentheses in the second calculation change the order of operations.

$$4 + 8 \times 3 = 28$$

because multiplication is completed before addition, but

$$(4 + 8) \times 3 = 36$$

because the addition within the parentheses must be completed before the multiplication.

❖

Example A-2. Use the proper order of operations to evaluate

$$5+3^2\times[(4+8)\times 3]-16\div\sqrt{64}.$$

Solution: Following the rules for operation priority yields

$5+3^2\times[(4+8)\times 3]-16\div\sqrt{64} =$ (first do operations inside parentheses)

$5+3^2\times[36]-16\div\sqrt{64} =$ (next do powers and roots)

$5+9\times[36]-16\div 8 =$ (then multiplication and division, left to right)

$5+324-2 =$ (then addition and subtraction, left to right)

327

❖

Time-Out to Think: Try the preceding examples on your calculator. Does your calculator follow the rules of order stated here? If not, what set of rules does your calculator use for the order of operations?

A.2 WORKING WITH FRACTIONS

Fractions are nearly everywhere: fractions of a dollar in counting change; fractions of a pound in buying groceries; fractional discounts like "10% off"; and so on. In general, fractions are used for the following three purposes.

- Fractions can express parts of a whole. For example, $2/3$ represents two parts of a whole divided into three parts, such as two pieces of a chocolate bar divided in thirds.
- Fractions can represent a comparison, or **ratio**, between two quantities. If one person weighs 100 pounds and another weighs 120 pounds, the ratio of their weights is 100 to 120; that is, the first person weighs $100/120$, or $5/6$, as much as the second person.
- Fractions can be used to represent *division* of one number by another. A fraction is its numerator divided by its denominator. For example,

$$\frac{10}{4} = 10 \div 4 = 2\frac{1}{2} = 2.5.$$

Three methods are widely used to express fractions: common fractions, decimals, and percentages. For example, one half may be expressed as $1/2$, 0.5, or 50%. We briefly review each method.

A.2.1 Common Fractions

Common fractions, or fractions in ratio form, are written

$$\frac{a}{b} \quad \text{or} \quad \frac{\text{numerator}}{\text{denominator}}.$$

Multiplying Fractions

Suppose that four slices remain from a pie that was cut in six pieces. If you take half the remaining pie, you take two slices. That is, taking $1/2$ of the remaining $4/6$ of the pie means taking $2/6$, or $1/3$, of the pie. This result demonstrates that multiplying common fractions requires multiplying both their numerators and denominators; in this case,

$$\frac{1}{2} \times \frac{4}{6} = \frac{1 \times 4}{2 \times 6} = \frac{4}{12} = \frac{1}{3}.$$

Multiplying common fractions: $\frac{a}{b} \times \frac{c}{d} = \frac{a \times c}{b \times d}$.

Study the following examples to be sure you understand multiplication of fractions, including the handling of negative fractions.

$$\frac{3}{4} \times \frac{6}{20} = \frac{3 \times 6}{4 \times 20} = \frac{18}{80}. \qquad \frac{3}{7} \times (-9) = \frac{3}{7} \times \frac{(-9)}{1} = \frac{3 \times (-9)}{7 \times 1} = -\frac{27}{7}.$$

$$5 \times \frac{3}{2} = \frac{5}{1} \times \frac{3}{2} = \frac{5 \times 3}{1 \times 2}. \qquad -\frac{2}{3} \times \left(-\frac{2}{3}\right) = \frac{-2 \times (-2)}{3 \times 3} = \frac{4}{9}.$$

Reciprocals

The reciprocal of x is $1/x$. The reciprocal of $\frac{a}{b}$ is $\frac{b}{a}$ $(a, b \neq 0)$.

Dividing fractions requires knowing how to find a **reciprocal**. Its definition is that, when a number is multiplied by its reciprocal, the result is 1. For example,

$$\frac{3}{4} \times \frac{4}{3} = 1, \quad \text{so } \frac{3}{4} \text{ and } \frac{4}{3} \text{ are reciprocals;}$$

$$9 \times \frac{1}{9} = 1, \quad \text{so 9 and } \frac{1}{9} \text{ are reciprocals;}$$

$$-7 \times \frac{1}{7} = 1, \quad \text{so } -7 \text{ and } \frac{1}{7} \text{ are reciprocals;}$$

$$0.25 \times 4 = 1, \quad \text{so 0.25 and 4 are reciprocals; and}$$

$$\frac{a}{b} \times \frac{b}{a} = 1, \quad \text{so } \frac{a}{b} \text{ and } \frac{b}{a} \text{ are reciprocals } (a \neq 0, b \neq 0).$$

Note that finding the reciprocal of any nonzero number simply requires interchanging the numerator and denominator.

Dividing Fractions

Imagine that you ride your bike a distance of 10 miles in a half hour. What is your average speed, in miles per hour? Your intuition probably tells you the answer: If your speed takes you 10 miles in $\frac{1}{2}$ hour, it would take you 20 miles in a full hour. Thus your average speed is 20 miles per hour. Mathematically, your average speed is the distance traveled divided by the time taken to travel the distance, which in this case is 10 (miles) divided by $\frac{1}{2}$ (hour).

$$10 \div \frac{1}{2} = \underbrace{10 \times \underbrace{\frac{2}{1}}_{\text{invert}}}_{\text{and multiply}} = 20.$$

Dividing is the same as multiplying by the reciprocal.

Note that the rule for dividing by a fraction is **invert and multiply**; that is, invert the fraction by replacing it with its reciprocal and then multiply. In general, we write this process as

$$\frac{a}{b} \div \frac{c}{d} = \frac{a}{b} \times \frac{d}{c} = \frac{a \times d}{b \times c}.$$

Study the following examples to be sure you understand the process.

$$\frac{3}{4} \div \frac{2}{5} = \frac{3}{4} \times \frac{5}{2} = \frac{15}{8}. \qquad 27 \div \frac{4}{3} = 27 \times \frac{3}{4} = \frac{27 \times 3}{1 \times 4} = \frac{81}{4}.$$

$$-\frac{2}{3} \div 9 = -\frac{2}{3} \times \frac{1}{9} = -\frac{2}{27}. \qquad -\frac{7}{10} \div \left(-\frac{100}{9}\right) = -\frac{7}{10} \times \left(-\frac{9}{100}\right) = \frac{-7 \times (-9)}{10 \times 100} = \frac{63}{1000}.$$

Equivalent Fractions

Fractions are equivalent if they represent the same number. For example,

$$\frac{1}{2} = \frac{2}{4} = \frac{13}{26} = \frac{1500}{3000}.$$

Multiplying or dividing the top (numerator) and bottom (denominator) of a fraction by the same nonzero number yields an equivalent fraction.

You can convert any fraction into an equivalent fraction by multiplying or dividing the top (numerator) and bottom (denominator) by the same nonzero number. Note that multiplying both the numerator and denominator by the same number is just a way of multiplying the entire fraction by 1. For example, let's rewrite the fraction $\frac{2}{3}$ with a denominator of 24. Changing the denominator from 3 to 24 requires multiplying by 8. We therefore must also multiply the numerator by 8.

$$\frac{2}{3} = \frac{2}{3} \times \underbrace{\frac{8}{8}}_{1} = \frac{16}{24}.$$

Thus $\frac{2}{3}$ and $\frac{16}{24}$ are equivalent fractions.

Symbolically, we express the methods for converting fractions to equivalent form as

$$\frac{a}{b} = \frac{a \times c}{b \times c} \quad \text{and} \quad \frac{a}{b} = \frac{a \div c}{b \div c}, \quad c \neq 0.$$

Again, because $c/c = 1$, this process is nothing more than multiplying or dividing by 1.

Finally, note that adding or subtracting the same number on the top and bottom of a fraction does *not*, in general, yield an equivalent fraction.

$$\frac{a}{b} \neq \frac{a+c}{b+c} \quad \text{and} \quad \frac{a}{b} \neq \frac{a-c}{b-c}.$$

Time-Out to Think: Why can't you form an equivalent fraction by adding the same number to both numerator and denominator? Try a few examples to convince yourself that the resulting fractions are not equivalent.

Fractions in Simplest Form

A fraction is in **simplest** (or **reduced**) form when the only natural number that is a factor of *both* the numerator and denominator is 1. For example, $2/4$ can be reduced to the simplest form $1/2$ by dividing both the numerator and denominator by 2.

Rewriting $9/12$ in simplest form requires recognizing that the numerator and denominator both are divisible by 3; carrying out this division yields

$$\frac{9}{12} = \frac{9 \div 3}{12 \div 3} = \frac{3}{4}.$$

Adding and Subtracting Fractions

Adding or subtracting fractions requires that the fractions have a common denominator. For example, adding or subtracting the following pairs of fractions is possible because each pair has a common denominator.

$$\frac{1}{4} + \frac{2}{4} = \frac{1+2}{4} = \frac{3}{4} \quad \text{and} \quad \frac{4}{5} - \frac{1}{5} = \frac{4-1}{5} = \frac{3}{5}.$$

When fractions have different denominators, multiplying their denominators is the easiest way to find a common denominator. Alternatively, the common denominator can be any number for which both denominators are factors.

For example, you can find a common denominator for $1/5$ and $2/7$ by multiplying the two denominators: $5 \times 7 = 35$. Converting $1/5$ to a fraction with a denominator of 35 requires multiplying by $7/7 = 1$, and converting $2/7$ to a fraction with a denominator of 35 requires multiplying by $5/5 = 1$:

$$\frac{1}{5} = \frac{1}{5} \times \frac{7}{7} = \frac{7}{35} \quad \text{and} \quad \frac{2}{7} = \frac{2}{7} \times \frac{5}{5} = \frac{10}{35}.$$

With a common denominator, $2/7$ clearly is larger than $1/5$ by $10/35 - 7/35 = 3/35$. We also can easily add the fractions now:

$$\frac{1}{5} + \frac{2}{7} = \frac{7}{35} + \frac{10}{35} = \frac{17}{35}.$$

Example A-3. Add $1/2$ and $1/3$.

Solution: Multiplying the denominators yields a common denominator: $2 \times 3 = 6$. Rewrite each fraction with this common denominator and then add.

$$\frac{1}{2} + \frac{1}{3} = \underbrace{\frac{3}{6} + \frac{2}{6}}_{\text{common denominator of 6}} = \frac{3+2}{6} = \frac{5}{6}$$

❖

Example A-4. Rewrite $3/4$ and $5/6$ with a common denominator. Which is larger and by how much?

Solution: You can find a common denominator by multiplying the two denominators: $4 \times 6 = 24$. However, as both 4 and 6 are factors of 12, you can use 12 as a simpler common denominator. To rewrite each fraction, multiply them by 1 in a form that converts the denominator to 12:

$$\frac{3}{4} = \frac{3}{4} \times \underbrace{\frac{3}{3}}_{1} = \frac{9}{12} \quad \text{and} \quad \frac{5}{6} = \frac{5}{6} \times \underbrace{\frac{2}{2}}_{1} = \frac{10}{12}.$$

Now subtract the fractions to find that $5/6$ is larger than $3/4$ by $1/12$:

$$\frac{5}{6}-\frac{3}{4}=\frac{10}{12}-\frac{9}{12}=\frac{1}{12}.$$

❖

Example A-5. Which is larger and by how much: $2/3$ or $3/5$?

Solution: Both fractions can be rewritten with a common denominator of 15:

$$\frac{2}{3}=\frac{2}{3}\times\frac{5}{5}=\frac{10}{15} \quad\text{and}\quad \frac{3}{5}=\frac{3}{5}\times\frac{3}{3}=\frac{9}{15}.$$

Thus $2/3$ is larger than $3/5$ by $1/15$:

$$\frac{2}{3}-\frac{3}{5}=\frac{10}{15}-\frac{9}{15}=\frac{1}{15}.$$

❖

Complex Fractions

In some fractions either the numerator or the denominator (or both) are themselves fractions and are called **complex fractions**. Such fractions can be readily simplified by remembering that fractions represent division of the numerator by the denominator. Study each of the following examples carefully to be sure that you understand the process of simplifying complex fractions.

$$\frac{\frac{2}{3}}{\frac{3}{4}}=\frac{2}{3}\div\frac{3}{4}=\underbrace{\frac{2}{3}\times\frac{4}{3}}_{\text{invert and multiply}}=\frac{2\times4}{3\times3}=\frac{8}{9}. \qquad \frac{\frac{3}{7}}{-8}=\frac{3}{7}\div(-8)=\underbrace{\frac{3}{7}\times\left(-\frac{1}{8}\right)}_{\text{invert and multiply}}=-\frac{3}{56}.$$

$$\frac{3}{\left(\frac{4}{9}\right)}=3\div\frac{4}{9}=\underbrace{3\times\frac{9}{4}}_{\text{invert and multiply}}=\frac{27}{4}=6\frac{3}{4}. \qquad \frac{-\frac{1}{2}}{-\frac{1}{3}}=-\frac{1}{2}\div\left(-\frac{1}{3}\right)=\underbrace{-\frac{1}{2}\times(-3)}_{\text{invert and multiply}}=\frac{3}{2}.$$

A.2.2 Decimal Equivalents

For fractions in **decimal form**, each successive column to the right of the decimal point has one-tenth the value of the previous column. The following table shows values and names for each of the first six columns right of the decimal point, both as common fractions and in decimal form.

Ones	Decimal Point	Tenths	Hundredths	Thousandths	Ten Thousandths	Hundred Thousandths	Millionths
1	.	$\frac{1}{10}$	$\frac{1}{100}$	$\frac{1}{1000}$	$\frac{1}{10,000}$	$\frac{1}{100,000}$	$\frac{1}{1,000,000}$
		0.1	0.01	0.001	0.0001	0.00001	0.000001

Reading a number in decimal form is easiest if you first write it (or think of it) in **expanded form**. For example, in expanded form the number 8.375 is

$$8.375 = (8\times1) + (3\times0.1) + (7\times0.01) + (5\times0.001).$$

From this point you can read the number by thinking of it in terms of common fractions, as in

$$\begin{aligned} 8.375 &= 8 + \frac{3}{10} + \frac{7}{100} + \frac{5}{1000} \\ &= 8 + \frac{300}{1000} + \frac{70}{1000} + \frac{5}{1000} \\ &= 8 + \frac{375}{1000} \\ &= \text{"eight and 375 one thousandths."} \end{aligned}$$

In general, converting a decimal into a common fraction simply means identifying the appropriate power of 10 in the denominator. If possible, the common fraction then may be simplified. Study each of the following examples to be sure you understand the procedure.

$$0.7 = \frac{7}{10}. \qquad 0.35 = \frac{35}{100} = \frac{35 \div 5}{100 \div 5} = \frac{7}{20}.$$

$$0.004 = \frac{4}{1000} = \frac{4 \div 4}{1000 \div 4} = \frac{1}{250}. \qquad 6.5625 = 6 + \frac{5625}{10,000} = 6 + \frac{5625 \div 625}{10,000 \div 625} = 6\frac{9}{16}.$$

Converting Common Fractions to Decimals

Converting a common fraction into its decimal equivalent simply involves the *division* implied by the common fraction. For example,

$$\frac{9}{16} = 9 \div 16 = 0.5625.$$

> **Time-Out to Think:** Use your calculator to confirm that 9/16 = 0.5625. Then confirm several more simple conversions, such as 1/2 = 0.5 and 3/4 = 0.75.

Some common fractions can't be written *exactly* in decimal form; rather, they have a repeating pattern of digits. For example, in decimal form $\frac{1}{3} = 0.333333\ldots$. Repeating digits often are indicated by drawing a bar over the pattern that repeats. For example,

$$\frac{1}{3} = 0.3333333\ldots = 0.\overline{3} \quad \text{and} \quad \frac{2}{7} = 0.285714285714\ldots = 0.\overline{285714}.$$

More often, decimals with repeating digits simply are rounded. In particular, most calculators don't use the bar notation; instead they round the fraction by showing only the decimal places that fit on the screen.

Most calculators internally keep track of several additional places beyond what can be shown on the screen.

> **Time-Out to Think:** Do long division (e.g., 1 ÷ 3) to convince yourself that $\frac{1}{3}$ is a repeating decimal. What happens when you convert $\frac{1}{3}$ to decimal form on your calculator? Why?

Irrational numbers can't be written as common fractions. Moreover, the decimal representation of an irrational number does not have a repeating pattern of digits. For example, consider the number that has digits appearing in the following pattern: 0.10110011100011110000.... Even though there is a clear pattern to the digits, no pattern of digits *repeats*, so this is an irrational number. Well-known irrational numbers include

$$\pi = 3.141592654\ldots,$$
$$e = 2.7182818284\ldots,$$
$$\sqrt{2} = 1.414213562\ldots, \text{ and}$$
$$\sqrt{3} = 1.732050808\ldots.$$

Note that fractions in decimal form make addition or subtraction particularly easy. For example, a quarter (the coin) represents ¼ of a dollar. A dime represents 1⁄10 of a dollar. If you have a quarter and a dime, you have ¼ + 1⁄10 of a dollar or, equivalently, \$0.25 + \$0.10 =\$0.35.

Example A-6. Subtract ⅝ – ⅖ by converting the common fractions to decimals.

Solution: Converting each fraction to a decimal makes the subtraction easy:

$$\frac{5}{8} - \frac{2}{5} = (5 \div 8) - (2 \div 5) = 0.625 - 0.4 = 0.225.$$

❖

A.2.3 Fractions as Percentages

A percentage is just another way to express a fraction. The word *percent* simply mean *per hundred* or *divided by one hundred*. For example, 19% means *nineteen per hundred*, or *nineteen divided by one hundred*:

% means divided by 100.

$$19\% = \frac{19}{100} = 0.19.$$

Percentages have been widely used since ancient times, when their earliest use probably was for levying taxes. The Roman Emperor Augustus (27 B.C.–A.D. 14) is said to have levied a 1% tax on proceeds from goods sold at auction.

Converting Percentage to Decimals or Common Fractions

To convert from percent to a decimal: Drop the % symbol and move the decimal point two places to the left.

Converting a percentage to a common fraction or decimal requires only remembering that the % sign means *divided by 100*. For example,

$$25\% = 25 \div 100 = \frac{25}{100} = 0.25; \qquad 86.2\% = 86.2 \div 100 = \frac{86.2}{100} = 0.862;$$

$$724\% = 724 \div 100 = \frac{724}{100} = 7.24; \qquad 0.1\% = 0.1 \div 100 = \frac{0.1}{100} = \frac{1}{1000} = 0.001.$$

Note that converting from percent to decimal form is easy: Simply drop the % symbol and move the decimal point two places to the left.

Converting Decimals to Percentages

100% = 100 ÷ 100 = 1.

Note that 100% means 100 ÷ 100 = 1. Therefore multiplying by 100% is just another way of multiplying by 1, so converting a decimal to a percentage simply requires multiplying by 100%. Study the following examples.

$$0.43 = 0.43 \times 100\% = 43\%. \qquad 0.003 = 0.003 \times 100\% = 0.3\%.$$

$$0.07 = 0.07 \times 100\% = 7\% \qquad 25 = 25 \times 100\% = 2,500\%.$$

$$1.02 = 1.02 \times 100\% = 102\%. \qquad -0.1 = -0.1 \times 100\% = -10\%.$$

Converting Common Fractions to Percentages

Converting a common fraction to a percentage requires two steps: First convert the common fraction to a decimal and then convert the decimal to a percentage, as shown in the following examples.

$$\frac{1}{2} = 0.5 = 50\%. \qquad \frac{5}{6} = 0.8333\ldots = 83.33\% \text{ (rounded)}.$$

$$\frac{3}{4} = 0.75 = 75\%. \qquad \frac{5}{3} = 1.666\ldots = 166.67\% \text{ (rounded)}.$$

A.3 POWERS AND ROOTS

Raising to powers and taking roots are two other common operations of arithmetic based on multiplication.

A.3.1 Basics of Powers and Roots

Raising to a power involves finding a product when a number is multiplied by itself some number of times. For example, 2^4 is read as "two to the fourth power," or "the fourth power of 2," and means 2 multiplied by itself four times:

$$2^4 = \underbrace{2 \times 2 \times 2 \times 2}_{4 \text{ times}} = 16.$$

The number that is multiplied, 2 in this case, is called the **base**. The number of times it is multiplied, 4 in this case, is called the **exponent**.

If you want to multiply 10 by itself six times, you are looking for "ten to the sixth power." The base is 10 and the exponent is 6, so it is it is written 10^6, which means

$$10^6 = \underbrace{10 \times 10 \times 10 \times 10 \times 10 \times 10}_{6 \text{ times}} = 1{,}000{,}000 \text{ (one million)}.$$

If we let the letter b represent the base and n represent the exponent (since n is the number of times that the base is multiplied), the general rule for raising to a power is

$$b^n = \underbrace{b \times b \times b \times \cdots \times b}_{n \text{ times}}.$$

The Zero and First Powers

Any number to the first power is itself: $b^1 = b$.

Any number to the zero power is 1: $b^0 = 1$ ($b \neq 0$).

What happens when the exponent is 0 or 1? We can answer this question by example, considering powers of 5. For each of the following powers of 5 note that we also show multiplication by 1, the reason for which should soon become clear.

$$5^3 = 1 \times \underbrace{5 \times 5 \times 5}_{3 \text{ times}} = 125; \quad 5^2 = 1 \times \underbrace{5 \times 5}_{2 \text{ times}} = 25; \text{ and } \quad 5^1 = 1 \times \underbrace{5}_{1 \text{ time}} = 5.$$

Note that the pattern clearly shows that 5 raised to the first power is itself. Generalizing, again using the symbol b to represent the base, we have

$$b^1 = 1 \times \underbrace{b}_{1 \text{ time}} = b.$$

Now continue the pattern of powers of 5 to the 0 power. The pattern shows that, for the 0 power, 1 should not be multiplied by 5 at all. Therefore

$$5^0 = 1 \qquad \text{(not multiplied by 5 any times)}.$$

That is, 5 raised to the zero power is 1. Generalizing, any nonzero number raised to the zero power is 1, or

$$b^0 = 1 \qquad (b \neq 0).$$

Fractions Raised to Powers

The rules for raising to a power directly apply to fractions. For example,

$$\left(\frac{3}{4}\right)^3 = \frac{3}{4}\times\frac{3}{4}\times\frac{3}{4} = \frac{3\times3\times3}{4\times4\times4} = \frac{3^3}{4^3} = \frac{27}{64}.$$

However, note that the power can be distributed over the numerator and denominator. In general,

$$\left(\frac{a}{b}\right)^n = \frac{a^n}{b^n}.$$

Finding a Root

Note that you can also get 4 by multiplying −2 by itself: (−2) × (−2) = 4. However, the square root symbol, $\sqrt{\ }$, is taken to mean only positive numbers.

The process of **finding a root** is the **inverse**, or opposite, of raising to a power. For example, finding the second root, or **square root**, of 4 means finding a number that, when raised to the second power, yields 4. Thus the square root of 4 is 2, as

$$2^2 = 2\times2 = 4.$$

Similarly, finding the fourth root of 625 means finding a number that, when raised to the fourth power, yields 625. Thus the fourth root of 625 is 5, as

$$5^4 = 625.$$

The word *radical* comes from the Latin word *radix*, which means "root."

The **radical sign**, $\sqrt{\ }$, is used to write roots. By itself a radical sign indicates a square root. Other roots are indicated by placing an **index** on the radical sign. For example,

$\sqrt{9}$ is the square root of 9; $\sqrt{9} = 3$, as $3^2 = 9$.

$\sqrt[3]{1000}$ is the third or cube root of 1000; $\sqrt[3]{1000} = 10$, as $10^3 = 1000$.

$\sqrt[4]{2401}$ is the fourth root of 2401; $\sqrt[4]{2401} = 7$, as $7^4 = 2401$.

$\sqrt[6]{64}$ is the sixth root of 64; $\sqrt[6]{64} = 2$, as $2^6 = 64$.

$\sqrt[n]{b}$ is read as the "*n*th root of *b*" and means the number that, raised to the *n*th power, yields *b*.

In general, $\sqrt[n]{b}$ is read as the "*n*th root of *b*" and means the number that, raised to the *n*th power, yields *b*. The following examples further illustrate the uses of powers and roots.

$5^2 = 5\times5 = 25.$	$\sqrt{9} = 3$ as $3^2 = 9.$
$7^4 = 7\times7\times7\times7 = 2401.$	$\sqrt[3]{1000} = 10$ as $10^3 = 1000.$
$(-6)^3 = (-6)\times(-6)\times(-6) = 216.$	$\sqrt[3]{-27} = -3$ as $(-3)^3 = 27.$
$\left(\frac{2}{3}\right)^4 = \frac{2^4}{3^4} = \frac{16}{81}.$	$\sqrt{\frac{1}{4}} = \frac{1}{2}$ as $\left(\frac{1}{2}\right)^2 = \frac{1}{2}\times\frac{1}{2} = \frac{1}{4}.$

A.3.2 Power Rules

Four simple rules make working with powers easy. We demonstrate each rule by example.

Multiplication Rule

Study the following examples in which we multiply powers of the same base.

$$2^2\times2^4 = (2\times2)\times(2\times2\times2\times2) = 2^6 = 2^{2+4}.$$

$$5^2\times5^3 = (5\times5)\times(5\times5\times5) = 5^5 = 5^{2+3}.$$

$$10^4\times10^1 = (10\times10\times10\times10)\times(10) = 10^5.$$

To multiply powers of the same base, *add* the exponents:
$b^n \times b^m = b^{n+m}$.

Note that, as shown in the last step of each preceding example, multiplying two powers of the same base simply requires *adding* their exponents. Again, note that this multiplication rule for powers applies only to powers *of the same base.*

Multiplication rule for powers of the same base: $b^n \times b^m = b^{n+m}$.

Division Rule

The following examples demonstrate that dividing powers of the same base requires *subtracting* the exponents.

$$\frac{5^5}{5^3} = \frac{5\times5\times5\times5\times5}{5\times5\times5} = 5\times5 = 5^2 = 5^{5-3}.$$

$$\frac{2^4}{2^7} = \frac{2\times2\times2\times2}{2\times2\times2\times2\times2\times2\times2} = \frac{1}{2\times2\times2} = \frac{1}{2^3} = 2^{-3} = 2^{4-7}.$$

Again, note that this division rule for powers applies only to powers *of the same base.*

To divide powers of the same base, *subtract* the exponents:
$\frac{b^m}{b^n} = b^{m-n}$

Division rule for powers of the same base: $\frac{b^m}{b^n} = b^{m-n}$.

The following examples further demonstrate the multiplication and division rules in action.

$$3^3 \times 3^3 = 3^{3+3} = 3^6 = 729. \qquad 10^5 \times 10^0 = 10^{5+0} = 10^5 = 100{,}000.$$

$$2^3 \times 2^4 \times 2^5 = 2^{3+4+5} = 2^{12} = 4096. \qquad \frac{2^6}{2^5} = 2^{6-5} = 2^1 = 2.$$

$$\frac{10^4}{10^2} = 10^{4-2} = 10^2 = 100. \qquad \frac{7^3}{7^1} = 7^{3-1} = 7^2 = 49.$$

$$\left(\frac{2}{3}\right)^3 \times \left(\frac{2}{3}\right)^2 = \left(\frac{2}{3}\right)^{3+2} = \left(\frac{2}{3}\right)^5 = \frac{2^5}{3^5} = \frac{32}{243}. \qquad \left(\frac{2}{3}\right)^3 \div \left(\frac{2}{3}\right)^2 = \left(\frac{2}{3}\right)^{3-2} = \left(\frac{2}{3}\right)^1 = \frac{2}{3}.$$

Powers of Powers Rule

The third rule applies to powers of powers. Again, we use examples to demonstrate the rule.

$$\left(2^2\right)^3 = 2^2 \times 2^2 \times 2^2 = (2\times2)\times(2\times2)\times(2\times2) = 2^6; \text{ note that } 2^6 = 2^{2\times3}.$$

$$\left(10^3\right)^4 = \underbrace{10^3 \times 10^3 \times 10^3 \times 10^3}_{\text{apply multiplication rule: add the exponents}} = 10^{3+3+3+3} = 10^{12}; \text{ note that } 10^{12} = 10^{3\times4}.$$

When a power is raised to another power, *multiply* the exponents:
$\left(b^n\right)^m = b^{n\times m}$.

Note that raising a power to another power simply requires multiplying the exponents.

Powers of powers rule: $(b^n)^m = b^{n\times m}$.

Combining Bases Rule

The fourth rule applies to multiplying or dividing two powers that have the *same exponent*. The following examples demonstrate that you may first multiply or divide the bases and then raise their product to the exponent.

$$2^2 \times 3^2 = (2\times3)^2 = 6^2 = 36.$$

$$2^2 \div 10^2 = \frac{2^2}{10^2} = \left(\frac{2}{10}\right)^2 = 0.2^2 = 0.04.$$

Again, note that two bases can be combined only when they have the same exponent.

Combining bases with same exponent rule:
$a^n \times b^n = (a \times b)^n$ and $a^n \div b^n = (a \div b)^n$.

Warning: Situations Without Rules

Be careful in applying the four rules. Remember that the multiplication and division rules work *only* when you are dealing with powers of a single base, and that the combining bases rule works only when the bases have the same exponent. No rules or "shortcuts" apply in other situations. Consider the following example in which the both the two bases are different and the two exponents are different.

$$2^3 \times 3^2 = (2 \times 2 \times 2) \times (3 \times 3) = 8 \times 9 = 72.$$

The only way to reach the correct answer of 72 is to carry out each operation step-by-step as shown. Two more examples with differing bases and exponents further demonstrate the step-by-step requirement when no special rules offer shortcuts.

$$5^2 \times 2^3 = (5 \times 5) \times (2 \times 2 \times 2) = 25 \times 8 = 400.$$

$$4^3 \div 5^4 = \frac{4^3}{5^4} = \frac{4 \times 4 \times 4}{5 \times 5 \times 5 \times 5} = \frac{64}{625}.$$

Similarly, there are no shortcuts for adding or subtracting powers. Instead, you must work through the addition or subtraction as shown in the following two examples.

$$2^6 + 2^2 = 64 + 4 = 68.$$

$$3^3 - 3^{-3} = 27 - \frac{1}{9} = 26\frac{8}{9}.$$

A.3.3 Negative Powers

What happens when a number is raised to a *negative* power, such as 2^{-3}? Note that, according to the multiplication rule for powers,

$$2^3 \times 2^{-3} = 2^{3+(-3)} = 2^0 = 1.$$

Because $2^3 = 8$, the preceding statement is true only if $2^{-3} = 1/8$, or $1/2^3$. Similarly, because $10^2 \times 10^{-2} = 10^{2-2} = 10^0 = 1$, 10^{-2} must be $1/10^2 = 1/100$. Let's try the same idea on a fraction raised to a power:

$$\left(\frac{3}{2}\right)^2 \times \left(\frac{3}{2}\right)^{-2} = \left(\frac{3}{2}\right)^{2-2} = \left(\frac{3}{2}\right)^0 = 1; \text{ as } \left(\frac{3}{2}\right)^2 = \frac{3^2}{2^2}, \left(\frac{3}{2}\right)^{-2} \text{ must be } \frac{2^2}{3^2} = \left(\frac{2}{3}\right)^2.$$

Generalizing, any number raised to a negative power is the *reciprocal* of the corresponding positive power. That is,

$$b^{-n} = \frac{1}{b^n} \quad \text{and} \quad \left(\frac{a}{b}\right)^{-n} = \left(\frac{b}{a}\right)^n.$$

Any number raised to a negative power is the *reciprocal* of the corresponding positive power:

$$b^{-n} = \frac{1}{b^n} \text{ and } \left(\frac{a}{b}\right)^{-n} = \left(\frac{b}{a}\right)^n.$$

The following examples further demonstrate the calculation of negative powers.

$$2^{-1} = \frac{1}{2^1} = \frac{1}{2}. \qquad -6^{-3} = -\frac{1}{6^3} = -\frac{1}{6\times 6\times 6} = -\frac{1}{216}.$$

$$5^{-2} = \frac{1}{5^2} = \frac{1}{25}. \qquad 10^{-6} = \frac{1}{10^6} = \frac{1}{10\times10\times10\times10\times10\times10} = \frac{1}{1{,}000{,}000}.$$

$$\left(\frac{1}{2}\right)^{-3} = \left(\frac{2}{1}\right)^3 = 2^3 = 8. \qquad \left(\frac{4}{7}\right)^{-3} = \left(\frac{7}{4}\right)^3 = \frac{7^3}{4^3} = \frac{343}{64}.$$

Multiplying and Dividing with Negative Powers

The four rules for multiplying and dividing powers apply directly to all powers, whether negative or positive. You should confirm each of the following examples.

$$2^4 \times 2^{-2} = 2^{4+(-2)} = 2^{4-2} = 2^2 = 4. \qquad 10^{-3}\times 10^{-1} = 10^{-3+(-1)} = 10^{-3-1} = 10^{-4} = 0.0001.$$

$$3^5 \div 3^{-2} = 3^{5-(-2)} = 3^{5+2} = 3^7 = 2187. \qquad 10^{-4} \div 10^{-7} = 10^{-4-(-7)} = 10^{-4+7} = 10^4 = 1000.$$

$$\left(\frac{3}{2}\right)^{-3} \times \left(\frac{3}{2}\right)^{2} = \left(\frac{3}{2}\right)^{-3+2} = \left(\frac{3}{2}\right)^{-1} = \frac{2}{3}. \qquad \left(\frac{1}{6}\right)^{-2} \div \left(\frac{1}{6}\right)^{2} = \left(\frac{1}{6}\right)^{-2-2} = \left(\frac{1}{6}\right)^{4} = \frac{1}{1296}.$$

$$\left(5^2\right)^{-2} = 5^{2\times(-2)} = 5^{-4} = \frac{1}{5^4} = \frac{1}{625}. \qquad \left(10^{-3}\right)^{-2} = 10^{(-3)\times(-2)} = 10^6 = 1{,}000{,}000.$$

A.3.4 Roots as Fractional Exponents

According to the powers of powers rule,

$$4^{\frac{1}{2}} \times 4^{\frac{1}{2}} = \left(4^{\frac{1}{2}}\right)^2 = 4^{\frac{1}{2}\times 2} = 4^1 = 4.$$

The result demonstrates that $4^{\frac{1}{2}}$ is a number that, when multiplied by itself, yields 4. That is, $4^{\frac{1}{2}}$ must be the *square root* of 4: $4^{\frac{1}{2}} = \sqrt{4} = 2$. In fact, any number with an exponent of one-half is a square root. For example,

Because the square root of 2 is irrational, it can only be approximated as a decimal; note that $1.41421^2 = 1.999989924 \approx 2$.

$$\sqrt{100} = 100^{\frac{1}{2}} = 10, \quad \sqrt{9} = 9^{\frac{1}{2}} = 3, \text{ and } \sqrt{2} = 2^{\frac{1}{2}} \approx 1.41421.$$

Similarly, according to the powers of powers rule,

$$8^{\frac{1}{3}} \times 8^{\frac{1}{3}} \times 8^{\frac{1}{3}} = \left(8^{\frac{1}{3}}\right)^3 = 8^{\frac{1}{3}\times 3} = 8^1 = 8.$$

Thus $8^{\frac{1}{3}}$ is a number that, when raised to the third power, yields 8. That is, $8^{\frac{1}{3}}$ must be the *cube root* of 8: $8^{\frac{1}{3}} = \sqrt[3]{8} = 2$. Generalizing, an exponent of $1/n$ always is the same as an nth root, as in the following examples.

$$32^{\frac{1}{5}} = \sqrt[5]{32} = 2. \qquad (-32)^{\frac{1}{5}} = \sqrt[5]{-32} = -2.$$

$$\sqrt[6]{15625} = 15625^{\frac{1}{6}} = 5. \qquad \sqrt[12]{4096} = 4096^{\frac{1}{12}} = 2.$$

$$\sqrt[3]{\frac{27}{64}} = \left(\frac{27}{64}\right)^{\frac{1}{3}} = \frac{3}{4}. \qquad \sqrt[7]{\frac{128}{2187}} = \left(\frac{128}{2187}\right)^{\frac{1}{7}} = \frac{2}{3}.$$

If the fractional power has a numerator other than 1, it can be rewritten by applying the power of powers rule, as in the following examples.

$$8^{\frac{2}{3}} = 8^{\frac{1}{3}\times 2} = \left(8^{\frac{1}{3}}\right)^2 = \left(\sqrt[3]{8}\right)^2 = 2^2 = 4.$$

$$4^{-\frac{5}{2}} = 4^{\frac{1}{2}\times(-5)} = \left(4^{\frac{1}{2}}\right)^{-5} = \left(\sqrt{4}\right)^{-5} = (2)^{-5} = \frac{1}{2^5} = \frac{1}{32}.$$

$$81^{\frac{3}{4}} = 81^{\frac{1}{4}\times 3} = \left(81^{\frac{1}{4}}\right)^3 = \left(\sqrt[4]{81}\right)^3 = (3)^3 = 27.$$

Generalizing the preceding results about fractional exponents yields the following.

Fractional exponent rules: $b^{\frac{1}{n}} = \sqrt[n]{b}$ and $b^{\frac{m}{n}} = \sqrt[n]{b^m} = \left(\sqrt[n]{b}\right)^m.$

PROBLEMS

1. **Order of Operations — Warmup.** Evaluate each of the following expressions using the rules for order of operations. Try each of these both with and without a calculator and be sure you get the same answer either way.
   - **a.** $12 \div 4 - 3$
   - **b.** $12 \times 3 \div 4$
   - **c.** $12 - 2 \times 3$
   - **d.** $(64 - 4) \times 6$
   - **e.** $48 \div (8 - 2)$
   - **f.** $24 + (12 - 4)$
   - **g.** $100 \div (30 - 5)$
   - **h.** $143 \div 11 + 4 \times 2$
   - **i.** $24 \times 2 - 8 + 6$

2. **Operations with Integers.** Follow the rules for addition, subtraction, and multiplication of integers and the usual order of operations in each of the following problems. Do each problem both with and without a calculator.
   - **a.** $18 - (-7) + (-4)$
   - **b.** $10 - [7 - (-7)]$
   - **c.** $21 - \{-[16 - (12 \times (-14))] + (-2)\}$

3. **Number Lines.** Draw a number line that extends from −10 to 10. Mark a dot at or near the location of the following numbers.
   - **a.** 1 **b.** −2 **c.** −9 **d.** 5.5
   - **e.** 0 **f.** −2.5 **g.** 3/2 **h.** π

4. **Reciprocals.** Find the reciprocals of each of the following numbers. Express each answer both as a common fraction and as a decimal (using a calculator if necessary).
   - **a.** $\frac{88}{91}$ **b.** $-\frac{537}{912}$ **c.** $\sqrt{2}$ **d.** $-\frac{3}{\pi}$
   - **e.** $\frac{3}{2}$ **f.** $-\frac{23}{41}$ **g.** 7.5 **h.** 0.45
   - **i.** 25 **j.** −3.7 **k.** 0.66 **l.** −4

5. **Fractions in Simplest Form.** Reduce each of the following fractions to its simplest form.
   - **a.** $\frac{8}{12}$ **b.** $\frac{28}{70}$ **c.** $\frac{25}{45}$ **d.** $\frac{693}{792}$
   - **e.** $\frac{537}{912}$ **f.** $-\frac{999}{37}$ **g.** $\frac{243}{81}$ **h.** $\frac{343}{16{,}807}$

6. **Equivalent Fractions.** Determine whether the following pairs of fractions are equivalent. Explain how you make each determination.
   - **a.** $\frac{3}{4}, \frac{15}{20}$ **b.** $-\frac{1}{8}, -\frac{8}{64}$ **c.** $\frac{72}{11}, \frac{731}{121}$ **d.** $\frac{31}{29}, \frac{43}{41}$
   - **e.** $\frac{5}{6}, \frac{10}{12}$ **f.** $\frac{993}{111}, \frac{331}{37}$ **g.** $\frac{360}{18}, \frac{600}{30}$ **h.** $-\frac{16}{3}, \frac{32}{6}$

7. **Common Denominators.** Rewrite each of the following pairs of fractions with the same common denominator; use the lowest possible common denominator in each case.
   - **a.** $\frac{2}{3}, \frac{1}{2}$ **b.** $\frac{8}{21}, \frac{16}{35}$ **c.** $\frac{6}{9}, \frac{18}{27}$
   - **d.** $\frac{2}{3}, \frac{6}{8}$ **e.** $-\frac{9}{49}, \frac{3}{14}$ **f.** $\frac{13}{144}, \frac{45}{120}$

8. **Comparing Fractions.** By finding a common denominator, determine which is the larger fraction in each of the following pairs, and by how much.
   - **a.** $\frac{5}{16}, \frac{11}{32}$ **b.** $\frac{4}{21}, \frac{17}{85}$ **c.** $-\frac{1}{2}, -\frac{6}{11}$
   - **d.** $\frac{7}{8}, \frac{11}{12}$ **e.** $\frac{2}{3}, \frac{6649}{10{,}000}$ **f.** $\frac{740}{987}, \frac{3}{4}$

9. **Practice with Fractions.** Evaluate each of the following expressions. Use your calculator only to check your work.
   - **a.** $\frac{4}{3} \times \frac{1}{2}$ **b.** $\frac{4}{3} + \frac{1}{2}$ **c.** $\frac{4}{3} \div \frac{1}{2}$
   - **d.** $\frac{4}{3} - \frac{1}{2}$ **e.** $\frac{7}{20} + \frac{3}{5}$ **f.** $\frac{12}{13} - \frac{1}{4}$

10. **More on Adding Fractions.** To show that you cannot add the same thing to the numerator and denominator of a fraction without changing the value of the fraction, do the following:
    - Write the fraction ½ in decimal form.
    - Write the fraction $\frac{1+3}{2+3}$ in decimal form.

    Is the answer the same in both cases? Why not?

11. Complex Fractions. Rewrite each of the following complex fractions as a common fraction reduced to its simplest terms.

a. $\dfrac{1/2}{1/2}$ b. $\dfrac{4/7}{2/5}$ c. $\dfrac{\left(\frac{3}{8}\right)}{10}$ d. $\dfrac{9}{\left(\frac{11}{16}\right)}$

e. $\dfrac{\left(\frac{9}{11}\right)}{16}$ f. $\dfrac{-7/11}{13/5}$ g. $\dfrac{-17/4}{-4/17}$

12. Fractions in Decimal Form. For each of the following numbers, (i) write the number in expanded form, (ii) write the number in words, and (iii) write the number as a common fraction. *Example:* 0.55. *Solution:* (i) $0.55 = (0 \times 1) + (5 \times 0.1) + (5 \times 0.01)$; (ii) 0.55 is read as "fifty-five hundredths"; (iii) $0.55 = 55/100 = 11/20$.

a. 0.003 b. 1000.7 c. −65.5 d. −85,700
e. 6005.4 f. 98.7 g. 0.124 h. 0.9999
i. 0.24816 j. 0.12345

13. Percentages to Common Fractions. Convert each of the following percentages first to a decimal and then to a common fraction reduced to simplest terms. *Example:* 10%. *Solution:* 10% = 0.1 = ⅒.

a. 25% b. 95% c. 2.5% d. −14%
e. −1.5% f. 200% g. −60% h. 100%
i. 0.04% j. 33% k. 7.5% l. 19.5%

14. Fractions to Percentages. Convert each of the following fractions to a percentage.

a. 8/2 b. 36/50 c. 130/200 d. 12/8
e. 0.72 f. 0.003 g. 2/5 h. −0.025
i. 17/20 j. 12/15 k. 30/25 l. 2/3
m. 0.995 n. 1.125 o. 3/99 p. −0.90

15. Working with Exponents. Simplify the following expressions as much as possible without using a calculator.

a. $2^4 \times 2^{10}$ b. $3^4 \times 3^2$ c. $5^0 \times 5^2$
d. $3^{12} \div 3^{10}$ e. $4^3 \div 4^1$ f. $10^8 \div 10^5$

16. Negative Exponents. Simplify the following expressions.

a. 7^{-2} b. 4^{-3} c. $2^{-2} \times 2^{-4}$ d. $3^{-1} \times 4^{-2}$

17. Fractional Exponents. Rewrite each of the roots as fractional powers. Express your answer in the simplest possible form without using a calculator.

a. $\sqrt{25}$ b. $\sqrt[4]{64}$
c. $\sqrt{81} \times \sqrt[3]{27}$ d. $\sqrt[3]{125} \div \sqrt{625}$

18. Powers and Roots. Simplify each of the following expressions without a calculator.

a. 4^3 b. $10^6 \times 10^{-8}$ c. $27^{2/3}$
d. 7^{-2} e. $4^{0.5}$ f. $10^{-5} \div 10^8$
g. $(10^6)^{1/2}$ h. $2^3 \times 3^3$ i. $144^{1/2}$
j. $125^{1/3}$ k. $3^{1/2} \times 3^2$ l. $16^{1/4} \div 16^{1/2}$

19. One and Zeros. The following operations involve ones and zeros in various ways. Evaluate each expression or state that the expression is undefined.

a. 4^1 b. 4^0 c. $4 \div 4$
e. 4×0 d. $4 \div 1$ f. $4 - 4$

20. Putting it All Together. Evaluate each of the following expressions. Be sure to follow the rules for order of operations.

a. $1000 - \sqrt[3]{27} \times \left\{5 + \left[4 \times \left(2 + 3^2\right) - 4\right]\right\} + 50$

b. $\sqrt{25} \times 2^5 + \left[13 - \left(4^3 + \sqrt[3]{8}\right)\right]$

c. $39 \div 13 + 2 \times 36 - 4 \div 2$

d. $2 \times \{5 + 3 \times [(7+1) - 56 \div 8]\} - \sqrt[5]{32}$

e. $\dfrac{24 - \sqrt{144} \times 2 + \left[33 - (6^2 - 3 \times 5)\right]}{\sqrt{9} \times \left[17 - (\sqrt{121} + 2^2)\right]}$

B REVIEW OF BASIC ALGEBRA

Among the most important mathematical tools are those of algebra. This appendix is designed to provide a brief review of the algebra skills that you will need in parts of the text. If you are already proficient in algebra, you may simply skim this appendix. If you need review, study it in depth and use it as a reference as you work through the text.

B.1 USES OF SYMBOLS AND EQUATIONS IN ALGEBRA

In arithmetic, we apply the operations of addition, subtraction, multiplication, and division to numbers. Algebra is similar, except that these operations are applied to symbols, such as n, x, y, or z, instead of individual numbers. The advantage of using symbols is that they can represent many different numbers, which allows generalization. With algebra, you can discover and use techniques that apply to a wide range of situations.

A meaningful combination of numbers, symbols, and arithmetic operations is called an **algebraic expression**. Unless told otherwise, you can assume that symbols in expressions stand for any real number. Here are a few examples of algebraic expressions:

$$25, \quad 3x, \quad \frac{2y+3}{4}, \quad 3a+4b, \quad \text{and} \quad 2n^2+3n+5.$$

Note that an expression, by itself, does not have much meaning. However, by using a mathematical connector, you can link two or more expressions to form mathematical sentences, or statements. Using the connectors shown in Table B-1, you can form two basic types of mathematical statements:

Table B-1. Mathematical Connectors

Connector	Meaning
$=$	is equal to
$\approx$	is approximately equal to
$<$	is less than
$\leq$	is less than or equal to
$>$	is greater than
$\geq$	is greater than or equal to

- equations, which use an equals sign (=); and
- inequalities, which use one of the "less than" ($<$, $\leq$) or "greater than" ($>$, $\geq$) signs.

We will be concerned only with equations. Here are a few examples of equations:

$$5 \times 5 = 25, \quad x + x = 2x, \quad p = 400 + 5n, \quad \text{and} \quad x + 12 = 2x.$$

Note that an equation always has two sides, connected by the equals sign and that each side is an expression. Some equations, such as the first one shown, involve only numbers; others include symbols as well, such as the other three shown.

Doing algebra primarily involves working with equations. Although the same rules apply to all types of equations, you should recognize that symbols are used in very different ways in different types of equations. In fact, the sample equations presented show three distinct types of equations with distinct uses of symbols. Let's investigate each type.

Identities

The first type of equation is called an **identity** because the expressions on both sides of the equals sign have the identical meaning. Identities make general statements that are always true. For example, adding a number to itself is always the same as multiplying the number by 2. That is, 3 + 3 is the same as 2 × 3; 12 + 12 is the same as 2 × 12; and so on. We state this fact in general by writing an identity, using the symbol x to represent all numbers:

$$x + x = 2 \times x = 2x, \qquad \text{for all numbers } x.$$

We discuss relations in greater depth in Chapter 10.

Relations

A second use of equations is to describe a *relation* between two (or more) quantities. For example, suppose that you have a job with a base salary of \$400 per week, and you receive \$15 per hour of overtime. Then your total paycheck is related to the number of hours of overtime you work as follows:

$$\text{paycheck} = \$400 + \left(15\frac{\$}{\text{hr}}\right) \times \text{(number of hours of overtime)}.$$

Symbols can be introduced to write the relation in a more compact form. For example, the symbol p can be used to represent the dollar amount of your paycheck, and n can represent the number of hours of overtime. Then the relation can be written as

$$p = \$400 + \left(15\frac{\$}{\text{hr}}\right) \times n, \quad \text{where} \begin{cases} n = \text{number of hours of overtime;} \\ p = \text{total paycheck in dollars.} \end{cases}$$

Note that we also explain the "code" for the symbols.

Time-Out to Think: Would it make any difference if we used the symbols x and y instead of n and p in the preceding equation? If the symbols don't matter, is it at all helpful to use n and p rather some other pair of symbols?

Each of the quantities in a relation is called a **variable** because it can assume varying values. In the paycheck relation, there are two variables: *the number of hours of overtime* that you work and your *total paycheck*. As the relation is written, the variable *total paycheck* depends on the variable *number of hours of overtime*. That is, if you choose a particular value for the *number of hours of overtime* worked, the relation will tell you the *total paycheck* amount.

Example B-1 Suppose that the amount of your weekly paycheck is given by the preceding equation. If you work 8 hours of overtime in a particular week, how much will you receive in your paycheck?

Solution: Simply substitute a value of 8 hours for the variable n. The relation then becomes

$$p = \$400 + \left(15\frac{\$}{\text{hr}}\right) \times 8 = \$520.$$

Working 8 hours of overtime will give you a paycheck of \$520.

❖

Equations with Unknowns

The third use of equations involves symbols as unknown quantities. For example, consider the question: *If John is 12 years older than Mary, when will John be twice as old as Mary?* Although we do not know Mary's age (it is an unknown), we can represent it with the symbol m. Because John is 12 years older than Mary, his age is always $m + 12$. To find when John's age will be twice Mary's age, or $2m$, we equate his age at any time ($m + 12$) with his age at the particular time when he is twice as old ($2m$), or

$$m + 12 = 2m.$$

Because the only unknown in the equation is m, or Mary's age, we can solve for it. Once we determine Mary's age, m, we know that John's age is $m + 12$.

Example B-2 The preceding problem can be solved algebraically, but for the moment, proceed by "trial and error." Substitute the following values for m in the equation: $m = 4, 8, 12, 16$. Do any of these values make the equation a true statement? What is the answer to the problem?

Solution: Try each of the values.

If $m = 4$, *then* $m + 12 = 2m$ *becomes* $4 + 12 = 2 \times 4$ *False!*

If $m = 8$, *then* $m + 12 = 2m$ *becomes* $8 + 12 = 2 \times 8$ *False!*

If $m = 12$, *then* $m + 12 = 2m$ *becomes* $12 + 12 = 2 \times 12$ *TRUE!*

If $m = 16$, *then* $m + 12 = 2m$ *becomes* $16 + 12 = 2 \times 16$ *False!*

The equation is true only when $m = 12$, where m represents Mary's age. Because John's age is $m + 12$, he will be 24 at that time. The answer to the original question is that John will be twice Mary's age when she is 12. ❖

When symbols are used as unknowns in equations, the message usually is "find all values of the unknown(s) that make the equation a true statement." The values of the unknown(s) that satisfy the equation are called **solutions**. Equations with unknowns do not necessarily have a single (unique) solution; they sometimes have no solution, and sometimes have more than one solution. For example,

$x + 2 = 5$ has one solution, $x = 3$, because $3 + 2 = 5$.

$x^2 = -1$ has no solution because the square of any real number is positive.

$x^2 = 4$ has two solutions, $x = 2$ and $x = -2$, because $2^2 = 4$ and $(-2)^2 = 4$.

B.1.2 Summary: Types of Algebraic Equations

To summarize, algebraic equations have at least three different uses, each of which uses symbols differently.

- **Identities** use symbols to represent all numbers (or a subset of all numbers); identities are equations that are always true.
- **Relations** use symbols to represent related quantities, or variables.
- **Equations with unknowns** use symbols to stand for unknown quantities; equations with unknowns may have no solution, one solution, or more than one solution.

Although all three types of equations have much in common and all are subject to the same algebraic manipulations, distinguishing between them will help you to interpret and solve problems.

B.2 FINDING EQUIVALENT EXPRESSIONS

Perhaps the most basic technique for working with algebraic problems involves converting one expression to an equivalent form. Just as the fraction 276/414 is more meaningful when rewritten as 2/3, the expression $(2x + 3x + x)$ is easier to work with when rewritten as $6x$. We say that $(2x + 3x + x)$ and $6x$ are equivalent forms of the same expression.

Identities provide the basic ground rules for manipulating expressions, because each side of the equals sign in an identity represents the same expression in a different form. Many algebraic manipulations can be done with the aid of just a few basic identities. Let's look at seven basic identities that are extremely useful.

1. The order in which we add two numbers does not matter (the **commutative law of addition**). For example, $3 + 5 = 5 + 3$. Algebraically, we can use symbols to write this identity as

$$a + b = b + a, \qquad \text{for all numbers } a \text{ and } b.$$

2. The order in which we multiply two numbers does not matter (the **commutative law of multiplication**). For example, $12 \times 4 = 4 \times 12$. We write this identity as

$$a \times b = b \times a, \qquad \text{for all numbers } a \text{ and } b.$$

3. We can regroup added numbers without changing the meaning (the **associative law of addition**). For example, $6 + (2 + 5) = (6 + 2) + 5$. The general form of the identity is

$$a + (b + c) = (a + b) + c, \qquad \text{for all numbers } a, b, \text{ and } c.$$

4. We can regroup multiplied numbers without changing the meaning (the **associative law of multiplication**). For example, $2 \times (3 \times 4) = (2 \times 3) \times 4$. In general,

$$a \times (b \times c) = (a \times b) \times c, \qquad \text{for all numbers } a, b, \text{ and } c.$$

5. We can *distribute* multiplication over a sum (the **distributive law**). For example, $4 \times (3 + 5) = (4 \times 3) + (4 \times 5)$. In general,

$$a \times (b + c) = (a \times b) + (a \times c), \qquad \text{for all numbers } a, b, \text{ and } c.$$

6. We can *distribute* division over a sum (no special name). For example, $(6 + 9)/3 = (6/3) + (9/3)$. The general rule is

$$\frac{a+b}{c} = \frac{a}{c} + \frac{b}{c}, \qquad \text{for all numbers } a, b, \text{ and } c; \text{ provided that } c \neq 0.$$

7. We can add fractions by forming a common denominator. For example,

$$\frac{2}{3} + \frac{3}{4} = \frac{(2 \times 4) + (3 \times 3)}{3 \times 4} = \frac{17}{12}.$$

The general rule is

$$\frac{a}{b} + \frac{c}{d} = \frac{(a \times d) + (b \times c)}{b \times d} = \frac{ad + bc}{bd}, \qquad \text{for all numbers } a, b, c, \text{ and } d, \text{ provided that } b \neq 0 \text{ and } d \neq 0.$$

Time-Out to Think: Substitute specific numbers into the preceding identities to convince yourself of their validity. Is the following statement an identity? Why or why not?

$$\frac{a}{b+c} \stackrel{?}{=} \frac{a}{b} + \frac{a}{c}.$$

Example B-3 Expand the expression $3a \times (2y + 3z)$.

Solution: Use identity 5 (the distributive law) to write this expression as

$$3a \times (2y + 3z) = (3a \times 2y) + (3a \times 3z) = 6ay + 9az.$$

Note that, in expanding the expression, you essentially found a new identity, $3a \times (2y + 3z) = 6ay + 9az$, which holds for all real numbers a, y, and z.

❖

> **Time-Out to Think:** Using $a = -3$, $y = 4$, and $z = -7$, check the identity found in the preceding example. Try several more sets of numbers a, y, and z to convince yourself that the example was done correctly.

Expanding and Factoring Expressions

The procedures of **expanding** and **factoring** expressions simply involve multiplication or division. Examples B-4 and B-5 illustrate these techniques.

Example B-4 Use the distributive law to expand the expression

$$(a + b)(c + d).$$

Solution: Begin by regarding the first term $(a + b)$ as a single term and distribute it across the term $(c + d)$. Recall that multiplication signs can be omitted by using the convention that $ab = a \times b$. The first step is to write

$$(a + b)(c + d) = (a + b)\,c + (a + b)\,d.$$

Now use the distributive law again to distribute c and d over the term $(a + b)$:

$$(a + b)\,c + (a + b)\,d = ac + bc + ad + bd.$$

In summary, the expression $(a + b)(c + d)$ can be written $ac + bc + ad + bd$.

Example B-5 Factor the expression $8xy + 12x$.

Solution: This time, use the distributive law in the opposite direction by noting that each of the terms in the expression has $4x$ as a factor:

$$8xy + 12x = (4x \times 2y) + (4x \times 3) = 4x\,(2y + 3).$$

Again, a new identity has been found in this process: $8xy + 12x = 4x(2y + 3)$.

Simplifying Expressions

Any term in an expression, such as $3xy$ or $4x^3$, is composed of two basic pieces: the number, called the **coefficient**, and the variables. For example, the coefficient is 3 in the term $3xy$, and the coefficient is 4 in the term $4x^3$. Terms that differ *only* in their coefficients, such as $3xy$ and $-7xy$ or $4x^3$ and $-2x^3$, are called **like terms** (or **similar terms**). An expression is in its **simplest form** when all its like terms are grouped together as completely as possible.

Example B-6 Simplify the expression

$$\frac{16xy + 2x^2 + 8xz + 7x^2}{4x}.$$

Solution: To simplify this expression, first group the like terms containing x^2 in the numerator:

$$\frac{16xy - 2x^2 + 8xz + 7x^2}{4x} = \frac{16xy + 8xz + (7x^2 - 2x^2)}{4x} = \frac{16xy + 8xz + 5x^2}{4x}.$$

Simplifying the expression further by using identity 6:

$$\frac{16xy + 2x^2 + 8xz + 7x^2}{4x} = \frac{16xy}{4x} + \frac{8xz}{4x} + \frac{5x^2}{4x} = 4y + 2z + 1.25x.$$

Because division by x is involved, this new identity holds only as long as $x \neq 0$.

Time-Out to Think: Using $x = -3, y = 4$, and $z = -7$, check the identities found in Examples B-5 and B-6. Try several more sets of numbers for x, y, and z to convince yourself that the examples were done correctly. Why isn't the identity in the last example true if $x = 0$?

B.3 WORKING WITH EQUATIONS

Just as expressions can be changed into other equivalent expressions, you can transform equations into other equivalent equations. The basic technique involves performing the same operations on *both sides* of the equals sign. The same rules of operation apply to identities, relations, and equations with unknowns.

B.3.1 Adding or Subtracting the Same Quantity from Both Sides

If we start with an identity, such as $6 \times 5 = 30$, we can form other identities simply by adding or subtracting the same quantity from *both sides*. For example, the following statements must also be true.

$$(6 \times 5) + 7 = 30 + 7. \qquad (6 \times 5) - 256 = 30 - 256.$$

$$(6 \times 5) + \pi = 30 + \pi \qquad (6 \times 5) + x^2 = 30 + x^2.$$

In general, adding or subtracting the same quantity from both sides of an equation leads to another equivalent equation. We use symbols to state this rule by writing:

If $x = y$, then $x + z = y + z$, for all numbers x, y, and z.

Let's see how this rule can help solve equations with unknowns. Recall the problem: If John is 12 years older than Mary, when will John be twice as old as Mary? Using the symbol x for Mary's age, recall that this problem led to the equation

$$x + 12 = 2x.$$

To solve for the unknown x, we need to isolate it on one side of the equation. We do so by subtracting x from both sides:

$$x + 12 \underbrace{-x}_{\substack{\text{subtract } x \\ \text{on left}}} = 2x \underbrace{-x}_{\substack{\text{subtract } x \\ \text{on right}}} \Rightarrow \quad 12 = x, \quad \text{or} \quad x = 12.$$

Thus the solution is $x = 12$ because that is the only value of the unknown x that makes the equation true. Because x represents Mary's age, the answer to the question is that Mary's age must be 12 and John's age must be 24.

Time-Out to Think: Let's be sure that this answer makes sense by thinking about the original question. According to the information given in the question, how old will John be when Mary is 12? Is it true that he will be twice Mary's age at that time?

The same rule is useful with relations. Recall the relation in which a paycheck depends on a base salary of \$400 per week, plus \$15 per hour of overtime. Using the symbol p for the amount of the paycheck, and n for the number of hours of overtime, this relation was written as

$$p = \$400 + \left(15 \frac{\$}{\text{hr}}\right) \times n, \quad \text{where} \begin{cases} n = \text{number of hours of overtime;} \\ p = \text{total paycheck in dollars.} \end{cases}$$

In this form, the relation gives the amount of the paycheck in terms of the number of hours of overtime, or p in terms of n. Suppose, instead, that you wanted to know how many hours of overtime would be needed to receive a particular amount in the paycheck. In that case, you would need to find a relation that gives n in terms of p. To transform the relation to n in terms of p, we must isolate the variable n on one side of the equation. First, we subtract 400 from both sides:

$$p \underbrace{-\$400}_{\text{subtract \$400 from left}} = \$400 + \left(15\frac{\$}{\text{hr}}\right) \times n \underbrace{-\$400}_{\text{subtract \$400 from right}} \Rightarrow$$

$$p - \$400 = \left(15\frac{\$}{\text{hr}}\right) \times n \quad \text{or} \quad \left(15\frac{\$}{\text{hr}}\right) \times n = p - \$400.$$

Note that n is not yet isolated, because it still is multiplied by \$15. To complete the process, we need a second rule.

B.3.2 Multiplying or Dividing Both Sides by the Same Quantity

Our second rule is that both sides of an equation can be multiplied or divided by the same *nonzero* quantity. For example, starting with an identity such as $6 \times 5 = 30$, the following statements are true. In each case, the original identity is multiplied or divided by a nonzero quantity.

$$(6 \times 5) \times 3 = 30 \times 3. \qquad (6 \times 5) \times a = 30 \times a.$$

$$\frac{6 \times 5}{10} = \frac{30}{10}. \qquad \frac{6 \times 5}{x^2 + y^2} = \frac{30}{x^2 + y^2}.$$

Why does the rule fail if we multiply or divide by zero? Recall that dividing by zero gives an undefined result, which destroys any information contained in the original equation. The problem with multiplying by 0 is that it could convert a false statement into a true one. For example, consider the false statement $2 + 2 = 3$. Multiplying both sides by zero, gives the statement $0 = 0$, which is true. Thus multiplying by zero also destroys the information in the original equation.

We can use this rule to finish the paycheck problem. The relation was left in the form

$$\left(15\frac{\$}{\text{hr}}\right) \times n = p - \$400.$$

To isolate n, we divide both sides of the equation by \$15 per hour, which gives the equivalent relation:

$$\frac{\left(15\frac{\$}{\text{hr}}\right) \times n}{15\frac{\$}{\text{hr}}} = \frac{p - \$400}{15\frac{\$}{\text{hr}}} \quad \Rightarrow \quad n = \frac{p - \$400}{15\frac{\$}{\text{hr}}}.$$

The variable n has been expressed in terms of p.

Example B-7 Suppose that you want a paycheck of \$700. Based on the preceding relation, with n expressed in terms of p, how many overtime hours will you need to work?

Solution: Substitute a value of \$700 for the variable p:

$$\text{if } p = \$700, \text{ then } n = \frac{\$700 - \$400}{15\frac{\$}{\text{hr}}} = 20 \text{ hours.}$$

You will need to work 20 hours of overtime to receive a paycheck of \$700. ❖

Time-Out to Think: The relation between paycheck and the number of hours of overtime has been expressed in two forms: one in which the paycheck is given in terms of the number of hours of overtime and the other in which the number of hours of overtime is given in terms of the paycheck amount. In words, describe the different types of questions that each form can answer.

This second rule also can be cast in general terms to show that it applies to all types of equations. If x, y, and z are any numbers and $z \neq 0$,

$$\text{if } x = y, \text{ then} \begin{cases} x \times z = y \times z \\ \quad \text{and} \\ \dfrac{x}{z} = \dfrac{y}{z}. \end{cases}$$

Example B-8 Solve the equation $\frac{3x}{4} - 2 = 10$ for the variable x.

Solution: The goal is to find all values of x that make the equation a true statement. First, isolate the term involving x by adding 2 to both sides:

$$\frac{3x}{4} - 2 + 2 = 10 + 2 \quad \Rightarrow \quad \frac{3x}{4} = 12.$$

Then, multiply both sides of the equation by $4/3$:

$$\frac{4}{3}\left(\frac{3x}{4}\right) = \frac{4}{3}(12) \quad \Rightarrow \quad x = 16.$$

The only value of x that satisfies the equation is $x = 16$. Check to be sure that it works.

❖

B.3.3 Other Operations on Both Sides of an Equals Sign

So far, we have shown two ways of transforming equations into equivalent equations:

- adding or subtracting the same quantity on both sides of the equals sign; and
- multiplying or dividing by the same thing (except zero) on both sides of the equals sign.

Other operations also transform equations legitimately. But how can you know whether using a particular operation on both sides is legitimate? As usual, the only way to know for sure is to create a deductive, mathematical proof. But, sometimes a few test cases can provide guidance. The easiest way to devise test cases is to start with statements that you know to be true; if an operation changes a true statement into a false statement, it is *not* legitimate. Examples will illustrate the point; be sure to notice the subtleties that sometimes come into play.

Example B-9 Use a few test cases to determine whether squaring both sides of an equation is legitimate.

Solution: To ensure a strong argument, try a variety of different test cases involving different operations as well as positive numbers and negative numbers.

Test case 1: If you square both sides of $4 \times 3 = 12$, is it still true? Check:

$$4 \times 3 = 12 \xrightarrow{\text{square both sides}} (4 \times 3)^2 = 12^2. \quad \text{Yes, it's still true.}$$

Test case 2: If you square both sides of $\sqrt{9} = 3$, is it still true? Check:

$$\sqrt{9} = 3 \xrightarrow{\text{square both sides}} (\sqrt{9})^2 = 3^2. \quad \text{Yes, it's still true.}$$

Test case 3: If you square both sides of $-7 + 7 = 0$, is it still true? Check:

$$-7 + 7 = 0 \xrightarrow{\text{square both sides}} (-7 + 7)^2 = 0^2. \quad \text{Yes, it's still true.}$$

Test case 4: Suppose that $x = -5$. Will this information be preserved if you square both sides?

$$x = -5 \xrightarrow{\text{square both sides}} x^2 = (-5)^2 = 25. \quad \text{Yes, but...}$$

The "but" refers to the fact that the equation $x^2 = 25$ actually has two solutions: –5 and 5. The original statement said only that $x = -5$; it did *not* say that x could also be 5. Thus, although information in the original statement ($x = -5$) is preserved, the new statement contains additional information ($x = 5$) that may or may not be correct!

What can you conclude? All four test cases checked out, although the last test case provides reason for caution: The operation of squaring both sides might add additional information that is not correct. These test cases do not *prove* that squaring both sides of an equation always is legitimate, but they certainly offer strong evidence of legitimacy.

❖

Example B-10 Suppose that you have an equation that reads $\sqrt{x} = 13$. Use the rule that allows squaring both sides of the equation to solve for x.

Solution: Simply square both sides of the equation:

$$\sqrt{x} = 13 \xrightarrow{\text{square both sides}} \left(\sqrt{x}\right)^2 = 13^2 = 169 \quad \Rightarrow \quad x = 169.$$

The only solution of the equation is $x = 169$.

❖

B.4 SUMMARY OF ALGEBRAIC TECHNIQUES

If you look back at what we have done, you will notice that the basic ideas of algebra can be summarized very concisely. Table B-2 presents this summary, along with some examples. If you remember that these ideas are the essence of algebra, you will find using algebra and learning new algebraic techniques relatively easy.

Table B-2. Summary of Algebraic Techniques

Algebra generally involves manipulating expressions or statements for one of three purposes:	
• to find new *identities* from old ones.	**Example:** Using the distributive law, $$x(y + z) = xy + xz,$$ you find that $$3x(2y + 3z) = 6xy + 9xz.$$
• to restate *relations* so that they are expressed in different terms.	**Example:** Given p in terms of n, as $$p = 400 + 15n,$$ you can rearrange to find n in terms of p: $$n = \frac{p - 400}{15}.$$
• to solve *equations with unknowns*.	**Example:** The equation $x + 12 = 2x$ contains an unknown, x. You can solve for x to find that $x = 12$.
Basically, only two general procedures are used to accomplish these purposes:	
1. transforming expressions, on either side of an equation, into equivalent expressions by using identities.	**Example:** You can expand the expression $$3x(2y + 3z)$$ into the equivalent expression $$6xy + 9xz.$$
2. transforming equations into equivalent equations by operating on both sides of the equals sign in the same way (as long as the operations preserve the information in the original statement).	**Example:** Adding 2 to both sides of $$\frac{3x}{4} - 2 = 10,$$ you find the equivalent equation $$\frac{3x}{4} = 12.$$

B.4.1 Graphing Relations

One of the best ways to interpret and visualize a relation is to draw its graph. In this section, we review methods for graphing relations. There are two basic approaches to graphing any relation.

- you can make a graph **point-by-point**; that is, create a table of points that satisfy the relation, and then plot and connect these points.
- you can make the graph using **analytical methods** by analyzing the properties of the relation. For example, drawing the graph of a linear relation by determining its slope and intercept (see Chapter 10) is an analytical method.

Because analytical methods can be rather involved, in this text we rely on the point-by-point method. If you choose enough points with appropriate spacing between them, this method always succeeds.

Recall the framework for graphing. A **number line** extends in both the positive and negative directions from the zero point. Every point on the number line is associated with a real number, and every real number has a unique

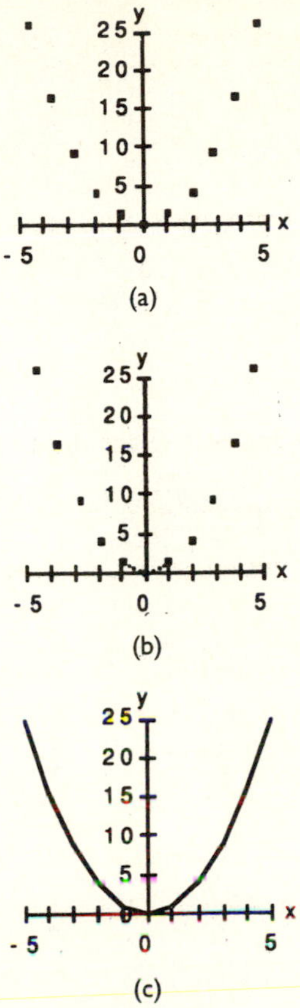

Figure B-1. The series shows the point-by-point method for graphing $y = x^2$. (a) Points in which x is an integer between –5 and 5 are plotted. (b) To clarify the graph, additional points (small dots) between $x = 0$ and $x = 1$ are added. (c) The graph is completed by drawing a smooth curve through the points.

point on the number line. To create a coordinate plane, draw two perpendicular number lines. Each of the number lines is called an **axis** (plural, **axes**). The intersection point of the two axes is called the **origin**. Normally, numbers increase to the right on the horizontal axis, and increase upward on the vertical axis.

Points in the coordinate plane are described by two numbers, called the **coordinates**, which give the horizontal and vertical distances between the point and the origin (which has coordinates (0, 0)). Coordinates are written in parentheses with the horizontal coordinate specified first; that is, in the form (*horizontal coordinate, vertical coordinate*). For example, you find the point (2, 3) by moving 2 units horizontally to the right and 3 units vertically upward from the point (0, 0). Similarly, you find the point (–3, 1) by moving –3 units (that is, 3 units to the left) horizontally and 1 unit vertically upward from (0, 0).

To keep the discussion of relations general and to minimize the amount of writing necessary, we discuss relations in which y changes with respect to x. Consider the graph of the relation $y = x^2$. The first step is to make a table of points:

x	–5	–4	–3	–2	–1	0	1	2	3	4	5
$y = x^2$	25	16	9	4	1	0	1	4	9	16	25

These points are plotted in Figure B-1a. Note that we have stretched the x-axis so that details of the graph can be seen more clearly. With only these points shown, however, we still don't know how they should be connected, especially near $x = 0$. Therefore several more points between 0 and 1 should be included:

x	–0.8	–0.6	–0.4	–0.2	–0.1	0	0.1	0.2	0.4	0.6	0.8
$y = x^2$	0.64	0.36	0.16	0.04	0.01	0	0.01	0.04	0.16	0.36	0.64

These additional points are shown in Figure B-1b. Now, the shape of the graph is much clearer, and the points clearly lie on a smooth curve as shown in Figure B-1c. This type of bowl-shaped curve is called a **parabola**.

Time-Out to Think: Note that the graph of $y = x^2$ is symmetric about the y-axis. That is, the left side looks like a mirror image of the right side. Can you explain why? Hint: would the graph of $y = x^3$ be symmetric? How about $y = x^4$?

As the preceding example shows, the point-by-point method often involves trial and error. In general, the following steps are useful in drawing graphs by the point-by-point method:

1. Create a table of a few points that you think will be representative of the relation. Unless the relation is not defined for negative values of x (such as $y = \sqrt{x}$), be sure that your table includes both positive and negative values of x.
2. Based on the points in the table, choose appropriate scaling for your axes (i.e., stretch or compress the axes, or show only a portion of the coordinate plane, if it will help show details of the graph). Plot the points.
3. Study the graph of the plotted points, looking for regions where you think that additional points might be needed to determine the precise shape. If necessary, make a table of additional points and plot them on the graph. Continue this process until the shape of the graph is clear.
4. Connect the points with a smooth curve.

Example B-11 Graph the relation $y = 5 - x^2$. Compare its graph to the graph of $y = x^2$.

Solution: Begin by making a table of points using integer values of x:

x	–5	–4	–3	–2	–1	0	1	2	3	4	5
$y = 5 - x^2$	–20	–11	–4	1	4	5	4	1	–4	–11	–20

These points are plotted in Figure B-2a. Note that the x-axis has been stretched so that the details of the graph can be seen more clearly. However, more detail seems to be needed near $x = 0$, so additional points should be included:

x	–0.8	–0.6	–0.4	–0.2	–0.1	0	0.1	0.2	0.4	0.6	0.8
$y = 5 - x^2$	4.36	4.64	4.84	4.96	4.99	5	4.99	4.96	4.84	4.64	4.36

The complete graph is shown in Figure B-2b. Note that this graph also is a parabola, but it is upside-down compared to the graph of $y = x^2$.

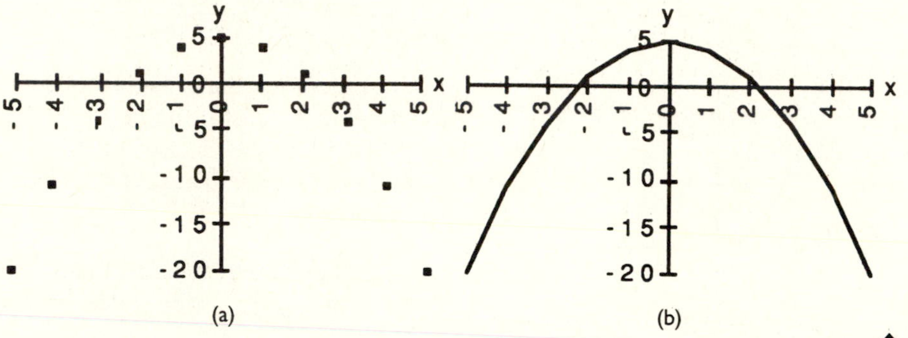

Figure B-2. (a) Selected points are shown for the graph of the relation $y = 5 - x^2$. (b) After plotting some additional points to clarify the graph near $x = 0$, the graph is completed.

❖

Example B-12 Graph the relation $y = 10^x$ (called an **exponential relation** because x appears in the exponent).

Solution: Again, begin by making a table of points, using integer values of x:

x	–4	–3	–2	–1	0	1	2	3	4
$y = 10^x$	0.0001	0.001	0.01	0.1	1	10	100	1000	10,000

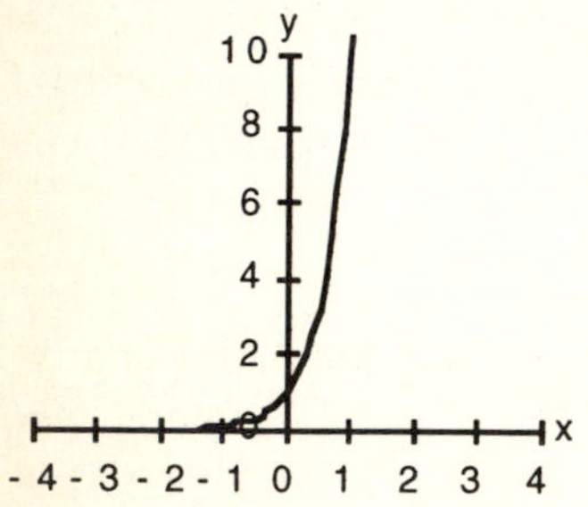

Figure B-3. The graph of the exponential relation $y = 10^x$ is shown.

This table has some important features. For example, y is never negative; thus, only the portion of the coordinate plane in which y is positive needs to be shown. Also, note that the values of y skyrocket as x slowly increases; therefore the graph should be drawn with a compressed scale for y. The result is shown in Figure B-3. Note that as x takes on larger and larger negative values, y gets closer and closer to zero. In fact, y will never reach zero; for example, when $x = -100$, $y = 10^{-100}$ — a very small, but positive, number.

❖

Example B-13 Graph the relation $y = 1/x$ (sometimes called a **reciprocal** or **inverse** relation because x appears in the denominator).

Solution: Begin by making a table of points using integer values of x.

x	–5	–4	–3	–2	–1	0	1	2	3	4	5
$y = 1/x$	–0.2	–0.25	–0.33	–0.5	–1	???	1	0.5	0.33	0.25	0.2

Even before drawing a graph, you should be aware of the problem around $x = 0$ because $y = 1/x$ is undefined there. Thus additional points should be included near $x = 0$:

x	–0.75	–0.5	–0.25	–0.1	–0.01	–0.001	0.001	0.01	0.1	0.25	0.5	0.75
$y = 1/x$	–1.33	–2	–4	–10	–100	–1000	1000	100	10	4	2	1.33

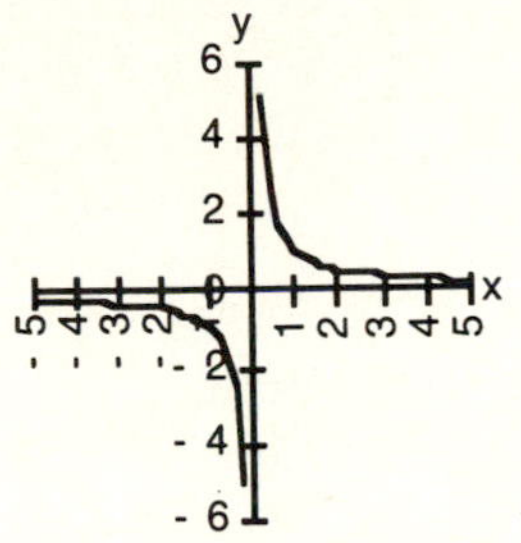

Figure B-4. The graph of the inverse relation $y = 1/x$ is shown.

The resulting graph is shown in Figure B-4. Note that the graph is discontinuous: It jumps from negative infinity to infinity at $x = 0$. Also note that each branch gets closer and closer to one of the axes without ever reaching it. We can describe this property for each of the four branches as follows: (1) the value of $y = 1/x$ approaches 0 as x approaches infinity; (2) the value of $y = 1/x$ also approaches 0 as x approaches negative infinity; (3) the value of $y = 1/x$ approaches infinity as x approaches zero from the positive side; and (4) the value of $y = 1/x$ approaches negative infinity as x approaches zero from the negative side.

❖

Using a Graphing Calculator

The only major drawback to the point-by-point method of graphing is that it can become very tedious. Fortunately, modern technology has made graphing much easier. For less than $100, you can buy a *graphing calculator* that, besides doing numerical calculations, can draw graphs of relations on a small screen. Although the details of graphing vary with the brand and model of the calculator, they all work in the same basic way:

1. You enter the relation that you want to graph in the form of an equation (e.g., $y = x^2$).
2. You enter numbers representing the portion of the x-axis that you want to display. For example, you might enter the range from $x = -5$ to $x = 5$ to get the same portion of the graph shown in Figure B-1. The calculator then automatically shows the appropriate range of values for the y-axis.
3. You instruct the calculator to draw the graph.

Graphing calculators usually have additional features that allow "zooming in" on certain portions of the graph, calculating values (called **roots**) where the graph crosses the x-axis, and obtaining other useful information.

Graphing with a Computer

If equipped with appropriate software, computers also can be used to draw graphs. Several software companies offer packages that graph almost any relation. Many schools have site licenses for some of the more popular packages, such as *Maple V*, *Mathematica*, *Derive*, or *MathCad*. Check your school's computing facilities to determine whether you can experiment with graphing with one of these or a similar program.

Graphing on a computer works essentially the same way as on graphing calculators: You enter the relation to be graphed and numbers representing the portion of the x-axis to be displayed, and the program does the rest. Like graphing calculators, computer programs have many additional features, including options for labeling graphs and adapting them for presentations.

Spreadsheet Graphing

You may also use *spreadsheet* software to graph relations. Spreadsheets are extremely common; virtually every business uses spreadsheets, as do most schools and many individuals. Often, specially priced versions for students are available.

A spreadsheet essentially allows you to make a table of (x, y) points; however, the program can calculate all the y values automatically, which saves a lot of work. Further, most spreadsheets have graphing capabilities that plot the resulting table of points.

Spreadsheets are arranged in **cells**; typically, *rows* of cells are labeled with numbers and *columns* by letters. Cells contain either numbers or formulas. To make a graph, you enter the x values in one set of cells and the relation you want to graph (a formula) in a second set of cells. For example, suppose that you want to graph the relation $y = 3x^3 - 2x^2 + 5$ and want to show the portion of the graph from $x = -5$ to $x = 5$. You begin by entering the value –5 in, say, cell B1. Next, you enter –4 in cell B2, –3 in cell B3, and so on until you reach 5 in cell B11 (this procedure can be automated within the spreadsheet program). Column B now contains x values for the table of points (Figure B-5a).

Next, you enter the corresponding y values in column C. In cell C1, you enter the formula $C1 = 3 \times (B1)^3 - 2 \times (B1)^2 + 5$. This formula instructs the spreadsheet to use the value from cell B1 as x in the relation $y = 3x^3 - 2x^2 + 5$. You then continue down column C, instructing each cell to use the value from the corresponding cell of row B in the relation. (Most spreadsheets have a "copy down" command that will copy the formula through the column of cells, automatically adjusting it to the corresponding cells of column B.) Figure B-5(a) shows the table with x values in column B and the appropriate formulas in column C. Once you have entered the formulas, the spreadsheet automatically calculates the y values; the resulting table of points, from which you can draw a graph, is shown in Figure B-5(b). Finally, the spreadsheet's graphing tools can be used to draw the graph, shown in Figure B-5(c).

Time-Out to Think: Check the points in the table shown in Figure B-5(b) by substituting *x*-values from column B into the relation. Confirm that the values are calculated properly.

	A	B	C
1		-5	=3*B1^3-2*B1^2+5
2		-4	=3*B2^3-2*B2^2+5
3		-3	=3*B3^3-2*B3^2+5
4		-2	=3*B4^3-2*B4^2+5
5		-1	=3*B5^3-2*B5^2+5
6		0	=3*B6^3-2*B6^2+5
7		1	=3*B7^3-2*B7^2+5
8		2	=3*B8^3-2*B8^2+5
9		3	=3*B9^3-2*B9^2+5
10		4	=3*B10^3-2*B10^2+5
11		5	=3*B11^3-2*B11^2+5

(a)

	A	B	C
1		-5	-420
2		-4	-219
3		-3	-94
4		-2	-27
5		-1	0
6		0	5
7		1	6
8		2	21
9		3	68
10		4	165
11		5	330

(b)

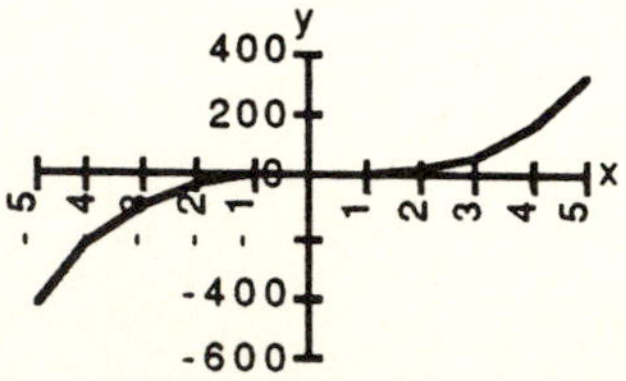

(c)

Figure B-5. The series shows the procedure for graphing the relation $y = 3x^3 - 2x^2 + 5$ between $x = -5$ and $x = 5$. (a) A spreadsheet is arranged in cells with rows numbered and columns labeled by letters. Values from –5 to 5 are entered in column B (cells B1 through B11). In column C the relation is entered as a formula; each cell in column C uses the corresponding cell of column B as x. (b) The spreadsheet automatically calculates the values for the formulas in column C; the calculated values are now shown. (c) The spreadsheet graphing tool gives a graph of the relation, as shown.

PROBLEMS

1. **Expanding Expressions.** Expand each of the following expressions.
   a. $3y(y + 2)$
   b. $x(y + 3)$
   c. $(3x + 4)(3x + 4)$
   d. $(3x + 4)(3x - 4)$
   e. $(p + q)^2$
   f. $(a + b + c)(4a - 3b)$

2. **Factoring Expressions.** Factor each expression by the term that follows the comma. *Example:* $8xy + 12xz - 4z$, $4x$. *Solution:* $8xy + 12xz - 4z = 4x(2y + 3z - 1)$
   a. $12x^2 - 30x$, $6x$
   b. $xy - 3y$, y
   c. $x^2 + 2x + 1$, $x + 1$
   d. $3y^4 - 27y^2$, $3y^2$

3. **Simplifying Expressions.** Simplify each of the following expressions or state that they are in simplest form.
   a. $x + x + x$
   b. $x + 2x + 3x$
   c. $x + y + 3x$
   d. $x(y+3)$
   e. $\dfrac{2x+4y}{2}$
   f. $\dfrac{4x}{2+4y}$

4. **Testing identities.** The following identities are false! Find values of the variables that show them to be false. Assume that w, x, y, and z are real numbers.
   a. $x - y = y - x$
   b. $x(y + z) = -xy + z$
   c. $x + yz = (x + y)(x + z)$
   d. $xyz = xy + xz$
   e. $\dfrac{x}{y} = \dfrac{y}{x}$
   f. $\dfrac{x+y}{w+z} = \dfrac{x}{w} + \dfrac{y}{z}$

5. **A Paycheck Relation.** Suppose that you receive a weekly salary of \$350 plus \$12 per hour for overtime. Your weekly paycheck amount is given by the following relation

$$p = \$350 + \left(12\frac{\$}{\text{hr}}\right) \times n, \quad \text{where} \begin{cases} n = \text{number of hours of overtime;} \\ p = \text{total paycheck in dollars.} \end{cases}$$

   a. Calculate your total paycheck for each of the following amounts of overtime: 0 hours, 2 hours; 12 hours; 6.5 hours.
   b. Solve the equation so that it becomes a relation for n in terms of p.
   c. Use the restated relation to calculate the number of hours of overtime needed to obtain each of the following amounts in your paycheck: \$374, \$400, \$450, \$500.

6. **Is It a Solution?** For each of the following equations a potential solution is given after the semicolon. Test the potential solution by substituting into the equation. Is it a solution?
   a. $6x - 15 = 3$; $x = 5$
   b. $x^2 - x = 3$; $x = 2$
   c. $7y^2 - 3 = 4y$; $y = -1$
   d. $4xy - 1 = 3xy$; $x = 3$, $y = 3$
   e. $125x^3 - 5x = 0$; $x = 1/5$

7. **Solving Equations.** Find the real number x that satisfies each of the following equations.
   a. $4x + 12 = 0$
   b. $6x - 10 = 32$
   c. $x^2 - 16 = 0$
   d. $\sqrt{x} = 9$
   e. $(x - 3)^2 = 4$
   f. $y = 3x - 16$, when $y = 2$

8. **Solving Word Problems.** Use the variable indicated to solve the following word problems for the unknown.
   a. Sue was 16 years old when Jerry was born. How old will Sue be when she is three times older than Jerry? (Let x be Jerry's age.)
   b. A rental car costs \$50 per day plus \$0.50 per mile. How far can you drive in a day with \$100? (Let x be the number of miles you drive.)
   c. The population of Tuba City has always been 1000 more than the population of Flute Town. One year the population of Tuba City also was three times the population of Flute Town. What was the population of both cities that year? (Let x be the population of Flute Town.)

9. **Graphing Parabolas.** Figure B-1 shows the graph of $y = x^2$, which is a bowl-shaped parabola that opens upward. Based on this graph, sketch the graphs of the relations
   a. $y = x^2 + 1$
   b. $y = x^2 - 1$
   c. $y = x^2 + 4$
   d. $y = -x^2$
   e. $y = -x^2 + 3$
   f. $y = -2x^2$

10. **Graphing Relations.** Using a method of your choice (e.g., the point-by-point method with pencil and paper, a graphing calculator, or a computer), graph each of the following relations.
   a. $y = 3 - x^3$
   b. $y = 4\sqrt{x} + 7$
   c. $y = 3/x$
   d. $y = 3x^3 - 2x^2 + 1$
   e. $y = 10^{0.1x}$
   f. $y = 4/(x+1)$
   g. $y = 0.1x^5$
   h. $y = 2^x$

C LOGARITHMS — NOT TO BE FEARED!

Many students dread logarithms even before their first encounter with them. Fortunately, the dire reputation of logarithms is undeserved; with a little practice, they are easy to understand and use. Thus, before beginning, we ask that you keep an open mind and believe that logarithms really are not to be feared. If you studied them before and had an unpleasant experience, forget the past and approach this subject as if you never saw logarithms before. If you have never studied logarithms, you might be pleasantly surprised by their simplicity and utility.

C.1 WHAT IS A LOGARITHM?

Perhaps the single biggest reason that many students have trouble with logarithms is that their funny name really does not describe how they are used. Logarithms were invented in the early 1600s by Scottish mathematician John Napier, in order to simplify calculations with large numbers. Napier did not think of his logarithms as exponents as we do today; instead, he derived them using ratios. Thus the word *logarithm* literally means "ratio numbers."

$\log_b a$ means: "the power to which we must raise b to get a."
or
$\log_b a$ means: "b to what power equals a?"

The statement "$\log_b a$" is read as "the logarithm base b of a." Here is a better translation:

$\log_b a$ means: "the power to which we must raise b to get a."

A less perfect, but perhaps easier to remember, translation is:

$\log_b a$ means: "b to what power equals a?"

For example, $\log_2 8$ should be read as "2 to what power equals 8?" Read in this way, you can see that $\log_2 8 = 3$ because $2^3 = 8$. Here are a few more examples:

- $\log_{10} 100$ means "10 to what power equals 100?"
  The answer is $\log_{10} 100 = 2$, because $10^2 = 100$.
- $\log_9 81$ means "9 to what power equals 81?"
  The answer is $\log_9 81 = 2$ because $9^2 = 81$.
- $\log_5 625$ means "5 to what power equals 625?"
  The answer is $\log_5 625 = 4$ because $5^4 = 625$.
- $\log_{10} 10^4$ means "10 to what power equals 10^4?"
  The answer is $\log_{10} 10^4 = 4$.

C.2 COMMON AND NATURAL LOGARITHMS

You can use any base when you work with logarithms. However, two bases are far more common than any others: base-10, or **common logarithms**; and base-e, or **natural logarithms** ($e \approx 2.718$). In this appendix we focus primarily on common logarithms, but all the properties that we discuss apply to logarithms to *any* base.

C.2.1 Common Logarithms

In the case of common logarithms (or simply *common logs*),

$\log_{10} x$ means: "10 to what power equals x?"

Powers of 10 therefore are especially easy to work with, as the following examples show.

- $\log_{10} 1000$ means: "10 to what power makes 1000?" Thus $\log_{10} 1000 = 3$ because $10^3 = 1000$.
- $\log_{10} 0.0001$ means: "10 to what power makes 0.0001?" Thus $\log_{10} 0.0001 = -4$ because $10^{-4} = 0.0001$.
- $\log_{10} 10^{23}$ means: "10 to what power makes 10^{23}?" The answer is $\log_{10} 10^{23} = 23$.
- $\log_{10} 10^{-6}$ means: "10 to what power makes 10^{-6}?" The answer is $\log_{10} 10^{-6} = -6$.
- $\log_{10} 1$ means: "10 to what power makes 1?" Thus $\log_{10} 1 = 0$ because $10^0 = 1$.

So far, all the examples have involved logs of an integer power of 10. However, you can take the log of *any* positive number. For example,

- $\log_{10} 30 = 1.477$ because $10^{1.477} = 30$;
- $\log_{10} 67{,}920 = 4.832$ because $10^{4.832} = 67{,}920$; and
- $\log_{10} 0.45 = -0.347$ because $10^{-0.347} = 0.45$.

Time-Out to Think: What key on *your* calculator do you use to find a common log? What key do you use to evaluate 10^x? Confirm each of the preceding examples on your calculator. (Note that the answers shown are rounded to the nearest thousandth.)

Although logs of negative numbers are not defined for the real numbers, they *are* defined if you use imaginary or complex numbers.

Recall that 10 raised to a positive power yields a number larger than 1 and that 10 raised to a negative power yields a number smaller than 1 but greater than 0. No power of 10 that equals 0 or a negative number. Therefore, logs of 0 and logs of negative numbers are not defined for real numbers.

Time-Out to Think: What happens if you try to take the log of 0 or a negative number on your calculator? Why?

Example C-1 Graph the relation $y = \log_{10} x$.

Solution: For each value of x, y is the power of 10 that equals x. Because raising 10 to a power and getting 0 or a negative result is impossible, only positive values of x can be used. A calculator can be used to make the following table (rounded to nearest hundredth).

x	1	2	3	4	5	6	7	8	9	10	100
$y = \log_{10} x$	0	0.30	0.48	0.60	0.70	0.78	0.85	0.90	0.95	1	2

Note that skipping from $x = 10$ to $x = 100$ in the table results in the relatively small change from $y = 1$ to $y = 2$. In fact, if $x = 1000$, $y = \log_{10} 1000 = 3$; if $x = 10{,}000$, $y = \log_{10} 10{,}000 = 4$; and so on. Additional points should be included for values of x less than 1.

x	0.9	0.7	0.5	0.2	0.1	0.01	0.001
$y = \log_{10} x$	-0.05	-0.15	-0.30	-0.70	-1	-2	-3

Note that y approaches negative infinity as x approaches zero. Plotting all the (x, y) pairs in the two tables results in the graph shown in Figure C-1.

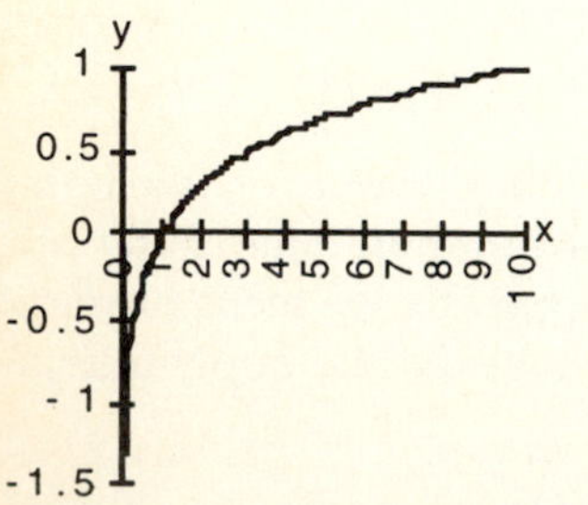

Figure C-1. The graph of $y = \log_{10} x$ rises slowly as x increases, and it approaches negative infinity as x approaches 0.

C.2.2 Natural Logarithms and a Note on Notation

Besides 10, the most commonly used base for logarithms is a special number, e. Logs with the base e are called *natural logs*. Like π, the number e is an irrational number; its value is $e \approx 2.718282\ldots$. A natural log, such as $\log_e 10$, asks the question: "e to what power equals 10?"

Because both common logs and natural logs are used frequently, shortcuts often are used in writing them. Generally, if no base is specified on a log it is taken to be a common log. That is, $\log x$ (with no base shown) is assumed to mean $\log_{10} x$. Natural logs are abbreviated "ln"; that is, $\log_e x$ is often written as $\ln x$. Most books and calculators follow these conventions.

> **Time-Out to Think:** Although you may never have heard of the number e, it is very important in mathematics — so important that nearly all calculators have a special key for e, just as they have a special key for π. Confirm that $e \approx 2.718282$ by pressing the key sequence for e^1 (e raised to the first power). Find the key on your calculator for natural logs. What is the natural log of e (ln e)? Why?

C.2.3 Working with Logarithms

Because logarithms imply questions about exponents, the rules for working with logarithms are analogous to those rules. Working with base-10 logs is easiest, and we begin by considering multiplication.

Recall that $\log_{10} 10 = 1$, $\log_{10} 100 = 2$, $\log_{10} 1000 = 3$, and $\log_{10} 10{,}000 = 4$; then consider the calculations:

$$\log_{10}(10^2 \times 10^2) = \underbrace{\log_{10}(10^4)}_{4} = \underbrace{\log_{10}(10^{2+2})}_{2+2} = \log_{10}(10^2) + \log_{10}(10^2); \text{ and}$$

$$\log_{10}(10^3 \times 10^1) = \underbrace{\log_{10}(10^4)}_{4} = \underbrace{\log_{10}(10^{3+1})}_{3+1} = \log_{10}(10^3) + \log_{10}(10^1).$$

These two calculations suggest that the logarithm of the product of two numbers is the sum of the individual logarithms. This *multiplication rule* is true in general:

$$\log_{10} xy = \log_{10} x + \log_{10} y.$$

The multiplication rule leads to an important insight about logs if we apply them to numbers written in scientific notation. As shown in Examples C-2 and C-3, simply knowing the common logs of numbers from 1 through 10, which always are between 0 and 1, is enough to find the common log of *any* number simply by adding or subtracting the appropriate integer. Before the advent of electronic calculators, logs usually were calculated in this way. Even today, this property of logs is extremely valuable to check answers found with calculators.

> **Time-Out to Think:** Why do the logs of all numbers between 1 and 10 fall between 0 and 1? *Hint:* What is $\log_{10} 1$? $\log_{10} 10$?

Example C-2 From $\log_{10} 3 = 0.477$ (which you can check on your calculator), find the logs of 30, 300, 0.3, and 0.03.

Solution: Use the multiplication rule for logarithms to find that

$$\log_{10} 30 = \log_{10}(3 \times 10^1) = (\log_{10} 3) + (\log_{10} 10^1) = 0.477 + 1 = 1.477;$$

$$\log_{10} 300 = \log_{10}(3 \times 10^2) = (\log_{10} 3) + (\log_{10} 10^2) = 0.477 + 2 = 2.477;$$

$$\log_{10} 0.3 = \log_{10}(3 \times 10^{-1}) = (\log_{10} 3) + (\log_{10} 10^{-1}) = 0.477 - 1 = -0.523; \text{ and}$$

$$\log_{10} 0.03 = \log_{10}(3 \times 10^{-2}) = (\log_{10} 3) + (\log_{10} 10^{-2}) = 0.477 - 2 = -1.523.$$

❖

Example C-3 You punch some buttons on your calculator and find that $\log_{10} 600 = 5.778$. Should you believe it?

Solution: First note that

$$\log_{10} 600 = \log_{10}(6 \times 100) = (\log_{10} 6) + (\log_{10} 100) = (\log_{10} 6) + 2.$$

Knowing that $\log_{10} 6$ is between 0 and 1, you expect the value of $\log_{10} 600$ to be between 0 + 2 = 2 and 1 + 2 = 3. Because the calculator gave an answer between 5 and 6, something went wrong!

❖

Division Rule

To obtain the division rule for logarithms, let's consider that

$$\log_{10} \frac{100}{10} = \log_{10} 10 = 1, \quad \text{and} \quad \log_{10} 100 - \log_{10} 10 = 2 - 1 = 1.$$

Note that the logarithm of the quotient of two numbers is the *difference* of the individual logarithms. This division rule also holds in general:

$$\log_{10} \frac{x}{y} = \log_{10} x - \log_{10} y.$$

Finally, we develop a rule for the logarithm of a power by applying the multiplication rule:

$$\log_{10} 3^2 = \log_{10} (3 \times 3) = \log_{10} 3 + \log_{10} 3 = 2 \log_{10} 3;$$

$$\log_{10} 3^4 = \log_{10} (3 \times 3 \times 3 \times 3) = \log_{10} 3 + \log_{10} 3 + \log_{10} 3 + \log_{10} 3 = 4 \log_{10} 3; \text{ and}$$

$$\log_{10} 3^n = \log_{10}\underbrace{(3 \times 3 \times \cdots \times 3)}_{n \text{ times}} = \underbrace{\log_{10} 3 + \log_{10} 3 + \cdots + \log_{10} 3}_{n \text{ times}} = n \log_{10} 3.$$

In general, the power rule is:

$$\log_{10} x^n = n \log_{10} x.$$

Table C-1 contains a summary of the three rules developed.

Table C- 1. Rules for Working with Exponents and Logs

	Exponents	Common Logs	Logs of Any Base
Multiplication	$b^m b^n = b^{m+n}$	$\log_{10} xy = \log_{10} x + \log_{10} y$	$\log_b xy = \log_b x + \log_b y$
Division	$\frac{b^m}{b^n} = b^{m-n}$	$\log_{10} \frac{x}{y} = \log_{10} x - \log_{10} y$	$\log_b \frac{x}{y} = \log_b x - \log_b y$
Powers	$(b^m)^n = b^{mn}$	$\log_{10} x^n = n \log_{10} x$	$\log_b x^n = n \log_b x$

C.2.4 A Note on Inverse Operations

Suppose that you start with any number and perform two consecutive operations on it. Suppose that after the second operation you have the same number with which you started. Then we say that the two operations are **inverses** of one another. (For example, operation 1, opening a door, and operation 2, closing a door, are inverses of each other.) Thinking about such inverse operations may help clarify powers, roots, exponents, and logs. Each of the following are pairs of inverse operations.

- *Addition and subtraction:* Starting with any number n, if you add 4 to n, and then subtract 4 from the result, you get n. The reverse, first subtracting 4 and then adding 4, also returns the starting value n. That is,

$$(n + 4) - 4 = n \quad \text{and} \quad (n - 4) + 4 = n.$$

- *Multiplication and division:* Starting with any number n (except 0), if you multiply n by 5 and then divide the result by 5, you get n. The reverse, first dividing and then multiplying 5, also returns the starting value n. That is,

$$(n \times 5) \div 5 = n \quad \text{and} \quad (n \div 5) \times 5 = n.$$

- *Powers and roots:* Starting with any number n, if you raise it to a power and then take the same root, you get n. The reverse order also works (subject to restrictions such as not taking the square root of a negative number). That is,

$$\sqrt[3]{n^3} = \left(n^3\right)^{1/3} = n \quad \text{and} \quad \left(\sqrt[4]{n}\right)^4 = n.$$

- *Logs and exponents:* As with the other examples, logs and exponents are inverse operations. If you take the log of a positive number n and then raise 10 to the power log n, you get n. Similarly, if you raise 10 to a power n and then take the log of the result, you also get n. That is,

$$10^{\log n} = n, \quad \text{and} \quad \log 10^n = n.$$

C.3 ALGEBRA WITH EXPONENTS AND LOG

Recall that operating on both sides of an equals sign is legitimate if the information in the original equation is preserved. In Appendix B we showed that the same quantity can be added or subtracted to both sides of an equation and that both sides can be multiplied or divided by the same nonzero quantity. We now discuss operations on equations that involve powers, roots, exponents, and logs.

Raising Both Sides to the Same Power (or Taking the Same Root)

In most cases, raising both sides of an equation to the same power or taking the same root of both sides (a root is a fractional power) is permissible However, as shown in Appendix B, care must be exercised because spurious solutions can be produced. Also, if one side of an equation is negative, you cannot take even roots (such as square roots) because even roots of negative numbers are not defined.

Example C-4 Solve the equation $x^3 = -64$ for x.

Solution: To isolate the unknown x, take the cube root of both sides (or, equivalently, raise both sides to the one-third power). Recalling that $(x^3)^{1/3} = x$, we have

$$\left(x^3\right)^{\frac{1}{3}} = (-64)^{\frac{1}{3}} \quad \Rightarrow \quad x = (-64)^{\frac{1}{3}} = \sqrt[3]{-64} = -4.$$

The only value of x that satisfies the equation is $x = -4$.

❖

Example C-5 The area of a square floor is 400 square feet. How long is each side of the room?

Solution: If you call the length of each side of the room x (which is unknown at the moment), the floor area is x^2. Because the area is 400 square feet, the room length must satisfy the equation $x^2 = 400$. You can solve this equation for x by taking the square root (or the one-half power) of both sides:

$$x^2 = 400 \text{ ft}^2 \quad \Rightarrow \quad (x^2)^{\frac{1}{2}} = \left(400 \text{ ft}^2\right)^{\frac{1}{2}} \quad \Rightarrow \quad x = \pm\sqrt{400 \text{ ft}^2} = \pm 20 \text{ ft}.$$

Note that taking the square root of the units ft^2 gives the units ft. Note also that two solutions exist, but the floor can have only one length. The negative solution ($x = -20$ ft) must be discarded, because a negative room length makes no sense. The remaining solution ($x = 20$ ft) says that the length of each side of the room is 20 feet.

❖

Making Both Sides an Exponent

Making both sides of an equation an exponent is always legitimate. That is,

if $x = y$, then $10^x = 10^y$, and

if $x = y$, then $b^x = b^y$, for any positive number b.

This rule helps you solve equations where a variable appears in a logarithm, as in Examples C-6 and C-7.

Example C-6 Solve $\log_{10} a = 3$ for a.

Solution: Make both sides an exponent of the base 10:

$$\log_{10} a = 3 \xrightarrow{\text{make both sides exponents}} 10^{\log_{10} a} = 10^3 \quad \Rightarrow \quad a = 1000.$$

The last step relies on the fact that $10^{\log_{10} a} = a$, which is a consequence of the inverse nature of exponents and logs. Alternatively, recall that $\log_{10} a$ means "the power of 10 that equals a." It follows that $10^{\log_{10} a}$ means "10 to the power of 10 that makes a," which must be a.

❖

Example C-7 Solve $\log_{10} 8x^3 = -3.9$ for x.

Solution: This problem requires several steps. First, to isolate the term $8x^3$, make both sides an exponent of 10:

$$10^{\log_{10} 8x^3} = 10^{-3.9} \quad \Rightarrow \quad 8x^3 = 10^{-3.9}.$$

Again, the property that $10^{\log_{10} x} = x$ has been used. Next divide both sides by 8 (or multiply both sides by 1/8):

$$\frac{1}{8}(8x^3) = \frac{1}{8}(10^{-3.9}) \quad \Rightarrow \quad x^3 = \frac{10^{-3.9}}{8}.$$

To solve for x, take the cube root of both sides:

$$(x^3)^{\frac{1}{3}} = \left(\frac{10^{-3.9}}{8}\right)^{\frac{1}{3}} \quad \Rightarrow \quad x = \frac{\left(10^{-3.9}\right)^{\frac{1}{3}}}{\sqrt[3]{8}} = \frac{\left(10^{-3.9}\right)^{\frac{1}{3}}}{2}.$$

All that remains is to apply the powers rule for exponents and use a calculator to find that

$$x = \frac{1}{2}\left(10^{\frac{-3.9}{3}}\right) = \frac{1}{2}\underbrace{\left(10^{-1.3}\right)}_{\text{use calculator to find value of 0.05}} = \frac{1}{2}(0.05) = 0.025.$$

After this much algebra, always check the solution. Substituting the solution ($x = 0.025$) into the original equation yields

$$\log_{10} 8x^3 = \log_{10} 8(0.025)^3 = \log_{10} 8(1.5625 \times 10^{-5}) = \log_{10} 1.25 \times 10^{-4} = -3.903.$$

The result checks almost perfectly; the difference (–3.903 instead of –3.9) is due to small rounding errors.

❖

Taking the Log of Both Sides

Taking the log of both sides of an equation is permissible, provided both sides are positive. That is,

if $x = y$ (x and y both positive), then $\log_{10} x = \log_{10} y$.

This rule allows you to solve for variables that appear in exponents.

Example C-8 Solve $2^x = 50$ for x.

Solution: Begin by taking the common log of both sides and then apply the power rule for logs:

$$\log_{10} 2^x = \log_{10} 50 \xrightarrow[\text{rule for logs}]{\text{apply power}} x \log_{10} 2 = \log_{10} 50.$$

To isolate the unknown x, divide both sides by $\log_{10} 2$ to obtain

$$x = \frac{\log_{10} 50}{\log_{10} 2} = \frac{1.699}{0.301} = 5.645.$$

A mental check helps to confirm the solution without even using a calculator. Because 50 is between 32 and 64, and because $2^5 = 32$ and $2^6 = 64$, the solution to $2^x = 50$ should be between 5 and 6, as it is ($x = 5.645$). A more precise calculator check can also be made. Substituting $x = 5.645$ into the original equation shows that $2^{5.645} = 50.04$. The result is not exactly 50 because of rounding errors.

❖

PROBLEMS

1. **Practice with Logs.** Translate each of the following expressions into words and then evaluate it (without a calculator). *Example*: $\log_2 4$. *Solution:* $\log_2 4$ means "2 to what power equals 4?" The answer is $\log_2 4 = 2$ since $2^2 = 4$.
   a. $\log_2 16$ b. $\log_5 125$
   c. $\log_{12} 144$ d. $\log_4 0.25$
   e. $\log_{10} 1{,}000{,}000{,}000$ f. $\log_{10} 10^{-6}$

2. **Common Logs and 10^x.** Without doing any calculations, determine whether the following statements are true or false and explain why. *Example:* $\log_{10} 2$ is between 0 and 1. *Solution:* The statement is true: $\log_{10} 2$ must be between $\log_{10} 1 = 0$ (because $10^0 = 1$) and $\log_{10} 10 = 1$ (because $10^1 = 10$).
   a. $\log_{10} 96$ is between 3 and 4.
   b. $\log_{10} 1{,}600{,}000$ is between 6 and 7.
   c. $\log_{10} (8 \times 10^9)$ is between 12 and 13.
   d. $\log_{10} 1/4$ is between 1 and 0.
   e. $\log_{10} 0.00045$ is between 6 and 5.
   f. $\log_{10} \pi$ is between 0 and 1.
   g. $10^{0.928}$ is between 10 and 100.
   h. $10^{3.334}$ is between 1000 and 10,000.
   i. $10^{-5.2}$ is between 100,000 and 1,000,000.
   j. $10^{-2.67}$ is between 0.001 and 0.01.

3. **Evaluating Logarithms.** Use $\log_{10} 5 = 0.699$ to evaluate each of the following numbers by using the rules of logarithms. Do not use your calculator.
   a. $\log_{10} 50$ b. $\log_{10} 5000$
   c. $\log_{10} 0.05$ d. $\log_{10} 25$
   e. $\log_{10} 0.20$ f. $\log_{10} 0.04$

4. **Algebra Practice.** Use the properties of logs and exponents to solve each of the following equations.
   a. Solve $\log_{10} x = 2$ for x.
   b. Solve $\log_{10} 2z^4 = -1.7$ for z.
   c. Solve $\log_{10} 5t^{-2} = 3.4$ for t.
   d. Solve $2^x = 125$ for x.
   e. Solve $4^{0.3y} = -5.6$ for y.
   f. Solve $251^{n/5} = 10$ for n.

5. **Word problems.** Use the properties of logs and exponents to solve the following practical problems.
   a. A population of bacteria grows according to the growth law $N = 10^t$, where N is the number of bacteria and t is the time measured in hours. After how many hours will the population equal 1550 bacteria?
   b. The number of cells in a tumor grows according to the growth law $N = 20 \times 10^t$, where N is the number of cells and t is the time measured in days. After how many days will the population of tumor cells equal 12,500?
   c. For an earthquake of magnitude M, the amount of energy released, E, is given by the formula $\log_{10} E = 4.4 + 1.5M$; the energy, E, has units of joules. How much energy is released by an earthquake with a magnitude of $M = 7.0$? What is the magnitude of an earthquake that releases $E = 10^{12}$ joules?

D USING YOUR CALCULATOR

A calculator is an indispensable tool when studying this text. In this section, we briefly review basic calculator operations.

D.1 CALCULATORS, COMPUTERS, AND BRAINS

Used properly, electronic calculators and computers can help you to improve your understanding of and intuition for mathematical concepts. They also offer tremendous computational power even to people with little formal mathematical training — power that was available only to professionals in the recent past. However, blind faith in calculators and computers inevitably leads to mistakes. Before reaching for a calculator or computer, always *think* about the problem you are trying to solve. If you obtain an unexpected answer, it may be a computational mistake; check your work carefully before accepting the unexpected. Use calculators and computers as extensions of your brain power, not as substitutes.

- **Calculator Rule #1:** Always use your intelligence with your calculator (or computer). You might accidentally press the wrong buttons, set up a calculation incorrectly, or otherwise foul up with an electronic device. Before accepting any answer, ask yourself: Does it make sense?

Example D-1. Suppose that, using your calculator, you obtain the result: $(3.28 \times 10^4) \times (2.4 \times 10^{-7}) = 2.3 \times 10^{15}$. Should you accept it?

Solution: In your head (without a calculator), you should note that $3.28 \times 2.4 \approx 3 \times 2 = 6$ and $10^4 \times 10^{-7} = 10^{4-7} = 10^{-3}$. Thus, you expect the answer to be near 6×10^{-3}. Clearly, 2.3×10^{15} is incorrect. You must have done something wrong with your calculator, so try again until you get an answer that makes sense.

Example D-2. Suppose that, using your calculator, you obtain the result: $\log_{10} 47 = 2.3$. Should you accept it?

Solution: Because 47 is between 10 and 100, and you know (from Appendix C) that $\log_{10} 10 = 1$ and $\log_{10} 100 = 2$, you *know* that $\log_{10} 47$ must lie between 1 and 2. Thus the answer of 2.3 is incorrect.

D.2 VARIATION AMONG CALCULATORS

All of the calculator work needed in this textbook can be accomplished with an inexpensive (typically less than $20) *scientific* calculator. If you want to draw graphs with your calculator, you can buy a somewhat more expensive and more powerful *graphing* calculator.

Both scientific and graphing calculators can perform all of the operations described in this Appendix. However, *your* calculator may have key labels that differ slightly from the labels we show. In addition, some calculators require pressing buttons in a different sequence than the one we describe to get a particular result.

As you study this Appendix, try each of the examples that we present. If you don't get a correct result by following our procedure, you will need to take a few moments to determine the correct procedure to use on *your* calculator. In many cases you can determine this procedure by trial and error; if that doesn't work, consult the calculator's instruction manual. If the procedure on your calculator differs from the one that we present, you should write it down in the margin so that you will be able to find it easily in the future.

Calculator Precision and Rounding

Most calculators can display up to nine decimal places, although they keep track of a few more decimal places internally. Although you should always round final answers appropriately, rounding during intermediate steps can lead to error. As a general rule, it is better to do a series of calculations in sequence on your calculator, rather than writing down intermediate answers and later rekeying them.

D .3 OPERATION PRIORITY

All calculators have some priority system for performing operations. Most calculators follow the operation priority rules presented in Appendix A (subsection A.1.3). For example, multiplication and division have higher priority than addition and subtraction. If you press the key sequence

[5] [+] [3] [×] [2] [=],

the calculator first multiplies $3 \times 2 = 6$, then adds the 5 to get 11. Be sure that you understand your calculator's priority system, especially when you are performing calculations that involve multiple steps.

D .3.1 The Parentheses Keys

Most calculators have keys to represent parentheses, allowing you to group operations.

Example D-3. Calculate $(5 + 3) \times 2$; then calculate $5 + (3 \times 2)$. Compare the answers.

Solution: For the first calculation enter

[(] [5] [+] [3] [)] [×] [2] [=]

yielding an answer of 16. For the second calculation enter

[5] [+] [(] [3] [×] [2] [)] [=]

yielding an answer of 11. Note that the latter calculation yields the same answer with or without the parentheses because it follows the standard priority rules.

In this appendix, we indicate specific calculator buttons to be pressed by one of two different notations, which have the same meaning. For example, to show how to enter the expression 2/3, we write:

[2] [÷] [3]

or

<2> <÷> <3>

D .3.2 Fractions

Enter fractions on your calculator simply by recognizing that fractions are a short-hand way to express division. Because most calculators display fractions in decimal form, you can use this technique to convert common fractions to decimal form easily.

Example D-4. Multiply $1/2 \times 3/4$ on your calculator.

Solution: Enter the fraction 1/2 as $1 \div 2$ and the fraction 3/4 as $3 \div 4$. That is, use the key sequence

<1> <÷> <2> <x> <3> <÷> <4> =

yields the answer, 0.375, which is 3/8 as expected.

Example D-5. Convert 2/3 to a decimal fraction using your calculator.

Solution: Simply do the division:

$$\boxed{2}\ \boxed{\div}\ \boxed{3} = 0.666666667.$$

Note that some calculators will show more or fewer digits than shown here.

Negative Numbers

Your calculator probably has a <+/–> key that can be used to enter negative numbers. For example, to enter the number –6 you first enter the 6 and then use the <+/–> key:

<6> <+/–>.

Your calculator should display the negative sign as you complete this key sequence.

Example D-6. Multiply –67 × –4 on your calculator.

Solution: Use the key sequence

<6> <7> <+/–> <x> <4> <+/–> =

which yields the answer, 268.

Second Functions

On most calculators, many keys can perform two (or more) operations. One operation is usually listed *on* the key, and the second is listed in small print above (or below) the key. Pressing the key performs the operation listed on it. To use **second function** in small print above the key, you must first press a button usually labeled <2nd> or <shift>: This button instructs the calculator to use the second function rather than the primary function written on the key.

For example, if the <$\sqrt{\ }$> has small print showing x^2 above it, the key can be used both to take square roots and to find the square. Because the square root is written *on* the key, simply pressing the key takes the square root. For example, the key sequence

<4> <$\sqrt{\ }$>

finds $\sqrt{4} = 2$. To square a number, you must use the second function key before pressing the button. For example, the key sequence

<4> <2nd> <$\sqrt{\ }$>

finds $4^2 = 16$. Note that the calculator ignores the square root written on the key, instead using the second function x^2 of the key.

Because calculators widely differ in which operations they assign directly to keys and which to second functions, we specify only operations in the remainder of this Appendix; for each operation we describe, you will need to determine whether using the second function key is necessary on *your* calculator.

D .4 SCIENTIFIC NOTATION

Your calculator should have a key, probably labeled *EXP* or *EE* (for *exponent* or *enter exponent*), for entering numbers in scientific notation. This key is used to enter the exponent on the power of 10; note that you do not have to enter the number 10. For example, you enter the number 3.5×10^6 using the key sequence

[3.5] [EE] [6] .

You can think of the *EE* key as meaning *times 10 to the power that I enter next*. For example, the key sequence

<7> <EE> <9>

means 7×10^9 because 9 is the key entered next after the *EE* key.

Negative Exponents

Just as with other negative numbers, you can enter negative exponents by first entering the digits of the exponent and then using the <+/–> key to make the exponent negative. For example, to enter 3.8×10^{-12} you should use the key sequence

[3.8] [EE] [12] [+/-] .

Special Care for Powers of 10

You must be especially careful when entering powers of 10 with the *EE* key. For example, because the number 10^8 is the same as 1×10^8, you enter it with the key sequence

[1] [EE] [8] .

Be careful to avoid the common mistake of entering the sequence

<10> <EE> <8>,

which means 10×10^8, or 10^9, rather than 10^8.

Switching Between Ordinary and Scientific Notation

Most calculators have a key which switches between scientific and ordinary notation. The label on this key varies significantly among calculators, but usually includes the letters *F* and *E* (for *f*loating point notation and *e*xponential notation, respectively). Common labels for this key include [F⇔E] and [FSE] . On some calculators you can simply press the same key used to enter scientific notation (e.g., *EE*). Still other calculators require a sequence of keys to switch between ordinary and scientific notation.

Time-Out to Think: Find the key for switching between scientific and ordinary notation on your calculator. Enter the number 5000 and use this key to convert it to scientific notation (5×10^3); then convert back to ordinary notation (5000).

Don't Speak in "Calculator Talk"

Most calculators show only the exponent when displaying scientific notation. For example, if your calculator display shows something like

3.765 09, 3.765E9, or 3.765^{09},

it means the number 3.765×10^9. Properly interpreting the calculator display requires knowing how to translate this "calculator talk."

As you are not a calculator, you should avoid speaking or writing in calculator talk. That is, if you are referring to the number 5×10^3, you should say so; never say 5E3 (or any of the other variations of "calculator talk").

Example D-7. Multiply $(4 \times 10^7) \times (-3.1 \times 10^{-12})$ on your calculator.

Solution: Use the key sequence

$$\underbrace{\langle 4\rangle\langle EE\rangle\langle 7\rangle}_{4\times10^7} \quad \langle\times\rangle \quad \underbrace{\underbrace{\langle 3\rangle\langle .\rangle\langle 1\rangle\langle +/-\rangle}_{-3.1}\langle EE\rangle\underbrace{\langle 1\rangle\langle 2\rangle\langle +/-\rangle}_{-12}}_{-3.1\times10^{-12}} \quad =$$

which yields the answer -1.24×10^{-4}.

D .5 POWERS AND ROOTS

A key that usually is labeled [y^x] or [x^y] is used for raising numbers to powers. For example, you enter the number 5^3 with the key sequence

[5] [y^x] [3] [=]

to obtain the answer, 125.

A key that usually is labeled [$\sqrt[x]{y}$] or [$\sqrt[y]{x}$] or [$x^{1/y}$] is used for taking roots. For example, to find the 4th root of 16 ($\sqrt[4]{16}$) you enter the key sequence

[16] [$\sqrt[x]{y}$] [4] [=]

to obtain the answer, 2. Note that, because $\sqrt[4]{16}$ is the same as $16^{1/4}$, or $16^{0.25}$, you also can use the key sequence

[16] [y^x] [.25] [=]

to obtain the same answer, $\sqrt[4]{16} = 16^{0.25} = 2$.

Example D-8. Use your calculator to find the answer to $5^7 \div \sqrt[3]{10}$.

Solution: Use the key sequence

$$\underbrace{\langle 5\rangle\langle y^x\rangle\langle 7\rangle}_{5^7} \quad \langle\div\rangle \quad \underbrace{\langle 1\rangle\langle 0\rangle\langle \sqrt[x]{y}\rangle\langle 3\rangle}_{\sqrt[3]{10}} \quad =$$

which yields the answer, 36,262.4.

Squares, Square Roots, and Reciprocals

Squares, square roots, and reciprocals all can be found with the keys for raising to powers and taking roots. For example, because 1/25 is the same as 25^{-1}, you can find the reciprocal of 25 by entering

[25] [y^x] [-1] [=]

to obtain the answer, 0.04.

However, most calculators have special keys that simplify these operations, usually labeled [x^2] for squaring, [$\sqrt{\ }$] for taking square roots, and [1/x] or [x^{-1}]

for finding reciprocals. Some calculators also have special keys for cubes $\boxed{x^3}$ and for cube roots $\boxed{\sqrt[3]{\ }}$.

Example D-9. Use your calculator to find the answer to $1/(5^2 \times \sqrt{7})$.

Solution: Use the key sequence

$$\underbrace{\langle 5\rangle\langle x^2\rangle}_{5^2} \quad \langle\times\rangle \quad \underbrace{\langle 7\rangle\langle\sqrt{x}\rangle}_{\sqrt{7}} \quad \underbrace{\langle 1/x\rangle}_{\text{take reciprocal}} \quad =$$

which yields the answer, 0.01512.

D .6 COMMON LOGS AND 10^x

You can find the base-10 logarithm (common logarithm) of any number by first entering the number and then using the $\boxed{\log}$ key. For example, find $\log_{10} 50$ with the key sequence

$$\boxed{50}\ \boxed{\log}\ \boxed{=},$$

yielding the answer 1.69897.

Use the $\boxed{10^x}$ key to raise 10 to any power (on some calculators this operation requires using the *inverse* key followed by the *log* key). For example, find $10^{1.69897}$ by entering the sequence

$$\boxed{1.69897}\ \boxed{10^x}\ \boxed{=},$$

which yields the answer, 49.9999995 (about 50, as you should expect because $\log_{10} 50 \approx 1.69897$). Note that you can also do this calculation with the $\boxed{y^x}$ key by entering

$$\boxed{10}\ \boxed{y^x}\ \boxed{1.69897}\ \boxed{=}\ .$$

Example D-10. Use your calculator to find the answer to $\log_{10} 25 + 10^{1.15}$.

Solution: Use the key sequence

$$\underbrace{\langle 2\rangle\langle 5\rangle\langle\log\rangle}_{\log_{10} 25} \quad \langle+\rangle \quad \underbrace{\langle 1\rangle\langle.\rangle\langle 1\rangle\langle 5\rangle\langle 10^x\rangle}_{10^{1.15}} \quad =$$

which yields the answer, 15.523.

Natural Logs and e^x

Most calculators also have special keys for natural (base-*e*) logarithms, usually labeled $\boxed{\ln}$, and for powers of *e*, usually labeled $\boxed{e^x}$; recall that $e \approx 2.718282846$ is an irrational number that is very important in mathematics.

Example D-11. Use your calculator to find the answer to $\log_e 10 - e^2$.

Solution: Use the key sequence

$$\underbrace{\langle 1\rangle\langle 0\rangle\langle \ln\rangle}_{\log_e 10} \quad \langle -\rangle \quad \underbrace{\langle 2\rangle\left\langle e^x\right\rangle}_{e^2} \quad =$$

which yields the answer, –5.08647.

D .7 TRIGONOMETRIC RATIOS

Your calculator should have keys for all of the standard trigonometric ratios including sine (sin), cosine (cos), and tangent (tan), as well as their inverse operations ($\sin^{-1}$, $\cos^{-1}$, and $\tan^{-1}$, respectively).

Degrees, Radians, and Grads

One complication arises when you use the trigonometric ratios: Because these ratios involve angles, and there is more than one system for measuring angles, you must always make sure your calculator is using the angle measurements.

Most calculators can be set to display angles either in *degrees*, *radians*, or *grads*. You're probably familiar with angle measurements in degrees. Radians are discussed in Chapter 11. Grads, which are common in engineering, do not appear in this textbook.

The label [DRG] is common for the key that switches the calculator mode between degrees, radians, and grads. Many calculators also show, somewhere in the display, a label of *deg*, *rad*, or *grad* to let you know which mode it is in.

Before using any trigonometric ratio, always put your calculator into the correct mode! Otherwise you will not get correct answers.

Entering π

Whenever you need to enter the number π, which is common when dealing with trigonometric ratios, you should use your calculator's π key. Because your calculator stores π to more decimal places than it can display, using this key is far more accurate than typing in an approximation to π.

Example D-12. Find the sine of 45° (sin 45°). Compare to the sine of 45 *radians* (sin 45).

Solution: To find sin 45°, *first* make sure that your calculator is in degree mode. then enter the key sequence

[45] [sin] [=],

which yields the answer, 0.707. To find sin 45 radians, switch your calculator to radian mode. Then, the same key sequence should yield the answer 0.851. Note that there is no relationship between sin 45° and sin 45 radians, so it is crucial that you set the calculator mode correctly.

Example D-13. Find the sine of $\pi/2$ (sin $\pi/2$).

Solution: Use the key sequence

<π> <÷> <2> <sin> =,

which yields the answer, 1.

D .8 OTHER KEYS

Modern calculators have many keys besides those that we have described in this Appendix. Among the most common keys are those for finding factorials, permutations, and combinations (discussed in Chapter 12), and those for entering statistical data and finding a mean or standard deviation (discussed in Chapter 14). If you have a graphing calculator or a programmable calculator, many more types of calculations are possible.

Because your calculator probably has so much untapped power, you probably will find it well worth your time to spend a few hours studying your calculator with the aid of its instruction manual.

D .9 PROBLEMS

1. This problem will help you understand the operation priority rules on your calculator. First do each of the following calculations without a calculator. Then, do them on your calculator, but without using the parentheses keys. Note that, to obtain correct answers, you may have to enter the calculations in a different order than shown. Once you have figured out how to get the correct answer on your calculator, write down the *exact sequence* of keys that you used.
   a. $(15 - 9) \div 3 \times (2 + 5)$
   b. $\frac{1}{3} + (\ \{[9 \times (8 - 2)] + 8\} \div 6)$
2. Repeat Problem 1, but this time with the aid of the parentheses keys. Again, write down the *exact sequence* of keys that you used.
3. Do the following calculation on your calculator: $10^9 \div 10^6$. Write down the exact sequence of keys that you used.
4. Do the following calculation on your calculator: $10^{-3} \times 10^5$. Write down the exact sequence of keys that you used.
5. Use your calculator for each of the following. In each case, write down the exact sequence of keys that you used in addition to stating the answer.
   a. $(8.92 \times 10^{-3}) \div (5.03 \times 10^{-6})$
   b. $(3.775 \times 10^4) \times (2.553 \times 10^5)$
   c. $(5.34 \times 10^{23}) \times 10^{-9}$
6. To make sure that you can switch into and out of scientific notation on your calculator, do the following.
   a. Enter the number 2000 on your calculator. Write down the exact sequence of keys that you must use on *your* calculator to get the number to switch to scientific notation (2×10^3).
   b. Now that the calculator is showing the number in scientific notation, write down the exact sequence of keys you must press to get it back to its original form (2000).
7. Use your calculator to answer each of the following. Round your answers to 3 decimal places.

a. 9^7	d. 76^3	g. $7.1^2 \times 10^{1/3}$
b. $\sqrt{2}$	e. $\sqrt[5]{1206}$	h. $2^{30} \times (3.6 \times 10^9)^{-1}$
c. $35^{\frac{3}{7}}$	f. $\frac{1}{3.77^5}$	i. $\frac{10^{1.77}}{\sqrt[3]{5}}$

8. Use your calculator to answer each of the following. Round your answers to 3 decimal places.

a. log 5	d. log 50	g. log 500
b. log 0.25	e. log 4,400	h. $10^{1.7}$
c. log 1	f. $10^{6.2}$	i. $10^{-0.23}$

9. Use your calculator to answer each of the following. Round your answers to 3 decimal places.
   a. ln 5
   b. ln 0.25
   c. ln 1
   d. ln 100
   e. ln 2.71828
   f. e^3
   g. ln 27.1828
   h. $e^{1.7}$
   i. $e^{-0.79}$
10. Use your calculator to answer the following. Round all answers to 3 significant figures.
   a. $(1.21 \times 10^{15}) \div 10^{42}$
   b. $(3.5 \times 10^{-4}) \times (20.97 + [1.44 \times 3.1])$
   c. $4^8 \times \sqrt[3]{10} \times \pi$
   d. $\frac{4\pi}{3}(6.1)^3$
11. With your calculator in degree mode, find each of the following:
   a. sin 10°
   b. sin 30°
   c. sin 90°
12. With your calculator in radian mode, find each of the following:
   a. cos 0
   b. cos (π/2)
   c. cos (π/4)

E ANSWERS TO SELECTED PROBLEMS

Chapter 2

6. a. i. The proposition in standard form.

ii. The subject is *biology courses* and the predicate is *science courses*.

iii.

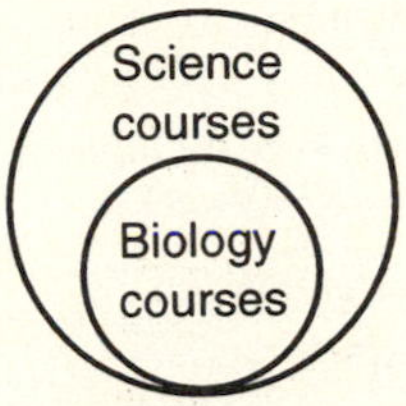

iv. Most people would agree that this proposition is unambiguously true.

c. i. Some police officers are people who are not tall people .

ii. The subject is *police officers* and the predicate is *tall people*.

iii.

iv. This is unambiguously true.

e. i. Some states are states without coastlines .

ii. The subject is *states* and the predicate is *states without coastlines*.

iii.

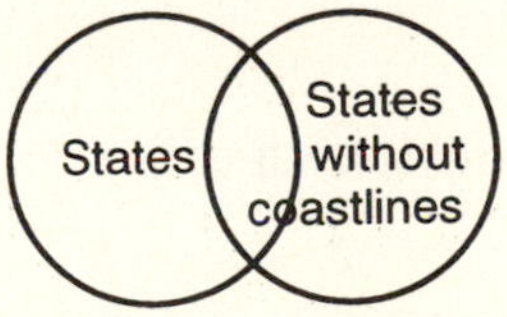

iv. This claim is unambiguous.

8. a. Both the propositions are true. The compound proposition is also true.

c. The first proposition is true, but the second is not. The compound proposition is true.

10. a. Seventy four people were surveyed.

c. Forty two people eat WonderCorn.

e. The number of people who eat WonderCorn *and* PrimePop is 18.

11. a. Twenty seven people read the WSJ only.

c. Sixty two people read the NYT.

e. Eight people read all three papers.

14. a. All candidates for governor are rich people.

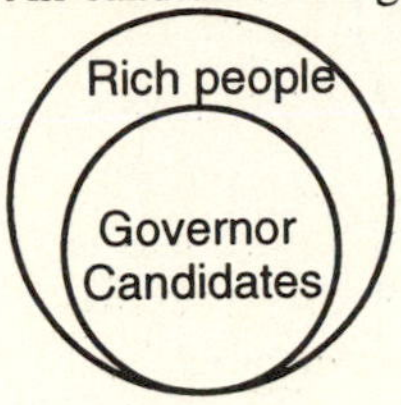

c. All columbines are blue flowers.

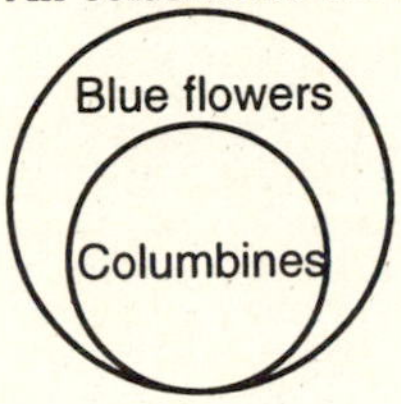

15. a. i. P1: All islands are tropical lands.

P2: All tropical lands are lands with jungles.

C: All islands are lands with jungles.

ii.

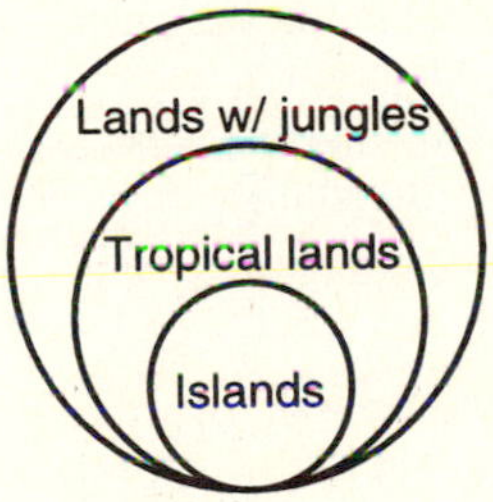

The argument is valid.

iii. The argument is not sound since P1 is false.

c. i. P1: All dairy products are products containing protein.

P2: No soft drinks are products containing protein.

C: No soft drinks are dairy products.

ii.

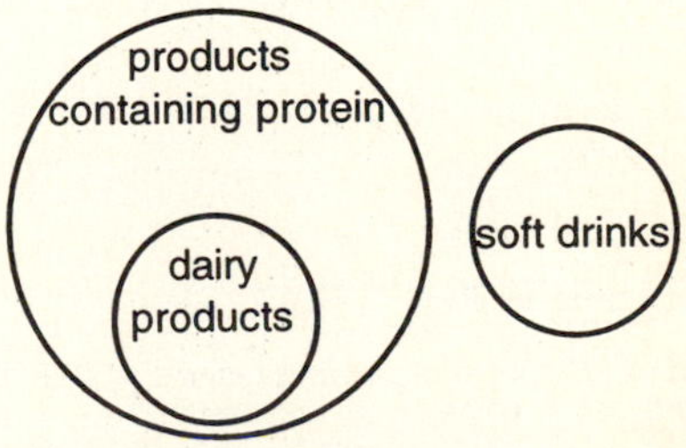

The argument is valid.

iii. The argument is sound. It is conceivable, however, to produce a soft drink containing protein. In this case, the second premise (P2) would cease to be true, and the argument would no longer be sound.

16. a. i. It does not need to be rephrased.

ii. 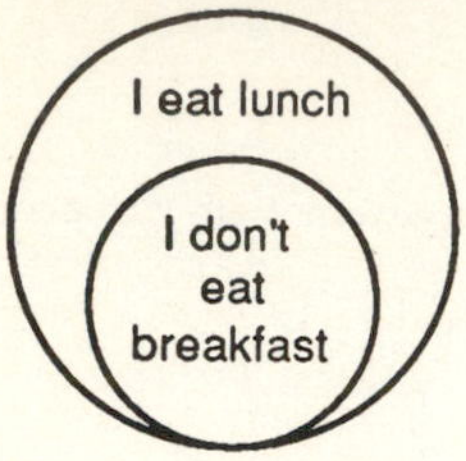

The argument is valid.

c. i. It does not need to be rephrased.

ii. 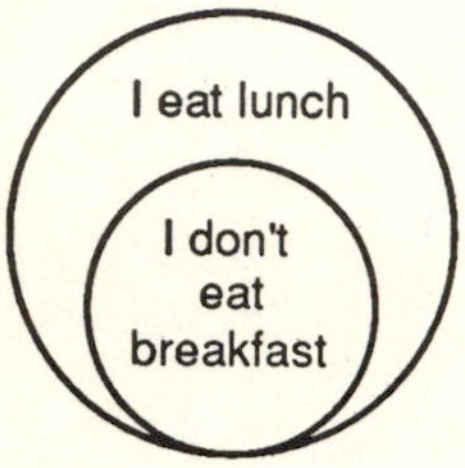

The argument is invalid.

iii. The argument is invalid because it denies the antecedent.

17. a. i. P1: If you live in the U.S., then you have the right to say anything at any time.

P2: If you yell "fire!" in a theater, then you are saying something.

C: If you live in the U.S., then you have the right to yell "fire!" in a theater.

ii. The argument is valid.

iii. The first premise (P1) is not true, so the argument is not sound.

19. a. It is true for all numbers that a x b = b x a.

b. It is not true for all numbers that a + b = b + a.

Chapter 3

19. a. The absolute change is 5. The relative change is 10%.

c. The absolute change is 1. The relative change is 0.1%.

e. The absolute change is –5,000. The relative change is –10%.

20. a. The total is 766.7.

c. The total is about 168.5.

23. a. The U.S. population increased by 20.5% between 1974 and 1995.

c. Between 1974 and 1989 the US defense budget increased by 333.33%; that is, the 1989 budget was over three times more than the 1974 budget.

e. The average student-teacher ratio in U.S. public schools decreased by 19% from 1974 to 1988.

Chapter 4

2. a. Apples sell for 50 cents/pound.

c. Tile installation costs $3.50/ft^2$.

e. There are 43,250 ft^2/acre.

3. a. Sam used 4 eggs for his omelet.

b. Troop 11 raised an average of $8/scout.

5. a. The swimming pool has a volume of 150 cubic meters (150 m^3).

c. The skyscraper has a volume of 25 million cubic feet (25 million ft^3).

6. a. At a cost of $0.50/lb, three pounds of apples would cost $1.50 ($0.50/lb x 3 lbs = $1.50).

c. There are 1,760 yards/mile (5,280 ft/mile ÷ 3 ft/yard = 1,760 yards/mile).

e. The amount earned would be $240/week (40 hr/week x $6/hr = $240/week).

8. a. There are 128 ounces/gallon.

9. a. There are 1,728 in^3/ft^3.

c. There are one million square meters in one square kilometer (10^6 m^2/km^2).

Chapter 5

1. a. The length of the hypotenuse is 10.

c. The length of the third side would be 14.7.

e. First, you must realizing that the length of the staircase (*l*) is the hypotenuse of the triangle whose other two sides are the height (*h*) between floors, and the base (*b*) of the staircase. Second, the length of a side is the square root of the difference between the hypotenuse squared and the length of the other side squared. Thus, the length of the base of the staircase (*b*) would be the square root of the difference between the length of the staircase (*l*) squared and the height between floors (*h*) squared ($b = \sqrt{l^2 - h^2}$).

8. The series sums to 2.

Chapter 6

1. a. The quotient is 10/x, x ≠ 0.

c. The difference is *a* – 12.

2. a. The ellipsis means 4 + 5.

c. The ellipsis means 10,000 + 100,000.

e. The next three entries would be $2^4 + 2^5 + 2^6$.

4. a. 53.

c. 121.

e. 1,443.

5. a. IV.

c. XLI.

e. CVI.

7. a. The number 75 in its expanded form is written as (7 x 10) + (5 x 1) and is read as *seventy five*.

c. The number 227 in its expanded form is written as (2 x 100) + (2 x 10) + (7 x 1) and is read as *two hundred twenty seven*.

e. The number 432,067 is written in its expanded form as (4 x 100,000) + (3 x 10,000) + (2 x 1000) + (0 x 100) + (6 x 10) + (7 x 1) and is read as *four hundred thirty two thousand sixty seven*.

9. a. 9.

c. 16.

10. a. 100_2.

c. 10111_2.

13. a. 10 is a factor of 2140.

c. 3 is a factor of 831.

e. 9 is a factor of 5436.

14. a. 23 is a prime number.

c. 67 is a prime number.

15. a. The factors of 10 are 1, 2, 5, and 10. The prime factorization is $10 = 5 \times 2$.

c. The factors of 16 are 1, 2, 4, 8, and 16. The prime factorization is $16 = 2^4$.

e. The factors of 32 are 1, 2, 4, 8, 16, and 32. The prime factorization is $32 = 2^5$.

16. a. $75 = 5^2 \times 3$.

c. $400 = 5^2 \times 2^4$.

e. $625 = 5^4$.

17. a. 2.3 is a rational number.

c. 3 is a natural number.

e. 100.1 is a rational number.

18. a. Multiplying three negative integers would result in a negative integer.

c. Multiplying a positive rational number by a positive irrational number would result in a positive irrational number.

e. Subtracting a negative rational number from a positive irrational number would result in a positive irrational number.

23. a. If I am 6 feet 2 inches tall, then I am 74 inches tall.

c. There are 8 furlongs/mile, so a 10.2 furlong race is 1.275 miles. This is almost 5 times shorter than a 6.2 mile road race, and over 20 times shorter than a 26 mile marathon.

e. A 6000 fathom deep sea trench is 36,000 ft deep. Thus, it is about 6.8 statute miles or 2.3 land leagues deep.

24. c. A 154 grain letter is 0.352 ounces.

25. a. A 12 fl oz can of soda is 21.648 in^3.

26. a. A centimeter is ten times larger than a millimeter.

c. A cubic meter is a million times larger than a cubic centimeter.

e. A gigasecond is 10^{15} times smaller than a yottasecond.

27. a. A mile is about 61% longer than a kilometer.

c. A quart is about 5.7% more voluminous than a liter.

28. a. 10 meters is 32.8 feet.

c. 20 gallons is 75.7 liters.

e. 150 lbs is 68.0 kilograms.

30. a. 50 K is –223 °C

c. The 500,00 K should say 500 K, which is 227 °C.

31. a. 0 °F is –17.8 °C.

c. 100 °C is 212 °F.

e. 70 °F is 21.1 °C.

32. a. 2.

c. 779.

e. 4.

g. 235.

33. a. 3/4 = 0.75.

c. 5/8 = 0.625.

e. 4/7 = 0.571.

34. a. i. 2,365.985

ii. 2,366.0

iii. 2,370

iv. 2,400

c. i. 6,000.000

ii. 6,000.0

iii. 6,000

iv. 6,000

Chapter 7

2. a. $5 \times 10^6 = 5{,}000{,}000$ = five million.

c. $-2 \times 10^{-2} = -0.02$ = negative two hundredths.

e. $1 \times 10^{-7} = 0.0000001$ = one ten millionth.

3. a. $600 = 6 \times 10^2$.

c. $50{,}000 = 5 \times 10^4$.

e. $0.0005 = 5 \times 10^{-4}$.

4. a. $1{,}000{,}000 = 1 \times 10^6$.

c. $45 \times 10^{-1} = 4.5 \times 10^0$.

e. $540 \times 10^6 = 5.4 \times 10^8$.

5. a. $2.2 \times 10^{-4} = 0.00022$.

c. $2 \times 10^{-1} = 0.2$.

e. $3.5 \times 10^4 = 35{,}000$.

8. a. $(3 \times 10^4) \times (8 \times 10^5) = (3 \times 8) \times (10^4 \times 10^5) = 24 \times 10^9 = 2.4 \times 10^{10}$.

c. $(9 \times 10^3) \times (5 \times 10^{-7}) = (9 \times 5) \times (10^3 \times 10^{-7}) = 45 \times 10^{-4} = 4.5 \times 10^{-3}$.

e. $(8 \times 10^{12}) \div (4 \times 10^9) = (8 \div 4) \times (10^{12} \div 10^9) = 2 \times 10^3$.

g. $(3.2 \times 10^{22}) \div (1.6 \times 10^{-14}) = (3.2 \div 1.6) \times (10^{22} \div 10^{-14}) = 2.0 \times 10^{36}$.

h. $(6 \times 10^{10}) - (5 \times 10^9) = (60{,}000{,}000{,}000) - (5{,}000{,}000{,}000) = 5{,}500{,}000{,}000 = 5.5 \times 10^{10}$.

i. $(9 \times 10^3) + (5 \times 10^9) = (9{,}000) + (5{,}000{,}000{,}000) = 5{,}000{,}009{,}000 \approx 5 \times 10^9$.

10. a. 10^{35} is 10^9, or 1 billion, times larger than 10^{26}.

c. 1 billion is 10^3, or 1 thousand, times larger than 1 million.

e. 2×10^{-6} is 10^3, or 1 thousand, times larger than 2×10^{-9}.

i. 250 million is 20 times smaller than 5 billion.

14. a. The total distance run by the 35,000 runners was 3.5×10^5 km.

17. a. The scale factor is 100,000.

c. The scale factor is 2,000,000.

18. a. The actual distance between two points 7.5 cm apart on a map of scale 1 to 24,000 is 1.8 km.

21. The greatest distance between Pluto and the Sun is 7.8×10^{-4} light-years. At this distance, it takes light from the Sun about 6.9 hours to reach Pluto.

Chapter 8

9. a. The average score on the exam was 77.5 ± 2.5 points.

c. The average person requires 3/8 ± 1/8 pound of chocolate each week to prevent scurvy.

10. a. The population of China is between 1 and 1.4 billion people.

12. a. There are 3 significant digits in 96.2 km / hr.

c. There are 2 significant digits in 0.00098 mm.

e. There are 3 significant digits in 401 people.

g. There are 6 significant digits in 1.00098 mm.

13. a. The implied range of uncertainty in a measurement of 241 kg is 240.5 to 241.5 kg.

c. The implied range of uncertainty in a measurement of 20.0 mm is 19.95 to 20.05 mm.

e. The implied range of uncertainty in a measurement of 27,000 students is 26,500 to 27,500 students.

21. a. $48.49 + 4.237 + 12.1 = 64.8$.

c. $(4.326 \times 10^{-6}) + (9.36478 \times 10^{-9}) = 4.335 \times 10^{-6}$.

e. $(8.599 \times 10^9) + (7.62 \times 10^7) = 8.675 \times 10^9$.

22. a $(1.3 \times 10^{21}) \times (4.1 \times 10^{-12}) = 5.3 \times 10^9$.

c. $(3.43 \times 10^{-7}) \times (5.661 \times 10^{-5}) = 1.94 \times 10^{-11}$.

e. $(4.448921 \times 10^{13}) \times (1 \times 10^1) = 4 \times 10^{14}$.

Chapter 9

1. a. 1996 interest payment will be \$63.

2. a. In 1995, 15.7% of the total receipts was devoted to paying off interest on the debt.

c. In 1995, 24.9% of the total receipts was devoted to Social Security.

3. With a debt of \$4.96 trillion for 1995 and 125 million laborers, the federal debt per laborer is \$39,680/laborer.

5. a. The interest rate was about 4.7%.

6. a. The deficit in 1996 would be \$139 billion. Note: Even if 1996 spending is held at 1995 levels, the interest on the debt paid in 1996 is more than that paid in 1995. Thus, even if spending levels remain constant, the debt will continue to grow.

8. a. The density of the rock is 16.7 g/cm^3.

c. The density of the sphere is 0.48 g/cm^3.

9. a. The mass of the granite would be 54 kg; its weight would be 119 lbs.

c. As long as the masses are the same, a granite rock will be about three times more voluminous than an iron rock.

10. a. The population density of the U.S. is abut 71 people per square mile (71 people/mi.^{2}).

11. a. The information density on the disk would be 2.8 gigabytes/cm^2.

16. a. You will travel 315 km in 3.5 hrs at 90 km/hr.

c. The average speed of an athlete in an Ironman Triathlon is about 25 mi/hr.

e. It would take the runner about 34 minutes to run 10 km.

17. a. The average acceleration is about 4.6 mi/hr per second.

c. Your average deceleration would be 0.125 mi/hr per min. This is 7.5 mi/hr^2, or 9.3×10^{-4} m/s^2.

18. a. After one second, the ball will be going up at 30 m/s.

c. The ball will reach its highest point four seconds after it is thrown.

19. a. The ball will hit the ground just after six seconds from the hit.

20. a. The penny is falling 10 m/s 1 second after it's dropped.

c. After 6 seconds, the penny is moving at 60 m/s.

21. a. The skydiver's speed after 5.0 seconds would be about 180 km/hr.

25. a. The kinetic energy of a 2 kg rock moving at 5m/s is 25 kg(m/s)^2, or 25 Joules (J)

c. 300 kilowatt-hours is 1.08×10^9 J.

26. a. 1250 kilowatt-hours is 4.5×10^9 J.

27. a. The average person needs 1.0×10^7 J/day, or 2.9 kilowatt-hours/day of energy.

Chapter 10

5. a. i. The first variable is the date or time of year. The second variable is the average high temperature.

ii. The domain is from January 1 to December 1. The range extends from 38 to 85 °F.

6. a. The pressure at an altitude of 5,000 ft. would be 25 in. The pressure at an altitude of 10,000 ft. would be 22 in. The pressure at an altitude of 15,000 ft. would be 18 in. The pressure at an altitude of 20,000 ft. would be 14 in. The pressure at an altitude of 25,000 ft. would be 12 in.

13. a. An 8¢ increase in the price of a gallon of gas should result in selling 1,200 gal./wk less gas. If the price for a gallon of gas decreases by 2¢, the sales should go up 300 gal./wk [(–150 (gal./wk)/¢) x (–2 ¢) = 300 gal./wk].

14. a. i. Time is the first variable, and the average price of a new car is the second variable.

ii. The initial value of the second variable is \$12,000, and the rate of change is \$1,200/yr.

iii. The average price of a new car = $12,000 + ($1,200/yr x change in time from today).

iv. The average cost of a new car 2.5 years from now will be $15,000.

15. a. The depth of the pool after 8 minutes is 20 inches. The water is 15 inches deep after 6 minutes, and it is 75 inches deep after 30 minutes.

16. a. Profit = $5/raffle ticket x n – $350, where n is the number of raffle tickets sold. The revenue increases by $5 for each ticket sold. Seventy tickets must be sold before the raffle can begin to make a profit.

17. a. The slope of the graph is 2, and the y-intercept is –6.

18. a. The equation for the line is $y = 3x + 1$.

c. The equation for the line is $y = 11/9x - 136/9$.

21. The value of the washing machine = $1000 – $50/yr x time. Thus, the value of the washing machine has depreciated to a value of $0 after 20 years.

Chapter 11

2. a. It takes three dimensions to describe the volume of the book.

c. An edge of the book is a line, so it is described by one dimension.

3. a. 1/3 of a circle is 120°.

c. 1/20 of a circle is 18°.

e. 1/90 of a circle is 4°.

4. a. 1° is 1/360 of a circle.

c. 15° is 1/24 of a circle.

e. 60° is 1/6 of a circle.

5. a. There are 21,600 minutes of arc in a full circle.

6. a. 26.5° = 1,590'= 95,400".

c. 15.0161° = 900.966' = 54,058.0".

e. 126.9971° = 7,619.826' = 457,189.6".

7. a. 10°15' = 10.25°.

c. 36°45'8" = 36.75222°.

e. 1°1'1" = 1.01694°

8. a. The unknown angle is 120° (180° – 60° = 120°).

c. The unknown angle is 50° (180° - 80° - 50° = 50°).

9. a. The slope of a 5 in 12 roof is 5/12, or 41.7%. Thus, the roof rises 4.17 feet over a 10 ft span.

c. The angle of a 6 in 6 roof is 45° from the horizontal. It is possible to have a 7 in 6 roof.

10. a. The unknown side is of length 9.

c. The unknown angle is 45°, and the unknown sides are of length 8.5.

11. a. The theater is 200 m from the bus stop, as the crow flies.

13. a. The area of the property is 0.85 acre.

c. The phone cable would only need to be 0.90 mi long if it ran across the field diagonally.

14. a. The cost for all of the flooring is $804.

c. The area of the grass is 4,300 m^2.

17. The circumference of Nearth is 36,000 km.

19. a. The pool holds 2,090 cubic meters (2,090 m^3).

21. a. The volume of the duct is 94,248 cubic inches. 131 square feet of paint is needed to paint the duct.

22. a. A ball with a radius of 2.25 inches has a volume of 47.7 cubic inches and a surface area of 63.6 square inches.

25. The water level would rise about 2 mm.

27. a. 1/3 of a circle = $(2/3)\pi$ radians =2.09 radians.

c. 1/20 of a circle = 0.314 radians.

e. 1/30 of a circle = 0.209 radians.

28. a. 1° = 0.0175 radians.

c. 10° = 0.175 radians.

e. 45° = 0.785 radians.

29. a. $\pi = 180°$.

c. $\pi/15 = 12°$.

e. $\pi/45 = 4°$.

30. a. i. You would walk 10.0 m, or $1/2\pi$ of the circumference.

ii. You would walk 25.0 m, or $1.25/\pi$ of the circumference.

iii. You would walk 31.4 m, or 1/2 of the circumference.

iv. You would walk 3.14 m, or 1/20 of the circumference.

31. a. The actual diameter of the sun is 1.31×10^6 km.

32. a. sin 27° = 0.454.

c. cos 245° = –0.423.

e. tan 0.7 = 0.842.

33. a. $\sin^{-1} 0 = 0°$.

c. $\cos^{-1} 0 = 90°$.

e. $\cos^{-1} 0.87 = 30°$.

35. a. The sun is 26° above the horizon, or 64° from overhead.

36. a. The missing boat is about 5700 feet from your location along the surface.

37. The star is 1.5×10^{15} km or about 150 light-years away.

Chapter 12

2. a. 7! = 5,040.

c. 7! / (4! x 3!) = 35.

e. 12! / 8!(12-8)! = 495.

3. a. ${}_9C_3 = 9!\,/\,3!(9-3)! = 84$. How many three-card combinations may be made from a hand with nine cards?

c. ${}_4P_2 = 4!\,/\,(4-2)! = 12$. How many ways can a judge pick first and second place winners out of four competitors?

4. a. i.

	Sedan	Station Wagon	Hatchback
Color 1	Car #1	Car #2	Car #3
Color 2	Car #4	Car #5	Car #6
Color 3	Car #7	Car #8	Car #9
Color 4	Car #10	Car #11	Car #12
Color 5	Car #13	Car #14	Car #15
Color 6	Car #16	Car #17	Car #18
Color 7	Car #19	Car #20	Car #21
Color 8	Car #22	Car #23	Car #24

ii.

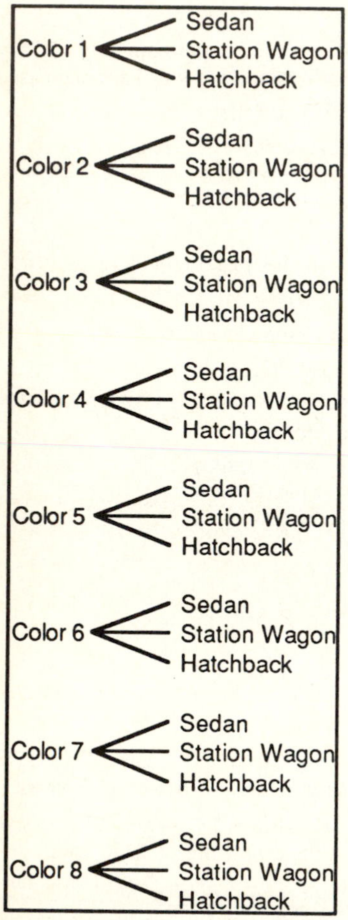

iii. 8 x 3 = 24 different cars.

5. a. i. 8 x 6 x 7 = 336 different ski/boot packages are possible.

ii. From each of the eight different kinds of skis would branch six different bindings. Seven different boots would branch from each binding/ski combination.

6. a. 10^4 different addresses can be formed.

7. a. Any six of the thirteen cook books may be arranged in 1,235,520 different ways.

8. a. 495 different subcommittees can be formed.

9. a. 72 different sundaes may be created.

c. The possible number of triple cones is 1,320.

21. a. Five vertices are needed to draw a network representation of Sleepy Waters. Eight edges are required.

c. The network for Sleepy Waters does not have an Euler circuit.

32. a. The cost of the spanning tree labeled (b) is 24.

Chapter 13

5. a. The probability of drawing a king is 4/52 or 7.7%.

c. The probability of drawing a spade is 1/4 or 25%.

7. c. The probability of getting three heads and one tail is 1/4 or 25%.

8. c. The probability of getting a total of 5 on a toss of two dice is 1/9 or 11%.

13. a.

Number of heads	Empirical Probability
0	0.260
1	0.495
2	0.245
3	0.000

15. a. The probability of drawing an ace on the first try is 4/52 or 7.7%.

c. The probability of the first four draws all being aces is $(4/52)^4$ or 0.0035%.

16. a. The probability of rolling two sixes is 1/36 or 2.8%.

c. The probability of purchasing three winning lottery tickets in a row is $(1/10)^3$ or 0.1%.

19. a. The probability of drawing an ace on the first try is 4/52 or 7.7%.

c. The probability of the first four draws all being aces is 24/6,497,400 or 0.00037%.

21. a. The probability of drawing either a 2 or a 4 on a single draw is 8/52 or 15.4%.

c. The probability of drawing either a king, an ace, or a diamond in a single draw is 19/52 or 36.5%.

22. a. The odds of winning would be 1 in 1.978 or slightly better than 50%.

23. a. The probability of getting a $2, $5, or a $10 winning lottery ticket is 12.2%. This is only slightly larger than the 12.0% probability of getting only a $2 winner.

24. a. The probability of Miami being hit by a hurricane any given year is 1/40 or 2.5%.

28. a. The expected number of boys is 2.

c. The probability of a family having 2 boys and 2 girls is 6/16 or 37.5%.

29. a. The probability of getting 25 heads in 50 tosses is 11.2%.

33. a. The most likely number of games won by the casino is 503,000.

c. The probability that the patrons will win more than 498,500 games is 0.15%.

37. a. There is an 8% chance that somebody in the class will have the same birthday as you.

39. a. The probability of not having a predictive dream on any given night in 9,999/10,000 or 99.99%.

43. a. The probability of the average American dying in a car wreck each year is about 1.6×10^{-4} or 0.016%.

c. The probability that the average American will not die in a car accident, over the next fifty years, is 99.2%.

48. a. If suspect A confesses, the minimum sentence he can expect is 1 yr. If suspect A maintains his innocence, the minimum sentence he can expect is 3 years. Confessing is the safest strategy for suspect A.

c. Either suspect could get a shorter sentence by switching to a guilty plea. Thus, the choice (A_2,B_2) is unstable.

Chapter 14

10. a.

Grade	Frequency	Rel. Freq.	Cum. Freq.
A	3	0.136	3
B	7	0.318	10
C	7	0.318	17
D	2	0.091	19
F	3	0.136	22
Total	**22**	**1.000**	**22**

16. a. In 1988 the average private college tuition rose the most compared to the consumer price index.

19. a. The mean of the set is 7.6.

20. a. The mean time is 10.92 s.

c. The median is between 11.0 and 11.1 seconds. The upper quartile = 11.1 s and the lower quartile = 10.8 s. The fastest time is 10.3 s and the slowest is 11.3 s.

21. a. The median height is 172 cm. The upper quartile is 188 cm and the lower quartile is 166 cm. The tallest person is 190 cm tall, and the height of the shortest person is 145 cm.

c. Since the mean of the heights is 172.6 cm, the variance = $[(0.6)^2 + (27.6)^2 + (6.6)^2 + (5.4)^2 + (17.4)^2 + (15.4)^2 + (3.6)^2] / (7 - 1) = 231\ \text{cm}^2$, and the standard deviation is 15.2 cm.

22. The mean for heat 1 is 10.06 s. The variance = $[(0.08)^2 + (0.05)^2 + (0.01)^2 + (0.03)^2 + (0.04)^2 + (0.09)^2] / (6 - 1) = 0.0039\ \text{s}^2$, and the standard deviation is 0.06 s.

29. a. The margin of error is 0.4%.

32. a. The 95% confidence range is between 50% and 58% of the population approving the President's performance.

Chapter 15

2. a. The population increase of Reno is linear because it is increasing by the same absolute amount each year.

c. The Austrian birth rate is decreasing exponentially. The decrease grows smaller each year because it is a fixed percentage of the diminishing population.

4. a. After 15 days you would have $163.84.

c. The first billion dollars will come on the 38th day.

6. a. The number of bacteria in the bottle after 50 minutes is $2^{50} = 1.13 \times 10^{15}$. The bottle is $1/2^{10}$, or one tenth of one percent full after 50 minutes of doubling.

7. a. The population will have increased by a factor of 2^8 or 256 times after 24 hours. In one week the population will have increased by 2^{56} or 7.2×10^{16} times the original population!

c. It will take 34.9 years for the population to triple.

8. a. The doubling time is approximately 10 years. After 3 years, the consumer price index will increase by approximately 1.23 times.

c. The doubling time is approximately 43.75 years. In 75 years, the population will have increased by approximately 3.28 times.

9. a. The half-life of the rabbit population is approximately 11.7 years. Thus, the rabbit population will fall to 50% its present size in approximately 11.7 years.

c. The population will fall to half its present size in approximately 233 years.

10. a. Its doubling time is exactly 4 days.

c. Its half-life is exactly 4 days.

11. a. The statement is true since the price has gone up 10% in one year and the rate of inflation is given as 10% per year.

c. The statement is false because the exact doubling time formula yields two different doubling times for a growth rate of 1% per month and 12% per year.

13. a. The exact doubling time is 10.2 years. This is within 2% of the approximate doubling time of 10 years.

c. The exact doubling time is 43.66 years. This is within 0.2% of the approximate doubling time of 43.75 years.

14 a. The exact half-life is 11.9 years. This is within 2% of the approximate half-life of 11.7 years.

c. The exact half-life is 231 years. This is within 1% of the approximate half-life of 233 years.

16. a. The population will increase by 2,040 people the first year. After 25 years, the population will be 1.538×10^5 people.

c. About 1,700 and 2,300 crimes will be committed in the years 2000 and 2010 respectively.

e. In one year, the population will drop to 9,646 people. The population will be down to 6,000 in about 14 years.

18. a. The cloth was painted about 2,100 years ago.

c. The meteorite formed about 3 billion years ago.

19. a. Q is the value of the portfolio, and $Q = \$300 \times (1.08)^t$, where t is the time in years since 1990. This model predicts Q to be $648 by the year 2000.

c. Presuming today is 1 March 1996, $Q = \$8{,}000 \times (0.9656)^t$, where t is the time in months since January 1990, and Q is the value of the computer at the given time. This model predicts Q to be $822 in June of 1995, and $394 in March of 1997. The model predicts the value of the computer will be down to $100 by June of the year 2000.

Chapter 16

4\. a. The frequency of the red light is 4.28×10^{14} Hertz.

c. The frequency of the gamma ray is 6.0×10^{20} Hertz.

e. The wavelength of the FM station is 2.8 meters.

8\. a. The loudness is 76.5 dB.

c. The loudness is –30 dB.

e. An 85 dB sound is 1.0×10^7 or 10 million times louder than a 15 dB sound.

9\. a. The absolute intensity of a jet at 30 meters is 100 watt/m^2.

c. The absolute intensity of a 62 dB sound is 1.6×10^{-6} watt/m^2.

10\. a. The siren will sound 90,000 times louder when it's 0.1 m from your ear than when it's 30 meters from your ear.

11\. a. 7.9×10^{11} joules of energy are released in a 5th magnitude earthquake.

c. 3.5×10^{16} joules of energy were released in 1985 Mexico City earthquake.

e. The energy released by a 1 megaton nuclear bomb is 6 times greater than that of a 7th magnitude earthquake.

13\. a. As the pH increases by one, the hydrogen ion concentration decreases by a factor of 10. This makes the solution more basic.

c. A solution with a pH of 8.5 has a hydrogen ion concentration of 1.9×10^{15} hydrogen ions per liter.

e. The pH is 1. This is an acid.

16\. a. The balance after 6 months and 1 year are $516.52 and $533.58 respectively. Thus, the annual yield for this account is 6.7%.

19\. a. At an APR of 8% compounded daily, it will take 13.7 years for a principal amount to triple.

c. It would take 5.8 years for a principle to grow by 50% at an APR of 7% compounded daily.

20\. a. The annuity will be worth $175,000 after 40 years. This is over seven times the $24,000 invested.

21\. a. You should deposit $330 per month.

24\. a. Your monthly payment would be $241.26.

26\. a. The gravitational force between the objects will be 1/9 the initial value.

28\. a. She is 28 years old when she returns.

Appendix A

1\. a. 0

c. 6

e. 8

g. 4

i. 46

2\. a. 21

c. 207

4\. a. 91/88 or 1.034

c. $1/\sqrt{2}$ or 0.707

e. 2/3 or 0.667

g. 1/7.5 or 0.133

i. 1/25 or 0.04

k. 1/0.66 or 1.515

5\. a. 2/3

c. 5/9

e. 179/304

g. 3/1 or 3

6\. a. Yes

c. No

e. Yes

g. Yes

7\. a. 4/6, 3/6

c. 18/27, 18/27

e. –18/98, 21/98

8\. a. 5/16 = 10/32 so, 5/16 < 11/32 by 1/32.

c. –1/2 = –11/22 and –6/11 = –12/22 so, –1/2 > –6/11 by 1/22.

e. 2/3 = 20,000/30,000 and 6649/10,000 = 19,947/30,000 so, 2/3 > 6649/10,000 by 53/30,000.

9\. a. 4/6 or 2/3

c. 8/3

e. 19/20

10\. 1/2 = 0.5; 4/5 = 0.8; No, because 0.5 π 0.8.

11\. a. 1

c. 3/80

e. 9/176

g. 289/16

12\. a. i) 0.003 = (0 x 1) + (0 x 0.1) + (0 x 0.01) + (3 x 0.001)

ii) Three one-thousandths

iii) 3/1000

c. i) –65.5 = –[(6 x 10) + (5 x 1) + (5 x 0.1)]

ii) Negative sixty five and five tenths

iii) –655/10 = –131/2

e. i) 6005.4 = (6 x 1000) + (0 x 100) + (0 x 10) + (5 x 1) + (4 x 0.1)

ii) Six thousand five and four tenths

iii) 60054/10 = 30027/5

g. i) 0.124 = (0 x 1) + (1 x 0.1) + (2 x 0.01) + (4 x 0.001)

ii) One hundred twenty four one-thousandths

iii) 124/1000 = 31/250

i. i) 0.24816 = (0 x 1) + (2 x 0.1) + (4 x 0.01) + (8 x 0.001) + (1 x 0.0001) + (6 x 0.00001)

ii) Twenty four thousand eight hundred sixteen one hundred-thousandths

iii) 24816/100000 = 1551/6250

13. a. 25% = 0.25 = 25/100 = 1/4
    c. 2.5% = 0.025 = 25/1000 = 1/40
    e. –1.5% = –0.015 = –15/1000 = –3/200
    g. –60% = –0.60 = –60/100 = –3/5
    i. 0.04% = 0.0004 = 4/10000 = 1/2500
    k. 7.5% = 0.075 = 75/1000 = 3/40

14. a. 400%
    c. 65%
    e. 72%
    g. 40%
    i. 85%
    k. 120%
    m. 99.5%
    o. 3%

15. a. 2^{14} = 16,384
    c. 5^2 = 25
    e. 4^2 = 16

16. a. $1/7^2$ = 1/49 = 0.020
    c. $2^{-6} = 1/2^6$ = 1/64 = 0.016

17. a. $25^{1/2}$ = 5
    c. $81^{1/2} \times 27^{1/3}$ = 9 x 3 = 27

18. a. 64
    c. 9
    e. 2
    g. 1000
    i. 12
    k. 15.588

19. a. 4
    c. 1

20. a. 915
    c. 73
    e. 2

Appendix B

1. a. $3y^2 + 6y$
   c. $9x^2 + 24x + 16$
   e. $p^2 + 2pq + q^2$

2. a. 2x – 5
   c. x +1

3. a. 3x
   c. 4x + y
   e. x + 2y

4. a. Any values except x = y.
   c. Any values except those that make x + y + z = 1 true.
   e. Any values except x = y.

5. a. For n = 0 hrs, p = $350
   For n = 2 hrs, p = $374
   For n = 12 hrs, p = $494
   For n = 6.5 hrs, p = $428
   c. For p = $374, n = 2 hrs
   For p = $400, n = 4.17 hrs
   For p = $450, n = 8.33 hrs
   For p = $500, n = 12.5 hrs

6. a. 15 ≠ 3, so x = 5 is not a solution.
   c. 4 ≠ -4, so y = -1 is not a solution.
   e. 0 = 0, so x = 1/5 is a solution.

7. a. x = -3
   c. x = ±4, that is x = +4 or x = -4.
   e. x = 5

8. a. 3x = x + 16 yr. Thus, x = 8 years and Sue will be 24 when she's three times older than Jerry.
   c. 3x = x + 1000 people. Thus, x = the population of Flute Town = 500 people and the population of Tuba City was 1,500 people that same year.

9. a.

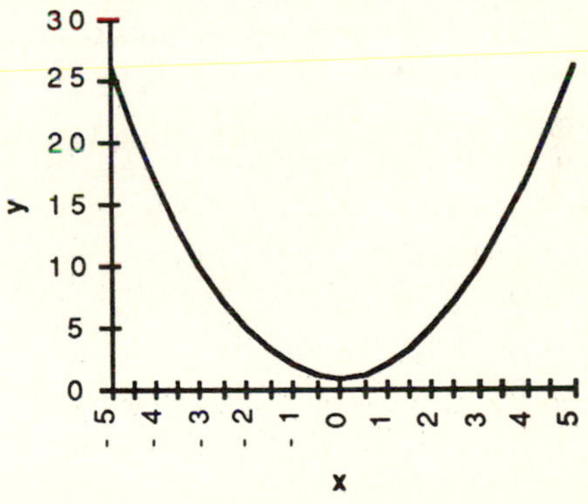

c.

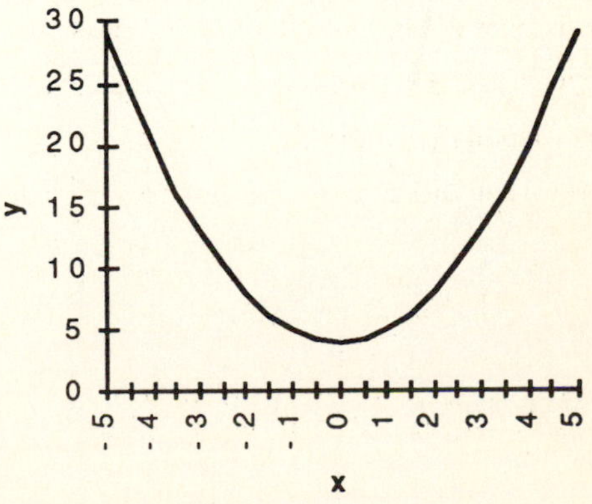

e.

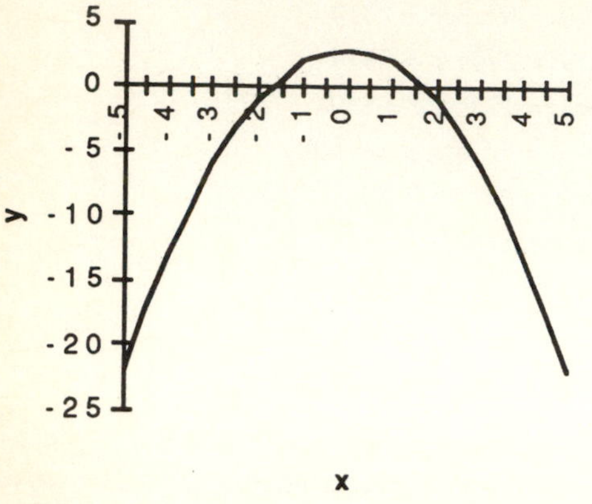

e.

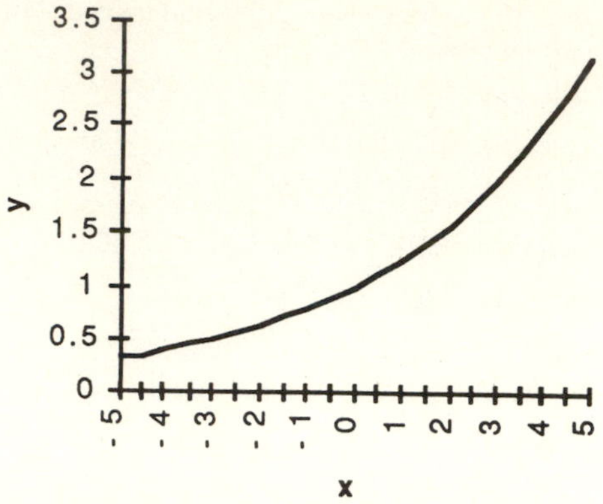

10. a.

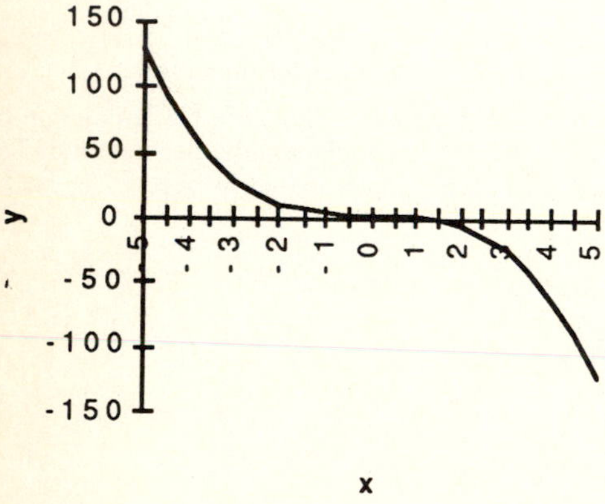

g.

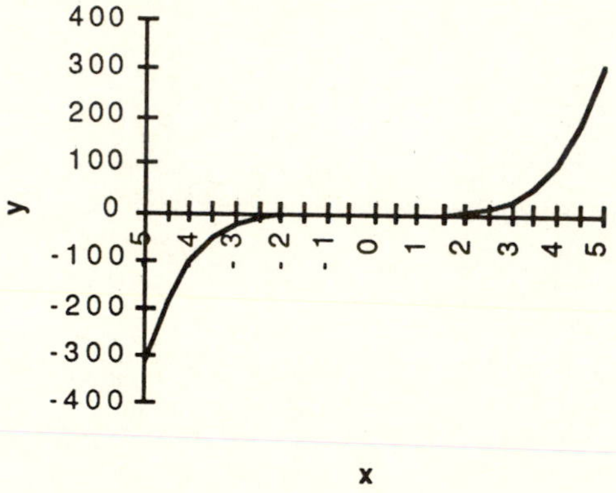

c.

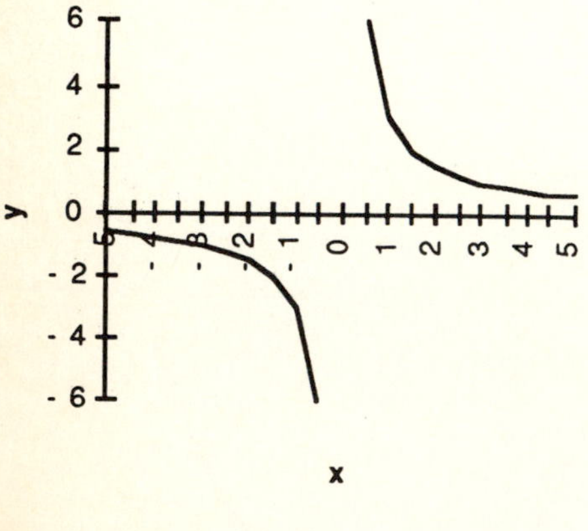

Appendix C

1. a. $\log_2 16$ means 2 to what power equals 16. In this case $2^4 = 16$, so the answer is 4.

   c. $\log_{12} 144$ means what power of 12 equals 144. Since 12 to the power of 2 is 144 the answer is 2.

   e. $\log_{10}1{,}000{,}000{,}000$ means 10 to what power equals a billion. Since 1 billion = 10^9, the answer is 9.

2. Recall that log x is assumed to mean $\log_{10} x$.

   a. False: log 100 = 2, which is less than 3, so log 96 must also be less than 3 and not between 3 and 4.

   b. True: log 1,000,000 = 6 and log 10,000,000 = 7 and 1,600,000 is between 1,000,000 and 10,000,000.

   c. False: $\log (8 \times 10^9) = \log 8 + \log 10^9 = \log 8 + 9 \times \log 10 = \log 8 + 9$ which is a number between 9 and 10 not 12 and 13.

   g. False: $10^0 = 1$ and $10^1 = 10$, so $10^{0.928}$ must be less than 10 and not between 10 and 100.

   i. False: $10^{-5.2}$ must be less than 1 and greater than 0, and thus not between 10^5 and 10^6.

3. a. log 50 = 1.699.

   c. log 0.05 = –1.301.

   e. log 0.20 = –0.699.

4\. a. $x = 10^2 = 100$.

c. $t = \left(\frac{10^{3.4}}{5} \right)^{-\frac{1}{2}} = 0.045 = 4.5 \times 10^{-2}$.

e. The only way to solve an equation with a variable in the exponent is to take the log of both sides, provided both sides of the equation are positive. Since both sides of the equation are not positive, this equation cannot be solved.

5\. a. t = 3.19 hours.

c. For M = 7.0, E = 7.9×10^{14} Joules. For E = 10^{12} Joules, M = 5.1.

Appendix D

1\. a. 14

b. 10.67

5\. a. 1.77×10^3

c. 5.34×10^{14}

7\. a. 4.783×10^6

c. 4.589

e. 4.133

g. 108.605

i. 34.436

8\. a. 0.699

c. 0.000

e. 3.643

g. 2.699

i. 0.589

9\. a. 1.609

c. 0.000

e. 1.000

g. 3.303

i. 0.454

10\. a. 1.21×10^{-27}

c. 4.44×10^5

11\. a. 0.174

c. 1

12\. a. 1

c. 0.707

Index

T

U